GEOMETRY *of* PDEs *and* MECHANICS

GEOMETRY
of
PDEs
and
MECHANICS

Agostino Prástaro

Dipartimento di Metodi e Modelli
Matematici per le Scienze Applicate
Università degli Studi di Roma "La Sapienza"

World Scientific
Singapore • New Jersey • London • Hong Kong

Published by

World Scientific Publishing Co. Pte. Ltd.

P O Box 128, Farrer Road, Singapore 912805

USA office: Suite 1B, 1060 Main Street, River Edge, NJ 07661

UK office: 57 Shelton Street, Covent Garden, London WC2H 9HE

Library of Congress Cataloging-in-Publication Data

Prástaro, Agostino.
Geometry of PDEs and mechanics / Agostino Prástaro.
760 p.; 22.5 cm.
Includes bibliographical references and index.
ISBN 9810225202
1. Mechanics--Mathematics. 2. Geometry, Algebraic.
3. Differential equations, Partial. 4. Mathematical physics.
I. Title.
QC125.2.P73 1996
530.1'55353--dc20 95-47858
CIP

British Library Cataloguing-in-Publication Data

A catalogue record for this book is available from the British Library.

This book is printed on acid-free paper.

Printed in Singapore by Uto-Print

To the memory of my father and mother

and

To my wife Paola and my sons Francesco and Alessandro

PREFACE

'Non inorgogliamoci troppo del perfetto rigore a cui crediamo
di essere oggi capaci di ridurre tanta parte della matematica
e non scartiamo quel che non ci pare del tutto rigoroso,
poiché domani si troveranno certo imperfezioni nella nostra perfezione
e da qualche geniale atto di intuizione, che non ha ancora i crismi del rigore,
si troveranno risultati impensabili'
Francesco Severi

This book is an introduction to Mechanics revisited with a modern geometric language. It is crowded, in some sense, around courses held at the University and researches in Mathematics and Mathematical Physics, that I developed until now. My principal interest in research is the geometrization of Physics and the study of geometry of partial differential equations. In fact, Physics is essentially geometry, so the most suitable language to use, in order to describe physical phenomena, is geometric. On the other hand, all Mathematics has been reformulated with an algebraic-geometric language by obtaining great results.
Any physical theory is characterized by a PDE. If one takes seriously this simple observation, we can consider that the true universe that characterizes a physical theory is the manifold that identifies the corresponding PDE. For example, the manifold that characterizes the General Relativity is not a 4-dimensional manifold M (space-time), but a 144-dimensional manifold, (Ein), that corresponds to the geometric structure $(Ein) \subset J\mathcal{D}^2(S^0_2 M)$ that characterizes Einstein equation (Ein). So, from this point of view, it becomes very important to study the geometric structures (manifolds) that describe PDEs. This structure is more important than the manifold of solutions, as it contains more information than the latter. But, the manifold that characterizes a PDE identifies also its set of solutions. So all the usual informations about PDEs can be recovered by studying the geometry of PDEs.
Thus, in this book we study the geometry of PDEs first. Boundary value problems, whether for regular solutions or singular ones are studied. We characterize also PDEs by means of some useful spectral sequences and study differential cobordisms in PDEs. These allow us to give a topological algebraic characterization whether of quantum solutions or classic ones. Then, the general geometric framework of PDEs, allows us to give a very elegant description of mechanics and field theory, and to open new prospects in these areas. Of course, the most suitable presentation of any physical theory is in a space-time framework. So, here, we start by introducing the Galilean space-time, and, in this panorama, we build the meaning of admissible motions and fields, by emphasizing the geometric role of dynamic equations. Quantizations of classical theories, have natural places in this picture. More precisely, we consider the

formal quantization of partial differential equations, first introduced in refs.[103,104], and further developed in refs.[105,106,108,109,111,112,113,114]. This is a procedure that allows us to associate to any PDE a measure space (quantum situs). This quantization becomes effective if on the quantum situs we recognize a (pre-)spectral measure (quantum spectral measure). Then it is proved that canonical quantization is a phenomenon governed by the geometry of PDEs. The geometrical structure of the quantum situs, $\Omega(E_\infty)$, of the infinity prolongation of a PDE E_k, is explicitly given. A structure of C_c^∞–manifold [56] is recognized on $\Omega(E_\infty)$. It is proved that formal quantization allows us to give a geometric description of quantum tunnel effects in PDEs. Applications to some important equations of the field theory are given. In the last chapter a theory of non-commutative manifolds (quantum manifolds) is proposed. Quantum manifolds are seen as locally convex manifolds where the model has the structure A^{m+n}, where A is a non-commutative algebra that satisfies some particular axioms (quantum algebra). For such manifolds we also formulate a geometry and a formal theory of PDEs. Furthermore, as we characterize quantum manifolds by means of some geometrical objects defined on the underlying topological manifold, we are able to develop a quantum theory simulating on microlevel what the gravitational theory makes on macrolevel one. In fact, we formulate a quantum physical theory where the dynamical variables (quantum structure) are defined on quantum manifolds, and are solutions of suitable PDEs. A general theory of quantum tunnel effects on such equations is obtained on the ground of cobordism theory and Morse theory. In this framework a model of quantum supergravity is given where the quantum structure interprets at the classic limit curvature, torsion, gravitino and electromagnetic fields as A-valued distributions on space-time.
We emphasize that in this book we have discussed some new results that are not yet published. For example, completely new results that are relative to the classifications of singular solutions by means of cobordisms in PDEs, and the geometry of quantum PDEs, and its applications to quantum field theory and quantum supergravity.
Roma, July 1995.

Agostino Prástaro

CONTENTS

1 - ALGEBRAIC GEOMETRY

1.1 - ALGEBRAIC COMPLEMENTS

The purpose of this section is to recall some fundamental definitions and results of Algebra that will be used in the next chapters. (For more details see also refs.[1,7,14,43, 49,53,79,80,86].)

DEFINITION 1.1 - *A* **matrix** *is a map:*

$$f : I_1 \times \cdots \times I_k \to \mathbf{R}, \quad f : (i_1, \cdots, i_k) \mapsto f_{i_1}{}^{i_2 i_3}{}_{\dots i_k} \in \mathbf{R}$$

where $I_1 \equiv \{1, \cdots, n_1\}, \cdots, I_k \equiv \{1, \cdots, n_k\}$. *We will denote the set of matrices with* k *indexes* $M_{n_1}{}^{n_2 n_3}{}_{\dots n_k}$, *where the lower and upper indexes are called* **covariant** *and* **controvariant** *respectively.*

EXAMPLE 1.1 - 1) $k = 2, n_1 = n, n_2 = m, I_1 = I \equiv \{1, \cdots, n\}, I_2 = J \equiv \{1, \cdots, m\}$,

$$f : I \times J \to \mathbf{R} \quad , \quad f : (i, j) \mapsto f_{ij} \, ,$$

where f_{ij} is the **generic element** of the matrix. In this case one can represent the matrix in the following way:

$$\begin{bmatrix} f_{11} & \cdots & f_{1m} \\ \cdots & \cdots & \cdots \\ f_{n1} & \cdots & f_{nm} \end{bmatrix} .$$

This is an example of 2×2 matrix with two covariant indexes.

2) $(f_i{}^j)$ belongs to the matrix-space $M_n{}^m$. The first index is covariant, the second one is controvariant. ■

PROPOSITION 1.1 - *The set of matrices of some type is a vector space on* $\mathbf{R}$.

EXAMPLE 1.2 - For example, in M_{nm} we have:

a) $(f_1 + f_2)_{ij} = (f_1)_{ij} + (f_2)_{ij}$ (addition)

b) $(\lambda f)_{ij} = \lambda f_{ij}$ (scalar multiplication)

The zero of M_{nm} is the matrix (0_{ij}), with $0_{ij} = 0 \in \mathbf{R}$; the **opposite** of (f_{ij}) is $(-f_{ij})$. In fact, for any index i, j one has: $f_{ij} - f_{ij} = 0$. ■

DEFINITION 1.2 - $V^* \equiv L(V; \mathbf{R})$. *$V^*$ is the space of linear applications $V \to \mathbf{R}$. We call V^* the* **dual** *of V.*

REMARK 1.1 - One has the following non-canonical isomorphisms.

$$\left\{ \begin{array}{l} V \cong M^n; \; V^* \cong M_n; \; L(V; W) \cong M_n^m; \; L(\mathbf{R}; W) \cong W \cong M^m \\ L(V; \mathbf{R}) \equiv V^* \cong M_n; \; L(V_1, V_2; W) \equiv M_{n_1 n_2}{}^m; \; L(V_1^*, V_2; W) \equiv M^{n_1}{}_{n_2}{}^m \\ L(V_1^*, V_2^*; W) \equiv M^{n_1 n_2 m}; \; L(V_1, V_2; W^*) \equiv M_{n_1 n_2 m}; \; L(V_1^*, V_2; W) \equiv M^{n_1}{}_{n_2 m} \\ L(V_1^*, V_2^*; W^*) \equiv M^{n_1 n_2}{}_m; \; L(V_1, V_2^*; W^*) \equiv M_{n_1}{}^{n_2}{}_m \\ L(V_1, \cdots, V_p; W) \equiv M_{n_1 \cdots n_p}{}^m . \end{array} \right\}$$

Here $L(V;W)$ denotes the space of all linear applications $V \to W$. We will also use the following notation $Hom_{\mathbf{K}}(V;W) \equiv L(V;W)$. Furthermore, $L(V_1,\cdots,V_p;W)$ denotes the space of p-linear applications $V_1 \times \cdots \times V_p \to W$.

DEFINITION 1.3 - *We set*

$$GL(n;\mathbf{R}) \equiv \{a \in M_n{}^n | \det a \neq 0\}.$$

We call $GL(n;\mathbf{R})$ *the* **general linear n-dimensional real group**.

THEOREM 1.1 - *The matrix that relates two different bases, in an n-dimensional vector space, belongs to* $GL(n;\mathbf{R})$.

DEFINITION 1.4 - *The* **tensor product** *between vector spaces V and W is*

$$V \bigotimes W \equiv < V \times W > / I,$$

where $< V \times W >$ *denotes the space of all linear combinations of couples* (v,w), *with* $v \in V, w \in W$. *I is the subspace in* $< V \times W >$ *generated by means of vectors of the types:*

$$\left\{\begin{array}{l} (v_1+v_2,w)-(v_1,w)-(v_2,w) \\ (\lambda v,w)-\lambda(v,w) \\ (v,w_1+w_2)-(v,w_1)-(v,w_2) \\ (v,\lambda w)-\lambda(v,w) \end{array}\right\}$$

Let us denote by $u \otimes v$ the equivalence class identified by $(u,v) \in < V \times W >$. *We say that $u \otimes v$ is the* **tensor product** *of u with v.*

THEOREM 1.2 - 1) *One has:* $\dim V \bigotimes W = nm$, *if* $\dim V = n, \dim W = m$.

2) *One has the following canonical isomorphisms:*

$$L(V,W) \cong V^* \bigotimes W; \quad L(V_1,V_2;W) \cong V_1^* \bigotimes V_2^* \bigotimes W.$$

DEFINITION 1.5 - *A* **tensor of the type** (r,s) *on* $V, \dim V = n$, *is a vector of*

$$T^r_s(V) \equiv V \bigotimes \cdots_r \cdots \bigotimes V \bigotimes V^* \bigotimes \cdots_s \cdots \bigotimes V^*.$$

We set $T^0_0(V) = \mathbf{R}$.

PROPOSITION 1.2 - *For any basis e_i in V one has the following linear representation:*

$$t = t^{i_1\cdots i_r}{}_{j_1\cdots j_s} e_{i_1} \otimes \cdots\cdots \otimes e_{i_r} \otimes \theta^{j_1} \otimes \cdots \otimes \theta^{j_s}$$

where $(t^{i_1\cdots i_r}{}_{j_1\cdots j_s}) \in M^{n\cdots_r\cdots n}{}_{n\cdots_s\cdots n}$. $\{\theta^j\}_{1\le j\le n}$ *is the* **dual basis** *of e_i defined by the condition:* $\theta^j(e_i) = \delta^j_i$, where (δ^j_i) the **identity matrix**. *One has:* $\dim T^r_s(V) = n^{r+s}$.

REMARK 1.2 - Of course one has $T_1^0(V) = V^*, T_0^1(V) = V$.

DEFINITION 1.6 - *A* **Euclidean space** *is an n-dimensional vector space V endowed with a map (***scalar product***) $g : V \times V \to \mathbf{R}$, such that: a) (***symmetry***) $g(v_1, v_2) = g(v_2, v_1)$; b) (***nondegeneracy***) $g(v,v) \neq 0, \forall v \neq 0, g(v,v) = 0 \Leftrightarrow v = 0$; c) (***positivity***) $g(v,v) > 0$, $\forall v \neq 0$.*

PROPOSITION 1.3 - 1) *If g is a Euclidean structure on the vector space V we get: $g \in T_2^0(V)$.*

2) *For any basis $\{e_i\}$ of V we have: $g = g_{ij}\theta^i \otimes \theta^j$. $(g_{ij}) \in M_{nm}$, or more precisely $(g_{ij}) \in GL(n; \mathbf{R})$.*

3) *There exists a basis $\{e_i\}$ of V such that $g_{ij} \equiv g(e_i, e_j) = \delta_{ij}$, (***orthonormal basis***).*

4) *One has $g_{ij} = g_{ji}$ (symmetry).*

5) *The inverse matrix (g^{ij}) of (g_{ji}) is a controvariant matrix; one has:*[1]

$$g^{ij} g_{jk} = \delta^i{}_k \, .$$

6) *One has the following canonical isomorphism:*

$$'g : V \to V^*, \quad 'g(v) \equiv g_v : V \to \mathbf{R} \, ,$$

where $g_v(u) = g(u,v)$, $\forall u \in V$. *The isomorphism $'g$ is represented by the matrix g_{ij}.*

$$v = (v^i) \mapsto (v_j = \sum_i v^i g_{ij}) = {}'g(v) \in V^*, (i = 1, 2, \cdots, n) \, , \forall v \in V \, .$$

For the inverse matrix g^{-1} we get:

$$\alpha \in V^* \Rightarrow (\alpha_j) \Rightarrow (v^i = \sum_i \alpha_j g^{ji}) = {}'g^{-1}(\alpha) \in V, (i = 1, 2, \cdots, n) \, .$$

REMARK 1.3 - In a Euclidean space (V, g) , by using the canonical identification $V \equiv V^*$, we can see each vector $v \in V$ as a vector of V^* too. Therefore, with respect to a basis $\{e_i\}$ of V , we can consider controvariant components (v^i) of v , and covariant ones, $v_j = g_{ji} v^i$, as two different aspects of only one entity.

THEOREM 1.3 - *With respect to an orthonormal basis the covariant and controvariant components coincide.*

DEFINITION 1.7 - 1) *A tensor $u \otimes v \in T_0^2(V)$ is called* **symmetric** *if it satisfies the following condition:*

$$u \otimes v = v \otimes u \, .$$

[1] We often use the convention of not putting the sum-symbol for summations between equal controvariant and covariant indexes.

2) *We denote the space of symmetric tensors of type* $(2,0)$ *by* $S_0^2(V)$.

3) *A tensor* $u\otimes v$ *is said to be* **skew-symmetric** *if it satisfies the following condition:*

$$u\otimes v=-v\otimes u\,.$$

4) *We denote the space of skew-symmetric tensors of type* $(2,0)$ *by* $\Lambda_0^2(V)$.

PROPOSITION 1.4 - 1) *One has the following canonical projections:*

$$(symmetrization):\left\{\begin{array}{l} s: T_0^2(V)\to S_0^2(V)\\ u_1\otimes u_2\mapsto \frac{1}{2}(u_1\otimes u_2+u_2\otimes u_1)\equiv u_1\odot u_2\end{array}\right\}$$

$$(skew\text{-}symmetrization):\left\{\begin{array}{l} a: T_0^2(V)\to \Lambda_0^2(V)\\ u_1\otimes u_2\mapsto \frac{1}{2}(u_1\otimes u_2-u_2\otimes u_1)\equiv u_1\wedge u_2\end{array}\right\}$$

2) *One has the following isomorphism :*

$$\boxed{\begin{array}{c} T_0^2(V)\cong S_0^2(V)\bigoplus\Lambda_0^2(V)\\ t\mapsto s(t)\bigoplus a(t)\end{array}}$$

3) *If* (V,g) *is a Euclidean structure, then* g *belongs to* $S_0^2(V)$.

DEFINITION 1.8 - *The symmetry properties can be extended to tensors of type* $(r,0)$ or $(0,r)$. $u_1\otimes\cdots\otimes u_r$ *is* **symmetric** *if* $u_1\otimes\cdots\otimes u_r=u_{\sigma(1)}\otimes\cdots\otimes u_{\sigma(r)}$, $\forall\sigma\in S_r$, *where* $S_r\equiv$ *group of permutations of* $(1,\cdots,r)$. *The space of symmetric tensors of type* $(r,0)$ *is denoted by* $S_0^r(V)$. *Similarly, we say that* $u_1\otimes\cdots\otimes u_r$ *is* **skew-symmetric** *if:* $u_1\otimes\cdots\otimes u_r=\epsilon(\sigma)u_{\sigma(1)}\otimes\cdots\otimes u_{\sigma(r)}$, $\forall\sigma\in S_r$, *where* $\epsilon(\sigma)$ *is the signature of the permutation* σ. *The space of skew-symmetric tensors of type* $(r,0)$ *is denoted by* $\Lambda_0^r(V)$.

PROPOSITION 1.5 - *One has the following canonical mappings:*

$$s:T_0^r(V)\to S_0^r(V),u_1\otimes\cdots\otimes u_r\mapsto u_1\odot\cdots\odot u_r\equiv\frac{1}{r!}\sum_{\sigma\in S_r}u_{\sigma(1)}\otimes\cdots\otimes u_{\sigma(r)}$$

and

$$a:T_0^r(V)\to \Lambda_0^r(V),u_1\otimes\cdots\otimes u_r\mapsto u_1\wedge\cdots\wedge u_r\equiv\frac{1}{r!}\sum_{\sigma\in S_r}\epsilon(\sigma)u_{\sigma(1)}\otimes\cdots\otimes u_{\sigma(r)}\,.$$

REMARK 1.4 - For $r > 2$, $S^r_0(V) \bigoplus \Lambda^r_0(V) \subset T^r_0(V)$.

REMARK 1.5 - (SYMMETRY GROUPS). 1) The **symmetry group** $GL(V)$ of a vector space V, i.e., the set of isomorphisms of V, is a group with respect to the composition law, and it is isomorphic to a group of matrices, if a basis has been fixed on V. More precisely one has the following: $GL(V) \cong GL(n; \mathbf{R}), \dim V = n$.

2) The symmetry group $O(V, g)$ of a Euclidean vector space (V, g) is the subgroup of $GL(V)$ that preserves the metric: $O(V, g) \equiv \{f \in GL(V) | S^0_2(f)(g) = g\}$. For any basis on V, one has the following isomorphism:

$$O(V, g) \cong O(n) \equiv \{(a_{ij}) \in GL(n; \mathbf{R}) | (\det(a_{ij}))^2 = 1\}.$$

We call $O(n)$ the **group of orthogonal matrices**.

REMARK 1.6 - The eigenvalues equation for a linear operator $A \in L(E)$, where (E, g) is a Euclidean n-dimensional vector space, can be written, with respect to a basis $\{e_i\}$ on E, in the following way:

$$(A^j_i - \lambda \delta^j_i) v^i = 0.$$

The above equation can also be written in the following way:

$$g_{jk}(A^j_i - \lambda \delta^j_i) v^i = 0,$$

hence

$$(A_{ik} - \lambda g_{ik}) v^i = 0 \tag{1.1}$$

where $A_{ik} \equiv g_{jk} A^j_i$. This means that we have used the canonical isomorphisms $L(E) \cong E^* \bigotimes E \cong E^* \bigotimes E^*$. Equation (1.1) admits solutions iff

$$\det[A_{ik} - \lambda g_{ik}] = 0 \tag{1.2}$$

This equation preserves the form even if we change basis: $\{e_i\} \to \{\bar{e}_i\}$. In fact,

$$\det[\bar{A}_{ik} B^i_r B^k_s - \lambda \bar{g}_{ik} B^i_r B^k_s] = 0 \Rightarrow \det[(\bar{A}_{ik} - \lambda \bar{g}_{ik}) B^i_r B^k_s] = 0$$

$$\Downarrow$$

$$(\det(B^i_r))^2 \det[\bar{A}_{ik} - \lambda \bar{g}_{ik}] = 0$$

where (B^i_r) is the matrix change of basis. As $\det(B^i_r) \neq 0 \Rightarrow \det[\bar{A}_{ik} - \lambda \bar{g}_{ik}] = 0$.

1.2 - AFFINE SPACES

A fundamental structure of Mathematics and Mechanics is that of affine space. In fact, the natural framework for classical Mechanics is the Galilean space-time. This is just an affine space (endowed with an additional structure).

DEFINITION 1.9 - (**Structure of affine space**). *The structure of affine space is given by the following:*

a) $M \equiv$ *set,* (**set of points**)*;*

b) $\mathbf{M} \equiv$ *vector space,* (**space of free vectors**)*;*

c) $\alpha \equiv \mathbf{M} \times M \to M \equiv$ **translation operation**, $\alpha : (v, p) \mapsto p' \equiv p + v$. *This action is* **transitive** *and* **without fixed points**. *This means that, if we consider the pointed space* [141] (M, O), *where* $O \in M$ *is the base point, we can then associate to any other point* $p \in M$ *a vector* $v \in \mathbf{M}$ *by means of a one-to-one mapping* $M \to \mathbf{M}$.

2) *The* **tangent space** *of* M *in* $p \in M$ *is the vector space:*

$$T_pM \equiv (\mathbf{M}, p) \cong \mathbf{M} .$$

The vectors of T_pM *are called* **applied vectors** *(in* p*).*

3) *If* $\mathbf{M}$ *is a Euclidean space, the corresponding affine space is also Euclidean.*

4) *We call* **dimension** *of the affine space* $(M, \mathbf{M}, \alpha)$ *the dimension of* $\mathbf{M}$.

PROPOSITION 1.6 - *For any point* $O \in M$ *we have the canonical identification:* $M \equiv \mathbf{M}$. *Furthermore, for any couple* $(O, \{e_i\})_{1 \leq i \leq n}$, *where* $\{e_i\}$ *is a basis of* $\mathbf{M}$, *we have the isomorphism:* $M \cong \mathbf{R}^n$. *We call* $(O, \{e_i\})$ *an* **affine frame**. *Then, to any affine frame it is associated with a coordinate system* $x^\alpha : M \to \mathbf{R}$, *where* $x^\alpha(p)$ *is the* α*-th component, in the basis* $\{e_i\}$, *of the vector* $p - O$.

THEOREM 1.4 - *Let* (x^α) *and* $(\bar{x}^\alpha)$ *be two coordinate systems corresponding to the following affine frames* $(O, \{e_i\})$ *and* $(\bar{O}, \{\bar{e}_i\})$ *respectively. These are related by an affine transformation:*

$$\bar{x}^\alpha = A^\alpha_\beta x^\beta + y^\alpha \tag{1.3}$$

where

$$(y^\alpha) \in \mathbf{R}^n, (A^\alpha_\beta) \in GL(n; \mathbf{R})$$

DEFINITION 1.10 - *We call* $A(n) \equiv Gl(n; \mathbf{R}) \times \mathbf{R}^n$ **affine group of dimension** n.

THEOREM 1.5 - *The symmetry group of an* n*-dimensional affine space is called the* **affine group** $A(M)$ *of* M . *One has the following non-canonical isomorphism:*

$$A(M) \cong A(n), \quad f \mapsto (f^\alpha_\beta, y^\alpha) ; \; f^\alpha \equiv x^\alpha \circ f = f^\alpha_\beta x^\beta + y^\alpha. \tag{1.4}$$

PROOF. Let us consider the following:
LEMMA 1.1 - *Let $f : M \to N$ be a differentiable mapping between affine spaces, $\dim M = m, \dim N = n$. The derivative of f is a mapping:*

$$Df : M \to J\mathcal{D}(M,N) \equiv \bigcup_{p,q \in M \times N} T_p^*M \bigotimes T_qN \cong M \times N \times \mathbf{M}^* \bigotimes \mathbf{N}.$$

If $\{x^\alpha\}$ are coordinates on M, and $\{y^j\}$ are coordinates on N, one has the following representation:

$$Df = \sum_{1 \le \alpha \le m; 1 \le \beta \le n} (\partial x_\alpha f^\beta) dx^\alpha \otimes \partial y_\beta \circ f$$

where
a)

$$f^\beta \equiv y^\beta \circ f,$$

b) *$dx^\alpha(p)$ is the natural canonical basis induced on T_p^*M by $\{x^\alpha\}$.*
c) *$\{\partial y_\beta(f(p))\}$ is the canonical natural basis induced on $T_{f(p)}N$ by $\{y^j\}$.*
LEMMA 1.2 - *If f is an affine mapping, we get: $Df(p) \equiv Df(p')$, $\forall p, p' \in M$. As a consequence we have the identification:*

$$Df \equiv \bar{f} \in L(\mathbf{M};\mathbf{N}) \cong \mathbf{M}^* \bigotimes \mathbf{N}.$$

Let us, now, consider an affine mapping: $f : M \to M$, we have the following commutative diagram:

$$\begin{array}{ccccc} & \mathbf{M} \times M & \overset{\alpha}{\to} & M & \\ {\scriptstyle \bar{f} \times f} & \downarrow & & \downarrow & {\scriptstyle f} \\ & \mathbf{M} \times M & \underset{\alpha}{\to} & M & \end{array}$$

For any coordinate system $\{x^\alpha\}$, we get: $\bar{f}^\beta_\alpha = (\partial x_\alpha f^\beta), f^\beta \equiv x^\beta \circ f$. By integration we get: $f^\beta = \bar{f}^\beta_\alpha x^\alpha + y^\beta$. Let us put: $f^\beta_\alpha = \bar{f}^\beta_\alpha$. Therefore the proof is complete. □
THEOREM 1.6 - (SOLUTIONS OF AFFINE EQUATIONS). *Let E and F be two vector spaces of finite dimension n. Let $A \in L(E;F)$ and $f \in F$. The following equation is called* **affine equation**

$$A(u) = f. \tag{1.5}$$

With respect to a basis $\{e_i\}$ of E and one $\{\epsilon_i\}$ of F, equation (1.5) can be written in the components x^k of $u = x^k e_k$ as follows:

$$A^j_k x^k = f^j \tag{1.6}$$

where $f = f^j \epsilon_j$. We can distinguish the following cases:

(a) $\det(A^j_k) \neq 0$. *The only solution of the linear associated equation* $A(u) = 0$ *is* $u = 0$. *The mapping* A *is also an isomorphism. Then there exists a unique solution of* (1.6):

$$u = A^{-1}(f). \tag{1.7}$$

(b) $\det(A^j_k) = 0$. *Let* $n-k$ *be the rank of* A. *Then the rank of the transpose mapping* $A^* : F^* \rightarrow E^*$ *is also* $n-k$. *Then equation* $A(u) = 0$ *has* k *linearly independent solutions* $\{u^{(1)}, \cdots, u^{(k)}\}$. *The same holds for* $A^*(\alpha) = 0$: $\{\alpha^{(1)}, \cdots, \alpha^{(k)}\}$. *Then equation* (1.5) *has solutions iff* f *satisfies the following condition:*

(1.8)(**consistency condition**) $$< f, \alpha^{(i)} >= 0, 1 \leq i \leq k.$$

If the above equations hold, equation (1.5) *has an infinite number of solutions:*

$$u = \sum_{1 \leq i \leq k} c_i u^{(i)} + u_p, \tag{1.9}$$

where u_p *is any vector of the space* $A^{-1}(f) \subset E$.

EXAMPLE 1.3 - Let the affine equation $A(u) = f$ be given by the following

$$\left\{ \begin{array}{r} x^1 + x^2 + x^3 = 1 \\ 2x^1 - x^2 + x^3 = 3 \\ x^1 - 2x^2 = 1 \end{array} \right\}. \tag{1.10}$$

Equation $A^*(\alpha) = 0$ is given by

$$\left\{ \begin{array}{r} \alpha_1 + 2\alpha_2 + \alpha_3 = 0 \\ \alpha_1 - \alpha_2 - 2\alpha_3 = 0 \\ \alpha_1 + \alpha_2 = 0 \end{array} \right\}.$$

We get $\det A^* = 0, \mathrm{rank} A^* = 2$. The general solution of $A^*(\alpha) = 0$ is $c(1, -1, 1)$. We have

$$< (1, 3, 1), (1, -1, 1) >= -1 \neq 0 .$$

Therefore, the consistency condition is not verified. As a consequence (1.10) has no solutions. ■

1.3 - DIFFERENTIAL MANIFOLDS

The globalization of the concept of n-dimensional vector space is that of n-dimensional manifold. Scope of this section is just to recall some fundamental definitions and results about differential manifolds that will be useful in the following. (See also refs.[18,22,24,25,50,57,61,90,98,118,120,149]).

DEFINITION 1.11 - 1) *A* **manifold**, *of* **dimension** n *and* **class** C^k, *is defined by the following:*

(a) $M \equiv$ *set;*

(b) $\{U_i\}_{i\in I}, \bigcup_i U_i = M$, **covering** *of* M *by means of subsets* U_i *of* M, *called* **open sets**;

(c)

$$(U_i, \phi_i) \equiv (\textbf{chart}) : \left\{ \begin{array}{l} \phi_i : U_i \to \Omega_i \subset \mathbf{R}^n \\ \Omega_i \equiv \textit{open subset of} \quad \mathbf{R}^n \\ \phi_i \equiv \textit{bijective mapping} \end{array} \right\}$$

(d) *the functions* $f_{ij} \equiv \phi_i \circ \phi_j{}^{-1} : \Omega_j \to \Omega_i$, *are of class* $C^k, 0 \leq k \leq \infty, \omega$, *where* C^ω *means real analytic class.*

2) *The functions* $x^\alpha \equiv pr^\alpha \circ \phi_i : U_i \to \mathbf{R}$, *where* $pr^\alpha : \mathbf{R}^n \to \mathbf{R}$ *is the projection on the* α*-th factor, are called* **coordinates**.

3) *If* $n = 1$, *the manifold is called a* **curve**.

4) *If* $n = 2$, *the manifold is called a* **surface**.

EXAMPLE 1.4 - 1) SPHERE $S^2 \subset \mathbf{R}^3$. We can obtain a covering by means of six open hemispheres.

2) TANGENT SPACE. $TM \equiv \bigcup_{p\in M} T_pM$, where $T_pM \equiv$ **tangent space** at M in p, (T_pM is the vector space of the velocities of all curves in M passing through p). One has the covering $\{\Theta_i\}_{i\in I}$, where $\Theta_i \equiv \pi^{-1}(U_i)$, with $\pi : TM \to M$ the canonical projection, and $\{U_i\}_{i\in I}$ is the covering of M that defines the structure of differentiable manifold on M . We will denote the corresponding coordinates on TM with $\{\breve{x}^\alpha, \dot{x}^\alpha\}$, such that $\breve{x}^\alpha \equiv x^\alpha \circ \pi$, $\dot{x}^\alpha(v(p)) = v^\alpha(p)$, if $v(p) = v^\alpha(p)\partial x_\alpha(p)$, where $\partial x_\alpha(p)$ is the velocity of the α-th coordinate-line passing through p . In fact, $\{\partial x_\alpha(p)\}_{1\leq\alpha\leq n}$ is a basis for T_pM . The change of coordinates on TM is given by:

$$\breve{x}^\alpha = f^\alpha(\breve{\bar{x}}^\beta) \, , \, \dot{x}^\alpha = (\partial \bar{x}_\beta . f^\alpha)\dot{\bar{x}}^\beta \, ,$$

where f^α are the functions giving the change of coordinates on M .

3) COTANGENT SPACE. $T^*M \equiv \bigcup_{p\in M}(T_pM)^*$. Here we can make considerations similar to the previous ones. Note that each coordinate system on M induces a natural basis $\{dx^\alpha(p)\}_{1\leq\alpha\leq n}$ in $(T_pM)^*$, where $dx^\alpha : M \to T^*M$ are the differentials of the coordinates x^α. $\{dx^\alpha(p)\}_{1\leq\alpha\leq n}$ results in the dual basis of $\{\partial x_\alpha(p)\}_{1\leq\alpha\leq n}$, i.e., $< dx^\alpha(p), \partial x_\beta(p) >= \delta^\alpha_\beta$.

4) SPACE OF DIFFERENTIAL q-FORMS ON M. $\Lambda^0_q M \equiv \bigcup_{p\in M} \Lambda^q(T_pM)^*$.
Note that if $q = n = \dim(M) \Rightarrow \dim(\Lambda^q(T_pM)^*) = 1$.

5) SUBMANIFOLDS. Let M be a differential manifold of dimension n. A **submanifold** of M is a subset $N \subset M$ that is a differential manifold with respect to the structure **induced** by M. More precisely, the charts $(\bar{U}, \bar{\phi})$ on N are obtained by the charts (U, ϕ) on M in the following way:

$$\bar{U} = U \bigcap N, \bar{\phi} = \phi|_{\bar{U}} .$$

If the manifold N is locally given by the equations $F^s = 0, 1 \leq s \leq p$, where $F^s : U \to \mathbf{R}$ are local numerical functions such that the rank of the $n \times p$ matrix $(\partial x_\alpha F^s)$ is p. Then, $\dim N = n - p$.

6) (LOCALLY TRIVIAL) FIBER BUNDLES. A **fiber bundle** is a map $\pi : W \to V$ between differential manifolds W and $V, \dim(W) = m + n, \dim(V) = n$. W is called **total space**, V **base** and $W_p \equiv \pi^{-1}(p) \subset W$, $\forall p \in V$, the **fiber** over p. Furthermore, if for any open set $U \subset V, \pi^{-1}(U) \cong U \times F$, where F is a differential manifold of dimension m (**type fiber**), then we say that the fiber bundle is **locally trivial**. In this way we can define a system of **fibered coordinates** on W, (x^α, y^j), $1 \leq \alpha \leq n$, $1 \leq j \leq m$, such that y^j are coordinates on F and x^α on U respectively. A **section** of a fiber bundle $\pi : W \to V$ is a map (eventually defined on open sets $U \subset V$ only) $s : V \to W$ such that $\pi \circ s = id_V$. A **vector fiber bundle** is a (locally trivial) fiber bundle such that the fiber over p is a vector space, $\forall p \in V$. $TM \to M, T^*M \to M$ and $\Lambda^0_q M \to M$ are examples of locally trivial vector fiber bundles. Sections of these are called **vector fields, differential 1-forms** and **differential q-forms** respectively.

☐ The **pull-back** f^*W of π : W$\to V$ by means of a differentiable mapping $f : N \to V$ of class C^k is defined by

$$f^*W \equiv \{(n, q) \in N \times W | f(n) = \pi(q)\} .$$

We have the following commutative diagram:

$$\begin{array}{cccc} & f^*W & \xrightarrow{f_*} & W & \\ \pi_* & \downarrow & & \downarrow & \pi \\ & N & \xrightarrow[f]{} & V & \end{array}$$

where π_* is the projection on the first factor and f_* is the projection on the second one. In particular, if f is an **immersion**, i.e., N is **diffeomorphic** to its image $f(N)$, (i.e., in correspondence one-to-one, and f is differentiable), then f_* is also an immersion, and f^*W can be identified to a submanifold of W that is a fiber bundle over $f(N) \subset V$, contained in the fiber bundle π : W$\to V$, (**subfiber bundle**).

☐ If in a fiber bundle π : W$\to V$, to any sections $s : M \to W$ and to any mapping $f : V \to V$, we can associate a mapping $\phi : W \to W$ such that $s' \equiv \phi \circ s \circ f$ is a section of π , i.e., $\pi \circ s' = id_V$, then we say that s' is the **pull-back** of s by means of (ϕ, f) . For example: (a) We can associate to any vector field $X : M \to TM$ its pull-back f^*X by means of a diffeomorphism $f : M \to M$. More precisely, we have: $f^*X = T(f^{-1}) \circ X \circ f$. If $X = X^\alpha \partial x_\alpha$ is the local representation of X then we have:

$$f^*X = (\partial x_\alpha f^\beta) X^\alpha \partial x_\beta .$$

(b) If $\alpha : M \to \Lambda_q^0 M$ is a differential q-form on M , then the pull-back of α is defined by means of a differentiable mapping $f : M \to M$. In fact we have: $f^*\alpha = \Lambda_q^0(f) \circ \alpha \circ f$. If $\alpha = \sum_{1 \le i_1 < \cdots < i_q \le n} \alpha_{i_1 \cdots i_q} dx^{i_1} \wedge \cdots \wedge dx^{i_q}$, then

$$f^*\alpha = j(f)^{i_1 \cdots i_q}_{j_1 \cdots j_q} \alpha_{i_1 \cdots i_q} dx^{j_1} \wedge \cdots \wedge dx^{j_q} ,$$

where $j(f)^{i_1 \cdots i_q}_{j_1 \cdots j_q}$ is the minor of the jacobian matrix of f obtained taking the rows $i_1 \cdots i_q$ and the columns $j_1, \cdots, j_q$. ■

DEFINITION 1.12 - *A* **volume form** *on a manifold M of dimension n is a section $\eta : M \to \Lambda_n^0 M$, such that $\eta(p) \in \Lambda^n(T_pM)^* \backslash \{0\}$, $\forall p \in M$.*

The local expression of η is the following: $\eta = f dx^1 \wedge \cdots \wedge dx^n$, with $f : U \to \mathbf{R}\backslash\{0\}$. If a volume form can be defined on M, then we say that M is **orientable**. Otherwise we say that M is **non-orientable**. If M is orientable, and it is chosen a volume form, then we say that M is **oriented**.

EXAMPLE 1.5 - 1) The sphere S^2 in $\mathbf{R}^2$ is orientable.

2) Examples of non orientable manifolds are the following.

(a) **Möbius band**: $B \equiv S^1 \times [-1,1]/\sim$. The equivalence relation $\sim$, expressed with respect to coordinates $(\phi, t), 0 \le \phi \le 2\pi, -1 \le t \le 1$, on $B \equiv S^1 \times [-1,1]$, is the following: $(0,t) \sim (2\pi, -t)$.

(b) **Klein bottle**: $K \equiv [0,1] \times [0,1]/\sim$. The equivalence relation $\sim$, expressed with respect to coordinates $(t, \tau), 0 \le t \le 1, 0 \le \tau \le 1$ on $K \equiv [0,1] \times [0,1]$, is the following: $(t,0) \sim (1-t,1), (0,\tau) \sim (1,\tau)$.

(c) **Projective plane**: $\mathbf{R}P^2 \equiv D^2/\sim$. The equivalence relation $\sim$ in the disk D^2 defined by using the following polar coordinates in D^2: $(\rho, \theta), 0 \le \rho \le 1, 0 \le \theta \le 2\pi$, is the following: $(1,\theta) \sim (1, \theta + \pi)$. A Möbius band can be considered a neighbourhood of the projective line $\mathbf{R}P^1 \subset \mathbf{R}P^2$. ■

THEOREM 1.7 - 1) *If $\phi : M \to M$ is a diffeomorphism of M then a volume form η is transformed by means of ϕ in the following one $\phi^*\eta = \det(j(\phi)).\eta$, where $(\phi^*\eta)(p) \equiv \Lambda^n(D\phi(p)^*)(\eta(\phi(p)))$, with $\Lambda^n(D\phi(p)^*) : \Lambda^n((T_{\phi(p)}M)^*) \to \Lambda^n((T_pM)^*)$, induced by the isomorphism $D\phi(p) : T_pM \to T_{\phi(p)}M$. $j(\phi)$ is the jacobian of ϕ .*

2) (VOLUME FORM OF A RIEMANN MANIFOLD). *Let (M, g) be a* **Riemann manifold,** *i.e., on M is defined a differentiable mapping $g : M \to S_2^0(M)$, such that*

$g(p)$, $\forall p \in M$, is a scalar product on the tangent space T_pM. Then on M we can also define a canonical volume form locally given by

$$\eta(g) = \sqrt{\det g_{ij}} dx^1 \wedge \cdots \wedge dx^n .$$

REMARK 1.7 - The expression in the above equation does not change with the change of coordinates.

THEOREM 1.8 - 1) *Let $D \subset M$ be a regular domain in an n-dimensional manifold M, such that $D \subset U$ is contained in an open set U of a chart (U, ϕ) of M. Then, we can define, for any volume form η on M a measure of D in the following way:*

$$\mu(D) = \int_D \eta = \int_{\phi(D)} f \circ \phi^{-1} dx^1 \cdots dx^n. \tag{1.11}$$

where $f dx^1 \wedge \cdots \wedge dx^n$ is the local expression of η in the chart (U, ϕ). The integral in (1.11) is the usual Riemann integral in a domain of $\mathbf{R}^n$.

2) If M is a Riemann manifold there is a canonical measure on M.

REMARK 1.8 - If D is not contained in a chart, we can extend the above result, by means of a proceeding of soldering by using numerical functions, $\{g_U\}$, (**partition of unity subordinate to the covering** $\{U\}$ on M). More precisely,

$$\int_D \eta = \sum_U \int_{D \bigcap U} g_U \eta, \sum_U g_U = 1, \operatorname{supp}(g_U) \subset U .$$

(For more informations see refs.[57,118,135,157].)

EXAMPLE 1.6 - 1) (MEASURE OF LINE). Let $\gamma : \mathbf{R} \to M$ be a line in a Riemann manifold (M, g). Then, we can induce a metric, i.e., a Riemann structure, on $\mathbf{R}$ by means of γ. More precisely: $\gamma^* g \equiv S^0_2(\gamma) \circ g \circ \gamma : \mathbf{R} \to S^0_2 \mathbf{R}$. As a consequence on $\mathbf{R}$ we recognize a canonical volume form

$$\eta(\gamma^* g) = \sqrt{g_{ij} \dot{x}^i \dot{x}^j} dt .$$

2) (AREA OF SURFACE). Let $\gamma : D \subset \mathbf{R}^2 \to M$ be a surface on a Riemann manifold (M, g). Similarly to previous example we can recognize on D a canonical volume form induced by γ. More precisely, we have

$$\eta(\gamma^* g) = \sqrt{EG - F^2} du \wedge dv ,$$

with

$$(\textbf{Gauss-symbols}) : \left\{ \begin{array}{l} E \equiv g_{ij}(\partial u.\gamma^i)(\partial u.\gamma^j) \\ F \equiv g_{ij}(\partial u.\gamma^i)(\partial v.\gamma^j) \\ G \equiv g_{ij}(\partial v.\gamma^i)(\partial v.\gamma^j) \end{array} \right\}$$

■

REMARK 1.9 - (SYMPLECTIC MANIFOLDS). On a vector space V, of even dimension $2n$, we can define a **symplectic form**, i.e., a mapping $\sigma : V \times V \to \mathbf{R}$, skew-symmetric, and non-degenerate. Then, (V, σ) is called a **symplectic space**. With respect to a basis $\{e_i\}$ of V, σ can be represented in the following way:

$$\sigma = \sigma_{ij} \theta^i \wedge \theta^j ,$$

where $\{\theta^i\}$ is the dual basis of $\{e_i\}$. Furthermore, σ_{ij} is a $2n \times 2n$ skew-symmetric matrix ($\sigma_{ij} = -\sigma_{ji}$), and non-degenerate, $\det(\sigma_{ij}) \neq 0$. If (V, σ) is a symplectic vector space, then there is a canonical isomorphism $V \to V^*, v \mapsto v \rfloor \sigma$, with $\sigma \in \Lambda^0_2(V)$. Moreover, we can find **canonical bases**, $\{\bar{e}_i\}$, i.e., such that $\bar{\sigma}_{ij} \equiv \sigma(\bar{e}_i, \bar{e}_j)$ assumes the following form:

$$(\bar{\sigma}_{ij}) = \begin{pmatrix} 0 & 1 & & & & & \\ -1 & 0 & 1 & & & \bigcirc & \\ & -1 & 0 & & & & \\ & & & \ddots & & & \\ & & & & 0 & 1 & \\ & \bigcirc & & & -1 & 0 & 1 \\ & & & & & -1 & 0 \end{pmatrix} .$$

The group of isomorphisms of (V, σ), (**symplectic isomorphisms**), is a subgroup $Sp(V, \sigma)$ of $GL(V)$, and it is called **symplectic group** of (V, σ). On a differential manifold X, of dimension $2n$, we can define a **symplectic structure**, i.e., a section σ of $\Lambda^0_2 X \setminus \{0\}$ on X, such that $\sigma(p) \in \Lambda^0_2(T_p X)$ is a symplectic structure on $T_p X$, $\forall p \in X$.

REMARK 1.10 - (**LIE GROUPS AND LIE ALGEBRAS**).(See also ref.[86].)

1) Important examples of differential manifolds are the Lie groups. A **Lie group** of dimension n and class C^k is a differential manifold G of dimension n that is a group, i.e., there is defined a **product** $G \times G \to G, (a, b) \mapsto ab$, for any $a, b \in G$, and a **unity** $e, ae = ea = a$, $\forall a \in G$, and any element $a \in G$ admits an **inverse** $a^{-1}, aa^{-1} = a^{-1}a = e$. Furthermore, the product mapping and taking the inverse, are differentiable of class $C^k, 0 \leq k \leq \infty, \omega$. An important example of Lie group is the general linear group $GL(n, \mathbf{R})$. In the following we give some other important examples.

2) (GLOBAL STRUCTURE FOR LIE GROUP). (a) Any Lie group G contains a **discrete number of unconnected components**; the component G_0 connected with the unity is an invariant subgroup, i.e., $aG_0a^{-1} = G_0$, $\forall a \in G$, and the other components are its cosets. In other words, there are as many unconnected components as are the elements of the quotient group $D \equiv G/G_0$ that is discrete.

(b) G is not necessary a (semi-)direct product [14] of the form $G = D \times_r G_0$, but this often happens. For example: $O(2n+1) = \mathbf{Z}_2 \times SO(2n+1), O(2n) = \mathbf{Z}_2 \times_r SO(2n)$.
(c) The connected component of the unity is the homeomorphic image (i.e., has the same structure of the differential manifold up the class) of a simply connected (i.e., without holes) covering group $\widetilde{G}_0, \pi : \widetilde{G}_0 \to G_0$, such that the kernel of π is a discrete group Z contained in the center $Z(\widetilde{G}_0)$ of $\widetilde{G}_0$, namely in the part of $\widetilde{G}_0$ where the product is commutative. Then, we have: $G_0 \cong \widetilde{G}_0/Z$.
3) (LIE ALGEBRA OF LIE GROUP). On the tangent space in the unity, T_eG, of a Lie group G, we can define a structure of Lie algebra, i.e., a bracket

$$[,] : T_eG \times T_eG \to T_eG, \quad (Z_1, Z_2) \mapsto [Z_1, Z_2],$$

such that the following relations are satisfied: (a)(anticommutativity)

$$[Z_1, Z_2] + [Z_2, Z_1] = 0, \quad \forall Z_1, Z_2 \in T_eG;$$

(b)(Jacobi identity)

$$[Z_1, [Z_2, Z_3]] + [Z_2, [Z_3, Z_1]] + [Z_3, [Z_1, Z_2]] = 0, \quad \forall Z_1, Z_2, Z_3 \in T_eG.$$

In fact, to each vector $Z \in T_eG$ we can associate a vector field $\sigma(Z) : G \to TG$ induced by the product of $G : \sigma(Z)(p) = D\phi_a(e)(Z) \in T_{\phi_a(e)}G = T_aG$. Then,

$$[Z_1, Z_2] = [\sigma(Z_1), \sigma(Z_2)](e) \in T_eG,$$

where $[\sigma(Z_1), \sigma(Z_2)] \equiv \mathcal{L}_{\sigma(Z_2)}\sigma(Z_1) \equiv$ **Lie derivative** of $\sigma(Z_1)$ with respect to $\sigma(Z_2)$. Namely,

$$\mathcal{L}_{\sigma(Z_2)}\sigma(Z_1) : G \to TG, \mathcal{L}_{\sigma(Z_2)}\sigma(Z_1)(a) = \lim_{\lambda \to 0} \frac{\psi_\lambda^*\sigma(Z_1)(a) - \sigma(Z_1)(a)}{\lambda},$$

where $\psi_\lambda^*\sigma(Z_1)(a) \equiv D(\psi_\lambda^{-1})(\psi_\lambda(a))(\sigma(Z_1)(\psi_\lambda(a)))$, with $\psi_\lambda : G \to G$, $\forall\lambda \in \mathbf{R}$, the 1-parameter group of diffeomorphisms of G generated by $\sigma(Z_2)$. In other words, ψ_λ satisfies the following system of ordinary differential equations: $\dot{\psi}_\lambda^\alpha = \sigma(Z_2)^\alpha, 1 \leq \alpha \leq n = \dim G$. Denote by $A(G)$, (or g, or $\mathbf{g}$) the Lie algebra of G. Then, if $\{Z_a\}_{1 \leq a \leq n}$ is a basis for $A(G)$, we can write

$$[Z_a, Z_b] = C_{ab}^c Z_c,$$

where $C_{ab}^c \in \mathbf{R}$ are called **structure constants** of G. They satisfy the following relations:

(antisymmetry)

$$C^c_{ab} = -C^c_{ba}\,,$$

(Jacobi identity)

$$C^k_{ls}C^s_{mn} + C^k_{ms}C^s_{nl} + C^k_{ns}C^s_{lm} = 0\,.$$

Vice versa one can prove that by assigning some constants satisfying the above conditions, they identify the connected component of the unity of a Lie group.
In the following we give some important examples of Lie algebras.
EXAMPLE 1.7 - (IMPORTANT LIE GROUPS (AND LIE ALGEBRAS) OF **K**-MATRICES, WITH $\mathbf{K} = \mathbf{R}, \mathbf{C}$).
1) $GL(n;\mathbf{K})$: all the $n \times n$ matrices with entries in $\mathbf{K}$ and with det $\neq 0$; $gl(n;\mathbf{K})$: all the $\mathbf{K} - n \times n$-matrices.
2) $SL(n;\mathbf{K})$: all the $n \times v$ matrices with entries in $\mathbf{K}$ and det $= 1$; $sl(n;\mathbf{K})$: all the $\mathbf{K} - n \times n$-matrices with tr $= 0$.
3) $D(n;\mathbf{K})$: all the diagonal, unitary $\mathbf{K} - n \times n$-matrices with det $\neq 0$; $d(n;\mathbf{K})$: all the diagonal $\mathbf{K} - n \times n$-matrices.
4) $T(n;\mathbf{K})$: all the upper triangular $\mathbf{K} - n \times n$-matrices with det $\neq 0$; $t(n;\mathbf{K})$: all the upper triangular $\mathbf{K} - n \times n$-matrices.
5) $T_0(n;\mathbf{K})$: all the upper-triangular, unit diagonal $\mathbf{K} - n \times n$-matrices; $t_0(n;\mathbf{K})$: all the strictly upper triangular $\mathbf{K} - n \times n$-matrices.
6) $U(n), \mathbf{K} = \mathbf{C}$: all the unitary $\mathbf{K} - n \times n$-matrices; $u(n)$: all the skew-hermitian $n \times n$-matrices with.
7) $SU(n), \mathbf{K} = \mathbf{C}$: all the unitary $n \times n$-matrices with det $= 1$; $su(n)$: all the skew-hermitian $n \times n$-matrices with tr $= 0$.
8) $O(n;\mathbf{K})$: all the orthogonal $\mathbf{K} - n \times n$-matrices; $o(n;\mathbf{K})$: all the skew-symmetric $\mathbf{K} - n \times n$-matrices.
9) $SO(n;\mathbf{K})$: all the orthogonal $\mathbf{K} - n \times n$-matrices with det $= 1$; $so(n) = 0(n;\mathbf{K})$.
10) $Sp(2n;\mathbf{K})$: all the unitary, symplectic $\mathbf{K} - 2n \times 2n$-matrices M (i.e., $M^tJM = J, t \equiv$ transpose,

$$J \equiv \begin{pmatrix} 0 & \mathbf{1} \\ -\mathbf{1} & 0 \end{pmatrix}, \quad \mathbf{1} \equiv \text{unit } n \times n \text{-matrix}$$

$$sp(2n;\mathbf{K}) = \text{all the skew-hermitian, symplectic } n \times n \text{ matrices}$$
$$(L^tJ + JL = 0)$$

11) $SU(p,q)$: all the pseudounitary, unimodular $n \times n$-matrices M, i.e.,

$$M^+IM = I, \quad I \equiv \begin{pmatrix} \mathbf{1}_p & 0 \\ 0 & -\mathbf{1}_q \end{pmatrix}$$

$su(p,q) =$ all the pseudo-skew-hermitian $n \times n$ matrices : $(L^+I + I)$, tr $= 0$.

Here $\mathbf{1}_p$ is the p-dimensional unit $n \times n$ matrix.
12) $O(p,q)$: pseudo-orthogonal (i.e., $M^t I M = I$) ; $o(p,q) = so(p,q)$: all the pseudo-skew-symmetric ($L^t I + IL = 0$) $n \times n$ matrices. ∎

4)(GENERAL DEFINITIONS RELATED TO LIE ALGEBRAS). (a) A **subspace** K of a Lie algebra L is a subspace L that is closed with respect to the commutation, namely $X, Y \in K \Rightarrow [X,Y] \in K$. (We write $[X,Y] \subset K$.)
(b) A subalgebra K of a Lie algebra L is called **invariant** if $[L,K] \subset K$.
(c) A Lie algebra L is a **direct sum** of two subalgebras: $L = K \bigoplus M$, if this happens for the underlying vector spaces and if $[M,K] = 0$.
(d) A Lie algebra L is a **semidirect sum** of two subalgebras: $L = K \times_r M$, if $[K,L] \subset K$.
☐ Alternatively, an invariant subalgebra $K \subset L$ and its complement $M \equiv L - K$, form a semidirect sum iff M is a subalgebra.
☐ The semi-direct sum becomes a direct sum iff both subalgebras are invariant.
(e) A Lie algebra L is **abelian** if $[X,Y] = 0$, $\forall X, Y \in L$.
(f) The **center** $Z(L)$ of a Lie algebra L is the abelian subalgebra such that $[Z(L), L] = 0$, Then, we can write: $L = Z(L) \bigoplus L_0$, where the subalgebra L_0 has no center.
(g) A Lie algebra is **semisimple** if it has not invariant abelian subalgebras.
(h) A Lie algebra is **simple** if it is semisimple and does not contain invariant subalgebras.
☐ One can prove that semisimple algebras are direct sum of simple ones.
(i)(CARTAN METRIC OF LIE ALGEBRA). On a Lie algebra L we can define a symmetric linear mapping:

$$k \in L^2_s(L) \cong L^* \odot L^* ,$$

such that in any basis we have the following matrix-representation:

$$k_{ab} = C^d_{ac} C^c_{bd} ,$$

where C^d_{ac} are the constant structures of L . If L is abelian, $k = 0$. Furthermore, k can be, in general, degenerate. k is not-degenerate, $\det k \neq 0$, iff L is semisimple.
(l) A Lie algebra is called **compact** if the corresponding Lie group is such that it can be covered by a finite number of open sets.
☐ We can prove that if L is compact one can write $L = \bigoplus_{1 \leq i \leq s} L_i$, where L_i are **irreducible compact** Lie algebras, where irreducible means simple or 1-dimensional.
☐ As a consequence, each compact Lie group G can be written as follows: $G = \prod_{1 \leq i \leq s} G_i$, where G_i are simple compact Lie groups or such that $G_i / Z_i \cong U(1)$, where Z_i is a discrete group.

(m) A simple compact Lie algebra, can be identified (faithful representation) with a **Cartan-form** of the Lie algebra such that its generators

$$(H_i, E_{\pm\alpha}), \alpha = (\alpha^{(1)}, \cdots, \alpha^{(l)})$$

($\alpha \equiv$ **roots**), satisfy the following commutation relations:

$$\begin{aligned} [H_i, H_j] &= 0 \\ [H_i, E_{\pm\alpha}] &= \pm\alpha^{(i)} E_{\pm\alpha} \\ [E_\alpha, E_{-\alpha}] &= \alpha^{(i)} H_i \\ [E_\alpha, E_\beta] &= N_{\alpha\beta} E_{\alpha+\beta} \\ i, j &= (1, \cdots, l), \alpha + \beta = 0\,, \end{aligned}$$

where H_i and E_α are orthonormal: $\mathrm{tr} H_i H_j) = \delta_{ij}, \mathrm{tr}(E_\alpha E_\beta) = \delta_{\alpha\beta}, E^+_\alpha = E_{-\alpha}$. The number l is called the **rank** of the algebra. The roots are $\frac{1}{2}(r - l), r = \dim A(G)$.

☐ The **Weyl group** of a simple compact Lie algebra is defined by $W \equiv N/C$, where N(= **normalizer**) is the maximum subgroup of the covering group $\widetilde{G}$ that preserves the Cartan algebra, and C (= **centralizer**) is the maximum subgroup of $\widetilde{G}$ that preserves a separated element of the Cartan algebra. Then, we can write $C = e^{i\alpha_s H_s}$.

☐ The roots α can be considered as vectors of a Euclidean vector space of dimension l. In fact, they can be expressed as linear combinations of some l ones (**primitive roots**):

$$\alpha = \pm \sum_{1 \leq i \leq l} n_i \alpha_i \quad , \quad n_i \in \mathbf{N} \bigcup \{0\} \quad , \quad \alpha_i \equiv \textbf{primitive roots}\,.$$

The matrix $C_{ij} \equiv 2 < \alpha_i, \alpha_j > / < \alpha_j, \alpha_j >, i, j = 1, \cdots, l$, is called matrix of **Cartan** (or matrix of Coxeter if we put on the denominator $< \alpha_i, \alpha_i >$).

1.4 - GRASSMANN MANIFOLDS

Grassmann manifolds are important objects in any geometric theory of PDEs. In fact the set of integral planes, at any point of a PDE, are examples of such structures. So, in this section we shall consider in some details the differential geometry of Grassmann manifolds.

DEFINITION 1.13 - 1) *The* **Grassmann manifolds**

$$G_{l,l+p}(E)$$

is the set of l-dimensional planes that pass through the origin $0 \in E$ in the $(l+p)$-dimensional Euclidean space E.
2) *If we take the oriented l-dimensional planes, then we get the* **oriented Grassmann manifold** $G^+_{l,l+p}(E)$.
3) *Complex and quaternionic Grassmann manifolds $\mathbf{C}G_{l,l+p}$ and $\mathbf{H}G_{l,l+p}$ can be similarly defined by replacing E by $\mathbf{C}^{l+p}$ and $\mathbf{H}^{l+p}$.*
REMARK 1.11 - In the following, unless otherwise stated, by a Grassmann manifold we shall mean a real Euclidean Grassmann manifold.
THEOREM 1.9 - *$G_{l,l+p}(E)$ is an analytic manifold of dimension lp.*
PROOF. Let us cover $G_{l,l+p}(E)$ with subsets U_α, where

$$U_\alpha \equiv \left\{ \begin{array}{l} \text{set of all } l\text{-dimensional planes whose projection onto a given} \\ l\text{-dimensional plane } \pi_\alpha \text{ is non-degenerate} \\ (\text{and preserves the orientation for } G^+_{l,l+p}(E)) \end{array} \right\}.$$

If we choose an orthonormal basis in E so that the plane π_α is given by the equations $x^{l+\mu} = 0, \mu = 1, 2, \cdots, p$, we see that the l-dimensional planes that are in the corresponding chart U_α are defined by the following equations:

$$x^{l+\mu} = \sum_{1\le j\le l} \xi^\mu_j x^j, \quad 1 \le \mu \le p, \xi^\mu_j \in \mathbf{R}.$$

So, the lp numbers ξ^μ_j can be taken as local coordinates on $G_{l,l+p}(E)$ in the neighbourhood of the point under consideration $(\xi^\mu_j(\pi_\alpha = 0))$. Furthermore, it is easy to verify that the functions change of coordinates are analytic ones. □
DEFINITION 1.14 - *We say that two planes $\tau, \tau' \in G_{l,l+p}(E)$ are* **close** *if the projection of each onto the other is non-degenerate.*
REMARK 1.12 - (**Another coordinate system on** $G_{l,l+p}(E)$). Fix in E an orthonormal basis $\{e_i\}_{1\le i\le l+p}$, such that the vectors $e_1, \cdots, e_l$ lie in the given plane π_α and the vectors $e_{l+1}, \cdots, e_{l+p}$ are orthogonal to it. We say that the plane π_α is **associated** with the basis $\{e_i\}_{1\le i\le l+p}$. Let us, now, consider the plane π associated with the orthonormal basis $\{f_i\}_{1\le i\le l+p}$. Let us write them in the basis $\{e_i\}$:

$$f_i = \Xi^j_i e_j, \tag{1.12}$$

where

$$\left(\Xi^j_i\right) = \left(\begin{array}{c|c} A & B \\ \hline C & D \end{array}\right)$$

where A, B, C, D are matrices respectively $l \times l, l \times p, p \times l, p \times p$.

$\square$ Let us assume $\{f_i\}$ be such that

$$f_i \bullet e_j = f_j \bullet e_i, \left\{ \begin{array}{ll} \text{if} & 1 \leq i,j \leq l, \\ \text{or} & l+1 \leq i,j \leq l+p \end{array} \right\}.$$

[This corresponds to decompose the matrices A and D into a symmetric matrix and an orthogonal matrix.]

$\square$ Furthermore assume that $f_i = e_i$ if $\pi = \pi_\alpha$.

$\square$ The correspondence between basis $\{e_i\}$ and plane π_α is assumed to be continuous. Then relation (1.12) assumes the form

$$f_i = \Xi_i^j e_j, \tag{1.13}$$

where

$$\left(\Xi_i^j\right) = \left(\begin{array}{c|c} 1+A & B \\ \hline C & 1+D \end{array}\right)$$

with $A_i^j = A_j^i$, $\forall i,j = 1,\cdots,l; D_i^j = D_j^i$, $\forall i,j = l+1,\cdots,l+p$. Now, taking into account that $\{f_i\}$ are orthonormal, we can represent A, C, D in terms of B , i.e., $l.p$-numbers: $\{B_i^j\}$. These form the local coordinates in the neighbourhood of the point π_α (where $B_i^j(\pi_\alpha) = 0$).

$\square$ The relation between this system of coordinates on $G_{l,l+p}(E)$ and the previous one is given by $B = \xi Q$, where Q is the unique symmetric $(l \times l)$-matrix such that $QQ^* = (I_l + \xi^* \xi)^{-1}$, where I_l is the $(l \times l)$-matrix identity. One has $Q = I_l$ if $\xi = 0$, and Q depends continuously on ξ .

In the following table we report some examples of Grassmann manifolds.

TAB.1.2 - **Examples of Grassmann manifolds**

l	$G_{l,p}(E)$	$G^+_{l,p}(E)$
0	$G_{0,p}(E) \cong pt$	$G^+_{0,p}(E) \cong S^0$
1	$G_{1,p+1}(E) \cong \mathbf{R}P^p$	$G^+_{1,p+1}(E) \cong S^p$
1	$\mathbf{C}G_{1,p+1} \cong \mathbf{C}P^p$	$\mathbf{H}G^+_{1,p+1} \cong \mathbf{H}^p$

THEOREM 1.10 - 1) *There is a canonical diffeomorphism (as smooth manifolds):*

$$G_{l,l+p}(E) \cong G_{p,l+p}(E), \quad F \mapsto F^\perp .$$

2) *Similarly,*

$$G^+_{l,l+p}(E) \cong G^+_{p,l+p}(E), \quad F \mapsto F^\perp .$$

3) *One has the following covering:* $\pi : G^{+}_{l,l+p}(E) \to G_{l,l+p}(E)$, *where* π *is defined by taking away the orientation of the planes. The fiber of* π *is* $\mathbf{Z}_2$.
4) *The Grassmann manifolds are compact and connected (except for* $G^{+}_{0,p} = S^0$ *).*
5) *One has the following:*

$$\pi_1(G^{+}_{l,l+p}(E)) = \left\{ \begin{array}{ccc} 0 & \textit{if} & (l,p) \neq (1,1) \\ \mathbf{Z}_2 & \textit{if} & (l,p) = (1,1) \end{array} \right\}, \quad \pi_1(G_{l,l+p}(E)) = \mathbf{Z}_2 \quad \textit{if} \quad l,p > 0\,,$$

where $\pi_1(-)$ *denotes the first homotopy group of* $(-)$*. (See ref.* [141] *and sections below).*
6) *A Grassmann manifold has a standard cell decomposition into* **Schubert cells***. More precisely, let* $E^0 \subset W^1 \subset \cdots \subset E, \dim E^i = i$, *be a filtration of* E *into subspaces. We define the numbers*

$$m_j(\pi) = \inf_m \{m : \dim(\pi \cap E^m) \geq j\}\,,$$

for an l*-dimensional plane* $\pi \subset E$ *passing from zero; (here* $m_1(\pi) < m_2(\pi) < \cdots < m_l(\pi)$*). Then,*

$$\left\{ \begin{array}{l} C_{m_1 \cdots m_l} = \{\pi \in G_{l,l+p}(E) | m_j(\pi) = m_j\}, m_1 < \cdots < m_l \\ \dim C_{m_1 \cdots m_l} = \displaystyle\sum_{s \leq j \leq l} m_j - \frac{l(l+1)}{2}. \end{array} \right\}$$

☐ *There is a single cell of dimension* 0*: this is* $C_{12\cdots l}$.
☐ *There is a single cell of dimension* lp*: this is* $C_{p+1\cdots p+l}$.
☐ $G_{l,l+p}(E)$ *decomposes into* $\binom{l+p}{l}$ *cells.*
☐ *The homology and cohomology* ([22, 149, 157], *sec.1.5) of Grassmann manifolds can be computed with the help of the cell decomposition. For example:*

$$H^k(G_{l,l+p}(E); \mathbf{Z}_2) \cong H_k(G_{l,l+p}(E); \mathbf{Z}_2) = \bigoplus_{N_k} \mathbf{Z}_2\,,$$

where N_k *is the number of Scubert cells of dimension* k .
☐ *One can see that*

$$H^k(\mathbf{C}G_{l,l+p}; G) \cong H_k(\mathbf{C}G_{l,l+p}; G) = \bigoplus_{N_k} G\,,$$

where G *is an Abelian group. All the cells have even dimension.*
REMARK 1.13 - The Grassmann manifolds play a central role in constructing the theory of characteristic classes. (See refs.[84,141] and below.) This is explained by the

fact that it is possible to construct a tautological vector bundle over them: π itself, considered as a linear l-dimensional space, is the fiber over the point $\pi \in G_{l,l+p}(E)$. Such a vector bundle is universal in the sense that it contains any vector bundle over a paracompact base as a subbundle.

THEOREM 1.11 - 1) (PLÜCKER COORDINATES). *Grassmann manifolds can be analytically embedded in* $\mathbf{R}P^{\binom{l+p}{l}-1}$ *in* **non-orientable** *case, and in* $S^{\binom{l+p}{l}-1}$ *in the* **orientable** *case.*

PROOF. The **Plücher coordinates** $P_{i_1\cdots i_l}, 1 \leq i_1 < \cdots < i_l \leq l+p$, of the l-dimensional plane $\pi \subset E$ are defined as the minors of order p in the $l \times (l+p)$-matrix whose rows are the coordinates of the vectors of a basis for the plane π in E. They are skew-symmetric in all pairs of indices and satisfy the Plücher relations:

$$P_{i_1\cdots i_{l-1}[i_l}P_{j_1\cdots j_l]} = 0\,.$$

Under a change of the basis in the plane π the Plücher coordinates are multiplied by the determinant of the transformation matrix. Thus, corresponding to every l-dimensional plane $\pi \subset E$, one can consider the collection of Plücher coordinates with

$$\sum_{i_1<\cdots<i_p} P^2_{i_1 i_2\cdots i_l} = 1\,.$$

This relation is one-to-one if we consider planes without orientation, and so this gives an embedding

$$G_{l,l+p}(E) \to \mathbf{R}P^{\binom{l+p}{l}-1}$$

and it is two-sheeted if we consider oriented planes, and so we get the embedding

$$G^+_{l,l+p}(E) \to S^{\binom{l+p}{l}-1}\,.$$

□

2) *The Plücher embedding is intrinsically defined by the following mapping*

$$\boxed{\begin{aligned} G^+_{l,l+p}(E) &\to \mathbb{P}(\Lambda^l(E)) \\ \pi &\mapsto \Lambda^l(\pi) \end{aligned}}$$

3) *One has the following embeddings:*

$$\boxed{\begin{aligned} &G_{l,l+p}(E) \to E^N, \text{with} \quad N = \left(\binom{l+p}{l} - 1\right)\binom{l+p}{l}/2 \\ &G^+_{l,l+p}(E) \to E^N, \text{with} \quad N = \binom{l+p}{l} \end{aligned}}$$

PROOF. This is obtained by considering the composition

$$\begin{array}{ccc} G_{l,l+p}(E) & \rightarrow & \mathbf{R}P^{\binom{l+p}{l}-1} \\ & \searrow & \downarrow \\ & & E^N \end{array}$$

and similarly for $G^+_{l,l+p}(E)$. □

EXAMPLE 1.8 - In the case $l = p = 2$ there is a simple Plücher relation

$$P_{12}P_{34} - P_{13}P_{24} + P_{14}P_{23} = 0 .$$

If we add to this equation the equation of the sphere $\sum_{i<j} P^2_{ij} = 1$, we see that $G^+_{2,4}(E)$ decomposes into the product of two spheres

$$G^+_{2,4}(E) \cong S^2 \times S^2$$

as a smooth manifold. In fact, if we put

$$\rho^1 = P_{12} + P_{34}, \quad \rho^2 = P_{13} - P_{24}, \quad \rho^3 = P_{14} + P_{23};$$
$$\sigma^1 = P_{12} - P_{34}, \quad \sigma^2 = P_{13} + P_{24}, \quad \sigma^3 = P_{14} - P_{23};$$

we see that the Plücher relation and the condition $\sum_{i<j} P^2_{ij} = 1$, are equivalent to $\sum(\rho^i)^2 = \sum(\sigma^i)^2 = 1$. ■

DEFINITION 1.15 - *The* **angles** $0 \leq \beta_1 \leq \cdots \leq \beta_l \leq \pi/2$ *between* $\pi, \tau \in G_{l,l+p}(E)$ *are the stationary values of the angle between a pair of vectors* $u \in \pi$ *and* $v \in \tau$.

PROPOSITION 1.7 - (PROPOSITION ON THE ANGLES). 1) *Let* P_π *and* P_τ *denote the orthogonal projections; let* $u_1, \cdots, u_l \in \pi$, *(resp.* $v_1, \cdots, v_l \in \tau$*), be orthonormal eigenvectors of the positive semidefinite symmetric operator* $P_\pi P_\tau : \pi \rightarrow \pi$ *(resp.* $P_\tau P_\pi : \tau \rightarrow \tau$*), then the angles* $\{\beta_i\}_{1\leq i\leq l}$ *are just the angles between the vectors* u_i *and* v_i *(possibly after renumbering and replacing some vectors by their negative ones). So, there are orthonormal bases* $\{u_i\}_{1\leq i\leq l} \subset \pi$ *and* $\{v_i\}_{1\leq i\leq l} \subset \tau$ *such that*

$$\boxed{< v_i, v_j >= \cos\beta_i\, \delta_{ij}\,, \quad \beta_i \in [0, \pi/2]}$$

2) *The* l *angles* $\{\beta_i\}_{1\leq i\leq l}$ *determine the relative position of the planes* π *and* τ *completely.*

DEFINITION 1.16 - *The* **distance** ρ *between the planes* π *and* τ *is*

$$\rho^2(\pi, \tau) = \sum_{1\leq i\leq l} \beta_i^2 .$$

THEOREM 1.12 - 1) *$G_{l,l+p}(E)$ is a Riemann manifold, where the metric g in coordinates*

$$\xi = (\xi_i^\mu)_{1 \le i \le l, 1 \le \mu \le p}$$

is given by

$$g_{\binom{\mu}{i}\binom{\nu}{j}} = \left((I_p + \xi\xi^*)^{-1}\right)_\mu^\nu \left((I_l + \xi^*\xi)^{-1}\right)_j^i .$$

In particular, at the point $\xi = 0$ one has $g_{\binom{\mu}{i}\binom{\nu}{j}} = \delta_\mu^\nu \delta_i^j$.
2) *The metric on $\mathbf{C}G_{l,l+p}$ is introduced in exactly the same way: the angles are taken in the sense of the Hermitian scalar product in $\mathbf{C}^{l+p}$* [102, 161]; *(the transformation is replaced by conjugation). Then, $\mathbf{C}^{l+p}$; with this metric is a Kähler manifold.* [101, 159]
3) *Similar considerations can be made for quaternionic Grassmann manifolds.*
4) *On the manifold $G_{l,l+p}^+(E)$ a Riemann structure can be induced by using the covering mapping*

$$G_{l,l+p}^+(E) \to G_{l,l+p}(E) .$$

5) *In the local coordinates $(B_i^j)_{1 \le i \le l, l+1 \le j \le l+p} \equiv (B_i^\alpha)_{1 \le i \le l, 1 \le \alpha \le p}$ the components of the metric tensor are given by*

$$g_{\binom{\alpha}{i}\binom{\beta}{j}} = \delta^{\alpha\beta}\delta_{ij} + \eta_i^\alpha \eta_j^\beta + O(\eta^4) .$$

6) *The curvature tensor* [18, 149] *can be expressed in terms of coordinates ξ. In particular for $\xi = 0$ all the Christoffel symbols* [18, 146] *vanish and for the curvature tensor one has:*

$$R_{\binom{\mu}{i}\binom{\nu}{j}\binom{\rho}{k}\binom{\lambda}{r}}|_{\xi=0} = \delta_i^j \delta_k^r (\delta_\rho^\mu \delta_\lambda^\nu - \delta_\lambda^\mu \delta_\rho^\nu) + \delta_\lambda^\mu \delta_\lambda^\rho (\delta_i^k \delta_j^r - \delta_i^r \delta_j^k) .$$

7) *The Grassmann manifold is a locally symmetric space: $\nabla R = 0$, where ∇ is the canonical absolute differential, (see refs.* [18, 98, 157] *and below).*
PROOF. In fact as

$$(\nabla_{\binom{\sigma}{h}} R)_{\binom{\mu}{i}\binom{\nu}{j}\binom{\rho}{k}\binom{\lambda}{r}}|_{\xi=0} = 0$$

that is, the curvature tensor [18, 149] is covariantly constant at the point 0, and hence, since the metric we have introduced is invariant $\Rightarrow \nabla R = 0$. □
8) *The sectional curvature* [57, 135, 149] *is given by*

$$\bar{\kappa}(X, Y) = \frac{\frac{1}{2}\mathrm{tr}(\Lambda_1 \Lambda_1^*) + \frac{1}{2}\mathrm{tr}(\Lambda_2 \Lambda_2^*)}{\mathrm{tr}(XX^*)\mathrm{tr}(YY^*) - (\mathrm{tr}(XY^*)^2)}$$

where $X = (x_i^\mu), Y = (Y_i^\mu)$ are two non-collinear tangent vectors to $G_{l,l+p}(E)$ at the point $\xi = 0$, written as $(l \times p)$-matrices, and $\Lambda_1 \equiv XY^* - YX^*, \Lambda_2 \equiv X^*Y - Y^*X$ are skew-symmetric matrices.

9) *The curvature of the Grassmann manifold is non-negative and does not exceed* 2.
10) *The Grassmann manifold is a Euclidean manifold, i.e., for the Ricci curvature* $R_{..}$ *one has:*

$$R_{..} = (l + p - 2)g ,$$

so the **scalar curvature** *is equal to* $(l + p.2)p$.
11) *In terms of local coordinates* $\xi = (\xi_i^\mu)$ *the equation for geodesics* ([18, 135, 149], *sec.4.8) on the Grassmann manifold* $G_{l,l+p}(E)$ *has the form:*

$$\ddot{\xi} - 2\dot{\xi}\xi^*(I + \xi\xi^*)^{-1}\dot{\xi} = 0 .$$

☐ *The curvature will be geodesic iff it forms a one-parameter family of l-planes in E such that the angles between any two nearly planes are proportional to a fixed collection of numbers* $\{a_i\}_{1 \le l \le l}, a_i \ge 0$.
☐ *Any two points on* $G_{l,l+p}(E)$ *can be pointed by a segment of geodesics length no greater than*

$$\min(\sqrt{l}, \sqrt{p})\frac{\pi}{2} .$$

The closed geodesics have been considered. In particular the length of any closed geodesic is $\sqrt{n}\pi$. *The shortest of them - of length* π *- are mutually congruent. In particular, the injectivity radius of the Grassmann manifold* $G_{l,l+p}(E)$ *is* $\frac{\pi}{2}$.
12) (**Relation between Plücher embeddings and Riemann structure**). *The embedding of a Grassmann manifold into Euclidean space using Plücher coordinates is isometric and equivariant, (that is, the isometry group of the Grassmann manifold is a subgroup of the motions of the Euclidean framework). The surfaces obtained will be minimal in the sphere, have second fundamental form of constant length and its Grassmann map will be homothetic and harmonic.*
PROOF. Let us consider, for example, the embedding $G^+_{l,l+p}(E) \to S^N \subset E, \dim E = N+1$, where $N = \binom{l+p}{l} - 1$. The Lie group of isometries of the manifold $G^+_{l,l+p}(E)$ as a symmetric space is $SO(l+p)$. Let $X = (x_{ij})_{1 \le i \le l, 1 \le j \le l+p}$ be the $(l \times (l+p))$ matrix of the components of the vectors $(x_1, \cdots, x_l)$ that span the l-plane $\pi \in G^+_{l,l+p}(E)$. We put

$$p_{i_1 \cdots i_l} = \det \begin{pmatrix} x_{1i_1} & \cdots & x_{1i_l} \\ \cdots & \cdots & \cdots \\ x_{li_1} & \cdots & x_{li_l} \end{pmatrix}$$

$$P_{i_1 \cdots i_l} = \frac{p_{i_1 \cdots i_l}}{\sqrt{\sum\limits_{j_1 < \cdots < j_l} (p_{j_1 \cdots j_l})^2}}, 1 \le i_1 < \cdots < i_l \le l + p.$$

Thus, the functions $\{P_{i_1 \cdots i_l}, 1 \le i_1 < \cdots < i_l \le l + p\}$ give the Plücher embedding. Let $U \in SO(l + p)$ be an element of the isometry group. Under the action of U the

l-plane π goes into some l-plane π' . We have:

$$X \mapsto XU^*, \quad p_{i_1\cdots i_l} \mapsto u_{i_1}^{j_1} u_{i_2}^{j_2} \cdots u_{i_l}^{j_l} p_{j_1\cdots j_l}\,,$$

where u_i^j are the elements of the matrix U . For two vectors p and q belonging to E we have

$$\begin{aligned}
< p, q > &= \sum_{i_1<\cdots<i_l} P_{i_1\cdots i_l} q_{i_1\cdots i_l} \\
&= \frac{1}{l!} \sum_{i_1\cdots i_l} P_{i_1\cdots i_l} q_{i_1\cdots i_l} \\
&= \frac{1}{l!} \sum_{i_1\cdots i_l} u_{i_1}^{k_1} u_{i_2}^{k_2} \cdots u_{i_l}^{k_l} u_{i_1}^{j_1} u_{i_2}^{j_2} \cdots u_{i_l}^{j_l} p_{k_1\cdots k_l} q_{j_1\cdots j_l} \\
&= \frac{1}{l!} \delta^{k_1 j_1} \cdots \delta^{k_l j_l} p_{k_1\cdots k_l} q_{j_1\cdots j_l} \\
&=< p, q >
\end{aligned}$$

Then, $P_{i_1\cdots i_l} \mapsto u_{i_1}^{j_1} \cdots u_{i_l}^{j_l} P_{j_1\cdots j_l}$ is an orthogonal transformation of E , that is, $SO(l+p) \subset SO(N)$, the Plücher embedding is equivariant. Hence, it is sufficient to investigate all its properties at one point. We take the point π_0 with $\xi_i^\mu = 0$. For a plane π close to π_0 we can say that it is spanned by the row- vectors of the matrix $(I_l; \xi)$. Hence, $P_{1\cdots l} = 1, P_{1\cdots \hat{k}\cdots ll+m} = (-1)^{l+k}\xi_k^m$ ("hat" means that the index k is discarded); $P_{1\cdots \hat{k}\cdots \hat{r}\cdots ll+ml+s} = (-1)^{k+r+1}(\xi_k^m \xi_r^s - \xi_r^m \xi_k^s), k < r <, m < s$. All the remaining $P_{i_1\cdots i_l}$ are $o(\xi^2)$. Then,

$$1/\sqrt{\sum_{i_1<\cdots<i_l} p_{i_1\cdots i_l}^2} = \Big[\sqrt{1 + \sum_{m,k} (\xi_n^m)^2 + o(\xi^2)}\Big]^{-1} = 1 - \frac{1}{2} \sum_{m,k} (\xi_n^m)^2 + o(\xi^2)\,.$$

Therefore,

$$\begin{aligned}
P_{1\cdots l} &= 1 - \frac{1}{2} \sum_{m,k} (\xi_n^m)^2 + o(\xi^2) \\
P_{1\cdots \hat{k}\cdots ll+m} &= (-1)^{l+k}\xi_k^m + o(\xi^2) \\
P_{1\cdots \hat{k}\cdots \hat{r}\cdots ll+ml+s} &= (-1)^{k+r+1}(\xi_k^m \xi_r^s - \xi_r^m \xi_k^s) + o(\xi^2) \\
P_{i_1\cdots i_l} &= 0(\xi^2)\,, \text{ for all the others.}
\end{aligned}$$

If we, now, compute the first and second fundamental form at zero and use equivariance, we easily prove 1), 3) and 5). Property 6) follows from 3). □

13) (**Grassmann manifolds as symmetric spaces**).

□ *A Grassmann manifold is a compact globally symmetric space.*

PROOF. The group $O(l+p)$ acts transitively on $G_{l,l+p}(E)$. $O(l)\times O(p)$ is the isotropy group of the point $\pi \in G_{l,l+p}(E)$, where $O(l)$ acts in the l-dimensional plane π and

$O(p)$ acts in its orthogonal complement. Thus, $G_{l,l+p}(E)$ is the homogeneous space $O(l+p)/O(l)\times O(p)$.

$$\boxed{G_{l,l+p}(E) \cong O(l+p)/O(l)\times O(p)}$$

Let $\{x^\alpha\}_{1\le\alpha\le l+p}$ be a coordinate system in E. Let π_0 be the plane characterized by the equations $x^{l+1} = \cdots = x^{l+p} = 0$. Then, the corresponding isotropy group is a matrix group of the form

$$\left(\begin{array}{c|c} U & 0 \\ \hline 0 & V \end{array}\right), \quad U \in O(l), V \in O(p)$$

Let $\sigma : O(l+p) \to O(l+p)$ be the involutory automorphism defined by

$$\sigma(A) = SAS^{-1}, \text{where} \quad S = \left(\begin{array}{c|c} -I_l & 0 \\ \hline 0 & I_p \end{array}\right).$$

This makes $G_{l,l+p}(E)$ into a symmetric space. □

□ *One has the following canonical decomposition of the Lie algebras*

$$\mathbf{g} = \mathbf{h} \bigoplus \mathbf{m},$$

where $\mathbf{g} = \nu(l+p)$ = *space of skew-symmetric* $(l+p)$*-matrices*

$$\mathbf{n} \equiv \nu(l) \bigoplus \nu(p) \equiv \left\{ \left(\begin{array}{c|c} U & 0 \\ \hline 0 & V \end{array}\right), U \in \nu(l), V \in \nu(p) \right\},$$

and

$$\mathbf{m} \equiv \left\{ \left(\begin{array}{c|c} 0 & -X^* \\ \hline X & 0 \end{array}\right), X = (l \times p) - \textit{matrix} \right\}.$$

The Lie algebra $\mathbf{g}$ *is the* lp*-dimensional tangent space to* $G_{l,l+p}(E)$.

□ *The* **scalar product**

$$< A, B >= \frac{1}{2}\mathrm{tr}(AB^*), A, B \in \mathbf{m}$$

is $O(p)\times O(l)$*-invariant and induces an invariant metric on* $G_{l,l+p}(E)$*; it is proportional to the restriction to* $\mathbf{m}$ *of the Killing metric on* $\nu(l+p)$.

$\square$ *The curvature operator has the form* $R(A,B) = -[A,B]$, *and the sectional curvature of* $G_{l,l+p}(E)$ *at the point* π_0 *on the 2-vector* $A \wedge C \subset \mathbf{m}$ *is*

$$\frac{<[A,B],[A,B]>}{<A,A><B,B>-(<A,B>)^2}.$$

If

$$A = \left(\begin{array}{c|c} 0 & -X^* \\ \hline X & 0 \end{array}\right), B = \left(\begin{array}{c|c} 0 & -Y^* \\ \hline Y & 0 \end{array}\right)$$

this is

$$\bar{\kappa}(A,B) = \frac{\frac{1}{2}\mathrm{tr}(\Lambda_1\Lambda_1^*) + \frac{1}{2}\mathrm{tr}(\Lambda_2\Lambda_2^*)}{\mathrm{tr}(XX^*)\mathrm{tr}(YY^*) - (\mathrm{tr}(XY^*))^2}$$

where $\Lambda_1 = XY^* - YX^*, \Lambda_2 = X^*Y - Y^*X$.

$\square$ *The rank of* $G_{l,l+p}(E)$ *- the dimension of a maximal totally geodesic flat submanifold - is equal to* $\min(l,p)$. *The corresponding submanifold is the flat torus* $e^{\mathbf{s}}$, *where* $\mathbf{s} \subset \mathbf{m}$ *can be taken to be*

$$\mathbf{s} \equiv \left\{ \left(\begin{array}{c|c} 0 & -X^* \\ \hline X & 0 \end{array}\right), X \equiv \left(\begin{array}{ccc|c} x_1 & & \bigcirc & \\ & \ddots & & \bigcirc \\ \bigcirc & & x_p & \end{array}\right), \quad x_j \in \mathbf{R} \right\}$$

with $p \leq l$ *(similarly when* $p > l$*).*

$\square$ *Similarly we get the results listed in the following table.*

*TAB.*1.3 – **Grassmann manifolds as symmetric spaces**

Grassmann manifold (*)	Symmetric Spaces
$G^+_{l,l+p}(E)$	$SO(l+p)/SO(l) \times SO(p)$
$\mathbf{C}G^+_{l,l+p}$	$SU(l+p)/SU(l) \times SU(p)$
$\mathbf{H}G^+_{l,l+p}$	$Sp(l+p)/Sp(l) \times Sp(p)$

(*) (All the Grassmann manifolds listed here are Einstein manifolds, i.e., the Ricci tensor is proportional to the metric tensor)

THEOREM 1.13 - (**Standard embeddings of Grassmann manifolds in Euclidean space**).

1) *Let* L *be a connected real non-compact semisimple Lie group. Let* $A(L)$ *be the Lie algebra of* L *and* K *a maximal compact subgroup, with Lie algebra* $A(K)$. *Let* $A(L) = A(K) \bigoplus \mathbf{p}$ *be a Cartan decomposition. Let* $E \in \mathbf{p}$ *such that* ad_E *is semisimple and*

has eigenvalues $\pm 1, 0$. *Finally, if* $H \subset K$ *is the subgroup such that* $H \equiv \{h \in K \ : ad_h E = E\}$, *then the map* $K \to ad_K E, \kappa \mapsto \mathbf{p}$ *induces an isometric equivariant embedding* $K/H \to \mathbf{p}$ *of a symmetric space in a Euclidean space. Symmetric spaces obtained in this way are called* **R-spaces**, *and the embedding described is a* **standard embedding**.

EXAMPLE 1.9 - The following are R-spaces: (a) All the classical compact symmetric spaces; (b) The projective Cayley plane. ■

☐ *(Properties of standard embeddings)*: The embeddings described above are: (a) isometric; (b) equivariant. Furthermore, the surface obtained (c) lies in a hypersphere and is minimal in this hypersphere; (d) is extrinsically symmetric (symmetry with respect to the normal space at any point maps it onto itself); (e) has a parallel second fundamental form; (f) is "tight", i.e., with least complete absolute curvature. ■

In the following table we give some other examples of R-spaces and standard embedding of Grassmann manifolds.

TAB.1.4 - **R-spaces and standard embeddings of Grassmann manifolds**

(L, K)	$G_{l,l+p}(E)$	$\mathbf{p}$
$(SL(p+l, \mathbf{C}), U(p+l))$	$\mathbf{C}G_{l,p+l}$	Sp.Hermit.zero trace $(l+p) \times (l+p)$-matrices over $\mathbf{C}$
$(SU^*(2p+2l), Sp(p+l))$	$\mathbf{H}G_{l,p+l}$	Sp.Hermit.zero trace $(l+p) \times (l+p)$-matrices over $\mathbf{H}$

EXAMPLE 1.10 - (LINE AND COLINES IN GRASSMANN MANIFOLDS OF $(m+n)$-DIMENSIONAL VECTOR SPACE W). A **straight line** in $G_{n,m+n}(W)$ is a subset $l \subset G_{n,m+n}(W)$ such that its image $\wp(l) \subset \mathbb{P}(\Lambda^n(W))$ by means of the Plücher mapping

$$\wp : G_{n,m+n}(W) \to \mathbb{P}(\Lambda^n(W))$$

is a projective line in $\mathbb{P}(\Lambda^n(W))$. A straight line in $G_{n,m+n}(W)$ can be considered a linear 1-parameter set of n-planes $p(t_1, t_2) \in G_{n,m+n}(W), t = [t_1 : t_2] \in \mathbf{R}P^1$, having $(n-1)$-dimensional intersection:

$$q \equiv \bigcap_t p(t) \in G_{n,m+n}(W) .$$

Furthermore, l can be identified with a couple (W_0, q) where: (a) $W_0 \subset$ is a subspace with $\dim W_0 = n+1$; (b) $q \subset W_0, \dim q = n-1$. More precisely, one has

$$l \equiv (W_0, q) \equiv l(W_0, q) = \{p \in G_{n,m+n}(W) | q \subset p \subset W_0\} .$$

The cone $Sp(p) \subset G_{n,m+n}(W)$ of all straight-lines passing through $p \in G_{n,m+n}(W)$ has dimension $m+n$: $\dim Sp(p) = m + m$. Moreover, for any subspace $W_0 \subset W, \dim W_0 = n+1$, one has an embedding

$$\mathbb{P}(W_0^*) \to G_{n,m+n}(W).$$

Thus, a line in $G_{n,m+n}(W)$ is image of a projective line by means of this embedding. Then, a **co-line** is the image of hyperplane in $\mathbb{P}(W_0^*)$ by means of this embedding, i.e., a set of n-planes $p \in G_{n,m+n}(W)$ such that are lying in some subspace $W_0 \subset W, \dim W_0 = n+1$, and form an hyperplane, i.e., have 1-dimensional intersection. Thus, one can see that a co-line is a linear $(n-1)$-parameter set of n-planes $p([t_1, \cdots, t_n]) \in G_{n,m+n}(W), [t_1 : \cdots : t_n] \in \mathbf{R}P^{n-1}$ having 1-dimensional intersection. Each co-line L can be identified with a pair $(W_0, w), w \in W_0, w \neq 0, \dim W_0 = n+1$, such that

$$L \equiv (W_0, w) \equiv L(W_0, w) = \{p \in G_{n,m+n}(W) | \mathbf{R}w \subset p \subset W_0\}.$$

Let $b \in W \bigotimes W$ be some non-degenerate tensor and let

$$B : W^* \to W, B(\alpha) = <\alpha, b>,$$

be the corresponding isomorphism. This isomorphism, together with the natural duality mapping

$$\mathcal{D}(p) : G_{n,m+n}(W) \to G_{m,m+n}(W^*), \quad p \mapsto Ann(p)$$

gives a b-**duality**

$$\mathcal{D}_b : G_{n,m+n}(W) \to G_{m,m+n}(W^*), \quad p \mapsto \Lambda^m(B)(Ann(p)).$$

Then, for any line $l \equiv l(W_0, q)$ one has

$$\mathcal{D}_b(l) = l(q^\perp, W_0^\perp), \text{where} \quad V^\perp \equiv \{v \in W | b(v, V) = 0\} \equiv b - \text{dual space of } V.$$

■

EXAMPLE 1.11 - (GRASSMANNIAN OF HORIZONTAL PLANES). A **connection** $\bar{\Gamma}$ on a fiber bundle $\pi : W \to M$ can be considered as a splitting, on the left, of the exact sequence

$$0 \to vTW \underset{\bar{\Gamma}}{\overset{i}{\rightleftarrows}} TW \xrightarrow{(T\pi,\beta)} \pi^*TM \to 0$$

with β the canonical projection $TW \to W$ and i the canonical inclusion. Let $\mathbb{H}(W)$ be the kernel of $\bar{\Gamma}$. Then, the sequence of vector bundles over W

$$0 \to \mathbb{H}(W) \to TW \xrightarrow{\bar{\Gamma}} vTW \to 0$$

is exact and one has

$$TW \cong vTW \bigoplus \mathbb{H}(W) .$$

We can associate to $\mathbb{H}(W)$ two fiber bundles: $\pi_H : \mathbb{H}(W) \to W$, and $\bar{\pi}_H : \mathbb{H}(W) \to M$, with $\bar{\pi}_H \equiv \pi \circ \pi_H$. In advance, the space of connections can be identified with the fiber bundle manifold $\underline{\mathbb{H}}(W) \equiv \bigcup_{p \in W} \underline{\mathbb{H}}(W)_p$, where $\underline{\mathbb{H}}(W)_p$ are Grassmann manifolds, $G_{n,n+q}(T_pW)$, of horizontal n-planes in $p \in W, n = \dim M, q = m - n, m = \dim W$. Locally, one has the trivialization $\underline{\mathbb{H}}(W)|_{\pi^{-1}(U)} \cong (U \times E) \times F$, where E is the type fiber of π and $F \cong \mathbf{R}^{nq} = M(n, q; \mathbf{R}) \equiv$ space of $n \times q$ real matrices. So the dimension of $\underline{\mathbb{H}}(W)$ is $nq + m$. A local coordiante system on $\underline{\mathbb{H}}(W)$ is $(\breve{x}^i, \breve{y}^h, \gamma^k_j), i, j = 1, \cdots, n; k, h = 1, \cdots, q$. A local change of chart is

$$\begin{aligned}
&\bar{\breve{x}}^i \mapsto \breve{x}^i,\\
&\bar{\breve{y}}^k \mapsto \breve{y}^k,\\
&\bar{\gamma}^k_i \mapsto \gamma^\beta_\alpha = \bar{\gamma}^k_i(\partial \breve{x}_\alpha . \bar{\breve{x}}^i)(\partial \bar{y}_k . x^\beta) + (\partial \bar{y}_k . y^\beta)(\partial \breve{x}_\alpha . \bar{y}^k).
\end{aligned}$$

■

REMARK 1.14 - (GRASSMANN MANIFOLDS AND HYPERGEOMETRIC FUNCTIONS).
In refs.[33,34] it is proved tha a generalized hypergeometric function $\Phi(\alpha, \epsilon, \tilde{\zeta})$ can be defined as a suitable function on the line bundle $\tilde{G}_{k,n}(\mathbf{R}^n)$ over the Grassmann manifold $G_{k,n}(\mathbf{R}^n)$, where

$$\tilde{G}_{k,n}(\mathbf{R}^n) \equiv \bigcup_{\zeta \in G_{k,n}(\mathbf{R}^n)} \Lambda^k(\zeta^*) ,$$

with $\alpha \equiv (\alpha_1, \cdots, \alpha_n) \in \mathbf{C}^n$, such that $\sum \alpha_j = n - k$, and $\epsilon \equiv (\epsilon_1, \cdots, \epsilon_n)$, where $\epsilon_j = \pm 1$, and $\tilde{\zeta} \in \tilde{G}_{k,n}(\mathbf{R}^n)$. Such hypergeometric function contains as a particular case the Gauss hypergeometric function $F(a, b, c; x)$, (see section 2.2).

1.5 - SPECTRAL SEQUENCES

In this section we will consider the techniques of spectral sequences that will be utilized in order to characterize PDEs. (See also refs.[49,80,141].)
Let us start by recalling some fundamental definitions and results on the theory of categories.

DEFINITION 1.17 - *A* **category** $\mathcal{C}$ *is a set of* **objects** C_i *and maps between these objects* $f_i : C_i \to C_j$ *(***morphisms***). One writes* $C_i \in Ob(\mathcal{C})$ *and* $f_i \in Hom_{\mathcal{C}}(C_i, C_j)$. *Further, a composition law of morphisms is given by*

$$Hom(Y,Z) \times Hom(X,Y) \to Hom(X,Z), \quad (g,f) \mapsto g \circ f \,.$$

Moreover, one has the following axioms:
(i) $Hom(X,Y) \bigcap Hom(X',Y') = \emptyset$ *except if* $X = X'$ *and* $Y = Y'$.
(ii) *The composition law of morphisms is associative, namely* $f \in Hom(X,Y), g \in Hom(Y,Z), h \in Hom(Z,W) \Rightarrow h \circ (g \circ f) = (h \circ g) \circ f$.
(iii) *For any object* $X \in Ob(\mathcal{C})$ *there exists in* $Hom(X,X)$ *a morphism* id_X *(identity of* X*) such that for any* $f \in Hom(X,Y)$ *and* $g \in Hom(Y,X)$ *one has*

$$f \circ id_X = f, \quad id_X \circ g = g \,.$$

An **isomorphism** *(or* **equivalence***) in a category* $\mathcal{C}$ *is a map* $f \in Hom(X,Y)$ *such that there is a morphism* $g \in Hom(Y,X)$ *such that* $g \circ f = id_X, f \circ g = id_Y$. *Then, one says that* X *and* Y *are* **isomorphic** *(or* **equivalent***).*

DEFINITION 1.18 - *Let two categories* $\mathcal{C}$ *and* $\mathcal{C}'$ *be given.*
1) *A* **covariant functor** *between the categories* $\mathcal{C}$ *and* $\mathcal{C}'$ *is a map* $T : \mathcal{C} \to \mathcal{C}$ *such that:*
(a) $T : Ob(\mathcal{C}) \to Ob(\mathcal{C})$;
(b) $T : Hom_{\mathcal{C}}(X;Y) \to Hom_{\mathcal{C}}(T(X);T(Y))$;
(c) $T(id_X) = id_{T(X)}$;
(d) $T(f \circ g) = T(f) \circ T(g)$.
2) *A* **controvariant functor** *is similarly defined except that conditions* (b) *and* (d) *are replaced by*
(b') $T : Hom_{\mathcal{C}}(X;Y) \to Hom_{\mathcal{C}}(T(Y);T(X))$;
(d') $T(g \circ f) = T(f) \circ T(g)$.

One can easily generalize the previous definitions to functors of several variables $T : \mathcal{C}_1 \times \cdots \times \mathcal{C}_n \to \mathcal{D}$, where $\mathcal{C}_i$ and $\mathcal{D}$ are categories. Let $\mathcal{C}$ be a category and $A \in Ob(\mathcal{C})$. If for any $X \in Ob(\mathcal{C}), Hom(A,X) \neq \emptyset$, then A is called a **total source**. If $Hom(A,X)$ has exactly one element, A is called an **initial object**. If $Hom(X,A) \neq \emptyset$, then A is called a **final object**. An object which is both initial and final is called a **zero object**. If a problem is reconduced to finding a final or initial object in a category, one says that it is a **universal problem**. The initial (resp. final) objects in a category are isomorphic.

EXAMPLE 1.12 - In the category of sets $\mathcal{E}ns$ extended to the empty set, this last set is an initial object. Any set with a single point is a final object. ∎

REMARK 1.15 - 1) If $\mathcal{C}$ is a category with zero objects, then $Hom_{\mathcal{C}}(X,Y)$ possesses a distinguished morphism $X \to 0 \to Y$ called **zero morphism** and written 0_{XY} or simply 0.
2) It is natural to define the concept of subcategory $\mathcal{C}_0$ of a category $\mathcal{C}$. Of particular importance between the subcategories of $\mathcal{C}$ are the **full subcategories** that is, those subcategories $\mathcal{C}_0$ of $\mathcal{C}$ such that $Hom_{\mathcal{C}_0}(X,Y) = Hom_{\mathcal{C}}(X,Y)$, $\forall X, Y \in Ob(\mathcal{C}_0)$. For example, the category of abelian groups is a full subcategory of the category of groups. But, the category of rings-with unit-elements, and ring-homomorphisms preserving unit-elements is simply a subcategory of the category of rings.
3) Given any category $\mathcal{C}$, we may form a new category, $\mathcal{C}^{opp}$, the category opposite to $\mathcal{C}$, such that $Ob(\mathcal{C}^{opp}) = Ob(\mathcal{C})$, and $Hom_{\mathcal{C}^{opp}}(X,Y) = Hom_{\mathcal{C}}(Y,X), \forall X, Y \in Ob(\mathcal{C}^{opp})$. Then, $\mathcal{C}^{opp}$ is a category with the same identity morphisms as $\mathcal{C}$, and if $\mathcal{C}$ has zero objects, then the same objects are zero objects of $\mathcal{C}^{opp}$. Moreover, $(\mathcal{C}^{opp})^{opp} = \mathcal{C}$.
4) We say that a functor $F : \mathcal{C} \to \mathcal{D}$ is **full**, if it maps $Hom_{\mathcal{C}}(X,Y)$ onto

$$Hom_{\mathcal{D}}(F(X), F(Y)), \forall X, Y \in Ob(\mathcal{C}).$$

Furthermore, we say that F is **faithful** if it maps $Hom_{\mathcal{C}}(X,Y)$ injectively to

$$Hom_{\mathcal{D}}(F(X), F(Y)).$$

Finally, F is a **full embedding** if F is full and faithful and one-to-one on objects. Notice then that $F(\mathcal{C})$ is a full subcategory of $\mathcal{D}$ (in general, $F(\mathcal{C})$ is not a category at all).
DEFINITION 1.19 - *We suppose that the set M of indices of the objects $A_i \in Ob(\mathcal{C})$ has a partial-ordering relation (reflexive and transitive) $\leq$ and that there are given morphisms $f_{ij} : A_i \to A_j$ for $i \leq j$ which satisfy the following conditions:*
(a) $f_{ii} = id_{A_i}$;
(b) $f_{jk} \circ f_{ik} = f_{ji}, i \leq j \leq k$. *Such a system $(A_i, f_{ij})_{i,j \in M}$ is called an* ***inductive family of objects.***
DEFINITION 1.20 - *Let us consider the category $\mathcal{C}_{\leq}^{A_i}$ whose objects are systems of morphisms $A_i \xrightarrow{u_i} X$ of $\mathcal{C}$ such that for $i \leq j$ we have a commutative diagram*

$$\begin{array}{ccc} A_i & \xrightarrow{u_i} & X \\ {\scriptstyle f_{ij}} \downarrow & \nearrow {\scriptstyle v_j} & \\ A_j & & \end{array} \tag{1.14}$$

and whose morphisms are the morphisms t with commutative diagrams

$$\begin{array}{ccc} A_i & \xrightarrow{u_i} & X \\ & {\scriptstyle v_i} \searrow & \downarrow {\scriptstyle t} \\ & & Y \end{array} \tag{1.15}$$

Thus $(A_i, f_{ij})_{i,j\in M}$ is an inductive family of objects in $\mathcal{C}$. Then, an initial object of the category $\mathcal{C}_{\leq}^{A_i}$ is called the **inductive** *(or* **direct***)* **limit** *of the family and we write $\varinjlim A_i$, namely $\varinjlim A_i$, consists of an object of $\mathcal{C}$ and of the system of morphisms $f_i : A_i \to \varinjlim A_i$, for which the diagrams (1.15) are commutative.*

DEFINITION 1.21 - *A* **projective family** $\{A_i, f_{ij}\}_{ij\in M}$ *in a category $\mathcal{C}$ is characterized by*

(a) $\{A_i\}_{i\in M}$ *objects of $\mathcal{C}$ with indices in the partially ordered set M;*

(b) f_{ij} *morphisms of $\mathcal{C}$ defined for every pair $i \geq j$ and satisfying the conditions*

$$f_{ii} = id_{A_i}, \quad f_{jk} \circ f_{ij} = f_{ik}, i \geq j \geq k\,.$$

To every family of this type we associate a category $\mathcal{C}_{\overline{A}_i}^{\leq}$ whose objects are systems of morphisms $u_i : X \to A_i$, such that for $i \geq j$ there is a commutative diagram

(1.16)
$$\begin{array}{ccc} & & A_i \\ & {}^{u_i}\nearrow & \\ X & & \downarrow \; f_{ij} \\ & {}_{u_j}\searrow & \\ & & A_j \end{array}$$

and whose morphisms are the morphisms f with commutative diagrams

$$\begin{array}{ccc} X & \xrightarrow{u_i} & A_i \\ t \downarrow & \nearrow v_i & \\ Y & & \end{array}$$

A final object (if it exists) of the category $\mathcal{C}_{\overline{A}_i}^{\leq}$ is called the **projective limit** *(or* **inverse limit***) of the family A_i which one writes $\varprojlim A_i$ and which consists of an object of $\mathcal{C}$ together with morphisms $f_i : \varprojlim A_i \to A_i$ which make the diagrams (1.16) commutative.*

REMARK 1.16 - 1) If $f \in Hom(\mathcal{C})$, then we say that f is a **monomorphism** (or **injective homomorphism**) if $fg = fh \Rightarrow g = h$, $\forall g, h \in Hom(\mathcal{C})$. Furthermore, we say that $f \in Hom(\mathcal{C})$ is an epimorphism (resp. **surjective homomorphism**) if $gf = hf \Rightarrow g = h$, $\forall g, h \in Hom(\mathcal{C})$. Then, $f \in Hom(\mathcal{C})$ is a monomorphism (resp. epimorphism) iff $f \in Hom(\mathcal{C}^{opp})$ is an epimorphism (resp. monomorphism).

2) Define the **kernel** of $\phi \in Hom_{\mathcal{C}}(X, Y)$, in a category $\mathcal{C}$ with zero morphism as a morphism $\mu : K \to X$ such that (i) $\phi\mu = 0$, (ii) if $\phi\psi = 0$ then $\psi = \mu\psi'$ and ψ' is unique. The dual concept (i.e., one in the category $\mathcal{C}^{opp}$) is that of **cokernel**.

3) Let $\{X_i\}_{i\in I}$ be a family of objects of the category $\mathcal{C}$, indexed by the set I. Then, a **product** (X, p_i) of the objects X_i is an object X, together with morphisms $p_i : X \to X_i$, called projections with the universal property: given any object Y and

morphisms $f_i : Y \to X_i$, there exists a unique morphism $f \equiv \{f_i\} : Y \to X$ such that $p_i f = f_i$. We write $X = \prod X_i$. The product identifies a functor $\prod : \mathcal{C}^I \to \mathcal{C}$. The dual concept is that of **coproduct** and we write $X = \coprod X_i$.

4) An **additive category** is a category $\mathcal{C}$ with zero object in which any two objects have a product and in which the morphism sets $Hom_{\mathcal{C}}(X, Y)$ are abelian groups such that the composition

$$Hom_{\mathcal{C}}(X,Y) \times Hom_{\mathcal{C}}(Y,Z) \to Hom_{\mathcal{C}}(X,Z)$$

is bilinear.

5) An **abelian category** is an additive category in which (i) every morphism has a kernel and a cokernel; (ii) every monomorphism is the kernel of its cokernel; every epimorphism is the cokernel of its kernel; (iii) every morphism is expressible as the composite of an epimorphism and a monomorphism.

EXAMPLE 1.13 - 1) The category of finite abelian groups is abelian.

2) The category of free abelian groups is additive but not abelian.

□ If the morphism α in the abelian category $\mathcal{C}$ is a monomorphism and an epimorphism, then it is an isomorphism.

DEFINITION 1.22 - 1) *Let $\mathcal{G}_A$ be the category of commutative left A-modules, where A is a ring. A* **0-sequence** *in $\mathcal{G}_A$ is a diagram*

$$L \xrightarrow{f} M \xrightarrow{g} N \tag{1.17}$$

with $L, M, N \in Ob(\mathcal{G}_A)$ and $f, g \in Hom(\mathcal{G}_A)$, such that $g \circ f = 0$, i.e., $\mathrm{im}(f) \subseteq \ker(g)$. $H \equiv \ker(g)/\mathrm{im}(f)$ is called the **homology** *of (1.17).*

2) *An* **exact sequence** *in $\mathcal{G}_A$ is a diagram like (1.17) such that $H = 0$. An exact sequence (1.17) is said to* **split** *if $\mathrm{im}(f) = \ker(g)$ is a direct summand of M.*

PROPOSITION 1.8 - [47]. *Let*

$$0 \to L \xrightarrow{f} M \xrightarrow{g} N \to 0 \tag{1.18}$$

be a sequence given in $\mathcal{G}_A$. Then, (1.18) is a split exact sequence iff there exist A-hmomorphisms $\bar{f} : M \to L$, and $\bar{g} : N \to M$ such that

$$\bar{f} \circ f = id_L, g \circ \bar{g} = id_N, f \circ \bar{f} + \bar{g} \circ g = id_M, g \circ f = 0.$$

So, a splitting of an exact sequence like (1.18) induces an isomorphism $M \cong L \bigoplus N$ defined by the following diagram:

$$\begin{array}{ccccc} L & & \longrightarrow & & M \\ & \searrow i_L & & & \| \\ & & L \bigoplus N & \cong & M \\ & \nearrow i_N & & & \| \\ N & & \longrightarrow & & M \end{array}$$

Further,

$$M \cong \ker(g) \bigoplus \bar{g}(N) \equiv \mathrm{im}(f) \bigoplus \bar{g}(N) .$$

In particular, any exact sequence like (1.18) in the subcategory $\mathcal{V} \subset \mathcal{G}_A$ of finite dimensional vector spaces splits. Of course, the splitting is not canonical. However, to fix the map $\bar{f}$ is equivalent to fix $\bar{g}$.

REMARK 1.17 - (HOMOTOPY FUNCTOR). 1) Let X be a topological space.[2] We set

$$\pi_0(X) \equiv X/\sim y \Leftrightarrow \exists \in C^0([0,1]), X) , \; f(0) = x , \; f(1) = y .$$

If $[x]$ is an equivalence class of $x \in X$ in $\pi_0(X)$, then it is called the 0-component of x in X. X is called **0-connected** (or path-connected) if $\pi_0(X)$ contains a single element, i.e. ,$\forall x \in X, [x] = X$. $\pi_0(X)$ identifies a covariant functor $\pi_0 : \mathcal{T}op \to \mathcal{E}ns$, between the category of topological spaces (and continuous mappings) and the category of sets. Let $P\mathcal{T}op$ be the category of pointed topological spaces, where the morphisms are the continuous maps that preserve the base points. If $X \in Ob(P\mathcal{T}op)$, then $\pi_0(X)$ has a base point ($\equiv [x_0]$).

2) The **loop functor** Ω is given by

$$\Omega : P\mathcal{T}op \to P\mathcal{T}op, \quad \Omega(Y, y_0) \equiv C^0((S^1, s_0), (Y, y_0)) \equiv \Omega Y ,$$

where $S^1 \equiv \{x \in \mathbf{R}^2 | \|x\| = 1\}$, and $s_0 = (1,0) \in \mathbf{R}^2$.

3) The **suspension functor** S is given by:

$$S : P\mathcal{T}op \to P\mathcal{T}op, \quad S(X, x_0) = (SX, *), SX \equiv [0,1] \times X/(\{0\} \times X \cup [0,1] \times \{x_0\}) ,$$

where $*$ is base point of SX .

4) The **homotopic equivalence functor** $[,]$ is given by

$$[,] : \mathcal{T}op \times \mathcal{T}op \to \mathcal{T}op, [,](X,Y) = [X,Y] \equiv C^0(X,Y)/\simeq ,$$

where $\simeq$ is the following equivalence relation in $C^0(X,Y)$:

$$f \simeq g \Leftrightarrow \exists F \in C^0([0,1] \times X, Y), F_0 = f, F_1 = g .$$

Then, one has the following natural equivalences:

---

[2] Recall that a **topological space** is a set X, with a set $\mathcal{T} \equiv \{U_i\}$ of subsets U_i of X, called **open sets**, such that $X \in \mathcal{T}$, $\emptyset \in \mathcal{T}$, $U_i \cap U_j \in \mathcal{T}$, $U_i \cup U_j \in \mathcal{T}$. A mapping $f: X \to Y$ between topological spaces $(X,\mathcal{T})$, $(Y,\overline{\mathcal{T}})$ is said to be **continuous** if $f^{-1}(\overline{U_i}) \in \mathcal{T}, \forall \overline{U_i} \in \overline{\mathcal{T}}$.

$(a)\quad [*;X] \leftrightarrow \pi_0(X), [f] \mapsto [f(*)];$

$(b)\quad [S^0, s_0; X, x_0] \leftrightarrow \pi_0(X, x_0), [f] \mapsto [f(-1)], (S^0 \equiv \{-1, 1\});$

$(c)\quad [SX, x; Y, y_0] \leftrightarrow [X, x_0; \Omega Y, \omega_0].$

5) By iteration one sets:

$$\Omega^n Y \equiv \Omega(\Omega^{n-1}Y), \Omega^0 Y \equiv Y;$$
$$S^n Y \equiv S(S^{n-1}Y), S^0 Y \equiv Y;$$
$$\pi_n(Y, y_0) \equiv \pi_0(\Omega^n Y, \omega_0) \equiv n\text{-th Hurewicz group}.$$

($\pi_1(Y, y_0)$ is usually called **fundamental group** (or **homotopy Poincaré group**)). One has the natural equivalences:

$$\pi_n(X, x_0) \simeq [S^0, +1; \Omega^n X, \omega_0] \simeq [S^1, s_0; \Omega^{n-1}X, \omega_0] \simeq \cdots \simeq [S^n, s_0; X, x_0].$$

Then, one has the covariant functor $\pi_n : PTop \to \mathcal{G} \equiv$ category of groups, for $n \geq 1$, and $PTop \to \mathcal{A} \equiv$ category of abelian groups, for $n \geq 2$, such that $\pi_n(f) \equiv (\pi_0 \circ \Omega^n)(f), \forall f \in Hom(PTop)$. A space X is said **n-connected** if $\pi_k(X, x) = 0, 0 \leq k \leq n$, $\forall x \in X$. For a contractible space (that is a space (X, x_0) such that id_X, and the constant map $X \to x_0$ belongs to the same equivalence class in $[X; X], \pi_k(X, x_0) = 0, n \geq 0$.

REMARK 1.18 - (FUNCTORS EXT AND TOR).

1)(EXTENSIONS OF Λ-MODULES). An **extension** of a Λ-module A by B is the following exact sequence:

$$0 \to B \to E \to A \to 0.$$

Thus, B is a sub-module of E, and $E/B \cong A$. The set $E(A, B)$ of equivalence classes of extensions of A by B identifies a bifunctor from the category of Λ-modules to the category of sets. It is controvariant in the first variable, and covariant in the second one. The functor $Ext_\Lambda(-, -)$ is a bifunctor to the category of abelian group, defined by

$$Ext_\Lambda(A, B) = \operatorname{coker}(\mu^* : Hom_\Lambda(P, B) \to Hom_\Lambda(R, B)),$$

where $0 \to R \to P \to A \to 0$ is a short exact sequence with P projective [49,88], (**projective presentation** of A), that induces the following exact sequence:

$$0 \to Hom_\Lambda(A, B) \overset{\epsilon_*}{\to} Hom_\Lambda(P, B) \overset{\mu_*}{\to} Hom_\Lambda(R, B).$$

☐ One has a canonical isomorphism:

$$E(A, B) \cong Ext_\Lambda(A, B).$$

☐ If P is projective $\Rightarrow Ext_\Lambda(P,B)=0, \forall\Lambda-\text{module}B$.
☐ If I is injective $\Rightarrow Ext_\Lambda(A,I)=0, \forall\Lambda-\text{module}A$.
☐ One has the following canonical isomorphisms:

$$Ext_\Lambda(\bigoplus_i A_i,B)\cong\prod_i Ext_\Lambda(A_i,B)$$
$$Ext_\Lambda(A,\prod_j B_j)\cong\prod_j Ext_\Lambda(A,B_j).$$

☐ If A is a finitely-generated abelian group such that

$$\left\{\begin{array}{c} Ext_{\mathbf{Z}}(A,\mathbf{Z})=0 \\ Hom_{\mathbf{Z}}(A,\mathbf{Z})=0 \end{array}\right\}\Rightarrow A=0\,.$$

☐ If A is a finite abelian group, one has:

$$Ext(A,\mathbf{Z})\cong A\,.$$

☐ (**Stein-Serre theorem for abelian groups**). If A is an abelian group of countable rank, then $Ext_{\mathbf{Z}}(A,\mathbf{Z})=0$ implies A free. (See below for the definition of free abelian group.)
2)(THE FUNCTOR TOR). Let A be a right Λ-module and B a left Λ-module. Set

$$Tor^\Lambda\equiv\ker(\mu_*:R\bigotimes_\Lambda B\to P\bigotimes_\Lambda B)\,,$$

where $0\to R\to P\to A\to 0$ is a projective presentation of A. Then, one has the following exact sequence:

$$0\to Tor^\Lambda(A,B)\to R\bigotimes_\Lambda B\to P\bigotimes_\Lambda B\to A\bigotimes_\Lambda B\to 0\,.$$

☐ If A (or B) is projective, then $Tor^\Lambda(A,B)=0$.
REMARK 1.19 - (COHOMOLOGY OF GROUPS). (See also refs.[7,17].)
☐ Let G be a group written multiplicatively. The **integral group ring** $\mathbf{Z}G$ of G is defined by:

$$\mathbf{Z}G\equiv\left\{\begin{array}{l}\sum_{x\in G}m(x)x|m:G\to\mathbf{Z} \\ \text{almost everywhere zero except on a finite number of elements of } G.\end{array}\right\}$$

The product in $\mathbf{Z}G$ is given by:

$$(\sum_{x\in G}m(x)x).(\sum_{y\in G}m'(y)y)=\sum_{x,y\in G}(m(x).m'(y))xy\,.$$

☐ One has the natural embedding:

$$i : G \to \mathbf{Z}G\,.$$

☐ (**Universal property**). Let R be a ring. To any function $f : G \to R$ with $f(xy) = f(x).f(y)$ and $f(1) = 1_R$, there exists a unique ring-homomorphism $f' : \mathbf{Z}G \to R$ such that $f' \circ i = f$, namely the following diagram is commutative:

$$\begin{array}{ccc} G & \xrightarrow{f} & R \\ {\scriptstyle i}\downarrow & \nearrow {\scriptstyle f'} & \\ \mathbf{Z}G & & \end{array}$$

☐ A (**left**) G-**module** is an abelian group A together with a group homomorphism $\sigma : G \to Aut(A)$. (Similarly for right G-module.) A (left) G-module is called **trivial** if the structure map $\sigma : G \to Aut(A)$ is trivial, i.e., $\sigma(x) = id_A$, $\forall x \in G$.

☐ Since $Aut(A) \subseteq End(A) \equiv R$ the universal property of the group ring yields a ring homomorphism $\sigma' : \mathbf{Z}G \to End(A)$, making A into a (left) module over $\mathbf{Z}G$.

☐ Conversely, if A is a (left) module over $\mathbf{Z}G$, then A is a (left) G-module, since any ring homomorphism takes invertible elements into invertible elements, and since the group elements in $\mathbf{Z}G$ are invertible. Thus, we need not retain any distinction between the concept of G-module and $\mathbf{Z}G$-module. One has the canonical mappings:

$$\begin{array}{l} G \to \mathbf{Z}, x \mapsto 1 \\ \epsilon : \mathbf{Z}G \to \mathbf{Z}, \epsilon(\sum_{x\in G} m(x)x) = \sum_{x\in G} m(x), (\textbf{augmentation}) \end{array}$$

We call $IG \equiv \ker(\epsilon)$ **augmentation ideal** of G.

☐ Let A, A', A'' be left G-modules. Let B, B', B'' be right G-modules. Then, set:

$$\begin{aligned} B \bigotimes_G A &\equiv B \bigotimes_{\mathbf{Z}G} A \\ Hom_G(A, A') &\equiv Hom_{\mathbf{Z}G}(A, A') \\ Tor_n^G(B, A) &\equiv Tor_n^{\mathbf{Z}G}(B, A) \\ Ext_G^n(A, A') &\equiv Ext_{\mathbf{Z}G}^n(A, A'). \end{aligned}$$

☐ The n-**th cohomology group** of G with coefficients in the left G-module A is defined by:

$$H^n(G, A) \equiv Ext_G^n(\mathbf{Z}, A)\,,$$

where $\mathbf{Z}$ is regarded as trivial G-module.

□ The n-**th homology group** of G with coefficients in the right G-module B is defined by:

$$H_n(G,A) \equiv Tor_n^G(B,\mathbf{Z}),$$

where $\mathbf{Z}$ is regarded as trivial G-module.

□ $H^n(G,-)$ and $H_n(G,-)$ are covariant functors.

□ $H^n(G,A)$ can be calculated as the n-th homology of the complex $Hom_G(\mathbb{P},A)$, where $\mathbb{P}$ is a G-projective resolution of the trivial (left) G-module $\mathbf{Z}$.

□ $H_n(G,A)$ can be calculated as the n-th homology of the complex $B\bigotimes_G \mathbb{P}$.

□ (PROPERTIES).

(a) If A is injective $\Rightarrow H^n(G,A)=0, n\geq 1$.

(b) If B is flat (in particular projective) $\Rightarrow H_n(G,B)=0, n\geq 1$.

(c) If $0\to A\to I\to A'\to 0$ is an injective presentation of A, then

$$H^{n+1}(G,A)\cong H^n(G,A'), n\geq 1.$$

(d) If $0\to B'\to P\to B\to 0$ is a projective (or flat) presentation of B, then

$$H_{n+1}(G,B)\cong H_n(G,B'), n\geq 1.$$

(e) $H^0(G,A)=Hom_G(\mathbf{Z},A)=A^G\equiv\{a\in A|x\circ a=a,\forall x\in G\}\equiv$ subgroup of invariant elements in A.

(f) $H_0(G,B)=B\bigotimes_G\mathbf{Z}=B_G\equiv B/B\circ IG=B/\{b(x-1)\quad|\quad b\in B, x\in G\}$.

(g) If A,B are trivial G-modules, then

$$\begin{aligned}&H^0(G,A)=A, H_0(G,B)=B;\\&H_1(G,B)\cong B\bigotimes IG/(IG)^2, H_1(G,\mathbf{Z})\cong \mathbf{Z}\bigotimes_G IG;\\&H^1(G,A)\cong Hom(H_1(G,\mathbf{Z}),A).\end{aligned}$$

(h) For a free group F one has

$$H^n(F,A)=0=H_n(F,B),\ \forall F-\text{modules}\quad A,B,n\geq 2.$$

REMARK 1.20 - (EXTENSIONS OF GROUPS).

□ If $0\to N\xrightarrow{i} G\xrightarrow{p} Q\to 0$ is an exact sequence of groups, then

$$\mathbf{Z}\bigotimes_N \mathbf{Z}G\cong\mathbf{Z}Q$$

as right G-modules.

☐ If

$$0 \to N \xrightarrow{i} G \xrightarrow{p} Q \to 0$$

is an exact sequence of groups, and if A is a left G-module, then

$$Tor_n^N(\mathbf{Z}, A) \cong Tor_n^G(\mathbf{Z}Q, A)\,.$$

☐ If

$$0 \to N \to G \to Q \to 0$$

is an exact sequence of groups, then

$$0 \to N_{ab} \xrightarrow{k} \mathbf{Z}G \bigotimes_G IG \xrightarrow{r} IQ \to 0$$

is an exact sequence of Q-modules, where

$$N_{ab} \equiv N/N' \equiv H_1(N, \mathbf{Z}) \cong \mathbf{Z} \bigotimes_N IN\,.$$

☐ Let

$$0 \to R \to F \to Q \to 0$$

be an exact sequence of groups, with F a free group, i.e., a free presentation of Q. Then

$$0 \to R_{ab} \xrightarrow{k} \mathbf{Z}Q \bigotimes_F IF \xrightarrow{r} IQ \to 0$$

is a Q-free presentation of IQ.

☐ Furthermore, for any Q-modules A, B and all $n \geq 3$, one has the following isomorphisms:

$$H_n(Q, B) \cong Tor_{n-1}^Q(B, IQ) \cong Tor_{n-2}^Q(B, R_{ab})$$
$$H^n(Q, A) \cong Ext_Q^{n-1}(IQ, A) \cong Ext_Q^{n-2}(R_{ab}, A).$$

☐ The equivalence classes $M(G, A)$ of an extension of the group G by the G-module A, i.e., the following exact sequence:

$$0 \to A \xrightarrow{i} E \to G \to 0$$

is identified (by means of a group isomorphism) with $H^2(G, A)$. The neutral element in the abelian group $M(G, A)$ is represented by the split extension

$$0 \to A \to A \times G \to G \to 0\,.$$

☐ If F is free, $M(F,A)$ consists of one element only, the class containing the semi-direct product.

☐ If G is finite and if A and B are finitely generated G-modules, then $H^n(G,A)$ and $H_n(G,B)$ are finitely generated.

☐ (**Universal coefficient theorem**).

(a) *Let G be a group and let C be an abelian group considered as a trivial G-module. Then, the following sequences are exact and natural, for any $n \geq 0$,*

$$0 \to H_n(G,\mathbf{Z}) \bigotimes C \to H_n(G,C) \to Tor(H_{n-1}(G),C) \to 0$$
$$0 \to Ext(H_{n-1}(G,\mathbf{Z}),C) \to H^n(G,C) \to Hom(H_n(G,\mathbf{Z}),C) \to 0.$$

Moreover, both sequences split by an unnatural splitting.

(b) *Let $G_i, i = 1,2$, be two groups and let $G = G_1 \times G_2$ be their direct product. Then, the following sequence is exact:*

$$0 \to \bigoplus_{p+q=n} H_p(G_1,\mathbf{Z}) \otimes H_q(G_2,\mathbf{Z}) \to H_n(G,\mathbf{Z})$$
$$\to \bigoplus_{p+q=n-1} Tor(H_p(G,\mathbf{Z}),H_q(G_2,\mathbf{Z})) \to 0$$

Moreover, the sequence splits by an unnatural splitting.

☐ There is a universal coefficient isomorphism

$$H^2(G,A) \cong Hom_{\mathbf{Z}}(H_2(G,\mathbf{Z}),A)$$

for any abelian group A.

REMARK 1.21 - (ABELIAN GROUPS).

Let $\alpha : A \to B$ be a homomorphism between abelian groups. Set

$$\mathrm{coim}(\alpha) \equiv A/\ker(\alpha) \cong \mathrm{im}(\alpha)$$
$$\mathrm{coker}(\alpha) \equiv B/\mathrm{im}(\alpha).$$

If the sequence

$$0 \to A' \stackrel{\alpha'}{\to} A \stackrel{\alpha''}{\to} A'' \to 0$$

is exact, then

$$B = \mathrm{im}(\alpha') = \ker(\alpha'') \quad \text{(is a subgroup of } A\text{)}$$
$$\cong A'$$
$$A/B \cong A''.$$

☐ The **direct product** of $\{A_\lambda\}_{\lambda\in\Lambda}$ is defined by:

$$\prod_{\lambda\in\Lambda} A_\lambda \equiv \{a : \Lambda \to \{A_\lambda\}_{\lambda\in\Lambda} | a(\lambda) \in A_\lambda\}.$$

In $\prod_{\lambda\in\Lambda} A_\lambda$ we define an addition:

$$(a+b)(\lambda) = a(\lambda) + b(\lambda) \in A_\lambda .$$

Then, $\prod_{\lambda\in\Lambda} A_\lambda$ becomes an abelian group. The elements $a_\lambda \equiv a(\lambda)$ are called the **components** of $a = \{a_\lambda\} \in \prod_{\lambda\in\Lambda} A_\lambda$. One has a canonical epimorphism $\prod_\lambda : \prod_{\lambda\in\Lambda} A_\lambda \to A_\lambda, a \mapsto a_\lambda$.

☐ The **direct sum** of $\{A_\lambda\}_{\lambda\in\Lambda}$ is the subgroup $\bigoplus_{\lambda\in\Lambda} A_\lambda \subset \prod_{\lambda\in\Lambda} A_\lambda$ defined by:

$$\bigoplus_{\lambda\in\Lambda} A_\lambda \equiv \Big\{a \in \prod_{\lambda\in\Lambda} A_\lambda | \text{supp}\,(a) = \text{finite, i.e., } a_\lambda = 0, \text{ for almost all } \lambda \in \Lambda\Big\} .$$

☐ If Λ is finite then: $\bigoplus_{\lambda\in\Lambda} A_\lambda \cong \prod_{\lambda\in\Lambda} A_\lambda$.

One has the monomorphism:

$$i_\nu : A_\nu \to \bigoplus_\lambda A_\lambda, \left\{ \begin{array}{l} i_\nu(a)_\mu = 0\,,\ \mu \neq \nu \\ i_\nu(a)_\nu = a \end{array} \right\}$$

☐ (**Universal property**).

(a) *Let* $\{\phi_\lambda : X \to A_\lambda\}_{\lambda\in\Lambda}$ *be a family of homomorphisms, then there exists a unique homomorphism* $\phi : X \to \prod_{\lambda\in\Lambda} A_\lambda$ *such that*

$$\phi(x) = \{\phi_\lambda(x)\}_{\lambda\in\Lambda}\,,\ \forall x \in X .$$

We write $\phi = \{\phi_\lambda\}$, *and call these* $\phi_\lambda = \prod_\lambda \phi$ *the* **components** *of* ϕ.

(b) *If* $\{\psi_\lambda : A_\lambda \to X\}$ *is a family of homomorphisms, then there exists a unique homomorphism* $\psi : \bigoplus_\lambda A_\lambda \to X$.

☐ *A short exact sequence*

$$0 \to A' \xrightarrow{\alpha'} A \xrightarrow{\alpha''} A'' \to 0$$

splits iff α' *(resp.* α''*) is one component of a direct sum (resp. product) representation* $A' \bigoplus A'' \cong A$. *In this case* A' *is a* **direct summand** *of* A. *In particular, if* $A_1 \subset A$ *is a subgroup we see that the exact sequence*

$$0 \to A_1 \xrightarrow{i} A \to A/A_1 \to 0$$

splits, (i.e., A_1 *is a direct summand) iff the inclusion map* i *has a left inverse:* $r : A \to A_1, r \circ i = id_{A_1}$.

☐ If A is an abelian group and $a \in A$, we define

$$i_a : \mathbf{Z} \to A \quad , \quad i_a(n) = na .$$

Then, i_a is the unique homomorphism such that $i_a(1) = a$.

A subset $B \subset A$ is said to be a **basis** of A if the family $\{i_a\}_{b \in B}$ is a direct sum representation:

$$\{i_b\} : \bigoplus_{b \in B} \mathbf{Z} \cong A .$$

Then, every element $a \in A$ has a unique representation as a finite linear combination of basis elements with integral coefficients:

$$a = \sum_{b \in B} n_b b, n \in \mathbf{Z} \text{ , almost all } \quad n_b = 0 .$$

If an abelian group has a basis it is said to be **free**.

An abelian group is free iff it is isomorphic to a direct sum of groups $\mathbf{Z}$. *Furthermore, every subgroup of a free abelian group is free.*

Every abelian group A *is isomorphic to a quotient of a free abelian group.*

PROOF. In fact, if Λ is any subset of A which generates A, then there is a (unique) homomorphism

$$\xi : \mathbf{Z}\Lambda \to A \, , \, \lambda \mapsto \lambda \, ,$$

where $\mathbf{Z}\Lambda$ is the *free abelian group generated by* Λ:

$$a \in \mathbf{Z}\Lambda \Leftrightarrow a = \sum_{\lambda \in \Lambda} a_\lambda \lambda, a_\lambda \in \mathbf{Z} \text{ , almost all } \quad a_\lambda = 0 .$$

Then, ξ is an epimorphism (as Λ generates A)

$$0 \to \ker(\xi) \to \mathbf{Z}\Lambda \xrightarrow{\xi} A \to 0$$

hence

$$A \cong \mathbf{Z}\Lambda / \ker(\xi) .$$

□

The quotient does necessitate to be free!!! It is free if it is a direct summand.

If F *is a free abelian group, then every short exact sequence*

$$0 \to A' \to A \xrightarrow{\alpha} F \to 0$$

splits. Hence, $A \cong A' \bigoplus F$.

Every finitely generated abelian group A *is a finite direct sum of cyclic subgroups* $\{C_j \subset A\}$:

$$(1.19) \qquad A = \bigoplus_{1 \le j \le k} C_j, C_j \cong \mathbf{Z}/\nu_j\mathbf{Z} \equiv \mathbf{Z}_{\nu_j}, \nu_j \in \mathbf{Z}, \nu_j \ge 0.$$

(a) *The partial sum*

$$T \equiv \bigoplus_{\nu_j > 0} C_j = \bigoplus_{\nu_j > 1} C_j$$

is called **torsion subgroup** *of* A. *It is a finite group and consists of all elements of* A *of finite order.*

(b) *The quotient*

$$A/T \cong \bigoplus_{\nu_j = 0} \mathbf{Z}$$

is called **free part** *of* A. *The number of summands* $\mathbf{Z}$ *in* A/T *is called the* **rank** *of* A. *It does not depend on the particular direct sum decomposition* (1.19). *In fact,* $\text{rank}(A) \equiv$ **maximal number of linearly independent elements in** A.

☐ The numbers $\nu_j > 1$ (**torsion coefficients** of A) that occur in (1.19) are not unique. However, they can be chosen as powers of prime numbers $\nu_j = p_j^{\rho_j}$, p_j prime, $\rho_j > 0$, and then they are unique (independent of the decomposition) up to permutation coefficients.

Two finitely generated abelian groups are isomorphic iff they have the same rank and the same system of torsion coefficients.

☐ If A is a finitely generated abelian group and, $A' \subset A$ is a subgroup, $\Rightarrow$ A' and A/A' are also finitely generated.

$$\text{rank}(A) = \text{rank}(A') + \text{rank}(A/A').$$

☐ *For arbitrary abelian groups* G *one can define a rank as follows:*

(a) *If* G *is free,* $\text{rank}(G) =$ *cardinality of a basis.*

(b) *Otherwise:* $\text{rank}(G) = \textit{supremum}\{\text{rank}(F)\}$, $F \in$ *set of free subgroups of* G. *Then*

$$\text{rank}(G) = \text{rank}(G') + \text{rank}(G/G'), \ \forall G' \subset G.$$

DEFINITION 1.23 - 1) *Let* $\mathcal{U}$ *be an abelian category. A* **differential object** *in* $\mathcal{U}$ *is a pair* (A, d) *where: (a)* $A \in Ob(\mathcal{U})$; *(b)* $d \in Hom_{\mathcal{U}}(A, A)$, *such that* $d^2 = 0$.

2) *Let* $\mathcal{D}(\mathcal{U})$ *be the* **category of differential objects** *in* $\mathcal{U}$. *We call* **homology** *the additive functor*

$$H : \mathcal{D}(\mathcal{U}) \to \mathcal{U}, \textit{given by} \quad (A, d) \mapsto H(A) \equiv \ker(d)/\text{im}(d) \equiv Z(A)/B(A),$$

where $Z(A) \equiv$ **set of cycles** *of* A, $B(A) \equiv$ **set of boundaries** *of* A, $H(A) \equiv$ **homology** *of* A.

DEFINITION 1.24 - 1) *A* **spectral sequence** *in* $\mathcal{U}$ *is a sequence of differential objects of* $\mathcal{U}$:

$$\mathbb{E} \equiv \{E_n, d_n\}, n = 1, 2, 3, \cdots,$$

such that

$$H(E_n, d_n) = E_{n+1}\ ,\ n = 1, 2, 3, \cdots .$$

2) *A* **morphism** $\phi : \mathbb{E} \to \mathbb{E}'$ *of spectral sequences is a sequence of morphisms* $\phi_n : E_n \to E'_n$ *of* $\mathcal{D}(\mathcal{U})$ *such that*

$$H(\phi_n) = \phi_{n+1}\ ,\ n = 1, 2, \cdots .$$

We write $\mathcal{S}(\mathcal{U})$ *for the* **category of spectral sequences** *in* $\mathcal{U}$.

PROPOSITION 1.9 - *The following propositions are equivalent:*

1) $\{E_n, d_n\}, n = 1, 2, 3, \cdots$, *is a spectral sequence of objects of* $\mathcal{D}(\mathcal{U})$.

2) $E_1 \in Ob(\mathcal{U})$ *is endowed with a tower of subobjects*

$$B_1 \subset B_2 \subset \cdots \subset B_n \subset \cdots \subset Z_n \subset \cdots \subset Z_2 \subset Z_1 \subset E_1 \tag{1.20}$$

such that

$$Z_n/Z_{n+1} \cong B_{n+1}/B_n\ ,\ \forall n. \tag{1.21}$$

PROOF. In fact if $\{E_n, d_n\}, n = 1, 2, 3, \cdots$, is a spectral sequence, we have $E_2 = Z_1/B_1$, where $B_1 \subset Z_1 \subset E_1$. On the other hand $E_3 = \bar{Z}_2/\bar{B}_2$, where $\bar{B}_2 \equiv \operatorname{im}(d_2) \subset \bar{Z}_2 \equiv \ker(d_2) \subset E_2$. So we can write $\bar{B}_2 = B_2/B_1, \bar{Z}_2 = Z_2/B_1$, for some subobject B_2 and Z_2 of Z_1 such that $B_1 \subset B_2 \subset Z_2 \subset Z_1$. Then we have $E_3 = (Z_2/B_1)/(B_2/B_1) \cong Z_2/B_2$. Iterating this process we have the filtration (1.20) . Furthermore, the following short exact sequence, induced by d_n:

$$0 \to Z_n/B_{n-1} \to Z_{n-1}/B_{n-1} \xrightarrow{d_n} B_n/B_{n-1} \to 0$$

gives rise to isomorphisms (1.21) . Conversely, starting with a tower of subobjects (1.20) we can define the sequence $\{E_n, d_n\}, n = 1, 2, 3, \cdots$, of objects of $\mathcal{D}(\mathcal{U})$, where $E_n \equiv Z_{n-1}/B_{n-1}$ and the differential d_n can be taken as a mapping $Z_{n-1}/B_{n-1} \to Z_{n-1}/B_{n-1}$ which has kernel Z_n/B_{n-1} and image B_n/B_{n-1} . Then we have

$$H(E_n, d_n) = (Z_n/B_{n-1})/(B_n/B_{n-1}) \cong Z_n/B_n \equiv E_{n+1}$$

So $\{E_n, d_n\}$ is a spectral sequence. □

DEFINITION 1.25 - 1) *Let* $\{E_n, d_n\}, n = 1, 2, 3, \cdots$, *be a spectral sequence and let* $B_1 \subset B_2 \subset \cdots \subset B_n \subset \cdots \subset Z_n \subset \cdots \subset Z_2 \subset Z_1 \subset E_1$ *be the corresponding filtration of* E_1 . *Then, we define the following subobjects of* E_1:

$$B_\infty \equiv \bigcup_n B_n\ ,\ Z_\infty \equiv \bigcap_n Z_n\ .$$

Furthermore put

$$E_\infty \equiv Z_\infty / B_\infty \,.$$

2) *A spectral sequence* $\{E_n, d_n\}$ **converges finitely** *if there exists* n_0 *such that*

$$E_{n_0} = E_{n_0+1} = \cdots = E_\infty \,.$$

3) *A spectral sequence* $\{E_n, d_n\}$ **collapses** *at the* N^{th} *term if the differentials* $d_n = 0$ *for* $n \geq N$ *. (Then* **converges finitely** *as* $E_N \cong E_{N+1} = \cdots = E_\infty$*.)*

4) *A spectral sequence* $\{E_n, d_n\}$ **converges** *to* $H \in Ob(\mathcal{U})$ *if there exists a filtration* F *of* H*:*

$$\cdots \subset F^{p+1}(H) \subset F^p(H) \subset F^{p-1}(H) \subset \cdots \subset H \; (\textbf{decreasing filtration})$$

or

$$\cdots \subset F^{p-1}(H) \subset F^p(H) \subset F^{p+1}(H) \subset \cdots \subset H \; (\textbf{increasing filtration})$$

such that

$$E_\infty \cong E_0(H, F) \equiv \left\{ \begin{array}{l} \bigoplus_{p\in\mathbf{Z}} F^p(H)/F^{p+1}(H)\,, \text{ if } F \text{ is decreasing,} \\ \bigoplus_{p\in\mathbf{Z}} F^p(H)/F^{p-1}(H)\,, \text{ if } F \text{ is increasing} \end{array} \right\},$$

where E_∞ *is the* **limit of the spectral sequences**.

EXAMPLE 1.14 - 1) **Spectral sequences of differential bigraded** R**-modules**, $R \equiv$ ring with unit.

☐ Let (A, d_1) and (B, d_2) be differential graded modules over R. Recall the definitions of the **tensor product** of differential graded modules over R:

$$(A \bigotimes_R B, D), (A \bigotimes_R B)^n \equiv \bigoplus_{p+q=n} A^p \bigotimes_R B^q, D(a \otimes b) = d_A(a) \otimes b + (-1)^{|a|} a \otimes d_B(b)\,.$$

☐ A **differential bigraded module** over R is a collection of R-modules, $\{E^{p,q}\}$, where p and q are integers together with an R-linear mapping $d : E^{\bullet,\bullet} \to E^{\bullet,\bullet}$, the **differential**, of bidegree $(s, 1-s)$, or $(-s, s-1)$, for some integer s, satisfying $d^2 = 0$. The **homology** of $(E^{\bullet,\bullet}, d)$ is :

(a) for differentials of bidegree $(s, 1-s)$:

$$H^{p,q}(E^{\bullet,\bullet}, d) = \frac{\ker(d : E^{p,q} \to E^{p+s,q-s+1})}{\operatorname{im}(d : E^{p-s,q+s-1} \to E^{p,q})}$$

(b) for differential of bidegree $(-s, s-1)$:

$$H^{p,q}(E^{\bullet,\bullet}, d) = \frac{\ker(d : E^{p,q} \to E^{p-s,q+s-1})}{\mathrm{im}(d : E^{p+s,q-s+1} \to E^{p,q})} .$$

☐ The **tensor product of differential bigraded modules** over R,

$$(E^{\bullet,\bullet}, d_E), (\bar{E}^{\bullet,\bullet}, d_{\bar{E}})$$

is a differential bigraded module $(E \bigotimes_R \bar{E}, D)$ with

$$\left\{ \begin{array}{l} (E \bigotimes\limits_R \bar{E})^{p,q} = \bigoplus\limits_{m+n=p, r+s=q} E^{m,r} \bigotimes\limits_R \bar{E}^{n,s} \\ D(e \otimes \bar{e}) = d_E(e) \otimes \bar{e} + (-1)^{r+m} e \otimes d_{\bar{E}}(\bar{e}) \\ e \in E^{m,r}, \bar{e} \in \bar{E}^{n,s}. \end{array} \right\}$$

☐ **A spectral sequence of differential bigraded modules** is a collection of differential bigraded modules $\{E_r^{\bullet,\bullet}, d_r\}$, where $r = 1, 2, \ldots$; the differentials are either all of bidegree $(-r, r-1)$, (for a spectral sequence of **homology type**), or all of bidegree $(r, 1-r)$, (for a spectral sequence of **cohomological type**), and

$$E_{r+1}^{p,q} \cong H^{p,q}(E_r^{\bullet,\bullet}, d_r) .$$

2) A **first quadrant spectral sequence** (of **cohomology type**) is a sequence of differential bigraded R-modules $\{E_r^{\bullet,\bullet}, d_r\}$, with differential of bidegree $(r, 1-r)$, such that $E_r^{p,q}$ are zero outside the first quadrant of the plane (p, q), and

$$E_{r+1}^{\bullet,\bullet} \cong H(E_r^{\bullet,\bullet}, d_r) = \frac{\ker(d_r : E_r^{p,q} \to E_r^{p+r,q-r+1})}{\mathrm{im}(d_r : E_r^{p-r,q+r-1} \to E_r^{p,q})} .$$

Such spectral sequence converges finitely. In fact, consider $E_r^{p,q}$ for $r > \max(p, q+1)$, here the differentials, d_r, become trivial: since $q+1-r < 0$, we have $E_r^{p+r,q-r+1} = \{0\}$ and so $\ker(d_r) = E_r^{p,q}$. Also $p - r < 0$ implies $E_r^{p-r,q+r-1} = \{0\}$ and $\mathrm{im}(d_r) = \{0\}$. Thus $E_{r+1}^{p,q} = E_r^{p,q}$ and, continuing in a similar fashion

$$E_{r+k}^{p,q} = E_r^{p,q}, \quad \text{for} \quad k \geq 0 .$$

We denote this common vector space just by $E_\infty^{p,q}$.
$\{E_r^{\bullet,\bullet}, d_r\}$ **converges** to a graded R-module $H^\bullet$ if $H^\bullet$ has a filtration $F^\bullet$ such that

$$\left\{ \begin{array}{l} E_\infty^{p,q} \cong (F^p H^{p+q} / F^{p+1} H^{p+q}) \equiv E_0^{p,q}(H^\bullet) \\ \text{where} \quad F^p H^r \equiv F^p H^\bullet \bigcap H^r \\ H^r \equiv \bigoplus\limits_{p+q=r} E_0^{p,q} . \end{array} \right\}$$

□ Similarly we can define **second quadrant spectral sequences**, $(p \leq 0, q \geq 0)$, **third quadrant spectral sequences**, $(p \leq 0, q \leq 0)$, and **fourth quadrant spectral sequences** $(p \geq 0, q \leq 0)$.

□ Let $\{E_r^{\bullet,\bullet}, d_r\}$ be a first quadrant spectral sequence of differential bigraded vector spaces. Suppose $E_2^{p,q} = \{0\}$ whenever p is even or q is odd. Then the spectral sequence collapses at E_2 term.

This, for example, can occur if the E_2-term is given by a tensor product $E_2^{p,q} = V^p \bigotimes W^q$ and $V^\bullet$ and $W^\bullet$ are graded vector spaces such that $V^\bullet$ is concentrated in odd dimensions and $W^\bullet$ is concentrated in even dimensions.

□ (**Zeeman's comparison theorem**). Suppose $\{E_r^{\bullet,\bullet}, d_r\}$ and $\{\bar{E}_r^{\bullet,\bullet}, \bar{d}_r\}$ are first quadrant spectral sequences of cohomology type (i.e., d_r and $\bar{d}_r$ of bidegree $(r, 1-r)$). Suppose $\{f_r\}$ is a morphism of spectral sequences. Emphasize that a **morphism** $f : \{E_r^{\bullet,\bullet}, d_r\} \to \{\bar{E}_r^{\bullet,\bullet}, \bar{d}_r\}$ between two spectral sequences is a sequence of homomorphisms $f_r : E_r^{\bullet,\bullet} \to \bar{E}_r^{\bullet,\bullet}, r = 1, 2, \cdots$ of bigraded objects, of bidegree $(0,0)$, such that f_r commutes with the differentials, $f_r \circ d_r = \bar{d}_r \circ f_r$, and each f_{r+1} is induced by f_r on homology, that is f_{r+1} is the composite

$$f_{r+1} : E_{r+1}^{\bullet,\bullet} \cong H(E_r^{\bullet,\bullet}, d_r) \underset{H(f_r)}{\to} H(\bar{E}_r^{\bullet,\bullet}, \bar{d}_r) \cong \bar{E}_{r+1}^{\bullet,\bullet} .$$

Let us, now, assume that the following diagram commutes with exact rows:

$$\begin{array}{ccccccccc}
0 & \to & E_2^{p,0} \bigotimes E_2^{0,q} & \to & E_2^{p,q} & \to & Tor_1^R(E_2^{p+1,0}, E_2^{0,q}) & \to & 0 \\
 & & {\scriptstyle f_2 \otimes f_2} \downarrow & & \downarrow {\scriptstyle f_2} & & \downarrow {\scriptstyle Tor(f_2, f_2)} & & \\
0 & \to & \bar{E}_2^{p,0} \bigotimes \bar{E}_2^{0,q} & \to & \bar{E}_2^{p,q} & \to & Tor_1^R(\bar{E}_2^{p+1,0}, \bar{E}_2^{0,q}) & \to & 0
\end{array}$$

Then any two of the following conditions imply the third:

I. $f_2 : E_2^{p,0} \to \bar{E}_2^{p,0}$ is an isomorphism for all p.

II. $f_2 : E_2^{0,q} \to \bar{E}_2^{0,q}$ is an isomorphism for all q.

III. $f_\infty : E_\infty^{p,q} \to \bar{E}_\infty^{p,q}$ is an isomorphism for all p, q.

□ (**Künneth spectral sequence**). Let $(K^\bullet, d_K)$ and $(L^\bullet, d_L)$ be differential graded modules over R with $K^\bullet$ flat. Then there is a spectral sequence with

$$E_2^{p,q} = \bigoplus_{s+t=q} Tor_R^p(H^s(K), H^t(L)) .$$

If K and L have differentials of degree $+1$, this is a second quadrant spectral sequence. If K and L have differentials of degree -1, this is a first quadrant spectral sequence. When $E_r^{\bullet,\bullet}$ converges, it does so to $H(K \bigotimes_R L)$ in either case.

3) **Spectral sequence of algebras** over R. If $H^\bullet$ is a graded R-module , then let $H^\bullet \bigotimes_R H^\bullet$ be the graded R-module given by

$$(H^\bullet \bigotimes_R H^\bullet)^n = \bigoplus_{p+q=n} H^p \bigotimes_R H^q \,.$$

$H^\bullet$ is a **graded algebra** if there is a mapping of graded vector spaces,

$$\phi : H^\bullet \bigotimes_R H^\bullet \to H^\bullet \,,$$

that is a R-linear mapping $\phi : H^p \bigotimes_R H^q \to H^{p+q}$ for all p and q . The mapping ϕ must satisfy the followng commutative diagram, namely the multiplication on $H^\bullet$ is associative:

$$\begin{array}{ccc} H^\bullet \bigotimes_R H^\bullet \bigotimes_R H^\bullet & \stackrel{\phi\otimes 1}{\to} & H^\bullet \bigotimes_R H^\bullet \\ {\scriptstyle 1\otimes\phi} \downarrow & & \downarrow {\scriptstyle \phi} \\ H^\bullet \bigotimes_R H^\bullet & \to & H^\bullet \end{array}$$

If the algebra $H^\bullet$ contains a unit element there is a mapping $\eta : R \to H^\bullet$, where R is the graded algebra combining the ring R in degree 0 and $\{0\}$ in higher degrees, such that one has the following commutative diagram:

$$\begin{array}{ccccc} R \bigotimes_R H^\bullet & \to & H^\bullet \bigotimes_R H^\bullet & \leftarrow & H^\bullet \bigotimes_R R \\ \downarrow & & \downarrow \phi & & \downarrow \\ H^\bullet & = & H^\bullet & = & H^\bullet \end{array}$$

□ Let $E^{\bullet,\bullet}$ be a bigraded R-module. $E^{\bullet,\bullet}$ is a **bigraded algebra** if there is a mapping of bigraded R-modules $\phi : E^{\bullet,\bullet} \bigotimes_R E^{\bullet,\bullet} \to E^{\bullet,\bullet}$, i.e., ϕ has components $\phi : E^{m,n} \bigotimes_R E^{r,s} \to E^{n+r,s+s}$. We assume further that ϕ satisfies the analogous conditions of associativity and having a unit.

EXAMPLE 1.15 - As an example of bigraded algebra, suppose $A^\bullet$ and $B^\bullet$ are graded algebras with ϕ and ψ their respective products. If we let $E^{p,q} \equiv A^p \bigotimes_R B^q$, we can define an algebra structure on $E^{\bullet,\bullet}$ by the composite:

$$\begin{array}{ccccc} A^p \otimes_R B^q \otimes_R A^r \otimes_R B^s & \stackrel{1\otimes T\otimes 1}{\to} & A^p \otimes_R A^r \otimes_R B^q \otimes_R B^s & \stackrel{\phi\otimes\psi}{\to} & A^{p+r} \otimes_R B^{q+s} \\ \| & & & & \| \\ E^{p,q} \otimes_R E^{r,s} & & \underset{\phi}{\to} & & E^{p+r,q+s} \end{array}$$

where $T(b \bigotimes a) = (-1)^{|a|} a \otimes b$.

□ A **differential graded algebra** over R , $(A^\bullet, d)$, is a graded algebra over R with a degree 1 linear mapping $d : A^\bullet \to A^\bullet$, which satisfies the Leibnitz rule:

$$d(a.a') = d(a).a' + (-1)^{|a|} a.d(a') \,.$$

□ A **differential bigraded algebra** over R, $(E^{\bullet,\bullet}, d)$, is a bigraded algebra over R with a total degree 1 mapping

$$d : \bigoplus_{p+q=n} E^{p,q} \to \bigoplus_{r+s=n+1} E^{r,s}$$

which satisfies the Leibniz rule: $d(e.e') = d(e).e' + (-1)^{p+q} e.d(e')$, when $e \in E^{p,q}$ and $e' \in E^{r,s}$.

EXAMPLE 1.16 - An example of differential bigraded algebra can be constructed from two differential graded algebras $(A^\bullet, d)$ and $(B^\bullet, d')$ by setting $E^{\bullet,\bullet} \equiv A^\bullet \bigotimes B^\bullet$ and defining the differential on $E^{\bullet,\bullet}$ by the formula $d(a \otimes b) = d(a) \otimes b + (-1)^{|a|} a \otimes d'(b)$.

$\{E_r^{\bullet,\bullet}, d_r\}$ is a **spectral sequence of algebras** if, for each r, $(E_r^{\bullet,\bullet}, d_r)$ is a differential bigraded algebra and furthermore, the product on $E_{r+1}^{\bullet,\bullet}$ is induced by the product of $E_r^{\bullet,\bullet}$ on homology. That is, the product on $E_{r+1}^{\bullet,\bullet}$, $\psi_{r+1} : E_{r+1} \bigotimes_R E_{r+1} \to E_{r+1}$ may be expressed as the composite:

$$\psi_{r+1} : E_{r+1} \bigotimes_R E_{r+1} \cong H(E_r) \bigotimes_R H(E_r) \xrightarrow{p} H(E_r \bigotimes_R E_r) \xrightarrow{H(\psi_r)} H(E_r) \cong E_{r+1}$$

where the map p is as in the Künneth theorem, that is given in the following short exact sequence:

$$0 \to \bigoplus_{r+s=n} H^r(A) \bigotimes_R H^s(B) \xrightarrow{p} H^n(A \bigotimes_R B) \to \bigoplus_{r+s=n-1} Tor_1^R(H^r(A), H^s(B)) \to 0$$

where $p([u] \otimes [v]) = [u \otimes v]$. In the case that $H^n(A)$ and $H^n(B)$ are projective R-modules for all n, the homomorphism p is an isomorphism.

Suppose $F^\bullet$ is a filtration of $H^\bullet$, a graded algebra with product ϕ. The filtration is said to be **stable** with respect to the product if $\phi(F^r(H^\bullet) \bigotimes F^s(H^\bullet)) \subset F^{r+s}(H^\bullet)$.

Thus a spectral sequence of algebras $\{E_r^{\bullet,\bullet}, d_r\}$ **converges** to $H^\bullet$ as a graded algebra if there is a stable filtration on $H^\bullet$ with $E_0^{\bullet,\bullet}(H^\bullet, F^\bullet)$ isomorphic as a bigraded algebra to the E_∞-term of the spectral sequence.

4) **Spectral sequences of differential bigraded vector spaces locally finite.**

□ Let $A^\bullet$ be a graded space over a field $\mathbf{K}$. We say that $A^\bullet$ is **locally finite** (or of **finite type**) if $\dim_{\mathbf{K}} A^n$ is finite for all n.

If $A^\bullet$ is locally finite, the **Poincaré series** for $A^\bullet$ is the formal power series

$$P(A^\bullet, t) = \sum_{n=0}^{\infty} (\dim_{\mathbf{K}} A^n) t^n .$$

One has:

(a) $A^\bullet$ is globally finite dimensional iff $P(A^\bullet, t)$ is a polynomial;
(b) $P(A^\bullet \bigotimes B^\bullet, t) = P(A^\bullet, t) \times P(B^\bullet, t)$, where the multiplication on the right is the Cauchy product of power series, that is

$$(\sum_{i=0}^{\infty} a_i t^i) \otimes (\sum_{j=0}^{\infty} b_j t^j) = \sum_{k=0}^{\infty} c_k t^k, c_k = \sum_{i=0}^{k} a_i b_{k-i} .$$

☐ The **Euler characteristic** for $A^\bullet$ is $\chi(A^\bullet) = P(A^\bullet, -1)$, where this expression makes sense.
If $A^\bullet = H^\bullet(M; \mathbf{Q})$ for M a finite dimensional manifold, this $\chi(A^\bullet)$ is the classical Euler-Poincaré characteristic.
☐ Let $E^{\bullet,\bullet}$ be a locally finite bigraded vector space, the **Poincaré -series** for $E^{\bullet,\bullet}$ is the formal series:

$$P(E^{\bullet,\bullet}, t) = \sum_{n=0}^{\infty} \dim_{\mathbf{K}}(\bigoplus_{p+q=n} E^{p,q}) t^n .$$

Let $\{E_r^{\bullet,\bullet}, d_r\}$ be a spectral sequence converging to $H^\bullet$ and that $E_2^{\bullet,\bullet}$ is locally finite. Then

$$P(E_2^{\bullet,\bullet}, t) \geq P(E_3^{\bullet,\bullet}, t) \geq \cdots \geq P(E_\infty^{\bullet,\bullet}, t) \geq P(H^\bullet, t) .$$

Furthermore, $P(H^\bullet, t) = P(E_i^{\bullet,\bullet}, t)$ for some finite i iff the spectral sequence collapses at the i-th term. Finally, if we let $\chi(E_r^{\bullet,\bullet}) = P(E_r^{\bullet,\bullet}, -1)$, whenever these expressions are meaningful, then $\chi(E_r^{\bullet,\bullet}) = \chi(H^\bullet)$ for all $r \geq 2$.
5) **Spectral sequences of $\Gamma^\bullet$-modules**, $\Gamma^\bullet \equiv$ graded algebra over R.
☐ Let $\Gamma^\bullet$ be a graded-algebra over R. A graded vector space $H^\bullet$ is a $\Gamma^\bullet$-module if there is a mapping of graded vector spaces $\psi : \Gamma^\bullet \bigotimes_R H^\bullet \to H^\bullet$ satisfying the usual module axioms; that is, the following diagram commutes with m, the multiplication on $\Gamma^\bullet$:

$$\begin{array}{ccc} \Gamma^\bullet \bigotimes \Gamma^\bullet \bigotimes H^\bullet & \rightarrow & \Gamma^\bullet \bigotimes H^\bullet \\ \downarrow m\otimes 1 & & \downarrow \psi \\ \Gamma^\bullet \bigotimes H^\bullet & \stackrel{\psi}{\rightarrow} & H^\bullet \end{array}$$

☐ Let $E^{\bullet,\bullet}$ be a bigraded vector space and $\Gamma^\bullet$ a graded algebra. We say that $\Gamma^\bullet$ **acts (vertically)** on $E^{\bullet,\bullet}$ if, for each $n \geq 0, E^{n,\bullet}$ is a $\Gamma^\bullet$-module; that is, there is a mapping $\psi_n : \Gamma^\bullet \bigotimes_R E^{n,\bullet} \to E^{n,\bullet}$.
☐ For a graded algebra $\Gamma^\bullet$ to **act on a spectral sequence**, $\{E_r^{\bullet,\bullet}, d_r\}$, we require that: (a) $\Gamma^\bullet$ acts on $E_r^{\bullet,\bullet}$ for each r; (b) the differentials d_r are $\Gamma^\bullet$-linear and; (c) the $\Gamma^\bullet$- action on $E_{r+1}^{\bullet,\bullet}$ is induced through homology from the action of $\Gamma^\bullet$ on $E^{\bullet,\bullet}$.
☐ Let $H^\bullet$ be a $\Gamma^\bullet$-module. We say that a **spectral sequence converges** to $H^\bullet$ as a $\Gamma^\bullet$- module if we have a spectral sequence converging to $H^\bullet$, on which $\Gamma^\bullet$ acts and

the filtration on $H^\bullet$ induces a $\Gamma^\bullet$-action on the associated bigraded space $E_0^{\bullet,\bullet}(H^\bullet)$ that is isomorphic to the $\Gamma^\bullet$-action on the E_∞-term of the spectral sequence.

6) **Spectral sequence of graded commutative** (or **skew-commutative**) **algebras.**

☐ A graded algebra $H^\bullet$ is said to be **graded commutative** if

$$x.y = (-1)^{pq} y.x, \quad \forall x \in H^p, \quad y \in H^q .$$

☐ A bigraded algebra $E^{\bullet,\bullet}$ is said to be **graded commutative** if

$$x.y = (-1)^{(r+s)(p+q)} y.x, \quad \forall x \in E^{p,q}, \quad y \in E^{r,s} .$$

If there is a spectral sequence converging to $H^\bullet$ as an algebra and the E_∞-term is a free bigraded skew-commutative algebra, then $H^\bullet$ is a free graded commutative algebra isomorphic to

$$TotE_\infty^{\bullet,\bullet} \equiv \bigoplus_{n\in\mathbf{Z}} \bigoplus_{p+q=n} E_\infty^{p,q} .$$

REMARK 1.22 - In the following we shall show how spectral sequences arise. The two principal sources of spectral sequences are filtered differential graded objects and exact couples.

DEFINITION 1.26 - (FILTERED DIFFERENTIAL GRADED OBJECTS). *A* **filtered differential graded object** *in $\mathcal{U}$ is an object $A \in Ob(\mathcal{U})$ such that:*

(a) $A \in Ob(\mathcal{D}(\mathcal{U}))$;

(b) *A is an object sum of subobjects of A: $A = \bigoplus_{n=0}^{\infty} A^n$;*

(c) *The differential $d : A \to A$ has a fixed degree: $|d| = s \in \mathbf{Z}$, (in the following we shall take $s = 1$);*

(d) *A has a* **filtration** *F and the differential d with respect to the filtration, that is, $d : F^p A \to F^p A$.*

If A is a filtered differential graded object in $\mathcal{U}$, then its homology $H(A)$ is filtered one: $F^p H(A) = \mathrm{im}(i_* : H(F^p A) \to H(A))$, where $i_* = H(i)$, and $i : F^p A \to A$ is the canonical inclusion.

THEOREM 1.14 - 1) *Let $\mathcal{D}(\mathcal{U}^{\mathbf{Z}})_f$ be the* **category of filtered differential graded objects** *in $\mathcal{U}$. Then one has a canonical functor:*

$$\bar{S} : \mathcal{D}(\mathcal{U}^{\mathbf{Z}})_f \to \mathcal{S}(\mathcal{U}^{\mathbf{Z}\times\mathbf{Z}}) .$$

2) *Moreover, if the filtration is* **bounded** *(or* **finite***), that is, for each dimension n, there are values $s = s(n)$ and $t = t(n)$, so that $\{0\} = F^s A^n \subset F^{s-1} A^n \subset \cdots \subset F^{t+1} A^n \subset F^t A^n = A^n$, then the spectral sequence converges to $H(A)$, that is*

$$E_\infty^{p,q} \cong F^p H^{p+q}(A) / F^{p+1} H^{p+q}(A) .$$

PROOF. 1) In fact, each filtered differential graded object $A \in Ob(\mathcal{U}^{\mathbf{Z}})$ determines a spectral sequence

$$\bar{S}(A) \equiv \{E_r^{\bullet,\bullet}, d_r\}\,,\ r = 1, 2, \cdots\,,$$

with d_r of bidegree $(r, 1-r)$ and

$$E_1^{p,q} \cong H^{p+q}(F^p A/F^{p+1}A)\,.$$

More precisely, set

$$E_r^{p,q} \equiv Z_r^{p,q}/(Z_{r-1}^{p+1,q-1} + B_{r-1}^{p,q})\,,$$

where

$$\left\{\begin{array}{l} Z_r^{p,q} = F^p A^{p+q} \bigcap d^{-1}(F^{p+r}A^{p+q+1}) \\ B_r^{p,q} = F^p A^{p+q} \bigcap d(F^{p-r}A^{p+q-1}) \\ Z_\infty^{p,q} = \ker d \bigcap F^p A^{p+q},\, B_\infty^{p,q} \\ \qquad = \operatorname{im} d \bigcap F^p A^{p+q}\,. \end{array}\right\}$$

One has the following tower of submodules: $B_0^{p,q} \subset B_1^{p,q} \subset \cdots \subset B_\infty^{p,q} \subset Z_\infty^{p,q} \subset \cdots \subset Z_1^{p,q} \subset Z_0^{p,q}$, and $d(Z_r^{p-r,q+r-1}) = B_r^{p,q}$. The differential d identifies a mapping $d : Z_r^{p,q} \to Z_r^{p+r,q-r+1}$ and a homomorphism $d_r : E_r^{p,q} \to E_r^{p+r,q-r+1}$ such that the following diagram commutes

$$\begin{array}{ccccc} & Z_r^{p,q} & \xrightarrow{d} & Z_r^{p+r,q-r+1} & \\ \pi_r^{p,q} & \downarrow & & \downarrow & \pi_r^{p+r,q-r+1} \\ & E_r^{p,q} & \underset{d_r}{\longrightarrow} & E_r^{p+r,q-r+1} & \end{array}$$

with $d_r d_r = 0$, (following from $dd = 0$). Furthermore, by some calculation we can see that

$$H(E_r^{p,q}, d_r) \equiv E_{r+1}^{p,q}\,.$$

2) The assumption that the filtration is bounded implies, for $r > s(p+q-1) - p$ and $r \geq p - t(p+q-1)$, that $Z_r^{p,q} = Z_\infty^{p,q}$ and $B_r^{p,q} = B_\infty^{p,q}$. Thus the spectral sequence converges . We have , $E_0^{p,q} = Z_0^{p,q}/(Z_{-1}^{p+1,q-1} + B_{-1}^{p,q})$, where we define $Z_{-1}^{p+1,q-1} = F^{p+1}A^{p+q}$ and $B_{-1}^{p,q} = d(F^{p+1}A^{p+q-1})$. So

$$\left\{\begin{array}{l} E_0^{p,q} = F^p A^{p+q} \bigcap d^{-1}(F^p A^{p+q+1})/F^{p+1}A^{p+q} + d(F^{p+q}A^{p+q-1}) \\ \qquad = F^{p+q}A^{p+q}/F^{p+1}A^{p+q}\,. \end{array}\right\}$$

Now the differential $d : F^pA^{p+q} \to F^{p+1}A^{p+q}$ induces the map $d : E_0^{p,q} \to E_0^{p,q-1}$ and so we have $E_1^{p,q} \cong H^{p+q}(F^pA/F^{p+q}A)$. Furthermore, by some calculation we can see that $E_\infty^{p,q} \cong F^pH^{p+q}(A,d)/F^{p+1}H^{p+q}(A,d)$. □

REMARK 1.23 - For the convergence we can give a weaker candition than the bounded filtration. Let us first introduce some definitions and results about filtrations of objects $A \in Ob(\mathcal{D}(\mathcal{U}^{\mathbf{Z}})_f)$:

A) The filtration is **exhaustive**, if $A = \bigcup_s F^sA$.

B) The filtration is **weakly convergent**, if for all $p, Z^p_\infty = \bigcap_p Z^p_r$, where $Z^p_r = F^pA \bigcap d^{-1}(F^{p+r}A)$, that is $F^pA \bigcap \ker d = \bigcap_r (F^pA \bigcap d^{-1}(F^{p+r}A))$. (Some simple conditions on the filtration imply that it is weakly convergent; for example:

1) the filtration is **bounded above**, that is for each n , there is a value $s(n)$ with, $F^{s(n)}A = \{0\}$;

2) $\bigcap_p F^pA = \{0\}$.)

C) The filtration is **convergent** (or **Hausdorff**) if it is weakly convergent and $\bigcap_p F^pH(A) = \{0\}$.

□ **(Cartan-Eilenberg)**. The following conditions are equivalent for a filtration F of a differential graded object $A \in Ob(\mathcal{D}(\mathcal{U}^Z)_f)$:

1) F is weakly convergent;

2)

$$\bigcap_{r\geq 1} \operatorname{im}(H(F^pA/F^{p+r}A) \to H(F^{p+1}A)) = \{0\};$$

3) For all p, the mappings induced by the filtration, $R^{p+1} \to R^p$, are monomorphisms, where

$$R^p = \bigcap_r \operatorname{im}(H(F^{p+r}A) \to H(F^pA)) .$$

THEOREM 1.15 - *Let $A \in Ob(\mathcal{D}(\mathcal{U}^{\mathbf{Z}})_f)$ such that the filtration is exhaustive and weakly convergent. Then there is spectral sequence with*

$$E_1^{p,q} \cong H^{p+q}(F^pA/F^{p+1}A),$$

which converges to $H(A)$.

PROOF. The construction is that of the above theorem. □

DEFINITION 1.27 - (EXACT COUPLES). *An* **exact couple** $\mathbb{E}C \equiv \{D, E; \alpha, \beta, \gamma\}$ *in $\mathcal{U}$ is an exact diagram of morphisms in $\mathcal{U}$ like the following:*

$$\begin{array}{ccc} D & \stackrel{\alpha}{\to} & D \\ \gamma \uparrow & & \downarrow \beta \\ E & = & E \end{array}$$

A **morphism** $\Phi : \mathbb{E}C \to \mathbb{E}C'$ *between two exact couples is a pair of morphisms* $(\kappa : D \to D', \lambda : E \to E')$ *such that* $\alpha'\kappa = \kappa\alpha, \beta'\kappa = \lambda\beta, \gamma'\lambda = \kappa\gamma$. *Let* $\mathcal{E}(\mathcal{U})$ *be the category of exact couples in* $\mathcal{U}$.

THEOREM 1.16 - *One has the canonical functor* $S : \mathcal{E}(\mathcal{U}) \to \mathcal{S}(\mathcal{U})$ *such that*

$$\mathbb{E}C \equiv \left\{ \begin{array}{ccc} D & \overset{\alpha}{\to} & D \\ {}_{\gamma}\uparrow & & \downarrow_{\beta} \\ E & = & E \end{array} \right\} \mapsto S(\mathbb{E}C) = (E_n, d_n), \forall \mathbb{E}C \in Ob(\mathcal{E}(\mathcal{U})),$$

where $E_1 = E, d_1 = \beta\gamma \Rightarrow$ *(intermediate exact couple)*

$$\begin{array}{ccc} D_2 & \overset{\alpha_2}{\to} & D_2 \\ {}_{\gamma_2}\uparrow & & \downarrow_{\beta_2} \\ E_2 & = & E_2 \end{array}$$

where $D_2 \equiv \alpha(D), E_2 \equiv H(E, d_1), \alpha_2$ *is induced by* α, β_2 *is induced by* $\beta\alpha^{-1}, \gamma_2$ *is induced by* $\gamma, d_2 = \beta_2\gamma_2$. *By induction*

$$E_n = H(E_{n-1}, d_{n-1}), d_n = \beta_n\gamma_n .$$

REMARK 1.24 - There exists a strong relation between the two above sources of spectral sequences. In the following points we shall show this relation.

1) Let B be an abelian category. Let $\mathcal{B}_f$ be the associated **category of filtered objects**

$$\cdots \subseteq B^{(p-1)} \subseteq B^{(p)} \subseteq \cdots \subseteq B, -\infty < p < \infty .$$

A morphism $\phi : B \to B'$ of filtered objects sends $B^{(p)}$ to $B'^{(p)}$ for all p.

2) Let $\mathcal{B}^{\mathbf{Z}}$ be the category of $\mathbf{Z}$-graded objects $\{B^p\}$ in $\mathcal{B}$. A **morphism** $\phi : \{B^p\} \to \{B'^p\}$, of degree $|\phi| = q$, of $\mathcal{B}^{\mathbf{Z}}$ is a morphism of $\mathcal{B}$ such that $\phi(B^p) \subset B'^{p+q}, \forall p \in \mathbf{Z}$.

3) Let $G_r : \mathcal{B}_f \to \mathcal{B}^{\mathbf{Z}}$ be the functor such that $G_r(B) = \{A^p\}_{p\in\mathbf{Z}}$, with $A^p \equiv B^{(p)}/B^{(p-1)}$.

4) The category of filtered objects in $\mathcal{D}(\mathcal{U}), \mathcal{D}(\mathcal{U})_f$, has objects $(C^{(p)}, d)$ filtered by subobjects of $\mathcal{D}(\mathcal{U}) : \cdots \subseteq C^{(p-1)} \subseteq C^{(p)} \subseteq \cdots \subseteq, -\infty < p < \infty$, and each $C^{(p)}$ is closed under the differential d on C , that is $dC^{(p)} \subseteq C^{(p)}$. The morphisms in $\mathcal{D}(\mathcal{U})_f$ are morphisms of $\mathcal{D}(\mathcal{U})$ that respect filtration and commute with differentials.

1) One has a functor

$$\bar{H} : \mathcal{D}(\mathcal{U})_f \to \mathcal{E}(\mathcal{U}^{\mathbf{Z}})$$

such that

$$\bar{H}(C^{(p)}, d) \equiv \left\{ \begin{array}{ccc} D^p & \overset{\alpha^p}{\to} & D^p \\ {}_{\gamma^p}\uparrow & & \downarrow_{\beta^p} \\ E^p & = & E^p \end{array} \right\},$$

where

$$D^p \equiv H(C^{(p)})$$
$$E^p \equiv H(C^{(p)}/C^{(p-1)})$$
$$|\alpha^p| = 1\,,\ |\beta^p| = 0\,,\ |\gamma^p| = -1$$

where the morphisms α, β, γ, are induced by the following short exact sequences:

$$0 \to C^{(p-1)} \to C^{(p)} \to C^{(p)}/C^{(p-1)} \to 0$$

and the associated homology exact diagram

$$\gamma \quad \begin{matrix} H(C^{(p-1)}) & \xrightarrow{\alpha} & H(C^{(p)}) \\ \uparrow & & \downarrow \\ H(C^{(p)}/C^{(p-1)}) & = & H(C^{(p)}/C^{(p-1)}) \end{matrix} \quad \beta$$

2) One has the functor:

$$E : \mathcal{D}(\mathcal{U})_f \to \mathcal{U}^{\mathbf{Z}}$$

such that

$$E(C^{(p)}, d) \equiv E^p \equiv H(C^{(p)}/C^{(p-1)}) .$$

3) One has the following factorization $E = H \circ G_r$, where H is the homology functor.
4) The homology functor H and G_r do not commute; that is $G_r \circ H \neq H \circ G_r = E$. In particular if $\mathcal{U} \equiv \mathcal{U}^{\mathbf{Z}}$, by considering differentials d that lower the degree by 1 we have the following functors:
(a)

$$\bar{H} : \mathcal{D}(\mathcal{U}^{\mathbf{Z}})_f \to \mathcal{E}(\mathcal{U}^{\mathbf{Z}\times\mathbf{Z}})$$

such that

$$(C^{(p)}, d) \mapsto \bar{H}(C^{(p)}, d) \equiv \left\{ \begin{matrix} D & \xrightarrow{\alpha} & D \\ \gamma \uparrow & & \downarrow \beta \\ E & = & E \end{matrix} \right\}$$

$D \equiv \{D^{p,q}\}, D^{p,q} \equiv H_q(C^{(p)}), E \equiv \{E^{p,q}\}, E^{p,q} \equiv H_q(C^{(p)}/C^{(p-1)})$, the bidegrees of α, β, γ, are the following: $|\alpha| = (1, 0)$, $|\beta| = (0, 0)$, $|\gamma| = (-1, -1)$. One has the exact sequence:

$$D^{p-1,q} \xrightarrow{\alpha} D^{p,q} \xrightarrow{\beta} E^{p,q} \xrightarrow{\gamma} D^{p-1,q-1} \xrightarrow{\alpha} D^{p,q-1}. \tag{1.22}$$

(b)

$$S : \mathcal{E}(\mathcal{U}^{\mathbf{Z}\times\mathbf{Z}}) \to \mathcal{S}(\mathcal{U}^{\mathbf{Z}\times\mathbf{Z}}) ,$$

such that

$$S(\mathbb{E}C \equiv \left\{ \begin{matrix} D & \xrightarrow{\alpha} & D \\ \gamma \uparrow & & \downarrow \beta \\ E & = & E \end{matrix} \right\}) = (E_n^{p,q}, d_n)$$

with $|\alpha_n| = (1,0)$, $|\beta_n| = (-n,0)$, $|\gamma_n| = (-1,-1)$, $|d_n| = (-n-1,-1)$. One has the following equivalence:

$$\bar{S} \cong S \circ \bar{H} .$$

REMARK 1.25 - In this case each object of $\mathcal{D}(\mathcal{U}^{\mathbf{Z}})$ is also a chain complex and each object of $\mathcal{D}(\mathcal{U}^{\mathbf{Z}})_f$ is also a filtered chain complex. Furthermore if the filtration of a chain complex C is finite, it is also **homologically finite**, that is for each q , there exist p_0, p_1 with $H_q(C^{(p)}) = 0$, for $p \leq p_0, H_q(C^{(p)}) = H_q(C)$, for $p \geq p_1$.

REMARK 1.26 - $\alpha : D \to D$ is **positively** (resp.**negatively**) **stationary** if given q , there exists p_0 such that $\alpha : D^{p,q} \cong D^{p+1,q}$, $p \geq p_0$, (resp. $\alpha : D^{p-1,q} \cong D^{p,q}, p \leq p_0$) . If α is both positively and negatively stationary it is **stationary**.

THEOREM 1.17 - 1) *If α is stationary the spectral sequence associated with exact couples* **converges finitely**.

2) *If the filtration of the chain complex $C \in Ob(\mathcal{D}(\mathcal{U}^{\mathbf{Z}})_f)$ is homologically finite, then:*

(i) *the induced filtration of $H(C)$ is finite;*

(ii) *the associated spectral sequence converges finitely to the graded object associated with $H(C)$, suitable filtered: $E^{p,q}_{\infty} \cong (G_r \circ H_q(C))_p$.*

PROOF. By using the exact sequence (1.22) with a fixed q , we directly have the results. □

EXAMPLE 1.17 - 1) (**Double complexes and spectral sequences**).

A) Let $(\mathbb{B}, \partial', \partial'')$ be a **double complex**, we have an anticommutative diagram

$$\begin{array}{cccccc} & B_{r,s} & \overset{\partial'}{\to} & B_{r-1,s} & \\ \partial'' & \downarrow & & \downarrow & \partial'' \\ & B_{r,s-1} & \underset{\partial'}{\to} & B_{r-1,s-1} & \end{array}$$

with $\partial'\partial' = \partial''\partial'' = \partial''\partial' + \partial'\partial'' = 0$, for each r, s . May be it is more convenient to replace anticommutative diagrams with commutative ones, by setting: $d' = \partial', d'' = (-1)^r \partial''$ on $B_{r,s}$:

$$\begin{array}{cccccc} & B_{r,s} & \overset{d'}{\to} & B_{r-1,s} & \\ d'' & \downarrow & & \downarrow & d'' \\ & B_{r,s-1} & \underset{d'}{\to} & B_{r-1,s-1} & \end{array}$$

We call d', d'' the **horizontal, vertical differentials** of $\mathbb{B}$ respectively. We say that the double complex $\mathbb{B}$ is **positive** if there exists n_0 such that $B_{r,s} = 0$ if $r < n_0$, or $s < n_0$. Set

$$Tot\mathbb{B} \equiv \text{graded module}$$

with

$$(Tot\mathbb{B})_n \equiv \bigoplus_{p+q=n} B_{p,q} .$$

As

$$\partial'((Tot\mathbb{B})_n) \subseteq (Tot\mathbb{B})_{n-1}$$
$$\partial''((Tot\mathbb{B})_n) \subseteq (Tot\mathbb{B})_{n-1}$$
$$\partial\partial = (\partial' + \partial'')(\partial' + \partial'')$$
$$= \partial'\partial' + \partial''\partial' + \partial'\partial'' + \partial''\partial'' = 0$$

we can consider

$$(Tot\mathbb{B}, \partial = (\partial' + \partial''))$$

a chain complex:

$$\partial : (Tot\mathbb{B})_n \to (Tot\mathbb{B})_{n-1}\,,\ \forall n\,.$$

Examples of double complexes. Given two chain complexes $\mathbf{C}, \mathbb{D}$ of right and left A-modules ($A \equiv \mathbf{Z}_2$-graded commutative algebra), we define the following double complexes:

(a)

$$\mathbb{B}^{\otimes} \equiv \{\mathbb{B}^{\otimes}_{p,q} \equiv C_p \bigotimes_A D_q, \partial', \partial''\}\,,$$

where

$$\partial'(c \otimes d) = \partial c \otimes d, \partial''(c \otimes d) = (-1)^p c \otimes \partial d, c \in C_p, d \in D_q\,,$$

(the sign $(-1)^p$ is inserted into the definition of ∂'' to guarantee that $\partial''\partial' + \partial'\partial'' = 0$).

(b)

$$\mathbb{B}^{Hom} \equiv \{\mathbb{B}^{Hom}_{p,q} \equiv Hom_A(C_{-p}, D_q), \partial', \partial''\}\,,$$

with

$$(\partial' f)(c) = (-1)^{p+q+1} f(\partial c), c \in C_{-p+1}, f \in Hom_A(C_{-p}, D_q),$$
$$(\partial'' f)(c) = \partial(f(c)), c \in C_{-p}, f \in Hom_A(C_{-p}, D_q).$$

The corresponding graded modules $Tot\mathbb{B}^{\otimes}, Tot\mathbb{B}^{Hom}$, are denoted by $Tot\mathbb{B}^{\otimes} \equiv \mathbf{C} \bigotimes_A \mathbb{D}, Tot\mathbb{B}^{Hom} \equiv Hom_A(\mathbf{C}, \mathbb{D})$ respectively, and identify equally chain complexes:

($\bar{a}$)

$$\left\{\begin{array}{l} \{\mathbf{C} \bigotimes_A \mathbb{D}, \partial^{\otimes}\} \\ \partial^{\otimes}(c \bigotimes d) = \partial c \otimes d + (-1)^p c \otimes \partial d \equiv \partial'(c \otimes d) + \partial''(c \otimes d)\,. \end{array}\right\}$$

($\bar{b}$)

$$\left\{\begin{array}{l} \{Hom_A(\mathbf{C}, \mathbb{D}), \partial^{Hom}\} \\ (\partial^{Hom} f)_{p,q} = (-1)^{p+q} f_{p+1,q}\partial + \partial f_{q,q+1} \\ f \equiv \{f_{p,q} \in Hom_A(C_{-p}, D_q)\}\,. \end{array}\right\}$$

One has a canonical isomorphism:

$$Hom_A(\mathbf{C} \bigotimes_A \mathbb{D}, \mathbb{E}) \cong Hom_A(\mathbf{C}, Hom_A(\mathbb{D}, \mathbb{E}))\,.$$

B) **Spectral sequences associated with double complexes.** To the double complex $(\mathbb{B}, \partial', \partial'')$ we can associate two spectral sequences:

(a)

$$(Tot\mathbb{B}, \partial) \Rightarrow (\text{filtration})\, {}_1F^p(Tot\mathbb{B})_n \equiv \bigoplus_{r+s=n, r\le p} B_{r,s}$$

$$\Downarrow$$

$$\{{}_1E_r^{p,q}, {}_1 d_r\}$$

with

$$\begin{aligned} {}_1E_0^{p,q} &= H_{q-p}(B_{p,0}, \partial'') \\ {}_1E_1^{p,q} &= H_p(H_{q-p}(\mathbb{B}, \partial''), \partial')\,. \end{aligned}$$

(b)

$$(Tot\mathbb{B}, \partial) \Rightarrow (\text{filtration})\, {}_2F^p(Tot\mathbb{B})_n \equiv \bigoplus_{r+s=n, s\le p} B_{r,s}$$

$$\Downarrow$$

$$\{{}_2E_r^{p,q}, {}_2 d_r\}$$

with

$$\begin{aligned} {}_2E_0^{p,q} &= H_{q-p}(B_{0,p}, \partial') \\ {}_2E_1^{p,q} &= H_q(H_{q-p}(\mathbb{B}, \partial'), \partial'')\,. \end{aligned}$$

If $\mathbb{B}$ is positive, then both the first and the second spectral sequences converge finitely to the graded object associated with $\{H_n(Tot\mathbb{B})\}$, suitably (finitely) filtered.

2) (**Filtration on topological spaces and spectral sequences**)

(A) (**Filtration and homology spectral sequences.**) Let X be a topological space with a filtration $\{X_q\}$:

$$\emptyset = X_{-1} \subset X_0 \subset X_1 \subset \cdots . \subset X_N = X\,,$$

where X_i are closed subspaces of X such that $\bigcup_i X_i = X$ and every compact subset of X is contained in some X_i. (For example, if X is a CW-complex then for X_i we may take the i-dimensional skeleton X^i of X.) Then for any Abelian group A, there are groups $E^r_{p,q}$, defined for $r \ge 0$ and all p, q (so that $E^r_{p,q} = 0$, for $p < 0$ and $q > N$), and homomorphisms $d^r : E^r_{p,q} \to E^r_{p-r,q+r-1}$ (so that: $d^r \circ d^r = 0$) such that:

(a) $\{E^r_{\bullet,\bullet}, d^r\}$ is a spectral sequence, $r \ge 1$.

(b) Set

$$ {}_pH_{p+q}(X;A) \equiv \mathrm{im}(H_{p+q}(X_p;A) \to H_{p+q}(X;A)) \,. $$

Then the subgroups

$$ {}_pH_n(X;A) \subset H_n(X;A) $$

form a filtration of $H_p(X;A)$:

$$ 0 = {}_{-1}H_n(X;A) \subset {}_0H_n(X;A) \subset \cdots \subset_N H_n(X;A) = H_n(X;A) $$

and one has a natural isomorphism

(•)

$$ E^{\infty}_{p,q} = \lim_{\to} E^r_{p,q} \cong \frac{{}_pH_{p+q}(X;A)}{{}_{p-1}H_{p+q}(X;A)}; $$

(c)

$$ E^0_{p,q} = C_{p+q}(X_p, X_{p-1}; A); $$

(d) In particular if A is a field , $A \equiv F$, one has the following isomorphism of F-vector spaces:

$$ H_n(X;\mathbf{Z}) \cong \bigoplus_{p+q=n} E^{\infty}_{p,q} \,. $$

PROOF. Let us set:

$$ \begin{aligned} Z^r_{p,q} &\equiv \mathrm{im}[j_* : H_{p+q}(X_p, X_{p-r}; A) \to H_{p+q}(X_p, X_{p-1}; A)] \\ B^r_{p,q} &\equiv \mathrm{im}[\partial : H_{p+q+1}(X_{p+r-1}, X_p; A) \to H_{p+q}(X_p, X_{p-1}; A)] \\ Z^{\infty}_{p,q} &\equiv \mathrm{im}[j_* : H_{p+q}(X_p; A) \to H_{p+q}(X_p, X_{p-1}; A)] \\ B^{\infty}_{p,q} &\equiv \mathrm{im}[\partial : H_{p+q+1}(X, X_p; A) \to H_{p+q}(X_p, X_{p-1}; A)]. \end{aligned} $$

The groups

$$ Z^r_{p,q} \,,\; B^r_{p,q} \,,\; Z^{\infty}_{p,q} \,,\; B^{\infty}_{p,q} $$

for fixed p and q are all subgroups of

$$ H_{p+q}(X_p, X_{p-1}; A) $$

and satisfy the following inclusion relations:

$$ \begin{aligned} 0 = B^1_{p,q} \subset B^2_{p,q} \subset \cdots \subset B^r_{p,q} \subset B^{r+1}_{p,q} \subset \cdots \\ \subset B^{\infty}_{p,q} \subset Z^{\infty}_{p,q} \subset \cdots \subset Z^{r+1}_{p,q} \subset \cdots \subset Z^1_{p,q} \end{aligned} $$

Set

$$ E^r_{p,q} \equiv Z^r_{p,q}/B^r_{p,q} \,, \text{ and } \quad E^{\infty}_{p,q} \equiv Z^{\infty}_{p,q}/B^{\infty}_{p,q} \,. $$

One has a natural isomorphism

$$(1.23) \qquad Z^r_{p,q}/Z^{r+1}_{p,q} \cong B^{r+1}_{p-r,q+r-1}/B^r_{p-r,q-r-1}.$$

This can be seen by considering the following commutative diagram

$$\begin{array}{ccc}
H_{p+q}(X_p, X_{p-r}; A) & \overset{j_{1*}}{\rightarrow} & H_{p+q}(X_p, X_{p-1}; A) \\
\downarrow & & \| \\
H_{p+q}(X_{p-1}, X_{p-r}; A) & \rightarrow H_{p+q}(X_p, X_{p-r}; A) \underset{j_{2*}}{\rightarrow} & H_{p+q}(X_p, X_{p-1}; A) \\
\downarrow & & \downarrow \partial \\
H_{p+q-1}(X_{p-r}, X_{p-r-1}; A) & = & H_{p+q-1}(X_{p-r}, X_{p-r-1}; A)
\end{array}$$

induced by the exact sequences of the triples

$$(X_p, X_{p-1}, X_{p-r}),\ (X_p, X_{p-r}, X_{p-r-1}).$$

The isomorphism (1.23) is induced by $\partial \circ j^{-1}{}_{2*}$. Then the isomorphism (1.23) allows us to define a homomorphism $d^r : E^r_{p,q} \to E^r_{p-r,q+r-1}$ by means of the following composition

$$\begin{array}{ccc}
Z^r_{p,q}/B^r_{p,q} & \rightarrow Z^r_{p,q}/Z^{r+1}_{p,q} \cong B^{r+1}_{p-r,q+r-1}/B^r_{p-r,q+r-1} \rightarrow & Z^r_{p-r,q+r-1}/B^r_{p-r,q+r-1} \\
\| & & \| \\
E^r_{p,q} & \underset{d^r}{\rightarrow} & E^r_{p,q}
\end{array}$$

From this definition it results that:

(a)

$$\ker(d^r) = Z^{r+1}_{p,q}/B^r_{p,q};$$

(b)

$$\operatorname{im}(d^r) = B^{r+1}_{p-r,q+r-1}/B^r_{p-r,q+r-1}.$$

Hence, $\operatorname{im}(d^r) \subset \ker(d^r)$, so $d^r \circ d^r = 0$. In other words, for fixed r, $\{E^r_{p,q}, d^r\}$ is a cochain complex. Furthermore

$$\ker(d^r)/\operatorname{im}(d^r) = (Z^{r+1}_{p,q}/B^r_{p,q})/(B^{r+1}_{p,q}/B^r_{p,q}) \cong Z^{r+1}_{p,q}/B^{r+1}_{p,q} \cong E^{r+1}_{p,q}.$$

This proves that $\{E^r_{p,q}, d^r\}$ is a spectral sequence, for $r \geq 1$. Now, we also have the natural inclusions:

$$0 = {}_{-1}H_n(X; A) \subset {}_0H_n(X; A) \subset \cdots \subset {}_NH_n(X; A) = H_n(X; A).$$

Since $\bigcup_i X_i = X$ we have

$$\{i_{p*}\} : \varinjlim H_n(X_p; A) \cong H_n(X; A).$$

But

$$\mathrm{im}\{i_{p*}\} = \bigcup_{p+q=n_p} H_{p+q}(X;A)\,.$$

So we can conclude that

$$H_{p+q}(X;A) = \bigcup_{r+s=p+q_r} H_{r+s}(X;A)\,.$$

The isomorphism (•) can be proved by considering the following commutative diagram

$$\begin{array}{ccccc} H_{p+q+1}(X,X_p;A) & & & \to & H_{p+q}(X_p,X_{p-1};A) \\ \downarrow \partial & & & & \| \\ H_{p+q}(X_{p-1};A) & \to & H_{p+q}(X_p;A) & \underset{-j_*}{\to} & H_{p+q}(X_p,X_{p-1};A) \\ & & & & \downarrow i_{2*} \\ & & & & H_{p+q}(X;A) \end{array}$$

induced by the exact sequences of pairs (X,X_p) and (X_p,X_{p-1}). In fact the isomorphism (•) is induced by

$$i_{2*}\circ j^{-1}{}_{*} : H_{p+q}(X_p,X_{p-1};A) \cong H_{p+q}(X;A).$$

Now, one can see that $Z^r_{p,q} = Z^\infty_{p,q}$ for $r > p$, and $B^\infty_{p,q} = \bigcup_{r\geq 1} B^r_{p,q}$. In fact, for $r > p$, we have

$$\begin{aligned} Z^r_{p,q} &\equiv \mathrm{im}[j_* : H_{p+q}(X_p,X_{p-r};A) \to H_{p+q}(X_p,X_{p-1};A)] \\ &= \mathrm{im}[i_* : H_{p+q}(X_p,\emptyset;A) \to H_{p+q}(X_p,X_{p-1};A)] \\ &= Z^\infty_{p,q}. \end{aligned}$$

Let $j_r : (X_{p+r-1},X_p) \to (X,X_p)$ be the inclusion. Since

$$\bigcup_i X_i = X\{j_{r*}\} : \lim_{\rightarrow} H_{p+q-1}(X_{p+r+1},X_p;A) \to H_{p+q+1}(X,X_p;A)$$

is an isomorphism. For each r, let

$$\partial : H_{p+q+1}(X_{p+r-1},X_p;A) \to H_{p+q}(X_p,X_{p-1};A)$$

be the boundary for the triple (X_{p+r-1},X_p,X_{p-1}). These induce a homomorphism

$$\bar{\partial} : \lim_{\rightarrow} H_{p+q+1}(X_{p+r-1},X_p;A) \to H_{p+q}(X_p,X_{p-1};A)\,.$$

Moreover, the diagram

$$\begin{array}{ccc}
\varinjlim H_{p+q+1}(X_{p+r-1}, X_p; A) & \xrightarrow{\bar{\partial}} & H_{p+q}(X_p, X_{p-1}; A) \\
\{j_{r*}\} \downarrow & & \| \\
H_{p+q+1}(X_{p+r-1}, X_p; A) & \underset{\bar{\partial}}{\rightarrow} & H_{p+q}(X_p, X_{p-1}; A)
\end{array}$$

commutes; hence $B^{\infty}_{p,q} = \mathrm{im}(\partial) = \mathrm{im}(\bar{\partial}) = \bigcup_{r\geq 1} B^r_{p,q}$. So, we have an epimorphism $E^r_{p,q} \to E^{r+1}_{p,q}$, for $r > p$ and $E^{\infty}_{p,q} = \varinjlim E^r_{p,q}$.

Finally, if $A = F$ is a field, then one has $\bigoplus_{p+q=n} E^{\infty}_{p,q}$ which is the adjoint of the F-vector space $H_n(X; A)$. So $H_n(X; \mathbf{Z}) \cong \bigoplus_{p+q=n} E^{\infty}_{p,q}$.

(B)(**Filtration and cohomology spectral sequence.**) Similarly we can prove that for any filtration

$$\emptyset = X_{-1} \subset X_0 \subset X_1 \subset \cdots . \subset X_N = X ,$$

of a topological space X and Abelian group A, there are groups $E^{p,q}_r$ defined for $r \geq 0$ and all p, q, (so that $E^{p,q}_r = 0$ for $p < 0$ and $q > N$) and homomorphisms $d_r : E^{p,q}_r \to E^{p+r,q-r+1}_r$, ($d_r \circ d_r = 0$) such that:

(a) $\{E^{p,q}_r, d_r\}$ is a spectral sequence for $r \geq 1$;

(b) $E^{p,q}_{\cdot} = C^{p,q}(X_p, X_{p-1}; A)$;

(c) Set

$$_pH^{p+q}(X; A) = \ker(H^{p+q}(X; A) \to H^{p+q}(X_p; A)) .$$

(The map $H^n(X; A) \to H^n(X_p; A)$ is induced by embedding $X_p \subset X$.) The subgroups $_pH^n(X; A) \subset H^n(X; A)$ define a filtration of $H^n(X; A)$:

$$0 = {}_NH^n(X; A) \subset \cdots \subset {}_0H^n(X; A) \subset {}_{-1}H^n(X; A) = H^n(X; A) .$$

One has the natural isomorphisms:

$$E^{p,q}_{\infty} \cong {}_{p-1}H^{p+q}(X; A)/{}_pH^{p+q}(X; A) .$$

Furthermore, one has

$$E^{p,q}_{\infty} = \varinjlim E^{p,q}_r .$$

(d) If $A = F$ is a field, one has the isomorphism of F-vector spaces

$$H^n(X; A) = \bigoplus_{p+q=n} E^{p,q}_{\infty} .$$

In the following we shall quote some fundamental results that allow to characterize topological spaces and fibrations by means of spectral sequences.

THEOREM 1.18 - (**Atiyah-Hirzebruch-Whitehead.**) *For any homology theory* $h_\bullet$ *and CW-complex* X *there is a spectral sequence*

$$\{E^r_{p,q}, d^r\}$$

with

$$E^2_{p,q} \cong H_p(X; h_q(.))$$

and converging to $h_\bullet(X)$, *where* $h_q(.)$ *are the coefficients groups of* h_q .

THEOREM 1.19 - (**The homology Leray-Serre spectral sequence.**) *Let* G *be an abelian group. Given a fibration,* $F \subset E \xrightarrow{p} B$, *where* B *is path-connected, there is a first quadrant spectral sequence* $\{E^r_{\bullet,\bullet}, d^r\}$ *with* $E^2_{p,q} \cong H_p(B; \mathcal{H}_q(F; G))$, *the homology of* B *with local coefficients in the homology of* F , *the fiber of* p , *and converging to* $H_\bullet(E; G)$. *Furthermore, this spectral sequence is natural with respect to fiber-preserving maps of fibrations.*

THEOREM 1.20 - (**The cohomology Leray-Serre spectral sequence.**) *Let* R *be a commutative ring with unit. Given a fibration,* $F \subset E \xrightarrow{p} B$, *where* B *is path-connected, there is a first quadrant spectral sequence of algebras,* $\{E_r^{\bullet,\bullet}, d_r\}$, *with* $E_2^{p,q} \cong H^p(B; \mathcal{H}(F; R))$, *the cohomology of* B *with local coefficients in the cohomology of* F , *the fiber of* p , *and converging to* $H^\bullet(E; R)$ *as an algebra. Furthermore, this spectral sequence is natural with respect to fiber-preserving maps of fibrations.*

PROPOSITION 1.10 - 1) *Suppose that the system of local coefficients on* B *determined by the fiber is simple; then, for a field* $\mathbf{K}$ *we have*

$$E_2^{p,q} \cong H^p(B; \mathbf{K}) \bigotimes_{\mathbf{K}} H^q(F; \mathbf{K}) .$$

2) *Suppose that the system of local coefficients is simple and that* B *and* F *are connected. Then*

$$E_2^{p,0} \cong H^p(B; H^0(F; \mathbf{K})) \cong H^p(B; \mathbf{K}); E_2^{0,q} \cong H^0(B; H^q(F; \mathbf{K})) \cong H^q(F; \mathbf{K}) .$$

3) (**Leray-Hirsh.**) *Given a fibration with* B *path-connected with the system of local coefficients on* B *induced by* F *simple, then, if* F *is totally non-homologuous to* 0 *in* E *with respect to a field* R ,

$$H^\bullet(E; \mathbf{K}) \cong H^\bullet(B; \mathbf{K}) \bigotimes H^\bullet(F; \mathbf{K})$$

as vector sapecs. (F *is* **totally non-homologuous to zero** *in* E *with respect to the ring* R *if this homomorphism*

$$i^* : H^\bullet(E; R) \to H^\bullet(F; R)$$

is onto. In the case of a trivial fibration, this holds for obvious reasons. In general F is totally non-homologuous to 0 in E with respect to $\mathbf{K}$ iff the spectral sequence $E_r(B, E, F)$ collapses at the E_2-term.)

4) Under the hypotheses of the above Leray-Hirsh theorem, if $H^q(F; \mathbf{K}) = 0$, if $q > 0$, or *$H^q(F; \mathbf{K}) = \mathbf{K}$, if $q = 0$, then one has the isomorphism of vector spaces $H^\bullet(E; \mathbf{K}) \cong H^\bullet(B; \mathbf{K})$. (This is for example the case where $p : E \to B$ is a $\mathbf{K}$-vector bundle, that is F is a $\mathbf{K}$-vector space.)*

DEFINITION 1.28 - *A* **spectrum** *is a sequence of spaces, E_n, with base point, together with weak equivalences $\epsilon_n : SE_n \to E_{n+1}$* or *adjointly $\epsilon'_n : E_n \to \Omega E_{n+1}$. The* **generalized cohomology theory** *associated to $\{E_n\}$ is denoted by $E^\bullet(X)$ and defined by*

$$E^n(X) \equiv [X, E_n]$$

for a space X. The **generalized homology theory** *associated to $\{E_n\}$ is denoted by $E_\bullet(X)$ and defined by*

$$E_n(X) = \pi_n(E_n \wedge X).$$

The **coefficients** *of the generalized theories determined by $\{E_n\}$ are the groups $E^\bullet(*) \cong E_\bullet(*) \equiv \pi_\bullet(E)$. (The analogue of the Steenrod algebra for the cohomology, $E^\bullet$, is the algebra $E^\bullet E, (E^\bullet E)_r = \lim_n [E^{n+r}, E^n]$.)*

REMARK 1.27 - E.Brown proved that the generalized cohomology functors, were representable, that is there is a space W_n for each n, such that $E^n(X) = [X, W_n]$.

THEOREM 1.21 - (**Atiyah-Hirzebruch spectral sequence.**) *Suppose $\{E_n\}$ is a spectrum and X a space. Then there is a spectral sequence with*

$$E_2^{p,q} \cong H^p(X; E^q(*))$$

converging to $E^\bullet(X)$. There is also a spectral sequence with

$$E^2_{p,q} \cong H_p(X; E_q(*))$$

converging to $E_\bullet(X)$.

2 - DIFFERENTIAL EQUATIONS (PDEs)

Here and in the following sections we shall consider differential manifolds of finite dimension and class C^∞.

DEFINITION 2.1 - 1) *Let* $f : V \to W$ *be a differentiable mapping between two manifolds* V *and* W. *Then, the* **derivative** *of* f *is the mapping*

$$Df : V \to \mathcal{D}(V,W) \equiv T^*V \bigotimes TW \equiv \bigcup_{(p,q)\in V\times W} T_p^*V \bigotimes T_qW \cong \bigcup_{(p,q)\in V\times W} Hom_{\mathbf{K}}(T_pV; T_qW)$$

such that

$$Df(p) \in T_p^*V \bigotimes T_{f(p)}W \cong Hom_{\mathbf{K}}(T_pV; T_{f(p)}W), \ \forall p \in V.$$

2) *The space* $\mathcal{D}(V,W)$ *is called the* **derivative space** *(of the first order) for the mappings* $V \to W$.

3) *If we have a fibered structure* $\pi : W \to V$ *between* W *and* V, *and if* f *is a section of* π, *then* Df *has values in a subspace of* $\mathcal{D}(V,W)$ *denoted with* $\mathcal{D}(W)$ *and defined by*

$$\mathcal{D}(W) \equiv d^*\mathcal{D}(V,W) = \bigcup_{p\in V}\{\bigcup_{q\in\pi^{-1}(q)} T_p^*V \bigotimes T_qW\}$$

where $d : V \subset V \times V$ *is the diagonal mapping.*

4) *By iteration one can define* $D^kf \equiv D(D^{k-1}f)$ *and the spaces* $\mathcal{D}^k(V,W)$ *and* $\mathcal{D}^k(W)$.

5) *Furthermore, in* $\mathcal{D}^k(W)$ (*as well as in* $\mathcal{D}^k(V,W)$) *one recognizes a subspace,* $J\mathcal{D}^k(W)$, *where all the points are derivatives of order k for some section* f *of* π *and at some point* $p \in V$. *More precisely, set:*

$$J\mathcal{D}^k(W) \equiv \{u \in \mathcal{D}^k(W) | \exists f \in C^k(W) : D^kf(\pi_k(u)) = u\}$$

where $C^k(W)$ *denotes the space of* C^k*–sections of the fiber bundle* $\pi : W \to V$, *and* $\pi_k : \mathcal{D}^k(W) \to V$ *is the canonical projection.* $J\mathcal{D}^k(W)$ *is called the* **jet-derivative space of order k for sections** *of* $\pi : W \to V$.

REMARK 2.1 - 1) Similarly we can define $J\mathcal{D}^k(V,W)$. Of course, one has the following diffeomorphism:

$$J\mathcal{D}^k(X,Y) \cong J\mathcal{D}^k(X \times Y)$$

for any differential manifold X and Y, and considering $X \times Y$ like a trivial fiber bundle over X: $\pi : X \times Y \to X$.

2) $J\mathcal{D}^k(W)$ can be identified with the **jet-space** of order k, $J^k(W)$ [97], on the fiber bundle $\pi : W \to V$, defined by:

$$J^k(W) \equiv \bigcup_{p \in V} J^k(W)_p; \quad J^k(W)_p \equiv \{[f]^k_p\}$$

where $[f]^k_p$ is an equivalence class of order k for sections at $p \in V$, i.e., two sections of π, f, f', defined in neighbourhoods of p, are equivalent in p to the order k if $D^k f(p) = D^k f'(p)$. So, $[f]^k_p$ is the set of all the sections of π that have a contact of order k at p. One has the following commutative diagram:

$$\begin{array}{ccccccc} & J^k(W) & \cong & J\mathcal{D}^k(W) & \subset & \mathcal{D}^k(W) & \\ {\scriptstyle j^k(f)} & \uparrow & & \uparrow D^k f & & \downarrow & {\scriptstyle \pi_k} \\ & V & = & V & = & V & \end{array}$$

where $j^k(f)$ is the k-jet of the mapping f, i.e., $j^k(f)(p) = [f]^k_p, \forall p \in V$. Of course, the immersion $J\mathcal{D}^k(W) \subset \mathcal{D}^k(W)$ allows to transfer on $J\mathcal{D}^k(W)$ the algebraic structure of $\mathcal{D}^k(W)$. In fact in this way we can consider derivative of mappings as "tensor fields", namely, we can represent them by means of algebraic formulas. For example, we can write the local expression of the derivative Df of a section $f : V \to W$ in the following algebraic form:

$$Df = \delta^j_i dx^i \otimes \partial x_j + (\partial x_i . f^j) dx^i \otimes \partial y_j \circ f$$

where $\{x^i, y^j\}$ are fibered coordinates on W, y^j are vertical coordinates, and $f^j \equiv y^j \circ f$. Then, $Df(p)$ is characterized by means of the 1-jet $j(f) = \{f^j, (\partial x_i . f^j)\}$. Similarly , for $D^2 f$ we have the following algebraic representation:

$$D^2 f = \delta^j_i dx^i \otimes \partial x_j + (\partial x_i . f^j) dx^i \otimes \partial y_j \circ Df + (\partial x_s \partial x_i . f^j) dx^i \otimes \partial y^s_j \circ Df .$$

Therefore, $D^2 f$ is characterized by means of the 2-jet

$$j^2(f) = \{f^j, (\partial x_i . f^j), (\partial x_s \partial x_i . f^j)\}$$

and so on.

One has an exact sequence of vector bundles over $J\mathcal{D}^k(W)$:

$$0 \to \pi^*_{k,0} S^0_k M \bigotimes vTW \stackrel{\mu}{\to} vTJ\mathcal{D}^k(W) \to \pi^*_{k,k-1} vTJ\mathcal{D}^{k-1}(W) \to 0 .$$

If $\pi : W \equiv E \to M$ is a vector bundle over M, we can define an exact sequence of vector bundles over M:

$$0 \to S^0_k M \bigotimes E \stackrel{\mu}{\to} J\mathcal{D}^k(E) \to J\mathcal{D}^{k-1}(E) \to 0$$

DEFINITION 2.2 - 1) *A* **system of partial differential equations** *(PDEs) of order k on the fiber bundle $\pi : W \to V$ is a sub-fiber bundle E_k of $J\mathcal{D}^k(W)$ over V:*

$$\begin{array}{ccc} E_k & \subset & J\mathcal{D}^k(W) \\ \downarrow & & \downarrow \pi_k \\ V & = & V \end{array}$$

2) *A* **linear system** *(resp.* **affine system***) of PDEs of order k on the vector fiber bundle, (resp. affine fiber bundle) $\pi : E \to V$ is a vector fiber bundle (resp. affine fiber bundle) E_k of the vector fiber bundle (resp. affine fiber bundle) $J\mathcal{D}^k(E) \to V$.*
3) *A system of* **ordinary differential equations** *(ODEs) is a PDE where* $\dim V = 1$.
REMARK 2.2 - 1) The local expression of E_k can be represented by local mumerical functions $\{F^i\}_{1\le i\le s}$, on $J\mathcal{D}^k(W)$:

$$(E_k) \qquad \{F^i = 0\,,\, 1 \le i \le s\}\,.$$

2) In the linear case F^i are linear functions with respect to the vertical coordinates of $J\mathcal{D}^k(E)$ relative to the projection $\pi_k : J\mathcal{D}^k(E) \to V$. Thus , if

$$\{x^i, y^j, y^j_i, \cdots, y^j_{i_1\cdots i_k}\}\,,\ 1 \le i, i_1, \cdots, i_k \le n\,,\ 1 \le j \le m$$

are fibered coordinates on $J\mathcal{D}^k(E)$ defined by

$$x^i(D^k f(p)) = x^i(p), y^j_{i_1\cdots i_r}(D^k f(p)) = (\partial x_{i_1} \cdots \partial x_{i_r}.f^j)(p), 0 \le r \le k$$

we get

$$F^i = \sum_{1\le j\le m\,,\,1\le i_1,\cdots,i_r\le n} A^{ii_1\cdots i_r}_j y^j_{i_1\cdots i_r}$$

where $A^{ii_1\cdots i_r}_j = A^{ii_1\cdots i_r}_j(x)$ are numerical functions of the coordinates $\{x^i\}$.
REMARK 2.3 - If $\{F^i : J\mathcal{D}^k(W) \to \mathbf{R}\}_{1\le i\le s}$ locally represents a PDE, $E_k \subset J\mathcal{D}^k(W)$, then, its h**-th prolongation**,

$$E_{k+h} \equiv J\mathcal{D}^h(E_k) \bigcap J\mathcal{D}^{k+h}(W) \subset J\mathcal{D}^{k+h}(W)$$

is locally characterized by the following functions:

$$\left\{\begin{array}{l} F^i \circ \pi_{k+h,k} : J\mathcal{D}^{k+h}(W) \to \mathbf{R} \\ (\partial x_\alpha.F^i) + (\partial y^{i_1\cdots i_r}_j.F^i) y^j_{i_1\cdots i_r\alpha} : J\mathcal{D}^{k+h}(W) \to \mathbf{R} \\ 1 \le i \le s\,,\, 1 \le \alpha \le n \end{array}\right\}$$

DEFINITION 2.3 - 1) *A* **regular solution** *of* $E_k \subset J\mathcal{D}^k(W)$ *is a (local) section* $f : U \subset V \to W$ *such that* $D^k f(U) \subset E_k$. *We say that* $E_k \subset J\mathcal{D}^k(W)$ *is* **determined, overdetermined, underdetermined** *if* codim $E_k = m$, $> m$ *and* $< m$ *respectively.*
2) *The* **symbol** *of* E_k *is the family of subspaces* $g_k \equiv \{g_{k,q}\}_{q \in E_k}$ *of the vector fiber bundle* $p^*_{k,0}(S^0_k V \bigotimes vTW)$*:*

$$g_k \equiv vTE_k \bigcap p^*_{k,0}(S^0_k V \bigotimes vTW)$$

where $p_{k,0}$ *is the restriction of* $\pi_{k,0} : J\mathcal{D}^k(W) \to W$ *to* E_k .
REMARK 2.4 - If E_k is locally characterized by the equations: $F^i = 0$, then $g_{k,q}$ is the space of vectors

$$v_q = X^j_{\gamma_1 \cdots \gamma_k}(q) \partial y_j^{\gamma_1 \cdots \gamma_k}(q)$$

such that

$$(\partial y_j^{\gamma_1 \cdots \gamma_k} . F^i)(q) X^j_{\gamma_1 \cdots \gamma_k}(q) = 0 \, .$$

REMARK 2.5 - (DIFFERENTIAL EQUATIONS AND DIFFERENTIAL OPERATORS).
1) In Mechanics, differential equations are frequently obtained as kernels of differential operators on suitable fibered manifolds. Therefore, it is useful to consider also differential equations by means of this point of view. (However, it is important to note that it is not always possible to globally express a differential equation as the kernel of some differential operator.) Thus, if Ω is an open fibered submanifold of $J\mathcal{D}^k(W)$ on M , $K : \Omega \to \mathbb{K}$ is a morphism of fibered manifold of locally constant rank, and s is a section of $\mathbb{K}$ on M that satisfies the condition $s(M) \subset K(\Omega)$, then

$$E_k = \ker_s K \equiv \{q \in \Omega | K(q) = s(\pi_k(q))\}$$

is a fibered submanifold of Ω on M , i.e., E_k is a PDE on W . Then we have the following exact sequence:

$$(2.1) \qquad 0 \to E_k \subset \Omega \underset{s \circ \pi_k}{\overset{K}{\rightrightarrows}} \mathrm{im}(K).$$

As a consequence, we have

$$E_{k+h} = \ker_{D^h s} K^{(h)}$$

where $K^{(h)}$ is the h-th prolongation of K . Furthermore,

$$g_k = \ker(\sigma_k(K))$$

where $\sigma_k(K)$ is the **symbol** of K given by the following morphism:

$$\sigma_k(K) \equiv vT(K) \circ \mu$$

fibered on $J\mathcal{D}^k(W)$, where $\mu : \pi^*_{k,0} S^0_k M \bigotimes vTW \to vTJ\mathcal{D}^k(W)$ is the canonical monomorphism.

$$(2.2) \qquad g_{k+h} = \ker(\sigma_{k+h}(K^{(h)})),$$

where $\sigma_{k+h}(K^{(h)})$ is the symbol of the h-th prolongation of K.
2) If $W \equiv E$ is a vector fiber bundle on M, then $\sigma_k(K)$ is identified with a morphism of vector fiber bundles on M: $\sigma_k(K)' \equiv K \circ \epsilon : S^0_k M \bigotimes E \to \mathbb{K}$, where $\epsilon : S^0_k M \bigotimes E \to J\mathcal{D}^k(E)$ is the canonical monomorphism. Then, the symbol of $E_k = \ker(K)$ is identified with $g_k = \ker(\sigma_k(K)')$. Furthermore, in the linear case we have that $\sigma_{k+h}(K^{(h)})$ is identified with a morphism

$$\sigma_{k+h}(K^{(h)})' : S^0_{k+h} M \bigotimes E \to S^0_h M \bigotimes E$$

of vector fiber bundles on M.
3) If $K^{(h)}$ is locally of constant rank, then E_{k+h} is a fibered submanifold

$$E_{k+h} \subset J\mathcal{D}^{k+h}(W), h \geq 0.$$

4) If $p_{k+h,k}$ is surjective, and $K^{(h)}$ is locally of constant rank, for $h \geq 0$, then g_{k+h} is a vector fiber bundle on E_k.
DEFINITION 2.4 - 1) *E_k is said to be* **regular** *if E_{k+r} is a fibered submanifold of $J\mathcal{D}^{k+r}(W)$, $\forall r \geq 0$.*
2) E_k is said to be **sufficiently regular** *if E_k is regular and the morphisms*

$$\pi_{k+r+s,k+r} : E_{k+r+s} \to E_{k+r}$$

have constant rank, for all $r, s \geq 0$.
REMARK 2.6 - If $(E_k)_{+1} \subset J\mathcal{D}^{k+1}(W)$ is a fibered submanifold of $J\mathcal{D}^{k+1}(W)$, then E_k has the same solutions of $(E_k)_{+1}$.
REMARK 2.7 - If $E_k \subset J\mathcal{D}^k(W)$ is sufficiently regular, then

$$E^{(s)}_{k+r} \equiv \pi_{k+r+s,k+r}(E_{k+r+s})$$

is a fibered submanifold of $J\mathcal{D}^{k+r}(W)$. Furthermore, one has the following finite decreasing filtration

$$E^{(s(r))}_{k+r} \subset \cdots \subset E^{(s)}_{k+r} \subset \cdots \subset E^{(1)}_{k+r} \subset E_{k+r} \subset J\mathcal{D}^{k+r}(W)$$

such that $E^{(t)}_{k+r} = E^{(s(r))}_{k+r}$, for $t \geq s(r) \in \mathbf{N}$. Furthermore, E_k and $E^{(s)}_{k+r}$ have the same solutions, $\forall r, s \geq 0$.

DEFINITION 2.5 - 1) **Formal compatibility**: *the map* $\pi_{k+r} : E_{k+r} \to M$ *is surjective,* $\forall r \geq 0$. *(If* E_k *is linear it is always formally compatible.)*

2) **Formal transitivity**: *the following maps are surjective* $\forall r \geq 0$, $\pi_{k+r,0} : E_{k+r} \to W$.

PROPOSITION 2.1 - 1) **Normal bundle** $N(i_k) \equiv F_0$ *of the embedding* $i_k : E_k \to J\mathcal{D}^k(W)$. *One has the following commutative and exact diagram of families of vector spaces over* E_k*:*

$$\begin{array}{ccccccccc} & & 0 & & 0 & & & & \\ & & \downarrow & & \downarrow & & & & \\ 0 & \to & g_k & \to & \pi^*_{k,0}(S^0_k M \bigotimes vTW) & \to & F_0 & \to & 0 \\ & & \downarrow & & \downarrow & & \Vert\wr & & \\ 0 & \to & vTE_k & \to & J\mathcal{D}^k(vTW) & \to & J\mathcal{D}^k(vTW)/vTE_k & \to & 0 \end{array}$$

2) *(***r-th prolongation** g_{k+r} *of the symbol of* E_k*). Set*

$$g_{k+r} \equiv vTE_{k+r} \cap \pi^*_{k+r,0}(S^0_{k+r} M \bigotimes vTW).$$

One has the following exact sequences of families of vector spaces over E_k*:*

$$0 \to g_{k+r} \to \pi^*_{k,0}(S^0_{k+r} M \bigotimes vTW) \to S^0_r M \bigotimes F_0;$$

$$0 \to g_{k+1} \to \pi^*_{k,0}(S^0_{k+1} M \bigotimes vTW) \xrightarrow{\psi} T^*M \bigotimes F_0 \to F_1;$$

$$0 \to \pi^*_{k+r,k} g_{k+r} \to vTE_{k+r} \to \pi^*_{k+r,k+r-1} vTE_{k+r-1};$$

If g_{k+1} *is a vector bundle over* E_k, *then* F_1 *is a vector bundle over* E_k. *One has:*

$$\dim F_1 = n \cdot \dim F_0 - m \frac{(n+k)!}{(k+1)!(n-1)!} + \dim g_{k+1}.$$

DEFINITION 2.6 - (**Formal integrability** of E_k). $E_k \subset J\mathcal{D}^k(W)$ *is formally integrable if for* $r \geq 0$:

(a) g_{k+r+1} *is a vector bundle over* E_k;

(b) $\pi_{k+r+1,k+r} : E_{k+r+1} \to E_{k+r}$ *is surjective.*

PROPOSITION 2.2 - 1) *The following propositions are equivalent:*

A) *Any prolongation* $E_{k+r}, 0 \leq r \leq h$, *is an affine subbundle of* $J\mathcal{D}^{k+r}(W)|_{E_{k+r-1}}$ *modeled on the vector bundle* $\pi^*_{k+r,k+r-1} g_{k+r}$ *over* E_{k+r-1};

B) g_{k+r-1} *is a vector bundle over* E_k *and* $E_{k+r+1} \to E_{k+r}$ *is surjective for* $0 \leq r \leq h$; *Further, if any one of these assertions is satisfied, then* $(E_{k+1})_{+r} = (E_k)_{+r+1}$.

2) (**First Spencer complex**). *Given a differential equation $E_k \subset J\mathcal{D}^k(W)$ one has the following complexes of vector spaces over E_k :*

$$g_m \xrightarrow{\delta^0} T^*M \bigotimes g_{m-1} \xrightarrow{\delta^1} \Lambda^0_2 M \bigotimes g_{m-2} \xrightarrow{\delta^2} \cdots$$
$$\to \Lambda^0_{m-k} M \bigotimes g_k \to \Lambda^0_{m-k+1} M \bigotimes S^0_{k-1} M \bigotimes vTW$$

where

$$g_{k+r-q} = \left\{ \begin{array}{rl} S^0_{k+r-q} M \bigotimes vTW & \text{if} \quad r-q<0 \\ 0 & \text{if} \quad k+r-q<0 \end{array} \right\}$$

The corresponding cohomology $\{H^{m-j,j}_q\}$, $q \in E_k$, *at* $(\Lambda^0_j M \otimes g_{m-j})_q$ *is called the* **Spencer cohomology** *of* E_k .

DEFINITION 2.7 - *We say that* E_k (or g_k) *is* r-**acyclic** *if* $H^{m,j} = 0$, $m \geq k, 0 \leq j \leq r$. *We say that* E_k (or g_k) *is* **involutive** *if* $H^{m,j} = 0$, $m \geq k, j \geq 0$. *We say that* E_k (or g_k) *is of* **finite type** *if* $\exists r \geq 0, g_{k+r} = 0$. *Note that the sequences*

$$0 \to g_{k+r} \to T^*M \bigotimes g_{k+r-1} \to \Lambda^0_2 M \bigotimes g_{k+r-2}$$

are exact for $r \geq 1$.

REMARK 2.8 - If g_k is involutive and of finite type, then we must have $g_k = 0$.

PROPOSITION 2.3 - (**Spencer families** of vector spaces associated to E_k). *One has the following families of vector spaces over* E_k *associated to a PDE* $E_k \subset J\mathcal{D}^k(W)$:

$$\begin{aligned} C^j &\equiv \{\Lambda^0_j M \bigotimes_{E_k} vTE_k\} / \delta(\Lambda^0_{j-1} M \bigotimes_{E_k} G_{k+1}); \\ G_{k+r} &\equiv g_{k+r} \text{ , if } E_k \subset J\mathcal{D}^k(W); \\ G_{k+r} &\equiv S^0_{k+r} M \bigotimes_{E_k} vTW \text{ , if } E_k = J\mathcal{D}^k(W) \\ C^j &\equiv C^j(W) \text{, if } E_k = J\mathcal{D}^k(W) \,. \end{aligned}$$

Note that $C^1 = J\mathcal{D}(E_k)/E_{k+1}$. *One has the following exact sequence of vector spaces over* E_k*:*

$$0 \to G_{k+2} \to S^0_2 M \bigotimes_{E_k} vTE_k \to T^*M \bigotimes_{E_k} C^1 \to C^2 \to 0 \,.$$

THEOREM 2.1 - 1) *If* $E_k \subset J\mathcal{D}^k(W)$ *is a PDE such that the map* $E_{k+1} \to E_k$ *is surjective and* g_{k+1} *is a vector bundle over* E_k , *then there is a morphism over* E_k*:*

$$\kappa(E_k) : E_{k+1} \to C^2 \equiv \{\Lambda^0_2 M \bigotimes vTE_k\} / \delta(T^*M \bigotimes g_{k+1})$$

such that one has the following exact sequence:

$$E_{k+2} \to E_{k+1} \underset{0\circ\pi_{k+1,k}}{\overset{\kappa(E_k)}{\rightrightarrows}} C^2$$

where $\kappa(E_k)$ has values in Spencer cohomology spaces:

$$H^{k,2} = \frac{\ker\{\delta : \Lambda_2^0 M \bigotimes g_k \to \Lambda_3^0 M \bigotimes S_{k-1}^0 M \bigotimes vTW\}}{\delta(T^*M \bigotimes g_{k+1})} \subset C^2$$

2) *If E_k is a PDE (resp. linear PDE) such that the map $E_{k+1} \to E_k$ is surjective, g_{k+1} is a vector bundle over E_k (resp. over M) and g_k is 2-acyclic then E_k is formally integrable.*

3) (**Criterion of formal integrability**). *Let E_k be a PDE (resp. linear PDE). Then, there is an integer $k_0 = k + h$, depending only on k, the dimension of M and dimension of W (resp. E) such that E_{k_0} is involutive and such that if g_{k+r+1} is a vector bundle over E_k (resp. over M) and $E_{k+r+1} \to E_{k+r}$ is surjective for $0 \le r \le h$, then E_k is formally integrable.*

REMARK 2.9 - Linear PDEs with constant coefficients and linear PDEs represented by linear bundles of geometric objects, in the sense of [93,96,97] and sections 2.7 and 4.8, such that g_{k+r} and E_{k+r} are vector bundles over M, $\forall r \ge 0$, are formally integrable (in fact g_{k+r} become 2-acyclic for r large enough).

PROPOSITION 2.4 - (**Family of vector spaces associated to E_k**). *Let us consider the following:*

$$F_r \equiv (\Lambda_r^0 M \bigotimes_{E_k} F_0)/\delta(\Lambda_{r-1}^0 M \bigotimes G_1'),$$
$$G_r' = \mathrm{im}(S_{k+r}^0 M \bigotimes_{E_k} vTW \to S_r^0 M \bigotimes_{E_k} F_0), \; 1 \le r \le n.$$

Then F_r are families of vector spaces over E_k. One has the short exact sequences over E_k:

$$0 \to g_{k+r} \to S_{k+r}^0 M \bigotimes_{E_k} vTW \to G_r' \to 0$$

One has:

$$H^{k+r}(g_k) = H^{r+1,s-1}(G_1'), \; \forall r \ge 0, \; \forall s \ge 2.$$

If g_k is involutive and g_{k+1} is a vector bundle over E_k, then F_r is a vector bundle over E_k for $r = 1, \cdots, n$.

THEOREM 2.2 - 1) *If* $E_k \subset J\mathcal{D}^k(W)$ *is PDE and if* g_{k+1} *is a vector bundle over* E_k *, then* F_1 *is a vector bundle over* E_k *, and a section* κ *of this vector bundle exists such that we have the exact sequence:*

$$E_{k+1} \to E_k \underset{0}{\overset{\kappa}{\rightrightarrows}} F_1 .$$

The application κ *is called the* **curvature map** *of* E_k *.*
2) *Let* $E_k \subset J\mathcal{D}^k(W)$ *be a regular system of order* k *on* $\pi : W \to M$ *. If* g_{k+r} *is a vector bundle over* E_k *and if* g_k *is 2-acyclic, then* $E^{(1)}_{k+r} = (E^{(1)}_k)_{+r}$ *.*
3) *Let* $E_k \subset J\mathcal{D}^k(W)$ *be a sufficiently regular PDE of order* k *on* $\pi : W \to M$ *, i.e.,* E_k *is a regular system such that the morphisms* $\pi_{k+r+s,k+r} : E_{k+r+s} \to E_{k+r}$ *have constant rank,* $\forall r, s \geq 0$ *. Then, there exist integers* $s \geq 0$ *and* $k_s \geq k$ *such that* $E_{k_s} \subset J\mathcal{D}^{k_s}(W)$ *is an involutive formally integrable PDE with the same solutions as* E_k *and such that:* $(E^{(s)}_{k_s})_{+r} = E^{(s)}_{k_s+r}, \forall r \geq 0$ *.*
REMARK 2.10 - If $\pi : W \to X$ is a fiber bundle we denote the sheaf of sections of π by $\underline{W}$. (For informations on sheaves see e.g., ref.[151]. Furthermore, for a fiber bundle morphirsm $A : J\mathcal{D}^k(W) \to K$, denote by $A^{(r)} : J\mathcal{D}^{k+r}(W) \to J\mathcal{D}^r(K)$ its r-th prolongation. Let

$$(\triangle) \cdots \to \underline{F_r} \overset{j_r}{\to} \underline{F_{r+1}} \overset{j_{r+1}}{\to} \underline{F_{r+2}} \to \cdots$$

be a sequence of linear differential operators of order $r, r+1, \cdots$ respectively. Let $\phi_r : J\mathcal{D}^r(F_r) \to F_{r+1}$ be the corresponding fiber bundles homomorphisms. Then we say that sequence $(\triangle)$ is **locally exact** if $\mathrm{im}(j_r) = \ker(j_{r+1})$, $\forall r$. We say that $(\triangle)$ is **formally exact** if the corresponding sequences of prolongated fibered morphisms

$$\cdots \to J\mathcal{D}^{r+(r+1)+s}(F_r) \overset{j_r{}^{(r+1+s)}}{\to} J\mathcal{D}^{r+1+s}(F_{r+1}) \overset{j_{r+1}{}^{(s)}}{\to} J\mathcal{D}^s(F_{r+2}) \to \cdots , \ \forall s \geq 0$$

are exact. We say that $(\triangle)$ is **exact** if it is locally and formally exact.
PROPOSITION 2.5 - (**Janet sequence of an involutive formally integrable linear PDE** $E_k \subset J\mathcal{D}^k(E)$).

$$(2.3) J^k(\Theta)(\mathbf{J-sequence}) \qquad 0 \to \Theta \to \underline{E} \overset{D}{\to} \underline{F_0} \overset{D_1}{\to} \underline{F_1} \overset{D_2}{\to} \cdots \overset{D_n}{\to} \underline{F_n} \to 0 .$$

There, $\Theta \equiv$ *sheaf solution of* $E_k \equiv \ker(\phi), \phi : J\mathcal{D}^k(E) \to J\mathcal{D}^k(E)/E_k \equiv F_0 . D_r \equiv \psi^r \circ D, \psi^r \equiv$ *canonic epimorphism* $\psi^r : J\mathcal{D}(F_{r-1}) \to F_r \equiv \mathrm{coker}(\psi^{(r-1)})^{(1)}, \phi = \psi^o, D = \phi \circ D^k . J^k(\Theta)$ *is formally exact. Furthermore,* $D_1 \circ D = 0$ *, the operator* D_1 *is the compatibility condition for* D *and express all the formal obstruction to solving the inhomogeneous equation* $D \cdot s = f, s \in \underline{E}, f \in \underline{F_0}$ *.*

EXAMPLE 2.1 - The d-Poincaré sequence

$$0 \to \Theta \to \underline{\Lambda^0_0 M} \xrightarrow{d} \underline{\Lambda^0_1 M} \xrightarrow{d} \cdots \to \underline{\Lambda^0_n M} \to 0$$

where Θ is the sheaf of functions locally constant on M, is an exact J-sequence associated to the exterior differential $d: JD(\Lambda^0_r M) \to \Lambda^0_{r+1} M$. ∎

DEFINITION 2.8 - (**Spencer operator**). *The Spencer operator is the first order differential operator* $\mathbf{D}: \underline{J\mathcal{D}^k(W)} \to \underline{T^*M \bigotimes J\mathcal{D}^{k-1}(vTW)}$ *defined by*

$$\mathbf{D} = D \circ \pi_{k,k-1} - (id_{J\mathcal{D}^{k-1}})^{(1)}.$$

It is easy to prove that $\ker(\mathbf{D}) = \{u \in \underline{J\mathcal{D}^k(W)} | u = D^k s, s \in \underline{W}\}$.

PROPOSITION 2.6 - (**Linear Spencer complexes of an involutive formally integrable linear PDE $E_k \subset J\mathcal{D}^k(E)$**).

1) *One has the following locally exact complex:*

$(2.4)(S^q_1(E))$

$$0 \to \underline{E} \xrightarrow{D^q} \underline{J\mathcal{D}^q(E)} \xrightarrow{\mathbf{D}} \underline{T^*M \bigotimes J\mathcal{D}^{q-1}(E)} \xrightarrow{\mathbf{D}} \underline{\Lambda^0_2 M \bigotimes J\mathcal{D}^{q-2}(E)} \xrightarrow{\mathbf{D}} \cdots$$
$$\to \underline{\Lambda^0_n M \bigotimes J\mathcal{D}^{q-n}(E)} \to 0$$

where $J\mathcal{D}^{q-r}(E) = 0$ *if* $r > q$,

$$\mathbf{D}(\omega \wedge u) = d\omega \wedge (\pi_{q,q-1} \circ u) + (-1)^j \omega \wedge \mathbf{D}, \omega \in \underline{\Lambda^0_j M}, u \in \underline{\Lambda^0_j M \bigotimes J\mathcal{D}^{q-j}(E)}.$$

2) $S^q_1(E)$ *(for* $q = k + r$*), gives by restriction to* $E_{k+r} \subset J\mathcal{D}^{k+r}(E)$ *the first linear Spencer complex of* E_k*:*

$$(2.5) S^q_1(\Theta) \quad \begin{array}{l} 0 \to \Theta \xrightarrow{D^{k+r}} \underline{E_{k+r}} \xrightarrow{\mathbf{D}} \underline{T^*M \bigotimes E_{k+r-1}} \xrightarrow{\mathbf{D}} \underline{\Lambda^0_2 M \bigotimes E_{k+r-2}} \to \cdots \\ \xrightarrow{\mathbf{D}} \underline{\Lambda^0_n M \bigotimes E_{k+r-n}} \to 0 \end{array}$$

where Θ = *sheaf of solutions of* E_k. $E_{k+r-s} = J\mathcal{D}^{k+r-s}(E)$ *if* $r < s \leq k + r, E_{k+r-s} = 0$ *if* $s > k + r$. *We say that* $S^q_1(\Theta)$ *is* **stable** *when* $r \geq n - 1$. $S^q_1(\Theta)$ *is always locally exact at* $\underline{E_{k+r}}$. *But in general* $S^q_1(\Theta)$ *is neither formally nor even locally exact.*

REMARK 2.11 - Note that the coboundary δ in the first Spencer complex is obtained by restriction on $-\mathbf{D}$. Furthermore, as g_{k+1} is involutive, the cohomology $H^{q-j,j}(E_k)$ of $S^q_1(\Theta)$ at $\Lambda^0_j M \bigotimes E_{q-j}$ is independent of $s = q - j \geq k + 1$, and is called the j**-th**

Spencer cohomology group $H^j(E_k)$ of E_k . The relation between δ and $S_1^q(\Theta)$ is given by the following commutative diagram:

(2.6)

$$\begin{array}{ccccccccc}
 & & & & 0 & & 0 & & 0 & \\
 & & & & \downarrow & & \downarrow & & \downarrow & \\
 & & 0 & \rightarrow & \underline{g_{k+r-1}} & \overset{-\delta}{\rightarrow} & \underline{T^*M \otimes g_{k+r}} & \cdots \overset{-\delta}{\rightarrow} & \underline{\Lambda_n^0 M \otimes g_{k+r-n+1}} & \rightarrow 0 \\
 & & \downarrow & & \downarrow & & \downarrow & & \downarrow & \\
0 \rightarrow & & \Theta & \overset{D^{k+r+1}}{\rightarrow} & \underline{E_{k+r+1}} & \overset{\mathbf{D}}{\rightarrow} & \underline{T^*M \otimes E_{k+r}} & \cdots \overset{\mathbf{D}}{\rightarrow} & \underline{\Lambda_n^0 M \otimes E_{k+r-n-1}} & \rightarrow 0 \\
 & & \| & & \downarrow & & \downarrow & & \downarrow & \\
0 \rightarrow & & \Theta & \overset{D^{k+r}}{\rightarrow} & \underline{E_{k+r}} & \overset{\mathbf{D}}{\rightarrow} & \underline{T^*M \otimes E_{k+r-1}} & \cdots \overset{\mathbf{D}}{\rightarrow} & \underline{\Lambda_n^0 M \otimes E_{k+r-n}} & \rightarrow 0 \\
 & & \downarrow & & \downarrow & & \downarrow & & \downarrow & \\
 & & 0 & & 0 & & 0 & & 0 &
\end{array}$$

If $E_k = J\mathcal{D}^k(W)$, the above commutative diagram has both rows and columns exact. The following is also called **first linear Spencer complex** of E_k

(2.7)$(S_1^q(E_k))$

$$0 \rightarrow \underline{E}_q \overset{\mathbf{D}}{\rightarrow} \underline{T^*M \bigotimes E_{q-1}} \overset{\mathbf{D}}{\rightarrow} \underline{\Lambda_2^0 M \bigotimes E_{q-2}} \overset{\mathbf{D}}{\rightarrow} \cdots \overset{\mathbf{D}}{\rightarrow} \underline{\Lambda_n^0 M \bigotimes E_{q-n}} \rightarrow 0,$$

with $q = k + r > k$, obtained from $(S_1^q(\Theta))$ by dropping the term Θ . Of course, $(S_1^q(E_k))$ is not more locally exact at E_q .

3) *There is a formally exact sequence* $S_2^k(\Theta)$ (**Second linear Spencer sequence**)*:*

$$(2.8)(S_2^k(\Theta)) \qquad 0 \rightarrow \Theta \rightarrow \underline{C^o} \overset{D^k}{\rightarrow} \underline{C^0} \overset{\mathbf{D}_1}{\rightarrow} \underline{C}^1 \rightarrow \cdots \rightarrow \underline{C}^n \rightarrow 0 .$$

We denote by $S_2^k(E)$ *the sequence* $S_2^k(\Theta)$, *when* $E_k = J\mathcal{D}^k(W)$. *We denote by* $J^k(\Omega)$ *the sequence obtain from* $J^k(\Theta)$ *by dropping the term* $\underline{E}$. *Moreover, the sequences* $S_2^k(\Theta), S_2^k(E), J^k(\Omega)$ *are related by the following commutative diagram in which the columns are exact and* $\phi_0, \cdots, \phi_n$ *are induced by* $\phi = \phi_0$.

$$\begin{array}{lccccccccccc}
 & & & 0 & & 0 & & 0 & & & 0 & \\
 & & & \downarrow & & \downarrow & & \downarrow & & & \downarrow & \\
(S_2^k(\Theta)) & 0 & \rightarrow & \Theta & \overset{D^k}{\rightarrow} & \underline{C^0} & \overset{\mathbf{D}_1}{\rightarrow} & \underline{C^1} & \overset{\mathbf{D}_2}{\rightarrow} & \cdots \overset{\mathbf{D}_n}{\rightarrow} & \underline{C}^n & \rightarrow 0 \\
 & & & \downarrow & & \downarrow & & \downarrow & & & \downarrow & \\
(S_2^k(E)) & 0 & \rightarrow & \underline{E} & \overset{D^k}{\rightarrow} & \underline{C}^0(E) & \overset{\mathbf{D}_1}{\rightarrow} & \underline{C}^0(E) & \overset{\mathbf{D}_2}{\rightarrow} & \cdots \overset{\mathbf{D}_n}{\rightarrow} & \underline{C}^n(E) & \rightarrow 0 \\
 & & & \downarrow & & \downarrow & & \downarrow & & & \downarrow & \\
(J^k(\Omega)) & 0 & \rightarrow & \Omega & \rightarrow & \underline{F}_0 & \overset{\mathcal{D}_1}{\rightarrow} & \underline{F_1} & \overset{\mathcal{D}_2}{\rightarrow} & \cdots \overset{\mathcal{D}_n}{\rightarrow} & \underline{F_n} & \rightarrow 0 \\
 & & & & & \downarrow & & \downarrow & & & \downarrow & \\
 & & & & & 0 & & 0 & & & 0 &
\end{array}$$

The sequence $J^k(\Omega)$ *is formally exact;* $S_2^k(E)$ *is exact and* $S_2^k(\Theta)$ *is formally exact. If* $S_2^k(\Theta)$ *is exact it is called the* **Spencer's resolution** *of the sheaf* Θ . *The cohomology* *of the sequence* $S_2^k(\Theta)$ *at* $\underline{C}^{r+1}$ *is the same as that of the sequence* $J^k(\Theta)$ *at* $\underline{F}_r$. *One has* $\dim \underline{C}^r = \dim \underline{C^r(E)} - \dim \underline{F}_r$. *If* $E_k \subset J\mathcal{D}^k(E)$ *is formally integrable involutive elliptic analytic equation of order* k *on* E, $S_2^k(\Theta)$ *is the Spencer's resolution of* Θ. *(For the definition of elliptic equation see below).*

4) *The sequence*

$$\underline{E} \xrightarrow{\mathcal{D}} \underline{F}_0 \xrightarrow{\mathcal{D}_1} F_1$$

is "minimum" that is to say , if there exists another sequence $\underline{E} \rightarrow \mathcal{D}\underline{F}_0 \xrightarrow{\mathcal{D}'_1} \underline{F}'_1$ *then there must exist an operator* $\mathcal{D}' : \underline{F_1} \rightarrow \underline{F}'_1$ *such that* $\mathcal{D}'_1 = \mathcal{D}' \circ \mathcal{D}_1$.

5) *The cohomology of the sequence* $S_2^k(\Theta)$ *at* $\underline{C}_{r+1}$ *is the same as that of the sequence* $J^k(\Omega)$ *at* $\underline{F}_r$.

6) $\dim C^r(E) = \dim C^r(E) - \dim F^r$.

DEFINITION 2.9 - *Let* $Char(E_k)_q$ *be the* **characteristic manifold** *at the point* $q \in E_k$, *that is*

$$Char(E_k)_q \equiv \left\{ \begin{array}{l} \alpha \in T_x^*M - \{0\} \\ \alpha^k \otimes vT_pW \cap g_{k,q} \neq \{0\} \subset T_x^*M - \{0\} \\ p \equiv \pi_{k,0}(q), x \equiv \pi_k(q) \end{array} \right\}$$

Any covector $\alpha \in Char(E_k)_q$ *is called* **characteristic covector** *for* E_k *at* q. E_k *is* **elliptic at the point** $q \in E_k$ *if* $Char(E_k)_q = \emptyset$. E_k *is* **elliptic** *if it is elliptic for any* $q \in E_k$.

REMARK 2.12 - $Char(E_k)_q$ is a conic manifold by construction.

REMARK 2.13 - Let $\phi : J\mathcal{D}^k(E) \rightarrow E'$ be a vector bundle morphism. Set $\mathcal{D} = \phi \circ D^k : \underline{E} \rightarrow \underline{E'}$. For any covector $\chi \in T^*M$, we define the map $\sigma_\chi(D) : \underline{E} \rightarrow \underline{E'}$ by composition:

$$\sigma_\chi(\mathcal{D}) : \underline{E} \rightarrow \underbrace{S_k^0 M \bigotimes E \underset{\sigma_k(\phi)}{\rightarrow}} \underline{E'}.$$

$\sigma_\chi(\mathcal{D})$ defines a linear map $A(x,\chi) \equiv \sigma_\chi(\mathcal{D})_x : E_x \rightarrow E'_x$, $\forall x \in M$. If $\underline{E} \rightarrow \mathcal{D}\underline{E'} \xrightarrow{\mathcal{D}'} \underline{E''}$ is a sequence, $\mathcal{D}' \circ \mathcal{D} = 0$, then $\sigma_\chi(\mathcal{D}') \circ \sigma_\chi(\mathcal{D}) = 0$, and

$$\underline{E} \xrightarrow{\chi} \underline{E'} \xrightarrow{\sigma_\chi(\mathcal{D}')} \underline{E''}$$

is a sequence. The coefficients of $A(x,\chi)$ belongs to $\mathbf{R}[\chi_1, \cdots, \chi_n] \equiv \mathbf{R}[\chi]$ and have coefficients in $C^\infty(M)$. If $E_k = \ker \phi$, one has that a non-zero characteristic covector $\chi \in T_x^*M$ is **characteristic** if $\sigma_\chi(\mathcal{D})$ is not injective, (in this case we say also that χ is a **characteristic covector** for $\mathcal{D}$). Furthermore, E_k is undertermined (determined) if $\forall x \in M, \exists \chi \in T_x^*M$, such that $\sigma_\chi(\mathcal{D}) : E_x \rightarrow E'_x$ is surjective (bijective). Otherwise $\mathcal{D}$ is **overdetermined**. If $\mathcal{D}$ is undertermined or determined then the morphism $\phi^{(r)} : J\mathcal{D}^{k+r}(E) \to J\mathcal{D}^{r}(E')$ are epimorphisms $\forall r \geq 0$ and $\mathcal{D}$ is formally integrable. Furthermore, $\mathcal{D}$ is involutive and its associated J-sequence is just:

$$0 \to \Theta \to \underline{E} \overset{\mathcal{D}}{\to} \underline{F}_0 \to 0\,.$$

However, $J^k(\Theta)$ is not always exact at $\underline{F}_0$. Finally, an operator $\mathcal{D}$ is of finite type iff its characteristic manifold is zero. (Note that an operator $\mathcal{D}$ is said to be of finite type if the corresponding equation is of finite type.)

For the calculations the following is useful.

PROPOSITION 2.7 - (**Criterion of involutiveness**)[46]. *Let $E_k \subset J\mathcal{D}^k(W)$ be a PDE on the fiber bundle $\pi : W \to M, \dim M = n.(g_k)_q, q \in E_k$, is involutive iff there exists a local coordinate system in which we have*

$$\dim(g_{k+1})_q = \sum_{i=0}^{n-1} \dim(g_k^{(i)}),\quad g_k^{(i)} = \{X \in (g_k)_q | X(v_1) = \cdots = X(v_i) = 0\}$$

where $\{v_1, \cdots, v_n\}$ is the natural basis of $T_{\pi_k(q)}M$ induced by the system of coordinates.

REMARK 2.14 - We call **characters** of g_k the integers:

$$\alpha_k^i \equiv \dim(g_k^{(i-1)}) - \dim(g_k^{(i)})\,.$$

For an involutive symbol g_k, we have:

a)

$$\dim(g_k) = \alpha_k^1 + \cdots + \alpha_k^n;$$

b)

$$\dim(g_{k+1}) = \alpha_k^1 + 2\alpha_k^2 + \cdots + n\alpha_k^n;$$

c)

$$\dim(g_{k+r}) = \alpha_k^1 + \cdots + \frac{(r+i-1)!}{r!(i-1)!}\alpha_k^i + \cdots + \frac{(r+n-1)!}{r!(n-1)!}\alpha_k^n\,.$$

Let us define the integers $\beta_k^i \equiv m\binom{k+n-i-1}{n-i} - \alpha_k^i, m = \dim \text{fiber} W$. Let $E_k \subset J\mathcal{D}^k(W)$ be an involutive PDE over a fiber bundle $\pi : W \to M$ with $\dim M = n, \dim \text{fiber} W = m$. One has:

d)

$$\alpha_k^n = \sum_{r=0}^{n} (-1)^r \dim C^r;$$

e)

$$m = \sum_{r=0}^{n} (-1)^r \dim C^r(W);$$

f)

$$\beta_k^n = \sum_{r=0}^{n} (-1)^r \dim F_r;$$

g)

$$\dim F_r = \sum_{i=r}^{n} \frac{i!}{(i-r)!r!} \beta_k^{n-i};$$

h)

$$\beta_k^0 = \dim F_0 - \sum_{i=1}^{n} \beta_k^i .$$

THEOREM 2.3 - (**Goldschmidt**)[39]. *Let $E_k \subset J\mathcal{D}^k(W)$ be an analytic, formally integrable PDE on $\pi : W \to M$. Then, given $q \in E_{k+r}$ there exists an analytic solution s of E_k over a neighbourhood of $x \equiv \pi_k(q) \in M$ such that $D^{k+r}s(x) = q$.*

REMARK 2.15 - Let us consider, now, the Cauchy problem for linear PDEs. Let M be a differentiable manifold of dimension n and class C^∞. Let X be an s-dimensional submanifold of M and $\rho : X \to M$ the canonical embedding. Let $\pi : F \to M$ be a vector bundle over M and let $\bar{\pi} : E \to X$ be a vector bundle over X. Set $E \bigotimes_X F \equiv E \bigotimes \rho^* F$. Then we can associate to the first linear Spencer complex $(S_1^q(\rho^* F))$ another complex $S_1^q(F)_\rho$ over X, such that the following sequence:

$$(S_1^q(F)_\rho) \to (S_1^q(\rho F^*)) \to 0$$

is exact. More precisely, one has the following commutative diagram:

$$\begin{array}{cccccccc}
0 \to & \underline{Z} & \overset{{}_\rho D^q}{\to} & \rho^* \underline{J\mathcal{D}^q(F)} & \overset{\mathbf{D}}{\to} & \underline{T^*X \otimes \rho^* J\mathcal{D}^{q-1}(F)} & \cdots \overset{\mathbf{D}}{\to} & \underline{\Lambda_s^0 X \otimes \rho^* J\mathcal{D}^{q-s}(F)} \to 0 \\
& \downarrow & & \downarrow \rho_q & & \downarrow \rho_{q-1} & & \downarrow \rho_{q-s} \\
0 \to & \underline{Z} & \overset{D^q}{\to} & \underline{J\mathcal{D}^q(Z)} & \overset{\mathbf{D}}{\to} & \underline{T^*X \otimes J\mathcal{D}^{q-1}(Z)} & \cdots \to & \mathbf{D}\underline{\Lambda_s^0 X \otimes J\mathcal{D}^{q-s}(Z)} \to 0 \\
& \downarrow & & \downarrow & & \downarrow & & \downarrow \\
& 0 & & 0 & & 0 & & 0
\end{array}$$

where $Z \equiv \rho^* F$. The map ${}_\rho D^q$ is induced from $D^q : \underline{F} \to \underline{J\mathcal{D}^q}(F)$ by $\rho : {}_\rho D^q(s \circ \rho) = (D^q s) \circ \rho$. $S_1^q(F)_\rho$ is exact at $\Lambda_i^0 X \bigotimes \rho^* J\mathcal{D}^{q-i}(F)$ for $i > 0$.

Let $E_k \subset J\mathcal{D}^k(F)$ be a linear PDE on $\pi : F \to M$. To $S_1^q(F)_\rho$ we can associate, by restriction and dropping the first term, the following sequence:

$$(2.9)(S_1^q(E_k)\rho) \qquad 0 \to \underline{\rho^* E_q} \overset{\mathbf{D}}{\to} \underline{T^*X \bigotimes \rho^* E_{q-1}} \overset{\mathbf{D}}{\to} \cdots \overset{\mathbf{D}}{\to} \underline{\Lambda_s^0 X \bigotimes \rho^* E_{q-s}} \to 0.$$

Let $H^{q,j}(E_k)_{\rho,x}$ (resp. $H^{q,j}(E_k)_{\rho(x)}$) be the cohomology of the complex $S_1^q(E_k)_\rho$ at $\underline{(\Lambda_j^0 \bigotimes \rho^* E_{q-j})}_x$ (resp. of complex $S_1^q(E_k)$ at $\underline{(\Lambda_j^0 \bigotimes E_{q-j})}_{\rho(x)}$, $x \in X$. Then one has a canonical map:

$$\rho_* : H^{q,j}(E_k)_{\rho(x)} \to H^{q,j}(E_k)_{\rho,x} \, , \; \forall x \in X, q \geq k, j \geq 0 \, .$$

The **Cauchy problem** for E_k is the study of the above maps with $j = 0$. Surjectivity of these maps corresponds to the existence theorems an injectivity to uniqueness theorems.
Note also that, to the commutative diagram (2.6) we can associate the following:

$$\begin{array}{ccccccccc}
 & 0 & & 0 & & 0 & & 0 & \\
 & \downarrow & & \downarrow & & \downarrow & & \downarrow & \\
0 \rightarrow & \underline{P_p} & \overset{-\delta}{\rightarrow} & \underline{T^*X \otimes P_{p-1}} & \overset{-\delta}{\rightarrow} & \underline{\Lambda^0_2 X \otimes P_{p-2}} & \cdots \overset{-\delta}{\rightarrow} & \underline{\Lambda^0_n X \otimes P_{p-n}} & \rightarrow 0 \\
 & \downarrow & \searrow & \downarrow & \searrow & \downarrow & & \downarrow & \\
0 \rightarrow & \underline{A_p} & \rightarrow & \underline{T^*X \otimes A_{p-1}} & \overset{\mathbf{D}}{\rightarrow} & \underline{\Lambda^0_2 X \otimes A_{p-2}} & \cdots \overset{\mathbf{D}}{\rightarrow} & \underline{\Lambda^0_n X \otimes A_{p-n}} & \rightarrow 0 \\
 & \downarrow & \searrow & \downarrow & \searrow & \downarrow & & \downarrow & \\
0 \rightarrow & \underline{A_{p-1}} & \overset{\mathbf{D}}{\rightarrow} & \underline{T^*X \otimes A_{p-2}} & \overset{\mathbf{D}}{\rightarrow} & \underline{\Lambda^0_2 X \otimes A_{p-3}} & \cdots \overset{\mathbf{D}}{\rightarrow} & \underline{\Lambda^0_n X \otimes A_{p-n-1}} & \rightarrow 0 \\
 & \downarrow & & \downarrow & & \downarrow & & \downarrow & \\
 & 0 & & 0 & & 0 & & 0 &
\end{array}$$

where the vertical lines are exact, and $P_s \equiv \rho^* g_p$, $A_s \equiv \rho^* E_s$. The inclined lines which define morphisms are also denoted by δ. So, we can define the Spencer families ${}_\rho C^j$ associated to a linear boundary value problem:

$$\begin{aligned}
&{}_\rho C^j \equiv \Lambda^0_j X \bigotimes \rho^* E_k / \delta(\Lambda^0_{j-1} X \bigotimes \rho^* g_{k+1}), \\
&({}_\rho C^0 \equiv \rho^* E_k),
\end{aligned}$$

and if $\delta : \Lambda^0_{j-1} X \bigotimes \rho^* g_{k+1} \rightarrow \Lambda^0_j X \bigotimes \rho^* g_k$ is of constant rank for $j \geq 1$, one has a **second Spencer sequence** for linear boundary value problem:

$$0 \rightarrow {}_\rho\underline{C^0} \overset{\mathbf{D}}{\rightarrow} {}_\rho\underline{C^1} \overset{\mathbf{D}}{\rightarrow} {}_\rho\underline{C^2} \overset{\mathbf{D}}{\rightarrow} \cdots \overset{\mathbf{D}}{\rightarrow} {}_\rho\underline{C^n} \rightarrow 0$$

where $\mathbf{D}$ is the operator canonically associated to the Spencer operator

$$\mathbf{D} : \underline{\Lambda^0_j X \bigotimes \rho^* E_{k+1}} \rightarrow \underline{\Lambda^0_{j+1} X \bigotimes \rho^* E_k}, j \geq 0 .$$

So we can translate geometric results on the formal theory of PDEs to boundary value problems ones, and to prove the following.
REMARK 2.16 - Let $N \subset \rho^* T^* M$ be the conormal bundle of X in M and $S^r_0 N$ denotes its r-th symmetric product. We say that X is **non-characteristic** for g_{k+1} if:

$$((S^{k+1}_0 N) \bigotimes \rho^* F) \cap \rho^* g_{k+1} = 0 .$$

If E_k is elliptic, then every hypersurface of $Y \subset M$ is non-characteristic.
THEOREM 2.4 - [40] *If g_{k+1} is involutive and X is non-characteristic for g_{k+1}, one has the existence and uniqueness for the Cauchy problem, i.e., there exists a unique*

section s of F defined on a neighbourhood U of X in Y such that $D^k s(x) \in E_k$, $\forall x \in U$ and $s|_X = s_0 \in C^{\omega}(F|_X)$, for some fixed s_0.

The following theorems relate the existence and unicity of analytical solutions to the boundary conditions.

THEOREM 2.5 - 1) (CARTAN-KÄHLER). *Let $E_k \subset J\mathcal{D}^k(W)$ be an analytical, involutive, formally integrable PDE, on the fiber bundle $\pi : W \to M, \dim M = n, \dim \text{fiber} W = m$. Then, there exists a unique solution $s \in C^{\infty}(W)$ that satisfies the following initial boundary conditions:*

(i) *$D^{k-1}s(x_0) = q$, where q is a fixed point on $\pi_{k,k-1}(E_k) \subset J\mathcal{D}^{k-1}(W)$ and $x_0 = \pi_{k-1}(q) \in M$;*

(ii) *For the α_k^i-parametric derivatives $(\partial x_{\alpha_1} \cdots \partial x_{\alpha_s}.s^j)(x)$ of class i, $i = 1, \cdots, n$, one has*

$$(\partial x_{\alpha_1} \cdots \partial x_{\alpha_s}.s^j)(x) = (f^j_{i\alpha_1 \cdots \alpha_s}.s^j)(x) = \text{fixed}, \ \forall x \in X$$

where X is a submanifold of M for x_0, locally characterized by the equations:

$$x^{i+1} = x_0^{i+1}, \cdots, x^n = x_0^n .$$

Note that

$$\alpha_k^i = \dim g_k^{(i-1)} - \dim g_k^{(i)}$$

with

$$g_k^{(i)} \equiv \{X \in (g_k)_{q \in E_k} | X(v_1) = \cdots = X(v_i) = 0\}$$

where $\{v_1, \cdots, v_n\} \equiv$ base of $T_{\pi_k}(q)M, g_k =$ symbol of E_k. For an involutive symbol g_k one has $\dim g_k = \alpha_k^1 + \cdots + \alpha_k^n, \dim g_{k+1} = \sum_{i=0}^{n-1} \dim(g_k^{(i)})$.

2) (CAUCHY-KOWALESKI). *Let $E_1 \subset J\mathcal{D}(W)$ be a PDE of the first order, involutive, formally integrable, and analytic on a fiber bundle $\pi : W \to M, \dim M = n, \dim \text{fiber} W = m$, such that $\pi_{1,0} : E_1 \to W$ is an epimorphism and*

$$\beta_1^1 = 0, \cdots, \beta_1^{n-1} = 0, \beta_1^n = m$$

where $\beta_1^i \equiv m - \alpha_1^i$. Then, there exists a unique solution $s = (s^1, \cdots, s^m)$ such that on the hypersurface $X \subset M$, locally characterized by the equations $x^n = const$, one has $s|_X = f$, for an assigned section $f : X \to W|_X$ of $W|_X$ over X.

2 The study of singular solutions of PDEs is obtained by considering on the manifold characterizing a PDE a suitable distribution (Cartan distribution).

REMARK 2.17 - Let M be a differential manifold of class C^{∞}. A **distribution** of dimension r on M is a vector subbundle $\mathbb{E} \subset TM$, such that $\dim \mathbb{E}_p = r$, where $\mathbb{E}_p$ is the fiber of $\mathbb{E}$ over $p \in M$. A distribution is said to be **involutive** if for any couple X, Y of vector fields on M, such that $X(p), Y(p) \in \mathbb{E}_p$, $\forall p \in M$, one has that $[X,Y] \equiv \mathcal{L}_X Y$ belongs also to $\mathbb{E}$. A submanifold $N \subset M$ is called **integral manifold** of $\mathbb{E}$ if $TN \subseteq \mathbb{E}$. An integral manifold N of $\mathbb{E}$ is called **maximal** if $\dim N = r$ =dimension of $\mathbb{E}$. One can prove [61,155] that if $\mathbb{E}$ is involutive for any point $p \in M$ passes a unique maximal integral manifold.

DEFINITION 2.10 - *The* **Cartan distribution** *on a PDE $E_k \subset J\mathcal{D}^k(W)$ is the sub-bundle $\mathbb{E}_k \subset TE_k$ generated by the tangent spaces to the graphs of the k-derivatives of sections of $\pi : W \to V$ that are solutions of E_k.*

REMARK 2.18 - $\mathbb{E}_k$ is the annulator of

$$C\Omega^1(E_k) \equiv \{\alpha \in \Omega^1(E_k) | (D^k s)^* \alpha = 0 \, , \, \forall s \in C^\infty(W)\}$$

where $\Omega^1(E_k)$ is the module of differential 1-forms on E_k.

DEFINITION 2.11 - *A (***singular***)* **solution** *of E_k is an integral manifold of the Cartan distribution $\mathbb{E}_k$.*

REMARK 2.19 - The regular solutions of E_k are also integral manifolds of $\mathbb{E}_k$. They diffeomorphically project on M. (See sections 2.6 and the following references [60,70,73,110,111,112].) Singular solutions of PDEs are also important in interpreting the meaning of the quantization of PDEs and tunnel effects in PDEs [103,107].

□ (INTEGRALS AND SYMMETRIES OF DISTRIBUTIONS).

DEFINITION 2.12 - *A smooth function $h \in C^\infty(M)$ on M is called an* **integral** *of the distribution $\mathbb{E}$ iff $h|_N = const.$ for any maximal integral manifold $N \subset M$.*

THEOREM 2.6 - 1) *Let $\mathbb{E} \subset TM$ be a completely integrable distribution. Then, a function $h \in C^\infty(M)$ is an integral of $\mathbb{E}$ iff $dh \in \Omega^1(\mathbb{E})$ or equivalently $X.h = 0 \, , \, \forall X \in C^\infty(\mathbb{E})$.*

2) *If $h_1, \cdots, h_k$ is a set of independent integrals of a completely integrable distribution, then submanifolds $N_c \equiv \{h_1 = c_1, \cdots, h_k = c_k\}, c_i \in \mathbf{R}$, $i = 1, \cdots, k, k =$ codim$(\mathbb{E})$, locally represent the set of all maximal integral manifolds of $\mathbb{E}$.*

3) *An* **infinitesimal symmetry** *of $\mathbb{E} \subset TM$ is a vector field $v \in C^\infty(TM)$ such that $L_v X \in C^\infty(\mathbb{E})$, $\forall X \in C^\infty(\mathbb{E})$. Let us denote by $\mathcal{S}ym(\mathbb{E})$ the space of infinitesimal symmetries of $\mathbb{E}$.*

4) *We call* **shufting symmetries** *of $\mathbb{E}$ the elements of the following quotient*

$$\mathcal{S}huf(\mathbb{E}) \equiv \mathcal{S}ym(\mathbb{E}) / \mathbf{C}har(\mathbb{E})$$

where $\mathbf{C}har(\mathbb{E})$ is the **characteristic distribution** *of $\mathbb{E}$, (see below).*

THEOREM 2.7 - (RELATION BETWEEN SYMMETRIES AND INTEGRALS OF DISTRIBUTIONS).

1) *Let* $\mathbb{E} \subset TM$ *be a completely integrable distribution. Assume that* $\mathbf{g} \subset Shuf(\mathbb{E})$ *is a* $k(=\operatorname{codim}(\mathbb{E}))$*-dimensional Lie subalgebra which is transversal to the distribution in the sense that the natural mapping* $Shuf(\mathbb{E}) \to T_aM/\mathbb{E}_a$ *is bijective at each point* $a \in M$. *Let* $\overline{X}_1, \cdots, \overline{X}_k$, *where* $\overline{X}_i = X_i \operatorname{mod} \mathbf{C}har(\mathbb{E}), X_i \in Sym(\mathbb{E})$ *be a basis of the Lie algebra* $\mathbf{g}$ *and* $\omega_1, \cdots, \omega_k$ *be a basis of the* $C^\infty(M)$*-module* $\Omega^1(\mathbb{E})$. *The transversality condition for the algebra* $\mathbf{g}$ *is equivalent to the requirement that the matrix* $w \equiv (\omega_i(X_j))$ *be non-degenerate at any point of the manifold. Therefore, one can choose another basis* $\omega'_1, \cdots \omega'_k$ *of the module* $\Omega^1(\mathbb{E})$ *such that*

$$\omega'_i(X_j) = v_{ij}\,,\ i,j = 1, \cdots, k. \tag{2.10}$$

Indeed, it is enough to set

$$\begin{pmatrix} \omega'_1 \\ \cdots \\ \omega'_k \end{pmatrix} = w^{-1} \begin{pmatrix} \omega_1 \\ \cdots \\ \omega_k \end{pmatrix}.$$

Assume that for the basis $\omega_1, \cdots, \omega_k$ *conditions* (2.10) *hold, and let* $C^l_{ij} \in \mathbf{R}, i,j = l, \cdots, k$, *be structure constants of the Lie algebra* $\mathbf{g}$, *i.e.,* $[\overline{X}_i, \overline{X}_j] = \sum_{1 \le l \le k} C^l_{ij} \overline{X}_l$, *for all* $i,j = 1, \cdots, k$. *Then, the following formula*
(2.11)(**Maurer-Cartan equations**)

$$\boxed{d\omega_l + \sum_{i,j} C^l_{ij}\omega_i \wedge \omega_j = 0, l = 1, \cdots, k}$$

holds.
2) *If the algebra* $\mathbf{g}$ *is commutative, a complete set of first integrals of the distribution can be found by quadratures:*

$$h_j(a) = \int_{[a_0,a]} \omega_j\,,\ j = 1, \cdots, k, a_0 \in M, \text{fixed point}.$$

PROOF. In fact, ω_j are closed and (locally), $\omega_j = dh_j$, for some smooth function $h_j \in C^\infty(M), j = 1, \cdots, k$. □

DEFINITION 2.13 - *A Lie algebra* $\mathbf{g}$ *is said to be* **solvable** *iff* $\mathbf{g}^{(i)} = 0$, *for some* i, *whereby* $\mathbf{g}^{(i)}$ *denote Lie subalgebras defined by induction:*

$$\mathbf{g}^{(i+1)} \equiv [\mathbf{g}^{(i)}, \mathbf{g}^{(i)}], \quad \mathbf{g}^{(1)} \equiv [\mathbf{g}, \mathbf{g}], \quad \mathbf{g}^{(0)} \equiv \mathbf{g}.$$

THEOREM 2.8 - 1) *Let* $\mathbb{E}$ *be a completely integrable distribution and* $\mathbf{g} \subset Shuf(\mathbb{E})$ *be a solvable Lie algebra transversal to* $\mathbb{E}$. *Then,* $\mathbb{E}$ *is integrable by quadratures,*

i.e., one can find a complete set of all integrals for $\mathbb{E}$ by integrating closed differential 1-forms and solving functional equations, i.e., to find functions defined by implicit formulas (e.g., $F(z_1, \cdots, z_n) = 0, \to z_1 = f(z_2, \cdots, z_{n-1})$).
2) (**Lie-Bianchi**). *If one knows a solvable k-dimensional Lie algebra of shufting symmetries of the k-th order ODE, then the general solution of the ODE can be found by quadratures.*
EXAMPLE 2.2 - 1)

$$y' = F(x,y) \Rightarrow \omega = dy - F(x,y)dy\,.$$

Let f be a generating function of some symmetry. In this case,

$$W = \omega(S_f) = f \Rightarrow \omega' = \frac{1}{f}\omega = \frac{1}{f}dy - \frac{F}{f}dx$$

is closed, i.e., f^{-1} is an integrating factor and the function

$$h(x,y) = \int_{((x_0,y_0),(x,y))} \omega'$$

is an integral. Note, that in this case Lie equation

$$\triangle_F(f) = X.f - (\partial y.F)f = 0$$

is equivalent to the condition $d\omega' = 0$.
2) (**Affine equation**).

$$y^{(k+1)} + A_k(x)y^{(k)} + \cdots + A_0(x)y = g(x) \tag{2.12}$$

A function $f = f(x)$ is a generating function for some symmetry of (2.12) iff $\triangle_F(f) \equiv f^{(k+1)}(x) + A_k(x)f^{(k)}(x) + \cdots + A_0(x)f(x) = 0$. Hence, an arbitrary solution of the corresponding homogeneous equation is a symmetry of the affine equation. As, the space of solutions of the linear equation is a commutative k-dimensional algebra, then the affine equation can be integrated by means of quadratures. So, if $f_1(x), \cdots, f_{k+1}(x)$ is a fundamental system of solutions of linear equation, then

$$S_f \equiv f(x)\partial y_0 + f^1(x)\partial y_1 + \cdots + f^{(k)}(x)\partial y_k$$

and

$$[S_{f_i}, S_{f_j}] = 0\,.$$

Since

$$\omega_0 \equiv dy_0 - dy_1 dx, \cdots, \omega_{k-1} \equiv dy_{k-1} - y_k dx, \omega_k \equiv dy_k - F dx$$

where

$$F(x, y_0, \cdots, y_k) = y(x) - A_0(x)y_0 - \cdots - A_k(x)y_k$$

one gets

$$\omega_i(S_f) = f^{(i)}, \text{and} \quad W = (\omega_i(S_{f_j}) = \begin{pmatrix} f_1 & \cdots & f_{k+1} \\ f_1' & \cdots & f_{k+1}' \\ \cdots & \cdots & \cdots \\ f_1^{(k)} & \cdots & f_{k+1}^{(k)} \end{pmatrix}$$

is a Wronski-matrix. Therefore, the set of differential 1-forms

$$\begin{pmatrix} \omega_0' \\ \cdots \\ \omega_k' \end{pmatrix} = W^{-1} \begin{pmatrix} \omega_0 \\ \cdots \\ \omega_k \end{pmatrix}$$

is a complete set of integrals $h_j(a) = \int_{[a_0,a]} \omega_j'$. ∎

THEOREM 2.9 - *Let* $\mathbf{g} \subset Shuf(\mathbb{E})$ *be an arbitrary transversal symmetry algebra of the distribution* $\mathbb{E}, \dim \mathbf{g} = \operatorname{codim}(\mathbb{E})$, *and let* $\mathcal{I} \subset \mathbf{g}$ *be some ideal of* $\mathbf{g}$. *We can choose a basis* $X_1, \cdots, X_k$ *in the algebra* $\mathbf{g}$ *such that* $X_{r+1}, \cdots, X_k, k - r = \dim \mathcal{I}$, *is a basis in the ideal* $\mathcal{I}$. *Moreover, let* $\omega_1, \cdots, \omega_k$ *be such differential* 1*-forms of* $\Omega^1(\mathbb{E})$ *that* $\omega_i(X_j) = \delta_{ij}, i, j = 1, \cdots, k$. *Denote by* $\mathbb{E}_{\mathcal{I}}$ *a distribution generated by the differential* 1*-forms* $\omega_1, \cdots, \omega_k$. *Then, one has the following:*

i) $\mathbb{E}_{\mathcal{I}}$ *is completely integrable.*

ii) $\mathcal{I} \subset Char(\mathbb{E}_{\mathcal{I}})$.

iii) $\mathbf{g}/\mathcal{I} \subset Shuf(\mathbb{E}_{\mathcal{I}})$.

iv) *By means of quadratures we can reduce the integrability problem for a completely integrable distribution with an arbitrary transversal Lie algebra to the case with simple one.*

EXAMPLE 2.3 - (EXAMPLES OF SYMMETRIES OF CARTAN DISTRIBUTIONS).

1) The Cartan distribution $\mathbb{E}_1$ on $J\mathcal{D}(\mathbf{R}^n, \mathbf{R}) \cong \mathbf{R}^{2n+1}$ is the usual contact distribution, that in coordinates $(q^1, \cdots, q^n, u, p_1, \cdots, p_n)$ on $\mathbf{R}^{2n+1}$ can be locally generated by the vector fields

$$X_k \equiv \partial q_k + p_k \partial u, \quad Y^k \equiv \partial p^k \, , 1 \leq k \leq n \, .$$

In the dual form it is generated by the following differential 1-form

$$\omega \equiv du - \sum_{k=1}^{n} p_k dq^k \, .$$

The manifold $V \equiv \{u(q^1, \cdots, q^n), p_1 = (\partial q_1.u), \cdots, p_n = (\partial q_n.u)\}$ is an integral manifold of $\mathbb{E}_1$.

Symmetries of the contact distribution $\mathbb{E}_1$ on $J\mathcal{D}(\mathbf{R}^n, \mathbf{R}) \cong \mathbf{R}^{2n+1}$ are for example

$$X_i \equiv \partial q_i \, , \; i = 1, \cdots, n, \quad Y \equiv \partial u \, .$$

But $\partial p^i \notin Sym(\mathbb{E}_1)$. In fact, for the vector field X_i we have

$$\phi_t : (q^1, \cdots, q^i, \cdots, q^n, u, p_1, \cdots, p_n) \mapsto (q^1, \cdots, q^i + t, \cdots, q^n, u, p_1, \cdots, p_n)$$

and therefore

$$\phi_t^* \omega = \omega \, .$$

A similar result holds for Y. Instead, for the vector field ∂p^i we get:

$$\phi_t : (q^1, \cdots, q^n, u, p_1, \cdots, p_n) \mapsto (q^1, \cdots, q^n, u, p_1, \cdots, p_i + t, \cdots, p_n)$$

and

$$\phi_t^* \omega = \omega - t dq^i \neq \omega \, , \; \text{if} \, t \neq 0 \, .$$

2) The Cartan distribution $\mathbb{E}_{k+1}$ associated to an ordinary differential equation $E_{k+1} \subset J\mathcal{D}^{k+1}(\mathbf{R}, \mathbf{R})$ of the $(k+1)$-th order

$$y^{(k+1)} = F(x, y, y^1, \cdots, y^{(k)})$$

is a 1-dimensional distribution on the manifold $\mathbf{R}^{k+2}$ endowed with the coordinates $(x, y \equiv y, \cdots, y_k \equiv y^{(k)})$. More precisely $\mathbb{E}_{k+1}$ is locally generated by the following vector field

$$X = \partial x + y_1 \partial y^0 + \cdots + y_k \partial y^{k-1} + F \partial y^k \, .$$

From the dual point of view $\mathbb{E}_{k+1}$ is locally identified by means of the following 1-forms

$$\omega_0 \equiv dy_0 - y_1 dx, \cdots, \omega_{k-1} \equiv dy_{k-1} - y_k dx \, , \; \omega_k \equiv dy_k - F(x, y_0, \cdots, y_k) dx \, .$$

2 The curve

$$V \equiv \{ y_0 = f(x), y_1 = f^1(x), \cdots, y_k = f^{(k)}(x) \}$$

is an integral manifold of $\mathbb{E}_{k+1}$ iff the following equation

$$f^{(k-1)} = F(x, f(x), \cdots, f^{(k)}(x))$$

holds. The vector field

$$X = \partial x + y_1 \partial y^o + \cdots + y_k \partial y^{k-1} + F \partial y^k$$

generates the distribution $\mathbb{E}_{k+1}$. On the other hand

$$[X, X] = 0 \in \mathbb{E}_{k+1} \Rightarrow X \in Sym(\mathbb{E}_{k+1}) \Rightarrow X \in Char(\mathbb{E}_{k+1}) .$$

So $Char(\mathbb{E}_{k+1})$ is generated by the vector field X .

$\boxed{2}$ The **shufting symmetries** of $\mathbb{E}_{k+1}$,

$$Shuf(\mathbb{E}_{k+1}) \equiv Sym(\mathbb{E}_{k+1})/Char(\mathbb{E}_{k+1})$$

has a unique representative

$$S_f = f\partial y_0 + (X.f)\partial y_1 + \cdots + X^k(f)\partial y_k$$

where $X^p \equiv X \cdots_p \cdots X$, and $f = f(x, y_0, \cdots, y_k)$ is any function called **generating function** of S .

PROOF. As the module $\mathbb{E}_{k+1}$ is generated by the vector field X we get

$$\partial x \equiv -y_1\partial y_0 - \cdots - y_k\partial y_{k-1} - F\partial y_k \bmod Char(\mathbb{E}_{k+1})$$

and hence any shufting symmetry $S \in Shuf(\mathbb{E}_{k+1})$ has a unique representative

$$S = \alpha_0(x, y_0, \cdots, y_k)\partial y_0 + \cdots + \alpha_k(x, y_0, \cdots, y_k)\partial y_k .$$

As $\mathbb{E}_{k+1}$ is in dual form generated by the differential 1-forms

$$\omega_0 \equiv dy_0 - y_1 dx, \cdots, \omega_k \equiv dy_k - F dx .$$

Let us calculate the Lie derivatives of $\omega_0, \cdots, \omega_{k-1}$ with respect to S . We get

$$\begin{aligned} \mathcal{L}_S\omega_i &= dS(y_i) - S(y_{i+1})dx \\ &= d\alpha_i - \alpha_{i+1}dx \\ &\equiv (X.\alpha_i - \alpha_{i+1})dx \bmod \Omega^1(\mathbb{E}_{k+1}) , \ i = 0, \cdots, k-1 . \end{aligned}$$

Therefore, $S \in Sym(\mathbb{E}_{k+1})$ iff $\alpha_{i+1} = X.\alpha_i, \forall i = 0, 1, \cdots, k-1$, and any shufting symmetry has the following representation: $S = S_f$, where $f = \alpha_0$. □

$\boxed{2}$ For any ordinary differential equation

$$y^{(k+1)} = F(x, y, \cdots, y^{(k)}) \tag{2.13}$$

there is an isomorphism

$$Shuf(\mathbb{E}_{k+1}) \cong \ker(\Delta_F)$$

where

$$\triangle_F \equiv X^{k+1} - \sum_{0 \le i \le k} (\partial y_i.F) X^i : C^\infty(\mathbf{R}^{k+2}) \to C^\infty(\mathbf{R}^{k+2}) .$$

PROOF. In fact, for the last differential 1-form ω_k one gets

$$\mathcal{L}_S \omega_k = d\alpha_k - (S.F)dx \equiv [X.\alpha_k - (S.F)]dx \bmod \Omega^1(\mathbb{E}_{k+1})$$
$$= [X^{k+1}.f - S_f.F]dx \bmod \Omega^1(\mathbb{E}_{k+1})$$

and hence $X^{k+1}.f - S_f.F = 0$. □

The Lie algebra structure and the set of all shufting symmetries of (2.13) is given by

$$[f, g] = \sum_{0 \le i \le k} [(X^i.f)(\partial y_i.g) - (X^i.g)(\partial y_i.f)] .$$

PROOF. Let $[f, g] = [S_f, S_g](y_0) \equiv S_{[f,g]}(y_0)$ or $[f, g] = S_f(g) - S_g(f)$. □

In the following table we resume some examples of symmetries.

TAB.2.2 - **Examples of symmetries**

vector field	gen.funct.	flow	name
$X = \partial x$	$f = -y_1$	$(x, y) \mapsto (x + t, y)$	translats. in x
$X = \partial y$	$f = 1$	$(x, y) \mapsto (x, y + t)$	translats. in y
$X = ax\partial x + by\partial y$ $a, b \in \mathbf{R}$	$f = by - axy_1$	$(x, y) \mapsto (e^{at}x, e^{bt}y)$	scale transfs.
$X = h(x)\partial y$	$f = h$	$(x, y) \mapsto (x, y + th(x))$	translats. in $h(x)$
$h(x) \in C^\infty(\mathbf{R})$ $X = -a\partial x + b\partial y$ $a, b \in C^\infty(\mathbf{R}^2)$	$f = ay_1 + b$		point transfs.

REMARK 2.20 - The characterization of singular solutions of PDEs $E_k \subset J\mathcal{D}^k(W)$ can be made by means of some suitable differential forms on $J\mathcal{D}^k(W)$. In fact, let us define the following 1-forms on $J\mathcal{D}^k(W)$, for any $f : J\mathcal{D}^{k-1}(W) \to \mathbf{R}$:

$$\chi_k(f) : J\mathcal{D}^k(W) \to T^* J\mathcal{D}^{k-1}(W), \quad \chi_k(f) = < \chi_k, df > .$$

Then, one has

$$\chi_k(f)(q)(X_q) = < df(\bar{q}), v(Y_{\bar{q}}) >$$

where $\bar{q} \equiv \pi_{k,k-1}(q) \in J\mathcal{D}^{k-1}(W), Y_{\bar{q}} \equiv T(\pi_{k,k-1}(X_q)) \in T_{\bar{q}}J\mathcal{D}^{k-1}(W)$, and $v(Y_{\bar{q}}) \equiv Y_{\bar{q}} - o(Y_{\bar{q}})$, where $o(Y_{\bar{q}})$ is the part of $Y_{\bar{q}}$ belonging to $L_q \subset T_{\bar{q}}J\mathcal{D}^{k-1}(W)$ determined by the point q. The operator

$$\underline{\chi}_k : F^{k-1}(W) \to \Omega^1(J\mathcal{D}^k(W)), \quad \underline{\chi}_k : f \mapsto \chi_k(f)$$

is a $C^\infty(M)$-linear differentiation over the mapping $\pi^*_{k,k-1}$:

$$\chi_k(fg) = f\chi_k(g)\,,\ \text{if } f \in C^\infty(M);$$
$$\chi_k(fg) = f\chi_k(g) + g\chi_k(f)\,,\ \text{if } f \in F^{k-1}(W).$$

THEOREM 2.10 - (METHASYMPLECTIC STRUCTURE ON CARTAN DISTRIBUTION). *On $J\mathcal{D}^k(W)$ one has a canonical $\pi^*_{k,0}(S^0_{k-1}M \bigotimes vTW)$-valued differential 2-form:*

$$\Omega_k : J\mathcal{D}^k(W) \to \pi^*_{k,0}(S^0_{k-1}M \bigotimes vTW) \bigotimes \Lambda^2(\mathbb{E}^*_k)$$

called **methasymplectic structure** *on the Cartan distribution* $\mathbb{E}_k$.
PROOF. In fact, for any function $f \in F^{k-1}(W)$, we define the following differential 2-form on $J\mathcal{D}^k(W)$:

$$\Omega_k(f) \equiv pr_{\Lambda^2(\mathbb{E}^*_k)} \circ d\chi_k(f) : J\mathcal{D}^k(W) \to \Lambda^2(\mathbb{E}^*_k)\,.$$

The properties of the operator $\underline{\chi}_k$ imply that Ω_k identifies a section Ω_k of the fiber bundle

$$\pi^*_{k,0}(S^0_{k-1}M \bigotimes vTW) \bigotimes \Lambda^2(\mathbb{E}^*_k) \to J\mathcal{D}^k(W)\,.$$

□

REMARK 2.21 - For any vectors

$$X \in T_pM, Y \in S^0_k(T_xM) \bigotimes vT_pW, \lambda \in S^{k-1}_0(T_xM) \bigotimes vT^*_pW, \pi(p) = x$$

one has the following formula

$$\Omega(\lambda)(X,Y) = < \lambda, X \rfloor \delta Y > \tag{2.14}$$

where $\delta : S^0_k(T_xM) \otimes vT_pW \to T^*_xM \otimes S^0_{k-1}(T_xM) \otimes vT_pW$ is the Spencer δ-operator.

☐ The degeneration subspace of $\Omega(\lambda)$ coincides with the first prolongation of the symbol

$$g_\lambda \equiv \ker(\lambda) = \left\{ \begin{array}{l} \gamma \in S^0_{k-1}(T_xM) \bigotimes vT_pW \\ <\gamma, \lambda>= 0 \end{array} \right\} \subset S^0_{k-1}(T_xM) \bigotimes vT_pW$$

namely

$$\ker(\Omega(\lambda)) = \{\theta \in S^0_{k-1}(T_xM) \bigotimes vT_pW \subset (\mathbb{E}_k)_q | \delta\theta \in T^*_xM \bigotimes g_\lambda\} .$$

In fact one has the following commutative and exact diagram:

$$\begin{array}{ccccc} & & 0 & & 0 \\ & & \downarrow & & \downarrow \\ 0 & \rightarrow & (g_\lambda)_{+1} & \rightarrow & S^0_k(T_xM) \otimes vT_pW \\ & & \downarrow \delta & & \downarrow \delta \\ 0 & \rightarrow & T^*_xM \otimes g_\lambda & \rightarrow & T^*_x \otimes S^0_{k-1}(T_xM) \otimes vT_pW \end{array}$$

EXAMPLE 2.4 - 1)(**Cotangent bundle** $W \equiv T^*M$).

☐ For any manifold M, there exists a differential mapping

$$\Phi : J\mathcal{D}(T^*M) \rightarrow \Lambda^2 T^*M$$

which has the following properties:

(a) Φ is a morphism of the vector bundles with base M, such that the following sequence is exact:

$$0 \rightarrow T^{*2}M \rightarrow J\mathcal{D}(T^*M) \xrightarrow{\Phi} \Lambda^2 T^*M$$

where $T^{*2}M$ is the cotangent bundle of order 2 on M.

(b) For every form ω on M, we get $d\omega = \Phi \circ D\omega$.

(c) The local expression of Φ is the following:

$$\begin{array}{rl} T^*M & :\{x^\alpha, \dot{x}_\alpha\} \\ J\mathcal{D}(T^*M) & :\{x^\alpha, \dot{x}_\alpha, \dot{x}_{\alpha\beta}\} \\ \Lambda^2 T^*M & :\{x^\alpha, y_{\alpha\beta}\} \end{array} \Rightarrow \left\{ \begin{array}{c} x^\alpha \circ \Phi = x^\alpha \\ y_{\alpha\beta} \circ \Phi = \frac{1}{2}\dot{x}_{[\alpha,\beta]} \end{array} \right\}$$

☐ The 2-form $\Theta : J\mathcal{D}(T^*M) \rightarrow \Lambda^2 J\mathcal{D}(T^*M)$ (called **Liouville form of order** 1 and **degree** 2), is defined by means of the following commutative diagram

$$\begin{array}{ccccc} & J\mathcal{D}(T^*M) & \xrightarrow{\Phi} & \Lambda^2 T^*M & \\ \Theta & \downarrow & & \downarrow & \theta^{(2)} = id_{\Lambda^2 T^*M} \\ & \Lambda^2 J\mathcal{D}(T^*M) & \underset{\Lambda^2(\phi_{1,0})}{\leftarrow} & \Lambda^2 T^*M & \end{array}$$

One has the following properties:
(a) Θ is a semibase form that vanishes on $T^{*2}M$;
(b) For every Pfaffian form ω on M, one has $d\omega = (D\omega)^*\Theta$. Namely, one has the following commutative diagram:

$$\begin{array}{cccc} & J\mathcal{D}(T^*M) & \stackrel{\Theta}{\rightarrow} & \Lambda^J\mathcal{D}(T^*M) \\ D\omega & \uparrow & & \downarrow \\ & M & \underset{d\omega}{\rightarrow} & \Lambda^2 T^*M \end{array}$$

(c) The local expression of Θ is the following:

$$\Theta = \sum_{1\leq i,j\leq n} \dot{x}_{\alpha\beta} dx^\alpha \wedge dx^\beta .$$

$\square$ The canonical form χ_1 on $J\mathcal{D}(T^*M)$ is given by means of the following commutative dagram:

$$\begin{array}{ccc} J\mathcal{D}(T^*M) & \rightarrow & vT(\pi_1^*T^*M)\bigotimes T^*J\mathcal{D}(T^*M) \\ \chi_1 \downarrow & & \wr\| \\ T^*M\bigotimes T^*J\mathcal{D}(T^*M) & = & T^*M\bigotimes T^*J\mathcal{D}(T^*M) \end{array}$$

The local espression of χ_1 is the following:

$$\chi_1 = dx^\alpha \otimes [d\dot{x}_\alpha - \dot{x}_{\alpha\beta} dx^\beta] .$$

$\boxed{2}$ A local section $s : M \rightarrow J\mathcal{D}(T^*M)$ is of the form $s = D\omega$ iff

$$(D\omega)^*\chi_1 = 0 .$$

$\boxed{2}$ To χ_1 there corresponds a 2-form locally written as

$$\eta = dx^\alpha \wedge [d\dot{x}_\alpha - \dot{x}_{\alpha\beta} d\dot{x}^\beta]$$

that can be globally written as follows:

$$\eta = \beta^* d\theta - \Theta$$

where $d\theta$ is the canonical symplectic form on T^*M, and β is the target mapping.
$\boxed{2}$ For every Pfaffian form ω on M one has

$$(D\omega)^*\eta = 0 , \ i.e., \quad d\omega = \omega^* d\theta = (D\omega)^*\Theta .$$

2 (**Torsion of a connection**). To a connection $\Gamma: T^*M \to J\mathcal{D}(T^*M)$ on T^*M we can associate the mapping

(2.15)(**torsion of** Γ) $$\tau : T^* \to \Lambda^2 T^*M, \quad \tau = \Phi \circ \Gamma.$$

The connection is said to be **symmetric** if it lifts T^*M into $T^{*2}M$. Such a connection my be defined locally by Christoffel symbols such that $\Gamma^i_{js} = \Gamma^i_{sj}$. Then, Γ is symmetric iff $\tau = 0$.

2 We can also define the **torsion 2-form** $\mathcal{T}$ by

$$\mathcal{T} \equiv \tau^* \theta^2 = \Gamma^* \Theta .$$

In the case of a linear connection, Γ, and consequently τ, are vector-bundle morphisms. The torsion $\mathcal{T}$ is then a section of the vector bundle $Hom(T^*M; \Lambda^2 T^*M) \cong TM \bigotimes \Lambda^2 T^*M$. So $\mathcal{T}$ is a tensor field of type $(1,2)$. Locally we can write

$$\mathcal{T} = \frac{1}{2} \sum_{1 \leq j,k \leq n} \mathcal{T}^i_{jk} \partial x_i \otimes dx^i \wedge dx^k$$

with

$$\mathcal{T}^i_{jk} = \Gamma^i_{jk} - \Gamma^i_{kj} .$$

2)(**Bundle of** s**-forms**): $\Lambda^s T^*M$. We have

$$\Phi_{(s)} : J\mathcal{D}(\Lambda^s T^*M) \to \Lambda^{s+1} T^*M .$$

For any s-form $\omega_{(s)}$ on M one has

$$d\omega_{(s)} = \Phi_{(s)} \circ D\omega_{(s)} = (D\omega_{(s)})^* \Theta_{(s)}$$

where the form $\Theta_{(s)}$ is defined by

$$\Theta_{(s)} = \Phi^*_{(s)} \theta^{(s+1)}$$

where $\theta^{(s+1)}$ is the Liouville form on $\Lambda^{s+1} T^*M$, i.e., the identity mapping. The form $\Theta_{(s)}$ may be called a Liouville form of order 1 and degree $s+1$.

2.2 - ORDINARY DIFFERENTIAL EQUATIONS (ODEs)

In this section we will consider in some detail the ordinary differential equations as they are the dynamical equations for systems of particles and rigid systems. Furthermore, we will show in the next section that ODEs are useful also to integrate PDEs. These are the central objects for continuous mechanics and field theory.

REMARK 2.22 - We say that a function $F(x^1,\cdots,x^n)$ satisfies to the **Lipschitz conditions** if for each variable one has

$$|F(x^1,\cdots x^{h'},\cdots,x^n)-F(x^1,\cdots,x^{h''},\cdots,x^n)|\leq A_h|x^{h'}-x^{h''}|,(h=1,\cdots,n)$$

for any $x^{h'},x^{h''}$ belonging to the interval of variation of x^h, and $A_h\in\mathbf{R}_+$.

THEOREM 2.11 - (**Cauchy-Lipschitz**). *Let $E_1\subset J\mathcal{D}(W),\pi:W\equiv\mathbf{R}\times\mathbf{R}^m\to\mathbf{R},(x,y^1,\cdots,y^m)\mapsto(x)$, be a system of ordinary differential equations written in the form*

$$(2.16)\qquad \boxed{\begin{aligned}\dot{y}^1&=f_1(x,y^1,\cdots y^m)\\ \cdots&=\cdots\\ \dot{y}^m&=f_m(x,y^1,\cdots,y^m)\end{aligned}}$$

Let $(y_0^1,\cdots,y_0^m)$ be the initial conditions. If we are able to find two positive numbers a and b such that in the domain D defined by means of the conditions $x_0-a\leq x\leq x_0+a,y_0^1-b\leq y^1\leq y_0^1+b,\cdots,y_0^m-b\leq y^m\leq y_0^m+b$, the functions $f_1,\cdots,f_m$ are continuous and satisfy the Lipschitz conditions then for the variables $y^1,\cdots y^m$, we can determine a positive number $\delta\leq a$ such that in the interval $(x_0,x_0+\delta)$ the system admits a unique solution that satisfies the initial conditions. The set of solutions of (2.16) can be identified with $\mathbf{R}^n$.

PROOF. See e.g. ref.[121]. □

EXAMPLE 2.5 - 1)(CIRCULAR FUNCTIONS).

Let us consider the following ODEs.

$$(2.17)(E_1)\qquad \begin{gathered}\dot{y}^1=y^2\;,\;\dot{y}^2=-y^1\\ +\text{initial conditions}:y^1(0)=0,y^2(0)=1\end{gathered}.$$

Solution: $\{y^1=\sin x,y^2=\cos x\}$.

Properties:

a) $(y^1)^2+(y^2)^2=1$. We can write this equation also in the following form: $\dot{y}^1=\sqrt{1-(y^1)^2},\dot{y}^2=-\sqrt{1-(y^2)^2}$. Therefore, the functions $\sin x$ and $\cos x$ can be considered as inverse functions of the two integrals:

$$x=\int\frac{dy^1}{\sqrt{1-(y^1)^2}},\quad x=-\int\frac{dy^2}{\sqrt{1-(y^2)^2}}.$$

b) $|y^1|\leq 1,|y^2|\leq 1$.

c) y^1, y^2 cannot be simultaneously zero and their zeros are all simple ones.
d) $y^1(\xi+\alpha) = y^2(\xi), y^2(\xi+\alpha) = -y^1(\xi)$, where α is the first positive zero of $y^2(x)$.
e) The functions y^1 and y^2 are periodic functions, with period 4α. (The even multiples of α are zeros of y^1, and odd ones are zeros of y^2.)
2)(ELLIPTIC FUNCTIONS)(JACOBI FUNCTIONS). Let us consider the following ODEs:

$$(2.18)(E_1) \quad \begin{array}{l} \dot{y}^1 = y^2y^3, \quad \dot{y}^2 = -y^1y^3, \quad \dot{y}^3 = -k^2y^1y^2, 0 \le k \le 1 (k \equiv \textbf{modulus}) \\ + \text{initial conditions} : y^1(0) = 0, y^2(0) = 1, y^3(0) = 1. \end{array}$$

Solution:

$$\begin{aligned} y^1(x) &\equiv \operatorname{sn} x = x - (1+k^2)\frac{x^3}{3!} + (1+14k^2+k^4)\frac{x^5}{5!} - \cdots, \\ y^2(x) &\equiv \operatorname{cn} x = 1 - \frac{x^2}{2!}x + (1+4k^2)\frac{x^4}{4!} - (1+44k^2+16k^4)\frac{x^6}{6!} + \cdots, \\ y^3(x) &\equiv \operatorname{dn} x = 1 - k^2\frac{x^2}{2!}x + k^2(4+k^2)\frac{x^4}{4!} - k^2(16+44k^2+k^4)\frac{x^6}{6!} + \cdots. \end{aligned}$$

The series have finite radius of convergence, so it is suitable to try to read the properties of the corresponding functions directly on (E_1).
Properties:
a) $\operatorname{sn} x^2 + \operatorname{cn} x^2 = 1, k^2 \operatorname{sn} x^2 + \operatorname{dn} x^2 = 1$.
PROOF.

$$(E_1) \Rightarrow \left\{ \begin{array}{l} \{2y^1\dot{y}^1 = 2y^1y^2y^3, 2y^2\dot{y}^2 = -2y^1y^2y^3\}, \\ \{2k^2y^1\dot{y}^1 = 2k^2y^1y^2y^3, 2y^3\dot{y}^3 = -2k^2y^1y^2y^3\}. \end{array} \right\}$$

By adding the first two between them and the other two also between them, we get:

$$2y^1\dot{y}^1 + 2y^2\dot{y}^2 = \frac{d}{dx}((y^1)^2 + (y^2)^2) = 0, \quad 2k^2y^1\dot{y}^1 + 2y^3\dot{y}^3 = \frac{d}{dx}(k^2(y^1)^2 + (y^3)^2) = 0.$$

Therefore, we have the two integrals: $(y^1)^2 + (y^2)^2 = const.$, $k^2(y^1)^2 + (y^3)^2 = const.$, that with the initial conditions give the above properties. □
b) From the first property we get that the functions oscillate between -1 and 1, and when one of these functions is zero, the other pass from the maximum 1 to the minimum -1.
c) From the second property we get that dn^2x oscillates between $1-k^2$ and 1, that implies that $\operatorname{dn} x$ is never zero.
d) If we introduce an angle ϕ, (**amplitude** of elliptic functions), such that $\operatorname{sn} x = \sin\phi$ (**sine amplitude**), that is zero for $x = 0$, and increases with x, we get $\operatorname{cn} x = \pm\cos\phi$ (**cosine amplitude**), $\operatorname{dn} x = \pm\sqrt{1-k^2\sin\phi} \equiv \Delta\phi$ (**delta amplitude**).

e) $\mathrm{sn}\,(\xi+k)=\frac{\mathrm{cn}\,\xi}{\mathrm{dn}\,\xi}$, $\mathrm{cn}\,(\xi+k)=-k'\frac{\mathrm{sn}\,\xi}{\mathrm{dn}\,\xi}$, $\mathrm{dn}\,(\xi+k)=\frac{k'}{\mathrm{dn}\,\xi}$, $k'\equiv\sqrt{1-k^2}$ (**complementary modulus**), where k is the first positive zero of $\mathrm{cn}\,x$.

f) The functions $\mathrm{sn}\,x$ and $\mathrm{cn}\,x$ are periodic with period $4k$, and the function $\mathrm{dn}\,x$ is periodic with period $2k$. The even multiples of k are zeros of $\mathrm{sn}\,x$ (and the maxima of $\mathrm{dn}\,x$); the odd multiples ones are zeros of $\mathrm{cn}\,x$ (and minima of $\mathrm{dn}\,x$).

g) For the elliptic functions we have addition theorems similar to circular functions ones.

h) The above functions can be considered as inverse functions of integrals of algebraic functions, (**elliptic integral**). For example, $y^1(x)\equiv\mathrm{sn}\,x$ is the inverse function of the integral

$$x=\int_{[0,y^1]}\frac{dy^1}{\sqrt{(1-(y^1)^2)(1-k^2(y^1)^2)}} \tag{2.19}$$

i) (Limit cases for the parameter k).

($k^2=0$). The elliptic functions reduce to circular functions: $\mathrm{sn}\,x=\sin x, \mathrm{cn}\,x=\cos x, \mathrm{dn}\,x=1$.

($k^2=1$). The elliptic functions reduce to hyperbolic functions: $\mathrm{sn}\,x=\tanh x, \mathrm{cn}\,x=\mathrm{dn}\,x=1/\cosh x$, that are not more periodic, ($k=\infty$). In fact, from (2.19) we get

$$x=\int_{[0,y^1]}\frac{dy^1}{1-(y^1)^2}=\log\sqrt{\frac{1+y^1}{1-y^1}}\,.$$

Hence,

$$\mathrm{sn}\,x=y^1=\frac{e^x-e^{-x}}{e^x+e^{-x}}\quad,\quad \mathrm{cn}\,x=\mathrm{dn}\,x=\frac{2}{e^x+e^{-x}}\,.$$

THEOREM 2.12 - (ELEMENTARY METHODS OF INTEGRATIONS).

1) (**Order reduction**). *Let us consider the following ODE:*

$$(2.20)(E_k\subset J\mathcal{D}^k(\mathbf{R}\times\mathbf{R}))\qquad y^{(n)}=f(x,y,y',\cdots,y^{(n-1)}).$$

This can be integrated by an iterative method, by using the immersions:

$$\begin{aligned}J\mathcal{D}^k(\mathbf{R}\times\mathbf{R})&\to J\mathcal{D}(J\mathcal{D}^{k-1}(\mathbf{R}\times\mathbf{R}))\\&\to J\mathcal{D}(J\mathcal{D}(J\mathcal{D}^{k-2}(\mathbf{R}\times\mathbf{R})))\\&\to\cdots\to J\mathcal{D}(\cdots J\mathcal{D}(\mathbf{R}\times\mathbf{R})\cdots)\,.\end{aligned}$$

In other words, by putting $u^1=y,u^2=y',\cdots,u^n=y^{(n-1)}$, we can write the equation (E_k) as a first order one:

$$(2.21)(\widetilde{E}_1\subset J\mathcal{D}(\mathbf{R}\times\mathbf{R}^n))\qquad \{\dot u^1=u^2,\dot u^2=u^3,\cdots,\dot u^n=f(x,u^1,\cdots,u^n)\}$$

that represents a vector field on $\mathbf{R}^n$.
2) (**Separation of variables**). *Let us consider the ODE:*

$$(2.22)(E_1 \subset J\mathcal{D}(\mathbf{R}\times\mathbf{R})) \qquad \dot{y} = A(x)B(y).$$

This can be written as follows:

$$dy/B(y) = A(x)dx .$$

Then, by integration, we get:

$$G[y] \equiv \int dy/B(y) = \int A(x)dx \equiv f(x) + const .$$

3) (**Linear equation**). *Let us consider the ODE:*

$$(2.23)(E_k \subset J\mathcal{D}^k(\mathbf{R}\times\mathbf{R})) \qquad y^{(n)} + a_1 y^{(n-1)} + \cdots + a_{n-1}y' + a_n y = 0,$$

with coefficients $a_i = a_i(x)$. *Then, every solution of* E_k *can be written in the form:*

$$y(x) = c_1 y_1(x) + \cdots + c_n y_n(x)$$

where $y_i(x)$ *are particular independent solutions of* E_k , *i.e., such that their Wronskian is* $\neq 0$ *in any point of the interval of definition of the equation.*
4) (**Affine equation**). *Let us consider the ODE:*

$$(2.24)(E_k \subset J\mathcal{D}^k(\mathbf{R}\times\mathbf{R})) \qquad y^{(n)} + a_1 y^{(n-1)} + \cdots + a_{n-1}y' + a_n y = h(x),$$

where the coefficients $a_i = a_i(x)$. *The general solution (i.e., the space of solutions) of this equation can be written as follows:*

$$y(x) = c_1 y_1(x) + \cdots + c_n y_n(x) + \eta(x)$$

where $\overline{y}(x) \equiv c_1 y_1(x) + \ldots + c_n y_n(x)$ *is the general solution of* $\overline{E}_k \equiv$ *linear equation associated to* E_k , *and* $\eta(x)$ *is a particular solution of* E_k *expressed by means of the* **fundamental system** $\{y_1(x), \cdots, y_n(x)\}$. *More precisely, we have:*

$$\eta(x) = \int_{[x_0,x]} K(x,\xi)h(\xi)d\xi$$

with

$$K(x,\xi) = \frac{1}{W(\xi)} \begin{bmatrix} y_1(\xi) & \cdots & \cdots & y_n(\xi) \\ \dot{y}_1(\xi) & \cdots & \cdots & \dot{y}_n(\xi) \\ \cdots & \cdots & \cdots & \cdots \\ y_1^{(n-2)}(\xi) & \cdots & \cdots & y_1^{(n-2)}(\xi) \\ y_1(x) & \cdots & \cdots & y_n(x) \end{bmatrix} .$$

EXAMPLE 2.6 - 1)(INCOMPLETE GAMMA FUNCTION). Let us consider the following ODE:

$$(2.25) \qquad x\dot{y} + (a - x)y = 1.$$

The general integral of the associated linear equation is $\overline{y} = Cx^{-a}e^{x}, C \in \mathbf{R}$. Let us find a particular solution of (2.25) of the type $\eta(x) = \gamma(x)x^{-a}e^{x}$. We get

$$\gamma(x) = \int_{[0,x]} t^{a-1}e^{-t}dt \equiv \gamma(a,x) \equiv \textbf{incomplete gamma function}.$$

Then, the general integral of (2.25) is

$$y = x^{-a}e^{x}(\gamma(a,x) + C).$$

2)(INCOMPLETE BETA FUNCTION). Let us consider the following ODE:

$$(2.26) \qquad x(1-x)\dot{y} + [p - (p+q)x]y = 1\,,\ p,q \in \mathbf{R}.$$

The general integral of the associated linear equation is $\overline{y} = C[x^{-p}(1-x)^{-q}], C \in \mathbf{R}$. Let us find a particular solution of (2.26) of the type $\eta(x) = \gamma(x)x^{-p}(1-x)^{-q}$. We get

$$\gamma(x) = \int_{[0,x]} t^{p-1}(1-t)^{q-1}dt \equiv Bx(p,q) \equiv \textbf{incomplete beta function}.$$

Then, the general integral of (2.26) is

$$y = x^{-p}(1-x)^{-q}(Bx(p,q) + C).$$

5)(**Transformation of non-linear equation into linear one**).(**Bernoulli equation**). Let us consider the following ODE:

$$(2.27)(E_1 \subset J\mathcal{D}(\mathbf{R}\times\mathbf{R})) \qquad \dot{y} + p(x)y + q(x)y^{\alpha} = 0\,,\ y > 0, \alpha \neq 1, \alpha \neq 0.$$

The above equation can be transformed into an affine one by means of the following transformation $z = y^{1-\alpha}$. In fact, we get: $\dot{z} + (1-\alpha)pz + (1-\alpha)q = 0$.

2 THEOREM 2.13 - (CONSERVATION LAWS AND FIRST INTEGRALS FOR ODEs).

1) *Let $E_k \subset J\mathcal{D}^k(\mathbf{R}\times\mathbf{R}^p)$ be a system of ordinary differential equations. A* **conservation law** *for E_k is a numerical function $f : J\mathcal{D}^k(\mathbf{R}\times\mathbf{R}^p) \to \mathbf{R}$ such that for*

any solution $s \equiv (s^1, \dots, s^p) : \mathbf{R} \to \mathbf{R}^p, Ds : \mathbf{R} \to E_k$, *one has* $d(s^*f) = 0$, *where* $s^*f \equiv f \circ D^k s : \mathbf{R} \to \mathbf{R}$.

If f *is a conservation law, then for any solution* s *of* E_k , *we get* $s^*f = const$. *The value of the constant depends on the particular solution.*

2) *A* **first integral** *of* $E_k \subset J\mathcal{D}^k(\mathbf{R}\times\mathbf{R}^p)$ *is a function* $f : J\mathcal{D}^{k-1}(\mathbf{R}\times\mathbf{R}^p) \to \mathbf{R}$ *such that for any solution* s *of* E_k *one has* $s^*f = const.$, *where* $s^*f \equiv f \circ D^{k-1}s : \mathbf{R} \to \mathbf{R}$.

Each first integral of E_k *is a conservation law.*

PROOF. In fact, f , identifies the function $\widetilde{f} \equiv f \circ \pi_{k,k-1} : J\mathcal{D}^k(\mathbf{R} \times \mathbf{R}^p) \to \mathbf{R}$. Then, one has $s^*\widetilde{f} \equiv \widetilde{f} \circ D^k s = f \circ \pi_{k,k-1} \circ D^k s = f \circ D^{k-1} s = const$. □

3) *Let* $f : J\mathcal{D}^{k-1}(\mathbf{R} \times \mathbf{R}^p) \to \mathbf{R}$ *be a first integral of* $E_k \subset J\mathcal{D}^k(\mathbf{R} \times \mathbf{R}^p)$, *then* f *identifies an ODE* $I_k \subset J\mathcal{D}^k(\mathbf{R} \times \mathbf{R}^p)$ *of order* k *such that*

$$I_k \supset E_k .$$

PROOF. Let us consider the following function ($\equiv$ prolongation of f):

$$F \equiv (\partial x.f) + (\partial y_{i(s)}.f)y^{i(s+1)} .$$

Then, equation $I_k \equiv \{F = 0\}$ is such that $I_k \supset E_k$. In fact, for any solution s of E_k one has $f \circ D^{k-1}s = const$. Then, we have

$$\begin{aligned}
&d(f \circ D^{k-1}s) = 0 \Rightarrow D((D^{k-1}s)^\alpha)(\partial\xi_\alpha.f) = 0\\
&\Downarrow\\
&(\partial x.(x \circ (D^{k-1}s)))(\partial x.f) + (\partial x.(y^i \circ (D^{k-1}s)))(\partial y_i.f) + \dots\\
&+ (\partial x.(y^{i(k-1)} \circ (D^{k-1}s)))(\partial y_{i(k-1)}.f) = 0\\
&\Downarrow\\
&(\partial x.f) + (\partial x.s^i)(\partial y_i.f) + \dots + (\partial x \dots_k \dots \partial x.s^i)(\partial y_{i(k-1)}.f) = 0.
\end{aligned}$$

This means that $F \circ D^k s = 0$, namely $I_k \supset E_k$. □

4) *Let* $(f^j) : J\mathcal{D}^{k-1}(\mathbf{R} \times \mathbf{R}^p) \to (\mathbf{R}^s)$ *be such that each function* f^j *is a first integral of* $E_k \subset J\mathcal{D}^k(\mathbf{R} \times \mathbf{R}^p)$ *and* $df^1 \wedge \dots \wedge df^s \neq 0$. *Then,* (f^j) *identifies an ODE of order* k, $I_k \subset J\mathcal{D}^k(\mathbf{R} \times \mathbf{R}^p)$, *such that* $I_k \supset E_k$. *In particular, if* $\dim(I_k) = \dim(E_k)$, *then* $I_k = E_k$.

PROOF. It is a natural consequence of the above theorem, taking into account that the functions f^j are independent. □

EXAMPLE 2.7 - The following system: $(E_1)\{\dot{y}^1 = 2(y^2)^3, \dot{y}^2 = -3(y^1)^5\}$, admits the following first integral $f = (y^1)^6 + (y^2)^4$. In fact, let us integrate the system by separation of variables. We get: $(y^1)^6 = -(y^2)^4 + const$. Therefore, f is a constant on the set of solutions. Note, also, that the first prolongation of f is zero on the equation. In fact, $F \equiv f^{(1)} = 6(y^1)^5\dot{y}^1 + 4(y^2)^3\dot{y}^2$. Therefore, by restriction of F on the given equation, we get $F = 12(y^1)^5(y^2)^3 - 12(y^2)^3(y^1)^5 = 0$. Hence, $E_1 \subset (I_0)_{+1}$.

THEOREM 2.14 - 1) (LINEAR ODEs OF SECOND ORDER IN **R**). *The equations of the type*

$$(2.28)(E_2 \subset J\mathcal{D}^2(\mathbf{R}\times\mathbf{R})) \qquad \ddot{y} + p_1(x)\dot{y} + p_2(x)y = 0$$

*can also be written in the following form (**adjoint form**):*

$$\frac{d}{dx}[p(x)\dot{y}] + P(x)y = 0\,,\; p(x) > 0. \tag{2.29}$$

(For the general properties of these equations see e.g. refs.[115, 138] and bibliography reported there.)

2) (LINEAR ODEs OF SECOND ORDER IN **C**). *Let us consider the above equation but in* **C** *instead of* **R**. *Denote such an equation as equation (2.29). Let us assume that the functions $p_i(x)$, i= 1,2 , are analytic regular in a domain $D \subset \mathbf{C}$, except for some points $x_1, x_2, \cdots$, where they can have some singularities. The **singular points** of (2.29) are by definition the singular points of the functions $p_i(x)$, $i = 1,2$. Then, we have the following propositions:*

(a) *The solutions of (2.29) can have as singular points only the singular points of (2.29).*

(b) *The solutions of (2.29), in a neighbourhood of an isolated singular point x_0 of (2.29) are represented as multivalued functions of the type:*

$$(I)\quad (x-x_0)^{r_1};\quad (II)\quad (x-x_0)^{r_1};\quad (III)\quad \log(x-x_0)\,.$$

There r_1 and r_2 are two suitable numbers belonging to **R**, **C** *such that $|r_1 - r_2| \notin \mathbf{N}$. More precisely, one has that in a neighbourhood U of a singular point x_0 of (2.29) we can have the following two possible representations of the general integral of (2.29):*

(A)

$$y(x) = C_1(x-x_0)^{r_1}\phi_1(x) + C_2(x-x_0)^{r_1}\phi_2(x)$$

where ϕ_i , $i = 1,2$, are suitable analytic functions in U and r_1, r_2 are suitable constants;

(B)

$$y(x) = C_1(x-x_0)^r\phi(x) + C_2[A\log(x-x_0) + \psi(x)]$$

where ϕ and ψ are suitable analytic functions in U and r and A are suitable constants.

(c) *(**Fuchs theorem**). Equation (2.29) admits a fundamental system of solutions represented in a neighbourhood of a singular point x_0 as in above theorem, where*

ϕ_1, ϕ_2 and ψ are functions having for $x = x_0$ at most a pole iff p_1 has at most a pole of order one at $x = x_0$, and p_2 has at most a pole of order two at $x = x_0$. If these conditions are verified, we say that x_0 is a **fuchsian singularity** *(or* **regular singularity***) of equation (2.29), (which is then called a fuchsian equation). As a consequence, if $x = x_0$ is a fuchsian point, then we have that (2.29) can be written in the following form:*

$$(2.30)(E_2 \subset J\mathcal{D}^2(\mathbf{R}\times\mathbf{R})) \qquad \ddot{y} + \frac{A(x)}{x-x_0}\dot{y} + \frac{B(x)}{(x-x_0)^2}y = 0$$

*where $A(x)$ and $B(x)$ are analytic functions, regular for $x = x_0$. (So they can be represented in Taylor-series: $A(x) = \sum_{0\le n\le\infty} a_n(x-x_0)^n, B(x) = \sum_{0\le n\le\infty} b_n(x-x_0)^n$.) The numbers r_1 and r_2 are solutions (**roots**) of the following equation:*

$$(2.31)(\textbf{indicial equation}) \qquad r^2 + [A(x_0)-1]r + B(x_0) = 0.$$

*Furthermore, the sum of all the roots of all the indicial equations (**characteristic exponents**) of a fully Fuchsian equation, with $n+1$ singular points is $n-1$:*

$$(2.32) \qquad \sum_{1\le h\le\infty} r_h{}^{(1)} + r_h{}^{(2)} = n-1.$$

PROOF. See e.g. ref.[123]. □

EXAMPLE 2.8 - 1)

$$(2.33)(\text{BESSEL EQUATION}) \qquad \ddot{y} + \frac{1}{x}\dot{y} + (1-\frac{\nu^2}{x^2})y = 0.$$

The above equation satisfies Fuchs condition at the point $x = 0$, but not at $x = \infty$, namely it is not a fully fuchsian equation.

2)

$$(2.34)(\text{LAGUERRE EQUATION}) \qquad \ddot{y} + \frac{\alpha+1-x}{x}\dot{y} + \frac{\lambda}{x^2}y = 0.$$

The above equation satisfies Fuchs condition at $x = 0$, but not at $x = \infty$, namely it is not a fully fuchsian equation.

3)

$$(2.35)(\text{LEGENDRE EQUATION}) \qquad \ddot{y} - \frac{2x}{(1-x)(1+x)}\dot{y} + \frac{\lambda}{(1-x)(1+x)}y = 0.$$

The above equation satisfies Fuchs conditions at $x = \pm 1$, and at $x = \infty$, namely it is a fully fuchsian equation.

4)(FULLY FUCHSIAN EQUATIONS WITH 3 SINGULAR POINTS). Let us consider the following equations:

(2.36)(singular points $\alpha_1, \alpha_2, \infty$)

$$\ddot{y} - \frac{h_1 x + h_2}{(x-\alpha_1)(x-\alpha_2)}\dot{y} + \frac{k_1 x^2 + k_2 x + k_3}{[(x-\alpha_1)(x-\alpha_2)]^2} y = 0.$$

(2.37)(singular points a, b, c)

$$\ddot{y} - \frac{2x^2 + \underline{h}_1 x + \underline{h}_2}{(x-a)(x-b)(x-c)}\dot{y} + \frac{\underline{k}_1 x^2 + \underline{k}_2 x + \underline{k}_3}{[(x-a)(x-b)(x-c)]^2} y = 0.$$

[As the point ∞ must be a regular point, we require that p_2 should have a zero of this order at ∞ and p_1 a zero of first order and such that $\lim_{x\to\infty} xp_1 = 2$.] The equation is fully determined by 5 parameters = 3 parameters $(a, b, c) + 6(= 3$ couples of roots of 3 singular points $(\alpha, \alpha'), (\beta, \beta'), (\gamma, \gamma')) - 4(3$ determinant equations$+(\alpha + \alpha' + \beta + \beta' + \gamma + \gamma' = 1)$. More precisely equation (2.37) can be written in the following form (**Papperitz form**)

$$(2.38) \qquad \left\{ \begin{array}{l} \ddot{y} + (\frac{A}{(x-a)} + \frac{B}{(x-b)} + \frac{C}{(x-c)})\dot{y} \\ + (\frac{A'}{(x-a)} + \frac{B'}{(x-b)} + \frac{C'}{(x-c)})\frac{1}{(x-a)(x-b)(x-c)} y = 0 \end{array} \right\}$$

with: $A = 1 - \alpha - \alpha', B = 1 - \beta - \beta', C = 1 - \gamma - \gamma', A' = (a-b)(a-c)\alpha\alpha', B = (b-a)(b-c)\beta\beta', C = (c-a)(c-b)\gamma\gamma'$. Each function, solution of (2.38) is called **function P of Riemann** and is denoted by the following symbol:

$$P \left\{ \begin{array}{cccc} a & b & c & \\ \alpha & \beta & \gamma & x \\ \alpha' & \beta' & \gamma' & \end{array} \right\}.$$

5)

(2.39)(HYPERGEOMETRIC GAUSS EQUATION)

$$x(1-x)\ddot{y} + [c - (a+b+1)x]\dot{y} - aby = 0.$$

This is a particular case of Papperitz equation with singular points $(0, 1, \infty)$. Except for the three cases: $c =$ integer, $c - a - b =$ integer, $a - b =$ integer, (where logarithm terms can appear) the above equation admits the following fundamental integrals in the neighbourhoods of the singular points:

$$\begin{array}{ll} (x=0) & y_1 = P_1(x), \quad y_2 = x^{1-c}P_2(x); \\ (x=1) & y_3 = P_3(1-x), \quad y_4 = (1-x)^{c-a-b}P_4(x); \\ (x=\infty) & y_5 = (\frac{1}{x})^a P_5(\frac{1}{x}), \quad y_6 = (\frac{1}{x})^b P_6(\frac{1}{x}); \end{array}$$

where $P_1, \ldots, P_6$ denote power series with positive integer exponents of the corresponding arguments, with radius of convergence defined, for each couple, by the following:

$$(x=0) \quad |x|<1; \qquad (x=1) \quad |1-x|<1; \qquad (x=\infty) \quad |1/x|<1,$$

so they have all radius of convergence 1. More precisely, one has:
$(x=0)$ (**hypergeometric series**)

$$P_1(x) \equiv F(a,b,c;x) \equiv 1 + \sum_{1\leq n\leq\infty} \frac{a(a+1)\ldots(a+n-1)b(b+1)\ldots.(b+n+1)}{1.2\ldots n.c(c+1)\ldots(c+n-1)} x^n$$
$$= \frac{\Gamma(c)}{\Gamma(a)\Gamma(b)} \sum_{1\leq n\leq\infty} \frac{\Gamma(a+n)\Gamma(b+n)}{\Gamma(c+n)} \frac{x^n}{n!},$$

where $\Gamma(x)$ is the gamma function $\equiv \frac{1}{x}\int_{[0,\infty]} e^{-t}t^x dt, (\Re(x)>-1)$.

$$P_2(x) \equiv F(a-c+1, b-c+1, 2-c; x).$$

6)(SPHERICAL FUNCTIONS). These are the hypergeometric functions such that between the three parameters a, b, c there is the relation: $2c = a+b+1$. These satisfy the following ODE:

$$(1-\xi)^2\ddot{y} - 2\xi\dot{y} + [\nu(\nu+1) - \frac{\mu^2}{1-\xi^2}]y = 0$$

with $\mu = \frac{1}{2}(a+b-1), \nu = \frac{1}{2}(-a+b-1)$. This can be obtained from equation (2.39) with the transformation $x = \frac{1-\xi}{2}$.
7)(HYPERGEOMETRIC POLYNOMIALS OR JACOBI POLYNOMIALS). These are the hypergeometric functions where the series stop, and reduce to polynomials. This happens iff at least one of the parameters a and b is an integer $-n \leq 0$. In the following we quote some particularly important cases.
7/1)(LEGENDRE POLYNOMIALS). $(a=-n, b=n+1, c=1)$. They satisfy the following ODE:

(2.40)(**Legendre equation**) $\qquad x(1-x)\ddot{y} + [1-2x]\dot{y} + n(n+1)y = 0.$

The **Legendre polynomials** are the unique polynomial solution of this equation:

$$P_n(x) \equiv F(-n, n+1, 1; \frac{1-x}{2}).$$

Therefore, the Legendre polynomials are **spherical polynomials** with $(\mu = 0, \nu = n)$.
8)(CONFLUENT HYPERGEOMETRIC EQUATION). This is the limit case of the hypergeometric equation of Gauss when two among the three singular points coincide, generating a non-fuchsian singular point. So, we have the following equation:

(2.41)(**confluent hypergeometric equation**) $\quad x\ddot{y} + (c - x)\dot{y} - ay = 0$

with singular points 0 and ∞. More precisely, by applying to equation (2.40) the transformation $x = z/b$, the equation becomes

$$z(1 - \frac{z}{b})\ddot{y} + [c - (1 + \frac{a+1}{b})z]\dot{y} - ay = 0$$

with singular points $(0, b, \infty)$.
Therefore, taking the limit $b \to \infty$, (and changing z with x for uniformity).
We define: **Kummer function** (or **confluent hypergeometric function**) the following:

$$\Phi(a, c; x) \equiv \lim_{b\to\infty} F(a, b, c; \frac{x}{b}) = \frac{\Gamma(c)}{\Gamma(a)} \sum_{0\le n\le\infty} \frac{\Gamma(a+n)}{\Gamma(c+n)} \frac{x^n}{n!}.$$

☐ For $a = c$ we get: $\Phi(a, a; x) = e^x$.
☐ Therefore, the general integral of the confluent hypergeometric equation is the following:

$$y = A\Phi(a, c; x) + Bx^{1-c}\Phi(a - c + 1, 2 - c; x)$$

with A and B arbitrary constants.
☐ One has the **Kummer relation**: $\Phi(a, c; x) = e^x \Phi(a - c, c; -x)$.
☐ The confluent hypergeometric equation coincides with the Laguerre equation by putting $c = \alpha + 1, a = -\lambda$. The unique polynomial solutions of this equation are the **Laguerre polynomials** $L_n^{(c-1)}(x)$ ($\equiv$ **confluent hypergeometric polynomials**), defined by the following:

$$L_n^{(\alpha)}(x) \equiv \binom{\alpha+n}{n} \Phi(-n, \alpha + 1; x).$$

☐ If $\alpha = \pm\frac{1}{2}$ the Laguerre polynomials are called **Hermite polynomials**:

$$H_n(x) \equiv L_n^{(\pm\frac{1}{2})}(x).$$

☐ In the case $c = 2a$ the hypergeometric confluent equation can be transformed into the **Bessel equation**:

$$x\ddot{y} + \dot{y} + (x - \frac{\nu^2}{x})y = 0. \tag{2.42}$$

In fact, substituting $a = \nu + \frac{1}{2}, c = 2\nu + 1$, into equation (2.41), namely we have $x\ddot{y} + (2\nu + 1 - x)\dot{y} - (\nu + \frac{1}{2})y = 0$. After the change of variables $y = \alpha(x)z$, with $\alpha(x) = e^{x/2}x^{-\nu} = e^{(x/2)-\nu \log x}$, we get: $x\ddot{z} + \dot{z} - (\frac{x}{4} + \frac{\nu^2}{x})z = 0$. Therefore, putting $x = 2i\xi$, we obtain equation (2.42).

☐ If we denote by $\mathcal{C}(a, c; x)$ the general integral of the hypergeometric confluent equation, and with $\mathcal{Z}_\nu(x)$ of the Bessel equation one, one has:

$$\mathcal{Z}_\nu(x) = x^\nu e^{-ix}\mathcal{C}(\nu + \frac{1}{2}, 2\nu + 1; 2ix).$$

☐ Each solution of the Bessel equation is called **cylindrical function.**

☐ We call **Bessel functions of first type** ones corresponding to the Kummer function (that is multivalued):

$$J_\nu(x) = \frac{1}{\Gamma(\nu+1)}(\frac{x}{2})^\nu e^{-ix}\Phi(\nu + \frac{1}{2}, 2\nu + 1; 2ix). \tag{2.43}$$

☐ We call **Bessel functions of third type** (or **Hankel functions**) (x and ν real numbers):

$$H_\nu^{(1)}(x) = -\frac{2i}{\sqrt{\pi}}(2x)^\nu e^{i(x-\nu\pi)}\Psi(\nu + \frac{1}{2}, 2\nu + 1; -2ix), \tag{2.44}$$

and

$$H_\nu^{(2)}(x) = -\frac{2i}{\sqrt{\pi}}(2x)^\nu e^{-i(x-\nu\pi)}\Psi(\nu + \frac{1}{2}, 2\nu + 1; 2ix), \tag{2.45}$$

where

$$\Psi(a, c; x) \equiv \frac{\Gamma(1-c)}{\Gamma(a-c+1)}\Phi(a, c; x) + \frac{\Gamma(c-1)}{\Gamma(a)}x^{1-c}\Phi(a - c + 1, 2 - c; x), \tag{2.46}$$

is a solution of the hypergeometric confluent equation.

☐ We call **Bessel functions of second type** (or **Neumann functions**, or **Weber functions**):

$$Y_\nu(x) \equiv \frac{1}{2i}(H_\nu^{(1)}(x) - H_\nu^{(2)}(x)) \equiv N_\nu(x). \tag{2.47}$$

Thus the general integral of the Bessel equation can be written as follows:

(a)

$$\mathcal{Z}_\nu(x) = AJ_\nu(x) + BJ_{-\nu}(x)$$

where A and B are arbitrary constants. (If $\nu = n =$ integer is $J_{-n}(x) = (-1)^n J_n(x)$ so the above representation of the general integral of the Bessel equation does not work anymore.)

(b)

$$\mathcal{Z}_\nu(x) = AJ_\nu(x) + BY_\nu(x)\,.$$

(c)

$$\mathcal{Z}_\nu(x) = AH_\nu^{(1)}(x) + BH_\nu^{(2)}(x)\,.$$

☐ We call **uniform Bessel functions**: $E_\nu(\xi) \equiv \xi^{-\nu/2} J_\nu(2\sqrt{\xi})\,.$

☐ We call **modified Bessel functions** that corresponding to purely imaginary values of x . In particular, we put:

$$I_\nu(\xi) \equiv e^{-\nu\pi i/2} J_\nu(e^{\pi i/2} x) = (\frac{x}{2})^\nu E_\nu(-\frac{x^2}{4}) = \sum_{0 \leq n \leq \infty} \frac{1}{\Gamma(\nu + n + 1)n!} (\frac{x}{2})^{1+2n}$$

$$K_\nu(x) = \frac{\pi}{2} \frac{I_\nu(x) - I_\nu(x)}{\sin(\nu\pi)}\,.$$

REMARK 2.23 - 1) If $\nu \notin \mathbf{N}$, any couple of the following functions

$$J_\nu(x), \quad J_{-\nu}(x), \quad Y_\nu(x), \quad H_\nu^{(1)}(x), \quad H_\nu^{(2)}(x)$$

is independent. Then, the general integral of the Bessel equation can be written as a linear combination of two of the above functions.

2) If $\nu = n \in \mathbf{N}$, any couple of the following functions

$$J_n(x), \quad Y_n(x), \quad H_n^{(1)}(x), \quad H_n^{(2)}(x)$$

is independent. Then, the general integral of the Bessel equation can be written as a linear combination of two of the above functions.

3) In the following table integral representations of some special functions are given. (See also refs.[31,143].)

TAB.2.3 - **Integral representation of some special functions**

a) $F(a,b,c;x)$	$=\frac{\Gamma(c)}{\Gamma(a)\Gamma(c-a)}\int_{[1,0]} t^{a-1}(1-t)^{c-a-1}(1-xt)^{-b}dt$ $(\Re(c)>\Re(a)>0)$
b) $F(a,b,c;1)$	$=\frac{\Gamma(c)\Gamma(c-a-b)}{\Gamma(c-a)\Gamma(c-b)}$
c) $\Phi(a,b,c;x)$	$=\frac{\Gamma(c)}{\Gamma(a)\Gamma(c-a)}\int_{[1,0]} e^{-xt}t^{a-1}(1+t)^{c-a-1}dt$ $=\frac{1}{\Gamma(a)}\mathcal{L}_x[t^{a-1}(1+t)^{c-a-1}]$
d) $J_\nu(x)$	$=\frac{1}{\sqrt{\pi}\Gamma(\frac{1}{2}+\nu)}(\frac{x}{2})^\nu\int_{[-1,1]} e^{ixz}(1-z^2)^{\nu-1/2}dz$ $(\Re(\nu)>-\frac{1}{2})$

2.3 - CHARACTERISTICS OF PDEs

In this section we will show that on a PDE we can recognize vector fields (characteristic fields) that allow, under suitable conditions, the effective construction of (singular) solutions. In this way, the integration of PDEs is reconducted to that of ODEs.

THEOREM 2.15 - *Let M be an n-dimensional manifold. Let*

$$\mathbb{D}\subset TM$$

a p-dimensional distribution on M, namely a vector subbundle of TM, with fibers vector spaces of dimension p.

1) Let $\Omega^1(M)_{\mathbb{D}}$ be the submodule of $\Omega^1(M)\equiv C^\infty(T^*M)$ defined by

$$\Omega^1(M)_{\mathbb{D}}\equiv\{\theta\in\Omega^1(M)|\theta\rfloor v=0,\forall v\in\mathbb{D}\}.$$

Then, $\Omega^1(M)_{\mathbb{D}}$ is locally a $C^\infty(M)$ free-module of dimension $n-p\equiv r$. Thus, there exist r local 1-forms $\theta^1,\dots,\theta^r$, such that $\forall\alpha\in\Omega^1(M)_{\mathbb{D}}$ has the following local representation:

$$\alpha=\sum_{1\le j\le r}\alpha_j\theta^j,\alpha_j\in C^\infty(M).$$

We say that $\mathbb{D}$ has r **rank**.

2) Let $\mathcal{I}(\mathbb{D})\subset\Omega^\bullet(M)$ be the ideal in $\Omega^\bullet(M)$ defined by

$$\mathcal{I}(\mathbb{D})\equiv\{\omega\in\Omega^\bullet(M)|\omega=\sum_p\omega_p,\omega_p\in\Omega^p(M),\omega_p(v_1,\dots,v_p)=0\,,\ \forall v_i\in\mathbb{D}\}.$$

Then, $\mathcal{I}(\mathbb{D})$ is locally generated by $\theta^1, \ldots, \theta^r$, namely we have locally, for any $\omega \in \mathcal{I}(\mathbb{D})$, the following representation $\omega = \sum_{1 \leq j \leq r} \beta_j \wedge \theta^j, \beta_j \in \Omega^\bullet(M)$. Furthermore, for any $v \in \mathbb{D}$ and $\omega \in \mathcal{I}(\mathbb{D})$ one has $v \rfloor \omega \in \mathcal{I}(\mathbb{D})$.
3) *Vice versa, if $\mathcal{I} \subset \Omega^\bullet(M)$ is an ideal locally generated by $r \equiv n - p$ independent differential 1-forms $\theta^1, \ldots, \theta^r$, then there exists a unique distribution $\mathbb{D}$ of dimension p on M such that $\mathcal{I} = \mathcal{I}(\mathbb{D})$.*
4) *In particular, let $\mathcal{I} \equiv < \Sigma >$ be the ideal generated by a set $\Sigma \equiv \{\omega^\alpha\}_{1 \leq \alpha \leq N}$ of differential forms. Let $\Omega^1(M)_\mathcal{I} \subset \Omega^1(M)$ be the smallest submodule such that $\Sigma \subset \Lambda^\bullet(\Omega^1(M)_\mathcal{I})$. $\Omega^1(M)_\mathcal{I}$ is a $C^\infty(M)$-module locally finite. The local dimension of $\Omega^1(M)_\mathcal{I}$ is the* **rank** *of $\mathcal{I}$. If $\theta \in \mathcal{I}$ is a differential 1-form, it also belongs to $\Omega^1(M)_\mathcal{I}$. Then, the distribution $\mathbb{D}$ associated to $\mathcal{I}$ is the kernel of $\Omega^1(M)_\mathcal{I}$:*

$$\mathbb{D} \equiv \{X \in C^\infty(TM) | X \rfloor \theta = 0 \, , \forall \theta \in \Omega^1(M)_\mathcal{I}\}.$$

Furthermore, one has

$$\mathbb{D} \equiv \{X \in C^\infty(TM) | x \rfloor \omega \in \mathcal{I} \, , \forall \omega \in \mathcal{I}\} = \{X \in C^\infty(TM) | x \rfloor \omega \in \mathcal{I} \, , \forall \omega \in \Sigma\}.$$

5) *In particular, if $\Sigma \equiv \{\omega\}$, then $\mathbb{D} \equiv \{X \in C^\infty(TM) | x \rfloor \omega = 0\}$, as $\omega = \sum_{1 \leq \alpha_1 < \ldots < \alpha_p \leq r} \omega_{\alpha_1} \ldots \alpha_p \theta^{\alpha_1} \wedge \ldots \wedge \theta^{\alpha_p}, \omega_{\alpha_1} \ldots \alpha_p \in \mathbf{R}$, where $\{\theta^j\}_{1 \leq j \leq r}$ is a local basis of $\Omega^1(M)_\mathcal{I}$. In this case, if $\{\psi^j\}_{1 \leq j \leq n}$ is a local basis for $\Omega^1(M)$, then the module $\Omega^1(M)_\mathcal{I}$ is locally generated by the following differential 1-forms:*

$$\gamma_{i_2 \ldots i_p} \equiv \psi^i \omega_{i i_1 \ldots i_p} \, , \, 1 \leq i_2 < \ldots < i_p \leq n$$

if

$$\omega = \sum_{p \geq 0 \, , \, 1 \leq i_1 < \ldots < i_p \leq n} \frac{1}{p!} \omega_{i_1 \ldots i_p} \psi^{i_1} \wedge \ldots \wedge \psi^{i_p}$$

is a local representation of ω.

EXAMPLE 2.9 - The differential 2-form $F = \frac{1}{2} \sum_{1 \leq i < j \leq n} F_{ij} \psi^i \wedge \psi^j$ has an associated space $\Omega^1(M)_\mathcal{I}$ one generated by 1-forms $\gamma_j = \sum_{1 \leq j \leq n} F_{ij} \psi^i$. The local dimension of this model, i.e., the rank of F, is the rank of the matrix (F_{ij}). [The rank of a square skew-symmetric matrix is an even number. In particular, the determinant of (F_{ij}) is zero if n=odd; this means that the rank of F is $< n$.] For example, let us consider the 2-form on $\mathbf{R}^3$, $F = A dx \wedge dy + B dy \wedge dz + C dz \wedge dx$, namely

$$(F_{ij}) = \begin{pmatrix} 0 & A & -C \\ -A & 0 & B \\ C & -B & 0 \end{pmatrix}.$$

The conditions on the components $(v^k)_{1\leq k\leq 3}$ of the vector fields $v \in \mathbb{D}$ are as follows:

$$
\begin{aligned}
v\rfloor\omega &= (v\rfloor\psi_1)\wedge\omega^1 + (-1)^{\deg\psi_1}\psi_1\wedge(v\rfloor\omega^1)\\
&\quad + (v\rfloor\psi_2)\wedge\omega^2 + (-1)^{\deg\psi_2}\psi_2\wedge(v\rfloor\omega^2)\\
&= (v\rfloor\psi_1)\wedge\omega^1\\
&\quad + (v\rfloor\psi_2)\wedge\omega^2\\
&\quad + (-1)^{\deg\psi_2}\psi_2\wedge[(A_2v^3 - A_3v^2)\theta^1 + (A_3v^1 - A_1v^3)\theta^2 + (A_1v^2 - A_2v^1)\theta^3] .
\end{aligned}
$$

Thus, $v\rfloor\omega \in \mathcal{I}$ iff

$$
\begin{aligned}
&(A_2v^3 - A_3v^2)\theta^1 + (A_3v^1 - A_1v^3)\theta^2 + (A_1v^2 - A_2v^1)\theta^3 = f\omega^1\\
&\Updownarrow\\
&A_2v^3 - A_3v^2\\
&= fa_1, A_3v^1 - A_1v^3 = fa_2\\
&A_1v^2 - A_2v^1 = fa_3 .
\end{aligned}
$$

Therefore, a system of 1-forms generating the module $\Omega^1(M)_{\mathcal{I}}$ is the following:

$$\{\omega^1, (A_2\theta^3 - A_3\theta^2)a_2 - (A_3\theta^1 - A_1\theta^3)a_1, (A_3\theta^1 - A_1\theta^3)a_3 - (A_1\theta^2 - A_2\theta^1)a_2\} .$$

Hence, the system $\{\omega^1, \omega^2\}$ has rank 3 if the above 1-forms are independent ones. Furthermore, $\Omega^1(M)_{\mathcal{I}}$ has rank 2, and is generated by the 1-forms θ^1, θ^2 iff $a_3 = 0, A_2a_2 + A_1a_1 = 0$.

DEFINITION 2.14 - (CHARACTERISTIC SYSTEM OF IDEAL $\mathcal{I} \subset \Omega^\bullet(M)$). *Let*

$$\mathcal{I} \equiv \bigoplus_{q\geq 0} \mathcal{I}^q \subset \Omega^\bullet(M), \dim(M) = n$$

be an exterior differential system on M. We define **characteristic system** *of $\mathcal{I}$ the Pfaffian differential ideal (i.e., formed by 1-forms), $Char(\mathcal{I}) \equiv \mathcal{I} < \Omega^1(M)_{\mathcal{I}} >$, identified with the smallest submodule $\Omega^1(M)_{\mathcal{I}} \subset \Omega^1(M)$ such that $\mathcal{I} \subset \Lambda(\Omega^1(M)_{\mathcal{I}})$. [Note that $\Omega^1(M)_{\mathcal{I}}$ is a locally finite $C^\infty(TM)$-module; the local dimension of $\Omega^1(M)_{\mathcal{I}}$ is called* **rank** *of $Char(\mathcal{I})$, an is called the* **class** *of $\mathcal{I}$.] A* **characteristic manifold** *of $\mathcal{I}$ is an integral manifold of dimension $(n-r)$ of the characteristic system $Char(\mathcal{I})$.*

REMARK 2.24 - 1) If $\mathcal{I}$ is generated by 1-forms $\Sigma \equiv \{\theta^1, \ldots, \theta^r\}$, namely Σ is a basis for $\Omega^1(M)_{\mathcal{I}}$, then $Char(\mathcal{I})$ is also generated by the same Σ. In fact,

$$\mathcal{I} =< \theta^1, \ldots, \theta^r >=< \Sigma >\Rightarrow Char(\mathcal{I}) =< \theta^1, \ldots, \theta^r, d\theta^1, \ldots, d\theta >\equiv< \Sigma, d\Sigma > .$$

Thus, we can write,

$$\omega = \sum_{1\leq j\leq r} (\alpha_j \wedge \theta^j) + (\beta_j \wedge d\theta^j), \forall\omega \in Char(\mathcal{I}), \forall\alpha_j, \beta_j \in \Omega^\bullet(M).$$

This means that $d\theta^i = \{\alpha_j = 0, \beta_j = \delta^i_j\}$.
2) Of course, the characteristic system of $\mathcal{I}$ is fully integrable. The corresponding distribution $\mathbf{C}har(\mathcal{I})$ is involutive. If $y^1, \cdots, y^r$ is a complete system of independent first integrals of $Char(\mathcal{I})$, then $\mathcal{I} =< \Sigma >$, with

$$\Sigma \equiv \{a_{i_1\dots i_s} dy^{i_1} \wedge \dots \wedge dy^{i_s}, 0 \leq s \leq r, a_{i_1\dots i_s} = a_{i_1\dots i_s}(y^1, \dots, y^r)\}.$$

Furthermore, if $\mathbb{D}$ is a distribution associated to $\mathcal{I}$, one has

$$\mathbf{C}har(\mathcal{I}) \subseteq \mathbb{D}, \; Char(\mathcal{I}) \supseteq \mathcal{I}.$$

3) Let $f : N \to M$ be an integral manifold of $\mathcal{I}, \dim(N) = m, \dim(M) = n$. Let r be the class of $\mathcal{I}$. Then, the manifold $\Sigma \equiv N \times \mathbf{R}^{n-r}$ is also an integral manifold with respect to the mapping $F : \Sigma \to M$, locally defined by

$$(y^1, \cdots, y^m, u^1, \cdots, u^{n-r}) \mapsto (f^1(y^\alpha), \dots, f^r(y^\alpha), u^1, \cdots, u^{n-r}) \equiv (x^1, \dots, x^n)$$

where $\{y^\alpha\}$ are local coordinates on N, $\{u^A\}$ are local coordinates on $\mathbf{R}^{n-r}$ and $\{x^i\}$ are local coordinates on M such that the subbundle $\Xi \subset T^*M$, that identifies the Pfaffian system $Char(\mathcal{I})$ is generated by $\{dx^1, \cdots, dx^r\}$. Furthermore, if N is a submanifold of M such that the matrix $(\partial y_\alpha . x^A)$ is of constant rank m, Σ is also in a neighbourhood of N a submanifold of M, of dimension $m + n - r$; then we can say that the submanifold Σ built by considering at each point of N the characteristic submanifold that passes through this point, is also an integral manifold.
EXAMPLE 2.10 - $M \equiv \mathbf{R}^3, \Sigma \equiv \{a(x^1, x^2) dx^1 \wedge dx^2\}$. The characteristic manifolds are the following:

$$x^1 = c^1, x^2 = c^2, \quad c^1, c^2 \in \mathbf{R}.$$

Each line in $\mathbf{R}^3$ is an integral manifold of $< \Sigma >$. Let us consider the line

$$\Gamma \equiv \{[0,1] \to \mathbf{R}^3, t \mapsto (x^1(t), x^2(t), x^3(t))\}.$$

The surface $F : \Sigma \equiv [0,1] \times \mathbf{R} \to \mathbf{R}^3, (t, \lambda) \mapsto (x^1(t), x^2(t), \lambda)$, is also an integral manifold of Σ.
DEFINITION 2.15 - *Let $\mathbb{D} \subset TM$ be a distribution. Then, to $\mathbb{D}$ we can associate a mapping*

$$\Omega_{\mathbb{D}} : \Omega^1(M)_{\mathbb{D}} \to C^\infty(\Lambda^2(\mathbb{D}^*)), \omega \mapsto \Omega(\omega)(X, Y) \equiv d\omega(X, Y), \; \forall X, Y \in C^\infty(\mathbb{D}).$$

We call $\Omega_{\mathbb{D}}$ the **methasymplectic structure** *of the distribution $\mathbb{D}$.*

THEOREM 2.16 - 1) *If we consider the following exact sequence:*

$$0 \to \mathbb{D} \to TM \to TM/\mathbb{D} \to 0$$

that induces the following one:

$$0 \leftarrow \mathbb{D}^* \leftarrow T^*M \leftarrow (TM/\mathbb{D})^* \leftarrow 0$$

we get the following identification:

$$\Omega^1(M)_{\mathbb{D}} \equiv C^\infty((TM/\mathbb{D})^*).$$

2) *Thus,* $\Omega_{\mathbb{D}}$ *can be identified with a vector space valued differential 2-form:*

$$\Omega_{\mathbb{D}} \in C^\infty(\Lambda^2(\mathbb{D}^*)) \bigotimes C^\infty(TM/\mathbb{D}).$$

3) *The* **characteristic distribution** $\mathbf{C}har(\mathcal{I})_a, \mathcal{I} = \mathcal{I}_{\mathbb{D}}$, *of the distribution* $\mathbb{D}$ *is the distribution of the subspaces of degeneration of the methasymplectic structure, i.e.,*

$$\mathbf{C}har(\mathcal{I})_a = \ker(\Omega_{\mathbb{D}})_a \equiv \bigcap_{\omega \in \Omega^1(M)_{\mathbb{D}}} \ker(\Omega_{\mathbb{D}})(\omega)_a\,, \ \forall a \in M\,.$$

DEFINITION 2.16 - *The* **characteristic vector fields** *of a PDE*

$$E_k \subset J\mathcal{D}^k(W), \quad \pi : W \to M$$

are vector field belonging to the **characteristic distribution**

$$\mathbf{C}har(E_k)$$

namely to the distribution corresponding to the **characteristic differential ideal** *of* E_k*,*

$$\mathbf{C}har(E_k) \equiv \mathbf{C}har(\mathcal{I}(\mathbb{E}_k))$$

where $\mathcal{I}(\mathbb{E}_k)$ *is the ideal generated by the Cartan distribution on* E_k.

PROPOSITION 2.8 - 1) *A vector field of the Cartan distribution* $\mathbb{E}_k$ *has the following local representation:*

$$v = X^\alpha[\partial x_\alpha + \sum_{0 \le s \le k-1} y^j_{\alpha_1 \ldots \alpha_s \alpha} \partial y_j^{\alpha_1 \ldots \alpha_s}] + Y^j_{\alpha_1 \ldots \alpha_k} \partial y_j^{\alpha_1 \ldots \alpha_k}, \tag{2.48}$$

where $X^\alpha, Y^j_{\alpha_1 \ldots \alpha_k} : J\mathcal{D}^k(W) \to \mathbf{R}$ *are the local numerical functions, such that the following equations are satisfied:*

$$X^\alpha[(\partial x_\alpha.F^I) + \sum_{0 \le s \le k-1} y^j_{\alpha_1 \ldots \alpha_s \alpha}(\partial y_j^{\alpha_1 \ldots \alpha_s}.F^I)] + Y^j_{\alpha_1 \ldots \alpha_k}(\partial y_j^{\alpha_1 \ldots \alpha_k}.F^I) = 0, \tag{2.49}$$

where F^I are numerical functions on $J\mathcal{D}^k(W)$ that locally define E_k .

PROOF. The Cartan distribution is defined by the vector fields on $J\mathcal{D}^k(W)$ that annihilate the following differential 1-forms that generate $\mathcal{I}(\mathbb{E}_k)$:

$$\{\omega_0^I \equiv df^I, \omega^j_{\alpha_1\dots\alpha_s} \equiv dy^j_{\alpha_1\dots\alpha_s} - y^j_{\alpha_1\dots\alpha_s\alpha}dx^\alpha\}\,.$$

As a consequence such vector fields are:

$$v = X^\alpha \partial x_\alpha + \sum_{0\leq s\leq k} Y^j_{\alpha_1\dots\alpha_s}\partial y_j^{\alpha_1\dots\alpha_s}$$

such that

$$\begin{aligned}&\{ <\omega_0^I, v>=0, <\omega^j_{\alpha_1\dots\alpha_s}, v>=0\}\\ &\Leftrightarrow \{X^\alpha(\partial x_\alpha.F^I) + \sum_{0\leq s\leq k} Y^j_{\alpha_1\dots\alpha_s}(\partial y_j^{\alpha_1\dots\alpha_s}.F^I)=0, Y^j_{\alpha_1\dots\alpha_s} - y^j_{\alpha_1\dots\alpha_s\alpha}X^\alpha = 0\}.\end{aligned}$$

Thus equations (2.48) and (2.49) must be verified. □

2) *The vertical vector fields $v = Y^j_{\alpha_1\dots\alpha_k}\partial y_j^{\alpha_1\dots\alpha_k}$ belong to $\mathbb{E}_k$ iff it belongs to the symbol g_k of E_k .*

PROOF. In fact, such vector fields must satisfy equations (2.49) with $X^\alpha = 0$, namely

$$Y^j_{\alpha_1\dots\alpha_k}(\partial y_j^{\alpha_1\dots\alpha_k}.F^I) = 0\,.$$

But these equations say that $v \in g_k$. □

2 3) *Solutions of E_k , namely integral manifolds (in general no maximal ones) of $\mathbb{E}_k$, are integral manifolds of $\mathbf{C}har(E_k)$, namely are characteristic manifolds.*

PROOF.The characteristic system $Char(\mathcal{I}(\mathbb{E}_k))$ is the following differential ideal:

$$Char(\mathcal{I}(\mathbb{E}_k)) =< \mathcal{I}(\mathbb{E}_k), d\mathcal{I}(\mathbb{E}_k) >=< \omega_0^I, \omega^j_{\alpha_1\dots\alpha_s}, dy^j_{\alpha_1\dots\alpha_s\gamma}\wedge dx^\gamma > \,.$$

Thus , if $f : N \to E_k$ is an integral manifold of $\mathbb{E}_k$, it is also an integral manifold of $\mathbf{C}har(E_k)$. □

4) *A vector field $X \in \mathbb{E}_k$ is characteristic, namely $X \in \mathbf{C}har(E_k)$, iff*

$$\mathcal{L}_X\omega \in Char(\mathcal{I}(\mathbb{E}_k))\,, \forall\omega \in \mathcal{I}(\mathbb{E}_k)\,.$$

PROOF. Assume that $X \in \mathbb{E}_k$ then $X\rfloor\omega \in \mathcal{I}(\mathbb{E}_k)$, $\forall\omega \in \mathcal{I}(\mathbb{E}_k) \to d(X\rfloor\omega) \in Char(\mathcal{I}(\mathbb{E}_k))$, $d\omega \in Char(\mathcal{I}(\mathbb{E}_k)) \to \mathcal{L}_X\omega = d(X\rfloor\omega) + X\rfloor d\omega \in Char(\mathcal{I}(\mathbb{E}_k))$.

Vice versa, if

$$\begin{aligned}&X \in Char(\mathcal{I}(\mathbb{E}_k))\\&\Downarrow\\&X \rfloor d\omega \in Char(\mathcal{I}(\mathbb{E}_k)),\\&d(X\rfloor\omega) \in Char(\mathcal{I}(\mathbb{E}_k))\\&\Downarrow\\&L_X\omega \in Char(\mathcal{I}(\mathbb{E}_k))\,.\end{aligned}$$

□

5) (a) *Let* $v \in \mathcal{S}ym(\mathbb{E}_k) \equiv$ **space of infinitesimal symmetries** *of* E_k. *An infinitesimal symmetry of* E_k *is a vector field on* E_k *such that*

$$\mathcal{L}_v X \in \mathbb{E}_k\,,\ \forall X \in \mathbb{E}_k\,.$$

Thus, $\mathcal{S}ym(\mathbb{E}_k)$ *results in a Lie algebra with respect to the* **Lie bracket***:* $[v,u] \equiv \mathcal{L}_v u$ *. Let* $\phi_t : E_k \to E_k, t \in \mathbf{R}$*, be the one-parameter group of transformations on* E_k *associated to* v*.* ϕ_t *is a* **symmetry** *of* E_k *in the sense that induces a transformation on the tangent space* TE_k *that conserves the Cartan distribution* $\mathbb{E}_k$ *, namely* $T(\phi_t) : \mathbb{E}_k \to \mathbb{E}_k$ *. Then , if* V *is a maximal integral manifold of* $\mathbb{E}_k$*, then so is* $\phi_t(V)$*.*
(b) $\phi_t(V) = V$ *iff* $\partial\phi \equiv v \in$ **C***har*(E_k) *.*
(c) **C***har*(E_k) *is an ideal of the Lie algebra* $\mathcal{S}ym(\mathbb{E}_k)$ *.*
6) *The* **characteristic equations** *of* E_k *are the ODEs that define the integral curves of the characteristic vector fields. More precisely, one has:*

$$\begin{aligned}\dot{x}^\alpha &= X^\alpha\\ \dot{y}^j_{\alpha_1\dots\alpha_s} &= y^j_{\alpha_1\dots\alpha_s\alpha}X^\alpha, 0 \le s \le k-1\\ \dots &= \dots\\ \dot{y}^j_{\alpha_1\dots\alpha_k} &= Y^j_{\alpha_1\dots\alpha_k}\end{aligned} \tag{2.50}$$

where the functions on the right are such that they satisfy equations (2.48), (2.49)*, and such that* $< v, dy^j_{\alpha_1\dots\alpha_s\gamma} \wedge dx^\gamma >$ *are linear combinations of* ω^I_0 *and* $\omega^j_{\alpha_1\dots\alpha_s}$*.*
THEOREM 2.17 - 1) *Every point* $\overline{q} \in (E_k)_{+1}$ *determines on* $q \equiv \pi_{k+1,k}(\overline{q}) \in E_k$ *,* **characteristics***, namely straight lines identified with 1-dimensional subspaces* $L_{\overline{q}} \cap L_{\overline{q}'} \subset T_q E_k$ *, where* $\overline{q}' \in (g_k)_{\overline{q}} \cap S_1(\overline{q})$ *, where* $S_1(\overline{q})$ *is the characteristic cone at* $\overline{q}$ *. Each characteristic line can be identified with a line on* $T_x M, x \equiv \pi_k(q) \in M$ *, that is also called* **bicharacteristic** *on* M *.*

2) *The dual version of the concept of characteristics is that of* **characteristic covectors** *that belong to* $(n-1)$*-dimensional subspaces of* $(L_{\overline{q}})^*$ *given by* $Ann(L_{\overline{q}} \cap L_{\overline{q}'})$. *As* $(L_{\overline{q}})^*$ *can be identified with* $(T_xM)^*$, $Ann(L_{\overline{q}} \cap L_{\overline{q}'})$ *can be identified with a* $(n-1)$*-dimensional subspace of* $(T_xM)^*$. *The set of such cocharacteristic planes is a submanifold* $Char_1(E_k)_{\overline{q}}$ *of* $G_{n-1,s}(T_qE_k) \equiv$ *Grassmannian of* $(n-1)$*-dimensional subspaces of* $T_qE_k, q \equiv \pi_{k+1,k}(\overline{q}), s = \dim T_qE_k$.

REMARK 2.25 - A generalization of the concept of characteristics can be obtained by considering i-**dimensional characteristic strips**. These are i-dimensional subspaces of $L_{\overline{q}} \cap L_{\overline{q}'} \subset T_qE_k$ where $\overline{q}' \in (g_k)_{\overline{q}} \cap S_i(\overline{q})$, and $S_i(\overline{q})$ is the i-characteristic cone at $\overline{q}$. These identify i-dimensional subspaces on T_xM . The dual aspect is obvious.

The following definition relates the concept of characteristic directions to that of characterisic submanifolds of M with respect to PDEs.

DEFINITION 2.17 - *Let* $\sigma \subset E_k$ *a* $(n-1)$*-dimensional integral manifold of* E_k . *We say that* σ *is* **characteristic** *if there exists a submanifold* $\overline{\sigma} \subset (E_k)_{+1}$ *of dimension* $(n-1)$ *such that* $\pi_{k+1,k}(\overline{\sigma}) = \sigma$ *and such that for any* $\overline{q} \in \overline{\sigma}$ *and for some* $q' \in (g_{k+1})_{\overline{q}} \cap S_1(\overline{q})$ *the corresponding characteristic line passing through* $q \equiv \pi_{k+1,k}(\overline{q}) \in \sigma$ *is tangent to* σ; *otherwise we say that* σ *is* **non-characteristic**. *Similarly , we can define characteristic manifolds of dimension* $(n-i)$ *of* E_k . *Similar extensions can be obtained by considering* $\sigma \subset E_{k+s}$, *where* $s \in \mathbf{Z}$, *if* $s < 0, E_{k+s} = \pi_{k,k+s}(E_k)$.

PROPOSITION 2.9 - *If* σ *is characteristic, the* $(n-1)$*-dimensional submanifold* $\pi_k(\sigma) \equiv N \subset M$ *is* (**bi**) **characteristic**.

PROOF. In fact, the tangent spaces of N have as annihilators characteristic covectors. □

THEOREM 2.18 - *Let* $E_k \subset J\mathcal{D}^k(W)$, *be a PDE, over the fiber bundle* $\pi : W \to M$. *Let us consider the restriction on* E_k *of the Cartan distribution on* $J\mathcal{D}^k(W)$ *and describe the characteristics of this distribution in the sense introduced above. For any point* $\overline{q} \in J\mathcal{D}^{k+1}(W)_x, x \in M$, *and any vector* $v \in T_xM$, *denote* $v(\overline{q}) \in L_{\overline{q}}$ *the vector such that* $T(\pi_{k,0})(v(\overline{q})) = v$. *Assume that* E_k *is such that* $\pi_{k+1,k} : (E_k)^{(1)} \to E_k, \pi_{k,k-1} : E_k \to J\mathcal{D}^{k-1}(W)$ *are smooth fiber bundles, where* $(E_k)^{(1)}$ *is the first prolongation of* E_k . *Then,* $Char(E_k)_q$ *is the set of vectors* $v(\overline{q})$, $v \in T_xM, \overline{q} \in (E_k)^{(1)}$, *such that* $v \rfloor \theta = 0$, *for any tensor* $\theta \in g_k(q) \subset S^k((T_xM)^*) \bigotimes vT_pW, p \equiv \pi_{k,0}(q)$. *Here* $Char(E_k)_q$ *denote the characteristic vectors of the Cartan distribution at the point* $q \in E_k$.

PROOF. See ref.[73]. □

DEFINITION 2.18 - *We say that a PDE* E_k *is* **degenerate at the point** $q \in E_k$ *if there is a subspace* $\Xi(q) \subset (T_xM)^*$ *such that* $g_k(q) \subset S^k(\Xi(q)) \bigotimes vT_pW$. *The subspace* $\Xi(q)$ *is called* **degeneration subspace**.

COROLLARY 2.1 - 1) $Char(E_k)_q \neq 0$ *iff* E_k *is a PDE degenerate at the point* $q \in E_k$.
2) *The subspace* $\Xi(q) = Ann(T(\pi_{k,0})(Char(E_k)_q))$ *is the degeneration subspace of* E_k *at the point* $q \in E_k$.
3) *A vector* $v(\overline{q}) \in (\mathbb{E}_k)_q$ *is a characteristic vector of the Cartan distribution iff* $v(\overline{q})$ *is a characteristic vector of* E_k, *namely some integral co-line* $L(W, w)$ *is characteristic.*
4) *Let* $w \in (\mathbb{E}_k)_q$ *be a characteristic vector of the Cartan distribution at the point* $q \in E_k$, *then* $w \in L_{\overline{q}}$, *for some element* $\overline{q} \in (E_k)^{(1)}$.
THEOREM 2.19 - *Let* $E_k \subset J\mathcal{D}^k(W)$ *be a PDE such that:*
(i) $\pi_{k+1,k} : (E_k)^{(1)} \to E_k, \pi_{k,k-1} : E_k \to J\mathcal{D}^{k-1}(W)$ *are smooth fiber bundles;*
(ii) $\Xi : q \mapsto \Xi(q)$, $\forall q \in E_k$, *is a smooth vector fiber bundle, where* $\Xi(q)$ *is a space of degeneration of* E_k *to the point* q.
$Char(E_k)_q$ *is a smooth distribution on* E_k *and the fiber* $Char(E_k)_q$ *coincides with* $\Xi(\overline{q})$, *for any point* $\overline{q} \in (S^0_{k+1}(T_xM) \bigotimes vT_pW) \cap (E_k)^{(1)}$, *and the solutions of* E_k *can be written with the method of the characteristics.*
DEFINITION 2.19 - *Integral manifolds,* P, *of dimension* $(n-1)$, *of the Cartan distribution on* $J\mathcal{D}^k(W)$ *will be called* **Cauchy data**.
REMARK 2.26 - Let us consider the mapping $\pi_{k,k-1} : P \to J\mathcal{D}^{k-1}(W)$ and denote by $\Sigma(l)$ the set of singular points q of this application with $\dim\ker((D\pi_{k,k-1})(q)) = l$. Let us put $\Sigma(P) = \cup_{l\geq 1}\Sigma(l)$.
From the above results we can describe the structure of the tangent planes P at the points $q \in \Sigma(P)$.
THEOREM 2.20 - *Let* $L \subset (\mathbb{E}_k)_q$, *be an involutive subspace and* $\dim L = n-1$. *Denote by* $\Xi(q) \subset (T_xM)^*$ *the following subspace:* $\Xi(q) \subset Ann(T(\pi_{k,0})(L))$. *Then,* L *is determined by the following,* $L \equiv (\overline{q}, \Xi, L_0)$ *where* $\overline{q} \in S^k((T_xM)^*) \bigotimes vT_pW, L_0 \subset S^k(\Xi) \bigotimes vT_pW$ *and* $L_0 = \ker(T(\pi_{k,k-1})_q|_L)$. *One has:* $\dim L_0 = l, \dim \Xi = l+1$, *if* L *is the tangent plane to the integral manifold* P *at the singular point* q *of type* $\Sigma(l)$.
Denote by $I_l(n-1)$ *the set of involutive* $(n-1)$*-dimensional subspaces in* $(\mathbb{E}_k)_q$ *of type* $\Sigma(l)$, *namely with* $\dim L_0 = l$, *and let* $\widetilde{I}_l(n-1)$ *the total space of a fiber bundle over* $G_{l+1,n}((T_xM)^*)$ *with fiber* $G_{l,r}(S^k(\Xi) \bigotimes vT_pW)$ *at the point* $\Xi \subset G_{l+1,n}((T_xM)^*)$, *where* $r = \dim(S^k(\Xi) \bigotimes vT_pW)$. *Then, one has two fiber bundles:*

$$\left\{ \begin{array}{l} \alpha : I_l(n-1) \to \widetilde{I}_l(n-1) \\ \alpha(\overline{q}, \Xi, L_0) = (\Xi, L_0) \end{array} \right\}$$

$$\left\{ \begin{array}{l} \beta : \widetilde{I}_l(n-1) \to G_{l+1,n}((T_xM)^*) \\ \beta(\Xi, L_0) = \Xi \end{array} \right\}.$$

Therefore,

$$\dim I_l(n-1) = m[\tbinom{n+k}{k+1} - \tbinom{l+k+1}{k+1} + l\tbinom{l+k}{k}] - l^2 - l - 1 + m.$$

If we denote by χ_l the codimension of $\Sigma(l)$-singularity, namely $\chi_l = \dim(I_0(n-1)) - \dim(I_l(n-1))$, *then we get*

$$\chi_l = m[\binom{l+k}{k} - (l - \frac{lk}{k+1}) - 1] + l^2 + l \,.$$

We can see [73] *that if $q \in P$ is the point where the application $\pi_{k,k-1} : P \to J\mathcal{D}^{k-1}(W)$ has a singularity of Thom-Bordman type $\Sigma(l)$ and $l \geq 3$, and $T_qP \in I_l(n-1)$ does not lie on the boundary of the regular cell $I_0(n-1)$. Then, if* $\dim(I_l(n-1)) > \dim(I_0(n-1))$, *some neighbourhood of the point q completely consists of singular points, that cannot be removed with little deformations. The solutions of PDEs with Cauchy data P are n-dimensional integral manifolds Q such that $Q \supset P$. By using the description of involutive subspaces, and the following immersion $T_qQ \supset T_qP$ to each point $q \in P$, we get*

(a) $L_0' \supset L_0$;

(b) $\Xi' \subset \Xi$. *If*

$$T_qQ = (\bar{q}, \Xi', L_0')$$

and

$$T_qP = (\bar{q}, \Xi, L_0) \,.$$

But

$$L_0 \subset S^k(\Xi) \bigotimes vT_pW \text{ and } L_0 \subset L_0' \subset S^k(\Xi') \bigotimes vT_pW \,.$$

Therefore, L_0 must be the degeneration space in $S^k(\Xi) \otimes vT_pW$ in the sense introduced above.

THEOREM 2.21 - *The Cauchy data $P \subset J\mathcal{D}^k(W)$ admits a smooth solutions at the point $q \in P$ (namely there is an n-dimensional integral manifold $Q \supset P$ and $q \in Q$) if*

$$L_0 = \ker(T(\pi_{k,k-1}|P)_q) \subset S^k(\Xi) \bigotimes vT_pW$$

is a degenerate subspace.

DEFINITION 2.20 - *Define $q \in \Sigma(l) \subset P$* **frozen singularity** *if L is a nondegenerate subspace in*

$$S^k(\Xi) \bigotimes vT_pW \,.$$

Then we have

THEOREM 2.22 - *There are no n-dimensional integral manifolds with Cauchy data and passing through frozen singularities.*

THEOREM 2.23 - (CLASSIFICATION OF AFFINE PDEs OF THE SECOND ORDER BY MEANS OF CHARACTERISTICS). *Set $M \equiv \mathbf{R}^2, E \equiv M \times \mathbf{R}$. Let us consider the following PDE:*

$$(2.51)(E_2) \qquad Au_{xx} + 2Bu_{xy} + Cu_{yy} + Du_x + Eu_y + Fu + G = 0$$

with A, B, C, D, E, F, G C^1-numerical functions on M. The symbol of E_2, $(g_2)_q \subset S^0_2(\mathbf{R}^2)$ is defined by the set of vectors $\theta = X_{\alpha_1\alpha_2} dx^{\alpha_1} \odot dx^{\alpha_2}$ such that $X_{\alpha_1\alpha_2}$ satisfy the following equations $X_{xx}A + X_{xy}2B + X_{yy}C = 0$. Thus a vector $v \in T_pM \cong \mathbf{R}^2$ is characteristic if for any element $\theta \in (g_2)_q$ one has $v \rfloor \delta(\theta) = 0$. Thus the following system must be satisfied:

$$AX_{xx} + 2BX_{xy} + CX_{yy} = 0, v^x X_{xx} + v^y X_{xy} = 0, v^x X_{xy} + v^y X_{yy} = 0\,.$$

This system has solutions if the determinant of the coefficients is zero:

$$det \begin{bmatrix} A & 2B & C \\ v^x & v^y & 0 \\ 0 & v^x & v^y \end{bmatrix} = A(v^y)^2 - 2Bv^x v^y + C(v^x)^2 = 0\,.$$

If the determinant is $\neq 0$ the system has no characteristic vectors at q. If the determinant is $= 0$ we can divide by $(v^x)^2$, and put $\frac{dy}{dx} = \frac{v^y}{v^x}$, we get the following equation:

(2.52)(**characteristic equation of** E_2) $$A(\frac{dy}{dx})^2 - 2B\frac{dy}{dx} + C = 0.$$

From this equation we get

$$\frac{dy}{dx} = \frac{B \pm \sqrt{\Delta}}{A}\,, \Delta \equiv B^2 - AC\,.$$

Therefore, we distinguish the following three cases:
(a) $\Delta > 0$ (**hyperbolic** PDE);
(b) $\Delta = 0$ (**parabolic** PDE);
(c) $\Delta < 0$ (**elliptic** PDE).
REMARK 2.27 - This equation can be written as follows:

(2.53)(**characteristic equation** of E_2) $$A(\phi_x)^2 - 2B(\phi_x)(\phi_y) + C(\phi_y)^2 = 0$$

where $\phi(x, y)$ is the equation of a curve in $\mathbf{R}^2$. This equation can also be written as follows:

(2.54)(**characteristic equation** of E_2) $$A(\frac{\phi_x}{\phi_y})^2 - 2B\frac{\phi_x}{\phi_y} + C = 0\,.$$

By considering that $\phi(x, y) = 0 \to d\phi = 0 \Rightarrow \phi_x dx + \phi_y dy = 0 \Rightarrow \frac{\phi_x}{\phi_y} = -\frac{dy}{dx}$. Then from the above equation we get equation (2.52).
REMARK 2.28 - In the hyperbolic case the characteristic lines that start from a point of $\mathbf{R}^2$ are two real lines; in the parabolic case the characteristic lines coincide;

and in the elliptic case are imaginary ones, namely from a point of $\mathbf{R}^2$ do not pass through real characteristic lines.
EXAMPLE 2.11 - 1) (STRING EQUATION).(HYPERBOLIC PDE). $u_{xx}-u_{yy}=0$.
The characteristic equations are: $\frac{dy}{dx}=\pm 1 \Rightarrow y=\pm x+const$.
2)(LAPLACE EQUATION). (ELLIPTIC PDE). $u_{xx}+u_{yy}=0$.
The characteristic equations are: $\frac{dy}{dx}=\pm i$.
3)(HEAT EQUATION).(PARABOLIC PDE). $u_{xx}-u_y=0$.
The characteristic equations are: $\frac{dy}{dx}=0 \Rightarrow y=const$.
4)(CASE $E\equiv \mathbf{R}^3\times\mathbf{R}$). Let $E_2\subset J\mathcal{D}^2(E)$ be the following equation:

$$(2.55)(E_2) \qquad Au_{xx}+Bu_{xy}+Cu_{zz}+Du_x+Eu_y+Fu_z+G=0.$$

Let $\phi(x,y,z)=0$ be the equation of a surface $\sigma\in\mathbf{R}^3$. Then, σ is characteristic if it satisfies the following equation:

(2.56)(**characteristic equation of** E_2) $\qquad A(\phi_x)^2+B(\phi_y)^2+C(\phi_z)^2=0.$

By using a transformation we can put this in the form $\epsilon_1(\phi_x)^2+\epsilon_2(\phi_y)^2+\epsilon_3(\phi_z)^2=0$, with $\epsilon_1,\epsilon_2,\epsilon_3=\pm 1,0$. Thus we can distinguish the following cases:
(I)(Elliptic). $\epsilon_1,\epsilon_2,\epsilon_3$ have all the same sign. Thus the solutions are only $\phi=const$. (There are not real characteristics.) EXAMPLES: (a) (Poisson Equation) $u_{xx}+u_{xy}+u_{zz}+f=0$; (b)(Laplace Equation) $u_{xx}+u_{xy}+u_{zz}=0$.
(II)(Hyperbolic). Two coefficients have the same sign and others have opposite sign. Then the characteristic surfaces are circular cones. For example, let us assume the following form

$$(2.57) \qquad (\phi_x)^2+(\phi_y)^2-(\phi_z)^2=0.$$

By considering that the derivatives are proportional to the components (α,β,γ) of the normal to the characteristic surfaces, we can write equation (2.57): $\alpha^2+\beta^2-\gamma^2=0$, that together with the condition $\alpha^2+\beta^2+\gamma^2=1$ gives $\gamma=\pm 1/\sqrt{2}$. Thus the characteristic surface passing through the point (x_o,y_0,z_0) is the cone of revolution with vertex in (x_o,y_0,z_0) and axis parallel to the z axis. EXAMPLE. (Vibrating membrane equation): $u_{xx}+u_{yy}-\frac{1}{v^2}u_{tt}=0$.
(III)(Parabolic). A coefficient is zero. Then we can write the following case:
(a) $(\phi_x)^2+(\phi_y)^2=0. \Rightarrow \phi_x=\phi_y=0\Rightarrow \phi=\phi(z)$. Thus the characteristics are the planes $z=const$. EXAMPLE: (Heat equation): $u_{xx}+u_{yy}-\frac{1}{D}u_t=0$. The surfaces $t=const$ are characteristic surfaces.
(b) $(\phi_x)^2-(\phi_y)^2=0. \Rightarrow$ The general integral is the following: $\phi=f(x\pm y,z)$, where f is an arbitary function. Thus the characteristics are the cylindric surfaces with generatrices parallel to the bisectors of the axes x,y.

(c) $(\phi_x)^2 = 0. \Rightarrow \phi = \phi(y,z)$. The characteristics are cylindric surfaces parallel to the x axis.

5) (SEMILINEAR PDEs OF THE SECOND ORDER). Set $W \equiv \mathbf{R}^n \times \mathbf{R}^m$ and let $E_2 \subset J\mathcal{D}^2(W)$ be locally given by

$$(2.58)E_2 \qquad A_j^{\alpha\beta} y^j_{\alpha\beta} + f(x^\alpha, y^j, y^j_{\alpha\beta}) = 0, \quad A_j^{\alpha\beta} : \mathbf{R}^n \to \mathbf{R}.$$

For simplicity we put $n = 2$ and $m = 1$. Thus the above equation can be written as:

$$(2.59) \qquad Az_{xx} + 2Bz_{xy} + Cz_{yy} + f(x,y,z,z_x,z_y) = 0.$$

One can classify these equations in four cases, following the fact that characteristics are real ones or not. Here, characteristics mean system of curves defined by the differential equation:

$$\frac{dy}{dx} = \frac{B \pm \sqrt{\Delta}}{A}, \; \Delta \equiv B^2 - AC$$

namely:

(I) ELLIPTIC PDE: $\Delta < 0$; (II) HYPERBOLIC PDE: $\Delta > 0$; (III) PARABOLIC PDE: $\Delta = 0$; (IV) MIXED PDE: This assume in particular different domains of the above types.

DEFINITION 2.21 - *We say* **canonical forms** *of the above three types the following ones:*

(a) *Elliptic:* $\Delta_2 z + f = 0$, *with* $\Delta_2 z \equiv z_{xx} + z_{yy}$;

(b) *Hyperbolic:* $z_{xy} + f = 0$;

(c) *Parabolic:* $z_{xx} + f = 0$;

(d) *Mixed:* $yz_{xx} + z_{yy} + f = 0$.

PROPOSITION 2.10 - *Any equation of the type* (2.59) *can be placed in the canonical form.*

PROOF. (Hyperbolic case). Let us assume that after integration of the charcteristic equations of the two previous systems we have the following form: $\phi(x,y) = const., \psi(x,y) = const$. Let us make the change of variables $\xi = \phi(x,y), \eta = \psi(x,y)$, thus from (2.59) we get an equation of the same type, with new coefficients $A_1, B_1 \ldots$. As in the plane (ξ,η) the characteristics will be represented by the equations $\xi = const., \eta = const.$, it follows that their equations should be reduced to the form $d\xi \wedge d\eta = 0$. Hence we shall have $A_1 = C_1 = 0$. Thus, by dividing by $2B_1$, the given equation reduces to the canonical form.

(Elliptic case). The characteristic equations will be of the form $\lambda(x,y) + i\mu(x,y) = const., \lambda(x,y) - i\mu(x,y) = const.$, where λ and μ are two real functions. On the other hand, the substitution $\xi = \lambda(x,y), \eta = \mu(x,y)$, should reduce equation (2.59)

to the form $(d\xi + id\eta)(d\xi - id\eta) = d\xi^2 + d\eta^2 = 0$, that implies $A_1 = C_1, B_1 = 0$. As a consequence, by dividing by A_1, equation assumes the canonical form.
(Parabolic case). Here there is only one system of double characteristics, that can be represented by the equation $\phi(x, y) = const$. Thus, by considering the change of variables, $\xi = \phi(x, y), \eta = y$, equation (2.59) will assume the form $d\xi^2 = 0$. Thus $B_1 = C_1 = 0$. Dividing by A_1 we will have the canonical form. □

EXAMPLE 2.12 - (SUBSONIC AND SUPERSONIC FLOW AROUND PLANE DOMAIN: FROM ELLIPTIC PDE TO HYPERBOLIC PDE).
1) Let D be a domain that moves with a stationary velocity U in air at rest. With respect to a frame attached to the domain D we have that the potential u of the velocity satisfies the following PDE:

$$(1 - M^2)u_{xx} + u_{yy} = 0$$

where $M \equiv$ **Mach number** $\equiv \frac{U}{v}$, with $v =$ sound velocity. We have the following velocities:
(a) $M < 1$: Subsonic flow (Elliptic PDE) : $B^2 - AC = M^2 - 1 < 0$;
In this case the flow lines around the domain D are regular enough.
(b) $M > 1$: Supersonic flow (Hyperbolic PDE) : $B^2 - AC = M^2 - 1 > 0$;
In this case the flow lines around the domain D are non-continuous (shock waves).
(c) $M = 1$: Transonic flow (Parabolic PDE) : $B^2 - AC = M^2 - 1 = 0$.
2) (TRICOMI PDE). For the transonic flow a more suitable equation is the following:

$$u_{xx} - xu_{yy} = 0 .$$

Now, $B^2 - AC = x \Rightarrow$ for $x < 0$ Tricomi equation is elliptic and for $x > 0$ it is hyperbolic.

2 THEOREM 2.24 - (INTEGRATION OF PDEs BY MEANS OF ODEs: THE METHOD OF THE CHARACTERISTICS).
1) (PDEs OF FIRST ORDER). *Let us consider the equation*

$$(2.60)(E_1 \subset J\mathcal{D}(M, \mathbf{R})) \qquad F(x^\alpha, y, y_\alpha) = 0,$$

that identifies the following exterior differential system on $J\mathcal{D}(M, \mathbf{R})$:

$$(2.61)(S) \qquad \{\omega_0 \equiv F, \omega_1 = dy - y_k dx^k\}.$$

A solution of E_1 *is identified by an* n*-dimensional integral manifold of* (S) *locally defined by a mapping*

$$U \subset \mathbf{R}^n \to \mathbf{R}^{2n+1} \equiv J\mathcal{D}(M, \mathbf{R}), \{x^\alpha\} \mapsto \{x^\alpha, u(x^\alpha), (\partial x_\alpha.u)\} .$$

The **closure** *of (S) is obtained by adding the following differential forms:*

$$(2.62)(dS) \qquad \{\omega_2 \equiv d\omega_0 \equiv dF = [(\partial x_\alpha.F) + y_\alpha(\partial y.F)]dx^\alpha, \omega_3 \equiv dy_\alpha \wedge dx^\alpha\}.$$

The **characteristic system** *of E_1 is the following*

$$(2.63) \qquad \boxed{\begin{aligned} \dot{x}^\alpha &= (\partial y^\alpha.F) \\ \dot{y} &= y_\alpha(\partial y^\alpha.F) \\ \dot{y}_\alpha &= -[(\partial x_\alpha.F) + y_\alpha(\partial y.F)] \end{aligned}}$$

namely, the characteristic vector field is the following vector field $v : E_1 \to TE_1$, represented by

$$v = (\partial y^\alpha.F)\partial x_\alpha + y_\alpha(\partial y^\alpha.F)\partial y - [(\partial x_\alpha.F) + y_\alpha(\partial y.F)]\partial y^\alpha .$$

This is a vector field on E_1 that is tangent to any solution of E_1. Let us define **bicharacteristics** *the projection of the characteristic curves on $W \equiv M \times \mathbf{R}$.*
EXAMPLE 2.13 - 1) (LINEAR EQUATIONS). $F \equiv v^k(x)y_k = 0$.
The characteristic vector field on E_1 is the following:

$$v = v^\alpha \partial x_\alpha + y_\alpha v^\alpha \partial y - [(\partial x_\alpha.v^k)y_k]\partial y^\alpha .$$

The corresponding vector field on M is $v_0 = v^\alpha \partial x_\alpha$. A hypersurface $X \subset M$ is characteristic for the equation if the characteristic vector field v_0 on M is tangent to the hypersurface X. [In fact, if X is defined by the equation $f = 0$, X is characteristic if $v^\alpha(\partial x_\alpha.f) = 0$. This equation shows that v_0 is tangent to X].
2) (QUASI AFFINE EQUATIONS). $F \equiv v^k(x)y_k - f = 0, v^k, f : W \to \mathbf{R}$.
The projection of the characteristic vector field on W is the following

$$v = v^\alpha \partial x_\alpha + f\partial y .$$

The corresponding characteristic equation is

$$(2.64) \qquad \boxed{\begin{aligned} \dot{x}^\alpha &= v^\alpha \\ \dot{y} &= f \end{aligned}}$$

The corresponding solutions are curves on $W \equiv \mathbf{R}^{n+1}$. If $f_I(x^\alpha, y) = c_I$, $1 \le I \le n$, are the general integral of (2.64) (the intersections of these surfaces generate the characteristic curves). The general solution of E_1 is the following

$$\phi(f_1(x^k, y), \dots, f_n(x^k, y)) = 0 \, , \; \phi : \mathbf{R}^n \to \mathbf{R}$$

where ϕ is an arbitrary function. Such solutions are hypersurfaces V. Furthermore, for any point pass a characteristic curve that is in V. Moreover, for any characteristic curve passes infinite solutions V.

EXAMPLE 2.14 - Let us consider some examples of equations of this type:

(a) $au_x + bu_y - c = 0, a, b \in \mathbf{R}$.

Characteristic equations: $\dot{x} = a, \dot{y} = b, \dot{u} = c$. The general integral of this system is the following: $cx - au = c_1; cy - bu = c_2$. The general integral is the following $\phi(cx - au, cy - bu) = 0$, where $\phi : \mathbf{R}^2 \to \mathbf{R}$ is an arbitrary function. These are cylindric surfaces having as generatrices the lines with director cosines (a, b, c).

(b) $2y(2 - x)u_x + (x^2 - y^2 + u - u_x)u_y + 2yu = 0$.

The characteristic system is the following: $\dot{x} = 2y(2 - x), \dot{y} = x^2 - y^2 + u^2 - 4x, \dot{u} = -2yu$. The general integral is the following: $\frac{x-2}{u} = c_1, \frac{x^2+y^2+u^2}{u} = c_2$. These are circles. The general solution is the following: $\phi(\frac{x-2}{u}, \frac{x^2+y^2+u^2}{u}) = 0$, where ϕ is an arbitrary function on $\mathbf{R}^2$.

3) (CAUCHY PROBLEM FOR PDEs OF FIRST ORDER). *Let*

$$(2.65)(E_1 \subset J\mathcal{D}(M, \mathbf{R})) \qquad F(x^\alpha, y, y_\alpha) = 0,$$

be a PDE of first order. Then, if $\sigma \subset E_1$ is a non-characteristic regular integral manifold of dimension $n - 1$, (namely the characteristic vectors passing for $q \in \sigma$, do not belong to the tangent space $T_q\sigma$), then there exists a regular solution passing through σ generated by the characteristic curves passing through σ.

PROOF. Let $\sigma \equiv Ds(N)$, where $N \equiv \pi_1(\sigma) \subset M$ is a submanifold of M of dimension $(n - 1)$, and let $s : \mathbf{R}^n \to \mathbf{R}$ defined in a neighbourhood of N. Then , if $x^i = x^i(u^\lambda), (u^\lambda) \in \Omega \subset \mathbf{R}^{n-1}$, $i = 1, \cdots, n$, are parametric equations for N, the condition on σ are the following (σ is an integral manifold of E_1):

$$\begin{aligned}
&\omega_0|_\sigma = 0 \Rightarrow H_0 \equiv F(x^i(u^\lambda), y(u^\lambda), y_\alpha(u^\lambda)) = 0,\\
&\omega_1|_\sigma = 0 \Rightarrow dy(u^\lambda) - y_\alpha(u^\lambda)dx^\alpha(u^\lambda) = 0\\
&\Downarrow\\
&H_\lambda \equiv (\partial u_\lambda.y) - y_\alpha(\partial u_\lambda.x^\alpha) = 0, 1 \leq \lambda \leq n - 1.
\end{aligned}$$

From the implicit function theorem, we have that if the jacobian $j \equiv \det(\partial y^\alpha.H_\mu) \neq 0, 0 \leq \mu \leq n - 1$, these equations have a differentiable solution $y_\alpha(u^\lambda)$ in the neighbourhood of any point of N. On the other hand, under our hypotheses, must be $j \neq 0$. In fact, as σ is a regular manifold the matrix $(\partial u_\lambda.x^\alpha)$ has rank $n - 1$, namely the map $\Omega \to M$ is an immersion. Furthermore, as σ is not characteristic, the characteristic vector field on M is not tangent to $\pi_1(\sigma) \equiv N \subset M$. This is the vector field $v = (\partial y^\alpha.F)\partial x_\alpha$. As the tangent space to N is generated by $e_\lambda \equiv (\partial u_\lambda.x^\alpha)\partial x_\alpha$, it follows that v belongs to T_pN iff

$$(\partial y^\alpha.F) = \mu^\lambda(\partial u_\lambda.x^\alpha)\,,\ 1 \leq \alpha \leq n. \tag{2.66}$$

This is a linear system of n-equations in $(n-1)$ functions $(\mu^1,\ldots,\mu^{n-1})$. As the matrix of the coefficients $(\partial u_\lambda.x^\alpha)$ has rank $(n-1)$, from the Rouché-Cappelli theorem it follows that the system (2.66) is compatible iff the matrix $\binom{(\partial u_\lambda.x^\alpha)}{(\partial y^\alpha.F)}$ has zero determinant. Thus the condition that σ is not characteristic assures that $j \neq 0$. Let us, now, $\{\phi_t\}_{t\in]-\epsilon,\epsilon[} \subset \mathbf{R}, \phi_t : E_1 \to E_1$, be the 1-parameter group of transformations generated by the characteristic vector field v. In local coordinates we have:

$$\begin{aligned}
&(a) \qquad x^\alpha = x^\alpha \circ \phi(t, x^i(u^\lambda), y(u^\lambda), y_\alpha(u^\lambda))\\
&(b) \qquad y = y \circ \phi(t, x^i(u^\lambda), y(u^\lambda), y_\alpha(u^\lambda))\\
&(c) \qquad y_\alpha = y_\alpha \circ \phi(t, x^i(u^\lambda), y(u^\lambda), y_\alpha(u^\lambda))
\end{aligned}$$

where ϕ is the unique solution of the characteristic system for any initial condition $(0, x^i, y, y_\alpha) \in \sigma$. From (a) and (b), by means of eliminitation of t, we have a differential function $y = y(x^\alpha)$ that represents a regular solution for the Cauchy problem. This condition is possible if $\bar{j} \equiv \det(\binom{(\partial u_\lambda.x^\alpha)}{(\partial t.x^\alpha)}) \neq 0$. On the other hand $(\partial t.x^\alpha) = (\partial y^\alpha.F)$. Thus $\bar{j} = j \neq 0$. □

EXAMPLE 2.15 - 1) (HAMILTON-JACOBI EQUATION).

$$E_1 \subset J\mathcal{D}(M,\mathbf{R}) \equiv T^*M \times \mathbf{R} : H(x^\alpha, (\partial x_\alpha.u)) = 0\,.$$

The projection of the characteristics on the cotangent bundle, by means of the canonical projection

$$J\mathcal{D}(M,\mathbf{R}) \to T^*M$$

gives the following system of equations:

$$\boxed{\begin{aligned} \dot{x}^\alpha &= (\partial p^\alpha.H)\\ \dot{p}_\alpha &= -(\partial x_\alpha.H)\end{aligned}}$$

The projections of these on M gives, after integration, the **bicharacteristic curves** of E_1 on M.

2) (EIKONAL EQUATION). $(\partial x_1.u)^2 + \ldots + (\partial x_n.u)^2 = 1$, where $u \equiv$ **optic distance**.

This is a particular case of Hamilton-Jacobi equation. The level surfaces of u are called **front waves**. The **caustic** is the set of critical values of the projection of an integral manifold on M. The characteristic equations for the eikonal equation are the following:

$$\boxed{\begin{aligned} \dot{x}^\alpha &= 2y_\alpha \\ \dot{y} &= 2 \\ \dot{y}_\alpha &= 0 \end{aligned}}$$

Therefore, the characteristic that starts from $q_0 \equiv (x^k_0, y_0, y_{k,0}) \in E_1$, (such that $\sum_{1\leq k\leq 3}(y_{k,0})^2 = 1$) is the line of parametric equations:

$$\{x^\alpha = 2y_{\alpha,0}(t-t_0) + x^\alpha_0, y = 2(t-t_0) + y_0, y_\alpha = y_{\alpha,0}\}.$$

By elimination of $(t-t_0)$ we get:

$$\{x^\alpha = y_{\alpha,0}(y-y_0) + x^\alpha_0, y_\alpha = y_{\alpha,0}\}. \tag{2.67}$$

The set of characteristic lines passing for $q_0 \equiv (x^k_0, y_0, y_{k,0}) \in E_1$, that project on the same point $p_0 \in W$, generates on W a conic surface. In fact from (2.67) and from E_1 we get

$$\sum_{1\leq k\leq 3}(x^k - x^k_0)^2 = (y-y_0)^2.$$

REMARK 2.29 - (INDEX OF SINGULAR POINTS OF VECTOR FIELDS).

1)DEFINITION 2.22 - *Let $v: U \to TU$ be a smooth vector field defined in a region U of the plane $\mathbf{R}^2$ with coordinates (x^1, x^2). Let $v = v^1\partial x_1 + v^2\partial x_2$ be the linear representation of v. In the plane $\mathbf{R}^2$ the system of coordinates defines an orientation and a euclidean structure. Let eliminate from U the set Σ of the singular points. Put $U' \equiv U \setminus \Sigma$. Let us define the following mapping:*

$$f: U' \to S^1, f(x) = \phi(x) = \arctan\frac{v_2(x)}{v_1(x)}. \tag{2.68}$$

This mapping is smooth as we have removed the singular points $(v_1 \neq 0, v_2 \neq 0)$. The differential

$$d\phi = \frac{v_2 dv_1 - v_1 dv_2}{v_1^2 + v_2^2}$$

is defined in all U'. Let define **index** *of an oriented closed curve $\Gamma \equiv \{\gamma: S_1 \to U'\}$ the following integral:*

$$\mathrm{ind}(\gamma) = \frac{1}{2\pi}\int_\Gamma d\phi.$$

2)THEOREM 2.25 - *The index of γ is equal to the sum of the indexes of the singular points that are inside the domain D with boundary Γ.*

PROOF. Assume that v has a finite number of singular points inside D only. Denote by $D' \equiv D \setminus \Sigma'$, where Σ' is the set of the singular points with a little circular neighbourhood. Then, the boundary $\partial D'$ of D' is given by: $\partial D' = \Gamma - \sum_i \Gamma_i$. So, we get:

$$0 = \int_{D'} d(d\phi) = \int_{\partial D'} d\phi = \int_{\Gamma} d\phi - \sum_i \int_{\Gamma_i} d\phi.$$

$$\Downarrow$$

$$0 = \text{ind}(\gamma) - \sum_i \text{ind}(\gamma_i) \Rightarrow \text{ind}(\gamma) = \sum_i \text{ind}(\gamma_i).$$

□

3)THEOREM 2.26 - *The index of γ is an integer:*

$$\text{ind}(\gamma) \in \mathbf{Z}\,.$$

4)(DEGREE OF MAPPING). The generalization of the concept of index on higher dimensional domains is given by the concept of degree of mapping.

(a) Let $f : M_1 \to M_2$ be a mapping between differentiable manifolds of dimension n. A point $x \in M_1$ is called **regular point** if $Df(x) : T_x M_1 \to T_{f(x)} M_2$ is a linear nonsingular application.

(b) Let us define the **degree** of $f : M_1 \to M_2$ in a regular point $x \in M_1$, the number

$$\deg_x f = \left\{ \begin{array}{lll} +1 & \text{if} \quad Df(x) & \text{conserves orientation} \\ -1 & \text{if} \quad Df(x) & \text{changes orientation} \end{array} \right\}$$

5)PROPOSITION 2.11 - (a) *The degree of $Df(x) \equiv A \in L(\mathbf{R}^n)$ does not depend on the point $x \in \mathbf{R}^n$ and one has*

$$\deg_x A = \text{sgn}\ \det(A) = (-1)^{m_-}$$

where m_- is the number of eigenvalues of A with negative real part.

(b) *The degree of a mapping $Df(x) \equiv A \in L(\mathbf{R}^n)$ of the restriction $f : S^{n-1} \subset \mathbf{R}^n \to S^{n-1} \subset \mathbf{R}^n$ is given by:*

$$\deg_x f = \deg A\,.$$

(c) *(a) Let $f : S^{n-1} \to S^{n-1}$ be a mapping that associates to each point its diametrically-opposite one. One has:*

$$\deg_x f = (-1)^n\,.$$

(d) *Let* $A \in L_{\mathbf{C}}(\mathbf{C}^n)$. *Then, one has:*

$$\deg_x A_{\mathbf{R}} = +1 .$$

(e) *Let* $y \in M_2$. y *is called* **regular value** *of* $f : M_1 \to M_2$ *if any* $x \in f^{-1}(y)$ *is regular point.*

7) THEOREM 2.27 - *Let* M_1 *and* M_2 *be two compact and connected* n*-dimensional manifolds.*

(a) *There exist regular values.*

(b) *The number of points of the inverse image of a regular value is finite.*

(c) *The following sum,* $\sum_{x\in f^{-1}(y)} \deg_x f$, *where* y *is a regular value, does not depend on the particular value* y *considered.*

(d) *The nonregular values in* M_2 *form a set of zero measure.*

8)DEFINITION 2.23 - *Define* **degree of a mapping** $f : M_1 \to M_2$ *the following number:*

$$\deg(f) \equiv \sum_{x\in f^{-1}(y)} \deg_x f .$$

EXAMPLE 2.16 - (a) The degree of the mapping $f : S_1 \to S_1, f(z) = z^n, n = 0, \pm 1, \pm 2$, is n: $\deg(f) = n$.

(b) The degree of $f : \mathbf{C}P \to \mathbf{C}P, f(z)$=polynomial of degree n, is: $\deg(f) = n$.

(c) Let $f : U' \to S^1$ be the mapping given in (2.68) by means of a vector field v in the domain U'. Let $\gamma : S^1 \to U'$ be a closed curve, and $h = f \circ \gamma : S^1 \to S^1$. Then, the index of γ defined above coincides with the degree of the mapping h: $\mathrm{ind}(\gamma) = \deg(h)$.

9)DEFINITION 2.24 - *The* **index of an isolated singular point** $p_0 \in D$ *of a vector field defined into a domain* $D \subset \mathbf{R}^n$ *is the degree of the corresponding application* $h : S^{n-1} \to S^{n-1}, h(x) = \frac{rv(x)}{|v(x)|}$, *where* $S^{n-1} \equiv \{x \in \mathbf{R}^n \| x - p_0| = r\}$.

EXAMPLE 2.17 - The index of the singular point $0 \in \mathbf{R}^n$ of $\dot{x} = -x$ is $(-1)^n$.

10)THEOREM 2.28 - *The sum* χ *of the indexes of the singular points of a vector field on a compact manifold of any dimension does not depend on the vector field, and it is a characteristic property of the manifold. This number coincides with the* **Euler characteristic** *of the manifold* [25, 50].

[In order to identify the Euler characteristic of a manifold, it is enough to identify the singular points of some ODE defined on this manifold.]

EXAMPLE 2.18 - (a) EULER CHARACTERISTIC OF S^n AND $\mathbf{R}P^n$.

$$\chi(S^n) = 2\chi(\mathbf{R}P^n) = 1 + (-1)^n .$$

PROOF. Let us consider the mapping $f : S^n \to \mathbf{R}P^n, p \mapsto r_p \equiv$ line identified by the vector $\overrightarrow{OP} \equiv p$. This is locally a diffeomorphism. The inverse mapping f^{-1} associates to r_p two points on S^n that are diametrically-opposite. So, any vector field on $\mathbf{R}P^n$ defines on S^n a field with double number of singular points. Furthermore, the index of each singular point of the sphere is the same as the index of the corresponding point on the projective space. In order to calculate $\chi(S^n)$, define the sphere with equation $x_0^2 + \ldots + x_n^2 = 1$ in $\mathbf{R}^{n+1}$ and consider the function $x_0 : S^n \to \mathbf{R}$. Let us consider the differential function $\dot{x} = \mathrm{grad}(x_0)$ on S^n and let us study its singular points. The vector field $\mathrm{grad}(x_0)$ is zero in two points: north pole $(\mathcal{N}) \equiv (x_0 = 1)$ and south pole $(\mathcal{S}) \equiv (x_0 = -1)$. By linearization of the differential equation at north pole and at south pole we get

$$\dot{\xi} = -\xi \,,\; \xi \in \mathbf{R}^n \cong T_{\mathcal{N}}S^n \,;\; \dot{\eta} = \eta \,,\; \eta \in \mathbf{R}^n \cong T_{\mathcal{S}}S^n \,.$$

Thus, the index in $\mathcal{N}$ is $(-1)^n$, and that in $\mathcal{S}$ is $(+1)^n$. Thus we get: $\chi(S^n) = 1+(-1)^n$. As a consequence we get that any vector field on a sphere of even dimension has at least one singular point. □

(b) EULER CHARACTERISTIC OF n-DIMENSIONAL TORUS.

$$T^n \equiv S^1 \times \ldots_n \ldots \times S^1 : \quad \chi(T^n) = 0 \,.$$

PROOF. A differential equation, without singular points, exists in any dimension on T^n. For example: $T^n \cong \mathbf{R}^n/\mathbf{Z}^n \Rightarrow \{\dot{\phi}_1 = \omega_1, \dot{\phi}_2 = \omega_2, \cdots, \dot{\phi}_n = \omega_n\}$. □

11) (RELATION BETWEEN MORSE FUNCTION AND CHARACTERISTIC VECTOR FIELDS).

Taking into account that any smooth manifold M admits a structure of Riemannian manifold, we can associate to any function $f : M \to \mathbf{R}$ a vector field, $v \equiv \mathrm{grad}(f) : M \to TM$. Then, to a critical point of a Morse function f (namely $Df(x) = 0$) there corresponds a zero of the vector field v. A **separatrix diagram** of the critical point x, is the collection of all the integral trajectories of the vector field having x as target or source. The trajectories of the first type go in an λ-dimensional disc D^λ corresponding to the local coordinates $x^1, \cdots x^\lambda$, and trajectories of the second type go in a disc $D^{n-\lambda}$, corresponding to the other coordinates and transverse to D^λ in the point of their intersection x.

EXAMPLE 2.19 - Let $\dim(M) = 2$. Let x be a critical point of $f : M \to \mathbf{R}$. Then x can be a maximum, minimum or saddle point. The theorem of the index of Poincaré [25,50] assures that if M is a 2-dimensional compact manifold and f is a Morse function, then the Euler characteristic of M, $\chi(M)$, is equal to $N_0 - N_1 + N_2$, where N_0 = number of minimum points, N_1 = number of saddle points and N_2 = number of maximum points of f.

THEOREM 2.29 - *Let $N = N_0 \times 0 \cup N_1 \times 1$ be a given Cauchy data for a PDE $E_k \subset J\mathcal{D}^k(W)$. Let $V \subset E_k$ be a solution that cobords N_0 with N_1, i.e., $\partial V = N$.*

Assume that V has an admissible Morse function (always possible) such that f is of the type $(\nu_0, \cdots, \nu_s)$, where ν_k denotes the number of critical points with index k. Then, V defines, and it is defined by, a characteristic vector field having $\nu_0 + \ldots + \nu_s$ zero points on V.

PROOF. In fact we can associate to the Morse function f its gradient $v \equiv \mathrm{grad}(f)$. Then, we can write $V = \bigcup_t \phi_t(N_0)$, if $\partial\phi = v$. $\square$

2.4 - AFFINE PDEs AND GREEN FUNCTIONS

In this section we give a geometric formulation for the Green functions. This allows us to express such functions by means of singular solutions of PDEs.

(A) - GREEN OPERATORS AND GREEN KERNELS

Let $\mathbf{K} \equiv \mathbf{R}$, or $\mathbf{C}$. Let E and F be $\mathbf{K}$-vector fiber bundles of finite rank and class C^r on a manifold M of locally finite dimension n and class C^r. Let K be a differential operator on M between E and F, of order $\leq k$ and class $C^h, k \leq h \leq r-k$. Let K^* be the corresponding transpose operator between $F' \equiv F^* \bigotimes \Lambda^0_n M$ and $E' \equiv E^* \bigotimes \Lambda^0_n M$. One has the following properties:

(a) The mapping $K \mapsto K^*$ is $\mathbf{K}$-linear. If $k = 0, K^*$ is the morphism $L(K; id) : F' \to E'$.

(b) The **symbol** of K defines the following isomorphism:

$$S^k_0 M \bigotimes Hom_{\mathbf{K}}(E, F) \to S^k_0 M \bigotimes Hom_{\mathbf{K}}(F', E')$$

given by

$$\alpha \otimes f \mapsto (-1)^k \alpha \otimes Hom(f, id).$$

(c) Let K_1, K_2 be two differential operators. We get:

$$(K_2 \circ K_1)^* = {K_1}^* \circ {K_2}^*.$$

(d) In particular, for trivial vector fiber bundles, $E \cong F \cong M \times \mathbf{K}$, and with respect to a fixed volume form η on M, one has $E' \cong F' \cong M \times \mathbf{K}$ and the mapping $K \mapsto K^*$ transforms scalar differential operators in scalar differential ones.

EXAMPLE 2.20 - $(\Delta^\alpha)^* = (-1)^{|\alpha|} \Delta^\alpha, \alpha \in \mathbf{N}^n$.

DEFINITION 2.25 - *Let K_1 be a differential operator between E and F. Let K_2 be another differential operator between F' and E'. Let the orders of these operators be $\leq k$. Define* **Green operator** *of (K_1, K_2) any differential operator G of order $\leq k-1$ (and class $C^h, h \geq 1$) between $E \bigotimes F'$ and $\Lambda^{n-1} T^* M \equiv \Lambda^0_{n-1}(M)$, such that*

$$< K_1(u), v > - < u, K_2(v) >= dG(u \otimes v).$$

THEOREM 2.30 - 1) *The existence of* G *implies that* $K_2 = K_1{}^*$. *Then, we say that* G *is a Green operator for* K_1 .

2) *Let* K *be a differential operator between* E *and* F *of order* $\leq k$, *class* $C^h, k \leq h \leq r - k$. *Then, there exists a Green operator of class* C^{h-k+1} *for* K *in each of the following cases:*

(a) $\mathbf{K} = \mathbf{R}$, $r = \infty$, $M =$ *differential manifold of dimension* n ;

(b) M *is isomorphic to an open set of* $\mathbf{K}^n$, *and the fiber bundles* E *and* F *are isomorphic to trivial fiber bundles;*

(c) *The operator* K *is of order* ≤ 1 .

3) *If* G_i *is a Green operator for* K_i , $i = 1, 2$, *there exists a unique Green operator* H *for* $K_2 \circ K_1$ *such that*

$$H(u \otimes v) = G_2(K_1(u) \otimes v) + G_1(u \otimes K^*(v)) .$$

4) *If* K *is of order* ≤ 1 *there exists a unique Green operator, it is of order* 0; *it can be identified with a morphism of class* C^h: $Hom(E \bigotimes F', \Lambda^0_{n-1}(M))$.

5) (GREEN's FORMULA). *Set* $\mathbf{K} \equiv \mathbf{R}$. *Let* M *be a separated and oriented manifold. Let* A *be a closed domain of* M . *Let us give to* A *(and hence to* ∂A*) the orientation induced by* M . *Let* $k \in \mathbf{N}, 2k \leq r$, *and let* K *be a differential operator on* M *between* E *and* F , *of order* $\leq k$ *and class* C^k . *Let* G *be a Green operator for* K . *Let* $u \in C^k_0(E)$ *and* $v \in C^k_0(F)$ *be such that* $\mathrm{supp}(u) \cap A \neq \emptyset \neq \mathrm{supp}(v) \cap A$. *Then, one has*

$$\int_A < K(u), v > - \int_A < u, K^*(v) >= \int_{\partial A} G(u \otimes v). \tag{2.69}$$

In particular , *if* $\partial A = \emptyset$, *one has*

$$\int_A < K(u), v >= \int_A < u, K^*(v) > . \tag{2.70}$$

EXAMPLE 2.21 - 1)(INFINITESIMAL TRANSFORMATIONS). Let K be a differential operator of order ≤ 1 defined by means of infinitesimal variation induced by a vector field ξ on M: $K(u) = \partial \widetilde{u}$, for $u : M \to E$ a section of a natural vector fiber bundle $\pi : E \to M$ on M, $\widetilde{u}_t \equiv \mathbb{B}(\phi_t{}^{-1}) \circ u \circ \phi_t$, $\mathbb{B} \equiv$ functor of the natural structure of E. Then, the Green operator of K is $G(u \otimes v) = \xi \rfloor < u, v >$. In fact, one has the following Green formula:

$$< \partial u, v > + < u, \partial v >= d(\xi \rfloor < u, v >) .$$

2)(EXTERIOR DIFFERENTIAL). If $E \equiv \Lambda^0_p(M), n = \dim M$, we can identify E' with $\Lambda^0_{n-p-1}(M)$. Thus , if $K \equiv d$ one has $K^* = (-1)^{p+1} d$. The Green operator is the exterior product:

$$G : \Lambda^0_p(M) \bigotimes \Lambda^0_{n-p-1}(M) \to \Lambda^0_{n-1}(M) .$$

3)(LAPLACIAN). $E \equiv M \times \mathbf{R}, r = \infty, M \equiv \mathbf{R}^n, F \equiv E$. $K \equiv \sum_{1 \leq i \leq n} \partial\xi_i \partial\xi_i$. Then one has $E' \equiv F' \equiv M \times \mathbf{R}$. The Green operator is the following:

$$G(u \otimes v) = v.\mathrm{grad}(u) - u.\mathrm{grad}(v) \,.$$

Furthermore, if A is a compact domain in $\mathbf{R}^n$, the Green formula can be written as follows:

$$\int_A (K.u)v\eta - \int_A u.(Kv)\eta = \int_{\partial A} [v.\mathrm{grad}(u) - u.\mathrm{grad}(v)]\rfloor\eta \,.$$

THEOREM 2.31 - (RELATION BETWEEN GREEN OPERATOR AND SYMBOL FOR LINEAR DIFFERENTIAL OPERATOR OF FIRST ORDER). *Let*

$$K. : C^\infty(E) \to C^\infty(F)$$

be a linear differential operator of first order between vector fiber bundles $\pi : E \to M$, *and* $\pi' : F \to M$ *on a differential manifold* M *of dimension* n . *Then, one has the following relation between the symbol* $\sigma(K)$ *of* K *and Green operator* G*:*

$$< \sigma_\lambda(K)(u), v >= \lambda \wedge G(u \otimes v) \,, \, \forall \lambda \in \Omega^1(M) \,.$$

PROOF. If $K_1. : C^\infty(E) \to C^\infty(F), K_2. : C^\infty(F') \to C^\infty(E')$ are two differential operators of the first order, such that the symbols $\sigma_\lambda(K_1)$ and $\sigma_\lambda(K_2)$, for any differential 1-form $\lambda \in \Omega^1(M)$ are skew-symmetric: $\sigma_\lambda{}^*(K_1) + \sigma_\lambda(K_2) = 0$, then the operator $\gamma(K_1, K_2) : C^\infty(E \bigotimes F') \to \Omega^n(M)$, given by $\gamma(K_1, K_2)(u \otimes v) =< K_1(u), v > - < u, K_2(v) >$ is a differential operator of the first oprder. The symbol of this operator determines the morphism $\omega : C^\infty(E \bigotimes F') \to \Omega^{n-1}(M)$ such that $\sigma_\lambda(\gamma(K_1, K_2))(u \otimes v) = \lambda \wedge \omega(u \otimes v)$. Thus, the operators $\gamma(K_1, K_2)$ and ω have the same symbol, and, hence, differ by an operator of zero order. By substituting, if necessary, the operator K_2 with $K_2 + K_2{}'$, where $K_2{}' \in Hom(C^\infty(F'), C^\infty(E'))$, we get for each operator $K_1 \in Diff_1(C^\infty(E), C^\infty(F))$, the operator $K_1{}^* \in Diff_1(C^\infty(F'), C^\infty(E'))$, that is the adjoint of the operator K_1 for which $\gamma(K_1, K_1{}^*) = d\omega$. Thus $\omega = G$. □

REMARK 2.30 - Let $\pi : E \to M$ and $\overline{\pi} : F \to N$ be two vector fiber bundles on differential manifolds M and N respectively. Set:

$$E \boxtimes F \equiv \bigcup_{(p,q) \in M \times N} E_p \bigotimes F_q$$

where E_p is the fiber (vector space) of E over $p \in M$ and F_q the fiber (vector space) of F over $q \in N$.

In the following table we resume some useful definitions, results and examples relative to the adjoint of differential operators of order k .

TAB.2.4 - **Formal adjoint of operator of order** k

Operator: L Relation $vLu - uL^*v = div(j(u,v))$
$L: C^\infty(E) \to C^\infty(F), L.s = \sum_U g_U (L.s)^j_U f_{U,j}(\dagger)$ $(L.s)^j_U \equiv \sum_{\lvert q\rvert \leq k} a_q(x)(D^q s)^j(x), D^q \equiv \frac{\partial^{\lvert q\rvert}}{\partial x_1^{q_1}\ldots\partial x_n^{q_n}} = \frac{\partial^{q_1+\ldots+q_n}}{\partial x_1^{q_1}\ldots\partial x_n^{q_n}}$ $L^*: C^\infty(F') \to C^\infty(E'), L^*(\phi\otimes\eta) = \sum_U g_U L^*(\phi\otimes\eta)_{U,i}\theta^i \otimes dx^1 \wedge\ldots\wedge dx^n$ $L^*(\phi\otimes\eta)_{U,i} \equiv \sum_{\lvert q\rvert\leq k}(-1)^{\lvert q\rvert} D^q(a_q\phi_i)$
Examples a) $L \equiv \nabla^2 \equiv \frac{\partial^2}{\partial x_1^2}+\ldots+\frac{\partial^2}{\partial x_n^2}, L^* = L \quad vLu - uL^*v = \mathrm{div}(v\,\mathrm{grad}\,u - u\,\mathrm{grad}\,v)$
b) $L \equiv \nabla^4 \equiv \nabla^2\nabla^2, L^* = L$ $J(u,v) = v\,\mathrm{grad}\,\nabla^2 u - u\,\mathrm{grad}\,\nabla^2 v + (\nabla^2 v)\mathrm{grad}\,u - (\nabla^2 u)\mathrm{grad}\,v$
c) $L \equiv \partial t - \nabla^2, L^* = -\partial t - \nabla^2$ $J(u,v) = uv\partial t - (v\,\mathrm{grad}_x\,u - u\,\mathrm{grad}_x\,v)$
d) $L \equiv \Box^2 \equiv \frac{\partial^2}{\partial t^2} - \nabla^2, L^* = L$ $J(u,v) = [v(\partial t.u) - u(\partial t.v)]\partial t - (v\,\mathrm{grad}_x\,u - u\,\mathrm{grad}_x\,v)$

($\dagger$) $\{g_U\}$= partition of unity subordinate to the covering $\{U\}$ of M

DEFINITION 2.26 - 1) *The* **Dirac kernel** $\mathbb{D}$ *of the vector fiber bundle* F *on* M *is the distributive kernel of local character*

$$\mathbb{D} \in C_0^\infty(F \boxtimes F')' \cong C_0^\infty(F' \boxtimes F)'$$

defined by

$$\mathbb{D}(f\otimes\alpha) = \int_M < f,\alpha >, \forall f \in C_0^\infty(F), \alpha \in C_0^\infty(F').$$

2) *Let* $K.: C^\infty(E) \to C^\infty(F)$ *be a differential operator of order* k *between the vector fiber bundles* $\pi: E \to M$, *and* $\pi': F \to M$. *The* **Green kernel** *of* K *is a kernel* $G \in C_0^\infty(E' \boxtimes F)' \cong C_0^\infty(F \boxtimes E')'$, *such that it satisfies the following equation:*

$$\widetilde{K} \bigotimes \mathbf{1}(G) = \mathbf{1} \bigotimes \widetilde{K}(G) = \mathbb{D},$$

where

(a) $\widetilde{K}: C^\infty(E')' \to C^\infty(F')'$ *is the distributional extension of* K;

(b) $\mathbf{1}\bigotimes\widetilde{K}: C_0^\infty(F\boxtimes E')' \to C_0^\infty(F\boxtimes F')', \mathbf{1}\bigotimes\widetilde{K}(\Xi)(f\otimes\alpha) =_f \Xi(K^*(\alpha))$;

(c) $\widetilde{K}\bigotimes\mathbf{1}: C_0^\infty(E'\boxtimes F)' \to C_0^\infty(F'\boxtimes F)', \widetilde{K}\bigotimes\mathbf{1}(\Xi)(\alpha\otimes f) = \Xi_f(K^*(\alpha))$;

(d) $\mathbb{D} \in C_0^\infty(F \boxtimes F')' \cong C_0^\infty(F' \boxtimes F)'$ *is the Dirac kernel of* F.

THEOREM 2.32 - *The Dirac kernel admits the following local representation:*

$$\mathbb{D}^j_{U,i}(x, x') = \delta^j_i \delta(x, x')$$

where $\delta(x, x')$ *is the Dirac function.*

PROOF. In fact we can write:

$$\mathbb{D}(f \otimes \alpha) = \sum_U \int_U g_U < f_U, \alpha_U >= \sum_U \int_{U \times U} g_{U \times U} f^i_U(x) \alpha_{U,j}(x') \delta^j_i \delta(x, x') d\mu(x')$$

$\mathbb{D}_{U,i}{}^j(x, x') = \delta^j_i \delta(x, x')$.

□

REMARK 2.31 - The relation between Dirac function written in Cartesian coordinates (x^k) and curvilinear coordinates (y^k) respectively is the following:

$$\delta(x^1 - \xi^1) \ldots \delta(x^n - \xi^n) = \frac{1}{|J|} \delta(y^1 - \eta^1) \ldots \delta(y^n - \eta^n), \; J \equiv \det(\partial y_i . x^j).$$

2) A point where J is zero is called **singular point** of the transformation $x^\alpha = x^\alpha(y^j)$. If the singular point p is determined by the equations $(y^1 = \eta^1, \cdots, y^k = \eta^k)$, then we call **ignorable coordinates** $y^{k+1}, \cdots, y^n$. In this case we have:

$$\delta(x^1 - \xi^1) \ldots \delta(x^n - \xi^n) = \frac{1}{|J^k|} \delta(y^1 - \eta^1) \ldots \delta(y^k - \eta^k),$$

where

$$J^k \equiv \int J dy^{k+1} \ldots dy^n.$$

EXAMPLE 2.22 - 1) Transformations from Cartesian coordinates (x, y) to polar ones (ρ, θ):

$$\{x = \rho \cos \theta, y = \rho \sin \theta\}.$$

The origin $\rho = 0$ is a singular point and θ is an ignorable coordinate. One has:

$$\text{(a)} \quad \delta(x - x_0)\delta(y - y_0) = \frac{1}{\rho} \delta(\rho - \rho_0)\delta(\theta - \theta_0);$$

$$\text{(b)} \quad \delta(x)\delta(y) = \frac{1}{2\pi\rho} \delta(\rho).$$

2) Transformations from Cartesian coordinates (x, y, z) to spherical ones (ρ, θ, ϕ): $\{x = \rho \sin \phi \cos \theta, y = \rho \sin \phi \sin \theta, z = \rho \cos \phi\}$, $J = \rho^2 \sin \phi$. The origin $\rho = 0$ is a singular point, and (θ, ϕ) are ignorable coordinates. Thus, we have:

$$\text{(a)} \quad \delta(x - x_0)\delta(y - y_0)\delta(z - z_0) = \frac{1}{\rho^2 \sin \phi} \delta(\rho - \rho_0)\delta(\theta - \theta_0)\delta(\phi - \phi_0);$$

$$\text{(b)} \quad \delta(x)\delta(y)\delta(z - z_0) = \frac{1}{2\pi \rho^2 \sin \phi} \delta(\rho - \rho_0)\delta(\phi - \phi_0).$$

The z-axis ($\phi = 0$) is a locus of singular points, θ is an ignorable coordinate. Thus, we have:

$$\delta(x)\delta(y)\delta(z) = \frac{1}{4\pi\rho^2}\delta(\rho).$$

PROPOSITION 2.12 - *Let $\pi : E \to M$ and $\bar{\pi} : F \to M$ be vector fiber bundles on M. Let $K. : C^\infty(E) \to C^\infty(F)$ be a linear continuous mapping. Then, there exists an associated linear application $\widetilde{K} : (C_0^\infty(E'))' \to (C_0^\infty(F'))'$ such that one has the following exact commutative diagram:*

$$\begin{array}{ccccc} 0 & \to & C^\infty(F) & \overset{j}{\to} & (C_0^\infty(F'))' \\ & & {\scriptstyle K}\uparrow & & \uparrow{\scriptstyle \widetilde{K}} \\ 0 & \to & C^\infty(E) & \underset{j}{\to} & (C_0^\infty(E'))' . \end{array}$$

More precisely, $\widetilde{K}$ is given by

$$\widetilde{K}(\omega)(\alpha \otimes \eta) = < \omega, K^*(\alpha \otimes \eta) > , \ \forall \omega \in (C_0^\infty(E'))', \alpha \otimes \eta \in C_0^\infty(F')$$

where K^* is the formal adjoint of K , namely, K^* is the mapping

$$K^* : C_0^\infty(F') \to C_0^\infty(E')$$

such that

$$\int_M < e, K^*(\alpha \otimes \eta) > = \int_M < K(e), \alpha > \eta \, , \ \forall \alpha \otimes \eta \in C_0^\infty(F'), e \in C_0^\infty(E) .$$

THEOREM 2.33 - *If*

(2.71)($\widetilde{A}$) $$\widetilde{K}.\omega = \widetilde{f}$$

is the distributional equation corresponding to the affine equation

(2.72)(A) $$K.e = f$$

then, any distributional solution $\omega \in (C_0^\infty(E'))'$ of (A) (namely, $\omega \in Sol(\widetilde{A})$) can be represented in the following way:

(2.73) $$\omega =_f G.$$

PROOF. We must prove that $\widetilde{K}({}_fG) = \widetilde{f}$, where $\widetilde{f} = j(f) \in (C_0^\infty(F'))'$. In fact, for any $\alpha \in C_0^\infty(F')$ we have:

$$\widetilde{K}({}_fG)(\alpha) =_f G(K^*(\alpha)) = (K \otimes 1)(G)(f \otimes \alpha) = \mathbb{D}(f \otimes \alpha) = \int_M < f, \alpha > = \widetilde{f}(\alpha) .$$

Vice versa, let $\omega \in Sol(\widetilde{A})$. Then, one has

$$\widetilde{K}(\omega)(\alpha) = \widetilde{f}(\alpha) = \int_M < f, \alpha >$$
$$= \mathbb{D}(f \otimes \alpha) = (K \otimes \alpha) = \int_M < f, \alpha >= \widetilde{f}(\alpha), \forall \alpha \in C_0^\infty(F') .$$

On the other hand $\widetilde{K}(\omega)(\alpha) = \omega(K^*(\alpha))$, $\forall \alpha \in C_0^\infty(F')$. Then, $\omega =_f G$. □

THEOREM 2.34 - (DISTRIBUTIONAL EXTENSION FOR GREEN FORMULA ON COMPACT MANIFOLDS WITH BOUNDARY).

1) *For linear differential operators of the first order one has:*

$$(2.74) \qquad \widetilde{K}(\omega)(\alpha) = \omega(K^*(\alpha)) + \widetilde{\sigma}_1(K)|_{\partial M}(\omega)(\alpha),$$

where $\omega \in (C_0^\infty(E'))', \alpha \in C_0^\infty(F')\widetilde{K}$ *is the distributional extension of* $K. : C^\infty(E) \to C^\infty(F)$, $\widetilde{\sigma}_1(K)|_{\partial M}(\omega)$ *is the distribution supported on* ∂M *identified by the symbol* $\sigma_1(K)$ *of* K.

PROOF. In fact, one has the following Green formula:

$$< K(e), \alpha > - < e, K^*(\alpha) >= d(< \sigma_1(K)(e), \alpha >) , \forall e \in C^\infty(E), \alpha \in C_0^\infty(F').$$

Then, taking into account the canonical immersion $j : C^\infty(E) \to (C_0^\infty(E'))'$, we get:

$$\widetilde{K}(j(e), \alpha) = \int_M < K(e), \alpha >$$
$$= \int_M < e, K^*(\alpha) > + \int_M d(< \sigma_1(K)(e), \alpha >)$$
$$= j(e)(K^*(\alpha)) + \int_{\partial M} < \sigma_1(K)(e), \alpha >$$
$$= j(e)(K^*(\alpha)) + \widetilde{\sigma}_1(K)|_{\partial M}(\omega)(\alpha)$$

where $\widetilde{\sigma}_1(K)|_{\partial M}(e) \in (C_0^\infty(F'))'$ is the distribution with support on ∂M. Then, we get:

$$\widetilde{K}(j(e))(\alpha) = j(e)(K^*(\alpha)) + \widetilde{\sigma}_1(K)|_{\partial M}(e)(\alpha).$$

On the other hand, as $C^\infty(E)$ is dense in $(C_0^\infty(E'))'$, for any $\omega \in (C_0^\infty(E'))'$ we can write

$$\widetilde{K}(\omega)(\alpha) = \widetilde{K}(\sum_s j(e_s))(\alpha)$$
$$= \sum_s \int_M < K(e_s), \alpha >$$
$$= \sum_s \int_M < e_s, K^*(\alpha) > + \sum_s \int_M d(< \sigma_1(K)(e_s), \alpha >)$$
$$= \sum_s j(e_s)(K^*(\alpha)) + \sum_s \int_{\partial M} < \sigma_1(K)(e_s), \alpha > .$$

Then, we have

$$j(\omega)(K^*(\alpha)) + \widetilde{\sigma}_1(K)_1|_{\partial M}(\omega)(\alpha) = j(\sum_s e_s)(K^*(\alpha)) + \sum_s \widetilde{\sigma}_1(K)_1|_{\partial M}(e_s)(\alpha).$$

□

2) *As particular cases we have:*
(a) *For the Green kernel G of K we have,* $\forall f \in C^\infty(F)$,

$${}_fG \in (C_0^\infty(E'))'$$

thus we get

$$\widetilde{K}({}_fG)(\alpha) = {}_fG(K^*(\alpha)) + \widetilde{\sigma}_1(K)|_{\partial M}({}_fG)(\alpha), \alpha \in C_0^\infty(F').$$

b) *If the Green kernel G is such that it can be identified with a section*

$$G \in C^\infty(E \boxtimes F)\,,$$

we can then write:

$$\int_M < K({}_fG), \alpha >= \int_M < {}_fG, K^*(\alpha) > + \int_{\partial M} < \sigma_1(K)({}_fG), \alpha > .$$

3) *The Green kernel satisfies the following equivalent equations:*

(a) $\widetilde{K}({}_fG) = {}_f\mathbb{D}, \forall f \in C^\infty(F)$

(b) ${}_fG(K^*(\alpha)) + \widetilde{\sigma}_1(K)|_{\partial M}({}_fG)(\alpha) = \int_M < f, \alpha >, \forall f \in C^\infty(F), \alpha \in C_0^\infty(F')$

(c) *If* $G \in C^\infty(E \boxtimes F)$, *the above equations become :*

(d) $K({}_fG) = {}_f\mathbb{D}, \forall f \in C^\infty(F)$

(e) $\int_M < {}_fG, K^*(\alpha) > + \int_{\partial M} < \sigma_1(K)({}_fG), \alpha >= \int_M < f, \alpha > .$

(B) - DIFFERENTIAL ALGEBRA AND GREEN KERNELS.

NOTATION. $\mathbf{N} = \{1, 2, 3, \dots\}$. Multi-index: $i \equiv (i_1, \dots, i_m) \in \mathbf{N}^m$; length of a multi-index: $|i| \equiv i_1 + \dots + i_m, i \in \mathbf{N}^m; i! = i_1! \dots i_m!, i \in \mathbf{N}^m; i \leq k \Leftrightarrow i_1 \leq k_1, \dots, i_m \leq k_m, i, k \in \mathbf{N}^m; (\mu) \equiv (0, \dots, 0, 1, \dots, 0), \mu \in \mathbf{N}^m$, where 1 is the μ-th position; $i+k = (i_1+k_1, \dots, i_m+k_m), i, k \in \mathbf{N}^m$. Polynomial coefficient corresponding to a multi-index $i \in \mathbf{N}^m : c(i) \equiv \frac{|i|!}{i!}$. Multinomial coefficient corresponding to multi-indexes $i, k \in \mathbf{N}^m, i \leq k : \binom{k}{i} \equiv \frac{k!}{i!(k-1)!}$.

DEFINITION 2.27 - 1) *Let $\mathcal{A}$ be a real, or complex, algebra with unity e, let*

$$D_1, \ldots, D_m$$

be differentiations of $\mathcal{A}$ that commute among them, namely, linear applications $\mathcal{A} : D_i \to D_i$ that satisfy the Leibnitz rule: $D_i(fg) = (D_i f)g + f(D_i g), f, g \in \mathcal{A}$. Define **differential algebra** *a couple $(\mathcal{A}, D)$, where $\mathcal{A}$ is an algebra and $D \equiv (D_1, \cdots, D_m)$ a set of differentiations of $\mathcal{A}$.*

2) *Let $\mathcal{A}[D]$ the* **polynomial algebra** *$P = \sum_i f^i D_i, f^i \in \mathcal{A}$, where i is a multi-index $i \equiv (i_1, \ldots, i_m) \in \mathbf{N}^m$. The degree of a polynomial P is the highest number $|i|$ such that $f^i \neq 0$. [The algebra $\mathcal{A}$ is identified with the subalgebra $\mathcal{A}[D]$ of all polynomials of order zero. The differentiations $D_1, \ldots, D_m$ are elements of the algebra $\mathcal{A}[D]$.]*

3) *A constant $c \in \mathcal{A}[D]$ is an element $c \in \mathcal{A}$ such that $D_r c = 0$, for all $r = 1, \ldots, m$. [The set of all constants defines the sub-algebra $\mathcal{K} \subset \mathcal{A}$. In particular, the element identity belongs to $\mathcal{K}$.]*

4) *The* **characteristic of the polynomial** *$P \in \mathcal{A}[D]$ is the element*

$$\chi(P) = \sum_i (-1)^{|i|} D_i f^i \in \mathcal{A}.$$

In particular,

$$\chi(D_\mu) = 0\,,\ \mu = 1, \ldots, m;$$
$$\chi(P \circ R) = R(\chi(P))\,,\ \forall P, R \in \mathcal{A}[D];$$
$$\chi(D_\mu \circ P) = 0, \mu = 1, \ldots, m\,,\ P \in \mathcal{A}[D].$$

Let us call **characteristic application** *of $\mathcal{A}[D]$ the application $\chi : \mathcal{A}[D] \to A$, given by $\chi : P \mapsto \chi(P)$.*

5) *Set*

$$\mathcal{A}[D]^p \equiv \left\{ \begin{array}{l} (P^{\mu_1 \cdots \mu_p} \equiv \sum_q \sum_{\nu_1 \ldots \nu_q} f^{\nu_1 \ldots \nu_q \mu_1 \ldots \mu_p} D_{\nu_1} \ldots D_{\nu_q}) \\ P^{\mu_1 \cdots \mu_p} \in \mathcal{A}[D], \mu_1, \ldots, \mu_p = 1, \ldots, m \end{array} \right\}.$$

(a) *Define the linear application $D\circ : \mathcal{A}[D]^p \to \mathcal{A}[D]^{p-1}, p = 1, \ldots, m$, where $D \circ P$ denotes the element of $\mathcal{A}[D]^{p-1}$ with components*

$$(D \circ P)^{\mu_1 \cdots \mu_{p+1}} = \sum_\mu D_\mu \circ P^{\mu\mu_1 \cdots \mu_{p-1}} \in \mathcal{A}[D].$$

In particular $D \circ P = 0$ if $P \in \mathcal{A}[D]^o$.

(b) *Define the linear application*

$$h : \mathcal{A}[D]^p \to \mathcal{A}[D]^{p+1},\ p = 0, \ldots, m-1$$

given by

$$h(P)^{\mu_1\dots\mu_{p-1}} = \sum_{r,s}(-1)^r c(p,r,s)\sum_{(\nu)}(\sum_{(\lambda)} D_{(\lambda)} f^{(\lambda)(\nu)[\mu_1\dots\mu_{p+1}]})D_{(\nu)}$$

where

$$r+s=0,\dots,Q-1, \quad Q = \text{degree of } P$$

$$c(p,r,s) = \frac{(p+1)(s+1)\dots(s+r+1)}{(p+s+1)\dots(p+s+r+1)} = (p+1)\frac{\binom{p+s}{p}}{\binom{p+s+r+1}{p}}$$

$$(\nu) = \nu_1\dots\nu_s, (\lambda) = \lambda_1\dots\lambda_r,$$

the brackets $[\dots]$ *denote the skew-symmetrization in the indexes* $\mu_1\dots\mu_{p+1}$, (e.g.,

$$f^{[\mu_1\dots\mu_{p+1}]} = \frac{1}{p+1}\sum_{k=1}^{p+1}(-1)^k f^{\mu_k\mu_1\dots\widehat{\mu}_k\dots\mu_{p+1}}$$

where the index $\widehat{\mu}_k$ *is omitted. In particular,* $h(P)=0, P\in\mathcal{A}[D]^m$, *or the order* q *of* P *is equal zero.*

THEOREM 2.35 - 1) (HOMOTOPY-FORMULA). *One has the following formula:*

$$D\circ h(P) + h(D\circ P) = P - \delta(p)\chi(P)$$

for any $P\in\mathcal{A}[D]^p, p=0,\dots,m$, *where* $\delta(p)=1$ *if* $p=0$ *and* $\delta(p)=0$ *if* $p\geq 1$. *More precisely, we get:*

$$\begin{aligned} (p=0)(\text{GREEN-FORMULA}) \quad & D\circ h(P) = P-\chi(P) \\ (p=1,\dots,m-1) \quad & D\circ h(P)+h(D\circ P) = P \\ (p=m) \quad & h(D\circ P) = P. \end{aligned}$$

PROOF. See ref.[166]. □

2)(CASE $p=0$). *Here* $c(0,r,s)=1$. *Thus, for any polynomial* $P=\sum_i f^i D_i\in\mathcal{A}[D]$ *one has the following formula:*

$$\sum_{\mu=1}^{m} D_\mu\circ h^\mu(P) = P-\chi(P), \tag{2.75}$$

where the differential polynomials

$$h(P)^\mu = \sum_{r,s}(-1)^r\sum_{(\nu)}(\sum_{(\lambda)} D_{(\lambda)} f^{(\lambda)(\nu)\mu})D_{(\nu)}\ , \ \mu=1,\dots,m$$

or

$$h(P)^\mu = \sum_i(\sum_k(-1)^{|k|}M(i,k,\mu)D_k f^{i+k+(\mu)})D_i\ , \ \mu=1,\dots,m$$

with

$$M(i,k,\mu) = \frac{c(i)c(k)}{c(i+k+\mu)}. \tag{2.76}$$

3) (GREEN FORMULA AND ADJOINT OPERATOR).
(a)(SCALAR CASE). *Let $P \in \mathcal{A}[D]$. Consider P as an operator $\mathcal{A} \to \mathcal{A}$ and define the adjoint operator $P^* : \mathcal{A} \to \mathcal{A}$ with the rule:*

$$P^*(\phi) = \chi(\phi P) = \sum_i (-1)^{|i|} D_i(\phi f^i) \in \mathcal{A}\,,\ \forall \phi \in \mathcal{A}.$$

If $\mathcal{A}$ is commutative then

$$P^* = \sum_i f^{*i} D_i$$

where

$$f^{*i} \equiv \sum_k (-1)^{|i+k|} \binom{i+k}{k} D_k f^{i+k}.$$

Then, (2.76) can be written as follows:

$$\phi P(\psi) - P^*(\phi)\psi = \sum_\mu D_\mu J^\mu$$

where

$$J^\mu \equiv h(\phi P)^\mu \psi = \sum_{i,k} (-1)^{|i|} M(i,k,\mu) D_i \phi f^{i+k+(\mu)} (D_k \psi),\ \mu = 1,\dots,m\,.$$

(b)(EXTENSION TO THE MATRICIAL CASE). *Let $P : \mathcal{A}^a \to \mathcal{A}^b$ be an operator. Then,*

$$\sum_\alpha \phi^\alpha P(\psi)_\alpha - \sum_\beta P^*(\phi)_\beta \psi^\beta = \sum_\mu D_\mu J^\mu$$

where

$$J^\mu \equiv \sum_{\alpha,\beta} \sum_{i,k} (-1)^{|i|} M(i,k,\mu) (D_i \phi^\alpha f_{\alpha\beta}^{i+k+(\mu)})(D_k \psi^\beta)$$

$$\mu = 1,\dots,m\,,\ \phi \in \mathcal{A}^a, \psi \in \mathcal{A}^b\,.$$

4) *One has the following exact complex of linear spaces:*

$$0 \to \mathcal{A}[D]^m \xrightarrow{D\circ} \mathcal{A}[D]^{m-1} \xrightarrow{D\circ} \dots \xrightarrow{D\circ} \mathcal{A}[D]^1 \xrightarrow{D\circ} \mathcal{A}[D]^o \xrightarrow{\chi} \mathcal{A} \to 0\,.$$

PROOF. In order to prove the exactness of the complex it is enough to note that for each element $f \in \mathcal{A}$ the characteristic $\chi(P)$ of the polynomial of zero order

$P = f \in \mathcal{A}[D]$ is equal to the element $f, (\chi(f) = f)$, and to use the homotopy-formula. □

5) *Let us consider the following complex:*

$$0 \to \mathbf{K} \to \mathcal{A}^m \stackrel{D}{\to} \mathcal{A}^{m-1} \stackrel{D}{\to} \dots \stackrel{D}{\to} \mathcal{A}^1 \stackrel{D}{\to} \mathcal{A}^o \to 0$$

where the linear applications $D : \mathcal{A} \to \mathcal{A}, p = 1, \dots, m$, *are defined by*

$$\phi \equiv (\phi^{\mu_1 \dots \mu_p}) \mapsto D\phi = ((D\phi)^{\mu_1 \dots \mu_{p-1}} = \sum_{\mu} D_\mu \phi^{\mu\mu_1 \dots \mu_{p-1}}).$$

This complex is not, in general, exact, and denote with $\mathcal{H}^p(\mathcal{A})$ *its spaces of homology,* $p = 0, \dots, m$.

DEFINITION 2.28 - 1) *A* **differential ideal** $\mathcal{I}$ *of a differential algebra* $\mathcal{A}$ *is an ideal that is closed with respect to the differentiations* D_μ, *namely, if*

$$D_\mu \phi \in \mathcal{I}, \forall \mu = 1, \dots, m, \phi \in \mathcal{I}.$$

2) *Let* $(\mathcal{A}, D)$ *be a differential algebra and* $\mathcal{I}$ *a differential algebra of* $(\mathcal{A}, D)$. *Then, the* **quotient differential algebra** $(\overline{\mathcal{A}}, \overline{\mathcal{D}})$ *is defined by*

(a) $\overline{\mathcal{A}} \equiv \mathcal{A}/\mathcal{I}$;

(b) $\overline{\mathcal{D}}_\mu[\phi] = \overline{\mathcal{D}}_\mu(\phi + \mathcal{I}) = D_\mu(\phi) + \mathcal{I} \equiv [D_\mu \phi] \in \overline{\mathcal{A}}, \forall [\phi] \in \overline{\mathcal{A}}$.

THEOREM 2.36 - 1) *One has the following exact sequence:*

$$0 \to \mathcal{I}[D] \to \mathcal{A}[D] \to \overline{\mathcal{A}}[\overline{\mathcal{D}}] \to 0.$$

2) *One has the following exact complexes* $\{\overline{\mathcal{A}}[\overline{\mathcal{D}}]^p, \overline{\mathcal{D}}\}, \{\mathcal{I}[D]^p, D\}$. *For these, the homotopy-formula holds.*

3) *One has the following commutative diagram:*

$$\begin{array}{ccccccccccccccc}
 & & 0 & & 0 & & 0 & & & & 0 & & 0 & & \\
 & & \uparrow & & \uparrow & & \uparrow & & & & \uparrow & & \uparrow & & \\
0 & \to & \overline{\mathcal{K}} & \to & \overline{\mathcal{A}}^m & \stackrel{\overline{\mathcal{D}}}{\to} & \to & \overline{\mathcal{A}}^{m-1} & \stackrel{\overline{\mathcal{D}}}{\to} & \dots & \stackrel{\overline{\mathcal{D}}}{\to} & \overline{\mathcal{A}}^1 & \stackrel{\overline{\mathcal{D}}}{\to} & \overline{\mathcal{A}}^o & \to 0 \\
 & & \uparrow & & \uparrow & & \uparrow & & & & \uparrow & & \uparrow & & \\
0 & \to & \mathcal{K} & \to & \mathcal{A}^m & \stackrel{D}{\to} & \to & \mathcal{A}^{m-1} & \stackrel{D}{\to} & \dots & \stackrel{D}{\to} & \mathcal{A}^1 & \stackrel{D}{\to} & \mathcal{A}^o & \to 0 \\
 & & \uparrow & & \uparrow & & \uparrow & & & & \uparrow & & \uparrow & & \\
0 & \to & \mathcal{L} & \to & \mathcal{I}^m & \stackrel{D}{\to} & \to & \mathcal{I}^{m-1} & \stackrel{D}{\to} & \dots & \stackrel{D}{\to} & \mathcal{I}^1 & \stackrel{D}{\to} & \mathcal{I}^o & \to 0 \\
 & & \uparrow & & \uparrow & & \uparrow & & & & \uparrow & & \uparrow & & \\
 & & 0 & & 0 & & 0 & & & & 0 & & 0 & &
\end{array}$$

where $\mathcal{L}$ denotes the sub-algebra of all the constants in $\mathcal{I}$, namely $\mathcal{L} \equiv \mathcal{K} \cap \mathcal{I}$, and $\overline{\mathcal{K}} \equiv \mathcal{K}/\mathcal{L}$. Denote $\mathcal{H}^p(\overline{\mathcal{A}}), \mathcal{H}^p(\mathcal{A})$ and $\mathcal{H}^p(\mathcal{I})$ as the homology spaces of the horizontal lines. As all the columns are exact, it follows that the connecting linear applications

$$\delta : \mathcal{H}^p(\overline{\mathcal{A}}) \to \mathcal{H}^{p-1}(\mathcal{I})$$

are defined by

$$\delta\phi = [D\phi] \in \mathcal{H}^{p-1}(\mathcal{I}), \; \forall[\phi] \in \mathcal{H}^p(\overline{\mathcal{A}}), \; p = 1, \dots, m$$

and the sequence of homology

$$\begin{aligned} &0 \to \mathcal{H}^m(\mathcal{I}) \to \mathcal{H}^m(\mathcal{A}) \to .\mathcal{H}^m(\overline{\mathcal{A}}) \xrightarrow{\delta} \mathcal{H}^{m-1}(\mathcal{I}) \to \cdots \\ &\to \mathcal{H}^1(\overline{\mathcal{A}}) \xrightarrow{\delta} \mathcal{H}^o(\mathcal{I}) \to \mathcal{H}^o(\mathcal{A}) \to \mathcal{H}^o(\overline{\mathcal{A}}) \to 0 \end{aligned}$$

is exact. Note that one has: $\mathcal{H}^m(\mathcal{I}) \cong \mathcal{H}^m(\mathcal{A}) = 0$. One has the following results:
(a) If $\mathcal{H}^p(\mathcal{A}) = 0$, for all $p = 1, \dots, m$, then $\mathcal{H}^p(\overline{\mathcal{A}}) \cong \mathcal{H}^p(\mathcal{I}), p = 2, \dots, m$, and

$$0 \to \mathcal{H}^1(\overline{\mathcal{A}}) \xrightarrow{\delta} \mathcal{H}^o(\mathcal{I}) \to \mathcal{H}^o(\mathcal{A}) \to \mathcal{H}^o(\overline{\mathcal{A}}) \to 0 \,.$$

In particular, $\delta : \mathcal{H}^1(\overline{\mathcal{A}}) \cong \mathcal{I} \cap D(\mathcal{A}^1)/D(\mathcal{I}^1)$.

(C) - INTEGRAL OPERATORS.

DEFINITION 2.29 - 1) *Let $\pi : E \to M$ and $\overline{\pi} : F \to M$ be vector fiber bundles on M. An* **integral operator**

$$K : C_0^0(E) \to C^0(F)$$

is a continuous linear mapping,

$$K \in \mathcal{L}(C_0^0(E); C^0(F))$$

that can be locally represented as follows:

$$K(\phi)_U^\beta = \int_U \sum_{1 \le \alpha \le m} \mathbb{K}_{U,\alpha}^\beta \phi^\alpha d\mu_U \,, \; 1 \le \beta \le \overline{m}$$

where μ_U is a regular measure on U, and $\mathbb{K}_{U,\alpha}^\beta : U \times U \to \mathbb{K}$ are continuous functions.
2) We say that an integral operator $K : C_0^0(E) \to C^0(F)$ is of class $C^r, (r \ge 0, \infty)$, if the functions $\mathbb{K}_{U,\alpha}^\beta$ are of class C^r.
3) Such an operator of class C^∞ is also called **regularizing**.
4) Set $I^r(E; F)$ for the space of integral operators of class C^r.

THEOREM 2.37 - (PROPERTIES OF INTEGRAL OPERATORS).

1) Let $K \in I^r(E;F)$. One has:
(a) $K(C_0^0(E)) \subset C^r(F)$;
(b) For any compact subset $L \subset M, K|_{C_0^0(E|_L)}$ is a continuous linear application between the spaces $C_0^0(E|_L)$ and $C^r(F)$.
2) Let M be a compact manifold and $K \in I^r(E;F)$. Then, one has:
(a) If r is finite, $K_r \equiv K|_{C^r(F)}$ is a compact operator between the Banach spaces $C^r(E)$ and $C^r(F)$; [A linear transformation $T: V_1 \to V_2$ is **compact** *(or* **completely continuous**) *if for any sequence $\{x_n\}$ with norms uniformly limited (i.e., $\|x_n\| < c$, for $c > 0$ and for all n), there exists a subsequence, $\{x_{n_j}\}$ and an element $y \in V_2$, such that $\{T(x_{n_j})\}$ converges to y].*
(b) If $E \equiv F$ and if I_r is the identity mapping in $C^r(E)$, the kernel $I_r - K_r$ has finite dimension, and the image of $I_r - K_r$ is a closed subspace of finite codimension in the space $C^r(F)$. (In other words $I_r - K_r$ is a Fredholm's operator [24].)
3) One has the following canonical isomorphism: $I^r(E;F) \cong C^r(E' \boxtimes F)$. The section $B: M \times M \to E' \boxtimes F$ corresponding to the integral operator K is called the **kernel section** *of K.*
4) For an integral operator we can define a number (the **trace** *of K), that, if it exists, is given by: $\mathrm{tr}(K) = \int_M \mathrm{tr}(B|_\triangle)$, where $\triangle \subset M \times M$ is the diagonal subset of $M \times M$ and B is the kernel section corresponding to K.*

EXAMPLE 2.23 - (INTEGRAL OPERATORS). 1) Any continuous function

$$\mathbb{K} : [a,b] \times [a,b] \to \mathbf{R}$$

defines a linear application

$$K : L[a,b] \to C^o([a,b]), f \mapsto K(f)(x) = \int_{[a,b]} \mathbb{K}(x,y) f(y) dy.$$

Here $\mathbb{K}(x,y)$ is the kernel of K.

PROOF. For any $\epsilon > 0$, there is a $\delta(\epsilon) > 0$ such that, for points $x_1, x_2, y_1, y_2 \in [a,b]$, results

$$|K(x_1,y_1) - K(x_2,y_2)| < \epsilon, \text{ when } \quad |x_1 - x_2| < \delta(\epsilon), |y_1 - y_2| < \delta(\epsilon).$$

Therefore,

$$\begin{aligned}|K(f)(x) - K(f)(t)| &= |\int_{[a,b]} [\mathbb{K}(x,y) - \mathbb{K}(t,y)] f(y) dy| \\ &\leq \int_{[a,b]} |\mathbb{K}(x,y) - \mathbb{K}(t,y)||f(y)| dy \leq \epsilon \int_{[a,b]} |f(y)| dy,\end{aligned}$$

and it follows that

$$\lim_{x \to t} K(f)(x) = K(f)(t).$$

□

2) If we consider the **C**-valued functions on $[a,b]$, then a continuous function $\mathbb{K}$: $[a,b] \times [a,b] \to \mathbf{C}$ defines a linear application

$$K : L^{\mathbf{C}}[a,b] \to C^o([a,b],\mathbf{C}), f \mapsto K(f)(x) = \int_{[a,b]} \mathbb{K}(x,y)f(y)dy.$$

Furthermore, to K there corresponds the **adjoint** K^* given by:

$$K^*(f)(x) = \int_{[a,b]} \overline{\mathbb{K}(x,y)}f(y)dy.$$

Taking into account that the scalar product in $L^{\mathbf{C}}[a,b]$ is given by

$$(f,g) = \int_{[a,b]} f(x)\overline{g}(x)dx,$$

we get

$$\begin{aligned}(K(f),g) &= \int_{[a,b]} \{\int_{[a,b]} \mathbb{K}(x,y)f(y)dy\}\overline{g}(x)dx \\ &= \int_{[a,b]} \{\int_{[a,b]} \mathbb{K}(x,y)\overline{g}(x)dx\}f(y)dy \\ &= (f,K^*(g)) \to (K(f),g) = (f,K^*(g)).\end{aligned}$$

□The previous equation uniquely determines the adjoint transformation K^*. In fact, if there exists another transformation M such that $(K(f),g) = (f,M(g))$, for all $f,g \in L^{\mathbf{C}}[a,b] \Rightarrow (f,(K^*-M)(g)) = 0$, $\forall f,g \in L^{\mathbf{C}}[a,b] \Rightarrow K^* = M$.
□We have:

$$\|K\| \leq (\int_{[a,b]}\int_{[a,b]} |\mathbb{K}(x,y)|^2 dxdy)^{\frac{1}{2}}.$$

PROOF. Let us choose f such that $\|f\| = 1$, then

$$\begin{aligned}|K(f)(x)|^2 &= |\int_{[a,b]} \mathbb{K}(x,y)f(y)dy|^2 \leq \int_{[a,b]} |K(x,y)|^2 dy \int_{[a,b]} |f(y)|^2 dy \\ &= \int_{[a,b]} |K(x,y)|^2 dy.\end{aligned}$$

Therefore,

$$\|K(f)(x)\|^2 \equiv \int_{[a,b]} |\mathbb{K}(f)(x)|^2 dx \leq \int_{[a,b]}\int_{[a,b]} |K(x,y)|^2 dxdy$$

and write

$$\|K\| = l.u.b.\{\|K(f)(x)\| \,|\, \|f\| = 1\} \leq (\int_{[a,b]}\int_{[a,b]} |\mathbb{K}(x,y)|^2 dxdy)^{\frac{1}{2}}.$$

□

□ A continuous kernel $\mathbb{K}(x,y)$ defines a compact transformation.

□ The integral transformation K is called **symmetric** if for all the elements f and g of $L^{\mathbf{C}}[a,b]$ one has

$$(K(f),g) = (f,K(g)).$$

(Thus K is symmetric iff $K = K^*$.)

□ A real kernel $\mathbb{K}(x,y)$ is called **symmetric** if $\mathbb{K}(x,y) = \mathbb{K}(y,x)$, for all x, y such that $a \leq x \leq b$, $a \leq y \leq b$.

□ A kernel $\mathbb{K}(x,y)$ defined on $L^{\mathbf{C}}[a,b]$ is called **Hermitian** iff

$$\mathbb{K}(x,y) = \overline{\mathbb{K}(y,x)}.$$

□ The integral transformation K is symmetric iff it has a symmetric kernel.

□ (a) The eigenvectors corresponding to different eigenvalues of a symmetric integral transformation K , are orthogonal ones.

(b) The number of different non-zero eigenvalues of a symmetric integral transformation K is numerable. For each non-zero eigenvalue there is at most a finite number of linearly independent eigenvalues.

(c) Let $\mathbb{K}(x,y)$ be a **separable kernel**, namely $\mathbb{K}(x,y) = \sum_{1\leq i,j\leq n} c_{ij}h_i(x)h_j(x)$, where $c_{ij} = c_{ji}$ and $h_i(x)$ are continuous functions on $[a,b]$. Let K be also symmetric. Then, there is a set of orthonormal vectors $f_1(x), \cdots, f_m(x) \in L^{\mathbf{C}}[a,b]$ and non-zero real scalars, $\lambda_1, \cdots, \lambda_m$, such that

$$K(f_i)(x) = \lambda_i f_i(x) \, , \; i = 1, \cdots, m.$$

If also $g \in L^{\mathbf{C}}[a,b]$ and it is $\perp f_i$ for all i , then $K(g) = 0$. The scalars λ_i are the non-zero eigenvalues of K and there is a finite number of linearly independent eigenvectors corresponding to each eigenvalue.

(d) Let $f_i(x)$, $i = 1, \cdots, m$, be a basis of orthonormal eigenvectors of a symmetric integral operator K corresponding to non-zero eigenvalues λ_i , $i = 1, \cdots, m$. If K has no more non-zero eigenvalues, then

$$\mathbb{K}(x,y) = \sum_{1\leq j\leq m} \lambda_j f_j(x) f_j(y).$$

(e) The symmetric kernel $\mathbb{K}(x,y)$ is **semidefinite** (namely $(K(f),f) \geq 0$, $\forall f \in L^{\mathbf{C}}[a,b]$) iff their eigenvalues are non-negative.

(f) If $\mathbb{K}(x,y)$ is a symmetric, positive and semi-definite, then: $\mathbb{K}(x,x) \geq 0$, for $a \leq x \leq b$. Furthermore,

$$\mathbb{K}(x,y) = \sum_{1 \leq j \leq m} \lambda_j f_j(x) f_j(y)$$

and the series converge absolutely and uniformly for $a \leq x \leq b$, $a \leq y \leq b$.

DEFINITION 2.30 - *The* **unity function of Heaviside** $H(x)$ *is the following piecewise continuous function:*

$$H(x) = \begin{Bmatrix} 1, & x > 0 \\ 0, & x < 0 \end{Bmatrix}.$$

THEOREM 2.38 - *For the derivative of* $H(x)$ *one has* $H'(x) = \delta(x)$.

PROOF. One has:

$$\begin{aligned} \int_{[-\infty,\infty]} H'(x)\phi(x)dx &= -\int_{[-\infty,\infty]} H(x)\phi'(x)dx \\ &= -\int_{[0,\infty]} \phi'(x)dx = \phi(0)\,,\ \forall \phi \in C_0^\infty(\mathbf{R})\,. \end{aligned}$$

Therefore, by comparison with $\int_{[-\infty,\infty]} \delta(x)\phi(x)dx = \phi(0)$, we see that the theorem holds. □

THEOREM 2.39 - *Let* f *be a piecewise continuous function having steps of height* $a_1, \ldots, a_n$, *at the points* $x_1, \cdots, x_m$ *respectively. Then, the function*

$$g(x) \equiv f(x) - \sum_{1 \leq k \leq m} a_k H(x - x_k)$$

is continuous. Then, we can define the distribution:

$$f(x) \equiv g(x) + \sum_{1 \leq k \leq m} a_k H(x - x_k).$$

EXAMPLE 2.24 - (EXAMPLES OF WEAK SOLUTIONS).

1)(DISTRIBUTIONAL (OR WEAK) SOLUTIONS OF WAVE EQUATION).

The function $H(x)$ allows us to express the distributional solutions of the following equation:

$$Lu \equiv u_{tt} - u_{xx} = 0.$$

A solution of this equation is $u(x,t) = f(x-t)$ where $f : \mathbf{R} \to \mathbf{R}$ is any function of class C^2. We can see that a distributional solution of the above equation is the following:

$$u(x,t) = H(x-t).$$

In fact, as $L^* = L$, we get $< u, L^*(\phi) >= \int_{x>t}(\phi_{tt} - \phi_{xx})dxdt$, where ϕ is a smooth function with compact support on $\mathbf{R}^2$ (**test function**) that is zero outside a limited region $D \subset \mathbf{R}^2$. From Tab.2.3 we get that for the operator $\square^2$ we can write

$$\int_D (v\square^2 u - u\square^2 v)dxdt = \int_{\partial D} \mathbf{n}.[(vu_t - uv_t)\partial t + u\,\mathrm{grad}_x v - v\,\mathrm{grad}_x u]d\sigma$$
$$= \int_{\partial D} [v(\partial q.u) - u(\partial q.v)]d\sigma$$

where $\partial q.\partial t = \mathbf{n}.\partial t, \partial q.\partial x_k = -\mathbf{n}.\partial x_k$, $1 \le k \le n$. For $n = 1$, we get

$$\int_D [v(u_{tt} - u_{xx}) - u(v_{tt} - v_{xx})]dxdt = \int_{\partial D} [v(\partial q.u) - u(\partial q.v)]d\sigma.$$

Then, one has:

$$< u, L^*(\phi) >= \int\int_{x>t} \phi(u_{tt} - u_{xx})dxdt + \int_{x=t} (\partial q.\phi)ds.$$

As $\partial q|_{x=t} \|$ line $x = t$, $\Rightarrow (\partial q.\phi) = (\partial s.\phi) \Rightarrow \int_{x=t}(\partial q.\phi)ds = \int_{x=t}(\partial s.\phi)ds = [\phi]_a^b = 0$, as ϕ is zero outside a finite region of the line $x = t$. Furthermore, taking into account that $u_{tt} = u_{xx} = 0$ for $x > t \Rightarrow< u, L^*(\phi) >= 0 \Rightarrow u(x,t) = H(x - t)$ is a weak solution of the wave equation. $\square$

$\square$Note also that discontinuity of the solution $u(x,t) = H(x - t)$ propagates on the line $x = t$ that is a characteristic line of the equation. Furthermore, we can prove that the unique curves that bring the discontinuity of the weak solutions are the characteristic lines, and the step of this discontinuity is constant. In fact, let γ be a curve in $\mathbf{R}^2$, (coordinates (x,t)), that separates the discontinuity of the weak solution. Let A_+ and A_- be the corresponding domains in $\mathbf{R}^2$. We have:

$$\begin{aligned}
0 &=< u, L^*(\phi) > \\
&= \int\int_{\mathbf{R}^2} u(\phi_{tt} - \phi_{xx})dxdt \\
&= \int\int_{A_+} u(\phi_{tt} - \phi_{xx})dxdt + \int\int_{A_-} u(\phi_{tt} - \phi_{xx})dxdt \\
&= \int_\gamma \{[u]_\pm(\partial q.\phi) - \phi[(\partial q.u)]_\pm\}ds,
\end{aligned}$$

where $[]_\pm$ represents the step through γ of the quantity within the brackets. As this equation holds for any ϕ, we get $[u]_\pm = [(\partial q.u)]_\pm = 0$. Therefore, ∂q is not tangent to the curve γ, this curve cannot separate discontinuities of u and its derivatives. On the other hand, if ∂q is tangent to γ, namely $\partial q.\mathbf{n} = 0 \Rightarrow \partial q = (\mathbf{n}.\partial t)\partial t - (\mathbf{n}.\partial x)\partial x \Rightarrow$

$0 = \partial q.\mathbf{n} = (\mathbf{n}.\partial t)^2 - (\mathbf{n}.\partial x)^2 \Rightarrow \mathbf{n}.\partial t = \pm\mathbf{n}.\partial x$. Therefore, ∂q is tangent to γ iff γ belongs to the family of curves $x \pm t = const$. These are the characteristic lines of the wave equation. Now, if γ is characteristic, then $\partial q = k\partial s$, $k =$ const. and we can write:

$$0 = \int_\gamma \{[u]_\pm(\partial q.\phi) - \phi[(\partial q.u)]_\pm\}ds\,.$$

By considering

$$[(\partial s.u)]_\pm = \partial s.[u]_\pm$$

it follows that

$$0 = \int_\gamma \{\partial s.(\phi[u]_\pm) - 2\phi(\partial s.[u]_\pm)\}ds\,.$$

As ϕ is zero outside a finite region of $\gamma \Rightarrow \phi[u]_\pm = 0$. Thus, we get

$$0 = \int_\gamma \phi(\partial s.[u]_\pm)ds.$$

As ϕ is an arbitrary test function, we have

$$\partial s.[u]_\pm = 0 \Rightarrow [u]_\pm = const.$$

□

2)(DISTRIBUTIONAL SOLUTIONS OF LAPLACE EQUATION). One can prove that the equation $\nabla^2 u = 0$ has no weak solutions with simple discontinuities in u and its first derivatives. In fact, any weak solution of Laplace equation is a C^∞ solution. [This is a general behaviour for elliptic PDEs.] The explicit calculation can be made following a way similar to the previous calculation and by utilizing the results reported in tab.2.3.]

(D) - GREEN FUNCTIONS FOR LINEAR DIFFERENTIAL OPERATORS.

Here we will show that to any singular solution of an affine PDE we can associate a kernel, that we will call generalized Green kernel of the singular solution.

REMARK 2.32 - In general, a Green function G is not uniquely determined. In fact, any solution u of the associated linear equation, $K.u = 0$, identifies kernels of the type $u \otimes \beta$, $\forall \beta \in C_0^\infty(F)'$, by means of the following commutative diagram:

$$\begin{array}{ccc} C_0^\infty(E' \boxtimes F)' & \cong & C_0^\infty(E')' \bigotimes C_0^\infty(F)' \\ \uparrow & & \uparrow j \\ C^\infty(E) \bigotimes C_0^\infty(F)' & \cong & C^\infty(E) \bigotimes C_0^\infty(F)' \\ \uparrow & & \uparrow \\ 0 & & 0 \end{array}$$

More precisely: ${}_f(u\otimes\beta)(K^*(\alpha))=0$, $\forall\alpha\in C_0^\infty(F')$, $\forall f\in C_0^\infty(F)$. In fact, one has:

$$\begin{aligned}{}_f(u\otimes\beta)(K^*(\alpha)) &=< u\otimes\beta, f\otimes K^*(\alpha)>\\ &=\beta(f).u((K^*(\alpha))\\ &=\beta(f).\int_M < K^*(\alpha),u>=\beta(f).\int_M < K(u),\alpha>=0\end{aligned}$$

as $K(u)=0$. Then, we also have: $1\otimes\widetilde{K}(G+u\otimes\beta)=\mathbf{1}\otimes\widetilde{K}(G)+\mathbf{1}\otimes\widetilde{K}(u\otimes\beta)=\mathbb{D}$. Thus, if G is a Green kernel of K, the following are also Green kernels of K: $G+u\otimes\beta$, where $K(u)=0$, $\forall\beta\in C_0^\infty(F)'$.

THEOREM 2.40 - 1) *A kernel* $G\in C_0^\infty(E'\boxtimes F)'\cong C_0^\infty(E')'\bigotimes C_0^\infty(F)'$ *can be identified with a section*

$$\underline{G}: M\times M\to J\mathcal{D}^\infty(E')^*\bigotimes\Lambda_n^0(M)\boxtimes J\mathcal{D}^\infty(F)^*\bigotimes\Lambda_n^0(M).$$

More precisely one has:

$$G(\alpha\otimes f)=\int_{M\times M}<\underline{G},D^\infty\alpha\otimes D^\infty f>.$$

2) *We have the following local representation:*

$$G(\alpha\otimes f)=\sum_U\sum_{|\gamma|,|\beta|\le\infty}\int_{U\times U}g_{U\times U}G_i^{\beta\gamma j}(x,x')(\partial_\gamma.\alpha_j)(x)(\partial_\beta.f^j)(x')d\mu(x)d\mu(x')$$

where $G_i^{\beta\gamma j}$ *are the local components of* $\underline{G}$, *and* $g_{U\times U}$ *is a partition of unity subordinate to the covering* $\{U\times U\}$ *of* $M\times M$.

3) *If* $\underline{G}$ *is identified with a section of* $(E'\boxtimes F)'$ *on* $M\times M$ *then* G *is identified with local functions* G_i^j *on* $M\times M$ *that are called* **Green functions**. *(For abuse of notation we shall also write* $\underline{G}=G$ *and we will say that* G *is a Green function.)*

THEOREM 2.41 - *If* G *is a Green function of the linear differential operator of order* k, $K.: C^\infty(E)\to C^\infty(F)$, *it must satisfy the following local equations:*

$$\sum_{|\gamma|\le k}K_i^{\gamma j}(x)(\partial_{x,\gamma}.G_u^i)(x,x')=\delta_u^j\delta(x,x').\tag{2.77}$$

PROOF. In fact, in local coordinates we must have:

$$\begin{aligned}0=\sum_U\int_{U\times U}g_{U\times U}\sum_{|\omega|,|\epsilon|\le k}&[G_u^{\omega\epsilon i}(x,x')(\partial_\epsilon.K^*(\alpha)_i)(x)(\partial_\omega.f^u)(x')\\ &-\alpha_i(x)f^i(x')]d\mu(x)d\mu(x').\end{aligned}$$

Now, if G is a Green function, the above expression can be written:

$$0 = \sum_U \int_{U\times U} g_{U\times U}[G^i_u(x,x')K^*(\alpha)_i(x)f^u(x') - \alpha_j(x)f^u(x')\delta^j_u\delta(x,x')]d\mu(x)d\mu(x') .$$

As this expression must hold for any f we can write:

$$G^i_u(x,x')K^*(\alpha)_i(x) = \alpha_j(x)\delta^j_u\delta(x,x') .$$

Taking into account that $K^*(\alpha)_i = \sum_{|\gamma|\leq k} K^{*\gamma j}_i(\partial_{x,\gamma}.\alpha_j)$, we can also write

$$\sum_{|\gamma|\leq k} G^j_u(x,x')K^{*\gamma i}_j(x)(\partial_{x,\gamma}.\alpha_i)(x) = \alpha_j(x)\delta^j_u\delta(x,x') .$$

From definition of adjoint of K and taking into account that the above equation holds for any α we get equations (2.77). □

REMARK 2.33 - Of course, the Green functions are not, in general, uniquely determined. In fact, if G^j_i are the local components of a Green function G,

$$\widehat{G}^j_i(x,x') \equiv G^j_i(x,x') + u^j(x)g_i(x') ,$$

where

$$\sum_{|\gamma|\leq k} K^{*\gamma j}_i(x)(\partial_{x,\gamma}.u^j)(x) = 0$$

and g_i are local arbitrary functions on M, are local components of Green functions of K.

THEOREM 2.42 - (FUNDAMENTAL SOLUTIONS (≡ GREEN FUNCTIONS) OF LINEAR ORDINARY DIFFERENTIAL OPERATORS).

1) *Set*

$$L_x \equiv \sum_{0\leq k\leq n} a_k(x)\frac{d^{n-k}}{d^{n-k}x}.$$

Define **fundamental solution** *any distributional solution E of the following equation:*

$$L_xE \equiv \sum_{0\leq k\leq n} a_k(x)\frac{d^{n-k}}{d^{n-k}x}E = \delta(x-t). \tag{2.78}$$

The general fundamental solution of L_x can be written in the form:

$$E = F(x,t) + \sum_{1\leq k\leq n} c_k(t)u_k(x), \tag{2.79}$$

where:
(a) $u_k(x)$ *are linearly independent solutions of* $L_x u = 0$;
(b) $F(x,t)$ *is a particular solution of* (2.78). *This can be written in the following form:*

$$F(x,t) = u(x)H(t-x) + v_t(x)H(x-t), \tag{2.80}$$

where $u(x)$ *is a fixed arbitrary solution of* $L_x u = 0$; $v_t(x)$ *is another solution of* $L_x u = 0$ *that satisfies the following initial conditions:*

$$\left\{ \begin{array}{l} v_t(t) = u(t) \\ v'_t(t) = u'(t) \\ \dots \\ v_t^{(n-2)}(t) = u^{(n-2)}(t) \\ v_t^{(n-1)}(t) = u^{(n-1)}(t) + \dfrac{1}{a_0(t)} \end{array} \right\}.$$

2) *Any fundamental solution of* L_x *can be identified with a* **weak solution**, *i.e.,* **regular distribution**, *(≡distribution that can be identified with a locally integrable function* $F(x,t)$: $F_t = \int F(x,t)\phi(x)dx$).
3) *Any fundamental solution of* L_x *identifies an integral operator that coincides with the Green function of* L_x. *(Thus, we can say that the space of the fundamental solutions of* L_x *, and the space of Green functions, without boundary conditions, coincide.)*
EXAMPLE 2.25 - 1) The fundamental solutions of $L_x \equiv \frac{d^2}{d^2x}$.
The general solution of the following equation:

$$\frac{d^2u}{d^2x} = 0 \tag{2.81}$$

is

$$u(x) = x\alpha + \beta\,,\ \alpha, \beta \in \mathbf{R}. \tag{2.82}$$

As α and β are constants we can write: $u(x) = x\alpha(t) + \beta(t)$. For $u(x)$ we can take the particular solution $u(x) = x - t$. Then, we have the following initial conditions:

$$u(t) = 0\,,\ v'_t(t) = u'(t) + 1 = 2.$$

Then, from (2.80) we get the following particular solution of (2.81):

$$u(t) = 0 = t\alpha + \beta\,,\ u'(t) = 2 = \alpha \Rightarrow \beta = -2t\,,\ \alpha = 2.$$

Therefore, we have the following solution:

$$v_t(x) = 2x - 2t = 2(x - t).$$

Then, by using (2.80) we get

$$\begin{aligned} F(x,t) &= (x-t)H(t-x) + 2(x-t)H(x-t) \\ &= (x-t)[H(t-x) + 2H(x-t)] = (x-t)[1 + H(x-t)]\,. \end{aligned}$$

Therefore, we get

$$\begin{aligned} E(x,t) &= F(x,t) + x\overline{\alpha}(t) + \overline{\beta}(t) = (x-t)H(x-t) + x(\alpha(t)+1) + \beta(t) - t \\ &= (x-t)H(x-t) + x\alpha(t) + \beta(t) = G(x,t). \end{aligned}$$

□

THEOREM 2.43 - *The problem of finding the Green kernel for a differential operator K can be reduced to the problem of finding the Green kernel of a Hermitian operator.*
PROOF. Let K be a differential operator. Let $H \equiv KK^+$. So H is a Hermitian operator. Let G_H its Green kernel: $\widetilde{H} \bigotimes \mathbf{1}(G_H) = \mathbf{1} \bigotimes \widetilde{H}(G_H) = \mathbb{D}$. Then, we can say that $G \equiv \mathbf{1} \bigotimes \widetilde{K^+}(G_H)$ is a Green kernel for K. In fact, we get: $\mathbf{1} \bigotimes \widetilde{K}(G) = \mathbf{1} \bigotimes \widetilde{K}(\mathbf{1} \bigotimes \widetilde{K^+}(G_H)) = 1 \bigotimes K\widetilde{K^+}(G_H) = \mathbf{1} \bigotimes \widetilde{H}(G_H) = \mathbb{D}$. □

THEOREM 2.44 - (SPECTRUM OF AN OPERATOR AND INVERTIBILITY). *Let A be a closed linear operator defined on a linear manifold D_A that we assume to be dense in a Hilbert space $\mathcal{H}$. Then, the operator $A - \lambda I, \lambda \in \mathbf{C}$ is also closed: $D_{A-\lambda I} = D_A$.*
1) (CLASSIFICATION OF POINTS $\lambda \in \mathbf{C}$ DEPENDING ON THE FACT THAT $A - \lambda I$ IS REGULAR OR SINGULAR).
(a) *A* **regular value** *(or* **regular point***) of A is a value $\lambda \in \mathbf{C}$ such that $A - \lambda I$ is regular. The set of regular values of A is called* **resolvent** *of A: $\mathcal{R}(A)$.*
(b) *The* **spectrum** *of A: $Sp(A) \equiv \mathbf{C}/\mathcal{R}(A)$. (Thus $\lambda \in Sp(A)$ iff $A - \lambda I$ is singular.)*
2) (CLASSIFICATION OF THE POINTS OF THE SPECTRUM DEPENDING ON THE WAY THAT OPERATOR $A - \lambda I$ IS SINGULAR).
(a) $\ker(A - \lambda I) \neq \{0\}$, *$\lambda$ is called an* **eigenvalue** *of A. The set of eigenvalues is called* **point spectrum** *of A: $Sp(A)_p$. [The set of eigenvectors corresponding to the same eigenvalue is a linear manifold, called* **eigenmanifold** *of λ. The dimension of this manifold is the* **multiplicity** *of λ .]*
(b) *If* $\ker(A - \lambda I) = \{0\} \Rightarrow (A - \lambda I)^{-1}$ *is unlimited,*

$$\operatorname{im}(A - \lambda I) \neq \mathcal{H}, \quad \overline{\operatorname{im}(A - \lambda I)} = \mathcal{H}$$

λ belongs to the **continuous spectrum** *of A: $Sp(A)_c$.*

(c) *If* $\ker(A-\lambda I)=\{0\}, \overline{\mathrm{im}(A-\lambda I)} \neq \mathcal{H}, \lambda$ *belongs to the* **residual spectrum** *of* A*:* $Sp(A)_r$*. The closed linear manifold* $(\mathrm{im}(A-\lambda I))^{\perp}$ *has positive dimension. This dimension is called the* **deficiency** *of* λ.

3) (PROPERTY OF THE SPECTRUM OF OPERATOR).

(i) *If* $\lambda \in Sp(A)_r$ *, deficiency* $(\lambda)=m \Rightarrow \lambda \in Sp(A^*)_p$ *, multiplicity*$(\lambda)=m$.

(ii) *If* A *is symmetric* $\Rightarrow Sp(A)_p \subset \mathbf{R}, Sp(A)_c \subset \mathbf{R}$.

(iii) *If* A *is selfadjoint* $\Rightarrow Sp(A) \subset \mathbf{R}, Sp(A)_r = \emptyset$*. Furthermore, one has:*

(a) $\lambda \in \mathcal{R}(A) \Leftrightarrow \mathrm{im}(A-\lambda I)=\mathcal{H}$;

(b) $\lambda \in Sp(A)_p \Leftrightarrow \overline{\mathrm{im}(A-\lambda I)} \neq \mathcal{H}$;

(c) $\lambda \in Sp(A)_c \Leftrightarrow \mathrm{im}(A-\lambda I) \neq \mathcal{H}$, but $\overline{\mathrm{im}(A-\lambda I)}=\mathcal{H}$.

4) (**Resolution of the equation** $A(u)=f$ **by means of the spectrum of** A). *Let us assume that* $Sp(A)_p$ *has the corresponding eigenvectors that form a basis for all the space. Then, we have the following representation of the general solution of* $A(u)=f$*:*

$$u=(\sum_n \frac{a_n}{\lambda_n} v_n)+\ker(A),$$

where a_n *are Fourier-coefficients of* f *in the basis of eigenvectors* $\{v_n\}$.

PROOF. Set $f=\sum_n a_n v_n, u=\sum_n b_n v_n$. Determine b_n in such a way to satisfy the equation. We get

$$A(u)=\sum_n b_n A(v_n)=\sum_n b_n \lambda_n v_n=\sum_n a_n v_n \Rightarrow \sum_n (b_n\lambda_n-a_n)v_n=0 \Rightarrow b_n=\frac{a_n}{\lambda_n}.$$

□

5) (**Spectral representation of a function** $f(A)$ **of an operator** A).

$$f(A)(u)=\sum_n a_n f(\lambda_n)v_n.$$

PROOF. In fact,

$$f(A)(u)=f(A)(\sum_n a_n v_n)=\sum_n a_n f(A)(v_n)=\sum_n a_n f(\lambda_n)v_n.$$

□

6) *In particular, the inverse* $(A-\lambda I)^{-1}$ *is given by:*

$$(A-\lambda I)^{-1}(u)=\sum_n (\frac{a_n}{\lambda_n-\lambda})v_n, \ \forall \lambda \neq \lambda_n \in Sp(A)_p.$$

Of course, $\lambda=\lambda_n$ *is admissible if* $a_n=0 \Leftrightarrow <v_n, f>=0$.

7) (**Resolution of the equation** $(A-\lambda)(u)=f$ **by means of the spectrum of** A).

(a) *If* $\lambda \in \mathcal{R}(A), \lambda \in Sp(A)_c \Rightarrow$ *there is a unique solution:*

$$u = (A - \lambda I)^{-1}(f) = \sum_n (\frac{a_n}{\lambda_n - \lambda}) v_n$$

if $f = \sum_n a_n v_n$, *with* $\{v_n\}$ *eigenvectors of* A.
(b) *If* $\lambda \in Sp(A)_p \Rightarrow$ *the general solution is:*

$$u = \sum_n (\frac{a_n}{\lambda_n - \lambda}) v_n + \ker(A - \lambda I)$$

iff $\lambda = \lambda_n$, and $a_n = 0 \Leftrightarrow < v_n, f >= 0$.
EXAMPLE 2.26 - 1) An ordinary differential operator has not residual spectrum.
2) *For selfadjoint ordinary differential operators, the eigenvectors generate the domains of the operators.*
3) $A_1 \equiv \frac{d}{dt}|_{D_1}, D_1 \equiv \{x \in \mathcal{L}_2^{(c)}(0,1)|x(t)$ *absolutely continuous,* $\dot{x} \in \mathcal{L}_2^{(c)}(0,1)\}$. *We get:*

$$\ker(A_1 - \lambda I) = \{x = Ce^{\lambda t} | \forall \lambda \in \mathbf{C} \Rightarrow Sp(A) = Sp(A)_p = \mathbf{C}\}.$$

4) $A_2 \equiv \frac{d}{dt}|_{D_2}, D_2 \equiv \{x \in D_1 | x(0) = 0\}$. *We get:*

$$\ker(A_2 - \lambda I) = \{0\} \Rightarrow Sp(A)_p = \emptyset.$$

The affine equation $(A_2 - \lambda I)(x) = f$ *has a unique solution for fixed* f. *In fact, we can build the Green function* G *that satisfies the following system:*

$$\frac{d}{dt}G(t,\tau) - \lambda G(t,\tau) = \delta(t - \tau), 0 \leq t, \tau \leq 1$$
$$G(t,\tau)|_{t=0} = 0.$$

We get:

$$G(t,\tau) = e^{\lambda(t-\xi)} H(t - \xi).$$

Therefore, the affine equation admits the unique solution

$$x(t) = \int_{[0,1]} G(t,\tau) f(\tau) d\tau = e^{\lambda t} \int_{[0,t]} e^{-\lambda\tau} f(\tau) d\tau.$$

The operator $(A_2 - \lambda I)^{-1}$ *is limited* $\Rightarrow Sp(A_2) = \emptyset$.
5) $A_3 \equiv \frac{d}{dt}|_{D_3}, D_3 \equiv \{x \in D_1 | x(0) = x(1)\}$. *We get:*

$$\ker(A_3 - \lambda I) = \{Ce^{\lambda t}, \lambda = 2\pi n i\} \Rightarrow \ker(A_3 - \lambda I) = \{0\} \Leftrightarrow \lambda \neq 2\pi n i.$$

The affine equation $(A_3 - \lambda I)(x) = f$ has a unique solution for fixed f. In fact, we can build the Green function G that satisfies the following system:

$$\frac{d}{dt}G(t,\tau) - \lambda G(t,\tau) = \delta(t-\tau)\,,\ 0 \le t,\tau \le 1$$
$$G(0,\tau) = G(1,\tau).$$

We get

$$G(t,\tau) = \frac{e^{-\lambda\tau}}{e^{-\lambda}-1}e^{\lambda t}H(\tau - t) + \frac{e^{-\lambda\tau}e^{\lambda(\tau-1)}}{e^{-\lambda}-1}e^{\lambda t}H(t-\tau).$$

Therefore, the unique solution is the folllowing:

$$x(t) = \int_{[0,1]} G(t,\tau)f(\tau)d\tau.$$

The operator $(A_3 - \lambda I)^{-1}$ is limited $\Rightarrow Sp(A_3) = Sp(A_3)_p = \{\lambda_n = 2n\pi i\} \cong \mathbf{Z}$.
6) $A_4 \equiv \frac{d}{dt}|_{D_4}, D_4 \equiv \{x \in D_1 | x(0) = x(1) = 0\}$. *We get:*

$$\ker(A_4 - \lambda I) = \{0\} \Rightarrow Sp(A)_p = \emptyset.$$

The affine equation $(A_4 - \lambda I)(x) = f$ has the unique solution that satisfies the condition in $x = 0$.

$$x(t) = e^{\lambda t}\int_{[0,t]} e^{-\lambda\tau}f(\tau)d\tau.$$

The condition at $x = 1$ can be satisfied iff

(2.83)(**consistency condition**) $$0 = \int_{[0,1]} e^{-\lambda\tau}f(\tau)d\tau.$$

This is a condition on f, that, therefore, cannot be arbitrary one. Thus, the range of $A_4 - \lambda I$ contains the functions $f(\tau)$ that are $\perp e^{-\lambda\tau}$. If f satisfies the consistency condition the solution of the affine equation is the same as example 4 $\Rightarrow (A_4 - \lambda I)^{-1}$ is limited $\Rightarrow$ any value of $\lambda \in Sp(A_4)_r$.
7) *Let $R \subset \mathbf{R}^2, (x,y)$, be the rectangular domain $R \equiv [0,a] \times [0,b]$. Consider the operator:*

$$L \equiv -\nabla^2|_D, D \equiv \{u \in \mathcal{L}_2(R) | u|_{\partial R} = 0\}.$$

We get

$$Sp(L)_p \equiv \{\lambda_{m,n} = (\frac{m\pi}{a})^2 + (\frac{n\pi}{b})^2, m,n = 1,2,\dots\}.$$

The orthonormalized eigenfunctions (form a complete system) are:

$$u_{m,n}(x,y) = \sqrt{\frac{4}{ab}}\sin(\frac{m\pi x}{a})\sin(\frac{n\pi y}{b}), m,n = 1,2,\dots.$$

PROOF. Let us consider the eigenvalue problem:

$$\left\{ \begin{array}{l} -\nabla^2 u = \lambda u \, , \; x \in R \\ u(0,y) = u(a,y) = 0 \, , \; 0 < y < b \\ u(x,0) = u(x,b) = 0 \, , \; 0 < x < a \end{array} \right\}$$

and integrate with the method of the separation of variables:

$$u(x,y) = X(x)Y(y) \Rightarrow \frac{\ddot{X}}{X} = -\frac{\ddot{Y}}{Y} - \lambda$$

$$\Downarrow$$

$$\left\{ \frac{\ddot{X}}{X} = -\mu = const \right\}, \{X(0) = X(a) = 0 \, , \; 0 < x < a\} \, ,$$

$$\left\{ \frac{\ddot{Y}}{Y} - \lambda = -\mu = const, Y(0) = Y(b) = 0, 0 < y < b \right\},$$

$$\Downarrow$$

$$\mu_m = (\frac{m\pi}{a})^2, X_m(x) = \sqrt{\frac{2}{a}} \sin(\frac{m\pi x}{a}), m = 1, 2, \ldots;$$

and therefore .... □

8) *Let* $R \subset \mathbf{R}^2, (r,\theta)$, *be the disc* R *of unitary radius and center* 0. *Let us consider the following operator:*

$$L \equiv -\nabla^2|_D, D \equiv \{u \in \mathcal{L}_2(R) | u|_{\partial R} = 0\}.$$

We get

$$Sp(L)_p \equiv \{\lambda_{n,k} = [\beta_k^{(n)}]^2, n = 0, \pm 1, \pm 2, \ldots, k = 1, 2, \ldots\}$$

where $\beta_k^{(n)}$ *are the positive roots of* $J_n(\sqrt{\lambda})$. *Furthermore,* $\lambda_{n,k} = \lambda_{-n,k}$. *The corresponding orthonormalized eigenfunctions (form a complete system) are:*

$$u_{n,k}(r,\theta) = \frac{e^{in\theta} J_n(\beta_k^{(n)} r)}{\sqrt{\pi} J_n(\beta_k^{(n)})}.$$

PROOF. We must consider the eigenvalue system:

$$\left\{ \begin{array}{l} -\frac{1}{r}(\partial r(r(\partial r.u))) - \frac{1}{r^2}(\partial\theta\partial\theta.u) = \lambda u \, , \; r < 1 \\ u(1,\theta) = 0 \, , \; -\pi < \theta < \pi \end{array} \right\}.$$

□

9) *Let* $R \subset \mathbf{R}^3, (r, \theta, \phi)$, *be the sphere of center* 0 *and unitary radius. Let us consider the following operator:*

$$L \equiv -\nabla^2|_D, D \equiv \{u \in \mathcal{L}_2(R) | u|_{\partial R} = 0\}.$$

We get

$$Sp(L)_p \equiv \{\lambda_{n,k} = [\beta_k^{(n+\frac{1}{2})}]^2, n = 0, 1, 2, \dots, k = 1, 2, \dots\}$$

where $\beta_k^{(n)}$ *are the positive roots of* $J_{n+\frac{1}{2}}(\sqrt{\lambda})$. *The corresponding orthonormalized eigenfunctions (form a complete system) are:*

$$u_{m,n,k}(r, \theta, \phi) = \frac{1}{r} J_{n+\frac{1}{2}}(\beta_k^{(n+\frac{1}{2})} r) Y_n^m(\theta, \phi).$$

PROOF. We must consider the eigenvalue system:

$$\left\{\begin{array}{l} -\dfrac{1}{r^2}(\partial r(r^2(\partial r.u))) + \dfrac{1}{r^2} S(u) = \lambda u \ , \ r < 1 \\ u(1, \theta, \phi) = 0 \\ S(u) \equiv -(\partial\theta(\sin\theta(\partial\theta.u))) - \dfrac{1}{\sin\theta}(\partial\phi\partial\phi.u) \end{array}\right\}.$$

We integrate the system by putting $u = R(u)Y(\theta, \phi)$. We get

$$\left\{\begin{array}{l} -\dfrac{d}{dr}(r^2\dfrac{dR}{dr}) - \mu R + \lambda r^2 R = 0 \\ S(Y) = \mu Y \end{array}\right\}.$$

From the second we get:

$$\left\{\begin{array}{l} \mu = n(n+1) \ , \ n = 0, 1, 2, \dots \\ Y = Y_n^m(\theta, \phi) = P_n^{|m|}(\cos\theta) e^{im\phi} \ , \ |m| \leq n \end{array}\right\}$$

and in combination with the first one, by putting $z = r\sqrt{\lambda}, Z = zR$, we get:

$$\left\{\begin{array}{l} -\dfrac{d}{dz}(z\dfrac{dZ}{dz}) + zZ - \dfrac{(n+\frac{1}{2})^2}{z} Z = 0 \\ R(1) = 0 \ , \ Z(\sqrt{\lambda}) = 0, \text{finite solution in } 0. \end{array}\right\}$$

The above equation is the Bessel equation of order $n + \frac{1}{2}$. The eigenvalues are determined by $J_{n+\frac{1}{2}}(\sqrt{\lambda}) = 0, n = 0, 1, 2, \dots$. □

THEOREM 2.45 - (GREEN FUNCTIONS EXPRESSED BY MEANS OF EIGENFUNCTIONS). *Let* K *be a Hermitian operator. Let* $\phi^{i(n)}, {\phi_i}^{*(n)}$, *be the eigenfunction solutions of the following eigenvalue equations:*

$$\sum_{|\gamma| \leq k} K_i^{\gamma j}(x)(\partial_\gamma.\phi^{i(n)})(x) = \lambda_n \phi^{j(n)}(x); \sum_{|\gamma| \leq k} K_i^{*\gamma j}(x)(\partial_\gamma.{\phi_j}^{*(n)})(x) = \lambda_n {\phi_i}^{*(n)}(x).$$

The eigenvalues λ_n are real and the same in both equations, and one has: $\phi_i{}^{*(n)} = \bar{\phi}^{j(n)}$. Furthermore, the functions $\phi^{j(n)}$ can be assumed orthonormal ones:

$$\int_M \phi_i{}^{*(n)}(x)\phi^{j(m)}(x) = \delta^{nm}\delta_i^j.$$

Then, we can write the Green function of K in the following way:

$$G_i^j(x, x') = \sum_n \frac{1}{\lambda_n}\phi_i{}^{*(n)}(x')\phi^{j(n)}(x). \tag{2.84}$$

PROOF. In fact, by using the completness condition, we get:

$$\sum_n \frac{1}{\lambda_n}\phi_i{}^{*(n)}(x') \sum_{|\gamma|\leq k} K_j^{\gamma s}(x)(\partial_{x,\gamma}.\phi^{j(n)})(x) = \sum_n \phi_i{}^{*(n)}(x')\phi^{s(n)}(x) = \delta_i^s\delta(x, x').$$

Therefore, we can conclude that equation (2.84) holds. □

REMARK 2.34 - (**GREEN FUNCTIONS AND BOUNDARY PROBLEMS**). Let $K : J\mathcal{D}^k(E) \to F$ be a morphism of vector fiber bundle on a manifold, M, of dimension n, such that it defines a linear differential operator of order k, $K. : C^\infty(E) \to C^\infty(F)$. Let $A \subset M$ be a domain of M and let ∂A be its boundary. Then, define **linear boundary conditions** a vector subbundle $B \subset J\mathcal{D}^k(E)|_{\partial A}$. B is locally defined by the following equations:

$$U^I[y] \equiv \sum_{1\leq s\leq p} A_j^{I(s)}y^j(s) + \ldots + A_j^{I(s)\alpha_1\ldots\alpha_k}y^j_{\alpha_1\ldots\alpha_k}(s) = 0$$

$$\Phi^{(s)J} = 0,$$

where $\Phi^{(s)J} : M \to \mathbf{R}$ are numerical functions that fully define the connected component Y_s of $\partial A = Y_1 \dot{\bigcup} \ldots \dot{\bigcup} Y_p$, $y^j_{\alpha_1\ldots\alpha_e}(s)$ are numerical functions on $J\mathcal{D}^k(E)|_{Y_s}$, and $A_j^{I(s)\alpha_1\ldots\alpha_e}$ are numerical functions on Y_s. For any linear boundary condition B we identify a subspace $\mathbf{\Phi} \subset C^\infty(E)$ defined by: $\mathbf{\Phi} \equiv \{s \in C^\infty(E) | D^k s|_{\partial A} : \partial A \to B\}$. Thus, we can identify a linear operator $L : \mathbf{\Phi} \to C^\infty(F)$ defined by the following commutative diagram:

$$\begin{array}{ccc} \mathbf{\Phi} & \xrightarrow{L} & C^\infty(F) \\ \downarrow & & \| \\ C^\infty(E) & \underset{K.}{\rightarrow} & C^\infty(F) \end{array}$$

Any section $s \in \mathbf{\Phi}$ satisfies the following local equations:

$$U^I[s] \equiv \sum_{1\leq s\leq p} A_j^{I(s)}s^j|_{Y_s} + \ldots + A_j^{I(s)\alpha_1\ldots\alpha_k}(\partial x_{\alpha_1}\ldots\partial x_{\alpha_k}.s^j)|_{Y_s} = 0\,.$$

2 THEOREM 2.46 - (GREEN FUNCTIONS AND BOUNDARY VALUES FOR ODEs).

Let us consider the differential operator L identified by the following operator of order k

$$(K.y)^j = \sum_{1\leq i\leq m,|\alpha|\leq k} p^j_{i\alpha} y^{i(\alpha)}\, , \; a \leq x \leq b$$

and the linear boundary conditions:

$$U^I[y] \equiv \sum_{1\leq j\leq m,|\alpha|\leq k} \alpha^I_{j\alpha} y^{j(\alpha)}(a) + \beta^I_{j\alpha} y^{j(\alpha)}(b)\, , \; \alpha, \beta \in \mathbf{R}.$$

1) *If the linear boundary problem $L.y = 0$ has trivial solutions only, then L has a Green function only that satisfies the following boundary values problem:*

$$\sum_{|\gamma|\leq k} K^{\gamma j}_i(x)(\partial_{x,\gamma}.G^i_u)(x,x') = 0\, ; \; U^I[G] = 0.$$

Such a function satisfies the following conditions:
(a) *$G^i_u(x,x')$ is continuous and has continuous derivatives with respect to x until the $(k-2)$-th order for all the values of x and x' in the interval $[a,b]$.*
(b) *For any $x' \in (a,b)$ the functions $G^i_u(x,x')$ have uniformly continuous derivatives of order k, with respect to x in each of the semi-intervals $[a,x')$ and $(x',b]$ and the derivatives of order $k-1$ satisfy the condition:*

$$G^i_u(x'+,x')^{(k-1)} - G^i_u(x'-,x')^{(k-1)} = 1/p^u_{ik}(x')$$

if $x = x'$. Furthermore, for any continuous function f on $[a,b]$ there is a solution of the following problem to the boundary values:

$$L.y = f, \text{or} \;\; \{(K.y)^j = f^j\, , \; U^I[y] = 0\}$$

that can be expressed by the following formula:

$$y^j(x) = \int_{[a,b]} G^j_i(x,x') f^i(x') dx'.$$

If the operator L has the Green function $G^j_i(x,x')$, the adjoint operator L^* has as Green function the following function:

$$\overline{G^j_i(x,x')}.$$

In particular, if L is selfadjoint, $L = L^*$, then $G_i^j(x,x') = \overline{G_i^j(x,x')}$, namely, the Green function is, in this case, a **hermitian kernel**.
2) If the boundary value problem $L.y = 0$ has nontrivial solutions, namely there exist m linearly independent solutions of the problem $L.y = 0$, then a generalized Green function $\widetilde{G}_i^j(x,x')$ exists with the properties (a) and (b) of an ordinary Green function, satisfies the boundary conditions as a function of x if $a < x' < b$, and it is a solution of the equation

$$(K.y)^j = - \sum_{1 \leq k \leq m} \bar{\nu}_{jk}(x')\phi^{jk}(x).$$

Here, $\{\nu_k\}_{1 \leq k \leq m}$ is a system of linearly independent solutions of the adjoint problem $L^*.y = 0$, and $\{\phi^k\}_{1 \leq k \leq m}$ is a system of ordinary continuous functions, biorthogonal to it, i.e., $< \phi^i, \nu_j >= 0, i \neq j$. Then,

$$y^j(x) = \int_{[a,b]} \widetilde{G}_i^j(x,x') f^i(x') d\mu(x')$$

is a solution of the boundary problem $L.y = f$ if the function f is continuous and satisfies the **consistency resolubility criterion**, *namely f is orthogonal to all ν_k. Finally, if $\widetilde{G}_0$ is a generalized Green function of L, then any other generalized Green function can be represented in the form:*

$$\widetilde{G}_i^j(x,x') = \widetilde{G_0}_i^j(x,x') + \sum_{1 \leq k \leq m} u_k^j(x)\psi_i^k(x'),$$

where $\{u_k\}_{1 \leq k \leq m}$ is a full system of solutions of $L.y = 0$, and $\psi^k(x')$ are arbitrary functions.

EXAMPLE 2.27 - 1) The Green function of the selfadjoint operator of the second order generated by the differential operator with real coefficients

$$K.y = \frac{d}{dx}[p(x)\frac{dy}{dx}] + q(x)y \ , \ a < x < b$$

and boundary conditions $y(a) = y(b) = 0$ has the form:

$$G(x,x') = \left\{ \begin{array}{ll} Cy_1(x)y_2(x'), & \text{if} \quad x \leq x' \\ y_1(x')y_2(x), & \text{if} \quad x > x' \end{array} \right\}$$

Here y_1 and y_2 are arbitrary independent solutions of the equation $K.y = 0$ that satisfy the following conditions: $y_1(a) = 0, y_2(b) = 0; C \equiv [p(x')W(x')]^{-1}$, where W is the Wronskian of y_1 and y_2. We can prove that C is independent of x'.

2) Let us consider the following system:

$$\frac{d^2u}{d^2x} = f \, , \; u(0) = u(1) = 0, \tag{2.85}$$

where f is an integrable function with compact support. Then, the solution of (2.85) can be written in the following way:

$$u(x) = \int_{[0,1]} [(x-t)H(x-t) - x(1-t)]f(t)dt. \tag{2.86}$$

3) (FUNDAMENTAL CAUSAL SOLUTION FOR THE INITIAL VALUE PROBLEM). Let us define such a fundamental solution that satisfies the following system:

$$L_xG(x,t) \equiv \sum_{0\le k\le n} a_k(x)\frac{d^{n-k}}{d^{n-k}x}G(x,t) = \delta(x-t)$$

$$G(0,t) = \frac{d}{dx}G(x,t)|_{x=0} = 0, \ldots, \frac{d^{n-k}}{d^{n-k}x}G(x,t) = |_{x=0} = 0.$$

The solution of such a problem is the following: $G(x,t) = v_t(x)H(x-t)$, where $v_t(x)$ is a solution of $L_xu = 0$, such that it satisfies the initial conditions: $v_t(t) = v_t'(t) = \ldots = v_t^{(n-2)}(t) = 0$, $v_t^{(n-1)}(t) = \frac{1}{a_0(t)}$. Then, the initial value problem:

$$L_xu = f, u(0) = u'(0) = \ldots. = u^{(n-1)}(0) = 0$$

admits the following solution:

$$u(x) = \int_{[0,x]} v_t(x)f(t)dt$$

with $v_t(x)$ as above.

4) (IMPULSIVE ANSWER IN ELECTRIC SYSTEMS). As an application of the above example, let us consider the case $n = 2$.

$$L_xu \equiv \sum_{0\le k\le 2} a_k(x)\frac{d^{2-k}}{d^{2-k}x} = f(t), t > 0, u(0) = u'(0) = 0. \tag{2.87}$$

Then, the Green function satisfies the following system:

$$\left\{ \begin{array}{c} L_xG(x,t) = \delta(x-t) \\ G(x,t) = 0 \, , \text{ for } x < t \\ G(0,t) = 0 \, , \text{ for } x > t \\ \frac{d}{dx}G(x,t)|_{x=0} = 0 \end{array} \right\}. \tag{2.88}$$

Then, we have for $x > t$, $G(x,t) = v_t(x)H(x-t)$, and the solution of (2.87) can be written

$$u(x) = \int_{[0,x]} G(x,t)f(t)dt. \tag{2.89}$$

☐ Note that we can also express this solution by means of a **step answer** $Z(x,t)$, namely we consider a weak solution $Z(x,t)$ that satisfies the following distributional system:

$$\left\{\begin{array}{l} L_x Z(x,t) = H(x-t) \\ \quad Z(0,t) = \frac{d}{dx}Z(x,t)|_{x=0} = 0\,,\ \text{for}\ x > t \\ \quad Z(x,t) = 0\,,\ \text{for}\ x < t \end{array}\right\} \tag{2.90}$$

In fact, as $\delta(x-t) = (\partial x.H(x-t)) = -(\partial t.H(x-t))$, it follows that by derivation of (2.90) with respect to t we get:

$$\begin{array}{l} L_x(\partial t.Z) = -\delta(x-t); \\ (\partial t.Z)(0,t) = (\partial t.\frac{d}{dx}Z)(x,t)|_{x=0} = 0\,,\ \text{for}\ x > t \\ (\partial t.Z)(x,t) = 0\,,\ \text{for}\ x < t \end{array}$$

By comparison with (2.88) we have that $G(x,t) = -(\partial t.Z)(x,t)$. Therefore, by using (2.89) we get

$$u(x) = -\int_{[0,x]} (\partial t.Z)f(t)dt = \int_{[0,x]} Z(x,t)f(t)dt - [Z(x,t)f(t)]_{t=0}^{t=t}$$

as $Z(t,t) = 0$

$$\Downarrow$$

$$u(x) = Z(x,0)f(0) + \int_{[0,x]} Z(x,t)f(t)dt.$$

Of course, this is equivalent to (2.89).

5) (HOMOGENEOUS BOUNDARY VALUE PROBLEM). Let us consider the operator:

$$L \equiv -(\frac{d^2}{d^2x} + k^2)\,,\ x \in [0,1]\,,\ u(0) = u'(1) = 0.$$

The corresponding Green function is the following:

$$G(x,t) = \frac{\sin kx \cos k(1-t)}{k\cos k}H(t-x) + \frac{\sin kt \cos k(1-x)}{k\cos k}H(x-t).$$

6) 2 (STEADY STATE DIFFUSION TEMPERATURE IN ROD OF UNITARY LENGTH). The system is described by the following equation:

$$-\frac{d}{dx}(k(x)\frac{d\theta}{dx}) = f(x)\,,\ \theta(0) = \theta(1) = 0, \tag{2.91}$$

where $k(x) =$ thermal conductivity, $\theta(x) =$ temperature, $f(x) =$ heat density generated by applying heat sources along the rod. We get:

$$G(x,t) = \frac{1}{J(w_2, w_1)}[w_1(x)w_2(t)H(t-x) + w_1(t)w_2(x)H(x-t)]$$

with

$$J(w_2, w_1) = \int_{[0,1]} \frac{1}{k(t)}dt.$$

We have:

$$\theta(x) = \int_{[0,1]} G(x,t)f(t)dt.$$

7) (STEADY STATE TEMPERATURE DIFFUSION IN CYLINDRICAL SURFACE). The system is described by the following equations:

$$\left\{\begin{array}{l} \frac{d}{dx}(x\frac{d\theta}{dx}) + x\theta = f(x) \\ x \in [a,b]\,,\ \theta(a) = \theta(b) = 0 \\ \text{with} \quad a > 0, (\text{in order to avoid the singularity at } x = 0) \end{array}\right\}.$$

In this case, the associated linear equation is the Bessel equation: $\frac{d}{dx}(x\frac{d\theta}{dx}) + (x - \frac{\nu^2}{x})u = 0$, with $\nu = 0$. So the fundamental integrals are expressed by means of cylindrical functions.

8) (TRANSVERSE FORCED VIBRATIONS OF STRESSED STRING: STEADY STATE). The system is described by the following system:

$$\ddot{u} + k^2 u = -f(x)\,,\ 0 \le x \le l\,,\ u(0) = u(l) = 0$$

where l is the length of the string. Its solution is:

$$u(x) = \int_{[0,l]} G(x,t)f(t)dt.$$

with

$$G(x,t) = \left\{\begin{array}{ll} \frac{\sin kt \sin k(l-x)}{k \sin kl}, & \text{for} \quad 0 \le t \le x; \\ \frac{\sin kx \sin k(l-t)}{k \sin kl}, & \text{for} \quad x \le t \le l \end{array}\right\}.$$

9) (NON-HOMOGENEOUS BOUNDARY VALUE PROBLEM). Let us consider the following system:

$$\left\{ \begin{array}{l} L.u \equiv -\frac{d}{dx}[p(x)\frac{du}{dx}] + q(x).u = f \ , \ x \in [0,1] \\ B_1(u) = a \ , \ B_2(u) = b \end{array} \right\}$$

where $B_1(u) = a$ involves only values at $x = 0$ and $B_2(u) = b$ involves values only at $x = 1$. Then, its solution is the following:

$$u = u_1 + u_2, \tag{2.92}$$

with

$$u_1(x) = \int_{[0,1]} G(x,t)f(t)dt; \tag{2.93}$$

and

$$\left\{ \begin{array}{l} u_2(x) = -p(1)\{u'(1)G(x,1) + u(1)(\partial t.G(x,t))|_{t=1}\} \\ \qquad + p(0)\{u'(0)G(x,0) + u(0)(\partial t.G(x,t))|_{t=0}\} \end{array} \right\} \tag{2.94}$$

where

$$G(x,t) = \frac{1}{J(w_2,w_1)}[w_1(x)w_2(t)H(t-x) + w_1(t)w_2(x)H(x-t)]$$

with:

$$w_1(x) = c_1v_1(x) + c_2v_2(x) \ , \text{ such that } B_1(w_1) = 0,$$
$$w_2(x) = d_1v_1(x) + d_2v_2(x) \ , \text{ such that } B_2(w_2) = 0,$$

where $v_1(x)$ and $v_2(x)$ are two linearly independent solutions of the homogeneous equation:

$$L_x u = 0 \ , \ J(w_2,w_1) = -p(t)W(w_1,w_2)$$

with $W(w_1,w_2)$ the Wronskian of the couple (w_1,w_2).

PROOF. First, let us find a solution of the type $u = u_1 + u_2$ such that u_1 satisfies the system with the homogeneous boundary conditions (namely $a = b = 0$), and u_2 is a solution of the problem:

$$\left\{ \begin{array}{l} L.u = 0 \ , \ x \in [0,1] \\ B_1(u) = a \ , \ B_2(u) = b \end{array} \right\}.$$

We can write $u_1(x) = \int_{[0,1]} G(x,t)f(t)dt$ as said above. Furthermore, in order to find u_2, consider the distributional extension of L. Taking into account that L is formally self-adjoint, we get:

$$(Lu, v) = (u, Lv) = \int_{[0,1]} u[-(pv')' + qv]dx = \int_{[0,1]} v[-(pu')' + qu]dx - p(uv' - vu')|_0^1.$$

Then, we have the following distributional representation of L:

$$Lu = f + p(1)[u(1)\delta'(x-1) + u'(1)\delta(x-1)] - p(0)[u(0)\delta'(x) + u'(0)\delta(x)]$$

where $u(0), u'(0), u(1), u'(1)$ are chosen in order to satisfy the non-homogeneous boundary conditions. Then, we can write:

$$Lu_2 = p(1)[u(1)\delta'(x-1) + u'(1)\delta(x-1)] - p(0)[u(0)\delta'(x) + u'(0)\delta(x)].$$

Therefore, taking into account that

$$L_x G(x,t) = -\delta(x,t), L_x(\partial t.G(x,t))|_{t=0} = -\delta'(x), L_x(\partial t.G(x,t))|_{t=1} = -\delta'(x-1)$$

we can write (2.92). □

10) (MODIFIED GREEN FUNCTIONS). Let us consider the following system:

$$(2.95) \qquad \left\{ \begin{array}{l} Lu = f \,,\; x \in [a,b] \\ B_1(u) = B_2(u) = 0 \end{array} \right\}.$$

Assume that the linear operator L, restricted to the submanifold defined by the boundary conditions, is not injective, namely the equation $Lu = 0$ also has nontrivial solutions). As a consequence, the inverse mapping L^{-1} is not univalued, hence does not exist a Green function. On the other hand, under the condition that the solutions $u_1(x)$ of the associated linear system:

$$Lu = 0 \,,\; x \in [a,b]\,;\; B_1(u) = B_2(u) = 0$$

satisfies the **consistency condition**: $\int_{[a,b]} u_1^2 dx = 1$, one can find a **modified Green function** G_M such that the solution of (2.95) can be written in the following:

$$(2.96) \qquad u(x) = \int_{[a,b]} G_M(\xi, x) f(\xi) d\xi + C u_1(x) \,, \text{ if } \int_{[a,b]} f(\xi) u_1(\xi) d\xi = 0.$$

The G_M satisfies the system:

$$(2.97) \qquad L_x G_M(x,\xi) = \delta(x-\xi) - u_1(x)u_1(\xi) \,,\; B_1(G_M) = B_2(G_M) = 0.$$

This system has a solution as $\int_{[a,b]}[\delta(x-\xi)-u_1(x)u_1(\xi)]u_1(x)dx = 0$. The construction of G_M is similar to that of the ordinary Green function, but is not uniquely determined as we can add $Cu_1(x)$ to a modified Green function. Furthermore, G_M will be symmetric, if the following condition is verified

$$\int_{[a,b]} G_M(\xi,x)u_1(x)dx = 0. \tag{2.98}$$

EXAMPLE 2.28 -

$$\left\{\begin{array}{l} -\ddot{u} = f(x)\,,\ x\in[0,l] \\ \dot{u}(0)=\dot{u}(l)=0 \end{array}\right\} \tag{2.99}$$

As the associated linear system admits a nontrivial solution $u = const$, we cannot build the ordinary Green function. But the modified Green function must satisfy the following system:

$$\left\{\begin{array}{l} -\dfrac{d^2G_M}{d^2x}(x,\xi) = \delta(x-\xi)-\dfrac{1}{l} \\ \dfrac{d}{dx}G_M|_{x=0} = \dfrac{d}{dx}G_M|_{x=l} = 0 \end{array}\right\} \tag{2.100}$$

and admits solution as $\int_{[0,l]}[\delta(x-\xi)-\frac{1}{l}]dx = 0$. In fact, one has

$$G_M(\xi,x) = \left\{\begin{array}{ll} A+Bx+\dfrac{x^2}{2l} & \text{for} \quad 0\le x<\xi \\ C+Dx+\dfrac{x^2}{2l} & \text{for} \quad \xi<x\le l \end{array}\right\}$$

The boundary conditions give $B=0, D=-1$. The continuity at $x=\xi$ implies $A+(\xi^2/2l) = C-\xi+(\xi^2/2l) \Rightarrow C = A+\xi$. The step condition on $\frac{d}{dx}G_M|_{x=\xi}$ gives $1-(\xi/l)+(\xi/l) = 1 \Rightarrow$ it is automatically satisfied. Therefore, we get:

$$G_M(x,\xi) = (A+\frac{x^2}{2l})H(\xi-x)+(A+\xi-x+\frac{x^2}{2l})H(x-\xi).$$

As there is an arbitrary constant A, we can give a more elegant expression to $G_M(x,\xi)$ by requiring that it should be symmetric (**symmetric modified Green function**). We get:

$$G_{MS}(x,\xi) = (\frac{2}{3}-\xi+\frac{x^2+\xi^2}{2l})H(\xi-x)+(\frac{l}{3}-x+\frac{x^2+\xi^2}{2l})H(x-\xi).$$

Furthermore, from the equation for $G_M(x,\xi)$ we get:

$$-\frac{d^2G_M}{d^2x}(x,\xi_1) = \delta(x-\xi_1)-\frac{1}{l}\,,\ \frac{d}{dx}G_M|_{x=0} = \frac{d}{dx}G_M|_{x=l} = 0,$$
$$-\frac{d^2G_M}{d^2x}(x,\xi_2) = \delta(x-\xi_2)-\frac{1}{l}\,,\ \frac{d}{dx}G_M|_{x=0} = \frac{d}{dx}G_M|_{x=l} = 0\,.$$

By combining these equations in the usual way, we get:

$$G_M(\xi_1,\xi_2) - G_M(\xi_2,\xi_1) - \frac{1}{l}\int_{[0,l]} G_M(x,\xi_2)dx + \frac{1}{l}\int_{[0,l]} G_M(x,\xi_1)dx = 0.$$

If we impose the condition: $\int_{[0,l]} G_M(x,\xi)dx = 0$, $\forall\xi$, then $G_M(x,\xi)$ becomes symmetric. The above condition gives

$$A = \frac{1}{l}(\frac{l^2}{3} + \frac{\xi^2}{2} - \xi l).$$

(E) - INTEGRAL EQUATIONS AND GREEN FUNCTIONS.

DEFINITION 2.31 - *Equations where the unknown function appears under the sign of integration are called* **integral equations**.

In the following table we resume some useful classifications of integral equations.

TAB.2.5 - **Types of integral equations**

Non-homogeneous Fredholm equation: $f(x) = y(x) + \int_{[a,b]} K(x,z)y(z)dz$
Homogeneous Fredholm equation: $y(x) = \int_{[a,b]} K(x,z)y(z)dz$
PARTICULAR TYPES OF FREDHOLM EQUATIONS: $K(x,z) \equiv \overline{K}(x,z)H(x-z)$ Non-homogeneous Volterra equation: $f(x) = y(x) + \int_{[a,x]} \overline{K}(x,z)y(z)dz$ Homogeneous Volterra equation: $0 = y(x) + \int_{[a,x]} \overline{K}(x,z)y(z)dz$
Transpose integral equation of: $\phi - K(\phi) = f : \psi - K^*(\psi) = g$ $K^* \equiv$ adjoint integral operator of K $(K(\phi),\psi) = (\phi, K^*(\psi)), K^*(\psi) = \int_{[a,b]} K(z,x)y(z)dz$

THEOREM 2.47 - 1) *The Fredholm equation of second type*

(2.101) $$\phi - K(\phi) = f \text{ , where } K(\phi) \equiv \int_{[a,b]} K(x,z)\phi(z)dz$$

has a unique solution given by the uniformly convergent series:

(2.102)(**Neumann series**) $$\phi(x) = f(x) + \sum_{1\le m\le\infty} \lambda^m K^m(f),$$

where

$$K^m(f) \equiv \int_{[a,b]} K_m(x,s)\phi(s)ds, K_m(x,s)$$

$$\equiv \int_{[a,b]} K_r(x,t)K_{m-r}(t,s)dt \text{ , } r \le m, m \ge 2,$$

$$K_2(x,s) = \int_{[a,b]} K(x,t)K(t,s)dt$$

where
(a) f *is square summable;*
(b) $\int_{[a,b]}\int_{[a,b]} K(x,t)^2 dx dt \equiv B^2 < \infty$; $|\lambda| B < 1$.
2) *A necessary condition that (2.101) has a solution is that the function f is orthogonal to any solution ψ of the associated transpose homogeneous equation.*
3) (**The Fredholm alternative**).
(a) *If the homogeneous equations*

$$\phi - \lambda K(\phi) = 0\,,\ \psi - \lambda K^*(\psi) = 0 \tag{2.103}$$

have only trivial solutions $\phi = \psi = 0$, then the non-homogeneous equations

$$\phi - \lambda K(\phi) = f\,,\ \psi - \lambda K^*(\psi) = g \tag{2.104}$$

have unique solutions ϕ and ψ respectively, for all the square summable functions f and g.
(b) *If the homogeneous equations have nontrivial solutions (eigenfunctions):*

$$\phi_1, \cdots, \phi_k$$

and

$$\psi_1, \cdots, \psi_k$$

then the number of these solutions, that are linearly independent, is the same for each equation. In such a case the non-homogeneous equation

$$\phi - \lambda K(\phi) = f$$

has a solution iff f is $\perp$ the eigenfunctions $\psi_i, (\psi_i, f) = 0$. Similarly , the other equation

$$\psi - \lambda K^*(\psi) = g$$

will have a solution iff $(\phi_i, f) = 0$.
4) (**Symmetric integral equations**). (a) *Each symmetric continuous kernel that is not identically zero has eigenvalues and eigenfunctions; their number is denumerable infinite iff the kernel is not degenerate. All the eigenvalues of a real symmetric kernel are also real.*
(b) *Each function f square summable is $\perp$ all the eigenfunctions ϕ_i of the symmetric kernel $K(x,t)$ iff $(K(x,t), f(t)) = 0$. The sequence of eigenfunctions $\{\phi_i\}$ is complete iff $(K(x,t), f(t)) = 0$, for any $f \neq 0$.*

5) (**Hilbert-Schmidt**). *Any function f that can be expressed in the form*

$$(K(x,t), g(t)) = f(x), \tag{2.105}$$

where g is a square summable function and K is an integral operator with symmetric kernel $K(x,t)$, has a representation absolutely and uniformly convergent:

$$f(x) = \sum_{1 \le i \le \infty} (\phi_i, f)\phi_i = \sum_{1 \le i \le \infty} \frac{(\phi_i, g)}{\lambda_i} \phi_i(x), \tag{2.106}$$

where ϕ_i are eigenfunctions of K.

THEOREM 2.48 - (EQUIVALENCE OF INTEGRAL AND DIFFERENTIAL EQUATIONS).

1) *Let us consider the following boundary value problem:*

$$\boxed{\begin{aligned} & Lu + \lambda \rho u = -f \,,\ x \in [0,1] \\ & L \equiv -\frac{d}{dx}[p(x)\frac{du}{dx}] + q(x) \\ & + \text{homogeneous boundary conditions} \\ & \rho > 0 \,,\ \lambda = const. \\ & f = \text{piecewise continuous function in } [0,1] \end{aligned}} \tag{2.107}$$

If the Green function $G(x,t)$ exists for the operator L, and the given boundary conditions, then the solution of (2.107) can be expressed in the following form:

$$u(x) = \int_{[0,1]} G(x,t)\rho(t)u(t)dt + g(x) \,,\ g(x) \equiv -\int_{[0,1]} G(x,t)f(t)dt. \tag{2.108}$$

Hence, the integral equation (2.108) is equivalent to the differential equation (2.107).

2) (a) *To the homogeneous equation*

$$Lu + \lambda \rho u = 0 + \text{homogeneous boundary conditions} \tag{2.109}$$

there corresponds the homogeneous integral equation:

$$u(x) = \lambda \int_{[0,1]} G(x,t)\rho(t)u(t)dt. \tag{2.110}$$

(b) *This last equation can be written in the more convenient form:*

$$z(x) = \lambda \int_{[0,1]} K(x,t)z(t)dt$$

after substituting

$$(2.111) \qquad z(x) \equiv u(x)\sqrt{\rho(x)}\,,\ K(x,t) \equiv G(x,t)\sqrt{\rho(x)\rho(t)}.$$

Equation (2.111) has the symmetric kernel, as L is selfadjoint.

3) (CONCLUSIONS).

(a) *If for a fixed λ, any solution of the linear equation (2.109) is identically zero, (λ is not eigenvalue of (2.109)); then the affine differential equation (2.107) has a unique solution for any choice of f).*

(b) (**Theorem of the alternative**). *If for some $\lambda = \lambda_i$, the linear equation (2.109) has nontrivial solutions, u_i, (λ_i is an eigenvalue of (2.109) corresponding to the eigenfunction u_i), the solution of the affine equation (2.107) for $\lambda = \lambda_i$ exists iff $\int_{[0,1]} \rho u_i f dx = 0$ for any sequences $\{\lambda_i\}$ with the associated eigenfunctions $\{u_i\}$, that form an infinite set of orthogonal functions that satisfy the following conditions:*

$$\int_{[0,1]} \rho u_i u_k dx = 0\,,\ i \neq k\,,\ \int_{[0,1]} \rho {u_i}^2 dx = 1.$$

If we can represent $g(x)$ with a piecewise continuous function $\phi(x)$ in the form:

$$g(x) = \int_{[0,1]} G(x,t)\phi(t)dt$$

then $g(x)$ can be developed in form of eigenfunctions:

$$g(x) = \sum_{1 \leq n \leq \infty} c_n u_n(x)\,,\ c_n \equiv \int_{[0,1]} g\rho u_n dx$$

that absolutely and uniformly converges.

EXAMPLE 2.29 -

1) (FORCED VIBRATIONS OF STRING, WITH FIXED ENDS, UNDER APPLIED PRESSURE $f(x,t), x \equiv$ space, $t \equiv$ time)

The system is described by the following:

$$Tu_{xx} = \rho u_{tt} - f(x,t)\,,\ u(0,t) = u(l,t) = 0$$

where $T \equiv$ tension of the string, $\rho \equiv$ mass density. For simplicity we put: $t = \rho = 1, F(x)e^{-i\omega t} = f(x,t)$. Then, we can write the system in the following way:

$$(2.112) \qquad u_{xx} - u_{tt} = -F(x)e^{-i\omega t}\,,\ u(0,t) = u(l,t) = 0.$$

Try to find solutions of the type $u(x,t) = X(x)e^{-i\omega t}$. From (2.112) we get:

$$(2.113) \qquad L(\omega).X \equiv \ddot{X} + \omega^2 X = -F(x)\,,\ X(0) = X(l) = 0.$$

The operator L, with these boundary conditions, is selfadjoint. If the associated linear equation

(2.114) $$L(\omega).Z = 0$$

has a nontrivial solution $Z_\omega(x)$, then from the above two equations (2.113) and (2.114) we get:

$$\begin{array}{ccc} \int_{[0,l]}(\ddot{Z}_\omega X - \ddot{X} Z_\omega)dx & = & \int_{[0,l]} Z_\omega F dx, \\ \| & & \| \\ 0 & = & (\dot{Z}_\omega X - \dot{x} Z_\omega)|_0^l. \end{array}$$

Then, in order to solve (2.112) it is necessary that

$$\int_{[0,l]} Z_\omega F dx = 0.$$

This is a condition on Z_ω that can be very restrictive if $Z_\omega \neq 0$. On the other hand (2.114) has a nontrivial solution when ω is one of the **natural frequencies** of the system, namely:

(2.115) $$L(\omega).Z \equiv \ddot{z} + \omega^2 Z = 0\,,\ Z(0) = Z(l) = 0.$$

By integrating this system we get that the natural frequencies (eigenvalues) are $\omega_n = \frac{\pi n}{l}$, with the corresponding eigenfunctions $Z_n(x) = C\sin(\frac{\pi n}{l}x)$. Hence, when ω coincides with a natural frequency of the system (**resonance situation**) the solution of Z is just $Z_n(x)$. In this case, the affine equation admits a solution iff:

(2.116) $$0 = \int_{[0,l]} Z_n F dx = \int_{[0,l]} \sin(\frac{\pi n}{l}x) F dx.$$

This means that if condition (2.116) is not verified the system has no vibrations with frequencies ω_n. Now, let us integrate (2.114) by series:

(2.117) $$u(x,t) = \sum_{1\le k\le\infty} a_k(t)\sin(\frac{\pi k}{l}x))\,,\ a_k(t) \equiv \frac{2}{l}\int_{[0,l]} u(x,t)\sin(\frac{\pi k}{l}x)dx.$$

We get

$$a_k(t) = \frac{lF_k}{k\pi}\int_{[0,t]} e^{-i\omega\tau}\sin(\frac{\pi k}{l}(t-\tau))d\tau\,,\ k = 1,2,\dots .$$

(a) If ω is not a natural frequency:

$$a_k(t) = \frac{F_k}{2\alpha_k}[e^{-i\omega t}(\frac{2\alpha_k}{{\alpha_k}^2 - \omega^2}) - \frac{e^{-i\alpha_k t}}{\alpha_k - \omega} - \frac{e^{i\alpha_k t}}{\alpha_k + \omega}]\,,\ \alpha_k \equiv \frac{\pi k}{l}.$$

(b) If $\omega = \frac{\pi n}{l}$, namely it is a natural frequency the result for $a_k(t)$ is the same when $k \neq n$, but for $k = n$ we get:

$$a_n(t) = \frac{F_n}{2i\omega}[\frac{\sin \omega t}{\omega} - te^{-i\omega t}].$$

Therefore, if ω coincides with a natural frequency, (2.117) will have a term that increases with the time (**resonance phenomenon**), except if

$$F_n = 0 \Leftrightarrow \int_{[0,l]} \sin(\frac{\pi n}{l}x) F dx = 0.$$

If this last circumstance is verified, the nth-term in the series does not contribute, and the solution does not increase in the time, (namely the phenomenom of the resonance does not appear).

Finally, note that if ω does not coincide with one of the natural frequencies (namely, we are not in resonance condition) $L(\omega)$ is an injective operator and hence admits a Green function $G(x, \xi)$. Then, the solution of (2.113) can be written in the following form:

$$X(x) = \int_{[0,l]} G(x, \xi) F(\xi) d\xi$$

where $G(x, \xi)$ is the Green function that satisfies the following system:

$$\ddot{G} + \omega^2 G = \delta(x - \xi)\,,\ G(0, \xi) = G(l, \xi) = 0.$$

Therefore, the solution of (2.112) becomes

$$u(x, t) = X(x)e^{-i\omega t}$$

that shows that the string vibrates with imposed frequency ω.

2)(BENDING OF COLUMN OF LENGTH l WITH AXIAL FORCE APPLIED TO THE ENDS). The system is described by the following:

$$-\ddot{u} - \lambda u = f(x), x \in [0, l]\,,\ u(0) = u(l) = 0. \tag{2.118}$$

For the corresponding linear system one has:

$$-\ddot{u} - \lambda u = 0\,,\ x \in [0, l]\,,\ u(0) = u(l) = 0 \tag{2.119}$$

there are nontrivial solutions iff λ is an eigenvalue of $-\frac{d^2}{dx^2}$, namely λ belongs to the **point spectrum** of $-\frac{d^2}{dx^2}$. This is given by:

$$\lambda_n = (\frac{n\pi}{l})^2, \text{with the corresponding eigenfunctions} \quad \phi_n(x) = \sqrt{\frac{2}{l}} \sin(\frac{n\pi}{l}x).$$

The sequence $\{\phi_n(x)\}$ is an orthonormal complete system and can be used to give a (**spectral**) representation of a solution of (2.118). In fact, by multiplying the differential equation (2.118) for $\phi_n(x)$ and by integrating from 0 and l, we get:

$$\sqrt{\frac{2}{l}}\int_{[0,l]} \ddot{u}\sin(\frac{\pi n}{l}x)dx - \lambda u_n = u_n \equiv < u, \phi_n > = \sqrt{\frac{2}{l}}\int_{[0,l]} u(x)\sin(\frac{\pi n}{l}x)dx,$$

$$f_n \equiv < f, \phi_n > = \sqrt{\frac{2}{l}}\int_{[0,l]} f(x)\sin(\frac{\pi n}{l}x)dx.$$

After integration by parts and by using the boundary conditions: $[(\frac{\pi n}{l})^2 - \lambda] = f_n, n = 1, 2, \ldots$ we get

$$u_n \equiv \frac{f_n}{(\frac{\pi n}{l})^2 - \lambda}$$

hence

$$u_n(x) = \sum_{1\leq n\leq\infty} \frac{f_n\phi_n}{(\frac{\pi n}{l})^2 - \lambda} = \sum_{1\leq n\leq\infty} \frac{2/l}{(\frac{\pi n}{l})^2 - \lambda}\sin(\frac{\pi n}{l}x)[\int_{[0,l]} f(x)\sin(\frac{\pi n}{l}x)dx].$$

Assume that we apply an axial force P to the ends of the column. Then, the resulting lateral bending is $u(x)$, solution of (2.118) with $\lambda = P/EI$, where $EI \equiv$ physical constant of the column, called **bending rigidity**. If $f = 0 \Rightarrow u(x) = 0$ except if P is such that $\lambda = P/EI = (\frac{\pi n}{l})^2$, then the bending results $u(x) = A\sin(\frac{\pi n}{l}x)$, where A is an arbitrary constant. The corresponding eigenfunctions $\phi_n(x) = \sqrt{\frac{2}{l}}\sin(\frac{n\pi}{l}x)$ are called **bending modes** and the relative loads $P_n = \frac{l^2\pi^2 EI}{l^2}$ are called **bending loads**. The corresponding minimum value $P_1 \equiv P_{cr}$ is called **critical bending load**. Note that $P_{cr} \to \infty$ as $l \to 0$, and $P_{cr} \to 0$ as $l \to \infty$.

22 THEOREM 2.49 - (MODIFIED GREEN FUNCTIONS IN EIGENVALUES BOUNDARY PROBLEMS).

Let us consider the following problem:

$$(2.120) \qquad L_s\phi - \lambda\phi = 0, \ a < x < b, \ B_a(\phi) = B_b(\phi) = 0.$$

This problem admits the following modified Green function:

$$(2.121) \qquad G_M(x, \xi|\lambda_1) = \lim_{\lambda\to\lambda_1} G_M(x, \xi|\lambda_1) = \sum_{2\leq n\leq\infty} \frac{\phi_n(x) - \overline{\phi_n(\xi)}}{\lambda_n - \lambda_1},$$

where λ_1 is an eigenvalue of L_s and ϕ_n are eigenfunctions. Furthermore, the sequence $\{\phi_n\}$ is complete with weight s, namely

$$\int_{[a,b]}\int_{[a,b]} s(x)s(\xi)|G_M|^2 dx d\xi < \infty.$$

(G)-ODEs OF SECOND ORDER WITH SINGULAR POINTS ON THE BOUNDARY AND GREEN FUNCTIONS).

DEFINITION 2.32 - (CLASSIFICATION OF SINGULAR POINTS). *Let us consider systems of the type:*

(2.122)

$$
\begin{array}{l}
L_s u - \lambda u \equiv Lu - \lambda u = 0\,,\ x \in [a,b] \\
L_s \equiv \frac{1}{s} L\,,\ L \equiv -\frac{d}{dx}[p(x)\frac{du}{dx}] + q(x) \\
s > 0\,,\ \lambda = const.
\end{array}
$$

☐ *Let* $\mathcal{H} \equiv \{u : [a,b] \to \mathbf{C} | \int_{[a,b]} s \mid u|^2 dx < \infty\}$. *The Schwarz inequality implies the existence of*

$$\int_{[a,b]} su\overline{v}dx$$

and

$$|\int_{[a,b]} s\, u\, \overline{v} dx| \leq [\int_{[a,b]} s|u|^2 dx]^{1/2} [\int_{[a,b]} s|v|^2 dx]^{1/2}.$$

Thus, $\mathcal{H}$ *is a Hilbert space with interior product*

$$< u, v >_s = \int_{[a,b]} s\, u\, \overline{v} dx$$

and norm

$$||u||_s \equiv < u, u >_s^{1/2} = [\int_{[a,b]} s|u|^2 dx]^{1/2}.$$

Then, $\mathcal{H}$ *is called the* **space of finite norm** s *on* (a,b).

☐ *For any differentiable function* u *and* v *we have:*

$$\overline{v} L_s(u) - u L_s(\overline{v}) = \frac{1}{s}\frac{d}{dx}[p(u\overline{v}' - u'\overline{v})], a < x < b.$$

☐ *For any* a_0, b_0 *such that* $a < a_0 < b_0 < b$ *we get*

$$
\begin{aligned}
(2.123)\qquad \int_{[a_0,b_0]} s[\overline{v} L_s(u) - u L_s(\overline{v})] dx &= \int_{[a_0,b_0]} [\overline{v} L(u) - u L(\overline{v})] dx \\
&= p(b_0) W(u, \overline{v}; b_0) - p(a_0) W(u, \overline{v}; a_0),
\end{aligned}
$$

where $W(u, \overline{v}; x) \equiv (u\overline{v}' - u'\overline{v})$ *is the Wronskian of* u *and* $\overline{v}$.

☐ *In order to extend this formula to the interval* (a,b) *we must require that the terms of the type* $\int_{[a_0,b_0]} s\overline{v} L_s(u) dx$ *have finite values. Define the following submanifold* $D \subset \mathcal{H}$:

$$
D \equiv \left\{
\begin{array}{l}
u : [a,b] \to \mathbf{C} | u \\
u' \text{absolutely continuous} \\
u, L_s(u) \quad \text{of finite norm } s \text{ on} \quad (a,b)
\end{array}
\right\} \subset \mathcal{H}.
$$

Then , if u , $v \in D \Rightarrow < L_s(u), v >_s = < u, L_s(v) >_s$ exists and we can write (2.123) as follows:

$$< L_s(u), v >_s = < u, L_s(v) >_s = \lim_{a_0 \to a, b_0 \to b} [p(b_0)W(u, \overline{v}; b_0) - p(a_0)W(u, \overline{v}; a_0)].$$

As the two limits can be taken separately, we have that both the following limits

$$\lim_{b_0 \to b} p(b_0)W(u, \overline{v}; b_0), \quad \lim_{a_0 \to a} p(a_0)W(u, \overline{v}; a_0)$$

exist. Denote such limits $p(b)W(u, \overline{v}; b), p(a)W(u, \overline{v}; a)$ respectively.

□*Let $\lambda \in \mathbf{C}$, let $\phi \in \ker(L_s - \lambda)$; then, as the coefficients in L are real, we also have:*

$$\phi \in \ker(L_s - \overline{\lambda})$$

thus

$$(\lambda - \overline{\lambda})s|\phi|^2 = \overline{\phi}L(\phi) - \phi L(\overline{\phi}) = \frac{d}{dx}[p(x)W(\phi, \overline{\phi}; x)]$$

and

$$2i(\mathrm{im}(\lambda)) \int_{[a,b]} s|\phi|^2 dx = p(b_0)W(\phi, \overline{\phi}; b_0) - p(a_0)W(\phi, \overline{\phi}; a_0)$$

if $\phi \in D$ one has:

$$2i(\mathrm{im}(\lambda)) \int_{[a,b]} s|\phi|^2 dx = p(b)W(\phi, \overline{\phi}; b) - p(a)W(\phi, \overline{\phi}; a). \tag{2.124}$$

□(**Weyl theorem**). *Let a be a regular point and b a singular point of the equation* (2.122). *We get:*

(a) *If for some value of λ any solution of* (2.122) *is of finite norm s on (a, b) then for any other value of λ , any solution is also of finite norm s on (a, b).*

(b) *For any λ with $\mathrm{im}(\lambda) \neq 0$, there exists at least a solution of finite norm s on (a, b). Then we can classify the singular points in the following way:*

(a) **limit-circle-type**: *If all the solutions u are of finite norm s , namely*

$$\int_{[a,b]} s|u|^2 dx < \infty \, , \forall \lambda.$$

(b) **limit-point-type**: *For $\mathrm{im}(\lambda) \neq 0$ there is exactly one solution (up to multiplicative constants) of finite norm s.*

In these cases we build the Green functions for intervals of regularity and take the limit versus the singular points.

EXAMPLE 2.30 - 1)

$$-\ddot{u} - \lambda u = 0 \, , \, 0 < x < \infty. \tag{2.125}$$

In this case $s(x) = p(x) = 1$. We get: $0 =$ regular point; $\infty =$ singular point. If $\lambda = 0$ the general solution $u(x) = Ax + B$. As no nontrivial solution is in $\mathcal{L}_2^{(c)}(0, \infty)$, and taking into account that: (i) if $\text{im}(\lambda) \neq 0$, the solution $e^{i\sqrt{\lambda}x}$ is the unique of finite norm; (ii) if $\text{im}(\lambda) = 0$, $\Re(\lambda) \geq 0 \Rightarrow$ no solution is of finite norm; (iii) if $\text{im}(\lambda) = 0$, $\Re(\lambda) < 0 \Rightarrow$ only one solution is of finite norm; conclude that we are in the case limit-point at $x = \infty$.

(b)

$$\text{Equation (2.125)}, \; -\infty < x < \infty.$$

The points $\pm\infty$ are both of limit-point type.

2)(BESSEL EQUATION OF ORDER 0 WITH PARAMETER λ).

(a)

$$-(xu')' - \lambda xu = 0, a < x < b, b > 0, b < \infty, s(x) = x, p(x) = x. \tag{2.126}$$

As $p(0) = 0$, the point $x = 0$ is a singular point. This result is of limit-circle type. In fact, if $\lambda = 0$, the functions $u_1(x) = 1, u_2(x) = \log x$, are independent solutions of the equation, as

$$\int_{[a,b]} xu_1^2 dx < \infty, \int_{[a,b]} xu_2^2 dx < \infty.$$

This agrees with the fact that the general solution, $\forall\lambda$, is $u(x) = AJ_0(x\sqrt{\lambda}) + BY_0(x\sqrt{\lambda})$ that is of finite norm s.

(b)

$$\text{Equation (2.126)}, \; -\infty < x < \infty, a < x < \infty, a > 0.$$

The singular point is $x = \infty$ and its result is of limit-point type. In fact, $\int_{[a,b]} xu_1^2 dx = \infty$, $\int_{[a,b]} xu_2^2 dx = \infty$. Furthermore, (i) if $\lambda \notin [0, \infty)$, the solution $H_0^{(1)}(x\sqrt{\lambda})$ is unique and is of finite norm s; if $\lambda \in [0, \infty)$, there are not solutions of finite norm s.

3)(LEGENDRE EQUATION).

$$-[(1-x)^2 u']' + \lambda u = 0, -1 < x < 1, s(x) = 1, p(x) = 1 - x^2.$$

As $p(1) = p(-1) = 0$, we have that ± 1 are singular points. These result in limit-circle type. In fact, for $\lambda = 0$ there are two independent solutions:

$$u_1(x) = 1, u_2(x) = \log(1+x) - \log(1-x).$$

On the other hand

$$\int_{[-1,l]} |u_1|^2 dx < \infty, \int_{[l,1]} |u_1|^2 dx < \infty, \int_{[-1,l]} |u_2|^2 dx < \infty, \int_{[l,1]} |u_2|^2 dx < \infty.$$

THEOREM 2.50 - (GREEN FUNCTIONS FOR ODEs WITH ONE SINGULAR POINT).

1) *Let us assume that the singular point b is singular. The Green function must satisfy the following system:*

(2.127)

$$LG - \lambda s G = \delta(x-\xi), x, \xi \in [a,b], L \equiv -\frac{d}{dx}[p(x)\frac{du}{dx}] + q(x)$$

$$B_a(G) \equiv \alpha_2 G|_{x=a} - \alpha_1 p(a)\frac{dG}{dx}|_{x=a} = 0, \alpha_1, \alpha_2 \in \mathbf{R}, \text{not both zero;}$$

$$\lim_{b_0 \to b}(B_{b_0}(G) \equiv \beta_1 G|_{x=b_0} + \beta_2 p(b_0)\frac{dG}{dx}|_{x=b_0}), \beta_1, \beta_2 \in \mathbf{R}, \text{not both zero}, b_0 < b.$$

The Green function for the regular problem on (a, b_0) is:

$$\hat{G} = A[\phi(x,\lambda)u(\xi,\lambda)H(\xi - x) + \psi(\xi,\lambda)u(x,\lambda)H(x-\xi)]$$

where ϕ and ψ are independent solutions of (2.127) that satisfy the initial conditions at the regular point a:

$$\phi(a,\lambda) = -\alpha_2 \ , \ \dot{\phi}(a,\lambda) = \frac{\alpha_1}{p(a)} \ , \ \psi(a,\lambda) = \alpha_1 \ , \ \dot{\psi}(a,\lambda) = \frac{\alpha_2}{p(a)}.$$

Any solution of (2.127) can be written in the form $u = \phi + m\psi$, up to multiplicative constants, where

$$m = -\frac{h\phi(b_0,\lambda) + p(b_0)\dot{\phi}(b_0,\lambda)}{h\psi(b_0,\lambda) + p(b_0)\dot{\psi}(b_0,\lambda)} \ , \ h \equiv \frac{\beta_1}{\beta_2}.$$

As $\hat{G}$ satisfies (2.127) we have the step conditions:

$$\frac{d\hat{G}}{dx}|_{x=\xi+} - \frac{d\hat{G}}{dx}|_{x=\xi-} = -\frac{1}{p(\xi)} = AW(\psi, u; \xi)$$

$$W(\psi, \phi + m\psi; \xi) = W(\psi, \phi; \xi) = \frac{\alpha_1^2 + \alpha_2^2}{p(\xi)} .$$

Therefore,

$$\hat{G} = -(\alpha_1^2 + \alpha_2^2)\left\{ \begin{array}{l} \psi(x,\lambda)[\phi(\xi,\lambda) + m\psi(\xi,\lambda)]H(\xi - x) \\ + \psi(\xi,\lambda)[\phi(x,\lambda) + m\psi(x,\lambda)]H(x-\xi)] \end{array} \right\}$$

where $m = m(h, b_0, \lambda)$. [The functions ϕ and ψ are integer functions of λ, instead m is a meromorphic function of λ (namely the singularities of m at finite are poles) [159].]

2) *(Case b = limit-point-type,* $\mathrm{im}(\lambda) \neq 0$*). In this case* $\lim_{b_0 \to b} m(b_0, h, \lambda) = m_b(\lambda) \equiv$ *limit point of* m*. Furthermore,* $\lim_{b_0 \to b}(u = \phi + m\psi) = u_b(x, \lambda) \equiv$ *limit point of* u *(unique solution of finite norm* s*). Therefore,* $\lim_{b_0 \to b} \hat{G}(x, \xi, \lambda) = G(x, \xi, \lambda)$ *is uniquely determined, are the unique solution of the system:*

$$\left\{ \begin{array}{l} LG - \lambda s G = \delta(x - \xi)\,,\; x, \xi \in [a, b] \\ L \equiv -\dfrac{d}{dx}[p(x)\dfrac{du}{dx}] + q(x) \\ B_a(G) = 0\,,\; \displaystyle\int_{[a,b]} s|G|^2 dx < \infty. \end{array} \right\}$$

3) *In order to build* G *for all* λ , with $\mathrm{im}(\lambda) \neq 0$, we can use the condition:

$$\frac{1}{2\pi i}\int_{\Gamma} G(x, \xi, \lambda) d\lambda = -\frac{\delta(x - \xi)}{s(x)}$$

where Γ *is a large circle in the complex plane. Therefore, this integral reduces to the calculus of the sum of residues (corresponding to the poles of* $m_b(\lambda)$ *and related to the point-spectrum) plus an integral on a part of the real axis (corresponding the cut of the branching of* $m_b(\lambda)$ *and related to the continuous spectrum). We get:*

$$\frac{1}{2\pi i}\int_{\Gamma} G(x, \xi, \lambda) d\lambda = -\sum_n \phi_n(x)\overline{\phi}_n(\xi) - \int \phi_\nu(x)\phi_\nu(\xi) d\nu. \tag{2.128}$$

If there is only the point-spectrum, we can prove that the eigenfunctions form an orthonormal complete set of weight s*.*

EXAMPLE 2.31 -

1)(BESSEL EQUATION OF ORDER ν WITH PARAMETER).

$$-(xu')' + \frac{\nu^2}{x}u - \lambda x u = 0\,,\; 0 < x < 1, \nu^2 \in \mathbf{R}_+. \tag{2.129}$$

Now we have $s(x) = x, p(x) = x$.

I) The point $x = 1$ is regular; the point $x = 0$ is singular: $\nu \geq 1$ limit-point type, $\nu < 1$ limit-circle type.

II) (CASE $\nu \geq 1$). We study the boundary value problem:

$$-(xu')' + \frac{\nu^2}{x}u - \lambda s u = 0\,,\; 0 < x < 1, \nu(1) = 0, \int_{[0,1]} x|\phi|^2 dx < \infty. \tag{2.130}$$

We have the following.

(a) The set of eigenfunctions of (2.130) forms a complete orthonormal set of weight s.

(b) The Green function is the following:

$$(2.131)\quad G = \frac{\pi}{2J_\nu(\sqrt{\lambda})}[J_\nu(x\sqrt{\lambda})Z_\nu(\xi\sqrt{\lambda})H(\xi - x) + J_\nu(\xi\sqrt{\lambda})Z_\nu(x\sqrt{\lambda})H(x-\xi)]\,.$$

(c) The normalized eigenfunctions are the following:

$$\phi_n(x) = \frac{\sqrt{2x}}{\dot{J}_\nu(\sqrt{\lambda_n})}J_\nu(x\sqrt{\lambda_n})$$

where $\sqrt{\lambda_n}$ are the roots of $J_\nu(x)$.
(d)(FOURIER-BESSEL REPRESENTATION FOR ARBITRARY FUNCTION).

$$(2.132)\qquad f(\xi) = 2\sum_{1\le n\le\infty}\frac{\sqrt{\xi}}{[\dot{J}_\nu(\sqrt{\lambda_n})]^2}\int_{[0,1]} f(x)(\sqrt{x})J_\nu(x\sqrt{\lambda_n})dx.$$

Observe that this formula holds also for $\nu > -1$. (See the case singular point of limit-circle type).
PROOF.
I) If $\lambda = 0$ we have two independent solutions: $u_1(x) = x^r, u_2(x) = x^{-\nu}$. One has:

$$\int_{[0,1]} s|u_1|^2dx < \infty, \int_{[0,1]} s|u_2|^2dx = \infty, \nu \ge 1; \int_{[0,1]} s|u_2|^2dx < \infty, \nu < 1.$$

II)(a) The system (2.132) can be transformed in the integral equation:

$$\phi(x) = \lambda\int_{[0,1]} G(x,\xi,0)s(\xi)\phi(\xi)d\xi.$$

Set

$$\mu = \frac{1}{\lambda}, \psi(x) = \phi(x)\sqrt{s}\,,\ K(x,\xi) = G(x,\xi,\lambda)\sqrt{s(x)s(\xi)}.$$

We have

$$\mu\psi(x) = \int_{[0,1]} K(x,\xi)\phi(\xi)d\xi.$$

As $\int_{[0,1]}\int_{[0,1]} |K(x,\xi)|^2d\xi dx < \infty$ one has that K is an integral operator of Hilbert-Schmidt. Therefore,...
II)(b) For im$(\lambda) \ne 0$ we have that $J_\nu(x\sqrt{\lambda}), Z_\nu(x\sqrt{\lambda})$ are independent solutions of (2.132). The first satisfies the condition of finite norm s, and the second one satisfies the boundary condition $x = 1$. Furthermore, as $\frac{dZ_\nu(x\sqrt{\lambda})}{dx}|_{x=1} = -\frac{2}{\pi x}$, namely $Z_\nu(x\sqrt{\lambda})$ satisfies an initial condition independent of λ, it follows that $Z_\nu(x\sqrt{\lambda})$ is an integer function of λ. Therefore, we can write:

$$G = A[J_\nu(x\sqrt{\lambda})Z_\nu(\xi\sqrt{\lambda})H(\xi - x) + J_\nu(\xi\sqrt{\lambda})Z_\nu(x\sqrt{\lambda})H(x-\xi)]$$

where $W(J_\nu, Z_\nu; \xi\sqrt{\lambda})A\sqrt{\lambda} = -\frac{1}{\xi}$. On the other hand, $W(J_\nu, Z_\nu; x) = -2\frac{J_\nu(\sqrt{\lambda})}{\pi x} \Rightarrow A = \frac{\pi}{2J_\nu(\sqrt{\lambda})}$.

II)(c) The non-normalized eigenfunctions are $J_\nu(x\sqrt{\lambda_n})$. In order to build the normalized eigenfunctions let us study the analytical properties of $G(x, \xi, \lambda)$. Note that $Z_\nu(x\sqrt{\lambda})$ is an integer function of λ. $J_\nu(x\sqrt{\lambda})$ is an analytical function, except for a cut on the positive real half axis. These branching points disappear from the quotient $\frac{J_\nu(x\sqrt{\lambda})}{J_\nu(\sqrt{\lambda})}$. Therefore, the only singularities of (2.132) are simple poles, the zeros λ_n of $J_\nu(\sqrt{\lambda})$, that belong to the real axis. The calculus of residues of these gives

$$R_{\lambda=\lambda_n} = -\frac{2}{[\dot{J}_\nu(\sqrt{\lambda_n})]^2} J_\nu(x\sqrt{\lambda_n})J_\nu(\xi\sqrt{\lambda_n}). \tag{2.133}$$

Therefore, we have

$$\frac{\delta(x-\xi)}{x} = \frac{\delta(x-\xi)}{\sqrt{x\xi}} = \sum_{1\le n\le\infty} \frac{2}{[\dot{J}_\nu(\sqrt{\lambda_n})]^2} J_\nu(x\sqrt{\lambda_n})J_\nu(\xi\sqrt{\lambda_n}).$$

II)(d) It is a direct consequence of the above formulas. □

2)(MELLIN TRANSFORMS).

(a) Let us consider the following system:

$$-(xu')' - \lambda\frac{u}{x} = 0\,,\ 0 < x < \infty.$$

Now we have $s(x) = 1/x, p(x) = x$.

I) The points $x = 0$ and $x = \infty$ are singular of limit-point type.

II) The Green function for $\lambda \notin [0, \infty)$ is the following:

$$G(x, \xi, \lambda) = \frac{i}{2\sqrt{\lambda}}[x^{-i\sqrt{\lambda}}\xi^{i\xi\sqrt{\lambda}}H(\xi - x) + \xi^{-i\sqrt{\lambda}}x^{i\xi\sqrt{\lambda}}H(x-\xi)]\,.$$

III) Spectral representation of $\delta(x - \xi)$:

$$x\delta(x-\xi) = \frac{1}{2\pi}\int_{[-\infty,\infty]} x^{-i\nu}\xi^{i\nu}d\nu\,,\ 0 < x, \xi < \infty\,.$$

IV) One has the following integral representation:

(2.134)(**inverse Mellin transform**)
$$f(\xi) = \frac{1}{2\pi}\int_{[-\infty,\infty]} \xi^{i\nu}F(\nu)d\nu,$$

where

(2.135)(**Mellin transform**)
$$F(\nu) = \int_{[0,\infty]} \frac{f(x)}{x}x^{-i\nu}dx\,.$$

V) 2 RELATION BETWEEN MELLIN TRANSFORM AND EXPONENTIAL FORM OF THE FOURIER TRANSFORM.

The transformation $y = \log x$ produces (2.134) and (2.135) in the exponential form:

$$f(\xi) = \frac{1}{2\pi}\int_{[-\infty,\infty]} e^{-i\nu\xi}[\int_{[-\infty,\infty]} f(x)e^{i\nu x}]d\nu.$$

In fact, the above change of variables transforms the above system in the following one: $-\ddot{y} - \lambda y = 0\,,\ -\infty < y < \infty$.

(b) (SINE MELLIN TRANSFORM). Let us consider the following system:

$$-(xu')' - \lambda\frac{u}{x} = 0\,,\ 0 < x < 1, u(1) = 0.$$

Now we have $s(x) = 1/x, p(x) = x$.

I) The point $x = 0$ is singular of limit-point type; the point $x = 1$ is regular one.

II) The Green function is the following:

(2.136)

$$G(x,\xi,\lambda) = \frac{i}{2\sqrt{\lambda}}[x^{-i\sqrt{\lambda}}(\xi^{i\sqrt{\lambda}} - \xi^{-i\sqrt{\lambda}})H(\xi - x) + \xi^{-i\sqrt{\lambda}}(x^{i\sqrt{\lambda}} - x^{-i\sqrt{\lambda}})H(x-\xi)]\,.$$

III) Spectral representation of $\delta(x-\xi)$:

(2.137) $$x\delta(x-\xi) = \frac{2}{\pi}\int_{[0,\infty]}[\sin(\nu\log x)][\sin(\nu\log\xi)]d\nu\,,\ 0 < x,\xi < \infty\,.$$

IV)(**Inverse sine Mellin transform**)

(2.138) $$f(\xi) = \frac{2}{\pi}\int_{[0,\infty]}[\sin(\nu\log x)]F(\nu)d\nu,$$

where

(2.139)(**sine Mellin transform**) $$F(\nu) = \int_{[0,1]}\frac{f(x)}{x}[\sin(\nu\log x)]dx.$$

IV) The transformation $y = \log x$ produces (2.138) and (2.139) in the form of sine Fourier transform:

(2.140) $$f(x) = (\frac{2}{\pi})^{1/2}\int_{[0,\infty]} F(\nu)\sin(\nu x)d\nu,$$

with

(2.141) $$F(\nu) = (\frac{2}{\pi})^{1/2}\int_{[0,\infty]} f(x)\sin(\nu x)dx.$$

(c) (COSINE MELLIN TRANSFORM). Let us consider the following system:

$$-(xu')' - \lambda\frac{u}{x} = 0\,,\, 0 < x < 1, \dot{u}(1) = 0.$$

By using considerations similar to previous ones, we get the cosine Mellin transform.
3)(HERMITE EQUATION).

$$-\ddot{u} + x^2u - \lambda u = 0\,,\, -\infty < x < \infty. \tag{2.142}$$

In this case $s(x) = p(x) = 1$.
I) Set $u(x) = ze^{x^2/2}$. Then, the above equation becomes:

$$-\ddot{z} - 2x\dot{z} - z - \lambda z = 0\,,\, -\infty < x < \infty. \tag{2.143}$$

The singular points $\pm\infty$ are both of limit-point type. In fact, for $\lambda = -1$, the functions

$$z_1 = 1\,,\, z_2 = \int_{[0,x]} e^{-t^2}\,dt$$

are independent solutions of (2.143). Then, the functions

$$u_1(x) = e^{x^2/2}\,,\, u_2(x) = e^{x^2/2}\int_{[0,x]} e^{-t^2}\,dt$$

are independent solutions of (2.142). Furthermore, $u_1(x)$ is of finite norm in $(-\infty, l)$ and in (l, ∞), for any l.
II) The set of eigenfunctions of (2.142) forms a complete set.
4)(BESSEL EQUATION OF ORDER 0 AND WITH PARAMETER $\lambda \in [0,\infty)$).
Let us consider the following system:

$$-(xu')' - \lambda xu = 0\,,\, 0 < x < \infty.$$

I) The points 0 and ∞ are singular points. $x = 0$ is of limit-circle type and $x = \infty$ is of limit-point type.
II) The Green function for $\lambda \notin [0,\infty)$ is the following:

$$G(x,\xi,\lambda) = \frac{\pi i}{2}[J_0(x\sqrt{\lambda})H_0^{(1)}(\xi\sqrt{\lambda})H(\xi - x) + J_0(\xi\sqrt{\lambda})H_0^{(1)}(x\sqrt{\lambda})H(x-\xi)]. \tag{2.144}$$

III) The spectral representation of δ is the following:

$$\frac{\delta(x-\xi)}{x} = \int_{[0,\infty]} \mu J_0(x\mu)J_0(\xi\mu)d\mu. \tag{2.145}$$

IV) One has the following representation:

(2.146)(**Hankel transform**) $$F(\mu) = \int_{[0,\infty]} xJ_0(x\mu)f(x)dx,$$

and

(2.147)(**inverse Hankel transform**) $$f(x) = \int_{[0,\infty]} \mu J_0(x\mu)F(\mu)d\mu.$$

5)(BESSEL EQUATION OF ORDER 0 AND WITH PARAMETER $\lambda \in (a,\infty)$). Let us consider the following system:

(2.148) $$-(xu')' - \lambda xu = 0, a < x < \infty, u(a) = 0.$$

I) The point $x = \infty$ is of the limit-point type.
II) The Green function for $\lambda \notin [0,\infty)$ is the following:

(2.149) $$G(x,\xi,\lambda) = \frac{\pi i}{2H_0^{(1)}(a\sqrt{\lambda})}\left[\begin{array}{l} Z_0(x\sqrt{\lambda})H_0^{(1)}(\xi\sqrt{\lambda})H(\xi - x) \\ + Z_0(\xi\sqrt{\lambda})H_0^{(1)}(x\sqrt{\lambda})H(x-\xi) \end{array}\right].$$

III) The spectral representation of δ is the following:

(2.150) $$\frac{\delta(x-\xi)}{x} = \int_{[a,\infty]} \mu \frac{Z_0(x\mu)Z_0(\xi\mu)}{J_0^2(a\mu) + N_0^2(a\mu)} d\mu.$$

IV) One has the following representation:

(2.151)(**Weber transform**) $$F(\mu) = \int_{[a,\infty]} xZ_0(x\mu)f(x)dx,$$

and

(2.152)(**inverse Weber transform**) $$f(x) = \int_{[0,\infty]} \frac{Z_0(x\mu)}{J_0^2(a\mu) + N_0^2(a\mu)}\mu d\mu.$$

6)(BESSEL EQUATION OF ORDER ν AND WITH PARAMETER $\lambda \in [0,\infty)$). Let us consider the following system:

(2.153) $$-(xu')' + (\frac{\nu^2 u}{x}) - \lambda xu = 0\,,\ 0 < x < \infty, \nu^2 \in \mathbf{R}_+.$$

I) The singular points $x = \infty$ (limit-point type), and $x = 0$ (limit-point type for $\nu \geq 1$, limit-circle type for $\nu < 1$).

II) The Green function is the following:

$$(2.154)\quad G(x,\xi,\lambda)=\frac{\pi i}{2}[J_\nu(x\sqrt{\lambda})H_\nu^{(1)}(\xi\sqrt{\lambda})H(\xi-x)+J_\nu(\xi\sqrt{\lambda})H_\nu^{(1)}(x\sqrt{\lambda})H(x-\xi)].$$

III) δ admits the following spectral representation:

$$(2.155)\qquad \frac{\delta(x-\xi)}{x}=\int_{[0,\infty]}\mu J_\nu(x\mu)J_\nu(\xi\mu)d\mu.$$

III) One has the following representation:

(2.156)(**Hankel transform of order** ν) $\quad F_\nu(\mu)=\int_{[0,\infty]}xJ_\nu(x\mu)f(x)dx,$

and

(2.157)(**inverse Hankel transform of order** ν) $\quad f(x)=\int_{[0,\infty]}\mu J_\nu(x\mu)F_\nu(\mu)d\mu.$

THEOREM 2.51 - (AFFINE EQUATIONS AND SINGULAR POINTS). *Let us consider the following system:*

$$(2.158)\qquad \boxed{\begin{array}{c} L_s u-\lambda u=f\ ,\ x\in[a,b]\\ L_s\equiv\frac{1}{s}L\ ,\ L\equiv-\frac{d}{dx}[p(x)\frac{du}{dx}]+q(x)\\ s>0,\lambda=const.,a=\text{regular},b=\text{singular},\\ f\ \text{is of finite norm}\ s\ \text{in}\ (a,b),\mathrm{im}(\lambda)\neq0\end{array}}$$

Assume the following boundary conditions.
(a) (regular point):

$$B_a(G)\equiv\alpha_2G|_{x=a}-\alpha_1p(a)\frac{dG}{dx}|_{x=a}=0\ ,\ \alpha_1,\alpha_2\in\mathbf{R},\text{not both zero;}$$

(b) (singular point): *Assume that the solutions should be of finite norm s.*
1) (LIMIT-POINT TYPE CASE). *There exists a unique solution of* (2.158) *given by the following:*

$$u(x)=\int_{[a,b]}G(x,\xi,\lambda)s(\xi)f(\xi)d\xi.$$

2) (LIMIT-CIRCLE TYPE CASE). *The Green function can be written in the following form:*

$$(2.159)\qquad G(x,\xi,\lambda)=\psi(x,\lambda)v(\xi,\lambda)H(\xi-x)+\psi(\xi,\lambda)v(x,\lambda)H(x-\xi)$$

where ψ is the nontrivial solution of the associated linear system that satisfies the boundary condition $B_a(\psi) = 0$ and v is a solution of finite norm s on (a, b) linearly independent of u such that

$$\lim_{b_0 \to b} p(b_0) W(v, \overline{v}; b_0) = 0. \tag{2.160}$$

For any function $v(x, \lambda)$ that satisfies (2.160) we have a different Green function. Furthermore, the unique solution of (2.158) is the following:

$$\begin{aligned} u(x) &= \int_{[a,b]} G(x, \xi, \lambda) s(\xi) f(\xi) d\xi \\ &= v(x, \lambda) \int_{[a,x]} \psi(\xi, \lambda) s(\xi) f(\xi) d\xi + \psi(x, \lambda) \int_{[x,b]} v(\xi, \lambda) s(\xi) f(\xi) d\xi. \end{aligned}$$

(F) EXAMPLES OF GREEN FUNCTIONS FOR PDEs.

2 REMARK 2.35 - (LAPLACE TRANSFORM AND LINEAR PDEs). A method that allows us to solve the integration problem of linear PDEs is that of the Laplace transform. Let us give, here, a sketch.

1)DEFINITION 2.33 - a) *Let $F(t)$ be a* **R** *or* **C**-*valued function defined on a curve l of the complex plane: $F : [a, b] \to \mathbf{K}$, then the* **Laplace transform** *of F is given by the following:*

$$f(x) = \mathcal{L}_x[F] \equiv \int_l e^{-xt} F(t) dt.$$

b) *In particular, if $l \equiv [0, \infty) \subset \mathbf{R}$, we have:*

$$f(x) = \mathcal{L}_x[F] \equiv \int_{[0,\infty)} e^{-xt} F(t) dt. \tag{2.161 (strict transform)}$$

c) *If $l \equiv (-\infty, \infty) \subset \mathbf{R}$, we have the following:*

$$f(x) = \mathcal{L}_x[F] \equiv \int_{(-\infty,\infty)} e^{-xt} F(t) dt. \tag{2.162 (two-sided transform)}$$

2)(EXAMPLES). a) $F \equiv 1$,

$$f(x) = \mathcal{L}_x[F] \equiv \int_{[0,\infty)} e^{-xt} dt = \frac{1}{x} - \frac{1}{x} \lim_{t \to \infty} e^{-xt}.$$

As

$$|e^{-xt}| = e^{-t\Re(x)}$$

$$\Rightarrow$$

$$\text{in the half plane} \left\{ \begin{array}{ll} \Re(x) > 0, & \mathcal{L}_x[1] = \dfrac{1}{x} \\ \Re(x) \leq 0, & \mathcal{L}_x[1] \quad \text{is divergent} \end{array} \right\}.$$

b)(**Gamma function**).

$$\Gamma(r) \equiv \int_{[0,\infty)} e^{-y} y^{r-1} dy \; , \; \Re(r) > 0.$$

Set $t = xt, x > 0$; we get:

$$\Gamma(r) = x^r \int_{[0,\infty)} e^{-xt} t^{r-1} dt \Rightarrow \frac{\Gamma(r)}{x^r} = \mathcal{L}_x[t^{r-1}] \; , \; \Re(r) > 0.$$

[This formula holds also for $x \in \mathbf{C}$ and $\Re(x) > 0$].

3) THEOREM 2.52 - *One has the following properties for the Laplace transform:*

a) *If the Laplace integral converges in a point $x_0 \in \mathbf{C}$, it converges also for $\Re(x) > \Re(x_0)$.*

b) *If the Laplace integral is absolutely convergent for $x = x_0$, namely*

$$\int_{[0,\infty)} e^{-xt} |F(t)| dt$$

is convergent, it is also absolutely convergent for $\Re(x) \geq \Re(x_0)$.

c) *If the Laplace integral converges to a point $x_0 \in \mathbf{C}$, it uniformly converges in any regular domain $\mathcal{D}(x_0, \eta)$.*

d) $\mathcal{L}_x[F'] = x\mathcal{L}_x[F] - F_0, F_0 \equiv \lim_{t \to 0+} F(t)$.

e) *$\mathcal{L}_x[F]$ is an analytical function of x, regular at least in the convergence half plane and such that*

$$\frac{d^n}{dx^n} \mathcal{L}_x[F] = (-1)^n \mathcal{L}_x[t^n F] \; , \; n = 1, 2, \ldots.$$

f) *Any Laplace transform is limited in any angular domain $\mathcal{D}(x_0, \eta)$, where x_0 is any point of convergence, and goes to zero if $x \to \infty$.*

g) (**Faltung theorem**). *Set*

$$R(t) \equiv \int_{[0,\infty)} F(\tau) G(t - \tau) d\tau \equiv F(t) * G(t).$$

We have:

$$\mathcal{L}_x[F * G] = \mathcal{L}_x[F].\mathcal{L}_x[G].$$

h) (**Riemann inversion formula**). *One has the following formula of inversion of the strict and bilater Laplace transforms:*

$$F(t) \equiv \frac{1}{2\pi i} \int_{[\xi - i\infty, \xi + i\infty]} e^{xt} f(x) dx.$$

4)(EXAMPLES). a)

$$\mathcal{L}_x[e^{ht}] = \mathcal{L}_{x-h}[1] = \frac{1}{x-h}\,,\ \Re(x) > \Re(h).$$

b)

$$\mathcal{L}_x[\text{const}] = \frac{x}{1+x^2}\,,\ \Re(x) > 0.$$

c)

$$\mathcal{L}_x[\sin t] = \frac{1}{1+x^2}\,,\ \Re(x) > 0.$$

d)(**Laplace transform of distribution**): Let $\phi \in C_0^\infty(M)' \equiv \{$ space of distributions on the manifold $M \equiv \mathbf{R}^n\}$, $C_0^\infty(M) \equiv \{f : M \to \mathbf{R} | \text{supp}(\phi) = \text{compact }\}$. Then, we have:

$$\mathcal{L}_x[\phi] = <\phi, e^{-sx}> .$$

For example: (i) $\mathcal{L}_x[H] = < H(x), e^{-sx} >= \int_{[0,\infty)} e^{-sx}dx = \lim_{a\to\infty} \int_{[0,a)} e^{-sx}dx = \frac{1}{s}$.
(ii) $\mathcal{L}_x[\delta] = < \delta(x), e^{-sx} >= (e^{-sx})|_{x=0} = 1$.
e)(**Heat equation**).

$$\left\{\begin{array}{l} u_{xx} - u_t = 0; u(0,t) = 0\,,\ t > 0 \\ u(x,0) = f(x)\,,\ x > 0 \end{array}\right\}$$

In order to integrate, we can apply the separation of variables method: $u(x,t) = X(x)T(t)$. The equation becomes:

$$\frac{\ddot{X}}{X} = \frac{\dot{T}}{T} = const \equiv -\mu^2 \Rightarrow \ddot{X} + \mu^2 X = 0\,,\ \dot{T} + \mu^2 T = 0$$

$$\Downarrow X = C(\mu)\sin(\mu x)\,,\ T = e^{-\mu^2 t}$$

$$\Downarrow$$

$$u(x,t|\mu) = e^{-\mu^2 t}C(\mu)\sin(\mu x).$$

The system also admits the following solution:

$$u(x,t) = \int_{[0,\infty)} u(x,t|\mu)d\mu = \int_{[0,\infty)} e^{-\mu^2 t}C(\mu)\sin(\mu x)d\mu,$$

where $C(\mu)$ is a suitable function such that $u(x,t)$ satisfies the boundary conditions. The first condition imposes that $\sin(\mu 0) = 0$, that is automatically satisfied. The second one gives $f(x) = \int_{[0,\infty)} C(\mu)\sin(\mu x)d\mu$. This means that

$$C(\mu) = \frac{2}{\pi}\int_{[0,\infty)} f(\xi)\sin(\mu\xi)d\xi$$

$$\Downarrow$$

$$u(x,t) = \frac{2}{\pi}\int_{[0,\infty)} e^{-\mu^2 t}\sin(\mu x)d\mu \int_{[0,\infty)} f(\xi)\sin(\mu\xi)d\xi\,.$$

We can also write:

$$u(x,t)=\frac{1}{\sqrt{\pi t}}\int_{[0,\infty)}\{e^{[-\frac{(x-\xi)^2}{4t}]}-e^{[-\frac{(x+\xi)^2}{4t}]}\}f(\xi)d\xi\,.$$

This means that the Green function of the problem can be written as follows:

$$G(x,t,x_0,t_0)=\frac{1}{\sqrt{\pi t(1-t_0)}}e^{-\frac{(x-x_0)^2}{4(t-t_0)}}\,,\ t>t_0.$$

f)(**String equation**).

$$\left\{\begin{array}{l} u_{xx}-\alpha^2u_{tt}=0\\ 0\le x\le l: u(x,0)=0; u_t(x,0)=0\\ t\ge 0: u(0,t)=0, u(l,t)=F(t)\end{array}\right\}.$$

Let us produce the Laplace transform of $u(x,t)$ with respect to the time coordinate:

$$\zeta(x,\xi)=\mathcal{L}_\xi[u]\equiv\int_{[0,\infty)}e^{-\xi t}u(x,t)dt.$$

Taking into account the properties on the Laplace transform, we get:

$$\zeta_{xx}-\alpha^2[\xi^2\zeta-\xi u(x,0)-u_{xt}(x,0)]=0.$$

This is an ordinary equation in $\zeta(x|\xi)$, where ξ is a parameter. Taking into account the boundary conditions, the above equation becomes:

$$\zeta_{xx}-\alpha^2\xi^2\zeta=0.$$

The general solution is the following:

$$\zeta(x|\xi)=A(\xi)\sinh(\alpha\xi x)+B(\xi)\cosh(\alpha\xi x).$$

There A and B must be determined in such a way that

$$\zeta(0|\xi)=\mathcal{L}_\xi[u(0,t)]=0\,,\ \zeta(l|\xi)=\mathcal{L}_\xi[u(l,t)]=\mathcal{L}_\xi[F].$$

We have

$$A(\xi)=\frac{\mathcal{L}_\xi[F]}{\sinh(\alpha\xi l)}\,,\ B(\xi)=0\Rightarrow\zeta(x|\xi)=\frac{\sinh(\alpha\xi x)}{\sinh(\alpha\xi l)}\mathcal{L}_\xi[F]$$

$$\Downarrow$$

$$u(x,t)=\mathcal{L}_\xi^{-1}[\zeta]=\int_{[0,\infty)}e^{\xi t}\zeta(x|\xi)d\xi.$$

THEOREM 2.53 - 1) (GREEN FUNCTIONS FOR ELLIPTIC BOUNDARY PROBLEMS). *Let*

$$(K.y)^j = \sum_{|\alpha|\leq k} a^j_{i\alpha}(x)(D^\alpha y^i)$$

be an elliptic operator and let $B_I[y] = 0$ *be a linear boundary condition. If the coefficients* $a^j_{i\alpha}$ *are smooth ones and* $L.y = 0$ *has trivial solutions only, there exists a Green function that satisfies the boundary conditions* $B_I[G] = 0$. *Furthermore, the corresponding problem* $L.y = f$ *admits a solution that can be represented in the following way:*

$$y^j(x) = \int_A G^j_i(x, x') f^i(x') d\mu(x').$$

The eigenvalue boundary problem $K.y = \lambda y$ *is equivalent to the integral equation*

$$y^j(x) = \lambda \int_A G^j_i(x, x') y^i(x') d\mu(x')$$

to which we can apply the Fredholm theory. Here, the Green function of the adjoint problem is $\overline{G^j_i(x, x')}$. *It follows, in particular, that the number of cigcnvalues is numerable at most, and there are not limit-points at the finite; the adjoint problem has complex-conjugate eigenvalues with the same multiplicity. The Green function* $G^j_i(x, x')$ *for an operator of the second order* K *with coefficients in a domain* A *with boundary* ∂A , *allows us to write the solution of the following system:*

$$K.y - f = 0, on \quad A \, , \, y|_{\partial A} = \phi$$

in the form

$$y^j(x) = \int_A G^j_i(x, x') f^i(x') d\mu(x') + \int_{\partial A} (\partial\nu_{x'}.G^j_i(x, x'))\phi^i(x') d\sigma_{x'}$$

If the associated linear system has nontrivial solutions we can determine a modified Green function exactly in the same way as made for ODEs.

EXAMPLE 2.32 - **Laplace operator**. One has:

$$\left\{ \begin{array}{l} G(x, x') = -\dfrac{\Gamma(n/2)}{2\pi^{n/2}(n-2)}|x - x'|^{2-n} + \gamma(x, x') \quad \text{if} \quad n > 2 \\ G(x, x') = \dfrac{1}{2\pi} \ln|x - x'| + \gamma(x, x') \quad \text{if} \quad n = 2 \end{array} \right\}$$

where $\gamma(x, x')$ is a harmonic function, (see section 3.3), in the domain such as to satisfy the boundary conditions.

2)(GREEN FUNCTIONS FOR PARABOLIC BOUNDARY PROBLEMS). Let L be a parabolic differential operator of order k identified by the following:

$$(K.y)^j = (\partial t.y^j) - \sum_{|\alpha|\leq k} a^j_{i\alpha}(t,x)(D_x{}^\alpha y^i)$$

plus homogeneous boundary conditions: $y^j(x,0) = 0; B_I[y] = 0$, where B_I are boundary operators with coefficients defined for $x \in \partial A$ and $t \geq 0$. The Green functions of L are functions $G^j_i(t,x,t'x')$ such that for $(t',x'), t > t'$ and $x' \in A$ satisfy $B_I[G] = 0$ beside the equations

$$(\partial t.G^j_i)(t,x,t'x')) - \sum_{|\alpha|\leq k} a^j_{s\alpha}(t,x)(D_x{}^\alpha .G^s_i)(t,x,t'x') = \delta^j_i \delta(t,x,t'x').$$

For operators with smooth coefficients and normal boundary conditions, there exists a unique solution of $K.y = 0$, and a Green function exists. Furthermore, the corresponding affine problem $K.y - f = 0$ that satisfies the same boundary conditions and initial condition $y(0,x) = \phi(x)$ has the following solution:

$$y^j(t,x) = \int_{[a,b]} dt' \int_A G^j_i(t,x,t',x') f^i(t',x') d\mu(x') + \int_A G^j_i(t,x,0,x')\phi^i(x')d\mu(x').$$

EXAMPLE 2.33 - (**Heat equation: Infinite rod without sources**). Let us consider the following problem with boundary conditions:

$$\left\{ \begin{array}{l} u_t - u_{xx} = 0 \,,\, -\infty < x < \infty, t > 0 \\ u(x,0) = f(x) \end{array} \right\}.$$

☐ The fundamental causal solution is the following:

$$G(x,t;x_0,t_0) = \frac{1}{\sqrt{4\pi(t-t_0)}} e^{-(x-x_0)^2/4(t-t_0)} \,,\; t > t_0.$$

☐ The solution is the following:

$$u(x,t) = \int_{(-\infty,\infty)} \frac{1}{\sqrt{4\pi t}} e^{-(x-x_0)^2/4t} f(x_0) dx_0.$$

If $f(x)$ is limited in $(-\infty,\infty)$ then the above integral converges and it is C^∞ for $t > t_0$ and for any x.

☐ In particular:

(a) If $f(x) = \delta(x-\xi) \Rightarrow u(x,t) = \frac{1}{\sqrt{4\pi t}} e^{-(x-\xi)^2/4t}$, is the causal fundamental solution corresponding to a unitary source at $(\xi, 0)$.

(b) If $f(x) = -\delta'(x) \Rightarrow u(x,t) = \frac{x}{\sqrt{16\pi t^3}} e^{-x^2/4t}$.

3)(GREEN FUNCTIONS FOR HYPERBOLIC BOUNDARY PROBLEMS).

Let M be a hyperbolic manifold, that is a metric that in suitable coordinate systems has the diagonal form $(1, -1, \ldots, -1)$ defined on M. Let $K. : C^\infty(E) \to C^\infty(F)$ be a linear differential operator of order k. Then, K admits two Green functions $G^{+}{}^{j}_{i}(x, x'), G^{-}{}^{j}_{i}(x, x')$ such that

$$\int_M G^{+}{}^{j}_{i}(x, x') f^i(x') d\mu(x') \quad (\text{resp.} \int_M G^{-}{}^{j}_{i}(x, x') f^i(x') d\mu(x'))$$

has support in the future (resp. past) of supp(f). Then, the **propagator**

$$\boxed{\widetilde{G}^j_i(x, x') \equiv G^{+}{}^{j}_{i}(x, x') - G^{-}{}^{j}_{i}(x, x')}$$

satisfies the following boundary problem (Cauchy problem):

$$(2.163) \qquad \boxed{\begin{aligned} &\sum_{|\gamma| \leq k} K^{\gamma j}_i(x)(\partial_{x,\gamma}.\widetilde{G}^i_u)(x, x') = 0 \\ &\widetilde{G}^j_i(x, x')|_{x^0 = x^{0\prime}} = 0 \\ &(\partial x'_0.\widetilde{G}^j_i)|_{x^0 = x^{0\prime}} = \frac{i}{2} \delta^j_i \delta(x, x')|_{x^0 = x^{0\prime}} \end{aligned}}$$

Furthermore, any distributional solution ν of the affine equation $K.u - f = 0$ can be written as follows:

$$\nu^j(x) = \int_M G^j_i(x, x') f^i(x') d\mu(x').$$

(G) - GEOMETRIC GREEN KERNELS FOR PDEs.

All the above considerations can be generalized also to non-linear PDEs, by introducing the concept of GREEN KERNELS FOR PDEs. (See also refs.[73,111]).

THEOREM 2.54 - (GREEN FUNCTIONS AND SINGULAR BOUNDARY PROBLEMS)

1) *Let $K : C^\infty(E) \to C^\infty(F)$ be a linear differential operator of order k. Let $E_k \equiv \ker_f K \subset J\mathcal{D}^k(E)$ be an affine equation, where $f \in C^\infty(F)$. Assume that $R \subset J\mathcal{D}^{k+s}(E)$ is an integral manifold of dimension n, $\pi_{k+s}|R : R \to M$ a proper application, and ${\omega_1}^{(k+s)}(R) = 0$, where ${\omega_1}^{(k+s)}(R)$ is the characteristic of Stiefel-Withney [80] of R. Then, we can associate to R a distribution $F[R]$ on E, $F[R] \in (C^\infty_0(E'))'$, $E' \equiv E^* \bigotimes \Lambda^0_n M$. Then, $F[R]$ is a distributive solution of E_k iff R is a mulivalued solution. Furthermore, the distributional solution of E_k can be written*

as follows: $\nu = G_f$, *where* G *is the* **Green kernel** *of* K , *namely* G *is the distributive kernel*

$$G \in C_0^\infty(E' \boxtimes F)' \cong C_0^\infty(E \boxtimes F')'$$

such that

$$(\widetilde{K} \otimes \mathbf{1})(G) = (\mathbf{1} \otimes \widetilde{K})(G) = \widetilde{\mathbb{D}}$$

where $\widetilde{K}$ *is the distributional extension of* K , *and*

$$\widetilde{\mathbb{D}} \in C_0^\infty(F \boxtimes F')'$$

is the **Dirac kernel** *of* F , *namely*

$$\widetilde{\mathbb{D}}(f \otimes \alpha) = \int_M < f, \alpha > , \forall f \in C_0^\infty(F), \alpha \in C_0^\infty(F').$$

2) *For any singular solution* R *of an affine PDE* $E_k \equiv \ker_f K \subset J\mathcal{D}^k(E)$, *that satisfies the boundary condition, i.e.,* $\partial R = R_1 \dot{\bigcup} R_2$, *where* R_1 *and* R_2 , *are such that:*

(i) $\pi_{k+s}|R : R \to M$ *is a proper application;*

(ii) ${\omega_1}^{(k+s)}(R) = 0$;

identifies a distributive kernel

$$G[R] \in C_0^\infty(E' \boxtimes F')'.$$

More precisely, let $F[R]$ *be the distribution associated to* R *by means of the above theorem. Then,* $G[R] \in C_0^\infty(E' \boxtimes F')'$ *is given by the following:*

$$G[R](\alpha \otimes \phi) = F[R](\alpha) \int_M < \phi, f > .$$

Define $G[R]$ *the* **generalized Green kernel** *of the singular solution* $R \subset E_{k+s}$ *that satisfies the boundary condition* $\partial R = R_1 \dot{\bigcup} R_2$.

3) *In particular, if* R *is the singular solution corresponding to the Green kernel* G *of* K , *namely* $F[R] = G_f$, *then the corresponding generalized Green kernel is defined by the following:*

$$G[R](\alpha \otimes \phi) = G_f(\alpha) \int_M < \phi, f > .$$

In this case we say that R *is the* **Green singular solution** *of the boundary problem* $\partial R = R_1 \dot{\bigcup} R_2$.

4) *In particular, if* L *is a singular solution corresponding to the Green function* $G_i^j(x,x')$ *of* K , *the corresponding generalized Green function is given by the following:*

$$G[L](\alpha\otimes\phi)=\int_M G_i^j(x,x')\alpha_j(x)f^i(x')d\mu(x)d\mu(x')\int_M \phi_i(x')f^i(x')d\mu(x').$$

Then, we call L *the* **Green solution** *of the boundary problem.*
5) *Let* $E_k=\ker_\chi K\subset J\mathcal{D}^k(W)$ *be a PDE given as kernel of a differential operator of order* k: $K: J\mathcal{D}^k(W)\to\mathbf{K}$, *with respect to a section* $C^\infty\chi: M\to K$ *of the fiber bundle* $\underline{\pi}:\mathbf{K}\to M$. *Then, for any section* $s\in C^\infty(W)$, *solution of* E_k , *and* $j[\chi]=\partial\widetilde{\chi}$, *where* $\widetilde{\chi}$ *is a deformation of* χ , *we can associate to* E_k *an affine equation* $E_k[s]=\ker_{j[\chi]}J[s]\subset J\mathcal{D}^k(s^*vTW)$, *where* $J[s]: C^\infty(s^*vTW)\to C^\infty(\chi^*vT\mathbf{K})$ *is a linear application. More precisely* $J[s]$ *is the linearized of* K *at the section* s. *Define* $E_k[s]$ **Jacobi equation** *of* E_k *at the solution* s. *Furthermore, define* **propagator** *of* E_k **at the section** s *the propagator of* $E_k[s]$.
6) *Assume that* $E_k\equiv\ker_f K\subset J\mathcal{D}^k(E)$ *is a PDE as given in the above theorem. Then, the propagator* $G[s]$ *identifies an integral manifold* (**quantum cobord**) R, *such that* $G[s]$ *satisfies to the boundary condition* $R_1\dot{\bigcup}R_2$, *then* $\partial R=R_1\dot{\bigcup}R_2$. *The set of such manifolds* R *is called* **quantum situs** *of* E_k *and denoted by* $\Omega(E_k)$.
PROOF. See refs.[73,109]. □

(H)-GREEN FORMULA AND CARTAN DISTRIBUTION OF PDEs.

In this section we will consider a generalization of the Green formula to non-linear PDEs. Let us resume the characterization of PDEs by means of Cartan distribution.

CARTAN DISTRIBUTION, CARTAN FORMS AND CONSERVATION LAWS OF PDEs.

Let $X\equiv(M,C(M))$ be an object of the category of differential equations [60,110,150], namely, M is a smooth manifold and $C(M)$ is an involutive distribution (**Cartan distribution**) of finite dimension n (**Cartan dimension** of X).[1]

If M is of finite dimension, then the equation is called **holonomic** or **maximally overdetermined**. The Cartan dimension coincides with the number of independent variables. Let (x^μ,u^i), $1\le\mu\le n, i\in I$, be a local chart adapted to X , then any vector field v of the Cartan distribution, i.e., $v\in C^\infty(C(M))$, admits the following local representation:

$$v=v^\mu\partial_\mu\,,\ \partial_\mu\equiv\partial x_\mu+\sum_{i\in I}A^i_\mu(x,u)\partial u_i$$

[1] In general M is not of finite dimension, but a Fréchet manifold. See section 5.1 for a detailed study of infinite dimensional manifolds.

where ∂_μ is a local basis (**standard basis**) for the Cartan vector fields. The involutiveness of $C(M)$ requires that

$$\partial_\mu.A^i_\nu = \partial_\nu.A^i_\mu \, , \, \mu \, , \, \nu = 1, \ldots, n, i \in I.$$

An n-dimensional smooth submanifold $V \subset M$ is **integral** iff its local representation

$$V = \{x, u | u^i = s^i(x), i \in I\}$$

satisfies the following contact equations:

$$(\partial x_\mu.s^i) = A^i_\mu(x, s) \, , \, \mu \, , \, \nu = 1, \ldots, n, i \in I.$$

An alternative way to characterizate an n-dimensional integral submanifold $V \subset X$ is to require that the following equations should be satisfied:

$$(\partial x_\mu.\phi)|_{u=s(x)} = (\partial_\mu.\phi)|_{u=s(x)}, \mu = 1, \ldots, n.$$

for any smooth function ϕ on an open set U where the chart is defined. Any vector field v on M has a unique local representation

$$v = \sum_{1 \leq \mu \leq n} v^\mu(x, u)\partial_\mu + \sum_{i \in I} \bar{v}^i(x, u)\partial u_i$$

where $v^\mu, \bar{v}^i$ are smooth functions on U. Any 1-form $\omega \in \Omega^1(M)$ has the unique local representation

$$\omega = \sum_{1 \leq \mu \leq n} \omega_\mu(x, u)dx^\mu + \sum_{i \in I} \bar{\omega}_i(x, u)\delta u^i$$

where

$$\delta u^i \equiv du^i - \sum_{1 \leq \mu \leq n} A^i_\mu(x, u)dx^\mu, i \in I.$$

and $\omega_\mu, \bar{\omega}_i$ are smooth functions on U. Furthermore, if $\phi : M \to \mathbf{R}$ is a smooth function on M , then the differential $d\phi$ of ϕ has the following local representation:

$$d\phi = \sum_{1 \leq \mu \leq n} (\partial_\mu.\phi)dx^\mu + \sum_{i \in I}(\partial u_i.\phi)\delta u^i.$$

Let $f : X \to Y$ be a morphism between two objects of the category of differential equations, hence f is a smooth application $f : M \to N$ that preserves the Cartan distributions of X and Y respectively. We say that f is a **Lie application**. Let

(x^μ, u^i), $1 \leq \mu \leq n, i \in I, (y^\mu, v^i)$, $1 \leq \mu \leq m, i \in J$, be local coordinates in M and N respectively. Let

$$\partial_\mu \equiv \partial x_\mu + \sum_{i \in I} A^i_\mu(x,u)\partial u_i \,, \; 1 \leq \mu \leq n$$
$$\bar{\partial}_\mu \equiv \partial y_\mu + \sum_{i \in J} B^i_\mu(y,v)\partial v_i \,, \; 1 \leq \mu \leq m$$

be the corresponding local expression of the standard basis of $C(M)$ and $C(N)$ respectively. Then, the local expression of f is given by the following functions:

$$y^\beta \circ f \equiv Y^\beta(x,u) \,, \; v^j \circ f \equiv V^j(x,u)$$

such that they satisfy the following equations:

$$(\partial_\mu . V^j) - \sum_{1 \leq \nu \leq m} B^j_\nu(Y,V)(\partial_\mu . Y^\nu) = 0 \,, \; 1 \leq \mu \leq n, j \in J.$$

Let $\mathcal{L}ie(C(M))$ be the set of all **Lie vector fields** on M:

$$v \in \mathcal{L}ie(C(M)) \Leftrightarrow [v,u] \in C^\infty(C(M)) \,, \; \forall u \in C^\infty(C(M)).$$

Then, $\mathcal{L}ie(C(M))$ is a subalgebra of $C^\infty(T(M))$ and $C^\infty(C(M))$ is its ideal. Let

$$\mathcal{S}ym(C(M)) \equiv \mathcal{L}ie(C(M))/C^\infty(C(M))$$

be the algebra of the **infinitesimal symmetries** of $X \equiv (M, C(M))$. The local expression of a Lie vector field is the following:

$$v = \sum_{1 \leq \mu \leq n} v^\mu(x,u)\partial_\mu + \sum_{i \in I} \bar{v}^i(x,u)\partial u_i$$

such that

$$(\partial_\mu . \bar{v}^i) - \sum_{j \in J} (\partial u_j . A^i_\mu)\bar{v}^j = 0 \,, \; \mu = 1, \ldots, n, i \in I.$$

where

$$\partial_\mu \equiv \partial x_\mu + \sum_{i \in I} A^i_\mu(x,u)\partial u_i \,, \; \mu = 1, \ldots, n$$

is a standard basis of the Cartan distribution $C(M)$. If ϕ_t is a 1-parameter group of Lie transformations of $X \equiv (M, C(M))$, then $v = \partial\phi \in \mathcal{L}ie(C(M))$. A **Cartan form** of X is a q-form $\omega \in \Omega^q(M)$ such that $\omega(a)(v_1, \ldots, v_q) = 0, \forall a \in M, v_i \in C^\infty(C(M))$. Let $C\Omega^q(M)$ be the linear space of all the Cartan forms on $X, q = 0, 1, \ldots$. Set

$C\Omega^0(M) = 0$. Note that, $C\Omega^q(M) = \Omega^q(M), q > n =$ Cartan dimension of X. If V is an integral manifold, $\omega|_V = i^*\omega = 0$, for any Cartan form ω, where $i : V \to M$ is the canonical immersion. If X is finite dimensional, this property characterizes the Cartan forms. The involutiveness of the Cartan distribution implies that the exterior differential defines a linear application $d : C\Omega^q(M) \to C\Omega^{q+1}(M), q = 0, 1, \ldots$. A **$q$-current** (or **current of dimension** q) is a q-form $\omega \in \Omega^q(M)$ such that $d\omega \in \Omega^{q+1}(M)$. Let $\mathcal{C}urr^q(C(M))$ be the linear space of the currents of dimension q of X. Exact forms and Cartan forms are current forms too; thus these are called **trivial currents**. The **space of the conservation laws (of dimension** q**) of** X is the following:

$$\mathcal{C}ons^q(C(M)) \equiv \frac{d^{-1}C\Omega^{q+1}(M)}{C\Omega^q(M) \bigoplus d\Omega^{q-1}(M)}, \; q = 0, 1, \ldots$$

where

$$d^{-1}C\Omega^{q+1}(M) \equiv \{\omega \in \Omega^q(M) : d\omega \in C\Omega^{q+1}(M)\}.$$

As $d(\omega|_V) = (d\omega)|_V$ for any submanifold V of M, we can consider the conservation law as a non trivially closed form on integral submanifolds. One has:

$$\mathcal{C}ons^0(C(M)) = \{f \in C^\infty(M, \mathbf{R}) : df \in C\Omega^1(M)\},$$

$$\mathcal{C}ons^n(C(M)) \equiv \frac{\Omega^n(M)}{C\Omega^n(M) \bigoplus d\Omega^{n-1}(M)},$$

$$\mathcal{C}ons^q(C(M)) = 0 \,,\; q = n+1, n+2, \ldots .$$

Any q-form $\omega \in \Omega^q(M)$ has the unique local representation

$$\omega = \sum_{k+r=q} \sum_{(i)_k} \sum_{(\mu)_r} \omega_{(i)_k;(\mu)_r} (\delta u^i)^k \wedge (dx^\mu)^r$$

where $(i_k) \equiv i_1 \ldots i_k \in I^k, (\mu)_r \equiv \mu_1 \ldots \mu_r \in (1, \ldots, m)^r, \omega_{(i)_k};\ (\mu)_r$ are smooth functions on $U \subset M$ skew-symmetric in the indexes $i_1, \cdots, i_k$, and $\mu_1, \ldots, \mu_r$ respectively;

$$(\delta u^i)^k \equiv \delta u^{i_1} \wedge \ldots \wedge \delta^{i_k}, (dx^\mu)^r \equiv dx^{\mu_1} \wedge \ldots \wedge dx^{\mu_r}.$$

Furthermore, any q-form $\omega \in \Omega^q(M)$ has the unique representation:

$$\omega = \sum_{\mu_1, \ldots, \mu_q} \omega_{\mu_1 \ldots \mu_q}(x, u) dx^{\mu_1} \wedge \ldots \wedge dx^{\mu_q} \bmod C\Omega^q(M).$$

One has:

$$d\omega = \partial\omega \bmod C\Omega^q(M)$$

where

$$\partial\omega = \sum_{\mu_0, \ldots, \mu_q} (\partial_{[\mu_0} .\omega_{\mu_1 \ldots \mu_q]}) dx^{\mu_0} \wedge \ldots \wedge dx^{\mu_q} \bmod C\Omega^q(M)\,.$$

The brackets [...] denote skew-symmetry of the indexes $\mu_0, \dots, \mu_q$. In particular,

$$\partial f = \sum_{1\leq\mu\leq n} (\partial_\mu . f) dx^\mu \, , \forall f \in C^\infty(M, \mathbf{R}),$$

$$\partial\omega = 0 \, , \forall\omega \in \Omega^n(M).$$

One has the following representations:

$$\mathcal{C}ons^0(C(M)) = \{f \in C^\infty(M, \mathbf{R}) : \partial f = 0\},$$

$$\mathcal{C}ons^q(C(M)) = \frac{\{\omega \in \Omega^q(M) : \partial\omega = 0\}}{\{\omega = \partial\theta, \theta \in \Omega^{q-1}(M)\}} \, , \, q = 0, 1, \dots m-1.$$

Let $\omega \in \mathcal{C}urr^q(C(M))$, $V \subset M$ an integral manifold. The restriction $\omega|_V$ is an exact q-form on the n-dimensional smooth manifold V:

$$\int_{\partial D} \omega|_V = \int_D d(\omega|_V) = \int_D (d\omega)|_V = 0$$

by means of Stokes theorem, for any compact domain $D \subset V$ with boundary ∂D step-smooth. If $f : M \to N$ is a Lie application between differential equations $X \equiv (M, C(M))$ and $Y \equiv (N, C(N))$, then the induced application $f^* : \Omega^q(N) \to \Omega^q(M), q = 0, 1, \dots$ transforms Cartan forms in Cartan forms, hence we have the following induced application:

$$f^* : \mathcal{C}ons^q(C(M)) \to \mathcal{C}ons^q(C(M)), q = 0, 1, \dots .$$

The Lie derivative with respect to the Lie vector fields $v \in \mathcal{L}ie(C(M))$, and infinitesimal symmetries $v \in \mathcal{S}ym(C(M))$, induces the following applications:

$$\mathcal{L}_v : C\Omega^q(M) \to C\Omega^q(M), q = 0, 1, \dots$$

$$\mathcal{L}_v : \mathcal{C}ons^q(C(M)) \to \mathcal{C}ons^q(C(M)) \, , \, q = 0, 1, \dots .$$

REMARK 2.36 - (SPECTRAL SEQUENCES FOR PDEs). A **q-form is p-Cartan** on

$$X \equiv (M, C(M))$$

if it is a q-form $\omega \in \Omega^q(M)$ such that $\omega(v_1, \dots, v_q) = 0$ when at least $q - p + 1$ of the fields $v_1, \dots, v_q \in C^\infty(TM)$ are Cartan ones. The linear space of the q-forms of p-Cartan on M is denoted with $C^p\Omega^q(M)$. One has:

$$C^0\Omega^q(M) = \Omega^q(M)$$

$$C^1\Omega^q(M) = C\Omega^q(M)$$

$$C^p\Omega^q(M) = 0, \quad \text{if} \quad p > q$$

$$C^{q-n}\Omega^q(M) = C^{p+1}\Omega^q(M), \quad \text{if} \quad q > n = \text{ Cartan dimension of } X$$

$$C^p\Omega^q(M) \supset C^{p+1}\Omega^q(M)$$

$$d : C^p\Omega^q(M) \to C^p\Omega^{q+1}(M) .$$

The de Rham complex:

$$0 \to \mathbf{R} \to \Omega^0(M) \to \Omega^1(M) \to \dots$$

of the differential equation $X \equiv (M, C(M))$ has the following filtration:

$$C^0\Omega^q(M) \supset C^1\Omega^q(M) \supset C^2\Omega^q(M) \supset \dots . \supset C^q\Omega^q(M) \supset 0$$

compatible with the exterior differentiation. As a consequence we have associated a spectral sequence:

$$\{E_r^{p,q}(M), d_r^{p,q}\}.$$

The operations $d\omega, v\rfloor\omega, \mathcal{L}_v\omega$ have the following properties on the spaces $C^p\Omega^q(M)$:

$$d : C^p\Omega^q(M) \to C^p\Omega^{q+1}(M)$$
$$\left\{\begin{array}{l} v\rfloor : C^p\Omega^q(M) \to C^{p-1}\Omega^{q-1}(M)\,,\ v \in C^\infty(TM) \\ v\rfloor : C^p\Omega^q(M) \to C^p\Omega^{q-1}(M), v \in C^\infty(C(M)) \end{array}\right\}$$
$$\left\{\begin{array}{l} \mathcal{L}_v : C^p\Omega^q(M) \to C^{p-1}\Omega^q(M) v \in C^\infty(TM) \\ \mathcal{L}_v : C^p\Omega^q(M) \to C^p\Omega^q(M)\,,\ v \in \mathcal{L}ie(C(M)) \end{array}\right\}$$
$$\left\{\begin{array}{l} [v]\rfloor : E_r^{p,q}(M) \to E_r^{p-1,q}(M) \\ [v] \in \mathcal{S}ym(C(M))\,,\ p, q = 0, 1, \cdots, r = 0, 1, \cdots \end{array}\right\}$$
$$\left\{\begin{array}{l} \mathcal{L}_{[v]} : E_1^{p,q}(M) \to E_1^{p,q}(M) \\ [v] \in \mathcal{S}ym(C(M)),\ p, q = 0, 1, \dots \end{array}\right\}.$$

One has the following formula:

$$\mathcal{L}_{[v]}[\omega] = d_1([v]\rfloor[\omega]) + [v]\rfloor(d_1[\omega]).$$

The linear application:

$$d_1 : E_1^{0,q}(M) \to E_1^{1,q}(M)$$

is called **characteristic** when $q = n-1$, and **variational** or **Euler-Lagrange** when $q = n$.

REMARK 2.37 - (REDUCTIONS OF PDEs). $Y \equiv (N, C(N))$ is a **sub-equation** of $X \equiv (M, C(M))$ if N is a submanifold of M and $C(N) = TN \cap C(M)|_N$, namely the application $Y \to X$ is a Lie immersion. Furthermore, if the Cartan dimension of Y is equal to the Cartan dimension of X, then Y is called a **reduction** of X. In this case any submanifold of Y is an integral manifold of X, and vice versa, any integral submanifold of X, that is in M, is an integral submanifold of Y. Let Y be a sub-differential equation of X. Then, for any point $a \in N$ there exists a standard chart $V \subset M$ with coordinates $(x^{\mu}, u^{i}, y^{\nu}, v^{j}), 1 \leq \mu \leq m =$ Cartan dimension of N , $1 \leq \nu \leq n, m+n =$ Cartan dimension of $M, i \in I, j \in J$, and basis Cartan fields

$$\partial_{\mu} \equiv \partial x_{\mu} + \sum_{i \in I} A^{i}_{\mu}(x,u,y,v)\partial u_{i} + \sum_{j \in J} B^{j}_{\mu}(x,u,y,v)\partial v_{j}, \mu = 1,\ldots,m,$$

$$\bar{\partial}_{\nu} \equiv \partial u_{\nu} + \sum_{i \in I} C^{i}_{\mu}(x,u,y,v)\partial u_{i} + \sum_{j \in J} D^{j}_{\mu}(x,u,y,v)\partial v_{j}, \nu = 1,\ldots,n,$$

such that

$$uN \cap V = \{x,u,y,v : y^{\nu} = 0, v^{j} = 0, \nu = 1,\ldots,n, j \in J\}$$

on $U \equiv N \cap V, \{x^{\mu}|_{U}, u^{i}|_{U}\}_{1 \leq \mu \leq m}$ are coordinates on U and

$$\partial_{\mu}|_{U} \equiv \partial x_{\mu} + \sum_{i \in I} a^{i}_{\mu}(x,u)\partial u_{i}\,,\ 1 \leq \mu \leq m\,,\ \text{with}\, a^{i}_{\mu}(x,u) \equiv A^{i}_{\mu}(x,u,0,0)$$

is a basis of Cartan fields for Y.

Let $\mathcal{I}_N$ be an ideal that defines N , i.e., the set of all the functions $f \in C^{\infty}(M,\mathbf{R})$ such that $f|_{N} = 0$. Then, a vector field $v \in C^{\infty}(TN)$ iff

$$v : \mathcal{I}_N \to \mathcal{I}_N$$

namely

$$(v.f)|_{N} = 0 \quad \text{if} \quad f|_{N} = 0.$$

Furthermore, $Y \equiv (N, C(N))$ is a reduction of $X \equiv (M, C(M))$ iff $v : \mathcal{I}_N \to \mathcal{I}_N$ for any Cartan vector field $v \in C^{\infty}(C(M))$.

Now, we are ready to state the following.

THEOREM 2.55 - (GREEN FORMULA FOR PDEs).

1) *Let $Y \equiv (N, C(N))$ be a reduction of $X \equiv (M, C(M))$. One has the following application:*

$$<,>: \mathcal{S}ym(C(N)) \times E^{p,q}_{1}(C(N)) \to E^{p-1,q}_{1}(C(N)), ([v],[\omega]) \mapsto < [v],[\omega] >= [v]\rfloor[\omega].$$

A similar expression holds for X.

2) (GREEN FORMULA). *One has the following duality:*

$$< \nu[v],[\omega] >=< [v], \nu^{*}[\omega] >$$

for all $[v] \in \mathcal{S}ym(C(M)|_N)$ and $[\omega] \in E_1^{p,q}(N), p \geq 1$, *where* ν *is the canonical application:*

$$\nu : \mathcal{S}ym(C(M)|_N) \to \mathcal{S}ym(C(N)|)$$
$$\nu^* : E_1^{p,q}(C(N)) \to E_1^{p,q}(C(M)|_N).$$

3) (GREEN FORMULA IN COORDINATES). *Let us suppose that* $\{F = 0\}$ *is realized as reduction of a* **trivial differential equation**, *i.e., any* m*-dimensional submanifold of* M *is integral one, where* m *is the Cartan dimension of* M. *Then, for any smooth function* $\phi \in C^\infty(N, \mathbf{R})^a$ *and* $\psi \in C^\infty(N, \mathbf{R})_s$ *there exists a smooth function* $h \in C^\infty(N, \mathbf{R})^s$ *such that one has the following equations:*

$$\sum_{1 \leq \sigma \leq s} \bar{F}_*(\phi)^\sigma \psi_\sigma - \sum_{1 \leq \alpha \leq a} \phi^\alpha \overline{F^*(\psi)_\alpha} = \sum_{1 \leq \mu \leq m} (\partial_\mu . h^\mu)$$

with

$$\bar{F}_*(\phi)^\sigma = \sum_{1 \leq \alpha \leq a} \sum_k (\partial w_\alpha^k . F^\sigma)|_N (\partial_k . \phi^\alpha)\,,\ \phi \in C^\infty(N, \mathbf{R})^a, \sigma = 1, \ldots, s;$$
$$\bar{F}^*(\psi)_\alpha = \sum_{1 \leq \sigma \leq s} \sum_k (-1)^{|k|} (\partial_k . (\psi_\sigma (\partial w_\alpha^k . F^\sigma)|_N), \psi \in C^\infty(N, \mathbf{R})_s, \alpha = 1, \ldots, a.$$

More precisely we take:

$$h^\mu = \sum_{\alpha, \sigma, i, k} (-1)^{|k|} M(i, k, \mu) \partial_i (\phi^\alpha)(\partial_k . (\psi_\sigma (\partial w_\alpha^{i_1 + k + (\mu)} . F^\sigma)|_N)\,,$$

where $\mu = 1, \ldots, m$, *and* $M(i, k, \mu)$ *is given in (2.76).*

EXAMPLE 2.34 - 1)(LAPLACE EQUATION): $(\triangle w = 0)$. We get:

$$a = s = 1, x = (x^\mu) \in \mathbf{R}^m, w = (w_k) \in \mathbf{R}_K, K \equiv \mathbf{N}^m,$$
$$F = \sum_{1 \leq \mu \leq m} w_{2(\mu)}, \bar{F}_*(\phi) = \sum_\mu \partial_\mu^2 \phi, \overline{F^*(\psi)} = \sum_\mu \partial_\mu^2 \phi, \phi, \psi \in C^\infty(N, \mathbf{R})$$
$$h^\mu = (\partial_\mu . \phi)\psi - \phi(\partial_\mu . \psi)\,,\ \mu = 1, \ldots, m.$$

Therefore, we get

$$\sum_\mu (\partial_\mu^2 . \phi)\psi - \phi(\sum_\mu \partial_\mu^2 . \psi) = \sum_\mu \partial_\mu ((\partial_\mu . \phi)\psi - \phi(\partial_\mu . \psi)).$$

If we take $\phi = u(x), \psi = v(x) \in C^\infty(\mathbf{R}^m, \mathbf{R})$ we recover the classical Green formula:

$$(\triangle u)v - u(\triangle v) = \nabla((\nabla u)v - u(\nabla v)).$$

2)(SYMMETRIES, CONSERVATION LAWS AND GREEN FORMULA). If ϕ is a symmetry, i.e., $\bar{F}_*(\phi) = 0$ and ψ is the characteristic of a conservation law, i.e., $\overline{F^*(\psi)} = 0$, then $h = h(\phi, \psi)$ is a conserved current (it can eventually be trivial) of the system $\{F = 0\}$, i.e., $\sum_\mu (\partial_\mu . h^\mu) = 0$. The Green formula relates symmetries with conservation laws of the considered differential equation.

3)(LINEAR CASE). We get:

$$F^\sigma(x, w) = \sum_{\alpha,k} a_\alpha^{\sigma,k}(x) w_k^\alpha, \sigma = 1, \ldots, s$$

$$\bar{F}_*(\phi)^\sigma = \sum_{\alpha,\ k} a_\alpha^{\sigma,k}(x)(\partial_k . \phi^\alpha), \sigma = 1, \ldots, s;$$

$$\bar{F}^*(\psi)_\alpha = \sum_{\sigma,\ k} (-1)^{|k|}(\partial_k . (\psi_\sigma a_\alpha^{\sigma,k}(x)), \alpha = 1, \ldots, a.$$

$$h^\mu = \sum_{\alpha,\sigma,i,k} (-1)^{|k|} M(i, k, \mu) \partial_i(\phi^\alpha)(\partial_k . (\psi_\sigma a_\alpha^{\sigma, i_1 + k + (\mu)}(x)), \mu = 1, \ldots, m.$$

where $\phi \in C^\infty(F)^a$, $\psi \in C^\infty(F)_s$. If, now, assume that ϕ and ψ depend only on x, we get the usual Green formula for the linear system:

$$(\psi, F(x, \partial x), \phi)_s - F^*(x, \partial x)\psi, \phi)_a = \nabla h$$

where $(.,.)_s$ and $(.,.)_a$ are interior products in the Euclidean spaces $\mathbf{R}^s$ and $\mathbf{R}^a$ respectively. Taking into account that any solution ϕ of the differential system $F.w = 0$ identifies a symmetry of this system: $w \mapsto w + t\phi$, we recognize also in the classic case a relation between symmetries and Green formula.

2.5 - SPECTRAL SEQUENCES IN PDEs

Let $\pi : W \to M$ be a fiber bundle, where M is an n-dimensional manifold and W is an $(n + m)$-dimensional manifold, both of class C^∞. Set

$$\Omega^p(J\mathcal{D}^\infty(W)) \equiv \lim \{\pi^*{}_{\infty,k} C^\infty(\Lambda^p T^* J\mathcal{D}^k(W))\}.$$

Similarly we can define $\Omega^p(E_\infty)$ for a PDE $E_k \subset J\mathcal{D}^k(W)$. Then the canonical connection on E_∞, identifies a bigradation on $\Omega^r(E_\infty)$

$$\Omega^r(E_\infty) \equiv \bigoplus_{p+q} \Omega^{p,q}(E_k)$$

where $\Omega^{p,q} E_\infty \equiv C^\infty(\Lambda^p v T^* E_\infty \bigotimes \Lambda^q \mathbf{H}_{E_k}^\perp)$. Since $\mathbf{H}_{E_k}$ is flat, d has only $(1,0)$ and $(0,1)$ components:

$$\left\{ \begin{array}{l} (1,0)\text{-component} : \delta : \Omega^{p,q}(E_\infty) \to \Omega^{p+1,q}(E_\infty) \\ (0,1)\text{-component} : (-1)^p \partial : \Omega^{p,q}(E_\infty) \to \Omega^{p,q+1}(E_\infty) \end{array} \right\}$$

$$(d^2 = 0) \to \delta^2 = \partial^2 = \delta\partial - \partial\delta = 0$$

We call $\{\Omega^{\bullet,\bullet}E_\infty;\delta,\partial\}$ the **variational bicomplex** of E_k.
The flatness of $u\in C^\infty(E_\infty)$ can be expressed using the variation bicomplex of E_k by one of the following equivalent conditions:
1)
$$u^*\Omega^{+,\bullet}(E_\infty)=0,\quad \Omega^{+,\bullet}(E_\infty)\equiv\bigoplus_{p\geq 0}\Omega^{p,\bullet}(E_\infty);$$

2)
$$u^*\Omega^{1,q}(E_\infty)=0,\text{for some}\quad q\in\mathbf{Z}(0,n-1)\equiv\text{set of integers in}\,[0,n-1];$$

3)
$$u^*\Omega^{1,0}(E_\infty)=0;$$

4)
$$du^*f=u^*\partial f\,,\,\forall f\in C^\infty(E_\infty;\mathbf{R}).$$

DEFINITION 2.34 - *The filtration* $F^p\Omega^\bullet(E_\infty)\equiv\bigoplus_{p'\geq p}\Omega^{p',\bullet}(E_\infty)$, *is compatible with* d *and induces a spectral sequence* $\{E_r^{\bullet,\bullet},d_r\}$ *that we call* ***$\mathcal{C}$*-spectral sequence** *of* $E_k\subset J\mathcal{D}^k(W)$.
REMARK 2.38 - In the following table we resume important meanings of some terms of $\mathcal{C}$-spectral sequence. In ref.[102] it is given a general method to find conservation laws for solutions of PDEs, $E_k\subset J\mathcal{D}^k(W)$, associated to symmetry (pseudo)groups of E_k. This method has been further developed in [77,78,104]. Note that such conservation laws could not be included in the space $E_1^{0,n-1}$. In fact, in $E_1^{0,n-1}$ there are the conservation laws that are shared by all the solutions of E_k.
REMARK 2.39 - Set

$$\mathcal{C}\Omega^\bullet(E_\infty)\equiv\{\omega\in\Omega^\bullet(E_\infty)|D^\infty s^*\omega=0\,,\,\forall s\in C^\infty(W),D^\infty s(M)\subset E_\infty\}.$$

One can see that,
$$\mathcal{C}\Omega^1(E_\infty)\wedge\Omega^{i-1}(E_\infty)=\mathcal{C}\Omega^i(E_\infty).$$

So $\mathcal{C}\Omega^\bullet(E_\infty)$ is an ideal in $\Omega^\bullet(E_\infty)$. Then the $\mathcal{C}$-spectral sequence $\{E_r^{\bullet,\bullet},d_r\}$ of E_k can be obtained also by considering $\Omega^\bullet(E_\infty)$ filtered by the ideal $\mathcal{C}\Omega^\bullet(E_\infty)$ and its powers

$$\mathcal{C}^p\Omega^\bullet(E_\infty),p\geq 0:\Omega^\bullet(E_\infty)\supset\mathcal{C}^0\Omega^\bullet(E_\infty)\supset\mathcal{C}\Omega^\bullet(E_\infty)\supset\cdots\supset\cdots$$

As usually in $E_r^{p,q}$, p denotes the filtration index and $p+q$ stands for the degree. Furthermore, the complex $\{\Omega^\bullet(E_\infty),d\}$ induces the complex $\{\bar{\Omega}^\bullet(E_\infty),\bar{d}\}$, where $\bar{\Omega}^\bullet(E_\infty)\equiv\Omega^\bullet(E_\infty)/\mathcal{C}\Omega^\bullet(E_\infty)$. We shall denote the correspondng cohomologies by $H^\bullet(E_\infty)$ and $\bar{H}^\bullet(E_\infty)$ respectively.

TAB.2.6 - **Meaning of some spectral terms of $\mathcal{C}$-spectral sequence**

$E_r^{\bullet,\bullet}$	Observations	Meaning
$E_1^{0,0}$	ODE	$E_1^{0,0}$ = sp.motion constants
$E_1^{0,n-1}$	$n = \dim M$	$E_1^{0,n-1}$ = sp.conservation laws
$E_1^{0,n}$	$n = \dim M$	$E_1^{0,n}$ = sp.Lagrangian densities
		$\{d_1^{0,0} < \omega >= 0\}$
		Euler-Lagrange eq. of $< \omega > \in E_1^{0,n}$

THEOREM 2.56 - 1) $\{E_r^{\bullet,\bullet}, d_r\}$ *converges to the de Rham cohomology algebra*

$$H^\bullet(E_\infty).$$

2) *For formally integrable PDE* $E_k \subset J\mathcal{D}^k(W)$ *one has*

$$H^\bullet(E_k) \cong H^\bullet(E_\infty)$$

3)

$$\begin{aligned} &E_0^{0,q} \cong \bar{\Omega}^q(E_\infty), d_0 = \bar{d}; \\ &E_0^{p,q} = 0 \quad \text{if} \quad q < 0, \quad \text{or} \quad q > n; \\ &E_1^{0,q} \cong \bar{H}^q(E_\infty). \end{aligned}$$

4) *If* $E_\infty \subset J\mathcal{D}^\infty(W)$ *is open then one has:*

$$\begin{aligned} &E_1^{p,q} = 0, p > 0, q \neq 0; \\ &E_1^{0,q} \cong E_\infty^{0,q}, q < n; \\ &E_2^{p,n} = E_\infty^{p,\infty}, p \geq 0; \\ &E_r^{p,q} = 0, 2 \leq r \leq \infty, \quad \text{if} \quad p > 0, q \neq n, \quad \text{or}\, p = 0, q > n; \\ H^q(E_\infty), = &\left\{ \begin{array}{ll} \bar{H}^q(E_\infty), & \text{if} \quad q < n \\ E_2^{q-n,n}, & \text{if} \quad q \geq n \end{array} \right\}. \end{aligned}$$

5) *If* $E_\infty = J\mathcal{D}^\infty(W) \Rightarrow H^q(J\mathcal{D}^\infty(W)) = H^q(W)$.

☐ (LERAY-SERRE SPECTRAL SEQUENCES OF PDEs). Let $E_k \subset J\mathcal{D}^k(W)$ be a PDE on a fiber bundle $\pi : W \to M$. The following theorems give a relation between the (co-)homology of Grassmannian bundles of integral planes $I(E_k)$ of E_k (resp. Grassmannian of oriented integral planes $I^+(E_k)$ of E_k), and spectral sequences. In particular we prove that cohomology Leray-Serre spectral sequence associated to the fibration $I(E_k) \to E_k$, (resp. $I^+(E_k) \to E_k$) converges to the cohomology of $I(E_k)$ (resp. $I^+(E_k)$). Furthermore, we find conditions under which the Leray-Serre spectral sequence of a PDE converges to the same space of $\mathcal{C}$-spectral sequence. (See also ref.[109].)

THEOREM 2.57 - *Let $E_k \subset J\mathcal{D}^k(W)$ be a PDE on a fiber bundle $\pi : W \to M$. Let G be an abelian group. There is a first quadrant spectral sequence $\{E^r_{\bullet,\bullet}, d^r\}$ with*

$$E^2_{p,q} \cong H_p(E_k; H_q(F_k; G))$$

the homology of E_k with local coefficients in the homology of F_k, the fiber of $I(E_k) \to E_k$, and converging to $H_\bullet(I(E_k); G)$. Furthermore, the spectral sequence is natural with respect to fiber preserving maps of fibrations.

2) *Let G be an abelian group, there is a first quadrant spectral sequence $\{{}^+E^r_{\bullet,\bullet}, d^r\}$ with ${}^+E^2_{p,q} \cong H_p(E_k; H_q(F^+_k; G))$, the homology of E_k with local coefficients in the homology of F^+_k, the fiber of $I^+(E_k) \to E_k$ and converging to $H_\bullet(I^+(E_k); G)$. Furthermore, this spectral sequence is natural with respect to fiber- preserving maps of fibrations.*

PROOF. Points 1) and 2) are direct consequences of homology Leray-Serre spectral sequences applied to the fibrations $I(E_k) \to E_k$, and $I^+(E_k) \to E_k$ respectively. □

THEOREM 2.58 - 1) *Let R be a commutative ring with unit. There is a first quadrant spectral sequence of algebras $\{E^{\bullet,\bullet}_r, d_r\}$ with $E^{p,q}_2 \cong H^p(E_k; H^q(F_k; R))$ the cohomology of E_k with local coefficients in the cohomology of F_k, the fiber of $I(E_k) \to E_k$ and converging to $H^\bullet(I(E_k); R)$ as an algebra. Furthermore, this spectral sequence is natural with respect to fiber-preserving map of fibrations.*

2) *Let R be a commutative ring with unit. There is a first quadrant spectral sequence of algebras $\{{}^+E^{\bullet,\bullet}_r, d_r\}$ with ${}^+E^{p,q}_2 \cong H^p(E_k; H^q(F^+_k; R))$ the cohomology of E_k with local coefficients in the cohomology of F^+_k, and converging to $H^\bullet(I^+(E_k); R)$, as an algebra. Furthermore, this spectral sequence is natural with respect to fiber-preserving map of fibrations.*

PROOF. Points 1) and 2) are direct consequences of cohomology Leray-Serre spectral sequences applied to the fibrations $I(E_k) \to E_k$ and $I^+(E_k) \to E_k$ respectively. □

DEFINITION 2.35 - *We shall call the above spectral sequence associated to*

$$E_k \subset J\mathcal{D}^k(W)$$

Leray-Serre spectral *sequences of E_k. We call also* (**oriented**) **Leray-Serre** (**co**-) **homology** *of E_k the algebras to which Leray-Serre spectral sequences of E_k converge. We shall denote the spectral terms by the following symbols:*

(a) *(non-oriented case)*

$$E^{\bullet,\bullet}_r(E_k), E^r_{\bullet,\bullet}(E_k);$$

(b) *(oriented case)*

$$ {}^{+}E_r^{\bullet,\bullet}(E_k), {}^{+}E^r_{\bullet,\bullet}(E_k). $$

THEOREM 2.59 - 1) *Assume that $H^q(F_k;R)$ (resp. $H^q(F_k^+;R)$) is simple. Then for $R = \mathbf{K}$ = field, we have:*

$$ E_2^{p,q}(E_k) \cong H^p(E_k;\mathbf{K}) \bigotimes H^q(F_k;\mathbf{K}); $$
$$ (resp. {}^{+}E_2^{p,q}(E_k) \cong H^p(E_k;\mathbf{K}) \bigotimes H^q(F_k^+;\mathbf{K})). $$

2) *If $H^q(F_k;R)$ (resp. $H^q(F_k^+;G)$) is simple and E_k and F_k (resp. F_k^+) are connected, then*

$$ E_2^{p,0}(E_k \cong H^p(E_k;R); E_2^{0,q}(E_k \cong H^q(F_k;R); $$
$$ (resp. {}^{+}E_2^{p,0}(E_k) \cong H^p(E_k;R); {}^{+}E_2^{0,q}(E_k) \cong H^q(F_k^+;R)). $$

3) *If $I(E_k)$ is path-connected $H^q(F_k;R)$; (resp. $H^q(F_k^+;R)$) is simple, $R \equiv \mathbf{K} \equiv$field and F_k (resp. F_k^+) is totally non-homologous to 0 in $I(E_k)$ (resp. $I^+(E_k)$) with respect to $\mathbf{K}$, then one has the following isomorphism of vector spaces:*

$$ H^\bullet(I(E_k);\mathbf{K}) \cong H^\bullet(E_k;\mathbf{K}) \bigotimes H^\bullet(F_k;\mathbf{K}); $$
$$ (resp. H^\bullet(I^+(E_k);\mathbf{K}) \cong H^\bullet(E_k;\mathbf{K}) \bigotimes H^\bullet(F_k^+;\mathbf{K})). $$

In these cases $E_r^{\bullet,\bullet}(E_k)$ (resp. ${}^{+}E_r^{\bullet,\bullet}(E_k)$) collapses.

4) *Under same hypotheses of point 3) and furthermore if $H^q(F_k;\mathbf{K}) = 0$, if $q > 0$, or $H^q(F_k;\mathbf{K}) = \mathbf{K}$, if $q = 0$, (resp. $H^q({F^+}_k;\mathbf{K}) = 0$, if $q > 0$, or $H^q({F^+}_k;\mathbf{K}) = \mathbf{K}$, if $q = 0$) then*

$$ H^\bullet(I(E_k);\mathbf{K} \cong H^\bullet(E_k;\mathbf{K}); $$
$$ (resp. H^\bullet(I^+(E_k);\mathbf{K}) \cong H^\bullet(E_k;\mathbf{K})). $$

THEOREM 2.60 - *If $E_k \subset J\mathcal{D}^k(W)$ is a formally integrable PDE on the fiber bundle $\pi : W \to M, \dim M = n, \dim W = n + m$, such that:*

(i) *$I(E_k)$ is path-connected;*

(ii) *$H^q(F_k;R)$ is simple with $R = \mathbf{R}$;*

(iii) *F_k is totally non-homologous to 0 in $I(E_k)$ with respect to $\mathbf{R}$;*

(iv) *$H^q(F_k;\mathbf{R}) = 0$, if $q > 0$, or $H^q(F_k;\mathbf{R}) = \mathbf{R}$, if $q = 0$. Then the cohomology Leray-Serre spectral sequence $E_r^{\bullet,\bullet}(E_k)$ of E_k and the $\mathcal{C}$-spectral sequence of E_k converge to same space.*

PROOF. In fact the first converges to $H^\bullet(E_k;\mathbf{R}) \cong H^\bullet(E_\infty)$ and the second also to $H^\bullet(E_\infty)$. □

REMARK 2.40 - A similar theorem also holds for oriented case ${}^{+}E_r^{\bullet,\bullet}(E_k)$.

2 Here let us make some digressions on the "system of local coefficients" that interest the meaning of "simple space".

DEFINITION 2.36 - 1) *A* **coordinate bundle** $\mathcal{B}$ *is a collection*

$$(p : B \to X, Y, G \subset Aut(Y), \{V_j\}_{j\in J}, \phi_j : V_j \times Y \to p^{-1}(V_j)), \ \forall j \in J$$

where p *is a surjective continuous mapping,* $\{V_j\}_{j\in J}$ *is an open set covering of* X *, and* ϕ_j *are homeomorphisms.*

☐ The mapping $\phi_{j,x} : Y \to p^{-1}(x)$ define $g_{ji} : V_i \cap V_j \to G, g_{ji}(x) \equiv \phi_{j,x}^{-1}\phi_{i,x} : Y \to Y$, such that

$$\left\{\begin{array}{c} g_{kj}(x)g_{ji}(x) = g_{ki}(x), x \in V_i \cap V : j \cap V_k \\ g_{ii}(x) = e \ , \ \forall x \in V_i \end{array}\right\}$$

☐ Two coordinate bundles $\mathcal{B}$ and $\mathcal{B}'$ are said to be **equivalent in the strict sense** if the union of the two sets of coordinate functions is a set of coordinate functions of a bundle with some X, B, Y and G.

☐ This is an equivalent relation. A **fiber bundle** is an equivalence class of coordinate bundles.

☐ A **bundle map** $h : \mathcal{B} \to \mathcal{B}'$ is defined by $h : B \to B'$ such that

(2.164)

$$\left\{\begin{array}{ll} h : Y_x \to Y_{x'} & \text{(homomorphism)} \\ \bar{h} : X \to X' & \text{(continuous map)} \end{array}\right\} \quad \begin{array}{ccccc} & B & \xrightarrow{h} & B' & \\ p & \downarrow & & \downarrow & p' \\ & X & \to & X' & \end{array} \quad \text{(commutative diagram)}$$

The map

$$\bar{g}_{kj} : V_j \cap \bar{h}^{-1}(V'_k) \to G \quad \text{mapping transformation}$$

defined by

$$\bar{g}_{kj}(x) = \phi'^{-1}_{k,x'}h_x\phi_{j,x} = \phi'_k h_x \phi_{j,x} : Y \to Y$$

is continuous. One has the following properties:

$$\left\{\begin{array}{c} \bar{g}_{kj}(x)g_{ki}(x) = \bar{g}_{ki}(x), x \in V_i \cap V_j \cap \bar{h}^{-1}(V'_k) \\ g_{lk}(\bar{h}(x))\bar{g}_{kj}(x) = \bar{g}_{lj}(x), x \in V_j \cap \bar{h}^{-1}(V'_k \cap V'_l) \end{array}\right\} \qquad (2.165)$$

☐ Let $X \subset X'$. Then, we denote by $\mathcal{B}_X \equiv \mathcal{B}'|_X$, where $\mathcal{B}'|_X \equiv (B \equiv p'^{-1}(X), p \equiv p'|_B, X, Y, \phi_j \equiv \phi'_j(V'_j \cap X) \times Y, G)$. Then, the inclusion $\mathcal{B} \subset \mathcal{B}'$ is a bundle map.

LEMMA 2.1 - *A bundle map* $h : \mathcal{B} \to \mathcal{B}'$ *is uniquely determined by condition* (2.165).

LEMMA 2.2 - *If* $\bar{h} : X \to X'$ *is* $1-1$ *and has a continuous inverse* $\bar{h}^{-1} : X' \to X$ *, then* h *has a continuous inverse* $h^{-1} : B' \to B$ *and* h^{-1} *is a map* $\mathcal{B}' \to \mathcal{B}$.

DEFINITION 2.37 - *Two coordinate bundles* $\mathcal{B}$ *and* $\mathcal{B}'$ *with* $X = X', Y = Y', G \subset G'$ *are equivalent iff there exist continuous maps* $\bar{g}_{kj} : V_j \cap V'_k \to G, j \in J, k \in J'$, *such that* (2.165) *are satisfied.*

LEMMA 2.3 - *Let* $\mathcal{B}$ *and* $\mathcal{B}'$ *be two coordinate bundles with* $X = X', Y = Y', G = G'$ *and* $\{V_j\} = \{V'_j\}$. *Then,* $\mathcal{B}$ *is equivalent to* $\mathcal{B}'$ *if there exist continuous functions* $\lambda_j : V_j \to G, \forall j \in J$, *such that* $g'_{ji}(x) = \lambda_j(x)^{-1} g_{ji}(x) \lambda_i(x), \forall x \in V_i \cap V_j$.

DEFINITION 2.38 - *A coordinate bundle is called a* **product bundle** *if there is just one coordinate neighbourhood* $V \equiv X$ *and the group* G *consists of the identity element* e *alone:* $G = \{e\}$.

THEOREM 2.61 - *If the group of a bundle consists of the identity element alone, then the bundle is equivalent to a product bundle.*

THEOREM 2.62 - (ENLARGING THE GROUP OF A BUNDLE). *If* H *is a closed subgroup of the topological group* G , *any bundle* $\mathcal{B}$ *with group* H *can also be considered as a bundle with group* G *(*G*-***image of** $\mathcal{B}$*).*

DEFINITION 2.39 - 1) *Let* H *and* K *be closed subgroups of a topological group* G. *Let* $\mathcal{B}$ *and* $\mathcal{B}'$ *be bundles with groups* H *and* K *respectively. We say that* $\mathcal{B}, \mathcal{B}$ *are* **equivalent in** G *(or* G*-equivalent) if the* G*-images of* $\mathcal{B}$ *and* $\mathcal{B}'$ *are equivalent.*

2) *In particular, if* $K = \{e\}$ *we say that the* H*-bundle* $\mathcal{B}$ *is* G-**equivalent to the product bundle**.

THEOREM 2.63 - (G-EQUIVALENCE TO PRODUCT BUNDLE). *Let* $\mathcal{B}$ *be a bundle with group* H *and coordinate transformation* $\{g_{ji}\}$. *Let* H *be a subgroup of* G. *Then,* $\mathcal{B}$ *is* G*-equivalent to the product bundle iff there exist maps* $\lambda_j : V_j \to G$ *such that* $g_{ji}(x) = \lambda_j(x)\lambda_i(x)^{-1}, \forall x \in V_i \cap V_j$. *(This is equivalent to saying that its group* $= \{e\}$*). (If* $H = G$ *we say that* $\mathcal{B}$ *is* **simple**.*)*

EXAMPLE 2.35 - Let $\mathcal{B}$ denote the twisted torus. The group H of $\mathcal{B}$ is a cyclic group of order 2. This bundle is not H-equivalent to the product bundle. But $\mathcal{B}$ is G-equivalent to the product bundle if G = full group of rotations of the circle Y. Thus, the twisted torus is not a product bundle, but it is equivalent to the product bundle in the full group of rotations.

REMARK 2.41 - In general, is not very significant to consider equivalence of bundles in a larger group unless this group is also a group of homomorphisms of the common fiber.

THEOREM 2.64 - (TRANSLATING A FIBER ALONG A CURVE ON THE BASE). 1) *Let* X *be an arcwise connected space, and let* $\mathcal{B}$ *be a bundle over* X *with fiber* Y *and group* G. *Let* $c : I \equiv [0,1] \to X$ *be a curve in* X *from* $x_0 \equiv c(0)$ *to* $x_1 \equiv c(1)$. *Let* $Y_t \equiv p^{-1}(c(t))$. *As one has the homeomorphism* $\phi_{j,x} : Y \to Y_x$ *we can choose an admissible map* $h_0 : Y \to Y_0$. *Let us regard* Y *as a bundle over a point* $\{pt\}$. *So* h_0

can be interpreted as a bundle map

$$\begin{array}{ccc} Y \times \{pt\} & \overset{h_0}{\rightarrow} & Y_0 \times \{x_0\} \\ \downarrow & & \downarrow \\ \{pt\} & \underset{\bar{h}_0}{\rightarrow} & \{x_0\} \end{array}$$

Thus, c can be considered a homotopy of $\bar{h}_0$, $c : I \times \{pt\} \rightarrow X$. The covering homotopy theorem provides a bundle map $h : I \times Y \rightarrow \mathcal{B}$ such that $h(0,y) = h_0(y)$ and $ph(t,y) = c(t)$. More precisely, one has

$$h(t,y) = \phi_{j,c(t)}(y) \in Y_{c(t)} \subset B.$$

Then, we have $h_t : Y \rightarrow Y_t$ which allows to define

$$h_t h_0^{-1} : Y_0 \rightarrow Y_t \tag{2.166}$$

that is a 1-parameter family of maps beginning with the identity and ending with a map $Y_0 \rightarrow Y_1$. We call (2.166) a translation of Y_0 along c into Y_1.
2) If the topology of the group G is totally disconnected, then the translation of fibers along curves is a unique operation. Then, we also have the following results:
(a) The homotopic curves c_1, c_2 from x_0 to x_0 define the same mapping $c^* : Y_1 \rightarrow Y_0$, and $(c_1c_2)^* = c_2^*c_1^*$.
(b) One has a mapping $\chi : \pi_1(X, x_0) \rightarrow Aut(Y_0)$ that is a group homomorphism.
(c) χ is determined up to its equivalence class under inner automorphisms of G (different choices of $h_0 : Y \rightarrow Y_0$). We call the **characteristic class of the bundle** *$\mathcal{B}$ the equivalence class of χ.*
(d) Let X be arcwise connected and arcwise locally connected. Let $\mathcal{B}$ and $\mathcal{B}'$ be two bundles over X having the same group G which is totally disconnected. Then, $\mathcal{B}$ and $\mathcal{B}'$ are the associated bundles, i.e., associated to equivalent principal bundles, if $\chi(\mathcal{B}) = \chi(\mathcal{B}')$.
(e) For a criterion of existence of bundles with prescribed χ see ref.[136].
(f) If X is arcwise connected, arcwise locally connected, and simply connected, and if G is totally disconnected, then any bundle over X with group G is equivalent to the product bundle.
DEFINITION 2.40 - 1) *A* **bundle of coefficients** *means a bundle of groups where the fiber is an abelian group, written additively, and the group of the bundle is totally disconnected.*

$$(B, X, Y, G)$$

where Y is the fiber, and G is the group.

2) *The bundle of coefficients is called* **simple** *if it is a product bundle.*

□ Such a bundle is determined up to an equivalence by its characteristic homomorphism $\chi : \pi_1(X) \to G$.

□ A bundle of coefficients is simple iff $\chi(\pi_1(X)) = \{e\}$, i.e., each closed path operator as the identity.

LEMMA 2.4 - *Let $E \to B$ be a fiber bundle over B with fiber E. Let $H^0(E, R)$ be a free R-module. Let us assume that a basis $\{\epsilon_i\}$ is given by $\{\epsilon_i = i^* e_i\}$, where $e_i \in H^0(E, R)$ and $i : F \to E$ is the natural inclusion. Then, the system of local coefficients induced by F on B is simple.*

PROOF OF LEMMA 2.4 - Let us consider the characteristic map $\chi : \pi_1(B, b_0) \to Aut(F), \chi(\gamma) : F \to F$. Let $\chi(\gamma)^* : H^0(F, R) \to H^0(F, R)$ be the induced mapping. We have $\chi(\gamma)^*(\epsilon_i) = \epsilon_i$. But, $\{\epsilon_i\}$ is a basis for $H^0(F, R)$; hence $\chi(\gamma)^* = 1 \Rightarrow \chi(\gamma) = id_F$. Then, recalling the above results we conclude that the system of local coefficients is simple. □

LEMMA 2.5 - *Let $E \to B$ be a fiber bundle over B with fiber F. Let $H^0(F, R)$ be* **simple**, *i.e., there exist elements $\{v_i\}, v_i \in H^0(F, R)$, that are* **simple system of generators** *for the R-algebra $H^0(F, R)$, i.e., the monomials $v_1^{h_1} v_2^{h_2} \cdots v_m^{h_m}, (h_i = 0, or\ \ 1, m \geq 0)$ form a basis for A. Let us assume that each $v_i = i^* e_i$, where $e_i \in H^0(E, R)$, with $i : F \to E$ the natural inclusion. Then, the system of local coefficients induced by F over B is simple.*

PROOF OF LEMMA 2.5 - In fact, under our hypotheses $H^0(F, R)$ results R-free; hence we can use the above Lemma. □

THEOREM 2.65 - *Take $E_k = J\mathcal{D}^k(W)$ be path-connected. Let $R = \mathbf{Z}_2$. Then one has the following isomorphisms:*

(a)

$$E_2^{p,q}(J\mathcal{D}^k(W)) \cong H^p(W; \mathbf{Z}_2) \bigotimes_{\mathbf{Z}_2} H^q(F_k; \mathbf{Z}_2)$$

(b)

$$E_2^{p,0}(J\mathcal{D}^k(W)) \cong H^p(W; \mathbf{Z}_2)$$

(c)

$$E_2^{0,q}(J\mathcal{D}^k(W)) \cong H^q(F_k; \mathbf{Z}_2)$$

(d) *Furthermore, if $\pi : W \to M$ is k-**non-homologous** to 0, that is F_k is totally non-homologous to 0 in $I_k(W) \equiv I_k(J\mathcal{D}^k(W))$, then one has the following isomorphism of vector spaces:*

$$H^\bullet(I(W); \mathbf{Z}_2) \cong H^\bullet(W; \mathbf{Z}_2) \bigotimes H^\bullet(F_k; \mathbf{Z}_2)\,.$$

In this case the Leray-Serre spectral sequence $E_r^{\bullet,\bullet}(J\mathcal{D}^k(W))$ collapses.

e) *Furthermore if* $H^q(F_k;\mathbf{Z}_2)=0$, *if* $q>0$, *or* $H^q(F_k;\mathbf{Z}_2)=\mathbf{Z}_2$, *if* $q=0$ *one has the isomorphism:*

$$H^\bullet(I(W);\mathbf{Z}_2)\cong H^\bullet(W;\mathbf{Z}_2).$$

PROOF. (a) Let us note, first, that $H^q(F_k;\mathbf{Z}_2)$ is simple. In fact we know [70] that the cohomology ring $H^\bullet(F_k;\mathbf{Z}_2)$ is isomorphic in dimension up to n to the ring $\mathbf{Z}_2[\omega_1^{(k)},\cdots,\omega_n^{(k)}]$ of polynomials in the generator $\omega_i^{(k)}$. So, as $\mathbf{Z}_2$ is a field, we get the following isomorphism:

$$E_2^{p,q}(J\mathcal{D}^k(W))\cong H^p(J\mathcal{D}^k(W);\mathbf{Z}_2)\bigotimes H^q(F_k;\mathbf{Z}_2).$$

Furthermore, as $J\mathcal{D}^q(W)\to J\mathcal{D}^{q-1}(W), q\geq 1$, are affine bundles, we also have the following canonical isomorphisms:

$$H^q(J\mathcal{D}^k(W);\mathbf{Z}_2)\cong H^q(J\mathcal{D}^{k-1}(W);\mathbf{Z}_2)\cong\cdots\cong H^q(W;\mathbf{Z}_2). \tag{2.167}$$

Therefore, we get the isomorphisms (a).
(b),(c),(d) and (e) can be similarly proved considering that $H^q(F_k;\mathbf{Z}_2)$ is simple and $J\mathcal{D}^k(W)$, as well as F_k are connected, and taking into account isomorphisms and the above theorems on spectral sequences. (For more details see ref.[111].) □

THEOREM 2.66 - *Take* $E_k=J\mathcal{D}^k(W)$ *be path-connected. Take* $R=\mathbf{Q}$. *Then one has the following isomorphisms:*

(a)

$$^+E_2^{p,q}(J\mathcal{D}^k(W))\cong H^p(W;\mathbf{Q})\bigotimes H^q(F_k^+;\mathbf{Q})$$

(b)

$$^+E_2^{p,0}(J\mathcal{D}^k(W))\cong H^p(W;\mathbf{Q})$$

(c)

$$^+E_2^{0,q}(J\mathcal{D}^k(W))\cong H^q(F_k^+;\mathbf{Q}).$$

(d) *Furthermore, if* $\pi: W\to M$ *is* k-**non-homologous** *to* 0, *that is* F_k^+ *is totally non-homologous to* 0 *in* $I^+{}_k(W)\equiv I^+{}_k(J\mathcal{D}^k(W))$, *then one has the following isomorphism of vector spaces:*

$$H^\bullet(I^+(W);\mathbf{Q})\cong H^\bullet(W;\mathbf{Q})\bigotimes H^\bullet(F_k^+;\mathbf{Q}).$$

In this case the Leray-Serre spectral sequence $^+E_r^{\bullet,\bullet}(J\mathcal{D}^k(W))$ *collapses.*
e) *Furthermore if* $H^q(F_k^+;\mathbf{Q})=0$, *if* $q>0$, *or* $H^q(F_k^+;\mathbf{Q})=\mathbf{Q}$, *if* $q=0$ *one has the isomorphism:*

$$H^\bullet(I^+(W);\mathbf{Q})\cong H^\bullet(W;\mathbf{Q}).$$

PROOF. (a) In this case $H^q(F_k^+;\mathbf{Q})$ is simple. In fact we know [70] that for the cohomology ring $H^\bullet(F_k^+;\mathbf{Q})$ we have the following isomorphisms:
(i) In all cases (except $n=2, m=1, k=2$), up to n:

$$H^\bullet(F_k^+;\mathbf{Q})\cong \mathbf{Q}[p_1{}^{(k)},\cdots,p_{1[l/2]}{}^{(k)}],\ \text{for } n=2l+1$$

or

$$H^\bullet(F_k^+;\mathbf{Q})\cong \mathbf{Q}[p_1{}^{(k)},\cdots,p_{1[l/2]}{}^{(k)},e^{(k)}],\ \text{for } n=2l$$

where $p_i{}^{(k)}$ coincide with the rational Pontryagin classes of the tautological bundle, while $e^{(k)}$ coincides with the Euler class.
(ii) If $n=1, m=1, k=2$, one has $H^1(F_k^+;\mathbf{Q})=0$, $H^2(F_k^+;\mathbf{Q})\cong \mathbf{Q}e^{(k)}\bigoplus \mathbf{Q}\alpha$.
Now we can apply the above theorems on spectral sequences to get the following isomorphism:

$${}^+E_2^{p,q}(J\mathcal{D}^k(W))\cong H^p(J\mathcal{D}^k(W);\mathbf{Q})\bigotimes H^q(F_k^+;\mathbf{Q})\,.$$

Then, taking into account the isomorphism induced by the affine structure $J\mathcal{D}^q(W)\to J\mathcal{D}^{q-1}(W)$, $q\geq 1$, we have the isomorphism (a).
(b), (c), (d) and (e) are similarly proved. □

THEOREM 2.67 - 1) *Take* $E_1 = J\mathcal{D}^1(W)$ *be a path-connected, first order PDE on the fiber bundle* $\pi: W\to M, n=\dim M, \dim W=n+m$. *Take* $R=\mathbf{Z}_2$ *(or* $R=\mathbf{Q}$*). Assume that* E_1 *is a formally integrable PDE such that* $\dim_{\mathbf{C}} Char_1(E_1)_q>0, q\in E_1$, *and* $Char_1(E_1)_q$ *does not lie in any hyperplane, where* $Char_i(E_1)_q$ *is the manifold of complex characteristics of at the point* q. *Let* $k_0\geq 0$ *be such that*

$$g_\eta^{(k_0)}(q)\equiv g^{(k)}(q)\bigcap S^k(\eta)\bigotimes vT_{\pi_{1,0}(q)}W$$

is 2-acyclic with respect to the Spencer δ*-cohomology for all the characteristic subspaces* η. *Here* $g(q)\subset T^*{}_{\pi_1(q)}M\bigotimes vT_{\pi_{1,0}(q)}W$ *is the symbol of* E_1 *at* q,

$$g^{(k)}(q)\subset S^{k+1}\subset T^*{}_{\pi_1(q)}M\bigotimes vT_{\pi_{1,0}(q)}W$$

its k*-th prolongation, and* η *is a characteristic subspace for all the system* E_1 *at* q, *that is a subspace of* $T^*{}_{\pi_1(q)}M$ *such that*

$$g_\eta(q)\equiv g(q)\bigcap\eta\bigotimes vT_{\pi_{1,0}(q)}W\neq 0.$$

The vector spaces $g_\eta^{(k)}(q)$ *defines families* $\xi_i^{(k)}$ *of vector spaces over* $Char_1(E_1)_q$. *Furthermore* $\xi_i^{(k_0)}$ *and* $\xi_i^{(k_0+1)}$ *are vector bundles over* $Char_1(E_1)_q$ *for all* $i, 1\leq i\leq n-1$, *then there is a number* $k_1\geq k_0$ *such that for all* $l\geq k_1$ *the embeddings:*

$$I(E_{l+1})_u\to I(J\mathcal{D}^{l+1}(W))_u, I^+(E_{l+1})_u\to I^+(J\mathcal{D}^{l+1}(W))_u, \pi_{l+1,1}(u)=q$$

induce an isomorphism of the corresponding cohomology algebras in dimension up to n. (See ref.[111].) Then we can conclude that under the above hypotheses $F_{l+1}(E_1)$ and ${F^+}_{l+1}(E_1)$, the fibers of $I(E_{l+1}) \to E_{l+1}$, and $I^+(E_{l+1}) \to E_{l+1}$ respectively, are simple, so we have the following isomorphisms:

(a)

$$E_2^{p,q}(E_{l+1}) \cong H^p(E_1; \mathbf{Z}_2) \bigotimes_{\mathbf{Z}_2} H^q(F_{l+1}(E_1); \mathbf{Z}_2)$$

(b)

$$E_2^{p,0}(E_{l+1}) \cong H^p(E_1; \mathbf{Z}_2)$$

(c)

$$E_2^{0,q}(E_{l+1}) \cong H^q(F_{l+1}(E_1); \mathbf{Z}_2)$$

(a′)

$${}^+E_2^{p,q}(E_{l+1}) \cong H^p(E_1; \mathbf{Q}) \bigotimes_{\mathbf{Q}} H^q(F^+_{l+1}(E_1); \mathbf{Q})$$

(b′)

$${}^+E_2^{p,0}(E_{l+1}) \cong H^p(E_1; \mathbf{Q})$$

(c′)

$${}^+E_2^{0,q}(E_{l+1}) \cong H^q(F^+_{l+1}(E_1); \mathbf{Q}) .$$

2) *Furthermore, if $F_{l+1}(E_1)$, (resp. $F^+_{l+1}(E_1)$) is totally non-homologous to 0 in $I(E_{l+1})$ (resp. $I^+(E_{l+1})$) then one has the following isomorphism of vector spaces:*

$$H^\bullet(I(E_{l+1}); \mathbf{Z}_2) \cong H^\bullet(E_1; \mathbf{Z}_2) \bigotimes H^\bullet(F_{l+1}(E_1); \mathbf{Z}_2);$$
$$(resp.\ H^\bullet(I^+(E_{l+1}); \mathbf{Q}) \cong H^\bullet(E_1; \mathbf{Q}) \bigotimes H^\bullet(F^+_{l+1}(E_1); \mathbf{Q})).$$

In these cases the Leray-Serre spectral sequences $E_r^{\bullet,\bullet}(E_{l+1})$, (resp.${}^+E_r^{\bullet,\bullet}(E_{l+1})$) collapse.

3) *Furthermore,*

$$\text{if} \quad H^q(F_{l+1}(E_1); \mathbf{Z}_2) = 0, \text{if } q > 0, \text{or } H^q(F_{l+1}(E_1); \mathbf{Z}_2) = \mathbf{Z}_2\ , \text{ if } q = 0$$
$$(\text{resp. } H^q(F^+_{l+1}(E_1); \mathbf{Q}) = 0)\ , \text{ if } q > 0, \text{or} \quad H^q(F^+_{l+1}(E_1); \mathbf{Q}) = \mathbf{Q}, \text{if } q = 0)$$

then one has the isomorphisms:

$$H^\bullet(I(E_{l+1}); \mathbf{Z}_2) \cong H^\bullet(E_1; \mathbf{Z}_2); (resp.\ H^\bullet(I^+(E_{l+1}); \mathbf{Q}) \cong H^\bullet(E_1; \mathbf{Q})).$$

4) *Assume that $W \equiv M \times \underline{W}$, where M is an affine n-dimensional manifold and $\underline{W}$ is an m-dimensional $\mathbf{R}$-vector space. We have the isomorphism $J\mathcal{D}^k(W) \cong M \times \underline{W}_k$,*

where $\underline{W}_k$ is an s-dimensional **R***-vector space. Then the corresponding cohomology Leray-Serre spectral sequence converges to $H^\bullet(W)$, (as well as the $\mathcal{C}$-spectral sequence ones).*

5) *If $E_k \subset J\mathcal{D}^k(W)$ is a formally integrable PDE on the above bundle $W \equiv M \times \underline{W}$, such that $E_k = M \times \underline{S}_k$, with $\underline{S}_k$ a* **R***-vector subspace of $\underline{W}_k$, then one has that the corresponding cohomology Leray-Serre spectral sequences converge to $H^\bullet(E_\infty)$ (as well as the $\mathcal{C}$-spectral sequence ones).*

☐ (ATIYAH-HIRZEBRUCH SPECTRAL SEQUENCES OF PDEs).

Here we shall prove that there is an important spectral sequence associated to any PDE E_k that converges to the homology of E_k. In fact we have the following.

THEOREM 2.68 - *Let $E_k \subset J\mathcal{D}^k(W)$ be a PDE on the fiber bundle $\pi : W \to M, \dim M = n, \dim W = n+m$. Let us consider the following decreasing filtration*

$$\cdots \subset F^{q+1}(E_k) \subset F^q(E_k) \subset \cdots \subset F^0(E_k) \equiv E_k$$

with $F^q(E_k) \equiv {E_k}^{(q)} \equiv \pi_{k+q,k}((E_k)_{+q}) \subset J\mathcal{D}^k(W)$. To this filtration corresponds a spectral sequence

$$\{\widehat{E}^r_{\bullet,\bullet}(E_k), \widehat{d}^r\}$$

that converges to $H_\bullet(E_k)$. Furthermore, if E_k is formally integrable then

$$\{\widehat{E}^r_{\bullet,\bullet}(E_k), \widehat{d}^r\}$$

collapses and

$$H_\bullet(E_k) = H_\bullet(E_\infty).$$

PROOF. Set

$$\begin{aligned}
X^q &\equiv F^q(E_k), \quad X \equiv E_k,\\
Z^r_{p,q} &\equiv \mathrm{im}[j_* : H_{p+q}(X^p, X^{p+r}) \to H_{p+q}(X^p, X^{p+1})],\\
B^r_{p,q} &\equiv \mathrm{im}[\triangle : H_{p+q+1}(X^{p-r+1}, X^p) \to H_{p+q}(X^p, X^{p+1})],\\
Z^\infty_{p,q} &\equiv \mathrm{im}[j_* : H_{p+q}(X^p) \to H_{p+q}(X^p, X^{p+1})],\\
B^\infty_{p,q} &\equiv \mathrm{im}[i_* : H_{p+q}(X^p) \to H_{p+q}(X)],\\
F_{p,q} &\equiv \mathrm{im}[i_* : H_{p+q}(X^p) \to H_{p+q}(X)]\,.
\end{aligned}$$

Then, $Z^r_{p,q}, B^r_{p,q}, Z^\infty_{p,q}$ and $B^\infty_{p,q}$, for fixed p and q are all subgroups of $H_{p+q}(X^p, X^{p+1})$ and satisfy the following inclusion relations:

$$0 = B^1 p,q \subset B^2 p,q \subset \cdots \subset B^r p,q \subset B^{r+1} p,q \subset \cdots \subset B^\infty p,$$
$$q \subset Z^\infty_{p,q} \subset \cdots \subset Z^{r+1}_{p,q} \subset Z^r_{p,q} \subset \cdots \subset Z^1_{p,q} \subset H_{p+q}(X^p, X^{p+1})\,.$$

Set

$$E^r_{p,q} \equiv Z^r_{p,q}/B^r_{p,q}.$$

One has the canonical isomorphism:

$$Z^r_{p,q}/Z^{r+1}_{p,q} \cong B^{r+1}_{p-r,q+r-1}/B^r_{p-r,q+r-1}.$$

One has the canonical differential:

$$d^r : E^r_{p,q} \to E^r_{p-r,q+r-1}$$

for all p, q , given by

$$\begin{array}{ccccc} Z^r_{p,q}/B^r_{p,q} & \to Z^r_{p,q}/Z^{r+1}_{p,q} \cong B^{r+1}_{p-r,q+r-1}/B^r_{p-r,q+r-1} \to & Z^r_{p-r,q+r-1}/B^r_{p-r,q+r+1} \\ \| & & \| \\ E^r_{p,q} & \stackrel{d^r}{\to} & E^r_{p-r,q+r-1} \end{array}$$

One has the following properties for d^r that allow to associate to $\{X^q\}$ the spectral sequence

$$\{E^r_{p,q}, d^r\} :$$

(a)

$$\ker d^r = Z^{r+1}_{p,q}/B^r_{p,q}$$

(b)

$$\mathrm{im}[d^r : E^r_{p+r,q-r+1} \to E^r_{p,q}] \cong B^{r+1}_{p,q}/B^r_{p,q}$$

(c) $\mathrm{im}\, d^r \subset \ker d^r$, so $d^r \circ d^r = 0$. So, for fixed r

$$\cdots \to E^r_{p+r,q-r+1} \stackrel{d^r}{\to} E^r_{p,q} \to E^r_{p-r,q+r-1} \to \cdots$$

is a chain complex.

(d)

$$\begin{aligned} \ker d^r/\mathrm{im} d^r &= (Z^{r+1}_{p,q}/B^r_{p,q})/(B^{r+1}_{p,q}/B^r_{p,q}) \\ &\cong Z^{r+1}_{p,q}/B^{r+1}_{p,q} \\ &= E^{r+1}_{p,q} \to E^{r+1}_{\bullet,\bullet} = H(E^r_{\bullet,\bullet}, d^r) \end{aligned}$$

$$Z^\infty_{p,q} = Z^r_{p,q}, \text{ for } r > p, B^\infty p, q = \bigcup_{r\geq 1} B^r p, q.$$

One has an epimorphism $E^r_{p,q} \to E^{r+1}_{p,q}$, for $r > p$, and

$$E^\infty_{p,q} = \lim E^r_{p,q}.$$

One has the inclusion relations:

$$(2.168)\qquad H_{p+q}(X)\supset\cdots\supset F_{p,q}\supset F_{p-1,q+1}\supset\cdots\supset F_{-1,p+q-1}=0.$$

There is a natural isomorphism $E^{\infty}_{p,q}\cong F_{p,q}/F_{p-1,q+1}$, for all p,q. Thus the groups $E^{\infty}p,q$ are the quotients of a filtration (2.168) of the graded **R**-module $H_\bullet(X)$. Thus one has that the spectral sequence $\{E^r_{p,q},d^r\}$ converges to the graded **R**-module $H_\bullet(X)$. So , if we put

$$\{\widehat{E}^r_{p,q},\widehat{d^r}\}\equiv\{E^r_{p,q},d^r\}$$

the theorem is proved. □

DEFINITION 2.41 - *We call the above spectral sequence associated to $E_k\subset J\mathcal{D}^k(W)$ the* **Atiyah-Hirzebruch spectral sequence** *of E_k.*

□(SPECTRAL SEQUENCES AND COBORDISM OF PDEs)[49,138,141,152]. Let us first consider some fundamental definitions and results on the cobordism of compact manifolds.

DEFINITION 2.42 - 1) *Two compact manifolds M_0,M_1 are called* **cobordant,** *and we write $M_0\sim M_1$, if there is a compact manifold W such that $\partial W\cong M_0\times 0\bigcup M_1\times 1$. We call W a* **cobordism** *from M_0 to M_1.*

2) *Two oriented manifolds (M_i,η_i), $i=0,1$, are* **cobordant** *if there is a compact oriented manifold (W,θ) and an oriented preserving diffeomorphism*

$$(\partial W,\theta_{\partial W})\cong(M_0\times 0,\eta_0)\bigcup(M_1\times 1,\eta_1).$$

PROPOSITION 2.13 - 1) *For each dimension n, the cobordism (resp. oriented cobordism) is an equivalence relation. The set of cobordism classes, (resp. oriented cobordism classes) is denoted by Ω_n (resp.$^+\Omega_n$); the cobordism class (resp. oriented cobordism class) of M is denoted by $[M]$, (resp.$^+[M]$).*

2) *The operation of disjoint union makes Ω_n and $^+\Omega_n$ into abelian groups.*

PROOF. 1) (a)(reflexivity). Let M be an n-dimensional compact manifold without boundary, then M cobords with itself. In fact, we can take $W\equiv M\times I, I\equiv[0,1]\subset\mathbf{R}$, then $\partial W\cong M\times 0\bigcup M\times 1$.

(b)(symmetry). If one has $\partial W\cong M_0\times 0\bigcup M_1\times 1$ one also has $\partial W\cong M_0\times 0\bigcup M_1\times 1$.

(c)(transitivity) If $\partial W\cong M_0\times 0\bigcup M_1\times 1, \partial W'\cong M_1\times 0\bigcup M_2\times 1$, then we also have that $\partial W''\cong\partial(W\bigcup W')\cong M_0\times 0\bigcup M_2\times 1$; so $M_0\sim M_2$.

2) First of all, diffeomorphic manifolds are cobordant. In fact if $M_0\cong M_1$ then $\partial(M\times I)\cong M_0\times 0\bigcup M_0\times 1\cong M_0\times 0\bigcup M_1\times 1$ so $M_0\sim M_1$. Furthermore, the zero element of the group is $[\emptyset]$. Note that any n-dimensional manifold that is the boundary of some compact manifold is cobording with $\emptyset$. In fact if $V=\partial W$, then

$[M]\bigcup[V] = [M\bigcup V]$. On the other hand $M \times I$ is an $(n+1)$-dimensional manifold that cobord M with itself and so we can consider the manifold $M \times I \bigcup W$, we have: $\partial(M \times I \bigcup V)\bigcup(M \times 1)$. So $M \bigcup V \sim M$, therefore $[M\bigcup V] = [M]$.
The inverse of $[M]$ is $[M]$. In fact, $[M]\bigcup[M] = [M\bigcup M] = [\emptyset]$, as the disjoint union of a set with itself is empty! The inverse of $[M,\eta]$ is $[M,-\eta]$. □
THEOREM 2.69 - *The operation of cartesian product of manifolds induces bilinear maps*

$$\Omega_i \times \Omega_j \to \Omega_{i+j} \quad (resp.^{+}\Omega_i \times^{+} \Omega_j \to^{+} \Omega_{i+j})$$

these pairings are associative and commutative $\alpha\beta = \beta\alpha$, *(resp. graded commutative* $\alpha\beta = (-1)^{i+j}\beta\alpha$*). Thus the sequence* $\Omega. \equiv (\Omega_0, \Omega_1, \Omega_2, \cdots)$, *(resp.*$^{+}\Omega. \equiv (^{+}\Omega_0, ^{+}\Omega_1, ^{+}\Omega_2, \cdots)$*), has the structure of a ring (resp. graded ring). This ring possesses a 2-sided identity element* $1 \in \Omega_0$ *. Furthermore as* $M_1 \times M_2$ *is isomorphic, as manifold (resp. oriented manifold), to* $M_2 \times M_1$, *(resp.* $(-1)^{mn} M_2 \times M_1$*), we have that* $\Omega.$ *is a commutative ring.*
In the following table we list the cobordism groups of n-dimensional (oriented) manifolds.
REMARK 2.42 - In the following we shall characterize the cobordism in PDEs by means of some characteristic numbers and Leray-Serre cohomology of PDEs. Let us first recall some definitions and results on characteristic classes.
Let $\pi : W \to M$ be a G-bundle over X. The set of characteristic classes of W is the $\mathbf{Z}_2$-graded sub A-module $Kar^\bullet(W;A) \subset H^\bullet(X;A)$ given by $Kar^\bullet(W;A) = \mathrm{im}(f^\bullet{}_W)$, where $f^\bullet{}_W : H^\bullet(BG;A) \to H^\bullet(X;A)$ is the homomorphism induced by the classifying map $f_W : X \to BG$. The elements $\chi^\bullet(W) \equiv f^\bullet{}_W(\chi) \in Kar^\bullet(W;A)$ for some $\chi^\bullet \in H^\bullet(BG;A)$ are called **characteristic classes** of W. **Real** and **integer characteristic classes** are those which take values in $H^\bullet(X;\mathbf{R})$ and $H^\bullet(X;\mathbf{Z})$ or $H^\bullet(X;\mathbf{Z}_2)$ respectively. For example, integer characteristic classes for the groups $G = U(n), O(n), Sp(n)$ are respectively called **Chern, Stiefel-Whitney** and **Pontrjagin** and denoted by $c_i(X), w_i(X)$ and $p_i(X)$ respectively [84,141]. One has the one-to-one correspondence $Kar^\bullet(W;A) \leftrightarrow H^\bullet(BG;A)$. So we can adopt the following identification $Kar^\bullet(W;A) \equiv H^\bullet(BG;A)$. Let G be a non-compact Lie group and let K be a maximal compact subgroup of K. One has the isomorphism $H^\bullet(BG;A) \cong H^\bullet(BK;A)$ as $BG \simeq BK$ (homotopically equivalent). So characteristic classes of G-bundles that admit reduction to K belong to $H^\bullet(BK;A)$. The algebra $H^\bullet(BG)$ has a basis $\{d_1, \cdots, d_k\}$ whose integrals over all cycles in BG of appropriate dimension are integer valued:

$$< d_i, c_i > \in \mathbf{Z}, d_i \in H^i(BG), c_i \in H_i(BG).$$

As a consequence we have that:

□ The algebra $H^\bullet(BG)$ has a multiplicative basis consisting of classes whose integrals over all cycles of the base manifolds of arbitrary G-bundles are integer valued. In fact, let $[f] \in [X, BG]$, and let $\phi : D \to X$ be the smooth map representing a closed oriented q-dimensional submanifold D into X. Let (d_i) be a basis of $H^\bullet(BG)$. Let $d'_i \equiv f^* d_i \in H^\bullet(X)$ be the corresponding characteristic class on X. Assume that d_i (as well as d'_i) has rank q. Then one has

$$\int_D d'_i = \int_D \phi^* d'_i = \int_D \phi^*(f^* d_i) = \int_{f\circ\phi(D)} d_i \in \mathbf{Z}.$$

□ For any characteristic class $\chi^q(W) \in H^q(X)$ of a fiber bundle $\pi : W \to X$ over X with structure group G, the integral $\int_D \alpha$, where $\alpha \in \chi^q(W)$ and D is a closed q-differentiable submanifold of dimension q of X, is a topological invariant of the bundle (and D). In fact let $\alpha, \bar{\alpha} \in \chi^q(W)$, so that $\alpha = \bar{\alpha} + d\beta$, for some $(q-1)$-differential form β on X. So we have

$$\int_D \alpha = \int_D \bar{\alpha} + \int_D d\beta = \int_D \bar{\alpha} + \int_{\partial D} \beta = \int_D \bar{\alpha}.$$

In the following tables we resume some examples of characteristic classes for G-bundles over CW-complexes [138].

2 EXAMPLE 2.36 - EXAMPLES OF CHARACTERISTIC CLASSES FOR G-BUNDLES OVER CW-COMPLEX X.

One has

$$H^\bullet(BG; A) \cong H^\bullet(BG) \bigotimes_{\mathbf{R}} A$$

1) $G = \mathbf{Z}$

$$H^1(BG) \cong \mathbf{R}; \quad H^q(BG) = 0, q > 1$$

2) $G = \mathbf{Z}_m, (m \neq \infty)$. Classifying space = lens space.

$$H^q(BG) = 0, q > 0$$

3) $G = \mathbf{Z} \bigoplus \cdots_n \cdots \bigoplus \mathbf{Z}$. Classifying space $= T^n =$ torus of dimension n. One has

$$H^\bullet(BG; A) = \text{exterior algebra in } n\text{-dimensional group}$$

4) $G = \mathbf{R}^k \times T^m$. Classifying space $= \mathbf{C}P^\infty \times \cdots_m \cdots \times \mathbf{C}P^\infty$. One has

$$H^\bullet(BG; A) = \mathbf{R}[t_1, \cdots, t_m], \quad t_i \in H^2(BG)$$

5) $G = U(m)$. Classifying space $= G_{\infty,m}(\mathbf{C}) \equiv \bigcup_{p\geq 0} G_{m+p,m}(\mathbf{C})$. One has

$$H^{\bullet}(BG; A) = \mathbf{R}[c_1, \cdots, c_m], \quad c_i \in H^{2i}(BG)$$

6) $G = SU(m)$. Classifying space $= G^{+}{}_{\infty,m}(\mathbf{C}) \equiv \bigcup_{p\geq 0} G^{+}{}_{m+p,m}(\mathbf{C})$. One has

$$H^{\bullet}(BG; A) = \mathbf{R}[c_2, \cdots, c_m], \quad c_i \in H^{2i}(BG)$$

7) $G = O(m)$. Classifying space $= G_{\infty,m}(\mathbf{R}) \equiv \bigcup_{p\geq 0} G_{m+p,m}(\mathbf{R})$. One has

$$H^{\bullet}(BG; A) = \mathbf{R}[w_1, \cdots, w_m], \quad w_i \in H^{i}(BG)$$

8) $G = SO(m)$. Classifying space $= G^{+}{}_{\infty,m}(\mathbf{R}) \equiv \bigcup_{p\geq 0} G^{+}{}_{m+p,m}(\mathbf{R})$. One has

$$H^{\bullet}(BG; A) = \left\{ \begin{array}{l} \mathbf{R}[c_2, \cdots, c_{2(n-1)}, \chi_n], m = 2n \\ \mathbf{R}[c_2, \cdots, c_{2n}], m = 2n+1 \\ c_{2i} \in H^{4i}(BG), \chi_n \in H^{2n}(BG) \end{array} \right\}$$

9) $G = S_p(m)$. Classifying space $= G_{\infty,m}(\mathbf{H}) \equiv \bigcup_{p\geq 0} G_{m+p,m}(\mathbf{H})$. One has

$$H^{\bullet}(BG; A) = \mathbf{R}[p_1, \cdots, p_m], \quad p_i \in H^{4i}(BG)$$

EXAMPLE 2.37 - (Examples of integer characteristic classes for G-bundles)
1)(CHERN). $G = U(n), A = \mathbf{Z}, H^{\bullet}(BG; A) = \mathbf{Z}[c_1, \cdots, c_n]$

$$\left\{ \begin{array}{l} c_i \in H^{2i}(BG; \mathbf{Z}); \quad c_0(\xi) = 1 \\ c_i(\xi \bigoplus \eta) = \displaystyle\sum_{k+j=i; 0\leq i\leq m+n} c_k(\xi) c_j(\eta); \\ \xi = U(m)\text{-bundle}, \eta = U(n)\text{-bundle} \\ c_i(\xi) = 0, i > m; c_i(\eta) = 0, i > n; \\ c(\xi \oplus \eta) = c(\xi) c(\eta), c(\xi) \equiv 1 + \displaystyle\sum_{i\geq 1} c_i(\xi) \end{array} \right\}$$

2)(STIEFEL-WHITNEY). $G = O(n), A = \mathbf{Z}_2, H^{\bullet}(BG; A) = \mathbf{Z}_2[w_1, \cdots, w_n]$

$$\left\{ \begin{array}{l} w_i \in H^{i}(BG; \mathbf{Z}_2); \quad w_0(\xi) = 1 \\ w_i(\xi \oplus \eta) = \displaystyle\sum_{k+j=i; 0\leq i\leq m+n} w_k(\xi) w_j(\eta); \\ \xi = 0(m)\text{-bundle}, \eta = 0(n)\text{-bundle} \\ w(\xi \oplus \eta) = w(\xi) w(\eta), w(\xi) \equiv 1 + \displaystyle\sum_{i\geq 1} w_i(\xi) \end{array} \right\}$$

3)(PONTRJAGIN). $G = S_p(n), A = \mathbf{Z}, H^\bullet(BG; A) = \mathbf{Z}[p_1, \cdots, p_n]$.

$$\left\{\begin{array}{l} p_i \in H^{4i}(BG; \mathbf{Z}); \quad p_0(\xi) = 1 \\ p_i(\xi \oplus \eta) = \sum\limits_{k+j=i; 0 \leq i \leq m+n} p_k(\xi) p_j(\eta) \\ \xi = S_p(m)\text{-bundle}, \eta = S_p(n)\text{-bundle} \\ p_i(\xi) = 0, i > m; p_i(\eta) = 0, i > n \\ p(\xi \oplus \eta) = p(\xi) p(\eta), p(\xi) \equiv 1 + \sum\limits_{i \geq 1} p_i(\xi). \end{array}\right\}$$

☐ Let $\pi : E \to X$ be a vector bundle of fiber dimension $n > 0$. An **orientation** for $\pi : E \to X$ is a function which assigns an orientation to each fiber E_x of E such that there exist a local coordinate system (U, h) with $x \in U$ and $h : U \times \mathbf{R}^n \to \pi^{-1}(U)$, so that for each fiber $F_{x'} = \pi^{-1}(x')$ over U the homomorphism $h_{x'} : \mathbf{R}^n \to F_{x'}$ is orientation preserving. (Note that the space $\mathbf{R}^n$ has a canonical orientation, corresponding to its canonical ordered basis.)

☐ (PROPERTIES OF THE ORIENTATION OF VECTOR BUNDLES). 1) Let $\pi : E \to X$ be a vector bundle similar to in above definition. Then the following propositions are equivalent:

(a) An orientation which is fixed on E ; (b) For any open coordinate U$\subset X$ there exist n linear independent local sections $s_1, \cdots, s_n : U \to \pi^{-1}(U)$; (c) For each fiber F_x there is assigned a preferred generator $u_x \in H^n(F_x, F_{0,x}; \mathbf{Z})$ such that there exists a neighbourhood U of x and a cohomology class $u \in H^n(\pi^{-1}(U), \pi^{-1}(U_0); \mathbf{Z})$ so that the restriction $u|(F_x, F_{0,x}) \in H^n(F_x, F_{0,x}; \mathbf{Z})$ is equal to u_x .

2) Let $\pi : E \to X$ be an oriented n-plane bundle. We have the following :

(a) $H^i(E, E_0; \mathbf{Z}) = 0$ for $i < n$, and $H^n(E, E_0; \mathbf{Z})$ contains one and only one cohomology class u whose restriction $u|(E_x, E_{0,x}) \in H^n(E_x, E_{0,x}; \mathbf{Z})$ is equal to the preferred generator u_x for every fiber E_x of E .

(b) $H^\bullet(E, E_0; \mathbf{Z})$ is a free $H^\bullet(E; \mathbf{Z})$-module on one generator u of degree n.

(c)(**Thom isomorphism**) . One has the canonical isomorphism : $H^{k+n}(E, E_0; \mathbf{Z}) \cong H^k(X; \mathbf{Z})$.

The **Euler class** of an oriented n-plane bundle $\pi : E \to X$ is the cohomology class $e(E) \in H^n(X; \mathbf{Z})$, which corresponds to the fundamental class $u \in H^n(E, E_0; \mathbf{Z})$ via the following composition of homomorphisms:

$$\begin{array}{ccc} H^n(E, E_0; \mathbf{Z}) & \xrightarrow{j} & H^n(X; \mathbf{Z}) \\ \underline{j} \downarrow & & \| \\ H^n(E; \mathbf{Z}) & \to & H^n(X; \mathbf{Z}) \end{array}$$

where $j(u) = e(E)$. The map $\underline{j}$ is induced by the inclusion $(E, \emptyset) \to (E, E_0)$.

☐ (PROPERTIES OF THE EULER CLASS.)
1) If $f_E : E \to E'$ is an orientation preserving bundle map over $f_M : M \to M'$, then $e(E) = f^*_M e(E')$.
2) If $\pi : E \to X$ is a trivial n-space, $n > 0$, then $e(E) = 0$.
3) If the orientation of E is reversed, then the Euler class $e(E)$ changes sign.
4) If the fiber dimension n is odd, then $e(E) + e(E) = 0$.
5) The natural homomorphism $H^n(X;\mathbf{Z}) \to H^n(X;\mathbf{Z}_2)$ carries the Euler class $e(E)$ to the topological Stiefel-Whitney class $w_n(E)$.
6) $e(E \bigoplus E') = e(E) \cup e(E')$.
7) $e(E \times E') = e(E) \times e(E')$.
8) Let $\pi : E \to X$ be a vector bundle for which $2e(E) \neq 0$. Then, it follows that E cannot split as the Whitney sum of two oriented odd dimensional vector bundles. As an example, let M be a smooth compact manifold. Assume that the tangent bundle TM is oriented, and that $e(TM) \neq 0$. Then, TM cannot admit any odd dimensional sub-vector bundle.

REMARK 2.43 - Characteristic classes can be expressed by means of connections. Recall that if $\pi : P \to X$ is a principal G-bundle over X, a principal connection can be defined by means of the Eheresmann differential 1-form over P, $\omega : P \to T^*P \bigotimes A(G)$, where $A(G)$ is the Lie algebra of G. The corresponding curvature is a $A(G)$-valued differential form on $U \subset X$: $A \equiv s^*\omega : U \subset X \to T^*X \bigotimes A(G), F \equiv s^*\Omega : U \subset X \to \Lambda^0_2\ X \bigotimes A(G)$. Then, we can obtain characteristic classes for G-bundles on X by using such differential forms and the elements of the Weyl algebra. More precisely one has.

☐ The **Weyl algebra** of G, $W(G)^\bullet$, is defined by:

$$W(G)^\bullet \equiv \bigoplus_{q\geq 0} W(G)^q \equiv S(A(G)^*)/G$$

where $S(A(G)^*)/G$ is the Ad-invariant symmetric algebra of $A(G)^*$. In other words, $W(G)$ is the algebra generated by all scalar-valued, symmetric, multilinear forms $\psi \in A(G)^* \odot \cdots_q \cdots \odot A(G)^*$ which are Ad-invariant, that is $\psi(gzg^{-1}, \cdots, gzg^{-1}) = \psi(z_1, \cdots, z_q)$, $\forall z_i \in A(G)$, $\forall g \in G$. $W(G)^\bullet$ is isomorphic to the algebra of Ad-invariant polynomials, $P(A(G)) \equiv \bigoplus_{q\geq 0} P(G)^q$, where $P(A(G))^q$ is the space of G-invariant polynomials of degree q on $A(G)$($\equiv$ G-invariant numerical functions, homogeneous of degree q, on $A(G)$). More precisely one has the following isomorphism:

$$W(G)^q \to P(G)^q; \phi \mapsto P(\phi)$$

where $P(\phi)(v) = \phi(v, \cdots_q \cdots, v)$, $\forall v \in A(G)$. The inverse isomorphism is given by $p \mapsto \phi(p)$, where

$$\phi(p)(v_1, \cdots, v_q) = \frac{1}{q!}\alpha_{i_1\cdots i_q}\xi^{i_1}_{(1}\cdots\xi^{i_q}_{q)}$$

if

$$p(v) = \frac{1}{q!}\alpha_{i_1\cdots i_q}\xi^{i_1}\cdots\xi^{i_q}$$

is the representation of P in a basis $\{e_i\}$ of $A(G)$, where $\alpha_{i_1\cdots i_q}$ are symmetric coefficients, and $v = \xi^i e_i, v_\lambda = \xi^i_\lambda e_i, (\lambda = 1, \cdots, q)$. This procedure does not depend on the choice of the basis (e_i) .

22 (REPRESENTATION OF CHARACTERISTIC CLASSES BY CONNECTIONS). One has a canonical map:
(CHERN-WEYL HOMOMORPHISM)

$$W(G)^q \to Kar^q(X; \mathbf{R}), \phi \mapsto c_\phi \in H^q(X)$$

where c_ϕ is the characteristic class defined by the following representative q-differential form:

$$\alpha_\phi(F)(X_1, \cdots, X_q) = \sum_{i_1<i_2,\cdots,i_{2m-1}<i_{2m}} \text{sgn}\phi[F(X_{i_1}, X_{i_1}), \cdots, F(X_{i_{2m}}, X_{i_{2m}})]$$

where $(X_1, \cdots, X_{2m})$ is an arbitrary $2m$-tuple of tangent vectors at any point of X , and σ is the permutation $\binom{1\cdots.2m}{i_1\cdots.i_{2m}}$. In fact, the Ad-invariance of ϕ implies that the scalar-valued form $\alpha_\phi(F)$ is well defined on the basis X. Furthermore, one can easily verify that $d\alpha_\phi(F) = 0$ and that another curvature connection F' determines $\alpha_\phi(F') = \alpha_\phi(F) + d\beta$ for some 1-form β on X.
The above construction can also be applied to obtain characteristic classes on vector fiber bundles associated to G-bundles.
□ 1) Let $\pi : E \to X$ be a $\mathbf{K}$-vector bundle ($\mathbf{K} = \mathbf{R}, \mathbf{C}, \mathbf{H}$). Let

$$\psi \in Hom(E^* \bigotimes E \times \cdots_q \cdots .E^* \bigotimes E, X \times \mathbf{K}).$$

Then, we can associate to ψ a morphism of vector bundles (over X):

$$\Psi : \Lambda^0_{d_1} X \bigotimes E^* \bigotimes E \times \cdots_q \cdots . \times \Lambda^0_{d_q} X \bigotimes E^* \bigotimes E \to \Lambda^0_{d_1+\cdots+d_q} X$$

given by

$$\Psi(\alpha_1 \otimes a_1, \cdots, \alpha_q \otimes a_q) = \alpha_1 \wedge \cdots \wedge \alpha_q \otimes \psi(a_1, \cdots, a_q).$$

The local expression of

$$\Psi(\theta_1, \cdots, \theta_q) \in C^\infty(\Lambda^0_{d_1+\cdots+d_q} X)$$

for

$$\theta_i \in C^\infty(\Lambda^0_{d_i} X \bigotimes E^* \bigotimes E)$$

is as follows

$$\Psi(\theta_1,\cdots,\theta_q)=\sum a^{\beta_1\cdots\beta_q}_{\alpha_1\cdots\alpha_q}\theta_1{}^{\alpha_1}_{\beta_1}\wedge\cdots\wedge\theta_q{}^{\alpha_q}_{\beta_q}$$

where the coefficients $a^{\beta_1\cdots\beta_q}_{\alpha_1\cdots\alpha_q}$ are local numerical functions on X and $\theta_i{}^{\alpha_i}_{\beta_i}$ are differential forms on X.
2) Let $\pi:E\to X$ be a $\mathbf{K}$-vector bundle over X with structure group G. One has the following.
(a) Ψ is a differential invariant iff $\Psi(\theta_1,\cdots,\theta_q)=\Psi(g\theta_1g^{-1},\cdots,g\theta_qg^{-1})$, where $g:U_i\bigcap U_j\to G$ are the coordinate transformations of E. This implies that the functions $a^{\beta_1\cdots\beta_q}_{\alpha_1\cdots\alpha_q}$ are constants $a^{\beta_1\cdots\beta_q}_{\alpha_1\cdots\alpha_q}\in\mathbf{K}$.
(b) If Ψ is an invariant function then

$$d\Psi(\theta_1,\cdots,\theta_q)=\sum_{1\le i\le q}(-1)^{d_1+\cdots+d_{i-1}}\Psi(\theta_1,\cdots,{}_\urcorner d\theta_i,\cdots,\theta_q)$$

where ${}_\urcorner d\theta_i$ is the covariant exterior derivative associated to any connection $\urcorner:E\to J\mathcal{D}(E)$ on E.
(c) Any connection $\urcorner:E\to J\mathcal{D}(E)$ on E and any invariant function Ψ identifies a closed differential form

$$\Psi(\Omega)=\Psi(\Omega,\cdots_q\cdots,\Omega):X\to\Lambda^o_{2q}X$$

where Ω is the curvature of

$$\urcorner:\Omega:X\to\Lambda^0_2X\bigotimes E^*\bigotimes E.$$

☐ The **Weyl algebra** of a $\mathbf{K}$-vector bundle $\pi:E\to X$ with structure group G is

$$W(G;E)^\bullet\equiv\bigoplus_{q\ge0}S^q(E\bigotimes E^*)/G\,.$$

☐ 1)The de Rham cohomology class of a form $\Psi(\Omega)=\Psi(\Omega,\cdots_q\cdots,\Omega)$, when $\Psi\in W(G;E)^\bullet$, does not depend on the choice of the connection of E.
2) We have an algebra homomorphism
(WEYL's HOMOMORPHISM)

$$w:W(G;E)^\bullet\to H^\bullet(X;\mathbf{K}),\Psi\mapsto[\Psi(\Omega)].$$

In the following we report some examples of characteristic classes expressed in terms of curvature of connections on vector fiber bundles with some structure group.

EXAMPLE 2.38 - (EXAMPLES OF CHARACTERISTIC CLASSES EXPRESSED BY MEANS OF CONNECTIONS ON **K**-VECTOR BUNDLES BY USING WEYL ALGEBRA)[2]

1)(REAL CHERN). $G = U(n), \mathbf{K} = \mathbf{C}$

$$c_j(E) = \frac{(-1)^j}{(2\pi i)^j j!}\psi_j(\Omega) = \frac{(-1)^j}{(2\pi i)^j j!}\sum_{\alpha,\beta} \delta^{\beta_1\cdots\beta_j}_{\alpha_1\cdots\alpha_j}\Omega^{\alpha_1}_{\beta_1}\wedge\cdots\wedge\Omega^{\alpha_j}_{\beta_j}; c(E) = \det(1 - \frac{1}{2\pi i}\Omega).$$

Note that the determinant is defined by the same expression as the determinant of numerical expressions, but with the ordinary products replaced by exterior products. Terms that correspond to elements of $W(G;E)$ are the following:

$$W(G;E) = \left\{ \begin{array}{l} \text{Chern character}: ch(E) = \mathrm{tr}(e^{-\frac{1}{2\pi i}\Omega}) \\ = \text{closed nonhomog.diff.1-form} \\ \text{Real k-th Chern character}: ch(E)_k = \mathrm{tr}(\frac{(-1)^k}{(2\pi i)^k k!}\Omega^k). \end{array} \right\}$$

REMARK 2.44 - If $E = F^c \equiv \mathbf{C}\bigotimes_{\mathbf{R}} F$, where F is a **R**-vector bundle, then

$$c_{2n+1}(F^c) = 0; p_j(F) = \sum_j p_j(E) \in H^\bullet(X);$$

2)(PONTRJAGIN). $G = S_p(n), \mathbf{K} = \mathbf{R}$.

$$p_j(E) = \frac{(-1)^j}{(2\pi)^{2j}(2j)!}\sum_{\alpha,\beta} \delta^{\beta_1\cdots\beta_j}_{\alpha_1\cdots\alpha_j}\Omega^{\alpha_1}_{\beta_1}\wedge\cdots\wedge\Omega^{\alpha_{2j}}_{\beta_{2j}}; p(E) = \det(1 - \frac{1}{2\pi i}\Omega).$$

REMARK 2.45 - If $E = F^c \equiv \mathbf{C}\bigotimes_{\mathbf{R}} F$, where F is a **R**-vector bundle, then

$$p_j(F) = (-1)^j c_{2j}(F^c) \in H^{4j}(X).$$

The Pontrjagin classes of a locally affine manifold vanish; (in fact we have a linear connection with zero curvature).

3)(EULER CLASS) $G = SO(n), \mathbf{K} = \mathbf{R}$.

$$e(E) = \left\{ \begin{array}{ll} 0 \text{ if } q = \text{odd} & \\ \frac{(-1)^q}{(2)^{2q}\pi^q q!}\sum_{\alpha,\beta}\epsilon^{\beta_2\cdots\beta_{2q}}_{\alpha_1\cdots\alpha_{2q-1}}\Omega^{\alpha_1}_{\beta_2}\wedge\cdots\wedge\Omega^{\alpha_{2q-1}}_{\beta_{2q}} & \text{if } q = \text{even} \end{array} \right\}.$$

[2] If $E{=}TX$ we say that the corresponding characteristic classes are of the manifold X. It is possible to give a representation of the Stiefel-Whitney classes by similar procedures outside the theory of the Weyl algebra. Then real characteristic classes are real images of integral ones.

Here Ω = curvature of Euclidean connection, and q = fiber dimension of E.

REMARK 2.46 - If E has the structure group reducible to $GL(q;\mathbf{C})$, then $e(E) = c_q(E)$, where the Chern class $c_q(E)$ is calculated for the complexificated of E.

The **Euler-Poincaré characteristic** of X (or **Euler number** of X) is defined by

$$\chi(X) = \sum_{k\geq 0}(-1)^k \dim H^k(X).$$

THEOREM 2.70 - (GAUSS-BONNET THEOREM). 1) *If X is a compact orientable differentiable manifold, the Euler-Poincaré characteristic is defined by the formula*

$$\chi(X) = \int_X e(TX)$$

where $e(TX)$ is expressed in terms of a Riemannian connection of the manifold.

2) *The Euler class of a locally Euclidean manifold vanishes. (In fact we have a Euclidean connection of zero curvature.)*

3) *If the differentiable manifold X has a Riemannian metric of constant sectional curvature k, one has that the Pontrjagin classes $p_j(X) = 0$ for $j > 0$. If, moreover, X is compact and orientable and has even dimension $2q$, $\chi(X)$ has the sign of k^q and vanishes iff $k = 0$.*

Let X be a closed, possibly disconnected, smooth n-dimensional manifold. We have a map

$$H^n(X;\mathbf{Z}_2) \to \mathbf{Z}_2; v \mapsto v[X] \equiv < v, \mu_X >$$

where μ_X is the unique fundamental homology class of X: $\mu_X \in H_n(X) \in \mathbf{Z}_2$. $v[X]$ is called the **Kronecker index** (or **characteristic number**) of the cohomology class of v belonging to $H^n(X;\mathbf{Z}_2)$.

In the following we report some important examples of Kronecker indexes.

EXAMPLE 2.39 - (EXAMPLES OF KRONECKER INDEXES OF COHOMOLOGY CLASSES ((≡ CHARACTERISTIC NUMBERS)). Denote the characteristic classes by $v \in H^n(M;\mathbf{Z}_2)$.

1)(ORIENTED MANIFOLDS).

(A)(EULER NUMBER OF MANIFOLD M):$v = e(TM), < e(TM), \mu_M > \equiv \chi(M)$.

(B)(STIEFEL-WHITNEY NUMBERS OF $O(n)$-BUNDLE $\{\pi : E \to M)$:

$$\left\{\begin{array}{l} v = w_1(E)^{r_1} \cdots w_n(E)^{r_n} \\ {w_1}^{r_1} \cdots {w_n}^{r_n}[E] \\ = \text{Stiefel-Whitney numbers of } E \\ (r_1, \cdots, r_n) = \text{non-negative integers} \\ r_1 + 2r_2 + \cdots nr_n = n \end{array}\right\}$$

(C)(STIEFEL-WHITNEY NUMBERS OF MANIFOLD M):

$${w_1}^{r_1} \cdots {w_n}^{r_n}[TM].$$

One has:

$$w_n[M] \equiv \chi(M) \bmod 2.$$

2)(CHERN NUMBERS OF n-DIMENSIONAL COMPLEX MANIFOLD M):

$$\begin{array}{l} v = c_{i_1}(TM) \cdots c_{i_r}(TM), I \equiv (i_1, \cdots, i_r) = \text{partition of } n \\ c_I[M] \equiv c_{i_1} \cdots c_{i_r}[M] = I\text{-th Chern number.} \end{array}$$

Then, M has $p(n)$ Chern numbers, where $p(n)$ = number of distinct partitions of n.
EXAMPLE 2.40 - (EXAMPLES OF CHERN NUMBERS).
1)($P^n(\mathbf{C})$:complex projective space). $c_{i_1} \cdots c_{i_r}[P^n(\mathbf{C})] = \binom{n+1}{i_1} \cdots \binom{n+1}{i_r}$.
2)(Complex 1-dimensional manifold M).$c_1[M]$ = Euler characteristic $= \chi(M)$.
3)(Complex 2-dimensional manifold M).$c_{11}[M], c_2[M]$ = Euler characteristic$= \chi(M)$.

2 THEOREM 2.71 - *Let $E_k \subset JD^k(W)$ be a k-order PDE on the fiber bundle $\pi : W \to M, \dim M = n$. Let (N_i, h_i), $i = 0, 1$, be a couple of Cauchy data for the system E_k, defined on E_{k+r}. Then one has a canonical map $\phi_{(N,h)} : H^{n-1}(I(E_{k+r}); \mathbf{Z}_2) \to \mathbf{Z}_2$ where $N \equiv N_0 \times 0 \bigcup N_1 \times 1$, h is the natural extension of h_i, $i = 0, \ldots, N$, and $H^{n-1}(I(E_{k+r}); \mathbf{Z}_2)$ is the $(n-1)$-Leray-Serre cohomology of E_k. Furthermore, in order for a solution to exist cobording N_0 and N_1 it is necessary that the map $\phi_{(N,h)}$ should be identically zero.*

PROOF. We know from the Pontrjagin-Thom theorem [138,141,142] that an $(n-1)$-dimensional manifold X is the boundary of a smooth compact n-dimensional manifold Y iff all the Stiefel-Whitney numbers of X are zero. Furthermore, two smooth closed $(n-1)$-dimensional manifolds belong to the same cobordism class iff all their corresponding Stiefel-Whitney numbers are equal. So, in order that there exists a solution L of E_k cobording N_0 and N_1 it is necessary that Stiefel-Whitney numbers of N are zero or equivalently that Stiefel-Whitney numbers of N_0 and N_1 are equal.

On the other hand, by means of the embedding $N \to E_{k+r}$ we can represent such numbers as $< h^*\omega, \mu_N >$, where $\omega \in H^{n-1}(I(E_{k+r}); \mathbf{Z}_2)$ and $\mu_N \in H_{n-1}(N; \mathbf{Z}_2)$ is a fundamental cycle of N. This completes the proof. □

REMARK 2.47 - Taking into account that Leray-Serre spectral sequence converges to $H^\bullet(I(E_{k+r}); \mathbf{Z}_2)$, we can conclude that above theorem gives a relation between Leray-Serre spectral sequences of PDEs and cobordism in PDEs.

REMARK 2.48 - (HYPERGEOMETRIC FUNCTIONS ON (SINGULAR) SOLUTIONS OF PDEs). By using the last remark in section 1.4, and taking into account the Grassmann bundle $I(E_k)$ of integral planes over a PDE $E_k \subset J\mathcal{D}^k(W)$, we can define a generalized hypergeometric function directly on E_k, and, hence, by restriction on each (singular) solution V of E_k. In this sense the classical theory of hypergeometric functions [31,146] can be generalized over PDEs.

2.6 - TUNNEL EFFECTS IN PDEs

COVERING SPACES

Singular solutions of PDEs can be understood in the covering spaces framework. In fact, we shall see that singular solutions of PDEs identify coverings of the base manifolds. So, we shall give, here, some fundamental definitions and results on covering spaces, that implicitly will be useful to characterize singular solutions of PDEs.

A **locally trivial covering space** is a locally trivial fibration $\pi : W \to M$ such that all its fibers are discrete. The cardinality of a fiber W_p over $p \in M$ is called the **multiplicity** of the covering at point p. This number is locally constant and for M connected also globally constant. If the multiplicity is constant and equal to n, we talk about an **n-fold covering**. Note that a locally covering map $\pi : W \to M$ is always locally homeomorphic, i.e. every $q \in W$ has an open neighbourhood V such that $\pi(V)$ is open in M and π defines a homeomorphism $V \cong \pi(V)$. More generally a **covering space** is a fibration $\pi : W \to M$ such that π is an open continuous map (images of open sets are open) and the fibers are discrete spaces. The points $q \in W$ at which the map π is not locally homeomorphic are called **branching points**.

EXAMPLE 2.41 - 1) The map $f : \mathbf{C} \to \mathbf{C}, f(z) = z^2$ is a double covering. The only branching point is 0.

2) The map $f : \mathbf{C} \setminus 0 \to \mathbf{C} \setminus 0$ (or $S^1 \to S^1$), $f(z) = z^n, n \geq 2$, gives an n-fold covering.

3) The map $f : \mathbf{R} \to S^1, f(x) = e^{ix}$, gives a covering with a countably infinite number of leaves.

4) The canonical projection $S^n \to P\mathbf{R}^n$ is a 2-fold covering.

REMARK 2.49 - (PROPERTIES OF PATH LIFTINGS ON COVERING SPACES).
1)(**Existence and uniqueness of path liftings**). If W is a covering space of M, then for every path $\gamma : [a,b] \to M$ in M and every $q_0 \in W$ over $\gamma(a)$, there is one and only one lifting $\breve{\gamma}$ of γ starting at q_0, i.e. $\breve{\gamma}(a) = q_0, \pi \circ \breve{\gamma} = \gamma$.
2)(**Lifting of homotopies over covering spaces**). Let W be a covering space of M. Let B be another topological space, $h : B \times [0,1] \to M$ a continuous map, and $\breve{h}_0 : B \to W$ a continuous lifting of h_0, i.e. the following diagram is commutative:

$$\begin{array}{ccc} B \times 0 & \overset{h_0}{\to} & W \\ \downarrow & & \downarrow \pi \\ B \times [0,1] & \overset{h}{\to} & M \end{array}$$

Then the map $\breve{h} : B \times [0,1] \to W, (b,t) \mapsto \breve{\gamma}_b(t)$, starting at $\overset{0}{h_0}(b)$, is continuous: the following diagram is commutative:

$$\begin{array}{ccc} B \times 0 & \overset{h_0}{\to} & W \\ \downarrow & & \| \\ B \times [0,1] & \overset{\breve{h}}{\to} & W \\ \| & & \downarrow \pi \\ B \times [0,1] & \overset{h}{\to} & M \end{array}$$

3)(**Monodromy lemma**). Let W be a covering space of M and $\alpha, \beta : [0,1] \to M$ two paths in M which are homotopic with fixed end points, i.e. there is a homotopy $h : [0,1] \times [0,1] \to M$, with $h_0 = \alpha, h_1 = \beta, h_t(0) = \alpha(0), h_t(1) = \alpha(1)$, $\forall t \in [0,1]$. If $\breve{\alpha}$ and $\breve{\beta}$ are liftings of α and β starting at the same point q_0, then they end at the same points as well: $\breve{\alpha}(1) = \breve{\beta}(1)$.
4)(**Behaviour of the groups of homotopy relative to covering spaces**). Let $\pi : (W, q_0) \to (M, p_0)$ be a covering map in the category of spaces with base point. Then the induced group homomorphism $\pi_\star : \pi_n(W, q_0) \to \pi_n(M, p_0)$ is an isomorphism for all $n > 1$ and a monomorphism for $n = 1$. Furthermore, if W is 0-connected and a locally trivial covering space with type fiber F, then the points of F are in a 1-1 correspondence with the cosets of the characteristic group $G(W, q_0)$ in $\pi_1(M, p_0)$, where $G(W, q_0) \equiv \pi_\star(\pi_1(W, q_0)) \subset \pi_1(M, p_0)$.
5) A loop α in (M, p_0) lifts to a loop $\breve{\alpha}$ in (W, q_0) iff $[\alpha] \in G(W, q_0)$.
6) If the characteristic subgroups of two covering spaces of (M, p_0) are contained in one another, the covering space with the smaller group canonically covers the other space, and the following diagram

$$\begin{array}{ccc} (B, b_0) & \to & (W, q_0) \\ \downarrow & & \| \\ (M, p_0) & \leftarrow & (W, q_0) \end{array}$$

is commutative.

REMARK 2.50 -

(LIFTABILITY CRITERION OF MAPS ON COVERING SPACES).

1) Let $\pi : (W, q_0) \to (M, p_0)$ be a covering mapping. Let $f : (B, b_0) \to (M, p_0)$ be a continuous map. In order that f should be liftable, i.e. there exists a unique map $\breve{f} : (B, b_0) \to (W, q_0)$ such that $\pi \circ \breve{f} = f$, it is necessary that

$$f_\star(\pi_1(B, b_0)) \subset G(W, q_0). \tag{2.169}$$

☐ SINGULARITIES OF DIFFERENTIABLE MAPS.

Let us consider here some fundamental definitions and results about singularities of differentiable manifolds. (See also refs. [3,50,81] and references quoted in [3].)

DEFINITION 2.43 - *Let M and N be manifolds of dimension m and n respectively. A point $x \in M$ is said to be a* **critical point** *(resp.* **regular point***) of the smooth map $f : M \to N$ if the rank of the derivative $Df(x) : T_xM \to T_{f(x)}N$ at point x is less than (resp. equal to) $min(\dim M, \dim N)$. A* **critical value** *(resp.* **regular value***) of f is the image of a critical point (resp. regular point). Let us denote by $\Sigma(f) \subset M$ the set of critical points of f and by $f(\Sigma(f)) \subset N$ the set of critical values of f.*

EXAMPLE 2.42 - 1) $M \equiv S^2$, N = 2-dimensional plane, $f : S^2 \to N$ vertical projection. The set of critical points is the equator and the set of critical values is S^1 in the plane N.

2)(**LAGRANGIAN SINGULARITIES**). A **symplectic structure** on a manifold V is a closed nondegenerate 2-form ω on V. The dimension of a symplectic manifold is necessarily even. (Darboux's theorem.) All the symplectic structures on manifolds of a fixed dimension are locally equivalent. So in a neighbourhood of every point every symplectic structure can be expressed in appropriate coordinates (x^α, y_α) , $1 \leq \alpha \leq n$, in the form $\omega = dy_\alpha \wedge dx^\alpha$. A submanifold $N \subset V$ is said to be **isotropic** if $\omega|_N = 0$. The dimension of an isotropic submanifold cannot be greater than n. The isotropic submanifolds of greatest possible dimension n are said to be **Lagrangian**. Let $\pi : W \to M$ be a **Lagrangian bundle**, i.e., a $2n$-dimensional fiber bundle on an n-dimensional manifold X such that W is equipped with a symplectic structure ω , and the fibers, W_x , are **Lagrangian submanifolds**: $\omega|_{W_x} = 0, \dim W_x = n$. A **casustic** is the set of critical values of a map (**Lagrangian map**) $p = \pi|_L : L \to X$, where L is a Lagrangian submanifold of W $(\omega|_L = 0, \dim L = n)$. An example of Lagrangian bundle is $T^*X \to X$, where T^*X is endowed with the canonical symplectic form $\omega = d\alpha$, where α is the Liouville 1-form on T^*X.

3)(**LEGENDRIAN SINGULARITIES**). A **contact form** is a 1-form χ defined on a differentiable manifold V of (2n+1) dimensions such that $\chi(p) \wedge d\chi(p)^n \neq 0$, $\forall p \in V$. This implies that rank$(d\chi) = 2n$ everywhere, (i.e. $d\chi$ is a presymplectic form on V). Thus the couple (V, χ) is called a **contact manifold**. The annihilator of $d\chi$ is 1-dimensional, and it is generated by the global vector field v such that $v \rfloor d\chi = 0, v \rfloor \chi = 1$, that is called **characteristic vector field** of the contact manifold (V, χ). If the orbits of v form a **nice foliation** (i.e. the space of its leaves V/v is a Hausdorff differentiable manifold, and the natural projection $\rho : V \to V/v$ is a submersion), we have a **symplectic reduction**, (i.e., there exists a symplectic form ω on V/v, such that $\rho^*\omega = d\chi$). Furthermore, if V is compact and if the orbits of v form a **regular foliation** (i.e. the leaves are the fibers of a submersion), then V is a principal circle bundle over the symplectic manifold V/v and χ is proportional to a connection form on V. A contact manifold (V, χ) whose foliation is regular is called a **regular contact manifold**. Let V be a $(2n+1)$-dimensional manifold, a **contact structure** (or **contact distribution**) on V is a field of hyperplanes ($\equiv 2n$-dimensional tangent planes) $\mathbb{B} \equiv \{B_p\}_{p \in V} \subset TV, B_p \equiv$ **contact planes**, satisfying the following condition of maximal non-integrability: there exists a **contact form** χ, locally generating the given field of hyperplanes,

$$< \chi(p), B_p >= 0, p \in V, \qquad (2.170)$$

(the 2-form $d\chi$ is nondegenerate on every plane where equation (2.170) is satisfied). This is equivalent to saying that χ is nondegenerate. More precisely, we say that a differential 1-form $\omega \in \Omega^1(M)$ is **nondegenerate** if the following conditions hold:
(i)

$$\mathbb{E} \equiv \bigcup_{p \in M} \mathbb{E}_p, \quad \mathbb{E}_p \equiv \ker(\omega(p)) \subset T_pM$$

is a smooth distribution, codim$(\mathbb{E}) = 1$. (We write $\mathbb{E} \equiv \ker(\omega)$.)
(ii) $d\omega(p)|_{\mathbb{E}_p}$, $\forall p \in M$, is a nondegenerate 2-form, i.e., the following implication hold: if $X \in \mathbb{E}_p, d\omega(X, Y) = 0$, $\forall Y \in \mathbb{E}_p \Rightarrow X = 0$.
NOTE. A differential 1-form ω is nondegenerate if the distribution $\mathbb{E} \equiv \ker(\omega)$ has no characteristic symmetries.
The contact form generating a contact structure is defined only locally and only up to multiplication by a function different from 0. If the contact form associated to a contact structure $\mathbb{B}$ is always the same (for any open set) we say that $\mathbb{B}$ is a **strictly contact distribution**. (Since nondegenerate skew-symmetric bilinear forms exist only on even-dimensional spaces, a contact structure can exist only on odd-dimensional manifolds.) Let $v : V \to TV$ be the characteristic vector field of

a contact manifold (V,χ). Then, as $v\rfloor\chi = 1$, the contact distribution $\mathbb{B}$ is supplementary to the annihilator: $TV = \mathbb{B}\bigoplus < v >$. Furthermore, if (V,χ) is a contact manifold, then $(\mathbb{B} \to V, d\chi|_{\mathbb{B}})$ is a symplectic vector bundle on V. (Darboux's theorem.) All the contact structures on a manifold of a fixed dimension are locally diffeomorphic. So all the contact forms on a manifold, in suitable coordinates, take the form $\chi = dz - y_\alpha dx^\alpha$. The set $J^1(V,1)_p$ of all contact elements touching V at a given point p$\in V$ is isomorphic with $\mathbb{P}T_p{}^*V$. In fact any element $\alpha \in \mathbb{P}T_p{}^*V$ identifies up to a real number, a 1-form $\chi(p)$, and therefore a contact element $\mathbb{B}_p$ as the annihilator of $\chi(p)$. The **manifold of contact elements** on V is the $(2\dim V - 1)$-dimensional manifold given by $\mathbb{P}T^*V \equiv \bigcup_{p\in V}\mathbb{P}T_p{}^*V$. This manifold has a natural contact structure. More precisely, on $J^1(V,1)$ we can define the following distribution: $\mathbb{E} \equiv \bigcup_{L\in J^1(V,1)}\mathbb{E}_L, \mathbb{E}_L \equiv T(\pi^{-1})(L)$, where π is the canonical projection $\pi : J^1(V,1) \to V$. Then, $\mathbb{E}$ is a contact distribution. Let (V,χ) be a contact manifold of dimension $2n+1$. Let $N \subset V$ be an h-dimensional manifold, $h < 2n$, such that $TN \subset \mathbb{B}$. (The contact distribution $\mathbb{B}$ on V is not completely integrable, but it may have integral submanifolds of dimension $h < 2n$.) Then TN is an isotropic subbundle of $\mathbb{B}$. (In fact let X, Y be vector fields on N, and therefore $\chi(X) = \chi(Y) = 0$. Then $[X,Y] \in TN$, hence we also have $\chi([X,Y]) = 0$. This implies $d\chi(X,Y) = 0$.) (From this we deduce that in fact $h \leq n$.) An **integral manifold** of a contact structure (V,χ) is a submanifold $N \subset V$ such that $i^*\chi = 0$, where $i : N \subset V$ is the canonical inclusion, or equivalently $T_pN \subset B_p$, $\forall p \in N$. The dimension of any integral manifold is $< \frac{1}{2}(2n+1)$. Integral manifolds of greatest possible dimension n, are said to be **Legendrian submanifolds** of the contact manifold (V,χ). If $X \subset V$ is a hypersurface of a contact manifold $V, \mathbb{P}T^*X$ is a Legendrian submanifold of $\mathbb{P}T^*V$. Let $\pi : W \to X$ be a **Legendrian bundle**, i.e., a $(2n+1)$-dimensional fiber bundle W on an $(n+1)$-dimensional manifold X such that W is equipped with a contact structure (identified by a contact form χ) while its fibers are **Legendrian submanifolds** $(\chi|_{W_x} = 0, \dim W_x = n)$. A **front** is the image of a map (**Legendrian map**) $p = \pi|_L : L \to X$, where $L \subset W$ is a **Legendrian submanifold** of W $(\chi|_L = 0, \dim L = n)$. Singularitiers of p are called **Legendrian singularities**. A front is an n-dimensional submanifold of W, (eventually with singularities). An example of Legendrian bundle is

$$J\mathcal{D}(M,\mathbf{R}) \cong T^*M \times \mathbf{R} \to M \times \mathbf{R}.$$

Every regular solution of $J\mathcal{D}(M,\mathbf{R})$ is a Legendrian submanifold. If N is a Legendrian submanifold of a contact manifold (V,χ), then TN is a Lagrangian subbundle of the symplectic vector bundle $(\mathbb{B}|_N, d\chi|_{\mathbb{B}|_N})$.

TAB.2.7 - **Examples of contact transformations on** (V, χ)

Name	Definition	$\phi^*\chi = \lambda_\phi\chi$
translation	$\phi : (x, u, p) \mapsto (x + c, u, p), c \in \mathbf{R}^n$	$\lambda_\phi = 1$
translation	$\phi : (x, u, p) \mapsto (x, u + c, p), c \in \mathbf{R}^n$	$\lambda_\phi = 1$
translation	$\phi : (x, u, p) \mapsto (x, u + h(x), p + (\partial x.h))$ $h \in C^\infty(\mathbf{R})$	$\lambda_\phi = 1$
scale tras.	$\phi : (x, u, p) \mapsto (e^{\lambda t}x, e^{\nu t}u, e^{(\nu-\lambda)t}p)$ $\lambda \in \mathbf{R}^n, t, \nu \in \mathbf{R}$	$\lambda_\phi = e^{\nu t}$
Legendre tras	$\phi : (x, u, p) \mapsto (p, u - px, -x)$	$\lambda_\phi = 1$

A **contact transformation** of $(V, \mathbb{B})$ is any symmetry $\phi : V \to V$ of the contact distribution. A **contact vector field** is any infinitesimal symmetry $X \in C^\infty(TV)$ of the contact distribution. A diffeomorphism $\phi : V \to V$ is a contact transformation of a strictly contact distribution iff $\phi^*\chi = \lambda_p\chi$, for some smooth function $\lambda_\phi \in C^\infty(V)$. Furthermore, a vector field $X \in C^\infty(TV)$ is a contact vector field iff $\mathcal{L}_X\chi = \lambda_X.\chi$. A contact transformation ϕ with $\lambda_\phi = 1$ is called a **strictly contact transformation.** A contact vector field X with $\lambda_X = 0$ is called an **Hamiltonian vector field.** The characteristic vector field v of the contact manifold (V, χ) is a strictly contact vector field. In fact $\mathcal{L}_v\chi = v\rfloor d\chi + d(v\rfloor\chi) = 0$. In particular, if $V = J\mathcal{D}(M, \mathbf{R})$ and $\chi = du - p_i dx^i$ is its Cartan form, one has $v = \partial u$. [$J\mathcal{D}(M, \mathbf{R})$ has a strictly contact distribution, i.e., Cartan distribution.] In table 2.7 we resume some example of contact transformations.

Any contact vector field X defines a **generating function** $f = \chi(X)$. Denote by X_f the contact vector field generated by a function $f \in C^\infty(M)$. Then, one has:

$$X_f = -\sum_{1\le i\le n}(\partial p^i.f)\partial x_i + \sum_{1\le i\le n}[(\partial x_i.f) + p_i(\partial u.f)]\partial p^i + [f - \sum_{1\le i\le n} p_i(\partial p^i.f)]\partial u.$$

The set of contact vector fields on (V, χ) is a Lie algebra with respect to the product $[X_f, X_g] = X_{[f,g]}$, where $[f, g]$ is the **Lagrange bracket** given by

$$\begin{aligned}[f, g] &= X_f(g) - v(f)g \\ &= \sum_{1\le i\le n}[(\partial x_i.f) + p_i(\partial u.f)](\partial p^i.g) \\ &\quad - [(\partial x_i.g) + p_i(\partial u.g)](\partial p^i.f) + f(\partial u.g) - g(\partial u.f)\,.\end{aligned}$$

So the set of all smooth functions on a strictly contact manifold is a Lie algebra with respect to the Lagrange bracket, i.e., the following conditions hold:

$$
\begin{aligned}
&\text{(i)} \quad [f,g] = -[g,f];\\
&\text{(ii)} \quad [\lambda_1 f_1 + \lambda_2 f_2, g] = \lambda_1[f_1,g] + \lambda_2[f_2,g];\\
&\text{(iii)} \quad [[f,g],h] + [[g,h],f] + [[h,f],g] = 0;\\
&\text{(iv)} \quad [f,gh] = g[f,h] + h[f,g] + v(f)gh = 0,
\end{aligned}
$$

where $f_1, f_2, f, g, h \in C^\infty(V), \lambda_1, \lambda_2 \in \mathbf{R}$.

4)(**SINGULAR SOLUTIONS IN PDEs OF FIRST ORDER ON MANIFOLDS**).

Let us denote by $\chi_1 : J\mathcal{D}(M,\mathbf{R}) \to T^*J\mathcal{D}(M,\mathbf{R})$ the canonical contact structure on $J\mathcal{D}(M,\mathbf{R})$. One has $\chi_1 = \bar{\pi}_1^* dz - pr_1{}^*\lambda, d\chi_1 = -(pr_1)^*\omega$, where λ is the Liouville form on $T^*M, \omega = d\lambda$ is the canonical form on T^*M, z is a coordinate on $\mathbf{R}$ and $\bar{\pi}_1 : J\mathcal{D}(M,\mathbf{R}) \to \mathbf{R}$, and $pr_1 : J\mathcal{D}(M,\mathbf{R}) \to T^*M$ are canonical projections. A Legendre submanifold of $J\mathcal{D}(M,\mathbf{R})$ is an n-dimensional ($n = \dim M$) submanifold $i : L \subset J\mathcal{D}(M,\mathbf{R})$ such that $i^*\chi_1 = 0$. A section $u : M \to J\mathcal{D}(M,\mathbf{R})$ of $\pi_1 : J\mathcal{D}(M,\mathbf{R}) \to M$ is of the form $u = Df$, for some numerical function $f : M \to \mathbf{R}$ iff $u^*\chi_1 = 0$. L can be represented, in a neighbourhood V of a point $q \in L$ as the image of Df for some differentiable function $f : M \to \mathbf{R}$, defined in a neighbourhood $U \equiv \pi_1(V)$ of $p \equiv \pi_1(q) \in M$, iff $\pi_1|_V : V \to M$ is a diffeomorphism, $(\mathrm{rank}(D(\pi_1|_V)) = n)$. This is equivalent to saying that L is transverse to the fiber ${\pi_1}^{-1}(p)$. For any point $q \in L \subset J\mathcal{D}(M,\mathbf{R})$ of a Legendrian submanifold $L \subset J\mathcal{D}(M,\mathbf{R})$, there exists an open neighbourhood $V \subset L, q \in V$, such that $pr_1(V) \subset T^*M$ is a Lagrangian submanifold of T^*M. Conversely, for each connected Lagrangian submanifold $\underline{L} \subset T^*M$, there exists a Legendrian submanifold $L \subset J\mathcal{D}(M,\mathbf{R})$ such that the map $pr_1 : L \to \underline{L}$ is a covering. A PDE of first order on an n-dimensional manifold M is a PDE of first order on the trivial fiber bundle $\pi : M \times \mathbf{R} \to M$, i.e., a subbundle E_1 of $J\mathcal{D}(M,\mathbf{R}) \to M$ of dimension $\geq 1+n$. A **singular solution** of E_1 is a Legendrian submanifold L such that $L \subset E_1$. A **singular point** of the solution L is a point $p \in L$ where $\pi_1|_L : L \to M$ is not a diffeomorphism (i.e. at p L is not transverse to the fiber ${\pi_1}^{-1}(x), x = \pi_1(q) \in M$). A **regular solution** of E_1 is a numerical function $f : M \to \mathbf{R}$ such that $Df(M) \subset E_1$. Any solution of E_1 identifies a Legendrian submanifold of T^*M by means of the natural projection $pr_1 : J\mathcal{D}(M,\mathbf{R}) \to T^*M$. The converse is in general false. However, there is a class of first order PDEs on M, where this condition is verified. These are the PDEs of Hamiltonian-Jacobi type. More precisely, a **PDE of Hamiltonian-Jacobi type** on an n-dimensional manifold M, is a first order PDE on the fiber bundle $\pi : M \times \mathbf{R} \to M$, identified by a submanifold C of the cotangent bundle $T^*M : E_1 = C \times \mathbf{R} \subset J\mathcal{D}(M.\mathbf{R}) \cong T^*M \times \mathbf{R}$.

Locally it can be written as follows: $F^A(x^\alpha, z_\alpha) = 0, A = 1, \cdots, r$, where (x^α, z, z_α) is a fibered coordinate system on $J\mathcal{D}(M, \mathbf{R})$. C is called the **characteristic manifold** of the system. A regular solution of such a system is a numerical function $f : M \to \mathbf{R}$ such that $df(M) \subset C$. Let s be a section of $T^*M \to M$. Then s identifies a solution of $E_1 = C \times \mathbf{R}$ iff

$$s : M \to C \quad \text{and} \quad s^*\lambda = df, \tag{2.171}$$

where λ is the canonical Liouville form on T^*M. In order that condition (2.171) should be satisfied it is necessary that $s^*\omega = 0$, where $\omega = d\lambda$. A singular solution of an Hamiltonian-Jacobi type PDE $E_1 = C \times \mathbf{R} \subset J\mathcal{D}(M, \mathbf{R})$ is identified by a Lagrangian submanifold $\underline{L}$ of $C \subset T^*M$. A necessary condition that such solution exists is that $\dim C \geq n = \dim M$. Furthermore, $\underline{L}$ is locally defined by means of a section $s : M \to T^*M$, (and therefore it is defined by a function $f : M \to \mathbf{R}$ such that $df = s$), iff $\pi|_{\underline{L}} : \underline{L} \to M$ is a diffeomorphism. (In this case the solution L is regular). A necessary condition for $\underline{L}$ to be defined by a cross section is that $\underline{L}$ be transversal to the vertical foliation $vT(T^*M)$. If $\underline{L} = s(M)$, where s is a section of T^*M over M, then we have the following isomorphism $T(T^*M)|_{\underline{L}} \cong vT(T^*M)|_{\underline{L}} \bigoplus T\underline{L}$ and $vT(T^*M)|_{\underline{L}}$ and $T\underline{L}$ are transversal Lagrangian subbundles of $T(T^*M)$. **Maslov characteristic classes** are obstructions to the transversality of two Lagrangian subbundles. So Maslov classes are the obstruction to represent $\underline{L}$ as the image of some section s of $T^*M \to M$ [147].

THEOREM 2.72 - 1) *(A) Let M and N be C^r manifolds, $r \geq 1$, without boundaries of dimension m and n respectively. If $y \in f(M) \subset N$ is a regular value of a C^r map $f : M \to N$, then $f^{-1}(y)$ is a C^r submanifold of M.*

(B) *Let M and N be C^r manifolds with boundaries, of dimension m and n respectively. If $y \in N - \partial N$ is a regular value for both f and $f|_{\partial M}$, then $f^{-1}(y)$ is a* **neat** *C^r submanifold of M, that is $\partial f^{-1}(y) = f^{-1}(y) \cap \partial M$ and $f^{-1}(y)$ is covered by charts (ϕ, U) of M such that $f^{-1}(y) \cap U = \phi^{-1}(\mathbf{R}^m)$.*

2) *(A) Let M and N be C^r manifolds without boundaries, of dimension m and n respectively. Let $f : M \to N$ be a C^r map,* **transverse** *to a C^r submanifold A of N without boundary, that is $T_{f(x)}A \bigoplus Tf(T_xM) = T_{f(x)}N$, $\forall x \in M$. Then $f^{-1}(A)$ is a C^r submanifold of M. Furthermore,* $\operatorname{codim}(f^{-1}(A)) = \operatorname{codim}(A)$.

(B) *Let M and N be C^r manifolds with boundaries, of dimension m and n respectively. Let $f : M \to N$ be a C^r map, transverse to a C^r submanifold A of N, such that either*

(i) *$\partial A \subset \partial N$ and $f|_{\partial M}$ is transverse to A; or*

(ii) *$A \subset N - \partial N, f$ and $f|_{\partial M}$ are transverse to both A and ∂A.*

Then $f^{-1}(A)$ is a C^r submanifold and $\partial f^{-1}(A) = f^{-1}(\partial A)$.

3) *(A) Let M be a compact Hausdorff C^r n-dimensional manifold, $2 \leq r \leq \infty$. Then there is a C^r embedding of M in $\mathbf{R}^{2n+1}$.*

(B) *Let M be a compact Hausdorff C^r n-dimensional manifold, $r \geq 1$. Then there is a* **neat** *C^r embedding into a halfspace of $\mathbf{R}^{2n+1}$. (A neat embedding is one whose image is a neat submanifold.)*

(C) *Let M be a compact manifold without boundary. Then M is diffeomorphic to a C^ω submanifold of Euclidean space.*

4) (**Bertini-Morse-Sard's theorem**). *The measure of the set of critical values $f(\Sigma(f))$ of any C^r map $f : M \to N, \dim M = m, \dim N = n$, such that $r > \max\{0, m-n\}$, has measure zero, that is for every chart (ϕ, U) of N, the set $\phi(U \cap f(\Sigma(f))) \subset \mathbf{R}^n$ has measure zero. (In particular, $f(\Sigma(f))$ has no interior.) The set of regular values of f, $N \setminus f(\Sigma(f))$, is* **residual** *(that is it contains the intersection of a countable family of dense open sets) and therefore is dense.*

5) *Let M be a compact manifold and $f : M \to N$ an analytic map (that is f is of class C^ω). If $\Sigma(f) \neq M, f^{-1}(f(\Sigma(f)))$ has measure zero.*

If f is merely C^∞, the above conclusion can be false.

DEFINITION 2.44 - 1) *A* **differentiable equivalence** *of differentiable maps is a commutative diagram whose vertical arrows are diffeomorphisms:*

$$\begin{array}{cccc} & M_1 & \xrightarrow{f_1} & N_1 & \\ h_1 & \downarrow & & \downarrow & h_2 \\ & M_2 & \xrightarrow[f_2]{} & N_2 & \end{array}$$

2) *Let $C^\infty(M, N)$ be the set of all smooth maps from M to N. Let us consider $Diff(M) \times Diff(N)$ the direct product of groups of diffeomorphisms of M and N respectively. $Diff(M) \times Diff(N)$ acts on $C^\infty(M, N)$ in the following way: $(h, k).f = k \circ f \circ h^{-1}, h \in Diff(M), k \in Diff(N), f \in C^\infty(M, N)$. This is a group action, in fact one has: $(h_1.h_2, k_1k_2).f = (h_1, k_1).((h_2, k_2).f)$. (This action is said to be* **left-right** *(the action of $Diff(M)$ is said to be* **right** *and the action of $Diff(N)$ to be* **left**.*) Two maps $f_1, f_2 \in C^\infty(M, N)$ are* **differentiably equivalent** *iff they belong to the same orbit of the left-right action.*

3) *A map $f \in C^\infty(M, N)$ is said to be* **differentiably stable** *(or more precisely* **left-right differentiably stable**, *or briefly simply* **stable**) *if*

$$(Diff(M) \times Diff(N))(f) \subset C^\infty(M, N)$$

is open.

4) *Two smooth map-germs, $f_x, g_x, x \in M, f, g \in C^\infty(M, N)$, are said to be* (**left-right**) **differentially equivalent** *if there are germs of diffeomorphisms of the source,*

h_x, and target space, k_y, such that

$$f_x = k_{gh^{-1}(x)} \circ g_{h^{-1}(x)} \circ {h_x}^{-1}.$$

The equivalence class of a germ at a critical point $x \in M$ is said to be a **singularity**.
5) *A smooth map-germ $f_x, x \in M, f \in C^\infty(M,N)$, at a point $x \in M$, is said to be (***left-right, differentiably***)* **stable** *if for a sufficiently small neighbourhood U of x there is a neighbourhood E of the map $f \in C^\infty(M,N)$ such that for any map $\widetilde{f}$ in E there is a point $\widetilde{x} \in M$ such that the germ $\widetilde{f}_x$ is equivalent to the germ f_x.*

EXAMPLE 2.43 - 1) (**Morse theory**). (a) The stable maps $M \to \mathbf{R}$ ($\dim M = n$), where M is a closed manifold M, form an everywhere dense set in $C^\infty(M,N)$.
(b) For a map f to be **stable** it is necessary and sufficient that the following two conditions be satisfied:
(i) The map f is stable at each point (in other words all the **critical points** $p \in M$ of the function are **nondegenerate** ($D^2 f(p) \neq 0$.)
(ii) All the critical values of the function f are distinct.
(c) A map $f : M \to \mathbf{R}$ is stable at a point x_0 iff there are neighbourhoods of $x_0 \in M$ and $y_0 \equiv f(x_0) \in \mathbf{R}$ in which coordinates $x^1, \cdots, x^m, y$ can be chosen so that the map can be written in one of the following $m+2$ forms:
(i) $y = x^1$;
(ii) $y = (x^1)^2 + \cdots + (x^k)^2 - (x^{k+1})^2 - \cdots - (x^m)^2, (k = 0, 1, \cdots, m)$.
2) (**Whitney**) Every smooth map of a compact two-dimensional manifold to a two-dimensional manifold can be approximated arbitrarily closed (for any number of derivatives) by a stable map.
3) (**Whitney's theorem**). A map of a two-dimensional manifold to a two-dimensional manifold is stable at a point iff the map can be described with respect to local coordinates (x^1, x^2) in the source and (y^1, y^2) in the target in one of the three terms:
(i) $y^1 = x^1, y^2 = x^2$ (**regular point**);
(ii) $y^1 = (x^1)^2, y^2 = x^2$ (**fold**);
(iii) $y^1 = (x^1)^3 + x^1 x^2, y^2 = x^2$ (**Whitney cusp**).

DEFINITION 2.45 - *Let $f : M \to N$ be a smooth map, $\dim M = m, \dim N = n$. The point $x \in M$ is said to be a* **point of class** Σ^i *for f, if $\dim(\ker(Df(x))) = i$. All the points of class Σ^i for f form a subset of M, called the set Σ^i for f and denoted by $\Sigma^i(f)$.*

EXAMPLE 2.44 - $M \equiv \mathbf{R}^2, N \equiv \mathbf{R}^2, f = (f^1 = (x^1)^3 + x^1 x^2, f^2 = x^2)$. All the critical points are of class Σ^1, while the non critical points are of class Σ^0, (Whitney cusp). The singularity at $(x^1 = 0, x^2 = 0)$ of the map $g \equiv f|_{\Sigma^1(f)} : \Sigma^1(f) \to \mathbf{R}^2$ is called the **cusp** of the map f.

DEFINITION 2.46 - *Let $f : M \to N$ be a smooth map,* $\dim M = m, \dim N = n$. *Let $x \in M$ be a point of M such that* $\mathrm{rank}(Df(x)) = r$. *We call* **coranks** *of f at x the differences* $m - r$ *and* $n - r$.

THEOREM 2.73 - 1) *The coranks are related to the dimension of the kernel i of $Df(x)$ by the formulas:*

$$m - r = i, \quad n - r = n - m + i.$$

2) (**Corank product formula**). *Let $f : M \to N$ be a smooth map such that the sets $\Sigma^i(f)$ are smooth submanifolds of source space. The codimension of the manifold is moreover equal to the product of the coranks:*

$$\dim M - \dim \Sigma^i(f) = (m - r).(n - r).$$

(If this number is negative, then the set is empty.)

DEFINITION 2.47 - (**Secondary singularities**). *Set*

$$\Sigma^{(i_1,i_2)}(f) \equiv \Sigma^{i_2}(f)\Sigma^{i_1}(f).$$

For any set of natural numbers $I \equiv (i_1, \cdots, i_k)$ the set $\Sigma^I(f)$ is defined inductively as follows. Let

$$\Sigma^I(f) \equiv \Sigma^{(i_1,\cdots,i_k)}(f) \subset M$$

be a smooth manifold. Then

$$\Sigma^{(i_1,\cdots,i_k,i_{k+1})}(f) = \Sigma^{i_{k+1}}(f|_{\Sigma^I(f)})$$

is the set of points where the kernel of the restriction of f to $\Sigma^I(f)$ has dimension i_{k+1}.

REMARK 2.51 - According to the definition the manifolds

$$M \supset \Sigma^{i_1}(f) \supset \Sigma^{(i_1,i_2)}(f) \supset \Sigma^{(i_1,i_2,i_3)}(f) \supset \cdots$$

are embedded one inside the other. Therefore, the kernels of the differentials of the restriction of f to these embedded submanifolds are also embedded one inside the other. Thus, the sequence of numbers $i_1, i_2, \cdots$ comprising the inedx I must be non-increasing: $m \geq i_1 \geq i_2 \geq i_3 \geq \cdots \geq 0$. If even one of these inequalities is not satisfied then the set Σ^I is empty. The set $\Sigma^I(f)$ need not be a manifold. However, there is the following important result that gives a criterion to recognize when $\Sigma^I(f)$ is a manifold.

THEOREM 2.74 - (**Bordman**)[12] *Let M and N be manifolds having dimensions m and n respectively. For each sequence $I \equiv (i_1, \cdots, i_k)$ of integers there exists a submanifold, not necessarily closed, $\Sigma^I \subset J\mathcal{D}^\infty(M,N)$ such that the following holds:*
1) $\Sigma^I = \emptyset$ *unless*

$$
\begin{aligned}
&(a) \quad i_1 \geq i_2 \geq \cdots \geq i_{k-1} \geq i_k \geq 0;\\
(2.172) \qquad &(b) \quad m \geq i_1 \geq m-n;\\
&(c) \quad \text{if} \quad i_1 = m-n, \text{then} \quad i_1 = i_2 = \cdots = i_k.
\end{aligned}
$$

2) Σ^I *has codimension* ν_I, *where ν_I is defined, for any sequence of integers $I \equiv (i_1, \cdots, i_k)$ with $i_1 \geq i_2 \geq \cdots \geq i_{k-1} \geq i_k \geq 0$ (we need consider only this case by (2.172)), by*

$$\nu_I \equiv (n-m+i_1)\mu_{i_1 i_2 \cdots i_k} - (i_1 - i_2)\mu_{i_2 i_3 \cdots i_k} - (i_2 - i_3)\mu_{i_3 i_4 \cdots i_k} - \cdots - (i_{k-1} - i_k)\mu_{i_k}$$

where μ_I is the number of sequences of integers $(j_1, \cdots, j_k)$ that satisfy
(a) $j_1 \geq j_2 \geq \cdots \geq j_k$;
(b) $i_r \geq j_r \geq 0$, *for all* $r (1 \leq r \leq k)$, *and* $j > 0$.
3) $\Sigma^I = \pi_{\infty,k}{}^{-1}(\Sigma^I_k)$, *where Σ^I_k is a submanifold of $J\mathcal{D}^k(M,N)$. The set Σ^I is empty unless I satisfies (2.172).*
4) *If $f : M \to N$ is a map such that $D^\infty f : M \to J\mathcal{D}^\infty(M,N)$ is transverse to Σ^I, (or equivalently $D^k f : M \to J\mathcal{D}^k(M,N)$ is transverse to Σ^I_k, then $\Sigma^I(f) \equiv (D^\infty f)^{-1}(D^\infty f(M) \cap \Sigma^I) = (D^k f)^{-1}(D^k f(M) \cap \Sigma^I_k)$ is a submanifold of M having codimension ν_I.*
5) *If I, j denotes the extended sequence $(i_1, i_2, \cdots, i_k, j)$,* we have

$$\Sigma^{I,j}(f) = \Sigma^j(f|_{\Sigma^I(f)}).$$

Also, when $I = \emptyset, \Sigma^j(f) = \{p \in M | \dim \ker(Df(p)) = j\}$.
6) *Any map $f : M \to N$ may be approximated in the fine C^∞ sense by a map $g : M \to N$ whose ∞-derivative $D^\infty g : M \to J\mathcal{D}^\infty(M,N)$ is transverse to all the submanifolds Σ^I.*

REMARK 2.52 - For any $r > 0$ and for a given map $f : M \to N$, transverse to each Σ^I, where the length of I is r we have the following stratifications:
(a)

$$J\mathcal{D}^r(M,N) = \bigcup_{I \equiv (i_1, \cdots, i_r)} \Sigma^I$$

(b)

$$M = \bigcup_{I \equiv (i_1, \cdots, i_r)} \Sigma^I(f).$$

In other words, $J\mathcal{D}^r(M,N)$, (resp. M), is the disjoint union of finitely many submanifolds Σ^I (resp. $\Sigma^I(f)$), where I runs on the sequence of length r.
REMARK 2.53 - 1) $\Sigma^I(f \times id_{\mathbf{R}}) = \Sigma^I(f) \times \mathbf{R}$.
2) Let $f : M \times \mathbf{R}^k \to N \times \mathbf{R}^k$ be a level-preserving map, then $\Sigma^I(f) \cap (M \times t) = \Sigma^I(f_t)$, for each I, $\forall t \in \mathbf{R}^k$.
From above we obtain a general procedure for making any map transverse by adding extra coordinates. In fact, let $f : M \to N$ be any map. Then, for sufficiently huge k, we can take a "homotopy" $f_t : M \to N, t \in \mathbf{R}^k$, with $f_0 = f$, such that the map $M \times \mathbf{R}^k \to J\mathcal{D}^r(M,N), r > 0$, given by $(p,t) \mapsto (D^r f_t)(p)$, is locally a projection. This map is certainly transverse to any submanifold of $J\mathcal{D}^r(M,N)$. Of course, if we want the transversality condition to be satisfied only for particular submanifolds, the number k has to be reduced.
EXAMPLE 2.45 - 1) Let X and Y be C^∞ manifolds of dimensions n and m respectively. Set

$$\begin{aligned}\Sigma_1{}^i &\equiv \bigcup_{(p,q)\in X\times Y} \Sigma_1{}^i{}_{(p,q)} \subset J\mathcal{D}(X,Y) \\ &\cong \bigcup_{(p,q)\in X\times Y} L(T_pX, T_qY)\,,\ i = 0,1,\cdots, s \equiv \min(n,m)\end{aligned}$$

where

$$\Sigma_1{}^i{}_{(p,q)} \equiv \{u \in J\mathcal{D}(X,Y) | u = Df(p), f(p) = q, \mathrm{rank}(Df(p)) = s - i\}.$$

$\Sigma_1{}^i$ are the orbits of the natural action of the group $Diff(X)\times Diff(Y)$ on $J\mathcal{D}(X,Y)$ and

$$\mathrm{codim}(\Sigma^i) = (n - s + i)(m - s + i).$$

One has
(**Whitney-Thom stratification**)

$$J\mathcal{D}(X,Y) = \bigcup_{i=0}^{s} \Sigma_1{}^i.$$

One has:

$$\partial\Sigma_1{}^i = \bar{\Sigma}_1^i - \Sigma_1^i = \bigcup_{j>i} \Sigma_1^j.$$

Then, let $f : X \to Y$ be a C^1 map. Then one has $(Df)^{-1}(\Sigma_1{}^i) = \Sigma^i(f) \subset X$.
☐ If $f : X \to Y$ is C^k-smooth for $k \geq 2$ and if the map $Df : X \to J\mathcal{D}(X,Y)$ is transversal to $\Sigma_1{}^i$ then $\Sigma^i(f)$ is a C^{k-1}-submanifold of codimension $(n - s + i)(m -$

$s+i$) in X. (Recall that the transversality condition means codim $[(Df)^{-1}(T_q{\Sigma_1}^i)] =$ codim $[T_q{\Sigma_1}^i]$, for all $q \in {\Sigma_1}^i$ and for all $p \in X$ such that $f(p) = q$.)

☐ If $\Sigma^i(f) = \emptyset$, for $i > 0$, then the map f is **regular**.

☐ If $\bigcup_{i>0} {\Sigma_1}^i(f) = X$, then f is **totally degenerate**. In this case f is a solution of the relation

$$\bigcup_{i>0} {\Sigma_1}^i \subset J\mathcal{D}(X,Y).$$

☐ If $\dim Y \geq \dim X$, then the regular maps $f : X \to Y$ are **immersions** defined by $\text{rank}(f) = n$.

☐ If $\dim Y \leq \dim X$, then the regular maps $f : X \to Y$ are **submersions** defined by $\text{rank}(f) = m$.

☐ A k**-mersion**, i.e., a C^1 map $f : X \to Y$ such that $\text{rank}(Df(p)) \geq k$, $\forall p \in X$, is a solution of a differential relation (k**-mersion relation**) $\mathcal{I}_k \subset J\mathcal{D}(X,Y)$ such that $\mathcal{I}_k = \bigcup_{i \leq s-k} {\Sigma_1}^i$.

☐ Every open subset in $J\mathcal{D}(X,Y)$ which is invariant under $Diff(X) \times Diff(Y)$ equals the k-mersion relation $\mathcal{I}_k$ for some $k = 0, 1, \cdots$.

2) A C^2 map $f : X \to Y$ is said to **fold along a submanifold** $X_0 \subset X$ if:

(a) f is regular on $X \setminus X_0$:

(b) the map $f|_{X_0} : X_0 \to Y$ is an immersion;

(c) the derivative $Df : X \to J\mathcal{D}(X,Y)$ is transversal to ${\Sigma_1}^1 \subset J\mathcal{D}(X,Y)$;

(d) $\Sigma^1(f) = X_0$.

☐ A C^2 map $f : X \to Y$ **folds along a submanifold** $X_0 \subset X$ if it satisfies an open relation $F_2 \subset J\mathcal{D}^2(X,Y)$ which is invariant under $Diff(Y)$ as well as under the diffeomorphisms of the pair (X, X_0). Furthermore, for $\dim X = \dim Y = n$, local coordinates $(x^1, \cdots, x^n)$ near each point $p \in X_0$ and some coordinates in Y near $f(p) \in Y$ exist such that the map f becomes

$$f : (x^1, \cdots, x^n) \mapsto ((x^1)^2, x^2, \cdots, x^n)$$

for X_0 given by $x^1 = 0$.

3) A map between equidimensional manifolds, $f : X \to Y$, is said to **ramify along a codimension 2 submanifold** $X_0 \subset X$ if in suitable local coordinates $(x^1, \cdots, x^n)$ near each point $p \in X_0$ and for some coordinates in Y near $f(p)$ the map f becomes

$$f : (x^1, \cdots, x^n) \mapsto (\Re(z^s), \Im(z^s), x^3, \cdots, x^n)$$

for $z = x^1 + \sqrt{-1}x^2$ and for some integer $s \geq 1$. (This s may be different for different components of X_0.)

4) The action of the group $Diff(X) \times Diff(Y)$ on $J\mathcal{D}^k(X,Y)$ may have infinitely many orbits for $k \geq 2$. In fact, the number of orbits is infinite for $k \geq 2$ and $\dim X \geq 2$ unless $k = 2$ and $\dim Y = 1$.

Any union of such orbits gives an invariant relation $R_k \subset J\mathcal{D}^k(X,Y)$.

Finally, we shall consider some concepts on the Morse functions and to see as they can be related to the procedure of recognizing the topology of manifolds.

REMARK 2.54 - A **critical point** of a numerical function $f : X \to \mathbf{R}$ on an n-dimensional manifold X belongs to $(df)^{-1}(Z^*) \subset X$, where $Z^* \subset T^*X$ is the n-dimensional submanifold of zeros. A critical point p of f is **non-degenerate** if df is transverse to Z^* at p; this is equivalent to saying that the Hessian $H(f)(p) \in L(T_pX)$ is non-degenerate. If all critical points of f are non-degenarate, f is called a **Morse function**. The **index** of a non-degenarate critical point p of $f : X \to \mathbf{R}$ is the index of $H(f)(p)$ that is the largest possible dimension of a subspace of T_pX on which $H(f)(p)$ is negative definite. If $p \in X$ is a non-degenerate critical point of $f : X \to \mathbf{R}$ with index k, then there is a local coordinate system in a neighbourhood U of p such that

$$f(x) = f(p) - \sum_{1 \leq i \leq k} (x^i)^2 + \sum_{1 \leq j \leq k+1} (x^i)^2, \forall x \in U.$$

We call a Morse function $f : X \to [a,b]$ **admissible** if:

(i) $\partial X = f^{-1}(a) \cup f^{-1}(b)$;

(ii) a and b are regular values.

DEFINITION 2.48 - 1) *A point is called* **regular** *for the function f on X if there is an open neighbourhood U on this point such that U is homeomorphic to $f^{-1}(a) \times I$, (where $I \equiv [0,1] \subset \mathbf{R}$ and $a = f(x)$).*

2) *The point $x \in X$ is called the* **bifurcation point** *of the function f if x is not regular for this function.*

EXAMPLE 2.46 - If $X = M$ is a smooth manifold and f is a smooth function on M, then any point that is regular for f, in the sense of the theory of smooth mappings, is also regular in the sense of the above definition. If $x \in M$ is nondegenerate critical point for the smooth function f, then x is a point of bifurcation. At the same time, if $x \in M$ is a degenerate critical point of f, then it must not be a point of bifurcation.

DEFINITION 2.49 - *The* **k-th type number** *of a Morse function $f : M \to \mathbf{R}$ is the number $\nu_k \equiv \nu_k(f)$ of critical points of index $k, 0 \leq k \leq n =$ $\dim M$. We say f has* **type** $(\nu_0, \cdots, \nu_n)$.

THEOREM 2.75 - 1) *A compact manifold has an admissible Morse function taking prescribed constant values on boundary components.*

2) Let $f : M \to [a,b]$ be a C^{r+1} map on a compact ∂-manifold, $1 \le r \le \omega$. Assume that f has no critical points and $f(\partial M) = \{a,b\}$. Then, there is a C^r diffeomorphism $F : f^{-1}(a) \times [a,b] \to M$ so that the diagram

$$\begin{array}{ccc} f^{-1}(a) \times [a,b] & \stackrel{F}{\to} & M \\ \downarrow & & \downarrow f \\ [a,b] & = & [a,b] \end{array}$$

commutes. In particular, all level surfaces of f are diffeomorphic.

3) Let M be a compact manifold and assume that $\partial M = A \cup B$, where A and B are disjoint closed sets. Assume that there exists a C^2 map $f : M \to \mathbf{R}$ without critical points such that $f(A) = 0, f(B) = 1$. Then M is diffeomorphic to both $A \times I$ and $B \times I$.

4) Let M be a compact n-dimensional manifold without boundary, admitting a Morse function $f : M \to \mathbf{R}$ with only 2 critical points. Then M is homeomorphic to the n-sphere S^n.

5) Let M be a compact manifold and $f : M \to [a,b]$ an admissible Morse function. Assume that f has a unique critical point z, of index k. Then there exists a k-cell (i.e. the image of an embedding $D^k \subset M$)$e^k \subset M - f^{-1}(b)$ such that $e^k \cap f^{-1}(a) = \partial e^k$, and there is a deformation retraction of M onto $f^{-1}(a) \cup e^k$.

6) Let $f : M \to [a,b]$ be an admissible Morse function having a unique critical point z, of index k. Then there exists an $(n-k)$-cell $e_*^{n-k} \subset M - f^{-1}(a)$ such that $e_*^{n-k} \cap f^{-1}(b) = \partial e_*^{n-k}$, and there is a deformation retraction of M onto $f^{-1}(b) \cup e_*^{n-k}$. Moreover e_*^{n-k} can be chosen so that e^k (of point 1) meets e_*^{n-k} only at z, and transversally. (We speak of these cells e^k and e_*^{n-k} as **dual** *to each other.)*

7) Let $f : M \to [a,b]$ be an admissible Morse function of type $(\nu_0, \cdots, \nu_n)$ on a compact manifold M. Assume f has just one critical value $c, a < c < b$. Then there are disjoint k-cells $e_i^k \subset M - f^{-1}(b)$, $1 \le i \le \nu_k, k = 0, \cdots, n$, such that $e_i^k = \cap f^{-1}(a) = \partial e_i^k$ and there is a deformation retraction of M onto $f^{-1}(a) \cup \{ \cup_{i,k} e_i^k \}$.

PROOF. See ref.[50]. □

Let us now give the fundamental theorem that describe quantum tunnel effects in PDEs.

22 THEOREM 2.76 - (TOPOLOGY TRANSITIONS AS QUANTUM TUNNEL EFFECTS IN PDEs).

Let $E_k \subset J\mathcal{D}^k(W)$ be a k-order PDE on the fiber bundle $\pi : W \to M, \dim M = n$. Let A_0 and A_1 be Cauchy hypersurfaces data of E_k. Let us assume that there are closed domains $N_i \subset A_i$, $i = 0,1$, such that there exists a (singular) quantum cobordism $L \in \Omega_s(E_k)$ between N_0 and N_1: $\partial L = N_0 \times 0 \cup N_1 \times 1$. Then there exists an admissible Morse function $f : L \to [a,b]$ such that:

(A) (**Simple quantum tunnel effect**). *If f has a critical point q of index k, then there exists a k-cell $e^k \subset L - N_1$ and an $(n-k)$-cell ${e_*}^{n-k} \subset L - N_0$ such that:*
(i) $e^k \cap N_0 = \partial e^k$;
(ii) ${e_*}^{n-k} \cap N_1 = \partial {e_*}^{n-k}$;
(iii) *there is a deformation retraction of L onto $N_0 \cup e^k$;*
(iv) *there is a deformation retraction of L onto $N_1 \cup {e_*}^{n-k}$;*
(v) ${e_*}^{n-k} \cap e^k = q; \quad {e_*}^{n-k} \pitchfork e^k$.
(B) (**Multi quantum tunnel effect**). *If f is of type $(\nu_0, \cdots, \nu_n)$, where ν_k denotes the number of critical points with index k, such that f has only one critical values $c, a < c < b$, then there are disjoint k-cells $e_i^k \subset L \setminus N_1$, and disjoint $(n-k)$-cells $(e_*)_i^{n-k} \subset L \setminus N_0$, $1 \le i \le \nu_k, k = 0, \cdots, n$, such that:*
(i) $e_i^k \cap N_0 = \partial e_i^k$;
(ii) ${e_*}_i^{n-k} \cap N_1 = \partial {e_*}_i^{n-k}$;
(iii) *there is a deformation retraction of L onto $N_0 \bigcup \{ \cup_{i,k} (e_*)_i^k \}$;*
(iv) *there is a deformation retraction of L onto $N_1 \bigcup \{ \cup_{i,k} (e_*)_i^{n-k} \}$;*
(v) $(e_*)_i^{n-k} \cap e_i^k = q_i; \quad (e_*)_i^{n-k} \pitchfork e_i^k$.
(C) (**No topology transition**). *If f has no critical point then $L \cong N_0 \times I$, where $I \equiv [0,1]$.*
PROOF. The proof can be conducted by considering the above results on the Morse functions. For details see ref.[111]. □

2.7 - COBORDISM GROUPS IN PDEs

In this section we characterize quantum cobordisms and integral cobordisms in PDEs by means of subgroups of usual cobordism groups. More precisely, we prove that integral cobordism groups can be expressed as extensions of quantum cobordism groups and the latter are extensions of subgroups of usual cobordism groups. Furthermore, a complete (co)homological characterization of integral cobordism and quantum cobordism is given. Applications to particular important classes of PDEs are considered. Finally, we give a complete characterization of integral and quantum cobordism by means of some suitable characteristic numbers. As an application, we relate integral cobordism to the term $E_1^{0,n-1}$, of the $\mathcal{C}$-spectral sequence, that represents the space of conservation laws for PDEs. This gives, also, a general method to associate in a natural way a Hopf algebra to $E_1^{0,n-1}$. This is useful in order to characterize PDEs by means of the associated algebraic structures.

(A) DIFFERENTIAL EQUATIONS OF SUBMANIFOLDS.

In this section we shall consider some further fundamental definitions and results on the geometry of PDEs that will be used in the following. (See also refs.[60,72,73,109] for related subjects.)

NOTATION. $\mathbb{P}$ denotes the projectivization functor. $\mathbf{R}P$, (resp. $\mathbf{C}P$) denotes the real (resp. complex) projective space. If H is a subgroup of G we write $H \lhd G$.

DEFINITION 2.50 - *Let W be a smooth manifold of dimension $m+n$. Set*

$$J^k_n(W) \equiv \bigcup_{p\in W} J^k_n(W)_p\,, \quad with \quad J^k_n(W)_p \equiv \{[N]^k_p\}$$

where $[N]^k_p$ is the set of all submanifolds of W of dimension n, that have in p a contact of order k. We call $J^k_n(W)$ the **space of k-jets** *of n-dimensional submanifolds of W.* One has the natural projections $\pi_{k,s} : J^k_n(W) \to J^s_n(W), s < n$. For any n-dimensional submanifold $N \subset W$ one has an embedding

$$j_k : N \to J^k_n(W)\,, \quad j_k : p \mapsto [N]^k_p.$$

The submanifold $N^{(k)} \equiv j_k(N) \subset J^k_n(W)$ is called the **k-jet extension** of N.

TAB.2.8 - PARTICULAR CASES OF $J^k_n(W)$	
Definition	Properties
$J^0_n(W) = W$	
$J^1_n(W)$	$J^1_n(W)_p \cong G_{n,m+n}(T_pW)$
$J^1_1(W)$	$J^1_1(W)_p \cong \mathbb{P}T_pW$
$J^1_{n-1}(W)$	$J^1_{n-1}(W)_p \cong \mathbb{P}T^*_pW$

Let $\Pi^0(W)$ denote the pseudogroup of local diffeomorphisms of the manifold W. Then, we have the following natural action

$$\Pi^0(W) \times J^k_n(W) \to J^k_n(W), \quad (\phi, [N]^k_n) \mapsto [\phi(N)]^k_{\phi(p)}.$$

We denote by $\phi^{(k)} : J^k_n(W) \to J^k_n(W)$ the action on $J^k_n(W)$ induced by $\phi \in \Pi^0(W)$, and call $\phi^{(k)}$ the **k-th prolongation** of ϕ. One has

$$\pi_{k,s} \circ \phi^{(k)} = \phi^{(s)} \circ \pi_{k,s}, \quad k > s$$

and

$$(\phi \circ \psi)^{(k)} = \phi^{(k)} \circ \psi^{(k)}, (id)^{(k)} = 1\,, \ \forall \phi, \psi \in \Pi^0(W).$$

Let $\Pi^k(W) \subset J\mathcal{D}^k(W,W)$ be the subbundle of $J\mathcal{D}^k(W,W)$ of k-jet derivative of local diffeomorphisms $W \to W$. We put

$$\Pi^k(W)_{(a,b)} \equiv \big\{[\phi]^k_a \in \Pi^k(W) | \phi(a) = b\big\}.$$

One has a natural pairing

$$\Pi^k(W)_{(b,c)} \times \Pi^k(W)_{(a,b)} \to \Pi^k(W)_{(a,c)}, \quad ([\psi]^k_b, [\phi]^k_a) \mapsto [\psi \circ \phi]^k_a$$

that defines a structure of Lie groups on $\Pi^k(W)_a \equiv \Pi^k(W)_{(a,a)}$. One has the natural epimorphism of Lie groups

$$1 \to \Pi^k(W)_a \to \Pi^{k-1}(W)_a \to \cdots \to \Pi^1(W)_a \equiv GL(T_aW) \to \Pi^0(W)_a \equiv 1, k \geq 1.$$

If $k \geq 2$ one has the following exact sequence

$$1 \to \Pi^{k+q,k}(W)_a \to \Pi^{k+q}(W)_a \to \Pi^k(W)_a \to 1.$$

This defines $\Pi^{k+q,k}(W)_a$ as the kernel of the bundle $\Pi^{k+q}(W)_a \to \Pi^k(W)_a$. One has the diffeomorphism

$$\Pi^{k+q,k}(W)_a \cong \bigoplus_{1\leq i\leq q} S^{k+i}(T^*_aW) \bigotimes T_aW.$$

The group $\Pi^k(W)_a$ acts on the space $J^k_n(W)_a$ in a natural way : $[\phi]^k_a([N]^k_a) = [\phi(N)]^k_a$. Under this action the kernel $\Pi^{k,k-1}(W)_a$ acts transitively on fibers :

$$\pi^{-1}_{k,k-1}(q) \subset J^k_n(W)_a \,, \ \forall q \equiv [N]^{k-1}_a \in J^k_n(W)_a, \pi_{k-1}(q) = a.$$

A stationary subgroup in $\Pi^{k,k-1}(W)_a$ of the element $u \in \pi^{-1}_{k,k-1}(q)$ under such action, $k \geq 2$, is

$$(Ann(T_aN)) \odot S^{k-1}(T^*_aW) \bigotimes T_aW \bigoplus S^k(T^*_aW) \bigotimes T_aN$$

where $(Ann(T_aN)) \subset T^*_aW$ is the annihilator of T_aN. Therefore, the fibers

$$\pi^{-1}_{k,k-1}(q), \quad k \geq 2$$

are homogeneous spaces with respect to the action of the connected abelian group $\Pi^{k,k-1}(W)_a$ and thus carry an affine structure with the associated vector sapce

$$S^k(T^*_aN) \bigotimes \nu_a, \quad \nu_a \equiv T_aW/T_aN.$$

The prolongations $\phi^{(k)}$ of any local diffeomorphism $\phi : W \to W$, for $k \geq 2$, preserves the affine structure described above. Jet bundles $\pi_{k,k-1} : J^k_n(W) \to J^{k-1}_n(W)$ are affine for $k \geq 2$. Here fiber $\pi^{-1}_{k,k-1}(q)$ over the point $q \equiv [N]^{k-1}_a \in J^k_n(W)_a$ are affine spaces with $S^k(T^*_aN) \otimes \nu_a$ as associated vector space. The prolongations $\phi^{(k)}$:

$J^k_n(W) \to J^k_n(W)$ of the local diffeomorphism $\phi \in \Pi^0(W)$ are affine authomorphisms for $k \geq 2$, and are linear collineations of Grassmannians $G_{n,m+n}(T_aW)$, for $k = 1$.
REMARK 2.55 - If $M \equiv \mathbf{R}^{n+m} \equiv \mathbf{R}^s$, we use the following notations [102]:

$$L^k_s \equiv \Pi^k(\mathbf{R}^s)_{(0,0)}, \quad L^{k+q,k}_s \equiv \Pi^{k+q,k}(\mathbf{R}^s)_0.$$

■

THEOREM 2.77 - (CARTAN DISTRIBUTION). *One has a canonical distribution on $J^k_n(W)$ (**Cartan distribution**), defined by*

$$\mathbb{E}^k_n(W) \equiv \bigcup_{q \in J^k_n(W)} \mathbb{E}^k_n(W)_q, \mathbb{E}^k_n(W)_q \equiv D\pi_{k,k-1}(q)^{-1}(L_q) \subset T_qJ^k_n(W)$$

where $L_q \subset T_{\pi_{k,k-1(q)}}J^{k-1}_n(W)$ is the subspace uniquely identified by $q \equiv [N]^k_a \in J^k_n(W)$. More precisely

$$L_q = T_{\pi_{k,k-1(q)}}N^{(k-1)}.$$

For any $u \in \pi^{-1}_{k+1,k}(q)$, $\mathbb{E}^k_n(W)_q$ can be represented in the following split way

$$\mathbb{E}^k_n(W)_q \cong T_q\pi^{-1}_{k,k-1}(\bar{q}) \bigoplus L_u \cong S^k(T^*_aN) \otimes \nu_a \bigoplus L_u, \bar{q} \equiv \pi_{k,k-1}(q) \in J^{k-1}_n(W).$$

The n-dimensional submanifolds V of the form $V = N^{(k)} \equiv j_k(N) \subset J^k_n(W)$, are the unique n-dimensional integral manifolds of $\mathbb{E}^k_n(W)$, such that $\pi_{k,0} : V \to W$ is an embedding.
THEOREM 2.78 - (DUAL DESCRIPTION OF CARTAN DISTRIBUTION). *The Cartan distribution of $J^k_n(W)$ coincides with the zeros of the differential 1-form*

$$\omega_f \equiv < f, \chi_k > \in \Omega^1(J^k_n(W)), \forall f \in C^\infty(\nu^*_k)$$

where

$$\nu_k \equiv \bigcup_{q \in J^k_n(W)} \nu_k(q), \quad \nu_k(q) \equiv T_{\pi_{k,k-1(q)}}J^{k-1}_n(W)/L_q$$

and

$$\chi_k \in \Omega^1(J^k_n(W)) \bigotimes C^\infty(\nu_k)$$

is the section on $J^k_n(W)$ defined by the following composition

$$T_qJ^k_n(W) \to T_{\pi_{k,k-1(q)}}J^{k-1}_n(W) \to T_{\pi_{k,k-1(q)}}J^{k-1}_n(W)/L_q.$$

*In fact $\mathbb{E}^k_n(W)_q = \ker(\chi_k(q))$. We call **Cartan forms** the differential 1-forms on $J^k_n(W)$, ω_f, for any $f \in C^\infty(\nu^*_k)$. One has the following short exact sequences*

$$0 \to S^{k-1}(T^*_aN) \otimes \nu_a \to \nu_k(q) \to \nu_{k-1}(\bar{q}) \to 0$$

and

$$0 \to \nu_{k-1}^*(\overline{q}) \to \nu_k^*(q) \to S^{k-1}(T_aN) \bigotimes \nu_a^* \to 0$$

with

$$\pi_k(q) = a, \overline{q} \equiv \pi_{k,k-1}(q), \forall q \in J_n^k(W).$$

There exists a canonical $S^{k-1}(T^*N) \bigotimes \nu$*-valued differential 2-form on the Cartan distribution on* $J_n^k(W)$. *(***methasymplectic structure** *on* $J_n^k(W)$*):*

$$\Omega_k : J_n^k(W) \to \Lambda^2(\mathbb{E}_n^k(W)^*) \bigotimes \nu_k .$$

More precisely,

$$\Omega_k(q)(f)(X,Y) = d\omega_f(X,Y), \forall q \in J_n^k(W), f \in C^\infty(\nu_k^*), X, Y \in \mathbb{E}_n^k(W)_q$$

where $\Omega_k(q)(f) = 0$, *for any section* $f \in C^\infty(\nu_{k-1}^*) \subset C^\infty(\nu_k^*))$. *Then, for any* $\lambda \in S^{k-1}(T_aN) \otimes \nu_a^*, q \in J_n^k(W)$ *one has*

$$\ker(\Omega_k(q)(\lambda)) = (g_\lambda)_{+1} = \left\{ \begin{array}{l} \text{first prolongation of} \\ g_\lambda = \{\theta \in S^k(T_a^*N) \otimes \nu_a | \delta\theta \in T_a^*N) \bigotimes g_\lambda\} \end{array} \right\}$$

with

$$g_\lambda \equiv \ker(\lambda) = \{\gamma \in S^{k-1}(T_a^*N) \bigotimes \nu_a | < \lambda, \gamma >= 0\}.$$

DEFINITION 2.51 - *We say that vectors* $X, Y \in \mathbb{E}_n^k(W)_q$ *are in* **involution** *if*

$$\Omega_k(q)(\lambda)(X,Y) = 0, \forall \lambda \in S^{k-1}(T_aN) \bigotimes \nu_a^*.$$

A subspace $P \subset \mathbb{E}_n^k(W)_q$ *is called* **isotropic** *if any vectors* $X, Y \in P$ *are in involution. We say that a subspace* $P \subset \mathbb{E}_n^k(W)_q$ *is a* **maximal isotropic subspace** *if* P *is not a proper subspace of any other isotropic subspace.*

THEOREM 2.79 - (STRUCTURE OF MAXIMAL ISOTROPIC SUBSPACES). *Any maximal isotropic subspace* $P \subset \mathbb{E}_n^k(W)_q$ *has the form*

$$P \equiv \{ < x \otimes \theta > | x \in Ann(U) \subset T_aN, \theta \in S^kU \bigotimes \nu_a \subset S^k(T_a^*N) \bigotimes \nu_a\}$$

where

$$U \equiv Ann(T(\pi_{k,0})(P)) \subset T_a^*N.$$

Now, let $P \subset \mathbb{E}_n^k(W)_q$ *be an isotropic subspace for which the mapping* $T(\pi_{k,0}) : P \to T_aN$ *is an isomorphism. Then* P *is a maximal isotropic subspace and* $P = L_u$ *for some* $u \in \pi_{k+1,k}^{-1}(q)$.

PROPOSITION 2.14 - *Let $N \subset W$ be an n-dimensional submanifold of W and let $N_0 \subset N$ be a submanifold in N. Set*

$$N_0^{(k)}(N) \equiv \{q \in J^k_n(W) | \pi_{k,k-1}(q) \in N_0^{(k-1)}, L_q \supset T_{\pi_{k,k-1}(q)} N_0^{(k-1)}\}$$

where

$$N_0^{(k-1)} \equiv \{[N]^{k-1}_a | a \in N_0\} \subset J^{k-1}_n(W).$$

Then the tangent planes to $N_0^{(k)}(N)$ coincide with the maximal involutive subspaces described in the previous theorem. Therefore, $N_0^{(k)}(N)$ is a maximal integral manifold of the Cartan distribution.

REMARK 2.56 - Let us assume that there exists a fiber bundle structure $\alpha : W \to M$, $\dim M = n$, then we get the natural embeddings

$$J\mathcal{D}^k(W) \cong J^k(W) \hookrightarrow J^k_n(W), \quad k \geq 0$$

and the following commutative diagram

$$\begin{array}{ccc} J^k(W) & \to & J^k_n(W) \\ {\scriptstyle \alpha_{k,k-1}} \downarrow & & \downarrow {\scriptstyle \pi_{k,k-1}} \\ J^{k-1}(W) & \to & J^{k-1}_n(W) \end{array}$$

Here $J\mathcal{D}^k(W)$ is the k-jet derivative space for section of α and $J^k(W)$ is the corresponding k-jet space [96]. For the methasymplectic structure one has

$$\Omega_k(q) \in \Lambda^2(\mathbb{E}^k_n(W)^*_q) \bigotimes S^{k-1}(T^*_b M) \bigotimes vT_a W$$

with $a \equiv \pi_{k,0}(q) \in W, b \equiv \pi_k(q) \in M$. If α is a trivial bundle $\alpha : W \equiv M \times F \to M$, then one has

$$J^k(W) \cong J^k(M, F)$$

and

$$\Omega_k(q) \in \Lambda^2(\mathbb{E}^k_n(W)^*_q) \bigotimes S^{k-1}(T^*_b M) \bigotimes T_f F, \quad \forall a \equiv (b, f).$$

■

DEFINITION 2.52 - *A* **PDE of order k of submanifolds of dimension n** *of the $(m+n)$-dimensional manifold W (resp. of sections of a fiber bundle $\alpha : W \to M$) is a smooth submanifold*

$$E_k \subset J^k_n(W) \quad (\text{resp. } E_k \subset J^k(W)).$$

A **prolongation** *of order $l \geq 0$ of $E_k \subset J^k_n(W)$ is defined by :*

$$(E_k)_{+l} \equiv J^l_n(E_k) \cap J^{k+l}_n(W).$$

In general $(E_k)_{+l}$ *is not a smooth submanifold in* $J^{k+l}_n(W)$. *The system* E_k *is called* **formally integrable** *if all prolongations* $(E_k)_{+l}, l \geq 0$, *are smooth submanifolds and projections*

$$\pi_{k+l+1,k+l} : (E_k)_{+(l+1)} \to (E_k)_{+l}, \quad l = 0, 1, ...$$

are smooth bundles. The **symbol** *of* E_k *at* $q \equiv [N]^k_a \in E_k$ *is defined by*

$$g_k(q) \equiv T_q E_k \cap T_q \pi^{-1}_{k,k-1}(\overline{q}) \subset S^k(T^*_a N) \bigotimes \nu_a$$

with $\overline{q} \equiv \pi_{k,k-1}(q), a = \pi_k(q)$. *If all prolongations* $(E_k)_{+l}, 0 \leq l \leq l_0$, *are smooth submanifolds, then their symbols*

$$g_{k+l}(q) \subset S^{k+l}(T^*_a N) \bigotimes \nu_a$$

at the point $u \in (E_k)_{+l}, \pi_{k+l,k}(u) = q \in E_k$, *are the* **l-th prolongations of the symbol** $g_k(q)$ *at* $q \in E_k$. *One has*

$$\delta(g_{k+l}(q)) \subset T^*_a N \bigotimes g_{k+l-1}(q), l = 1, 2, ...$$

where δ *is the* δ**-Spencer operator***:*

$$\delta : \Lambda^r(T^*_a N) \bigotimes S^t(T^*_a N) \bigotimes \nu_a \to \Lambda^{r+1}(T^*_a N) \bigotimes S^{t-1}(T^*_a N) \bigotimes \nu_a.$$

Then one has

$$g_{k+l}(q) \equiv T_u((E_k)_{+l}) \cap T_u(\pi^{-1}_{k+l,k+l-1}(\overline{u})), \quad \overline{u} \equiv \pi_{k+l,k+l-1}(u)$$

if $(E_k)_{+l}$ *are smooth submanifolds in* $J^{k+l}_n(W)$. *Therefore, at each point* $q \in E_k$ *one has the* δ**-Spencer complex***:*

$$0 \to g_{k+l}(q) \xrightarrow{\delta} (T^*_a N) \bigotimes g_{k+l-1}(q) \xrightarrow{\delta} \cdots \xrightarrow{\delta} \Lambda^{r+1}(T^*_a N) \bigotimes g_{k+l-r-1}(q) \xrightarrow{\delta}$$
$$\cdots \xrightarrow{\delta} \Lambda^n(T^*_a N) \bigotimes g_{k+l-n}(q) \to 0$$

where $g_{k+t}(q) = S^{k+t}(T^*_a N) \otimes \nu_a$ *if* $t < 0$ *is defined. The cohomology of this complex at the term* $\Lambda^i(T^*_a N) \otimes g_{k+j}(q)$ *is denoted by* $H^{j,i}(E_k)_q$ *and called* **Spencer** δ **- cohomology** *of* E_k *at* $q \in E_k$. *We shall say that* E_k *is r-***acyclic** *if* $H^{j,i}(E_k)_q = 0$, *for all* $0 \leq i \leq r, j \geq 0, q \in E_k$.

THEOREM 2.80 - *Let* $E_k \subset J^k_n(W)$ *be a 2-acyclic PDE such that*

$$\pi_{k+1,k} : (E_k)_{+1} \to E'_k, \pi_{k,0} : E_k \to M$$

are smooth bundles. Then E_k is a formally integrable system.

DEFINITION 2.53 - *The* **Cartan vector space** *of $E_k \subset J^k_n(W)$ at q is*

$$(\mathbb{E}_k)_q \equiv \mathbb{E}_k(W)_q \cap T_q E_k.$$

We say that $q \in E_k$ is a **regular point** *if in some neighbourhood $U \subset E_k$ of this point the function $q' \mapsto \dim(\mathbb{E}_k)_{q'}$ is constant, $\forall q' \in U$. Otherwise, the point $q \in E_k$ is called* **singular**. *If all points $q \in E_k$ are regular then*

$$\mathbb{E}_k \equiv \bigcup_{q \in E_k} (\mathbb{E}_k)_q$$

is a distribution on E_k called **Cartan distribution** *of E_k. A* **Cartan connection** *on E_k is an n-dimensional sub-distribution $\mathbb{H} \subset \mathbb{E}_k$ such that*

$$T(\pi_{k,k-1})(H_q) = L_q\ ,\ \forall q \in E_k.$$

Let us assume that $(E_k)_{+1} \to E_k$ is a smooth subbundle of $J^{k+1}_n(W) \to J^k_n(W)$. Then any section $H : E_k \to (E_k)_{+1}$ is called a **Bott connection**. *We call* **curvature** *of the Cartan connection $\mathbb{H}$ on $E_k \subset J^k_n(W)$ the field of geometric objects*

$$\begin{aligned}\Omega_{\mathbb{H}} : E_k \to & \Lambda^2(\mathbb{H}^*) \bigotimes \left[S^{k-1}(T_a N) \bigotimes \nu^*_a / Ann(g_{k-1})\right]^* \\ & \cong \Lambda^2(T^* N) \bigotimes \left[S^{k-1}(T_a N) \bigotimes \nu^*_a / Ann(g_{k-1})\right]^*\end{aligned}$$

defined by restriction of the methasymplectic structure on $\mathbb{H}$. [Note that we can consider the quotient $[S^{k-1}(T_a N) \bigotimes \nu^*_a / Ann(g_{k-1})]$, in the above expression of $\Omega_{\mathbb{H}}-$ map, as if $\lambda \in S^{k-1}(T_a N) \bigotimes \nu^*_a$ is such that belongs to $Ann(g_{k-1}(\pi_{k,k-1}(q))), q \in E_k, \pi_{k,0}(q) = a$, then one has $\Omega_{\mathbb{H}}(q)(\lambda) = 0$].

PROPOSITION 2.15 - *A Cartan connection $\mathbb{H}$ is a Bott one iff $\Omega_{\mathbb{H}} = 0$.*

PROPOSITION 2.16 - *A Cartan connection $\mathbb{H}$ gives a splitting of the Cartan distribution over E_k*

$$\mathbb{E}_k \cong g_k \bigoplus \mathbb{H}.$$

Two Cartan connections $\mathbb{H}, \mathbb{H}'$ on E_k identify a field of geometric objects λ on E_k called **soldering form**

$$\lambda \equiv \lambda_{\mathbb{H},\mathbb{H}'} : E_k \to \mathbb{H}^* \bigotimes g_k, \lambda(q) \in T^*_a N \bigotimes g_k(q)$$

One has

$$\Omega_{\mathbb{H}'} = \Omega_{\mathbb{H}} + \delta\lambda; \left\{ \begin{array}{l} \textbf{(Bianchi identity)}\quad \delta\Omega_{\mathbb{H}} = 0 \\ \Omega_{\mathbb{H}}(q)(\mathrm{mod}\quad \delta(T^*_a N \bigotimes g_k(q)) \in H^{k-1,2}(E_k)_q \end{array} \right\}$$

We call such δ-cohomology class of $\Omega_{\mathbb{H}}(q)$ the **Weyl tensor** *of E_k at $q \in E_k$: $W_k(q) \equiv [\Omega_{\mathbb{H}}(q)]$. Then, there exists a point $u \in (E_k)_{+1}$ over $q \in E_k$ iff $W_k(q) = 0$. Assume that g_{k+1} is a vector bundle over $E_k \subset J^k_n(W)$. Then if the Weyl tensor W_k vanishes the projection $\pi_{k+1,k} : (E_k)_{+1} \to E_k$ is a smooth affine bundle. If g_{k+l} are vector bundles over $E_k \subset J^k_n(W)$ and $W_{k+l} = 0, l \geq 0$, then E_k is formally integrable. If the system $E_k \subset J^k_n(W)$ is of finite type, i.e., $g_{k+l}(q) = 0$, $\forall q \in E_k, l \geq l_0$, then $W_{k+l} = 0, 0 \leq l \leq l_0$, is a sufficient condition for integrability.*

PROOF. In fact in such conditions a unique point $u \in \pi^{-1}_{k+l_0,k+l_0-1}(q)$, over $q \in (E_k)_{+(l_0-1)}$ defines a subspace $L_u \subset T_u[(E_k)_{+(l_0-1)}]$ and hence we get a distribution $L : u \mapsto L_u$ on $(E_k)_{+(l_0-1)}$. This distribution is integrable iff $W_{k+l_0} = 0$ (Frobenius theorem). □

DEFINITION 2.54 - *Let $E_k \subset J^k_n(W)$ be a PDE. A* **regular solution** *of E_k is a smooth submanifold $N \subset W$ (resp. section $s : M \to W$) such that $N^{(k)} \subset E_k$ (resp. $j_k(s)(M) \subset E_k$). [Such solutions are integral manifolds of the Cartan distribution $\mathbb{E}_k$ of E_k that can be projected on W (resp. M) without singularities.] A* **singular solution** *of E_k is an n-dimensional integral manifold, N, of the Cartan distribution $\mathbb{E}_k$ of E_k, such that the set $\Sigma(N)$ of the points $q \in N$ where $\ker(D(\pi_{k,0})(q)) \neq 0$ is nowhere dense in N.* **Singular points** *of such solutions are the points of $\Sigma(N)$. We will denote by $Sol(E_k)$ the set of solutions of E_k.*

PROPOSITION 2.17 - *Given a Cartan connection $\mathbb{H}$ on E_k , for any regular solution, $N^{(k)} \subset E_k$, we identify a section*

$$ {}_{\mathbb{H}}\nabla \in C^\infty(T^*N \bigotimes g_k) $$

called **covariant differential** *of $\mathbb{H}$ at the solution N. Furthermore, for any vector field $X : N \to TN$ we get a section*

$$ {}_{\mathbb{H}}\nabla X \in C^\infty(g_k|_{N^{(k)}}). $$

PROOF. In fact N identifies another Cartan connection on $N^{(k)}, TN^{(k)}$, and therefore the soldering form

$$ {}_{\mathbb{H}}\nabla(q) \in T^*_a N \bigotimes g_k(a) \, , \, \forall q \in [N]^k_a \in N^{(k)} \subset E_k. $$

□

DEFINITION 2.55 - *Let $E_k \subset J^k_n(W)$ be a PDE. A* **symmetry** *of E_k is a local diffeomorphism*

$$ \phi : W \to W \quad \textit{such that} \quad \phi^{(k)}(E_k) \subset E_k. $$

An **infinitesimal symmetry** *of E_k is a vector field $X \in C^\infty(TW)$ such that $X^{(k)} = \partial\phi^{(k)}$, where $\phi^{(k)}_t : J^k_n(W) \to J^k_n(W), t \in \mathbf{R}$, is the 1-parameter group* *of transformations on $J^k_n(W)$ corresponding to the flow ϕ_t associated to X on W. In particular, if $E_k \subset J^k(W)$, where $\alpha : W \to M$ is a fiber bundle, a local authomorphism of α, defined by the following commutative digram*

$$\begin{array}{ccc} W & \stackrel{\phi}{\rightarrow} & W \\ \alpha \downarrow & & \downarrow \alpha \\ M & \underset{\overline{\phi}}{\rightarrow} & M \end{array}$$

is a symmetry of E_k if $\phi^{(k)}(E_k) \subset E_k$. Furthermore, a projectable vector field on W, $X \in \widehat{C}^\infty(TW)$, defined by the following commutative diagram

$$\begin{array}{ccc} W & \stackrel{X}{\rightarrow} & TW \\ \alpha \downarrow & & \downarrow T\alpha \\ M & \underset{\overline{X}}{\rightarrow} & TM \end{array}$$

where $\overline{X} \in C^\infty(TM)$, is an infinitesimal symmetry of E_k if $X^{(k)}$ is tangent to E_k. Let $E_k \subset J^k_n(W)$ be a PDE. Then the linear PDE $R_k \subset J^k(TW)$ defined by

$$R_k \equiv \left\{[X]^k_a, X \in C^\infty(TW) | X^{(k)}(q) \in T_qE_k, \pi_k(q) = a\right\}$$

is called the **infinitesimal Lie equation** *of E_k. The set of regular solutions of R_k is the set of infinitesimal symmetries of E_k. If there exists a fiber bundle structure $\alpha : W \to M$ and $E_k \subset J^k(W)$ then one has*

$$R_k \equiv \{[X]^k_a, X \in \widehat{C}^\infty(TW) | X^{(k)}(q) \in T_qE_k, \pi_k(q) = a\}.$$

The set of regular solutions of R_k is the set of infinitesiml symmetries of E_k.
THEOREM 2.81 - *The pseudogroup G of symmetries of E_k is the set of solutions of a Lie equation $\mathcal{R}_k \subset \Pi^k(W)$ such that $J^k(G) \subseteq \mathcal{R}_k$, where $J^k(G)$ is the k-prolongation of the groupoid associated to G. The equality holds if $\mathcal{R}_k$ is completely integrable. The linearized equation R_k of $\mathcal{R}_k$ is the infinitesimal Lie equation of E_k.*

(B) QUANTUM AND INTEGRAL COBORDISM IN PDEs.

In this section we define quantum cobordism groups and integral cobordism groups and relate them to the usual cobordism groups. These results are the extension of those given in refs.[108,111]. Here we completely solve the problem to determine the integral cobordism group and the quantum cobordism group of any PDE.
Let $R \subset J^k_n(W)$ be an integral manifold, $\dim R = p, p \leq n-1$. We denote by $\Sigma(R) \subset R$ the set of all singular points of the mapping $\pi_{k,0} : R \to W$, and

$$\Sigma_l(R) \equiv \left\{q \in R | \dim \ker(\pi_{k,0}|_R)_{*,q} = l\right\} \subset \Sigma(R).$$

Set

$$L_{0,q} \equiv \ker(\pi_{k,0}|_R)_{*,q} \subset T_qR.$$

Then, one has

$$L_{0,q} \subset S^k(\Xi_q) \bigotimes \nu_a, \tag{2.173}$$

for some subspace $\Xi_q \subset L_q^*$. We will consider only the minimal subspace Ξ_q with the property (2.173).

DEFINITION 2.56 - *An integral manifold $R \subset J_n^k(W)$ will be called* **admissible** *if the following properties hold.*

(i) *$\Sigma(R) \subset R$ has no open subsets and has no* **frozen singularities**, *i.e., for any $q \in \Sigma(R), L_{0,q}$ is a degenerate subspace in $S^k(\Xi_q) \bigotimes \nu_a$, with respect to the exterior 2-form. $\Omega(q)(\lambda) \in C^\infty(\Lambda^2(\mathbb{E}_n^k(W)_q^*))$*[73]. *Recall that*

$$\ker \Omega(q)(\lambda) = \{\theta \in S^k(T_a^*N) \bigotimes \nu_a | \delta\theta \in T_a^*N \bigotimes g_\lambda\} \cong g_\lambda^{(1)}$$

where g_λ is a symbol space of tensor $\lambda \in S^{k-1}(T_aN) \bigotimes \nu_a^*$:

$$g_\lambda \equiv \{\gamma \in S^{k-1}(T_a^*N) \bigotimes \nu_a| < \lambda, \gamma >= 0\}.$$

[*Recall* [73] *that there are no smooth $(p+1)$-dimensional integral manifolds containing integral submanifolds of dimension p passing through frozen singularities.*]

(ii) *There is a vector bundle $e : E \to R$ such that the fibers $E_q \supset \Xi_q$ for any point $q \in R$, and $E_q \subset L_q^*, \dim E_q \leq l$.*

(iii) *The family $L_0 : q \mapsto L_{0,q}$ over $\Sigma(R)$ can be prolonged to some subbundle $L_0 \subset W \bigotimes \nu$.*

DEFINITION 2.57 - *Let $E_k \subset J_n^k(W), k \geq 0$, be a k-order PDE,* $\dim W = n + m$. *Let $N_i \hookrightarrow E_k$, $i = 1, 2$, be two s-dimensional compact smooth closed admissible integral manifolds. Then we say that they are* **E_k-cobordant** *if there exists an $(s+1)$-dimensional compact smooth manifold V , also integral, such that $\partial V = N_1 \bigcup N_2$. Then we write $N_1 \sim_{E_k} N_2$.* [3]

DEFINITION 2.58 - *Let X be a diffiety* [152], *i.e., an object in the category of differential equations of dimension n. (For example $X = E_\infty \to M, \dim M = n$.) Let*

---

[3] More precisely, we can say that if $f_i{:}P_i{\to}E_k$, $i{=}1,2$, are C^∞ mappings that represent s-dimensional integral manifolds N_i, $i{=}1,2$, respectively, then $N_1{\sim_{E_k}}N_2 \Leftrightarrow \exists$ a C^∞ mapping $f{:}R{\to}E_k$ such that R is an $(s{+}1)$-dimensional smooth manifold with $\partial R{=}P_1\dot{\cup}P_2$, $f|_{P_i}{=}f_i$, $i{=}1,2$, $f(R){\equiv}V \subset E_k$ is an $(s{+}1)$-dimensional integral smooth submanifold of E_k. Furthermore, note that if $N{=}\partial V$ is nonorientable, V cannot be simply connected. In fact, a compact nonorientable n-manifold without boundary cannot be embedded in a simply connected $(n{+}1)$-manifold [50].

$N_i \hookrightarrow X$, $i = 1, 2$, *be two s-dimensional compact smooth closed admissible integral manifolds. Then we say that they are* X-**cobordant** *if there exists an* $(s+1)$-*dimensional compact smooth integral manifold* $V \hookrightarrow X$ *such that* $\partial V = N_1 \dot{\bigcup} N_2$. *Then we write* $N_1 \sim_X N_2$.

CONVENTION. The empty set $\emptyset$ will be regarded as an s-dimensional smooth compact admissible integral manifold for all $s \geq 0$.

LEMMA 2.6 - $\sim_{E_k}$ *(resp.* $\sim_X$ *) is an equivalence relation. We write*

$$\Omega_s^{E_k} \quad (resp.\ \Omega_s^X)$$

for the set of all E_k*-cobordisms (resp.* X*-cobordisms) classes* $[N]_{E_k}$ *(resp.* $[N]_X$ *) of closed s-dimensional compact smooth admissible integral submanifolds of* E_k*, (resp.* X*),* $0 \leq s \leq n-1$*. The operation of taking disjoint union* $\dot{\bigcup}$ *defines a sum* $+$ *on* $\Omega_s^{E_k}$ *(resp.* Ω_s^X*) such that* $\Omega_s^{E_k}$ *(resp.* Ω_s^X*) becomes an abelian group,* $n - s > 0$*. The class* $[\emptyset]_{E_k}$ *(resp.* $[\emptyset]_X$*) defines the zero element. For* $k = 0$ *one has* $\Omega_s^{E_0} = \underline{\Omega}_s(E_0)$, *where* $E_0 \subset J_n^0(W) \equiv W$.

DEFINITION 2.59 - *We call* $\Omega_s^{E_k}$, *(resp.* Ω_s^X*), the* **integral s-cobordism group** *of* E_k*,(resp.*X*).*

THEOREM 2.82 - *If* $N = \partial V$, *where* N *is an s-dimensional admissible integral submanifold of* E_k *and* V *is an* $(s+1)$*-dimensional integral submanifold of* E_k , *then* $[N]_{E_k} = [\emptyset]_{E_k}$, *i.e.,* N *cobords with* $\emptyset$.

PROOF. The proof is similar to that for ordinary cobordism. □

THEOREM 2.83 - *In* $\Omega_s^{E_k}$ *one has that the inverse of* $[N]_{E_k}$ *is just* $[N]_{E_k}$.

PROOF. Let us calculate $[N]_{E_k} + [N]_{E_k}$. We can have two cases: (a) $[N]_{E_k} = [\emptyset]_{E_k}$. Then $[N]_{E_k} + [N]_{E_k} = [\emptyset]_{E_k} = 0$. (b) There exists a non-empty s-dimensional integral submanifold N_0 that cobords with N by means of an $(s+1)$-dimensional integral manifold $V \subset E_k$. (This is possible as we consider only admissible integral manifolds.) Then $[N]_{E_k} + [N]_{E_k} = [N]_{E_k} + [N_0]_{E_k} = [N \dot{\bigcup} N_0]_{E_k} = [\partial V]_{E_k} = [\emptyset]_{E_k} = 0$. □

THEOREM 2.84 - *There exists a natural action of the symmetry group* G *of* E_k *on* $\Omega_s^{E_k}$ *by means of isomorphisms. Furthermore, with respect to the reduced equation* E_k/G *one has the following short exact sequence*

$$1 \to G \to G \to Aut(\Omega_s^{E_k/G}) \to 1.$$

PROOF. Let ϕ be a symmetry for $E_k \subset J_n^k(W)$, i.e., a Lie diffeomorphism of $J_n^k(W)$ such that $\phi(E_k) \subseteq E_k$. Then , if $[N]_{E_k} \in \Omega_s^{E_k} \Rightarrow \phi_*[N]_{E_k} = [\phi(N)]_{E_k}$. In fact, one can distinguish two cases: (a) $[N]_{E_k} = [\emptyset]_{E_k}$, i.e., $N = \partial V$, for some $(s+1)$-dimensional integral submanifold V of E_k. Then, $\phi(V)$ is yet an $(s+1)$-dimensional integral submanifold of E_k. Furthermore, $\partial\phi(V) = \phi(\partial V) = \phi(N)$, as diffeomorphic manifolds have diffeomorphic boundaries [48]. Hence, $\phi_*[N]_{E_k} = [\phi(N)]_{E_k} = [\partial\phi(V)]_{E_k} = [\emptyset]_{E_k} = [N]_{E_k}$. So the natural action of ϕ on $\Omega_s^{E_k}$ preserves $[\emptyset]_{E_k}$. (b) There exists a nonempty s-dimensional integral submanifold N_0 of E_k such that $N_0 \in [N]_{E_k}$. Then there exists an $(s+1)$-dimensional integral submanifold V of E_k such that $\partial V = N \dot{\bigcup} N_0$. Then we get

$$\partial\phi(V) = \phi(N) \dot{\bigcup} \phi(N_0) \Rightarrow \phi(N_0) \in [\phi(N)]_{E_k} \in \Omega_s^{E_k}.$$

Furthermore, as

$$\begin{aligned} \phi_*([N_1]_{E_k} + [N_2]_{E_k}) &= \phi_*([N_1 \dot{\bigcup} N_2]_{E_k}) = [\phi(N_1) \dot{\bigcup} \phi(N_2)]_{E_k} \\ &= [\phi(N_1)]_{E_k} + [\phi(N_2)]_{E_k} \\ &= \phi_*[N_1]_{E_k} + \phi_*[N_2]_{E_k} \end{aligned}$$

we get that ϕ_* is a group homomorphism. Furthermore, taking into account also that $\ker(\phi_*) = \{0\}$, and that ϕ_* is surely surjective we get that ϕ_* is the isomorphism identity on $\Omega_s^{E_k}$. The last part of the theorem came from the previous one in a natural way. □

THEOREM 2.85 - *One has a canonical homomorphism*

$$j_s : \Omega_s^{E_k} \to \Omega_s, \quad \textit{given by} \quad j_s : [N]_{E_k} \mapsto [N].$$

The morphism j_s is not, in general, injective.

COROLLARY 2.2 - *If two s-dimensional admissible integral submanifolds $N, N' \subset E_k$ belong to the same integral cobordism class $N' \in [N]_{E_k}$, then their Stiefel-Whitney characteristic numbers are equal. The converse is, in general, a false statement.*

COROLLARY 2.3 - *If two s-dimensional admissible integral submanifolds $N, N' \subset E_k$ are diffeomorphic, then they are not necessarily integral cobordant.*

COROLLARY 2.4 - *If two s-dimensional admissible integral submanifolds $N, N' \subset E_k$ are homotopic, they are not necessarily integral cobordant.*

DEFINITION 2.60 - *An $(s+1)$-**quantum cobord** of a PDE $E_k \subset J_n^k(W), k \geq 0$, is an $(s+1)$-dimensional smooth integral manifold $V \hookrightarrow J_n^k(W)$ such that ∂V is an s-dimensional admissible integral submanifolds of E_k. The set of $(s+1)$-quantum cobords of E_k is denoted by $\Omega(E_k)_{s+1}$. An n-quantum cobord is simply called **quantum cobord**. Furthermore, we call $\Omega(E_\infty)_n \equiv \Omega(E_k)$ **quantum situs** of E_k. The **classic limit** of $\Omega(E_k)_{s+1}$ is the subset $\Omega(E_k)_{s+1,c} \subset \Omega(E_k)_{s+1}$ of $(s+1)$-quantum cobords contained in E_k. In particular, $\Omega(E_\infty)_{n,c} = Sol(E_k)$ is the set of solutions of E_k.*

THEOREM 2.86 - *The $(s+1)$-quantum cobordism, (in particular quantum cobordism), is an equivalence relation in the set of s-dimensional integral submanifolds of E_k (in particular in the set of hypersurface Cauchy data on E_k). The set of s-quantum cobordism classes is denoted by $\Omega_s(E_k)$. The operation of disjoint union makes $\Omega_s(E_k)$ into an abelian group.*

DEFINITION 2.61 - *We call $\Omega_s(E_k)$ the* **s-quantum cobordism group** *of E_k. (In particular $\Omega_{n-1}(E_k)$ is the* **quantum cobordism group** *of E_k.)*

PROPOSITION 2.18 - *One has a natural surjective mapping*

$$\sigma : \Omega(E_k)_{s+1} \to \Omega_s(E_k), L \mapsto [N]_{\overline{E_k}}$$

where $[N]_{\overline{E_k}}$ is an equivalence class in $\Omega_s(E_k)$ such that L cobords some elements of $[E]_{\overline{E_k}}$.

THEOREM 2.87 - *The classic limit $\Omega(E_k)_{s+1,c} \subset \Omega(E_k)_{s+1}$ identifies by means of the map σ a subgroup $\Omega(E_k)_c$* (**classic limit** *of the s-quantum cobordism group of E_k). The fiber $\Omega(E_k)_{s+1,[N]_{\overline{E_k}}}$ on $[N]_{\overline{E_k}} \in \Omega_s(E_k)$ is the* **quantum fluctuation** *of the set of classic solutions $(\Omega(E_k)_{s+1,c})_{[N]_{\overline{E_k}}}$. In particular, if $(\Omega(E_k)_{s+1,c})_{[N]_{\overline{E_k}}}$ is reduced to a point, i.e., there is a unique $(s+1)$-dimensional integral submanifold L of E_k that cobords any two s-dimensional integral manifolds of $[N]_{\overline{E_k}}$, then $(\Omega(E_k)_{s+1})_{[N]_{\overline{E_k}}}$ is the* **quantum fluctuation** *of L.*

THEOREM 2.88 - *Let us introduce in $\Omega(E_k)_{s+1}$ the following equivalence relation*

$$V \sim V' \Leftrightarrow V, V' \in \Omega(E_k)_{s+1,[N]_{\overline{E_k}}}, \quad \text{for some} \quad [N]_{\overline{E_k}} \in \Omega_s(E_k).$$

One has the following bijective mapping

$$\Omega(E_k)_{s+1}/\sim \quad \cong \quad \Omega_s(E_k)$$

that identifies a structure of abelian group on $\Omega(E_k)_{s+1}/\sim$.

THEOREM 2.89 - *One has the following identification*

$$\Omega(E_k)_{s+1,c}/\sim \quad \cong \Omega_s^{E_k}.$$

2 THEOREM 2.90 - (EXTENSION THEOREM FOR INTEGRAL AND QUANTUM COBORDISM GROUPS).

The integral cobordism groups $\Omega_s^{E_k}, 0 \leq s < n$, are extensions of the corresponding quantum cobordism groups $\Omega(E_k)_s, 0 \leq s < n$. One has the following exact commu-

tative diagrams

$$\begin{array}{ccccccccc} & & & & & & 0 & & \\ & & & & & & \downarrow & & \\ & & & & & & K_s(E_k) & & \\ & & & & & & \downarrow & & \\ 0 & \to & K_s^{E_k} & \to & \Omega_s^{E_k} & \xrightarrow{\widehat{\bar{i}}_s} & \Omega_s(E_k) & \to & 0 \\ & & & & & & \downarrow & & \\ & & & & & & \widehat{\Omega}_s(E_k) & & \\ & & & & & & \downarrow & & \\ & & & & & & 0 & & \end{array}$$

for $0 \leq s \leq n-1$. *So we can write*

$$\Omega_s^{E_k} \in H^2(\Omega_s(E_k), K_s^{E_k}) \cong Hom_{\mathbf{Z}}(H_2(\Omega_s(E_k), \mathbf{Z}), K_s^{E_k})$$
$$\Omega_s(E_k) \in H^2(\widehat{\Omega}_s(E_k), K_s(E_k)) \cong Hom_{\mathbf{Z}}(H_2(\widehat{\Omega}_s(E_k), \mathbf{Z}), K_s(E_k)).$$

PROOF. In fact we have the following commutative diagram

$$\begin{array}{ccccccc} & & & & & & 0 \\ & & & & & & \downarrow \\ & & & & & & K_s(E_k) \\ & & & & & & \downarrow \\ 0 & \to & K_s^{E_k} & \to & \Omega_s^{E_k} & \xrightarrow{\widehat{\bar{i}}_s} & \Omega_s(E_k) \\ & & & & & & \downarrow \widehat{\bar{j}}_s \\ & & & & & & \Omega_s \end{array}$$

where

$$K_s^{E_k} = \left\{ \begin{array}{l} [N]_{E_k} \in \Omega_s^{E_k} | N = \partial V, \text{for some} \\ (s+1)\text{-dimensional integral submanifold} \\ V \subset J_n^k(W) \end{array} \right\}$$

and

$$K_s(E_k) = \big\{[N]_{\overline{E_k}} \in \Omega_s(E_k) | N = \partial V, \text{for some } (s+1)\text{-dimensional manifold} \quad V\big\}.$$

Moreover, we put

$$\widehat{\Omega}_s(E_k) \equiv \Omega_s(E_k)/K_s(E_k), \quad \widehat{\Omega}_s^{E_k} \equiv \Omega_s^{E_k}/K_s^{E_k}.$$

It is easy to see that the map $\widehat{\bar{i}}_s : [N]_{E_k} \mapsto [N]_{\overline{E_k}}$ is surjective. □

DEFINITION 2.62 - *We call* $\widehat{\Omega}_s^{E_k}$ **reduced s-integral cobordism group** *of* E_k *, and* $\widehat{\Omega}_s(E_k)$ **reduced s-quantum cobordism group** *of* E_k. *Furthermore, we say* *that* E_k *is* **s-integral cobording reducible**, *(resp.* **s-quantum cobording reducible**), *if* $H^2(\Omega_s(E_k), K_s^{E_k}) = 0$, *(resp.* $H^2(\widehat{\Omega}_s(E_k), K_s^{E_k}) = 0$).

COROLLARY 2.5 - *The reduced s-integral cobordism group of* E_k *is isomorphic to the corresponding s-quantum cobordism group:*

$$\widehat{\Omega}_s^{E_k} \cong \Omega_s(E_k), 0 \leq s \leq n-1.$$

THEOREM 2.91 - (STRUCTURE OF THE REDUCED QUANTUM COBORDISM GROUPS). *The reduced s-quantum cobordism groups of* E_k, $\widehat{\Omega}_s(E_k), 0 \leq s \leq n-1$, *are abelian torsion finite groups of the form*

$$\widehat{\Omega}_s(E_k) \cong \mathbf{Z}_2 \bigoplus \cdots_p \cdots \bigoplus \mathbf{Z}_2.$$

Furthermore, Stiefel-Whitney numbers completely characterize $\widehat{\Omega}_s(E_k)$. *More precisely, we have:*

$$\widehat{\Omega}_s(E_k) = \left\{ \begin{array}{l} [N] \in \Omega_s \\ \exists\, X \equiv \text{s-dimensional integral submanifold of } \quad E_k \\ w[X] = w[N] \end{array} \right\}$$

PROOF. We shall use the following lemmas.

LEMMA 2.7 - (PONTRJAGIN-THOM). *A closed n-dimensional manifold* V, *belonging to the category of smooth differentiable manifolds is cobordant in this category iff the Stiefel-Whitney numbers*

$$w_{i_1}[V], \cdots, w_{i_s}[V]$$

are all zero, where $i_1 + \ldots + i_s = n$ *is any partition of* n. *[Recall that the Stiefel-Whitney cohomology classes* $w_i \in H^i(V; \mathbf{Z}_2)$ *and the Stiefel-Whitney numbers are obtained by means of the canonical integration homomorphism:* $[V] : H^i(V; \mathbf{Z}_2) \to \mathbf{Z}_2$.*]*

PROOF OF LEMMA 2.7 - Let M be an n-dimensional manifold. Let $\tau : M \to G_{n,N} \equiv B0_n$ be the classifying mapping for the tangent bundle TM, where $M \subset \mathbf{R}^N$, and $N \to \infty$. Let $\omega \in H^\bullet(B0_n; \mathbf{Z}_2)$ be a characteristic class mod 2. We set $\omega(N) \equiv \tau^*\omega$. The **stable characteristic classes** $\omega \in H^\bullet(B0_n)$ are defined by means of the restriction

$$\omega \equiv \lambda^*\overline{\omega}, \quad \overline{\omega} \in H^\bullet(BO_{n+1}), \quad \lambda : B0_n \to BO_{n+1}.$$

Similarly we can introduce the notions of stable characteristic classes for

$$BS0_n, BU_n, BSp_n.$$

Let $M = \partial V$. Then, one has $\omega(M) = \tau_M^* \omega = \tau_M^* \lambda^* \overline{\omega}$, $\overline{\omega}(V) = \tau_V^* \overline{\omega}$. Let j denote the immersion $M \to V$. We get

$$TV|_M = j^* TV = TM \bigoplus \mathbf{1}, \quad \omega(M) = \tau_M{}^* \lambda^* \overline{\omega} = j^* \tau_V^* \overline{\omega}.$$

As $j_*[M] = 0$, we get $< j^* \tau_V^* \overline{\omega}, [M] > = < \tau_M{}^* \overline{\omega}, j_*[M] > = 0$. □

LEMMA 2.8 - (PONTRJAGIN-THOM). *A closed n-dimensional manifold, belonging to the category of oriented differentiable manifolds is cobordant in this category iff both its Stiefel-Whitney number and its Pontrjagin number are zero. [Recall that the Pontrjagin numbers are defined only if the dimension $n = k4$.]*

PROOF OF LEMMA 2.8 - The proof in the oriented case is similar to the above one, by substituting the $\mathbf{Z}_2$-homologies with the $\mathbf{Q}$-homologies and taking into account that in orientable case the equality $j_*[M] = 0$ in $H_n(V)$ is true in rational homology. □

LEMMA 2.9 - (STRUCTURE OF THE COBORDISM GROUPS Ω_n AND ${}^+\Omega_n$). *The cobordism group Ω_n is a finite abelian torsion group of the form:*

$$\Omega_n \cong \mathbf{Z}_2 \bigoplus \cdots_p \cdots \bigoplus \mathbf{Z}_2$$

where p is the number of non-dyadic partitions of n. (A partition $(i_1, \ldots, i_r)$ of n is non dyadic if none of the i_β are of the form $2^s - 1$.) Two smooth closed n-dimensional manifolds belong to the same cobordism class iff all their corresponding Stiefel-Whitney numbers are equal. Furthermore, ${}^+\Omega_n$ is a finitely generated abelian group of the form

$${}^+\Omega_n \cong \mathbf{Z} \bigoplus \cdots \bigoplus \mathbf{Z} \bigoplus \mathbf{Z}_2 \bigoplus \cdots \bigoplus \mathbf{Z}_2$$

where infinite cyclic summands occur only if $n \equiv 0 \pmod 4$. Two smooth closed oriented n-dimensional manifolds belong to the same cobordism class iff all their corresponding Stiefel-Whitney and Pontrjagin numbers are equal.

PROOF OF LEMMA 2.9 - The first part of the theorem follows from the above considerations on the group and from the Thom-Pontrjagin theorem. In fact, if $[N] = [\emptyset]$, all the Stiefel-Whitney numbers of N are zero. Furthermore, if $[N] \neq [\emptyset]$, then there exists a nonempty n-dimensional manifold $N_0 \in [N] \Rightarrow \exists$ $(n+1)$-dimensional manifold V such that $\partial V = N_0 \dot{\bigcup} N$. Hence, we get $[N_0 \dot{\bigcup} N] = [\partial V] = [\emptyset] = 0$. Therefore, for all the Stiefel-Whitney class w we should have $0 = w[N_0 \dot{\bigcup} N] = w[N_0] + w[N]$. Taking into account that $w[N_0], w[N] \in \mathbf{Z}_2$, we conclude that V cobords N_0 with N if N_0 has the same Stiefel-Whitney numbers of N.

The second part of the lemma can be similarly proved. □

Now, theorem is a direct consequence of above theorem, taking into account that $\Omega_s(E_k)$ is a subgroup of the corresponding cobordism group Ω_s :

$$\widehat{\Omega}_s(E_k) \lhd \Omega_s \cong \mathbf{Z}_2 \bigoplus \cdots_r \cdots \bigoplus \mathbf{Z}_2 \ , r \geq p.$$

Moreover, $K_s(E_k) = \cap \ker(w[-])$, where $w[-] : \Omega_s(E_k) \to \mathbf{Z}_2$ are the mappings defined by means of the Stiefel-Whitney characteristic classes; the intersection is made on all the kernel of these maps. Finally, one has

$$\widehat{\bar{j}}_s[N]_{\overline{E_k}} = \widehat{\bar{j}}_s[N']_{\overline{E_k}} \Leftrightarrow w[N]_{\overline{E_k}} = w[N']_{\overline{E_k}}.$$

Therefore, $\widehat{\Omega}_s(E_k) \cong \mathrm{im}(\widehat{\bar{j}}_s)$ is completely characterized by means of Stiefel-Whitney numbers. □

THEOREM 2.92 - (STRUCTURE OF THE QUANTUM COBORDISM GROUPS). *$\Omega_s(E_k)$ is an extension of a finite abelian torsion group of the form*

$$\mathbf{Z}_2 \bigoplus \cdots_p \cdots \bigoplus \mathbf{Z}_2.$$

One has the following exact commutative diagram

$$\begin{array}{ccccccccc} & & & & & & 0 & & \\ & & & & & & \downarrow & & \\ & & & & & & K_s(E_k) & & \\ & & & & & & \downarrow & & \\ 0 & \to & K_s^{E_k} & \to & \Omega_s^{E_k} & \xrightarrow{\widehat{\bar{i}}_s} & \Omega_s(E_k) & \to & 0 \\ & & & & & & \downarrow & & \\ & & & & & & \mathbf{Z}_2 \bigoplus \cdots_p \cdots \bigoplus \mathbf{Z}_2 & & \\ & & & & & & \downarrow & & \\ & & & & & & 0 & & \end{array}$$

So we can write

$$\begin{aligned} \Omega_s(E_k) &\in H^2(\mathbf{Z}_2 \bigoplus \cdots_p \cdots \bigoplus \mathbf{Z}_2, K_s(E_k)) \\ &\cong Hom_{\mathbf{Z}}(H_2(\mathbf{Z}_2 \bigoplus \cdots_p \cdots \bigoplus \mathbf{Z}_2, \mathbf{Z}); K_s(E_k)) \\ &\cong \bigoplus_{q \in \{1,\dots,r\}} Hom_{\mathbf{Z}}(\mathbf{Z}_2 \bigotimes \mathbf{Z}_2; K_s(E_k)) \end{aligned}$$

where $r = r(p)$. One has

$$\mathrm{rank}(\Omega_s(E_k)) = \mathrm{rank}(K_s(E_k)).$$

We can write

$$K_s(E_k) = \big\{[N]_{\overline{E_k}} | w[N] = 0\big\}.$$

PROOF. It is a direct consequence of the above results. Note that one has

$$H_2(\mathbf{Z}_2 \bigoplus \cdots_p \cdots \bigoplus \mathbf{Z}_2, \mathbf{Z}) \cong \mathbf{Z}_2 \bigotimes_{\mathbf{Z}} \mathbf{Z}_2 \bigoplus \cdots_r \cdots \bigoplus \mathbf{Z}_2 \bigotimes_{\mathbf{Z}} \mathbf{Z}_2.$$

This can be seen by considering that for any finite cyclic group G of order n one has

$$H_0(G, \mathbf{Z}) = \mathbf{Z}; H_{2s+1}(G, \mathbf{Z}) = \mathbf{Z}_n; H_{2s}(G, \mathbf{Z}) = 0 = H_s(G, \mathbf{Z}), s > 0.$$

Furthermore, one uses the Künneth theorem for groups, that allows to write the unnatural isomorphism

$$\begin{aligned} H_s(G_1 \bigoplus G_2, \mathbf{Z}) \cong & \bigoplus_{p+q=s} H_p(G_1, \mathbf{Z}) \bigotimes_{\mathbf{Z}} H_q(G_2, \mathbf{Z}) \\ + & \bigoplus_{p+q=s-1} Tor^{\mathbf{Z}}(H_p(G_1, \mathbf{Z}), H_q(G_2, \mathbf{Z})) \end{aligned}$$

for any two groups G_1 and G_2 , and one considers that $Tor^{\mathbf{Z}}(A, B) = 0$ for projective, (e.g, free) $\mathbf{Z}$-module A (or B). Furthermore, recall that for an arbitrary abelian group G one can define rank as follows:

$$\text{rank}(G) \equiv \sup\{\text{rank}(F)\}_{F \in \mathcal{F}(G)}$$

where $\mathcal{F}(G)$ is the set of free subgroups of G. Furthermore, rank(F) = cardinality of a base. □

COROLLARY 2.6 - *If E_k is s-integrable reducible, (resp. s-quantum reducible), one has the isomorphism:*

$$\Omega_s^{E_k} \cong K_s^{E_k} \bigoplus \Omega_s(E_k), (\textit{resp. } \Omega_s(E_k) \cong K_s(E_k) \bigoplus \widehat{\Omega}_s(E_k)).$$

THEOREM 2.93 - (COHOMOLOGY STRUCTURE OF $\Omega_s^{E_k}$ AND $\Omega_s(E_k)$). *One has the following isomorphisms*

$$\Omega_s^{E_k} \cong H^2(\Omega_s(E_k), f_s)(u_s) \quad ; \quad \Omega_s(E_k) \cong H^2(\mathbf{Z}_2 \bigoplus \cdots \bigoplus \mathbf{Z}_2, h_s)(v_s)$$

where

$$\begin{aligned} u_s &\in H^2(\Omega_s(E_k), H_2(\Omega_s(E_k), \mathbf{Z})) \\ v_s &\in H^2(\mathbf{Z}_2 \bigoplus \cdots \bigoplus \mathbf{Z}_2, H_2(\mathbf{Z}_2 \bigoplus \cdots \bigoplus \mathbf{Z}_2, \mathbf{Z})) \end{aligned}$$

and f_s and h_s are uniquely determined mappings

$$f_s : H_2(\Omega_s(E_k), \mathbf{Z}) \to K_s^{E_k} \quad ; \quad h_s : H_2(\mathbf{Z}_2 \bigoplus \cdots \bigoplus \mathbf{Z}_2, \mathbf{Z}) \to K_s(E_k).$$

PROOF. We shall use the following lemma.

LEMMA 2.10 - *For any extension of abelian groups*

$$v \in H^2(G, A) \cong Hom_{\mathbf{Z}}(H_2(G, \mathbf{Z}); A)$$

there exists a universal cohomology class $u \in H^2(G, H_2(G, \mathbf{Z}))$ *and a unique mapping* $f : H_2(G, \mathbf{Z}) \to A$, *such that* $v = H^2(G, j)(u)$.

PROOF OF LEMMA 2.10 - This lemma is a direct consequence of the following. □

LEMMA 2.11 - (YONEDA)[81]. *Let* $\mathcal{C}$ *be a category and* $A \in Ob(\mathcal{C})$, *and let* $h_A : \mathcal{C} \to \mathcal{S} \equiv$ *category of sets, be the covariant functor represented by* A , *i.e.,* $h_a \equiv Hom_{\mathcal{C}}(A, -)$. *Let* $u_A \in h_A(A)$ *be the identity map* $A \to A$. *Let* $T : \mathcal{C} \to \mathcal{S}$ *be an arbitrary covariant functor. For any* $v \in T(A)$, *there is a unique natural transformation* $\phi : h_A \to T$ *such that* $\phi(u_A) = v$. *Then, if* $\mathcal{F}$ *is the category of functors* $\mathcal{C} \to \mathcal{S}$, *one has* $Hom_{\mathcal{F}}(h_A, T) \cong T(A)$, *i.e.,* h_A *is "freely generated" by* u_A.

Then from the above considerations we get the proof of theorem. □

EXAMPLE 2.47 - In order to consider some examples of reduced s-quantum cobordism groups, let us first consider some examples of ordinary cobordism groups.

1) Any smooth closed compact n-dimensional manifold N always cobords with an odd number of copies of itself:

$$N \sim N \dot{\bigcup} \cdots_{2s+1} \cdots \dot{\bigcup} N. \tag{2.174}$$

Furthermore, any smooth closed compact n-dimensional manifold N cobords with an even number of copies of itself $N \sim N \dot{\bigcup} \cdots_{2s} \cdots \dot{\bigcup} N$ iff $N \in [\emptyset]$. ■

2) In the category of compact smooth 1-dimensional manifolds

$$\Omega_1 = 0.$$

In fact, any possible smooth 1-dimensional closed compact manifold can be obtained as disjoint union of copies of S^1. Now, we can have manifolds like the following:

$$N \equiv S^1 \dot{\bigcup} \cdots_{2s+1} \cdots \dot{\bigcup} S^1, s \geq 0, N_0 \equiv S^1 \dot{\bigcup} \cdots_{2s} \cdots \dot{\bigcup} S^1, s \geq 1.$$

[Of course in the category of singular 1-dimensional closed manifolds the situation should be different as, for example, we should consider also manifolds like

$$(S^1, *) \dot{\bigcup} (S^1, *)/\{*\} \dot{\bigcup} \{*\}.]$$

From the above example we get that manifolds like N belong to the same cobordism class of S^1. Furthermore, manifolds like N_0 cobords with $\emptyset$. On the other hand $S^1 \in [\emptyset]$, as $S^1 = \partial D^2$, where D^2 is the 2-dimensional disk. Therefore $\Omega_1 = 0$. ■

3) One has the following $\Omega_2 \cong \mathbf{Z}_2$. In fact, it is possible to prove (Rohlin) that any 2-dimensional closed compact smooth manifold is or $\mathbf{R}P^2$+handles or K^2+handles, where K^2 is the Klein bottle. Now, as $\chi(\mathbf{R}P^2) = 1$ it follows that $\mathbf{R}P^2 \notin [\emptyset] \in \Omega_2$, hence $\Omega_2 \neq 0$. On the other hand $K^2 = \partial V$, where V is a suitable strip, hence $K^2 \in [\emptyset] \in \Omega_2$. Then, we can conclude that $\Omega_2 \cong \mathbf{Z}_2$. ∎

4) In the following we give some examples of calculated cobordism groups.

$$\Omega_0 \cong \mathbf{Z}_2 \quad ; \quad \Omega_1 \cong 0$$
$$\Omega_2 \cong \mathbf{Z}_2 \quad ; \quad \Omega_3 = 0$$
$$\Omega_4 = \mathbf{Z}_2 \bigoplus \mathbf{Z}_2 \quad ; \quad {}^{+}\Omega_s \cong 0, s = 1,2,3,6,7 \quad \text{and} \quad s > 11$$
$${}^{+}\Omega_s \cong \mathbf{Z}, s = 0,4 \quad ; \quad {}^{+}\Omega_s \cong \mathbf{Z}_2, s = 5,11$$
$${}^{+}\Omega_8 \cong \mathbf{Z} \bigoplus \mathbf{Z} \quad ; \quad {}^{+}\Omega_9 \cong \mathbf{Z}_2 \bigoplus \mathbf{Z}.$$

∎

5) Taking into account, now, that $\widehat{\Omega}_s(E_k)$ are subgroups of Ω_s we have the following.

$$\widehat{\Omega}_s(E_k) = 0 \,, \text{ if } s = 1,3;$$
$$\widehat{\Omega}_s(E_k) = 0, \text{or} \quad \widehat{\Omega}_s(E_k) = \mathbf{Z}_2, \text{if} \quad s = 0,2;$$
$$\widehat{\Omega}_4(E_k) = 0, \text{or} \quad \widehat{\Omega}_4(E_k) = \mathbf{Z}_2, \text{or} \quad \widehat{\Omega}_4(E_k) = \mathbf{Z}_2 \bigoplus \mathbf{Z}_2.$$

Of course, when there are more that one possibilities it is necessary to look more to the structure of the particular PDE in order to decide about. ∎

In order to characterize reduced quantum cobordisms and their classic limits by means of subgroups of suitable homology groups, let us consider some useful definitions and results on the ordinary cobordism groups.

DEFINITION 2.63 - *Let B be a closed differential connected manifold and let $\xi \equiv \{p : E \to B, F \equiv \mathbf{R}^n\}$ be a vector bundle over B with fiber $\mathbf{R}^n$ and structure group $G = O(n), SO(n), U(n), \ldots$. Let $\tilde{E} \to B$ be the subbundle of ξ defined by the vectors in the fibers with length ≤ 1. The fiber F' of $\tilde{E}$ is $F' = D^n \subset \mathbf{R}^n$. The boundary $\partial\tilde{E}$ is a fiber bundle with fiber S^{n-1}. The* **Thom complex** *of the vector bundle ξ is the quotient complex $M(\xi) \equiv \tilde{E}/\partial\tilde{E}$, where $\partial\tilde{E}$ is reduced to a point. [So $M(\xi)$ is the compactified to a point of $E \quad : \quad M(\xi) \equiv E \bigcup\{\infty\}$.]*

LEMMA 2.12 - *One has the natural isomorphisms:*

$$\phi : H_i(B; A) \to H_{n+i}(M(\xi); A), \phi : H^i(B; A) \to H^{n+i}(M(\xi); A)$$
$$z \mapsto \phi(z) \equiv p^* z (\operatorname{mod} \partial\tilde{E}), i \geq 0, n = \dim F$$

where $A = \mathbf{Z}_2$ if $G = O(n), A = \mathbf{Z}, \mathbf{Q}, \mathbf{Z}_p$ if $G = SO(n), \ldots$. More precisely, $\phi = D_{\tilde{E}} \circ D_B$, where D_X are the following Poincaré duality operators:

$$D_B : H_q(B) \to H^{m-q}(B), \dim B = m,$$
$$D_{\tilde{E}} : H^{m-q}(\tilde{E}) \cong H^{m-q}(B) \to H_{n+m-(m-q)}(\tilde{E}, \partial\tilde{E}) \cong H_{q+n}(M(\xi)), q > 0.$$

One has a **fundamental class** *in the* **cohomology of Thom** *of* ξ, *i.e.,* $\phi(1) \in H^n(M(\xi))$. *Furthermore, the following identifications hold:*

$$M(\xi)/B \equiv \xi, \quad M(\xi)\backslash B \cong \{*\}$$

where B *is identified with a submanifold of* $M(\xi)$ *by means of the zero section.*
DEFINITION 2.64 - *We call* **class** $\omega_i \in H^i(B; \mathbf{Z}_2)$ *an element*

$$\phi^* s_q^i(\phi(1)), \text{ where } \quad \phi : H^q(B; \mathbf{Z}_2) \to H^{n+q}(M(\xi); \mathbf{Z}_2), \text{ and } \quad s_q^i$$

are Steenrod squares [133]

$$s_q^i : H^n(M(\xi); \mathbf{Z}) \to H^{n+i}(M(\xi); \mathbf{Z}_2).$$

PROPOSITION 2.19 - *The group* $O(n)$ *contains a subgroup of diagonal matrices* $D(n) \subset O(n)$ *like*

$$\begin{bmatrix} \pm1 & 0 & 0 & 0 \\ 0 & \pm1 & 0 & 0 \\ \bullet & \bullet & \bullet & \bullet \\ 0 & 0 & 0 & \pm1 \end{bmatrix}.$$

One has

$$D(n) \cong \mathbf{Z}_2 \times \cdots \times \mathbf{Z}_2.$$

Furthermore, one has a canonical mapping between classifying spaces

$$BD(n) \cong \mathbf{R}P_1^\infty \times \cdots_s \cdots \times \mathbf{R}P_n^\infty \xrightarrow{i} BO(n)$$

and the induced cohomology mapping , $i^* : H^\bullet(BO(n); \mathbf{Z}_2) \to H^\bullet(BD(n); \mathbf{Z}_2)$. *One can see that* i^* *is a monomorphism and that* $\mathrm{im}(i^*)$ *is the set of symmetric polynomials in* $\chi_1, ..., \chi_n$ *, where* $0 \neq \chi_i \in H^1(\mathbf{R}P_i^\infty; \mathbf{Z}_2)$. *Moreover, the Stiefel-Whitney classes are elementary symmetric polynomials*

$$i^*(w_q) = \sum_{i_1 < \ldots < i_q} \chi_{i_1} \cdots \chi_{i_q}.$$

The mapping $j : BSO(n) \to BO(n)$ *induces an epimorphism*

$$j^\bullet : H^\bullet(BO(n); \mathbf{Z}_2) \to H^\bullet(BSO(n); \mathbf{Z}_2).$$

Furthermore, $\ker(j^\bullet)$ *is generated, as ideal, by the element* $\omega_1 \in H^1(BO(n); \mathbf{Z}_2)$. *Now, let* $M(\xi)$ *be the Thom complex of the universal bundle* ξ *on* $BO(n)$. *Let* $f : BO(n) \to M(\xi)$ *be the natural embedding. Then, for the induced mapping* $f^\bullet$:

$H^\bullet(M(\xi);\mathbf{Z}_2) \to H^\bullet(BO(n);\mathbf{Z}_2)$ *we get that* $f^\bullet$ *is a monomorphism and* $\mathrm{im}(f^\bullet)$ *is the set of polynomials in* ω_i *divisible by* $\omega_n \in H^n(BO(n);\mathbf{Z}_2)$, *where* $i^*\omega_n = \chi_1...\chi_n$. *Furthermore,*

$$f^\bullet\phi(1) = \omega_n, f^\bullet\phi(\omega_i) = s^i_q(\omega_n) = \omega_i\omega_n.$$

In general,

$$f^\bullet\phi(x) = x\omega_n.$$

Similar results can be obtained for $H^\bullet(BSO(n);\mathbf{Z}_2)$.

EXAMPLE 2.48 - If $B = BG$, with $G = O(n), SO(n), U(n/2), SU(n/2), Sp(n/2)$, and if ξ is the universal bundle with fiber $\mathbf{R}^n$, the Thom complex $M(\xi)$ is denoted by $MO(n), MSO(n), MU(n/2), MSp(n/4)$.

EXAMPLE 2.49 - If $G = e$, the universal bundle ξ is $\xi \equiv BG \times \mathbf{R}^n = \{*\} \times \mathbf{R}^n$. One has $Me = S^n$. In particular, $SO(1) = e, MSO(1) = S^1$. ■

EXAMPLE 2.50 - $O(1) = \{\pm 1\}, BO(1) = \mathbf{R}P^\infty$ (or $\mathbf{R}P^N$ for big N). The universal bundle $p : E \to \mathbf{R}P^N$, fiber $F = \mathbf{R}$, is like the normal bundle of $\mathbf{R}P^N$ into $\mathbf{R}P^{N+1}$. The corresponding bundle $\tilde{E}$ is of fiber $D^1 = I$ is a Mobius strip. The corresponding boundary $\partial\tilde{E}$ is a sphere S^N , covering $\mathbf{R}P^N$. So the Thom complex is given by $M(\xi) = MO(1) = \tilde{E}/\partial\tilde{E} = \mathbf{R}P^{N+1} \supset \mathbf{R}P^N = B$. ■

EXAMPLE 2.51 - For $G = SO(2)$ similarly one has $MSO(2) = MU(1) = \mathbf{C}P^{N+1} \supset \mathbf{C}P^N = B, N \to \infty$. ■

EXAMPLE 2.52 - The fundamental class is a base element of the groups: $u = \phi(1) \in H^1(S^1;\mathbf{Z})$, for $MSO(1) = S^1; u = \phi(1) \in H^1(\mathbf{R}P^\infty;\mathbf{Z}_2)$, for $MO(1) = \mathbf{R}P^\infty; u = \phi(1) \in H^2(\mathbf{C}P^\infty;\mathbf{Z}_2)$, for $MSO(2) = \mathbf{C}P^\infty$. These spaces are Eilenberg-MacLane complexes [133] of the type $K(\mathbf{G},n)$, for $n = 1,2; \mathbf{G} = \mathbf{Z},\mathbf{Z}_2$; the element $u = \phi(1)$ coincides with the fundamental element of the complex $K(\mathbf{G},n)$. ■

LEMMA 2.13 - *The Thom complexes* $M(\xi)$ *are simply connected for* $n > 1$. *Their homotopy groups are reported in the following table.*

TAB.2.9 - **Homotopy groups of** $M(\xi)$	
$\pi_j(M(\xi))$	Conditions
0	$1 \leq j < n$
$\mathbf{Z}_2$	$j = n$, non-orientable fiber bundle
$\mathbf{Z}$	$j = n$, orientable fiber bundle

PROOF OF LEMMA 2.13 - The cell-decomposition of $M(\xi)$ is obtained from that of B by means of the multiplication of one cell (fiber). $B \supset \sigma^i \mapsto \phi(\sigma^i) = p^{-1}(\sigma^i) = \sigma^{n+i}$. Furthermore, there exists a 0-dimensional cell $\sigma^0 \subset M(\xi)$, obtained by reduction of $\partial\tilde{E}$ to a point. Thus, $\pi_j(M(\xi)) = 0$, for $j < n$, as cells do not exist in these dimensions. Let us assume that in B there exists only one 0-dimensional cell, (if B is connected, we can reduce the situation to this case). Then, in $M(\xi)$ there exists only one n-dimensional cell, (i.e., the fiber over a point). Thus $\pi_n(M(\xi))$ is cyclic. If the fiber bundle is not orientable, in the basis B there exists a loop (that can be assimiled to a cell σ^1) that inverts the orientation of the fiber. Its reciprocal image $p^{-1}(\sigma^1_j) = \phi(\sigma^1) = \sigma^{n+1}$ is a cell in $M(\xi)$ such that $\partial\sigma^{n+1} = 2\sigma^n$. If the fiber bundle is orientable, the boundaries of all the cells $p^{-1}(\sigma^1_j)$ in the complex $M(\xi)$ are zero. Then, the cycle $[\sigma^n]$ is of infinite order. So taking into account that $H_n(M(\xi)) = \pi_n(M(\xi))$ we complete the proof. □

THEOREM 2.94 - *The cobordism groups are canonically isomorphic to the stable homotopy groups. See the following table.*

TAB.2.10 - **Cobordism gs.vs.isomorphic stable homotopy gs.**	
Cobordism group	Stable homotopy group
Ω_s	$\pi_{n+s}(MO(n))$
$^+\Omega_s$	$\pi_{n+s}(MSO(n))$
Ω^U_s	$\pi_{n+s}(MU(n/2))$
Ω^{SU}_s	$\pi_{n+s}(MSU(n/2))$
Ω^{Sp}_s	$\pi_{n+s}(MSp(n/4))$

PROOF. Let us consider a closed s-dimensional manifold $M \subset \mathbf{R}^{n+s}$, where $s < n-1$. All the immersions $M \subset \mathbf{R}^{n+s}$ are isotopic and the normal bundle of M in $\mathbf{R}^{n+s}$ does not depend on the immersion. Let us denote ν this normal bundle. Let us

$$\begin{array}{ccc} M & \rightarrow & BO(n) \\ \uparrow & & \uparrow \\ \nu & \rightarrow & \xi \end{array}$$

denote the relation with the universal bundle of fiber $\mathbf{R}^n$. The image of the fiber bundle ν covers $\tilde{E}$. Prolong this application to all the complementary of the neighbourhood, in such a way that all this complementary should be mapped in a cell $\sigma^0 \in M(\xi)$ obtained by contraction of $\partial\tilde{E}$. So we get the application $f : S^{n+s} \rightarrow M(\xi)$. This mapping is t-regular on the submanifold $BO(n) \subset M(\xi)$ and $f^{-1}(BO(n)) = M$. The notion of t-regularity along $BO(n) \subset M(\xi)$ means that in any point $x \in f^{-1}(BO(n))$ the image of the tangent space T_xS^{n+s} by means of the linear mapping $Df(x)$ is trasversal to the tangent plane $T_{f(x)}BO(n) \subset T_{f(x)}M(\xi) : T_{f(x)}M(\xi) \cong Df(x)(T_xS^{n+s}) \bigoplus T_{f(x)}BO(n)$. Let us now consider the cobord V of dimension $s+1$ such that: (i) $\partial V = M_1 \dot{\bigcup} M_2$, (ii) $V \subset \mathbf{R}^{n+s} \times I$, (iii) $M_1 \subset \mathbf{R}^{n+s} \times \{0\}$, (iv) $M_2 \subset \mathbf{R}^{n+s} \times \{1\}$, (v) V meets transversally the boundary. By respecting the previous construction for the bundle ν normal to $V \subset \mathbf{R}^{n+s} \times I$, we get a homotopy $S^{n+s} \times I \rightarrow M(\xi)$. Then we have built the homomorphism $\Omega_s \rightarrow \pi_{n+s}(MO(n)), s < n-1$. Similarly we can build the homomorphism $^+\Omega_s \rightarrow \pi_{n+s}(MSO(n)), s < n-1$. Let us now prove that the above correspondence is an isomorphism. Let $a \in \pi_{n+s}(MO(n))$, be represented by a mapping $f : S^{n+s} \to MO(n)$. After an eventual little deformation, we can assume that f is t-regular along the submanifold $BO(n) \subset MO(n)$. The reciprocal image $f^{-1}(BO(n)) = M$ is a non-singular differential submanifold $M \subset \mathbf{R}^{n+s} \subset S^{n+s}$. The image by means of Df of n-plane normal to M in $\mathbf{R}^{n+s}$ is trasversal to $BO(n)$. We can make this image normal to $BO(n)$ everywhere along $BO(n)$ by using an elementary deformation of the mapping; we can also fully reduce the complementary of the neighbourhood of the manifold M in S^{n+s} to a point σ^0 obtained by contraction of $\partial\tilde{E}$ in $MO(n)$. So the theorem is proved for Ω_s. For $^+\Omega_s$ the proof can be similarly deduced. □

DEFINITION 2.65 - *A k-***cycle** *of M is a couple (N, f) where N is a k-dimensional closed (orientable) manifold and $f : N \to M$ is a differentiable mapping. A* **period** *of the k-form ω on M, on the orientable cycle (N, f) is the integral*

$$\int_N f^*\omega.$$

*We call $(k+1)$-***strip** *of M a couple (V, F) where V is a $(k+1)$-dimensional (oriented) manifold with boundary $\partial V \equiv N$ and $F : V \to M$ is a differentiable mapping.*

REMARK 2.57 - For any orientable cycle (N, f), $\dim N = k$, the period of an exact form $\omega = d\alpha$ is zero:

$$\int_N f^*\omega = 0.$$

In fact,

$$\int_N f^*\omega = \int_N f^*d\alpha = \int_N d(f^*\alpha) = \int_{\partial N} f^*\alpha = 0, \quad \text{as} \quad \partial N = 0.$$

Furthermore, if the orientable cycle (N, f) is the boundary of a strip (V, F), i.e., $N = \partial V, F|_N = f$, then the period of a closed k-form on this cycle is zero: $\int_N f^*\omega = 0$. In fact, if $\partial(V, F) = (N, f)$ and $d\omega = 0$ we have

$$\int_N f^*\omega = \int_V d(F^*\omega) = \int_V F^*(d\omega) = 0.$$

Moreover, if the period of a closed form ω on all orientable cycles are zero then the form ω is exact. Finally a k-cycle (N, f) is the boundary of a $(k+1)$-strip (V, F) if for all their periods on all the closed k-forms are zero. These results are the meaning of the **de Rham theorem** that says that for any manifold M one has a canonical nondegenerate bilinear mapping

$$H_p(M; \mathbf{R}) \times H^p(M; \mathbf{R}) \to \mathbf{R}.$$

Then, the associated homomorphisms

$$H_p(M;\mathbf{R}) \to H^p(M;\mathbf{R})^*, \quad H^p(M;\mathbf{R}) \to H_p(M;\mathbf{R})^*$$

are injective, In particular, if $H_p(M;\mathbf{R})$, and hence also $H^p(M;\mathbf{R})$, is finite dimensional, (e.g., if M is a finite CW-complex), then the above homomorphisms become isomorphisms.

REMARK 2.58 - Let M be a compact symmetric space of a connected compact Lie group G. Then: (a) any invariant differential form on M is closed; (b) any closed differential form on M is cohomologous to an invariant form; (c) a nonzero invariant differential form is never cohomologous to zero. ■

EXAMPLE 2.53 - (TORUS) : $T^n \equiv \mathbf{R}^n/\Gamma$, where Γ is a lattice with integer numbers in $\mathbf{R}^n$, generated by n linearly independent vectors. The torus is a compact abelian Lie group. Let $x^1, ..., x^n$ be Euclidean coordinates in $\mathbf{R}^n$. Then, all the form $dx^{i_1} \wedge ... \wedge dx^{i_k}$ are invariant (for translations) on $\mathbf{R}^n$. So these forms define the invariant forms on T^n. Therefore, a form $\omega = \omega_{i_1...i_k} dx^{i_1} \wedge \cdots_s \wedge dx^{i_k}$ is invariant on T^n iff $\omega_{i_1...i_k}(x+y) = \omega_{i_1...i_k}(x)$, i.e., $\omega_{i_1...i_k} = const$. So any invariant form on T^n is a linear combination with constant coefficients of exterior products of the forms $dx^1, ..., dx^n$. As a consequence the cohomology ring $H^\bullet(T^n)$ is an exterior algebra $\Lambda(e_1, ..., e_n)$ with generators $e_1, ..., e_n$ of degree 1. Here e_i is the cohomology class of dx^i. ■

DEFINITION 2.66 - *A* **group of cycles** (N,f) *of an n-dimensional manifold M is the set of formal sums $\sum_i (N_i, f_i)$, where (N_i, f_i) are cycles of M. The quotient of this group by the cycles equivalent to zero, i.e., the boundaries gives the* **bordism groups** $\underline{\Omega}_s(M)$. *We define* **relative bordisms** $\underline{\Omega}_s(X,Y)$, *for any pair of manifolds $(X,Y), Y \subset X$. Similarly we define the* **bordisms** ${}^+\underline{\Omega}_s(X)$, *and* **oriented bordisms** ${}^+\underline{\Omega}_s(X,Y)$, *for oriented pair of cycles $Y \subset X$.*

PROPOSITION 2.20 - *One has:*

$$\underline{\Omega}_s(*) \cong \Omega_s \quad ; \quad {}^+\underline{\Omega}_s(*) \cong {}^+\Omega_s.$$

For bordisms, the theorem of invariance of homotopy is valid. Furthermore, for any CW-pair $(X,Y), Y \subset X$, one has the isomorphisms :

$$\underline{\Omega}_s(X,Y) \cong \underline{\Omega}_s(X/Y), \quad s \geq 0.$$

REMARK 2.59 - One has a natural group-homomorphism

$$\underline{\Omega}_s(X) \to H_s(X;\mathbf{Z}_2).$$

This is an isomorphism for $s = 1$. Note that for contractible manifolds, $H_s(X) = 0$, for $s > 0$, but $\underline{\Omega}_s(X)$ cannot be trivial for any $s > 0$. So, in general, $\underline{\Omega}_s(X) \neq H_s(X)$.
THEOREM 2.95 - *A cycle $x \in H_s(M;\mathbf{Z}_2)$, $\dim M = n + s$, is realized by means of a closed s-dimensional submanifold $N \subset M$ iff there exists a mapping $f : M \to MO(n)$ such that $f^*u = Dx$, where $u \in H^n(MO(n);\mathbf{Z}_2)$ is a fundamental class and $D : H_s(M;\mathbf{Z}_2) \to H^n(M;\mathbf{Z}_2)$ is the Poincaré duality operator.*[4]
Let M be an $(n+s)$-dimensional oriented manifold. A cycle $x \in H_s(M;\mathbf{Z})$ is realized by means of a closed oriented submanifold $N \subset M$ iff there exists a mapping $f : M \to MSO(n)$ such that $f^*u = Dx$. A cycle $x \in H_s(M;\mathbf{Z})$ is realized by means of a closed oriented submanifold $N \subset M$ of trivial normal bundle (i.e., defined by means of a family of nonsingular equations $\psi_1 = 0, \cdots, \psi_k = 0$ in M) iff there exists a mapping $f : M \to Me \cong S^n$ such that $f^*u = Dx$. (Similar theorems hold in the cases of realizations of cycles by means of submanifolds with normal bundles endowed with structural groups $U(n/2), SU(n/2)$, etc. A mapping $M \to MU(n/2), M \to MSU(n/2), M \to MSp(n/4)$ generates such restrictions.) In some cases the complexes $MO(s)$, $MSO(s)$ are Eilenberg-MacLane complexes of type $K(\mathbf{G}, n)$. (The following table resume such cases.)

TAB.2.11 - $MO(s)$ **and** $MSO(s)$ $\mathbf{K(G, n)}$**-complexes**

$MO(1) = \mathbf{R}P^\infty \cong K(\mathbf{Z}_2, 1)$	$\pi_j = 0, j > 1$
$MSO(1) = Me \cong S^1 \cong K(\mathbf{Z}, 1)$	$\pi_j = 0, j > 1$
$MSO(2) = \mathbf{C}P^\infty \cong K(\mathbf{Z}, 2)$	$\pi_j = 0, j \neq 2$

Here the element $\phi(1) = u \in H^n(MG)$ coincides with the fundamental class of the complex $K(\mathbf{G}, n)$.
PROOF. Let us prove these theorems for $G = O(n)$. Let $N \subset M$ be a submanifold. The normal bundle defines the mapping

$$\begin{array}{ccc} M & \xrightarrow{f} & MO(n) \\ \cup & & \cup \\ N & \rightarrow & BO(n) \end{array}$$

the complement of the neighbourhood of N in M fully reduces to a point σ^0 by means of the contraction of $\partial\tilde{E}$ in the construction of $M(\xi)$. Then we get $f^*u = D[N]$. On the other hand, if we assign a mapping $f : M \to MO(n)$ **t-regular** along $BO(n) \subset MO(n)$, the reciprocal image of $N \equiv f^{-1}(BO(n))$ is such that $f^*u = D[N]$. $\square$

[4] D is generated by means of the de Rham theorem on the duality between the singular homology space $H_p(X)$ and the singular cohomology space $H^p(X)$.

THEOREM 2.96 - *Any cycle $x \in H_n(M;\mathbf{Z}_2), \dim M = n+1$, can be realized with a closed submanifold. Furthermore, any cycle $x \in H_n(M;\mathbf{Z}), n+1 \leq \dim M \leq n+2$, can be realized with an orientable closed submanifold. If $s < n/2$, for any cycle $x \in H_s(M;\mathbf{Z}), \dim M = n$, there exists a $\lambda \neq 0$ such that the cycle λx is represented by an s-dimensional submanifold $N \subset M$.*

THEOREM 2.97 - *For any cycle $x \in H_s(M;\mathbf{Z})$ there exists an $\lambda \neq 0$ such that λx is the image of an s-dimensional manifold $M, \phi : M \to X, \phi_*[M] = x$. One has the natural epimorphism $\underline{\Omega}_s^{SO}(X,Y) \bigotimes \mathbf{Q} \to H_s(X,Y;\mathbf{Q})$.*

PROOF. We use the immersion $X \subset \mathbf{R}^{N+s}$ and consider a ∂-manifold $U \supset X$ that reduces to $X, U \sim X$. The cycle $\lambda x \in H_s(U) \cong H_s(X)$ can be realized by a submanifold by means of the application $f : (U,\partial U) \to MSO(N)$, where ∂V reduces to a point and $f^*u = D[N]$. □

DEFINITION 2.67 - (MILNOR). *Let O_∞ denote the union of the orthogonal groups $O_1 \subset 0_2 \subset O_3 \subset ...$ in the fine topology. Then, we recognize that this infinite orthogonal group O_∞ acts on the space X. It follows that each 0_k acts on X. Hence there is a bundle of geometric objects of type X (fiber) on every manifold V. A homotopy class of cross-sections of the bundle of geometric objects with fiber X over V is called an* **X-structure** *on V. A manifold V together with an X-structure on V is called an* **X-manifold**. *If V is an X-manifold, then so is ∂V. Given any closed X-manifold V one can define a second X-manifold $-V$ so that $\partial(V \times I) \cong V \dot{\bigcup}(-V)$. Thus one can define a cobordism group for the class of X-manifolds. The resulting group will be denoted by $\underline{\Omega}_n^X$ and called the* **n-th X-cobordism group**.

THEOREM 2.98 - (MILNOR-NOVIKOV). *A* **closed weakly complex manifold** *V, i.e., a manifold with O_∞/U_∞-structure, where O_∞/U_∞ denotes the union of the spaces $0_2/U_1 \subset O_4/U_2 \subset O_6/U_3 \subset \cdots_s \subset O_\infty/U_\infty$ in the fine topology, with O_∞ acting on O_∞/U_∞ in the usual way, is the boundary of a weakly complex manifold iff its Chern numbers $c_{i_1}...c_{i_n}[V]$ are all zero. [Note that an O_∞/U_∞-structure on V determines a preferred U_∞-bundle over V. Hence, Chern classes are defined.] It follows that*

$$\underline{\Omega}_s^{O_\infty/U_\infty} = 0\,,\ \text{for}\, n = 2n+1; \quad \underline{\Omega}_s^{O_\infty/U_\infty} = \text{ free abelian group for } \quad n = 2k.$$

THEOREM 2.99 - *The graded group $\underline{\Omega}_\bullet^{O_\infty/U_\infty}$ has a natural ring structure, making it into a polynomial ring $\mathbf{Z}[Y_2, Y_4, Y_6, ...]$ with one generator in each even dimension: $\underline{\Omega}_\bullet^{O_\infty/U_\infty} \cong \mathbf{Z}[Y_2, Y_4, Y_6, ...]$. (More generally, one could use any subgroup G of O_∞ instead of U_∞.) The following seems to be particularly interesting:*

$$1 \subset Sp_\infty \subset SU_\infty \subset U_\infty \subset SO_\infty \subset O_\infty.$$

Note that

$$\underline{\Omega}_\bullet^{O_\infty/O_\infty} \simeq \Omega_\bullet, \quad \underline{\Omega}_\bullet^{O_\infty/SO_\infty} \simeq^+ \Omega_\bullet.$$

EXAMPLE 2.54 - To any sphere an O_∞/U_∞-structure can be given although only S^2 and S^6 possess quasi-complex structures. ■

EXAMPLE 2.55 - $\underline{\Omega}_\bullet^{O_\infty/1} \cong \underline{\Omega}_\bullet^{O_\infty}$. An O_∞-structure on V is a trivialization of the tangent O_∞-bundle of V (the **stable tangent bundle**). Manifolds which admit such a structure are called **π-manifolds**. One has the isomorphism

$$\underline{\Omega}_\bullet^{O_\infty} \cong \pi_{n+m}(S^m)$$

with m large. ■

THEOREM 2.100 - (THOM). *One has the following isomorphism*

$$\underline{\Omega}_\bullet^X \cong \pi_\bullet(MX)$$

where MX is the Thom space of X , i.e., $MX \equiv TX \bigcup\{\infty\}$.

EXAMPLE 2.56 -

$$\underline{\Omega}_\bullet^{O_\infty} \cong \mathbf{Z}_2[Y_2, Y_4, Y_6, ...]$$

$$\underline{\Omega}_\bullet^{U_\infty} \cong \mathbf{Z}[Y_2, Y_4, Y_6, ...]$$

$$\underline{\Omega}_\bullet^{SO_\infty}/\mathcal{I} \cong \mathbf{Z}[Y_4, Y_8, ...]$$

■

DEFINITION 2.68 - *A* **singular X-manifold** *in the space Y is a continuous map (***singular simplex***) $f : M \to Y$, where M is a closed X-manifold. Two singular X-manifolds $(M, f), (M', f')$ in Y are called* **X-cobordant** *if there is a pair (W, g) such that W is a compact X-manifold with boundary $\partial W \cong M \bigcup(-M'), ((-M')$ in the sense of negative in the cobordism group) and g is a continuous map $g : W \to Y$ such that $g|_M = f, g|_{M'} = f'$.*

THEOREM 2.101 - *For the group $\underline{\Omega}_n^X(Y)$ of X-***cobordism classes** *of singular n-dimensional X-manifolds in Y one has the following isomorphism :*

$$\underline{\Omega}_n^X(Y) \cong MX_n(Y) \cong \pi_n(MX \wedge Y^+)$$

where $Y^+ \equiv Y \bigcup\{\infty\} \equiv Y/\emptyset$.

REMARK 2.60 - A **spectrum** is a sequence of spaces $\{E_n\}$, with base point, together with weak equivalences $\epsilon_n : SE_n \to E_{n+1}$ or adjointly $\epsilon'_n : E_n \to \Omega E_{n+1}$, where S is the suspension functor, and Ω is the loop functor. The **generalized cohomology** associated to $\{E_n\}$ is denoted by $h^\bullet(X)$ and defined by $h^n(X) \equiv [X, E_n]$ for a space X. The **generalized homology theory** associated to $\{E_n\}$ is denoted by $h_\bullet(X)$ and defined by $h_n(X) \equiv \pi_n(E_n \wedge X)$. The **coefficients** of the generalized theories determined by $\{E_n\}$ are groups: $h^\bullet(*) \cong h_\bullet(*) = \pi_\bullet(E)$. ■

COROLLARY 2.7 - *One has the following isomorphism*

$$\underline{\Omega}_n^X(Y) \cong h_n(Y^+)$$

where $h_\bullet(Y^+)$ is the homology induced by the spectrum MX.

COROLLARY 2.8 - *If Y is any space like in the above theorem, then one has a spectral sequence $\{E^r_{p,q}, d^r\}$ with $E^2_{p,q} \cong H_p(Y; h_q(*))$ and converging to $h_\bullet(Y)$.*

PROOF OF COROLLARY 2.8 - It is a direct consequence of the above theorem taking into account that MX is a spectrum. □

COROLLARY 2.9 - *Furthermore, there is a spectral sequence $\{E^{p,q}_r\ ,\ d_r\}$ with $E^{p,q}_2 \cong H^p(Y, h^q(*))$ converging to $h^\bullet(Y)$.*

PROOF OF COROLLARY 2.9 - From the above considerations and taking into account the Atiyah-Hrizebruch theorem the proof is obtained. □

THEOREM 2.102 - (D.QUILLEN [119]). 1) *One has the canonical isomorphism*

$$\underline{\Omega}_p(X) \cong \bigoplus_{r+s=p} H_r(X; \mathbf{Z}_2) \bigotimes_{\mathbf{Z}_2} \Omega_s.$$

In particular, as $\Omega_0 = \mathbf{Z}_2$ and $\Omega_1 = 0$, we get

$$\underline{\Omega}_1(X) \cong H_1(X; \mathbf{Z}_2).$$

2) *For the relative cobordism one has the following isomorphism:*

$$\underline{\Omega}_s(X, Y) \cong H_s(X, Y; MO).$$

Furthermore, if (X, Y) is a CW-pair one also has the following isomorphisms:

$$\underline{\Omega}_p(X, Y) \cong \bigoplus_{r+s=p} H_r(X, Y; \mathbf{Z}_2) \bigotimes_{\mathbf{Z}_2} \Omega_s$$

and the natural evaluation homomorphism $e : \underline{\Omega}_s(X, Y) \to H_s(X, Y; \mathbf{Z}_2)$ which sends the class represented by $f : (V, \partial V) \to (X, Y)$ into $f_*[V, \partial V]$ is an epimorphism. In particular, one has:

$$\Omega_1(X, Y) \cong H_1(X, Y; \mathbf{Z}_2).$$

If (X, Y) is a finite CW-complex one has:

$$\underline{\Omega}_p(X/Y) \cong \bigoplus_{r+s=p} H_r(X/Y; \mathbf{Z}_2) \bigotimes_{\mathbf{Z}_2} \Omega_s\,, \quad p \neq 0.$$

In particular, one has

$$\Omega_1(X, Y) \cong H_1(X/Y; \mathbf{Z}_2).$$

COROLLARY 2.10 - *For each $x \in H^m(X, Y; \mathbf{Z}_2)$ and partition ω of $m - n$ one has a* **generalized Stiefel-Whitney number** *which is defined for a map $f : (V, \partial V) \to$*

$(X,Y), \dim V = n$, *by* $\{w_\omega(TX) \cup f^*(x)\}[V, \partial V]$. *Since the class* $x \bigotimes w_\omega$ *form a base of* $H^*((X/Y) \times BO, \mathbf{Z}_2)$, *the associated numbers give a complete set of invariants.*
It is important to note that we can characterize manifolds on which are defined PDEs as manifolds with X-structure in the sense of J.Milnor. In fact we have the following.
THEOREM 2.103 - *Any PDE* $E_k \subset J^k_n(W)$ *identifies an* X*-structure on* W, *where* X *is a homogeneous space.*
PROOF. We have necessitate to introduce some definitions and lemmas.
DEFINITION 2.69 - *A* **super-bundle of geometric objects** [101, 102] *is a 3-plet* $\mathcal{S} \equiv (\mathcal{P}, \mathcal{B}; \mathbb{B})$, *where* (i) $\mathcal{P} \equiv (P, X, \pi_P)$ *is a fiber bundle over* X (**total bundle**), (ii) $\mathcal{B} \equiv (B, X, \pi_B)$ *is a fiber bundle over* X (**basic bundle**), (iii) $\mathbb{B}$ *is a covariant functor* $\mathbb{B} : \mathcal{C}(\mathcal{B}) \to \mathcal{C}(\mathcal{P})$, *where* $\mathcal{C}(\mathcal{B})$ *(resp.* $\mathcal{C}(\mathcal{P})$*) is the category whose objects are open subbundles of* $\mathcal{B}$ *(resp. open subundles of* $\mathcal{P}$*) and whose morphisms are the local fiber bundle authomorphisms between these objects; (if* $\mathcal{B}$ *is a principal fiber bundle then one requires that the morphisms of* $\mathcal{C}(\mathcal{B})$ *are local principal fiber bundles authomorphisms), such that* (a) *if* $B|_U \in Ob(\mathcal{C}(\mathcal{B})) \Rightarrow \mathbb{B}(B|_U) = \pi_P^{-1}(U) \in Ob(\mathcal{C}(\mathcal{P}))$, (b) *if* $f \in Hom(\mathcal{C}(\mathcal{B}))$, *with* $f \equiv (f_B, f_X) : B|_U \to B|_{U'}, \mathbb{B}(f) \in Hom(\mathcal{C}(\mathcal{P}))$, *and satisfies* (c) $\pi_P \circ \mathbb{B}(f) = f_X \circ \pi_B$; (d) *if* $B|_U \in Ob(\mathcal{C}(\mathcal{B})), U \subset U' \Rightarrow \mathbb{B}(f)|_{\pi_P^{-1}(U')} = \mathbb{B}(f|_{B|_{U'}})$.
A **reduced** *Lie super-bundle of geometric objects of order* k *over a manifold* X [104] *is a super-bundle of geometric objects* $\mathcal{S} \equiv (\mathcal{P}, \mathcal{B}; \mathbb{B})$, *such that:* (i) $\mathbb{B}$ *depends only on the sub-category* $\mathcal{C}_B(X) \subset \mathcal{C}(X)$, *image of the natural projection functor* $R : \mathcal{C}(\mathcal{B}) \to \mathcal{C}(X)$; (ii) *The* s*-isotropy pseudogroup* G_s *associated to any section* s *of* $\mathcal{P}$:

$$G_s \equiv \left\{ f \equiv (f_P, f_X) \in G_P | f_P \circ s \circ f_X^{-1} \equiv f^* s = s \right\} \subset G_P$$

is a Lie pseudogroup of order k *on* X, *where*

$$G_P \equiv \mathbb{B}(Hom(\mathcal{C}(\mathcal{B}))) \subset Hom(\mathcal{C}(\mathcal{P})).$$

LEMMA 2.14 - *A reduced Lie superbundle of geometric objects of order* k *on* X *identifies a natural action of* $\mathcal{R}_k$ *(*$\equiv$ *Lie equation defining* G_s*) on* P $\gamma_k : \mathcal{R}_k \times_X P \to P$ *with the inverse* $\gamma_k^{-1} : P \times_X \mathcal{R}_k \to P$ *such that the following diagrams commute*

$$\begin{array}{ccc} \mathcal{R}_k \times_X P & \overset{\gamma_k}{\to} & P \\ {\scriptstyle pr_1} \downarrow & & \downarrow {\scriptstyle \pi} \\ \mathcal{R}_k & \to & X \end{array}$$

$$\begin{array}{ccc} \mathcal{R}_k \times_X \mathcal{R}_k \times_X P & \xrightarrow{id_{\mathcal{R}_k} \times \gamma_k} & \mathcal{R}_k \times_X P \\ {\scriptstyle id_{\mathcal{R}_k} \times \gamma_k} \downarrow & & \downarrow {\scriptstyle \gamma_k} \\ \mathcal{R}_k \times_X P & \xrightarrow[\gamma_k]{} & X \end{array}$$

$$\begin{array}{ccc} P & \xrightarrow{(id_{\gamma_k}\circ\pi, id_P)} & \mathcal{R}_k \times_X P \\ \| & & \downarrow \gamma_k \\ P & = & P \end{array}$$

Furthermore, a natural associative action of $C^\infty(\mathcal{R}_k)$ on $C^\infty(P)$ is defined by means of the following commutative diagram

$$\begin{array}{ccc} \mathcal{R}_k \times_X P & \xrightarrow{\gamma_k} & P \\ (f_k, s) \uparrow & & \uparrow f_k(s) \\ X & \xleftarrow[f^{-1}]{} & X \end{array}$$

Moreover, any solution ρ of $\mathcal{R}_k$ can be lifted on P in a fiber bundle diffeomorphism $(\widehat{\rho}, \rho)$, such that the following diagram

$$\begin{array}{ccccc} P & \xrightarrow{(D^k \rho\circ\pi, id_P)} & \mathcal{R}_k \times_X P & \xrightarrow{\gamma_k} & P \\ \pi \downarrow & & \downarrow & & \downarrow \pi \\ X & \xrightarrow{D^k \rho} & \mathcal{R}_k & \xleftarrow{\overline{p}_k} & X \\ \| & & & & \| \\ X & & \xrightarrow[\rho]{} & & X \end{array}$$

is commutative. On the other hand, a k-order Lie equation $\mathcal{R}_k \subset \Pi^k(X)$ on X identifies a reduced Lie super-bundle of geometric objects of order k on X with isotropy equation $\mathcal{R}_k$. Furthermore, a natural epimorphism $\Phi : \Pi^k(X) \to P$ and section s of P exist such that $\mathcal{R}_k = \ker_s \Phi$; namely the following sequence

$$0 \to \mathcal{R}_k \to \Pi^k(X) \overset{\Phi}{\underset{s\circ p_k}{\rightrightarrows}} P$$

is exact.

PROOF OF LEMMA 2.14 - In fact γ_k is given by $\gamma_k(D^k f(x), s(x)) = (\mathbb{B}(f) \circ s \circ f^{-1})(f(x))$. On the other hand, given $\mathcal{R}_k \subset \Pi^k(X)$ we can canonically obtain a reduced Lie superbundle of geometric objects of order k given by: (a) basic bundle

$$B \equiv \bigcup_{x\in X} B_x, \quad B_x \equiv \mathcal{R}_k \cap \overline{p}_k^{-1}(x),$$

where $\overline{p}_k : \Pi^k(X) \to X$ is the target projection of $\Pi^k(X)$. B is a principal fiber bundle with structure group

$$G \equiv \mathcal{R}_k \cap \overline{p}_{k,0}^{-1}(x, x)$$

where $\overline{p}_{k,0} \equiv (p_k, \overline{p}_k) : \Pi^k(X) \to X \times X$, where $p_k : \Pi^k(X) \to X$ is the source projection of $\Pi^k(X)$. (b) total bundle

$$P \equiv \bigcup_{x\in X} P_x, P_x \equiv p_k^{-1}(x)/G$$

where P is a bundle of homogeneous spaces with typical fiber

$$F \equiv p_{k,0}^{-1}(x,x)/G.$$

(c) covariant functor $\mathbb{B} : \mathcal{C}(\mathcal{B}) \to \mathcal{C}(\mathcal{P})$ given by: (i) $\mathbb{B}(B|_U) = P|_U$; (ii) (note that $Hom(\mathcal{C}(\mathcal{B})) \cong \Theta =$ set of invertible sections of $\mathcal{R}_k \equiv \{f_k\}$),

$$\mathbb{B}(f_k)([D^k g(p)]) = [D^k(f \circ g \circ f^{-1})(f(p))].$$

Furthermore, one can see that

$$C^\infty(\mathcal{R}_k) = \{f \in Hom(\mathcal{C}(\mathcal{B})) | f_k(s) = s\}$$

where s is a section of P, identified by means of the image in P of $\mathcal{R}_k \cap \overline{p}_k^{-1}(x)$ by means of the natural projection $\Phi : \overline{p}_k^{-1}(x) \to P$. Finally we can construct an epimorphism $\Phi : \Pi^k(X) \to P$ that extends the previous Φ by defining $\Phi \circ f_k = f_k^{-1}(s)$, $\forall f_k \in C^\infty(\Pi^k(X))$. Furthermore, one can see that $f_k \in C^\infty(\mathcal{R}_k)$ iff $f_k(s) = s$. □

LEMMA 2.15 - *For the linearized R_k of the Lie equation $\mathcal{R}_k \subset \Pi^k(W)$ of $E_k \subset J^k_n(W)$ we can write $R_k = \ker(\mathcal{L}s)$ where $\mathcal{L}s : J\mathcal{D}^k(TW) \to s^*vTP$ is the Lie derivative associated to s [93], where $\pi_P : P \to W$ is the total bundle associated to $\mathcal{R}_k$.*

LEMMA 2.16 - *Let $\mathcal{R}_k \subset \Pi^k(W)$ be the Lie equation of a PDE $E_k \subset J^k_n(W)$. Then $\mathcal{R}_k$ is formally integrable iff its defining section s:$W \to P$ verifies $\dim(s^*P_1)$ analytic integrability conditions that constitute a natural system*

$$B_1(s|c) \equiv \ker_c I \subset J\mathcal{D}(P)$$

where: (a) $\pi_P : P \to W \leftarrow B : \pi_B$ is the superbundle structure associated to $\mathcal{R}_k$; (b)

$$P_1 \equiv \Lambda^0_1(W) \bigotimes vTP/\delta(N_1), N_1 \equiv \mathrm{im}(\sigma_1(vT(\Phi^{(1)})))$$

in the following exact sequence of vector bundles over W:

$$0 \to M_{k+1} \to S^0_{k+1}W \bigotimes TW \xrightarrow{\sigma_1(vT(\Phi^{(k)}))} S^0_1 W \bigotimes vTP$$

where M_k is the symbol of R_k; (c) c is a suitable section $c : P \to P_1$; (d) I is the affine natural epimorphism $I : J\mathcal{D}(P) \to P_1$ over P given by

$$I(q) = pr(q - a)\,,\ \forall q \in J\mathcal{D}(P)$$

where pr is the natural projection $T^*W \bigotimes vTP \to P_1$ and $a = \Phi^{(1)}(f_{k+1}(s))$, where $f_k \in C^\infty(\Pi^k(W))$, is such that $\Phi(f_k(x)) = \pi_{1,0}(q) \in P$. [The natural section locally depends on a finite number of constants called **structure constants**.] Furthermore, $B_1(s|c) \subset J\mathcal{D}(P)$ is analytic and formally integrable iff the section c is a solution of a differential equation

$$J(c) \subset J\mathcal{D}(P_1), (\textbf{Jacobi conditions}).$$

PROOF OF LEMMA 2.16 - Let $\mathcal{R}_k$ be formally integrable with a 2-acyclic symbol M_k. Set $B_1 \equiv \mathrm{im}(\Phi^{(1)}) \subset J\mathcal{D}(P)$. As B_1 coincides with the kernel of I it is a fibered submanifold of $J\mathcal{D}(P)$, so we have the following commutative exact diagram of natural affine bundles over W

$$\begin{array}{ccccccccc}
0 & \to & N_1 & \to & T^*W \bigotimes vTP & \to & P_1 & \to & 0 \\
 & & \cdot & & \cdot & & \| & & \\
0 & \to & B_1 & \to & J\mathcal{D}(P) & \xrightarrow{I} & P_1 & \to & 0 \\
 & & \downarrow & & \downarrow & & \downarrow\uparrow 0 & & \\
 & & P & = & P & = & P & &
\end{array}$$

where $0 : P \to P_1$ is the zero section. We also have the following commutative and exact diagram of affine bundles

$$\begin{array}{ccccccccccc}
0 & \to & M_{k+1} & \to & S^0_{k+1}W \bigotimes TW & \xrightarrow{\sigma_1(vT(\Phi^{(1)}))} & TW \bigotimes vTP & \xrightarrow{\tau} & P_1 & \to & 0 \\
 & & \cdot & & \cdot & & \cdot & & \| & & \\
0 & \to & \mathcal{R}_{k+1} & \to & \Pi^{k+1}(W) & \xrightarrow{\Phi^{(1)}} & J\mathcal{D}(P) & \xrightarrow{I} & P_1 & & \\
 & & & & \downarrow & & \downarrow & & & & \\
0 & \to & \mathcal{R}_k & \to & \Pi^k(W) & \xrightarrow[\Phi]{} & P & & \to & 0 &
\end{array}$$

(Note that the upper horizontal lines in the above two diagrams denote the sequence of vector bundles corresponding to the affine ones that are in vertical correspondence and separated by a point.) As a consequence we have an exact sequence

$$0 \to \mathcal{R}_{k+1} \to \mathcal{R}_k \underset{0}{\overset{k}{\rightrightarrows}} p_k^*(s^*P_1).$$

Now, for any $f_{k+1} \in C^\infty(\Pi^{k+1}(W))$ over $f_k \in C^\infty(\mathcal{R}_k)$ we have

$$k \circ f_k = \tau(f_{k+1}(Ds) - Ds) = f_k(I \circ Ds) - I(Ds).$$

Therefore

$$k \circ f_k = 0 \quad \text{iff} \quad f_k(I \circ Ds) = I \circ Ds, \quad \text{whenever} \quad f_k(s) = s.$$

The above condition is satisfied iff a natural section $c : P \to P_1$ exists such that $I \circ Ds = c \circ s$. As the symbol M_k of $\mathcal{R}_k$ is 2-acyclic, we have the following commutative diagram in which the upper row is an exact sequence of natural vector bundles over P:

$$\begin{array}{ccccccccc}
0 & \to & N_2 & \to & S_2^0 W \bigotimes vTP & \to & TW \bigotimes P_1 & \to & P_2 & \to & 0 \\
 & & \cdot & & \cdot & & \cdot & & & & \\
0 & \to & B_2 & \to & J\mathcal{D}^2(P) & \to & J\mathcal{D}(P_1) & & & & \\
 & & \downarrow & & \downarrow & & \downarrow & & & & \\
 & & P & = & P & = & P & & & &
\end{array}$$

where

$$N_2 \equiv \mathrm{im}(\sigma_2(vT(\Phi^{(2)}))), P_2 \equiv \Lambda_2^0 W \bigotimes vTP/\delta(\Lambda_1^0 W \bigotimes N_1).$$

We shall proceed by steps: (a) We are able to construct a natural morphism $J : J\mathcal{D}(P_1) \to P_2$; (b) The restriction of $J \circ J\mathcal{D}(c) : J\mathcal{D}(P) \to P_2$ to $B_1(s|c)$ is the curvature map k of $B_1(s|c)$; (c) The condition for the surjectivity of $B_1(s|c)_{+1} \to B_1(s|c)$ by means of the exact sequence

$$B_1(s|c)_{+1} \to B_1(s|c) \xrightarrow{k} P_2$$

is given by $J(c)$:

$$k \circ Ds = J \circ J\mathcal{D}(c) \circ Ds = 0$$

that, for fixed s , gives a condition on c. □

Then theorem follows from the above considerations by considering $\mathcal{R}_k$ as the Lie equation of E_k. □

REMARK 2.61 - By means of the bove theorems, we see that we can consider X-manifolds, (in the sense of Milnor), where the X-structure is defined by means of differential equations, and for such class of manifolds we can consider cobordism groups $\underline{\Omega}_s^X$ or $\underline{\Omega}_s^X(Y)$. Then, by using the above, we can calculate these groups by means of homotopy groups and homology groups. ■

The following important theorem allows us to characterize reduced s-quantum cobordism groups as subgroups of homological groups.

THEOREM 2.104 - (HOMOLOGICAL STRUCTURE OF THE REDUCED QUANTUM GROUPS). *One has the following:*

$$\widehat{\Omega}_s(E_k) \triangleleft h_s(J_n^k(W)^+).$$

PROOF. In fact one has the following short exact sequence

$$0 \to \ker(l_s) \to \Omega_s(E_k) \xrightarrow{l_s} \underline{\Omega}_s^O(J_n^k(W))$$

where

$$\ker(l_s) = \left\{ \begin{array}{l} [N]_{\overline{E_k}} \in \Omega_s(E_k) \\ N = \partial V, \text{for some } (s+1)\text{-dimensional submanifold} \\ V \subset J_n^k(W) \end{array} \right\}$$

Now, we can recognize a natural isomorphism $K_s(E_k) \cong \ker(l_s)$. In fact any smooth $(s+1)$-dimensional manifold V such that $N = \partial V, N \subset E_k$, can be embedded in J_n^k(W) if $\dim(J_n^k(W)) \geq 2(s+1)+1$ [50]. But this condition is always verified if $n \geq 1$ and $m \geq 1$. These last conditions are always verified in our paper! We then get also the following isomorphism : $\Omega_s(E_k)/\ker(l_s) \cong \widehat{\Omega}_s(E_k)$. Now, taking into account Corollary 2.7 we also get the isomorphism $\underline{\Omega}_s^O(J_n^k(W)) \cong h_s(J_n^k(W)^+)$. As a consequence we have the following short exact sequence

$$0 \to \widehat{\Omega}_s(E_k) \to h_s(J_n^k(W)^+)$$

and therefore the proof is obtained. □

THEOREM 2.105 - (QUANTUM COBORDISM GROUPS AND RELATIVE COBORDISM GROUPS). *Let $E_k \subset J_n^k(W)$ be a PDE. The relations between quantum cobordism groups of E_k and relative cobordism groups are given by the following commutative diagrams*

$$\begin{array}{ccccccc} & & 0 & & & & \\ & & \downarrow & & & & \\ 0 \to & & K_s(E_k) & \to & \Omega_s(E_k) & \to & \Omega_s \cong H_s(*; MO) \\ & & \downarrow & & \| & & \uparrow \\ 0 \to & & K_s(J_n^k(W), E_k) & \to & \Omega_s(E_k) & \to & \underline{\Omega}_s(J_n^k(W), E_k) \cong H_s(J_n^k(W), E_k; MO) \\ & & \downarrow & & & & \\ & & 0 & & & & \end{array}$$

with $0 \leq s \leq n-1$. Furthermore, if $(J_n^k(W), E_k)$ is a CW-pair, we get the following commutative diagrams

$$\begin{array}{ccccccc} & & 0 & & & & \\ & & \downarrow & & & & \\ 0 \to & & K_s(E_k) & \to & \Omega_s(E_k) & \to & \Omega_s \\ & & \downarrow & & \| & & \uparrow \\ 0 \to & & K_s(J_n^k(W), E_k) & \to & \Omega_s(E_k) & \to & \underline{\Omega}_s(J_n^k(W), E_k) \\ & & \downarrow & & & & \|\wr \\ & & 0 & & & & \bigoplus_{p+q=s} H_p(J_n^k(W), E_k; \mathbf{Z}_2) \otimes_{\mathbf{Z}_2} \Omega_q \end{array}$$

As a consequence, we get the following canonical isomorphism

$$\widehat{\Omega}_s(E_k) \cong \Omega_s(E_k)/K_s(J^k_n(W), E_k).$$

PROOF. Taking into account Theorem 2.104 and the fact that if $\partial V = N_0 \dot{\bigcup} N_1$, with $N_0, N_1 \subset E_k$, (integral submanifolds), we can embed V into $J^k_n(W)$, (see the proof of the above theorem). Hence, we get easily the proof. □

REMARK 2.62 - The above theorem proves that to refer $\Omega_s(E_k)$ to relative cobordism groups $\underline{\Omega}_s(J^k_n(W), E_k)$ does not increase the information on $\Omega_s(E_k)$, with respect to referring to Ω_s.

There are important classes of differential relations where it is possible to give some further general informations of the structure of integral and quantum cobordism groups. For this, let us give the following definition that generalizes previous one given by Gromov [44] for holonomic situation.

DEFINITION 2.70 - *A PDE $E_k \subset J^k_n(W)$ satisfies the* **(s)-homotopy principle**, *(or E_k is of* **(s)-h-type***), if for any compact smooth $(s+1)$-dimensional submanifold $V \subset E_k$, $0 < s+1 \leq n$, such that $N = \partial V$ is an admissible integral s-dimensional submanifold, there exists an $(s+1)$-dimensional integral submanifold X of E_k with $\partial X = N$, $X \simeq_N V$ (X homotopic to V with relation to N in the class of $(s+1)$-dimensional smooth submanifolds of E_k).*

THEOREM 2.106 - *If E_k is a PDE that satisfies the (s)-h-principle then*

$$\Omega^{E_k}_s \lhd \underline{\Omega}_s(E_k) \cong \bigoplus_{p+q=s} H_p(E_k; \mathbf{Z}_2) \otimes_{\mathbf{Z}_2} \Omega_q.$$

In particular, for $s = 1$ one has

$$\Omega^{E_k}_1 \lhd H_1(E_k; \mathbf{Z}_2).$$

PROOF. Under our hypotheses one has the following short exact sequence

$$0 \to \ker(\gamma_s) \to \Omega^{E_k}_s \xrightarrow{\gamma_s} \underline{\Omega}_s(E_k)$$

where γ_s is given by $\gamma_s : [N]_{E_k} \mapsto [N]_{\underline{E_k}}$, where $[N]_{\underline{E_k}}$ is the equivalence class of N in $\underline{\Omega}_s(E_k)$. One has

$$\ker(\gamma_s) = \{[N]_{E_k} \in \Omega^{E_k}_s | N = \partial V, V \subset E_k, \dim V = s+1\}.$$

On the other hand, from our assumption there must exist an integral manifold X such that $X \simeq_N V$, $X \subset E_k$, $\partial X = N$, and $\dim X = s+1$. So we have

$$\gamma_s : [N]_{E_k} = [N]_{\underline{E_k}} = 0.$$

Hence, $\ker(\gamma_s) = 0$, and the homomorphism γ_s is injective. Then, $\Omega_s^{E_k}$ must be a subgroup of $\underline{\Omega}_s^{E_k}$. In particular, if $s = 1$ one has $\Omega_0 \cong \mathbf{Z}_2$, and $\Omega_1 = 0$. Hence, $\underline{\Omega}_1(E_k) \cong H_1(E_k; \mathbf{Z}_2)$. □

COROLLARY 2.11 - *If E_k is contractible to a point one has*

$$\Omega_s^{E_k} \lhd \Omega_s, \quad s \geq 0.$$

PROOF. In fact, from the above theorem we get

$$\begin{aligned} \Omega_s^{E_k} \lhd \underline{\Omega}_s &\cong H_0(E_k; \mathbf{Z}_2) \otimes_{\mathbf{Z}_2} \Omega_s \\ &\cong \mathbf{Z}_2 \otimes_{\mathbf{Z}_2} \Omega_s \\ &\cong \Omega_s \end{aligned}$$

as $H_p(E_k; \mathbf{Z}_2) = 0$ for $p > 0$, and $H_0(E_k; \mathbf{Z}_2) \cong \mathbf{Z}_2$. □

EXAMPLE 2.57 - If $E_k \subset J_n^k(W)$ satisfies the e-principle [30] it also satisfies the $(n-1)$-h-principle. So the example given in ref.[110] about differential relations of immersions, Lagrangian immersions, and Legendrian immersions fit the above theorem. ■

The above considerations can also be simulated to obtain more informations on the quantum cobordism. More precisely, we can give the following.

DEFINITION 2.71 - *A PDE $E_k \subset J_n^k(W)$ satisfies the* **quantum (s)-homotopy principle**, *(or E_k is of* **q(s)-h-type***), if for any compact smooth $(s+1)$-dimensional submanifold $V \subset J_n^k(W)$, $0 < s+1 \leq n$, such that $N = \partial V$ is an admissible integral s-dimensional smooth submanifold of E_k, there exists an integral $(s+1)$-dimensional submanifold $X \subset J_n^k(W)$, with $\partial X = N, X \simeq_N V$, ($X$ homotopic to V with relation to N in the class of $(s+1)$-dimensional smooth submanifolds of $J_n^k(W)$).*

THEOREM 2.107 - *If E_k is a PDE that satisfies the q(s)-h-principle then one has*

$$\Omega_s(E_k) \cong \mathbf{Z}_2 \bigoplus \cdots_p \cdots \bigoplus \mathbf{Z}_2 \lhd \Omega_s.$$

PROOF. We shall prove that the homomorphism $\hat{\bar{j}}_s : \Omega_s(E_k) \to \Omega_s$ is injective. Indeed, let us assume that $[N]_{\overline{E_k}} \in K_s(E_k)$. Then, there exists an $(s+1)$-dimensional compact smooth manifold V such that $\partial V = N$. We know (see proof in Theorem 2.104) that we can embed V into $J_n^k(W)$ to a manifold Y with $\partial Y = N$. Furthermore, as E_k satisfies the q(s)-h-principle, we have $Y \simeq_N X$, where $X \subset J_n^k(W)$ and it is an integral submanifold of dimension $s+1$. Therefore, $[N]_{\overline{E_k}} = [\emptyset]_{\overline{E_k}} = 0$. Namely, $K_s(E_k) = 0$. □

COROLLARY 2.12 - *If E_k satisfies the q(s)-h-principle, $\Omega_s^{E_k}$ is an extension of an abelian finite torsion group of the form $\mathbf{Z}_2 \bigoplus \cdots_p \cdots \bigoplus \mathbf{Z}_2$, i.e., one has the following short exact sequence:*

$$0 \to K_s^{E_k} \to \Omega_s^{E_k} \to \mathbf{Z}_2 \bigoplus \cdots_p \cdots \bigoplus \mathbf{Z}_2 \to 0.$$

So we can write

$$\begin{aligned}\Omega_s^{E_k} &\in H^2(\mathbf{Z}_2 \bigoplus \cdots_p \cdots \bigoplus \mathbf{Z}_2; K_s^{E_k})\\ &\cong Hom_{\mathbf{Z}}(H_2(\mathbf{Z}_2 \bigoplus \cdots_p \cdots \bigoplus \mathbf{Z}_2; \mathbf{Z}); K_s^{E_k})\\ &\cong \bigoplus_{q\in\{1,\dots,r\}} Hom_{\mathbf{Z}}(\mathbf{Z}_2 \bigotimes \mathbf{Z}_2; K_s^{E_k}).\end{aligned}$$

EXAMPLE 2.58 - In ref.[109] we have considered PDEs over "regular" fiber bundles. Recall that we have called regular fiber bundle one $\pi : W \to M$ such that if it admits a cross-section s, defined on a closed subset $X \subset M$, then there exist sections, defined on an open neighbourhood of X in M, that extend s. There, it is also proved that for Cauchy hypersurfaces in such PDEs, $E_k \subset J\mathcal{D}^k(W)$, the corresponding statistical set $\Omega(N_1, N_2)$ is nonempty. This means, with the language of the cobordism, that in such cases $\Omega_{n-1}(E_k) \triangleleft \mathbf{Z}_2$. Therefore, we can have $\Omega_{n-1}(E_k) = \mathbf{Z}_2$, or $\Omega_{n-1}(E_k) = 0$. ■

EXAMPLE 2.59 - Let us consider the d'Alembert equation

$$(2.175)(d'A) \qquad F \equiv uu_{xy} - u_x u_y = 0$$

in the open subset $C_2 \equiv u^{-1}(\mathbf{R} \setminus 0) \subset J\mathcal{D}^2(\mathbf{R}^2, \mathbf{R})$, where (d'A) defines a subbundle of $J\mathcal{D}^2(\mathbf{R}^2, \mathbf{R}) \to \mathbf{R}^2$. (For a geometric study of the set of solutions of $(d'A)$ see ref.[113].) Of course, we can apply the (1)-h-principle to $J\mathcal{D}^2(\mathbf{R}^2, \mathbf{R})$. So we have

$$\Omega_1(d'A) \triangleleft \Omega_1 = 0 \quad \Rightarrow \quad \Omega_1(d'A) = 0.$$

Hence, from Theorem 2.90 we get that $\widehat{\Omega}_1^{(d'A)} = 0$. This means that $K_1^{(d'A)} = \Omega_1^{(d'A)}$, i.e., $\Omega_1^{(d'A)} \to \Omega_1(d'A)$ is the zero homomorphism. ■

EXAMPLE 2.60 - It is, often, workable to use PDEs reduced with respect to their symmetry groups. For example, let $E_k \subset J\mathcal{D}^k(\mathbf{R}, \mathbf{R})$ be a k-order linear ordinary differential equation

$$(2.176) \qquad \sum_{0\leq i\leq k} a_{k-i}(x) y^{(k-i)} = 0.$$

Then, for any point $q \in E_k$, (initial condition), passes only one solution, (integral curve). So we get

$$\mathcal{S}ol(E_k)_c \cong E_k/\mathbf{R}$$

where $\mathbf{R}$ acts on E_k by means of the following action map

$$\gamma : \mathbf{R} \times E_k \to E_k, (\lambda, q) \mapsto \gamma_q(\lambda)$$

where $\gamma_q(\lambda)$ is the value at λ of the unique integral curve passing through the initial condition q. Note that E_k is a linear submanifold of $J\mathcal{D}^k(\mathbf{R},\mathbf{R}) \cong \mathbf{R}^{k+2}$ of dimension $k+1$. So one has the following identification

$$E_k \cong \mathbf{R}^{k+1}, \quad Sol(E_k) \cong \mathbf{R}^{k+1}/\mathbf{R} \cong \mathbf{R}^k.$$

As a consequence, we get, after the identification of initial conditions by means of the symmetry group $G \equiv GL(\mathbf{R}^k)$ of E_k , the following isomorphism $\Omega_0^{E_k/G} \cong \mathbf{Z}_2$. So the homomorphism j_0 in Theorem 2.90 is now an isomorphism : $j_0 : \Omega_0^{E_k/G} \cong \mathbf{Z}_2 \cong \Omega_0$. Of course, one has also $\Omega_0(E_k/G) \cong \mathbf{Z}_2$. Note that in this case on E_k we can apply the q(0)-h-principle. So we should have $\Omega_0(E_k) \triangleleft \Omega_0 \cong \mathbf{Z}_2$. On the other hand the following sequence $\Omega_0(E_k) \to \Omega_0(E_k/G) \to 0$ must be exact. Hence, we also have $\Omega_0(E_k) \cong \Omega_0 \cong \mathbf{Z}_2$. One has the following exact commutative diagram:

$$\begin{array}{ccccccccc} 0 & \to & K_0^{E_k} & \to & \Omega_0^{E_k} & \to & \mathbf{Z}_2 & \to & 0 \\ & & & & \downarrow & & \| & & \\ & & & & \Omega_0^{E_k/G} & \cong & \mathbf{Z}_2 & & \end{array}$$

■

LEMMA 2.17 - *If $E_k \subset J_n^k(W)$ is a PDE of q(s)-h-type, then it is also of q(r)-h-type, for $0 \leq r \leq s$.*

THEOREM 2.108 - *If W is s-connected and any closed compact s-dimensional smooth manifold can be embedded into W, up to bordisms, then one has the following isomorphism*

$$\Omega_s^{J_n^\infty(W)} \cong \Omega_s.$$

PROOF. In fact, if N_0 and N_1 are two s-dimensional integral manifolds of $J_n^\infty(W)$ such that there exists an $(s+1)$-dimensional smooth manifold V with $\partial V = N_0 \bigcup N_1$, then we can embed V into $J_n^\infty(W)$ onto an s-dimensional smooth submanifold $X \subset J_n^\infty(W)$ with $\partial X = N_0 \dot{\bigcup} N_1$. Then, taking into account that $\pi_{\infty,0} : J_n^\infty(W) \to W$, is an affine bundle and, hence $\pi_{\infty,0}$ is a homotopy equivalence, we can deform any s-dimensional submanifold of $J_n^\infty(W)$ to integral one. In other words $J_n^\infty(W)$ satisfies the (s)-h-principle. Therefore, the homomorphism $\widehat{\overline{j}}_s : \Omega_s(J_n^\infty(W)) \cong \Omega_s^{J_n^\infty(W)} \to \Omega_s$ is injective. On the other hand, as any smooth s-dimensional manifold can be embedded into W, hence it determines an integral manifold of $J_n^\infty(W)$. So we see that $\widehat{\overline{j}}_s$ is an epimorphism, hence an isomorphism too. □

THEOREM 2.109 - *Let $E_k \subset J_n^k(W)$ be a PDE. Then $\Omega_s(E_\infty)$ is a finite group of the form*

$$\Omega_s(E_\infty) \cong \mathbf{Z}_2 \bigoplus \cdots_p \cdots \bigoplus \mathbf{Z}_2 \triangleleft \Omega_s, 0 \leq s < n, (\text{i.e., } K_s(E_\infty) = 0).$$

PROOF. It follows directly from the above considerations that E_∞ satisfies the q(s)-h-principle; then from Theorem 2.106 follows that $\Omega_s(E_\infty) \triangleleft \Omega_s$. $\square$

THEOREM 2.110 - *Let $E_k \subset J^k_n(W)$ be a PDE. Then $\Omega_s^{E_\infty}$ is an abelian group such that one has a short exact sequence*

$$0 \to K_s^{E_\infty} \to \Omega_s^{E_\infty} \to \mathbf{Z}_2 \bigoplus \cdots_p \cdots \bigoplus \mathbf{Z}_2 \to 0$$

with

$$K_s^{E_\infty} \equiv \left\{ \begin{array}{l} [N]_{E_\infty} \quad N = \partial V \\ V = (s+1)\text{-dimensional integral submanifold of } J^\infty_n(W) \end{array} \right\}$$

So we can write

$$\begin{aligned} \Omega_s^{E_\infty} &\in H^2(\mathbf{Z}_2 \bigoplus \cdots_p \cdots \bigoplus \mathbf{Z}_2, K_s^{E_\infty}) \\ &\cong Hom_{\mathbf{Z}}(H^2(\mathbf{Z}_2 \bigoplus \cdots_p \cdots \bigoplus \mathbf{Z}_2, \mathbf{Z}), K_s^{E_\infty}). \end{aligned}$$

PROOF. It follows from Theorem 2.92 and the above theorem. $\square$

In order to characterize more integral and quantum cobordism groups, we shall emphasize the fact that on such cobordism manifolds there are characteristic vector fields. This allows us to relate the cobordism in PDEs to the cobordism with vectors [138]. More precisely, we have the following important theorem.

2 THEOREM 2.111 - (CHARACTERISTIC NUMBERS FOR CLASSES IN INTEGRAL AND QUANTUM COBORDISM GROUPS). *If N_0 and N_1 are two s-dimensional admissible integral submanifolds of E_k and $N_0 \bigcup N_1 = \partial V$, for some integral submanifold of E_k, (or $J^k_n(W)$), then one has*

$$\chi(N_0) = \chi(N_1)\,, \quad w[N_0] = w[N_1].$$

In particular, $\chi(\partial V) = 0$. Furthermore, one has the following exact commutative diagram

$$\begin{array}{ccccccc}
 & & & & & & 0 \\
 & & & & & & \downarrow \\
 & & & & 0 & \to & K_s^{\uparrow} \\
 & & & & \uparrow & & \downarrow \\
0 & \to & K_s^{\uparrow}(E_k) & \to & \Omega_s(E_k) & \to & \Omega_s^{\uparrow} \\
 & & & & \widehat{i}_s \uparrow & & \downarrow j_s^{\uparrow} \\
0 & \to & K_s^{\uparrow E_k} & \to & \Omega_s^{E_k} & \to & \Omega_s \\
 & & & & \uparrow & & \downarrow \\
 & & & & K_s^{E_k} & \to & 0 \\
 & & & & \uparrow & & \\
 & & & & 0 & &
\end{array}$$

where $\Omega_s^{\uparrow}$ *(resp.*$^{+}\Omega_s^{\uparrow}$*) are the* **Reinhart cobordism groups**, *(resp.* **Reinhart oriented cobordism groups***), by taking into account the existence of a vector field* v *(***Reinhart vector field***), on any manifold* V, $\partial V = N_0 \dot{\bigcup} N_1$, *such that* v *is interior normal on* N_0 *and exterior normal on* N_1.

PROOF. We shall use the following lemmas.

LEMMA 2.18 - [50] *Let* (X, A) *be a pair and* F *a field. The* **homological Euler characteristic** *of* (X, A) *is*

$$\xi'(X,A) \equiv \sum_{0 \leq k \leq \infty} (-1)^k \lambda_k(X, A; F)$$

where

$$\lambda_k(X, A; F) \equiv \dim_F H_k(X, A; F).$$

If $\lambda_k < \infty$ *these are called the* **Betti numbers** *of* (X, A). *When* X *is a compact manifold and* A *is a compact submanifold then* $\xi'(X, A)$ *is finite and is independent of the field* F. *Furthermore, if* $f : M \to [a, b]$ *is an admissible Morse function on a compact manifold* M , *of type* $(\nu_0, \ldots, \nu_n)$, *then one has*

$$\sum_{0 \leq k \leq m} (-1)^{k+m} \nu_k \geq \sum_{0 \leq k \leq m} (-1)^{k+m} \beta_k , \quad 0 \leq m \leq n$$
$$\sum_{0 \leq k \leq n} (-1)^k \nu_k = \sum_{0 \leq k \leq n} (-1)^k \beta_k \equiv \xi'(M, f^{-1}(a))$$

where $\beta_k \equiv \dim_F H_k(M, f^{-1}(a); F)$. *In particular, if* $f^{-1}(a) = \emptyset$ *one has*

$$\sum_{0 \leq k \leq n} (-1)^k \nu_k = \sum_{0 \leq k \leq n} (-1)^k \beta_k \equiv \xi'(M).$$

If $f^{-1}(a) = f^{-1}(b) = \emptyset$, *i.e.* $\partial M = \emptyset$, *and* $\dim M = 2s + 1$, *then one has* $\xi'(M) = 0$. *In general, if* $\partial M = \emptyset$ *we have* $\chi'(M) = \chi(M)$. *Finally, one also has*

$$\nu_k \geq \beta_k.$$

LEMMA 2.19 - *If* $M \in [\emptyset] \in \Omega_n$, *i.e.,* $M = \partial V$, *where* V *is an* $(n + 1)$*-dimensional manifold, then the Euler-Poincaré characteristic of* M *is even:* $\chi(M) = 2m$.

PROOF OF LEMMA 2.19 - Let $n = 2s + 1$. Then, from the above theorem, we have $\chi(M) = 0$. Let $n = 2s$. Let us consider $W = V \bigcup_M V$. (W is the manifold obtained by gluing two copies of V along the submanifold M [50]). Now, taking into account that $\chi(X \bigcup_L Y) = \chi(X) + \chi(Y) - \chi(L)$, if $L = X \bigcap Y$, we get $0 = \chi(W) = 2\chi(V) - \chi(M)$. □

LEMMA 2.20 - *If two s-dimensional compact smooth closed manifolds N_0, N_1 are cobordant, then their Euler characteristic belong to the same congruence class in* $\mathbf{Z}_2$.

PROOF OF LEMMA 2.20 - Let $N_0 \bigcup N_1 = \partial V$. Then, $2m = \chi(\partial V) = \chi(N_0)+\chi(N_1)$.

□

LEMMA 2.21 - *If $M = \partial V$ one has:*

$$\text{If} \quad \dim M = 2s+1 \Rightarrow \chi(V) = 0; \qquad \text{If} \quad \dim M = 2s \Rightarrow \chi(V) = \frac{1}{2}\chi(M).$$

REMARK 2.63 - $\chi(M)$ is an example of non-stable characteristic class. The Stiefel-Whitney classes $w_q \in H^q(M;\mathbf{Z}_2)$ with all polynomials in w_q of dimension n, and the Pontrjagin classes $p_q \in H^{4q}(M;\mathbf{Q})$, with all polynomials in p_q of dimension n, (for $n = 4k$), are the full family of stable characteristic numbers for Ω_n and ${}^+\Omega_n$. ■

LEMMA 2.22 - (REINHART) *$N' \in [N]_\uparrow \in \Omega_s^\uparrow$, (resp. $N' \in [N,\eta]_\uparrow \in {}^+\Omega_s^\uparrow$), iff N' and N have the same Stiefel-Whitney (resp. Stiefel-Whitney, Pontrjagin) and Euler numbers, (resp.* (a)*($n \neq 4k+1$), the same Euler numbers;* (b)*($n = 4k+1$), the manifold M with $\partial M = V \bigcup(-V')$ has even Euler number. [This only depends on V and V' and not on the choice of M].) Furthermore, the following commutative diagram, where the horizontal lines are exact, relate the groups $\Omega_s^\uparrow$ and ${}^+\Omega_s^\uparrow$ with Ω_s and ${}^+\Omega_s$ respectively.*

$$\begin{array}{ccccccccc} 0 & \to & K_s^\uparrow & \to & \Omega_s^\uparrow & \to & \Omega_s & \to & 0 \\ & & \uparrow & & \uparrow & & \uparrow & & \\ 0 & \to & {}^+K_s^\uparrow & \to & {}^+\Omega_s^\uparrow & \to & {}^+\Omega_s & \to & 0 \end{array}$$

Here, $K_{2k+1}^\uparrow = 0$, and $K_{2k}^\uparrow$ is free cyclic. The generator is the class of the sphere S^{2k}. Moreover, ${}^+K_{4k-1}^\uparrow = 0$, and ${}^+K_{4k+1}^\uparrow$ is cyclic of order 2, and ${}^+K_{2k}^\uparrow$ is free cyclic. In each case, the generator is the class of the sphere. One has

$${}^+\Omega_s^\uparrow \cong {}^+K_s^\uparrow \bigoplus {}^+\Omega_s.$$

Now, let us return to the proof of our theorem. If $N_0 \bigcup N_1 = \partial V$, where N_0, N_1, and V are integral manifolds of E_k, (or $J_n^k(W)$), there must exist a Cartan vector field v tangent to V that is characteristic for $N_0 \bigcup N_1$, i.e., is of Reinhart type. Then, we can apply Reinhart's theorem that requires just $\chi(N_0) = \chi(N_1)$, other than $w[N_0] = [N_1]$. □

COROLLARY 2.13 - *If $E_k \subset J_n^k(W)$ is a PDE, then*

$$K_s^{E_k} = \{[N]_{E_k} | w[N] = \chi(N) = 0\}.$$

PROOF OF COROLLARY 2.13 - It follows directly from the commutative diagram in the above theorem. □

EXAMPLE 2.61 - Let us consider again d'Alembert equation. Any closed 1-dimensional integral manifold N of $(d'A)$ is of the type

$$N = N_0 \dot{\bigcup} \cdots \dot{\bigcup} N_s$$

where N_i is any closed integral curve in $(d'A)$ obtained by means of a smooth mapping

$$f_i : S^1 \to J\mathcal{D}^2(\mathbf{R}^2, \mathbf{R}), \quad f_i^* \omega_\alpha = 0$$

where $\omega_\alpha, 0 \leq \omega_\alpha \leq 3$, are the differential 1-forms on $J\mathcal{D}^2(\mathbf{R}^2, \mathbf{R})$ generating the Cartan distribution of $(d'A)$. More precisely,

$$\{\omega_\alpha\} \equiv \left\{ \begin{array}{l} \omega_0 \equiv du u_{xy} + u du_{xy} - du_x u_y - u_x du_y \\ \omega_1 \equiv du - u_x dx - u_y dy \\ \omega_2 \equiv du_x - u_{xx} dx - u_{xy} dy \\ \omega_3 \equiv du_y - u_{yx} dx - u_{yy} dy \end{array} \right\}.$$

Then, the necessary condition that

$$N \in [N']_{(d'A)}$$

where

$$N' = N'_0 \dot{\bigcup} \cdots \dot{\bigcup} N'_r$$

is another closed integral 1-dimensional submanifold of $(d'A)$, is that

$$w[N] = w[N'], \quad \chi(N) = \chi(N').$$

On the other hand, one has

$$w[N] = w[N'] = \chi(N) = \chi(N') = 0.$$

So there are no obstructions of cobord-type to obtain integral manifolds in $(d'A)$ cobording N with N'. Furthermore, as all the closed 1-dimensional integral manifolds of (d'A) satisfy

$$w[N] = w[N'] = \chi(N) = \chi(N') = 0$$

then

$$K_1^{(d'A)} = \Omega_1^{(d'A)}.$$

This agree with the results of Example 2.59. ■

The conditions contained in Theorem 2.111 are also sufficient to characterize quantum cobordism in some particularly important classes of PDEs. In fact we have the following theorem.

THEOREM 2.112 - (CHARACTERISTIC NUMBERS IDENTIFYING CLASSES IN QUANTUM COBORDISM GROUPS). *If a PDE $E_k \subset J^k_n(W)$ satisfies the q(s)-h-principle, then two s-dimensional admissible integral submanifolds N_0, N_1 quantum cobord, i.e., $N_1 \in [N_0]_{\overline{E_k}}$, iff they have the same Stiefel-Whitney numbers and Euler number.*

PROOF. If $V \subset E_k$ is an integral manifold such that $\partial V = X_0 \dot{\bigcup} X_1$. Then, X_0 and X_1 must necessarily have the same Stiefel-Whitney numbers and Euler number. Conversely , if X_0 and X_1 are two admissible s-dimensional integral manifolds having the same Stiefel-Whitney numbers and Euler number, then there exists an (s+1)-dimensional manifold V such that $\partial V = X_0 \dot{\bigcup} X_1$ and having a Reinhart vector field. On the other hand, we can imbed V into $J^k_n(W)$ to an (s+1)-dimensional manifold N and, (taking into account that E_k is of q(s)-h-type), to deform N to integral one Y , with $\partial Y = X_0 \dot{\bigcup} X_1$. Then, there also exists a Reinhart vector field on Y which is also a Cartan vector field. □

THEOREM 2.113 - *Let $E_k \subset J^k_n(W)$ be a PDE. If E_k admits an s-dimensional connected closed compact integral manifold X that integral (or quantum) cobords with $N_1 \dot{\bigcup} \cdots \dot{\bigcup} N_p \equiv N$, where N_i , $i = 1, \dots, p$, are some other s-dimensional connected closed compact integral manifolds of E_k, $\partial V = X \dot{\bigcup} N$, then one of the following conditions must be verified:*

$$\text{If } \ (w[X] = \chi(X) = 0)$$
$$\Downarrow$$
$$\left\{ \begin{array}{c} w[N_1] = \ldots = w[N_{2r}] = 1 = \chi[N_1] = \ldots = \chi[N_{2r}] \\ w[N_{2r+1}] = \ldots = w[N_p] = 0 = \chi[N_{2r+1}] = \ldots = \chi[N_p] \end{array} \right\}$$

and

$$\text{If } \ (w[X] = \chi(X) = 1)$$
$$\Downarrow$$
$$\left\{ \begin{array}{c} w[N_1] = \ldots = w[N_{2r+1}] = 1 = \chi[N_1] = \ldots = \chi[N_{2r+1}] \\ w[N_{2r+2}] = \ldots = w[N_p] = 0 = \chi[N_{2r+2}] = \ldots = \chi[N_p] \end{array} \right\}.$$

In both cases V will be an integral manifold of dimension $s+1$ with p singular points. The above conditions are also sufficient for the quantum cobording case, if E_k is of q(s)-h-type.

PROOF. In fact, in order that an integral cobording class $[X]_{E_k}$, or quantum cobording class $[X]_{\overline{E_k}}$, should be represented in the form $[X]_{E_k} = [N_1]_{E_k} + \ldots + [N_p]_{E_k}$, or $[X]_{\overline{E_k}} = [N_1]_{\overline{E_k}} + \dots + [N_p]_{\overline{E_k}}$, where N_i, $i = 1, \dots, p$, are given as in theorem, it is necessary that (a) or (b) be verified, depending on the characteristic numbers of X. Furthermore, if E_k is of q(s)-h-type, conditions (a), (b) are also sufficient in quantum cobordism case, as we can apply the above theorem. $\square$

DEFINITION 2.72 - *If* $\widehat{\Omega}_s(E_k) = 0$ *we say that* E_k *is* **reduced-q(s)-cobording free**. *Furthermore, if* $\Omega_s(E_k) = 0$ *we say that* E_k *is* **q(s)-cobording free**.

PROPOSITION 2.21 - *If* E_k *is reduced-q(s)-cobording free we have*

$$\Omega_s(E_k) \cong K_s(E_k).$$

This is equivalent to saying that all the classes $[N]_{\overline{E_k}} \in \Omega_s(E_k)$, *have zero Stiefel-Whitney numbers. If* E_k *is also of q(s)-h-type, we have* $\Omega_s(E_k) = 0$. *So in this case* E_k *results of q(s)-cobording are also free. Hence,* $\Omega_s^{E_k} = K_s^{E_k}$. *So we have the following implication:*

$$\textit{reduced q(s)-cobording free} + \textit{q(s)-h-type} \Rightarrow \textit{q(s)-cobording free.}$$

PROPOSITION 2.22 - *If* E_k *is q(s)-cobording free we have*

$$\Omega_s^{E_k} \cong K_s^{E_k}.$$

This is equivalent to saying that all the classes $[N]_{E_k} \in \Omega_s^{E_k}$, *have zero Stiefel-Whitney numbers and Euler numbers.*

Let us give, now, a full characterization of integral and quantum cobordism groups by means of suitable characteristic numbers.

DEFINITION 2.73 - 1) *We define* **space of characteristic** q**-forms** *on*

$$E_k \subset J_n^k(W), q = 1, 2, \cdots$$

the following

$$Ch\Omega^q(E_k) \equiv \{\beta \in \Omega^q(E_k) | \beta(\zeta_1, \cdots, \zeta_q)(p) = 0, \zeta_i(p) \in \mathbf{C}har(E_k)_p\,, \forall p \in E_k\}.$$

Set $Ch\Omega^0(E_k) = 0$.

2) *We define* **space of Cartan** q**-forms** *on* $E_k \subset J_n^k(W), q = 1, 2, \cdots$ *the following*

$$C\Omega^q(E_k) \equiv \{\beta \in \Omega^q(E_k) | \beta(\zeta_1, \cdots, \zeta_q)(p) = 0, \zeta_i(p) \in (\mathbb{E}_k)\,, \forall p \in E_k\}.$$

Set $C\Omega^0(E_k) = 0$.

3) *We define* **space of** p**-characteristic** q**-forms** *on* $E_k \subset J_n^k(W), q = 1, 2, \cdots$ *the following*

$$Ch^p\Omega^q(E_k) \equiv \left\{ \begin{array}{l} \beta \in \Omega^q(E_k) | \beta(\zeta_1, \cdots, \zeta_q) = 0 \\ \textit{if at least } q - p + 1 \textit{ of the fields } \zeta_1, \cdots, \zeta_q \textit{ are characteristic} \end{array} \right\}$$

4) *We define* **space of p-Cartan q-forms** *on* $E_k \subset J^k_n(W), q = 1, 2, \cdots$ *the following*

$$C^p\Omega^q(E_k) \equiv \left\{ \begin{array}{l} \beta \in \Omega^q(E_k) | \beta(\zeta_1, \cdots, \zeta_q) = 0 \\ \textit{if at least } q - p + 1 \textit{ of the fields } \zeta_1, \cdots, \zeta_q \text{ are Cartan} \end{array} \right\}$$

PROPOSITION 2.23 - *If the fiber dimension of* $\mathbf{C}har(E_k)_p$ *is* s *one has*

$$Ch\Omega^q(E_k) = \Omega^q(E_k), \quad q > s.$$

PROOF. In fact, the maximum number of linearly independent characteristic vector fields is s. So, if $\alpha \in \Omega^q(E_k), q > s$, then $\alpha(\zeta^1, \cdots, \zeta^q) = 0$, for all characteristic vector fields ζ_i. □

PROPOSITION 2.24 - *If the fiber dimension of* $\mathbb{E}_k$ *is* r *one has*

$$C\Omega^q(E_k) = \Omega^q(E_k), \quad q > r.$$

PROPOSITION 2.25 - *If* $k = \infty$ *one has:*

$$\begin{aligned} &Ch\Omega^q(E_k) = C\Omega^q(E_\infty) \\ &Ch^p\Omega^q(E_k) = C^p\Omega^q(E_\infty) \\ &C\Omega^q(E_\infty) = \Omega^q(E_\infty) = Ch\Omega^q(E_k), \quad q > n. \end{aligned}$$

PROOF. In fact, for $k = \infty$ one has $\mathbf{C}har(E_k) = \mathbb{E}_\infty$, and $r = s = n$. □

THEOREM 2.114 - $\alpha \in Ch\Omega^q(E_k)$, *iff* $\alpha|_V = 0$, *for all the characteristic integral manifolds of* E_k.

PROOF. If α is a characteristic q-form it follows that $\alpha|_V = 0$ for any characteristic integral manifold $V \subset E_k$. Conversely, let us assume that $\alpha|_V = 0$ on all the characteristic integral manifolds of E_k. Then, as $\mathbf{C}har(E_k)$ is an involutive distribution and E_k is of finite dimension, it follows that for any point $p \in E_k$ passes a unique maximal charactersitic integral manifold N of $\mathbf{C}har(E_k)$. Then, all the possible q-dimensional characteristic integral planes for p are contained in $G_{q,s}(T_pN)$, where $s = \dim T_pN$. Hence, under our assumptions it follows that β is characteristic. □

COROLLARY 2.14 - *If* $E_\infty \subset J^k_n(W)$, *then* $\mathbf{C}har(E_\infty) = \mathbb{E}_\infty$, *and* $Ch\Omega^q(E_\infty) = C\Omega^q(E_\infty)$. *Furthermore, even if for any* $p \in E_\infty$, *one has an infinity number of maximal integral manifolds (of dimension* n*) passing for* p, *one has that all these integral manifolds have at* p *the same tangent space* $(\mathbb{E}_\infty)_p$. *Hence, a differential* q*-form on* E_∞ *is Cartan iff it is zero on all the integral manifolds of* E_∞.

THEOREM 2.115 - *One has the following natural differential complex:*

$$\begin{aligned} &0 \to Ch\Omega^1(E_k) \xrightarrow{d} Ch\Omega^2(E_k) \xrightarrow{d} \cdots \\ &\to Ch\Omega^s(E_k) \xrightarrow{d} Ch\Omega^{s+1}(E_k) \xrightarrow{d} \cdots \to Ch\Omega^r(E_k) \xrightarrow{d} 0 \end{aligned} \tag{2.177}$$

where s = fiber dimension of $\mathbf{C}har(E_k)$ *, and* $r = \dim E_k$ *, with* $k \leq \infty$*. In particular, if* $k = \infty$ *we can write the above complex by putting* $Ch\Omega^q(E_\infty) = C\Omega^q(E_\infty)$.

PROOF. Let us consider the exterior differential $d : \Omega^q(E_k) \to \Omega^{q+1}(E_k)$, and restrict d to $Ch\Omega^q(E_k)$. Then, for any $\alpha \in Ch\Omega^q(E_k)$ and $\zeta_i \in \mathbf{C}har(E_k)$, we get

$$d\alpha(\zeta_1, \cdots, \zeta_q) = \frac{1}{q+1} \sum_{0 \leq \lambda \leq q} (-1)^\lambda \alpha(\zeta_1, \cdots, \hat{\zeta}_\lambda, \cdots, \zeta_q)$$
$$+ \frac{1}{q+1} \sum_{\lambda<\mu} (-1)^{\lambda+\mu} \alpha([\zeta_\lambda, \zeta_\mu], \zeta_1, \cdots, \hat{\zeta}_\lambda, \cdots, \hat{\zeta}_\mu, \cdots, \zeta_q).$$

As $\alpha \in Ch\Omega^q(E_k)$ and $\mathbf{C}har(E_k)$ is involutive, we get that each term in both the above summations is zero. Therefore, $d\alpha(\zeta_1, \cdots, \zeta_q) = 0 \Rightarrow d\alpha \in Ch\Omega^{q+1}(E_k)$. Similar results apply if $k = \infty$. □

COROLLARY 2.15 - *One has*

$$d : Ch^p\Omega^q(E_k) \to Ch^p\Omega^{q+1}(E_k), \quad k \leq \infty.$$

In particular, for $k = \infty$ *we can write*

$$d : C^p\Omega^q(E_\infty) \to C^p\Omega^{q+1}(E_\infty).$$

COROLLARY 2.16 - *One has the following filtration compatible with the exterior differential:*

$$Ch^0\Omega(E_k) \equiv \Omega^q(E_k) \supset Ch^1\Omega^q(E_k) \equiv Ch\Omega^q(E_k) \supset Ch^2\Omega^q(E_k) \supset \cdots$$
$$\supset Ch^q\Omega^q(E_k) \supset 0$$

for $k \leq \infty$*. As a consequence we have associated a spectral sequence (***characteristic spectral sequence of*** E_k*)*

$$\{E_r^{p,q}(E_k), d_r^{p,q}\}.$$

In particular, if E_∞ *is the infinity prolongation of a PDE* $E_k \subset J^k_n(W)$ *above spectral sequence applied to* E_∞ *coincides with the* C*-spectral sequence of* E_k *previously considered.*

□ *Of particular importance is the following spectral term:*

$$E_1^{0,q}(E_k) = \frac{\Omega^q(E_k) \cap d^{-1}Ch\Omega^{q+1}(E_k)}{d\Omega^{q-1}(E_k) \oplus Ch\Omega^q(E_k)}$$

that for $k = \infty$ *can also be written*

$$E_1^{0,q}(E_k) = \frac{\Omega^q(E_k) \cap d^{-1}C\Omega^{q+1}(E_k)}{d\Omega^{q-1}(E_k) \oplus C\Omega^q(E_k)}.$$

PROPOSITION 2.26 - *Set*

$$\bar{\Omega}^q(E_k) = \Omega^q(E_k)/Ch\Omega^q(E_k), \quad k \leq \infty.$$

Then, one has the following differential complex associated to E_k *(***bar de Rham complex** *of* E_k*):*

$$0 \to \bar{\Omega}^0(E_k) \xrightarrow{\bar{d}} \bar{\Omega}^1(E_k) \xrightarrow{\bar{d}} \bar{\Omega}^2(E_k) \xrightarrow{\bar{d}} \cdots \to$$
$$\bar{\Omega}^q(E_k) \xrightarrow{\bar{d}} \cdots \to \bar{\Omega}^s(E_k) \xrightarrow{\bar{d}} 0\,.$$

We call **bar de Rham cohomology** *of* E_k *the corresponding homology* $\bar{H}^q(E_k)$*. One has the following canonical isomorphism:*

$$E_1^{0,q}(E_k) \cong \bar{H}^q(E_k)\,, \quad k \leq \infty.$$

PROOF. Let us explicitly prove only the above isomorphism.

$$\begin{aligned}\bar{H}^q(E_k) &= \frac{\ker(\bar{d} : \bar{\Omega}^q(E_k) \to \bar{\Omega}^{q+1}(E_k))}{\bar{d}(\Omega^{q-1}(E_k))} \\ &= \frac{\ker(\bar{d} : \frac{\Omega^q(E_k)}{Ch\Omega^q(E_k)} \to \frac{\Omega^{q+1}(E_k)}{Ch\Omega^{q+1}(E_k)})}{\bar{d}(\frac{\Omega^{q+1}(E_k)}{Ch\Omega^{q-1}(E_k)})} \\ &= \frac{\Omega^q(E_k) \cap d^{-1}(Ch\Omega^{q+1}(E_k))}{d\Omega^{q-1}(E_k) \oplus Ch\Omega^q(E_k)} \\ &= E_1^{0,q}(E_k)\end{aligned}$$

□

DEFINITION 2.74 - *Let* $E_k \subset J_n^k(W)$ *be a PDE. We call* **bar singular chain complex** *of* E_k *the chain complex*

$$\{\bar{C}_p(E_k; \mathbf{R}), \bar{\partial}\}$$

where $\bar{C}_p(E_k; \mathbf{R})$ *is the vector space of formal linear combinations, with coefficients in* $\mathbf{R}$, $\sum \lambda_i c_i$, *where* c_i *is a singular p-chain* $f : \Delta^p \to E_k$ *that extends on a neighbourhood* $U \subset \mathbf{R}^{p+1}$, *such that* f *on* U *is differentiable and* $Tf(\Delta^p) \subset \mathbb{E}_k$. *Let*

$$\{\bar{C}^p(E_k; \mathbf{R}) \equiv Hom_{\mathbf{R}}(\bar{C}_p(E_k; \mathbf{R}), \mathbf{R}), \bar{\delta}\}$$

be the corresponding dual complex and $\bar{H}^p(E_k; \mathbf{R})$ *the associated homology spaces (***bar singular cohomology** *of* E_k*).*[5]

[5] Similar definitions can be done with coefficients into an abelian group G. For example, other than G=$\mathbf{R}$, it is important to take G=$\mathbf{Z}$,$\mathbf{Z}_2$.

THEOREM 2.116 - (BAR DE RHAM THEOREM FOR PDEs). *One has a natural bilinear mapping*

$$<,>: \bar{C}_p(E_k;\mathbf{R}) \times \bar{C}^p(E_k;\mathbf{R}) \to \mathbf{R}$$

such that

$$(2.178)(\textbf{bar Stokes formula}) \qquad <\bar{\delta}\alpha, c> + (-1)^p < \alpha, \bar{\partial} c> = 0.$$

One has the canonical isomorphism

$$\bar{H}^p(E_k;\mathbf{R}) \cong Hom_{\mathbf{R}}(\bar{H}_p(E_k;\mathbf{R}),\mathbf{R}) \equiv \bar{H}_p(E_k;\mathbf{R})^*.$$

One has the following nondegenerate mapping

$$<,>: \bar{H}_p(E_k;\mathbf{R}) \times \bar{H}^p(E_k;\mathbf{R}) \to \mathbf{R}.$$

Hence one has the following short exact sequence

$$0 \to \bar{H}_p(E;\mathbf{R}) \to \bar{H}^p(E_k;\mathbf{R})^*.$$

2 *This means that if* c *is a closed bar singular* p*-chain* ($\bar{\partial} c = 0$) *of* E_k, c *is the boundary of a singular* $(p+1)$*-chain* c' *of* E_k ($\bar{\partial} c' = c$) , *iff* $< c, \alpha >= 0$, *for all the* $\bar{\delta}$*-closed bar singular* p*-cochains* α *of* E_k. *Furthermore, if* α *is a* $\bar{\delta}$*-closed bar singular* p*-cochain of* E_k, α *is* $\bar{\delta}$*-exact,* ($\alpha = \bar{\delta}\beta$) *iff* $< c, \alpha >= 0$, *for all the* $\bar{\partial}$*-closed bar singular* p*-chains* c *of* E_k.

PROOF. The proof can be deduced by simulating standard one for ordinary manifolds and ordinary singular (co)homology [141, 157]. Let us explicitly prove only that $\bar{H}^p(E_k : \mathbf{R}) \cong (\bar{H}_p(E_k;\mathbf{R}))^*$. In fact, as $\bar{C}_p(E_k : \mathbf{R})$ is a free complex over $\mathbf{R}$ we can use the universal coefficient theorem in cohomology [49] and say that there is the following unnatural isomorphism of $\mathbf{R}$-vector spaces:

$$\bar{H}^p(E_k;\mathbf{R}) \cong Hom_{\mathbf{R}}(\bar{H}_p(E_k;\mathbf{R}),\mathbf{R}) \bigoplus Ext^1_{\mathbf{R}}(\bar{H}_{p-1}(E_k;\mathbf{R})).$$

As $\mathbf{R}$ is an injective $\mathbf{R}$-module, one has $Ext^1_{\mathbf{R}}(\bar{H}_{p-1}(E_k;\mathbf{R})) = 0$. □

THEOREM 2.117 - *One has the following canonical isomorphism:*

$$\bar{H}^p(E_\infty;\mathbf{R}) \cong \bar{H}^p(E_\infty).$$

PROOF. Let us consider the following homomorphism of $\mathbf{R}$-vector spaces

$$j_p : \bar{\Omega}^p(E_\infty) \to \bar{C}^p(E_\infty : \mathbf{R})$$

$$[\omega] \mapsto j_p[\omega], j_p[\omega](c \equiv \sum_i a^i u_i) = \sum_i a^i (\int_{u_i} \omega) = \sum_i a^i (\int_{\triangle^p} u_i^* \omega)$$

where $\int_{\triangle^p} u_i^*\omega$ are the standard Riemann integrals, and $a^i \in \mathbf{R}$. Then, j_p passes to quotient giving the following homomorphism

$$j_p^* : \bar{H}^p(E_\infty) \to \bar{H}^p(E_\infty; \mathbf{R}) \cong Hom_{\mathbf{R}}(\bar{H}_p(E_\infty; \mathbf{R}); \mathbf{R}).$$

Let us see that j_p^* are isomorphisms. Let us first prove that j_p^* are injective. Assume that there exist $0 \neq [\alpha] \in \ker(j_p^*)$. Therefore, we have

$$\int_c \alpha = j_p^*[\alpha][c] = 0\,,\ \forall [c] \in \bar{H}_p(E_\infty : \mathbf{R}).$$

From the bar de Rham theorem for PDEs, we see that α should be $\bar{\delta}$-exact, in contradiction with the assumption that $[\alpha] \neq 0 \in \bar{H}^p(E_\infty)$. Therefore, j_p^* are injective homomorphisms, $p \geq 0$. Let us, now, prove that j_p^* are surjective. This proof is more complex. In fact, we shall refer to the technique of sheaves. In fact, the homomorphisms j_p can be defined for arbitrary open sets in E_∞ , and yield presheaf homomorphisms

$$\{\bar{\Omega}^p(U); \rho_{U,V}\} \xrightarrow{j_p} \{\bar{C}^p(U; \mathbf{R}); \rho_{U,V}\}, \quad (p \geq 0)$$

which commute, by the bar Stokes formula for PDEs, with the bar coboundary homomorphisms

$$\{\bar{\Omega}^p(U); \rho_{U,V}\} \xrightarrow{\bar{d}} \{\bar{\Omega}^{p+1}(U; \mathbf{R}); \rho_{U,V}\}, \quad (p \geq 0)$$

and

$$\{\bar{C}^p(U); \rho_{U,V}\} \xrightarrow{\bar{\delta}} \{\bar{C}^{p+1}(U; \mathbf{R}); \rho_{U,V}\}, \quad (p \geq 0).$$

Thus, the induced homomorphisms of the associated sheaves form a commutative diagram

$$\begin{array}{ccccccccc} 0 \to & \mathcal{R} & \to & \bar{\Lambda}_0^0(E_\infty) & \to & \bar{\Lambda}_1^0(E_\infty) & \to & \bar{\Lambda}_2^0(E_\infty) & \to \cdots \\ & \| & & \downarrow j_0 & & \downarrow j_1 & & \downarrow j_2 & \\ 0 \to & \mathcal{R} & \to & \bar{C}^0(E_\infty; \mathbf{R}) & \to & \bar{C}^1(E_\infty; \mathbf{R}) & \to & \bar{C}^2(E_\infty; \mathbf{R}) & \to \cdots \end{array} \tag{2.179}$$

where $\mathcal{R}$ is the sheaf of real functions on E_∞.

Let us give, now, the following lemma.

LEMMA 2.23 - *Let $\{S_U; \rho_{U,V}\}$ be a presheaf on a manifold M such that whenever the open set U is expressed as union $\bigcup_\alpha U_\alpha$ of open sets in M, then whenever there is an element $f_\alpha \in S_{U_\alpha}$ for each α such that $\rho_{U_\alpha \cap U_\beta, U_\alpha} f_\alpha = \rho_{U_\alpha \cap U_\beta, U_\alpha} f_\beta$ for all α and β, then there exists $f \in S_U$ such that $f_\alpha = \rho_{U_\alpha U} f$ for each α. Let*

$$(S_M)_0 \equiv \{s \in S_M : \rho_{m,M}(s) = 0, \quad \text{for all } m \in M\}.$$

Then, the sequence

$$0 \to (S_M)_0 \to S_M \xrightarrow{\gamma} \Gamma(S) \to 0$$

is exact.

Let us now consider the following commutative diagram of cochain complexes with exact rows:

$$\begin{array}{ccccccc} & 0 & \to & \bar{\Omega}^\bullet(E_\infty) & \xrightarrow{\bar{1}} & \Gamma(\underline{\bar{\Lambda}_0^0(E_\infty)}) & \to 0 \\ & & & \downarrow j_\bullet & & \downarrow \bar{3} & \\ 0 \to & \bar{C}^\bullet(E_\infty;\mathbf{R})_0 & \to & \bar{C}^\bullet(E_\infty;\mathbf{R}) & \xrightarrow{\bar{2}} & \Gamma(\underline{\bar{C}^\bullet(E_\infty;\mathbf{R})}) & \to 0 \end{array}$$

where $\Gamma(\underline{X})$ denotes the space of sections of the sheaf $\underline{X}$, and

$$\bar{C}^\bullet(E_\infty;\mathbf{R})_0 \equiv \{s \in \bar{C}^\bullet(E_\infty;\mathbf{R}) \quad | \quad \rho_{u,E_\infty}(s) = 0, \forall u \in E_\infty\}$$

The cochain map $\bar{1}$ induces the isomorphisms $\bar{H}^p(E_\infty) \cong \bar{H}^p(E_\infty;\mathcal{R})$. Furthermore, $\bar{2}$ induces the isomorphisms $\bar{H}^p(E_\infty;\mathbf{R}) \cong \bar{H}^p(E_\infty;\mathcal{R})$. In fact, the lower short exact sequence in the above commutative diagram induces the following long exact sequence

$$\cdots \to \bar{H}^{q-1}(E_\infty;\mathcal{R}) \to H^q(\bar{C}^\bullet(E_\infty;\mathbf{R})_0) \to \bar{H}^q(E_\infty;\mathbf{R}) \to \bar{H}^q(E_\infty;\mathcal{R}) \to \cdots$$

Thus, if we prove that

$$H^q(\bar{C}^\bullet(E_\infty;\mathbf{R})_0) = 0\,, \forall q \tag{2.180}$$

it follows that there are canonical isomorphisms

$$\bar{H}^q(E_\infty;\mathbf{R}) \cong \bar{H}^q(E_\infty;\mathcal{R}).$$

The proof of (2.180) goes as follows. It is trivially satisfied for $q < 0$ since in this range the modules of the cochain complex $\bar{C}^\bullet(E_\infty;\mathbf{R})_0$ are all zero. The $\mathbf{R}$-vector space $\bar{C}^\bullet(E_\infty;\mathbf{R})_0$ is also zero since the presheaf $\{\bar{C}^0(U;\mathbf{R});\rho_{U,V}\}$ is complete, i.e., it satisfies the conditions of the above lemma and also the following condition is satisfied:

Whenever f and g in $\bar{C}^0(U;\mathbf{R})$ are such that $\rho_{U_\alpha,U} f = \rho_{U_\alpha,U} g$ for all α, then $f = g$.

Thus $H^0(\bar{C}^0(E_\infty;\mathbf{R})_0) = 0$. It remains to prove (2.180) for $q \geq 1$. Let $\mathcal{U} \equiv \{U_i\}$ be an arbitrary open cover of E_∞. Let $\bar{C}^\bullet_{\mathcal{U}}(E_\infty;\mathbf{R})$ be the cochain complex consisting of $\mathbf{R}$-vector spaces $\bar{C}^p_{\mathcal{U}}(E_\infty;\mathbf{R})$ of bar singular cochains α, defined only on those bar singular p-simpleces whose ranges lie in elements of $\mathcal{U}$. Each element of $\bar{C}^p(E_\infty;\mathbf{R})$

determines an element of $\bar{C}^p_{\mathcal{U}}(E_\infty;\mathbf{R})$ by means of suitable restriction. Then the homomorphisms $j_{\mathcal{U}}:\bar{C}^p(E_\infty;\mathbf{R})\to\bar{C}^p_{\mathcal{U}}(E_\infty;\mathbf{R})$ yield a surjective cochain map

$$j_{\mathcal{U}}:\bar{C}^\bullet(E_\infty;\mathbf{R})\to\bar{C}^\bullet_{\mathcal{U}}(E_\infty;\mathbf{R}).$$

The kernels of the homomorphisms $j_{\mathcal{U}}$ form a cochain complex $K^\bullet_{\mathcal{U}}$ such that

$$0\to K^\bullet_{\mathcal{U}}\to\bar{C}^\bullet(E_\infty;\mathbf{R})\to\bar{C}^\bullet_{\mathcal{U}}(E_\infty;\mathbf{R})\to 0 \tag{2.181}$$

is an exact sequence of cochain complexes. We can prove that $j_{\mathcal{U}}$ induces isomorphisms of the cohomology spaces. The proof can be conducted by simulating one for the ordinary singular cohomology; (see e.g., ref.[157]). It follows from the long exact cohomology sequence for (2.181) that

$$H^q(K^\bullet_{\mathcal{U}})=0\,,\ \forall q. \tag{2.182}$$

So, now, let $q\ge 1$ and let $f\in\bar{C}^q(E_\infty;\mathbf{R})_0$ such that $\bar{\delta}f=0$. Then, by definition of $\bar{C}^q(E_\infty;\mathbf{R})_0$, there is an open cover $\mathcal{U}$ of E_∞ consisting of sufficiently small open sets that $f\in K^q_{\mathcal{U}}$. It follows from (2.182) that there exists $g\in K^{q-1}_{\mathcal{U}}\subset\bar{C}^{q-1}(E_\infty;\mathbf{R})_0$ such that $dg=f$, that proves (2.180). Finally, in order to prove that $\overline{3}$ induces isomorphism on cohomology we shall use the following lemmas.

LEMMA 2.24 - *Let $\mathcal{H}$ and $\widetilde{\mathcal{H}}$ be cohomology theories on a manifold M with coefficients in sheaves of κ-modules over M, where κ is a fixed principal ideal domain (e.g., $\kappa=\mathbf{R}$). Then, there exists a unique isomorphism $\mathcal{H}\to\widetilde{\mathcal{H}}$.*

LEMMA 2.25 - *Assume that $\mathcal{H}$ is a cohomology theory for M with coefficients in sheaves of κ-modules over M. Let*

$$0\to\mathcal{S}\to\mathcal{C}_0\to\mathcal{C}_1\to\mathcal{C}_2\to\cdots \tag{2.183}$$

be a fine resolution of the sheaf $\mathcal{S}$. [6]

---

[6] An exact sheaf sequence

$$0\to\mathcal{A}\to\mathcal{C}_0\to\mathcal{C}_1\to\mathcal{C}_2\to\cdots \tag{2.184}$$

is called a **resolution** of the sheaf $\mathcal{A}$. The resolution (2.184) is called **fine** (resp. **torsionless**) if each of the sheaves $\mathcal{C}_i$ is fine (resp. torsionless). Recall that a sheaf $\mathcal{S}$ over M is said to be **fine** if for each locally finite cover $\{U_i\}$ of M by open sets there exists for each i an endomorphism l_i of $\mathcal{S}$ such that: (a) $\mathrm{supp}(l_i)\subset U_i$; (b) $\sum_i l_i=id$. Here, by $\mathrm{supp}(l_i)$ (the support of l_i) we mean the closure of the set of points in M for which $l_i|_{\mathcal{S}_m}$ is nonzero. We shall call $\{l_i\}$ **a partition of unity** for $\mathcal{S}$ subordinate to the cover $\{U_i\}$ of M. Furthermore, a sheaf of κ-modules is said to be **torsionless** if each stalk X is a torsionless κ-module, i.e., there is no nonzero element $x\in X$ for which there exists a nonzero element $\lambda\in\kappa$ such that $\lambda x=0$.

Then, there are canonical isomorphisms

$$H^q(M, \mathcal{S}) \cong H^q(\Gamma(\mathcal{C}^\bullet)), \forall q. \tag{2.185}$$

Then, $\overline{3}$ induces isomorphisms on cohomology as an application of above lemmas to the homomorphism (2.179) of fine torsionless resolutions of $\mathcal{R}$. Thus, from the uniqueness of the isomorphism between sheaf cohomology theories, $\overline{3}$ induces the identity isomorphism of $\bar{H}^p(E_\infty; \mathcal{R})$. It follows that j_p^* is a canonical isomorphism for each integer p. □

REMARK 2.64 - 1) Similar to the classical case, we can also define the relative (co)homology spaces $\bar{H}^p(E_k, X; \mathbf{R})$ and $\bar{H}_p(E_k, X; \mathbf{R})$, where $X \subset E_k$ is a bar singular chain.

2) One has the following exact sequence:

$$\begin{aligned} &\cdots \to \bar{H}_p(X; \mathbf{R}) \to \bar{H}_p(E_k; \mathbf{R}) \to \bar{H}_p(E_k, X; \mathbf{R}) \to \bar{H}_{p-1}(X; \mathbf{R}) \to \cdots \\ &\bar{H}_0(X; \mathbf{R}) \to \bar{H}_0(E_k; \mathbf{R}) \to \bar{H}_0(E_k, X; \mathbf{R}) \to 0\,. \end{aligned}$$

3) One has the following isomorphisms:

$$\begin{aligned} &\bar{H}_p(E_k, *; \mathbf{R}) \cong \bar{H}_p(E_k; \mathbf{R}), \quad p > 0 \\ &\bar{H}_0(E_k, *; \mathbf{R}) = 0 \end{aligned}$$

if E_k is arcwise connected.

DEFINITION 2.75 - 1) *A G-***singular p-dimensional integral manifold** *of $E_k \subset J_n^k(W)$, is a bar singular p-chain V with $p \leq n$, and G an abelian group, such that $V \subset E_k$.*

2) *A G-***singular p-dimensional quantum manifold** *of $E_k \subset J_n^k(W)$ is a bar singular p-chain $V \subset J_n^k(W)$, with $p \leq n$, and G an abelian group, such that $\partial V \subset E_k$.*

REMARK 2.65 - Smooth p-dimensional integral manifolds and smooth p-dimensional quantum manifolds of $E_k \subset J_n^k(W)$, are particular cases of the above singular ones respectively. Let us denote ${}^G\Omega_{p,s}^{E_k}$ and ${}^G\Omega_{p,s}(E_k)$ the corresponding cobordism groups in the singular case. Let us denote also by ${}^G[N]_{E_k}^s$ and ${}^G[N]_{\overline{E_k}}^s$ the equivalence classes in these groups respectively.

In the following, for simplicity, we will only consider the case $G = \mathbf{R}$, and we will omit the apex G in the above symbols.

THEOREM 2.118 - *Let us assume that $E_k \subset J_n^k(W)$ is a formally integrable PDE.*

1) *As $\pi_\infty : E_\infty \to E_k$ is surjective, one has the following short exact sequence of chain complexes:*

$$\begin{aligned} &\bar{C}_\bullet(E_\infty; \mathbf{R}) \to \bar{C}_\bullet(E_k; \mathbf{R}) \to 0 \\ &\bar{C}^\bullet(E_\infty; \mathbf{R}) \leftarrow \bar{C}^\bullet(E_k; \mathbf{R}) \leftarrow 0\,. \end{aligned}$$

These induce the following homomorphisms of vector spaces:

$$\bar{H}_p(E_\infty;\mathbf{R}) \to \bar{H}_p(E_k;\mathbf{R})$$
$$\bar{H}^p(E_\infty;\mathbf{R}) \leftarrow \bar{H}^p(E_k;\mathbf{R})$$

2) *One has the following isomorphisms:*

$$\Omega^{E_k}_{p,s} \cong \bar{H}_p(E_k;\mathbf{R}), \quad k \le \infty$$
$$\Omega_{p,s}(E_k) \cong \bar{H}_p(J^k_n(W), E_k;\mathbf{R})$$

THEOREM 2.119 - 1) *One has the following exact sequences of vector spaces:*

$$\Omega^{E_k}_{n-1,s} \to \Omega^{J^k_n(W)}_{n-1,s} \to \Omega_{n-1,s}(E_k) \to \Omega^{E_k}_{n-2,s} \to \cdots$$
$$\to \Omega^{E_k}_{0,s} \to \Omega^{J^k_n(W)}_{0,s} \to \Omega_{0,s}(E_k) \to 0\,.$$

2) *Therefore, one has unnatural splittings:*

$$\Omega_{p,s}(E_k) \cong \Omega^{J^k_n(W)}_{p,s} \times \Omega^{E_k}_{p-1,s}$$
$$\Omega^{J^k_n(W)}_{p,s} \cong \Omega^{E_k}_{p,s} \times \Omega_{p,s}(E_k)$$

3) *One has a natural homomorphism*

$$\pi_{\infty,k*} : \Omega^{E_\infty}_{p,s} \to \Omega^{E_k}_{p,s}$$

DEFINITION 2.76 - 1) *We call* **singular integral characteristic numbers** *of a p-dimensional $\bar{\partial}$-closed singular integral manifold $N \subset E_k \subset J^k_n(W)$ the numbers*

$$i[N] \equiv < N, \alpha > \in \mathbf{R}$$

where α is a closed bar singular p-chain of E_k.
2) *We call* **singular quantum characteristic numbers** *of a p-dimensional $\bar{\partial}$-closed singular integral manifold $N \subset E_k \subset J^k_n(W)$, the numbers*

$$q[N] \equiv < N, \alpha > \in \mathbf{R}$$

where α is a closed bar singular p-cochain of $J^k_n(W)$.

THEOREM 2.120 - 1) *$N' \in [N]^s_{E_k} \Leftrightarrow N'$ and N have equal singular integral characteristic numbers: $i[N'] = i[N]$.*
2) *$N' \in [N]^s_{\overline{E_k}} \Leftrightarrow N'$ and N have equal singular quantum characteristic numbers: $q[N'] = q[N]$.*

PROOF. It follows from the bar de Rham theorem that one has the following short exact sequences:

$$0 \to \Omega_{p,s}^{E_k} \to \bar{H}^p(E_k;\mathbf{R})^*$$
$$0 \to \Omega_{p,s}(E_k) \to \bar{H}^p(J_n^k(W),E_k;\mathbf{R})^* .$$

□

THEOREM 2.121 - (RELATION BETWEEN SINGULAR INTEGRAL COBORDISM GROUPS AND HOMOLOGY). *One has the following exact commutative diagram:*

$$\begin{array}{ccccccc} & & & & 0 & & \\ & & & & \downarrow & & \\ 0 \to K\bar{H}_p(E_k;\mathbf{R}) & \to & \Omega_{p,s}^{E_k} & \to & \widehat{\bar{H}}_p(E_k;\mathbf{R}) & \to 0 \\ & & & & \downarrow & & \\ & & & & H_p(E_k;\mathbf{R}) & & \end{array}$$

where

$$K\bar{H}_p(E_k;\mathbf{R}) \equiv \{[N]_{E_k}^s | N=\partial V, V = \text{singular } p\text{-chain in } E_k\}$$
$$\equiv \{[N]_{E_k}^s | <[\alpha]|[N]_{E_k}^s>=0, \forall[\alpha] \in H^p(E_k;\mathbf{R}\} .$$

We call $s[N] \equiv <[\alpha]|[N]_{E_k}^s> \equiv$ **singular characteristic numbers** *of* $[N]_{E_k}^s$.

THEOREM 2.122 - (RELATION BETWEEN SINGULAR QUANTUM COBORDISM GROUPS AND HOMOLOGY). *One has the following exact commutative diagram:*

$$\begin{array}{ccccccc} & & & & 0 & & \\ & & & & \downarrow & & \\ 0 \to K\bar{H}_p(J_n^k(W),E_k;\mathbf{R}) & \to & \Omega_{p,s}(E_k) & \to & \widehat{\bar{H}}_p(J_n^k(W),E_k;\mathbf{R}) & \to 0 \\ & & & & \downarrow & & \\ & & & & H_p(J_n^k(W),E_k;\mathbf{R}) & & \end{array}$$

where

$$K\bar{H}_p(J_n^k(W),E_k;\mathbf{R}) \equiv \{[N]_{\overline{E_k}}^s | N=\partial V, V = \text{ singular } p\text{-chain in } J_n^k(W)\}$$
$$\equiv \{[N]_{\overline{E_k}}^s | <[\alpha]|[N]_{\overline{E_k}}^s>=0, \forall[\alpha] \in H^p(J_n^k(W),E_k;\mathbf{R}\} .$$

We call $s_q[N] \equiv <[\alpha]|[N]_{\overline{E_k}}^s> \equiv$ **singular characteristic numbers** *of* $[N]_{\overline{E_k}}^s$.

THEOREM 2.123 - (RELATION BETWEEN INTEGRAL COBORDISM GROUPS AND SINGULAR INTEGRAL COBORDISM GROUPS). *The integral cobordism group* $\Omega_p^{E_k}, 0 \le p \le n-1$, *is an extension of a subgroup* $\hat{\Omega}_{p,s}^{E_k}$ *of the singular integral cobordism group* $\Omega_{p,s}^{E_k}$.

PROOF. In fact, one has a canonical group-homomorphism $j_p : \Omega_p^{E_k} \to \Omega_{p,s}^{E_k}$, that generates the following exact commutative diagram:

$$\begin{array}{ccccccc} & & & & 0 & & \\ & & & & \downarrow & & \\ 0 \to K_{p,s}^{E_k} & \to & \Omega_p^{E_k} & \to & \hat{\Omega}_{p,s}^{E_k} & \to 0 & \\ & & & {\scriptstyle j_p}\searrow & \downarrow & & \\ & & 0 & \to & \Omega_{p,s}^{E_k} & \to & \bar{H}_p(E_k;\mathbf{R}) \to 0 \end{array}$$

where $K^{E_k}_{p,s} \equiv \ker(j_p)$ and $\hat{\Omega}^{E_k}_{p,s} \equiv \Omega^{E_k}_p / K^{E_k}_{p,s}$. Furthermore, $K^{E_k}_{p,s}$ can be characterized by means of characteristic numbers. In fact we get

$$\begin{aligned} K^{E_k}_{p,s} &= \{[N]^s_{E_k} | \exists \\ &\quad (p+1)\text{-dimensional singular integral submanifold } V \subset E_k, \text{ with } \partial V = N\} \\ &= \{[N]^s_{E_k} | i[N] = 0 \text{ for all singular integral characteristic numbers}\}\,. \end{aligned}$$

□

THEOREM 2.124 - (RELATION BETWEEN QUANTUM COBORDISM GROUPS AND SINGULAR QUANTUM COBORDISM GROUPS). *The quantum cobordism group $\Omega_p(E_k), 0 \leq p \leq n-1$, is an extension of a subgroup $\hat{\Omega}_{p,s}(E_k)$ of the singular quantum cobordism group $\Omega_{p,s}(E_k)$.*

PROOF. In fact, one has a canonical group homomorphism $\bar{j}_p : \Omega_p(E_k) \to \Omega_{p,s}(E_k)$, hence one has the following exact commutative diagram:

$$\begin{array}{ccccccc} & & & 0 & & & \\ & & & \downarrow & & & \\ 0 \to K_{p,s}(E_k) \to & \Omega_p(E_k) & \to & \hat{\Omega}_{p,s}(E_k) & \to 0 & & \\ & & j_p \searrow & \downarrow & & & \\ & 0 & \to & \Omega_{p,s}(E_k) & \to & \bar{H}_p(J^k_n(W), E_k; \mathbf{R}) \to 0 & \end{array}$$

where

$$K_{p,s}(E_k) \equiv \{[N]^s_{\overline{E_k}} | q[N] = 0 \text{ for all singular quantum characteristic numbers}\}.$$

□

DEFINITION 2.77 - *Set*

$$\mathcal{I}(E_k)^p \equiv \frac{\Omega^p(E_k) \cap d^{-1}(C\Omega^{p+1}(E_k))}{d\Omega^{p-1}(EE_k) \oplus \{C\Omega^p(E_k) \cap d^{-1}(C\Omega^{p+1}(E_k))\}}$$

REMARK 2.66 - For $k = \infty$ one has

$$\mathcal{I}(E_\infty)^p \cong E_1^{0,q}(E_\infty) \cong \bar{H}^q(E_\infty).$$

THEOREM 2.125 - *Let us assume that $\mathcal{I}(E_k)^p \neq 0$. One has a natural group homomorphism:*

$$\begin{aligned} &j_p : \Omega^{E_k}_p \to (\mathcal{I}(E_k)^p)^* \\ &[N]_{E_k} \mapsto j_p([N]_{E_k}), \quad j_p([N]_{E_k})([\alpha]) = \int_N \alpha \equiv < [N]_{E_k}, [\alpha] > \,. \end{aligned}$$

We call $i[N] \equiv< [N]_{E_k}, [\alpha] >$ **integral characteristic numbers** *of* N *for all* $[\alpha] \in \mathcal{I}(E_k)^p$. *Then a necessary condition that* $N' \in [N]_{E_k}$ *is the following*

$$i[N'] = i[N]\,,\ \forall[\alpha] \in \mathcal{I}(E_k)^p. \tag{2.186}$$

2 *The above condition is also sufficient for* $k = \infty$ *in order to identify elements belonging to the same singular integral cobordism classes of* $\Omega_{p,s}^{E_\infty}$. *In fact, one has the following exact commutative diagram:*

$$\begin{array}{ccccc}
 & & & 0 & \\
 & & & \downarrow & \\
 & \Omega_p^{E_\infty} & \xrightarrow{i_p} & \Omega_{p,s}^{E_\infty} \equiv \bar{H}_p(E_\infty; \mathbf{R}) & \\
 & \downarrow j_p & & \downarrow & \\
0 \rightarrow & (\mathcal{I}(E_\infty)^p)^* & \rightarrow & \bar{H}^p(E_\infty; \mathbf{R})^* & \rightarrow 0
\end{array}$$

PROOF. Let us assume that there is a $0 \neq [N]_{E_k} \in \ker(j_p)$. This is equivalent to saying that

$$\int_N \alpha \equiv< [N]_{E_k}, [\alpha] >= 0\,,\ \forall[\alpha] \in \mathcal{I}(E_k)^p. \tag{2.187}$$

Now, let us prove the following characterization of $\ker(j_p)$.

$$\ker(j_p) = \{[N]_{E_k}\} = \left\{ \begin{array}{l} (a)\ N \text{ is non-orientable} \\ (b) \left\{ \begin{array}{l} N \text{ is orientable and boundary of admissible} \\ \text{integral manifold contained in } E_k \end{array} \right\} \end{array} \right\}$$

In fact, let $N \subset E_k$ be an orientable p-dimensional compact closed integral submanifold of E_k , then for the Poincaré duality theorem, one has

$$H^p(N) \cong H^p(N; \mathbf{R}) \cong H_p(N; \mathbf{R}) \cong H^{p-p}(N; \mathbf{R}) = H^0(N; \mathbf{R}) \cong H_0(N; \mathbf{R}) \cong \mathbf{R}^r$$

where r is the number of connected components of N. Then, one has

$$\int_N \alpha = \alpha_1 s^1 + \cdots + \alpha_q s^q, \forall 0 \neq [\alpha] \equiv (\alpha_1, \cdots, \alpha_q) \in \hat{H}^p(N) \cong \mathbf{R}^q$$
$$(s^1, \cdots, s^r) = [N] \in H^p(N; \mathbf{R}) \cong \mathbf{R}^r$$

where $\hat{H}^p(N)$ is the canonical image of $\mathcal{I}(E_k)^p$ into $H^p(N)$ and hence $q \leq r$. So, condition (2.187) implies, in particular, that

$$[\alpha] = (1, 0, \cdots, 0) \Rightarrow s^1 = 0$$
$$\cdots$$
$$[\alpha] = (0, \cdots, 0, 1) \Rightarrow s^q = 0\,.$$

So, condition (2.187) implies that

$$[N] = (0, \cdots_q \cdots, 0, s^{q+1}, \cdots, s^r) \in \mathbf{R}^r.$$

On the other hand, $[N] \in H^p(N; \mathbf{R})$ must coincide with the image $\underline{[N]} \otimes 1$ of the fundamental class $\underline{[N]} \in H_p(N; \mathbf{Z})$ by means of the canonical homomorphism

$$H_p(N; \mathbf{Z}) \to H_p(N; \mathbf{Z}) \otimes_{\mathbf{Z}} \mathbf{R} \cong H_p(N; \mathbf{R}).$$

Taking into account that

$$\underline{[N]} = (1, \cdots_r \cdots, 1) \in \mathbf{Z}^r$$

we see that if

$$[N] = (0, \cdots_q \cdots, 0, s^{q+1}, \cdots, s^r)$$

it cannot coincide with $\underline{[N]} \otimes 1$. So, the assumption that there is some

$$0 \neq [N]_{E_k} \in \ker(j_p),$$

with N orientable, is absurd. On the other hand, if N is not orientable one has $H_0(N; \mathbf{Z}) = \mathbf{Z}^r$ and $H_p(N; \mathbf{Z}) = 0$. So,

$$H_0(N; \mathbf{R}) \cong H_0(N; \mathbf{Z}) \otimes_{\mathbf{Z}} \mathbf{R} \cong \mathbf{R}^r$$

and

$$H_p(N; \mathbf{R}) \cong H_p(N; \mathbf{Z}) \otimes_{\mathbf{Z}} \mathbf{R} = 0.$$

Therefore, the Poincaré duality does not work more on $\mathbf{Z}$ and $\mathbf{R}$.[7]
Furthermore, as $H_p(N; \mathbf{R}) = 0$, we get that $\int_N \alpha = 0$, $\forall [\alpha] \in \mathcal{I}(E_k)^p$. Hence , if N is not orientable $[N]_{E_k} \in \ker(j_p)$. $\square$

COROLLARY 2.17 - *For any $k \leq \infty$ one has the following exact commutative diagram:*

$$\begin{array}{ccccccc} & & & & 0 & & \\ & & & & \downarrow & & \\ 0 \to \overline{K_p^{E_k}} & \to & \Omega_p^{E_k} & \to & \overline{\Omega_p^{E_k}} & \to 0 \\ & & & j_p \searrow & \downarrow & & \\ & & & & (\mathcal{I}(E_k)^p)^* & & \end{array}$$

[7] One could see that the Poincaré duality works for non-orientable manifolds on $\mathbf{Z}_2$: $H_q(N;\mathbf{Z}_2) \cong H^{p-q}(N;\mathbf{Z}_2)$, $\dim N = p$. But these (co)homology spaces are not related to $H_q(N;\mathbf{R})$ and $H^q(N;\mathbf{R})$ respectively.

Therefore, we can write

$$\overline{K_p^{E_k}} \equiv \{[N]_{E_k} | < [\alpha], [N]_{E_k} >= 0, \forall[\alpha] \in \mathcal{I}(E_k)^p\}$$

$$N' \in \overline{[N]_{E_k}} \in \overline{\Omega_p^{E_k}} \Leftrightarrow \int_{N'} \alpha = \int_N \alpha \,, \forall[\alpha] \in \mathcal{I}(E_k)^p$$

DEFINITION 2.78 - *Set*

$$\mathcal{Q}_n^k(W)^p \equiv \mathcal{I}(J_n^k(W))^p.$$

COROLLARY 2.18 - *Let us assume that $\mathcal{Q}_n^k(W)^p \neq 0$. One has a natural group homomorphism*

$$\begin{aligned} &\bar{j}_p : \Omega_p(E_k) \to (\mathcal{Q}_n^k(W)^p)^* \\ &[N]_{\overline{E_k}} \mapsto \bar{j}_p([N]_{\overline{E_k}}) \\ &\bar{j}_p([N]_{\overline{E_k}})([\alpha]) = \int_N \alpha \equiv < [N]_{\overline{E_k}}, [\alpha] > \end{aligned}$$

We call $q[N] \equiv < [N]_{\overline{E_k}}, [\alpha] >$ **quantum characteristic numbers** *of N, for all $[\alpha] \in \mathcal{Q}_n^k(W)^p$. Then, a necessary condition that $N' \in [N]_{\overline{E_k}}$ is that*

$$q[N'] = q[N] \,, \forall[\alpha] \in \mathcal{Q}_n^k(W)^p. \tag{2.188}$$

From the above theorem we also get the following important criterion.

THEOREM 2.126 - (CRITERION FOR THE CONDITION (2.186) TO BE SUFFICIENT). *Let us assume that $E_k \subset J_n^k(W)$ is such that all its p-dimensional admissible compact closed integral submanifolds are orientable and $\mathcal{I}(E_k)^p \neq 0$. Then,* $\ker(j_p) = 0$, *i.e.,*

$$N' \in [N]_{E_k} \Leftrightarrow \int_{N'} \alpha = \int_N \alpha \,, \forall[\alpha] \in \mathcal{I}(E_k)^p.$$

In particular, for $k = \infty$, one has

$$\Omega_p^{E_\infty} \cong \Omega_{p,s}^{E_\infty}$$

as $\mathcal{I}(E_\infty)^p \cong \bar{H}^p(E_\infty)$.

PROOF. The proof is a direct consequence of what was considered in the proof of the above theorem. □

REMARK 2.67 - Let us emphasize that we can have $\mathcal{I}(E_k)^p \neq 0$ even if E_k is p- homologically trivial, i.e., $H^p(E_k; \mathbf{R}) = 0$. This, for example, happens if E_k is contractible to a point.

COROLLARY 2.19 - *Under the same hypotheses of the above theorem one has*

$$N' \in [N]_{\overline{E_k}} \Leftrightarrow \int_{N'} \alpha = \int_N \alpha \,, \quad \forall[\alpha] \in \mathcal{Q}_n^k(W)^p.$$

REMARK 2.68 - In the above criterion $\Omega_p^{E_k}$ (resp. $\Omega_p(E_k)$) does not necessarily coincides with the oriented version of the integral (resp. quantum) cobordism groups. In fact, the Möbius band is an example of nonorientable manifold B with $\partial B \cong S^1$, that, instead, is an orientable manifold.

REMARK 2.69 - The oriented version of integral and quantum cobordism can be similarly obtained by substituting the group Ω_s with $^+\Omega_s$. We will not go in to details.

(C) COBORDISM AND CONSERVATION LAWS OF PDEs

In this section, as an application, we relate integral cobordism to the spectral term $E_1^{0,n-1}$ of the $\mathcal{C}$-spectral sequence, that represents the space of conservation laws of PDEs, in such a way to represent the space of conservations laws of a PDE into an Hopf algebra.

DEFINITION 2.79 - *We define* **conservation law** *of a PDE* $E_k \subset J_n^k(W)$, *any differential* $(n-1)$*-form* β *belonging to the following quotient space:*

$$\mathcal{C}ons(E_k) \equiv \frac{\Omega^{n-1}(E_\infty) \cap d^{-1}C\Omega^n(E_\infty)}{C\Omega^{n-1}(E_\infty) \bigoplus d\Omega^{n-2}(E_\infty)}$$

where $\Omega^q(E_\infty), q = 0, 1, 2, \dots$, *is the space of differential* q*-forms on* E_∞,

$$\begin{aligned} C\Omega^q(E_\infty) &\equiv \{\beta \in \Omega^q(E_\infty) \quad | \quad \beta(\zeta_1, \dots, \zeta_q) = 0, \zeta_i(p) \in \mathbf{E}_\infty, \forall p \in E_\infty\} \\ &\equiv \textit{space of all } \textbf{Cartan q-forms} \textit{ on } \quad E_\infty, q = 1, 2, \dots \end{aligned}$$

and

$$C\Omega^0(E_\infty) \equiv 0 \quad ; \quad C\Omega^q(E_\infty) \equiv \Omega^q(E_\infty), \quad \textit{for} \quad q > n, \Omega^{-1}(E_\infty) = 0.$$

Thus a conservation law is an $(n-1)$*-form on* E_∞ *non-trivially closed on the (singular) solutions of* E_k.

The space of conservation laws of E_k can be identified with the spectral term $E_1^{0,n-1}$ of the $\mathcal{C}$-spectral sequence associated to E_k. One can see that locally we can write

$$\mathcal{C}ons(E_k) = \frac{\{\omega \in \Omega^{n-1}(E_\infty) | \partial\omega = 0\}}{\{\omega = \partial\theta | \theta \in \Omega^{n-2}(E_\infty)\}}$$

where

$$\partial\omega = \sum_{\mu_0, \dots, \mu_{n-1}} (\partial_{[\mu_0}\omega_{\mu_1 \dots \mu_{n-1}]}) dx^{\mu_0} \wedge \dots \wedge dx^{\mu_{n-1}}$$

with

$$\omega = \sum_{\mu_1, \dots, \mu_{n-1}} \omega_{\mu_1 \dots \mu_{n-1}}(x^\mu, y^j) dx^{\mu_1} \wedge \dots \wedge dx^{\mu_{n-1}} \mathrm{mod}\, C\Omega^{n-1}(E_\infty)$$

and

$$\partial_\mu \equiv \partial x_\mu + \sum_{i\in I} A^i_\mu(x,y)\partial y_i \ , \ \mu = 1,\ldots,n$$

basis Cartan fields of $\mathbb{E}_\infty$, where $\{x^\mu, y^j\}_{1\leq\mu\leq k, j\in I}$ are adapted coordinates to $\mathbb{E}_\infty$.
THEOREM 2.127 - *One has the canonical isomorphism:*

$$\mathcal{I}(E_\infty)^{n-1} \cong \mathcal{C}ons(E_\infty).$$

So that integral numbers of E_∞ can be considered as conserved charges of E_k.
THEOREM 2.128 - *One has the following monomorphism of vector spaces*

$$j : E_1^{0,n-1} \to \mathbf{R}^{\Omega^{E_\infty}_{n-1}}. \tag{2.189}$$

Then $E_1^{0,n-1}$ can be identified with a subspace of $\mathbf{R}^{\Omega^{E_\infty}_{n-1}}$, where

$$E^{0,n-1} \equiv \mathrm{im}(j) = \{\phi \in \mathbf{R}^{\Omega^{E_\infty}_{n-1}} | \exists \beta \in E_1^{0,n-1}, \quad \phi([N]_{E_\infty}) = \int_N \beta|_N\}.$$

PROOF. In fact, to any conservation law $\beta : E_\infty \to \Lambda^0_{n-1}(E_\infty)$ we can associate a function

$$j(\beta) \equiv \phi : \Omega^{E_\infty}_{n-1} \to \mathbf{R}, \quad \phi([N]) = \int_N \beta|_N.$$

This definition makes sense as it does not depend on the representative used for $[N]_{E_\infty}$. In fact, if β is a conservation law, then $\forall V \in \Omega(E_\infty)_c$, with $\partial V = N_0 \dot{\bigcup} N_1$, we have

$$\int_{\partial V} \beta|_{\partial V} = \int_V d\beta|_V = 0 \Rightarrow \int_{N_0} \beta|_{N_0} = \int_{N_1} \beta|_{N_1}.$$

Furthermore, the mapping j is injective. Indeed one has

$$\ker(j) = \left\{ \begin{array}{l} \beta \in E_1^{0,n-1} \\ \int_N \beta|_N = 0 \\ \text{for all } (n-1)\text{-dimensional integral submanifolds of} \\ E_\infty \end{array} \right\} \tag{2.190}$$

So $\ker(j)$ fits in the zero-class $[0] \in \mathcal{C}ons(E_k)$. □

REMARK 2.70 - Note that one has the following short exact sequence

$$0 \to \mathbf{R}^{\Omega^{E_\infty}_{n-1,s}} \xrightarrow{i_*} \mathbf{R}^{\Omega^{E_\infty}_{n-1}}$$

where i_* is the mapping $i_* : \phi \mapsto \phi \circ i$, $\forall \phi \in \mathbf{R}^{\Omega^{E_\infty}_{n-1,s}}$, and i is the canonical mapping defined in the following commutative diagram:

$$\begin{array}{ccccc} & \mathbf{R} & = & \mathbf{R} & \\ i_*(\phi) & \uparrow & & \uparrow & \phi \\ & \Omega^{E_\infty}_{n-1} & \underset{i}{\rightarrow} & \Omega^{E_\infty}_{n-1,s} & \end{array}$$

So any function on $\Omega^{E_\infty}_{n-1,s}$ can be identified with a function on $\Omega^{E_\infty}_{n-1}$.

EXAMPLE 2.62 - Let us consider $E_k \equiv J\mathcal{D}^k(\mathbf{R},\mathbf{R})$. Therefore $E_\infty \equiv J\mathcal{D}^\infty(\mathbf{R},\mathbf{R})$. Then, from Theorem 2.108 we get

$$\Omega_0^{E_\infty} = \Omega_0^{E_k} \cong \Omega_0 \cong \mathbf{Z}_2 \quad \Rightarrow \quad \mathbf{R}^{\Omega_0^{E_\infty}} \cong (\mathbf{R})^{\mathbf{Z}_2} \cong \mathbf{R}^2.$$

On the other hand the Cartan distribution $\mathbb{E}_\infty$ corresponding to the infinity prolongation E_∞ of E_k is a 1-dimensional distribution on $J\mathcal{D}^\infty_n(W)$, endowed with coordinates $(x, y, y^{(i)}), i \in \mathbf{N}$, generated by the following vector field $\partial_x = \partial x + \sum_{i\in\mathbf{N}} y^{(i)} \partial y_{(i)}$. Then,

$$\mathcal{C}ons(E_k) = E_1^{0,0} \cong \left\{ f(x, y^{(i)}) | (\partial_x . f) = 0 \right\} \cong \mathbf{R}.$$

Therefore, j is the canonical monomorphism $\mathbf{R} \to \mathbf{R}^2$. ■

EXAMPLE 2.63 - Let us consider equation E_k/G as given in Example 2.60, where $\Omega_0^{E_k/G} \cong \mathbf{Z}_2$. Thus, we get the following exact sequence

$$0 \to E_1^{0,0}(E_k/G) \to \mathbf{R}^2.$$

So, $E_1^{0,0}$ is a subspace of $\mathbf{R}^2$, i.e., 0, $\mathbf{R}$ or $\mathbf{R}^2$. ■

EXAMPLE 2.64 - If $E_k \subset J^k_n(W)$ is a q(n-1)-cobording free PDE, one has that $E_1^{0,0}$ is a subspace of $\mathbf{R}^{K^{E_k}_{n-1}}$. ■

By means of Theorem 2.128 we are able to represent $E_1^{0,n-1}$ by means of a Hopf algebra. In the following Ω can be considered indifferently as one of the previously considered "cobordism groups".

LEMMA 2.26 - *Denote by* $\mathbf{K}\Omega$ *the free* $\mathbf{K}$*-module generated by* Ω. *Then,* $\mathbf{K}\Omega$ *has a natural structure of* $\mathbf{K}$*-bialgebra* (**group K-bialgebra**). *(Here* $\mathbf{K} = \mathbf{R}, \mathbf{C}$.*)*

PROOF OF LEMMA 2.26 - In fact define the following multiplication on the free $\mathbf{K}$-module $\mathbf{K}\Omega$:

$$(\sum_{x\in\Omega} a_x x)(\sum_{y\in\Omega} b_y y) = \sum_{z\in\Omega} (\sum_{xy=z} a_x b_y) z.$$

Then, $\mathbf{K}\Omega$ becomes a ring. The map

$$\eta_{\mathbf{K}\Omega} : \mathbf{K} \to \mathbf{K}\Omega, \quad \eta_{\mathbf{K}\Omega}(\lambda) = a1$$

where 1 is the unit in Ω, making $\mathbf{K}\Omega$ a $\mathbf{K}$-algebra. Furthermore, if we define $\mathbf{K}$-linear maps

$$\triangle : \mathbf{K}\Omega \to \mathbf{K}\Omega \bigotimes_{\mathbf{K}} \mathbf{K}\Omega\,, \quad \triangle(s) = s \bigotimes s$$

and

$$\epsilon : \mathbf{K}\Omega \to \mathbf{K}\,, \quad \epsilon(s) = 1$$

then $(\mathbf{K}\Omega, \triangle, \epsilon)$ becomes a $\mathbf{K}$-coalgebra. □

LEMMA 2.27 - *The dual linear space* $(\mathbf{K}\Omega)^*$ *of* $\mathbf{K}\Omega$ *can be identified with the set*

$$\mathbf{R}^{\Omega} \equiv Map(\Omega, \mathbf{K})$$

where the dual $\mathbf{K}$*-algebra structure of* $\mathbf{K}\Omega$ *is given by*

$$\left\{ \begin{array}{l} (f+g)(s) = f(s) + g(s) \\ (fg)(s) = f(s)g(s) \\ (af)(s) = af(s)\,,\ \forall f, g \in Map(\Omega, \mathbf{K}), s \in \Omega, a \in \mathbf{K} \end{array} \right\}$$

LEMMA 2.28 - $A \equiv Map(\Omega, \mathbf{K})$ *has a natural structure of* $\mathbf{K}$*-bialgebra* $(\mu, \eta, \triangle, \epsilon)$, *with*

$$\begin{array}{ll} (a) & \mu : A \bigotimes_{\mathbf{K}} A \to A\,, \quad \mu(f \bigotimes g) = f.g; \\ (b) & \eta : \mathbf{K} \to A\,, \quad \eta(\lambda)(s) = \lambda\,,\ \forall s \in \Omega; \\ (c) & \triangle : A \to A \bigotimes_{\mathbf{K}} A\,, \quad \triangle(f)(x, y) = f(xy); \\ (d) & \epsilon : A \to \mathbf{K}\,, \quad \epsilon(f) = f(1). \end{array}$$

LEMMA 2.29 - $\mathbf{K}\Omega$ *has a natural structure of* $\mathbf{K}$*-Hopf algebra.*

PROOF OF LEMMA 2.29 - Define the $\mathbf{K}$-linear map

$$S : \mathbf{K}\Omega \to \mathbf{K}\Omega\,, \quad S(x) = x^{-1}\,,\ \forall x \in \Omega.$$

Then, $(1 * S)(x) = xS(x) = xx^{-1} = 1 = \epsilon(x)1 = \eta \circ \epsilon(x), x \in \Omega$ and S is the antipode of $\mathbf{K}\Omega$ so that $\mathbf{K}\Omega$ becomes a $\mathbf{K}$-Hopf algebra. □

LEMMA 2.30 - $A \equiv Map(\Omega, \mathbf{K})$ *has a natural structure of* $\mathbf{K}$*-Hopf algebra.*

PROOF OF LEMMA 2.30 - In fact one can define the antipode

$$S(f)(x) = f(x^{-1})\,,\ \forall f \in A, x \in \Omega.$$

It satisfies the equalities: $\mu(1 \otimes S)\triangle = \mu(S \otimes 1)\triangle = \eta \circ \epsilon$. □

THEOREM 2.129 - *The space of conservation laws* $E_1^{0,n-1}$ *of a PDE identifies in a natural way a* $\mathbf{K}$*-Hopf algebra.*

PROOF. It is an immediate consequence of Theorem 2.128 and the above lemmas, and taking into account the following commutative diagram

$$\begin{array}{ccc} \mathbf{R}^{\Omega_{n-1}^{E_\infty}} \times \mathbf{R}^{\Omega_{n-1}^{E_\infty}} & \rightarrow & \mathbf{R}^{\Omega_{n-1}^{E_\infty}} \\ \uparrow & & \uparrow \\ E^{0,n-1} \times E^{0,n-1} & \rightarrow & < E^{0,n-1} > \end{array}$$

where $< E^{0,n-1} >$ is the Hopf subalgebra of $\mathbf{R}^{\Omega_{n-1}^{E_\infty}}$ generated by $E^{0,n-1}$. We denote by f_α the image of the conservation law $\alpha \in E_1^{0,n-1}$ into $\mathbf{R}^{\Omega_{n-1}^{E_\infty}}$. So in $< E^{0,n-1} >$ we have the following product

$$< E^{0,n-1} > \times < E^{0,n-1} > \rightarrow < E^{0,n-1} >, \quad (f_\alpha, f_\beta) \mapsto f_\alpha . f_\beta .$$

Furthermore, we can explicitly write

$$\begin{aligned} &\overline{\eta} : \mathbf{K} \rightarrow < E^{0,n-1} >, \quad \overline{\eta}(\lambda)(s) = \lambda \\ &\triangle : < E^{0,n-1} > \rightarrow < E^{0,n-1} > \bigotimes_{\mathbf{K}} < E^{0,n-1} >, \quad \triangle(f)(x,y) = f(xy) \\ &\epsilon : < E^{0,n-1} > \rightarrow \mathbf{K}, \epsilon(f) = f(1) \\ &S : < E^{0,n-1} > \rightarrow < E^{0,n-1} >, S(f)(x) = f(x^{-1}). \end{aligned}$$

So the proof is complete. □

DEFINITION 2.80 - *If* $< E^{0,n-1} > \cong \mathbf{R}^{\Omega_{n-1}^{E_\infty}}$, *we say that* E_k *is* **wholly Hopf-cobording**.

THEOREM 2.130 - *If* $\Omega_{n-1}^{E_\infty}$ *is trivial, then* E_k *is wholly Hopf-cobording. Furthermore, in such cases one has* $E_1^{0,n-1} = \mathbf{R}$.

PROOF. In fact in such cases $\mathrm{coker}(j) \equiv \mathbf{R}^{\Omega_{n-1}^{E_\infty}} / E^{0,n-1} = 0$. □

EXAMPLE 2.65 - If $E_\infty = J_n^\infty(W)$, with $n = 2, 4$, then E_k is wholly Hopf-cobording. In fact, in such cases, we have $\Omega_{n-1}^{E_\infty} \cong \Omega_{n-1} = 0$. ■

3 - MECHANICS

3.1 - STRUCTURE OF GALILEAN SPACE-TIME

In the last 30 years many literature has been produced on the geometry of Mechanics. Let us quote the following excellent fundamental books by Abrham & Marsden [2], Libermann & Marle [64] and Misner, Thorne & Wheeler [85]. Mechanics assumes a particular simple formulation if it is formulated with respect to some space-time manifold. In classical mechanics, the space-time interested is the so-called Galilean space-time. In fact, in this way it is possible to naturally recognize absolute objects and others that, instead, depend on frames. Thus, scope of this section is to give some fundamental axioms for the Galilean space-time structure, and to describe in this framework the motion of bodies and frames. (See also ref.[100].)

DEFINITION 3.1 - *The* **Galilean space-time structure** *is defined by the couple*

$$(\mathcal{G}, g)$$

where

(a) $\mathcal{G}$ *is the following structure (***fiber bundle space-time***)*

$$\mathcal{G} \equiv \{\tau : M \to T\}$$

with M = *4-dimensional affine manifold (***space-time***); the corresponding affine structure is the following:* $(M, \mathbf{M}, \alpha)$;

(b) T = *1-dimensional affine space (***time***); the corresponding affine structure is the following:* $(T, \mathbf{T}, \beta)$;

(c) τ = *surjective affine mapping, of constant rank* 1, *such that to each point* $p \in M$ *associates its time* $\tau(p) \in T$. *Put*

$$\mathbf{S} = \ker(\underline{\tau}) \in M,$$

where $\underline{\tau} = D\tau$; D *is the symbol of derivative. Furthermore, define*

$$g : M \to vS^0_2(M) \equiv M \times S^0_2(\mathbf{S}), \quad g(p) = (p, \underline{g}), \quad \forall p \in M,$$

where $\underline{g}$ *is a Euclidean structure on* $\mathbf{S}$. g *is called* **vertical metric field**.

REMARK 3.1 - In the following we will identify T with $\mathbf{R}$, namely we will assume that an origin for times, and a time-measure unity, have been chosen. In this way we will recognize on the space-time a canonical form

$$\sigma \equiv d\tau : M \to T^*M \equiv M \times \mathbf{M}^*.$$

An **adapted coordinate system** on M is a system of coordinates

$$\{x^\alpha\} \equiv \{x^0, x^1, x^2, x^3\}$$

such that

$$dx^0 = \sigma,$$

hence

$$\partial x_k(p) \in vTM \cong M \times \mathbf{S}, \quad k \in (1,2,3).$$

In other words, the coordinate lines $x_{\alpha,p}$ that pass for $p \in M$ are valued in $M_{\tau(p)} \equiv \tau^{-1}(p) \subset M$ only for $k = (1,2,3)$.

REMARK 3.2 - An **event** is, by definition, a point of the space-time M. The events that have the same time $M_t \equiv \tau^{-1}(t) \subset M, t \in T$, identify a 3-dimensional affine space M_t. In fact, as the mapping τ has constant rank 1, we have

$$\dim M_t = \dim M - \dim T = 3, \quad \forall t \in T.$$

As the **time mapping**:

$$\tau : M \to \mathbf{R}$$

is affine, its derivative does not depend on the events. Thus, we can write $\underline{\tau} \equiv D\tau(p), \forall p \in M$. Then, for any two points p and p' on M_t we have

$$\underline{\tau}(v \equiv p - p') = D\tau(p)(v) = \tau(p') - \tau(p) = 0.$$

Therefore, M_t is an affine space with associated vector space $\mathbf{S}$.

DEFINITION 3.2 - *A* **motion** *is a section of* $\mathcal{G}$, *i.e., a mapping* $m : T \to M$ *such that* $\tau \circ m = id_T$. *The* **velocity** $\dot{m}$ *of a motion* $m : T \to M$ *is the first derivative of* m, *i.e.,*

$$\dot{m} \equiv Dm.$$

REMARK 3.3 - Recall that

$$\forall t \in T, Dm(t) \in L(T_tT; T_{m(t)}M) \cong L(\mathbf{R}; \mathbf{M}) \cong \mathbf{M},$$

namely $Dm(t)$ is identified with the image of the derivative of m for $1 \in \mathbf{R}$.

$$1 \in \mathbf{R} : Dm(t)(1) \equiv \overline{m}(t) \in \mathbf{M}.$$

Thus, the application $\dot{m}$ is defined by the following:

$$\dot{m}(t) = (m(t), \overline{m}(t)) \in M \times \mathbf{M} = TM, \forall t \in T.$$

It is important to note that $\dot{m}$ takes always values into an affine subspace $J\mathcal{D}(M)$ of TM. More precisely we get:

$$\dot{m} : T \to J\mathcal{D}(M) \equiv M \times \mathbf{I} \subset TM,$$

where

$$\mathbf{I} \equiv \{v \in \mathbf{M} | < \underline{\tau}, v >= 1\} \subset \mathbf{M}.$$

This can be seen by derivation of the diagram that defines a motion:

$$\begin{array}{ccc} \mathbf{R} & \stackrel{m}{\rightarrow} & M \\ & id_{\mathbf{R}} \searrow & \downarrow \tau \\ & & \mathbf{R} \end{array}$$

namely

$$\begin{array}{ccc} \mathbf{R} \equiv T_t\mathbf{R} & \xrightarrow{Dm(t)} & T_{m(t)}M \equiv \mathbf{M} \\ & id_{\mathbf{R}} \searrow & \downarrow \\ & & \mathbf{R} \equiv T_t\mathbf{R} \end{array} \quad {}_{D\tau(m(t)) \equiv \underline{\tau}} \; .$$

As a consequence, we get:

$$< \underline{\tau}, \overline{m}(t) >= 1$$

as $1 \mapsto Dm(t)(1) \equiv \overline{m}(t)$.

DEFINITION 3.3 - *We call $J\mathcal{D}(M)$ the* **space of velocity of motions**.

$J\mathcal{D}(M)$ is an affine space. In fact, $\mathbf{I}$ is an affine space, with associated vector space $\mathbf{S}$. In fact, if $v, v' \in \mathbf{I}$ one has:

$$< \underline{\tau}, v - v' >=< \underline{\tau}, v > - < \underline{\tau}, v >= 1 - 1 = 0.$$

So, $v - v' \in \mathbf{S}$. Let us, now, introduce the following fundamental concept.

DEFINITION 3.4 - *A* **frame** *is a mapping:*

$$\psi : T \times M \to M$$

such that

a)

$$\psi_t : M \to M_t$$

b)

$$\psi_{t'} \circ \psi_t : M \to M_{t'}$$

c)

$$\psi_{t|M_t} = id_{M_t}.$$

PROPOSITION 3.1 - *With respect to a frame we have the following splitting:*

$$M \cong T \times S_\psi,$$

where $S_\psi \equiv M/\sim$, with $\sim$ the equivalence relation in M defined by the following:

$$p \sim p' \Leftrightarrow p' = \psi(t, p), \quad t = \tau(p').$$

In other words, S_ψ is the set of flow-lines of ψ.

REMARK 3.4 - A particularly important category of frames is that of **inertial frames**. For these, the flow-lines are parallel lines, namely they have constant velocities: $\dot{\psi} = const.$ Emphasize that

$$\dot{\psi}(p) = D_1\psi(\tau(p), p).$$

As

$$D_1\psi(\tau(p), p) = D\psi_p(\tau(p))$$

it follows that $\dot{\psi}(p) \in J\mathcal{D}(M)$. In fact, $\psi_p : T \to M$ is a motion.

DEFINITION 3.5 - 1) *For any vector field $X : M \to TM$ on the Galilean space-time we can define its* **absolute differential***:*

$$\begin{array}{ccc} M & \stackrel{DX}{\rightarrow} & T^*M \bigotimes T(TM) \\ \nabla X \downarrow & & \downarrow id_{T^*M}\otimes\Gamma\equiv\widehat{\Gamma} \\ T^*M \bigotimes TM \cong T^*M \bigotimes X^*vTM & \rightarrow & T^*M \bigotimes vT(TM) \end{array}$$

where

$$\begin{aligned} X^*vT(TM) &= \bigcup_{p\in M} vT_{X(p)}(TM) \\ &\cong \bigcup_{p\in M} vT_{X(p)}(M \times \mathbf{M}) \cong \bigcup_{p\in M} T_{X(p)}\mathbf{M} \cong \bigcup_{p\in M} \mathbf{M} = M \times \mathbf{M} = TM \end{aligned}$$

then, we have

$$X^*vT(TM) \cong TM.$$

2) *The* **covariant derivative** *of X with respect to another vector field Y is the following:*

$$\nabla_Y X \equiv Y \rfloor \nabla X : M \to TM,$$

where $\rfloor$ is the symbol of contraction between linear forms on $\mathbf{M}$ and vectors of $\mathbf{M}$.

$$\begin{aligned} \rfloor : \mathbf{M}^* \times \mathbf{M} \to \mathbf{R} \\ (\alpha, v) \mapsto v\rfloor\alpha = \alpha(v). \end{aligned}$$

PROPOSITION 3.2 - *In coordinates we have:*

$$\boxed{\begin{aligned} \nabla X &= [(\partial x_\alpha X^\beta) + \Gamma_{\alpha}{}^{\beta}_{\gamma} X^\gamma] dx^\alpha \otimes \partial x_\beta \\ \nabla_Y X &= Y^\alpha[(\partial x_\alpha X^\beta) + \Gamma_{\alpha}{}^{\beta}_{\gamma} X^\gamma]\partial x_\beta. \end{aligned}}$$

DEFINITION 3.6 - 1) *The* **acceleration** $\ddot{m}$ *of the motion* $m : T \to M$ *is the covariant derivative of* $\dot{m}$ *with respect to* $\dot{m}$.

$$\ddot{m} = \nabla_{\dot{m}}\dot{m}.$$

We shall also use the following notation:

$$\ddot{m} = \frac{\delta \dot{m}}{\delta t} = \frac{\nabla \dot{m}}{dt}.$$

2) *The* **acceleration of a frame** ψ *is the covariant derivative of* $\dot{\psi}$ *with respect to* $\dot{\psi}$:

$$\ddot{\psi} = \nabla_{\dot{\psi}}\dot{\psi}.$$

THEOREM 3.1 - *One has the following formulas:*

$$\nabla g = \nabla \sigma = 0, \quad T = 0\,,$$

where T *is the canonical torsion of the canonical connection* Γ *defined by:*

$$T(X,Y) \equiv \nabla_X Y - \nabla_Y X - [X,Y]$$

for any vector field X *and* Y *on* M. *The bracket* $[X,Y]$ *is called* **Lie bracket**. *Thus, for any numerical function* $f : M \to \mathbf{R}$ *we get:*

$$[X,Y]\cdot f = X\cdot(Y\cdot f) - Y\cdot(X\cdot f).$$

We have used the fact that any vector field is equivalent to a mapping

$$X : f \mapsto X\cdot f \equiv < df, X >= X^{\alpha}(\partial x_{\alpha} f),$$

where

$$X = X^{\alpha}\partial x_{\alpha}$$

is the representation in coordinates of X. *Furthermore, we have:*

$$[X,Y] = \mathcal{L}_X Y.$$

PROOF. It is a direct consequence of the definitions of ∇, g, σ. (For detailed proofs see ref.[100].) □

THEOREM 3.2 - *In coordinates one has the following expressions for the connection coefficients:* $\Gamma_{\alpha}{}^{\gamma}_{\beta}$ *of* Γ:

$$\Gamma^{j}_{k_i} = [ki,s]g^{sj}, [ki,s] \equiv \frac{1}{2}[(\partial x_k g_{is}) + (\partial x_i g_{sk}) - (\partial x_s g_{ki})] \tag{3.1}$$

The symbols $[ki, s]$ are called **Christoffel symbols**. *One has the following properties:*

(3.2) $$\Gamma^0_{\alpha\beta} = \Gamma^0_{\beta\alpha} = 0; \Gamma^k_{\beta\alpha} = \Gamma^k_{\alpha\beta}; \widehat{\Gamma}^k_{ij} = -\Gamma^k_{ji};$$

where $\widehat{\Gamma}$ is the canonical connection defined on the cotangent space T^*M. More precisely one has:

$$\widehat{\Gamma} : T(T^*M) \to vT(T^*M).$$

PROOF. It follows from direct calculations in coordinates of the following equations $\nabla g = 0, \nabla \sigma = 0$ and $T = 0$. □

DEFINITION 3.7 - *A* **Galilean transformation** *is a diffeomorphism $f : M \to M$ that preserves the Galilean space-time structure. More precisely we have the following:*

(a) *f is a fiber transformation of $\mathcal{G}$, i.e., one has the following commutative diagram:*

$$\begin{array}{ccccccccc} TM \equiv M \times \mathbf{M} & \to & M & \xrightarrow{\tau} & T & \xleftarrow{\beta} & TT & \equiv T \times \mathbf{T} \\ {\scriptstyle T(f)=(f,\overline{f})} \uparrow & & f \uparrow & & f_T \uparrow & & & \uparrow {\scriptstyle T(f_T)=(f_T,\overline{f}_T)} \\ TM \equiv M \times \mathbf{M} & \to & M & \xrightarrow{\tau} & T & \underset{\beta}{\leftarrow} & TT & \equiv T \times \mathbf{T} \end{array}$$

(b) $\overline{f}|_{\mathbf{S}} \in O(\mathbf{S})$, *($\overline{f}$ is called the* **associated linear Galilean transformation***).*

(c) $\overline{f}_{\mathbf{T}} = id_{\mathbf{T}}$ *(f_T is a rigid transformation of T).*

The set of Galilean transformations is a group: the **Galilean group G**.

The set of linear Galilean transformations is a group: the **linear Galilean group *LG***.

PROPOSITION 3.3 - *For any coordinate system $\{x^\alpha\}$ one has the following morphisms:*

$$LG \subset GL(4; \mathbf{R})$$

$$\mathbf{G} \cong LG \times \mathbf{R}^4 \subset A(4) \equiv GL(4; \mathbf{R}) \times \mathbf{R}^4 .$$

REMARK 3.5 - 1) For any adapted coordinates system $\{x^\alpha\}$ one has the following isomorphisms:

(3.3) $$LG \cong LG(4),$$

where $LG(4)$ is the group of 4×4 matrices of the type

(3.4)

$$\left[\begin{array}{c|ccc} 1 & 0 & 0 & 0 \\ \hline h^1 & & & \\ h^2 & & (B^j_i) & \\ h^3 & & & \end{array}\right]$$

with $h \equiv (h^1, h^2, h^3) \in \mathbf{R}^3$ and $(B_i^j) \in O(3)$.

$$\mathbf{G} \cong L\mathbf{G}(4) \times \mathbf{R}^4 \subset A(4) = GL(4) \times \mathbf{R}^4.$$

More precisely, one has:

$$f^\alpha \equiv x^\alpha \circ f = A^\alpha_\beta x^\beta + b^\alpha, \quad \forall f \in \mathbf{G}, (b^\alpha) \in \mathbf{R}^4, (A^\alpha_\beta) \in L\mathbf{G}(4).$$

2) If the Galilean space-time is oriented, namely if we consider the structure $(\mathcal{G}, g, \eta)$, with η a constant section of the fiber bundle $\pi' : v\Lambda^0_3 M \to M$, then condition (b) in the above definition is substituted by the following one:

$$(\bar{b}) \qquad \underline{f}|_{\mathbf{S}} \in SO(\mathbf{S}),$$

and $\mathbf{G}$ (resp. $L\mathbf{G}$) is substituted by the following group: $S\mathbf{G} \equiv$ **special Galilean group** (resp. $SL\mathbf{G} \equiv$ **linear special Galilean group**). Furthermore, for any adapetd coordinates system, we have the following isomorphisms:

$$(a') \quad SL\mathbf{G} \cong M(4) \equiv \text{group of Euclidean rigid motions},$$
$$(b') \quad S\mathbf{G} \cong M(4) \times \mathbf{R}^4.$$

One can prove that $L\mathbf{G}(4)$ (resp. $M(4)$) is the structure group of the principal fiber bundle of the linear orthogonal frames (resp. oriented linear orthogonal frames) adapted on M [101,102].

PROPOSITION 3.4 - *If ψ is a frame one has the following isomorphism:*[1]

$$M \times \mathbf{I} \equiv J\mathcal{D}(M) \cong vTM \equiv M \times \mathbf{S}, \quad (p, v) \mapsto (p, v_\wedge \equiv v - \bar{\psi}),$$

where $\bar{\psi}$ is the free part of the velocity of the frame; and its inverse application is the following:

$$(p, v) \mapsto (p, v + \bar{\psi})$$

DEFINITION 3.8 -

$$J\mathcal{D}^2(M) \subset TTM,$$

where

$$J\mathcal{D}^2(M)$$

[1] A frame identifies a connection on the fiber bundle space-time $\tau : M \to T$, i.e., a splitting $TM \cong vTM \oplus \langle \dot{\psi} \rangle$.

is the **space of second derivative of motions**. *$J\mathcal{D}^2(M)$ is the subspace of TTM identified by means of the following equations:*

$$\left\{\begin{array}{l} \ddot{x}^0 = 0 \\ \widehat{\dot{x}}^0 = \dot{\widehat{x}}^0 = 1 \\ \widehat{\dot{x}}^k = \dot{\widehat{x}}^k \end{array}\right\}$$

One has:

$$\boxed{J\mathcal{D}^2(M) \cong M \times \mathbf{I} \times \mathbf{S}}$$

REMARK 3.6 - The acceleration of a motion $m : T \rightarrow M$ is an application

$$\ddot{m} : T \rightarrow vTM,$$

defined by the following commutative diagram:

$$\begin{array}{ccccc} & T & = & T & \\ D^2m & \downarrow & & \downarrow & \ddot{m} \\ & M \times \mathbf{I} \times \mathbf{S} & \rightarrow & M \times \mathbf{S} & \\ & \| & & \| & \\ & J\mathcal{D}^2(M) & \underset{\Gamma|}{\rightarrow} & vTM & \\ & & & \cap & \\ & \cap & & TM & \\ & & & \uparrow & \\ & TTM & \underset{\Gamma}{\rightarrow} & vTTM & \end{array}$$

(a) Γ is the canonical connection on the affine space-time.
(b) $\Gamma|$ is the restriction of Γ on $J\mathcal{D}^2(M)$.
(c) $J\mathcal{D}^2(M) \subset TTM$ is the canonical inclusion given by: $(p, u, w) \mapsto (p, u, u, w)$.
(d) $vTM \subset TM$ is the canonical inclusion given by: $(p, w) \mapsto (p, w)$.
(e) $vTTM \rightarrow TM$ is the canonical projection given by: $(p, v, w) \mapsto (p, w)$.

PROPOSITION 3.5 - *The free part of the acceleration coincides with the the free part of the derivative of the free part of the velocity.*

PROOF. In fact, we have

$$\ddot{m}(t) = \Gamma|[D^2m(t)] = \Gamma|[m(t), \overline{\dot{m}}(t), \overline{\ddot{m}}(t)] = (m(t), \overline{\ddot{m}}(t)).$$

Therefore, the free part of $\ddot{m}(t)$ is just $\overline{\ddot{m}}(t)$. In particular, we get:

$$D^2m = D(Dm) = D(m, \overline{\dot{m}}) = (Dm, \overline{D\dot{m}})$$

where: $Dm = (m, \overline{\dot{m}}) : T \to J\mathcal{D}(M) \cong M \times \mathbf{I}$; thus $m : T \to M$ and $\overline{\dot{m}} : T \to \mathbf{I}$ is the free part of the velocity. In the following we will write:

$$D^2 m = (m, \overline{\dot{m}}, \overline{\dot{m}}, \overline{\ddot{m}}) : T \to J\mathcal{D}^2(M) \subset TTM$$
$$\Gamma| \circ D^2 m = (m, \overline{\ddot{m}}) : T \to vTM \cong M \times \mathbf{S}.$$

So

$$\overline{\ddot{m}} = \overline{D\dot{m}}\,.$$

□

THEOREM 3.3 - 1) *An* **adapted basis** *on* **M** *is a basis* $\{e_\alpha)_{0\leq\alpha\leq 3}$, *such that:*

$$e_0 \in \mathbf{I}, \quad \text{and} \quad e_k \in \mathbf{S}.$$

1) *With respect to an adapted basis* $\{e_\alpha\}_{0\leq\alpha\leq 3}$ *we have the following representation of the free part* $\overline{\dot{m}}$ *of the velocity:*

$$\overline{\dot{m}}(t) = e_0 + \overline{\dot{m}}^k(t) e_k.$$

2) *With respect to an adapted basis we have the following representation of the free part of the acceleration:*

$$\overline{\ddot{m}}(t) = \ddot{m}^k(t) e_k.$$

3) *If the adapted basis* $\{e_\alpha\}$ *is induced by an adapted system of coordinates*

$$\{x^\alpha\}_{0\leq\alpha\leq 3}, \quad e_\alpha = \partial x_\alpha(p),$$

by means of the identification $T_pM \cong \mathbf{M}$, *we get the following representation of the free part of the velocity:*

$$\overline{\dot{m}}(t) = \partial x_0(m(t)) + \dot{m}^k \partial x_k(m(t)) \tag{3.5}$$

and the following representation of the free part of acceleration:

(3.6)
$$\overline{\ddot{m}}(t) = [\Gamma^k_{0\,0}(m(t)) + 2\Gamma^k_{0\,j}(m(t))\dot{m}^j(t) + \Gamma^k_{i\,j}(m(t))\dot{m}^i(t)\dot{m}^j(t) + \ddot{m}^k(t)]\partial x_k(m(t)).$$

PROOF. With respect to an adapted coordinate system, we have the following representation of Γ:

$$\begin{Bmatrix} \hat{\hat{x}} \circ \Gamma = \hat{\hat{x}}^\alpha \\ \hat{\dot{x}}^\alpha \circ \Gamma = \hat{\dot{x}}^\alpha \\ \dot{\hat{x}}^\alpha \circ \Gamma = 0 \\ \ddot{x}^\alpha \circ \Gamma = \Gamma^\alpha_{\gamma\delta} \hat{\dot{x}}^\gamma \dot{\hat{x}}^\delta + \ddot{x}^\alpha \end{Bmatrix}$$

A vector $(v, w) \in T_{(p,u)}TM \cong T_pM \bigoplus T_u\mathbf{M}$. The natural basis on $T_{(p,u)}TM$ induced by the adapted coordinate system $\{x^\alpha\}$ on M is the following: $\{\partial\widehat{x}_\alpha, \partial\dot{x}_\alpha\}$. Then, we get:

$$\begin{aligned} (v,w) &= v^\alpha\partial\widehat{x}_\alpha + w^\gamma\partial\dot{x}_\gamma \\ \Gamma(v,w) &= v^\alpha\Gamma(\partial\widehat{x}_\alpha) + w^\gamma\partial\dot{x}_k = v^\alpha\Gamma^\gamma_\alpha \circ v\partial\dot{x}_\gamma + w^\gamma\partial\dot{x}_\gamma \\ &= (v^\alpha v^\delta\Gamma^\gamma_{\alpha\delta} + w^\gamma)\partial\dot{x}_\gamma\,. \end{aligned}$$

□

THEOREM 3.4 - (EXPRESSION OF THE VELOCITY OF AN OBSERVED MOTION PARAMETRIZED BY MEANS OF THE FUNCTION LENGTH OF CURVE).

If an observed motion $m_\psi : \mathbf{R} \to S_\psi$ is parametrized by means of the following function:

$$s(t) = \int_{[0,t]} \sqrt{g_{ij}\dot{x}^i(t')\dot{x}^j(t')}\,dt', \tag{3.7}$$

then one has the following commutative diagram:

$$\begin{array}{ccc} \mathbf{R} & \overset{m_\psi}{\rightarrow} & S_\psi \\ \| & & \uparrow\ \widetilde{m}_\psi \\ \mathbf{R} & \underset{s}{\rightarrow} & \mathbf{R} \end{array}$$

Thus the velocity of $\widetilde{m}_\psi$ is unitary:

$$|\dot{\widetilde{m}}_\psi| = 1;$$

and one has:

$$\dot{m}_\psi = \dot{s}\dot{\widetilde{m}}_\psi \circ s.$$

Furthermore, if we put $\dot{m}_\psi = v$ and $\dot{\widetilde{m}}_\psi = \mathbf{T}$, we can write the following formulas:

$$\boxed{\begin{aligned} v &= \dot{s}\mathbf{T} \circ s \\ \dot{s} &= \sqrt{g_{ij}\dot{x}^i\dot{x}^j} \end{aligned}} \tag{3.8}$$

PROOF. By derivation of the compose application $m_\psi = \widetilde{m}_\psi \circ s$ we get $\dot{m}_\psi(t) = T(\widetilde{m}_\psi)_{s(t)}(\dot{s}(t))$. Furthermore, taking into account that $\dot{s}(t) \in T_{s(t)}\mathbf{R} \cong \mathbf{R}$ we can represent $\dot{s}(t)$ in the canonical basis $1 \in \mathbf{R}$ and write $\dot{s}(t) = \dot{s}(t) \cdot 1$. Now, observing that $T(\widetilde{m}_\psi)_t$ is a linear application, we get:

(3.9). $$\dot{m}_\psi(t) = \dot{s}(t)T(\widetilde{m}_\psi)_{s(t)}(1) = \dot{s}(t) \cdot \dot{\widetilde{m}}_\psi(t)\,.$$

From (3.8) we get

$$\dot{s}(t) = \sqrt{g_{ij}\dot{x}^i(t)x^j(t)} = |\dot{m}_\psi(t)|.$$

Finally, from (3.9) we complete the proof. In fact, we get:

$$|\dot{m}_\psi(t)| = |\dot{s}(t)||\dot{\tilde{m}}(t)|$$

hence

$$|\dot{\tilde{m}}_\psi(t)| = 1.$$

□

PROPOSITION 3.6 - *With respect to an adapted coordinate system* $\{x^\alpha\}$, *we can write* $\mathbf{T} : \mathbf{R} \to TS_\psi$, *in the following form:*

$$\boxed{\mathbf{T} = T^k \circ s\,\partial x_k \circ m_\psi}$$

where $T^k : \mathbf{R} \to \mathbf{R}$ *is defined by the following:*

$$g \circ m_\psi(\mathbf{T} \circ s, \partial x_i \circ m_\psi) = \mathbf{T}^k \circ s\, g_{ki} \circ m_\psi$$

such that $\{\partial x_i\}$ *is the basis on the tangent space of* S_ψ: $\partial x_i : S_\psi \to TS_\psi$, *identified by* $\{x^\alpha\}$. *We get:*

$$\mathbf{T}^k = \frac{d\tilde{m}^k_\psi}{ds} : \mathbf{R} \to \mathbf{R}$$

$$\dot{m}^k_\psi = \frac{dm^k_\psi}{dt} = \dot{s}\frac{d\tilde{m}^k_\psi}{ds} \circ s = \dot{s}(\mathbf{T}^k \circ s)\,.$$

PROPOSITION 3.7 - (ACCELERATION OF THE OBSERVED MOTION IN CURVILINEAR COORDINATES).

$$\boxed{\begin{aligned} & a = \nabla_v v : \mathbf{R} \to TS_\psi \\ & a = [\ddot{m}^i_\psi + \dot{m}^i_\psi \dot{m}^k_\psi \Gamma^j_{i\,k} \circ m_\psi]\partial x_j \circ m_\psi \\ & \quad = [\ddot{s}T^j \circ s + (\dot{s})^2 \frac{dT^i}{ds} \circ s + (\dot{s})^2 T^i \circ s T^k \circ s \Gamma^j_{i\,k} \circ m_\psi]\partial x_j \circ m_\psi \end{aligned}}$$

REMARK 3.7 - (GENERALIZED FRENET FORMULAS). Let us use the invariant length of curve

$$s(t) = \int_{[t_0,t]} ds = \int_{t_0,t]} \sqrt{g_{\alpha\beta}\dot{x}^i\dot{x}^j}\,dt$$

to parametrize a curve $x^i = x^i(t)$ of a Riemannian manifold (M, g). So, we can write $x^i = x^i(s)$. Then, the unit vector $\mathbf{T}$ tangent to the curve, can be written

$$\mathbf{T} = \mathbf{T}^i(t)\partial x_i(t)$$

with

$$\mathbf{T}^i = \frac{\dot{x}^i}{\sqrt{g_{ij}\dot{x}^i\dot{x}^j}} = \frac{1}{\sqrt{g_{ij}\dot{x}^i\dot{x}^j}}\frac{dx^i}{dt}.$$

By covariant derivation of the following expression:

$$g_{ij}\mathbf{T}^i\mathbf{T}^j = 1$$

we get

$$0 = g_{ij}\mathbf{T}^i\frac{\delta T^j}{\delta s}.$$

Therefore, the following vector field, $\frac{\delta T^j}{\delta s}\partial x_j$ is $\perp$ $\mathbf{T}$ along the curve. Define **principal normal vector** as follows:

(3.10)(**I Generalized Frenet Formula**)

$$\boxed{\mathbf{N} = \frac{1}{\kappa}\frac{\delta \mathbf{T}}{\delta s}}$$

where κ is the **curvature** of the curve. Hence,

$$\kappa^2 = g_{ij}\frac{\delta T^i}{\delta s}\frac{\delta T^j}{\delta s}.$$

If the Riemann manifold M has dimension 3, it is useful to define another unit vector field, $\mathbf{B}$, along the curve, that is $\perp$ two of the above ones. Thus, $(\mathbf{T}, \mathbf{N}, \mathbf{B})$ is an orthonormal basis for T_pM. In fact, we can choose $\mathbf{B}$ such that

(3.11)(**II Generalized Frenet Formula**)

$$\boxed{\mathbf{B} = \frac{1}{\tau}(\frac{\delta \mathbf{N}}{\delta s} + \kappa \mathbf{T})}$$

where τ is the invariant called **torsion of the curve** and it is given by the following local formula:

$$\tau = \epsilon^{ijk}\mathbf{T}_i\mathbf{N}_j\frac{\delta \mathbf{T}^k}{\delta s}.$$

Then, we get the following.

THEOREM 3.5 - (FRENET FORMULAS). *For a curve in a 3-dimensional Riemann manifold M, one has the following formulas:*

$$\boxed{\begin{aligned}\frac{\delta \mathbf{T}}{\delta s} &= k\mathbf{N}\\ \frac{\delta \mathbf{T}}{\delta s} &= \tau\mathbf{B} - k\mathbf{T}\\ \frac{\delta \mathbf{B}}{\delta s} &= -\tau\mathbf{N}\end{aligned}} \tag{3.12}$$

where $\tau \equiv$ **torsion**, *and* $\sigma \equiv \frac{1}{\tau} \equiv$ **torsion radius**. *The* **osculating plane** *to the curve at a point P is that containing* $\mathbf{T}$ *and* $\mathbf{N}$. *The* **normal plane** *is that containing* $\mathbf{B}$ *and* $\mathbf{N}$. *The* **rectifying plane** *is that containing* $\mathbf{T}$ *and* $\mathbf{B}$.

PROOF. In fact, as

$$\mathbf{T} \perp \mathbf{N} \Rightarrow g_{ij}\mathbf{T}^i\mathbf{N}^j = 0.$$

By covariant derivation, we get:

$$\begin{gathered} g_{ij}\mathbf{T}^i\frac{\delta \mathbf{N}^j}{\delta s} + g_{ij}\frac{\delta \mathbf{T}^i}{\delta s}\mathbf{N}^j = 0\\ \Downarrow\\ g_{ij}\mathbf{T}^i\frac{\delta \mathbf{N}^j}{\delta s} + g_{ij}\kappa\mathbf{N}^i\mathbf{N}^j = 0\\ \Downarrow\\ g_{ij}\mathbf{T}^i\frac{\delta \mathbf{N}^j}{\delta s} + \kappa g_{ij}\mathbf{T}^i\mathbf{T}^j = 0\\ \Downarrow\\ g_{ij}\mathbf{T}^i(\frac{\delta \mathbf{N}^j}{\delta s} + \kappa\mathbf{T}^j) = 0\,. \end{gathered}$$

Therefore $\mathbf{B} = \frac{1}{\tau}(\frac{\delta \mathbf{N}}{\delta s} + \kappa\mathbf{T})$ is $\perp$ $\mathbf{T}$. From relation $g_{ij}\mathbf{N}^i\mathbf{N}^j = 1$, by means of the covariant derivation, we get

$$g_{ij}\mathbf{N}^i\frac{\delta \mathbf{N}^j}{\delta s} = 0 \Rightarrow g_{ij}\mathbf{N}^i(\frac{\delta \mathbf{N}^j}{\delta s} + \kappa\mathbf{T}^j) = 0.$$

Therefore, we have: $\mathbf{B} \perp \mathbf{N}$. Then, we can write:

$$\mathbf{B}^i = \epsilon^{ijk}\mathbf{T}_j\mathbf{N}_k.$$

By covariant derivation of this last relation, we get:

$$\begin{aligned}\frac{\delta \mathbf{B}^i}{\delta s} &= \epsilon^{ijk}\frac{\delta \mathbf{T}_j}{\delta s}\mathbf{N}_k + \epsilon^{ijk}\mathbf{T}_j\frac{\delta \mathbf{N}_k}{\delta s}\\ &= \kappa\epsilon^{ijk}\mathbf{N}_j\mathbf{N}_k + \epsilon^{ijk}\mathbf{T}_j\frac{\delta \mathbf{N}_k}{\delta s}\\ &= \kappa\epsilon^{ijk}\mathbf{N}_j\mathbf{N}_k + \epsilon^{ijk}\mathbf{T}_j[\tau\mathbf{N}_k - \kappa\mathbf{T}_k]\,.\end{aligned}$$

As ϵ^{ijk} is skew-symmetric, it follows:

$$\boxed{\frac{\delta \mathbf{T}_i}{\delta s} = \tau \epsilon^{ijk} \mathbf{T}_j \mathbf{B}_k}$$

As $(\mathbf{T}, \mathbf{N}, \mathbf{B})$ is an orthonormal basis for $T_p M$, we get

$$\frac{\delta \mathbf{T}_i}{\delta s} = \tau \mathbf{B}_i.$$

□

THEOREM 3.6 - *We have the following representation of the acceleration:*

$$\boxed{a = \ddot{s}\mathbf{T} + (\dot{s})^2 \kappa \mathbf{N}} \tag{3.13}$$

PROOF. By using Frenet formulas, we have:

$$\begin{aligned} a &= (\ddot{s} T^j \circ s)\partial x_j \circ m_\psi + (\dot{s})^2 [\frac{dT^j}{ds} + T^i \circ s T^k \circ s \Gamma^j_{i\,k} \circ m_\psi] \partial x_j \circ m_\psi \\ &= \ddot{s}\mathbf{T} + (\dot{s})^2 [\frac{dT^j}{ds} + T^i \circ s \frac{dx^k}{ds} \Gamma^j_{i\,k} \circ m_\psi] \partial x_j \circ m_\psi \\ &= \ddot{s}\mathbf{T} + (\dot{s})^2 k \mathbf{N}. \end{aligned}$$

□

THEOREM 3.7 - (CORIOLIS THEOREM). *The acceleration a_a of a motion can be expressed in terms of relative acceleration a_r, with respect to a frame ψ in the following way:*

$$a_a = a_r + a_\tau + a_c,$$

*where a_τ is the acceleration of the frame (**draging acceleration**), and a_c is an additive term, called **Coriolis acceleration** or **deviation acceleration**.*

PROOF. With respect to a coordinate system $\{x^\alpha\}$ adapted to a frame ψ, (i.e., $\partial x_0 = \dot{\psi}$), we have $\ddot{\psi} \equiv a_\tau = \Gamma^k_{00} \partial x_k$. Therefore, from (3.8) we have: $a_a - a_\tau - a_r = 2\Gamma^k_{0i} v^i \partial x_k \equiv a_d$. □

DEFINITION 3.9 - (RIGID MOTIONS). 1) *We define* **motion of a continuum system** *a 1-parameter group of transformations of the space-time:*

$$\phi : \mathbf{T} \times M \to M.$$

2) *A motion* $\phi : \mathbf{T} \times M \to M$ *of a continuum system is said to be* **rigid** *if* $\forall \lambda \in \mathbf{T}$, *one has:*

$$
\boxed{
\begin{aligned}
&(a) \quad \phi_\lambda \quad \textit{is an affine transformation} \\
&(b) \quad \phi_\lambda^* g \equiv g_\lambda = g \\
&(c) \quad \phi_\lambda^* \eta \equiv \eta_\lambda = \eta
\end{aligned}
}
$$

THEOREM 3.8 - *An affine motion preserving the orientation is rigid iff for any* $p, p' \in M_t$, *the distance* $d(p,p')$, *is preserved.*

PROOF. Let ϕ_λ be a rigid motion. The distance of two points $p, p' \in M_t$ is given by the relation:

$$d(p,p') = |\mathbf{s}| = |p' - p| = \sqrt{g(\mathbf{s},\mathbf{s})}.$$

On the other hand, we have:

$$
\begin{aligned}
d(\phi_\lambda(p), \phi_\lambda(p')) &= \sqrt{g(T(\phi_\lambda)(\mathbf{s}), T(\phi_\lambda)(\mathbf{s}))} \\
&= \sqrt{\phi_\lambda^* g(\mathbf{s},\mathbf{s}) + R_\lambda(\mathbf{s})} = \sqrt{g(\mathbf{s},\mathbf{s})} = d(p,p').
\end{aligned}
$$

Here, $R_\lambda(\mathbf{s})$ is an infinitesimal of order superior to $\phi_\lambda^* g(\mathbf{s},\mathbf{s})$, as $\mathbf{s} \to 0$. As we have assumed that each mapping ϕ_λ is affine, we must have $R_\lambda(\mathbf{s}) = 0$.

The converse proof is similarly made by the converse. □

REMARK 3.8 - The **space** $S_\phi \equiv M/\sim$ of a rigid motion, $\phi : \mathbf{T} \times M \to M$ of a continuum system is an affine space with the associated vector space $\mathbf{S}$:

$$TS_\phi \cong S_\phi \times \mathbf{S}.$$

REMARK 3.9 - From the above considerations it is easy to see that a rigid flow can be identified with a 1-parameter subgroup of the special Galilean group $S\mathbf{G}$.

DEFINITION 3.10 - *Let* $\phi : \mathbf{T} \times M \to M$ *be a motion of a continuum system. A frame* ψ *is* **compatible** *with* ϕ *if the following is verified:*

$$\psi_{t+\lambda} \circ \phi_\lambda = \phi_\lambda \circ \psi_t, \quad \forall t \in T, \lambda \in \mathbf{T}. \tag{3.14}$$

THEOREM 3.9 - *If a frame* ψ *is compatible with respect to a motion* ϕ, *one has a 1-parameter group of transformations on the observed space*

$$S_\psi : \phi_\psi : \mathbf{T} \times S_\psi \to S_\psi$$

defined by

$$\phi_{\psi,\lambda}([p]_\psi) = [\phi_\lambda(p)]_\psi.$$

In other words one has the following commutative diagram:

$$\begin{array}{ccccc} & \mathbf{T}\times M & \stackrel{\phi}{\rightarrow} & M & \\ id_{\mathbf{T}}\times\psi/ & \downarrow & & \downarrow & \psi \\ & \mathbf{T}\times S_{\psi} & \underset{\phi_{\psi}}{\rightarrow} & S_{\psi} & \end{array}$$

PROOF. In fact, if ψ is compatible with ϕ the definition of the mapping ϕ_{ψ} does not depend on the representative used. In fact, if $p' \in [p]_{\psi}$, namely $p' = \psi_t(p)$, for some $t \in T$, we get: $\phi_{\lambda}(p') \in [\phi_{\lambda}(p)]_{\psi}$; hence we can find a time $t'' = t + \lambda$, such that $\phi_{t''}(\phi_{\lambda}(p)) = \phi_{\lambda}(p')$. We have:

$$\psi_{t+\lambda}(\phi_{\lambda}(p)) = \phi_{\lambda}(\psi_t(p)) = \phi_{\lambda}(p').$$

□

REMARK 3.10 - If a frame ψ is not compatible with a motion ϕ, it can happen that for any point $[p]_{\psi} \in S_{\psi}$ pass different curves. In fact, for any point $p \in M$ the following commutative diagram holds:

$$\begin{array}{ccccc} & \mathbf{T} & \rightarrow & M & \\ \phi_{p,\psi} & \downarrow & & \downarrow & \psi/ \\ & S_{\psi} & \rightarrow & S_{\psi} & \end{array}$$

that defines a curve in S_{ψ} passing through $[p]_{\psi}$ in correspondence of $\lambda = 0$. If $p' \in [p]_{\psi}$ this happens also for the curve $\phi_{p',\psi} : \mathbf{T} \rightarrow S_{\psi}$ given by $[p]_{\psi}$ for $\lambda = 0$. The condition that two curves $\phi_{p,\psi}$ and $\phi_{p',\psi}$ coincide, is just the compatibility condition.

THEOREM 3.10 - *If ϕ is a rigid motion and ψ is a compatible rigid frame, the induced motion ϕ_{ψ} on S_{ψ} preserves the metric g_{ψ} canonically induced by means of the following commutative diagram:*

$$\begin{array}{ccccc} & vS^0_2M & \rightarrow & S^0_2(S_{\psi}) & \\ & \| & & \| & \\ & M\times\mathbf{S}^*\odot\mathbf{S}^* & \rightarrow & S_{\psi}\times\mathbf{S}^*\odot\mathbf{S}^* & \\ (id_M,\bar{g})=\mathbf{g} & \uparrow & & \uparrow & \mathbf{g}_{\psi}=(id_{S_{\psi}},\bar{g}) \\ & M & \underset{\psi/}{\rightarrow} & S_{\psi} & \end{array}$$

So, we have:

$$\phi^*_{\psi,\lambda}\mathbf{g}_{\psi} = \mathbf{g}_{\psi}.$$

PROOF. We have:

$$\begin{aligned}\phi^*_{\phi,\lambda}\mathbf{g}_{\phi}([p]_{\psi}) &= S^0_2(\phi_{\psi,\lambda}) \circ g_{\psi} \circ \phi_{\psi,\lambda}([p]_{\psi}) \\ &= S^0_2(\phi_{\psi,\lambda})(\mathbf{g}_{\psi}(\phi_{\psi,\lambda}([p]_{\psi}))) \\ &= ([p]_{\psi}, S^0_2(vT(\phi_{\lambda})(\bar{g})) \\ &= ([p]_{\psi}, \bar{g}) = \mathbf{g}_{\psi}([p]_{\psi}).\end{aligned}$$

□

REMARK 3.11 - Under the same hypotheses of the above theorem, if ϕ is a rigid motion, then ϕ_ψ preserves the orientation of S_ψ.

DEFINITION 3.11 - *A* **rigid body** *is a continuum system such that its admissible flows are rigid.*

THEOREM 3.11 - *A rigid body can be identified with a 3-dimensional Euclidean space, $\mathcal{B}$. In the following we will consider a rigid body from this point of view, i.e., as belonging, at any instant, to the space S_ψ.*

PROOF. Let $\phi : \mathbf{T} \times M \to M$ be a rigid motion. Denote by S_ϕ the corresponding space of flow-lines. If the motion rigid is regular, S_ϕ is a 3-dimensional Riemann manifold. Furthermore, as ϕ is rigid, one has the identification, at each instant $t \in T$, of S_ϕ with M_t. As M_t is a Euclidean 3-dimensional manifold, it follows that S_ϕ is also a 3-dimensional Euclidean manifold. Furthermore, taking into account that the motion of a rigid body must also be rigid, if ϕ' is another rigid motion of such a system, we get the following canonical diffeomorphism:

$$\begin{array}{cccccc} & T \times S_\phi & \cong & M & (t,[p]_\phi) \mapsto & \psi_\phi(t,p) \\ (id_T, j) & \| & & \| & & \downarrow \\ & T \times S_\phi & = & T \times S_{\phi'} & & (t,[\psi_\phi(t,p)]_{\phi'}) \end{array}$$

where ψ_ϕ is the frame associated to ϕ. Thus, $j_t : S_\phi \cong S_{\phi'}$ is a canonical diffeomorphism $\forall t \in T$. In other words, all the Euclidean 3-dimensional spaces $S_\phi, S_{\phi'}, \cdots$ are canonically diffeomorphic at each instant. Now, let ψ be an inertial frame, compatible with ϕ. We obtain the following canonical application:

$$\begin{array}{rccc} & M & \overset{\psi/}{\to} & S_\psi \\ & \| & & \| \\ f: & T \times S_\phi & \to & S_\psi \end{array}$$

For any $t \in T, f_t : S_\phi \to S_\psi$ is also a diffeomorphism between S_ϕ and S_ψ. Then, we can see that S_ϕ move inside $S_\psi, \forall t \in T$, by means of the application $f_t : S_\phi \to S_\psi$. We can, therefore, represent a rigid body as a Euclidean 3-dimensional space B that at each instant change its position inside S_ψ. □

THEOREM 3.12 - *An oriented rigid body has* 6 *degrees of freedom.*

PROOF. In fact, at each instant any point $\Omega \in S_\phi$ of S_ψ can be represented by 3 coordinates (y^j), as S_ψ is 3-dimensional). The orientation of S_ϕ in S_ψ can be obtained by choosing a 3-vector $\epsilon_1 \wedge \epsilon_2 \wedge \epsilon_3$, where $\{\epsilon_i\}_{1 \leq i \leq 3}$ is a basis defined on the tangent space $T_\Omega S_\phi \equiv \mathbf{S} \cong T_\Omega S_\psi$. Therefore, the vectors can be represented, with respect to a basis $\{e_k\}_{1 \leq k \leq 3}$ of $\mathbf{S}$, by means of the following relation:

$$\epsilon_i = \sum_{1 \leq k \leq 3} y_i^k e_k.$$

Observe that $y_i^k \in \mathbf{R}$ are 9 numbers, but the conditions of orthonormality add $3+3=6$ further conditions. So we have $9-6=3$ numbers to add to the previous 3; thus, we necessitate 6 numbers in order to identify an oriented rigid body. □

REMARK 3.12 - A point $p \in S_\phi$ of a rigid body is identified by the following expression written in the space of the free vectors $\mathbf{S}$:

$$(P-O)(t) = (\Omega - O)(t) + \sum_{1 \le k \le 3} \xi^k \epsilon_k(t),$$

where $\{\xi^k\}$ are three numbers that do not depend on the time.

DEFINITION 3.12 - (a) *The* **displacement** *of a continuum body is a mapping*

$$u : \mathbf{T} \times M \to TM, u(\lambda, p) = (p, \bar{u}_\lambda(p)),$$

where $\bar{u}_\lambda(p) \equiv p' - p \in \mathbf{M}$, *and* $p \in M_{\tau(p)}$, *and such that* $p' = \phi_\lambda(p)$, *with* ϕ_λ *the motion that generates the displacement.*

(b) *A* **rigid displacement** *is a displacement such that it preserves the distance between equal-time events and the spatial orientation of the Euclidean 3-dimensional manifold* $M_t, \forall t \in T$.

PROPOSITION 3.8 - (a) *A motion is rigid if any displacement is also rigid.*

(b) *With respect to an inertial frame, a displacement* $u : \mathbf{T} \times M \to TM$ *is represented by an application* $u_\psi : \mathbf{T} \times T \times S_\psi \to TS_\psi$. *One has the following commutative diagram:*

$$\begin{array}{ccc} \mathbf{T} \times M & \xrightarrow{u} & TM \\ \| & & \downarrow \; T(\psi/) \\ \mathbf{T} \times T \times S_\psi & \to_{u_\psi} & TS_\psi \end{array}$$

(c) *With respect to a frame, a rigid displacement is defined by the displacements of 3 non-aligned points. Therefore, it is identified by* $9-3=6$ *(* $9=$ *components of the displacements vectors,* $3=$ *rigidity conditions).*

(d) *A rigid displacement does not change the geometrical dimension of a figure and its orientation.*

EXAMPLE 3.1 - (**Examples of rigid displacements**).

1)(SPATIAL TRANSLATION). It is represented by a vector $\bar{u} \in \mathbf{S}$. Thus, the set of translations can be identified with $\mathbf{S}$.

2)(SPATIAL ROTATION). It is the case where there exists a fixed line a (**rotation axis**). Then, any point belonging to a plane $\perp a$ describes the same angle ϕ (**rotation angle**). We call **rotation vector** the vector $r \equiv \phi u$, where u is the unit vector identified by a. A rotation is identified by 5 parameters.

3)(SPHERICAL DISPLACEMENT). In this case there is a fixed point C, called **center**.

□ (**Euler Theorem**). Any spheric displacement is of rotation one, with rotation axis passing through C.

2 THEOREM 3.13 - FUNDAMENTAL RELATION IN THE KINEMATICS OF RIGID BODY). *Let $v_p, v_\Omega \in \mathbf{S}$ be the velocities of two points P and Ω of a rigid body that moves with respect to an inertial frame. Then, one has the following relation:*

$$v_p = v_\Omega + \omega \times \overrightarrow{\Omega P} \tag{3.15}$$

where ω is the **spin**, *(or* **angular velocity***) of the flow (See section 4.1). ω does not depend on the position.*

PROOF. The spatial components of a rigid flow can be written as follows:

$$\phi^k(\lambda) = y^k(\lambda) + A^k_j \xi^j + A^k_0 x^0,$$

with

$$(A^\alpha_\beta) = \left(\begin{array}{c|ccc} 1 & 0 & 0 & 0 \\ \hline A^1_0 & & & \\ A^2_0 & & (A^i_j) & \\ A^3_0 & & & \end{array}\right)$$

where $(A^i_j) \in SO(3), \lambda \in \mathbf{T}$. So, by derivation with respect to λ we get:

$$v^k \equiv \dot{\phi}^k = \dot{y}^k + \dot{A}^k_j \xi^j + \dot{A}^k_0 x^0.$$

Then, taking into account that the matrix (A^k_j) is expressed by means of three parameters (e.g., Euler angles), we see that $(\dot{A}^k_j)$ can also be expressed by means of these parameters. Then, for two different elements P and Q of a rigid body, we get

$$\begin{aligned} v^k_P &= \dot{y}^k + \dot{A}^k_j \xi^j_P + \dot{A}^k_0 x^0 \\ v^k_Q &= \dot{y}^k + \dot{A}^k_j \xi^j_Q + \dot{A}^k_0 x^0 \end{aligned}$$

$$v^k_P - v^k_Q = \dot{A}^k_j(\xi^j_P - \xi^j_Q)$$

Furthermore, we can also write

$$v^k_P - v^k_Q = \Omega^k_j \chi^j_{PQ}, \tag{3.16}$$

where

$$(\Omega^k_j) \equiv \begin{pmatrix} 0 & \omega^3 & -\omega^2 \\ -\omega^3 & 0 & \omega^1 \\ \omega^2 & -\omega^1 & 0 \end{pmatrix}$$

and χ^j_{PQ} are the transformed of $\xi^j_P - \xi^j_Q$ by means of the matrix (A^k_j). Of course, expression (3.16) is equivalent to $v_p - v_Q = \omega \times \overrightarrow{QP}$. □

COROLLARY 3.1 - (**Poisson formulas**).

$$\boxed{\begin{aligned} &\dot{\epsilon}_k = \omega \times \epsilon_k, 1 \leq k \leq 3 \\ &\omega = \frac{1}{2}[\epsilon_1 \times \dot{\epsilon}_1 + \epsilon_2 \times \dot{\epsilon}_2 + \epsilon_3 \times \dot{\epsilon}_3] \end{aligned}} \tag{3.17}$$

COROLLARY 3.2 - *If the motion of one body is referred to an inertial frame with base point O, and to a rigid frame with base point Ω, one has the following formulas:*

$$v_a = v_r + v_\tau = v_r + v_\Omega + \omega \times \overrightarrow{\Omega P}$$

$$a_a = a_r + a_\tau + a_c, \quad a_c = 2\omega \times v_r, a_\tau = \dot{v}_\Omega + \dot{\omega} \times \overrightarrow{\Omega P} - \omega^2 \overrightarrow{QP}$$

where Q is the point on the instantaneous axis of rotation identified by P.

EXAMPLE 3.2 - (**Examples of instantaneous distributions of velocities in rigid motion**).

1)(TRANSLATION). In this case one has $v_P = \tau \in \mathbf{S}$. This is verified iff $\omega = 0$.

2)(ROTATION). There exists a line a (**instantaneous axis of rotation**) such that $v_P = 0, \forall P \in a$. This condition is verified iff there is a point Ω with $v_\Omega = 0$.

3)(INSTANTANEOUS PLANE MOTION). Any velocity v_P is parallel to the same plane σ. This happens iff $v_P \cdot \omega = 0$, for any point P. The **center of instantaneous rotation** A_t is the point intersection of σ with the instantaneous axis of rotation.

THEOREM 3.14 - *One has the following canonical decomposition:*

$$\boxed{v_p = \tau + \omega \times \overrightarrow{A_t P}} \tag{3.18}$$

where A_t is the instantaneous center of rotation in the plane $\sigma \perp \tau$ passing through P, and τ is a translation parallel to ω.

PROOF. Start from the relation:

$$v_p = v_\Omega + \omega \times \overrightarrow{\Omega P} \Rightarrow v_p \cdot \omega = v_\Omega \cdot \omega \Rightarrow \tau \equiv v_p \cdot \omega \frac{\omega}{|\omega|}.$$

As τ is a vector that does not depend on P, it represents a translation. Set ${v_p}^{\perp} \equiv v_p - \tau$. Then, ${v_p}^{\perp}$ is a rotation. In fact, we must prove that there is a line r having the points with zero velocity with respect to τ. Now, let σ be a plane through $P \perp \omega$. Then, on the line $s \subset \sigma$, passing for P, and $\perp {v_p}^{\perp}$, there is a unique point A_t such that:

$$0 = v_{A_t} = {v_p}^{\perp} + \omega \times \overrightarrow{PA_t} \Rightarrow {v_p}^{\perp} = -\omega \times \overrightarrow{PA_t} \Rightarrow {v_p}^{\perp} = \omega \times \overrightarrow{A_tP}.$$

Now, let $O \in r$. We have

$$\begin{aligned} v_O &= v_\Omega + \omega \times \overrightarrow{\Omega O} \Rightarrow v_O \\ &= v_\Omega + \omega \times \overrightarrow{\Omega A_t} \omega \times \overrightarrow{A_t O} \\ &\Downarrow \\ v_O &= v_\Omega + \omega \times \overrightarrow{\Omega A_t} \\ &= v_{A_t}; \\ {v_O}^{\perp} &= v_O - \tau \\ &= {v_{A_t}}^{\perp} = 0 \\ &\Downarrow \\ v_O - \tau &= {v_O}^{\perp}. \end{aligned}$$

This proves that ${v_p}^{\perp}$ is a rotation. Note also that for physical dimensions we put also:

$$\tau = \frac{v_p \cdot \omega}{|\omega|}.$$

□

THEOREM 3.15 - (INSTANTANEOUS DISTRIBUTION OF THE ACCELERATION IN RIGID MOTION). *One has the following formula:*

$$a_p \equiv \nabla_{v_p} v_p : T \times S_\psi \to TS_\psi.$$

Furthermore, we get

$$a_p = (\dot{v}_p^j + \Gamma^j_{i\,k} {v_p}^i {v_p}^k) \partial x_j \tag{3.19}$$

with

$$\begin{aligned} v_p &= v_\Omega + \omega \times \overrightarrow{\Omega P} \\ \dot{v}_p &= \dot{v}_\Omega + \dot{\omega} \times \overrightarrow{\Omega P} - \omega^2 \overrightarrow{QP} \end{aligned}$$

EXAMPLE 3.3 - (**Examples of rigid motions**).
1) (TRANSLATION). The flow-lines are straight lines.
2)(ROTATION). There is a fixed line a, (**rotation axis**). The flow-lines are circles with centers belonging to a. For an element of the rigid body distant R from a one has the following formulas:

$$\boxed{\begin{aligned} s &= \theta R \\ |v_p| &= |\dot{s}| = |R\dot{\theta}| \\ |\omega| &= \dot{\theta} \\ a &\perp \omega \end{aligned}} \tag{3.20}$$

For the acceleration one has the following:

$$\boxed{a_T = R\ddot{\theta}, \quad a_N = R\dot{\theta}^2.} \tag{3.21}$$

3)(SPHERICAL MOTION). In this case there is a fixed point Ω. The surfaces described, (in the spaces of the inertial frame, and body respectively), by the motion of the instantaneous axis of rotation are called **Poinsot cones**.[2]
The orientation of the body is made by means of the **Euler angles** (ϕ, ψ, θ). More precisely, if (Ω, e_k) and (Ω, ϵ_k) are affine frames in S_ψ, where (e_k) and (ϵ_k) are respectively equioriented orthonormal bases in $\mathbf{S}$, the first fixed, i.e., joint with the inertial frame, and the second fixed joint with the rigid body, then one has:

$$\left\{\begin{aligned} \phi &= \widehat{e_1 \overrightarrow{N}} \\ \psi &= \widehat{\overrightarrow{N} \epsilon_1} \\ \theta &= \widehat{e_3 \epsilon_3} \end{aligned}\right\}$$

where $\overrightarrow{N}$ is the unitary vector identified by the intersection line of the planes (e_1, e_2) and (ϵ_1, ϵ_2), and oriented in such a way that $(e_3, \epsilon_3, \overrightarrow{N})$ is a basis for $\mathbf{S}$ with the

[2] We say that the **body cone** (fixed in the body frame), rolls without slipping on the **space cone**. It is to be emphasized that if two rigid surfaces with one moving on the other having at least a contact of first order we can talk of **slipping velocity** in the contact points as the draging velocity, in the relative motion, of these points. Furthermore, if there are contact points where the slipping velocities are zero, then the instantaneous rotation axis a_t passes through these points. So we say that the surfaces roll one on the other **without slipping**. Moreover, we say that the motion is a **pure rolling** if the instantaneous rotation axis a_t belongs to the instantaneous common tangent plane π_t. We say that the motion is **pivoting** one if $a_t \perp \pi_t$.

same orientation of (e_k). The line passing for Ω and with direction $\overrightarrow{N}$ is called **line of nodes**. Let $(x^k) \equiv (x, y, z)$ be the coordinates associated to the inertial frame (**space coordinates**), and let $(\xi^k) \equiv (\xi, \eta, \zeta)$ be coordinates associated to the frame attached to the body (**body coordinates**). One has the following relation:

$$x^k = A^k_j \xi^j,$$

where A^k_j is the matrix change of basis in $\mathbf{S}$, defined by the following relation:

$$A^k_j \equiv D_s{}^k C_r{}^s B_j{}^r \equiv \begin{bmatrix} A_1{}^1 & A_1{}^2 & A_1{}^3 \\ A_2{}^1 & A_2{}^2 & A_2{}^3 \\ A_3{}^1 & A_3{}^2 & A_3{}^3 \end{bmatrix}$$

with

$$D \equiv \begin{bmatrix} \cos\phi & \sin\phi & 0 \\ -\sin\phi & \cos\phi & 0 \\ 0 & 0 & 1 \end{bmatrix}$$

$$C \equiv \begin{bmatrix} 1 & 0 & 0 \\ 0 & \cos\theta & \sin\theta \\ 0 & -\sin\theta & \cos\theta \end{bmatrix}$$

$$B \equiv \begin{bmatrix} \cos\psi & \sin\psi & 0 \\ -\sin\psi & \cos\psi & 0 \\ 0 & 0 & 1 \end{bmatrix}$$

where $D =$ rotation around z-axis, $C =$ rotation around $\overrightarrow{N}$, and $B =$ rotation around ζ-axis. So, we have:

$$A \equiv \begin{bmatrix} \cos\psi\cos\phi - \cos\theta\,\sin\phi\sin\psi & \cos\psi\sin\phi + \cos\theta\,\cos\phi\sin\psi & \sin\theta\sin\psi \\ -\sin\psi\cos\phi - \cos\theta\,\sin\phi\cos\psi & -\sin\psi\sin\phi + \cos\theta\,\cos\phi\cos\psi & \sin\theta\cos\psi \\ \sin\theta\sin\phi & -\sin\theta\cos\phi & \cos\theta \end{bmatrix}$$

and

$$D^{-1} = \begin{bmatrix} \cos\phi & -\sin\phi & 0 \\ \sin\phi & \cos\phi & 0 \\ 0 & 0 & 1 \end{bmatrix}$$

$$C^{-1} = \begin{bmatrix} 1 & 0 & 0 \\ 0 & \cos\theta & -\sin\theta \\ 0 & \sin\theta & \cos\theta \end{bmatrix}$$

$$B^{-1} = \begin{bmatrix} \cos\psi & -\sin\psi & 0 \\ \sin\psi & \cos\psi & 0 \\ 0 & 0 & 1 \end{bmatrix}$$

and

$$A^{-1} = \begin{bmatrix} \cos\phi\cos\psi - \sin\phi\sin\psi\cos\theta & -\cos\phi\sin\psi - \sin\phi\cos\psi\cos\theta & \sin\phi\sin\theta \\ \sin\phi\cos\psi + \cos\phi\sin\psi\cos\theta & -\sin\phi\sin\psi + \cos\phi\cos\psi\cos\theta & -\cos\phi\sin\theta \\ \sin\psi\sin\theta & \cos\psi\sin\theta & \cos\theta \end{bmatrix}.$$

As a consequence, one has the following expression of the angular velocity:

$$\omega = \dot{\phi}e_3 + \dot{\theta}N + \dot{\psi}\epsilon_3.$$

The expressions of ω in the bases e_k and ϵ_k respectively are the following:

$$\begin{aligned} \omega &= (\dot{\theta}\cos\phi + \dot{\psi}\sin\theta\sin\phi)e_1 \\ &+ (\dot{\theta}\sin\phi - \dot{\psi}\sin\theta\cos\phi)e_2 \\ &+ (\dot{\phi} + \dot{\psi}\cos\theta\,)e_3 \\ \omega &= (\dot{\phi}\sin\theta\sin\psi + \dot{\theta}\cos\psi)\epsilon_1 \\ &+ (\dot{\phi}\sin\theta\cos\psi - \dot{\theta}\sin\psi)\epsilon_2 \\ &+ (\dot{\phi}\cos\theta\, + \dot{\psi})\epsilon_3. \end{aligned}$$

PROOF. Let us, first, prove the representation of ω with respect to e_k. We have:

$$\begin{aligned} \omega &= \omega^i e_j, \omega^j \equiv \omega\cdot e_j, \\ \omega^1 &\equiv \omega\cdot e_1 = \dot{\theta}N\cdot e_1 + \dot{\psi}\zeta\cdot e_1, \\ \zeta\cdot e_1 &= x^1 = A^1_3 = \sin\phi\sin\theta \\ &= \dot{\theta}\cos\phi + \dot{\psi}\sin\theta\sin\phi, \\ \omega^2 &\equiv \omega\cdot e_2 = \dot{\theta}N\cdot e_2 + \dot{\psi}\zeta\cdot e_2, \\ \zeta\cdot e_2 &= x^2 = A^2_3 = -\sin\theta\cos\phi \\ &= \dot{\theta}\sin\phi - \dot{\psi}\sin\theta\cos\phi \\ \omega^3 &\equiv \omega\cdot e_3 = \dot{\psi}\cos\theta\, + \dot{\phi}\,. \end{aligned}$$

The representation of ω in the basis ϵ_k can similarly be proved. More precisely, we have:

$$\begin{aligned} \bar{\omega}^k &\equiv \omega\cdot e_k \\ &\Downarrow \\ \omega^1 &\equiv \omega\cdot\xi = \dot{\phi}e_3\cdot\epsilon_1 + \dot{\theta}\cos\psi = \dot{\phi}\sin\theta\sin\psi + \dot{\theta}\cos\psi, \\ e_3\cdot\epsilon_1 &= x^3 = A^3_1 = \sin\theta\sin\psi. \\ \omega^2 &\equiv \omega\cdot\eta = \dot{\theta}e_3\cdot\epsilon_2 - \dot{\theta}\sin\psi = \dot{\phi}\cos\psi\sin\theta - \dot{\theta}\sin\psi, \\ e_3\cdot\epsilon_2 &= x^3 = A^3_2 = \cos\psi\sin\theta. \\ \omega^3 &\equiv \omega\cdot\zeta = \dot{\phi}\cos\theta + \dot{\psi}. \end{aligned}$$

□

REMARK 3.13 - (PRECESSIONS). 1) A rigid spherical motion is a **precession** if the instantaneous rotation axis is in the same plane of two other lines p and f through the center Ω. We call p **precession axis** if it is fixed in the inertial frame. Furthermore, we call f **figure axis** if it is fixed in the rigid body-frame.
2) A **direct precession** has $\widehat{\omega_p\omega_f} \leq \pi/2$, where ω_p and ω_f are the components of ω along p and f respectively. Otherwise, we say that the precession is **retrograd**.
3) A **regular precession** has $|\omega_f| =$ constant, or $|\omega_p| =$ constant, i.e., $|\omega| =$ constant.
4) One has the following properties:
(i) In a precession the Poinsot cones are of revolution iff $\frac{|\omega_f|}{|\omega_p|} =$ constant.
(ii) In particular, regular precession Poinsot cones are of revolution.
(iii) In a precession f describes a cone of revolution with axis $= p$. This is called **precession cone**.
5)(PLANE MOTION). In this case, the flow-lines are plane-curves belonging to planes parallel to a same plane σ. Then, the instantaneous centre of rotation A_t describes in σ, with respect to the inertial frame, a curve called **base**, and, with respect to the rigid body-frame, another curve, called **roulette**. These curves are called **polar curves**. They are tangent, at any instant, at the point A_t.

3.2 - ONE-BODY DYNAMICS

Let n be an integer: $n \in \mathbf{N}$.
DEFINITION 3.13 - *A* **constraint of order** $0 \leq n \leq 2$ *on the Galilean space-time* M *is a submanifold of* $J\mathcal{D}^n(M)$, *where*

$$\begin{aligned} J\mathcal{D}^0(M) &\equiv M \\ J\mathcal{D}^1(M) &= J\mathcal{D}(M) = M \times \mathbf{I} \\ J\mathcal{D}^2(M) &= M \times \mathbf{I} \times \mathbf{S} \end{aligned}$$

where

$$\mathbf{I} = \{v \in \mathbf{M} | < \underline{\tau}, v >= 1\}, \quad \mathbf{S} \equiv \ker \underline{\tau} \subset \mathbf{M}.$$

In particular, a constraint of order 0 *is a subbundle of* $\mathcal{G}$; *this is called* **time-independent** *if it is diffeomorphic to* $T \times Y$, *where* Y *is an* m*-dimensional manifold,* $1 \leq m \leq 3$; *otherwise it is called* **time-dependent constraint**.
DEFINITION 3.14 - *A* **Newtonian mechanical system** *is defined by the following:*

$$NMS = (\mathcal{G}, N(\mathcal{G})),$$

where

(a) $\mathcal{G}$ *is the fiber-bundle space-time.*
(b) $N(\mathcal{G}) \subset J\mathcal{D}^2(M)$ *is a ODE of second order on* $\mathcal{G}$, *called* **Newton equation**, *that is defined by means of the following differential operator:*

(3.22)(**Newton operator**) $$\mathcal{N} \equiv \mathcal{F} - \mu\Gamma : J\mathcal{D}^2(M) \to vTM,$$

where

$$\mathcal{F} : J\mathcal{D}^2(M) \to vTM,$$

is the **operator force**, $\mu \in \mathbf{R}$, *is called* **mass**, *and* Γ *is the canonical connection on* M. *Set:*[3]

$$N(\mathcal{G}) = \ker \mathcal{N} \subset J\mathcal{D}^2(M).$$

We call **dynamic motion** *a section of* $\mathcal{G}$, $m : T \to M$, *such that*

$$D^2m : T \to N(\mathcal{G}) \subset J\mathcal{D}^2(M).$$

THEOREM 3.16 - *The following propositions are equivalent:*
(a) m *is a dynamic motion ;*
(b) $\mathcal{N} \circ D^2m = 0$;
(c) $F - \mu\ddot{m} = 0$, *with* $F = \mathcal{F} \circ D^2m$. *In adapted coordinates we have:*

$$\mu(\Gamma^i_{00} \circ m + 2\Gamma^j_{0j} \circ m\, \dot{m}^j + \Gamma^i_{jk} \circ m\, \dot{m}^j \dot{m}^k + \ddot{m}^i) = F^i.$$

PROOF. The proof is obtained by a direct calculation. □

Let $\mathbb{K}$ be a trivial vector fiber bundle of geometric objects on M: $\mathbb{K} \cong M \times \overline{\mathbb{K}}$.

DEFINITION 3.15 - *A* **Newtonian physical entity** f *is a mapping*

$$f : J\mathcal{D}^2(M) \to \mathbb{K}$$

fibered on M. *In other words, one has the following commutative diagram:*

$$\begin{array}{ccc} M & \cong & M \\ \uparrow & & \uparrow \\ J\mathcal{D}^2(M) & \xrightarrow{f} & \mathbb{K} \\ \downarrow & & \downarrow \pi_{\mathbf{K}} \\ T & \cong & T \end{array}$$

[3] The operator force can be considered coming from a connection on the space-time, i.e., connection on the vector bundle $TM \to M$, by means of the following commutative diagram:

$$\begin{array}{ccc} TTM & \to & vTTM \\ \cup & & \downarrow \\ J\mathcal{D}^2(M) & \to & vTM \end{array}$$

In this way the concept of force assumes a full geometric meaning.

Furthermore, f is **reducible to the k-order**, $0 \leq k < 2$, *if the following diagram is commutative, for a suitable derivative operator $\widehat{f} : J\mathcal{D}^k(M) \to \mathbb{K}$:*

$$\begin{array}{ccccc} & J\mathcal{D}^2(M) & \xrightarrow{f} & \mathbb{K} & \\ p_{2,k} & \downarrow & & \uparrow & \hat{f} \\ & J\mathcal{D}^k(M) & \cong & J\mathcal{D}^k(M) & \end{array}$$

where $p_{2,k}$ is the canonical projection.

REMARK 3.14 - The evaluation of a physical entity f at the motion m is the section

$$f.m \equiv f \circ D^2 m : T \to \mathbb{K}$$

of $\pi_{\mathbb{K}} : \mathbb{K} \to T$. If f is reducible to $\hat{f}$, at the order k, then $f.m = \hat{f} \circ D^k m : T \to \mathbb{K}$. In fact, one has the following commutative diagram:

$$\begin{array}{ccccc} & T & = & T & \\ D^2 m & \downarrow & & \downarrow & f\cdot m \\ & J\mathcal{D}^2(\mathcal{G}) & \xrightarrow{f} & \mathbb{K} & \\ p_{2,k} & \downarrow & & \| & \\ & J\mathcal{D}^k(\mathcal{G}) & \xrightarrow{\hat{f}} & \mathbb{K} & \\ D^k m & \uparrow & & \uparrow & \hat{f}\cdot m \\ & T & = & T & \end{array}$$

REMARK 3.15 - By means of a frame ψ we can define the observed $\mathbb{K}_\psi$ of a fiber bundle of geometric objects $(\mathbb{K}, M, \pi; \mathbf{B}\mathbb{K})$ setting $\mathbb{K}_\psi \equiv \mathbf{B}\mathbb{K}(S_\psi)$ where $S_\psi \equiv M/\sim$ is the 3-dimensional manifold induced by ψ on M. In this way, to any Newtonian physical entity $f : J\mathcal{D}^2(M) \to \mathbb{K}$ we can associate the corresponding observed entity f_ψ:

$$f_\psi \equiv \pi_\psi \circ f : J\mathcal{D}^2(M) \to \mathbb{K}_\psi,$$

where $\pi_\psi : \mathbb{K} \to \mathbb{K}_\psi$ is the canonical projection. Therefore, for any motion m, one has:

$$f_\psi \cdot m = f_\psi \circ D^2 m = \pi_\psi \circ f \circ D^2 m : T \to \mathbb{K}_\psi.$$

The following diagram is commutative:

$$\begin{array}{ccccccc} & J\mathcal{D}^2(M) & \xrightarrow{\quad f_\psi \quad} & & & \mathbb{K}_\psi & \\ & \| & & & & \| & \\ & J\mathcal{D}^2(M) & \xrightarrow{f} & \mathbb{K} & \xrightarrow{\pi_\psi} & \mathbb{K}_\psi & \\ D^2 m & \uparrow & & \uparrow \; f\cdot m & & \uparrow & f_\psi\cdot m \\ & T & \cong & T & \cong & T & \end{array}$$

DEFINITION 3.16 - *Let a numerical function $\mathcal{T} : J\mathcal{D}(M) \to \mathbf{R}$ be given on $J\mathcal{D}(M)$. Define* **Legendre transformation,** *(associated to $\mathcal{T}$), the following fiber morphism on M:*

$$\sigma(\mathcal{T}) : J\mathcal{D}(M) \to vT^*M$$

given by composition:

$$\begin{array}{ccc} J\mathcal{D}(M) & \xrightarrow{vd\mathcal{T}} & vT^*(J\mathcal{D}(M)) \cong M \times \mathbf{I} \times \mathbf{S}^* \\ {\scriptstyle \sigma(\mathcal{T})} \downarrow & & \downarrow {\scriptstyle \pi} \\ vT^*M & = & M \times \mathbf{S}^* \end{array}$$

One has

$$\sigma(\mathcal{T})(p,u) = (\pi \circ v\overline{d\mathcal{T}}(p,u)),$$

where vd is the symbol of vertical differentiation with respect to the projection $J\mathcal{D}(M) \to M$.

PROPOSITION 3.9 - *One has the following:*

(a) *If X is a field of velocities on M, the differential 1-form α, obtained by means of $\sigma(\mathcal{T})$, that corresponds to X is the following:*

$$\alpha = \sigma(\mathcal{T}) \circ X : M \to vT^*M.$$

In coordinates one has:

$$\alpha = (\partial \dot{x}_k \mathcal{T}) \circ X dx^\alpha.$$

(b) *If $\mathcal{T}$ is the* **kinetic energy**

$$\mathcal{T}(p,v) = \frac{1}{2}\mu g(p)(v_\wedge, v_\wedge),$$

then, $\sigma(\mathcal{T})$ defines an isomorphism:

$$\sigma(\mathcal{T}) : J\mathcal{D}(M) \Rightarrow vT^*M \cong M \times \mathbf{S}.$$

In this way the Legendre transformation induces a canonical fiber isomorphism $\mathring{\sigma}(\mathcal{T})$ on M given by the composition:

$$\begin{array}{ccc} vT(J\mathcal{D}(M)) & \xrightarrow{\mathring{\sigma}(\mathcal{T})} & vT^*(J\mathcal{D}^*(M)) \\ vT(\sigma(\mathcal{T})) \downarrow & & \| \\ vT(vT^*M) & \to & vT^*(J\mathcal{D}(M)) \end{array}$$

One has:

$$\overset{\circ}{\sigma}(\mathcal{T})(p,u,v) = (p,u,\overline{\overset{v}{D}\,\sigma(\mathcal{T})(p,u)(v)}).$$

Therefore, Newton equation is equivalent to **Legendre equation** $\mathcal{L}(\mathcal{G})$ *given by:*

$$\mathcal{L}(\mathcal{G}) = \ker \mathcal{L} \subset J\mathcal{D}^2(M),$$

where

$$\mathcal{L} \equiv \overset{\circ}{\sigma}(\mathcal{T}) \circ \mathcal{N}$$

is a derivative operator of the second order called **Legendre operator**. *The physical entity*

$$\widetilde{\mathcal{F}} \equiv \overset{\circ}{\sigma}(\mathcal{T}) \circ \mathcal{F} : J\mathcal{D}^2(M) \to vT^* J\mathcal{D}(M)$$

is called **Legendre force**.
(c) *In adapted coordinates Legendre equation* $\mathcal{L}(\mathcal{G})$ *is as follows:*

$$(\partial^2 \dot{x}_i \dot{x}_s \cdot \mathcal{T}) \circ \dot{m}\{[\mu(\Gamma^i_{00} \circ m + 2\Gamma^i_{0j} \circ m\, \dot{m}^j + \Gamma^i_{jk} \circ m\, \dot{m}^j \dot{m}^k + \ddot{m}^i)] - F^i\} = 0.$$

As

$$\mathcal{T} = \frac{1}{2}\mu \widehat{g}_{ij}(\dot{x}^i - \widehat{\dot{\psi}}^i)(\dot{x}^j - \widehat{\dot{\psi}}^j),$$

where $\widehat{g}_{ij}$ *and* $\widehat{\dot{\psi}}^i$ *are respectively the liftings of* g_{ij} *and* $\dot{\psi}^i$ *on* $J\mathcal{D}(M)$. *We have:*

$$(\partial \dot{x}_i \partial \dot{x}_s \cdot \mathcal{T}) = \mu g_{is}\,.$$

As a consequence, we get:

$$g_{ik} \circ m\mu[\Gamma^i_{00} \circ m + 2\Gamma^i_{0j} \circ m\, \dot{m}^j + \Gamma^i_{je} \circ m\, \dot{m}^j \dot{m}^e + \ddot{m}^i] - F_k = 0,$$

where $F_k = g_{ik}F^i$; $\mathcal{L}(\mathcal{G})$ *represents the covariant expression of Newton equation.*
DEFINITION 3.17 - *One has the following useful definitions:*
(a) *A* **conservative force** *is a force* $\widetilde{\mathcal{F}} : J\mathcal{D}^2(M) \to vT^*M$ *such that there exists a function* $f : M \to \mathbf{R}$ *such that gives the following commutative diagram:*

$$\begin{array}{ccc} J\mathcal{D}^2(\mathcal{G}) & \overset{\widetilde{\mathcal{F}}}{\rightarrow} & vTM \\ p_{2,0}\downarrow & & \downarrow {}'g \\ M & \underset{-vdf}{\rightarrow} & vT^*M \end{array}$$

where $'g$ *is the canonical isomorphism.*

(b) *The conservative force is called* **time-independent** *if*

(3.23) $$\partial_\psi f = 0,$$

where $\partial_\psi f$ *is the* **time-derivative** *with respect to a frame* ψ. *One has:*

(3.24) $$\partial_\psi f = \dot{\psi} \rfloor Df.$$

Equation (3.23), *in a coordinate system adapted to the frame, is written in the following way:*

$$\partial_\psi f = (\partial x_0 \cdot f).$$

REMARK 3.16 - For a conservative system we can introduce the following numerical function on $J\mathcal{D}(\mathcal{G})$ (**Lagrangian**):

$$L \equiv \mathcal{T} - f : J\mathcal{D}(\mathcal{G}) \to \mathbf{R}.$$

Then, by means of the variational calculus (see section 4.8) we get the Euler-Lagrange equation,

$$\mathcal{E}[m] \cdot \sigma = 0,$$

that, in adapted coordinates, has the following form:

(3.25)(**Lagrange equation**) $$\boxed{\frac{d}{dt}(\partial \dot{x}_k \cdot L) - (\partial x_k \cdot L) = 0.}$$

This equation, written in coordinates adapted to a frame, is equivalent to Newton equation. In fact, taking into account that

$$\mathcal{T} = \frac{1}{2}\mu g_{ij}\dot{x}^i\dot{x}^j,$$

we get [4]

$$\begin{aligned}(\partial x_k \cdot L) &= \frac{1}{2}\mu(\partial x_k \cdot g_{ij})\dot{x}^i\dot{x}^j - (\partial x_k \cdot f)\\ &= -\frac{1}{2}\mu[\widehat{\Gamma}^s_{ik} g_{sj} + \widehat{\Gamma}^s_{jk} g_{is}]\dot{x}^i\dot{x}^j - (\partial x_k \cdot f),\\ (\partial \dot{x}_k \cdot L) &= \mu g_{ik}\dot{x}^i,\\ \frac{d}{dt}(\partial \dot{x}_k \cdot L) &= \mu g_{ik}\ddot{x}^i + \mu(\partial x_r . g_{ik})\dot{x}^r\dot{x}^i\end{aligned}$$

$$\Downarrow$$

$$\mu[-\widehat{\Gamma}^t_{ji}\dot{x}^i\dot{x}^j + \ddot{x}^t] = F^t$$

$$\Downarrow$$

$$\mu[\Gamma^t_{ji}\dot{x}^i\dot{x}^j + \ddot{x}^t] = F^t.$$

□

[4] Note that from $\nabla_X g = 0$ it follows that $(\partial x_s . g_{ij}) = -g_{rj}\widehat{\Gamma}^r_{si} - g_{ir}\widehat{\Gamma}^r_{sj}$.

REMARK 3.17 - From the variational calculus (see section 4.8), we get that Euler-Lagrange equation is equivalent to **Hamilton equation**:[5]

$$\boxed{\dot{m}^*(\xi \rfloor \Omega) = 0,}$$

where

$$\begin{aligned} \Omega &= d\Theta, \forall \xi \in C^\infty(vTJ\mathcal{D}(M)), \\ \Theta &= \theta \bar{\wedge} {p_0}^* \eta_0 : J\mathcal{D}(M) \to T^*J\mathcal{D}(M) \\ &\equiv \textbf{1-form of Cartan} \\ \eta_0 &\equiv \text{volume 1-form on} \quad T. \\ \theta &: J\mathcal{D}(M) \to T^*J\mathcal{D}(M) \bigotimes p_{1,0}^* TM \\ < \theta(q), \xi > &= L(q)T(p_0)(\xi) + \sigma(L)(q)(< \omega_1(q), \xi >) \\ \omega_1 &: J\mathcal{D}(M) \to T^*J\mathcal{D}(M) \otimes p_{1,0}^* vTM. \\ < \omega_1(q), \xi > &= T(p_{1,0})(\xi) - T(M) \circ T(p_0)(\xi). \end{aligned}$$

In adapted coordinates we get:

$$\begin{aligned} \omega_1 &= -\dot{x}^j dx^0 \otimes \partial x_j + \delta_i^j dx^i \otimes \partial x_j \\ \theta &= (\partial \dot{x}_j \cdot L) dx^j - H dx^0, H = (\partial \dot{x}_j \cdot L)\dot{x}^j - L = \textbf{Hamiltonian}. \\ \Theta &= p_j dx^j - H dx^0 \equiv \theta, p_j = (\partial \dot{x}_j \cdot L) \, . \end{aligned}$$

Therefore, Hamilton equation becomes:

$$\boxed{\begin{aligned} (\partial x_j \cdot H) \circ \dot{m} &= -\dot{p}_j \\ (\partial p^j \cdot H) \circ \dot{m} &= \dot{m}^j \end{aligned}}$$

where $\dot{p}_j \equiv d(p_j \circ \dot{m})$.

REMARK 3.18 - (POISSON ALGEBRA). We have the following important vector fields defined on $J\mathcal{D}(M)$:

(a) **Locally Hamiltonian vector fields**:

$$\mathcal{L}_\xi \Omega = 0 \, .$$

[5] Let W be a vector space of dimension n and $u \in W^* \otimes W$, $\beta \in \Lambda_n^0(W)$, then we set $u \bar{\wedge} \beta \equiv (\alpha \otimes \xi) \bar{\wedge} \beta = \alpha \wedge (\xi \rfloor \beta)$.

(b) **Hamiltonian vector fields**:

$$\xi \rfloor \Omega = df.$$

As

$$0 = \mathcal{L}_\xi \Omega = d(\xi \rfloor \Omega) + \xi \rfloor (d\Omega)$$
$$\Downarrow$$
$$\xi \rfloor \Omega = df \quad \text{(locally)}.$$

(c) The space of **observables** $\mathcal{P}$ is the space of 0-forms f on $J\mathcal{D}(M)$ such that:

$$\boxed{\xi \rfloor \Omega = df}$$

for a suitable vector field $\xi : J\mathcal{D} \to TJ\mathcal{D}$. In $\mathcal{P}$ we recognize a Lie algebra (**Poisson algebra**):

$$\{\mathcal{P}, (,)\},$$

where the bracket is defined by the following:

$$(f_i, f_j) = \xi_i \rfloor df_j = \xi_i \rfloor \xi_j \rfloor \Omega.$$

In fact, one has:

$$(f_1, f_2) = -(f_2, f_1)$$
$$((f_1, f_2), f_3) + ((f_2, f_3), f_1) + ((f_3, f_1), f_2) = 0 \quad \textbf{(Jacobi identity)}$$

(d) If $df = \xi \rfloor \Omega$, we have the following representation in adapted coordinates:

$$\boxed{\begin{aligned} \xi &= -(\partial p^j \cdot f)\partial x_j + (\partial x_j \cdot f)\partial p^j \\ \eta &= -(\partial p^j \cdot g)\partial x_j + (\partial x_j \cdot g)\partial p^j \\ \Omega(\xi, \eta) &= (f, g) \\ &= (\partial x_j \cdot f)(\partial p^j \cdot g) - (\partial p^j \cdot f)(\partial x_j \cdot g) \end{aligned}}$$

(e) If $i_t : J\mathcal{D}(M)_t \to J\mathcal{D}(M)$ is the canonical inclusion, $\forall t \in T$, and α_t is the canonical 1-form on T^*M_t (**Liouville form**), then we have[6]

$$i_t^* \Theta = \sigma(T)^* \alpha_t$$

[6] $\alpha_t : T^*M_t \to T^*T^*M_t$; $\alpha_t(\beta) = T^*(\pi)(\beta), \pi : T^*M_t \to M_t$.

with

$$\sigma(\mathcal{T})' \equiv \bar{\pi} \circ \sigma(\mathcal{T}) \circ \dot{J}_t,$$

where $\dot{J}_t : \mathcal{JD}(M)_t \to T(M)$ is the canonical inclusion and $\bar{\pi} : T^*M \to T^*M_t$ is the canonical projection. One has the following commutative diagram:

$$\begin{array}{ccccccc}
\mathbf{R} \times vT^*M & \rightarrow & vT^*M & \leftarrow & T^*M_t & & \\
{\scriptstyle \sigma(L)} \uparrow & & & & \| & & \\
\mathcal{JD}(M) & \overset{i_t}{\leftarrow} & \mathcal{JD}(M)_t & \overset{\sigma(L)_t}{\rightarrow} & T^*M_t & \overset{\alpha}{\rightarrow} & T^*T^*M_t \\
\Theta \downarrow & & \downarrow i_t^*\Theta & & & & \| \\
T^*\mathcal{JD}(M) & \underset{T^*(i_t)}{\rightarrow} & T^*\mathcal{JD}(M)_t & & & \underset{T^*(\sigma(L)_t)}{\leftarrow} & T^*T^*M_t
\end{array}$$

(f) As $d\alpha_t$ is the **standard symplectic form** on T^*M_t,[7]
the 2-form $i_t^*\Omega \equiv \omega_t$ defines a symplectic structure on $\mathcal{JD}(M)_t$. The space

$$C^\infty(\mathcal{JD}(M)_t, \mathbf{R})$$

of the numerical functions on $\mathcal{JD}(M)_t$ is a Lie algebra with respect to the Poisson bracket, defined by means of ω_t.
(g) The mapping $\mathcal{P} \to C^\infty(\mathcal{JD}(M)_t, \mathbf{R})$ that associates to the 0-form $f \in \mathcal{P}$ its restriction on $\mathcal{JD}(\mathcal{G})_t$, is an homomorphism of Lie algebras.
(h) There exists a unique vector field ζ on $\mathcal{JD}(\mathcal{G})$ identified by the following conditions:

$$\boxed{\begin{array}{c} \zeta \rfloor \Omega = 0 \\ < \zeta, p_0^*\eta_0 > = 1 \end{array}}$$

Then, ζ is called **Euler vector field**, and it has the following representation in adapted coordinates:

$$\boxed{\zeta = \partial x_0 + (\partial p^j \cdot H)\partial x_j - (\partial x_j \cdot H)\partial p^j} \qquad (3.26)$$

(i) A motion m is dynamic iff:

$$\zeta \circ \dot{m} = D\dot{m},$$

---

[7] On T^*M_t, ($\dim(T^*M_t)$=6), there exists a canonical symplectic structure: $\sigma = dx^i \wedge d\dot{x}_i$. This can be obtained as the differential, $\sigma = d\alpha_t$, of the **Liouville, 1-form** $\alpha_t = \dot{x}_i dx^i$.

namely iff the following diagram

$$\begin{array}{cccc} & TJ\mathcal{D}(M) & \overset{D\dot{m}}{\leftarrow} & T \\ \zeta & \uparrow & & \| \\ & J\mathcal{D}(M) & \underset{\dot{m}}{\leftarrow} & T \end{array}$$

is commutative.

(l) A function $f : J\mathcal{D}(M) \to \mathbf{R}$ belongs to $\mathcal{P}$ iff

$$\zeta \cdot f = 0.$$

PROOF. In fact, $f \in \mathcal{P} \Rightarrow \exists \zeta$ such that $\zeta \rfloor \Omega = df$. From (3.26) we get:

$$\zeta \rfloor \Omega = (\partial p^j \cdot f) dp_j + (\partial x_j \cdot f) dx^j + \{(\partial x_j \cdot H)(\partial p^j \cdot f) - (\partial p^j \cdot H)(\partial x_j \cdot f)\} dx^0$$

hence

$$(\partial x_0 \cdot f) = (\partial x_j \cdot H)(\partial p^j \cdot f)(\partial p^j \cdot H)(\partial x_j \cdot f)$$

that proves $\zeta \cdot f = 0$. The converse can be easily proved conversely. □

(m) In particular, $H \in \mathcal{P}$ iff in adapted coordinates one has:

$$(\partial x_0 \cdot H) = 0,$$

i.e., the Hamiltonian is a first integral of the motion. One has

$$\boxed{\begin{aligned} \xi_H \rfloor \Omega &= dH, \\ \xi_H &= -(\partial p^j \cdot H)\partial x_j + (\partial x_j \cdot H)\partial p^j \end{aligned}}$$

Then, for any $f \in \mathcal{P}$ we get

$$\boxed{(\partial x_0 \cdot f) = (H, f)}$$

DEFINITION 3.18 - (DYNAMIC SYMMETRIES OF ONE-BODY).

(a) *A* **dynamic symmetry** *is a fiber-transformation of $\mathcal{G}$ that induces a transformation of the dymanic equation into itself. Therefore, any dynamic symmetry is an* *element of* **G**, *hence the* **group of dynamic symmetries** G *is a subgroup of* **G**: $G \subset \mathbf{G}$. *The following diagram resumes the situation.*

$$\begin{array}{ccccccc}
 & & (\mathcal{N}) & \xrightarrow{J\mathcal{D}^2(f)|} & (\mathcal{N}) & & \\
 & & \downarrow & & \downarrow & & \\
TTM & \supset & J\mathcal{D}^2(M) & \xrightarrow{J\mathcal{D}^2(f)} & J\mathcal{D}^2(M) & \subset & TTM \\
\downarrow & & \downarrow & & \downarrow & & \downarrow \\
TM & \supset & J\mathcal{D}(M) & \xrightarrow{J\mathcal{D}(f)} & J\mathcal{D}(M) & \subset & TM \\
\downarrow & & \downarrow & & \downarrow & & \downarrow \\
M & = & M & \xrightarrow{f} & M & = & M \\
 & & {\scriptstyle\tau}\downarrow & & \downarrow{\scriptstyle\tau} & & \\
 & & T & \xrightarrow[f_T]{} & T & &
\end{array}$$

(b) *A dynamic motion is* (f_T, f)*-symmetric,* $(f_T, f) \in G$ *if:*

$$f \circ m \circ f_T^{-1} = m.$$

The symmetry group of a motion is denoted by G_m *and called the* m**-isotropy group**: $G_m \subset G$.
(c) *A state* D^2m *is* (f_T, f)*-symmetric if:*

$$J\mathcal{D}^2(f_T, f) \circ D^2m \circ f_T^{-1} = D^2m.$$

PROPOSITION 3.10 - *One has the following properties:*
(a)

$$G_{D^2m} = G_m \subset G.$$

(b) *Iff* m *is a dynamic motion, then*

$$m' = f \circ m \circ f_T^{-1}(m' \quad \text{related to} \quad m), \quad (f_T, f) \in G$$

is dynamic one. Thus, in the set of the solutions of the dynamic equation, G *generates a partition in classes, an the dynamic motions belonging to the same class are considered equivalent.*
DEFINITION 3.19 - *Let* $\mathcal{K} \equiv (\mathbb{K}, M, \pi_k; \mathbf{B}\mathbb{K})$ *be a vector fiber bundle of geometric objects. A* **physical isomorphism** *of* $\mathcal{K}$ *is an isomorphism of vector fiber bundle* $(\bar{\psi}, f)$ *of* $\pi_{\mathbf{K}}$ *with* $f \in \mathbf{G}$. *The set of physical isomorphisms of* $\mathcal{K}$ *is denoted by* $\mathcal{G}_{\mathbb{K}}$ *and is called the* **physical group** *of* $\mathcal{K}$.
PROPOSITION 3.11 - *One has*

$$\mathcal{G}_{\mathbf{K}} \cong \mathbf{G} \times Hom_M(\pi_{\mathbf{K}}, \pi_{\mathbf{K}})$$

with

$$(\bar{\psi}, f) \mapsto (f, g \equiv \bar{\psi} \circ \mathbf{B}\mathbb{K}(f^{-1}))$$

and its inverse

$$(f, g) \mapsto (f, \bar{\psi} \equiv g \circ \mathbf{B}\mathbb{K}(f)).$$

DEFINITION 3.20 - *Two sections k and k' of $\pi_{\mathbf{K}}$ are $(\bar{\psi}, f)$-related if there exists $(\bar{\psi}, f) \in \mathcal{G}_{\mathbf{K}}$ such that*

$$k' = \bar{\psi} \circ k \circ f^{-1}.$$

We get thc following cases:

$$(3.27) \qquad \begin{array}{ll} (a) & \bar{\psi} \equiv \mathbf{B}\mathbb{K}(f) :\Rightarrow k' = f^* k, \\ (b) & f_M = id_M :\Rightarrow k' = \bar{\psi} \circ k. \end{array}$$

DEFINITION 3.21 - *Let $K : J\mathcal{D}^2(\mathcal{G}) \to \mathbb{K}$ be a physical entity. A $\mathbb{K}$-symmetry is an element of $(f_T, f) \in G$, such that there exists a fiber-bundle isomorphism of $\mathbb{K}$ such that $\bar{\psi} \circ K = K \circ \mathcal{D}^2((f_T)f)$; namely the following diagram*

$$\begin{array}{cccc} & J\mathcal{D}^2(\mathcal{G}) & \xrightarrow{K} & \mathbb{K} \\ J\mathcal{D}^2(f_T, f) & \downarrow & & \downarrow \bar{\psi} \\ & J\mathcal{D}^2(\mathcal{G}) & \xrightarrow[K]{} & \mathbb{K} \end{array}$$

is commutative. We will call $\mathbb{K}$**-group** *the corresponding group, and we will denote it by $K(G)$. Furthermore, denote $K(\mathbf{G})$ the canonical image of $K(G)$ inside $\mathbf{G}$: $K(\mathbf{G}) \subset G \subset \mathbf{G}$.*

PROPOSITION 3.12 - 1) *Let m and m' be two arbitrary dynamic solutions f-related, with $(f_T, f) \in G$. If K is a physical entity, we get that $K \cdot m$ is $(f_T, f; \bar{\psi})$-related to $K \cdot m'$ for a fixed choice of $\bar{\psi} \in Hom(\pi_{\mathbf{K}}, \pi_{\mathbf{K}})$ iff $(f_T, f; \bar{\psi}) \in K(G)$.*
2) *Furthermore, if $\bar{\psi} = \mathbf{B}\mathbb{K}(f)$, we get:*

$$K \cdot m' = f^* K \cdot m.$$

DEFINITION 3.22 - *Let us consider a dynamic motion m and a physical entity K. The (m, K)*-**isotropy group** *is defined by the following:*

$$K_m(G) \equiv \{(f_T, f; \bar{\psi}) \in K(G) | (f_T, f) \in G_m\}.$$

PROPOSITION 3.13 - *One has the following properties:*
(a) *If m is a dynamic motion and $(f_T, f; \bar{\psi}) \in K_m(G)$ we get:*

$$K \cdot m = \bar{\psi} \circ (K \cdot m) \circ f^{-1}.$$

(b) *The following condition*

$$K^m(G) \equiv \{(f_T, f; \bar{\psi}) \in K(G) | K \cdot m = \bar{\psi} \circ (K \cdot m) \circ f^{-1}\} = K_m(G)$$

is verified iff K is an application fully elastic, i.e., depends on the point of the base. Otherwise we get:

$$K_m(G) \subset K^m(G).$$

PROPOSITION 3.14 - *The derivative operator Γ, and hence also $\mu\Gamma, \mu \in \mathbf{R}$, is invariant for any affine transformation of M. Therefore, the image of the invariance group of Γ inside $\mathbf{G}$ is just $\mathbf{G}$:*

$$\Gamma(\mathbf{G}) \equiv \mathbf{G}.$$

PROOF. In fact, it is easy to verify the commutativity of the following diagram

$$\begin{array}{ccccc} & T^2(M) & \xrightarrow{\Gamma} & vTTM & \\ T^2(f) & \uparrow & & \uparrow & vTT(f) \\ & T^2(M) & \underset{\Gamma}{\rightarrow} & vTTM & \end{array}$$

for any affine transformation of M. □

REMARK 3.19 - The image of the group of invariance of the Newton operator in $\mathbf{G}$ is contained in that of a force F. More precisely, we get

$$\mathcal{N}(\mathbf{G}) \subset F(\mathbf{G}) \cap G \subset G \subset \mathbf{G} \equiv \Gamma(\mathbf{G}).$$

Therefore, $\mathcal{N}(\mathbf{G})$ is a group contained in G.

DEFINITION 3.23 - *One has the following definitions.*

(a) *A physical entity K is a* **constant of motion** $m : T \rightarrow M$ *if*

$$\nabla K \cdot m \equiv \Gamma_{\mathbb{K}} \circ D(K \cdot m) = 0,$$

where $\Gamma_{\mathbb{K}} : T\mathbb{K} \rightarrow vT\mathbb{K}$ is the canonical connection of $\mathbb{K} \rightarrow M$.

(b) *If K is a constant of the motion, for any dynamic motion m, then K is called* **dynamic constant**.

PROPOSITION 3.15 - *If $K \cdot m$ is a numerical function (0-form) then the above definition is equivalent to the following:*

$$dK \cdot m = 0.$$

THEOREM 3.17 - (NOETHER THEOREM). *Let a physical entity $K : J\mathcal{D}^2(M) \rightarrow \mathbb{K}$ be given and let m be a dynamic motion. Furthermore, let K be invariant with* *respect to a deformation $\widetilde{m}$ of m and with respect to a flow ϕ on M, (with induced flow $\bar{\phi}$ on T and with infinitesimal generator $\bar{\xi}$)*

$$\widetilde{K} \cdot m \equiv \mathbb{K}(\phi_t^{-1}) \circ (K \circ \dot{\widetilde{m}}_t) \circ \bar{\phi}_t = K \circ \dot{m}.$$

Then, we get

$$\boxed{\partial \widetilde{K} \cdot m = 0} \tag{3.28}$$

In particular, if $K \equiv L \equiv$ Lagrangian, and the deformation of m comes from an infinitesimal transformation of **G***, then this last preserves either the Lagrangian, or Lagrange equations; thus it can be considered as an infinitesimal dynamic symmetry. Furthermore, equation (3.28) gives the following dynamic constants:*

$$\boxed{K \cdot m = [(L \circ \dot{m})\bar{\xi}^0 + (\partial \dot{x}_j \cdot L) \circ \dot{m}(\widetilde{X})^j \circ m]} \tag{3.29}$$

where $\widetilde{X} \circ m$ is the vertical vector field $T \to m^* vTM$, along the motion m, corresponding to the deformation of m.

EXAMPLE 3.4 - Galilean group is of dimension 10. Therefore, we can find at most 10 motion constants (functionally independent) that come from the symmetry of the Lagrangian. The situation is resumed in the following table.

TAB.3.2 - **Dynamic symmetries and dynamic constants**

Symmetry	N^*	Conservation law $K.m$
spatial translation	3	momentum: $p_k = (\partial \dot{x}_k \cdot L)$
spatial rotation	3	angular momentum: $m_{ij} = [x_i(\partial \dot{x}_j \cdot L) - x_j(\partial \dot{x}_i \cdot L)]$
time-translation	1	Lagrangian: L Hamiltonian: $H = (\partial \dot{x}_i \cdot L)\dot{x}^i - L$
$(f^0 = x^0, f^j = \lambda x^0)$	3	$p_j \cdot x^0$

(*) N= number of parameters

In adapted coordinates a Galilean transformation $f \in$ **G** is given by:

$$f^\alpha \equiv x^\alpha \circ f = A^\alpha{}_\beta x^\beta + b^\alpha,$$

with $(A^{\alpha}{}_{\beta})$ a matrix of the type (3.4).

THEOREM 3.18 - (HAMILTONIAN FORM OF THE NOETHER THEOREM).

1) *Let* $\Psi \equiv (\phi_t)_{t\in \mathbf{R}}$ *be a 1-parameter subgroup of* $\mathbf{G}$ *such that*

$$J\mathcal{D}(\phi_t)^*\Theta = \Theta, \quad \forall \lambda \in \mathbf{R}$$

we get:

I) Ψ *is a 1-parameter subgroup of* G.

II) *Hamiltonian vector field is associated to* Ψ*:*

$$\xi = \partial\phi$$

III) *A motion-constant is associated to* Ψ*:*

$$\dot{m}^*(\xi\rfloor\Theta)$$

that in adapted coordinates can be written as follows:

$$\boxed{\dot{m}^*(\xi\rfloor\Theta) = (\partial\dot{x}_j \cdot L) \circ \dot{m}(\partial\phi)^j \circ m - H \circ \dot{m}(\partial\bar{\phi}^0)}$$

where $(\bar{\phi}_t)_{t\in\mathbf{R}}$ *is the 1-parameter subgroup of transformations on* T, *corresponding to* Ψ, *namely one has the following commutative diagram:*

$$\begin{array}{ccccc}
 & J\mathcal{D}(M) & \overset{J\mathcal{D}(\phi_t)}{\longrightarrow} & J\mathcal{D}(M) & \\
p_{1,0} & \downarrow & & \downarrow & p_{1,0} \\
 & M & \underset{\phi_t}{\longrightarrow} & M & \\
\tau & \downarrow & & \downarrow & \tau \\
 & T & \underset{\bar{\phi}_t}{\longrightarrow} & T &
\end{array}$$

2) *Let* H *be a subgroup of* $\mathbf{G}$ *such that*

$$J\mathcal{D}(\phi_a)^*\Theta = \Theta, \forall \phi_a \in H.$$

We get the following.

I) H *is a subgroup of* G*:* $H \subseteq G$.

II) *There is a homomorphism of Lie algebras*

$$\psi : A(H) \to \mathcal{P}, \psi : \xi \mapsto \widehat{\xi}$$
$$\widehat{\xi}\rfloor\Omega = -d(\widehat{\xi}\rfloor\Theta) \quad (\Leftrightarrow \mathcal{L}_{\widehat{\xi}}\Theta = 0)\,.$$

So, for any vector of the Lie algebra of H, the corresponding vector field on $J\mathcal{D}(M)$ is a Hamiltonian vector field, and the associated 0-form is the following:

$$f_\xi \equiv -\widehat{\xi}\rfloor\Theta : J\mathcal{D}(M) \to \mathbf{R}.$$

III) *For any $\xi \in A(G)$ and $\phi_\alpha \in H$, there corresponds a motion-constant:*

$$\beta_{(a,\xi)} \equiv J\mathcal{D}(\phi_a^{-1})^* f_\xi - f_{ada}(\xi)\,.$$

IV) *For any $\xi, \eta \in A(G)$ we get the following motion-constants.*

$$\gamma_{(\xi,\eta)} \equiv \mathcal{L}_{\widehat{\xi}} f_\xi + f_{[\xi,\eta]} = (f_\xi, f_\eta) + f_{[\xi,\eta]}.$$

REMARK 3.20 - (MANY-BODIES DYNAMICS). In order to describe a system having N particles, we must consider the fiber bundle

$$W \equiv M \times_T \cdots \times_T M \to T.$$

A section m of this fiber bundle is a **motion** for the system of N particles, and it is represented by N-motions $m_i, i = 1, \cdots, N$, one for each particle:

$$m = (m_1, \cdots, m_N).$$

Furthermore, the space of velocity of motions is the following:

$$J\mathcal{D}(M \times_T \cdots \times_T M) \cong J\mathcal{D}(M) \times_T \cdots \times_T J\mathcal{D}(M).$$

Then, the velocity of the motion, m, for a system of N particles, splits into N velocities, for each particle, according to the following commutative diagram:

$$\begin{array}{ccc} J\mathcal{D}(M \times_T \cdots \times_T M) & \cong & J\mathcal{D}(M) \times_T \cdots \times_T J\mathcal{D}(M) \\ {\scriptstyle \dot{m}} \nwarrow & & \nearrow {\scriptstyle (\dot{m}_1, \cdots, \dot{m}_p)} \\ & T & \end{array}$$

Similarly, for the space of k-derivative for motions, $k \geq 1$, we get the following commutative diagram:

$$\begin{array}{ccc} J\mathcal{D}^k(M \times_T \cdots \times_T M) & \cong & J\mathcal{D}^k(M) \times_T \cdots \times_T J\mathcal{D}^k(M) \\ {\scriptstyle D^k m} \nwarrow & & \nearrow {\scriptstyle (D^k m_1, \cdots, D^k m_p)} \\ & T & \end{array}$$

Then, dynamic equations can be represented as submanifolds

$$E_2 \subset J\mathcal{D}^2(M \times_T \cdots \times_T M) \cong J\mathcal{D}^2(M) \times_T \cdots \times_T J\mathcal{D}^2(M).$$

If $E_2 \neq {}_1E_2 \times \cdots \times {}_NE_2$, where $E_i, i = 1, \cdots, N$, is the image of E_2 in $J\mathcal{D}^2(M)$ by means of the i-th projection $J\mathcal{D}^2(M \times_T \cdots \times_T M) \to J\mathcal{D}^2(M)$, we say that the N particles interact among them; otherwise we say that they do not interact.

REMARK 3.21 - (CONSTRAINTS AND FRICTION). If a body is constrained by a manifold X, the constraint reaction ${}^vF$ has, in general, a non-zero tangential component ${}^vF_T$. In a dynamic situation one has the following relation between tangential and normal components of ${}^vF$:

$${}^vF_T = -\phi_d|{}^vF_N|\frac{v}{|v|},$$

where v is the velocity of the body (that is, of course, tangential to the constraint). ϕ_d is called **dynamic friction coefficient**. In static situations, we have, instead, the following inequality:

$$\frac{|{}^vF_T|}{|{}^vF_N|} \leq \phi_s,$$

where ϕ_s is called **static friction coefficient**. The cone, with vertex on a point of the constraint, and with angle $\alpha = \arctan\phi_s$, is called **static friction-cone**. The constraint is called **without friction** if ${}^vF_T = 0$.

3.3 - IMPORTANT FORMULAS

PROPOSITION 3.16 - (IMPORTANT SYSTEMS OF COORDINATES ON RIEMANN MANIFOLDS). *Let (M, g) be a Riemann structure, with $\dim M = n$. One has the following formulas locally characterizing M in some important systems of coordinates.*

(a) (**Orthonormal coordinates**).

$$g_{ij} = \delta_{ij}, \sqrt{\det g_{ij}} = 1, \quad \Gamma^k_{ij} = 0.$$

(b) ($\dim M = 3$. **Spherical coordinates**: (ρ, θ, ϕ)).

Metric field:

$$(g_{ij}) = \begin{pmatrix} 1 & 0 & 0 \\ 0 & \rho^2 & 0 \\ 0 & 0 & \rho^2\sin^2\theta \end{pmatrix}, \sqrt{\det g_{ij}} = \rho^2\sin\theta.$$

Connection coefficients:

$$\Gamma^1_{22} = -\rho,$$
$$\Gamma^2_{21} = \Gamma^2_{12} = \frac{1}{\rho},$$
$$\Gamma^1_{33} = -\rho \sin^2\theta,$$
$$\Gamma^2_{33} = -\rho \sin\theta\cos\theta$$
$$\Gamma^3_{31} = \Gamma^3_{13} = \frac{1}{\rho},$$
$$\Gamma^3_{32} = \Gamma^3_{23} = \cot\theta$$

(c) ($\dim M = 3$. **Cylindrical coordinates**: (ρ, θ, z)).
Metric field:

$$(g_{ij}) = \begin{pmatrix} 1 & 0 & 0 \\ 0 & \rho^2 & 0 \\ 0 & 0 & 1 \end{pmatrix}, \sqrt{\det(g_{ij})} = \rho.$$

Connection coefficients:

$$\Gamma^1_{22} = -\rho,$$
$$\Gamma^2_{12} = \Gamma^2_{21} = \frac{1}{\rho}.$$

PROPOSITION 3.17 - (IMPORTANT DIFFERENTIAL OPERATORS).
1) (STAR OPERATOR).

$$* : C^\infty(\Lambda^0_p M) \to C^\infty(\Lambda^0_{n-p} M), \quad * : \alpha \mapsto *\alpha \equiv \bar{\alpha} \rfloor \eta.$$

So one has the following commutative diagram:

$$\begin{array}{ccccc} & M & \overset{\bar{\alpha}}{\to} & \Lambda^p_0 M & \\ \alpha & \downarrow & & \| & \Lambda^p(g') \\ & \Lambda^0_p M & \equiv & \Lambda^0_p M & \end{array}$$

with $\eta : M \to \Lambda^0_n M \equiv$ *canonical volume form on* M.

$$\eta = \pm\sqrt{\det(g_{ij})}dx^1 \wedge \cdots \wedge dx^n.$$

2)(VECTOR PRODUCT). ($\dim M = 3$).

$$u \times v = \frac{1}{\sqrt{\det g_{ij}}}[(u_2v_3 - u_3v_2)e_1 + (u_3v_1 - u_1v_3)e_2 + (u_1v_2 - u_2v_1)e_3]$$
$$= \epsilon^{l_1 l_2 k} u_{l_1} v_{l_2} e_k,$$

where ϵ^{ijk} *is the Ricci tensor defined by the following:*

$$\epsilon^{l_1 l_2 k} = \left\{ \begin{array}{ll} 0 & \textit{if at least two indexes are equal} \\ \pm\dfrac{1}{\sqrt{\det g_{ij}}} & \textit{if all indexes are different} \end{array} \right\}$$

Furthermore, one has

$$u \times v \equiv (\underline{u} \wedge \underline{v}) \rfloor \bar{\eta},$$

where

$$\underline{u} \equiv g' \circ u : M \to T^*M$$
$$\underline{v} \equiv g' \circ v : M \to T^*M$$
$$\bar{\eta} \equiv \Lambda^n(g') \circ \eta : M \to \Lambda^n(TM)\,.$$

One has the following properties:

$$\begin{aligned}
&(a) \quad u \times (v \times w) = (u \cdot w)v - (u \cdot v)w \\
&(b) \quad (u \times v) \times w = (u \cdot w)v - (v \cdot w)u \\
&(c) \quad u \cdot (v \times w) = -(u \times w) \cdot v\,.
\end{aligned}$$

3) (CODIFFERENTIAL).

$$\delta : C^\infty(\Lambda^0_p M) \to C^\infty(\Lambda^0_{p-1} M), \quad \delta \equiv *d*\,.$$

In coordinates we get:

$$(\delta\alpha)_{i_2\cdots i_p} = \nabla_i {\alpha^i}_{i_2\cdots i_p} = \frac{1}{\sqrt{\det g_{hk}}} \partial x_i (\sqrt{\det(g_{hk})} {\alpha^i}_{i_2\cdots i_p}).$$

One has the following relation with the Lie derivative:

$$\delta \underline{X} = *\mathcal{L}_{\overline{X}} \eta$$

4) (LAPLACIAN).

$$\Delta : C^\infty(\Lambda^0_p M) \to C^\infty(\Lambda^0_p N), \quad \Delta \equiv (d\delta + \delta d).$$

The space $\Delta^p(M) \equiv \ker(\Delta)$ *is called* **space of harmonic** p**-forms**. *One has the following splitting* (**Hodge theorem**)*:*

$$C^\infty(\Lambda^0_p M) = dC^\infty(\Lambda^0_{p-1} M) \bigoplus \delta C^\infty(\Lambda^0_{p+1} M) \bigoplus \Delta^p(M).$$

5) (GRADIENT).

$$\text{grad} : C^\infty(T^0_0 M) \to C^\infty(T^1_0 M), \text{grad}(f) \equiv g'^{-1} \circ df = [g^{ij}(\partial x_j \cdot f)]\partial x_i.$$

One has the following properties:

$$\begin{aligned}
&(a) \quad \text{grad}(f_1 + f_2) = \text{grad}\, f_1 + \text{grad}\, f_2 \\
&(b) \quad \text{grad}(f_1 \cdot f_2) = f_2 \text{grad}\, f_1 + f_1 \text{grad}\, f_2
\end{aligned}$$

6) (ROTOR). ($\dim M = 3$).

$$\operatorname{rot} : C^\infty(T_0^1 M) \to C^\infty(T_0^1 M),$$

$$\operatorname{rot} u \equiv \overline{*d\underline{u}} = \frac{1}{\sqrt{\det g_{ij}}}\{[(\partial x_2 \cdot u_3) - (\partial x_3 u_2)]\partial x_1$$

$$+ [(\partial x_3 \cdot u_1) - (\partial x_1 \cdot u_3)]\partial x_2 + [(\partial x_1 \cdot u_2) - (\partial x_2 \cdot u_1)]\partial x_3\}$$

$$= \epsilon^{l_1 l_2 k}(\partial x_{l_1} . \dot{\phi}_{l_2})\partial x_k .$$

One has the following properties:

$$(a) \quad \operatorname{rot}(u_1 + u_2) = \operatorname{rot} u_1 + \operatorname{rot} u_2$$

$$(b) \quad \operatorname{rot} \operatorname{grad} f = 0$$

$$(c) \quad \operatorname{rot}(f \cdot u) = f \cdot \operatorname{rot} u + \operatorname{grad} f \times u$$

$$(d) \quad \operatorname{grad}(u \cdot v) = \frac{1}{2}[\operatorname{rot} u \times v + \operatorname{rot} v \times u + g' \circ (u \rfloor \mathcal{L}_v f + v \rfloor \mathcal{L}_u g)] .$$

7) (DIVERGENCE).

$$\operatorname{div} : C^\infty(T_0^1 M) \to C^\infty(T_0^0 M),$$

$$\operatorname{div} u \equiv *d*\underline{u} = \delta\underline{u} = \frac{1}{\sqrt{\det g_{ij}}}[\partial x_i(\sqrt{\det g_{hk}} u^i)] = (\partial x_i \cdot u^i) + \Gamma^i_{ij} u^j$$

with

$$\Gamma^i_{hk} \equiv \frac{1}{2} g^{ij}[(\partial x_h \cdot g_{jk}) + (\partial x_k \cdot g_{jh}) - (\partial x_j \cdot g_{hk})].$$

One has the following properties:

$$(a) \quad \operatorname{rot}(u \times v) = \frac{1}{2}(u \operatorname{div} v - v \operatorname{div} u) + [u, v]$$

$$(b) \quad \operatorname{div}(u + u') = \operatorname{div} u + \operatorname{div} u'$$

$$(c) \quad \operatorname{div}(u \times v) = v \cdot \operatorname{rot} u - u \cdot \operatorname{rot} v$$

$$(d) \quad \operatorname{div}(f \cdot u) = f \cdot \operatorname{div} u + \operatorname{grad} f \cdot u$$

$$(e) \quad \operatorname{div} \operatorname{grad} f = \Delta f = g^{ij}(\partial x_i \partial x_j \cdot f).$$

PROPOSITION 3.18 -
(PHYSICAL COMPONENTS OF GEOMETRIC OBJECTS).
The physical components of a vector field are the components of the vector fields with respect to unitary bases.

$$(x^\alpha) \Rightarrow \{\partial x_\alpha\}, \{dx^\alpha\} : \text{natural basis}$$

$$\{e_\alpha \equiv \frac{\partial x_\alpha}{|\partial x_\alpha|} = \frac{\partial x_\alpha}{\sqrt{g_{\alpha\alpha}}}\} \equiv \text{orthonormal basis}$$

$$\Downarrow$$

$$\{\theta^\alpha \equiv \frac{dx^\alpha}{|dx^\alpha|} = \frac{dx^\alpha}{\sqrt{g^{\alpha\alpha}}}\} \equiv \text{dual basis.}$$

Let $v = v^{\alpha}\partial x_{\alpha}$ *be a vector field locally represented in the basis* $\{\partial x_{\alpha}\}$; *we get:*

$$v = v^{\alpha}\partial x_{\alpha} = v^{\alpha}\sqrt{g_{\alpha\alpha}}\frac{\partial x_{\alpha}}{\sqrt{g_{\alpha\alpha}}} = v^{<\alpha>}e_{\alpha}$$
$$\underline{v} = v_{\alpha}dx^{\alpha} = v_{\alpha}\sqrt{g^{\alpha\alpha}}\frac{dx^{\alpha}}{\sqrt{g^{\alpha\alpha}}} = v_{<\alpha>}\theta^{\alpha}$$

with $v^{<\alpha>} \equiv v^{\alpha}\sqrt{g_{\alpha\alpha}}, v_{<\alpha>} \equiv v_{\alpha}\sqrt{g^{\alpha\alpha}}$. *In an orthogonal basis one has:*

$$g_{\alpha\alpha} = \frac{1}{g^{\alpha\alpha}}$$

then, we can also write:

$$v^{<\alpha>} \equiv \frac{v^{\alpha}}{\sqrt{g^{\alpha\alpha}}}, v_{<\alpha>} \equiv \frac{v_{\alpha}}{\sqrt{g_{\alpha\alpha}}}.$$

More generally, for tensor fields we get:

$$T_{<\alpha_1\cdots\alpha_k>}{}^{<\beta_1\cdots\beta_n>} = T_{\alpha_1\cdots\alpha_k}{}^{\beta_1\cdots\beta_n}\sqrt{g^{\alpha_1\alpha_1}\cdots g^{\alpha_k\alpha_k}}\sqrt{g_{\beta_1\beta_1}\cdots g_{\beta_n\beta_n}}.$$

In an orthogonal basis we get:

$$T_{<\alpha_1\cdots\alpha_k>}{}^{<\beta_1\cdots\beta_n>} = \frac{T_{\alpha_1\alpha_k}{}^{\beta_1\cdots\beta_n}}{\sqrt{g_{\alpha_1\alpha_1}\cdots g_{\alpha_k\alpha_k}}\sqrt{g^{\beta_1\beta_1}\cdots g^{\beta_n\beta_n}}}.$$

EXAMPLE 3.5 - 1) (**Physical components of derivatives**). Physical components of $A^j \equiv T^{1j}/j$ in cylindrical coordinates (ρ, θ, z).

$$\begin{aligned} A^j = T^{1j}/j &\equiv (\partial x_j \cdot T^{1j}) + \Gamma^1_{jk}\Gamma^{kj} + \Gamma^j_{jk}T^{1k} \\ &= (\partial\rho \cdot T^{\rho\rho}) + (\partial\theta \cdot T^{\rho\theta}) + (\partial z \cdot T^{\rho z}) - \rho T^{\theta\theta} + \frac{1}{\rho}T^{\rho\rho}. \end{aligned}$$

As

$$\begin{aligned} A^1 &= A^{\rho} = A^{<\rho>} \\ T^{\rho\rho} &= T^{<\rho\rho>} \\ T^{\rho z} &= T^{<\rho z>} \\ T^{\rho\theta} &= \frac{T^{<\rho\theta>}}{\rho} \\ T^{\theta\theta} &= \frac{T^{<\theta\theta>}}{\rho^2} \end{aligned}$$

we get

$$A^{<\rho>} = (\partial\rho \cdot T^{<\rho\rho>}) + \frac{1}{\rho}(\partial\theta \cdot T^{<\rho\theta>}) + (\partial z \cdot T^{<\rho z>}) + \frac{1}{\rho}(T^{<\rho\rho>} - T^{<\theta\theta>})$$

2)(Divergence: $\operatorname{div} v$ ($\dim M = 3$).
(Cartesian coordinates):$(\partial x \cdot v^{<x>}) + (\partial y \cdot v^{<y>}) + (\partial z \cdot v^{<z>})$.
(Cylindrical coordinates):$(\partial \rho \cdot v^{<\rho>}) + \frac{1}{\rho} v^{<\rho>} + \frac{1}{\rho}(\partial \theta \cdot v^{<\theta>}) + (\partial z \cdot v^{<z>})$.
(Spherical coordinates): $(\partial \rho \cdot v^{<\rho>}) + \frac{1}{\rho} v^{<\rho>} + \frac{1}{\rho}(\partial \theta \cdot v^{<\theta>}) + \frac{1}{\rho} v^{<\theta>} \cot \theta + \frac{1}{\rho \sin \theta}(\partial \phi \cdot v^{<\phi>})$.
3)(Rotor): $\omega = \operatorname{rot} v$ ($\dim M = 3$).
(Cartesian coordinates): $\omega^{<x>} = (\partial y \cdot v^{<z>}) - (\partial z \cdot v^{<y>})$, etc.
(Cylindrical coordinates):

$$\begin{aligned} \omega^{<\rho>} &= \frac{1}{\rho}(\partial \theta \cdot v^{<z>}) - (\partial z \cdot v^{<\theta>}) \\ \omega^{<\theta>} &= (\partial z \cdot v^{<\rho>}) - (\partial \rho \cdot v^{<z>}) \\ \omega^{<z>} &= (\partial \rho \cdot v^{<\theta>}) + \frac{1}{\rho} v^{<\theta>} - \frac{1}{\rho}(\partial \theta \cdot v^{<\rho>}) . \end{aligned}$$

(Spherical coordinates):

$$\begin{aligned} \omega^{<\rho>} &= \frac{1}{\rho}(\partial \theta \cdot v^{<\phi>}) + \frac{1}{\rho} v^{<\phi>} \cot \theta - \frac{1}{\rho \sin \theta}(\partial \phi \cdot v^{<\theta>}) \\ \omega^{<\theta>} &= \frac{1}{\rho \sin \theta}(\partial \phi \cdot v^{<\rho>}) - \frac{1}{\rho} v^{<\phi>} - (\partial \rho \cdot v^{<\phi>}) \\ \omega^{<\phi>} &= (\partial \rho \cdot v^{<\theta>}) + \frac{1}{\rho} v^{<\theta>} - \frac{1}{\rho}(\partial \theta \cdot v^{<\rho>}). \end{aligned}$$

THEOREM 3.19 - 1) *If* $\dim M = 3$ *the condition that a vector field* v *be* **locally conservative**, *i.e.,* $d\underline{v} = 0$, *is equivalent to the following one:*

$$\operatorname{rot} v = 0.$$

PROOF. In fact,

$$\operatorname{div} \operatorname{rot} v = \star d \star \star d\underline{v} = \star d^2 \underline{v}(-1)^{n-1} = 0.$$

On the other hand, $\overline{\star ddf} = \overline{\star d^2 f} = 0$. □

2) *If* $\dim M = 3$ *the condition that a vector field* v *be* **locally solenoidal**, *i.e., there exists a local vector field* u *such that* $v = \operatorname{rot} u$, *is equivalent to the following one:*

$$\operatorname{div} v = 0.$$

We call u *the* **vector potential** *of* v.

PROOF. If $v = \operatorname{rot} u \Rightarrow \operatorname{div} v = \operatorname{div} \operatorname{rot} u = 0$. □

3) *If* v *is a solenoidal vector field, its vector potential* u *is determined up to a scalar field* f.

$$u' = u + \operatorname{grad} f.$$

4)(*Stokes*).

$$\int_D d\omega = \int_{\partial D} \omega.$$

5) *If ω is a closed p-form ($d\omega = 0$), then for any compact domain B, of dimension $(p+1)$, such that $\partial B = D \dot{\bigcup} D'$ we get:*

$$\int_D \omega = \int_{D'} \omega.$$

6) *If $\underline{F} : M \to T^*M$ is an exact differential 1-form, $\underline{F} = df$, then we get:*

$$\int_\gamma \underline{F} = \int_{\gamma'} \underline{F} = f(a) - f(b)$$

where γ and γ' are two curves in M with the same ends.

3.4 - FUNDAMENTAL THEOREMS OF DYNAMICS

THEOREM 3.20 - (MOMENTUM THEOREM). *The variation of the free part of momentum of the observed motion of one body, in a time interval $\Delta t \equiv [0, t]$, is equal to the corresponding impulse:*

$$I[0,t] \equiv \int_0^t F dt \equiv (\int_0^t F^j dt) e_j,$$

where $\{e_j\}_{1 \leq j \leq 3}$ is a fixed basis of $\mathbf{S}$:

$$\bar{p}_\psi(t) - \bar{p}_\psi(0) = I[0,t].$$

PROOF. One has:

$$\bar{p}_\psi = \mu \ddot{m}_\psi = \overline{f}_\psi \Rightarrow \bar{p}_\psi = \dot{p}^j e_j = F^j e_j \Rightarrow \int_{[0,t]} \dot{p}^j dt = \int_{[0,t]} F^j dt.$$

□

THEOREM 3.21 - *The polar moment of the momentum of one body, with respect to the same pole, is zero.*
PROOF. One has:

$$\Omega_c \equiv \overrightarrow{CP} \times p_\psi.$$

By time-derivation, we get:

$$\dot{\Omega}_c = \dot{\overrightarrow{CP}} \times p_\psi + \overrightarrow{CP} \times \dot{p_\psi} = v_p \times p_\psi + \overrightarrow{CP} \times \dot{p_\psi}\,.$$

□

THEOREM 3.22 - (TOTAL MOMENT AND ITS PROPERTIES).
1) *Let*

$$M(A) \equiv \sum_{i=1}^{k} \overrightarrow{AP_i} \times w_i$$

be the moment of N *applied vectors* (P_i, w_i). *If* B *is another pole we get the following relation:*

$$M(B) \equiv M(A) + W \times \overrightarrow{AB},$$

where

$$W \equiv \sum_{i=1}^{N} w_i.$$

The total moment is independent of the pole iff: $W = \sum_i w_i = 0$.
2) *If* r *is an oriented line, and* $A, B \in r$, *from the above theorem we get:*

$$M(B) \cdot u = M(A) \cdot u \equiv M_r,$$

where u *is the unit vector associated to* r. *To this common value of the moment along* r *is given the name of* **axial moment along** r:

$$M_r \equiv M(B) \cdot u, \forall B \in r.$$

Then, one has:

$$M_r \equiv \sum_{i=1}^{N} (M_r)_i, (M_r)_i \equiv \overrightarrow{AP_i} \times w_i \cdot u, \forall A \in r.$$

3) *The axial moment of an applied vector* (P, w) *with respect to an oriented line, is* 0 *iff* $w \| \sigma$, *where* σ *is the plane that contains* r *and passes through* P.
4) *One has the following formula for the axial moment:*

$$M_r = \pm b |w^\pi|, \tag{3.30}$$

where $b \equiv$ **arm** *of the axial moment:*

$$w = w^\pi + w^r.$$

(The sign + or − according to the orientation around r *as identified by the vector* w^π.*) In fact, we have:*

$$M_r = \overrightarrow{AP} \times w \cdot u = \overrightarrow{AP} \times (w^\pi + w^r) \cdot u = \overrightarrow{AP} \times w^\pi \cdot u.$$

Taking into account that $\overrightarrow{AP} = \overrightarrow{AH} + \overrightarrow{HP}$ *we get:*

$$M_r = \overrightarrow{AH} \times w^\pi \cdot u = \pm b|w^\pi|.$$

Note that the polyhedron defined by the vectors $(\overrightarrow{AH}, w^\pi, u)$ *has volume given by* M_r.

5) *From* (3.30) *we get*

$$M = \pm b|w|u.$$

6) *The* **total moment of a couple**, *i.e., two applied vectors* $(p_i, w_i), i = 1, 2$, *such that*

$$|w_1| = |w_2| = |w|, \quad w_1 \| w_2, \quad w_1 = -w_2,$$

is constant with respect to a pole.

$$M = M(A) = M(B) \equiv \textbf{moment of the couple}.$$

7) *The scalar product* $M(A) \cdot W$ *is independent of the particular chosen pole. Furthermore, if one has:*

$$W = \sum_{i=1}^{N} w_i \neq 0,$$

there exists a line $c \| W$ (**central axis**) *such that* $M(C) \| W$ *if* $C \in c$. *Moreover,* $M(C)$ *does not depend on the particular point* $c \in C$, *and one has:*

$$|M(C)| = \inf_A |M(A)|.$$

PROOF. In fact, by the condition

$$M(C) = \lambda F$$

one has

$$\epsilon^{j_1 j_2 k}(\sum_{i=1}^{N}(x_i - y)_{j_1}(F_i)_{j_2}) = \lambda F^k, [(y) = \text{coordinates of} \quad C]$$

$$\Downarrow$$

$$\epsilon^{j_1 j_2 k}[\sum_{i=1}^{N}(x_i)_{j_1}(F_i)_{j_2} - y_{j_1} F_{j_2}] = \lambda F^k$$

$$\Downarrow$$

$$\epsilon^{j_1 j_2 k} y_{j_1} F_{j_2} = a^k,$$

where

$$a^k \equiv \lambda F^k + \sum_{i=1}^{N} \epsilon^{j_1 j_2 k}(x_i)_{j_2}(F_i)_{j_2}.$$

This is a system of 3 linear equations in the variables (y_1, y_2, y_3) that represent the coordinates of C. □

REMARK 3.22 - All the above theorems can be applied, in particular, to a set of applied forces F_i applied to N bodies placed in the points p_i.

THEOREM 3.23 - *Let* $F : J\mathcal{D}^2(M) \to vTM$ *be a force. For any motion* m *we identify, with respect to a frame* ψ, *a differential* 1*-form on* T*:*

$$\mathcal{L} : T \to T^*T.$$

PROOF. In fact, we get the following commutative diagram:

$$\begin{array}{ccccc} & J\mathcal{D}^2(M) & \xrightarrow{F} & vTM \overset{'g}{\cong} vT^*M & \\ D^2m & \uparrow & & \downarrow & T^*(m_\psi) \\ & T & \xrightarrow[\mathcal{L}\equiv D^2m^*\underline{F}]{} & T^*T & \end{array}$$

where

$$\underline{F} \equiv {}'g \circ F : J\mathcal{D}^2(M) \to vT^*M.$$

□

DEFINITION 3.24 - 1) *We call* **work of** F **in the motion** $m : T \to M$, *the differential* 1*-form*

$$\mathcal{L} \equiv (D^2m)^*\underline{F}$$

on T, *(with respect to* ψ*).*

2) *We call* **work of the force** F **in the motion** $m : T \to M$, **in the time-interval** $\Delta t \equiv [0, 1]$, *the following integral:*

$$\boxed{\mathcal{L}[0,t] \equiv \int_0^t \mathcal{L}} \tag{3.31}$$

PROPOSITION 3.19 - *In adapted coordinates we get:*

$$\mathcal{L} = (F^j g_{jk}) \cdot m\,\dot{m}^k dt = \underline{F} \cdot v dt,$$

where

$$F \circ D^2m = F^j \cdot m\,\partial x_j \cdot m.$$

So we can write:

$$\mathcal{L}[0,t] = \int_0^t (F^j g_{jk}) \cdot m\, \dot{m}^k dt.$$

By using the formula $\dot{m}_\psi = \dot{s}\mathbf{T} \cdot s$, referred to a frame ψ, and taking into account that in coordinates adapted to the frame one has $\dot{m}_\psi = \dot{m}^k \partial x_k \cdot m$, we have:

$$(3.32) \qquad \boxed{\mathcal{L} = (g_{jk} F_\psi{}^j T^k) \cdot m\dot{s} dt =< \underline{F}_\psi, \mathbf{T} > \dot{s}\, dt \equiv F_{\psi T} \dot{s}\, dt}$$

where $F_{\psi T}$ is the component of F_ψ along $\mathbf{T}$.

REMARK 3.23 - In the above formulas, for abuse of notation, we have noted g_{ij} the metric on S_ψ induced by the vertical metric on M. More precisely, we have the following commutative diagram:

$$\begin{array}{ccccccc} T & \leftarrow & M & \leftarrow & vTM & \cong & M \times \mathbf{S} \\ \| & & \| & & \| & & \| \\ T & \leftarrow & T \times S_\psi & \leftarrow & vT(T \times S_\psi) & \cong & T \times TS_\psi \end{array}$$

that induces the following commutative one:

$$\begin{array}{ccc} vTM \times_M vTM & \xrightarrow{g} & M \times \mathbf{R} \\ \|_j & & \| \quad {\scriptstyle (\tau \times \psi / \times id_{\mathbf{R}}) \equiv \phi} \\ T \times TS_\psi \times_{S_\psi} TS_\psi & \rightarrow & T \times S_\psi \times \mathbf{R} \end{array}$$

that defines $g_\psi \equiv \phi \circ g \circ j^{-1}$. So,

$$g_{\psi,t} \equiv (\psi / \times id_{\mathbf{R}}) \circ g \circ j_t^{-1} : TS_\psi \times TS_\psi \rightarrow \mathbf{R}, \forall t \in T,$$

is a metric on S_ψ. If the frame is rigid, we can then define g_ψ independent of the time.

REMARK 3.24 - With respect to an additional frame, we can split the absolute velocity $\dot{m}$ (or $\dot{m}_\psi$) into relative $\dot{m}_r$ and draging one $\dot{m}_\tau$ respectively. So, we can talk about **relative work**, $\mathcal{L}_r$, and **draging work**, $\mathcal{L}_\tau$ respectively:

$$\mathcal{L}_a = \mathcal{L}_r + \mathcal{L}_\tau,$$

with

$$\mathcal{L}_r = (F^j g_{jk}) \cdot m\, \dot{m}_r^k dt$$
$$\mathcal{L}_\tau = (F^j g_{jk}) \cdot m\, \dot{m}_\tau^k dt\,.$$

DEFINITION 3.25 - 1) *We call* **power of a force**, *with respect to a frame ψ, the following mapping:*

$$w : J\mathcal{D}^2(M) \rightarrow M \times \mathbf{R},$$

defined by means of the following commutative diagram:

$$\begin{array}{lccc} & J\mathcal{D}^2(M) & \overset{w}{\rightarrow} & M \times \mathbf{R} \\ {\scriptstyle (F,\pi_{2,1})} & \downarrow & & \\ & vTM \times_M J\mathcal{D}(M) & & \| \\ {\scriptstyle 'g\times\psi/} & \downarrow & & \\ & vT^*M \times_M vTM & \underset{<,>}{\rightarrow} & M \times \mathbf{R}. \end{array}$$

In other words we get:

$$w(D^2 m) =< \underline{F}_1 \dot{m}_\psi >\equiv F \cdot \dot{m}_\psi.$$

We have the following definitions:

$$w > 0: \quad \textbf{motive force}$$

$$w < 0: \quad \textbf{resistive force}$$

A **dissipative force** *has $w < 0$, for any admissible motion.*
EXAMPLE 3.6 - 1) The weight force has a character of motive force or resistive one according to the direction of its velocity.
2) The Coriolis forces have zero power with respect to a moving frame:

$$F_s \circ \dot{m}_{\wedge\psi} = (-2\mu\omega_r \times v_r) \cdot v_r = 0.$$

3) The constraint reaction has zero power with respect to a frame ψ identified by the constraint. In fact, in this case the velocity is tangent to the constraint. Furthermore, the constraint reactions are dissipative ones, with respect to ψ.
THEOREM 3.24 - *The work is an exact differential form, i.e., there exists a zero form*

$$\mathcal{T}_\psi \cdot m : T \rightarrow \mathbf{R},$$

such that

$$\mathcal{L} = d(\mathcal{T}_\psi \cdot m).$$

More precisely, $\mathcal{T}_\psi \cdot m$ can be factorized in the following form:

$$\begin{array}{lcccl} & J\mathcal{D}(M) & \overset{\mathcal{T}_\psi}{\rightarrow} & \mathbf{R} & \\ {\scriptstyle D^2 m} & \uparrow & & \uparrow & {\scriptstyle \mathcal{T}_\psi \cdot m} \\ & T & = & TT & \end{array}$$

with respect to any motion $m : T \rightarrow M$.
PROOF. Define $\mathcal{T}_\psi$ as follows:

$$\mathcal{T}_\psi(\dot{m}(t)) = \frac{1}{2}\mu \dot{m}_{\wedge\psi}(t) \cdot \dot{m}_{\wedge\psi}(t) \in \mathbf{R}.$$

Let us, now, calculate the differential $\mathcal{T}_\psi \cdot m$:

$$d(\tau_\psi \cdot m) = d(\frac{1}{2}\mu\dot{m}_{\wedge\psi} \cdot \dot{m}_{\wedge\psi}) = d(\frac{1}{2}\mu g_{ij} \cdot m\,\dot{m}^i\dot{m}^j)$$
$$= \frac{1}{2}\mu[(\partial x_k \cdot g_{ij})\dot{m}^k\dot{m}^j\dot{m}^i + 2g_{ij}\ddot{m}^i\dot{m}^j]dt.$$

Now, taking into account the following:

$$(\partial x_k \cdot g_{ij}) = -g_{sj}\widehat{\Gamma}^s_{ki} - g_{is}\widehat{\Gamma}^s_{kj},$$

we get

$$\begin{aligned} d(\mathcal{T}_\psi \cdot m) &= \frac{1}{2}\mu[(-g_{sj}\widehat{\Gamma}^s_{ki} - g_{is}\widehat{\Gamma}^s_{kj})\dot{m}^k\dot{m}^i\dot{m}^j + 2g_{ij}\ddot{m}^i\dot{m}^j]dt \\ &= \frac{1}{2}\mu[-g_{sj}\widehat{\Gamma}^s_{ki}\dot{m}^k\dot{m}^i\dot{m}^j - g_{js}\widehat{\Gamma}^s_{ki}\dot{m}^k\dot{m}^j\dot{m}^i + 2g_{ij}\ddot{m}^i\dot{m}^j]dt \\ &= \frac{1}{2}\mu[-2g_{sj}\widehat{\Gamma}^s_{ki}\dot{m}^k\dot{m}^i\dot{m}^j + 2g_{sj}\ddot{m}^s\dot{m}^j]dt \\ &= \mu\dot{m}^j[-g_{sj}\widehat{\Gamma}^s_{ki}\dot{m}^k\dot{m}^i + g_{sj}\ddot{m}^s]dt \\ &= \mu g_{sj}\dot{m}^j[-\widehat{\Gamma}^s_{ki}\dot{m}^k\dot{m}^i + \ddot{m}^s]dt \\ &= D^2m^*\underline{F} = \mathcal{L}. \end{aligned}$$

□

DEFINITION 3.26 - *We call* $\mathcal{T}_\psi$ **kinetic energy** *of one body, with respect to a frame* ψ.

THEOREM 3.25 - (WORK THEOREM). *We have the following:*

(3.33) $$\boxed{\mathcal{T}_\psi \cdot m(t) - \mathcal{T}_\psi \cdot m(0) = \mathcal{L}[0,t]}$$

REMARK 3.25 - If one body is on a constraint without friction, we get:

$$\mathcal{T}_\psi \cdot m(t) - \mathcal{T}_\psi \cdot m(0) = \mathcal{L}_{active}[0,t].$$

If there is friction, we get

$$\mathcal{T}_\psi \cdot m(t) - \mathcal{T}_\psi \cdot m(0) \leq \mathcal{L}_{active}[0,t].$$

THEOREM 3.26 - (WORK FOR CONSERVATIVE FORCES). *For a conservative force, one has:*

(3.34) $$\boxed{\mathcal{L} = -m^*df}$$

where f is the potential of the force. The work made by a conservative force in the time-interval $\Delta t \equiv [0,t]$ is given by

$$L[0,t] \equiv -(f(p_0) - f(p_1)),$$

where f is the potential, $p_1, p_0 \in S_\psi$, and does not depend on the particular motion considered, but depends only on the end-points p_1, p_0. In particular, if $p_1 = p_0$, we get

$$\mathcal{L}[0,t] = 0.$$

PROOF. In fact, for a conservative force we get:

$$\begin{aligned} F &= -\operatorname{grad} f \\ &\Downarrow \\ \mathcal{L} &= (F^j g_{jk}) \circ m\, \dot{m}^k dt \\ &= -(\partial x_k \cdot f) \circ m \frac{dx^k}{dt} dt \\ &= -m^* df\,. \end{aligned}$$

□

THEOREM 3.27 - (CONSERVATION OF ENERGY). *The total energy*

$$E \equiv \mathcal{T}_\psi + f$$

of one body, under conservative force, is constant.

REMARK 3.26 - If on one body a conservative force plus dissipative one are acted, then one has the following inequality:

$$E \leq E_0,$$

where E_0 is the energy at the initial time.

DEFINITION 3.27 - *A position P_e is called* **stable** *if for any couple of numbers $\epsilon_1, \epsilon_2 \in \mathbf{R}^+$, it is possible to determine two other numbers $\delta_1, \delta_2 \in \mathbf{R}^+$ such that $|\overrightarrow{P_0 P_e}| < \delta_1$, and $v_0 < \delta_2$, for the motion $\overrightarrow{OP}(t)$ corresponding to the initial conditions $\overrightarrow{OP_0}, v_0$, we get:*

$$|\overrightarrow{P_e P}(t)| < \epsilon_1, v(t) < \epsilon_2, \forall t \geq t_0.$$

PROPOSITION 3.20 - *A position of equilibrium P_e is stable if*

$$\lim_{p_0 \to p_e, v_0 \to 0} d_1(\overrightarrow{OP_0}, v_0) = 0, \qquad \lim_{p_0 \to p_e, v_0 \to 0} d_2(\overrightarrow{OP_0}, v_0) = 0,$$

where

$$d_1 \equiv d_1(\overrightarrow{OP_0}, v_0)$$
$$\equiv \sup\{|\overrightarrow{P_0P}(t)| : t \geq t_0\},$$
$$d_2 \equiv d_2(\overrightarrow{OP_0}, v_0)$$
$$\equiv \sup\{|\overrightarrow{P_0\dot{P}}(t)| : t \geq t_0\}.$$

THEOREM 3.28 - (Dirichlet theorem). *If the total force is conservative, any point of relative minimum, for the potential, is a point of stable equilibrium.*

THEOREM 3.29 - (Liapunov theorem). *If the total force is conservative, any critical point of the potential, such that it is not of relative minumum, is of unstable equilibrium.*

THEOREM 3.30 - 1) *The dynamic equation for N bodies are the following:*

$$\mu_i a_i = {}^aF_i(p,v,t) + {}^vF_i, \{i = 1, \cdots, N\}$$

where ${}^aF$ = active force, ${}^vF$ =constraint force.

2) *The static equations for N bodies are the following:*

$${}^vF_i = -{}^aF_i(p,a,t)\,.$$

Furthermore, one has:

$$\sum_{i=1}^{N} F_i = 0; \quad \sum_{i=1}^{N} M_i = 0.$$

DEFINITION 3.28 - *We call* **inertial force** *(acting on the i-th particle) the force*

$${}^mF_i = -\mu_i a_i.$$

We also write ${}^mM_i(A)$ to denote the corresponding moment with respect to A.

THEOREM 3.31 - *One has the following equations:*

$$(3.35)\qquad \boxed{\begin{aligned} {}^mF &\equiv \sum_{i=1}^{N} {}^mF_i = -\underline{\dot{P}} = -\mu\, a_G \\ {}^mM(A) &\equiv \sum_{i=1}^{N} {}^mM_i(A) = -\dot{\Omega}(A) - v_A \times \underline{P} \end{aligned}}$$

If the pole A has velocity $\| v_G$ we get:

$${}^mF = -\dot{P} = -\mu\, a_G; \quad {}^mM(A) = -\dot{\Omega}(A).$$

DEFINITION 3.29 - *We say the two systems of forces* $(P_i, F_i)_{1\leq i\leq N}, (P'_j, F'_j)_{1\leq j\leq N}$ *are* **equivalent** *if the following conditions are satisfied:*

$$F \equiv \sum_i F_i = \sum_j F'_j \equiv F'; \quad M(A) \equiv \sum_i M_i(A) = \sum_j M'_j(A), \forall A.$$

In particular, the system of forces is **reducible to zero** *if*

$$M(A) = F = 0.$$

PROPOSITION 3.21 - *1) A system of forces such that* $F \cdot M \neq 0$ *cannot be equivalent to a unique force or unique couple.*
2) The most general system of forces, such that $F \cdot M \neq 0$*, is equivalent to a force and a couple. More precisely, we get:*

$$(P_i, F_i) \Leftrightarrow ((C, F), M(C)),$$

where $C \in c \equiv$ *central axis of the system of forces, and* $F = \sum F_i$*. The couple is on a plane passing for* $C \perp c$ *and has moment* $M(C)$*.*
EXAMPLE 3.7 - 1) A system of concorde forces is equivalent to a unique force. In fact we have

$$F \cdot M = 0, \quad \text{and} \quad F \neq 0.$$

A system of concorde forces $(P_i, F_i)_{1\leq i\leq N}$ is equivalent to a unique force $F \equiv F_1 + \cdots + F_N$ applied in a point $C \in c \equiv$ central axis. For example, if $N = 2$, we get:

$$|\overrightarrow{CP_1}||F_1| = |\overrightarrow{CP_2}||F_2|.$$

Therefore, c is parallel both to F_1, F_2 and C is the point that satisfies the above equation. In particular a system of weight forces is equivalent to a unique force applied in the center of mass.
2) If the motion of a system is translatory with respect to an inertial frame, the system of draging forces

$$(P_i, -\mu_i a_\tau)_{1\leq i\leq N},$$

is equivalent to a unique force $(G, -\mu a_\tau)$ applied in the center of mass.
3) A **plane system of forces** is one $(P_i, F_i)_{1\leq i\leq N}$ such that each force is parallel to the unique plane σ that contains the points $(P_i)_{1\leq i\leq N}$. This system is reducible to unique force if $F \neq 0$, or to unique couple if $F = 0$. In fact, if we take as pole a point $A \in \sigma$ one has

$$M(A) \cdot F = 0.$$

THEOREM 3.32 - (CARDINAL EQUATIONS OF DYNAMICS)

1) (I form) *One has the following equations:*

$$(3.36)\qquad \boxed{\begin{aligned} &\sum_{i=1}^{N} \mu_i a_i = {}^{e}F \\ &\sum_{i=1}^{N} \overrightarrow{AP_i} \times \mu_i a_i = {}^{e}M(A) \end{aligned}}$$

2) (II form) *The above equations can also be written as follows:*

$$(3.37)\qquad \boxed{\begin{aligned} \mu\, a_G = \dot{Q} = {}^{e}F = {}^{a,e}F + {}^{r,e}F \\ \dot{\Omega}(A) + v_A \times \mu\, v_G = {}^{e}M(A) = {}^{a,e}M(A) + {}^{r,e}M(A) \end{aligned}}$$

If A is such that $v_A \| v_G$, or $v_A = 0$, *then we have:*

$$(3.38)\qquad \boxed{\begin{aligned} \mu\, a_G = \dot{Q} = {}^{e}F \\ \dot{\Omega}(A) = {}^{e}M(A) \end{aligned}}$$

PROOF. 1) It is a direct consequence of $\sum_{i=1}^{N} F_i = 0$.

2) It is sufficient to use the following equations for the inertial forces and their moments:

$${}^{m}F = -\dot{P} = -\mu\, a_G, \ {}^{m}M(A) = -\dot{\Omega}(A) - v_A \times Q.$$

□

REMARK 3.27 - In fact, the cardinal equations are not the dynamic equations for a system of N particles. These, instead, are described by the set of Newton equations of each particle under the action of all the forces, considering also that coming by the action of the other particles.

THEOREM 3.33 - (CONSERVATION THEOREMS).

1) *If for any* $t \in T$, *one has* ${}^{e}F \cdot u = 0$, *where* u *is a unitary vector joint with the space of the inertial frame, then we get the following conservation law:*

$$v_G \cdot u = constant \Rightarrow Q \cdot u = constant\,.$$

PROOF. From the first cardinal equation we get:

$$\mu\, a_G \cdot u = {}^{e}F \cdot u = 0 \Rightarrow \frac{d}{dt}(\mu v_G \cdot u) = 0.$$

□

2) *As a consequence of the above theorem we get the following.*
(i) *If* ${}^eF \perp \sigma$, *where* σ *is a plane joint with the inertial frame, then we get the following conservation law:*

$$v_G \cdot u_i = const., i = 1, 2 \Rightarrow \underline{P} \cdot u_i = const.,$$

where u_i *are two unitary linearly independent unitary vectors belonging to* σ.
(ii) *If* ${}^eF = 0$ *one has:*

$$v_G \cdot u = constant, i = 1, 2, 3 \Rightarrow Q \cdot u_i = constant, i = 1, 2, 3$$

where $u_i, i = 1, 2, 3$, *are three linearly independent unitary vectors. In this case the motion of* G *is rectilinear and uniform.*

PROOF. Let u be a unitary vector, and ${}^eM_b \equiv {}^eM(T) \cdot u \equiv$ axial moment with respect to a line passing for T and concord with u. If ${}^eM_b = 0$ for any t, we get: $\Omega_b =$ constant, where $\Omega_b \equiv \Omega(T) \cdot u$.
Furthermore, from the second cardinal equation we get: $\dot{\Omega}(T) \cdot u = {}^eM(T) \cdot u = {}^eM_b = 0$. □

EXAMPLE 3.8 - (**Motion of body on horizontal plane without friction**). As ${}^eF \cdot u_i = 0, i = 1, 2$, where u_i are two unitary vectors belonging to the plane σ, we get:

$$v_G \cdot u_i = \text{constant} .$$

If the initial condition is with v_G in the vertical direction, the possible motions of G are on the vertical line passing through the initial position. Therefore, any internal force cannot produce motions of G in horizontal directions. Furthermore, if the initial direction of v_G is not vertical, the motion of the projection of G on σ is rectilinear and uniform, and along the vertical axis passing through G we get:

$${}^eM(G) = 0 \Rightarrow \Omega(G) = \text{constant}$$

DEFINITION 3.30 - *The* **work of a system of forces** (P_i, F_i) *is given by the following differential* 1*-form on* T:

$$\mathcal{L} \equiv \sum_{i=1}^{N} \mathcal{L}_i,$$

where

$$\mathcal{L}_i \equiv (D^2 m_i)^* \underline{F}_i = F_i \cdot v_i dt.$$

The corresponding work in the time-interval $\Delta t \equiv [0, t]$, *is the following integral:*

$$\int_0^t \mathcal{L} = \sum_{i=1}^N \int_0^t \mathcal{L}_i.$$

REMARK 3.28 - If the system of forces is reducible to zero, the work, in general, is not zero. The usual machines are founded on this property.

THEOREM 3.34 - (PROPERTIES OF THE WORK FOR A SYSTEM OF FORCES).

1) *The work of a rigid motion is given by the following:*

$$\mathcal{L} = (F \cdot v_\Omega + M_\Omega \cdot \omega)dt.$$

2) *The work of the weight forces is the following:*

$$\mathcal{L} = \mu\, g\, v_G dt.$$

DEFINITION 3.31 - *The* **power of a system of forces** *is defined by the following:*

$$w = \sum_{i=1}^N F_i \cdot v_i.$$

DEFINITION 3.32 - *A system of forces is called* **conservative** *if the following conditions are verified:*

$$F_i = -\mathrm{grad}_i f \quad , \quad i = 1, \cdots, N.$$

Here the scalar function f *is called the* **potential of the system**.

THEOREM 3.35 - *In order for a system of forces to be conservative it is enough that each force is conservative.*

EXAMPLE 3.9 - 1) A system of constant forces is conservative. This is, e.g., the case of weight forces. Then considering the z-coordinate along the direction of g we get the corresponding potential:

$$f = \sum_{i=1}^N \mu_i g z_i + c = \mu g z_G + c.$$

2) A system of central forces is conservative:

$$F_{ij} = \phi_{ij}(\rho_{ij})\mathrm{vers}\overrightarrow{P_iP_j}, i \neq j \Rightarrow f = -\sum_{i=1}^N \int_{\rho_{ij}(0)}^{\rho_{ij}(t)} \phi_{ij}(\rho'_{ij})d\rho'_{ij} + c,$$

where the sum is extended to each combination of indexes i and j ($i \neq j$) in such a way that each couple of internal forces is counted once. In particular for the Newtonian (or Coulombian) forces we get:

$$f = k_\Omega \sum_{i=1}^{N} \frac{\mu_i \mu_j}{\rho_{ij}} + c\,.$$

3) For an elastic force we get:

$$f = -\frac{1}{2} \sum_{i=1}^{N} c_{ij} \rho_{ij}^2 + c\,.$$

4) The draging force of a frame in uniform rotatory motion around an axis (z-axis) of an inertial frame, has the following potential:

$$f = \frac{1}{2}\omega_\tau^2 \sum_{i=1}^{N} \mu_i \delta_i^2 + c = \frac{1}{2}\omega_\tau^2 \sum_{i=1}^{N} \mu_i(\xi_i^2 + \eta_i^2) + c,$$

where $\delta_i =$ distance of i-th particle from z-axis.

DEFINITION 3.33 - *The* **virtual velocities** *(resp.* **virtual displacements***) of one body E in the position P at the instant t, are all the possible velocities (resp. displacements) that E can have taking account of the constraints, considered independent on the time.*

PROPOSITION 3.22 - *If N is a two-side constraint, then the virtual velocities (resp. virtual displacements) span the tangent space $T_P N$. Furthermore, if N is a one-side constraint, then for the virtual velocities (resp. virtual displacements) we get:*

$$v \in T_P S_\psi \Leftrightarrow v \cdot n \geq 0,$$

where n is the unitary vector normal to the constraint.

PROPOSITION 3.23 - *If the body is rigid all the virtual velocities (resp. virtual displacements) are also rigid. We get:*

$$(i) \quad \text{virtual velocity} : v_p = v_\Omega + \omega \times \overrightarrow{\Omega P},$$

$$(ii) \quad \text{virtual displacement} : \delta\overrightarrow{OP} = \delta\overrightarrow{O\Omega} + \delta r \times \overrightarrow{\Omega P},$$

where v_Ω and ω have virtual characters, and $\delta\overrightarrow{O\Omega} = v_\Omega \delta t \equiv$ virtual displacement of Ω, $\delta r \equiv \omega \delta t \equiv$ virtual rotation vector.

DEFINITION 3.34 - *We define* **virtual work** *of a force F the differential 1-form on T:*

$${}_v\mathcal{L} \equiv (D^2 m)^* \underline{F} = F \cdot v dt,$$

where the motion m is such that its velocity v is virtual, and vdt is the corresponding virtual displacement.
2) Similarly, define the **virtual work** for a system of forces $\{F_i\}_{i\in 1,..,N}$, the following differential 1-form on T:

$$_v\mathcal{L} \equiv \sum_{i=1}^{N}(D^2m_i)^*\underline{F}_i = (\sum_{i=1}^{N} F_i \cdot v_i)dt.$$

PROPOSITION 3.24 - 1) *The relation*

$$(\spadesuit) \qquad {}^v_v\mathcal{L} = 0 \quad (resp. \quad {}^v_v\mathcal{L} \geq 0),$$

characterizes the relations that a two-side constraint (resp. one-side cosntraint) without friction can produce on one body (resp. on boundary positions).
2) *For a rigid body we get*

$$(3.39) \qquad _v\mathcal{L} = F \cdot v_\Omega dt + M(\Omega) \cdot \delta r,$$

where v_Ω is the virtual velocity of Ω and δr is the virtual rotation vector.
DEFINITION 3.35 - **Perfect constraints** *are those which satisfy conditions in* $(\spadesuit)$.
THEOREM 3.36 - (THEOREM OF THE WORK). *One has the following:*

$$dT = \mathcal{L}^i + \mathcal{L}^e$$
$$dT = \mathcal{L}^a + \mathcal{L}^r,$$

where

$\mathcal{L}^i \equiv$ work of internal forces, $\quad \mathcal{L}^e \equiv$ work of effective forces,

$\mathcal{L}^a \equiv$ work of active forces, $\quad \mathcal{L}^r \equiv$ work of reaction forces.

PROPOSITION 3.25 - 1) *If the system is perfectly holonomic and the constraints are time-independent, we get:*

$$dT = \mathcal{L}^a, \qquad \mathcal{L}^v = 0.$$

Furthermore, if the active forces are conservative ones, from the above equation we get the theorem of conservation of energy.
2) (**Principle of virtual works**). *For a system with perfect constraints, a configuration C is of equilibrium if the following conditions are verified:*
(i) *The configuration C is compatible with the constraint at any instant;*
(ii) *For any virtual displacement, relative to C, and at any instant t, the virtual work $_v\mathcal{L}^a(0)$ of the system of active forces, corresponding to zero velocity, satisfies the following conditions:*

$$_v\mathcal{L}^a = 0 \quad \text{if } C \text{ is an ordinary configuration}$$
$$_v\mathcal{L}^a \leq 0 \quad \text{if } C \text{ is a boundary configuration.}$$

3) (**Torricelli's principle**). *For a system with perfect constraints, if the system of active forces is only the weight forces, the configurations of equilibrium are that where the center of mass does not lower for any virtual displacement.*

THEOREM 3.37 - 1) (KÖENIG THEOREM). *The kinetic energy for N particles can be written as follows:*

$$\boxed{\mathcal{T} = \frac{1}{2}\mu\, v_G^2 + \mathcal{T}^G}$$

where

$$\mathcal{T}^G \equiv \frac{1}{2}\sum_{i=1}^{N}\mu_i v_{i,G}^2,$$

is the kinetic energy of the motion relatively to the center of mass.

2) (GENERALIZED KÖENIG THEOREM). *The kinetic energy for a system of N particles can be written in the following way:*

$$\boxed{\mathcal{T} = \frac{1}{2}\mu_\Omega^2 + \mathcal{T}^\Omega + v_\Omega \cdot \mu v_{G,r}}$$

where

$$\mathcal{T}^\Omega \equiv \frac{1}{2}\sum_{i=1}^{N}\mu_i v_{i,r}^2,$$

is the kinetic energy of the motion relatively to the point Ω.

3) *In particular, if $\Omega \equiv G, v_{G,r} = 0$ we recover the Köenig theorem.*

THEOREM 3.38 - *The condition $\mathcal{T} = 0$ is equivalent to say that the velocity of each element of the system is zero.*

PROOF. In fact, each term in $\mathcal{T}$ is non-negative. □

THEOREM 3.39 - (THEOREM OF THE WORK).

$$\mathcal{T} - \mathcal{T}_0 = {}^i\mathcal{L}[0,t] + {}^e\mathcal{L}[0,t].$$

Furthermore, if the system of forces is conservative one, we get:

$$\mathcal{T} - \mathcal{T}_0 = -(f - f_0),$$

and the total energy, $E \equiv T + f$, is constant: $E =$ constant. Finally, if the dissipative forces are present, we have the following inequality:

$$E \leq E_0.$$

THEOREM 3.40 - (CARDINAL EQUATION IN STATICS). *In static conditions one has the following equations:*

$$\boxed{\begin{aligned} {}^{e}F &= {}^{a,e}F + {}^{r,e}F = 0 \\ {}^{e}M(A) &= {}^{a,e}M(A) + {}^{r,e}M(A) = 0 \end{aligned}} \tag{3.40}$$

3.5 - LAGRANGIAN MECHANICS FOR PERFECT HOLONOMIC SYSTEMS

Let a system of N bodies be given. Let $\phi^k = 0, 1 \leq k \leq s$, be the constraint equations, where ϕ^k are numerical functions:

$$\phi^k : S_\psi \times \cdots_N \cdots \times S_\psi \times T \to \mathbf{R}.$$

This is a **rheonomic system**, i.e., the constraints are time-dependent. Let

$$\{F_i\}_{1 \leq i \leq N}$$

be a system of forces. One has the following.

PROPOSITION 3.26 - 1) *The configuration space $V_t, \forall t \in T$, is a submanifold of $S_\psi \times \cdots \times S_\psi$ of dimension $n = 3N - s$, if the rank of the jacobian $\mathcal{J}^k_\alpha \equiv (\partial x_\alpha \cdot \phi^k)$ is s. Denote with $q^1, \cdots, q^n$ the corresponding coordinates on V and we will call them* **Lagrangian coordinates** *of the system. The generic point P_i is identified by a vector:*

$$\overrightarrow{OP_i} = \overrightarrow{OP_i}(q^1, \cdots, q^n, t), 1 \leq i \leq N.$$

2) The velocity of P_i can be expressed by means of the Lagrangian coordinates in the following way:

$$v_i = (\partial q_k \overrightarrow{OP_i})\dot{q}^k + (\partial t \cdot \overrightarrow{OP_i}),$$

where $\dot{q}^k \equiv$ **Lagrangian components** *of the velocity of i-th particle. If the constraint is time-independent (***sclerhonomic system***) we get:*

$$v_i = (\partial q_k \cdot \overrightarrow{OP_i})\dot{q}^k.$$

3) *The virtual velocity is given by the following:*

$$v_i = (\partial q_k \cdot \overrightarrow{OP_i})\dot{q}^k.$$

4) *The kinetic energy, expressed in Lagrangian coordinates, is given by the following:*

$$\mathcal{T} = \frac{1}{2}G_{hk}\dot{q}^k\dot{q}^h + \frac{1}{2}H_k\dot{q}^k + \frac{1}{2}b,$$

where

$$G_{hk} \equiv \sum_{i=1}^{N} m_i(\partial q_h \cdot \overrightarrow{OP_i}) \cdot (\partial q_k \cdot \overrightarrow{OP_i})$$

$$H_k \equiv \sum_{i=1}^{N} 2m_i(\partial t \cdot \overrightarrow{OP_i}) \cdot (\partial q_k \cdot \overrightarrow{OP_i})$$

$$b \equiv \sum_{i=1}^{N} m_i(\partial t \cdot \overrightarrow{OP_i}) \cdot (\partial t \cdot \overrightarrow{OP_i})$$

If the constraint is time-independent $\mathcal{T}$ is a quadratic form:

$$\mathcal{T} = \frac{1}{2}G_{hk}\dot{q}^k\dot{q}^h.$$

5) (**Lagrange equations**). *The dynamic equations can be expressed in the following form:*

$$\sum_{i=1}^{N} m_i a_i \cdot (\partial q_k \cdot \overrightarrow{OP_i}) = {}^a Q_k,$$

where

$${}^aQ_k \equiv \sum_{i=1}^{N} F_i \cdot (\partial q_k \cdot \overrightarrow{OP_i})$$

are the **Lagrangian components of the system of forces**
6) *The Lagrange equations can be expressed in the following alternative form:*

$$\boxed{\frac{d}{dt}(\partial \dot{q}_k \cdot \mathcal{T}) - (\partial q_k \cdot \mathcal{T}) = {}^aQ_k} \tag{3.41}$$

7) *For conservative systems the above equations can also be written in the following form:*

$$\boxed{\frac{d}{dt}(\partial \dot{q}_k \cdot L) - (\partial q_k \cdot L) = 0} \tag{3.42}$$

where $L \equiv \mathcal{T} - f \equiv$ **Lagrange function**.

8) *For systems that are only partially conservative, Lagrange equations can be expressed as follows:*

$$\frac{d}{dt}(\partial \dot{q}_k \cdot L) - (\partial q_k \cdot L) = {}^aQ'_k,$$

where ${}^aQ'_k$ is the Lagrangian component for the system of non-conservative forces, and $L \equiv \mathcal{T} - f$ is the Lagrange function for conservative forces.

PROOF. The proof is a direct consequence of the Newton equations and the concepts of virtual displacements. In fact, from the dynamic equations:

$$m_i a_i - F_i = 0, 1 \leq i \leq N,$$

after scalar multiplication for any virtual velocity, and summing over all $i \in \{1, \cdots, N\}$, we get:

$$\sum_{i=1}^{N}(m_i a_i - F_i) \cdot v_i = 0 \Rightarrow \sum_{i=1}^{N}(m_i a_i - F_i) \cdot (\partial q_k \cdot \overrightarrow{OP_i})\dot{q}^k = 0.$$

As this equation must be verified for any $\dot{q}^k$, we must have:

$$\sum_{i=1}^{N}(m_i a_i - F_i) \cdot (\partial q_k \cdot \overrightarrow{OP_i})\dot{q}^k = 0 \Rightarrow \sum_{i=1}^{N} m_i a_i \cdot (\partial q_k \cdot \overrightarrow{OP_i}) \equiv {}^aQ_k,$$

with ${}^vF_i \cdot (\partial q_k \cdot \overrightarrow{OP_i}) = 0$. Furthermore,

$$(\partial q_k \cdot \mathcal{T}) = \sum_{i=1}^{N} m_i v_i \cdot (\partial q_k \cdot v_i), (\partial \dot{q}_i \cdot \mathcal{T}) = \sum_{i=1}^{N} m_i v_i \cdot (\partial \dot{q}_k \cdot v_i).$$

In addition, we have:

$$(\partial \dot{q}_k \cdot v_i) = (\partial q_k \cdot \overrightarrow{OP_i}) \Rightarrow (\partial \dot{q}_k \cdot \mathcal{T}) = \sum_{i=1}^{N} m_i v_i \cdot (\partial q_k \cdot \overrightarrow{OP_i})$$

$$\Downarrow$$

$$\frac{d}{dt}(\partial \dot{q}_k \cdot \mathcal{T}) = \sum_{i=1}^{N} m_i a_i \cdot (\partial q_k \cdot \overrightarrow{OP_i}) + \sum_{i=1}^{N} m_i v_i \cdot \frac{d}{dt}(\partial q_k \cdot \overrightarrow{OP_i})$$

and

$$\frac{d}{dt}(\partial q_k \cdot \overrightarrow{OP_i}) = (\partial q_k \cdot \frac{d}{dt}(\overrightarrow{OP_i})) = (\partial q_k \cdot v_i).$$

As a consequence, we get:

$$(\partial \dot{q}_k \cdot \mathcal{T}) = \sum_{i=1}^{N} m_i a_i \cdot (\partial q_k \cdot \overrightarrow{OP_i}) + \sum_{i=1}^{N} m_i v_i \cdot (\partial q_k \cdot v_i).$$

Hence,

$$\frac{d}{dt}(\partial\dot{q}_k \cdot \mathcal{T}) - (\partial q_k \cdot \mathcal{T}) = \sum_{i=1}^{N} m_i a_i (\partial q_k \cdot \overrightarrow{OP_i}).$$

This can be expressed as follows:

$$\frac{d}{dt}(\partial\dot{q}_k \cdot \mathcal{T}) - (\partial q_k \cdot \mathcal{T}) = {}^{a}Q_k.$$

Finally, if the force is conservative, we get:

$$Q_k = -(\partial q_k \cdot f).$$

□

THEOREM 3.41 - *An alternative way to write Lagrange equation is the following:*

$$G_{hs}\ddot{q}^h + [(\partial q_k \cdot G_{hs}) - \frac{1}{2}(\partial q_s \cdot G_{hk})]\dot{q}^h\dot{q}^k + (\partial t \cdot G_{hs})\dot{q}^h$$
$$+\frac{1}{2}\{(\partial t \cdot H_s) + [(\partial q_k \cdot H_s) - (\partial q_s \cdot H_k)]\dot{q}^k\} - \frac{1}{2}(\partial q_s \cdot b) = {}^{a}Q_s.$$

PROOF. It is a direct consequence of Lagrange equation, taking into account the expression for the kinetic energy. □

COROLLARY 3.3 - *For sclerhonomic system, with n-dimensional configuration manifold V (q^k), Lagrange equation can be expressed in the following pseudo-Newton form:*

$$\boxed{\ddot{q}^k + G^k_{ij}\dot{q}^i\dot{q}^j = Q^j}$$

where G^k_{ij} are the connection coefficients of the Levi-Civita connection on the Riemann manifold (V, G), where $G = G_{rs}dq^r \otimes dq^s$, with $\mathcal{T} = \frac{1}{2}G_{rs}\dot{q}^r\dot{q}^s$ the kinetic energy of the system. Furthermore, one has

$$Q^j = G^{hj}Q_h.$$

PROOF. It follows directly from the above theorem, taking into account the following relation:

$$\boxed{G^j_{ir} = [(\partial q_i . G_{rh}) - \frac{1}{2}(\partial q_h . G_{ri})]G^{hj}}$$

□

COROLLARY 3.4 - *Admissible motions for free sclerhonomic system are identifiable with geodesics of the Riemann manifold* (V, G) *as defined in the above corollary.*
THEOREM 3.42 - (THEOREM OF THE WORK). *For sclerhonomic systems we get:*

$$ {}^a\mathcal{L} = {}^aQ_k\dot{q}^k \tag{3.43}$$

and

$$ d\mathcal{T} = {}^a\mathcal{L}. \tag{3.44}$$

THEOREM 3.43 - (STATICS FOR PERFECT HOLONOMIC SYSTEMS). *A configuration* $(q^k(\circ))$ *is of equilibrium one iff the following equation is verified:*

(3.45)(**Lagrange equation in statics**) $\qquad {}^aQ_k(q^k(\circ), \dot{q}^k = 0) = 0.$

PROOF. From Lagrange equation, it follows that condition (3.45) is necessary as in an equilibrium configuration one has $\dot{q}^k = 0$, hence $\mathcal{T} = 0$. The sufficience follows from the existence and uniqueness theorem of solutions for ODE's. □
THEOREM 3.44 - (THE INTEGRAL OF ENERGY). *If* L *does not explicitly depend on the time, we get the following conservation law:*

$$H \equiv (\partial\dot{q}_k \cdot L)\dot{q}^k - L = const.$$

Furthermore, if the constraints are time-independent, we have

$$H = \mathcal{T} + f,$$

where H *represents the total energy of the system.*
PROOF. As

$$\frac{dL}{dt} = (\partial q_k \cdot L)\dot{q}^k + (\partial\dot{q}_k \cdot L)\ddot{q}^k + (\partial t \cdot L), \tag{3.46}$$

from the Lagrange equation we get:

$$[\frac{d}{dt}(\partial\dot{q}_k \cdot L) - (\partial q_k \cdot L)]\dot{q}^k = 0.$$

Hence,

$$\dot{q}_k\frac{d}{dt}(\partial\dot{q}_k \cdot L) + \ddot{q}^k(\partial\dot{q} \cdot L) - \frac{dL}{dt} + (\partial t \cdot L) = 0.$$

As a consequence, we have:

$$\frac{d}{dt}[(\partial\dot{q}_k \cdot L)\dot{q}^k - L] + (\partial t \cdot L) = 0.$$

Finally, if the constraints are time-independent, $\mathcal{T}$ is a quadratic form, and we get:

$$(\partial \dot{q}_k \cdot L)\dot{q}^k = (\partial \dot{q}_k \cdot \mathcal{T})\dot{q}^k = 2\mathcal{T}.$$

From (3.46) we have:

$$H = 2\mathcal{T} - (\mathcal{T} - f) = \mathcal{T} + f.$$

□

REMARK 3.29 - (CRITERION FOR STABILITY OF CONSERVATIVE PERFECT HOLONOMIC SYSTEM WITH TIME-INDEPENDENT CONSTRAINTS). The point $p \equiv (x_0^1, \cdots, x_0^n) \in M$, is an extremum (minimum or maximum) for the function $f(x^1, \cdots, x^n)$ iff the following conditions are satisfied:

$$Df(p) = D^2 f(p) = \cdots = D^{m-1} f(p) = 0, D^m f(p) \neq 0, m = even.$$

Furthermore, if the m-th form

$$H(q) \equiv H_{j_1 \cdots j_m} q^{j_1 \cdots j_m} \equiv (\partial x_{j_1} \cdots \partial x_{j_m} \cdot f)(p) q^{j_1} \cdots q^{j_m}$$

is positive (semi)definite (resp. negative (semi)definite) p is a point of (relative) minimum (resp. point of (relative) maximum) for f. Furthermore, if $H(q)$ is indefinite, then p_0 is a saddle point. In particular, if $m = 2$, setting

$$A \equiv [H_{ij}], A_{n-1} \equiv \begin{bmatrix} H_{11} & \cdots & H_{1n-1} \\ \cdots & \cdots & \cdots \\ H_{n-11} & \cdots & H_{n-1n-1} \end{bmatrix}, \cdots, A_1 \equiv H_{11},$$

we have a system of submatrices such that

$$A \supset A_{n-1} \supset \cdots \supset A_1.$$

Then, we have the following:

a) If

$$\det A > 0, \det A_{n-1} > 0, \cdots, \det A_1 > 0 \Rightarrow H \quad \text{is positive definite}$$

(namely p is a point of minimum);

b) If

$$\det A \begin{Bmatrix} > 0, & n = \text{even} \\ < 0, & n = \text{odd} \end{Bmatrix}, \det A_{n-1} \begin{Bmatrix} < 0, & n = \text{even} \\ > 0, & n = \text{odd} \end{Bmatrix} \Rightarrow H \text{ is negative definite}$$

(namely p is a point of maximum).

For example, if $(n = 2, m = 2)$ we have:

$$
\begin{aligned}
(H_{ij}) &= \begin{bmatrix} (\partial x \partial x.f) & (\partial x \partial y.f) \\ (\partial y \partial x.f) & (\partial y \partial y.f) \end{bmatrix} \\
&\Rightarrow \det(H_{ij}) \\
&= (\partial x \partial x.f)(\partial y \partial y.f) - (\partial x \partial y.f)^2 \\
&\equiv \det A\,.
\end{aligned}
$$

Therefore, one has the following:

$$
\begin{aligned}
&\det A > 0, \left\{ \begin{array}{l} \det A_1 \equiv (\partial x \partial x.f) > 0, \text{minimum;} \\ \\ \det A_1 < 0, \text{maximum} \end{array} \right\} \\
&\det A < 0, \text{saddle point;} \\
&\det A = 0, \text{it is not possible to decide}\,.
\end{aligned}
$$

REMARK 3.30 - In the following we resume classifications for the quadratic forms:

$$
(3.47)\ \textbf{(quadratic form)} \qquad \sum_{i,j=1}^{n} a_{ij}\lambda^i\lambda^j, \quad \lambda^i, \lambda^j \in \mathbf{R}.
$$

This will be useful in all applications concerning the stability of dynamic systems.

Classification of quadratic forms

Set

$$
A_n \supset A_{n-1} \supset A_{n-2} \supset \cdots \supset A_1
$$

the principal submatrices.

☐ Positive semidefinite:

$$
\begin{aligned}
&\sum a_{ij}\lambda^i\lambda^j \geq 0 \\
&\det A_n \geq, \det A_{n-1} \geq 0, \cdots, \det A_1 \geq 0
\end{aligned}
$$

☐ Negative semidefinite:

$$
\begin{aligned}
&\sum a_{ij}\lambda^i\lambda^j \leq 0 \\
&\det A_n \, \{\geq 0,\ n = \text{even}; \leq 0,\ n = \text{odd}\} \\
&\det A_{n-1} \, \{\leq 0,\ n = \text{even}; \geq 0,\ n = \text{odd}\} \\
&\ldots
\end{aligned}
$$

☐ Positive:

$$
\begin{aligned}
&\sum a_{ij}\lambda^i\lambda^j > 0 \\
&\det A \neq 0; \det A_n > 0, \cdots, \det A_1 > 0
\end{aligned}
$$

☐ Negative:

$$\begin{aligned}
&\sum a_{ij}\lambda^i\lambda^j < 0\\
&\det A \neq 0\\
&\det A_n\,\{> 0,\ n = \text{odd}; < 0,\ n = \text{odd}\}\\
&\det A_{n-1}\,\{< 0,\ n = \text{even}; > 0,\ n = \text{odd}\}\\
&\ldots
\end{aligned}$$

☐ Indefinite: None of above cases:

$$\begin{aligned}
&\text{Criterion 1} : \exists \det A_p < 0, (p = \text{even})\\
&\text{Criterion 2} : \exists \det A_{p_1} < 0, \det A_{p_2} > 0, p_1, p_2 = \text{odd}\\
&\text{Criterion 3} : \exists a_{ii} > 0, a_{jj} < 0, i \neq j\,.
\end{aligned}$$

☐ (QUADRATIC FORMS AND CONICS). If the quadratic form

$$a_{11}{\lambda_1}^2 + 2a_{12}\lambda_1\lambda_2 + a_{22}{\lambda_2}^2, \tag{3.48}$$

is **positive** (resp. **negative**) then the equation:

$$a_{11}x^2 + 2a_{12}xy + a_{22}y^2 = k > 0 \quad (\text{resp.}\, k < 0) \tag{3.49}$$

represents in the plane (xy) an ellipse with centre at O. If (3.48) is indefinite, and $\det a_{ik} \neq 0$, then (3.49) represent a hyperbola ($\forall x \neq 0$), with centre at O and with asymptots given by the following equations:

$$a_{11}x^2 + 2a_{12}xy + a_{22}y^2 = 0\,.$$

☐ (QUADRATIC FORMS AND QUADRICS). If the quadratic form

$$a_{11}{\lambda_1}^2 + a_{22}{\lambda_2}^2 + a_{33}{\lambda_3}^2 + 2a_{12}\lambda_1\lambda_2 + 2a_{13}\lambda_1\lambda_3 + 2a_{23}\lambda_2\lambda_3 \tag{3.50}$$

is positive (resp. negative) any equation of the type:

$$a_{11}x^2 + 2a_{12}xy + a_{22}y^2 + 2a_{23}yz + a_{33}z^2 + 2a_{13}xz = K > 0\,(\text{resp.}\, K < 0) \tag{3.51}$$

represents in cartesian coordinates (x, y, z) an ellipsoid with centre at the origin. Furthermore, if (3.50) is indefinite, and if $\det a_{ik} \neq 0$, then (3.51) represents a hyperboloid.

2 THEOREM 3.45 - (LINEARIZED EQUATIONS FOR CONSERVATIVE SYSTEMS).

1) The linearized Lagrange equation for conservative systems is the following:

$$(3.52) \qquad \sum_{m=1}^{n} A_{mk}(e)\ddot{\nu}^m + B_{mk}(e)\dot{\nu}^m + C_{mk}(e)\nu^m = 0, 1 \leq k \leq n,$$

where:

$$\begin{aligned}
A_{mk}(e) &= G_{mk}(e),\\
B_{mk}(e) &= \{(\partial q_m \cdot G_{hk})(e) - (\partial q_n \cdot G_{mk})(e) - (\partial q_k \cdot G_{mh})(e)\dot{q}^h(e)\\
&\quad + \frac{1}{2}[(\partial q_k \cdot H_m) - (\partial q_m \cdot H_k)](e) + (\partial t \cdot G_{mk})(e)\},\\
C_{mk}(e) &= (\partial q_m \cdot G_{hk})(e)\ddot{q}^k(e) + [(\partial q_m \partial q_s \cdot G_{hk})(e)]\\
&\quad - \frac{1}{2}(\partial q_m \partial q_k \cdot G_{hk})(e)]\dot{q}^s(e)\dot{q}^h(e)\\
&\quad + \{(\partial t \partial q_m \cdot G_{kh})(e) + (\partial q_m \partial q_h H_k)(e) - (\partial q_k \partial q_m \cdot H_h)(e)\}\dot{q}^h(e)\\
&\quad + \frac{1}{2}(\partial t \partial q_m \cdot H_k)(e) - \frac{1}{2}(\partial q_m \partial q_k \cdot) + (\partial q_m \partial q_h \cdot f)(e).
\end{aligned}$$

Here (e) is the point on the equation on which one wants linearize:

$$(e) \equiv (q^h(e), \dot{q}^k(e), \ddot{q}^k(e).$$

2) The general solution of (3.52) is the following:

$$(3.53) \qquad \nu^m = \sum_{i=1}^{2n} C_i \xi_i^m e^{\lambda_i t},$$

where $C_i, 1 \leq i \leq 2n$, are arbitrary constants, and ξ^m are solutions of the following system:

$$X_{mk}(\lambda)\xi^m = 0,$$

that is compatible iff one has

$$\Delta \equiv \det(X_{mk}(\lambda)) = 0,$$

where

$$X_{mk}(\lambda) \equiv A_{mk}\lambda^2 + B_{mk}\lambda + C_{mk}.$$

3) The roots of Δ can be real, complex or imaginary ones. Let us distinguish these cases.

a)

$$\lambda_j = \mu_j \pm i\omega_j,$$

then:

$$\nu^m = \sum_{j=1}^{n} e^{\mu_j t}\widetilde{C}_j\sqrt{({k_j}^m)^2 + ({\epsilon_j}^m)^2}\cos(\omega_j t + \delta_j + \phi_j) \tag{3.54}$$

where

$$\xi_j^m = {k_j}^m \pm {\epsilon_j}^m f, \tan\delta_j = \frac{{\epsilon_j}^m}{{k_j}^m},$$

and

$$\widetilde{C}_j, \phi_j = \text{arbitrary constants.}$$

We can have two possibilities:

(i) $\mu_j < 0 \Rightarrow$ *damped harmonic oscillations,* (**stable configuration**)

(ii) $\mu_j > 0 \Rightarrow$ *non-damped harmonic oscillations,* (**instable configuration**).

b)

$$\lambda_j = \pm i\omega_j, (\mu_j = 0),$$

then

$$\nu^m = \sum_{j=1}^{n} \widetilde{C}_j\sqrt{({k_j}^m)^2 + ({\xi_j}^m)^2}\cos(\omega_j t + \delta_j + \phi_j)$$

(harmonic oscillation (**stable motion**)).

c)

$$\lambda_j = \mu_j, (\omega_j = 0)$$

then:

$$\nu^m = \sum_{j=1}^{n} \widetilde{C}_j e^{\mu_j t}.$$

Thus, we have two possibilities:

(i) $\mu_j < 0 \Rightarrow$ (*damped oscillation* **stable motion**),

(ii) $\mu_j > 0 \Rightarrow$ (*non-damped oscillation* **instable motion**)

PROOF. 1) Let us consider Lagrange equation. After linearization we get:

$$(\partial{\xi_m}^{\alpha} \cdot \phi_k)(e){\nu^m}_{\alpha} = 0, ({\xi_\alpha}^m) \equiv (q^k, \dot{q}^k, \ddot{q}^k), ({\nu^m}_{\alpha}) \equiv (\nu^k, \dot{\nu}^k, \ddot{\nu}^k)$$

where $\phi_k \equiv$ functions that define Lagrange equation. Let us consider a solution of the type:

$$\nu^m = \xi^m e^{\lambda t}, 1 \le m \le n,$$

where $\xi^m e\lambda$ are constants. We get:

$$\xi^m(A_{mk}\lambda^2 + B_{mk}\lambda + C_{mk}) = 0. \tag{3.55}$$

Set:

$$X_{mk}(\lambda) \equiv A_{mk}\lambda^2 + B_{mk}\lambda + C_{mk}.$$

We have that ξ^m are solutions of the system:

$$X_{mk}(\lambda)\xi^m = 0, \tag{3.56}$$

if $\Delta \equiv \det(X_{mk}(\lambda)) = 0$. This is an algebraic equation of degree $2n$. Let $\lambda_1, \lambda_2, \cdots,$ be roots of the equation $\Delta = 0$; then, for any $\lambda_i, 1 \leq i \leq 2n$, we get a set of solutions ξ_i^m, of equation (3.55), hence a set of solutions of equation (3.52):

$$\nu^m = \xi_i^m e^{\lambda_i t}, 1 \leq i \leq 2n.$$

Therefore, the general solution of equation (3.52) can be written as given in (3.54). The roots of Δ can be real, complex or imaginary ones. Then, taking into account that:

$$e^{\pm i\omega t} = \cos\omega t \pm i \sin\omega t,$$

we get

$$\nu^m = \sum_{j=1}^{n} e^{\mu_i t}[(k_j{}^m a_j - \epsilon_j{}^m b_j)\cos\omega_j t - (k_j{}^m b_j + \epsilon_j{}^m a_j)\sin\omega_j t].$$

Therefore, after elementary calculations we have (3.54). □

THEOREM 3.46 - *If the configuration $q^k(e)$ is of equilibrium, then equation (3.52) becomes the following one:*

$$A_{mk}\ddot{\nu}^m + a_{mk}\nu^m = 0, 1 \leq k \leq n, \tag{3.57}$$

where $a_{mk} = (\partial q_m \partial q_k \cdot f)(e)$. Furthermore, if $(q^k(e))$ is a position of stable equilibrium, we get

$$\mathcal{T} = \frac{1}{2}A_{kh}\dot{q}^k\dot{q}^h, f \cong a_{hk}(q^k - q^k(e))(q^h - q^h(e))$$

and the quadratic form $\mathcal{T}$ and f are positive definite. Therefore, the solution of $\Delta = 0$ is imaginary. As a consequence, the general solution of equation (3.57) can be written as follows:

$$\nu^m = \sum_{j=1}^{n} \widetilde{C}_j \lambda_j{}^m \cos(\omega_j t + \phi_j + \delta_j).$$

DEFINITION 3.36 - 1) *An* **ignorable Lagrangian coordinate** *is one* q^k *such that for conservative systems results:*

$$(\partial q_k \cdot L) = 0.$$

Otherwise we say that q^k *is an* **non-ignorable Lagrangian coordinate**.
2) *A* **steady state motion** *is one such that the following conditions are satisfied:*
(i) *Any non-ignorable coordinate* q^k *is such that*

$$q^k = const, (\Leftrightarrow \dot{q}^k = 0).$$

(ii) *The velocity* $\dot{q}^k$ *of any ignorable coordinate* q^k *is such that*

$$\dot{q}^k = const.(\Leftrightarrow q^k = \dot{q}^k(t_0)(t - t_0) + q^k(t_0)).$$

THEOREM 3.47 - *The momentum of an ignorable coordinate is constant.*

$$q^k \textit{ignorable coordinate} \Leftrightarrow (\partial \dot{q}_k \cdot L) \equiv p_k = const.$$

PROOF. It is a direct consequence of Lagrange equation:

$$\frac{d}{dt}(\partial \dot{q}_k \cdot L) = 0.$$

□

THEOREM 3.48 - (CRITERION TO RECOGNIZE STEADY STATES). *For a system with time-independent constraints, the steady states are solutions of the following equations:*

$$(\partial q_r \cdot L) = 0.$$

PROOF. From Lagrange equation we get that for non-ignorable coordinates, in a steady state one has:

$$\{\dot{q}^k = 0 \quad , \quad \dot{\tilde{q}}^k = const.\}$$

where $\tilde{q}^k$ are ignorable coordinates. As a consequence we get:

$$\begin{aligned}
\frac{d}{dt}(\partial \dot{q}_k \cdot L) &= \frac{d}{dt}(\partial \dot{q}_k \cdot T) \\
&= \frac{d}{dt}(\partial \dot{q}_k \cdot [G_{ij}\dot{q}^i\dot{q}^j + G_{ij}\underline{\dot{q}}^i\underline{\dot{q}}^j + G_{ij}\dot{q}^i\underline{\dot{q}}^j]) \\
&= \frac{d}{dt}(\partial \dot{q}_k \cdot G_{ij}\dot{q}^i\dot{q}^j) \\
&\Downarrow \\
(\partial q_k \cdot L) &= 0.
\end{aligned}$$

□

THEOREM 3.49 - (STABILITY OF STEADY STATE).
The study of stability around the steady state motion is obtained by linearization of Lagrange equation around such a configuration. The discussion is similar to the above made for a generic configuration.

3.6 - RIGID BODY DYNAMICS

(A) CENTER OF MASS FOR CONTINUUM SYSTEM

THEOREM 3.50 - 1) *If the body is continuous, with mass distribution* ρ *and, at* $t \in T$, *occupies the domain* D, *then the vector position of the center of mass* $\overrightarrow{OG}$ *is given by the following formula:*

$$\overrightarrow{OG} = \frac{1}{\mu}\int_D \overrightarrow{OP}\rho\eta = x^i_G \partial x_i = \{\frac{1}{\mu}\int_D x^j \rho\sqrt{\det g_{ij}}dx^1 \wedge dx^2 \wedge dx^3\}\partial x_i$$

where $\eta = \sqrt{\det g_{ij}}dx^1 \wedge dx^2 \wedge dx^3$ *is the canonical volume form.*
2) *In particular, if* $\rho =$ *constant, we get*

$$x^i_G = \frac{1}{vol(D)}\int_D x^j \sqrt{\det g_{ij}}dx^1 \wedge dx^2 \wedge dx^3.$$

PROPOSITION 3.27 - (PROPERTIES OF CENTER OF MASS).
1) *If the body is plane, then the center of mass is in this plane.*
2) *If the body has a plane of symmetry, then the center of mass is in this plane.*
3) *If a body, with constant mass-density* ρ, *has some holes, then the center of mass coincides with one of the corresponding body without holes, but where the regions occupied by the holes have mass-density* $-\rho$.

EXAMPLE 3.10 - 1) CENTER OF MASS OF HOMOGENEOUS SEGMENT. G is in the middle point of the segment.
2) CENTER OF MASS OF ARC OF CIRCLE.

$$x_G = \frac{R\sin\theta_0}{\theta_0}, \theta = 2\theta_0 \equiv \text{angle-arc}.$$

PROOF. Taking into account that

$$x = R\cos\theta\,, y = R\sin\theta \Rightarrow \dot{x} = -\dot{\theta}R\sin\theta, y = \dot{\theta}R\cos\theta\,,$$

we get:

$$\begin{aligned} x_G &= \frac{1}{R\theta}\int_{[0,t]} R\cos\theta\,(\sqrt{\dot{x}^2+\dot{y}^2})dt \\ &= \frac{1}{R\theta}\int_{[0,t]} R^2\cos\theta\dot{\theta}dt = \frac{R}{\theta}\int_{[-\theta_0,\theta_0]}\cos\theta dt \\ &= \frac{R\sin\theta_0}{\theta_0} \end{aligned}$$

□

3) CENTER OF MASS OF CIRCULAR SECTOR.

$$x_G = \frac{2}{3}\frac{R\sin\theta_0}{\theta_0}.$$

PROOF. Taking into account that

$$vol(D) = \int_{[0,R]}\int_{[-\theta_0,\theta_0]} \rho d\rho \wedge d\theta = \frac{R^2\theta}{2},$$

we get:

$$\begin{aligned} x_G &= \frac{2}{R^2\theta}\int_D x\rho d\rho \wedge d\theta \\ &= \frac{2}{R^2\theta}\int_D \rho^2 \cos\theta d\rho \wedge d\theta \\ &= \frac{2}{R^2\theta}\int_{[0,R]} \rho^2 d\rho \int_{[-\theta_0,\theta_0]} \cos\theta d\theta \\ &= \frac{2}{3}\frac{R\sin\theta_0}{\theta_0}. \end{aligned}$$

□

(B) MOMENTS OF INERTIA OF CONTINUUM SYSTEMS.

THEOREM 3.51 - 1) *If the body is continuum, its moment of inertia J_a with respect to a line a is given by the following:*

$$J_a = \frac{1}{\mu}\int_D \rho\delta^2\eta,$$

where δ is the function distance from a.

2) *If the body has constant mass density ρ we can also write:*

$$J_a = \rho j_a, j_a \equiv \int_D \delta^2\eta \equiv \text{ moment of inertia of figure.}$$

3)(HUYGHENS THEOREM) *The moments of inertia with respect to two parallel axes a and b are related by the following:*

$$\boxed{J_a = J_b + \mu({d_a}^2 - {d_b}^2)}$$

where d_a, d_b are the respective distances of a and b from a parallel axis passing through the center of mass.

EXAMPLE 3.11 - 1) HOMOGENEOUS SEGMENT. The moments of inertia of figure, with respect to axes $\perp$ the segment, and passing through its middle point, are all equal to $j_\zeta = \frac{l^3}{12}$, where $l =$ length of the segment.

PROOF.

$$j_\zeta = \int_{[-l/2,l/2]} s^2 ds = \frac{l^3}{12}\,.$$

□

2) ARC OF HOMOGENEOUS CIRCLE. The moments of inertia of figure are given by the following:

$$j_\xi = \frac{R^3}{2}(\theta - \sin\theta), \quad j_\zeta = 2\theta_0 R^3 (1 - \frac{\sin^2\theta_0}{\theta_0 2})$$
$$j_\eta = j_\zeta - j_\xi\,.$$

PROOF.

$$\begin{aligned} j_\xi &= \int_D y^2 \eta \\ &= \int_D R^2 \sin^2\theta(\sqrt{\dot{x}^2 + \dot{y}^2})dt \\ &= \int_D R^2 \sin^2\theta(\sqrt{R^2\dot{\theta}^2})dt \\ &= R^3 \int_{[-\theta_0,\theta_0]} \sin^2\theta d\theta \end{aligned}$$

On the other hand, as $\sin\theta = \pm\sqrt{\frac{1-\cos^2\theta}{2}}$, we get:

$$\begin{aligned} \int_{[-\theta_0,\theta_0]} \sin^2\theta d\theta &= \frac{1}{2}\int_{[-\theta_0,\theta_0]} (1 - \cos 2\theta)d\theta = \frac{1}{4}\int_{[-2\theta_0,2\theta_0]} (1 - \cos x)dx \\ &= \frac{1}{4}(4\theta_0 - \int_{[-2\theta_0,2\theta_0]} d\sin x) = \frac{2\theta_0 - \sin 2\theta_0}{2}. \end{aligned}$$

In order to calculate j_ζ, it is simpler to apply the theorem of Huyghens, after calculating the moment of inertia with respect to an axis $\perp$ the plane of the arc, and passing through the center O of the circle which belongs to the arc. □

3) HOMOGENEOUS DISC. The central moments of inertia of figure are given as follows:

$$j_\xi = j_\eta = \frac{1}{2}j_\zeta = \frac{\pi R^4}{2}.$$

PROOF.

$$j_\zeta = \int_D \rho^2 \eta = \int_D \rho^2 \rho d\rho \wedge d\theta = \int_{[0,R]} \int_{[0,2\pi]} \rho^3 d\rho d\theta = \frac{\pi R^4}{2}.$$

□

4) HOMOGENEOUS RECTANGULAR PARALLELEPIPED. The central moments of inertia of figure are the following:

$$j_\xi = \frac{abc}{12}(b^2+c^2), \quad j_\eta = \frac{abc}{12}(a^2+c^2), \quad j_\zeta = \frac{abc}{12}(a^2+b^2).$$

5) HOMOGENEOUS CIRCULAR CYLINDER. The central moments of inertia of figure are the following:

$$j_\xi = j_\eta = \pi R^2 h[\frac{R^2}{4} + \frac{h^2}{12}], \; j_\zeta = \frac{\pi h R^4}{2}.$$

PROOF. Let us use a cylindrical coordinate system.

$$\begin{aligned}
j_\zeta &= \int_D \rho^2 \eta = \int_D \rho^2 \rho d\rho \wedge d\theta \wedge d\zeta = \int_{[-h/2,h/2]} d\zeta \int_{[0,2\pi]} d\theta \int_{[0,R]} \rho^3 d\rho = \frac{h\pi R^4}{2}. \\
j_\xi &= \int_D (\zeta^2 + \eta^2)\rho d\rho \wedge d\theta \wedge d\zeta \\
&= \int_{[-h/2,h/2]} \zeta^2 d\zeta \int_{[0,2\pi]} d\theta \int_{[0,R]} \rho d\rho + \int_{[-h/2,h/2]} d\zeta \int_{[0,2\pi]} \sin^2\theta d\theta \int_{[0,R]} \rho^3 d\rho \\
&= \frac{\pi R^2 h^3}{12} + \frac{\pi h R^4}{4}.
\end{aligned}$$

□

6) HOMOGENEOUS SPHERE. The central moments of inertia of figure are the following:

$$j_\xi - j_\eta = j_\zeta = \frac{8\pi R^5}{15}.$$

PROOF. Let us use a spherical coordinate system.

$$\begin{aligned}
j_\xi &= \int_D (\zeta^2+\eta^2)\rho^2 \sin\phi d\rho \wedge d\theta \wedge d\phi \\
&= \int_D \rho^4 \sin^2\theta \sin^3\phi d\rho d\theta d\phi + \int_D \rho^4 \cos^2\phi \sin\phi d\rho \wedge d\theta \wedge d\phi \\
&= \frac{R^5\pi}{5} \int_{[0,\pi]} \sin^3\phi d\phi + \frac{R^5 2\pi}{5} \int_{[0,\pi]} \cos^2\phi \sin\phi d\phi.
\end{aligned}$$

Taking into account the following

$$(a) \quad \int_{[0,\pi]} \sin^3 \phi d\phi$$
$$= \int_{[0,\pi]} \sin^2 \phi \sin \phi d\phi$$
$$= -\int_{[0,\pi]} (1 - \cos^2 \phi) d \cos \phi$$
$$= -\int_{[-1,1]} (1 - x^2) dx = \frac{4}{3};$$
$$(b) \quad \int_{[0,\pi]} \cos^2 \phi \sin \phi d\phi$$
$$= -\int_{[0,\pi]} \cos^2 \phi d \cos \phi$$
$$= -\int_{[-1,1]} x^2 dx = \frac{2}{3}.$$

We complete the proof. □

THEOREM 3.52 - (ANGULAR MOMENTUM AND MOMENT OF INERTIA TENSOR).

1) *The angular momentum* Ω *of a rigid body in instantaneous spherical motion (around the point* A*) can be written as follows:*

$$\boxed{\underline{\Omega} = < \underline{J}, \omega > \in \mathbf{S}^*} \qquad (3.58)$$

where ω *the angular velocity and*

$$\underline{J} \in S_2^0(\mathbf{S}) \equiv \mathbf{S}^* \odot \mathbf{S}^*$$

is a symmetric tensor of type $(0,2)$*, called the* **moment of inertia tensor** *of the rigid body with respect to* A*. Thus,* $\underline{J}$ *can be represented by a symmetric* 3×3*-matrix*

$$\underline{J} = J_{rp} \theta^r \odot \theta^p,$$

with

$$(J_{rp}) = \begin{bmatrix} A & -D & -F \\ -D & B & -E \\ -F & -E & C \end{bmatrix}.$$

Here, $\{\theta^r\}$ *is the dual basis of a basis* $\{\epsilon_r\}$ *of* $\mathbf{S}$*. By using the canonical isomorphism* $\mathbf{S} \cong \mathbf{S}^*$*, we can also write:*

$$\overline{\Omega} = \underline{J}(\omega) \in \mathbf{S}, \qquad (3.59)$$

where:

$$\underline{J} \in L(\mathbf{S}) \cong \mathbf{S}^* \bigotimes \mathbf{S}, \underline{J} \equiv \textbf{moment of inertia tensor}$$

$$\underline{J} \equiv (id_{\mathbf{S}^*} \otimes {'g}^{-1}) \circ \underline{J} .$$

2) *With respect to a coordinate system we have:*

$$\underline{\Omega} = \Omega_r \theta^r$$

$$\Omega_r = J_{rp}\omega^p$$

$$\overline{\Omega} = \Omega^k \epsilon_k, \quad \Omega^k = \Omega_r g^{kr} = J_{rp} g^{kr} \omega^p = J_p{}^k \omega, J_p{}^k \equiv J_{rp} g^{kr} .$$

3) *If u is a unitary vector passing through the point A, the invariant (≥ 0)*

$$J_u \equiv J(u) \equiv < \underline{J}, u \otimes u >= J_{rp} u^r u^p$$

represents the moment of inertia with respect to the axis identified by u and passing through A.

4) *The quadric of the symmetric tensor $\underline{J}$ is called the* **inertia ellipsoid**. *Its axes are called* **principal axes passing through** *A. In particular, if $A \equiv G$ is the center of mass, then the principal axes are called* **central axes**.

5) *The corresponding kinetic energy is the following:*

$$\mathcal{T} = \frac{1}{2} < \underline{J}\omega \otimes \omega >= \frac{1}{2} < \underline{\Omega}, \omega >= \frac{1}{2} J_{rp} \omega^r \omega^p$$

6) *With respect to a cartesian system (ϵ_k) we get:*

$$\Omega_r = \Omega^r, J_{rp} = J_r{}^p .$$

PROOF. The proof is standard and can be obtained by direct computations. □

THEOREM 3.53 - *The principal axes passing through A are identified by means of the eigenvectors of the moment of inertia tensor, with respect to A. Then the corresponding moments of inertia are the relative eigenvalues. Furthermore, if ω is parallel to a principal axis, then the corresponding angular momentum is parallel to ω:*

$$\Omega = \lambda \omega, \quad \lambda \in \mathbf{R}.$$

THEOREM 3.54 - (GENERALIZED KÖENIG THEOREM FOR RIGID BODY). *For a rigid body one has the following expressions for the kinetic energy.*

1)

$$\mathcal{T} = \frac{1}{2}\mu v_A^2 + \frac{1}{2} < \underline{\Omega}, \omega > + \mu v_A \cdot (\omega \times \overrightarrow{AG})$$

where A is any point of the body.

2)

$$\mathcal{T}^A = \mathcal{T}^G + \frac{1}{2}\mu(\omega \times \overrightarrow{AG})^2$$

PROOF. 1) From the generalized Köenig theorem we get:

$$\mathcal{T} = \frac{1}{2}\mu v_A^2 + \mathcal{T}^A + v_A \cdot \mu v_{G,r}.$$

On the other hand, we have:

$$v_G = v_A + \omega \times \overrightarrow{AG} \Rightarrow v_{G,r} = v_G - v_A = \omega \times \overrightarrow{AG}.$$

□

2) Let us consider the Köenig theorem

$$\mathcal{T} = \frac{1}{2}\mu v_G^2 + \mathcal{T}^G,$$

and from the fundamental relation of kinematics of rigid bodies we have:

$$v_G = v_A + \omega \times \overrightarrow{AG} \Rightarrow v_G^2 = v_A^2 + (\omega \times \overrightarrow{AG})^2 + 2v_A \cdot (\omega \times \overrightarrow{AG}).$$

By using this relation in the Köenig theorem, we have:

$$\mathcal{T} = \mathcal{T}^G + \frac{1}{2}\mu v_A^2 + \frac{1}{2}\mu(\omega \times \overrightarrow{AG})^2 + \mu v_A \cdot (\omega \times \overrightarrow{AG}).$$

Therefore, from the generalized Köenig theorem, we get:

$$\begin{aligned}\mathcal{T} &= \frac{1}{2}\mu v_A^2 + \mathcal{T}^A + \mu v_A \cdot (\omega \times \overrightarrow{AG}) \\ &= \mathcal{T}^G + \frac{1}{2}\mu v_A^2 + \frac{1}{2}\mu(\omega \times \overrightarrow{AG})^2 + \mu v_A \cdot (\omega \times \overrightarrow{AG}) \\ \mathcal{T}^A &= \mathcal{T}^G + \frac{1}{2}\mu(\omega \times \overrightarrow{AG})^2.\end{aligned}$$

□

REMARK 3.31 - (GENERAL EXPRESSION OF THE CARDINAL EQUATIONS FOR DYNAMICS OF RIGID BODY). The dynamics of rigid body can be described by the following ODE's:

$$\begin{aligned} \mu\, a_G &= {}^e F \\ \dot{\Omega}(A) + \mu\, v_A \times v_G &= {}^e M(A) \end{aligned} \tag{3.60}$$

where:

$$\dot{\Omega}(A) = \overline{J}^{(A)}(\dot{\omega}) + \omega \times \overline{J}^{(A)}(\omega) = J^{(A)}{}_r{}^s \dot{\omega}^r \epsilon_s + J^{(A)}{}_r{}^s \omega^r \omega \times \epsilon_s$$

is the velocity of the angular momentum.

$$(J^{(A)}{}_r{}^s) = \begin{bmatrix} J^{(A)}{}_1{}^1 & J^{(A)}{}_1{}^2 & J^{(A)}{}_1{}^3 \\ J^{(A)}{}_2{}^1 & J^{(A)}{}_2{}^2 & J^{(A)}{}_2{}^3 \\ J^{(A)}{}_3{}^1 & J^{(A)}{}_3{}^2 & J^{(A)}{}_3{}^3 \end{bmatrix} = \begin{bmatrix} A & -D & -F \\ -D & B & -E \\ -F & -E & C \end{bmatrix}.$$

The second cardinal equation can be written in the following way:

$$\begin{aligned} & J^{(A)}{}_r{}^1 \dot{\omega}^r + J^{(A)}{}_r{}^3 \omega^r \omega_2 - J^{(A)}{}_r{}^2 \omega^r \omega_3 \\ & + \omega_1 (v_A \cdot \overrightarrow{AG}) - (v_A \cdot \omega)(\overrightarrow{AG})_\xi = M_\xi \\ & J^{(A)}{}_r{}^2 \dot{\omega}^r + J^{(A)}{}_r{}^1 \omega^r \omega_3 - J^{(A)}{}_r{}^3 \omega^r \omega_1 \\ & + \omega_2 (v_A \cdot \overrightarrow{AG}) - (v_A \cdot \omega)(\overrightarrow{AG})_\eta = M_\eta \\ & J^{(A)}{}_r{}^3 \dot{\omega}^r + J^{(A)}{}_r{}^2 \omega^r \omega_1 - J^{(A)}{}_r{}^3 \omega^r \omega_1 \\ & + \omega_3 (v_A \cdot \overrightarrow{AG}) - (v_A \cdot \omega)(\overrightarrow{AG})_\zeta = M_\zeta \end{aligned} \tag{3.61}$$

If $(A; \epsilon_1, \epsilon_2, \epsilon_3)$ identifies principal axes for A, we have:

$$(J^{(A)}{}_r{}^s) = \begin{bmatrix} A & 0 & 0 \\ 0 & B & 0 \\ 0 & 0 & C \end{bmatrix}.$$

Then the second cardinal equations can be written in the following form: (**Euler equations**):

$$\begin{aligned} \dot{\omega}_\xi A - (B - C)\omega_\eta \omega_\zeta + \omega_\xi (v_A \cdot \overrightarrow{AG}) - (v_A \cdot \omega)(\overrightarrow{AG})_\xi &= M_\xi \\ \dot{\omega}_\eta B - (C - A)\omega_\zeta \omega_\xi + \omega_\eta (v_A \cdot \overrightarrow{AG}) - (v_A \cdot \omega)(\overrightarrow{AG})_\eta &= M_\eta \\ \dot{\omega}_\zeta C - (A - B)\omega_\xi \omega_\eta + \omega_\zeta (v_A \cdot \overrightarrow{AG}) - (v_A \cdot \omega)(\overrightarrow{AG})_\zeta &= M_\zeta \end{aligned} \tag{3.62}$$

THEOREM 3.55 - (POINSOT-MOTION). *We say that a rigid body motion is of* **Poinsot type** *if there is a fixed element* A *of the body, and* $M(A) = 0$. *The dynamic of Poinsot type is completely described by the following ODE's:*

$$\boxed{\begin{aligned}\dot{\omega}_\xi A - (B - C)\omega_\eta\omega_\zeta &= 0\\ \dot{\omega}_\eta B - (C - A)\omega_\zeta\omega_\xi &= 0\\ \dot{\omega}_\zeta C - (A - B)\omega_\xi\omega_\eta &= 0\end{aligned}} \tag{3.63}$$

Then, we gave the following properties.

1) *In a Poinsot-motion there are the following conservation laws:*[8]

$$\Omega(A) = const., \quad \mathcal{T} = const.$$

With these two integrals it is possible to integrate the above equations completely in terms of elliptic functions.

2) *The inertia ellipsoid, with respect to* A, *is in contact with two fixed planes,* π *and* π', *that have the same constant distance*

$$\delta = \frac{\sqrt{2\chi\mathcal{T}}}{|\Omega(A)|} = const.$$

from A, *and are* $\perp\Omega(A)$. *Furthermore,*

$$|\omega| = \frac{1}{k}|\overrightarrow{AP_t}| = \frac{1}{k}|\overrightarrow{AP'_t}|, \ k = \sqrt{\frac{\chi}{2\mathcal{T}}} = const.,$$

where P_t *and* P'_t *are the contact points of the inertia ellipsoid on the planes* π *and* π'.

3) *The Poinsot motions that are also rotation, are those around a principal axis of inertia. These are uniform ones. This can happen iff at the initial instant the rotation axis coincides with a principal axis of inertia.*

4) *In a generic Poinsot-motion the instantaneous rotation axis describes respectively in the inertial frame, and the body-frame, two cones* (**Poinsot cones**).[9]

[8] The line indicating the direction of $\Omega(A)$ is called **invariable line**.

[9] These cones are called respectively space and body cones. More precisely the space cone is fixed in the inertial frame, and the body cone is fixed in the body frame. Then, we can also say that in a Poinsot motion one has a rolling without slipping of one cone on the other. Equivalently we can say that the ellipsoid of inertia, corresponding to the rigid body, rolls without slipping on the invariable planes. In the space frame one sees a rotation (or precession) of the angular velocity ω about the angular momentum Ω. In general, the motion of the inertia ellipsoid on the invariable planes is not of pure rolling not pivoting one. In fact, the instantaneous rotation axis does not belong to the invariable planes neither is orthogonal to these planes.

These cones intersect respectively on the invariable planes π, π', (resp. on the inertia ellipsoid) two curves: **herpolhodes** *(resp.* **polhodes***) that are symmetric with respect to A. Each polhode rolls without slipping on the corresponding herpolhode. If the inertia ellipsoid is of revolution around the ζ-axis, the Poinsot-motions are* **regular precessions** *with ζ as figure-axis, and the line parallel to $\Omega(A)$ as precession-axis. Then, the Poinsot cones result of revolution ones.*

5) *If the inertia ellipsoid is a sphere, all the Poinsot-motions are uniform rotations.*

PROOF. 1) The conservation laws follow directly from the second cardinal equation and from the theorem of the work.

2) Let P_t and P'_t be the points where the inertia ellipsoid intersects the instantaneous rotation-axis. We have: $\overrightarrow{AP_t} = -\overrightarrow{AP'_t} = k\omega$. Let the inertia ellipsoid have the following equation: $F \equiv A\xi^2 + B\eta^2 + C\zeta^2 - \chi = 0$, with respect to the principal axes passing for A. We have $\overrightarrow{AP_t} = k\omega^k\epsilon_k = \xi\epsilon_1 + \eta\epsilon_2 + \zeta\epsilon_3$. Therefore, $k^2[A(\omega^1)^2 + B(\omega^2)^2 + C(\omega^3)^2] = \chi \Rightarrow k^2 2T = \chi \Rightarrow k = \sqrt{\frac{\chi}{2T}} = constant$. Furthermore Ω is $\perp$ the inertia ellipsoid in the points P_t and P'_t. In fact, a vector $v = v^k\partial\xi_k$ is tangent to the inertia-ellipsoid iff $v.F = 0 \Rightarrow v^k(\partial\xi_k.F) = 0$. This equation in the points P_t and P'_t can be written $2k[v^1 A\omega^1 + v^2 B\omega^2 + v^3 C\omega^3] = 0$. This means that the vector $\Omega(A) = A\omega^1\epsilon_1 + B\omega^2\epsilon_2 + C\omega^3\epsilon_3$ is $\perp$ the inertia ellipsoid in the points P_t and P'_t. Then, we recognize, also, that the planes π and π' are always parallel to the same fixed plane. Furthermore, their distance δ from A is also constant. In fact,

$$\delta = \overrightarrow{AP_t}.\frac{\Omega}{|\Omega|} = \sqrt{\frac{\chi}{2T}}\omega.\frac{\Omega}{|\Omega|} = \sqrt{\frac{\chi}{2T}}\frac{2T}{|\Omega|} = \frac{\sqrt{2T\chi}}{|\Omega|} = constant.$$

Therefore, the planes π and π' are at rest.

3), 4), 5) can be directly proved. Let us explicitly prove point 4). Thus, we will prove that, under our hypotheses, there exist two axes passing through A, that are respectively joint with the inertial frame and the body-frame, and are on the same plane with the axis for A parallel to ω. We will show that these axes are respectively ζ and an axis p passing for $A \parallel \Omega(A)$. In order to state that the motion is a precession it is enough to show that any vector $\perp$ the plane identified by the axes p, ζ is $\perp\omega$. In fact, let us consider a vector $x = Z^1\epsilon_1 + Z^2\epsilon_2 + Z^3\epsilon_3$, and impose $\perp\zeta$, i.e., $\perp\epsilon_3 \Rightarrow Z^3 = 0$. Impose, also, that $x\perp p \Rightarrow 0 = \omega_1 AZ^1 + \omega_2 BZ^2$. As $A = B \Rightarrow 0 = \omega_1 Z^1 + \omega_2 Z^2$. Let us, now, verify that $\omega.x = 0$. In fact, $\omega.x = \omega_1 Z^1 + \omega_2 Z^2$. In order to prove that the precession is regular, we will show that the components of ω on p and ζ respectively are constant. In fact, from the Euler equation we get $\dot{\omega}_\zeta C = 0 \Rightarrow \omega_\zeta = constant$.

Furthermore, from the conservation of kinetic energy we have:

$$\text{constant} = \mathcal{T} = \frac{1}{2}(A[\omega_\xi^2 + \omega_\eta^2] + C\omega_\zeta^2) \Rightarrow \omega_\xi^2 + \omega_\eta^2 = \text{constant}$$

$$\Downarrow$$

$$|\omega| = \text{constant} \Rightarrow \cos(\widehat{\zeta\omega}) = \frac{\omega_\zeta}{|\omega|} = \text{constant}$$

$$\Downarrow$$

$$\widehat{\zeta\omega} = \text{constant}$$

and

$$\cos(\widehat{p\omega}) = \frac{\Omega.\omega}{|\Omega(A)||\omega|} = \text{constant}$$

$$\Downarrow$$

$$\widehat{p\omega} = \text{constant} \Rightarrow \widehat{p\zeta} = \text{constant}.$$

This last relation assures that the **precession cone** (with axis p and generatrix ζ) is of revolution. Furthermore, also the Poinsot cones (the space cone with axis p, and body cone with axis ζ) are of revolution. □

4 - CONTINUUM MECHANICS

4.1 - FLOW

Kinematics of a continuum medium is described by the concept of flow, that, partially, we have described in section 3.1. Here, we will further characterize this fundamental structure, by showing important geometric objects that can be associated to any flow.

DEFINITION 4.1 - *A* **flow** *is a 1-parameter group of transformations of the Galilean space-time:*

$$\phi : \mathbf{T} \times M \to M$$

i.e., $\{\phi_\lambda\}_{\lambda \in \mathbf{T}}$ *is a group of diffeomorphisms of* M.

REMARK 4.1 - If at the initial instant $t = t_0$ we consider that the domain D of the continuous medium, is with boundary ∂D, locally given by the parametric equations $x^k = x^k(u^1, u^2)$, then at the instant $t + \lambda$, the domain $D_\lambda = \phi_\lambda(D)$, will be with boundary ∂D_λ given by the following parametric equations:

$$\phi^k_\lambda = \phi^k_\lambda(x^1(u^1, u^2), x^2(u^1, u^2), x^3(u^1, u^2), x^0) \equiv f^k(u^1, u^2; x^0, \lambda).$$

PROPOSITION 4.1 - *A flow* ϕ *is equivalent to a frame.*

PROOF. If ψ is a frame, we get the following associated flow:

$$\boxed{\phi_\psi(\lambda, p) = \psi(\tau(p) + \lambda, p)}$$

Vice versa, if ϕ is a flow, its frame is the following:

$$\boxed{\psi_\phi(t, p) = \phi(t - \tau(p), p)}$$

□

DEFINITION 4.2 - *To any continuum system, with flow* ϕ, *it is associated the frame* **body-frame** *defined by* ϕ.

2) *The* **space** *of a continuum system with flow* ϕ *is the set* S_ϕ *of its flow-lines:*

$$S_\phi \equiv M/\sim$$

where $\sim$ *is the equivalence relation induced in* M *by* ϕ:

$$p \sim p' \Leftrightarrow p' = \phi(\lambda, p), \lambda \in \mathbf{T}.$$

REMARK 4.2 - 1) For a regular flow, (and we will always assume flows of such a type), S_ϕ results in a Riemannian 3-dimensional manifold, with metric

$$g_\phi : S_\phi \times T \to S_2^0(S_\phi),$$

induced by the vertical metric g on M.
2) In particular, if the flow is rigid, S_ϕ results in a Euclidean 3-dimensional affine space.
In the following table we consider some important fields of geometric objects associated to a flow.[1]

TAB.4.2 - **Geometric objects ass.flow** ϕ

Name	Definition
velocity	$\dot{\phi} = \partial\phi : M \to JD(M)$
acceleration	$\ddot{\phi} = \nabla_{\dot{\phi}}\dot{\phi} : M \to vTM$
inf.strain velocity	$\dot{e} = \frac{1}{2}\mathcal{L}_{\dot{\phi}}g : M \to vS_2^0(M)$
spin	$s = \frac{1}{2}\mathrm{rot}\dot{\phi} : M \to v\Lambda_2^0(M)$
gradient of strain	$F = vD_2\phi : \mathbf{T} \times M \to v\mathcal{D}(M, M)$
strain	$U_\lambda = \phi_\lambda^* g : M \to vS_2^0(M)$
volume strain	$V_\lambda = \phi_\lambda^* \eta : M \to v\Lambda_3^0(M)$
inf.volume strain vel.	$\dot{v} = \mathcal{L}_{\dot{\phi}}\eta : M \to v\Lambda_3^0(M)$
displacement vector	$u(\lambda, p) = \phi(\lambda, p) - p$

☐ For rigid flows one has $\dot{e} = 0$ and $\overline{F}_\lambda \in O(\mathbf{S})$.
☐ For isocoric flows one has $\det F = \pm 1, V_\lambda = \eta, \dot{v} = 0$.
REMARK 4.3 - (EXTENSION AND EXPANSION OF FLOW).
I)(EXTENSION)
1) Let us define **extension** the following mapping:

$$\boxed{\begin{gathered} e : \mathbf{T} \times \mathbf{M} \times_{\mathbf{T}} \mathbf{M} \to \mathbf{R} \\ e(\lambda; p, p') \equiv \frac{d(\phi_\lambda(p), \phi_\lambda(p'))}{d(p, p')} - 1 \end{gathered}}$$

[1] Compare also with standard formulations of Continuum Mechanics, (see e.g., refs.[32,51,148]).

2) We can exprime the extension by means of the strain tensor U_λ. We have:

$$e(\lambda; p, p') = \sqrt{\frac{U_\lambda(s,s)}{g(s,s)}} - 1$$

where $s \equiv p' - p \in \mathbf{S}$.
3) We can also exprime the extension by means of the displacement vector:

$$\Delta u : \mathbf{T} \times \mathbf{M} \times_{\mathbf{T}} \mathbf{M} \to \mathbf{S}$$
$$\Delta u(\lambda; p, p') \equiv u(\lambda, p) - u(\lambda, p')$$

In fact, from the relation $s_\lambda = s - \Delta u$ and taking into account the bilinearity of g, we get:

$$e(\lambda; p, p') = \sqrt{1 - 2\frac{g(s, \Delta u)}{g(s,s)} + \frac{g(\Delta u, \Delta u)}{g(s,s)}} - 1$$

4) Note that, for $\lambda \in \mathbf{T}$ and $p \in M$, Δu defines a vertical vector field on M:

$$u_\lambda \equiv (\Delta u)_{\lambda, p'} : M \to vTM \cong M \times \mathbf{S}$$

that we call also **displacement vector**.
5) Then, we define **infinitesimal strain**, for a displacement $u_\wedge : M \to vTM$, the following field of geometric objects:

$$e \equiv \frac{1}{2}\mathcal{L}_{u_\wedge} g$$

Its local representation is the following:

$$e_{ij} \equiv \frac{1}{2}(u_{i/j} + u_{j/i}).$$

6) Then, we will say **infinitesimal extension in the direction** l the following scalar:

$$\boxed{< l \otimes l, e > \equiv e_{ij} l^i l^j}$$

II)(EXPANSION)

7) We similarly define, for any $\lambda \in \mathbf{T}$, and compact 3-dimensional vertical domain D, i.e., $D \subset M_t, t \in T$, **expansion**

$$\boxed{v(\lambda, D) = \frac{\int_D (V_\lambda - \eta)}{\int_D \eta}}$$

8) We define **infinitesimal expansion**

$$\boxed{v \equiv \mathcal{L}_{u_\wedge} = \mathrm{div} u_\wedge}$$

In coordinates we get

$$\boxed{v = u^i_{/i}}$$

Thus we also have:

$$\boxed{v = \mathrm{tr}(e) = e_{ij} g^{ij}}$$

that relates the infinitesimal expansion with the infinitesimal extension. For isocoric transformations we get $\mathrm{tr}(e) = 0$. In the following table we resume some important examples of infinitesimal transformations.

TAB.4.3 - **Examples of infinitesimal transformations**

1) uniform expansion	$u^i = cx^i, c \in \mathbf{R}$ $e_{ij} = c\delta_{ij}, v = 3c$
2) simple extension	$u^i = cl^i l_j x^j, c \in \mathbf{R}$ $e_{ij} = cl_i l_j, v = c$
3) shear strain	$u^i = 2cl_i m_j y^j, c \in \mathbf{R}$ $e_{ij} = c(l_i m_j + l_j m_i), v = 2cl_i m^i$

4.2 - STRESS TENSOR AND MOMENT OF STRESS TENSOR

DEFINITION 4.3 - 1) *The* **stress** $P(n)$ *in the direction* n, *in a continuous body, is the pressure on a surface, normal to* n.
2) *The* **stress tensor** *is a section:*

$$P : M \to vT_2^0 M \cong M \times T_2^0(\mathbf{S})$$

of the fiber bundle $vT_2^0 M \to M$.
3) *Then the stress in the direction* n *can be calculated by the following:*

$$P(n) = \overline{P\rfloor n} : M \to vTM \cong M \times \mathbf{S}.$$

If the surface is given by the equations $s(x,t) = \text{constant}$, *then the components* $n_j(x,t)$ *of* n *are given by:*

$$n_j(x,t) = (g^{\alpha\beta}(\partial x_\alpha \cdot s)(\partial x_\beta \cdot s))^{-1/2}(\partial x_j \cdot s)$$

where

$$n^k(x,t) = n_j(x,t)g^{jk}.$$

PROPOSITION 4.2 - 1) *The stress tensor* P *has the following representation in co-ordinates:*

$$P = P_{ij}dx^i \otimes dx^j.$$

Then, we get:

$$P(n) = P_i{}^j n^i \partial x_j$$

where

$$P_i{}^j = P_{ik}g^{kj}.$$

One has:

$$P(n)^j = P_i{}^j n^i; P(n)_k = P_{ik}n^i.$$

$$P(-n) = -P(n).$$

$$P(n) = P(n)^j \partial x_j; \quad P(n)_k dx^k.$$

REMARK 4.4 - In general we can write

$$P = -pg + \wp$$

where p is a numerical function on M, called **isotropic pressure** and $\wp$ is the **deviatory part**, (i.e., non-isotropic), of P. This splitting is useful for fluids, where in static situation we have $\wp = 0$ and the pressure becomes purely isotropic.

DEFINITION 4.4 - 1) *The* **moment** *(or* **couple***)* $m(n)$ *of the stress tensor in the direction* n *(with respect to a fixed point* O*), is the moment, with respect to* O*, of the pressure* $P(n)$*.*
2) *The* **stress moment** *is a section*

$$m : M \to vT^0_2 M$$

of the fiber bundle $vT^0_2 M \to M$*.*
3) *Then, the stress moment in the direction* n *can be written as follows:*

$$m(n) = \overline{m\rfloor n} : M \to vTM.$$

PROPOSITION 4.3 - 1) *If the stress moment* m *has the following representation in coordinates:*

$$m = m_{ij} dx^i \otimes dx^j$$

we get

$$m(n) = m(n)^j \partial x_j$$

with

$$m(n)^j = m_i{}^j n^i.$$

We have:

$$m(-n) = -m(n)\,.$$

2) $m(n)$ *is identified with the following vertical vector field:*

$$\widehat{m}(n) \equiv \overline{*m(n)} : m \to v\Lambda^2_0 M$$

and vice versa

$$m(n) = \overline{*\underline{\widehat{m}(n)}} : M \to vTM.$$

In coordinates we get:

$$\widehat{m}(n) = m^{ji}(n)\partial x_i \wedge \partial x_j,$$

with

$$2m^{ij}(n) = \epsilon^{jir} m(n)_r, m(n)_k = \epsilon_{kpr} m^{pr}(n).$$

3) *Furthermore,* m *identifies the following vertical vector field:*

$$\widehat{m} : \overline{*m} : M \to vTM \bigotimes v\Lambda^2_0 M.$$

In coordinates we get:

$$\widehat{m} = m^{kpr} \partial x_k \otimes \partial x_p \wedge \partial x_r$$

with

$$2m^{kpr} = \epsilon^{epr} m^k_e, m^k_e = \epsilon_{epr} m^{kpr}.$$

We have:

$$m^{kpr} = -m^{krp}.$$

$$m(n)_q = \epsilon_{pqr} m^{pr}(n) = \epsilon_{qpr} m^{kpr} n_k.$$

THEOREM 4.1 - (EULER EQUATIONS FOR CONTINUUM SYSTEMS). *The cardinal equations for continuum systems are the following:*

$$\boxed{\begin{aligned} \frac{d}{dt}\int_V v^i d\mu &= \int_{\partial V} P^i(n) d\sigma + \int_V B^i d\mu \\ \frac{d}{dt}\int_V \epsilon_{jk}{}^i x^j v^k d\mu &= \int_{\partial V} (m^i(n) + \epsilon_{jk}{}^i x^j P^k(n)) d\sigma + \int_V (l^j + \epsilon_{jk}{}^i x^j B^k) d\mu \end{aligned}}$$

where B = body force, $P(n)$ = stress in the direction normal to ∂V, l = body moment, $m(n)$ = stress moment in the direction normal to ∂V.
THEOREM 4.2 - *The normal component N of the stress $P(n)$ in the direction n is the following:*

$$N = P(n) \cdot n = P_{ij} n^i n^j.$$

DEFINITION 4.5 - (THE STRESS QUADRIC). *The* **stress quadric** *is defined by the following equation:*

$$N = P_{ij} n^i n^j, \quad g_{ij} n^i n^j = 1.$$

THEOREM 4.3 - 1) *There exist three principal directions of the stress, and three principal stresses, $P_\alpha, \alpha = 1, 2, 3$. These act on the principal planes.*
2) *On the principal planes the shear stresses are zero, and the normal stresses are constant.*
3) *The stress tensor P satisfies the following equation:*

$$P^3 - IP^2 + IIP - III = 0$$

where I, II, III are the following three invariants associated to P:

$$\begin{aligned} I &= P^m_m \\ II &= \frac{1}{2}(I^2 - P^n_m P^m_n) \\ III &= |P^n_m|. \end{aligned}$$

PROOF. We shall use the following lemma.

LEMMA 4.1 - (**Cayley-Hamilton theorem**). *Any endomorphism $A \in L(V)$ of an n-dimensional vector space V satisfies its* **characteristic equation**

$$\chi(A) = 0$$

where

$$\chi(A) \equiv \sum_{j=0}^{n} (-1)^j \mathrm{tr}(\Lambda^j(A))\lambda^{n-j}$$

is the **characteristic polynomial** *of A.*
In particular, if $\dim V = 3$ one has

$$\chi(A) \equiv \chi(\lambda) = \lambda^3 - I\lambda^2 + II\lambda - III$$

with

$$I = \mathrm{tr}(A),\ II = \mathrm{tr}(\Lambda^2(A)),\ III = \mathrm{tr}(\Lambda^3(A)) = \det(A).$$

□

EXAMPLE 4.1 - In the following we report some important examples of stress.

$$\text{uniaxial-stress}: \begin{bmatrix} P_1 & 0 & 0 \\ 0 & 0 & 0 \\ 0 & 0 & 0 \end{bmatrix} \quad \text{biaxial-stress}: \begin{bmatrix} P_1 & 0 & 0 \\ 0 & P_2 & 0 \\ 0 & 0 & 0 \end{bmatrix}$$

$$\text{triaxial-stress}: \begin{bmatrix} P_1 & 0 & 0 \\ 0 & P_2 & 0 \\ 0 & 0 & P_3 \end{bmatrix} \quad \text{shear stress}: \begin{bmatrix} 0 & P_{xy} & 0 \\ P_{xy} & 0 & 0 \\ 0 & 0 & 0 \end{bmatrix}$$

An invariant condition for pure shear stress is the following:

$$I = 0 \quad , \quad III = 0 \quad , \quad P_{xy} = (-II)^{1/2}.$$

4.3 - LOCAL DYNAMIC EQUATIONS

DEFINITION 4.6 - 1) *The* **flux** $N^{\cdots j \cdots k}_{\cdots i \cdots}$ *of a tensor entity with* **concentration** $C^{\cdots j \cdots}_{\cdots i \cdots}$ *is the transfer velocity of this entity across a surface.*
2) *The flux N, with respect to an inertial frame, can be written as follows:*

$$N = J + \widetilde{J}\,,$$

where:

a) J *is the* **molecular flux** *(or* **diffusion flux***), that flows with respect to the body frame of the continuum system;*
b)

$$\widetilde{J} \equiv N - J \equiv \textbf{convective flux} \equiv C \otimes v$$

where v is the velocity of the continuum system

$$\widetilde{J}^{\cdots j \cdots k}_{\cdots i \cdots} = C^{\cdots j \cdots}_{\cdots i \cdots} v^k .$$

3) *The* **generation velocity of a tensor entity** *with concentration $C^{\cdots j \cdots}_{\cdots i \cdots}$ is denoted by*

$$\mathcal{R}^{\cdots j \cdots}_{\cdots i \cdots}.$$

In the following table we report some important fluxes of tensor entities.

TAB.4.4 - **Examples of fluxes of tensor entities**

Entity	$C^{\cdots j \cdots}_{\cdots i \cdots}$	$J^{\cdots j \cdots k}_{\cdots i \cdots}$	$\bar{J}^{\cdots j \cdots k}_{\cdots i \cdots}$	$\mathcal{R}^{\cdots j \cdots}_{\cdots i \cdots}$
Mass	ρ	0	ρv^j	0
Momt.	ρv^j	$-P^{ij}$	$\rho v^i v^j$	ρB^j
Energy	$\rho(e + \frac{1}{2}v^2)$	$q^j - P^{ij} v_i$	$\rho(e + \frac{1}{2}v^2) v^j$	$\rho B^j v_j$
Angular momt.	$\rho v^j x^k \epsilon^i_{jk}$	$-m^{is} - P^{js} x^k \epsilon^i_{jk}$	$\rho v^j x^k \epsilon^i_{jk} v^s$	$\rho B^j \epsilon^i_{jk} x^k + \rho l^i$

THEOREM 4.4 - (LOCAL DYNAMIC EQUATIONS). *One has the following relations between $C, J, \widetilde{J}$ and $\mathcal{R}$:*

$$\boxed{\frac{\delta C}{\delta t} = -C \operatorname{div} v - \operatorname{div} J + \mathcal{R}}$$

or in coordinates:

$$\boxed{\frac{\delta C^{\cdots j \cdots}_{\cdots i \cdots}}{\delta t} = -C^{\cdots j \cdots}_{\cdots i \cdots} v^k{}_{/k} - J^{\cdots j \cdots k}_{\cdots i \cdots /k} + \mathcal{R}^{\cdots j \cdots}_{\cdots i \cdots} = 0} \tag{4.1}$$

PROOF. The general statement of balance equation can be expressed by means of the following formula:

$$\int_V (\partial t.C^{\cdots j \cdots}_{\cdots i \cdots}) d\mu = -\int_{\partial V} N^{\cdots j \cdots k}_{\cdots i \cdots .} n_k d\sigma + \int_V \mathcal{R}^{\cdots j \cdots}_{\cdots i \cdots} d\mu$$

and from the divergence theorem we get:

$$\int_V (\partial t . C^{\cdots j \cdots}_{\cdots i \cdots}) d\mu = -\int_V N^{\cdots . j \cdots . k}_{\cdots i \cdots / k} d\mu + \int_V \mathcal{R}^{\cdots j \cdots}_{\cdots i \cdots} d\mu$$

Thus we have:

$$\begin{aligned}
\int_V (\partial t . C^{\cdots j \cdots}_{\cdots i \cdots}) d\mu &= -\int_V (\widetilde{J}^{\cdots . j \cdots . k}_{\cdots i \cdots / k} + J^{\cdots . j \cdots . k}_{\cdots i \cdots / k}) d\mu + \int_V \mathcal{R}^{\cdots j \cdots}_{\cdots i \cdots} d\mu \\
&= -\int_V [(C^{\cdots j \cdots}_{\cdots i \cdots} v^k)_{/k} + J^{\cdots . j \cdots . k}_{\cdots i \cdots / k}] d\mu + \int_V \mathcal{R}^{\cdots j \cdots}_{\cdots i \cdots} d\mu \\
&= -\int_V [C^{\cdots . j \cdots .}_{\cdots i \cdots / k} v^k + C^{\cdots j \cdots}_{\cdots i \cdots} v^k{}_{/k} + J^{\cdots . j \cdots . k}_{\cdots i \cdots / k}] d\mu \\
&+ \int_V \mathcal{R}^{\cdots j \cdots}_{\cdots i \cdots} d\mu
\end{aligned}$$

$$\Downarrow$$

$$\int_V [(\partial t \cdot C^{\cdots j \cdots}_{\cdots i \cdots}) + C^{\cdots . j \cdots .}_{\cdots i \cdots / k} v^k] d\mu = \int_V [-C^{\cdots j \cdots}_{\cdots i \cdots} v^k{}_{/k} - J^{\cdots . j \cdots . k}_{\cdots i \cdots / k} + \mathcal{R}^{\cdots j \cdots}_{\cdots i \cdots}] d\mu .$$

If $\lambda^{\cdots i \cdots}_{\cdots j \cdots}$ is a constant entity, we can write

$$\int_V [(\frac{\delta C^{\cdots j \cdots}_{\cdots i \cdots}}{\delta t}) + C^{\cdots j \cdots}_{\cdots i \cdots} v^k{}_{/k} + J^{\cdots . j \cdots . k}_{\cdots i \cdots / k} - \mathcal{R}^{\cdots j \cdots}_{\cdots i \cdots}] \lambda^{\cdots i \cdots}_{\cdots j \cdots} d\mu = 0.$$

From the arbitrariness of λ (4.1) follows. □

REMARK 4.5 - The general transfer-equation can also be written as follows:

$$\boxed{(\partial t \cdot C^{\cdots j \cdots}_{\cdots i \cdots}) = -N^{\cdots . j \cdots . k}_{\cdots i \cdots / k} + \mathcal{R}^{\cdots j \cdots}_{\cdots i \cdots}}$$

or

$$\boxed{(\partial t \cdot C) = -\mathrm{div}\, N + \mathcal{R}}$$

THEOREM 4.5 - 1) *If $C = \rho$ we get the* **continuity equation**:

$$\boxed{\frac{\delta t}{\delta t} = -\rho v^k{}_{/k}}$$

or

$$\boxed{(\partial t \cdot \rho) + (\rho v^k)_{/k} = 0}$$

2) *If* $C = \rho v$ *we get the* **motion equation**:

$$\boxed{\frac{\delta}{\delta t}(\rho v^i) = -\rho v^i v^k{}_{/k} + P^{ik}{}_{/k} + \rho B^i}$$

or

$$\boxed{P^{ki}{}_{/k} + \rho(B^i - a^i) = 0}$$

and also

$$\boxed{\rho\frac{\delta v^i}{\delta t} = P^{ik}{}_{/k} + \rho B^i \equiv -g^{ij} p_{/j} + \wp^{ik}{}_{/k} + \rho B^i}$$

3) *If* $C \equiv$ *concentration of the moment of momentum, from the equation of the momentum, we get the* **angular momentum equation**:

$$\boxed{m^{ji}{}_{/j} + \rho l^i + \epsilon^{ijk} P_{jk} = 0}$$

or in bivectorial form, by putting $l_i \equiv \epsilon_{ijk} l^{jk}$, *we have:*

$$\boxed{m^{kij}{}_{/k} + \rho l^{ij} + P^{[ij]} = 0} \tag{4.2}$$

REMARK 4.6 - If the body-couples are absent ($l = 0$), and also the stress-moment is absent ($m = 0$), then the stress tensor is symmetric ($P^{[ij]} = 0$):

$$P^{ij} = P^{ji}.$$

PROOF. This follows directly from equation (4.2). □

REMARK 4.7 - For the symmetric part of the stress-tensor we get the following equation:

$$\boxed{P^{(jk)}{}_{/j} - m^{(ij)k}{}_{/ij} + \rho(B^k - l^{jk}_{/j} - a^k) = 0.}$$

THEOREM 4.6 - *During the flow the mass-density changes according to the following rule:*

$$\rho(t_0) = J_\lambda \rho(t), \tag{4.3}$$

with $\lambda = t - t_0$, *where* J_λ *is the jacobian of the flow.*
PROOF.

$$\phi^*_\lambda \eta = J_\lambda \eta, \quad \lambda = t - t_0.$$

From the conservation of mass we get:

$$\rho(t_0)\eta = \rho(t)\phi^*_\lambda \eta = \rho(t) J_\lambda \eta\,.$$

Then, we have the (4.3). □

REMARK 4.8 - For isocoric flow we get: $J_\lambda = 1, \forall \lambda \in \mathbf{R}$.

THEOREM 4.7 - 1) *If* $C = \rho(e + \frac{1}{2}v^2)$ *we get the* **energy equation**

$$\boxed{\frac{\delta}{\delta t}[\rho(e + \frac{1}{2}v^2)] = -e^k{}_{/k} + n^k B_k}$$

where

$$e^k \equiv \rho(e + \frac{1}{2}v^2)v^k + q^k - P^{ki}v_i \equiv \textbf{energy flux}$$
$$n^k \equiv \rho v^k \equiv \textbf{mass flux.}$$

2) *The energy equation can also be written in the following way:*

$$\boxed{\rho\frac{\delta}{\delta t}(e + \frac{1}{2}v^2) = -q^k{}_{/k} + (P^{kj}v_j)_{/k} + \rho B^i v_j.}$$

3) *An alternative form for the energy equation is the following:*

$$\rho\frac{\delta e}{\delta t} = P^{(ij)}\bar{e}_{ij} - m^{rij}\omega_{ij/r} - q^i{}_{/i} + \rho\bar{h} \tag{4.4}$$

where $\bar{h}$ *is the* **body energy source**.

PROOF. This equation can, of course, be proved starting from the general transfer equation. However, here, we want to prove it using the fundamental laws of the thermodynamics.

First law of thermodynamics:The **velocity of change of kinetic energy** $\bar{h}$ and of **internal energy** $\bar{e}$ equals the **work** W made on the body by the external forces per unit time plus the sum of the other energies U_α, per unit time. (Many types of energies can be considered, caloric, chemical, electric, etc.):

$$\frac{\delta \bar{h}}{\delta t} + \frac{\delta \overline{E}}{\delta t} = W + \sum_\alpha U_\alpha. \tag{4.5}$$

In the following formula we will consider only thermic one, denoted by Q. Thus, we have:

$$\overline{H} = \frac{1}{2}\int_V \rho v_i v^i d\mu, \quad \underline{E} = \int_V \rho e d\mu$$
$$W = \int_V \rho(B^j v_j - e^{js}\omega_{js})d\mu + \int_{\partial V}(P^{ij}v_j - m^{ijs}\omega_{js})h_i d\sigma$$
$$Q = \int_V \rho\bar{h}d\mu - \int_{\partial V} q^i n_i d\sigma.$$

Then, by means of Stokes theorem, we get:

$$\frac{\delta\overline{H}}{\delta t} = \int_V [\rho a^p v_p d\mu + \frac{1}{2}v^p v_p \frac{\delta}{\delta t}(\rho d\mu)]$$
$$\frac{\delta\overline{E}}{\delta t} = \int_V [\rho\frac{\delta e}{\delta t}d\mu + e\frac{\delta}{\delta t}(\rho d\mu)]$$
$$W = \int_V [P^{rp}{}_{/r}v_p + P^{rp}v_{p/r} - m^{rpq}{}_{/r}\omega_{pq} - m^{rpq}\omega_{pq/r} + \rho(B^p v_p - l^{pq}\omega_{pq})]d\mu$$
$$Q = \int_V (-q^p{}_{/p} + \rho\bar{h})d\mu$$

where $q = -q^i{}_{/i} + \rho\bar{h}$ is the heat per unit volume. It is also useful to define the heat per unit mass: $\bar{q} = -\frac{1}{\rho}q^i{}_{/i} + \bar{h}$. Of course $Q = \int_v \rho\bar{q}d\mu$. If we write $v_{p/r} = \dot{e}_{pr} + \omega_{pr}$ in W, we get:

$$W = \int_V [(P^{rp}{}_{/r} + \rho B^p)v_p - (P^{[pq]} + m^{rpq}{}_{/r} + \rho l^{pq})\omega_{pq} + (P^{(rp)}\dot{e}_{pr} - m^{rpq}\omega_{pq/r})]d\mu.$$

Therefore, from (4.5), we have:

$$\int_V [\rho\frac{\delta e}{\delta t} - P^{(pr)}\dot{e}_{pr} + m^{prq}\omega_{pq/r} + q^p{}_{/p} - \rho\bar{h}]d\mu$$
$$= \int_V (-\frac{1}{2}v^p v_p - e)\frac{\delta}{\delta t}(\rho d\mu) + \int_V v_p(P^{rp}{}_{/r} + \rho B^p - \rho a^p)d\mu$$
$$- \int_V \omega_{pq}(P^{[pq]} + m^{rpq}{}_{/r} + \rho l^{pq})d\mu = 0.$$

If the energy is balanced for little arbitrary elements of body, we get the conservation of energy. □

REMARK 4.9 - The **Fourier law** for heat flux transferred by **conduction** is the following: $q = -\lambda \operatorname{grad}\theta$ or, in coordinates, $q^j = -\lambda g^{ik}(\partial x_k \cdot \theta)$. However, the heat can also be transferred by **convection** (in this case is called **sensible heat**), and for **radiation**. In solids heat can only be transferred by conduction and radiation.

PROPOSITION 4.4 - 1)(NON-POLAR CASE). *In non-polar case the energy equation becomes:*

$$\rho\frac{\delta e}{\delta t} = P^{pq}\dot{e}_{pq} - q^p{}_{/p} + \rho\bar{h}, \tag{4.6}$$

or

$$\rho\frac{\delta e}{\delta t} = \mathcal{W} + q$$

where

$\mathcal{W} \equiv P^{pq}\dot{e}_{pq} \equiv$ **stress power** *(or* **dissipation function***)*.

PROOF. In fact, in this case we get: $m^{rij} = 0$ and $P^{(ij)} = P^{ij}$. □

2) (POTENTIAL ENERGY). *If the body forces are conservative, i.e.,*

$$B_p = -(\partial x_p \cdot u)$$

we can define the **potential energy**

$$U = \int_V \rho u d\mu.$$

In such a case, we have the following:

a) *The equation of energy conservation becomes:*

$$\frac{\delta\bar{h}}{\delta t} + \frac{\delta\overline{E}}{\delta t} + \frac{\delta U}{\delta t} = \int_{\partial V} P^{ij}v_j d\sigma + Q.$$

PROOF. In fact, the power can be written in the following way:

$$W = \int_V \rho(\partial x_j \cdot u)v^j d\mu + \int_{\partial V} P^{ij}v_j d\sigma$$

hence

$$\frac{\delta U}{\delta t} = \frac{\delta}{\delta t}\int_V \rho\, u\, d\mu = \int_V \rho(\frac{\delta u}{\delta t})d\mu = \int_V \rho(\partial x_p \cdot u)v^p d\mu.$$

Then, we get

$$W = \int_{\partial V} P^{ij}v_j d\sigma - \frac{\delta U}{\delta t}.$$

□

b) *Conservation of mechanic energy (in adiabatic process): If the body is isolated, does not contain heat-sources, and the surface-stress is* $\perp$ *the surface velocity, we get:*

$$\overline{H} + E + U = const.$$

3) (EXPRESSION OF INTERNAL ENERGY BY MEANS OF STRAIN ENERGY AND DISSIPATION ENERGY). *We can consider the stress tensor splitted into two parts:*

$$P^{ij} = {}_{\overline{D}}P^{ij} + {}_{D}P^{ij}, \tag{4.7}$$

where ${}_{\overline{D}}P^{ij}$ $\equiv$ **hyperelastic stress** $\equiv$ *reversible part, and* ${}_{D}P^{ij}$ $\equiv$ **dissipative stress**
$\equiv$ *irreversible part.* ${}_{\overline{D}}P^{ij}$ *comes from a potential* $\mathbf{d}(\partial a_k \cdot x^i) \equiv$ **strain energy function***:*

$$\rho\frac{\delta \mathbf{d}}{\delta t} = {}_{\overline{D}}P^{ij}\dot{e}_{ij}. \tag{4.8}$$

The velocity of internal energy can be written in the following way:

$$\frac{\delta E}{\delta t} = \frac{\delta \overline{D}}{\delta t} + D + Q$$

where

$$\overline{D} \equiv \textbf{total strain energy} = \int_V \rho\, \mathbf{d}\, d\mu$$

$$D \equiv \textbf{total dissipation power} = \int_V {}_{D}P^{ij}\dot{e}_{ij} d\mu.$$

PROOF. From (4.5), (4.3), we have:

$$\int_V \rho\frac{\delta e}{\delta t} d\mu = \int_V \rho\frac{\delta \mathbf{d}}{\delta t} d\mu + \int_V {}_{D}P^{ij}\dot{e}_{ij} d\mu + \int_V (-q^j{}_{/i} + \rho\bar{h}) d\mu.$$

$\square$

REMARK 4.10 - 1) From the splitting $P = -pg + \wp$ we get that the stress power is given by the following:

$$\boxed{\mathcal{W} = -pv^i{}_{/i} + \wp^{ij}\dot{e}_{ij}}$$

In particular, if $\wp = 0$, we get:

$$\mathcal{W} = \rho\frac{\delta}{\delta t}\mathbf{d}.$$

We can also write:

$$\mathcal{W} = -p\frac{1}{d\mu(t)}\frac{\delta}{\delta t}(d\mu).$$

2) (**Piola's stress**). *The Piola's stress is defined by the following:*

$$_{\overline{D}}P_i^R \equiv \rho_0(\partial B_i^R \cdot \mathbf{d})$$

where $B_R^i \equiv (\partial a_R \cdot x^i)$. *Furthermore, the Piola's stress satisfies equation* (4.8).
PROOF. a) In fact, we get:

$$P^{Ri} = JB_j^R P^{ji}.$$

Furthermore, taking into account that $J = \rho_0/\rho$, we have:

$$P^{ji} = \frac{\rho}{\rho_0}P^{Rj}(\partial a_R \cdot x^i).$$

On the other hand

$$_{\overline{D}}P^{ij}\dot{e}_{ij} = {}_{\overline{D}}P^{ij}v_{j/i} = \frac{\rho}{\rho_0}{}_{\overline{D}}P^{Rj}(\partial a_R \cdot x^i)v_{j/i} = \frac{\rho}{\rho_0}{}_{\overline{D}}P_i^R\frac{\delta}{\delta t}(B_R^j).$$

Therefore, by using Piola's stress we get (4.8). □
b) *One has:*

$$_{\overline{D}}P_j^i = \rho(\partial a_R \cdot x^i)\cdot(\partial(\partial a_R \cdot x^j)\cdot \mathbf{d}).$$

c) *In the particular case where* $\mathbf{d} = d(J)$ *we get:*

$$\begin{aligned} (i) &\quad {}_{\overline{D}}P^{ij} = -p_0\delta^{ij} \\ (ii) &\quad \rho\frac{\delta \mathbf{d}}{\delta t} = -p_0 v^i{}_{/j} \\ (iii) &\quad \frac{\delta \overline{D}}{\delta t} = -\int_V p_0\frac{\delta(d\mu)}{\delta t} \end{aligned}$$

where p_0 *is the isobaric pressure:*

$$p_0 = -\rho_0(\partial J \cdot \mathbf{d}).$$

PROOF. In fact, one has:

$$_{\overline{D}}P_j^i = \rho(\partial a_R \cdot x^i)(\partial J \cdot \mathbf{d})(\partial(\partial a_R \cdot x^j)\cdot J) = \rho(\partial a_R \cdot x^j)(\partial J \cdot \mathbf{d})J(\partial x_j a^R).$$

□

REMARK 4.11 - (LOCAL EQUATIONS FOR CONTINUUM SYSTEMS IN STATICS). If there is no motion ($v = 0$) dynamic equations give the corresponding static equations:

$$\boxed{\begin{aligned} (\partial t.\rho) &= 0 \\ P^{ik}_{/k} + \rho B^i &= 0 \\ m^{ji}_{/j} + \rho l^i + \epsilon^{ijk} P_{jk} &= 0 \\ \rho(\partial t.e) &= -q^k_{/k} \end{aligned}}$$

4.4 - THERMODYNAMICS OF CONTINUUM MEDIA

We report in the following tables some fundamental informations and definitions that are useful in Continumm Mechanics.

TAB.4.5 - **Important thermal conductibilities**

Systems	$\lambda \equiv$ thermal conductibility
Monoatomic gas	$\lambda = \frac{5}{2} C_v \eta, \quad C_p = C_v =$ specific heats
Polyatomic gas	$\lambda = (C_p + \frac{5}{4}\frac{R}{M})\eta, \quad \eta =$ viscosity, $M =$ mole
Liquids	$\lambda = 2.80(\frac{\widetilde{N}}{\widetilde{V}})^{2/3} v_s$
	$v_s =$ low frequency sound velocity $= \sqrt{\frac{C_p}{C_v}(\partial\rho \cdot P)_\theta}$
	$(\partial\rho \cdot P)_\theta =$ isothermic compressibility
	$\widetilde{V} = M/\rho, \quad \widetilde{N} =$ Avogadro number $= 6,023 \cdot 10^{23}$
Solids	$\frac{\lambda}{\lambda_e \theta}$, $L =$ Lorentz number, $\lambda_e =$ electric cond.

($\alpha \equiv \lambda/\rho C_p \equiv$ **thermal diffusivity**)

TAB.4.6 - **Important parameters in rheology**

Name	Definition
Apparent viscosity	$\eta = \wp_{12}/\dot{e}$ ($\dot{e} =$ inf.velocity strain)
Coefficient of normal stress	$\psi_{ii} = \wp_{ii}/\dot{e}^2$
Coeff.difference coeff.nor.stress	$\psi_{ij} = (\wp_{ii} - \wp_{jj})/\dot{e}^2$
Trouton (elongational) viscosity	$\eta_e = -\frac{\wp_{11}}{\dot{e}_{22}} = -\frac{\wp_{11}}{\dot{e}_{33}} = -\frac{2\wp_{11}}{\dot{e}_{11}}$

TAB.4.7 - **Important adimensional numbers**

Name	Definition
Reynolds	$R = \rho v l/\eta$, (ρ = mass density)
Strouhal	$S = v\tau/l$, (v = velocity, l = length)
Froude	$F = v^2/\rho g$, (τ = time, g = gravity acc.)

DEFINITION 4.7 - 1) *A* **thermodynamic state** *is characterized by a set of independent parameters*

$$(s, c_\alpha), \quad \alpha = 1, \cdots, n$$

that identify the internal energy:

$$e = e(s, c_\alpha, a^i)$$

where a^i *identifies the* i*-th material particle,* s *is the* **specific entropy**, *(with physical dimension* $[s]$ = *energy /temperature mass), and* c_α *are electric or mechanical parameters.*

2) *The system is said to be* **thermodynamically homogeneous** *if the following equations are satisfied:*

$$(\partial a_\alpha \cdot e) = 0.$$

3) *As a consequence we give the following definitions:*

$$\textbf{Temperature}: \theta \equiv (\partial s \cdot e).$$
$$\textbf{Thermodynamic stress}: t^\alpha \equiv (\partial c^\alpha \cdot e).$$
$$\textbf{Thermodynamic pressure}: p \equiv t^1 = -(\partial(\frac{1}{\rho}) \cdot e).$$
$$\textbf{Chemical potentials}: p^\alpha \equiv (\partial e^\alpha \cdot e), e_\alpha \equiv c_\alpha.$$

c_α = *concentration of* α*-component*

THEOREM 4.8 - *One has the following equations:*

$$(\textbf{Gibbs equations}): de = \theta ds + t^\alpha dc_\alpha$$
$$(\textbf{Generalized Gibbs equations}): \frac{\delta e}{\delta t} = \theta \frac{\delta s}{\delta t} + t^\alpha \frac{\delta c_\alpha}{\delta t}.$$

Here we have considered the following functional dependence:

$$\theta = \theta(s, c^\alpha, a^i), \quad t^\alpha = t^\alpha(s, c^\alpha, a^i).$$

Of course we can also assume

$$s = s(\theta, c^\alpha, a^i), \quad t^\alpha = t^\alpha(\theta, c^\alpha, a^i) \Rightarrow e = e(\theta, c^\alpha, a^i).$$

DEFINITION 4.8 - *We give also the following definitions for a material point:*

$$\begin{aligned}
&\textbf{Thermodynamic curve}: s = s(\lambda), \quad c^\alpha = c^\alpha(\lambda);\\
&\textbf{Isotropic thermodynamic curve}: s = constant;\\
&\textbf{Isothermic thermodynamic curve}: \theta = constant;\\
&\textbf{Specific heats}: C \equiv \frac{\delta q}{\delta\theta} = \frac{1}{\dot\theta}(\dot e - t^\alpha \dot c_\alpha);\\
&\textbf{Latent heats}: \left\{ \begin{array}{l} \widehat{e}^\alpha \equiv \dfrac{\delta q}{\delta t^\alpha} = \dfrac{1}{\dot t^\alpha}(\dot e - t^\beta \dot c_\beta) \\ \widehat{m}_\alpha \equiv \dfrac{\delta q}{\delta C_\alpha} = \dfrac{1}{\dot c_\alpha}(\dot e - t^\beta \dot c_\beta) \end{array} \right\}
\end{aligned}$$

THEOREM 4.9 - *For fixed* a^α, *a thermodynamic curve depends on the time:* $\lambda = \lambda(t)$. *Then, one has:*

$$\begin{aligned}
C &= \frac{1}{d\theta}(de - t^\beta dc_\beta) = \theta(\partial\theta \cdot s)\\
\widehat{e}_\alpha &= \frac{1}{dt^\alpha}(de - t^\beta dc_\beta) = \theta(\partial t_\alpha \cdot s)\\
\widehat{m}_\alpha &= \frac{1}{dc_\alpha}(de - t^\beta dc_\beta) = \theta(\partial c_\alpha \cdot s).
\end{aligned}$$

DEFINITION 4.9 - *One has the following important definitions:*

Specific heat at $c^\alpha = constant: C_c \equiv (\partial\theta \cdot e)_{c^\alpha}.$

Specific heat at $t^\alpha = constant: C_t = C_c + [(\partial c_\beta \cdot e)_{\theta, c_\alpha} - t^\beta](\partial\theta \cdot c_\beta)_t$

$$\gamma \equiv C_t / C_c.$$

THEOREM 4.10 - (ENTROPY EQUATION). *One has the following entropy equations:*

$$\rho\theta\frac{\delta s}{\delta t} = P^{(ij)}\dot e_{ij} - m^{rij}\omega_{ij/r} - q^j{}_{/i} + \rho\bar h - \rho t^\alpha \frac{\delta(c_\alpha)}{\delta t}$$

or

$$\rho\theta\frac{\delta s}{\delta t}=g_{ik}P^{(ij)}v^{k}{}_{/j}+m^{ri}{}_{j}v^{j}{}_{/ir}-q^{j}{}_{/i}+\rho\bar{h}-\rho t^{\alpha}\frac{\delta(c_{\alpha})}{\delta t}. \tag{4.9}$$

PROOF. It is a direct consequence of the generalized Gibbs equations, after elimination of $\dot{e}$ in (4.4). □

THEOREM 4.11 - *The stress tensor and moment of stress tensor have the following explicit expressions:*

$$\begin{aligned}P^{(ij)}&={}_{D}P^{(ij)}+\rho g^{ri}[(\partial B^{R}_{r}\cdot e)B^{j}_{R}+(\partial x^{:RL}_{r}\cdot e)x^{j}_{:RL}]\\ m^{(ri)}{}_{j}&={}_{D}m^{(ri)}{}_{j}+\rho(\partial x^{:RL}_{i}\cdot e)B^{i}_{R}B^{r}_{L}\end{aligned} \tag{4.10}$$

where

$$x^{j}_{:RL}\equiv(B^{j}_{R})_{/m}B^{m}_{L}$$

and

$$\begin{aligned}{}_{D}P^{(ij)}&\equiv\rho g^{ri}({}_{D}t^{R}_{r}B^{j}_{R}+{}_{D}t^{RL}_{r}x^{j}{}_{:RL})\\ {}_{D}m^{(ri)}{}_{j}&=\rho{}_{D}t^{RL}_{j}B^{i}_{R}B^{r}_{L}\end{aligned} \tag{4.11}$$

are the dissipative parts. The reversible parts come from a potential, i.e., the **internal energy density**.

PROOF. We have:

$$v^{r}{}_{/q}=B^{R}_{q}\frac{\delta}{\delta t}(B^{r}_{R})$$

hence

$$\begin{aligned}v^{q}{}_{/pr}&=[\frac{\delta}{\delta t}(B^{q}_{R})B^{R}_{p}]_{/r}=[\frac{\delta}{\delta t}(B^{q}_{R})]_{/r}B^{R}_{p}+\frac{\delta}{\delta t}(B^{q}_{R})x^{R}{}_{:pr}\\ &=\frac{\delta}{\delta t}[B^{q}{}_{R/m}B^{m}_{L}]B^{L}_{r}B^{R}_{p}+\frac{\delta}{\delta t}(B^{q}_{R})x^{R}{}_{/pr}\end{aligned}$$

where $x^{R}{}_{:pr}=(B^{R}_{p})_{/r}$. Therefore, setting $x^{q}{}_{:RL}\equiv(B^{q}_{R})_{/m}B^{m}{}_{L}$, we can write:

$$v^{q}{}_{/pr}=\frac{\delta}{\delta t}(x^{q}{}_{:RL})B^{R}_{p}B^{L}_{r}+x^{R}{}_{/pr}\frac{\delta}{\delta t}(B^{q}_{R}).$$

Then, the entropy equations (4.9) become:

$$\begin{aligned}\rho\theta\frac{\delta s}{\delta t}&=[g_{qr}P^{(ri)}B^{R}_{j}+m^{rp}{}_{q}x^{R}{}_{/pr}]\frac{\delta}{\delta t}(B^{q}_{R})\\ &+m^{rp}{}_{q}B^{R}_{p}B^{L}_{r}\frac{\delta}{\delta t}(x^{q}{}_{:RL})+q^{p}{}_{/p}+\rho h-\rho t^{\alpha}\frac{\delta c_{\alpha}}{\delta t}.\end{aligned} \tag{4.12}$$

These equations suggest to choose the following forms for the constitutive equations of the internal energy:

$$e = e(s, B^p_R, x^p{}_{:RL}, c_r).$$

Thus, setting:

$$t^R_p \equiv (\partial B^R_p \cdot e), \quad t^{RL}_p \equiv (\partial x^{:RL}_p \cdot e), \quad t^\gamma \equiv (\partial c^\gamma \cdot e)$$

we can write equation (4.11) in the following form:

$$\theta \frac{\delta s}{\delta t} = {}_Dt^R_q \frac{\delta}{\delta t}(B^q_R) + {}_Dt^{RL}_q \frac{\delta}{\delta t}(x^q{}_{:RL}) + \frac{1}{\rho} q^j{}_{/i} + h - t^\gamma \dot{c}_\gamma$$

where

$$\begin{aligned}
\rho_D t^R_q &\equiv g_{qr} P^{(rk)} B^r_k + m^{rp}{}_q x^R{}_{/pr} - \rho t^R_q \\
&\Downarrow \\
P^{(ij)} &= \rho g^{ik}(t^R_k - \frac{1}{\rho} m^{sr}{}_k x^R{}_{/rs} + {}_Dt^R_k) B^j_R \\
\rho_D t^{RL}_q &\equiv m^{rp}{}_q B^R_p B^L_r - \rho t^{RL}{}_q \\
&\Downarrow \\
m^{(ij)}{}_q &= \rho(t^{RL}_q + {}_Dt^{RL}_q) B^j_R B^i_L.
\end{aligned}$$

Note that we cannot determine $m^{[ij]}{}_q$ as the coefficient of $m^{ij}{}_q$ is symmetric in the indexes ij. Now, taking into account that $B^k_R B^R_r = \delta^k_r$ it follows that

$$x^R{}_{/rs} = -x^k{}_{:LM} B^R_k B^L_r B^k_s.$$

Therefore, we get:

$$\begin{aligned}
P^{(ij)} &= {}_DP^{(ij)} + \rho g^{ri}[(\partial B^R_r \cdot e) B^j_R + (\partial x^{:RL}_r \cdot e) x^j_{:RL}] \\
m^{(ij)}{}_q &= {}_Dm^{(ij)}{}_q + \rho(\partial x^{:RL}_q \cdot e) B^j_R B^i_L.
\end{aligned}$$

□

THEOREM 4.12 - 1) *The contribution to the entropy comes only from the dissipative part of the stress and momentum-stress.*

PROOF. 1) In fact, we can also write equation (4.9) as follows:

$$\rho\theta \frac{\delta s}{\delta t} = g_{ik} P^{(ij)} v^k{}_{/j} + m^{(ri)}{}_j v^j{}_{/ir} - q^j{}_{/i} + \rho \bar{h} - \rho t^\alpha \frac{\delta c_\alpha}{\delta t}$$

where $P^{(ij)}$ and $m^{(ri)}{}_j$ are defined in equations (4.10) and (4.11) respectively.

2) The dynamic equations do not allow one to know the skew-symmetric part of the stress tensor. □

THEOREM 4.13 - (NON-POLAR CASE).

$$\textbf{(Stress-tensor)}: \quad P^{(ij)} = \rho g^{ri}(\partial B^R_r \cdot e)B^j_R + {}_DP^{(ij)}$$
$$\textbf{(Entropy equation)}: \quad \rho\theta\frac{\delta s}{\delta t} = {}_DP^{(ij)}\dot{e}_{ij} - q^j{}_{/i} + \rho\bar{h} - \rho t^\alpha\frac{\delta c_\alpha}{\delta t}.$$

If the reversible part of the stress is purely hydrostatic, we can write:

$$P^i_j = -p\delta^i_j + {}_DP^i_j, \tag{4.13}$$

where $p = -(\partial c \cdot e)_s \equiv$ *thermodynamic pressure,* $c \equiv 1/\rho \equiv$ *specific volume. On the other hand*

$$P^i_j = -\widehat{p}\delta^i_j + \wp^i_j, \tag{4.14}$$

where $\widehat{p}$ *is the mechanical pressure* $= -(1/3)P^i_i$, *and* $\wp^i_j \equiv$ *deviatory stress* ($\wp^i_i = 0$). *From these two equations* (4.13) *and* (4.14) *we get:*

$${}_DP^i_j = -(\widehat{p} - p)\delta^i_j + \wp^i_i.$$

This equation assures that the part of the mechanical pressure that differs from the thermodynamic pressure contributes to the dissipation.

REMARK 4.12 - Usually the thermodynamic pressure and the mechanical pressure are assumed to be the same. In Stokesian fluids this assumption is called "Stokes hypothesis".

THEOREM 4.14 - (NON-DISSIPATIVE CASE) (${}_DP^{ij} = {}_Dm^{ij} = 0$).

$$P^{(ij)} = \rho g^{ri}[(\partial B^R_r \cdot e)B^j_R + (\partial x^{:RL}_r \cdot e)x^j_{:RL}];$$
$$m^{(rj)}{}_j = \rho(\partial x^{:RL}_j \cdot e)B^i_R B^r_L\,.$$

Furthermore, if e *is independent of* $x^p_{:RL}$ *we get the non-polar case:*

$$P^{ij} = \rho g^{ri}(\partial B^R_r \cdot e)B^j_R, \quad m^{ri}{}_j = 0.$$

The Piola stress results:

$$P^k_q = \rho_0(\partial B^R_q \cdot e)_s. \tag{4.15}$$

Therefore, in the non-dissipative case we get a **stress potential***:*

$$\Sigma \equiv \rho_0 e. \tag{4.16}$$

A relation such as in (4.16) *is called a* **stress-strain relation**.

THEOREM 4.15 - (GLOBAL PRODUCTION OF ENTROPY).

$$\mathbf{S} \equiv \int_V \rho s d\mu \equiv \textbf{global entropy}$$

$$\frac{\delta \mathbf{S}}{\delta t} + \int_{\partial V} \frac{q^p}{\theta} d\sigma_p = \int_V (\widehat{\triangle} + \frac{\rho \bar{h}}{\theta}) d\mu$$

$$\theta \widehat{\triangle} = {}_D P^{(ij)} \dot{e}_{ij} + {}_D m^{(rj)} \omega_{ij/r} - \rho t^\alpha \dot{c}_\alpha - q^j (\partial x_i \cdot (\log \theta)).$$

Then, we can say that the global production of energy is given by the following:

$$(a)\, \textbf{entropy flux}: \quad q^p{}_{/\theta} \equiv J^p(s)$$

$$(b)\, \textbf{Entropy production}: \quad \mathcal{R}(s) \equiv (\frac{\widehat{\triangle}}{\rho} + \frac{\bar{h}}{\theta}), \text{(in the body)}$$

where $\triangle \equiv \widehat{\triangle}/\rho$ *is called velocity of thermodynamic dissipation, given by:*

$$\triangle_M + \triangle_\theta.$$

Here, $\triangle_M$ *is the* **mechanic part**, *defined by:*

$$\triangle_M \equiv \frac{1}{\rho}[{}_D P^{(ij)} \dot{e}_{ij} + {}_D m^{(rj)} \omega_{ij/r} - \rho t^\alpha \dot{c}_\alpha]$$

and $\triangle_\theta$ *is the* **thermic part**, *defined by:*

$$\triangle_\theta \equiv -\frac{1}{\rho} q^i (\partial x_i \cdot (\log \theta)).$$

PROOF. We have

$$\frac{\delta \mathbf{S}}{\delta t} = \frac{\delta}{\delta t} \int_V \rho s d\mu = \int_V \rho \frac{\delta s}{\delta t} d\mu = \int_V \frac{1}{\theta} q^i{}_{/i} d\mu = \int_V [(\frac{q^j}{\theta})_{/i} + \frac{q^j (\partial x_i \cdot \theta)}{\theta^2}] d\mu$$

$$= \int_{\partial V} \frac{q^j}{\theta} d\sigma_i + \int_V q^j \frac{(\partial x_i \cdot \theta)}{\theta^2} d\mu.$$

□

THEOREM 4.16 - (CLAUSIUS-DUHEM EQUATION).

1) (**Isothermic case**):

$${}_D P^{(ij)} \dot{e}_{ij} + {}_D m^{(rj)i} \omega_{ij/r} - \rho t^\alpha \dot{c}_\alpha \geq 0 \ (\textbf{Planck inequality}).$$

2) (**Static case**):

$$-q^j(\partial x_i \cdot \theta) \geq 0 \text{ (\textbf{Fourier inequality})}.$$

3) (**Global equation of Clausius-Duehm**)(Second law of Thermodynamics):

$$\frac{\delta \mathbf{S}}{\delta t} + \int_{\partial V} \frac{q^i}{\theta} d\sigma_i \geq \int_V \frac{\rho \bar{h}}{\theta} d\mu.$$

If there are no **internal sources of heat** *we get:*

$$\frac{\delta \mathbf{S}}{\delta t} \geq - \int_{\partial V} \frac{q^j}{\theta} d\sigma_i\,. \tag{4.17}$$

(FUNDAMENTAL PRINCIPLE OF THERMODYNAMICS). *Therefore, in an adiabatic process the entropy must increase:*

$$\frac{\delta \mathbf{S}}{\delta t} \geq 0.$$

Equation (4.17) *can be written in the following local form:*

$$\rho \frac{\delta s}{\delta t} \geq \frac{q}{\theta} + \frac{q^j}{\theta^2} \theta_{/i}.$$

THEOREM 4.17 - *In a system in thermic equilibrium, i.e.,*

$$(a)(\textbf{Local equilibrium}): \frac{\delta s}{\delta t} = 0$$

$$(b)(\textbf{Global equilibrium}): \frac{\delta \mathbf{S}}{\delta t} = 0$$

we get:

$$P^{(ij)} = \frac{\rho}{\rho_0} g^{rj}[(\partial B_r^R \cdot \Sigma) B_R^j + (\partial x_r^{:RL} \cdot \Sigma) x^j_{:RL}]$$

$$m^{(ij)}{}_q = \frac{\rho}{\rho_0} (\partial x_q^{:RL} \cdot \Sigma) B_R^j B_L^i$$

where $\Sigma = \rho_0 e$, *is the* **stress potential**.

PROOF. In fact, in this case we have a non-dissipative situation. □

DEFINITION 4.10 - (IMPORTANT THERMODYNAMIC FUNCTIONS).

1) **Local functions**:

(a) $f \equiv e - \theta s \equiv$ **free energy density** (*or* **Helmholtz density**)

(b) $h \equiv e - t^\alpha c_\alpha \equiv$ **enthalpy density**

(c) $g \equiv h - \theta s \equiv$ **free enthalpy density** (*or* **Gibbs function**)

2) **Global entities:**

$$(a) \quad F \equiv \int_V \rho f d\mu \equiv \textbf{free energy}$$

$$(b) \quad H \equiv \int_V \rho h d\mu \equiv \textbf{enthalpy}$$

$$(c) \quad G \equiv \int_V \rho g d\mu \equiv \textbf{free enthalpy}.$$

THEOREM 4.18 - (GIBBS RELATIONS). *We have the following equations, assuming that the thermodynamic state should be* (s, c_α):

1)

$$\begin{array}{ll} (a) & \frac{\delta e}{\delta t} = \theta \frac{\delta s}{\delta t} + t^\alpha \frac{\delta c_\alpha}{\delta t} \\ (b) & \frac{\delta f}{\delta t} = -s \frac{\delta \theta}{\delta t} + t^\alpha \frac{\delta c_\alpha}{\delta t} \\ (c) & \frac{\delta h}{\delta t} = \theta \frac{\delta s}{\delta t} - c_\alpha \frac{\delta t^\alpha}{\delta t} \\ (d) & \frac{\delta g}{\delta t} = -s \frac{\delta \theta}{\delta t} - c_\alpha \frac{\delta t^\alpha}{\delta t} \end{array}$$

Of course one also has the following equations:

$$\begin{array}{ll} (a) & de = \theta ds + t^\alpha dc_\alpha \\ (b) & df = -s d\theta + t^\alpha dc_\alpha \\ (c) & dh = \theta ds - c_\alpha dt^\alpha \\ (d) & dg = -s d\theta - c_\alpha dt^\alpha \end{array}$$

DEFINITION 4.11 - *We call* **velocity of accumulation of elastic energy**:

$$\dot{e}|_{s=const.} = \dot{f}|_{\theta=const.} = t^\alpha \dot{c}_\alpha.$$

3) (ALTERNATIVE FORMS FOR THE CLAUSIUS-DUEHM INEQUALITY).

$$\begin{array}{ll} (a) & \theta \frac{\delta s}{\delta t} - \frac{\delta e}{\delta t} + \frac{1}{\rho}[P^{(ij)} \dot{e}_{ij} - m^{rij} \omega_{ij/r}] - \frac{q^p}{\rho \theta} \theta_{/p} \geq 0 \\ (b) & s \frac{\delta \theta}{\delta t} + \frac{\delta f}{\delta t} - \frac{1}{\rho}[P^{(ij)} \dot{e}_{ij} - m^{rij} \omega_{ij/r}] + \frac{q^p}{\rho \theta} \theta_{/p} \leq 0 \end{array}$$

PROOF. From the relation:

$$\rho\frac{\delta s}{\delta t} \geq \frac{q}{\theta} + \frac{q^j}{\theta^2}\theta_{/i}$$

we get:

$$\rho\dot{s} \geq -\frac{q^p{}_{/p}}{\theta} + \frac{q^p}{\theta^2}\theta_{/p} + \rho\frac{\bar{h}}{\theta}.$$

Furthermore, from

$$\rho\dot{e} = P^{(ij)}\dot{e}_{ij} - m^{rij}\omega_{ij/r} - q^i{}_{/i} + \rho\bar{h}$$

we have:

$$-\frac{q^p{}_{/p}}{\theta} = \rho\frac{\dot{e}}{\theta} - \frac{P^{(ij)}\dot{e}_{ij}}{\theta} + \frac{m^{rij}\omega_{ij/r}}{\theta} - \frac{\rho\bar{h}}{\theta}.$$

By substitution, we get (a). In order to prove (b) we use (a) and 1(a),(b). □

4) (ALTERNATIVE FORMS FOR THE DISSIPATION OF THE THERMODYNAMIC VELOCITY).

$$\begin{aligned}
\Delta &\equiv \Delta_M + \Delta_\theta \\
&= \theta\frac{\delta s}{\delta t} - \frac{\delta e}{\delta t} + \frac{1}{\rho}[{}_DP^{(ij)}\dot{e}_{ij} + {}_Dm^{(rj)}\omega_{i_{j/r}}] - \frac{q^i}{\rho}(\partial x_i(\log\theta)) \\
\Delta &\equiv \Delta_M + \Delta_\theta \\
&= -(\frac{\delta t}{\delta t} + s\frac{\delta\theta}{\delta t}) + \frac{1}{\rho}({}_DP^{(ij)}\dot{e}_{ij} + {}_Dm^{(rj)i}\omega_{ij/r}) - \frac{q^j}{\rho}(\partial x_j(\log\theta))
\end{aligned}$$

5) (OTHER FORMS FOR THE CLAUSIUS-DUHEM INEQUALITY).

$$\begin{aligned}
&(a) \quad \theta\frac{\delta s}{\delta t} - \frac{\delta e}{\delta t} + \frac{1}{\rho}{}_DP^{(ij)}\dot{e}_{ij} \geq 0 \\
&(b) \quad \frac{1}{\rho}{}_DP^{(ij)}\dot{e}_{ij} - \frac{\delta f}{\delta t} \geq 0.
\end{aligned}$$

REMARK 4.13 - 1) If we make the assumption that the constitutive relations do not explicitly depend on $\dot{\theta}$, we get:

$$\begin{aligned}
&(a) \quad s = -(\partial\theta \cdot f) \\
&(b) \quad e = f - \theta(\partial\theta \cdot f) \\
&(c) \quad \Delta_M = -\dot{f}|_{\theta=const.} + \frac{1}{\rho}[{}_DP^{(ij)}\dot{e}_{ij} + {}_D\, m^{(rj)i}\omega_{ij/r}]\,.
\end{aligned}$$

PROOF. a) Let us assume as independent variables (θ, c_α). Therefore, from theorem 3(b) we get:

$$[(\partial\theta \cdot f) + s]\dot{\theta} + \dot{f}|_{\theta=const.} - \frac{1}{\rho}[P^{(ij)}\dot{e}_{ij} - m^{rij}\omega_{ij/r}] + \frac{q^p}{\rho\theta}\theta_{/p} \leq 0.$$

As this equation must be independent of $\dot{\theta}$, we get:

$$s = -(\partial\theta \cdot f).$$

b) From equation (4.10) we get:

$$e = f + \theta s$$

hence, by substituting s into the above equation, we have:

$$e = f - \theta(\partial\theta \cdot f).$$

c) It is a direct consequence of the above formulas. □

REMARK 4.14 - In thermomechanics one usually takes as independent variables (θ, c_α).

THEOREM 4.19 - *Assuming* $e = e(\theta, V, \theta_{/i})$, *where* $V \equiv 1/\rho \equiv$ *specific volume, we get the following.*

1) **Dissipation function**:

$$w = -\rho p\dot{V}.$$

2) *As*

$$(\partial\theta^{/i} \cdot e) = 0, \quad (\partial\theta^{/i} \cdot f) = 0$$

we get

(*a*) $e = e(\theta, V)$

(*b*) $p = -(\partial V \cdot e) = -(\partial V \cdot f)$

(*c*) **Velocity of accumulation of elastic energy** :

$\dot{e}|_{s=const.} = \dot{f}|_{\theta=const.} = -p\dot{V}$

(*d*) **Velocity of mechanic dissipation for non-polar fluids** :

$\Delta_M = 0$

(*e*) **Velocity of mechanic dissipation for polar fluids** :

$\Delta_M = {}_D m^{(rij)}\omega_{ij/r}$

3) **Second law of Thermodynamics (Clausius-Duehm law).**

(*a*) **Polar fluids** : $-\frac{1}{\rho}{}_D m^{(rij)}\omega_{ij/r} + \frac{q^p}{\rho\theta}\theta_{/p} \leq 0$

(*b*) **Non-polar fluids** : $\frac{q^p}{\rho\theta}\theta_{/p} \leq 0$

4) **Gibbs relations**

$$\dot{e} = \theta\dot{s} - p\dot{V}, \quad \dot{f} = -s\dot{\theta} - p\dot{V}.$$

PROOF. 1) From the rheological equations we get:

$$-p\delta^{ij}\dot{e}_{ij} = -p(\dot{e}_{11} + \dot{e}_{22} + \dot{e}_{33}) = -pv^k{}_{/k}.$$

Then, by using the continuity equation we have:

$$-p\delta^{ij}\dot{e}_{ij} = \frac{p}{\rho}\frac{\delta\rho}{\delta t} = pV\frac{\delta}{\delta t}(\frac{1}{V}) = -\frac{p}{V}\dot{V} = -p\rho\dot{V}.$$

2) We have:

$$\dot{f}|_{\theta=const.} = (\partial V \cdot f)\dot{V} + (\partial\theta^{/i} \cdot f)\frac{\delta}{\delta t}(\theta_{/i}).$$

Then, by considering the Clausius-Duehm inequality we get:

$$\dot{f}|_{\theta=const.} - \frac{1}{\rho}[{}_DP^{(ij)}\dot{e}_{ij} + {}_Dm^{rij}\omega_{ij/r}] + \frac{q}{\rho\theta}\theta_{/p} \leq 0\,,$$

$$((\partial V \cdot f) + p)\dot{V} + (\partial\theta^{/i} \cdot f)\frac{\delta}{\delta t}(\theta_{/i}) - \frac{1}{\rho}{}_Dm^{rij}\omega_{ij/r} + \frac{q^p}{\rho\theta}\theta_{/p} \leq 0.$$

Now, by considering the indepedence on $\dot{V}$ and $\frac{\delta}{\delta t}(\theta_{/i})$, we complete the proof. Points 3 and 4 are direct consequences of the above statements. □

4.5 - RHEOLOGICAL CLASSIFICATION OF MATERIALS

TAB.4.8 - **Rheological classification of materials**

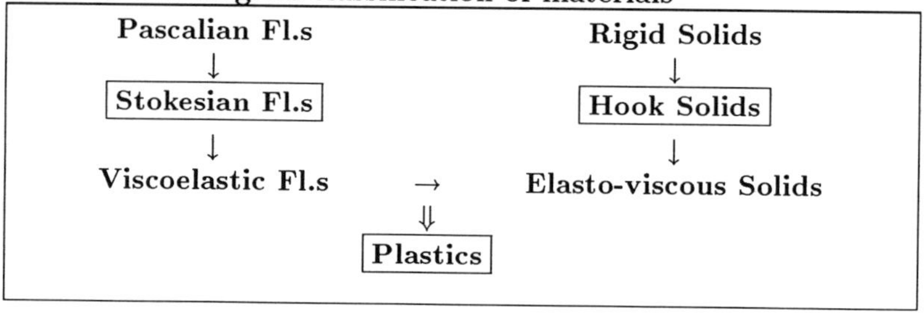

(A) - PASCALIAN FLUIDS (NONVISCOUS FLUIDS).

1) *Rheological equation:*

$$P^{jj} = -pg^{jj}$$

THEOREM 4.20 - 1) (**Euler equation**). *The motion equation is the following:*

$$g^{ij}p_{/j} = \rho(-\frac{\delta v^i}{\delta t} + B^i) \tag{4.18}$$

2) (**Bernoulli equation**). *In a gravitational field, and in steady state, we get the following constant of the motion along a flow-line:*

$$p + \frac{\rho}{2}v^2 + \rho g z = \text{constant}.$$

PROOF. In a gravitational field we have: $B^i = -gz_{/\alpha}g^{\alpha i}$. Then from (4.18), we get:

$$[p + \rho(\frac{1}{2}v^2 + gz)]_{/i}v^i = 0.$$

□

THEOREM 4.21 - 1) (**Energetic Euler equation**). *The energy equation is the following:*

$$\rho\frac{\delta e}{\delta t} = -pg^{jj}\dot{e}_{ij} - m^{rij}\omega_{ij/r} - q^i{}_{/i} + \rho\bar{h} \tag{4.19}$$

2) (**Bernoulli equation**). *In steady state, and in non-polar case,* $\bar{h} = 0, \rho =$ *constant, we have the following constant of the motion along a flow-line:*

$$(\rho e + p)v^k + q^k = \text{constant}.$$

PROOF. In fact, from (4.19) we get

$$[(\rho e + p)v^k + q^k]_{/k} = 0.$$

□

(B) STOKESIAN FLUIDS (VISCOUS FLUIDS).

1) *Rheological equations:*

$$P^{ij} = -pg^{ij} + \wp^{ij}, \quad \wp^{ij} = \alpha_0 g^{ij} + \alpha_1\dot{e}^{ij} + \alpha_2\dot{e}^{im}\dot{e}^j_m$$

where α_i *are polynomial functions of invariants* $I_{\dot{e}}, II_{\dot{e}}, III_{\dot{e}}$ *of* $\dot{e}^{ij}$ *and* $\alpha_0 = 0$ *if* $\dot{e}^{ij} = 0$.

2) $\frac{1}{2}\alpha_1 \equiv$ **apparent viscosity**.

3) $\alpha_2 \equiv$ **transverse viscosity** *(or* **viscosity coefficient of normal stress***)*.

4) *Internal energy:*

$$e = e(\rho, \theta).$$

REMARK 4.15 - If the material is submitted to adiabatic strain, we can write:

$$\rho = \rho(\theta) \Rightarrow e = e(\theta).$$

THEOREM 4.22 - 1) **Energy equation**.

$$\boxed{\rho C_v \frac{\delta \theta}{\delta t} = -\theta(\partial\theta \cdot p)_\rho v^k{}_{/k} - q^k{}_{/k} + \wp^{kj} v_{i/k}}$$

PROOF. From $e = e(\rho, \theta)$ we get:

$$\begin{aligned} \frac{\delta e}{\delta t} &= \frac{\delta \theta}{\delta t}(\partial\theta \cdot e)_\rho + \frac{\delta \rho}{\delta t}(\partial\rho \cdot e)_\theta = C_v \frac{\delta \theta}{\delta t} + \frac{1}{\rho} v^k{}_{/k}[\theta(\partial\theta \cdot p)_\rho - P] \\ C_v &= (\partial\theta \cdot e)_p \\ (\partial\rho \cdot e)_\theta &= -\frac{1}{\rho^2}[\theta(\partial\theta \cdot p)_\rho - p] \end{aligned}$$

and taking into account continuity equation, we get the proof. □

2) **Dissipation function**.

$$w = -\rho p \dot{V} + \wp^{ij} \dot{e}_{ij}.$$

3) *Assuming*

$$\begin{aligned} f &= f(\theta, V, \dot{e}^{ij}) \\ P^{ij} &= -p(\theta, V, \theta_{/i}) g^{ij} + \wp^{ij}(\theta, V, \dot{e}^{ij}, \theta_{/i}), \\ q^i &= q^i(\theta, V, \dot{e}^{ij}, \theta_{/i}) \end{aligned}$$

we get:

$$
\begin{aligned}
&(a) \quad (\partial \dot{e}_{ij} \cdot f) = 0 \\
&(b) \quad p = -(\partial V \cdot f) = -(\partial V \cdot e) \\
&(c) \quad (\partial \theta_i{}^{/i} \cdot p) = 0.
\end{aligned}
$$

(d) **Velocity of accumulation of elastic energy** :

$$\dot{e}|_{\rho=const.} = \dot{f}|_{\theta=const.} = -p\dot{V}$$

(e) **Gibbs relations** : $\left\{ \begin{aligned} \dot{f} &= -s\dot{\theta} - p\dot{V} \\ \dot{e} &= \theta\dot{s} - p\dot{V} \end{aligned} \right\}$

(f) **Velocity of mechanical dissipation** : $\Delta_M = \frac{1}{\rho}[\wp^{ij}\dot{e}_{ij} + {}_D m^{rij}\omega_{ij/r}]$

(g) **Second law of Thermodynamics (Clausius-Duehm inequality)** :

$$\frac{1}{\rho}[\wp^{ij}\dot{e}_{ij} + {}_D m^{rij}\omega_{ij/r}] - \frac{1}{\rho}q^i(\partial x_i \cdot (\log\theta)) \geq 0\,,$$

isothermic case : $\Delta_M \geq 0$,

static case : $\Delta_\theta \geq 0$.

Assuming:

$$
\begin{aligned}
f &= f(\theta, V) \\
P^{ij} &= -p(\theta, V)g^{ij} + \wp^{ij}(\theta, V, \dot{e}^{ij}) \\
q^j &= q^j(\theta, V, \theta_{/i})
\end{aligned}
$$

we similarly get: $\Delta_M \geq 0, \Delta_\theta \geq 0$; *then it follows, that in* **non-polar case** *the* **velocity of mechanical dissipation** *is the following:*

$$\wp^{ij}\dot{e}_{ij} = \wp^{ij}v_{i/j} \geq 0.$$

THEOREM 4.23 - 1) *The dissipative part of the stress, and non-dissipative one, are given by the following:*

$$
\begin{aligned}
{}_D P^{ij} &= \alpha_0 g^{ij} + \alpha_1 \dot{e}^{ij} + \alpha_2 \dot{e}^{im}\dot{e}^j_m, \\
{}_{\overline{D}}P &= -p_0 g^{ij}.
\end{aligned}
$$

PROOF. In fact, ${}_D P^{ij} = 0$ if $\dot{e}_{ij} = 0$. □

2) *Relation between thermodynamic pressure* p *and mechanic pressure* $\widehat{p}$:

$$3(p - \widehat{p}) = 3\alpha_0 + \alpha_1 I_{\dot{e}} + \alpha_2({I_{\dot{e}}}^2 - 2II_{\dot{e}}).$$

In the static case $\dot{e} = 0 = \alpha_0(0,0,0)$, where $I_{\dot{e}} = II_{\dot{e}} = III_{\dot{e}} = 0 \Rightarrow p = \widehat{p}$.

PROOF. Recall that the mechanic pressure $\widehat{p}$ is defined by the following:

$$\widehat{p} = -\frac{1}{3}P_i^i = -\frac{1}{3}I_P$$
$$P_j^i = -\widehat{p}\delta_j^i + \wp_j^i, (\wp_i^i = 0)$$
$$P_j^i = -p\delta_j^i + {}_DP_j^i.$$

By subtracting one from the other, we get:

$${}_DP_j^i = (p - \widehat{p})\delta_j^i + \wp_j^i.$$

For $i = j$ and by using point 1) for ${}_DP^{ij}$ we obtain the proof. □

3) *For an isotropic incompressible fluid one has:*

$$3(p - \widehat{p}) = -2\alpha_2 II_{\dot{e}}.$$

Therefore, $p = \widehat{p}$ *iff the constitutive equations are quasi-linear in* $\dot{e}_j^i$, *i.e.,* $\alpha_2 = 0$.

REMARK 4.16 - The concept of **isotropic tensor** T is such that its components do not change after a rigid rotation (in the vector space **S** where the tensor is defined). For example:

1) If T is a tensor of type $(1,0)$ (i.e., vector) T cannot be isotropic.
2) If T is a tensor of type $(2,0)$, it is isotropic if $T^{ij} = cg^{ij}$, where

$$c = \frac{1}{n}\sum_{1\leq k\leq n} T_k^k, \quad (n = \dim E).$$

THEOREM 4.24 - (LIMIT CASES FOR ENERGY EQUATION).

1) (**Ideal Gas**).

$$\rho C_v \frac{\delta\theta}{\delta t} = \lambda(\theta_{/k})^{/k} - pv^k{}_{/k}$$

or

$$\rho C_p \frac{\delta\theta}{\delta t} = \lambda(\theta_{/k})^{/k} + \frac{\delta p}{\delta t}.$$

PROOF. In this case $(\partial\theta \cdot p)_V = p/\theta$ and the viscosity is zero. □

2) (**Fluid with constant pressure**).

$$\rho C_p \frac{\delta\theta}{\delta t} = \lambda(\theta_{/k})^{/k} + \wp^{ij}\dot{e}_{ij}.$$

3) (**Incompressible fluid**).

$$\rho C_p \frac{\delta\theta}{\delta t} = \lambda(\theta_{/k})^{/k} + \wp^{ij}\dot{e}_{ij}.$$

PROOF. $C_p = C_v$. □

4) (**Rigid body**).

$$\rho C_p(\partial t \cdot \theta) = \lambda(\theta_{/k})^{/k}.$$

PROOF. $v^i = 0$. □

EXAMPLE 4.2 - (EXAMPLES OF STOKESIAN FLUIDS).

1) **Newtonian fluids**. Rheological equation:

$$(\text{I Form}): \quad \wp^{ij} = \mu I_{\dot{e}} g^{ij} + 2\eta \dot{e}^{ij}.$$

$$(\text{II Form}): \quad \wp^{ij} = (k - \frac{2}{3}\eta) I_{\dot{e}} g^{ij} + 2\eta \dot{e}^{ij}$$

where η = shear viscosity (or viscosity), k = body viscosity, $\mu = (k - \frac{2}{3}\eta)$.
With respect to the general expression for Stokesian fluids, we get:

$$\alpha_0 = \mu I_{\dot{e}}, \quad \alpha_1 = 2\eta, \quad \alpha_2 = 0.$$

In a Newtonian fluid we can consider the stress splitted into four parts:

$$P^{ij} = P_0{}^{ij} - \beta_0(\frac{\rho - \rho_0}{\rho_0}) g^{ij} + k I_{\dot{e}} g^{ij} + 2\eta(\dot{e}^{ij} - \frac{1}{3} I_{\dot{e}} g^{ij})$$

where:
(a) $P_0{}^{ij} \equiv$ static stress;
(b) $-\beta_0[(\rho - \rho_0)/\rho_0] J g^{ij} \equiv$ elastic stress that the body generates in order to contrast with the volume change;
(c) $k I g^{ij} \equiv$ viscous stress that the body generates to contrast with the volume change;
(d) $2\eta(\dot{e}^{ij} - \frac{1}{3} I_{\dot{e}} g^{ij}) \equiv$ viscous stress that the body generates to contrast with the form change.

2 (**Navier-Stokes equation**). The motion equation for Newtonian fluids is the following:

$$(\partial t \cdot v^k) + v^k{}_{/i} v^i = -\frac{1}{\rho} g^{ki} P_{/i} + 2\frac{\eta}{\rho} \dot{e}^{ki}_{/i} + B^k.$$

2) **Non-Newtonian Fluids**. Here we can consider all the models for Stokesian fluids that can be described by means of the rheological equations as follows:

$$\wp^{ij} = \alpha(y) \dot{e}^{ij}, \quad y \equiv \frac{1}{2} \dot{e}^n_m \dot{e}^m_n$$

or

$$\wp^{ij} = \alpha(\tau^2) \dot{e}^{ij}, \quad \tau^2 = \frac{1}{2} P^m_n P^n_m.$$

In the following table we reported some important phenomelogical models.

TAB.4.9 - **Important models of non-Newtonian fluids**

Name	Definition
Power	$\alpha(\tau^2) = 2m^{1/n}(\tau^2)^{-(1-n)/2n}$ $m = m(\theta), n = n(\theta)$
Ellis	$\frac{2}{\alpha(\tau^2)} = \frac{1}{\eta_0} + m^{-1/n}(\tau^2)^{(1-n)/2n}$
Sisko	$\alpha(\tau^2) = 2\eta_\infty + 2m^{1/n}(\tau^2)^{-(1-n)/2n}$ $(n < 1)$
Reiner-Philippoff	$\alpha = 2\eta_\infty + \frac{2(\eta_0 - \eta_\infty)}{1+\tau^2/G}$
Reiner (struct.stab.)	$\frac{2}{\alpha} = \frac{1}{\eta_\infty} - (\frac{1}{\eta_\infty} - \frac{1}{\eta_0})e^{-\tau^2/x^2}$

EXAMPLE 4.3 - (EXAMPLES OF NEWTONIAN FLOWS).

1) **Newtonian fluids: adimensional dynamic equation with constant mass density and in gravitational field.** Set: $D \equiv$ characteristic length of the system; $V \equiv$ average velocity; $g \equiv$ gravity acceleration. Introduce the following adimensional entities: $\bar{x}^i \equiv x^i/D$, $\bar{v}^i \equiv v^i/V$, $\bar{p} = (p - p_0)/\rho r^2$, $\bar{t} = tr/D$, $\bar{g}^k = g^k/g$, $\bar{\theta} = \frac{\theta - \theta_0}{\theta_1 - \theta_0}$. Then, define the following adimensional numbers:

$$R_e \equiv \frac{\rho v D}{\eta} = \text{Reynold}$$

$$F_r \equiv \frac{v^2}{gD} = \text{Frank}$$

$$P_r \equiv \frac{C_p \eta}{\lambda} = \text{Prandtl}$$

$$B_r \equiv \frac{\eta v^2}{\lambda(\theta_1 - \theta_0)} = \text{Brindeman}$$

Then, we can write the dynamic equations in the following way:

$$(a)\textbf{Continuity equation}: \bar{v}^k_{/k} = 0$$

$$(b)\textbf{Motion equation}: \frac{\delta \bar{v}^i}{\delta \bar{t}} = \frac{2}{R_e}\dot{\bar{e}}^{ki}_{/i} - g^{ki}\bar{p}_{/k} + (\frac{1}{F_r})\bar{g}^k$$

$$(c)\textbf{Energy equation}: \frac{\delta \bar{\theta}}{\delta \bar{t}} = \frac{1}{R_e P_r}(\bar{\theta}_{/k})^{/k} + \frac{B_r}{R_e P_r}\bar{\phi}.$$

2) **Newtonian fluids: dynamic equations for incompressible fluids with constant density in gravitational field and free convection.** In this case we can omit the dissipation-viscosity in the energy equation, as this contribution is

not important in the free convection. Let us introduce the following adimensional number:

$$G_r \equiv \frac{g\rho^2 B(\theta-\theta_0)D^3}{\eta^2} = \textbf{Groskof number}.$$

Then, dynamic equations can be written in the following form:

$$\begin{aligned} \bar{v}^k_{/k} &= 0 \\ \frac{\delta \bar{v}^k}{\delta \bar{t}} &= 2\bar{\dot{e}}^{ki}_{/i} - \bar{\theta} G_r \bar{g}^k \\ \frac{\delta \bar{\theta}}{\delta \bar{t}} &= \frac{1}{P_r}(\bar{\theta}_{/k})^{/k} \end{aligned}$$

with $\bar{\theta} \equiv \frac{\theta-\theta_0}{\theta_1-\theta_0}$, $\bar{v}^k \equiv \frac{v^k D\rho}{\eta}$, $\bar{t} \equiv \frac{t\eta}{\rho D^2}$.

3) **Trouton viscosity** (or **elongational viscosity**). Let us consider a fluid in the form of a thin rod, under elongational flow. (In order for such a situation to be possible, it is necessary that the fluid-viscosity should be about 10^6 poise.) Let us consider in cartesian coordinates the following expression for the infinitesimal velocity strain:

$$\dot{e}^{ij} = \frac{1}{2}\dot{\gamma}\begin{bmatrix} 2 & 0 & 0 \\ 0 & -1 & 0 \\ 0 & 0 & -1 \end{bmatrix}.$$

Then, we get: $I_{\dot{e}} = 0$. We have:

$$\wp^{ij} = 2\eta\dot{e}^{ij}.$$

Hence,

$$\begin{aligned} \wp_{11} &= 2\eta\dot{\gamma} \\ \wp_{22} &= \wp_{33} = -\eta\dot{\gamma} = -\frac{1}{2}\wp_{11} \\ &\Downarrow \\ P_{ij} &= -\widehat{p}\delta_{ij} + \wp_{ij} \\ &\Downarrow \end{aligned}$$

$$(\text{mechanic pressure}) = -\frac{1}{3}\mathrm{tr}(P) = -\frac{1}{3}I_P = \widehat{p}(\text{as} \quad \mathrm{tr}(\wp) = 0).$$

On the other hand

$$P_{ij} = -p\delta_{ij} + {}_D P_{ij},\ (p \equiv \text{thermodynamic pressure}).$$

Now, recalling that

$$
{}_DP_{ij} = (p - \widehat{p})\delta_{ij} + \wp_{ij}, e_DP_{ij} = \wp_{ij}
$$

$$
\Downarrow
$$

$$
p = \widehat{p}, (\text{as in this case} \quad \alpha_2 = 0).
$$

If the radial forces are absent $\Rightarrow$: $P_{22} = P_{33} = 0$, we then get:

$$
0 = P_{22} = -\,p = \eta\dot{\gamma}
$$

$$
\Downarrow
$$

$$
p = -\eta\dot{\gamma} = -\frac{1}{2}\wp_{11}
$$

$$
\Downarrow
$$

$$
P_{11} = -p + \wp_{11} = +\frac{1}{2}\wp_{11} + \wp_{11} = \frac{3}{2}\wp_{11} = 3\eta\dot{\gamma}.
$$

Note that the superficial tension σ produces a radial stress $P_{22} = \sigma/R$, where R is the radius of the cylinder. However, this component is very small, so we can, in general, omit it.

DEFINITION 4.12 - *Define* **Trouton viscosity** η_e *by means of the following:*

$$
P_{11} = \eta_e\dot{e}_{11} \Rightarrow \eta_e = 3\eta.
$$

4) **Newtonian fluid: steady state laminar flow in circular pipe**. Let us assume: (a) the mass density is constant: $\rho =$ constant; (b) the pipe is horizontal; (c) the pipe is very long (L); (d) there is not flow for $t < 0$; for $t = 0$ it is applied a gradient of pressure: $\frac{P_0 - P_L}{L}$; (e) the field of velocity has the following structure:

$$
v_r = v_\theta = 0, \quad v_z = v_z(r,t). \tag{4.20}
$$

Then the motion equation and continuity equations give the following equation:

$$
\rho(\partial t \cdot v_z) = \frac{P_0 - P_L}{L} + \eta\frac{1}{r}[\partial r \cdot (r(\partial r \cdot v_z))]
$$

plus the boundary conditions:

$$
\begin{aligned}
v_z(r,0) &= 0, \text{with} \quad 0 \leq r \leq R \\
v_z(0,t) &= \text{finite} \\
v_z(R,t) &= 0.
\end{aligned}
$$

As a consequence we get:

$$
v_z(r,t) = \frac{(P_0 - P_L)R^2}{4\eta L}[(1 - (\frac{r}{R})^2) - 8\sum_{n=1}^{\infty}\frac{J_0(\alpha_n\frac{r}{R})}{\alpha^3{}_nJ_1(\alpha_n)}e^{-\alpha_n^2\tau}],
$$

where

$$\tau \equiv \eta t/\rho R^2$$

$$\alpha_n = \text{ zeros of Bessel function of order } \quad 0, J_0$$

$$J_i \equiv \text{Bessel functions, with } \quad i \geq 0\,.$$

PROOF. Let us introduce the following adimensional variables:

$$\phi \equiv \frac{v_z}{(P_0 - P_L)R^2} 4\eta L, \quad \xi \equiv \frac{r}{R}, \quad \tau \equiv \frac{\eta t}{\rho R^2}.$$

Therefore, equation (4.20) can be written in the following adimensional form:

$$(\partial\tau \cdot \phi) = 4 + \frac{1}{\xi}(\partial\xi \cdot (\xi(\partial\xi \cdot \phi)))$$

plus the following boundary conditions:

$$\phi(\tau = 0) = 0, \quad \phi(\xi = 1) = 0, \quad \phi(\xi = 0) = \text{finite}, \quad \lim_{\tau \to \infty} \phi = \phi_\infty(\xi)(\text{steady state}).$$

Note that $\phi_\infty(\xi)$ satisfies the following equation:

$$0 = 4 + \frac{1}{\xi}\frac{d}{d\xi}(\xi \frac{d\phi_\infty}{d\xi}),$$

plus the following boundary conditions:

$$\phi_\infty(\xi = 1) = 0.$$

As a consequence

$$\phi_\infty(\xi) = 1 - \xi^2 \quad (\textbf{Poiseuille velocity}).$$

Therefore, let $\phi(\xi, \tau) = \phi_\infty(\xi) - \widehat{\phi}(\xi, \tau)$. Hence, $\widehat{\phi}$ must satisfy the following equation:

$$(\partial\tau \cdot \widehat{\phi}) = \frac{1}{\xi}\partial\xi \cdot (\xi(\partial\xi \cdot \widehat{\phi}))$$

plus the following boundary conditions:

$$\widehat{\phi}(\tau = \infty) = \phi_\infty, \quad \widehat{\phi}(\xi = 0) = \text{finite}, \quad \widehat{\phi}(\xi = 1) = 0.$$

Let us find a solution of the form $\phi(\xi,\tau)=\Xi(\xi)T(\tau)$:

$$\frac{1}{T}\frac{dT}{d\tau}=\frac{1}{\Xi}\frac{1}{\xi}\frac{d}{d\xi}(\xi\frac{d\Xi}{d\xi})$$

$$\Downarrow$$

$$\left\{\begin{array}{r}\frac{dT}{d\tau}=-\alpha^2 T\\ \frac{1}{\xi}\frac{d}{d\xi}(\xi\frac{d\Xi}{d\xi})+\alpha^2\Xi=0,\alpha^2=const.\end{array}\right\}$$

$$\Downarrow$$

$$\text{Solutions}:\left\{\begin{array}{l}T=c_0e^{-\alpha^2\tau}\\ \Xi=c_1J_0(\alpha\xi)+c_2Y_0(\alpha\xi)\\ c_0,c_1,c_2=const.\\ J_0,Y_0\equiv\text{Bessel functions of order 0.}\end{array}\right\}$$

By considering the boundary conditions, we get:

$$\widehat{\phi}(\xi=0)=\text{finite}\Rightarrow\text{as}\quad\lim_{\xi\to 0}Y_0(\alpha\xi)=-\infty\Rightarrow c_2=0$$

$$\widehat{\phi}(\xi=1)=0\Rightarrow J_0(\alpha)=0.$$

As $J_0(\alpha)$ is an oscillating function, it has many zeros α_n. Hence, we have many solutions $\Xi_n=C_{1n}J_0(\alpha_n)$, with $n=1,2,\cdots,n$. Then,

$$\widehat{\phi}(\xi,\tau)=\sum_{1\leq n\leq\infty}B_ne^{-\alpha_n\tau^2}J_0(\alpha_n\xi),\quad\text{with}\quad B_n\equiv c_0c_{1n}.$$

Now, impose the condition:

$$\lim_{\tau\to\infty}\widehat{\phi}(\xi,\tau)=\phi_\infty(\xi)=1-\xi^2;1-\xi^2=\sum_{1\leq n\leq\infty}B_nJ_0(\alpha_n\xi).$$

Multiplying both sides for $\xi J_0(\alpha_m\xi)$ and integrate

$$\int_{[0,1]}J_0(\alpha_m\xi)(1-\xi^2)\xi d\xi=\sum_{1\leq n\leq\infty}B_n\int_{[0,1]}J_0(\alpha_n\xi)J_0(\alpha_m\xi)\xi d\xi.$$

From the orthogonality of Bessel functions

$$4\frac{J_1(\alpha_m)}{\alpha_m^3}=B_m\frac{1}{2}[J_1(\alpha_m)]^2$$

it follows that

$$\frac{8}{J_1(\alpha_m)\alpha_m^3} = B_m.$$

□

5) **Newtonian fluid: form of the surface of rotating liquid**. Let us consider cylindric coordinates: (r, ϕ, z), and let us assume the following structure of the field velocity:

$$v_z = v_r = 0, \quad v_\phi = v_\phi(r).$$

Let us assume the following structure for the pressure:

$$p = p(r, z)\,.$$

Then, we can write the motion equations, in the followimg form:

$$\begin{aligned}
(r\text{-component}): &\quad \rho\frac{v_{\phi^2}}{r} = (\partial r \cdot p)\\
(\phi\text{-component}): &\quad 0 = \eta\, \partial r \cdot (\frac{1}{r}\partial r \cdot (r v_\phi))\\
(z\text{-component}): &\quad 0 = -(\partial z \cdot p) - \rho g.
\end{aligned}$$

We have:

$$v_\phi = \frac{1}{2}c_1 r + \frac{c_2}{r}, \quad (c_1, c_2 = const.).$$

As v_ϕ cannot be infinite, put $c_2 = 0 \Rightarrow v_\phi = \frac{1}{2}c_1 r$. Furthermore, from the boundary condition

$$v_\phi(R) = \Omega R \Rightarrow c_1 = 2\Omega$$

this means that any matherial element of the fluid moves as a rigid body:

$$v_\phi(r) = \Omega r.$$

From the other two dynamic equations, we get

$$(\partial r \cdot p) = \rho\Omega^2 r, \quad (\partial z \cdot p) = -\rho g.$$

On the other hand,

$$dp = (\partial r \cdot p)dr + (\partial z \cdot p)dz.$$

After integration we get:

$$p = -\rho g z + \frac{1}{2}\rho\Omega^2 r^2 + c.$$

The constant c can be determined by means of the following boundary condition:

$$p(r=0, z=z_0) = p_0$$
$$\Downarrow$$
$$p_0 = -\rho g z_0 + c\,.$$

Then, we get:

$$p - p_0 = -\rho g(z - z_0) + \frac{\rho\Omega r^2}{2}. \tag{4.21}$$

The free surface can be determined by imposing $p = p_0$ in equation (4.21). We get:

$$0 = -\rho g(z - z_0) + \frac{\rho\Omega^2 r^2}{2}$$
$$\Downarrow$$
$$z - z_0 = (\frac{\Omega^2}{2g})r^2.$$

This is the equation of a parable in the plane (z, r). As a consequence, the free surface is a paraboloid of rotation around the z-axis.
6) **Melt-spinning of cylindric rod**[94,95]. Let us assume the following structure for the velocity field:

$$v_z = v_z(z,t), \quad v_r = v_r(r,t), \quad v_\phi = 0.$$

Then, we have the following equations.
a) **Local equations**:

$$\left\{\begin{array}{l} \text{(continuity equation)}: \\ \frac{1}{r}\partial r(rv_r) + (\partial z \cdot v_z) = 0 \\ \text{(motion equation)}: \\ \left\{\begin{array}{l} \rho v_r(\partial r \cdot v_r) = (\partial r \cdot P_{rr}) + (\partial z \cdot P_{rz}) + \dfrac{P_{rr} - P_{\phi\phi}}{r} - \rho(\partial t \cdot v_r) \\ \rho v_z(\partial z \cdot v_z) = (\partial r \cdot P_{zr}) + (\partial z \cdot P_{zz}) + \dfrac{1}{r}P_{zr} + \rho\rho - \rho(\partial t \cdot v_z) \end{array}\right\} \\ \text{(energy equation)}: \\ \rho C_p[(\partial z \cdot \theta)v_z] = -\frac{\lambda}{r}\partial r \cdot (r(\partial r \cdot \theta)) - \rho C_p(\partial t.\theta) \end{array}\right\}$$

b) **Global equations**:

$$\left\{\begin{array}{l}
\text{(continuity equation)}: \\
W = \rho v_z A(z) \\
W = \int\limits_0^{R(z)} 2\pi r \rho v_z dr = \rho v_z 2\pi \int\limits_0^{R(z)} r dr = \rho v_z \pi R^2(z) = \rho v_z A(z) \\
\text{(motion equation)}: \\
(\partial z \cdot F) = (\partial t \cdot W) + \partial z \cdot (v_z W) - A\rho g + 2\pi R P_t \\
F(z) \equiv \int\limits_0^{R(z)} 2\pi r P_{zz} dr \\
\text{(energy equation)}: \\
\rho C_p[(\partial t \cdot <\theta>) + v_z(\partial z \cdot <\theta>)] = 2\sqrt{\frac{\pi}{A}} h(\theta_\infty - \theta_s) \\
\theta_s \equiv \text{(surface temperature)} \\
<\theta_{(z,t)}> \equiv \frac{1}{A} \int\limits_0^{R(z)} 2\pi r \theta dr.
\end{array}\right\}$$

c) **Steady state**.

Local equations:

$$\left\{\begin{array}{l}
\text{continuity equation}: \frac{1}{r}(\partial r \cdot (r \cdot v_r)) + (\partial z \cdot v_z) = 0 \\
\text{motion equations}: \left\{\begin{array}{l}
\rho v_r(\partial r \cdot v_r) = (\partial r \cdot P_{rr}) + (\partial z \cdot P_{rz}) + \frac{P_{rr} - P_{\phi\phi}}{r} \\
\rho v_z(\partial z \cdot v_z) = (\partial r \cdot P_{zr}) + (\partial z \cdot P_{zz}) + \frac{1}{r} P_{zr} + \rho\rho
\end{array}\right\} \\
\text{energy equation}: \rho C_p(\partial z \cdot \theta) v_z = -\frac{\lambda}{r} \partial r \cdot (r(\partial r \cdot \theta))
\end{array}\right\}$$

Global equations:

$$\left\{\begin{array}{l} \text{continuity equation}: W = \rho v_z A = const. \\ \text{strain equation}: (\partial z \cdot F) = (\partial z \cdot (v_z W)) - A\rho g + 2\pi R P \\ \text{energy equation}: \partial z \cdot <\theta> = \dfrac{2\sqrt{\pi A}}{W C_p} h(\theta_\infty - \theta_s) \\ \text{spinning tension}: F(z) = F_L + F_z, F_L \equiv \text{ends-force} \\ <\theta(z)> = <\theta(0)> + \int\limits_0^z \dfrac{2\pi A(z)}{W c_p} \lambda(z)(\theta_\infty - \theta_s(z)) dz \\ v_z(z) = W/\rho A(z), \\ v_r(R) = -\dfrac{1}{2}\sqrt{\dfrac{A(z)}{\pi}}(\partial z \cdot v_z). \end{array}\right\}$$

Let us assume:

$$F_L \gg F(z), \quad <\theta(z)> = \theta(z) = \theta_s(z).$$

Then, the global equations become:

Non-steady state :

$$\left\{\begin{array}{l} \text{continuity equation}: W = \rho v_z A(z) \\ \text{motion equation}: (\partial z \cdot F) = (\partial t \cdot W) \\ \text{energy equation}: \rho c_p[(\partial t \cdot \theta) + v_z(\partial z \cdot \theta)] = 2\sqrt{\dfrac{\pi}{A}} h(\theta_\infty - \theta) \end{array}\right\}$$

Steady state :

$$\left\{\begin{array}{l} \text{continuity equation}: W = \rho v_z A(z) = const. \\ \text{motion equation}: \dfrac{dF}{dz} = 0 \\ \text{energy equation}: \dfrac{d\theta}{dz} = \dfrac{2\sqrt{\pi A}}{W c_p} h(A)(\theta_\infty - \theta) \end{array}\right\}$$

d)(**Newtonian fluid**):

□ (Non-steady state).

(Local equations):

$$\left\{\begin{array}{l}\frac{1}{r}(\partial r\cdot(rv_r))+(\partial z\cdot v_z)=0\\ \rho[(\partial r\cdot v_r)v_r+(\partial t\cdot v_r)]=(\partial r\cdot P_{rr})+\frac{P_{rr}-P_{\phi\phi}}{r}\\ \rho[(\partial z\cdot v_z)v_z+(\partial t\cdot v_z)]=(\partial z\cdot P_{zz})+\rho g\\ \rho C_p[(\partial t\cdot\theta)+(\partial z\cdot\theta)v_z]=-\frac{\lambda}{r}(\partial r\cdot(r(\partial r\cdot\theta)))\\ \qquad P_{ij}=\begin{bmatrix}-p+2\eta(\partial r\cdot v_r) & 0 & 0\\ 0 & -p+2\eta\frac{v_r}{r} & 0\\ 0 & 0 & -p+2\eta\end{bmatrix}\end{array}\right\}$$

(Global equations):

$$\left\{\begin{array}{r}W=\rho v_zA(z)\\ (\partial z\cdot F)=(\partial t\cdot W)\\ \rho c_p[(\partial t\cdot\theta)+v_z(\partial z\cdot\theta)]=2\sqrt{\frac{\pi}{A}}h(\theta_\infty-\theta)\\ F=<\eta_e>A(\partial z\cdot v_z)-s\sqrt{\pi A},(s\equiv\text{surface tension})\\ <\eta_e>=\frac{1}{A}\int_0^R 2\pi r\eta_e dr.\end{array}\right\}$$

☐ (Steady state):

(Local equations)

$$\left\{\begin{array}{l}\frac{1}{r}\partial r\cdot(rv_r)+(\partial z\cdot v_z)=0\\ \rho[(\partial r\cdot v_r)v_r]=(\partial r\cdot P_{rr})+\frac{P_{rr}-P_{\phi\phi}}{r}\\ \rho[(\partial z\cdot v_z)v_z]=(\partial z\cdot P_{zz})+\rho g\\ \rho C_p(\partial z\cdot\theta)v_z=-\frac{\lambda}{r}\partial r(r(\partial r\cdot\theta))\\ \qquad P_{ij}=\begin{bmatrix}-p+2\eta(\partial r\cdot v_r) & 0 & 0\\ 0 & -p+2\eta\frac{v_r}{r} & 0\\ 0 & 0 & -p+2\eta(\partial z\cdot v_z)\end{bmatrix}.\end{array}\right\}$$

(Global equations)

$$\left\{\begin{array}{l} W = \rho v_z A(z) = const. \\ \dfrac{dF}{dz} = 0 \\ \dfrac{d\theta}{dz} = \dfrac{2\sqrt{\pi A}}{WC_p} h(A)(\theta_\infty - \theta) \\ \dfrac{dA}{dz} = -A\dfrac{\rho F_L}{W\eta_e(\theta)} \\ F_L = \dfrac{\int_{A_0}^{A_L} 2\sqrt{\pi A}\lambda(A)dA/A}{\int_{\theta_0}^{\theta_L} \frac{\rho C_p d\theta}{(\theta_\infty - \theta)\eta_e(\theta)}} \end{array}\right\}$$

7) **Steady state shear flow for Newtonian fluid in cone-plate system.** Let us assume the following structure for the velocity field:

$$v_r = v_\theta = 0, \quad v_\phi = v_\phi(r,\theta).$$

Then we have the following equations: Motion equations:

$$\left\{\begin{array}{r} -\rho\dfrac{v^2{}_\phi}{r} = -(\partial r \cdot p), (r\text{-component}) \\ -\rho\cot\theta\dfrac{v^2{}_\phi}{r} = -\dfrac{1}{r}(\partial\theta \cdot p), (\theta\text{-component}) \\ 0 = -(\dfrac{1}{r^2}\partial r \cdot (r^2 \wp_{r\phi}) + \dfrac{1}{r}(\partial\theta \cdot \wp_{\theta\phi}) + \dfrac{1}{r}\wp_{r\phi} + 2\cot\theta\dfrac{\wp_{\theta\phi}}{r}), (\phi\text{-component}) \end{array}\right\}$$

c) **(Creeping flow)**: Shear flow, such that $v^2{}_\phi \cong 0$. Then, we have:

$$v_\phi(r,\theta) = rf(\theta)$$
$$\Downarrow$$

(i) The angular velocity $\frac{v_\phi}{r}$ is independent of r.
(ii) $\wp_{r\phi} = 0$.
(iii) (ϕ-component of motion equation):

$$\frac{d\wp_{\theta\phi}}{d\theta} = -2\wp_{\theta\phi}\cot\theta.$$

For integration, we get:

$$\wp_{\theta\phi} = c_1/\sin^2\theta.$$

Let us calculate the constant c_1 by means of the boundary conditions: $\theta = \frac{\pi}{2}$, where the torque M is known:

$$M = \int_0^{2\pi}\int_0^R \wp_{\theta\phi}|_{\theta=\frac{\pi}{2}} r^2 dr d\phi = 2\pi(\frac{R^3}{3})(\frac{c_1}{\sin^2\frac{\pi}{2}}).$$

Thus, we get:

$$\wp_{\theta\phi} = 3M/2\pi R^3 \sin^2\theta.$$

Therefore, we have the following equation for $v_{\phi/r}$.

$$-\eta\sin\theta\frac{d}{d\theta}(\frac{v_{\phi/r}}{\sin\theta}) = \frac{3M}{2\pi R^3\sin^2\theta}.$$

By integrating by separations of variables, one has:

$$\frac{v_\phi}{r} = \frac{3M}{4\pi R^3\eta}[\cot\theta + \frac{1}{2}(\log\frac{1+\cos\theta}{1-\cos\theta})\sin\theta] + c_2.$$

Furthermore, we must have $c_2 = 0$, as $v_\phi(\pi/2) = 0$. Therefore, we get:

$$\frac{v_\phi}{r} = \frac{3M}{4\pi R^3\eta}[\cot\theta + \frac{1}{2}(\log\frac{1+\cos\theta}{1-\cos\theta_0})\sin\theta_0]\,. \tag{4.22}$$

Rewrite this formula for $\theta_0 = \pi - \theta_1$ and for $v_\phi = r\Omega\sin\theta_0$. We get:

$$\Omega\sin\theta_0 = \frac{3M}{4\pi R^3\eta}[\cot\theta_0 + \frac{1}{2}(\log\frac{1+\cos\theta}{1-\cos\theta_0})\sin\theta_0]. \tag{4.23}$$

This allows us to determine η from M. By eliminating M between (4.22) and (4.23), we get:

$$\frac{v_\phi}{r} = \Omega\sin\theta_0[\frac{\cot\theta + \frac{1}{2}(\log\frac{1+\cos\theta}{1-\cos\theta})\sin\theta}{\cot\theta_0 + \frac{1}{2}(\log\frac{1+\cos\theta_0}{1-\cos\theta_0})\sin\theta_0}].$$

For $\theta = \theta_0 \cong \pi/2, \theta_0 \equiv \pi - \theta_1$, one has:

$$\frac{v_\phi}{r} \cong \Omega\frac{\cos\theta}{\sin\theta_0} \cong \Omega(\frac{\frac{\pi}{2}-\theta}{\frac{\pi}{2}-\theta_0}).$$

EXAMPLE 4.4 - EXAMPLES OF FLOWS FOR STOKESIAN FLUIDS.

1) **Poiseuille flow in isothermic incompressible fluid.**

(Steady state).

Let us consider cylindric coordinates: $(x^1, x^2, x^3) = (r, \phi, z)$. Let us assume the following structure for the physical components of the velocity:

$$v_r = v_\phi = 0, \quad v_z = u(r).$$

As a consequence we get:

$$[\dot{e}_{ij}] = \frac{1}{2}(\partial r \cdot u)\begin{bmatrix} 0 & 0 & 1 \\ 0 & 0 & 0 \\ 1 & 0 & 0 \end{bmatrix} = \frac{1}{2}G\begin{bmatrix} 0 & 0 & 1 \\ 0 & 0 & 0 \\ 1 & 0 & 0 \end{bmatrix}$$

where $G \equiv (\partial r \cdot u) =$ **shear velocity**. Then, we have the following invariants associated to $\dot{e}_{ij}$:

$$\begin{aligned} I &= \dot{e}^m_m = 0 \\ II &= \frac{1}{2}(I^2 - \dot{e}^n_m \dot{e}^m_n) = -\frac{1}{4}G^2 \\ III &= |\dot{e}^n_m| = 0\,. \end{aligned}$$

Hence, α_1 and α_2 are functions of r. So we can write:

$$\begin{aligned} P^{ij} &= -g^{ij}p + \alpha_1 \dot{e}^{ij} + \alpha_2 \dot{e}^{im}\dot{e}^j_m \\ &\Downarrow \\ \wp_{rr} &= \wp_{zz} = \frac{1}{4}\alpha_2 G^2 \\ \wp_{\phi\phi} &= \wp_{r\phi} = \wp_{\phi r} = \wp_{\phi z} = \wp_{z\phi} = 0 \\ \wp_{rz} &= \wp_{zr} = \frac{1}{2}\alpha_1 G. \end{aligned}$$

Furthermore, the motion equations become:

$$\left\{ \begin{aligned} (\partial r \cdot p) &= \frac{1}{r}\frac{d}{dr}(r\wp_{rr}) \\ (\partial z \cdot p) - \rho g &= \frac{1}{r}\frac{d}{dr}(r\wp_{rz}) \end{aligned} \right\} \tag{4.24}$$

plus the following boundary conditions:

$$\lim_{r \to 0} \wp_{rz} = \text{finite}, \quad v_z(R) = 0.$$

Taking into account that

$$\partial z \cdot (\partial r \cdot p) = \partial r \cdot (\partial z \cdot p) = 0 \Rightarrow (\partial z \cdot p) = \text{function of } z \text{ and } G$$

and by using (4.24), after integration, we get

$$p = (P + \rho g)z + f(r), \quad \wp_{rz} = \frac{Pr}{2}, \quad P \equiv (\partial z \cdot p) - \rho g$$

where f is an arbitrary function of r. As

$$\wp_{rz}(r) = \frac{P_r}{2} = \frac{1}{2}\alpha_1(r)\frac{du(r)}{dr}$$

by integration we get:

$$u = -P \int_r^R \frac{r\,dr}{\alpha_1(r)}, \quad (u(R) = 0 : (\text{stik condition})).$$

Then, we have the following expression for the volume flow rate:

$$Q = 2\pi \int_0^R u(r) r\,dr \Rightarrow -2\pi P \int_0^R r\,dr \int_r^R \frac{\xi\,d\xi}{\alpha_1(\xi)} \Rightarrow Q = -P\pi \int_0^R \frac{r^3\,dr}{\alpha_1(r)}.$$

Set

$$B \equiv -\frac{PR}{2}, \quad F \equiv \frac{4Q}{\pi R^3},$$

Then we can write:

(4.25)**(Hogan-Poiseuille law)** $$F = 8B \int_0^1 \xi^3 d\xi / \alpha_1(\xi),$$

where $\xi = \frac{r}{R} \equiv$ adimensional coordinate.

□ For Newtonian fluids one has $\alpha_1 = 2\eta = const.$, then we have $F = \frac{1}{\eta}B$ or the usual formula:

(4.26)**(Classical Hogan-Poiseuille law)** $$P = -\frac{8Q\eta}{\pi R^4}.$$

Moreover, from equation (4.26) we get the following equation for $f(r)$:

$$\frac{df(r)}{dr} = \frac{1}{4r}\frac{d}{dr}(r\alpha_2(r)G^2).$$

Thus, after integration, we can determine f up to a constant. As a consequence, p can also be determined up to an arbitrary constant. (This is a general situation for incompressible fluids.)

2) **Stokesian fluids: couette flow: Weissenberg effect**. Let us consider the steady state isothermic rotational shearing flow of an incompressible fluid between coaxial cylinders. Let the exterior cylinder be at rest. Take cylindric coordinates, and assume the following structure for the velocity field:

$$v_r = v_z = 0, \quad v_\theta = v_\theta(r) \equiv u(r).$$

Then, we get:

$$(\dot{e}_{ij}) = \frac{1}{2}G\begin{bmatrix} 0 & 1 & 0 \\ 1 & 0 & 0 \\ 0 & 0 & 0 \end{bmatrix}, \quad G = -\frac{du}{dr}.$$

The corresponding invariants are:

$$I \equiv \dot{e}_k^k = 0$$
$$II \equiv \frac{1}{2}(I - \dot{e}_m^n \dot{e}_n^m) = -\frac{1}{4}G^2$$
$$III = |\dot{e}_m^n| = 0\,.$$

Then, for the rheological coefficients, we have:

$$\alpha_1 = \alpha_1(0, II, 0) = \alpha_1(r), \quad \alpha_2 = \alpha_2(0, II, 0) = \alpha_2(r).$$

The corresponding stress tensor is determined by the following:

$$\wp_{rr} = \wp_{\theta\theta} = \frac{1}{4}\alpha_2(r)G^2$$
$$\wp_{r\theta} = \wp_{\theta r} = \frac{1}{2}\alpha_1(r)G$$
$$\wp_{zz} = \wp_{rz} = \wp_{zr} = \wp_{\theta z} = \wp_{z\theta} = 0.$$

The dynamic equations are the following:

$$\left\{\begin{array}{l} \rho r(\frac{du}{dr})^2 = -(\partial r \cdot p) + \frac{1}{r}\frac{d}{dr}(r\wp_{rr}) - \frac{1}{r}\wp_{\theta\theta} \\ \frac{1}{r^2}\frac{d}{dr}(r^2\wp_{r\theta}) = 0 \\ (\partial z \cdot p) = -\rho g \end{array}\right\} \tag{4.27}$$

plus boundary conditions:

$$v_\theta(R_2) = 0, \quad v_\theta(R_1) = R_1\Omega$$

where $\Omega \equiv$ angular velocity of the interior cylinder.
Then we get:

$$\wp_{r\theta} = \frac{c}{r^2}, \quad c = \text{arbitrary constant}, \tag{4.28}$$

and

$$p = -\rho g z + f(r), \quad f(r) = \text{arbitrary function}\,. \tag{4.29}$$

Then, from (4.29), (4.27) and (4.28), we get:

$$(\frac{d^2 v_\theta}{dr^2})(\frac{dv_\theta}{dr})b(r) + a(r)(\frac{dv_\theta}{dr})^2 = \frac{df}{dr},$$

with

$$a(r) \equiv \frac{1}{2}\alpha_2(r), \quad b(r) \equiv \frac{1}{4}\frac{d\alpha_2}{dr} - \rho r.$$

☐ Let us, now, specialize on the case where α_1 and α_2 are constants. This is equivalent to considering a little shear strain velocity. We have:

$$\wp_{r\theta} = \frac{c}{r^2} = \frac{1}{2}\alpha_1 G = -\frac{1}{2}\alpha_1\frac{dv_\theta}{dr}$$
$$\Downarrow$$
$$\frac{dv_\theta}{dr} = -\frac{2c}{\alpha_1}\frac{1}{r^2} \equiv \frac{A}{r^2}.$$

Then we get

$$v_\theta = -\frac{A}{r} + B. \tag{4.30}$$

By using the boundary conditions we also get:

$$v_\theta(R_1) = v_0$$
$$v_\theta(R_2) = 0$$
$$\Downarrow$$
$$A = \frac{R_2 R_1}{R_1 - R_2}v_0$$
$$B = \frac{R_1}{R_1 - R_2}v_0.$$

Let us, now, consider the situation of the fluid at $z = z_0$, where the atmospheric pressure is present. From equation (4.29), we have:

$$p_0 = -\rho g z + f(r).$$

Let $z = z(r)$ be the equation of the free surface. By derivation we get:

$$0 = \rho g\frac{dz}{dr} + \frac{df}{dr}.$$

Hence,

$$\frac{dz}{dr} = \frac{1}{\rho g}\frac{df}{dr}. \tag{4.31}$$

From (4.27), we have:

$$(\partial r \cdot p) = f'(r) = \rho r(\frac{dv_\theta}{dr})^2 + \frac{1}{r}\frac{d}{dr}(r\wp_{rr}) - \frac{1}{r}\mathcal{P}_{\theta\theta}.$$

Then, by using equation (4.26), we have:

$$f'(r) = \rho r(\frac{dv_\theta}{dr})^2 + \frac{1}{2}\alpha_2(\frac{dv_\theta}{dr})(\frac{d^2v_\theta}{dr^2}).$$

Taking into account equations (4.30) and (4.31), we get:

$$f'(r) = \frac{A^2}{r^3}(\rho - \frac{\alpha_2}{r^2})$$

$$\Downarrow$$

$$\frac{dz}{dr} = \frac{A^2}{\rho g r^3}(\rho - \frac{\alpha_2}{r^2}).$$

If $(\rho - \frac{\alpha_2}{R_1{}^2}) < 0 \Rightarrow \frac{dz}{dr} < 0 \Rightarrow$ therefore the fluid climbs on the internal cylinder. This effect is called **Weissenberg effect**. In the Newtonian fluids $\alpha_2 = 0$ and this effect cannot be observed, as $\frac{dz}{dr} > 0$. By integration of (4.31), we have

$$z = \frac{4\gamma - 2\beta r^2}{r^4}, \quad \text{with} \quad \gamma \equiv \frac{A^2\alpha_2}{\rho g} \quad \text{and} \quad \beta \equiv \frac{A^2}{g}.$$

Thus, if $4\gamma - 2\beta R_2{}^2 > 0$ we have that z decreases from R_1 to R_2.

(C) ELASTIC MATERIALS (HOOK-SOLIDS).

DEFINITION 4.13 - *We call* **Hook-solids** *materials where the stress is uniquely determined by the strain. Furthermore, we made the following constitutive hypotheses:*

$$\begin{aligned} f &= f(\theta, B^i_k) \\ P^{ij} &= P^{ij}(\theta, B^i_k, \theta_{/i}) \\ q^i &= q^i(\theta, B^i_k, \theta_{/i}),\ (B^i_k) \equiv (\partial a_k \cdot x^i) = \textit{gradient strain tensor.} \end{aligned}$$

THEOREM 4.25 - 1) *We have the following equations:*

$$\left\{ \begin{aligned} P^{ij} &= \rho B^i_k(\partial B^k_n \cdot f)_\theta g^{hj} \\ 0 &= (\partial\theta_{/s} \cdot P^{ij}) \end{aligned} \right\}, \quad \left\{ \begin{aligned} f &= f(\theta, B^i_k) \\ P^{ij} &= P^{ij}(\theta, B^i_k) \\ q^i &= q^i(\theta, B^i_k, \theta_{/i}) \end{aligned} \right\} \tag{4.32}$$

2) *The velocity of accumulation of elestic energy is the following:*

$$\dot{f}|_{\theta=const.} = \dot{e}|_{s=const.} = \frac{P^{ij}v_{i/j}}{\rho}. \tag{4.33}$$

3) *The velocity of mechanic dissipation is zero:*

$$\Delta_M = 0.$$

4) *The velocity of thermic dissipation is non-negative:*

$$\Delta_\theta \geq 0.$$

[*This implies that, if we use the Fourier law, then the tensor conductivity must be positive.*]

5) *The Gibbs relations become:*

$$\boxed{\begin{aligned} \dot{f} &= -s\dot{\theta} + \frac{P^{ij}v_{i/j}}{\rho} \\ \dot{e} &= \theta\dot{s} + \frac{P^{ij}v_{i/j}}{\rho} \end{aligned}}$$

PROOF. In fact, we have:

$$\dot{f}|_{\theta=const.} = (\partial B_i^k \cdot f)_\theta \frac{\delta B_k^i}{\delta t} = (\partial B_i^k \cdot f)_\theta v^i{}_{/j} B_k^j. \tag{4.34}$$

Therefore, from the inequality of Clausius-Duehm, we have:

$$(\partial B_i^k \cdot f)_\theta v^i{}_{/j} B_k^j - \frac{1}{\rho} P^{ij} v_{i/j} + \frac{q^i}{\rho}(\partial x_i \cdot \log\theta) \leq 0$$

$$\Downarrow$$

$$[(\partial B_i^k \cdot f)_\theta B_k^j - \frac{P^{ij}}{\rho}] v_{i/j} + \frac{q^i}{\rho}(\partial x_i \cdot \log\theta) \leq 0.$$

As P^{ij} does not depend on $v_{i/j}$ and as this expression must be true for any process, we get (4.32). As f does not depend on $\theta_{/i}$ we get (4.33). Then, from (4.34) and (4.31) $\Rightarrow$ (4.33), and so on. □

THEOREM 4.26 - *We can write the rheological equation for a Hookean-solid in the following form:*

$$\boxed{P^{ij} = \frac{1}{2}\lambda K_1 g^{ij} + \mu(B^{ij} - g^{ij})}$$

where λ, $\mu \equiv$ **Lamé coefficients**, (μ *is also called* **rigidity modulus**).

$$K_1 \equiv g_{ij}(B^{ij} - g^{ij})$$

$$B^{ij}(x,t) = (\partial a_r \cdot x^i)(\partial a_s \cdot x^j) g^{rs}(a) \equiv \textbf{Finger strain tensor}.$$

The tensor

$$U^{ij} \equiv B^{ij} - g^{ij}$$

is also called **Ulmansi strain tensor**.

DEFINITION 4.14 - *We have the following definitions:*

1) (**hydrostatic**) **compression modulus**:

$$k \equiv \lambda + \frac{2}{3}\mu, \quad (1/k \equiv \textbf{compression coefficient}).$$

2) **Young modulus** (*or* **extension modulus**):

$$E \equiv \frac{qk\mu}{3k+\mu}.$$

3) $1/E \equiv$ **extension coefficient**.

4) **Poisson ratio**:

$$\sigma = \frac{1}{2}(3k - 2\mu)/(3k+\mu).$$

REMARK 4.17 - 1) For infinitesimal strain, the rheological equation for Hookean solids can be written as follows:

(4.35)(**Hooke's equation**)

$$\boxed{P^{ij} = \lambda \operatorname{tr}(e) g^{ij} + 2\mu e^{ij}}$$

where $e = e_{ij}dx^i \otimes dx^j = \frac{1}{2}(u_{i/j} + u_{j/i})dx^i \otimes dx^j$ is the **infinitesimal strain tensor**.

2) Then, for static conditions, the displacement vector u satisfies the following equations:

$$\boxed{(\frac{\lambda}{2} g^{qp} g^{jk} + \mu g^{qj} g^{pk})(u_{p/qk} + u_{q/pk}) + \rho B^j = 0}$$

3) E and σ can be measured in a little rod stressed along the principal axis z, with pressure p. In fact we have:

$$E = p/e_{zz}, \quad \sigma = -e_{xx}/e_{zz}.$$

Furthermore, we get:

$$\mu = E/2(1+\sigma), \quad k = E/3(1-\sigma), \quad \lambda = E\sigma/(1-2\sigma)(1+\sigma).$$

EXAMPLE 4.5 - EXAMPLES OF ELASTIC DEFORMATIONS

Let us consider the following constitutive equations:

a) (**Finger equation**: isotropic materials).

$$P_{ij} = 2\frac{\rho}{\rho_0}[I_2(\partial I_2 \cdot \bar{e}) + I_3(\partial I_3 \cdot \bar{e})]g_{ij} + (\partial I_1 \cdot \bar{e})B_{ij} - I_3(\partial I_2 \cdot \bar{e})C_{ij}.$$

where

$$C_{ij} = \text{Caundy tensor}; C_{ij}B^{jp} = \delta_i{}^p$$

$$(C_{ij} \equiv (\partial x_i \cdot a^m)(\partial x_j \cdot a^n)g_{mn})$$

$$I_j \equiv \text{invariant of} \quad (B_{ij})(\equiv \text{Finger tensor})$$

$$\Downarrow$$

$$I_1 \equiv B^{ij}B_{ij}, \quad I_2 \equiv \frac{1}{2}[I_1{}^2 - B_{ij}B^{ij}], \quad I_3 \equiv |B_{ij}|.$$

b) (**Rivlin's equations**: isotropic incompressible materials).

$$P_{ij} = -pg_{ij} + 2(\partial I_1 \cdot \bar{e})B_{ij} - e(\partial I_2 \cdot \bar{e})C_{ij}$$

where $p =$ isotropic pressure; (for incompressible materials $I_3 = 1$).
Now, let us make the following hypotheses: (i) The elastic deformation is so strong that the classic theory of Hooke-Cauchy does not work. (ii) The materials are incompressible ones b) ($\Rightarrow$ we can use Rivlin's equation). (iii) The strain is in equilibrium with the stress applied. Then the motion equations become:

$$P^{ij}{}_{/j} = 0. \tag{4.36}$$

Let us consider the following system.
(I) (**Elastic thin rod**). Let us assume that the rod should be deformed in the x direction with ratio $\lambda_{(1)}$. As the material is incompressible, we have:

$$\left\{ \begin{array}{l} x = x^1 = \lambda_{(1)}a^1 \\ y = x^2 = \lambda_{(2)}a^2 \\ z = x^3 = \lambda_{(3)}a^3 \end{array} \right\}, \quad \lambda_{(1)}\lambda^2{}_{(2)} = \lambda_{(1)}\lambda^2{}_{(3)} = 1.$$

We have the following geometric objects.

$$\textbf{Cauchy tensor}: C_{ij}(x) = \begin{bmatrix} 1/\lambda^2{}_{(1)} & 0 & 0 \\ 0 & \lambda_{(1)} & 0 \\ 0 & 0 & \lambda_{(1)} \end{bmatrix}$$

$$\textbf{Finger tensor}: B_{ij}(x) = \begin{bmatrix} \lambda^2_{(1)} & 0 & 0 \\ 0 & 1/\lambda_{(1)} & 0 \\ 0 & 0 & 1/\lambda_{(1)} \end{bmatrix}$$

$$I_1 = \lambda^2{}_{(1)} + 2/\lambda_{(1)}, \quad I_2 = 2\lambda_{(1)} + 1/\lambda^2{}_{(1)}, \quad I_3 = 1.$$

The stress is homogeneous as C_{ij} does not depend on the position. From the Rivlin's equation we get:

$$P_{\widehat{xx}} = -p + 2(\lambda^2{}_{(1)}(\partial I_1 \cdot \bar{e}) - \frac{1}{\lambda^2{}_{(1)}}(\partial I_2 \cdot \bar{e}))$$
$$P_{yy} = P_{zz} = -p + 2(\frac{1}{\lambda_{(1)}}(\partial I_1 \cdot \bar{e}) - \lambda_{(1)}(\partial I_2 \cdot \bar{e}))$$
$$P_{xy} = P_{yx} = P_{xz} = P_{zx} = P_{yz} = P_{zy} = 0.$$

Of course these stresses satisfy the equilibrium conditions (4.36) as they are independent of the position. As the boundary is $\parallel x$, we have

$$P_{yy} = P_{zz} = 0$$
$$\Downarrow$$
$$p = 2(\frac{1}{\lambda_{(1)}}(\partial I_1 \cdot \bar{e}) - \lambda_{(1)}(\partial I_2 \cdot \bar{e}))$$
$$P_{xx} = 2(\lambda^2{}_{(1)} - \frac{1}{\lambda_{(1)}})((\partial I_1 \cdot \bar{e}) + \frac{1}{\lambda_{(1)}}(\partial I_2 \cdot \bar{e})).$$

The **tension**, calculated on the non-deformed section of the rod, is the following:

$$F = P_{xx}\lambda_{(2)}\lambda_{(3)} = \frac{1}{\lambda_{(1)}}P_{xx} = 2(\lambda_{(1)} - \lambda^2{}_{(1)})((\partial I_1 \cdot \bar{e}) + \frac{1}{\lambda_{(1)}}(\partial I_2 \cdot \bar{e})).$$

(II) (**Stressed elastic strip**).

$$\left\{\begin{array}{l} x = x^1 = \lambda_{(1)}a^1 \\ y = x^2 = \lambda_{(2)}a^2 \\ z = x^3 = \lambda_{(3)}a^3 \end{array}\right\}, \quad \lambda_{(2)} = 1, \lambda_{(1)}\lambda_{(3)} = 1.$$

One has the following associated geometric objects:

$$C_{ij} = \begin{bmatrix} 1/\lambda^2_{(1)} & 0 & 0 \\ 0 & 1 & 0 \\ 0 & 0 & \lambda^2_{(1)} \end{bmatrix}$$

$$B_{ij} = \begin{bmatrix} \lambda^2_{(1)} & 0 & 0 \\ 0 & 1 & 0 \\ 0 & 0 & 1/\lambda^2_{(1)} \end{bmatrix}$$

$$I_1 = I_2 = \lambda^2{}_{(1)} + 1 + 1/\lambda^2{}_{(1)}, \quad I_3 = 1\,.$$

Thus the strain is homogeneous, (does not depend on x) and it is of pure shear. From the Rivlin's equations we get:

$$P_{xx} = -p + 2[\lambda^2{}_{(1)}(\partial I_1 \cdot \bar{e}) - \frac{1}{\lambda^2{}_{(1)}}(\partial I_2 \cdot \bar{e})]$$
$$P_{yy} = -p + 2[(\partial I_1 \cdot \bar{e}) - (\partial I_2 \cdot \bar{e})]$$
$$P_{zz} = -p + 2[\frac{1}{\lambda^2{}_{(1)}}(\partial I_1 \cdot \bar{e}) - \lambda^2{}_{(1)}(\partial I_2 \cdot \bar{e})]$$
$$P_{xy} = P_{yx} = P_{xz} = P_{zx} = P_{yz} = P_{zy} = 0.$$

As the stress are homogeneous, the equilibrium conditions (4.36) are satisfied. Furthermore, as in the direction z do not act forces, we have

$$P_{zz} = 0$$
$$\Downarrow$$
$$P = 2[\frac{1}{\lambda^2{}_{(1)}}(\partial I_1 \cdot \bar{e}) - \lambda^2{}_{(1)}(\partial I_2 \cdot \bar{e})]$$
$$\Downarrow$$
$$P_{xx} = 2[\lambda^2{}_{(1)} - \frac{1}{\lambda^2{}_{(1)}}][(\partial I_1 \cdot \bar{e}) + (\partial I_2 \cdot \bar{e})].$$

Therefore, the tension, with respect to the non-deformed section, is the following:

$$F = P_{xx}\lambda_{(2)}\lambda_{(3)} = \frac{1}{\lambda_{(1)}}P_{xx} = 2(\lambda_{(1)} - \frac{1}{\lambda^3{}_{(1)}})[(\partial I_1 \cdot \bar{e}) + (\partial I_2 \cdot \bar{e})].$$

(III) (**Torsion of rod**: **Poynting effect**).

$$r = x^1 = a^1, \quad \theta = x^2 = a^2 + ka^3, \quad z = x^3 = a^3$$

where k = **torsion angle per unity length of the rod**. Then, one has the following associated geometric objects:

$$(C_{ij}) = \begin{bmatrix} 1 & 0 & 0 \\ 0 & 1 & -kr \\ 0 & -kr & 1 + k^2r^2 \end{bmatrix}; (B_{ij}) = \begin{bmatrix} 1 & 0 & 0 \\ 0 & 1 + k^2r^2 & kr \\ 0 & kr & 1 \end{bmatrix}$$
$$P_{rr} = -p + 2[(\partial I_1 \cdot \bar{e}) - (\partial I_2 \cdot \bar{e})]$$
$$P_{\theta\theta} = -p + 2[(1 + k^2r^2)](\partial I_1 \cdot \bar{e}) - (\partial I_2 \cdot \bar{e})]$$
$$P_{zz} = -p + 2[(\partial I_1 \cdot \bar{e}) - (1 + k^2r^2)(\partial I_2 \cdot \bar{e})]$$
$$P_{\theta z} = P_{z\theta} = 2kr[(\partial I_1 \cdot \bar{e}) + (\partial I_2 \cdot \bar{e})]$$
$$P_{r\theta} = P_{\theta r} = P_{rz} = P_{zr} = 0$$
$$I_1 = I_2 = 3 + k^2r^2, I_3 = 1.$$

Note that the stress is non-homogeneous. The condition of equilibrium (4.36) gives:

$$P_{\theta\theta} = \partial r \cdot (r P_{rr})$$

$$\Downarrow$$

$$(\partial r \cdot P_{rr}) = 2k^2 r(\partial I_1 \cdot \bar{e})\,.$$

Assuming that the radius R of the cylinder does not change, hence there are no radial pressures, we can integrate the last equation:

$$P_{rr} = -2k^2 \int_r^R r(\partial I_1 \cdot \bar{e}) dr$$

$$\Downarrow$$

$$-p = P_{rr} - 2[(\partial I_1 \cdot \bar{e}) - (\partial I_2 \cdot \bar{e})]$$

$$P_{zz} = P_{rr} - 2[(\partial I_1 \cdot \bar{e}) - (\partial I_2 \cdot \bar{e})] + 2[(\partial I_1 \cdot \bar{e}) - (\partial I_2 \cdot \bar{e})] - 2k^2 r^2 (\partial I_2 \cdot \bar{e})$$

$$\Downarrow$$

$$P_{zz} = P_{rr} - 2k^2 r^2 (\partial I_2 \cdot \bar{e})$$

$$\Downarrow$$

$$P_{zz} = 2k^2 \int_R^r r(\partial I_1 \cdot \bar{e}) dr - 2k^2 r^2 (\partial I_2 \cdot \bar{e}).$$

Then, we get the following expression for the axial pressure:

$$N = 2\pi \int_0^R r P_{rr} dr = 2\pi [2k^2 \int_0^R r \int_R^r r(\partial I_1 \cdot \bar{e}) dr - 2k^2 \int_0^R r^3 (\partial I_2 \cdot \bar{e}) dr]$$

$$N = -2\pi k^2 \int_0^R r^3 [(\partial I_1 \cdot \bar{e}) + 2(\partial I_2 \cdot \bar{e})] dr.$$

2 If the general pressure N does not concern the end of the rod, the shear stress produces the breaking off of the rod. The torsion couple M is given by the following expression:

$$M = 2\pi \int_0^R r^2 P_{\theta z} dr = 4\pi K \int_0^R r^3 [(\partial I_1 \cdot \bar{e}) + (\partial I_2 \cdot \bar{e})] dr.$$

(IV) (**Torsion of thin rod, having any section, for little deformations**).

Let us assume the following structure of the displacement vector u, in orthogonal cartesian coordinates:

$$\left\{\begin{array}{l} u_x = -\tau z y \\ u_y = \tau z x \\ u_z = \tau\psi(x,y) \end{array}\right\}, \left\{\begin{array}{l} \tau \equiv \dfrac{d\phi}{dz} \\ \phi \equiv \textbf{rotation angle} \\ \psi \equiv \textbf{torsion function} \end{array}\right\}.$$

$$\Downarrow$$

$$\left\{\begin{array}{l} x \equiv x^1 = a^1 - \tau z y \\ y \equiv x^2 = a^2 + \tau z x \\ z \equiv x^3 = a^3 + \tau\psi(x,y) \end{array}\right\}$$

The infinitesimal strain tensor

$$e \equiv \frac{1}{2}\mathcal{L}_{u_\wedge} g = \frac{1}{2}(u_{i/j} + u_{j/i})dx^i \otimes dx^j$$

assumes the following form:

$$e_{ij} = \begin{bmatrix} 0 & 0 & (1/2)\tau(\partial x \cdot \psi - y) \\ 0 & 0 & (1/2)\tau(\partial y \cdot \psi + x) \\ (1/2)\tau(\partial x \cdot \psi - y) & (1/2)\tau(\partial y \cdot \psi + x) & 0 \end{bmatrix}.$$

Furthermore, the torsion is an isocoric deformation, namely it is of pure shear. We define **extension $e(l)$ in a point, in the direction l**: $e(l) = e_{ij}l^i l^j$. We define **expansion** $\theta = g^{ij}e_{ij}$. Assuming the rheological Hooke-equation, the stress tensor has the following structure: $P_{ij} = \lambda \mathrm{tr}(e)\delta_{ij} + 2\mu e_{ij}$. As $\mathrm{tr}(e) = 0$, we get:

$$(P_{ij}) = \begin{bmatrix} 0 & 0 & 2\mu\tau(\partial x \cdot \psi - y) \\ 0 & 0 & 2\mu\tau(\partial y \cdot \psi + x) \\ 2\mu\tau(\partial x \cdot \psi - y) & 2\mu\tau(\partial y \cdot \psi + x) & 0 \end{bmatrix}.$$

Let us introduce the function

$$\chi = \chi(x,y)$$

such that

$$(\partial y \cdot \chi) = \frac{1}{2}(\partial x \cdot \psi - y)$$

$$(\partial x \cdot \chi) = -\frac{1}{2}(\partial y \cdot \psi + x)$$

Then, the stress tensor can also be written as follows:

$$(P_{ij}) = \begin{bmatrix} 0 & 0 & 2\mu\tau(\partial y \cdot \chi) \\ 0 & 0 & -2\mu\tau(\partial x \cdot \chi) \\ 2\mu\tau(\partial y \cdot \chi) & -2\mu\tau(\partial x \cdot \chi) & 0 \end{bmatrix}.$$

□ In equilibrium conditions the torsion function ψ must satisfy the following equation:

$$\Delta\psi \equiv \psi_{xx} + \psi_{yy} = 0, \quad + \text{boundary conditions},$$

or, by means of the function χ we get:

$$\Delta\chi \equiv \chi_{xx} + \chi_{yy} + 1 = 0, \quad + \text{boundary conditions}.$$

If the section is simply connected, the boundary condition is $\chi|_{\partial D} = 0$; otherwise the boundary condition is $\int_{\partial D}(\partial n \cdot \chi)d\sigma = -S$, where $S \equiv$ area enclosed in the section.
PROOF. In fact, in equilibrium condition we have $(\partial x_j \cdot P_{ij}) = 0$, hence $(\partial x \cdot P_{zx}) + (\partial y \cdot P_{zy}) = 0$. Then, by using the expression for P, we get the above condition on ψ or χ.
In order to obtain the boundary conditions, let us consider that as the rod is thin, the lateral external forces must be negligible with respect to internal ones. Hence, set on the sides $P^{ij}n_j = 0$, with $n_z = 0 \Rightarrow$ (by using equation for χ)

$$(\partial y \cdot \chi)n_x - (\partial x \cdot \chi)n_y = 0$$

with

$$n_x = -\frac{dy}{ds}, \quad n_y = \frac{dx}{ds}$$

$$\Downarrow$$

$$(\partial x \cdot \chi)dx + (\partial y \cdot \chi)dy \equiv d\chi = 0$$

$$\Downarrow$$

$$\chi|_{\partial D} = const.$$

If the section is simply connected, we can put $\chi = 0$. Otherwise, $\partial D = \bigcup_{i\in\{1,\cdots,s\}} \Gamma_i$, where each Γ_i is a closed curve, χ will have different constant values for each Γ_i. We can put $\chi = 0$ only on one of these, e.g., Γ_1, but the other values χ_i remain to be determined. As $u_z = \tau\psi(x, y)$ is a univalued function of the coordinates, we can impose $\int_{\Gamma_i} d\psi = 0$. Then, taking into account the relations between the functions ψ and χ we get:

$$\begin{aligned} 0 = \int_{\partial D} d\psi &= \int_{\partial D} [(\partial x\psi)dx + (\partial y\psi)dy] \\ &= -2\int_{\partial D} [(\partial x\chi)dy - (\partial y\chi)dx] - 2\int_{\partial D} (xdy - ydx) \end{aligned}$$

hence, we have the imposed condition. □

In the following table we give some examples of functions χ for particular sections.

TAB.4.10 - **Examples of functions χ for particular sections**

Section-form	χ
Disc of radius R	$\chi = \frac{1}{4}(R^2 - x^2 - y^2)$
Circular sector: $(R_2 < R_1)$	$\chi = \frac{1}{4}(R_2^2 - r^2)$

Let us define **torsion rigidity**: $c = 4\mu \int_D (\text{grad}\,\chi)^2 d\sigma$. In the following table we give the explicitly expressions of c in correspondence to the homotopy properties of the sections.

TAB.4.11 - **Torsion rigidity and homotopy**

$\pi_1(D)$	c
$\pi_1(D) = 0$	$c = 4\mu \int_D \chi dx dy$
$\pi_1(D) \neq 0$	$c = 4\mu[\sum_i \chi_i S_i + \int_D \chi dx dy]$
Disc of radius R	$c = \frac{1}{2}\mu\pi R^4$
Elliptic-section: (a, b)	$c = \pi\mu a^3 b^3/(a^2 + b^2)$
Equilater triangle	$c = \sqrt{3}\mu a^4/80$
Circular sector: (R_1, R_2)	$c = \frac{1}{2}\mu\pi(R_2^4 - R_1^4)$
Arbitrary section(+)	$c = \frac{1}{3}\mu L h^3$

(+)(h= thickness, L= perimeter)

The above formulas can be obtained directly, taking into account the following.

$$(\text{grad}\,\chi)^2 = \text{div}(\chi \text{grad}\,\chi) - \chi \triangle \chi = \text{div}(\chi \text{grad}\,\chi) + \chi u.$$

☐ The free energy of a rod for torsion, in equilibrium conditions, is the following:

$$\boxed{E = \frac{1}{2}\int c z^2 dz}$$

In fact, by using the following equation:

$$\frac{\rho}{\rho_0}\frac{\delta e}{\delta t} = P_{ij}\dot{e}^{ij}$$

in condition of equilibrium, we get

$$\frac{\rho}{\rho_0}e = P_{ij}\frac{1}{2}(v_{i/j} + v_{j/i}) = P_{ij}e^{ij} = P_{xz}e_{xz} + P_{yz}e_{yz} = (P_{xz}^2 + P_{yz}^2)/\mu\,.$$

By using the expression of P in terms of χ we get

$$\frac{\rho}{\rho_0} e4\mu\tau^2[(\partial x\chi)^2 + (\partial y\chi)^2] = 4\mu\tau^2(\text{grad }\chi)^2.$$

By integration, we get the proof. □

□ Let the rod be fixed to one end, and an applied external forces be exerted on the other end in such a way to produce a couple of moment M that generates the torsion around the axis of the rod. Then, the torsion-angle, in equilibrium conditions, is constant along the rod, and one has:

$$\tau = M/c.$$

In fact, in equilibrium conditions the free energy of the rod must be minimum. The total energy of the rod is $E+U$, where U = potential energy generated by the action of external forces. Taking into account the above integral expression of E, and $\tau = \frac{d\phi}{dz}$, by a variation with respect to ϕ we get:

$$\delta(E+U) = \delta(\frac{1}{2}\int cz^2 dz) + \delta U = \int c\frac{d\phi}{dz}\frac{d\delta\phi}{dz}dz + \delta U = 0.$$

By integration by parts we get

$$-\int c\frac{d\tau}{dz}\delta\phi dz + \delta U + [cz\delta\phi] = 0.$$

The last term is the difference of the integration limit-values, namely at the ends of the rod. One of this is fixed, hence $\delta\phi = 0$. We know that $\delta U = -M\delta\phi$ hence

$$\int c\frac{d\tau}{dz}\delta\phi dz + [\delta\phi(-M + cz)] = 0.$$

As the variation $\delta\phi$ is arbitrary, we must have:

$$c\frac{d\tau}{dz} = 0, -M + cz = 0.$$

□

(V) (**Plane bending of thin rod for infinitesimal deformations**).
In the process of bending there are some regions of the rod that are compressed and other ones that are expanded. Between these two regions there is a zone that is neither compressed nor expanded. This zone is called **neutral surface**.

□ The bending of a thin rod is a simple extension (or compression), namely the unique non-zero component of the stress tensor is P_{zz}, where z is the axis of the rod. Furthermore, the infinitesimal strain Ulmansi tensor can be written as follows:

$$(e_{ij}) = \frac{\chi}{R}\begin{bmatrix} -\sigma & 0 & 0 \\ 0 & -\sigma & 0 \\ 0 & 0 & 1 \end{bmatrix}.$$

Then, assuming the rheological equation for Hooke's solids, we get the following expression for the stress tensor:

$$(P_{ij}) = \begin{bmatrix} 0 & 0 & 0 \\ 0 & 0 & 0 \\ 0 & 0 & E\chi/R \end{bmatrix}.$$

The neutral surface is determined by the condition:

$$\int_D \chi d\sigma = 0.$$

Thus, for a homogeneous system the neutral surface must cross the center of mass of the section of the rod. In fact, the considered strain must be of pure bending, without extension or compression. Then, the total forces acting on the section must be zero, i.e., $\int_D P_{zz} d\sigma = 0$. Thus, taking into account the expression of P we get the result.

□ The displacement vector is given by the following:

$$(u_x, u_y, u_z) = (-\frac{1}{2R}\{z^2 + \sigma(x^2 - y^2)\}, -\sigma xy/R, xz/R).$$

In fact, we must integrate the following PDE:

$$\left\{ \begin{array}{ll} (\partial x.u_z) + (\partial z.u_x) = 0, & (\partial z.u_z) = \frac{x}{R} \\ (\partial y.u_x) + (\partial x.u_y) = 0, & (\partial x.u_x) = -\frac{\sigma}{R}x \\ (\partial z.u_y) + (\partial y.u_z) = 0, & (\partial y.u_y) = -\frac{\sigma}{R}x \end{array} \right\} + \text{boundary conditions} .$$

By successive derivation we get:

$$\Delta u_x = -\frac{1}{R}, \quad \Delta u_y = 0, \quad \Delta u_z = 0.$$

From $\Delta u_y = 0$, integrating with separation of variables, we get:

$$\begin{aligned} u_y &= C_1C_3C_5xyz + C_1C_4C_5xz + C_2C_3C_5yz \\ &+ C_2C_4C_5z + C_1C_3C_6xy + C_1C_4C_6x + C_2C_3C_6y + C_2C_4C_6 . \end{aligned}$$

The condition $(\partial y.u_y) = -\frac{\sigma}{R}x$ implies

$$u_y(x)u_y(z)(\partial y.u_y) = -\frac{\sigma}{R}x, u_y(x)u_y(z)C_2 = -\frac{\sigma}{R}x$$

hence

$$u_y(z) = 1, u_y(x) = x, C_1 = 1, C_2 = 0, C_3 = -\frac{\sigma}{R}x, C_5 = 0, C_6 = 1.$$

$$\Downarrow$$

$$u_y(y) = -\frac{\sigma}{R}y + C_4$$

$$\Downarrow$$

$$u_y(x,y,z) = -\frac{\sigma}{R}xy + C_4 x.$$

Similarly, we get:

$$u_z(x,y,z) = \frac{1}{R}xz + C_{12}x.$$

Furthermore, for the component u_x, the following equation must be verified:

$$(\partial y.u_x) = -(\partial x.u_y) = \frac{\sigma}{R}y + C_4$$

$$\Downarrow$$

$$(\partial y.u_x) = \frac{\sigma}{R}y + C_4$$

$$(\partial z.u_x) = -(\partial x.u_z) = -\frac{z}{R} - C_{12}$$

$$\Downarrow$$

$$(\partial z.u_x) = -\frac{z}{R} - C_{12}$$

$$\Downarrow$$

$$\begin{aligned} du_x &= (\partial x.u_x)dx + (\partial y.u_x)dy + (\partial z.u_x)dz \\ &= -\frac{\sigma}{R}xdx + (-\frac{\sigma}{R}y + C_4)dy + (-\frac{z}{R} - C_{12})dz \end{aligned}$$

$$\Downarrow$$

$$u_x = -\frac{\sigma}{2R}x^2 - \frac{\sigma}{2R}y^2 + C_4 y - \frac{z^2}{2R} - C_{12}z + C_{13}.$$

The condition $u_i(0) = 0$ imposes $C_{13} = 0$. Furthermore, from

$$(\partial y.u_x) = \frac{\sigma}{R}y + C_4 \equiv \frac{1}{2R}\{-2\sigma y\} - C_4 \Rightarrow C_4 = 0.$$

Finally, from

$$(\partial x.u_z) = \tfrac{z}{R} + C_{12} \equiv \tfrac{1}{2R}\{2z\} + C_{12} \Rightarrow C_{12} = 0. \qquad \square$$

REMARK 4.18 - From the expressions of the displacements, we see that points initially on a section $z = z_0$ will be transformed, after deformation, in the following ones: $\{z = z_0 + u_z = z_0(1 + \frac{x}{R})\}$. Namely the section remains plane, but will be rotated by an angle relative to the initial position. (However, the form of the section change.)

☐ The energy per unit length of the bent rod is the following:

$$E = \frac{1}{2}\frac{E}{R^2}I_y$$

where $R =$ curvature radius of the neutral surface, (for a thin rod R can be identified with the corresponding curvature radius of the line that corresponds to the rod), $I_y \equiv \int_D x^2 d\sigma$ = moment of inertia of the section around the y-axis $\equiv$ **bending moment**. In fact, the free energy per unit volume is the following:

$$\frac{1}{2}P_{ij}e^{ij} = \frac{1}{2}P_{xz}e_{xz} = \frac{1}{2}\frac{Ex^2}{R^2}.$$

By integration on a section we get our formula.
Furthermore, the total moment of the forces with respect to the y-axis is the following:

$$M_y = \frac{E}{R}I_y.$$

In fact, as the force $P_{xz}d\sigma = \frac{xE}{R}d\sigma$ acts on the direction z on the element $d\sigma$ of surface of the section, the moment with respect to the origin is $xP_{xz}d\sigma$. Therefore, the total moment $M_y = \frac{E}{R}\int_D x^2 d\sigma = \frac{E}{R}I_y$.
(VI) (**Deformation of thin rod (torsion+bending)**).

☐ The deformation of a rod can be characterized by means of a vector $\mathbf{\Omega}$ (**deformation vector**) that represented in the local basis $(\mathbf{T}, \mathbf{N}, \mathbf{B})$ of the line representing the rod, has the following form:

$$\mathbf{\Omega} = \frac{1}{R}\mathbf{B} + \Omega_\zeta \mathbf{T}$$

where Ω_ζ is the component of pure torsion, and $\frac{1}{R}$ is the component of pure bending. We define, also, **rotation vector** one ϕ such that

$$\mathbf{\Omega} = \frac{d\phi}{ds}.$$

☐ The elastic energy, per unit length of the rod, can be written as follows:

$$E = \frac{1}{2}\{E(I_1\Omega_\xi^2 + I_2\Omega_\eta^2) + c\Omega_\zeta^2\}$$

where (ξ, η, ζ) are the coordinates with respect to the principal axes of inertia of the section, and I_1, I_2 are the corresponding moments of inertia (axes ξ and η).

□ The moment of the forces acting on a section is the following:

$$M = (M_\xi, M_\eta, M_\varsigma)$$
$$= (EI_1\Omega_\xi, EI_2\Omega_\eta, c\Omega_\varsigma)\,.$$

EXAMPLE 4.6 - For a circular rod one has:

$$E = \frac{1}{2}\{\frac{EI}{R^2} + c\Omega_\varsigma^2\}, \quad M = \frac{EI}{R}\mathbf{B} + c\Omega_\varsigma\mathbf{T}.$$

□ (EXPRESSION OF THE MOMENT BY MEANS OF THE STRESS ACTING ON THE SECTION $\perp\mathbf{T}$). We have:

$$M = (M_\xi, M_\eta, M_\varsigma)$$
$$= (-\int_D P_{\varsigma\varsigma}\eta d\sigma, \int_D P_{\varsigma\varsigma}\xi d\sigma, \int_D (P_{\xi\varsigma}\eta - P_{\eta\varsigma}\xi)d\sigma)\,.$$

In the case of pure bending one has $P_{\xi\varsigma} - P_{\eta\varsigma} = 0$.
The expression of $P_{\varsigma\varsigma}$ in terms of curvature is the following:

$$P_{\varsigma\varsigma} = \frac{E}{R}(B_\eta\xi - B_\xi\eta).$$

□(NON-PLANE BENDING). The vector displacement for non-plane bending is the following:

$$(u_x, u_y, u_z) = (-\frac{1}{R}z^2, -2\sigma xy/R, z(x+y)/R).$$

In fact, u is the sum of two plane bendings, respectively in the planes (z, x) and (z, y). The infinitesimal strain tensor of Ulmansi is the following:

$$(e_{ij}) = \begin{bmatrix} 0 & -\frac{\sigma}{R}y & -\frac{1}{R}z \\ -\frac{\sigma}{R}y & -\frac{2\sigma}{R}x & \frac{1}{2R}z \\ -\frac{1}{2R}z & \frac{1}{2R}z & \frac{x+y}{R} \end{bmatrix}$$

and the stress tensor becomes

$$\left\{\begin{array}{l} P_{11} = \dfrac{E\sigma}{R(1+\sigma)}(x + \dfrac{y}{1-\sigma}) \\ P_{22} = \dfrac{E\sigma}{R(1+\sigma)}(\dfrac{y}{1-2\sigma} - x) \\ P_{33} = \dfrac{E}{R}(x+y) \\ P_{12} = -\dfrac{E\sigma}{R(1+\sigma)}\dfrac{y}{R} \\ P_{13} = \dfrac{Ez}{2R(1+\sigma)} \\ P_{23} = \dfrac{Ez}{2R(1+\sigma)} \end{array}\right\}$$

The moment of the forces acting on the section $\perp z$-axis, with respect to the origin O is the following:

$$\begin{aligned} M &= (M_x, M_y, M_z) \\ &= (-\frac{E}{R}(I_{12}(z) + I_{11}(z)), \frac{E}{R}(I_{12}(z) + I_{22}(z)), \frac{EA(z)z}{2(1+\sigma)R}(y_G(z) + x_G(z))) \,. \end{aligned}$$

The form of the section changes with the bending. If in the plane (x, y) the boundary of the section has the equation $x_0 = f(y_0), \{x = x_0 + u_x, y = y_0 + u_y\}$, after the deformation we will have the equation $\{x = f(y_0) + u_x, y = y_0 + u_y\}$. Thus, by eliminating y_0, we have the implicit equation $F(x, y) = 0$ of the deformed boundary.
EXAMPLE 4.7 -

DEFORMATION OF CIRCULAR SECTION AFTER NON-PLANE BENDING.
If in the initial condition the boundary of the section has equation $x_0^2 + y_0^2 = R^2$, after deformation this boundary has equation:

$$x^2 + y^2 + \frac{4\sigma^2}{R^2}x^2y^2 + \frac{4\sigma}{R}xy^2 + \frac{2z_0^2}{R}x + \frac{z_0^4}{R^2} - R^2 = 0$$

that is the equation of a quadric. Furthermore, the z-coordinate becomes

$$z = \frac{Rz_0}{R - (x+y)}.$$

For little displacements we can omit the terms of third and fourth orders, and take $\frac{x+y}{R} \ll 1$. Then, the section in the plane (x, y) reduces to a conic (ellipse):

$$x^2 + y^2 + \frac{2z_0^2}{R}x + \frac{z_0^4}{R^2} - R^2 = 0$$

and

$$z = z_0.$$

□(EQUILIBRIUM EQUATIONS FOR ROD). We have the following equations:

$$\frac{d\mathbf{F}}{ds} = -B, \quad \frac{dM}{ds} = F \times \mathbf{T}$$

where $F_i = \int_D P_{i\varsigma} d\sigma$ = force acting on the section, B = external forces acting on the rod per unit length, M = moment of the forces acting on the section with respect to a point O.
□(EQUILIBRIUM EQUATIONS FOR ROD WITH FORCES CONCENTRATED ON ISOLATED POINTS).

1)

$$F = \text{constant along the axis}, \quad M = F \times \mathbf{r} + const.$$

where $\mathbf{r}$ is the position vector of a generic point.

2) If the concentrated forces are absent, and the rod is bent by means of concentrated moments, then one has:

$$F = \left\{ \begin{array}{l} = \text{constant (along the axis)} \\ M \text{ is discontinuous in the points where the couples are applied} \end{array} \right\}$$

(D) VISCOELASTIC MATERIALS.

DEFINITION 4.15 - **Viscoelastic materials** *are those where the stress tensor is determined by the history of the strain before the time considered. In other words P is an application of the type*

$$P : C^\infty(C) \to C^\infty(vT^0_2 M).$$

EXAMPLE 4.8 - 1) (**Linear viscoelasticity**):

$$\boxed{\begin{array}{l} P^{ij}(x,t) = -p(x,t)g^{ij}(x,t) + \wp^{ij}(x,t) \\ \wp^{ij}(x,t) = 2 \int\limits_{-\infty}^{t} \mu(t-t')(\partial a_m \cdot x^i)(\partial a_n \cdot x^j)\dot{e}^{mn}(x,t)dt' \end{array}}$$

where μ is called **relaxation function**. We call **elastic modulus**:

$$G = \int\limits_0^\infty G(\lambda)d\lambda$$

where

$$\mu(t) = \int\limits_0^\infty G(\lambda)e^{-t/\lambda}d\lambda = \mathcal{L}[G](t) = \text{Laplace transform of } G.$$

In the following table we report some important examples of relaxation functions.

TAB.4.12 - **Examples of relaxation functions**

Material	μ
Maxwell-fluid	$\mu(t) = \frac{\eta}{\lambda_0}e^{-t/\lambda_0}$
	(λ_0 = relaxation time)
Oldroyd-fluid	$\mu(t) = \frac{\eta}{\lambda_0}[(1 - \frac{\mu_0}{\lambda_0})e^{-t/\lambda_0} + \mu_0\delta(t)$
Newtonian-fluid	$\mu(t) = \eta\delta(t)$

THEOREM 4.27 - (STEADY FLOW FOR LINEAR VISCOELASTICITY). *A material can have a steady flow if $\mu(t)$ is such that the following integrals exist:*

$$\int_0^\infty \mu(t)dt = \int_0^\infty \lambda G(\lambda)d\lambda = \frac{\wp_{13}}{\dot{\gamma}} \equiv \eta \ (\textbf{apparent viscosity})$$

$$\int_0^\infty t\mu(t)dt = \int_0^\infty \lambda^2 G(\lambda)d\lambda = \frac{\wp_{33}}{\dot{\gamma}^2} \equiv \psi_{33} \ (\textbf{normal stress coeff.})$$

with $\dot{\gamma} =$ **shear strain**. *If these integrals do not exist the viscoelastic material is solid and cannot have a steady flow without breaking-off.*

PROOF. Let us consider the case of flow in a pipe with circular section. Let us use cylindrical coordinates (r, θ, z) and the following velocity field

$$\left\{ \begin{array}{c} v_r = v_\theta = 0 \\ t \leq 0 \Rightarrow v_z = 0 \\ t > 0 \Rightarrow v_z = u(r,t) \end{array} \right\}$$

Then, the corresponding flow is the following:

$$\left\{ \begin{array}{l} r = x^1 = a^1 \\ \theta = x^2 = a^2 \\ z = x^3 = a^3 + \int_{t'}^{t} u(a', z)dz \end{array} \right\}$$

Then, one has

$$(\dot{e}^{ij})(a,t') = \frac{\dot{\gamma}}{2}(r,t') \begin{bmatrix} 0 & 0 & 1 \\ 0 & 0 & 0 \\ 1 & 0 & 0 \end{bmatrix}$$

$$\dot{\gamma}(r,t') \equiv (\partial r \cdot u)(r,t') \equiv \text{shear strain.}$$

The non-zero physical components of $\wp$ are the following:

$$\wp_{rz} = \wp_{zr} = \int_0^t \mu(t-t')(\partial r \cdot u)(r,t')dt'$$

$$\wp_{zz} = 2\int_0^t \int_{t'}^t \mu(t-t')(\partial r \cdot u)(r,t')(\partial r \cdot u)(r,\tau)d\tau dt' .$$

LEMMA 4.2 - *Assume that there exists the following limit:*

$$\lim_{t\to\infty} f(t).$$

Then, for the following Laplace transform

$$\mathcal{L}[f](s) = \int_0^\infty e^{-st} f(t)dt,$$

we get

(4.37) $$\lim_{s\to\infty} s\mathcal{L}[f](s) = \lim_{t\to\infty} f(t).$$

By applying the above lemma we get:

$$s\mathcal{L}[\wp_{rz}](r,s) = \mathcal{L}[\mu](s)[s\frac{d\mathcal{L}[\mu]}{dr}(r,s)]$$

$$\Downarrow$$

$$\lim_{t\to\infty} \wp_{rz}(r,t) = \lim_{s\to\infty} s\mathcal{L}[\wp_{rz}](r,s)$$

$$= (\lim_{t\to\infty} \mathcal{L}[\mu](s))(\lim_{s\to\infty} s\frac{d\mathcal{L}}{dr}[u](r,s))$$

$$\Downarrow$$

$$\lim_{t\to\infty} \wp_{zr}(r,t) = \int_0^\infty \mu(t)dt\frac{du_\infty(r)}{dr}.$$

Hence,

$$\wp_{13}/\dot{\gamma} = \eta.$$

(Note that for viscoelastic solids it is not necessary that there exists the integral $\int_0^\infty \mu(t)dt$ as for solid materials $\frac{du_\infty}{dr} = 0$.) Let us consider, now,

$$\wp_{zz}(r,t) = 2\int_0^t (\partial r \cdot u)(r,t')dt' \int_t^{t'} \mu(t-t')(\partial r \cdot u)(r,\tau)d\tau.$$

Assume $\mu(t) = \int_0^\infty G(\lambda)e^{-t/\lambda}d\lambda$. We get:

$$\wp_{zz}(r,t) = 2\int_0^\infty G(\lambda)d\lambda \int_0^t e^{-(t-t')/\lambda}(\partial r \cdot u)(r,t')dt' \int_0^{t'} e^{-(t-z)/\lambda}(\partial r \cdot u)(r,\tau)d\tau$$

$$\Downarrow$$

$$\wp_{zz}(r,t) = 2\int_0^\infty G(\lambda)I_\lambda(r,t)d\lambda.$$

For the convolution theorem of the Laplace transform, we have:

$$s\mathcal{L}[I_\lambda](r,s) = \frac{\lambda}{1+\lambda s} s\mathcal{L}[\sigma_\lambda](r,s).$$

By using (4.37) we get:

$$\lim_{t\to\infty} I_\lambda(r,t) = \lambda \lim_{t\to\infty} \sigma_\lambda(r,t)$$

hence

$$\lim_{t\to\infty} \wp_{zz}(r,t) = 2\int_0^\infty G(\lambda)[\lim_{t\to\infty} I_\lambda(r,t)]d\lambda = 2\int_0^\infty G(\lambda)[\lambda \lim_{t\to\infty} \sigma_\lambda(r,t)]d\lambda.$$

On the other hand,

$$\lim_{t\to\infty} \sigma_\lambda(r,t) = [\frac{du_\infty}{dr}(r)]^2 \lambda$$

hence

$$\lim_{t\to\infty} \wp_{zz}(r,t) = [\frac{du_\infty}{dr}(r)]^2 2\int_0^\infty G(\lambda)\lambda^2 d\lambda.$$

On the other hand

$$\int_0^\infty \lambda^2 G(\lambda)d\lambda = \int_0^\infty t\mu(t)dt \equiv \epsilon.$$

Hence,

$$\lim_{t\to\infty} \wp_{zz}(r,t) = 2\psi_{33}[\frac{du_\infty}{dr}(r)]^2 = 2\epsilon\dot{\gamma}^2$$

$$\frac{\wp_{zz}}{\dot{\gamma}^2} = \psi_{33}, \psi_{33} = 2\epsilon.$$

□

THEOREM 4.28 - (OLDROYD FLUID IN CIRCULAR PIPE: STEADY STATE).
1) *The steady state flow is similar to those of Newtonian fluid.*
2) *In physical components, one has the following expression for the stress tensor:*

$$(P^{ij}) = \begin{bmatrix} -p & 0 & G\eta \\ 0 & -p & 0 \\ G\eta & 0 & -p - 2G^2\eta(\lambda_0 - \mu_0) \end{bmatrix}$$

where $G \equiv$ shear strain. For the normal stress differences we get:

$$P_{33} - P_{22} = -2G^2\eta(\lambda_0 - \mu_0)$$
$$P_{33} - P_{11} = P_{33} - P_{22} \, .$$

PROOF. 1) From the rheological equation we have:

$$\wp^{ij} = 2\int_{-\infty}^{t} \mu(t-t')(\partial a_m \cdot x^i)(\partial a_n \cdot x^j)\dot{e}^{mn}(t')dt'.$$

In steady state we have:

$$\left\{\begin{array}{l} x^1 = a^1 \\ x^2 = a^2 \\ x^3 = a^3 + v(a^1)(t-t^0) \end{array}\right\}$$

Hence,

$$\dot{e}^{ij} = \frac{1}{2}G\begin{bmatrix} 0 & 0 & 1 \\ 0 & 0 & 0 \\ 1 & 0 & 0 \end{bmatrix}.$$

For $t^0 = t'$, we have:

$$\wp^{33} = 2\int_{-\infty}^{t} \mu(t-t')[(\partial a_1 \cdot x^3)(\partial a_3 \cdot x^3)\dot{e}^{13}(a,t') + (\partial a_3 \cdot x^3)(\partial a_1 \cdot x^3)\dot{e}^{31}(a,t')]dt'$$

$$= G(r)\int_{-\infty}^{t} \mu(t-t')2[(\partial a_1 \cdot x^3)(\partial a_3 \cdot x^3)]dt'.$$

As $(\partial a_1 . x^3) = G(t-t') \Rightarrow (\partial a_1 . x^3) = 1$. Hence,

$$\begin{aligned}
\wp^{33} &= 2G^2\int_{-\infty}^{t} \mu(t-t')(t-t')dt' \\
&= -2G^2\int_{\infty}^{0} \mu(s)sds \\
&= 2G^2\frac{\eta}{\lambda_0}\int_{0}^{\infty} [(1-\frac{\mu_0}{\lambda_0})e^{-s/\lambda_0} + \mu_0\delta(s)]sds \\
&= 2G^2\frac{\eta}{\lambda_0}[(1-\frac{\mu_0}{\lambda_0})\int_{0}^{\infty} e^{-s/\lambda_0}sds + \mu_0\int_{0}^{\infty} s\delta(s)ds].
\end{aligned}$$

Here, $s \equiv t-t'$. Let us calculate, now, the two integrals:

$$\int_{0}^{\infty} e^{-s/\lambda_0}sds = \lambda_0^2\int_{0}^{\infty} e^{-x}xdx = \lambda_0^2\Gamma(2)$$

where $s/\lambda_0 = x$ and where $\Gamma(x)$ is the gamma function. This satisfies the relation $\Gamma(x+1) = x\Gamma(x)$. Hence:

$$\Gamma(2) = \Gamma(1) = 1$$

$$\Downarrow$$

$$\int_0^\infty e^{-s/\lambda_0} s ds = {\lambda_0}^2.$$

Furthermore,

$$\int_0^\infty s\delta(s) ds = 0.$$

Therefore, we get:

$$\wp^{33} = 2G^2 \frac{\eta}{\lambda_0}(1 - \frac{\mu_0}{\lambda_0})\lambda_0^2 = 2G^2\eta(\lambda_0 - \mu_0).$$

Similar relations also give:

$$\wp^{11} = 0, \quad \wp^{13} = G\eta.$$

The motion equation, in steady state, gives:

$$(\partial r \partial r \cdot v_z) + \frac{1}{r}(\partial r \cdot v_z) = \frac{1}{\eta}(\partial z \cdot p). \tag{4.38}$$

This equation coincides with the corresponding one for Newtonian fluids. Taking into account the boundary conditions:

$$v_z(R) = 0, \quad v_z(0) = \text{finite}$$

we get the following solution:

$$\left\{ \begin{aligned} v_r &= 0 \\ v_\phi &= 0 \\ v_z(r) &= \frac{\Delta p}{4\eta l}(R^2 - r^2) \end{aligned} \right\} \tag{4.39}$$

In fact from (4.38) we get:

$$(\partial z \cdot p) = const.$$

hence

$$(\partial r \cdot \partial r \cdot v_z) + \frac{1}{r}(\partial r \cdot v_z) = \frac{const.}{\eta}$$

Therefore, we can write $(\partial z \cdot p) = \Delta p/l$, and as a consequence we get (4.39).

THEOREM 4.29 - (OLDROYD FLUID IN CIRCULAR PIPE: PULSANT FLOW). *Let us give the following assumptions:* a) *for* $t \leq 0$ *the material is at rest, and without deformation;* b) *at* $t = 0$ *it is applied a gradient of pulsant pressure with little amplitude and such that we can consider the linear viscoelasticity;* c) *the velocity field has the following structure:*

$$v_r = v_\phi = 0, \quad v_z = u(r,t).$$

Then, the flow has the following structure:

$$\left\{ \begin{array}{l} r \equiv x^1 = a^1 \\ \phi \equiv x^2 = a^2 \\ z \equiv x^3 = a^3 + \int_{t'}^{t} u(a^1,\tau)d\tau \end{array} \right\}$$

The physical components of the stress are the following:

$$\wp_{rz} = \wp_{zr} = \int_0^t \mu(t-t')(\partial r \cdot u)(r,t')dt'$$

$$\wp_{zz} = 2\int_0^t \int_{t'}^t \mu(t-t')(\partial r \cdot u)(r,t')(\partial r \cdot u)(r,\tau)d\tau dt' .$$

The other components are zero. The motion equation gives the following equation for u:

$$\rho(\partial t \cdot u)(r,t) = p(t) + \frac{1}{r}\partial r \cdot [r\int_0^t \mu(t-t')(\partial r \cdot u)(r,t')dt']$$

$$+\text{boundary conditions} \left\{ \begin{array}{c} u(r,t) = 0, \text{with} \quad t \leq 0 \\ (\partial r \cdot u)(0,t) = 0 \\ u(R,t) = 0 \\ p(t) = \left\{ \begin{array}{r} 0, t < 0 \\ P_0 e^{i\omega t}, (P_0 \in \mathbf{R}_+), t > 0 \end{array} \right\} \end{array} \right\}$$

The solution is obtained by means of the Laplace transform:

$$\mathcal{L}[u](r,s) = \frac{P_0}{\rho s(s-i\omega)}[1 - \frac{J_0(\alpha r)}{J_0(\alpha R)}]$$

where $\alpha^2 = \frac{\rho s}{\mathcal{L}[\mu](s)}$, J_0 = Bessel of the first type. The inverse transform is obtained by means of the Heaviside's representation:

$$u(r,t) = \frac{P_0}{i\omega\rho}[1 - \frac{J_0(kr)}{J_0(kR)}]e^{i\omega t} - \frac{2P_0}{\rho R}\sum_{n,m=1}^{\infty} \frac{\mathcal{L}[\mu](s_{nm})J_0(kr)e^{s_{nm}t}}{s_{nm}(s_{nm} - i\omega)J_1(kR)(\frac{3R}{\beta_1} + \frac{\beta_n}{R}\frac{d\mathcal{L}[\mu]}{ds})}$$

where

$$k^2 = -\rho i\omega/\mathcal{L}[\mu](i\omega)$$
$$\beta_n = n\text{-}th \text{ positive zero of } J_0(x)$$
$$s_{nm} = m\text{-}th \text{ zero of } \frac{\mathcal{L}[\mu](s)}{s} + \frac{\rho R^2}{\beta_n{}^2} = 0.$$

Assuming that the sum are distinct and that $R_e(s_{nm}) < 0$, we can write:

$$u(r,t) = \frac{P_0}{i\omega\rho}[1 - \frac{J_0(kr)}{J_0(kR)}]e^{i\omega t}.$$

Therefore, the volume flow rate is given by the following:

$$Q(t) = 2\pi\int_0^R ru(r,t)dr = \frac{\pi P_0 R^2}{i\omega\rho}[1 - \frac{2J_1(kR)}{kRJ_0(kR)}]e^{i\omega t}.$$

REMARK 4.19 - (NON-LINEAR VISCOELASTICITY).
The following rheological equation (KBKZ-rheological equation), (KBKZ ≡ Kaye, Bernstein, Kearsly, Zopas) gives an example of non-linear viscoelasticity.

$$P_{ij}(x,t) = -p(x,t)g_{ij}(x) + \wp_{ij}(x,t)$$
$$\wp_{ij}(x,t) = \int_{-\infty}^{t} [(\partial I_B \cdot U)B_{ij}(t_0) - (\partial I_C \cdot U)C_{ij}(t_0)]dt_0$$

where

$$U = U(I_B, I_C, t - t_0)$$
$$I_C \equiv \mathrm{tr}(C^{ij})$$
$$I_B \equiv \mathrm{tr}(B^{ij})$$
$$B^{ij} = \text{Finger tensor}$$
$$C^{ij} = \text{Cauchy tensor.}$$

THEOREM 4.30 - (BARUS EFFECT: ELASTIC RECOVERY OF POISEUILLE FLOW)[96].
1) *We make the following assumptions:* a) *The flow is isothermic and incompressible;* b) *The ratio $L/d \to \infty$;* c) *The inertial effects can be omitted: $\rho\bar{u}d/\eta \ll 1$, where $\rho =$*

mass density, η = *characteristic viscosity,* $\bar{u}$ = *average velocity;* d) *The gravity, body forces, and superficial tension can be ignored:* $\rho\bar{u}^2\bar{t}/\eta \ll 1$, *where* $\bar{t}$ = *characteristic time of the fluid;* e) $\bar{u}, \bar{t}/d \gg 1$. *The length of the perturbation at the end of the pipe is* $\cong d$, *therefore there is a sudden change from the steady state flow to the state of zero stress and atmospheric pressure;* f) *Make the following hypotheses on the KBKZ-equation:*

$$\left\{\begin{array}{l}(\partial I_C \cdot U) = 0 \\ (\partial I_B \cdot U) \equiv U' \cong U^1(0) + (\partial I_B \cdot U')\Delta I_B.\end{array}\right\} \tag{4.40}$$

*Then outside the pipe, the section of the material increases with respect to pipe one (***Barus effect***). More precisely one has:*

$$\frac{D_e}{d} = \frac{D}{d} + 0.1, \quad \frac{D}{d} = \left[\frac{\int_0^{D/2} (N_1 + G - \frac{\tau^2}{G}) r\,dr}{\int_0^{D/2} G\, r\, dr}\right]^{1/6} = \frac{1}{\lambda} = \frac{r}{r'}, \tag{4.41}$$

where

$$N_1 \equiv P_{11} - P_{22}, \quad \tau \equiv P_{12}$$

$$\textbf{(elastic modulus)} \quad G \equiv \int_0^\infty (\partial I_D \cdot U) ds, s \equiv t - t_0.$$

2) *If*

$$(\partial\xi \cdot G) \ll (\partial\xi \cdot \tau) \quad (\partial\xi \cdot G) \ll (\partial\xi \cdot N_1)$$

it follows:

$$\xi \equiv 2r'/d$$

$$\frac{D}{d} = [2\int_0^1 (1 + \frac{N_1}{\bar{g}} - \frac{\tau^2}{\overline{G}^2})\xi d\xi]^{1/6}$$

$$\overline{G} \equiv 2\int_0^1 G(\xi)\xi d\xi \equiv \textbf{average sectional value of the elastic modulus.}$$

PROOF. The problem can be divided in the following three:
1)**Viscosimetric flow (steady state)**. Assume the following structure for the velocity field in cylindrical coordinates (r, ϕ, z):

$$(v^i) = (0, 0, \dot{\gamma} r).$$

Then, the corresponding flow becomes:

$$\left\{\begin{array}{l} r = a^1 \\ \phi = a^2 \\ z = \dot{\gamma} a^1 (t - t_0) + a^3 \end{array}\right\}$$

We then have the following geometric objects:

$$\text{train-gradient}: \left\{\begin{array}{l} (B^i_j) \equiv (\partial a_j \cdot x^j) = \begin{bmatrix} 1 & 0 & 0 \\ 0 & 1 & 0 \\ \dot{\gamma}(t - t_0) & 0 & 1 \end{bmatrix} \equiv F^{(v)-1} \\ (B^{-1}{}^i_j) \equiv (\partial x_j \cdot a^i) = \begin{bmatrix} 1 & 0 & 0 \\ 0 & 1 & 0 \\ -\dot{\gamma}(t - t_0) & 0 & 1 \end{bmatrix} \equiv F^{(v)} \end{array}\right\}$$

$$\text{Finger tensor}: (B^{ij}) = \begin{bmatrix} 1 & 0 & \dot{\gamma}(t - t_0) \\ 0 & 1 & 0 \\ \dot{\gamma}(t - t_0) & 0 & [1 + \dot{\gamma}^2 (t - t_0)^2] \end{bmatrix}$$

$$\text{Cauchy tensor}: (C^{ij}) = \begin{bmatrix} [\frac{1 - \dot{\gamma}^2}{(t - t_0)^2}] & 0 & -\dot{\gamma}(t - t_0) \\ 0 & 1 & 0 \\ -\dot{\gamma}(t - t_0) & 0 & 1 \end{bmatrix}$$

The stress tensor is the following:

$$P_{ij} = -\rho g_{ij} + \begin{bmatrix} \wp_{11} & 0 & \wp_{13} \\ 0 & \wp_{22} & 0 \\ \wp_{31} & 0 & \wp_{33} \end{bmatrix}$$

where

$$\wp_{11} = \int_0^\infty [(\partial I_B \cdot U) + (\partial I_C \cdot U)(1 - (\dot{\gamma})^2)] ds$$

$$\wp_{22} = \int_0^\infty [(\partial I_B \cdot U) + (\partial I_C \cdot U)] ds$$

$$\wp_{33} = \int_0^\infty [(\partial I_B \cdot U)(1 + \dot{\gamma}^2 s^2) + (\partial I_C \cdot U)] ds$$

$$\wp_{31} = \int_0^\infty [(\partial I_B \cdot U)\dot{\gamma} s + (\partial I_C \cdot U)\dot{\gamma} s] ds \equiv \tau = \eta \dot{\gamma}$$

Therefore, we have:

$$N_1 \equiv \wp_{33} - \wp_{11} = \int_0^\infty [(\partial I_B \cdot U) + (\partial I_C \cdot U)](\dot{\gamma} s)^2 ds$$

$$N_2 \equiv \wp_{11} - \wp_{22} = -\int_0^\infty (\partial I_C \cdot U)(\dot{\gamma} s)^2 ds.$$

Let us assume the following equation

$$(\partial I_C \cdot U) = 0, \tag{4.42}$$

i.e., $|N_2| \ll N_1$.

2) Let us consider that for $t = 0$, there is a sudden change in the system:

$$\left\{ \begin{array}{l} t < 0 : F^i_j = F^{(v)i}{}_\alpha (F_0)^\alpha_j, \\ t > 0 : F^j_j = \delta^j_j \end{array} \right\}$$

$$F^{(v)}{}^i_\alpha = (\partial x_\alpha \cdot a^i) = \begin{bmatrix} 1 & 0 & 0 \\ 0 & 1 & 0 \\ -\dot{\gamma} s & 0 & 1 \end{bmatrix}, \quad s \equiv (t' - t_0), \dot{\gamma} \equiv \frac{du}{dr'}.$$

$$(F_0)^\alpha_j \equiv (\partial x_j \cdot x^\alpha).$$

3) $(t > 0)$: Let us assume $P_{ij} = 0$. From the rheological equation we get:

$$p(F_0)^i_\alpha (F_0)_\beta{}^j g^{\alpha\beta} = \int_0^\infty U'(I_B, s) B^{(v)ij}(s) ds$$

(as $(F_0)^{\alpha i}$ does not depend on the time),

$$I_B \equiv \mathrm{tr}[(F_0{}^{-1})^{i\alpha} (B^{(v)})^{\beta r} (F_0{}^{-1})^{\omega j} g_{B\omega} g_{\alpha\gamma}]$$

$$U' \equiv (\partial I_B \cdot U)$$

$$p \equiv \text{unknown pressure}$$

$$(B^{(v)ij}) = \begin{bmatrix} 1 & 0 & \dot{\gamma} s \\ 0 & 1 & 0 \\ \dot{\gamma} s & 0 & [1 + \dot{\gamma}^2 s^2] \end{bmatrix}.$$

Set

$$\Delta I_B \equiv \mathrm{tr}\, B_{ij} - \mathrm{tr}\, B_{ij}{}^{(v)} = \mathrm{tr}[(F_0{}^{-1})_{i\alpha} B^{(v)}{}_{\beta\gamma} (F_0{}^{-1})_{\omega j} g^{\gamma\omega} g^{\alpha\beta} - (3 + \dot{\gamma}^2 s^2)]$$

Therefore, by using the Taylor representation for U' in terms of I_B, we get:

$$U' = U'(3+\dot{\gamma}^2 s^2, s) + (\partial I_B \cdot U')\Delta I_B + \frac{1}{2}(\partial I_B \partial I_B \cdot U')(\Delta I_B)^2 + \cdots$$

and by considering the first term only, (namely little expansions), we get from the conservation of the volume ($\det j = 0$) and from (4.42):

$$p^3 = G(G^2 + N_1, G - \tau^2), [G(s) \equiv U'(3+\dot{\gamma}^2 s^2, s)].$$

From (4.42) we have:

$$(4.43)\quad (F_0)^{j\alpha}(F_0)^{\beta j} g_{\alpha\beta} = p^{-1}\begin{bmatrix} G & 0 & \tau \\ 0 & G & 0 \\ \tau & 0 & G+N_1 \end{bmatrix}, \quad p = G(1+\frac{N_1}{G} - (\frac{\tau}{G})^2)^{1/3}.$$

If $(F_0)^{i\alpha}$ is considered independent of the coordinate z, we can prove that the stress with axial symmetry (without torsion) is the following:

$$\left\{\begin{array}{l} r' = \lambda r \\ \phi' = \phi \\ z' = \dfrac{z}{\lambda^2} + g(r), \lambda \in \mathbf{R} \end{array}\right\}$$

Hence:

$$(F_0{}^{ij}) = \begin{bmatrix} \lambda & 0 & 0 \\ 0 & 1 & 0 \\ g^1 & 0 & 1/\lambda^2 \end{bmatrix}, (F_0)^{j\alpha}(F_0)^{j\beta} g_{\alpha\beta} = \begin{bmatrix} \lambda^2 & 0 & \lambda g' \\ 0 & 1 & 0 \\ g\lambda & 0 & g'^2 + 1/\lambda^4 \end{bmatrix}.$$

This equation is not compatible with (4.43) as G, N_1, τ are independent. Therefore, the assumption $P_{ij} = 0$, for $t > 0$ is too restrictive. Thus, let us assume that $P_{ij} = 0$, but $P_{zz} \neq 0$ for $t > 0$. Moreover, as we require that the rod should be free, we must have

$$(4.44)\qquad 2\pi \int_0^{D/2} P_{zz} r\, dr = 0.$$

Therefore, we have:

$$\begin{bmatrix} 0 & 0 & 0 \\ 0 & 0 & 0 \\ 0 & 0 & P_{zz} \end{bmatrix} = -p \begin{bmatrix} 1 & 0 & 0 \\ 0 & 1 & 0 \\ 0 & 0 & 1 \end{bmatrix} + \int_0^\infty U'(s)[(F_0{}^{-1})^{i\alpha} B^{(r)\gamma B} (F_0{}^{-1})^{\omega j} g_{\beta\omega} g_{\alpha\gamma}] ds.$$

We can also write:

$$\begin{bmatrix} 0 & 0 & 0 \\ 0 & 0 & 0 \\ 0 & 0 & P_{zz}/\lambda^4 \end{bmatrix} = -p \begin{bmatrix} \lambda^2 & 0 & \lambda g' \\ 0 & 1 & 0 \\ g'\lambda & 0 & g'^2 + \frac{1}{\lambda^4} \end{bmatrix} + \int_0^\infty U'(s) \begin{bmatrix} 1 & 0 & \dot{\gamma}s \\ 0 & 1 & 0 \\ \dot{\gamma}s & 0 & [1+\dot{\gamma}^2 s^2] \end{bmatrix} ds.$$

Hence,

$$p\lambda^2 = G$$

$$P\lambda g' = \tau$$

$$N_1 + G - \frac{P_{zz}}{\lambda^4} = p[\frac{1}{\lambda^4} + g'^2].$$

By means of the elimination of p and g', we get:

$$N_1 + G - \frac{P_{zz}}{\lambda^4} = \frac{G}{\lambda^6} + \frac{\tau^2}{G}.$$

By integration and using (4.44), we get (4.41). □

COROLLARY 4.1 - (BARUS EFFECT FOR MAXWELL-FLUIDS). *Let us assume*

$$U'(s) = \frac{\eta}{\lambda^2} e^{-s/\lambda}.$$

Then we have the following.

1) *The maximum value of D/d is given by the following formula:*

$$D/d \cong [2 \int_0^1 (1 + (\frac{N_1}{2\tau})^2)\xi d\xi]^{1/6}.$$

2) *If $\frac{N_1}{\tau^2} = const.$, we have:*

$$\frac{D}{d} = (1 + f_w^2/2)^{1/6}, \quad f_w \equiv \frac{(N_1)_w}{2\tau_w}, \quad (w \equiv \textit{wall index}).$$

PROOF. In this case we have:

$$N_1 = 2\eta\dot{\gamma}^2\lambda_0, \tau = \eta\dot{\gamma}, G = \frac{\eta}{\lambda_0}$$

hence

$$\frac{N_1}{2\tau} = \dot{\gamma}\lambda_0, \ \frac{\tau}{G} = \dot{\gamma}\lambda_0$$

$$\Downarrow$$

$$\frac{N_1}{2G} = (\frac{\tau}{G})^2 = (\frac{N_1}{2\tau})^2 = 0,$$

$$\Downarrow$$

$$1 + \frac{N_1}{G} - \frac{\tau^2}{G^2} = 1 + 2(\frac{\tau}{G})^2 - (\frac{\tau}{G})^2 = 1 + (\frac{\tau}{G})^2 = 1 + \frac{N_1}{2G}.$$

□

COROLLARY 4.2 - (BARUS EFFECT FOR OLDROYD-MAXWELL FLUIDS).
Let us assume

$$U'(s) = \frac{\eta}{\lambda_0}[(1 - \frac{\mu_0}{\lambda_0})e^{-s/\lambda_0} + \mu_0\delta(s)].$$

We get:

$$D_e/d = D/d + 0.1$$
$$D/d = (1 + \frac{f_w^2}{2})^{1/6}$$
$$f_w = \frac{N_{1w}}{2\tau_w} = \dot{\gamma}\lambda_0.$$

PROOF. In fact, one has

$$N_1 = 2\dot{\gamma}^2\eta(\lambda_0 - \mu_0) = const.$$
$$G = \eta/\lambda_0 = const.$$
$$\tau = \dot{\gamma}\eta\frac{\lambda_0 - \mu_0}{\lambda_0} = const.$$

Note that for classic Oldroyd fluids $\tau = \dot{\gamma}\eta$. But in practice: $\frac{\lambda_0 - \mu_0}{\lambda_0} \cong 1$ with $\lambda_0 \gg \mu_0$. □

REMARK 4.20 - The KBKZ-equation gives for θ, τ and N_1 the same expressions as those coming from rheological equation for linear viscoelasticity. In fact, we get:

$$\wp_{31} = \int_0^\infty \dot{\gamma} s G(s) ds$$

$$\theta = \wp_{33} - \wp_{11}/\dot{\gamma}^2 = \int s^2 G(s) ds$$

$$\Downarrow$$

$$U'(3 + \dot{\gamma}^2 s^2, s) = G(s) \Rightarrow \mu(t) = \int_0^\infty e^{-t/s} G(s) ds\,.$$

Then G is exactly the elastic modulus: $G(s) = \int_0^\infty G(s)ds$.

THEOREM 4.31 - (OLDROYD-FLUID IN CONE-PLATE SYSTEM).
1) *Let us consider spherical coordinates:*

$$(x^1, x^2, x^3) \equiv (r, \theta, \phi).$$

We assume the following structure for the velocity:

$$(v^i) = (0, 0, \Omega(\theta, t)), \quad (g_{ij}) = \begin{bmatrix} 1 & 0 & 0 \\ 0 & r^2 & 0 \\ 0 & 0 & r^2\sin^2\theta \end{bmatrix}.$$

Hence, one has the following flow:

$$\left\{\begin{array}{l} r \equiv x^1 = a^1 \\ \theta \equiv x^2 = a^2 \\ \phi \equiv x^3 = a^3 + \int\limits_{\infty}^{t} \Omega(a^2, t')dt' \end{array}\right\}$$

As a consequence, we get:

$$(\dot{e}^{ij}) = \frac{r^{-2}}{2}(\partial\theta.\Omega)\begin{bmatrix} 0 & 0 & 0 \\ 0 & 0 & -1 \\ 0 & -1 & 0 \end{bmatrix}.$$

The corresponding physical components are the following:

$$\dot{e}^{ij} = (g^{ii}g^{jj})^{-1/2}\dot{e}^{ij} = -\frac{G}{2}\begin{bmatrix} 0 & 0 & 0 \\ 0 & 0 & 1 \\ 0 & 1 & 0 \end{bmatrix}, G \equiv \sin\theta(\partial\theta \cdot \Omega).$$

2) *If we assume that Ω is constant in time, we get the following (shear flow):*

$$\phi = a^3 + \int\limits_{\infty}^{t} \Omega(a^2, t')dt' = a^3 + \Omega(a^2)(t - t_0).$$

The corresponding stress tensor becomes:

$$P^{ij}(r,\theta) = \begin{bmatrix} -p & 0 & 0 \\ 0 & -p & -\eta G \\ 0 & -\eta G & -p + 2\eta G^2(\lambda_0 - \mu_0)/\sin\theta \end{bmatrix}$$

and the normal stress differences are the following:

$$N_1 = N_2 = P_{33} - P_{11} = P_{33} - P_{11} = \frac{2\eta G^2(\lambda_0 - \mu_0)}{\sin\theta}.$$

REMARK 4.21 - On the plate ($\theta = \frac{\pi}{2}$), we have $N_1 = 2\mu G^2(\lambda_0 - \mu_0)$. This gives a way to measure the elastic parameters difference: $\lambda_0 - \mu_0$.

(D1) HYPOELASTIC MATERIALS.

DEFINITION 4.16 - *We call* **hypoelastic materials** *ones such that their stress tensors satisfy the following equation:*

$$\mathcal{L}_{\dot{\phi}}P = \Phi(\dot{e}, P) \tag{4.45}$$

where Φ is any functional operator of $\dot{e}$ and P; $\dot{\phi}$ is the velocity of flow on the Galilean space-time.

REMARK 4.22 - Equation (4.45) can be written in coordinates adapted to an inertial frame, in the following way:

$$\boxed{\mathcal{D}_t P^{ij} \equiv (\partial t + \mathcal{L}_{\dot{\phi}})P^{ij} = \Phi^{ij}(\dot{e}^{ij}, P^{ij})}$$

where $\mathcal{D}_t \equiv \partial t + \mathcal{L}_{\dot{\phi}}$ is called **convective derivative.**

REMARK 4.23 - For such materials the memory is only for states infinitesimally near to the present one.

EXAMPLE 4.9 - (**Hypoelastic solids of degree** 0) These are characterized by the following equation (**Truesdell equation**):

$$\boxed{\mathcal{D}_t P^{ij} = 2\mu\dot{e}^{ij} + \lambda\dot{e}^k_k P^{ij}}$$

where λ and μ are called **Lamé coefficients.**

THEOREM 4.32 - (PARALLELEPIPED OF HYPOELASTIC MATERIAL OF DEGREE 0, UNDER HYDROSTATIC PRESSURE (OR TENSION)). *We have the following formula:*

$$\frac{\rho}{\rho_0} = (\frac{1 + \frac{3}{2}\frac{\lambda}{\mu} - \frac{P_0}{\mu}}{1 + \frac{3}{2}\frac{\lambda}{\mu} - \frac{P}{\mu}})^{3/2}.$$

PROOF. We have:

$$\dot{e}^i_j = \frac{1}{3}\gamma\delta^i_j, \quad \gamma(t) \equiv \frac{3k_0}{1 + k_0 t}, \qquad k_0 \in \mathbf{R}.$$

From the continuity equation we get:

$$\frac{d\rho}{dt} + \gamma\rho = 0.$$

Hence

$$\rho \equiv \frac{\rho_0}{(1 + k_0 t)^3}. \tag{4.46}$$

From the Truesdell equation we get:

$$\mathcal{D}_t P^{ij} = \frac{dP^{ij}}{dt} - \frac{2k_0}{1 + k_0 t}P^{ij}$$

and putting $P^{ij} = pg^{ij}$ one has:

$$-\frac{dp}{dt} + \frac{2k_0}{1+k_0t}p = \frac{2\mu k_0}{1+k_0t} + \frac{3\lambda k_0}{1+k_0t}.$$

By eliminating t with (4.46) we have:

$$\frac{d\log(p)}{dp} = \frac{1}{\frac{2}{3}\mu + \lambda\frac{2}{3}p}.$$

□

THEOREM 4.33 - *Let us consider the simple tension, in the z-direction. Let us assume the following:*

$$\dot{e}^{ij} = k\begin{bmatrix} -a & 0 & 0 \\ 0 & -a & 0 \\ 0 & 0 & 1 \end{bmatrix}$$

and that there is not acceleration. Then we have that the flow is given by the following:

$$\left\{\begin{array}{l} x = x^1 = a^1(1 - a_0k_0t) \\ y = x^2 = a^2(1 - a_0k_0t) \\ z = x^3 = a^3(1 + k_0t) \end{array}\right\}$$

iff the following equations are verified:

$$-\frac{1}{a_0}\frac{x^1 - a^1}{a^1} = -\frac{1}{a_0}\frac{x^2 - a^2}{a^2} = \frac{x^3 - a^3}{a^3} = k_0t$$

$$k(t) = \frac{k_0}{1+k_0t}$$

$$a(t) = a_0\frac{1+k_0t}{1-k_0a_0t}$$

where

$$\begin{aligned} k_0 &= \textit{extension velocity per length} \\ a_0 &= \textit{tansverse contraction velocity per length} \\ &= -\frac{(x^1 - a^1)/a^1}{(x^3 - a^3)/a^3} \\ &= -\frac{(x^2 - a^2)/a^2}{(x^3 - a^3)/a^3} \end{aligned}$$

As

$$\mathcal{D}_tP^{ij} = \begin{bmatrix} \frac{d}{dt}P_{xx} + 2kaP_{xx} & 0 & 0 \\ 0 & \frac{d}{dt}P_{yy} + 2kaP_{yy} & 0 \\ 0 & 0 & \frac{d}{dt}P_{zz} - 2kP_{zz} \end{bmatrix}$$

Truesdell equations become:

$$
\begin{aligned}
&(a): \quad P_{xx} = P_{yy} \\
&(b): \quad \frac{d}{dt}P_{xx} + 2kaP_{xx} = -2ka(\mu+\lambda) + \lambda k \\
&(c): \quad \frac{d}{dt}P_{zz} - 2kP_{zz} = 2k(\mu-\lambda) + \lambda k
\end{aligned}
$$

Set $\epsilon \equiv k_0 t, a_0 \equiv \sigma =$ **Poisson ratio**. *Therefore, equations (b) and (c) become:*

$$
\begin{aligned}
&(\tilde{a}): \quad (1+\epsilon)(1-\sigma\epsilon)\frac{dQ}{d\epsilon} + 2\sigma(1+\epsilon)Q = -\frac{2\sigma}{1-2\sigma}(1+\sigma)\epsilon \\
&(\tilde{b}): \quad (1+\epsilon)(1-\sigma\epsilon)\frac{dI}{d\epsilon} - 2(1-\sigma\epsilon)T = \frac{2}{1-\epsilon\sigma}[(1-\sigma) - \sigma(\epsilon(1+\sigma)+2\sigma)]
\end{aligned}
$$

where

$$
\begin{aligned}
&Q \equiv \frac{P_{xx}}{\mu} = \frac{P_{yy}}{\mu} \\
&T \equiv \frac{P_{zz}}{\mu}
\end{aligned}
$$

(Q and T *are adimensional stresses*)

$$
+ \textit{ initial conditions}(t = t_0) \begin{Bmatrix} T = T_0 \\ Q = Q_0 \end{Bmatrix}
$$

By integrating $(\tilde{a})$ *and* $(\tilde{b})$, *we get:*

$$
\begin{aligned}
Q &= (1-\sigma\epsilon)^2 Q_0 - \frac{2\sigma}{1-2\sigma}[(\frac{1-\sigma\epsilon}{1+\sigma})^2 \log(\frac{1-\sigma\epsilon}{1+\epsilon}) + \frac{\epsilon(2-\sigma\epsilon+\sigma^2\epsilon)}{2(1+\sigma)}] \\
T &= (1+\epsilon)^2 T_0 - \frac{4\sigma^3}{1-2\sigma}[(\frac{1+\epsilon}{1+\sigma})^2 \log(\frac{1+\epsilon}{1-\sigma\epsilon}) + \frac{\epsilon(2+\epsilon-6\sigma^2-5\sigma^2\epsilon)}{(1+\sigma)(1-2\sigma)}].
\end{aligned}
$$

From the equation of the flow, we get:

$$
\frac{V}{V_0} = (1-\sigma\epsilon)^2(1+\epsilon)
$$

where

$V \equiv xyz \equiv$ *volume at the deformation* ϵ

$V_0 \equiv a^1 a^2 a^3 \equiv$ *volume before the deformation.*

(D2) PLASTICS.

DEFINITION 4.17 - *The* **plastics** *are materials that for little values of deformation have behaviour similar to elastic materials, but for higher ones are viscous materials.*

This behaviour is formalized by the following equations:

$$\wp^{ij} = \begin{cases} \dot{e}^{ij} = 0, & \text{if } \frac{1}{2}\wp_i^k \wp_k^i \le P_y^2 \\ \dfrac{2\eta_p \dot{e}^{ij}}{1 - [\frac{P_y}{\sqrt{\frac{1}{2}\wp_i^k \wp_k^i}}]}, & \text{if } \frac{1}{2}\wp_i^k \wp_k^i > P_y^2 \end{cases}$$

$$(\text{or} \quad \wp^{ij} = 2[\eta_p + \frac{P_y}{\sqrt{2\dot{e}_i^k \dot{e}_k^i}}]\dot{e}^{ij})$$

The above equations are also referred to as **von Nises conditions of the yield point**, *and we call* P_Y **yield stress**. *Furthermore, we call* η_p **plastic viscosity** *(or* **mobility**).

REMARK 4.24 - The apparent viscosity is given by the following:

$$\eta = \frac{\eta_p}{1 - [\frac{P_y}{\sqrt{\frac{1}{2}\wp_i^k \wp_k^i}}]}.$$

THEOREM 4.34 - (TRACTION OF SOLID THREAD). 1) *We can characterize the system by means of the following geometric objects:*

$$(v_i)\begin{bmatrix} v_r(r) \\ 0 \\ v_z(z) \end{bmatrix}; \quad (\dot{e}^{ij}) = \begin{bmatrix} (\partial r \cdot v_r) & 0 & 0 \\ 0 & \frac{v_r}{r} & 0 \\ 0 & 0 & (\partial z \cdot v_z) \end{bmatrix}.$$

2) *The system is governed by the following ODE's:*
(Motion equations):

$$\frac{dF}{dz} = 2\pi R(z)P_z; P_z \equiv \text{Traction}$$

(Rheological equation):

$$F(z) = \frac{dv_z}{dz}\{3\eta_p A(z) + \frac{4\sqrt{2}}{5\Delta v}\pi P_y[(\frac{5}{4\pi}A(z) + \frac{1}{l^2})^{1/2} - \frac{1}{l}]\}$$

(Continuity equation):

$$W \equiv \rho v_z(z)A(z)$$

and the following boundary conditions:

$$\left\{ \begin{array}{l} A(0) = A_0 \\ v_z(0) = v_0 \\ F(0) = F_0 \end{array} \right\}.$$

PROOF. From the rheological equation we have:

$$F = \int_0^{R(z)} 2\pi r P_{zz} dr = 2\pi(3\eta_p \int_0^{R(z)} r(\partial z v_z) dr + \frac{P_y}{2} \int_0^{R(z)} r \frac{1}{\sqrt{I}} (\partial z \cdot v_z) dr)$$

with $I \equiv \dot{e}^k_s \dot{e}^s_k$. Therefore, we get:

$$F = 3\eta_p A(z)(\partial z \cdot v_z) + \frac{2\pi}{\sqrt{2}} P_y (\partial z \cdot v_z) \int_0^{R(z)} \frac{r}{\sqrt{I}} dr.$$

From the continuity equation we get

$$I = (\Delta v)^2 (\frac{5}{4} r^2 + \frac{1}{\rho^2}).$$

Hence, by integration we get explicit expression for the rheological equation. □

3) (DIFFERENTIAL EQUATION FOR F). *F must satisfy the following ODE:*

$$(4.47) \qquad \left\{ \begin{array}{l} (F')^3 F = -[\overline{m}(F')^2 + \sqrt{\overline{t}(F')^2 + \overline{h}} - \overline{\omega}] F'' \\ + \text{boundary conditions} F(0) = F_0 \end{array} \right\}$$

where

$$\overline{\omega} \equiv lh\overline{s}$$

$$h \equiv \frac{4\sqrt{2}}{5} \frac{\pi P_Y}{\Delta v}$$

$$\overline{s} \equiv \frac{4\pi W P_z^2}{\rho l}$$

$$\overline{t} \equiv 2h^2 (\frac{5W^2 P_z^2}{\rho^2})$$

$$\overline{h} \equiv 4h^2 s^2$$

$$\overline{m} = 2(3\eta_p \frac{W}{\rho}).$$

PROOF. It directly follows from the equation of motion and rheological equation after eliminating v_z. □

4) *The solution of the above equation can be written as the inverse function $F = F(z)$, of the following:*

$$z = z(F) = \int_{F_0}^{F} \frac{dF}{t(F)}$$

where $t = t(F)$ *is obtained in the following way:*

$$(4.48)\quad \frac{1}{2}(F^2 - F_0^2) = \overline{m}(t_0 - t) - \frac{\overline{\omega}}{2}(\frac{1}{t^2} - \frac{1}{t_2^0}) - \frac{1}{2\sqrt{\bar{t}}}\log(\frac{\bar{h} + 2\bar{t}t^2 - 2t^2}{\bar{h} + l\bar{t}t_0^2 - 2t_0^2\sqrt{\bar{t}(\bar{t} + \bar{h}t_0^{-2})}}).$$

PROOF. Set $t(F(z)) = F'(z) \Rightarrow F''(z) = \frac{dt}{dF}\frac{dF}{dz} = \frac{dt}{dF}t = t't$. Then, equation (4.47) becomes:

$$t^3 F + (\overline{m}t^2 + \sqrt{\bar{t}t^2 + \bar{h}} - \overline{\omega})t't = 0.$$

Hence, a first solution is $t = 0 \Rightarrow \frac{dF}{dz} = 0 \Rightarrow F = const.$ This solution is, now, non-integrable. Thus, we have

$$t^2 F + (\overline{m}t^2 + \sqrt{\bar{t}t^2 + \bar{h}} - \overline{\omega})t' = 0.$$

Set

$$g(t) \equiv \overline{m}t^2 + \sqrt{\bar{t}t^2 + \bar{h}} - \overline{\omega}$$
$$h(t) \equiv t^2.$$

We get

$$\frac{dt}{dF} = -\frac{h(t)}{g(t)}F.$$

Set

$$H(t) \equiv -(\frac{h(t)}{g(t)})^{-1}$$
$$\Downarrow$$
$$H(t)dt = FdF\,.$$

By integration, we have:

$$s(t) \equiv \int_{t_0}^{t} H(t)dt = \frac{1}{2}(F^2 - F_0^2) \equiv G(F).$$

5) *One has the following formulas:*

$$A(z) = \frac{1}{4\pi P_z^2}(\frac{dF}{dz})^2, \quad P_{zz}(z) = \frac{F(z)}{A(z)}.$$

6) *The energy equation becomes:*

$$\rho C_p[(\partial t \cdot \theta) + v_z(\partial z \cdot \theta)] = 2\sqrt{\frac{\pi}{A}}\lambda(\theta_\infty - \theta)\,.$$

THEOREM 4.35 - (PLASTIC COUETTE FLOW). *One has the following geometric objects:*

$$\textbf{velocity field}:\left\{\begin{array}{l} v_r = 0 \\ v_\theta = r\omega(r) \\ v_z = u(r) \end{array}\right\}$$

$$\Downarrow$$

$$(\dot{e}_{ij}) = \begin{bmatrix} 0 & \frac{1}{2}(r\omega' - \omega) & \frac{1}{2}u' \\ \frac{1}{2}(r\omega' - \omega) & 0 & 0 \\ \frac{1}{2}u' & 0 & 0 \end{bmatrix}.$$

Then, the stress tensor has the following structure:

$$(P^{ij}) = \begin{bmatrix} -p & p_{r\theta} & p_{rz} \\ p_{r\theta} & -p & 0 \\ p_{rz} & 0 & -p \end{bmatrix}$$

with

$$p = p(z), \quad p_{r\theta} = P_{\theta r}(r), \quad P_{rz} = P_{zr}(r).$$

The motion equations are the following:

$$\left\{\begin{array}{l} P_{r\theta} = \dfrac{\beta}{\xi^2} \\ P_{rz} = \dfrac{pR}{2}(\dfrac{\xi^2 - \lambda^2}{\xi}) \\ p = \dfrac{dp}{dz} - \rho g \end{array}\right\}$$

with $\beta, \lambda =$ *constant and* $\xi \equiv r/R$. *Then, the plastic flow is realized in the region* $k < \xi < 1, k \equiv R_1/R$, *where it is satisfied the von Nises-condition; that now can be written in the following way:*

$$\xi^6 - [2\lambda^2 + (\frac{2P_Y}{pR})^2]\xi^4 + \lambda^4\xi^2 + (\frac{2\beta}{pR})^2 > 0.$$

4.6 - RHEOPTICS

In this section we will study the interaction continuum medium-electromagnetic field. Deformation induces unisotropy in stressed materials. (For a detailed study of the concept of unisotropy, and its applications to constitutive characterizations of materials see refs.[97,100].)

(A) Maxwell equations in unisotropic materials.

Let us write Maxwell equations in tensor-form:

$$(4.49)\qquad \boxed{\begin{array}{ll}(a) & B^k{}_{/k} = 0 \\ (b) & D^k{}_{/k} = \rho 4\pi \\ (c) & \epsilon^{ijs} E_{j/s} = \frac{1}{c}(\partial t \cdot B^i) \\ (d) & \epsilon^{ijs} H_{j/s} - \frac{1}{c}(\partial t \cdot D^i) = \frac{4\pi}{c} I^i \end{array}}$$

with $\epsilon^{ijs} \equiv$ Ricci tensor and $\rho, c =$ invariant. Then, one has the following continuity equation:

$$(4.50)\qquad (\partial t \cdot \rho) + I^k{}_{/k} = 0$$

and the following equation of balance of electric charge, inside a compact region:

$$(\frac{de}{dt} \equiv \frac{d}{dt}\int_V \rho dv) + (\int_{\partial V} I^k{}_{n_k} d\sigma \equiv \mathcal{I}) = 0.$$

PROOF. In fact from (4.49)(d), as div rot $= 0$, we get (taking div on both sides):

$$I^k{}_{/k} = -\frac{1}{4\pi}(\partial t \cdot D^k)_{/k} = -\frac{1}{4\pi}\partial t \cdot (D^k{}_{/k})$$

and by using (4.49)(b), we get (4.50). The global balance equation is obtained from (4.50), after integration and by using Stokes theorem. □

REMARK 4.25 - 1) To the above equations the following constitutive equations (**material equations**) must be added:

$$(4.51)\qquad \boxed{\begin{array}{lll}(a) & I^k = \sigma^{ki} E_i, & \sigma^{ki} \equiv \textbf{conductivity tensor} \\ (b) & D^k = \epsilon^{ki} E_i, & \epsilon^{ki} \equiv \textbf{dielectric tensor}\,(\epsilon^{ki} = \epsilon^{ik}) \\ (c) & B^k = \mu^{ki} H_i, & \mu^{ki} \equiv \textbf{magnetic permeability tensor} \end{array}}$$

We have the following definitions:

(4.52)

$$(a)\quad w_e \equiv \frac{1}{8\pi} E^k D_k \equiv \textbf{electric energy density}$$

$$(b)\quad w_m \equiv \frac{1}{8\pi} B^k H_k \equiv \textbf{magnetic energy density}$$

$$(c)\quad w \equiv w_e + w_m \equiv \textbf{electromagnetic energy density}$$

$$(d)\quad S_i \equiv \frac{c}{4\pi}\epsilon_{kji} E^j H^i \equiv \textbf{Poynting vector}$$

THEOREM 4.36 - (ENERGY EQUATION FOR ELECTROMAGNETIC FIELD).

$$\frac{dw}{dt} = -S^j{}_{/j}$$

DEFINITION 4.18 - (PRINCIPAL DIELECTRIC AXES). *These are identified by means of the eigenvectors of the tensor* ϵ_{ij}*, at any point of the material. Hence, they are determined by the following equations:*

$$(\epsilon_{ij} - \epsilon g_{ij})L^j = 0$$

PROPOSITION 4.5 - *With respect to the orthonormal basis, identified with the principal dielectric axes, we can write:*

$$\begin{aligned}(\epsilon_{ij}) &= \begin{bmatrix} \epsilon_1 & 0 & 0 \\ 0 & \epsilon_2 & 0 \\ 0 & 0 & \epsilon_3 \end{bmatrix} \\ (D_i) &= (\epsilon_1 E^1, \epsilon_2 E^2, \epsilon_3 E^3) \\ w_e &= \frac{1}{8\pi}[\epsilon^1(E^1)^2 + \epsilon^2(E^2)^2 + \epsilon^3(E^3)^2].\end{aligned}$$

REMARK 4.26 - (AXIAL DISPERSION). As ϵ_{ij} is a function of the frequency, we have that the principal axes change with the frequency. Of course, if the electromagnetic wave is monocromatic there is not axial dispersion.

THEOREM 4.37 - (STRUCTURE OF THE MONOCROMATIC PLANE WAVE IN A NON-MAGNETIC ($B^i = H^i$) UNISOTROPIC MATERIAL).

1) *If a monocromatic plane wave propagates in a non-magnetic unisotropic material in the direction* **k**, *the vectors* (D, H, S) *are orthogonal among themselves. The same happens for the vectors* (E, H, S). *Furthermore, they are rotated one with respect to the other by an angle* α. *Therefore, the energy flux* S *is not along* **k**, *(as, instead happens in isotropic materials), but describes with* **k** *an angle* α.

PROOF. Let us consider the following expressions for the electromagnetic fields:

$$\begin{aligned}E^i &= E^i_0 e^{i(k^i x_i - \omega t)} \\ H^i &= H^i_0 e^{i(k^i x_i - \omega t)} \\ D^i &= D^i_0 e^{i(k^i x_i - \omega t)} \\ B^i &= B^i_0 e^{i(k^i x_i - \omega t)}.\end{aligned}$$

Then, from Maxwell equations, we get:
(4.53)

$$(a) \quad \frac{c}{\omega}\epsilon^{ijk}k_jH_k = -D^i$$
$$(b) \quad \frac{c}{\omega}\epsilon^{ijk}k_jE_k = B^k = H^k.$$

□

2) *We define* **refraction index vector***:*

$$\mathbf{n} = \frac{c}{\omega}\mathbf{k}.$$

Furthermore, define **radius vector s** *such that*

$$\mathbf{s} \parallel S \quad , \quad <\mathbf{n}, \mathbf{s}> = 1.$$

Then, the expressions of the electromagnetic fields in terms of **n** *and* **s** *are the following:*

$$\left\{\begin{array}{ll} (a) & H^i = \epsilon^{ijk}n_jE_k \\ (b) & D^i = -\epsilon^{ijk}n_jH_k \\ (c) & S^i = \frac{c}{4\pi}[E^2n^i - (E^kn_k)E^i] \\ (d) & s^iH_i = s^iE_i = 0 \\ (e) & H^i = \epsilon^{ijk}s_jD_k \\ (f) & E^i = -\epsilon^{ijk}s_jH_k \end{array}\right\}$$

5) *A wave that propagates in unisotropic material has, in general, an* **elliptic polarization***, i.e., the vectors D and E oscillate in a plane $\perp\mathbf{k}$ and describe an ellipse.*
PROOF. Let us consider some lemmas.
LEMMA 4.3 - *The components (n_i) of* **n** *must satisfy the following equation:*

(4.54)(**Fresnel equation**)

$$|\epsilon_{jk} - n^2g_{kj} + n_kn_j| = 0$$

For n^2 the above equation represents a quadric, called **surface of indexes** *(or* **surface of wave vector***).*
PROOF OF LEMMA 4.3 - Eliminate H^i in equations (4.53)(a) and (4.53)(b). We have:

(4.55)
$$D^i = n^2E^j - (n_kE^k)n^i.$$

Taking into account the following relation:

$$D^i = \epsilon^{it} E_t$$

we get

$$\epsilon^{it} E_t - n^2 E^i + (n_k E^k) n^i = 0$$

with $E_t = g_{ts} E^s$. Hence,

$$\epsilon^{it} g_{ts} E^s - n^2 E^i + n_s E^s n^i = 0$$

and by contraction with g_{ij}, we have:

$$(\epsilon_{js} - n^2 g_{js} + n_s n_j) E^s = 0.$$

The condition of compatibility gives:

$$|\epsilon_{js} - n^2 g_{js} + n_s n_j| = 0.$$

□

2 Let us consider the following **rule of change**: An equation that holds for $(E, \mathbf{n}, \epsilon_{ijk})$ can also be written for $(B, \mathbf{s}, \epsilon^{-1}_{ijk})$ by using the correspondence:

$$\left\{ \begin{array}{c} E \Leftrightarrow D \\ \mathbf{n} \Leftrightarrow \mathbf{s} \\ \epsilon_{ijk} \Leftrightarrow \epsilon^{-1}_{ijk} \end{array} \right\}$$

LEMMA 4.4 - *The components s_i of $\mathbf{s}$ must satisfy the following equations:*

(4.56)(**Fresnel equation**)

$$\boxed{|\mathbf{s}^2 g_{ik} - s_i s_j - \epsilon^{-1}_{ik}| = 0}$$

Therefore, we can say that the surface vector identifies, in a palne $\perp\mathbf{n}$, an ellipse having the axes that represent the refraction indexes

$$a = \sqrt{\epsilon_1} = n_1, \quad b = \sqrt{\epsilon_2} = n_2$$

in the directions of these axes. Thus these directions represent the directions of D^i. The conditions are similar for the radial surface vector. In fact, from (4.55) we get, taking a frame along the axis $\mathbf{n}$, that the transverse components $D_\alpha, \alpha = 1, 2$, have the following expressions:

$$D_\alpha = n^2 E_\alpha, \quad D_\alpha = \epsilon_{\alpha\beta} E^\beta$$

hence

$$\epsilon_{\alpha\beta}E^\beta - n^2 E_\alpha = 0$$
$$\Downarrow$$
$$\epsilon_{\alpha\beta}E^\beta - n^2 g_{\alpha\beta}E^\beta = 0$$
$$\Downarrow$$
$$(\epsilon_{\alpha\beta} - n^2 g_{\alpha\beta})E^\beta = 0$$

Therefore, in the plane $\sigma \perp \mathbf{n}$ the tensor $\epsilon_{\alpha\beta}$ $(\alpha\beta = 1,2)$ has eigenvectors E^β and eigevalues n^2. These eigenvectors are $\perp$ each other and identify the principal axes of the tensor $\epsilon_{\alpha\beta}$. Thus, with respect to these axes $(\epsilon_{\alpha\beta})$ can be written as follows:

$$(\epsilon_{\alpha\beta}) = \begin{bmatrix} {n_1}^2 & 0 \\ 0 & {n_2}^2 \end{bmatrix}.$$

So, the ellipse of equation

$$\epsilon_{\alpha\beta}x^\alpha x^\beta = 1 \Leftrightarrow \epsilon_1 {x_1}^2 + \epsilon_2 {x_2}^2 = 1$$

has the axes

$$a = \sqrt{\epsilon_1} = n_1, \quad b = \sqrt{\epsilon_2} = n_2.$$

□

DEFINITION 4.19 - 1) *A material is called (optically)* **uniaxial** *if*

$$(\epsilon_{ik}) = \begin{bmatrix} \epsilon_\perp & 0 & 0 \\ 0 & \epsilon_\perp & 0 \\ 0 & 0 & \epsilon_\parallel \end{bmatrix}$$

with $\epsilon_\perp \neq \epsilon_\parallel$. *The direction of* $\epsilon_\parallel$ *is called* **optic axis**. *The other two directions are arbitrary. In particular, we say* **negative uniaxial** *if* $\epsilon_\perp > \epsilon_\parallel$ *and* **positive uniaxial** *if* $\epsilon_\perp < \epsilon_\parallel$.[2]
2) *A material is called* **biaxial** *if:*

$$(\epsilon_{ik}) = \begin{bmatrix} \epsilon_1 & 0 & 0 \\ 0 & \epsilon_2 & 0 \\ 0 & 0 & \epsilon_3 \end{bmatrix}$$

with $\epsilon_1 < \epsilon_2 < \epsilon_3$.
THEOREM 4.38 - *In uniaxial materials we get the effect of the double refraction.*
PROOF. In fact, the tangential component $n_\parallel$ has a different value from $n_\perp$. □

[2] If $\epsilon_\perp = \epsilon_\parallel$ the material is isotropic.

DEFINITION 4.20 - 1) *We define* **optic-mechanical effects** *the unisotropy induced by means of stress tensor.*

2) *We define* **rheoptic equations** *as those which give a relation between ϵ_{ij} and $\dot{e}_{ij}$.*

REMARK 4.27 - In general, we can write: $\epsilon_{ij} = \epsilon g_{ij} + \overline{\epsilon}_{ij}$. In the following table we resume some important examples of rheoptic equations.

TAB.4.13 - **Examples of rheoptic equations**

Name	ϵ_{ij}
Isotropic	$\epsilon_{ij} = \epsilon g_{ij}$
Elastic	$\epsilon_{ij} = \epsilon g_{ij} + a_1 e_{ij} + a_2 g_{ij}$
	$a_1(\omega), a_2(\omega) =$ elasto-optic consts.
Newtonian	$\epsilon_{ij} = \epsilon g_{ij} + 2\lambda_1 e_{ij} + 2\lambda_2 \omega_{ij}$
Linear viscoelastic	$\overline{\epsilon}^{ij}(x,t) = 2\int_{[-\infty,t]} m(t-t')(\partial a_\beta . x^i)(\partial a_\alpha . x^j)$
	$\dot{e}^{\alpha\beta}(a,t')dt'$
	$m(t) \equiv$ rheoptic relaxation function
	PARTICULAR CASES
Newton:	$m(t) = \overline{\eta}\delta(t)$
Maxwell:	$m(t) = \frac{\overline{\eta}}{\overline{\lambda}_0} e^{-t/\overline{\lambda}_0}$
Oldroyd:	$m(t) = \frac{\overline{\eta}}{\overline{\lambda}_0}[(1 - \frac{\overline{\mu}_0}{\overline{\lambda}_0})e^{-t/\overline{\lambda}_0} + \overline{\mu}_0 \delta(t)]$

TAB.4.14 - **Examples of rheoptic PDEs**

Name	ϵ_{ij}
Rheoptic PDEs	$\mathcal{P}\overline{\epsilon}^{ij}(x,t) = 2\mathcal{Q}\dot{e}^{ij}(x,t)$
	$\mathcal{P} \equiv 1 + \overline{\lambda}_0 \frac{\delta}{\delta t} + \overline{\lambda}_1 (\frac{\delta}{\delta t})^2 + \cdots + \overline{\lambda}_N (\frac{\delta}{\delta t})^{N+1}$
	$\mathcal{Q} \equiv \overline{\mu}(1 + \overline{\mu}_0 \frac{\delta}{\delta t} + \overline{\mu}_1 (\frac{\delta}{\delta t})^2 + \cdots + \overline{\mu}_N (\frac{\delta}{\delta t})^{N+1})$
	PARTICULAR CASES
Maxwell:	$(1 + \overline{\lambda}_0 \frac{\delta}{\delta t})\overline{\epsilon}^{ij}(x,t) = 2\overline{\eta}\dot{e}^{ij}(x,t)$
Oldroyd:	$(1 + \overline{\lambda}_0 \frac{\delta}{\delta t})\overline{\epsilon}^{ij}(x,t) = 2\overline{\eta}(1 + \overline{\mu}_0 \frac{\delta}{\delta t})\dot{e}^{ij}(x,t)$
Newton:	$\overline{\epsilon}^{ij}(x,t) = 2\overline{\eta}\dot{e}^{ij}(x,t)$

THEOREM 4.39 - (RHEOPTIC IN POISEUILLE FLOW FOR OLDROYD FLUID).

1) *The dielectric tensor is the following:*

$$(\epsilon^{ij}) = \begin{bmatrix} \epsilon_0 & 0 & 0 \\ 0 & \epsilon_0 & \overline{\eta}\dot{\gamma} \\ 0 & \overline{\eta}\dot{\gamma} & a \end{bmatrix}$$

with $a \equiv \epsilon_0 + 2\overline{\eta}\dot{\gamma}^2(\overline{\lambda}_0 - \overline{\mu}_0)$, *where* ϵ_0 *is a dielectric constant for isotropic material. Note that in the Newtonian case,* $\overline{\lambda}_0 - \overline{\mu}_0 = 0$, *one has*

$$(\epsilon^{ij}) = \begin{bmatrix} \epsilon_0 & 0 & 0 \\ 0 & \epsilon_0 & \overline{\eta}\dot{\gamma} \\ 0 & \overline{\eta}\dot{\gamma} & \epsilon_0 \end{bmatrix}$$

2) *The eigenvectors and the eigenvalues of* (ϵ^{ij}) *are the following:*
(Eigenvalues):

$$\left\{ \epsilon_i = \epsilon_0 + \overline{\epsilon}_i, (i = 1, 2, 3), \left\{ \begin{array}{c} \overline{\epsilon}_1 = 0 \\ \overline{\epsilon}_{2,3} = \overline{\eta}\dot{\gamma}^2(\overline{\lambda}_0 - \overline{\mu}_0) \mp \overline{\eta}\dot{\gamma}[1 + \dot{\gamma}^2(\overline{\lambda}_0 - \overline{\mu}_0)^2]^{1/2} \end{array} \right\} \right\}$$

(Eigenvectors):

$$\left\{ \begin{array}{l} (L_i^{(1)}) = (1, 0, 0) \\ (L_i^{(2,3)}) = \left(0, \overline{\eta}\dot{\gamma}/[\overline{\epsilon}_{2,3}{}^2 + (\overline{\eta}\dot{\gamma})^2]^{1/2}, \overline{\epsilon}_{2,3}/[\overline{\epsilon}_{2,3}{}^2 + (\overline{\eta}\dot{\gamma})^2]^{1/2}\right) \\ L_1^{(2,3)} = 0 \\ L_2^{(2,3)} = 1/\{1 + \{\dot{\gamma}(\overline{\lambda}_0 - \overline{\mu}_0) \mp [1 + \dot{\gamma}^2(\overline{\lambda}_0 - \overline{\mu}_0)^2]^{1/2}\}^2\}^{1/2} \\ L_3^{(2,3)} = \dfrac{\dot{\gamma}(\overline{\lambda}_0 - \overline{\mu}_0) \mp [1 + \dot{\gamma}^2(\overline{\lambda}_0 - \overline{\mu}_0)^2]^{1/2}}{\{1 + \{\dot{\gamma}(\overline{\lambda}_0 - \overline{\mu}_0) \mp [1 + \dot{\gamma}^2(\overline{\lambda}_0 - \overline{\mu}_0)^2]^{1/2}\}^2\}^{1/2}} \end{array} \right\}$$

3) *The direction of maximum unisotropy is* $L^{(3)}$. *This is characterized by the angle* θ:

$$\cos\theta = L_3^{(3)} = \frac{\dot{\gamma}(\overline{\lambda}_0 - \overline{\mu}_0) + [1 + \dot{\gamma}^2(\overline{\lambda}_0 - \overline{\mu}_0)^2]^{1/2}}{\{1 + \{\dot{\gamma}(\overline{\lambda}_0 - \overline{\mu}_0) + [1 + \dot{\gamma}^2(\overline{\lambda}_0 - \overline{\mu}_0)^2]^{1/2}\}^2\}^{1/2}}.$$

4) *One has the following limit values:*

$$\begin{array}{ll} (a): & \lim_{\dot{\gamma}\to\infty} \cos\theta = 1 \Rightarrow \theta = 0; \\ (b): & \lim_{\dot{\gamma}\to 0} \cos\theta = 1/\sqrt{2} \Rightarrow \theta = 45; \\ (c): & \lim_{(\overline{\lambda}_0 - \overline{\mu}_0)\to 0} \cos\theta = 1/\sqrt{2} \Rightarrow \theta = 45; \\ (d): & \lim_{(\overline{\lambda}_0 - \overline{\mu}_0)\to\infty} \cos\theta = 1 \Rightarrow \theta = 0. \end{array}$$

5) *The birefringence is calculated by means of the variation of refraction indexes:*

$$\Delta n_\alpha \equiv \frac{1}{2n_\alpha} \triangle \epsilon_\alpha \simeq \frac{1}{2n_0} \triangle \epsilon_\alpha, \Delta\epsilon_\alpha \equiv \epsilon_\alpha - \epsilon_0, \alpha = 1, 2, 3.$$

6) *The birefringence of light that propagates in direction* x^1 *is the following:*

$$\Delta n_e \equiv n_{\|} - n_{\perp} = \Delta n_3 - \Delta n_2 = \frac{1}{2n_0}[\bar{\epsilon}_3 - \bar{\epsilon}_2] = \frac{1}{n_0}\bar{\eta}\dot{\gamma}[1 + \dot{\gamma}^2(\bar{\lambda}_0 - \bar{\mu}_0)^2]^{1/2}$$

where $n_{\|}$ *and* $n_{\perp}$ *are the refraction indexes in the direction* $\|$ *and* $\perp$ *the maximum birefringence respectively.*

7) *The birefringence in direction* x^2 *is*

$$\Delta n_3 - \Delta n_1 = n_{\|} - n_{\perp}.$$

4.7 - MULTICOMPONENT CONTINUUM SYSTEMS

As mass transfer and chemical reactions are also of interest, we would like to discuss multicomponent continuum systems here. The complete description of transport phenomena (mass, momentum and energy) in multicomponent systems is obtained by means of PDEs that can be resumed in the following:

$$\left\{\begin{array}{ll} (a) & \text{Continuity equation} \\ (b) & \text{Motion equation} \\ (c) & \text{Energy equation} \\ (d) & \text{Thermal state equation} : p = p(\rho, \theta, x_{A_i}) \\ (e) & \text{Caloric state equation} : e = e(\rho, \theta, x_{A_i}) \end{array}\right\}$$

In the following we shall consider in some details such PDEs.

DEFINITION 4.21 - *One has the following types of diffusions:*

(a) **ordinary diffusion**: *it is generated by a variation of concentration of one component* A *in* B;

(b) **pressure-diffusion**: *it is generated by a gradient of pressure.*

(c) **thermal diffusion**: *it is generated by a gradient of temperature.*

(d) **forced diffusion**: *it is generated by external forces that have different actions on each component.*

In the following table we report some useful definitions of concentration.

TAB.4.15 - **Definitions of concentrations**

Name	Definition
mass	ρ_{A_i}
mole	$c_{A_i} \equiv \rho_{A_i}/M_{A_i}$
mass-fraction	$\omega_{A_i} \equiv \rho_{A_i}/\rho$
mole-fraction	$\chi_{A_i} \equiv c_{A_i}/c$
mass-average velocity	$v \equiv (\sum_{1\leq j\leq n} \rho A_j v_{A_j})/\sum_{1\leq j\leq n} \rho A_j$
mole-average velocity	$\overset{*}{v} \equiv (\sum_{1\leq j\leq n} cA_j v_{A_j})/\sum_{1\leq j\leq n} cA_j$

(c= total mole density)

REMARK 4.28 - ρv (resp. cv) represents the local velocity to which the mass (resp. mole) cross the unit surface $\perp v$.

THEOREM 4.40 - 1) *The velocity of diffusion of i-th component with respect to v is:* $v_{A_j} - v$.

2) *The velocity of diffusion of i-th component with respect to $\overset{*}{v}$ is:* $v_{A_j} - \overset{*}{v}$.

Recall that the flux with respect to a frame is: $N = J + \widetilde{J}$, where J = mole-flux, $\widetilde{J}$ = convective flux. Then, in the following table we report some useful classifications of fluxes.

TAB.4.16 - **Classification of fluxes**

mass-flux	$nA_j \equiv \rho A_j v_{A_j}$
mass-flux relative to v	$jA_j \equiv \rho A_j (v_{A_j} - v)$
mass-flux relative to $\overset{*}{v}$	$\overset{*}{j}A_j \equiv \rho A_j (v_{A_j} - \overset{*}{v})$
mole-flux	$JA_j \equiv cA_j (v_{A_j} - v)$
mole-flux relative to v	$JA_j \equiv cA_j (v_{A_j} - v)$
mole-flux relative to $\overset{*}{v}$	$\overset{*}{j}A_j \equiv c_{A_j} (v_{A_j} - \overset{*}{v})$

THEOREM 4.41 - *For binary systems (A, B) one has the following relations:*

$$\rho = \rho_A + \rho_B, \quad c = c_A + c_B, \quad 1 = \chi_A + \chi_B, \quad 1 = \omega_A + \omega_B,$$

$$\chi_A M_A + \chi_B M_B = M, \quad \chi_A = \frac{(\omega_A/M_A)}{\frac{\omega_A}{M_A} + \frac{\omega_B}{M_B}}, \quad \omega_A = \frac{\chi_A M_A}{\chi_A M_A + \chi_B M_B},$$

$$\frac{\omega_A}{M_A} + \frac{\omega_B}{M_B} = 1/M, \quad d\chi_A = \frac{d\omega_A}{M_A M_B (\frac{\omega_A}{M_A} + \frac{\omega_B}{M_B})^2}, \quad d\omega_A = \frac{M_A M_B d\chi_A}{(\chi_A M_A + \chi_B M_B)^2}.$$

In the following table we report some useful relations for binary systems.

TAB.4.17 - **Fundamental relations for binary system**

Entity	absolute	relative to v and $\overset{*}{v}$
mass-flux	$n_A+n_B=\rho v$	$j_A+j_B=0$ $\overset{*}{j}_A+\overset{*}{j}_B=\rho(v-\overset{*}{v})$
mole-flux	$N_A+N_B=c\overset{*}{v}$	$J_A+J_B=c(\overset{*}{v}-v)$ $\overset{*}{j}_A+\overset{*}{j}_B=0$
flux(N_A)	$n_A=N_A M_A$	$J_A=N_A-\omega(N_A+\frac{M_B}{M_A}N_B)$ $\overset{*}{j}_A=N_A-\chi_A(N_A+N_B)$
flux(n_A)	$N_A=n_A/M_A$	$j_A=n_A-\omega_A(n_A+n_B)$ $\overset{*}{j}_A=n_A-\chi_A(n_A+n_B\frac{M_A}{M_B})$
flux((v,j_A))	$n_A=j_A+\rho_A v$	$J_A=j_A/M_A$ $\overset{*}{j}_A=j_A\frac{M}{M_B}$
flux(($\overset{*}{v}$))	$\overset{*}{j}_A$	$N_A=\overset{*}{j}_A+c_A\overset{*}{v}$ $J_A=\overset{*}{j}_A\frac{M_B}{M}\overset{*}{j}_A=J_A M_A$

(flux(-) $\equiv$ flux by means of (-))

In the following table we resume the first Fick diffusion law for binary systems.

TAB.4.18 - **First Fick law for binary systems**

Flux	Gradient	Diffusion flux (or molecular flux)
n_A	$\omega_{A/j}$	$n^j_A-\omega_A(n^j_A+n^j_B)=-\rho\mathcal{D}_{AB}\omega^{/j}_A$ (+)
N_A	$\chi_{A/j}$	$N^j_A-\chi_A(N^j_A+N^j_B)=-c\mathcal{D}_{AB}\chi^{/j}_A$ (+)
j_A	$\omega_{A/j}$	$j^j_A=-\rho\mathcal{D}_{AB}\omega^{/j}_A$
$\overset{*}{j}_A$	$\chi_{A/j}$	$\overset{*j}{j}_A=-c\mathcal{D}_{AB}\chi^{/j}_A$
J_A	$\chi_{A/j}$	$J^j_A=-(\frac{c^2}{\rho})M_A M_B\mathcal{D}_{AB}\chi^{/j}_A$
$\overset{*}{J}_A$	$\omega_{A/j}$	$\overset{*j}{j}_A=-(\frac{\rho^2}{cM_A M_B})\mathcal{D}_{AB}\omega^{/j}_A$
$c(v_A-v_B)$	$\chi_{A/j}$	$c(v^j_A-v^j_B)=-c(\frac{1}{\chi_A\chi_B})\mathcal{D}_{AB}\chi^{/j}_A$

(+)(In the first term there is the convective flux
and in the second diffusion one)

DEFINITION 4.22 - *We say that a system is with* **reaction** *if in the continuum system there is a chemical reaction. Furthermore, systems with reaction are called*

homogeneous systems *if the reaction is in the body of the system. In this case the production is usually denoted by the formula:* $r_A = \kappa''{}_n c_A^n$, *(n = order of reaction). Moreover,* **heterogeneous systems** *are those with reaction, where the reaction is limited to a region (e.g., on the boundary). In these cases one has the boundary conditions:* $|N_A|_{\partial V} = \kappa''{}_n c_A^n|_{\partial V}$.[3]

THEOREM 4.42 - (CONTINUITY EQUATION). *For the component* A_i*:*

$$(\partial t.\rho_{A_i}) + n^j_{A_i/j} = r_{A_i}, (i = 1, 2, \cdots, n)$$

For all the system:

$$(\partial t\rho) + (\rho v^j)_{/j} = 0$$

Other forms of such equations are the following:

$$(a_1): \quad (\partial t.c_{A_i}) + N^j_{A_i/j} = R_{A_i}, (i = 1, 2, \cdots, n)$$

$$(b_1): \quad (\partial t\rho c) + (c\overset{*}{v}{}^j)_{/j} = \sum_i R_{A_i}$$

where $n_{A_i} = \rho\omega_{A_i}v + j_{A_i}$, with $j_{A_i} = j^{(x)}_{A_i} + j^{(p)}_{A_i} + j^{(g)}_{A_i} + j^{(i)}_{A_i}$, where

(concentration flux) :

$$j^{(x)j}_{A_i} = \frac{1}{\rho R\theta} \sum_{1 \leq A_j \leq n} M_{A_i} M_{A_j} D_{A_i A_j} [\sum_{1 \leq A_k \leq n, A_k \neq A_j} (\partial x_{A_k}.\overline{G}_{A_j})_{\theta,p,\chi_s} \chi^{/j}_{A_k}]$$

(pressure flux)(Dufours effect) :

$$j^{(p)j}_{A_i} = \frac{1}{\rho R\theta} \sum_{1 \leq A_j \leq n} M_{A_i} M_{A_j} D_{A_i A_j} [\chi_{A_j} M_{A_j} (\frac{\overline{V}_{A_j}}{M_{A_j}} - \frac{1}{\rho}) p^{/j}]$$

(forced flux) :

$$j^{(g)j}_{A_i} = -\frac{c^2}{\rho R\theta} \sum_{1 \leq A_j \leq n} M_{A_i} M_{A_j} D_{A_i A_j} [\chi_{A_j} M_{A_j} (B^j_{A_j} - \sum_{1 \leq A_k \leq n} \frac{\rho_{A_k}}{\rho} B^j_{A_k})]$$

(thermal flux)(Sorel effect) :

$$j^{(\theta)j}_{A_i} = -D^\theta_{A_i} (\log\theta)^{/j}$$

[3] A good reference for problems of mass transfer and chemical reactions is the book by G.Astarita [4].

where $B^j \equiv$ *external forces;* $\overline{G}_{A_j} \equiv$ *Gibbs free energy* $\equiv$ *mole partial free enthalpy;* $\overline{V}A_j \equiv$ *mole partial volume;* $D_{A_iA_j} \equiv$ *diffusion coefficient. These have the following properties:*

$$\left\{ \begin{array}{l} D_{A_iA_i} = 0 \\ \sum\limits_{1 \leq A_j \leq n} [M_{A_i}M_{A_n}D_{A_iA_n} - M_{A_i}M_{A_j}D_{A_iA_n}] = 0 \\ \text{for} \quad n > 2 D_{A_iA_j} \neq D_{A_jA_i} \end{array} \right\}$$

THEOREM 4.43 - (CONTINUITY EQUATIONS IN TERMS OF D_{AB} FOR BINARY SYSTEMS)

1)

$$\left\{ \begin{array}{l} \textbf{(Mass continuity equation)} : (\partial t.\rho_A) + (\rho_A v^j)_{/j} = (\rho D_{AB}\omega^{/j})_{/j} + r_A \\ \textbf{(Mole continuity equation)} : (\partial t.c_A) + (c_A \overset{*}{v}{}^j)_{/j} = (\rho D_{AB}\chi^{/j})_{/j} + R_A \end{array} \right\}$$

2) (CONTINUITY EQUATIONS IN PARTICULAR SYSTEMS).

(a) *Liquid solutions with constant temperature and pressure, and low concentration:* $(\rho, D_{AB} =$ *const.).*

$$(\partial t.c_A) + (v^i c_{A/j} g_i^j) = D_{AB}(\triangle_2 c_A) + R_A$$

(b) *Gas at low density with constant temperature and pressure:* $(c, D_{AB} =$ *const.):*

$$(\partial t.c_A) + (\overset{*}{v}{}^i c_{A/j} g_i^j) = D_{AB}(\triangle_2 c_A) + R_A - \frac{c_A}{c}(R_A + R_B)$$

(c) **Second Fick diffusion law** *(solids, liquids or gas with steady molecular antidiffusion):*

$$(\partial t.c_A) = D_{AB}(\triangle_2 c_A).$$

THEOREM 4.44 - (GENERAL EXPRESSION FOR THE MASS FLUX IN BINARY SYSTEMS).

1)

$$\begin{aligned} j_A^j &= -j_B^j \\ &= -(\frac{c^2}{\rho R\theta})M_A^2 M_B D_{AB}\chi_A[(\partial x_A.\frac{\overline{G}_A}{M_A})_{\theta,p}\chi_A^{/j} + \frac{\rho_B}{\rho}(B_A^j - B_B^j) + (\frac{\overline{V}_A}{M_A} - \frac{1}{\rho})p^{/j}] \\ &+ D_A^\theta(\log\theta)^{/j}. \end{aligned}$$

This expression can be written in an alternative form by using the relation

$$(d\overline{G}_A)_{\theta,p} = R\theta d(\log q_A)$$

and by defining thermal diffusion ratio:

$$\kappa_\theta = \frac{\rho}{c^2 M_A M_B}(\frac{D_A^\theta}{D_{AB}})$$

One has:

$$j_A^j = -j_B^j = -(\frac{c^2}{\rho R\theta})M_A M_B D_{AB}[(\partial \log x_A . \log a_A)_{\theta,p}\chi_A^{/j} - \frac{M_A \rho_B \chi_A}{\rho R\theta}(B_A^j - B_B^j)$$
$$+ \frac{M_A \chi_A}{R\theta}(\frac{\overline{V}_A}{M_A} - \frac{1}{\rho})p^{/j} + \kappa_\theta(\log\theta)^{/j}]$$

with $a_A \equiv$ **concentration activity**.

2) *In the case of ordinary diffusion the above formula becomes:*

$$j_A^j = -j_B^j = -(\frac{c}{\rho})^2 M_A M_B D_{AB}(\partial \log x_A . a_A)_{\theta,p}\chi_A^{/j}.$$

Then, by comparison this formula with that in table 4.16 we conclude that D_{AB} *and* $\mathcal{D}_{AB}$ *are the same coefficient only for ideal solutions, i.e., when the activity is proportional to the mole fraction. [In general, one uses* $\mathcal{D}_{AB}$ *as this does not require knowledge on the concentration data.]*

THEOREM 4.45 - (ORDINARY DIFFUSION IN MULTICOMPONENT CASES AT LOW DENSITY).

$$j_{A_i}^j = (\frac{c}{\rho})^2 \sum_{1\le A_j \le n} M_{A_i} M_{A_j} D_{A_i A_j} \chi_{A_j}^{/j}.$$

As $D_{A_i A_j}$ *are functions of the concentration, the above formula is not workable. However, we can prove the following equation (***Sefan-Maxwell equation***):*

$$\chi_{A_i}^{/j} = \sum_{1\le A_j \le n} \frac{c_{A_i} c A_j}{c^2 \mathcal{D}_{A_i A_j}}(v^j A_j - v_{A_i}^j) = \sum_{1\le A_j \le n} \frac{1}{c\mathcal{D}_{A_i A_j}}(\chi_{A_i} N_{A_j}^j - \chi_{A_j} N_{A_i}^j).$$

There the coefficients $\mathcal{D}_{A_i A_j}$ *can be assumed independent of the concentration.*

DEFINITION 4.23 - (EFFECTIVE BINARY DIFFUSION FOR A_i IN SOLUTION).

$$N_{A_i}^j = -c\overline{\mathcal{D}}_{A_i A_m} \chi_{A_i}^{/j} + \chi_{A_i} \sum_{1\le A_j \le n} N_{A_j}^j. \tag{4.57}$$

THEOREM 4.46 - (RELATION BETWEEN $\overline{\mathcal{D}}_{A_i A_m}$ AND $\mathcal{D}_{A_i A_j}$).

$$\boxed{\frac{1}{c\overline{\mathcal{D}}_{A_i A_m}} = \frac{\sum_{1\le A_j \le n} \frac{1}{c\mathcal{D}_{A_i A_j}}(\chi_{A_j} N_{A_i}^j - \chi_{A_i} N_{A_j}^j)}{N_{A_i}^j - \chi_{A_i} \sum_{1\le A_j \le n} N_{A_j}^j}.}$$

PROOF. This formula can be obtained by using the expression of $\chi_{A_i}^{/j}$ in Stefan-Maxwell equation one obtained for $N_{A_i}^j$ from (4.57). □

In the following table we resume some expressions of the mass diffusivity by means of pressure and temperature.

TAB.4.19 - **Mass diffusivity in binary systems**

State	$\mathcal{D}_{AB}$
Low pressure gas	$p\mathcal{D}_{AB}/(p_{c_A}p_{c_B})^{1/3}(\theta_{c_A}\theta_{c_B})^{5/12}(\frac{1}{M_B}-\frac{1}{M_A})^{1/2}=a(\theta/\sqrt{\theta_{c_A}\theta_{c_B}})^b$ a, b =const. (non-polar gas: $a = 2.745 \times 10^{-4}, b = 1.823$) (non-polar gas+$H_2O$: $a = 3.640 \times 10^{-4}, b = 2.334$)
Liquids	hydrodynamic equation $\mathcal{D}_{AB} = \kappa\theta\frac{u_A}{F_A}$ spherical particles:κ = gas-constant, $\frac{u_A}{F_A}$ = mobility $F_A = 6\pi\eta_B u_A R_A(\frac{2\eta_B+\beta_{AB}R_A}{3\eta_B+\beta_{AB}R_A})$ η_B = solvent viscosity R_A = particle-radius β_{AB} = draging friction coefficient Particular cases ($\beta_{AB} = \infty$):$\frac{\mathcal{D}_{AB}\eta_B}{\kappa\theta} = \frac{1}{6\pi R_A}$ (Stokes-Einstein equation) ($\beta_{AB} = 0$):$\frac{\mathcal{D}_{AB}\eta_B}{\kappa\theta} = \frac{1}{4\pi R_A}$ (Self-diffusion):$\frac{\mathcal{D}_{AA}\eta_A}{\kappa\theta} = \frac{1}{2\pi}(\frac{\widetilde{N}}{\widetilde{V}_A})^{1/3}$

2 THEOREM 4.47 - (MOTION EQUATION FOR MULTICOMPONENT SYSTEMS).

$$\frac{\delta}{\delta t}(\rho v^i) = -\rho v^i v^k{}_{/k} + P^{ik}{}_{/k} + \sum_{1\leq A_j\leq n} \rho_{A_j} B^i A_j$$

An alternative form is the following:

$$\rho\frac{\delta v^i}{\delta t} = P^{ik}{}_{/k} + \sum_{1\leq A_j\leq n} \rho_{A_j} B^i A_j$$

If we consider the total momentum flux

$$\phi^{ij} = \rho v^i v^j - P^{ij}$$

we can write the motion equation also in the following alternative form:

$$\frac{\delta}{\delta t}(\rho v^i) = -\phi^{ij}{}_{/j} + \sum_{1 \leq A_j \leq n} \rho_{A_j} B^i_{A_j}$$

PROOF. It follows directly for the general equations for continuum systems. The last expression for the motion equation is obtained by taking into account that the total flux N is the sum of the diffusional and convective ones. □

2 THEOREM 4.48 - (ENERGY EQUATION FOR MULTICOMPONENT SYSTEMS).

$$\frac{\delta}{\delta t}\rho(e + \frac{1}{2}v^2) = -\rho(e + \frac{1}{2}v^2)v^k{}_{/k} - q^k{}_{/k} + (P^{kj}v_j)_{/k} + \sum_{1 \leq A_j \leq n} n^j_{A_i} B_{A_{ij}}$$

An alternative form is the following:

$$\rho\frac{\delta}{\delta t}(e + \frac{1}{2}v^2) = -q^k{}_{/k} + (P^{kj}v_j)_{/k} + \sum_{1 \leq A_j \leq n} \rho_{A_j} B^k_{A_j} v_{kA_j}$$

THEOREM 4.49 - (TOTAL FLUX OF ENERGY).

$$e^i = \rho(e + \frac{1}{2}v^2)v^i + q^k - P^{ij}v_j$$

PROOF. In fact, the convective flux is $\widetilde{J} = \rho(e + \frac{1}{2}v^2)$ and the diffusion flux is $J = q - P\rfloor v$. □

THEOREM 4.50 - (EXPRESSION FOR q).

$$q = q_{(c)} + q_{(d)} + q_{(x)} + q_{(v)}$$

where

(a) $q^i_{(c)} = -\lambda g^{ik}\theta_{/k}$

(*Fourier law* $\equiv$ *thermal conduction flux*)

(b) $q_{(d)} = \sum_{1\leq A_i \leq n} \frac{\overline{H}_{A_i}}{M_{A_i}} j_{A_i}$

$\equiv$ *interdiffusion* ($\overline{H}_{A_i}$ = *mole partial enthalpy of* A_i)

(c) $q^i_{(x)} \equiv$ (*Dufour flux*)

(*thermal diffusion: very complicated, but not too important*)

(d) $q^i_{(r)} \equiv$ *radiant flux*

When the following terms

$$\rho\frac{1}{2}v^2, \quad P^{kj}v_j, q^i_{(x)}, \quad q^i_{(r)}$$

are neglected, the energy flux is simply:

$$e^k = -\lambda g^{ik}\theta_{/k} + \sum_{1\leq A_i \leq n} \overline{H}_{A_i} N^k_{A_i}$$

THEOREM 4.51 - (ALTERNATIVE FORM FOR ENERGY EQUATION). *By using the above expression for the energy flux, we can write energy equation in the following alternative form:*

$$\frac{\delta}{\delta t}\rho(e + \frac{1}{2}v^2) = -e^k{}_{/k} + \sum_{1\leq A_j \leq n} n^k_{A_j} B_{kA_j}$$

Furthermore, in terms of C_p *the energy equation becomes:*

$$\begin{aligned}\rho\widehat{C}_p\frac{\delta\theta}{\delta t} &= -q^k{}_{/k} - v_{j/k}\mathcal{P}^{jk} \\ &+ \sum_{1\leq A_j \leq n} j^k_{A_j} B_{kA_j} \\ &+ (\partial \log\theta . \log\widehat{V})_{p,\chi_{A_i}}\frac{\delta p}{\delta t} + \sum_{1\leq A_j\leq n} \overline{H}_{A_j}[J^k_{A_j/k} - RA_j]\end{aligned}$$

where $\widehat{C}_p \equiv$ *specific heat at constant pressure for each unit mass,* $\widehat{V} \equiv$ *specific volume.*
DEFINITION 4.24 -
(INTERPHASE TRANSPORT FOR MULTICOMPONENT SYSTEMS).
These problems can be of two types:
(a) *mass transfer across the boundary* ∂V *of two phases;*
(b) *mass transfer across porous boundary* ∂V.
In the following table we report some important entities associated to problems specified in the above definition.

TAB.4.20 - **Transport interphase characterization**

Definition
boundary mole flux $\overset{*}{j}_A\lvert_{\partial V} = N_A\lvert_{\partial V} - \chi_{A_0}(N_A\lvert_{\partial V} + N_B\lvert_{\partial V}) = \overline{\kappa}_x \triangle \chi_A$ $\overset{*}{j}_B\lvert_{\partial V} = N_B\lvert_{\partial V} - \chi_{B_0}(N_A\lvert_{\partial V} + N_B\lvert_{\partial V}) = \overline{\kappa}_x \triangle \chi_A$ $\overset{*}{j}_A\lvert_{\partial V} = -\overset{*}{j}_B\lvert_{\partial V} = \overline{\kappa}_x \triangle \chi_A$ $\overline{\kappa}_x \equiv$ binary coeff.mass transfer in one phase $\triangle\chi_A \equiv$ difference mole fraction boundary- internal points
binary coeff. mass transfer in one phase at low velocity mass transfer $\kappa_x = \lim_{N_A\lvert_{\partial V} \to 0, N_B\lvert_{\partial V} \to 0} \frac{N_A\lvert_{\partial V} - \chi_{A_0}(N_A\lvert_{\partial V} + N_B\lvert_{\partial V})}{\triangle\chi_A}$
mole velocity of addition of A (or B) on the global surface ∂V $\mathcal{W}_A^{(m)} \equiv -\int_{\partial V} N_A^j d\sigma_j, \mathcal{W}_B^{(m)} \equiv -\int_{\partial V} N_B^j d\sigma_j$

THEOREM 4.52 - *One has the following results.*
1)

$$\mathcal{W}_A^{(m)} - \chi_{A_0}(\mathcal{W}_A^{(m)} + \mathcal{W}_B^{(m)}) = \kappa_x A \triangle \chi_A.$$

2)(RELATION BETWEEN (h, λ) AND $(\kappa_x, \mathcal{D}_{AB})$ IN BINARY SOLUTION FLOWING IN PIPE OF LENGTH L WITH POROUS BOUNDARY SURROUNDED BY A BINARY SOLUTION).
(4.58)

$$\boxed{\begin{array}{ll} (a) & h = \dfrac{1}{\pi DL(\theta_0 - \theta_1)} \displaystyle\int_{[0,L]} \int_{[0,2\pi]} (\lambda(\partial r.\theta)|_{r=R}) R d\phi dz \\ (b) & \kappa_x = \dfrac{1}{\pi DL(\chi_{A_0} - \chi_{A_1})} \displaystyle\int_{[0,L]} \int_{[0,2\pi]} (c\mathcal{D}_{AB}(\partial r.\chi_A)|_{r=R}) R d\phi dz \end{array}}$$

with $\chi_{A_0} \equiv$ *mole fraction at the boundary,* $\chi_{A_1} \equiv$ *mole fraction inside the pipe.*

PROOF. For the global heat we have:

$$Q = -\int_V q_j d\sigma^j = \int_V \lambda(\partial x_j.\theta) d\sigma^j.$$

On the other hand $h = \frac{Q}{A\triangle\theta}$, hence we have (4.58)($a$). Let us, now, consider the mole flux:

$$\begin{aligned}\mathcal{W}_A^{(m)} - \chi_{A_0}(\mathcal{W}_A^{(m)} + \mathcal{W}_B^{(m)}) &= -\int_{\partial V} [N_A^j - \chi_{A_0}(N_A^j + N_B^j)] d\sigma_j \\ &= \int_{\partial V} c\mathcal{D}_{AB} \chi_A^{/j} d\sigma_j\,.\end{aligned}$$

On the other hand

$$\kappa_x = \frac{\mathcal{W}_A^{(m)} - \chi_{A_0}(\mathcal{W}_A^{(m)} + \mathcal{W}_B^{(m)})}{A\,\triangle\,\chi_A}.$$

Therefore, we get (4.58)(b). □

TAB.4.21 - **Analogy between heat and mass transfer**

Name	θ	χ_A
diffusion	$\alpha = \frac{\lambda}{\rho}\widehat{C}_p$	$\mathcal{D}_{AB}$
diffusion	$\beta = -\frac{1}{\rho}(\partial\theta.\rho)_{p,\chi_A}$	$\zeta = -\frac{1}{\rho}(\partial\chi_A.\rho)_{p,\theta}$
flux	$q_{(c)}$	$\overset{*}{j}_A = N_A - \chi_{A_0}(N_A + N_B)$
transfer v.	Q	$\mathcal{W}_A^{(m)} - \chi_{A_0}(\mathcal{W}_A^{(m)} + \mathcal{W}_B^{(m)})$
transfer coeff.	$h = \frac{Q}{A\triangle\theta}$	$\kappa_x = \frac{\mathcal{W}_A^{(m)} - \chi_{A_0}(\mathcal{W}_A^{(m)} + \mathcal{W}_B^{(m)})}{A\triangle\chi_A}$
Equal adimensional numbers	$R_e = \frac{Dv\rho}{\eta} = \frac{Dv}{\nu}$ $Fr = \frac{v^2}{gD}$ L/D	R_e Fr L/D
Different adimensional numbers	$N_u = \frac{hD}{\lambda}$ $P_r = \frac{\widehat{C}_p\eta}{\lambda} = \frac{\nu}{\alpha}$ $G_r = \frac{D^3\rho^3 g\beta\triangle\theta}{\eta^2}$ $S_t = \frac{N_u}{R_e P_r} = \frac{h}{\rho\widehat{C}_p}$	$N_u = \frac{\kappa_x D}{c\mathcal{D}_{AB}}$ $S_c = \frac{\eta}{\rho\mathcal{D}_{AB}} = \frac{\nu}{\mathcal{D}_{AB}}$ $G_r = \frac{D^3\rho^3 g\zeta\triangle\chi_A}{\eta^2}$ $S_t = \frac{N_u}{R_e P_r} = \frac{\overline{\kappa}_x}{cv}$
special comb. adim. numbers.	$P_{e'} = R_e P_r = \frac{Dv\rho\widehat{C}_p}{\lambda}$ $j_H = N_u R_e^{-1} P_r^{-1/3}$ $= \frac{h}{\rho\widehat{C}_p}(\frac{\widehat{C}_p\eta}{\lambda})^{2/3}$	$P_{e'} = R_e P_r = \frac{Dv}{\mathcal{D}_{AB}}$ $j_D = N_u R_e^{-1} S_c^{-1/3}$ $= \frac{\overline{\kappa}_x}{cv}(\frac{\eta}{\rho\mathcal{D}_{AB}})^{2/3}$

REMARK 4.29 -
(ANALOGY BETWEEN HEAT AND MASS TRANSFER)
In the above table we have given some analogies between heat and mass transfer (at low velocity). In fact, we can translate any correlation of mass transfer in similar relations for mass transfer, simply by changing the adimensional groups reported in that table.

THEOREM 4.53 - (VELOCITY OF MOMENTUM, HEAT AND MASS TRNSFER AT THE BOUNDARY).

1) *Force of the fluid at the boundary in the direction of the principal flux:*

$$F_z \equiv -\overline{f}A\frac{1}{2}\rho v^2|_{\partial V} + (\mathcal{W}_A^{(m)}M_A + \mathcal{W}_B^{(m)}M_B)v_z|_{\partial V}$$

2) *Velocity of thermal conduction at the boundary:*

$$Q \equiv \overline{h}A\triangle\theta + (\mathcal{W}_A^{(m)}\overline{H}_A|_{\partial V} + \mathcal{W}_B^{(m)}\overline{H}_B|_{\partial V}) + (\mathcal{W}_A^{(m)}M_A + \mathcal{W}_B^{(m)}M_B)(\widehat{\Phi}_{\partial V} + \frac{v^2|_{\partial V}}{2}).$$

3) *Diffusion velocity of component A at the boundary:*

$$\mathcal{W}_A^{(m)} = \overline{\kappa}_x A\triangle\chi_A + (\mathcal{W}_A^{(m)} + \mathcal{W}_B^{(m)})\chi_{A_0}$$

where $\overline{f}, \overline{h}$ *and* $\overline{\kappa}_x$ *are* **transfer coefficients** *that, in general, depend on the transfer velocity.* $\widehat{\Phi}_{\partial V} \equiv$ **potential energy on the boundary**, $v|_{\partial V} \equiv$ **average flux mass velocity**.

THEOREM 4.54 - (MACROSCOPIC DYNAMIC EQUATIONS FOR MULTICOMPONENT SYSTEMS). *In the following we shall assume that the boundary is stationary.*

1) (GLOBAL CONTINUITY EQUATIONS).

$$(mass): \left\{ \begin{array}{l} (\partial t.M_{A_i}) = -\triangle W_{A_i} + W_{A_i}^{(m)} + \widehat{r}_{A_i} \\ (\partial t.M) = -\triangle W + W^{(m)} \end{array} \right\}$$

$$(mole): \left\{ \begin{array}{l} (\partial t.\mathcal{M}_{A_i}) = -\triangle \mathcal{W}_{A_i} + \mathcal{W}_{A_i}^{(m)} + R_{A_i} \\ (\partial t.\mathcal{M}) = -\triangle \mathcal{W} + \mathcal{W}^{(m)} + \sum_{1\le A_i\le n} R_{A_i} \end{array} \right\}$$

where

$$M_{A_i} \equiv \int_V \rho_{A_i} dv \equiv \text{ mass of component } \quad A_i$$

$$\triangle W_{A_i} \equiv W^2_{A_i} - W^1_{A_i} \equiv \rho^2_{A_i} < v_2 > S_2 - \rho^1_{A_i} < v_1 > S_1$$

$\equiv$ *difference of velocity mass flux at the surfaces* $1-2$

$$W^{(m)}_{A_i} \equiv \int_{S(m)} n^j \rho_{A_i} v A_j d\sigma$$

$\equiv$ *velocity mass flux across the surfaces at the boundary*

(it is positive if it is added, negative otherwise)

$$\widehat{r}_{A_i} \equiv \int_V r_i dv \equiv \text{velocity of production of } A_i\text{-th component in the reaction}$$

2) (GLOBAL MOTION EQUATIONS).

$$(\partial t.P^j) = \int_{\partial V} P^{jk} n_k d\sigma - \int_{\partial V} \rho v^j v^k n_k d\sigma + F^j_e - F^j_t$$

where

$$P^j \equiv \int_V \rho v^j dv \equiv \text{total momentum}$$

$$F^j_e \equiv \text{total external force}$$

$$F^j_t \equiv \text{mass transfer forces}$$

3) (GLOBAL ENERGY EQUATIONS).

$$\partial t.(\int_V \rho[e + \frac{1}{2}v^2]dv) = \int_{\partial V} P^{jk} v_j n_k d\sigma - \int_{\partial V} q^k n_k d\sigma - \int_{\partial V} \rho[e + \frac{1}{2}v^2]v^k d\sigma + \int_V B^k_{A_i} n_{A_i k}$$

PROOF. 1) Let us consider the macroscopic equation of change with

$$C \equiv \rho_{A_i}, \quad N \equiv \rho_{A_i} v^k_{A_i}, \quad \mathcal{R} \equiv v_{A_i}.$$

We have:

$$\begin{aligned}\partial t.(\int_V \rho_{A_i} dv) &= -\int_{\partial V} \rho_{A_i} v^k_{A_i} n_k d\sigma + \int_V r_{A_i} dv \\ &= -\int_{S(m)} \rho_{A_i} v^k_{A_i} n_k d\sigma + \int_{S_1} \rho_{A_i} v^k_{A_i} n_k d\sigma - \int_{S_2} \rho_{A_i} v^k_{A_i} n_k d\sigma \\ &\quad + \int_V r_{A_i} dv\,.\end{aligned}$$

Furthermore, in order to obtain the total continuity equation it is necessary to sum these terms and take into account that $\sum_{1\leq A_i\leq n}\widehat{r}_{A_i}=0$.
2) Similarly we can prove this equation by taking

$$\begin{aligned} C &\equiv \rho \\ N &\equiv -P+\rho v\otimes v \\ \mathcal{R} &\equiv \sum_{1\leq A_i\leq n} B_{A_i}\rho_{A_i}-F_t. \end{aligned}$$

3) Finally, here we take

$$\begin{aligned} C &\equiv \rho[e+\frac{1}{2}v^2] \\ N &\equiv q-P\rfloor v+\rho[e+\frac{1}{2}v^2]v \\ \mathcal{R} &\equiv \sum_{1\leq A_i\leq n} B^k_{A_i} n_{A_i k}. \end{aligned}$$

□

EXAMPLE 4.10 - (DRY-SPINNING). Let us assume: (i) The system has axial symmetry and the section is circular; (ii) The concentration of the polymeric component and solvent one are represented by the mass fractions ω_p and ω_s respectively: $\omega_p+\omega_s=1$; (iii) ρ and C_p are constant.
LOCAL EQUATIONS:
A) CONTINUITY EQUATION

$$\begin{aligned} &(a1):(\text{solution}) \quad \frac{1}{r}\partial r.(rv_r)+(\partial z.v_z)=0 \\ &(a2):(\text{solvent}) \quad (\partial t.\omega_s)+v_z(\partial z.\omega_s)=\frac{1}{r}\partial r.(r\mathcal{D}_{ps}(\partial r.\omega_s)). \end{aligned}$$

PROOF. Take into account the continuity equation

$$(\partial t.\rho)+(\rho v^i)_{/i}=0$$

and the continuity equation for the solvent

$$(\partial t.\rho_s)+{n_s}^i{}_{/i}=0.$$

As $\rho_s=\omega_s\rho$ and $n_s=\rho\omega_s v+j_s$, if we assume that in the continuum system the unique contribution to j_s comes from the diffusion for concentration, we get

$$j^j_s=-j^j_p=-\rho\mathcal{D}_{ps}\omega_s^{/j}.$$

□

B) MOTION EQUATION

$$\left\{\begin{aligned} &\rho[(\partial t.v_z)+v_r(\partial r.v_z)+v_z(\partial z.v_z)]=(\partial z.P_{zz})+\frac{1}{r}(r\partial r.P_{zz})+\rho g\\ &\rho[(\partial t.v_r)+v_r(\partial r.v_r)+v_z(\partial z.v_r)]=(\partial r.P_{rr})+(\partial z.P_{rz})+\frac{P_{rr}-P_{\phi\phi}}{r}\end{aligned}\right\}$$

C) ENERGY EQUATION

$$\rho C_p[(\partial t.\theta)+v_z(\partial z.\theta)]=\frac{\lambda}{r}\partial r.(r(\partial r.\theta))-(\frac{\overline{H}_s}{M_s}-\frac{\overline{H}_p}{M_p})\frac{1}{r}\partial r.(r\mathcal{D}_{ps}(\partial r.\omega_s)).$$

The term $(\frac{\overline{H}_s}{M_s}-\frac{\overline{H}_p}{M_p})\frac{1}{r}\partial r.(r\mathcal{D}_{ps}(\partial r.\omega_s))$ is caused by the interdiffusion polymer-solvent and can often be ignored.

RHEOLOGICAL EQUATIONS IN NEWTONIAN CASE

$$(P_{ij})=\begin{bmatrix} -p+2\eta(\partial r.v_r) & 0 & \eta[(\partial r v_z)+(\partial z v_r)]\\ 0 & -p+2\eta\frac{v_r}{r} & 0\\ \eta[(\partial r v_z)+(\partial z v_r)] & 0 & -p+2\eta(\partial z.v_z)\end{bmatrix}$$

GLOBAL EQUATIONS. Let us assume flat profile for the average mass velocity:

$$v_{s,z}=v_{p,z}=v_z=v_z(z,t).$$

A) CONTINUITY EQUATIONS

$$\left\{\begin{aligned} &(\text{polymer}): && \partial t.(\frac{W_p}{v_z})+(\partial z.W_p)=0\\ &(\text{solvent}): && \partial t.(\frac{W_s}{v_z})+(\partial z.W_s)+2\pi RM_sN_{s,0}=0\\ &(\text{solution}): && \partial t.(\frac{W}{v_z})+(\partial z.W)+2\pi RM_sN_{s,0}=0\end{aligned}\right\}$$

with

$$W_p\equiv A\rho v_z<\omega_p>$$
$$<\omega_p>\equiv\frac{1}{A}\int_{[0,R]}2\pi r\omega_p dr$$
$$W_s\equiv A\rho_s v_z<\omega_s>$$
$$<\omega_s>\equiv\frac{1}{A}\int_{[0,R]}2\pi r\omega_s dr$$
$$W\equiv A\rho v_z, A\equiv 2\pi Rz.$$

B) MOTION EQUATION

$$(\partial z.F)=(\partial t.W)+\partial z.(v_zW)-A\rho g+2\pi RP_f$$

with $F \equiv \int_{[0,R]} 2\pi r P_{zz} dr = A < P_{zz} > \equiv$ **spinning tension**.
c) ENERGY EQUATION

$$\rho C_p[(\partial t. < \theta >) + v_z(\partial z. < \theta >)] = 2\sqrt{\frac{\pi}{A}}[h(\theta_\infty - \theta_0) - L_s N_{s,0}]$$

with

$$\begin{aligned} < \theta > &\equiv \frac{1}{A}\int_{[0,R]} 2\pi r\theta dr \equiv \textbf{average temperature} \\ \infty &\equiv \textbf{air index} \\ 0 &\equiv \textbf{boundary index} \\ h &\equiv \textbf{heat transfer coefficient} \\ L_s &\equiv \textbf{vaporization heat of solvent.} \end{aligned}$$

4.8 - VARIATIONAL FIELD THEORY

The fundamental fields of physics, i.e., electromagnetic, gravitational and nuclear fields, all obey the dynamic equations that are of variational type. The main purpose of this section is to give a geometric formulation of global variational calculus in the same spirit of all the book, i.e., emphasizing the role of the dynamic PDE. We will give some examples that have great physical interest.

REMARK 4.30 - (THE CONCEPT OF COVARIANCE). The concept of **covariance** is fundamental for a theory of physical field. A field is considered a section s of fiber bundle $\pi : W \to M$. To say that s is covariant it means that for any local diffeomorphism ϕ of the base M we can calculate the pull-back $\phi^* s$ of s by means of ϕ:

$$\phi^* s \equiv \mathbb{B}(\phi) \circ s \circ \phi^{-1}$$

where $\mathbb{B}(\phi)$ is a local application on W, canonically associated to ϕ on M, such that the following diagram is commutative:

$$\begin{array}{ccccc} & W|_U & \xrightarrow{\mathbf{B}(\phi)} & W|_{\overline{U}} & \\ \pi & \downarrow & & \downarrow & \pi \\ & U & \xrightarrow[\phi]{} & \overline{U} & \end{array}$$

If this circumstance is verified, we say that $\pi : W \to M$ is a **natural fiber bundle** (or **fiber bundle of geometric objects**). We write: $(\pi : W \to M, \mathbb{B})$. In the previous sections we have just meet examples of such a structure. In the following table we resume some important examples.

TAB.4.22 - **Examples of fiber bundles of geometric objects**

Name	Definition
Tangent bundle	$(\pi : TM \to M, \mathbb{B} \equiv T)$
Cotangent bundle	$(\pi : T^*M \to M, \mathbb{B} \equiv T^*)$
Bundle of tensors	$(\pi : T^r_s M \to M, \mathbb{B} \equiv T^r_s)$
Bundle of derivative	$(\pi_1 : J\mathcal{D}(W) \to M)$
of sections	$\mathbb{B}^{(1)} \equiv J\mathcal{D}(-) \equiv T^*(-) \otimes T(\mathbb{B}(-))$
of natural bundle	$\pi_1 \circ Ds = id_M$
$(\pi : W \to M, \mathbb{B})$	$\phi^* Ds = (T^*(\phi) \otimes T\mathbb{B}(\phi^{-1})) \circ Ds \circ \phi$

A generalization of this concept is that of **super-fiber bundle of geometric objects** introduced in refs.[101,102], and considered also in section 2.7. A physical field is fully covariant iff it is a **superfield**, that is a section of such a structure. Recall that "superbundles" and "superfields" in the category of supermanifolds are particular examples of these structures [112,113,114].

DEFINITION 4.25 - *A* **structure of superbundle of geometric objects** *is given by two fiber bundles* $W \overset{\pi_W}{\to} M \overset{\pi_B}{\leftarrow} B$ *over the same base* M *and a covariant functor* $\mathbb{B} : \mathcal{C}(B) \to \mathcal{C}(W)$, *where* $\mathcal{C}(B)$, *(resp.* $\mathcal{C}(W)$*) is the category whose objects are open subbundles of* B *(respectively, open subbundles of* W*) and whose morphisms are the local fiber bundle authomorphisms between those objects such that:*
i) *if* $B|_U \in Ob(\mathcal{C}(B)) \Rightarrow \mathbb{B}(B|_U) = \pi^{-1}{}_W(U) \in Ob(\mathcal{C}(W))$;
ii) *if* $f \in Hom(\mathcal{C}(B))$ *with* $f \equiv (f_B, f_M) : B|U \to B|U'$, *then* $\mathbb{B}(f) \in Hom(\mathcal{C}(W))$ *and satisfies:*
(a) $\pi_W \circ \mathbb{B}(f) = f_M \circ \pi_W$;
(b) *if* $B|U \in Ob(\mathcal{C}(B)), \overline{U} \subset U \Rightarrow \mathbb{B}(f)|^{-1}_{\pi_W}(\overline{U}) = \mathbb{B}(f|_{B|_{\overline{U}}})$.
W *is called the* **total bundle** *and* B *the* **base bundle**. *A section of* π_W *is called a* **superfield of geometric objects**.

NOTE. G-**structures** (that are reductions of principal bundles of k-referments, $0 \leq k \leq \infty$, on a manifold [104,110]), can be considered as particular cases of superbundles of geometric objects. Kaluza-Klein theories are related to those structures.

In the physical applications it is important to consider the following.

THEOREM 4.55 - *Let* $(P, M, \pi; G)$ *be a principal fiber bundle with structure group* G. *Let* $W \equiv P \times F/G$ *be a fiber bundle associated to* P *with fibre* F. *Then, there exists a canonical covariant functor* $\mathbb{B}$ *such that* $(P, W; \mathbb{B})$ *is a superbundle of geometric objects if we restrict the category* $\mathcal{C}(P)$, *where* $\mathbb{B}$ *is defined, to the subcategory* $\mathcal{C}(P)$, *where* $Hom(P|U, P|U')$ *is the set of fibered diffeomorphisms* $P|U \to P|U'$, *such that the isomorphism on the structure group is the identity.*

PROOF. See ref.[102]. □

Here we consider the problem of existence of fully covariant classic and quantum Yang-Mills gauge fields and spinor fields on gravitational instantons. (For Yang-Mills equation see below.)

COROLLARY 4.3 - *A gauge theory is fully covariant iff it can be placed in a structure of superbundle of geometric objects.*

Recall that instantons are non-singular classical solutions of the Euclidean pure Yang-Mills equations with non-trivial topology. Similarly, gravitational instantons can be defined as four-dimensional Einstein-manifolds. If these solutions are closed (i.e., compact and without boundary), they are called compact instantons. Now, the problem that we shall consider is the following: When can we globally define a fully covariant spinor field on a gravitational instanton? As it is well known (see e.g. refs.[8,101,102]), there are some topological obstructions in order to globally define a spin-structure on a manifold M. This can be shortly resumed by requiring that the second Stiefel-Whitney class $w_2 \in H^2(M; \mathbf{Z}_2)$ should be zero. Recall that w_2 is given by $w_2 = \mu(\beta)$, where μ is the canonical connecting homomorphism in the Serre short exact sequence [141]:

$$(4.59) \qquad 0 \to H^1(M; \mathbf{Z}_2) \to H^1(SO(M); \mathbf{Z}_2) \xrightarrow{i^*} H^1(SO(M); \mathbf{Z}_2) \xrightarrow{\mu} H^2(M; \mathbf{Z}_2)$$

where: (a) $SO(M)$ is the total space of the principal bundle of oriented orthonormal frames on M; (b) For any topological space X, $H^1(X; \mathbf{Z}_2) \cong Hom(\pi_1(X); \mathbf{Z}_2)$, where $\pi_1(X)$ is the fundamental homotopy group of X; (c) $H^2(M; \mathbf{Z}_2)$ is the second cohomology group of M with coefficients in $\mathbf{Z}_2$; (d) β is the generator of $H^1(SO(M); \mathbf{Z}_2) \cong \mathbf{Z}_2$. We say that M is a spin manifold if the fiber bundle $SO(M)$ admits a non-trivial double covering $\pi : \widetilde{SO(M)} \to SO(M)$ over M. On the other hand the double covering of $SO(n)$ are classified by $H^1(SO(n); \mathbf{Z}_2) \cong \mathbf{Z}_2$, so other than the trivial covering $SO(n) \times \mathbf{Z}_2$ there is only one non-trivial double covering of $SO(n)$, usually denoted by $Spin(n)$. Furthermore, the double covering of $SO(M)$ are classified by $H^1(SO(M); \mathbf{Z}_2)$. So, M is a spin-manifold iff $\pi : SO(M) \to M$ is represented by a class $\theta \in H^1(SO(M); \mathbf{Z}_2)$ such that $i_*(\theta) = \beta$. But, for the exactness of (4.59) this condition is satisfied iff $\mu(\beta) = w_2$.

In the following table we report some examples of gravitational instantons. (References for these can be found in refs.[8,9,10,19,26,36,37,47,48,68,89,122,164].) In some cases it is also denoted if they are spin manifolds. In order to circumvent the topological obstruction ($w_2 \neq 0$), a good way is to consider generalized spin structures (see e.g.[101,102]). In fact, a less restrictive requirement is obtained by introducing the generalized spin structures, with structure group $Spin^G(4) \equiv Spin(4) \times_{\mathbf{Z}_2} G \equiv Spin(4) \times G/\sim$, where G is any Lie group with $\mathbf{Z}_2 = \{e, a\}$ in its centre; the equivalence relation $\sim$ is $(x, g) \sim (-x, ag)$. (One has the canonical projection: $Spin^G(4) \to SO(4), [x, g] \mapsto \Lambda(x)$, with $\Lambda : Spin(4) \to SO(4)$.) Then, in ref.[8] it is proved that an $SO(4)$-bundle ξ over a 4-dimensional manifold M, is covered by $Spin^G(4)$ bundle iff

$$\theta.(\xi) = 0, \tag{4.60}$$

with $\theta.$ the canonical morphism $\theta. : SO(4)[M] \to H^3(M; \pi_1(G))$, where $SO(4)[M]$ is the set of equivalence classes of principal $SO(4)$-bundles over M. In particular, if $\pi_1(G) = 0$ one has an exact sequence:

$$Spin^G(4)[M] \to SO(4)[M] \to 0$$

so $Spin^G(4)$-structures always exist.

TAB.4.23 - **Examples of gravitational instantons**

Name	Compact	Spin	Kahlerian
Schwarzschild (euclid.cont.)	NO		
Kerr (euclid.cont.)	NO		
Taub-NUT families	NO		
Eguchi-Hansen families	NO		
Page-Taub-NUT	NO		
$S^4 \subset \mathbf{R}^5$ (4-sphere) (+)	YES	YES	NO
$P^2(\mathbf{C})$ (complex proj.plane)	YES	NO	YES
$S^2 \times S^2$ (metric product) (++)	YES	NO	YES
$S^2 \to S^2$ ($\diamond$)	YES	NO	YES
$V_{2n} \subset P^2(\mathbf{C})$(†)	YES	YES	YES
PJ^+ (asymptotic twistor space)			YES

(+) Analytic continuation of de Sitter space

(++) Product of two 2-spheres each with radius $\Lambda^{-1/2}$ can be regarded as a limiting case of the Schwarzschild-de Sitter solution of Einstein equations with positive term: $R_{\alpha\beta} - \Lambda g_{\alpha\beta} = 0, \Lambda > 0$

($\diamond$) This is a limit case of the Kerr-de Sitter solution

(†) 4-dim. algebraic submanifold of $P^3(\mathbf{C})$

Then, we shall consider the group $Spin^G(4) \equiv G \times_{\mathbf{Z}_2} Spin(4)$, where G is any simple connected, simple Lie group which contains $\mathbf{Z}_2$ in its center. Physically $G/\mathbf{Z}_2$ represents the internal symmetry group. Then, a principal bundle of $Spin^G(4)$-referments can be built: $(P, M, \pi; Spin^G(4))$. So, a $Spin^G(4)$-spinor field is a section of a vector bundle $E \equiv P \times F/Spin^G(4)$ associated to P via a representation $r : Spin^G(4) \to Aut(F)$ of $Spin^G(4)$ in a vector space F. However, we have not, in general, a canonical way to associate to any symmetry $f \equiv (f_P, f_M)$ of a $Spin^G$-gravitational instanton (that is a gravitational instanton plus the principal bundle P), a fiber bundle transformation (f_E, f_M) of E over f_M. Therefore, if we define a spinor field in this way we have not obtained, in general, a fully covariant entity: The existence of $Spin^G$-structures on a gravitational instantons does not guarantee that any $Spin^G$ spinor field is fully covariant.

REMARK 4.31 - However, the physically important result is that any time we have the existence of $Spin^G$-structures we can build spinor-fields which are fully covariant. This is obtained if we consider the fiber bundle E over M having as fiber $E_x, x \in M$, the vector space $E_x \equiv \bigoplus_{p,q\geq 0} T^p{}_q(V \otimes_{\mathbf{R}} C(T_xM, g(x)))$, where V is a vector space over $\mathbf{R}$ or $\mathbf{C}$ such that its group of authomorphisms is G and $C(T_xM, g(x))$ is the Clifford algebra on the tangent space T_xM endowed with the inner product $g(x)$:

$$E \equiv \bigcup_{x\in M} E_x, \quad E_x \equiv \bigoplus_{p,q\geq 0} T^p{}_q(V \bigotimes_{\mathbf{R}} C(T_xM, g(x))) \equiv \bigoplus_{p,q\geq 0} T^p{}_q Clif^G(M, g).$$

One can see [101,102] that E is a vector bundle associated to P that for any symmetry $f \equiv (f_P, f_M)$ of $Spin^G$-gravitational instanton, associates a fiber transformation (f_E, f_M) of E over f_M. (As we have a natural inclusion $T^p{}_q Clif^G(M, g) \to E$, for any $p, q \geq 0$, we can also consider, for example, spinor fields sections of $Clif^G(M, g)$ only.)

If V is a complex vector space we can also consider the fiber bundle

$$Clif^G(M, g)_{\mathbf{C}} \equiv \bigcup_{x\in M} V \bigotimes_{\mathbf{C}} C(T_x{}^{\mathbf{C}}M, g^{\mathbf{C}}(x))$$

where $T_x{}^{\mathbf{C}}M$ is the complexificated tangent space and $g^{\mathbf{C}}(x)$ is the complexificated product. In particular, if $V = \mathbf{C}$ and $G = U(1)$, one has

$$Clif^c(M) \equiv Clif^{U(1)}(M, g)_{\mathbf{C}} = \bigcup_{x\in M} C(T^{\mathbf{C}}{}_xM, g^{\mathbf{C}}(x)).$$

Then, $Clif^G(M, g)_{\mathbf{C}}$ is also a fiber bundle that admits as structure group the group $Spin^G$ by means of the twisted adjoint representation.

REMARK 4.32 - Now, it is interesting to note that every time on a $Spin^G$-instanton a section ω of the fiber bundle $T^*M \bigotimes TM \to M$ exists such that $\omega(x)$ is a skew linear involution of $T_xM, x \in M$, we can canonically split the bundle $Clif^G(M,g)$ into the tensor product of two subbundles:

$$Clif^G(M,g) \cong S(M)^* \bigotimes S(M)^G$$

where

$$\begin{aligned} S(M)^G &\equiv V \bigotimes S(M) \\ S(M) &\equiv \Lambda(TM)_1 = \bigcup_{x\in M} (T_xM)_1 \\ (T_xM)_1 &\equiv \ker(\omega(x) - id_{T_xM}) \end{aligned}$$

$S(M)^G$ is called **generalized spin bundle**. For the fiber bundle $Clif^G(M)_{\mathbf{C}}$ we have similar results if on M a field J is defined such that $J(x)$ is a complex structure on T_xM and $g(x)$ is a non-degenerate hermitian metric. In fact, in these cases, one has the fiber bundle isomorphism :

$$Clif^G(M,g)_{\mathbf{C}} \cong S(M)^*{}_{\mathbf{C}} \bigotimes S(M)^G_{\mathbf{C}}, \tag{4.61}$$

where

$$\begin{aligned} S(M)_{\mathbf{C}} &\equiv \Lambda(T^{\mathbf{C}}M)_1 \\ (T^{\mathbf{C}}M)_1 &\equiv \bigcup_{x\in M} (T_x{}^{\mathbf{C}}M)_1 \\ (T_x^{\mathbf{C}}M)_1 &\equiv \ker(\omega(x) - id_{T_x{}^{\mathbf{C}}M}), \\ S(M)^G_{\mathbf{C}} &\equiv V \bigotimes_{\mathbf{C}} S(M)_{\mathbf{C}} \end{aligned}$$

where $\omega(x)$ is the involution canonically induced by $J(x)$. The spinor fields that are usually used in physics can be identified with sections of $S(M)^G$ or $S(M)^G{}_{\mathbf{C}}$ over M. We shall call such spinor fields **pure** $Spin^G$**-spinor fields**. It is important to note that the fully covariance for such objects is not generally assured for any $Spin^G$-gravitational instanton as in general the representations (4.60) or (4.61) cannot be conducted in a canonical way. In fact, the above mentioned structures ω or J are not canonically defined.

However, there are some manifolds where one can canonically recognize such fields: These are the so-called almost-Hermitian manifolds and almost-Kählerian manifolds; (for more details see e.g. refs.[101,102,161]). So, we have the following important

THEOREM 4.56 - *On almost Hermitian* $Spin^G$*-instantons and almost Kählerian* $Spin^G$*-instantons exist fully covariant pure* $Spin^G$*-spinor fields.*

2 EXAMPLE 4.11 - In table 5.23 we have reported also the existence of Kählerian structure on some gravitational instantons. So, on $P^2(\mathbf{C})$ we can recognize fully covariant $Spin^G$-spinor fields, where G is a simply connected, simple Lie group having $\mathbf{Z}_2$ in its center. For example we can take $G = SU(2n), n = 1, 2, 3, \cdots$, but not $SU(2n+1)$. In fact, condition (4.59) does not hold for these last groups, but valid for $SU(2n)$. So, the $SU(5)$-model of Georgi and Glashow [35] cannot be considered! In fact, $P^2(\mathbf{C})$ is a mathematical "counter-example" to the model of Georgi and Glashow.

For simplicity we shall consider only $Spin^c$-spinor fields. We have the following important theorem.

THEOREM 4.57 - *Any gravitational $Spin^c$-instanton admits fully covariant quantum spinor fields.*

PROOF. In fact, we can prove (see ref.[104] and chapter 5) that Dirac equation $(D) \subset J\mathcal{D}(E)$, $\quad E \equiv Clif^c(M)$, on any 4-dimensional (pseudo-)Riemannian manifold M with metric g and volume form $\mu \equiv *1$, admits a natural Dirac-quantization. Furthermore, we have that the quantum full covariance is satisfied because the complexificated Clifford bundle identifies a superbundle of geometric objects. As a particular case we have fully covariant quantum spinor fields on any gravitational $Spin^c$-instanton. □

COROLLARY 4.4 - *On almost Hermitian $Spin^c$-instantons and almost Kählerian $Spin^c$-instantons exist fully covariant quantum pure $Spin^c$-spinor fields.*

PROOF. This is obtained by restriction on the Dirac operator Dir on the subbundle $S(M)_{\mathbf{C}}$ and by using the above theorem. □

THEOREM 4.58 - *Any gravitational instanton M admits fully covariant quantum Yang-Mills fields.*

PROOF. The proof is similar to previous one by considering results of ref.[104], and the ones reported in Chapter 5. □

The above theorems can also be reformulated in the category of supermanifolds [112,113,114]. In fact we have the following important theorem:

THEOREM 4.59 - *A manifold M of dimension $2^{N-1}(m+n)$ that admits an almost superstructure of type $(m,n;N)$ [23, 112] is also a spin manifold if $N \geq 3$.*

PROOF. In fact such a manifold has the Stiefel-Whitney class w_i zero for $i < 2^{N-1}-1$. So in order for $w_2 = 0$ one must have $N \geq 3$. □

THEOREM 4.60 - *A gravitational instanton of type $(m,n;N)$ is a gravitational instanton that is a manifold of dimension $2^{N-1}(m+n)$ that admits an almost superstructure of type $(m,n;N)$.*

1) *Any gravitational instanton of type $(m,n;N)$ admits fully covariant quantum spinor fields.*

2) *On any almost Hermitian or almost Kählerian gravitational instanton of type $(m, n; N)$, fully covariant quantum pure spinor fields exist.*

2 Let us, now, consider the formulation of the global variational calculus.

DEFINITION 4.26 - 1) *A* **Lagrangian** *of the first order on the fiber bundle $\pi : W \to M$ is a numerical function*

$$L : J\mathcal{D}(W) \to \mathbf{R}. \tag{4.62}$$

2) *The* **action integral** *associated to a Lagrangian L, at the section s of π, on the compact domain $A \subset M$, and to the volume form η of M is the following integral:*

$$I_A[s] = \int_A (L \circ Ds)\eta. \tag{4.63}$$

3) *For any deformation $\widetilde{s} : [0,1] \times M \to W$ of s, i.e., the partial application $\widetilde{s}_\lambda : M \to W, \lambda \in \mathbf{R}$, is a section of π such that $\widetilde{s}_0 = s$, with $\widetilde{s}_\lambda|_{\partial A} = s|_{\partial A}$, the action integral defines a numerical function*

$$I_A[\widetilde{s}] : \mathbf{R} \to \mathbf{R},\ I_A[\widetilde{s}](\lambda) = \int_A (L \circ D\widetilde{s}_\lambda)\eta. \tag{4.64}$$

We say that s is an **extremal** *if for any deformation of s the function (4.64) has a critical point at $\lambda = 0$.*

THEOREM 4.61 - 1) *A section s of the fiber bundle*

$$\pi : W \to M, \dim M = n, \dim W = n + m$$

is an extremal for the Lagrangian $L : J\mathcal{D}(W) \to$ **R** *iff it is a solution of a second order PDE $(EL) \subset J\mathcal{D}^2(W)$:* **Euler-Lagrange equation** *associated to L. More precisely, in local coordinates*

$$(x^\alpha, y^j, y^j_\alpha, y^j_{\alpha\beta})$$

on $J\mathcal{D}^2(W)$, (EL) assumes the following form:

$$\boxed{(\partial y_j.L) \circ Ds - (\partial x_\alpha((\partial y^\alpha_j.L) \circ Ds)) = 0, j = (1, \cdots, m)} \tag{4.65}$$

PROOF. Let us derive (4.64) with respect to λ, and take $\lambda = 0$. Taking into account that the domain and η do not depend on λ, we get

$$\frac{d}{d\lambda}(I_A[\widetilde{s}])|_{\lambda=0} = \int_A \frac{d}{d\lambda}((L \circ D\widetilde{s}_\lambda)\eta)|_{\lambda=0} = \int_A \frac{d}{d\lambda}(L \circ D\widetilde{s}_\lambda)|_{\lambda=0}\eta.$$

On the other hand

$$\frac{d}{d\lambda}(L \circ D\widetilde{s}_\lambda)|_{\lambda=0} =< vdL, \partial D\widetilde{s} >$$

where vdL is the vertical differential of L, (i.e., the differential of L with respect to the vertical vectors of $\pi_1 : J\mathcal{D}(W) \to M$). Note that $\partial D\widetilde{s} : M \to Ds^* vTJ\mathcal{D}(W)$. In fact,

$$\partial D\widetilde{s}(p) = \frac{d}{d\lambda}(D\widetilde{s}_\lambda)|_{\lambda=0}(p) \in vT_{Ds(p)}J\mathcal{D}(W).$$

On the other hand

$$D\widetilde{s}(p,\lambda) = (D\widetilde{s}_\lambda)(p) \in J\mathcal{D}(W)_p, \quad \forall \lambda \in \mathbf{R}$$

as

$$\widetilde{s}_\lambda : M \to W, \quad \forall \lambda \in \mathbf{R}, \quad \pi \circ \widetilde{s}_\lambda = id_M.$$

Thus,

$$\frac{d}{d\lambda}(D\widetilde{s}_\lambda)(p) \in vT_{D\widetilde{s}_\lambda}(p)J\mathcal{D}(W)$$

and for $\lambda = 0$ we have

$$\frac{d}{d\lambda}(D\widetilde{s}_\lambda)|_{\lambda=0}(p) \in vT_{Ds(p)}J\mathcal{D}(W) \subset Ds^* vTJ\mathcal{D}(W).$$

On the other hand we have the isomorphism

$$Ds^* vTJ\mathcal{D}(W) \cong J\mathcal{D}(s^* vTW), \quad \partial D\widetilde{s}(p) \mapsto D(\partial\widetilde{s})(p).$$

As a consequence, we have in coordinates

$$\partial D\widetilde{s} = \nu^j \partial y_j + (\partial x_\alpha . \nu^j)\partial y_j^\alpha$$

if

$$\nu = \partial\widetilde{s} = \nu^j \partial y_j.$$

Furthermore, we have

$$vdL = (\partial y_j . L)dy^j + (\partial y_j^\alpha . L)dy_\alpha^j.$$

Hence,

$$< vdL, \partial D\widetilde{s} >= \nu^j(\partial y_j . L) \circ Ds + (\partial x_\alpha . \nu^j)(\partial y_j^\alpha . L) \circ Ds. \tag{4.66}$$

Moreover, $< vdL, \partial D\widetilde{s} >$ defines a linear differential operator of first order:

$$\mathcal{P}[s] : C^\infty(s^* vTW) \to C^\infty(M; \mathbf{R}), \nu \mapsto \mathcal{P}[s].\nu \equiv < vdL, \partial\nu > .$$

Let us denote the formal adjoint of $\mathcal{P}[s]$ with

$$\mathcal{E}[s] : C^\infty(\Lambda^0_n M) \to C^\infty(\Lambda^0_n M \bigotimes s^* vTW).$$

Taking into account the relation

$$(\mathcal{P}[s].\nu)\eta = \mathcal{E}[s].\eta + d(\sigma_1(\mathcal{P}[s])\nu\rfloor\eta)$$

where $\sigma_1(\mathcal{P}[s])$ is the symbol of $\mathcal{P}[s]$, by means of Stokes theorem, we have:

$$\begin{aligned}\frac{d}{d\lambda}(I_A[\widetilde{s}])|_{\lambda=0} &= \int_A < vdL, \partial D\widetilde{s} > = \int_A (\mathcal{P}[s].\nu)\eta = \int_A \mathcal{E}[s].\eta + \int_A d(\sigma_1(\mathcal{P}[s])\nu\rfloor\eta)\\ &= \int_A \mathcal{E}[s].\eta + \int_{\partial A} \sigma_1(\mathcal{P}[s])\nu\rfloor\eta = \int_A \mathcal{E}[s].\eta\end{aligned}$$

as $\nu = \partial\widetilde{s}|_{\partial A} = 0$. Finally, taking into account that ν is arbitrary we get $\mathcal{E}[s] = 0$. In order to obtain the expression in coordinates of this equation, we must calculate one for $\mathcal{E}[s]\eta = (\mathcal{P}[s].\nu)\eta - d(\sigma_1(\mathcal{P}[s])\nu\rfloor\eta)$.
From (4.66) we get:

$$(4.67) \qquad (\mathcal{P}[s].\nu)\eta = [\nu^j(\partial y_j.L)\circ Ds + (\partial x_\alpha.\nu^j)(\partial y^\alpha_j.L)\circ Ds]dx^1\wedge.\cdots dx^n.$$

Furthermore, taking into account that the local expression of the symbol $\sigma_1(\mathcal{P}[s])$ is the following:

$$\sigma_1(\mathcal{P}[s]) = (\partial y^\alpha_j.L)\dot{x}_\alpha \otimes y^j$$

and hence
(4.68)

$$\left\{\begin{array}{ll} (a) & \sigma_1(\mathcal{P}[s]).\nu = (\partial y^\alpha_j.L)\dot{x}_\alpha\nu^j\\ (b) & \sigma_1(\mathcal{P}[s]).\nu\rfloor\eta = (-1)^\alpha(\partial y^\alpha_j.L)\nu^j dx^1\wedge\cdots\wedge\widehat{dx^\alpha}\wedge\cdots\wedge dx^n\\ (c) & (\sigma_1(\mathcal{P}[s]).\nu\rfloor\eta) = (\partial x_\alpha((\partial y^\alpha_j.L)\nu^j))dx^1\wedge\cdots\wedge dx^n\end{array}\right\}$$

(the symbol ˆ denotes the absence). Hence, we get (4.65). □

THEOREM 4.62 - (HAMILTONIAN FORM OF EULER-LAGRANGE EQUATION).
1) *Associated to the Lagrangian L we can define the following application (***Legendre transform***)*

$$\sigma(L) : J\mathcal{D}(W) \to TM \bigotimes vT^*W$$

such that the following diagram is commutative:

$$\begin{array}{ccc} J\mathcal{D}(W) & \xrightarrow{\sigma(L)} & TM\bigotimes vT^*W\\ vdL\downarrow & & \uparrow\\ vT^*J\mathcal{D}(W) & \xrightarrow[j^*]{} & \pi^*_{1,0}(TM\bigotimes vT^*W)\end{array}$$

The representation in coordinates of $\sigma(L)$ is the following:

(4.69)
$$\boxed{\dot{x}^\alpha \otimes \dot{y}_j \circ \sigma(L) = (\partial y_j^\alpha.L),\ \alpha = (1, \cdots, n), j = (1, \cdots, m)}$$

Define **generalized momenta**

(4.70)
$$p_j^\alpha \equiv (\partial y_j^\alpha.L), \quad \alpha = (1, \cdots, n), j = (1, \cdots, m).$$

2) On $J\mathcal{D}(W)$ one recognizes a vTW-valued canonical 1-form ω_1. More precisely, one has:

$$\left\{\begin{array}{l} \omega_1 : J\mathcal{D}(W) \to T^*J\mathcal{D}(W) \bigotimes vTW \\ \omega_1(q) \in T_q^*J\mathcal{D}(W) \bigotimes vT_{\pi_{1,0}(q)}W \cong L(T_qJ\mathcal{D}(W); vT_{\pi_{1,0}(q)}W) \\ < \omega_1(q), \zeta >= T(\pi_{1,0})(\zeta) - T(s)[T(\pi_1)(T(\pi_{1,0})(\zeta))] \end{array}\right\}$$

where s is any section π, defined in the neighbourhood of $p \equiv \pi_1(q) \in M$ such that $Ds(p) = q$. In coordinates ω_1 can be written as follows:

(4.71)
$$\boxed{\omega_1 = -y_\alpha^k dx^\alpha \otimes \partial y_k + \delta_j^k \otimes \partial y_k}$$

3) To any Lagrangian $L : J\mathcal{D}(W) \to \mathbf{R}$, we can associate a TM-valued 1-form θ on $J\mathcal{D}(W)$. More precisaly, we have:

$$\left\{\begin{array}{l} \theta : J\mathcal{D}(W) \to T^*J\mathcal{D}(W) \bigotimes TM \\ \theta(q) \in T_q^*J\mathcal{D}(W) \bigotimes T_{\pi_1(q)}M \cong L(T_qJ\mathcal{D}(W); T_{\pi_1(q)}M) \\ < \theta(q), \zeta >= \frac{1}{n}L(q)T(\pi_1)(\zeta) + \sigma(L)(q)(\omega_1(q)(\zeta)). \end{array}\right\}$$

In coordinates θ can be written as follows:

$$\theta = \{[-y_\alpha^k(\partial y_k^\omega.L) + \frac{1}{n}L(q)]dx^\alpha + (\partial y_k^\omega.L)dy^k\} \otimes \partial x_\omega.$$

*4) To any Lagrangian $L : J\mathcal{D}(W) \to \mathbf{R}$, we associate a TM-valued 1-form Θ (**Cartan form**) on $J\mathcal{D}(W)$. More precisely, we have the following:*

$$\Theta : J\mathcal{D}(W) \to T^*J\mathcal{D}(W) \bigotimes TM, \quad \Theta = \theta \bar{\wedge} \pi_1^*\eta = (\alpha \otimes \xi) \bar{\wedge} \pi_1^*\eta \equiv \alpha \wedge (\xi \rfloor \pi_1^*\eta).$$

In coordinates Θ can be written as follows:

$$(4.72)\quad \left\{\begin{array}{l} (a)\quad \Theta = (\partial y_j^{\alpha}.L)dy^j \wedge dx^1 \wedge \cdots \wedge \widehat{dx^{\alpha}} \wedge \cdots \wedge dx^n - H dx^1 \wedge \cdots \wedge dx^n \\ (b)\quad H = \mathrm{tr}(T_{\alpha}{}^{\beta}) \equiv y_{\alpha}^k(\partial y_k^{\beta}.L) - \frac{1}{n}L\delta_{\alpha}{}^{\beta} = y_{\alpha}^k(\partial y_k^{\alpha}.L) - L \\ H \equiv \textbf{(Hamiltonian)}. \end{array}\right\}$$

Of course, if one uses (as it is frequently made) the formula

$$H = -\mathrm{tr}(T_{\alpha}{}^{\beta})$$

we have

$$H = -y_{\alpha}^k(\partial y_k^{\alpha}.L) + L.$$

5) *By means of generalized momenta, the Hamiltonian can be written:*

$$(4.73)\qquad H = L - y_{\alpha}^k p_k^{\alpha}.$$

6) *The extremal sections of L are solutions of the following equation:*

(4.74)**(Hamilton equation)**

$$\boxed{(Ds)^*(\xi \rfloor \Omega) = 0, \qquad \Omega \equiv d\Theta}$$

for any vector field ξ on $J\mathcal{D}(W)$. The expression in coordinates of this equation is the following:

$$(4.75)\qquad \boxed{(-1)^{\alpha+1}y_{\alpha}^j = (\partial p_{\alpha}^j.H), \quad (-1)^{\alpha+1}p_{j\alpha}^{\alpha} = -(\partial y_j.H)}$$

PROOF. Let us only show how one arrives at equation (4.74) starting from the action integral. First, note that the action integral can be written in the following way:

$$(4.76)\qquad I_A[\widetilde{s}] = \int_A (L \circ D\widetilde{s})\eta = \int_A (D\widetilde{s})^*\Theta,$$

where $(D\widetilde{s})^*\Theta$ is defined by the following commutative diagram:

$$\begin{array}{ccccc} & J\mathcal{D}(W) & \xrightarrow{\Theta} & \Lambda_n^0 J\mathcal{D}(W) & \\ \widetilde{Ds_{\lambda}} & \uparrow & & \downarrow & \Lambda_n^0(\widetilde{Ds_{\lambda}}) \\ & M & \xrightarrow[(\widetilde{Ds})^*\Theta]{} & \Lambda_n^0 M & \end{array}$$

Hence, we can write:

$$\frac{d}{d\lambda}I_A[\widetilde{s}]|_{\lambda=0} = \int_A \partial(L\circ D\widetilde{s})\eta = \int_A \partial((D\widetilde{s})^*\Theta). \tag{4.77}$$

On the other hand, as for any differential form α one has $\mathcal{L}_X\alpha = d(x\rfloor\alpha) + X\rfloor d\alpha$, and taking into account that $\mathcal{L}_X\alpha = \partial\widetilde{\alpha}$, where $\widetilde{\alpha}$ is the deformation of α by means of the flow associated to X: $\widetilde{\alpha}_\lambda \equiv \phi_\lambda{}^*\alpha \equiv \Lambda^o_p(\phi_\lambda)\circ\alpha\circ\phi_\lambda$, we get:
(4.78)

$$\left\{\begin{aligned}\frac{d}{d\lambda}I_A[\widetilde{s}]|_{\lambda=0} &= \int_A \partial((D\widetilde{s})^*\Theta) = \int_A d(X\rfloor(D\widetilde{s})^*\Theta) + \int_A X\rfloor d((D\widetilde{s})^*\Theta)\\ &= \int_{\partial A} X\rfloor(D\widetilde{s})^*\Theta + \int_A X\rfloor d((D\widetilde{s})^*\Theta)\\ &= \int_A X\rfloor d((D\widetilde{s})^*\Theta)\\ &= \int_A X\rfloor(Ds)^*d\Theta = \int_A (Ds)^*(X\rfloor\Omega)\end{aligned}\right\}$$

Therefore,

$$\frac{d}{d\lambda}I_A[\widetilde{s}]|_{\lambda=0} = 0 \Leftrightarrow (Ds)^*(X\rfloor\Omega) = 0.$$

The proof of the local equations (4.75) are obtained by means of the following change of coordinates on $J\mathcal{D}(W)$:

$$(x^\alpha, y^j, y^j_\alpha) \to (x^\alpha, y^j, p^j_\alpha)$$

and developing the direct calculation in coordinates for $(Ds)^*(\xi\rfloor\Omega) = 0$. □

THEOREM 4.63 - (SYMMETRY OF LAGRANGIAN AND CONSERVATION LAWS). 1) *Let a Lagrangian* $L : J\mathcal{D}(W) \to \mathbf{R}$ *and 1-parameter group of transformations* ϕ *of the fiber bundle* $\pi : W \to M$ *be given. We say that* π *is* **invariant** *for* ϕ*, if the following diagram is commutative:*

$$\begin{array}{ccccccc} & M & \leftarrow & W & \leftarrow & J\mathcal{D}(W) & \xrightarrow{L} & \mathbf{R}\\ \phi_{\lambda,M} & \downarrow & & \downarrow\phi_\lambda & & \downarrow J\mathcal{D}(\phi_\lambda) & & \|\\ & M & \leftarrow & W & \leftarrow & J\mathcal{D}(W) & \underset{L}{\to} & \mathbf{R}\end{array}$$

The mapping $J\mathcal{D}(\phi_\lambda)$ *is the* **natural lifting** *of* ϕ_λ *on* $J\mathcal{D}(W)$*. In coordinates we have:*

$$\left\{\begin{aligned} x^\alpha\circ J\mathcal{D}(\phi_\lambda) &= \phi^\alpha_{\lambda,M}\circ\pi_1\\ y^j\circ J\mathcal{D}(\phi_\lambda) &= \phi^j_\lambda\circ\pi_{1,0}\\ y^k_j\circ J\mathcal{D}(\phi_\lambda) &= (\partial x_\gamma\phi^{-1\alpha}_{M,\lambda})(\partial x_\alpha\phi^k_\lambda) + (\partial x_\alpha s^j)(\partial x_\gamma\phi^{-1\alpha}_{M,\lambda})(\partial y_j\phi^k_\lambda)\end{aligned}\right\} \tag{4.79}$$

Here $J\mathcal{D}(\phi_\lambda)$ is a 1-parameter group of transformations of $J\mathcal{D}(W)$. The corresponding vector field $Y^{(1)} : J\mathcal{D}(W) \to TJ\mathcal{D}(W)$ is called **natural lifting** *of $Y \equiv \partial\phi : W \to TW$. If Y has the local expression $Y = \xi^\alpha \partial x_\alpha + Y^j \partial y_j$, then $Y^{(1)}$ has the following local expression:*

$$(4.80) \qquad \boxed{Y^{(1)} = \xi^\alpha \partial x_\alpha + Y^j \partial y_j + [-y^k_\alpha(\partial x_\gamma \xi^\alpha) + (\partial x_\gamma Y^k) + y^j_\gamma(\partial y_j Y^k)]\partial y^\alpha_j}$$

The Lagrangian L is invariant for ϕ_λ iff

$$\boxed{\begin{aligned} Y^{(1)}.L &= \xi^\alpha(\partial x_\alpha L) + Y^j(\partial y_j L) \\ &\quad + [-y^k_\alpha(\partial x_\gamma \xi^\alpha) + (\partial x_\gamma Y^k) + y^j_\gamma(\partial y_j Y^k)](\partial y^\alpha_j L) = 0 \end{aligned}}$$

2) If the Lagrangian L is invariant for ϕ_λ, then there exists a differential $(n-1)$-form β on $J\mathcal{D}(W)$ such that

$$d[(Ds)^*\beta] = 0$$

for any extremal s of L. We say that β is a **conservation law** *of L. In coordinates β assumes the following form:*

$$(4.81) \qquad \left\{ \begin{aligned} \beta &= Z \rfloor \eta \\ Z &= Z^\alpha \partial x_\alpha \\ Z^\alpha &\equiv [L \circ Ds.\xi^\alpha + ((\partial y^\alpha_j.L) \circ Ds)\rfloor (D\tilde{s})^j_\alpha] \\ \partial\phi_M &= \xi^\alpha \partial x_\alpha : M \to TM \end{aligned} \right\}$$

PROOF. The proof can be obtained by direct calculations. □

REMARK 4.33 - These considerations can be generalized to Lagrangians of higher order $k > 1$. In fact, in Chapter 2 we have given a general formulation of variational calculus for systems constrained by PDEs. (See also refs.[60,77,78,102,104,107, 111,112,113,114]).

In the following we give some important examples of variational equations.

EXAMPLE 4.12 - 1)(RIGID BODY, COBORDISM GROUPS AND CONSERVATION LAWS). The configuration manifold of a rigid body can be identified with $\mathbf{R}^3 \times SO(3)$. Then, the dynamic equation can be identified with a second order ODE

$$E_2 \subset J\mathcal{D}^2(\mathbf{R}, \mathbf{R}^3 \times SO(3))$$

where E_2 is a $19 - 6 = 13$-dimensional submanifold of $J\mathcal{D}^2(\mathbf{R}, \mathbf{R}^3 \times SO(3))$. Let us assume, now, that the rigid body is free. Then, the dynamic equation can be obtained as Euler-Lagrange equation of the following Lagrangian function:

$$L = \mathcal{T} = \frac{1}{2}\mu v_G^2 + \frac{1}{2}I_{ij}(G)\omega^i\omega^j : J\mathcal{D}^2(\mathbf{R}, \mathbf{R}^3 \times SO(3)) \to \mathbf{R}$$

where μ is the total mass, v_G is the velocity of the center of mass G, $I(G)$ is the moment of inertia tensor with respect to G, and ω is the angular velocity, that we expressed by means of the Euler angles. As L is invariant for any spatial translation, from the Noether theorem one has the conservation of the total momentum P. Hence, from the first cardinal equation we get that the possible motions of the center of mass G are rectilinear and uniform. Now, let us note that E_2 splits into two different ODEs:

$$E_2 \cong (E_2)_1 \times_{\mathbf{R}} (E_2)_2$$

where $(E_2)_1 \subset J\mathcal{D}^2(\mathbf{R}, \mathbf{R}^3)$ is the 7-dimensional manifold that just describes the motion of the center of mass, and $(E_2)_2 \subset J\mathcal{D}^2(\mathbf{R}, SO(3))$ is a 7-dimensional manifold that describes the motion of the rigid body about the center of mass. As for the 0-cobordism group one has $\Omega_0 \cong \mathbf{Z}_2$, curves in $(E_2)_2$ representing motions, could at the beginning, be segment of lines (non-periodic motions) or loops (periodic motions). In order to decide which of these possibilities happens, let us consider again the conservation laws of (E_2) and $(E_2)_2$. Note that the Lagrangian function of E_2 is invariant for spatial rotations about the center of mass and time translations. Hence, the Noether theorem assures the existence of the following conservation laws:

$$(4.82) \qquad \left\{ \begin{array}{l} \Omega(G)_j = c_j,\ 1 \leq j \leq 3 \\ \text{(components of the total angular momentum)} \\ \mathcal{T} = c_4 \ \text{(kinetic energy)} \end{array} \right\}$$

On the other hand $(E_2)_2$ is also the Euler-Lagrange equation of the Lagrangian function $L = \frac{1}{2}I_{ij}(G)\omega^i\omega^j$, that admits the above conservation laws. In fact, these are functions on $T(SO(3))$, and one has the following exact commutative diagram:

$$\begin{array}{ccccc} (E_2) & \subset & J\mathcal{D}^2(\mathbf{R}, \mathbf{R}^3 \times SO(3)) & & \\ \downarrow & & \downarrow & & \\ (E_2)_2 & \subset & J\mathcal{D}^2(\mathbf{R}, SO(3)) & = & J\mathcal{D}^2(\mathbf{R}, SO(3)) \\ \downarrow & & \downarrow & & \downarrow \\ 0 & & 0 & & J\mathcal{D}(\mathbf{R}, SO(3)) \\ & & & & \downarrow \\ & & & & T(SO(3)) \\ & & & & \downarrow \\ & & & & 0 \end{array}$$

Then, the characteristic vector field v of $(E_2)_2$ identifies a (bicharecteristic) vector field $u : T(SO(3)) \to TT(SO(3))$. The admissible motions are the corresponding integral curves of u. $T(SO(3))$ is a 6-dimensional orientable manifold, and conditions (4.82) identify a 2-dimensional submanifold $V_c \subset T(SO(3))$. So any admissible motion is identified with a curve on such a surface. Furthermore, assuming $c_4 > 0$, it follows that the bicharacteristic vector field u on V_c cannot have singular points. Moreover, as V_c is a level surface for T, it follows that V_c is compact and orientable. Now, as the torus T^2 is the unique connected bidimensional compact orientable manifold that admits a tangent vector field without singular points, it follows that $V_c = T^2 \dot{\bigcup} \cdots \dot{\bigcup} T^2$. Then, we can choose suitable angular coordinates $\phi_1, \phi_2(\mathrm{mod})2\pi$ on T^2 such that $u|_{T^2}$ can be written $\dot{\phi}_1 = \omega_1(c)$, $\dot{\phi}_2 = \omega_2(c)$. As ω_1 and ω_2 are not, in general, commensurable entities, the resulting motion of the rigid body is not, in general, periodic one.

2)(GEODESIC EQUATION).

Let (M, g) be a 4-dimensional (pseudo-) Riemannian manifold, i.e., a smooth section g of the fiber bundle $S^0_2 M \to M$ is defined on M, such that there are local coordinates on M in which g assumes the diagonal form

$$(g_{\alpha\beta}) = \begin{bmatrix} 1 & 0 & 0 & 0 \\ 0 & \pm 1 & 0 & 0 \\ 0 & 0 & \pm 1 & 0 \\ 0 & 0 & 0 & \pm 1 \end{bmatrix}$$

where the sign $+$ holds if M is Riemannian and $-$ if M is pseudo-Riemannian (e.g., in Relativity the space-time is pseudo-Riemannian). Such a manifold admits a canonical connection (**Levi-Civita connection**) characterized by the following properties:

$$(4.83) \qquad \boxed{\left\{ \begin{array}{c} \nabla_X g = 0 \\ T(X,Y) \equiv \nabla_X Y - \nabla_Y X - [X,Y] = 0 \end{array} \right\}}$$

for any vector field X, Y on M. Here T is the **torsion**. Equations (4.83) characterize in coordinates the connection coefficients:

$$(4.84) \qquad \left\{ \begin{array}{c} \widetilde{\Gamma}^{\alpha}_{\beta\lambda} = -\Gamma^{\alpha}_{\beta\lambda}, \ \Gamma^{\alpha}_{\beta\lambda} = \Gamma^{\alpha}_{\lambda\beta}, \ \Gamma^{\alpha}_{\beta\lambda} = \frac{1}{2} g^{\alpha\omega}[\beta\lambda, \omega] \\ [\beta\lambda, \omega] \equiv (\partial x_\beta g_{\lambda\omega}) + (\partial x_\lambda g_{\omega\beta}) - (\partial x_\omega g_{\beta\lambda}) \end{array} \right\}$$

where $\widetilde{\Gamma}^{\alpha}_{\beta\lambda}$ are the connection coefficients on T^*M and $\Gamma^{\alpha}_{\beta\lambda}$ on TM respectively. If the metric is pseudo-Riemannian, on the tangent space T_pM we recognize a cone (**light cone**) by means of the following equation:

$$0 = g(p)(v,v) \Rightarrow g_{\alpha\beta}(p) v^\alpha v^\beta = 0.$$

The vectors inside the light cone, $g(p)(v,v) > 0$, are called **time-like**, and that outside the light cone, $g(p)(v,v) < 0$, are called **space-like**. The vector along the light cone, $g(p)(v,v) = 0$, are called **light-like**. Then, a **relativistic frame** on M is a pair (ψ, τ), where: (i) ψ is a 1-parameter group of transformations of M: $\psi : \mathbf{R}\times M \to M$, such that its velocity $\dot{\psi} \equiv \partial\psi : M \to TM$, is a time-like, or light-like vector field (i.e., $g(p)(\dot{\psi}(p), \dot{\psi}(p)) \geq 0, \forall p \in M$); (ii) $\tau : M \to \mathbf{R}$ is a numerical function (**proper time** of the relativistic frame) such that $< d\tau, \dot{\psi} >= 1$. A relativistic frame allows us to represent M as a product: $M \cong \mathbf{R} \times S_\psi$, where $S_\psi \equiv M/\sim$ (**proper space** of the relativistic frame), where $\sim$ is the equivalence relation in M induced by the frame-flow. If the latter is "regular" S_ψ is a 3-dimensional Riemannian manifold. If the coordinates (x^α) on M are **adapted** to ψ, i.e., $\dot{\psi} = \partial x_0$, then (x^i) identify coordinates on S_ψ, and the corresponding metric g_ψ induced by g has components that can be identified with $g_{ij} \equiv g(\partial x_i, \partial x_j)$.

REMARK 4.34 - (MINKOWSKI SPACE-TIME). If M is an affine 4-dimensional space, where the metric can be written (**Minkowski metric**):[4]

$$(g_{\alpha\beta}) = \begin{bmatrix} c^2 & 0 & 0 & 0 \\ 0 & -1 & 0 & 0 \\ 0 & 0 & -1 & 0 \\ 0 & 0 & 0 & -1 \end{bmatrix}$$

where c is the light velocity, then (G, g) is called **Minkowski space-time**. This is the space-time of the Special Relativity. Then, the differential ds, of the invariant length s of curve can be written:

$$ds = c\sqrt{1 - \frac{{u_\wedge}^2}{c^2}}dt\,.$$

Hence, if $u^\alpha = \frac{dx^\alpha}{dt} \equiv \dot{x}^\alpha$ are the velocity components of a motion in Minkowski space-time, we can write:

$$u^\alpha \equiv \dot{x}^\alpha = \frac{dx^\alpha}{ds}\dot{s} = \frac{dx^\alpha}{ds}c\sqrt{1 - \frac{{u_\wedge}^2}{c^2}}.$$

As $u^0 = 1$, we also have:

$$1 = u^0 \equiv \dot{x}^0 = \frac{dx^0}{ds}c\sqrt{1 - \frac{{u_\wedge}^2}{c^2}}$$

$$\Downarrow$$

$$\frac{dx^0}{ds} = \frac{1}{c}(1 - \frac{{u_\wedge}^2}{c^2})^{-1/2}.$$

[4] Usually one puts c=1.

Define **relativistic mass** of one-body

$$\mu \equiv \mu_0 c \frac{dx^0}{ds} = \mu_0 (1 - \frac{{u_\wedge}^2}{c^2})^{-1/2}$$

where $\mu_0 \in \mathbf{R}_+$. As for $u_\wedge = 0 \Rightarrow \mu = \mu_0$, μ_0 is called **rest-mass** of the one-body. Furthermore, define **relativistic momentum**

$$p = \mu u \Rightarrow p^\alpha = \mu \dot{x}^\alpha$$

As a consequence we get

$$\begin{aligned} p^0 &= \mu = \textbf{relativistic mass} \\ p^k &= \mu \dot{x}^k = \mu \frac{dx^k}{ds} c \sqrt{1 - \frac{{u_\wedge}^2}{c^2}} = \mu_0 c \frac{dx^k}{ds} \end{aligned}$$

Then, one can reformulate, mutatis mutandi, the geometric description of the Newton equation, in relativistic form. More precisely, the **Newton relativistic equation**, is the following second order ODE:

$$E_2 = \ker(\mathcal{F} - \mu\Gamma) \subset J\mathcal{D}^2_\bullet(\mathcal{M}) \subset J\mathcal{D}^2(\mathcal{M})$$

where $\mathcal{M}$ is the fiber bundle $\mathbf{R} \times M \to \mathbf{R}$, $J\mathcal{D}^2_\bullet(\mathcal{M})$ is the derivative space of the second order for mappings $\mathbf{R} \to M$, such that their velocity is time-like, and μ is the relativistic mass of one body, and $\mathcal{F} : J\mathcal{D}^2_\bullet(\mathcal{M}) \to TM$, is the **force operator**. In coordinates the relativistic Newton equation has the following form:

$$F^\alpha = \mu[\Gamma^\alpha_{\beta\gamma} \dot{x}^\beta \dot{x}^\gamma + \ddot{x}^\alpha] = 0$$

2 The **geodesic equation** is a second order ODE $E_2 \subset J\mathcal{D}^2(\mathbf{R}, M)$ that can be written in the following form:

$$\frac{\delta v}{\delta t} \equiv \nabla_v v = 0 \qquad (4.85)$$

where v is the velocity for curves in M. If (M,g) is pseudo-Riemannian, and if we want to consider motion-geodesics, then we must add the constraint that v is time-like or light-like. Equation (4.85) in coordinates has the following form:

$$\frac{\delta}{\delta s}\left(\frac{dx^l}{ds}\right) \equiv \frac{d^2x^l}{ds^2} + \Gamma^l_{ik}\frac{dx^i}{ds}\frac{dx^k}{ds} = 0. \tag{4.86}$$

This equation can be considered the Euler-Lagrange equation of the following action integral:

$$I_{[t_0,t_1]}[x(t)] = \int_{[t_0,t_1]} \sqrt{e g_{ij}\dot{x}^i\dot{x}^j}\,dt$$

where e is the indicator ($e = \pm 1$). This integral measures the length of an arc that pass for the points $P_0(t = t_0)$ and $P_1(t = t_1)$. We call **geodesic** a curve solution of equation (4.86). So a geodesic is a curve where the above action integral is stationary. Furthermore, for any initial condition, i.e., for any point $p \in M$ and vector $v \in T_pM$, pass a unique geodesic. For two points of M, instead, can pass more geodesics: this depends on the topology of M.

EXAMPLE 4.13 - 1) In a Euclidean space any straight line is a geodesic. So for two points passes a unique geodesic, and the arc of curve with the minimum distance relating two points is a geodesic arc (i.e., the segment identified by the two points). 2) The geodesics on S^2 are the great circles of S^2. So, for two generic points P and Q on the sphere ($P \neq \pm Q$), passes a unique geodesic. In the case that $P = -Q$, any meridian is a geodesic. Furthermore, P and Q are related by two geodesic arcs that, for generic points, are of different lengths; longer one does not minimize the length. In fact, a geodesic arc minimizes distances iff the arc has unit speed.

□ A vector field $A(t)$ along a curve $x^i = x^i(t)$ is said to be **autoparallel** if

$$\frac{\delta A}{\delta t} = 0 \Leftrightarrow \frac{\delta A_i}{\delta t} \equiv \frac{dA_i}{dt} - \Gamma^l_{ik}A_l\dot{x}^k = 0 \tag{4.87}$$

Taking into account that

$$\frac{\delta A^i}{\delta t} = \frac{\delta}{\delta t}(g^{ij}A_j) = g^{ij}\frac{\delta A_i}{\delta t},$$

equation (4.87) can be written in the following form:

$$\frac{\delta A^i}{\delta t} \equiv \frac{dA^i}{dt} - \Gamma^i_{jk}A^j\dot{x}^k = 0$$

REMARK 4.35 - (PROPERTIES OF GEODESICS).

(a) If $A^i \equiv \frac{dx^i}{ds}$ is the unitary velocity of a geodesic, it follows that A propagates with parallelism along this curve.

(b) The module of a vector that propagates with parallelism does not change.

(c) The angle θ between two vector fields A and B, that propagate with parallelism remains constant along the propagation.

(d) A vector that propagates along a closed curve, does not necessarily reproduces the original one.

(e) We call **geodesic coordinates**, passing for a point $p \in M$, the coordinates having geodesics as coordinate lines.

3)(KLEIN-GORDON EQUATION).[5]

Let (M, g) be a 4-dimensional (pseudo-) Riemannian manifold. Let, $\pi : E \equiv M \times \mathbf{R} \to M$ be a trivial vector fiber bundle with fibre $\mathbf{R}$. Klein-Gordon equation is the kernel (KG) of the following constant rank epimorphism:

$$\mathcal{KG} \equiv (\Box + \kappa) : J\mathcal{D}^2(E) \to M \times \mathbf{R},$$

where $\Box$ is the Laplacian for scalar fields: $\Box \equiv \delta d$, and κ is a numerical function on M. Thus, (KG) is a vector subbundle of $J\mathcal{D}^2(E)$, (i.e., a linear equation) locally characterized by the following numerical functions on $J\mathcal{D}^2(E)$:

$$\boxed{F \equiv g^{\alpha\beta} y_{\alpha\beta} - \Gamma^{\alpha}_{\beta\lambda} g^{\lambda\beta} y_{\alpha} + \kappa y}$$

where $\{x^\alpha, y, y_\alpha, y_{\alpha\beta}\}$ are local coordinates on $J\mathcal{D}^2(E)$, $\Gamma^{\alpha}_{\beta\lambda}: M \to \mathbf{R}$ are the connection coefficients of the Levi-Civita connection on (M, g). One can prove [104] that (KG) is an involutive formally integrable PDE, that is also completely integrable. (Note, also, that (KG) is an elliptic equation if g is Riemannian.) Then, we can apply the Cartan-Kähler theorem in order to state the existence and unicity of solutions C^ω for the Cauchy problem. Furthermore, (KG) is the Euler-Lagrange equation for the following Lagrangian:

$$L \equiv \frac{1}{2}\sqrt{|g|}\{g^{\mu\nu} y_\mu y_\nu - \kappa y^2\} : J\mathcal{D}(E) \to \mathbf{R}.$$

The corresponding generalized momenta are the following: $p^\alpha \equiv (\partial y^\alpha . L) : J\mathcal{D}(E) \to \mathbf{R}$. For any section $s \equiv (id_M, f) \in C^\infty(E), p^\alpha$ identifies a function $\pi^\alpha \equiv p^\alpha \circ Ds$:

[5] Here we consider only neutral scalar fields. In order to consider charged scalar fields we should take $E \equiv M \times \mathbf{C}$.

$M \to \mathbf{R}$ locally defined on M and given by $\pi^\alpha = \sqrt{|g|}g^{\alpha\nu}(\partial x_\nu.f)$. In particular, with respect to a (physical) frame on M, we can distinguish a **time coordinate** x^0 in $\{x^\alpha\}$ and to consider the conjugate field $\pi \equiv \pi^0$ to $s : \pi = \sqrt{|g|}g^{o\nu}(\partial x_\nu.f)$. If $\{x^\alpha\}$ is an orthonormal system of coordinates, one has $\pi = (\partial x_0.f))$.

$\square$ The Jacobi operator associated to (KG) coincides with F (as (KG) is linear):

$$J[s].\nu \equiv (\square + \kappa).\nu = g^{\alpha\beta}\nu_{\alpha\beta} - \Gamma^\alpha_{\beta\lambda}g^{\lambda\beta}\nu_\alpha + \kappa\nu.$$

In the minimally coupled case, i.e.,$\kappa = m^2 \in \mathbf{R}_+$, where m is the mass of the field quanta, the corresponding propagator is given by the following formula:

$$\widetilde{G}(x) = \theta(x^0)G^{(+)}(x) - \theta(-x^0)G^{(-)}(x)$$

such that

$$G^{(+)}(x) + G^{(-)}(x) = i\Delta(x)$$
$$G^{(+)}(x) - G^{(-)}(x) = \Delta^{(1)}(x)$$

where

$$\Delta(x) = -\frac{1}{2\pi}\epsilon(x^0)[\delta(x^2) - \frac{mJ_1(m\sqrt{x^2})}{2\sqrt{x^2}}\theta(x^2)]$$
$$\Delta^{(1)}(x) = \frac{m}{4\pi\sqrt{x^2}}Y_1(m\sqrt{x^2})\theta(x^2) + \frac{m}{2\pi^2\sqrt{-x^2}}K_1(m\sqrt{-x^2})\theta(-x^2)$$

where J_n and Y_n are Bessel functions, K_n is a modified Bessel function, and $\epsilon(x^0) \equiv \frac{x^0}{|x^0|}$ is the signature of x^0.

4)(MAXWELL EQUATION). Let (M, g) be as before. Set $E \equiv T^*M$. Maxwell equation is the kernel $(M) \subset J\mathcal{D}^2(E)$ of the following rank constant epimorphism: $\mathcal{M} : J\mathcal{D}^2(E) \to E$, given by $\mathcal{M} \circ D^2A = \delta dA$. With respect to a fibered coordinates system $\{x^\alpha, \dot{x}_\alpha, \dot{x}_{\alpha\beta}, \dot{x}_{\alpha\beta\lambda}\}$ on $J\mathcal{D}^2(E)$ the local expression of (M) is characterized by the following functions:

$$\boxed{F_\alpha \equiv g^{\beta\gamma}(\dot{x}_{[\alpha\beta]\gamma} + \Gamma^\beta_{\beta\gamma}\dot{x}_{[\alpha\beta]} + \Gamma^\beta_{\beta\alpha}\dot{x}_{[\gamma\delta]})}$$

where $\Gamma^\beta_{\alpha\gamma}$: $M \to \mathbf{R}$ are the coefficients defined above. One can prove [104] that (M) is an involutive, formally integrable PDE, as well as completely integrable. (Furthermore, if g is Riemannian, (M) is elliptic.) Thus, we can apply the Cartan-Kähler theorem to state the existence and unicity of solutions C^ω for the Cauchy problem. Finally, (M) is a variational equation with Lagrangian:

$$L \equiv -\frac{1}{2}\sqrt{|g|}g^{\mu\alpha}g^{\nu\beta}y_{\mu\nu}y_{\alpha\beta} : J\mathcal{D}(E) \to \mathbf{R}.$$

In orthonormal coordinates the generalized momentum of A has the following expression: $\pi_\mu = (\partial x_0 . A_\mu)$.

$\square$ The Jacobi operator of (M) coincides with F_α as (M) is linear:

$$(J[A].\nu)_\alpha = g^{\beta\gamma}(\nu_{[\alpha\beta]\gamma} + \Gamma^{\epsilon}_{\beta\gamma}\nu_{[\alpha\beta]} + \Gamma^{\epsilon}_{\beta\alpha}\nu_{[\gamma\delta]}).$$

5)(DIRAC EQUATION). Let (M,g) be as before, and let $\eta = *1$ be the volume form canonically associated to g. Let E be the **complexificated Clifford bundle** on M, defined by the following:

$$E \equiv \bigcup_{p\in M} E_p, E_p \equiv \bigoplus_{s\geq 0} T^s_0(\mathbf{C} \bigotimes T_pM)/\mathbf{I}_p$$

where $\mathbf{I}_p$ is the subspace (two-sided ideal) of $\bigoplus_{s\geq 0} T^s_0(\mathbf{C} \bigotimes T_pM)$ generated by the elements of the type $v \otimes v - g^c(v,v)\mathbf{1}$, with $\mathbf{1}$ the unity of $\mathbf{K}$ and g^c the scalar product on $\mathbf{C} \bigotimes T_pM$ defined by

$$g^c(z_1 \otimes v_1, z_2 \otimes v_2) = z_1 z_2 \otimes g(v_1, v_2).$$

Thus, each fiber E_p becomes an algebra (**complexificated Clifford algebra**), namely it is possible to define a product (**Clifford product**) of two elements of E_p.[101,102] Furthermore, the Levi-Civita connection on M lifts one on E, **Levi-Civita connection on** E. Then, Dirac equation is the kernel $(D) \subset J\mathcal{D}(E)$ of a constant rank epimorphism: $\mathcal{D} : J\mathcal{D}(E) \to Clif^c(M), \mathcal{D}.\psi = \breve{\pi} \circ \breve{\nabla}\psi + m\psi$, where $\breve{\nabla}$ is the absolute differential on E associated to the Levi-Civita connection on (M,g) and $\breve{\pi}$ is the application $T^*M \bigotimes E \to E$ induced by the Clifford product. Furthermore, $m \in \mathbf{R}$ is the **mass**. The local expression of (D), with respect to local coordinates $\{x^\alpha, z^{i_1\cdots i_s}, z^{i_1\cdots i_s}{}_\alpha\} \equiv \{x^\alpha, z^A, z^A_\alpha\}$ on $J\mathcal{D}(E)$ is the following:

$$\boxed{F^A \equiv \gamma^A_{\alpha B} g^{\alpha\beta} z^B_\beta + [\gamma^A_{\beta C} g^{\alpha\beta} \Gamma^C_{\alpha B} + m\delta^A_B] z^B = 0}$$

where

$$\Gamma^C_{\alpha A} \equiv \Gamma^{j_1\cdots j_s}_{\alpha i_1\cdots i_s}$$

are the connection coefficients on E and

$$\gamma^A_{\beta C} \equiv \gamma^{l_1\cdots l_s}_{\beta j_1\cdots j_s}$$

are the components of $a(\partial x_\beta \otimes \partial x_{j_1} \cdots \partial x_{j_s})$, where a is the Clifford multiplication. One can prove [104] that (D) is an involutive, formally integrable PDE, as well as completely integrable. (Moreover, if g is Riemannian then (D) is an elliptic equation.) Thus, we can apply the Cartan-Kähler theorem to state the existence and unicity of C^ω solutions for the Cauchy problem.

☐ (D) is a variational PDE with Lagrangian:

$$L \equiv -(\psi, \gamma^\mu(\partial_\mu.\psi) + m\psi).$$

In orthonormal coordinates the generalized momentum of ψ becomes:

$$\pi_B = (\partial x_0.\psi_B) = -\psi_A \gamma^{oA}_B.$$

☐ The Jacobi operator of (D) coincides with F^A as (D) is linear:

$$J[s].\nu^A \equiv \gamma^A_{\alpha B} g^{\alpha\beta} \nu^B_\beta + [\gamma^A_{\beta C} g^{\alpha\beta} \Gamma^C_{\alpha B} + m\delta^A_B]\nu^B.$$

6)(EINSTEIN EQUATION). Let (M, g) be as before and $E \subset S^0_2 M$ an open sub-bundle of non-degenerate metrics $(\det(g_{\alpha\beta})) \neq 0)$ on M. The Levi-Civita connection has the corresponding **curvature** given by the following:

$$\begin{aligned} R^{\cdot}_{\cdots}(g) &= R^\omega_{\alpha\lambda\beta} dx^\alpha \wedge dx^\lambda \otimes dx^\beta \otimes \partial x_\omega \\ R^\omega_{\alpha\lambda\beta} &= (\partial x_\alpha \Gamma^\omega_{\lambda\beta}) - (\partial x_\lambda \Gamma^\omega_{\alpha\beta}) \\ &+ \Gamma^\omega_{\alpha\gamma}\Gamma^\gamma_{\lambda\beta} - \Gamma^\omega_{\lambda\gamma}\Gamma^\gamma_{\alpha\beta} \end{aligned}$$

where the connection coefficients are given in (4.84). $R^\omega_{\alpha\lambda\beta}$ is called **Riemann Christoffel tensor**.

REMARK 4.36 - (PROPERTIES OF RIEMANN-CHRISTOFFEL TENSOR AND ASSOCIATED ENTITIES).

(A) For any vector field A on M we have:

$$A_{\alpha/\beta\gamma} - A_{\alpha/\gamma\beta} = R^\omega_{\alpha\beta\gamma} A_\omega.$$

One has the following properties:

$$\boxed{\begin{aligned} R^\omega_{\alpha\beta\gamma} &= -R^\omega_{\alpha\gamma\beta} \\ R^\omega_{\alpha\beta\gamma} + R^\omega_{\beta\gamma\alpha} + R^\omega_{\gamma\alpha\beta} &= 0 \\ R^\omega_{\omega\alpha\beta} &= 0. \end{aligned}}$$

(B) To the Riemann-Christoffel tensor one associates the **curvature tensor**:

$$R_{\alpha\beta\gamma\delta} \equiv g_{\omega\alpha}R^{\omega}_{\beta\gamma\delta}$$
$$= \frac{1}{2}[(\partial x_{\beta}\partial x_{\gamma}.g_{\alpha\delta}) + (\partial x_{\alpha}\partial x_{\delta}.g_{\beta\gamma}) - (\partial x_{\beta}\partial x_{\delta}.g_{\alpha\gamma})^2 - (\partial x_{\alpha}\partial x_{\gamma}.g_{\beta\delta})]$$
$$+ g^{\omega\epsilon}([\beta\gamma,\epsilon][\alpha\delta,\omega] - [\beta\delta,\epsilon][\alpha\gamma,\omega]).$$

One has the following properties

$$R_{\alpha\beta\gamma\delta} = -R_{\beta\alpha\gamma\delta}$$
$$R_{\alpha\beta\gamma\delta} = -R_{\alpha\beta\delta\gamma}$$
$$R_{\alpha\beta\gamma\delta} = R_{\gamma\delta\alpha\beta}$$
$$R_{\alpha\beta\gamma\delta} + R_{\alpha\delta\beta\gamma} + R_{\alpha\gamma\beta\delta} = 0$$

The number of independent components of $R_{\alpha\beta\gamma\delta}$ are $\frac{1}{12}n^2(n^2-1)$.
(C) To the curvature $R^{\cdot}_{\cdot\cdot\cdot}(g)$ one associates the **Ricci tensor**

$$R_{..}(g) = R^{\lambda}_{\alpha\lambda\beta}dx^{\alpha}\otimes dx^{\beta} \equiv R_{\alpha\beta}dx^{\alpha}\otimes dx^{\beta} = R_{\beta\alpha}dx^{\alpha}\otimes dx^{\beta}$$

and the **Ricci scalar curvature**

$$R(g) = R^{\lambda}_{\alpha\lambda\beta}g^{\alpha\beta}$$

In the following table we report some important relations for $R_{\alpha\beta}$.

TAB.4.24 - **Important relations for $R_{\alpha\beta}$**

	$R_{\alpha\beta} \equiv R^{\lambda}_{\alpha\lambda\beta} = -(\partial x_{\beta}\cdot\Gamma^{\lambda}_{\lambda\alpha}) + (\partial x_{\lambda}\cdot\Gamma^{\lambda}_{\beta\alpha}) + \Gamma^{\lambda}_{\lambda\omega}\Gamma^{\omega}_{\beta\alpha} - \Gamma^{\omega}_{\lambda\beta}\Gamma^{\lambda}_{\omega\alpha}$
	$R^{\omega}_{\alpha\beta\gamma/\delta} + R^{\omega}_{\alpha\delta\gamma/\beta} + R^{\omega}_{\alpha\delta\beta/\gamma} = 0$
Bianchi identities	$R_{\alpha\beta\gamma\delta/\omega} + R_{\beta\alpha\delta\omega/\gamma} + R_{\alpha\beta\gamma\delta} = -R_{\alpha\beta\delta\gamma}$ $R_{\alpha\beta\gamma\delta} = R_{\gamma\delta\alpha\beta}$ $R_{\alpha\beta\gamma\delta} + R_{\alpha\delta\beta\gamma} + R_{\alpha\gamma\beta\delta} = 0$

(D) The **invariant Riemann curvature** is the following scalar:

$$K = \frac{R_{\alpha\beta\gamma\delta}A^{\alpha}A^{\gamma}B^{\beta}A^{\delta}}{(g_{\alpha\gamma}g_{\beta\delta} - g_{\alpha\delta}g_{\beta\gamma})A^{\alpha}A^{\gamma}B^{\beta}B^{\delta}}$$

where A and B are any two vectors of the tangent space in a point $p \in M$. If $K = 0$ we say that the manifold is **flat**. We have the following properties:

☐ M is flat iff $R_{\alpha\beta\gamma\delta} = 0$.

☐ In a flat space the parallelism condition is independent of the curve.

☐ (**Shur theorem**). If $\dim M > 2$, and $K = K(x^{\alpha}) \Rightarrow K = constant$, (**space with constant curvature**).

Einstein equation is the kernel $(Ein) \equiv \ker \mathcal{E} \subset J\mathcal{D}^2(E)$, where $\mathcal{E}$ is the following differential operator $\mathcal{E} : J\mathcal{D}^2(E) \to S^0_2 M, \mathcal{E} \circ D^2 g = R..(g) - R(g)g - \chi T(g)$, where $R..(g)$ is the Ricci tensor of g and $R(g)$ is the Ricci scalar curvature of g; $\chi \in \mathbf{R}, T = T(g)$ is a fixed section of $S^0_2 M \to M$, for any $g \in C^{\infty}(E)$ such that $\operatorname{div} T = 0$. The local expression of (Ein) is as follows:

$$\begin{aligned} G_{\alpha\beta} &\equiv R_{\alpha\beta} - \frac{1}{2}R_{\gamma\delta}g^{\gamma\delta}g_{\alpha\beta} = \chi T_{\alpha\beta} \\ R_{\alpha\beta} &\equiv R^{\lambda}_{\alpha\lambda\beta} = -(\partial x_{\beta} \cdot \Gamma^{\lambda}_{\lambda\alpha}) + (\partial x_{\lambda} \cdot \Gamma^{\lambda}_{\beta\alpha}) + \Gamma^{\lambda}_{\lambda\omega}\Gamma^{\omega}_{\beta\alpha} - \Gamma^{\omega}_{\lambda\beta}\Gamma^{\lambda}_{\omega\alpha} \\ \Gamma^{\lambda}_{\nu\omega} &= \frac{1}{2}g^{\lambda\alpha}[(\partial x_{v} \cdot g_{\omega\alpha}) + (\partial x_{\omega} \cdot g_{\alpha v}) - (\partial x_{\alpha} \cdot g_{v_{\omega}})], \end{aligned}$$

where $T = T_{\alpha\beta}dx^{\alpha} \otimes dx^{\beta}$. We can prove that (Ein) is an involutive, formally integrable PDE. Then, we can use Cartan-Kähler theorem to state the existence and unicity of $C^{\omega}(E)$ solutions for the Cauchy problem.

☐ The Jacobi operator of Einstein equation has the following local components:

$$\begin{aligned} (J[g].\nu)_{\alpha\beta} = {} & (\partial y^{\omega\delta}.(G[g]_{\alpha\beta} - \chi T_{\alpha\beta}))\nu_{\omega\delta} + (\partial y^{\omega\delta\epsilon}.(G[g]_{\alpha\beta} - \chi T_{\alpha\beta}))\nu_{\omega\delta\epsilon} \\ & + (\partial y^{\omega\delta\epsilon\kappa}.(G[g]_{\alpha\beta} - \chi T_{\alpha\beta}))_{\omega\delta\epsilon\kappa} \end{aligned}$$

7)(YANG-MILLS EQUATION). Let M be as before. Let G be a semi-simple Lie group and denote by $A(G)$ its Lie algebra. Let $(P, M, \pi; G)$ a principal fiber bundle on M with structure group G. A **principal connection** on P, is a connection on P,

$\Gamma : TP \to vTP$, that is invariant by the natural action, $\phi : G \times P \to P$, of G on P, namely the following diagram is commutative for any $a \in G$:

$$\begin{array}{ccccc} & TP & \xrightarrow{\Gamma} & vTP & \\ T(\phi_a) & \downarrow & & \downarrow & vT(\phi_a) \\ & TP & \xrightarrow[\Gamma]{} & vTP & \end{array}$$

A principal connection on P can be identified with a section A of an affine fiber bundle $C(P)$ on M that is an open subbundle of the vector bundle $T^*M \bigotimes E(P) \to M$ where $E(P)$ is the vector bundle of G-invariant vector fields on P. The vector bundle associated to $C(P)$ is $T^*M \bigotimes A(P)$, with $A(P) \equiv P \times A(G)/G$, where G acts on $A(G)$ by means of the adjoint representation

$$ad : G \to L(A(G)), \quad ad(a)(v) = D(j_a)(e)(v)$$

where $j_a : G \to G, j_a(b) = a^{-1}ba$. To any principal connection A we can associate its **curvature**

$$R(A) \equiv dA + \frac{1}{2}[A, A] : M \to \Lambda_2^0 M \bigotimes A(P)$$

where the bracket $[A, A]$ is defined by the following:

$$[A, A] \equiv [A^i \otimes Z_i, A^j \otimes Z_j] = 2A^i \wedge A^j \otimes [Z_i, Z_j] = 2A^i \wedge A^j \otimes C_{ij}^k Z_k$$

where $\{Z_i\}$ is a basis for $A(G)$ and C_{ij}^k are the structure constants of G.
The operation that has allowed to associate to A its curvature, can be generalized to any $A(P)$-valued differential r-form. Let us denote such an operation by ${}_Ad$ and call it **absolute differential** associated to the connection A. So if $\alpha : M \to \Lambda_r^o M \bigotimes A(P)$, we get

$${}_Ad\alpha : M \to \Lambda_{2r}^0 M \bigotimes A(P).$$

Set $E \equiv C(P)$. Yang-Mills equation is a PDE of second order on E given as the kernel (YM) of the following constant rank epimorphism:

$$\mathcal{IM} : J\mathcal{D}^2(E) \to T^*M \bigotimes A(P), \quad \mathcal{IM} \circ D^2 A = {}_AdR(A)$$

where ${}_AdA \equiv R(A)$ is the curvature of A, and ${}_Ad$ is the absolute differential induced by the connection A. Let $\{x^\alpha, y_\alpha^H, y_{\alpha\beta}^H, y_{\alpha\beta\gamma}^H\}$ be a fibered coordinate system on $J\mathcal{D}^2(E)$. Note that $0 \le \alpha, \beta, \gamma \le n-1, 1 \le H \le \dim A(G) = m$. Then, the local expression of (YM) is the following:

$$\begin{array}{|l|} \hline F^{\alpha\beta H} \equiv g^{\alpha\delta} g^{\beta\gamma} F_{\delta\gamma}^H = g^{\alpha\delta} g^{\beta\gamma}[(\partial x_\delta . A_\gamma^H) - (\partial x_\gamma . A_\delta^H) + C_{BC}^H A_\delta^B A_\gamma^C] \\ \mathbf{Y}_\omega^H \equiv g_{\alpha\omega}[(\partial x_\beta . F^{\alpha\beta H}) + C_{BC}^H A_\beta^C F^{\alpha\beta B}] = 0 \\ C_{BC}^H \equiv \text{ structure constants of } G \\ \hline \end{array}$$

We can prove [104] that (YM) is an involutive, formally integrable PDE, as well as completely integrable on the fiber bundle $\pi : E \equiv C(P) \to M$. Thus, we can use the Cartan-Kähler theorem to state the existence and unicity of C^ω solutions of (YM) for the Cauchy problem.

$\square$ The Jacobi operator of (YM) has the following local components:

$$(J[A]_{x'}\nu)^H_\omega = (\partial y^\alpha_\kappa.\mathbf{Y}^H_\omega)\nu^\kappa_\alpha + (\partial y^{\alpha\gamma}_\kappa.\mathbf{Y}^H_\omega)\nu^\kappa_{\alpha\gamma} + (\partial y^{\alpha\gamma\delta}_\kappa.\mathbf{Y}^H_\omega)\nu^\kappa_{\alpha\gamma\delta}.$$

8)(EINSTEIN-YANG-MILLS EQUATION). Let us consider a gravitational field coupled with gauge one. Let $(P, M, \pi; G)$ be a principal fiber bundle on M as before. Set $E \equiv E_0 \times C(P)$ where E_0 is the fiber bundle of the non-degenerate metrics on M, and $C(P)$ is the fiber bundle of the principal connections on E. Einstein-Yang-Mills equation is PDE $(EYM) \subset J\mathcal{D}^2(E)$ have the following local expression:

$$\begin{array}{l}
F^{\alpha\beta H} \equiv g^{\alpha\delta}g^{\beta\gamma}F^H_{\delta\gamma} = g^{\alpha\delta}g^{\beta\gamma}[(\partial x_\delta.A^H_\gamma) - (\partial x_\gamma.A^H_\delta) + C^H_{BC}A^B_\delta A^C_\gamma] \\
\mathbf{Y}^H_\omega \equiv g_{\alpha\omega}[(\partial x_\beta.F^{\alpha\beta H}) + C^H_{BC}A^C_\beta F^{\alpha\beta B}] - J^H_\omega = 0 \\
(C^H_{BC} \equiv \text{structure constants of } G) \\
\mathbf{F}^H_{\lambda\alpha\beta} \equiv (\partial x_\lambda.F^H_{\alpha\beta}) + C^H_{AB}A^A_\lambda F^B_{\alpha\beta} = 0 \\
\mathbf{G}_{\alpha\beta} \equiv R_{\alpha\beta} - \frac{1}{2}R_{\gamma\delta}g^{\gamma\delta}g_{\alpha\beta} - \chi T_{\alpha\beta} = 0 \\
R_{\alpha\beta} \equiv R^\lambda_{\alpha\lambda\beta} = -(\partial x_\beta \cdot \Gamma^\lambda_{\lambda\alpha}) + (\partial x_\lambda \cdot \Gamma^\lambda_{\beta\alpha}) + \Gamma^\lambda_{\lambda\omega}\Gamma^\omega_{\beta\alpha} - \Gamma^\omega_{\lambda\beta}\Gamma^\lambda_{\omega\alpha} \\
\Gamma^\lambda_{\nu\omega} = \frac{1}{2}g^{\lambda\alpha}[(\partial x_v \cdot g_{\omega\alpha}) + (\partial x_\omega \cdot g_{\alpha v}) - (\partial x_\alpha \cdot g_{v\omega})] \\
\mathbf{T}_\beta \equiv \nabla_\beta T^\alpha_\beta = 0, \mathbf{J}^H \equiv \nabla_\alpha J^{\alpha H} = 0.
\end{array}$$

Equation (EYM) is analytic, involutive, formally integrable and completely integrable. Then, Cauchy problem admits a unique solution of class C^ω (Cartan-Kähler theorem).

9)(INCOMPRESSIBLE FLUIDS). The dynamic equation for continuum system, on the Galilean space-time, can be considered as the kernel of the following differential operator of order k:

$$\mathcal{D} : J\mathcal{D}^k(W) \to \mathbf{K} \equiv T^0_0 M \times_M vTM \times_M T^0_0 M$$

where (a) $W \equiv J\mathcal{D}(\mathcal{G}) \times_M \Pi \times_M \Theta$, $\mathcal{G} \equiv \{\tau : M \to T\}$ is the affine bundle that represents the Galilean space-time M on the space of time T, $\Pi \equiv M \times \mathbf{R}$ is the pressure space, $\Theta \equiv M \times \mathbf{R}$ is the temperature space;

(b)

$$\mathcal{D}.s \equiv \mathcal{D} \circ D^k s = \mathrm{div}(\mathcal{F}.s) - \mathcal{J}.s$$

with

$$\mathcal{F}.s \equiv (\mathcal{M}.s, \mathcal{T}.s, \mathcal{E}.s) : M \to J\mathcal{D}(\mathcal{G}) \times_M S^2_0 M \times_M TM$$

the total flux at the configuration s ($\mathcal{M}.s \equiv$ mass flux, $\mathcal{T}.s \equiv$ momentum flux, $\mathcal{E}.s \equiv$ energy flux), and

$$\mathcal{J}.s \equiv (0, \mathcal{B}.s, \mathcal{W}.s) : M \to \mathbf{K}$$

total current at the configuration s,

$$(0 \equiv \text{ mass current}, \mathcal{B}.s \equiv \text{body forces current}, \mathcal{W}.s \equiv \text{power forces current}).$$

A configuration s is given by $s \equiv (v, p, \theta)$, where v is the velocity field, p is the pressure and θ is the temperature.[99,109]

(A) EULER EQUATION. The geometric structure of Euler equation (E) is given in [99,100,109]. Recall some results. (E) is the kernel of a constant rank differential operator

$$\begin{gathered}
\mathcal{E} \equiv (\mathcal{E}^1, \mathcal{E}^2) : J\mathcal{D}(W) \to \mathbf{K} \equiv T^0_0 M \times_M vTM \\
\mathcal{E}^1.s = \mathcal{E}^1.(v,p) = \mathrm{div}(\rho v), \mathcal{E}^2.s = \mathcal{E}^2.(v,p) = \mathrm{div}(\rho v \otimes v + pg) - \rho B
\end{gathered}$$

where $s = (v,p), \rho =$ mass density $(=$ constant$)$, $B =$ body forces density, $g =$ metric on M. By means of a coordinate system: $\{x^\alpha\} = \{x^0, x^1, x^2, x^3\}$ adapted to an inertial frame ψ, we can write (E) in the following form:

$$\begin{aligned}
\Gamma^j_{jk} v^k + (\partial x_i.v^i) = 0 \\
\rho[\Gamma^j_{ik} v^i v^k + (\partial x_0.v^j) + v^i(\partial x_i.v^j)] + (\partial x_i.p)g^{ij} - \rho B^j = 0
\end{aligned}$$

where Γ^j_{ik} are numerical functions on M that represent the spatial components of the canonical connection on M. Then, we can prove that (E) is formally integrable. Furthermore, if the forces are conservatives with respect to ψ, one has the existence of analytic solutions for any initial condition, i.e., any point $q \in (E)_{+h}$. Then, we can use the Cartan-Kähler theorem to state the existence and unicity of analytic solutions for the Cauchy problem.

□ The Jacobi operator $J[s]$ of (E) has the following components:

$$\begin{array}{l}
\Gamma^j_{jk}\nu^k + (\partial x_i.\nu^i) = J[s]^\bullet \\
\rho[2\Gamma^j_{ik}v^k + (\partial x_i.v^j) - (\partial v_i.(\rho B^j))]\nu^i + [\rho\delta^j_k - (\partial v^o_k.(\rho B^j))](\partial x_0.\nu^k) \\
+ [\rho v^i\delta^j_k - (\partial v^i_k.(\rho B^j))](\partial x_i.\nu^k) + (\partial x_i.\mu)g^{ij} = J^j[s]
\end{array}$$

where $((\nu^j,\mu) \Leftrightarrow (v^j,p), 1 \leq j \leq 3)$.

(B) NAVIER-STOKES EQUATION. Let us consider the non-isothermal Navier-Stokes equation (NS) [99,100,109]. (NS) results in a subbundle of $J\mathcal{D}^2(W)$, where

$$W \equiv J\mathcal{D}(W) \times_M \Pi \times_M \Theta.$$

Let us assume that the internal energy depends on the temperature θ only, and that $C_p \equiv (\partial\theta.e) = constant$. Assume also that the thermal conductivity ϕ and viscosity χ are constants. Then, we can write (NS), in coordinates adapted to an inertial frame in the following way: $(NS) = \{A, B, C\}$, where:

$$\begin{array}{l}
(A) \quad \Gamma^j_{jk}v^k + (\partial x_i.v^i) = 0 \\
(B) \quad \rho[\Gamma^j_{ik}v^k v^i + (\partial x_0.v^j) + v^i(\partial x_i.v^j)] + (\partial x_i.p)g^{ij} \\
- 2\chi\{\Gamma^i_{ik}v^s(\partial x_s.g^{kj}) + \Gamma^j_{ik}v^s(\partial x_s.g^{ki}) + v^s(\partial x_i.\partial x_s.g^{ij}) \\
- \Gamma^i_{ik}g^{js}(\partial x_s.v^k) - \Gamma^i_{ik}g^{ks}(\partial x_s.v^j) \\
- 2\Gamma^j_{ik}g^{is}(\partial x_s.v^k)] - (\partial x_i.g^{is})(\partial x_s.v^j) - g^{js}(\partial x_i.\partial x_s.v^i) \\
- (\partial x_i.\partial x_s.v^j)g^{is}\} - \rho B^j = 0 \\
(C) \quad \rho C_p[(\partial x_0.\theta) + v^k(\partial x_k.\theta)] \\
- \phi[\Gamma^j_{ji}(\partial x_s.\theta)g^{is} + (\partial x_s.\theta)(\partial x_i.g^{is}) + (\partial x_s\partial x_i.\theta)g^{is}] \\
- 2\chi[\Gamma^j_{ik}v^k v^p g_{js}(\partial x_p.g^{is}) - \Gamma^j_{ik}v^k(\partial x_p.v^s)g_{js}g^{ip} \\
- \Gamma^j_{ik}v^k(\partial x_j.v^i) + v^p(\partial x_i.v^k)(\partial x_p g^{is})g_{ks}] \\
- (\partial x_i.v^k)(\partial x_p.v^s)g_{ks}g^{ip} - (\partial x_i.v^k)(\partial x_k.v^i)] = 0
\end{array}$$

One can prove that (NS) is involutive but not formally integrable. In fact, $(NS)_{+1} \to (NS)$ is not surjective. Thus, C^∞ solutions s for any initial condition $q \in (NS)_{+h}$ do not exist. However, we can prove [109] that there are integers $r \geq 0, s \geq 0$, such that

$$(NS)^{(s)}_{+r} \equiv \pi_{2+r+s,2+r}((NS)_{+(r+s)}) \subset J\mathcal{D}^{2+r}(W)$$

is an involutive, formally integrable PDE with the same solutions of (NS) such that

$$((NS)^{(s)}_{+r})_{+t} = (NS)^{(s)}_{+(r+t)}, \forall t \geq 0.$$

Thus, in the following we can consider as Navier-Stokes equation just

$$\overline{(NS)} \equiv (NS)^{(s)}_{+r} \subset J\mathcal{D}^{2+r}(W).$$

To such an equation we can apply the Cartan-Kähler theorem and state the existence and unicity of analytic solutions for the Cauchy problem.

$\square$ The Jacobi operator $J[s]$ of (NS) has the following components:

$$
\begin{aligned}
&(a)\quad \Gamma^j_{jk}\nu^k + (\partial x_i.\nu^i) = J[s]^\bullet\\
&(b)\quad \nu^s\{2\rho\Gamma^i_{sk}v^k + (\partial x_s.v^j) - 2\chi[\Gamma^i_{ik}(\partial x_s.g^{kj}) + \Gamma^j_{ik}(\partial x_s.g^{ki}) - (\partial v_s.\rho B^j)\\
&+ (\partial x_i.\partial x_s.g^{ij})]\} + (\partial x_0.\nu^i)\{\rho\delta^j_i\\
&- (\partial v_0.\rho B^j)\} + (\partial x_s.\nu^k)\{v^s\delta^j_k - 2\chi[-\Gamma^i_{ik}g^{j^s} - \Gamma^i_{ip}g^{p^s}\delta^j_k\\
&- 2\Gamma^i_{ik}g^{i^s} - (\partial x_i.g^{i^s})\delta^j_k] - (\partial v^s_k.\rho B^j)\} + (\partial x_i.\mu)g^{ij}\\
&+ (\partial x_i.\partial x_s.v^k)\{2\chi[g^{j^s}\delta^i_k - g^{i^s}\delta^j_k] = J[s]^j\\
&(c)\quad \nu^s\{\rho C_p(\partial x_s.\theta) - 2\chi[[\Gamma^j_{ik}(\partial x_s.g^{pi})v^k g_{jp} - \Gamma^j_{is}(\partial x_p.v^k)g_{kq}g^{qp}\\
&- \Gamma^j_{is}(\partial x_j.v^i) + (\partial x_i.v^k)(\partial x_s.g^{pi})g_{kp}]\}\\
&+ (\partial x_s.\nu^k)\{-2\chi[-\Gamma^j_{iq}v^q g_{kj}g^{i^s} - 2(\partial x_p.v^i)g_{ki}g^{sp}\\
&- 2(\partial x_k.v^s) - \Gamma^s_{ik}v^i + v^p(\partial x_p.g^{i^s})\\
&g_{ki}]\} - \Gamma^s_{ik}v^i + v^p(\partial x_p.g^{i^s})g_{ki}]\} + (\partial x_0.\nu^i)\rho C_p + (\partial x_k.\epsilon)\{2\rho C_p v^k\\
&- \phi[\Gamma^j_{ji}g^{ki} + (\partial x_i.g^{i^k})]\} - (\partial x_i.\partial x_s.\epsilon)\phi g^{i^s} = J[s]^\blacksquare
\end{aligned}
$$

with $((\nu^j, \mu, \epsilon) \Leftrightarrow (v^j, p, \theta), 1 \leq j \leq 3)$.

5 - QUANTUM FIELD THEORY

Some fundamental structures in quantum field theory (e.g., quantum situs), require expanding the point of view adopted until now, that is centered on finite-dimensional manifolds, and to consider also infinite dimensional manifolds. So scope of this section is to study the differential geometry of locally convex manifolds. This can be conducted, for many respects, similar to that for finite dimensional manifolds, but, of course, there are some important differences, that must be emphasized.

REMARK 5.1 - (SPACE OF MULTILINEAR MAPPINGS). Let E and F be locally convex topological vector spaces (locally convex spaces). We will denote by $L^m(E;F), m \geq 0$, the space of m-linear mappings $E^m \to F$. We set $L^0(E;F) \equiv F, L^1(E;F) \equiv L(E;F)$. We will denote by $\mathcal{L}^m(E;F)$ the subspace of $L^m(E;F)$ of all continuous m-linear mappings $E^m \to F$. Set

$$\mathcal{L}^0(E;F) \equiv F, \quad \mathcal{L}^1(E;F) \equiv \mathcal{L}(E;F).$$

One has the canonical isomorphism

$$\theta^{m,n} : L^m(E; L^n(E;F)) \cong L^{m+n}(E; L^m(E;F)).$$

Then, $L^m(E;F)$, and as a consequence $\mathcal{L}^m(E;F)$, has a natural structure of locally convex space with respect to the topology of pointwise convergence (that coincides with that induced by the product topology on $F^{(E^n)}$; (see also refs.[56,59,126,144]). With respect to each family $\mathcal{G}$ of bounded subsets of E such that their union is dense in E, we can recognize on $\mathcal{L}^m(E;F)$ a topology, $\Lambda_{\mathcal{G}}$: that of uniform convergence. If $\mathcal{G} \subset \mathcal{G}'$ it follows that $\Lambda_{\mathcal{G}}$ is coarser than $\Lambda_{\mathcal{G}'}$. Furthermore, the locally convex space

$$\mathcal{L}^{m+n}(E;F)_{\mathcal{G}} \equiv (\mathcal{L}^{m+n}(E;F); \Lambda_{\mathcal{G}})$$

is a subspace of the locally convex space $\mathcal{L}^m(E; \mathcal{L}^n(E;F)_{\mathcal{G}})_{\mathcal{G}}$. Moreover, we have that the canonical isomorphism $\theta^{m,n}$ induces a canonical isomorphism

$$\mathcal{L}^m(E; \mathcal{L}^n(E;F)_{\mathcal{G}})_{\mathcal{G}} \cong \mathcal{L}^{m+n}(E;F)_{\mathcal{G}}$$

in the following two cases: (a) E is a metrizable locally convex space and F is an arbitrary locally convex space and $\mathcal{G}$ is a collection of bounded subsets of E which contains all the compact sets; (b) E is a metrizable and barrelled locally convex space, F is an arbitrary locally convex space and $\mathcal{G}$ any cover of E by bounded sets.

EXAMPLE 5.1 - 1)(CONTINUOUS CONVERGENCE). Let E and F be locally convex spaces, then the topology Λ_c on $\mathcal{L}^m(E;F)$, called **continuous convergence**,

is the coarsest topology which makes the evaluation map $e : E^m \times \mathcal{L}^m(E;F) \to F$ continuous. So Λ_c is the topology induced by the compact open topology on $C^0(E^m;F)$. Furthermore, the canonical linear isomorphism $\theta^{m,n}$ induces an isomorphism $\mathcal{L}^m(E;\mathcal{L}^n(E;F)_c)_c \cong \mathcal{L}^{m+n}(E;F)_c$ of locally convex space.
2)(STRONG TOPOLOGY). Let E and F be locally convex spaces. The **strong topology** Λ_s is that induced by $\mathcal{G}$, where $\mathcal{G}$ is the set of subsets of E that are bounded, convex and balanced.
□ Let $\mathcal{H}^m(E;F)_{\mathcal{G}}, n \in \mathbf{N}$, be locally convex space defined inductively as follows:

$$\mathcal{H}^0(E;F)_{\mathcal{G}} \equiv F; \mathcal{H}^m(E;F)_{\mathcal{G}} \equiv \mathcal{L}(E;\mathcal{H}^{m-1}(E;F)_{\mathcal{G}})_{\mathcal{G}}, \quad \text{if} \quad m \geq 1.$$

One has the following
1)

$$\mathcal{H}^m(E;\mathcal{H}^n(E;F)_{\mathcal{G}})_{\mathcal{G}} \cong \mathcal{H}^{m+n}(E;F)_{\mathcal{G}};$$

2) There exists a canonical injective map

$$\psi^m : \mathcal{H}^m(E;F)_{\mathcal{G}} \to L^m(E;F)$$

defined inductively:

$$\psi^0 = id_F, \psi^m(u) = \theta^{1,m-1}(\psi^{m-1} \circ u), \quad \forall u \in \mathcal{H}^m(E;F)_{\mathcal{G}}.$$

So the elements of $\mathcal{H}^m(E;F)_{\mathcal{G}}$ are considered m-linear mappings that are called $\mathcal{G}$-**hypocontinuous**;
3) The locally convex space $\mathcal{L}^m(E;F)_{\mathcal{G}}$ is a subspace of the locally convex space $\mathcal{H}^m(E;F)_{\mathcal{G}}$.
4) If $\mathcal{G} \subset \mathcal{G}'$, then $\mathcal{H}^m(E;F)_{\mathcal{G}'}$ is a linear subspace of $\mathcal{H}^m(E;F)_{\mathcal{G}}$, and the inclusion is continuous.
5) Let $\mathcal{G} \equiv \mathcal{G}_p, \mathcal{G}_k, \mathcal{G}_{\bar{k}}, \mathcal{G}_b$, be the collection of all respectively finite, compact, precompact, bounded subsets of E. The corresponding topologies $\Lambda_{\mathcal{G}}$ on $\mathcal{L}^m(E;F)$ are finer than the preceding. Λ_p is the coarsest and Λ_b is the finest of all $\mathcal{G}$-topologies on $\mathcal{L}^m(E;F)$. Then one has:

$$\mathcal{H}^m(E;F)_b \subset \mathcal{H}^m(E;F)_{\bar{k}} \subset \mathcal{H}^m(E;F)_k \subset \mathcal{H}^m(E;F)_b$$

and the inclusion map are continuous. The space $\mathcal{H}^m(E;F)_{\mathcal{G}}$ consists just of those m-linear maps from E^m into F which are **separately continuous** (its topology is induced from $F^{(E^m)}$).
6)

$$\mathcal{H}^m(E;F)_{\mathcal{G}} \cong \mathcal{L}^m(E;F)_{\mathcal{G}}$$

if one of the following conditions is satisfied:
(a) E is metrizable locally convex space, and $\mathcal{G}$ is a collection of bounded subsets of E which contains all compact sets;
(b) E is metrizable and baralled locally convex space, and $\mathcal{G}$ is a cover of E by bounded sets.
REMARK 5.2 - (DIFFERENTIABLE MAPPINGS). Let $f : X \subset E \to F$ be a map between locally convex spaces E and F, where X is an open set in E. The map f is called **weakly differentiable** (or of class C^1_w) at a point $x \in E$, if in some neighbourhood U of x, we have the following representation of f:

$$f(x+th) = f(x) + Df(x)(h) + Rf_x(th), \forall (x,h) \in U \times E, t \in \mathbf{R}, x+th \in U;$$

where

$$\left\{ \begin{array}{ll} (a) & Df(x) \in L(E;F), (Df(x) \equiv \textbf{weak derivative of } f \textbf{ at } x) \\ (b) & \lim_{t\to 0} \frac{Rf_x(th)}{t} = 0 \in F \\ & \text{where the limit is with respect to the topology } \Lambda_p \text{ on } L(E;F) \end{array} \right\}$$

If $Df(x) \in \mathcal{L}(E;F)$ we say that $Df(x)$ is the **Gateaux-Levy derivative** of f at x. (In this case we say that f is of class C^1_{GL} at x.)
A function $f : X \subset E \to F$ is of **class** C^1_Λ, with respect to a topology Λ on $\mathcal{L}(E;F)$, if the corresponding Gateaux-Levy derivative $Df : X \to \mathcal{L}^m(E;F)_\Lambda$ is continuous.
EXAMPLE 5.2 - 1) Let E and F be normed spaces. A map $f : X \subset E \to F$ of class C^1_Λ, where Λ is the norm-topology on $\mathcal{L}(E;F)$, is called **Fréchet differentiable**, i.e.

$$\lim_{h\to 0} \frac{Rf_x(h)}{||h||} = 0.$$

2) Let us denote by C^1_c and C^1_p the classes C^1_Λ when $\Lambda = \Lambda_c$ and $\Lambda = \Lambda_p$ respectively. If E is a Banach space or a (F)-space one has $C^1_c = C^1_p$. If E is a finite dimensional space, then all the classes C^1_Λ coincide and we simply write C^1.
□ A function $f : X \subset E \to F$ is called **weakly k-times differentiale** (or of class C^k_w) if there exist functions $D^q f : X \to L^q(E;F), q = 0, 1, \cdots, k$, such that $D^0 f = f$, and for each $x \in X$, each $h \in E$, and $q = 0, 1, \cdots, k-1$, we have

$$\lim_{t\to 0} \frac{D^q f(x+th) - D^q f(x)}{t} = D^{q+1} f(x)(h^{(q+1)})$$

where $h^{(q+1)} \equiv (h, \cdots, h) \in E^{q+1}$, the limit exists with respect to the topology Λ_p of pointwise convergence in $L^q(E;F)$. If $f : X \subset E \to F$ is weakly k-times differentiable, one has the following:

1) For every $q \leq k$, f is weakly q-times differentiable and $D^q f : X \to L^q(E;F)_p$ is weakly $(k-q)$-times differentiable, furthermore, for every $s \leq k-q$ we have $D^s(D^q f) = D^{s+q} f$.
2) If $D^k f : X \to L^k(E;F)_p$ is weakly s-times differentiable, then f is weakly $(k+s)$-times differentiable and $D^s(D^k f) = D^{k+s} f$.
3) If $D^k f : X \to L^k(E;F)_p$ is continuous, then $D^r f : X \to L^r_s(E;F), 0 \leq r \leq k$, where $L^r_s(E;F)$ is the **space of symmetric r-linear mappings $E \to F$**.
4) Let $D^k f : X \to L^k(E;F)_p$ be continuous. Let $(x,h) \in X \times E$ such that $[x, x+h] \subset X$. Then, we have the following **Taylor's expansion**:

$$\left\{ \begin{array}{l} f(x+h) = \sum_0^k \frac{1}{s!} D^s f(x)(h^{(s)}) + R_k f(x,h) \\ R_k f(x,h) = \frac{1}{(k-1)!} \int_0^1 (1-t)^k (D^k f(x+th) - D^k f(x))(h^{(k)}) dt \end{array} \right\}$$

Furthermore, one has: (a) the remainder $R_k f(x,h)$ satisfies

$$\lim_{t \to 0} \frac{R_k f(x,th)}{t^k} = 0, \forall\, h \in E;$$

(b) For every $s \leq k-1$, and for every choice of $(h_1, \cdots, h_s) \in E^s$ one has:

$$\left\{ \begin{array}{l} (I) \quad (D^s f(x+h) - D^s f(x))(h_1, \cdots, h_s) = \int_0^1 (D^{s+1} f(x+th))(h, h_1, \cdots, h_s) dt; \\ (II) \quad (\frac{D^s f(x+th) - D^s f(x)}{t} - D^{s+1} f(x+th)(h))(h_1, \cdots, h_s) \\ \quad = \int_0^1 (D^{s+1} f(x+\tau t h) - D^{s+1} f(x))(h, h_1, \cdots, h_s) d\tau \end{array} \right\}$$

for every $t \in [0,1]$.
We say that $f : X \subset E \to F$ is **weakly smooth** (or of class C^∞_w) if it is of class $C^k_w, \forall\, k \geq 0$. We say that f is **weakly analytic** (or of class C^ω_w) if it is C^∞_w and $D^k f : X \to L^k(E;F)_p$ is continuous, $\forall\, k \geq 0$. Then if f is weakly analytic we have the following Taylor's expansion:

$$f(x+h) = \sum_0^\infty \frac{1}{s!} D^s f(x)(h^{(s)}).$$

5) If $f : X \subset E \to F$ is weakly k-times differentiable and $D^k f : X \to L^k(E;F)_p$ is continuous, we cannot assert that $D^s f : X \to L^k(E;F)_p$ is also continuous. However,

if $E \equiv \mathbf{R}^m$ (or E is a finite dimensional vector space), for a map $f : X \to F$ that is weakly k-times differentiable with

$$D^k f : X \to L^k(E;F)_p \cong \mathcal{L}^k(E;F) \cong F^{m^k}$$

continuous, we can also say that $D^s f : X \to F^{m^s}, 0 \le s \le k$, is continuous. So $f : X \subset \mathbf{R}^m \to F$ is of **class** C^k if it is weakly k-times differentiable and if $D^k f : X \to F^{m^k}$ is continuous.

☐ Let $\mathcal{G}$ be a cover of E consisting of bounded sets. A function $f : X \subset E \to F$ is said to be **differentiable of class** $C^k_{\mathcal{G}}$ if it is weakly k-times differentiable and such that for $s \in \mathbf{N}, 0 \le s \le k$, the following two conditions are satisfied:

$$\left\{ \begin{array}{ll} (a) & D^s f(x) \in \mathcal{H}^s(E;F)_{\mathcal{G}} \\ (b) & D^s f : X \to \mathcal{H}^s(E;F)_{\mathcal{G}} \text{ is continuous} \end{array} \right\}$$

Let $f : X \subset E \to F$ be a map of class $C^k_{\mathcal{G}}, k \ge 1$. Then, we have the following:
(I) For each $x \in X$, each $s \le k-1$, and each $S \in \mathcal{G}$, the limit

$$\lim_{t \to 0} \frac{D^s f(x+th) - D^s f(x)}{t} = D^{s+1} f(x)(h)$$

exists in the locally convex space $\mathcal{H}^s(E;F)_{\mathcal{G}}$, uniformly, with respect to $h \in S$.
(II) For $s \le k$: (a) f is of class $C^s_{\mathcal{G}}$; (b) $D^s f : X \to \mathcal{H}^s(E;F)_{\mathcal{G}}$ is of class $C^{k-s}_{\mathcal{G}}$; (c) for each $r \le k-s, D^r(D^s f) = D^{r+s} f : X \to \mathcal{H}^{r+s}(E;F)_{\mathcal{G}} = \mathcal{H}^r(E;\mathcal{H}^s(E;F)_{\mathcal{G}})_{\mathcal{G}}$.
(III) If $D^k f : X \to \mathcal{H}^k(E;F)_{\mathcal{G}}$ is of class $C^s_{\mathcal{G}}$, then f is of class $C^{k+s}_{\mathcal{G}}$.
(IV) Let $C^k(X;F)_{\mathcal{G}}$ denote the linear subspace of F^X. The mapping

$$D^s : C^k(X;F)_{\mathcal{G}} \to C^{k-s}(X;\mathcal{H}^s(E;F)_{\mathcal{G}})_{\mathcal{G}}$$

is linear.

EXAMPLE 5.3 - 1) $C^k(X;F)_b \subset C^k(X;F)_{\bar{k}} \subset C^k(X;F)_k \subset C^k(X;F)_p$.
2) In the case where E has finite dimension, all these function spaces coincide with the space $C^k(X;F)$.
3) If E and F are normable locally convex spaces then $C^k(X;F)_b$ is the space of all functions from X to F which are Fréchet differentiable of class C^k.

☐ Let Λ be a topology on $\mathcal{L}^k(E;F)$. A function $f : X \subset E \to F$ is said to be differentiable of class C^k_Λ if f is weakly k-times differentiable and such that for each $s \le k$ the following two conditions are satisfied:

$$\left\{ \begin{array}{ll} (a) & D^s f(X) \subset \mathcal{L}^s(E;F) \\ (b) & D^s f : X \to \mathcal{L}^s(E;F)_\Lambda \quad \text{is continuous} \end{array} \right\}$$

Let $f : X \to F$ be a mapping of class C^k_Λ. Then, the limit

$$\lim_{t\to 0} \frac{D^s f(x+th) - D^s f(x)}{t} = D^{s+1} f(x)(h)$$

exists in $\mathcal{L}^m(E;F)_\Lambda, \forall\, s \leq k-1$.
Let $C^k(X;F)_\Lambda$ be the linear subspace of F^X of all functions $X \to F$ of class C^k_Λ. We say that $f : X \to F$ is of class C^∞_Λ if it is of class C^k_Λ for $k \geq 0$. The mapping

$$D^k : C^k(X;F)_\Lambda \to C^0(X;\mathcal{L}^k(E;F)_\Lambda)$$

is linear, for $k \geq 0$.
EXAMPLE 5.4 - 1) $C^{k+1}(X;F)_c \subset C^k(X;F)_d$.
2) If E is a Banach space and F a locally convex space, or E is a (F)-space and F a normable locally convex space, then all C^∞_Λ coincide. In this case the difference between Gateaux-Levy and Fréchet differentiability disappears and we may speak of a function differentiable of class C^∞ without reference to any convergent structure Λ: $f \in C^\infty(E;F)$ iff there exist continuous functions $D^k f : X \to \mathcal{L}^k(E;F)_p, k = 0,1,2,\cdots$, such that $D^{k+1} f$ is the Gateaux-Levy derivative of $D^k f$.
DEFINITION 5.1 - 1) *A* **locally convex manifold of class** $C^k_w, 1 \leq k$, *is a topological space* M *locally homeomorphic to open subsets of a locally convex space* E *(over* $\mathbf{K}$*), such that on* M *a* **maximal** C^k**-atlas** *is defined. A* C^k_w*-atlas on* M *is defined to be an open covering* $\bigcup U_a = M$ *and a collection of homeomorphisms* $\mu_a : U_a \to \mu_a(U_a) \subset E$ *such that the mapping*

$$\mu_a \circ \mu_b^1 : \mu_b(U_a \cap U_b) \to \mu_a(U_a \cap U_b)$$

is of class C^k_w *for any* a *and* b*. Then we say that* M *is* **modeled on** E *and* $\mathbf{K}$ *is its* **field of numeros**.
2) *A continuous function* $f : M \to N$ *between locally convex manifolds of* **class** C^k_w, *is said to be of class* C^k_w *if for any point* $x \in M$*, there exists a local chart* (U,μ) *in* x *and a local chart* (V,β) *in* $y = f(x)$ *such that the map* $\beta \circ f \circ \mu^{-1} : \mu(f^{-1}(V)) \to \beta(V)$ *is a map of class* C^k_w*. If* f *is a one-to-one homeomorphism we say that* f *is a* **diffeomorphism of class** C^k_w*. If* E *has finite dimension,* s*, then we can introduce a basis* $\{e_a\}, 1 \leq a \leq s$*, and identify* E *with* $\mathbf{K}^s, E \equiv \mathbf{K}^s$*. Then we locally recognize on any locally convex manifold modeled on* E *a system of coordinates* $(x^1,\cdots,x^s)$ *by using the* $\mathbf{K}$*-valued functions* $x^i = pr^i \circ \mu : U \to \mathbf{K}$*, where* (U,μ) *is a chart in* M *and* pr^i *the projection on the* i*-th factor. In this case we say that* M *is of dimension* s.
A curve of class C^k_w (or equivalently a curve of class C^k) in the locally convex manifold M, passing through a point $p \in M$ is defined to be a C^k_w mapping $c : I \subset \mathbf{K} \to M$,

where I is an open subset of $\mathbf{K}$ with $0 \in I$, and $c(0) = p$. Two curves c and c' passing through p are said to be equivalent if $\mu \circ c(t)$ and $\mu \circ c'(t)$ have the same derivative at $t = 0$, where (U, μ) is some chart in a neighbourhood of p.

DEFINITION 5.2 - 1) *A* **tangent vector** *to M at a point p is defined to be an equivalence class of curves passing through p.*

2) *The set of tangent vectors at a point p is called the* **tangent space** *and denoted by T_pM. (T_pM has the structure of locally convex space.)*

3) *The* **cotangent space** *at $p \in M$ is the dual, T_p^*M, of T_pM: $T_p^*M \equiv L_{\mathbf{K}}(T_pM; \mathbf{K})$. The* **strong cotangent space at** *$p \in M$ is the topological dual, $T_p'M$ of T_pM. (We shall consider $T_p'M$ endowed with the strong topology.)*

REMARK 5.3 - Let M and W be locally convex manifolds of class C_w^k. Let $\pi : W \to M$ be a surjective C_w^k mapping. Then (W, M, π) is called a C_w^k **bundle**. Moreover, if there exists an open covering (U_a) of M and C_w^k diffeomorphisms $\mu_a : \pi^{-1}(U_a) \to U_a \times N$, with N a locally convex manifold such that $\mu_{a,p} \circ \pi^{-1}(p)$ is diffeomorphic to N and for two open sets U_a, U_b such that $U_a \cap U_b \neq \emptyset$, the map $\mu_{ab,p} \equiv \mu_{b,p} \circ \mu_{a,p}{}^{-1}$, maps N diffeomorphically into N, then (W, M, π) is called a C_w^k (**locally trivial fiber bundle**). (Linear, affine and algebraic fiber bundles can be similarly defined imitating definitions on finite dimensional manifolds.) A **section** of π is a C_w^k map $s : M \to E$ such that $\pi \circ s = id_M$ (s can only be defined locally and is then called a **local section**).

EXAMPLE 5.5 - **Tangent bundle** $TM \equiv \bigcup_{p \in M} T_pM$, **cotangent bundle** $T^*M \equiv \bigcup_{p \in M} T_p^*M$, their tensor products

$$T_s^rM = TM \bigotimes_{\mathbf{K}} \cdots_r \cdots \bigotimes_{\mathbf{K}} TM \bigotimes_{\mathbf{K}} T^*M \bigotimes_{\mathbf{K}} \cdots \bigotimes_{\mathbf{K}} T^*M$$

and their symmetric and antisymmetric subbundles S_s^rM, Λ_s^rM, are examples of locally trivial fiber bundles. A section respectively of TM, T^*M and T_s^rM is called **vector field**, **differential form**, and **tensor field of type** (r, s) respectively. A section of Λ_0^rM is called an r-**vector field** and a section of Λ_s^0M is called an **differential** s-**form**. We have the pairing

$$<,>: T^*M \times TM \to M \times \mathbf{K}.$$

DEFINITION 5.3 - 1) *The* **derivative space** *$D(M, N)$ is the locally convex manifold defined by*

$$D(M, N) \equiv L_{\mathbf{K}}(TM; TN) \equiv \bigcup_{(p,q) \in M \times N} L_{\mathbf{K}}(T_pM; T_qN).$$

2) *The* **strong derivative space** $\mathcal{D}(M,N)$ *is the locally convex manifold defined by*

$$\mathcal{D}(M,N) \equiv \mathcal{L}_{\mathbf{K}}(TM;TN) \equiv \bigcup_{(p,q)\in M\times N} \mathcal{L}_{\mathbf{K}}(T_pM;T_qN).$$

THEOREM 5.1 - *One has the following exact sequence of vector bundles over* $M\times N$*:*

$$0 \to T'M\widehat{\bigotimes}_{\mathbf{K}} TN \equiv \bigcup_{(p,q)\in M\times N} T'_pM\widehat{\bigotimes}_{\mathbf{K}} T_qN \to \mathcal{D}(M,N)$$

where $\widehat{\otimes}$ *denotes completed tensor product.*

THEOREM 5.2 - *If* M *and* N *are modeled on Fréchet spaces, such that the model of* M *is a nuclear space, one also has the following exact sequence of vector bundles over* $M \times N$*:*

$$0 \to T'M\widehat{\bigotimes}_{\mathbf{K}} TN \to \mathcal{D}(M,N) \to 0.$$

This is, in particular, the case when M *and* N *are manifolds of finite dimension. In this case* $\mathcal{D}(M,N) = D(M,N)$.

REMARK 5.4 - Particular cases are:

$$\left\{\begin{array}{ll} (a) & D(\mathbf{K},N) \cong \mathbf{K}\times TN \\ (b) & D(M,\mathbf{K}) \cong \mathbf{K}\times T^*M \end{array}\right\}$$

The derivative of a map $f: M \to N$ of class C^1_w is a section $Df: M \to D(M,N)$ of the fiber bundle $\pi_1 : D(M,N) \to M$, where Df denotes the weak derivative of f. If f is of class C^1_{GL} then one has $Df: M \to \mathcal{D}(M,N)$. Furthermore, if f is of class C^1_Λ then one has

$$Df: M \to \mathcal{D}(M,N)_\Lambda \equiv \bigcup_{(p,q)\in M\times N} \mathcal{L}_{\mathbf{K}}(T_pM;T_qN)_\Lambda.$$

Furthermore we have the following diffeomorphisms:

$$\left\{\begin{array}{ll} (c) & D(M,N_1\times\cdots\times N_m) \cong D(M,N_1)\times\cdots\times D(M,N_m) \\ (d) & D(M_1\times\cdots\times M_m,N) \cong D(M_1,N)\times\cdots\times D(M_m,N) \end{array}\right\}$$

The second diffeomorphism allows one to interpret the **partial derivatives**:

$$D_if : M_1\times\cdots\times M_m \to D(M_i,N), \quad i=1,\cdots,m,$$

as

$$D_if = pr_i \circ Df.$$

The first diffeomorphism interprets the **component derivations**

$$Df^i : M \to D(M, N_i), \quad Df^i \equiv pr_i \circ Df \quad i = 1, \cdots, m.$$

The **velocity** of a map $f : \mathbf{K} \to N$ (a **curve** of N) is

$$\dot{f} : \mathbf{K} \to TN, \quad \dot{f} = pr_2 \circ Df.$$

Similarly the **differential** of a map $f : M \to \mathbf{K}$ is

$$df : M \to T^*M, \quad df = pr_2 \circ Df.$$

For a map $f : \mathbf{K} \times M \to N$ we can define its **infinitesimal variation**

$$\partial f : M \to TN, \quad \partial f = D_1 f \circ j,$$

where j is the inclusion map $j : M \to \mathbf{K} \times M$ given by $p \mapsto (0, p)$. If f is a deformation of a section s of the fiber bundle $\pi : W \to M$ such that $f_0 = s$, then $\partial f : M \to s^*vTW$, where vTW is the vertical tangent bundle of W and s^*vTW is the pull-back of vTW with respect to s.
A vector field X of class C^1_w on M identifies a mapping $X : F(p) \to F(p)$, where $F(p)$ is the commutative algebra of C^1_w functions defined in the neighbourhoods of $p \in M$, given by $X : f \mapsto X.f =< df, X >$, such that $X(\mu f + \beta g) = \mu.Xf + \beta.Xg, \beta, \mu \in \dot{f}$, and $X.(fg) = (X.f)g + f(X.g)$.
DEFINITION 5.4 - *$\pi : W \to M$ is a fiber bundle between locally convex manifolds, the* **derivative space of sections** *$D(W)$ (resp.* **strong derivative space of sections** *$\mathcal{D}(W)$), of π is defined by means of the following commutative diagram:*

$$\begin{array}{ccc}
\mathcal{D}(W) \equiv d^*\mathcal{D}(M, W) & \rightarrow & \mathcal{D}(M, W) \\
\bigcap & & \bigcap \\
D(W) \equiv d^*D(M, W) & \rightarrow & D(M, W) \\
 & & \downarrow \\
\downarrow & & M \times W \\
 & & \downarrow d \\
M & \underset{d}{\rightarrow} & M \times M
\end{array}$$

where d is the diagonal map.
One has the following fiber structures:

$$\pi_{1,0} : D(W) \to W \quad \text{and} \quad \pi_1 : D(W) \to M.$$

The derivative Ds of a section s of π is a section of π_1. If $\pi_i : W_i \to M (i = 1, 2)$ are vector bundles over M and s_i are sections of π_i, the Leibnitz rule $D(s_1 \otimes s_2) = (Ds_1) \otimes s_2 + s_1 \otimes Ds_2$ holds.

DEFINITION 5.5 - **Derivative spaces of higher orders** *(resp.* **strong derivative spaces of higher order**) *can be defined as:*

$$D^k(M,N) \equiv D(D^{k-1}(M,N)), \quad (\text{resp.}\, \mathcal{D}^k(M,N) \equiv \mathcal{D}(\mathcal{D}^{k-1}(M,N))).$$

THEOREM 5.3 - *If M and N are modeled on Fréchet spaces, such that the model of M is a nuclear space, one has the following canonical isomorphism of vector bundles over $\mathcal{D}^{k-1}(M,N)$:*

(5.1)

$$\begin{array}{ccc} \mathcal{D}^k(M,N) & \cong & T'M\widehat{\check{\bigotimes}}_{\mathbf{K}}T\mathcal{D}^{k-1}(M,N) \\ \downarrow & & \downarrow \\ \mathcal{D}^{k-1}(M,N) & \cong & \mathcal{D}^{k-1}(M,N) \end{array}$$

where $\check{\bigotimes}$ denotes the pull-back with respect to the diagonal map $M \to M \times M$.

REMARK 5.5 - Natural structures of fiber bundles can be recognized on $D^k(M,N)$:

$$\begin{array}{ccc} M & = & M \\ {\scriptstyle \pi_k}\uparrow & & \uparrow{\scriptstyle pr_1} \\ D^k(M,N) & \overset{\pi_{k,k-1}}{\to} D^{k-1}(M,N) \to \cdots \to D^o(M,N) = & M \times N \\ {\scriptstyle \bar{\pi}_k}\downarrow & & \downarrow{\scriptstyle pr_2} \\ N & = & N \end{array}$$

If $\pi : W \to M$ is a fiber bundle, the k**-derivative space of sections** of π is a subbundle

$$D^k(W) \subset D^k(W,M)$$

over M. Furthermore, in $D^k(W)$ we can distinguish some further subbundles that are also useful:

holonomic k-jet-derivative space :

$JD^k(W) \equiv \{u \in D^k(W) \,|\, \exists\, s \in C^k_w(W), D^k s(p) = u\}$

semiholonomic k-jet derivative space :

$\bar{J}D^k(W)$

sesquiholonomic k-jet-derivative space :

$\check{J}D^k(W)$

The definitions of these last two spaces are similar to the ones for finite dimensional case in ref.[117]. One has:

$$JD^k(W) \subset \check{J}D^k(W) \subset \bar{J}D^k(W) \subset D^k(W).$$

Then we can define **k-derivative connections**, on the fiber bundle $\pi : W \to M$ as sections

$$\rceil : D^{k-1}(W) \to D^k(W)$$

of the fiber bundle $\pi_{k,k-1} : D^k(W) \to D^{k^{-1}}(W)$. Similarly we have also

$$\textbf{\textit{k}-holonomic connections}: \quad \rceil : JD^{k-1}(W) \to JD^k(W)$$
$$\textbf{\textit{k}-semiholonomic connections}: \quad \bar{\rceil} : \bar{J}D^{k-1}(W) \to \bar{J}D^k(W)$$
$$\textbf{\textit{k}-sesquiholonomic connections}: \quad \check{\rceil} : \check{J}D^{k-1}(W) \to \check{J}D^k(W)$$

At the first order a holonomic connection can be identified with the usual connection on the fiber bundle π, that is a splitting on the left of the following exact sequence of vector bundles over W:

$$0 \to vTW \to TW \to \pi^*TM \to 0.$$

(See also refs.[98,102,114].) (Similar considerations can be made for $\mathcal{D}^k(M, N)$.) Finally, note that a C^{∞}_{A} vector field X tangent on M, is not, in general, integrable, i.e., there does not exist a 1-parameter group of diffeomorphisms $\phi : I \times M \to M, I \subset \mathbf{K}$, such that $\partial\phi = X$. However, if $X : M \to TM$ is integrable, and $\pi : W \to M$ is a bundle of geometric objects, we can associate to a section of π its Lie derivative $L_X s$ given by $L_X s = \partial\widetilde{s}$, where $\widetilde{s}$ is the deformation of s by means of the 1-parameter group ϕ associated to X,

$$\widetilde{s} : I \times M \to W, \quad \widetilde{s}_t \equiv \mathbb{B}({\phi_t}^{-1}) \circ s \circ \phi_t,$$

where $\mathbb{B}$ is the natural functor characterizing the bundle of geometric objects $\pi : W \to M$. For example if $W = TM$, we have the usual Lie bracket $L_X Y = [X, Y]$.
A **locally convex Lie group of class C^k_w**, is simultaneously a group and a locally convex manifold such that the mapping

$$G \times G \to G, (a, b) \mapsto ab^{-1}$$

is C^k_w. The set of left-invariant vector fields on G forms a Lie algebra $A(G)$ over $\mathbf{K}$.
A **principal bundle**, in the category of locally convex manifolds, is just defined as for the case of manifolds of finite dimension; the same holds for the definitions of principal connections [98,102].
A **derivative** PDE of order k on the fiber bundle $\pi : W \to M$ is a subbundle E_k of $D^k(W)$, over M. Similarly we can define (holonomic) PDEs of order k as well as non-holonomic PDEs.

EXAMPLE 5.6 - Derivative, holonomic and nonholonomic k-connections on a fiber bundle $\pi : W \to M$ define synonymous PDEs of order k on W respectively.

REMARK 5.6 - Besides manifolds modeled on locally convex spaces, we shall also use manifolds modeled on Banach spaces; similar considerations hold. In particular, we have the following.

THEOREM 5.4 - *If M and N are manifolds modeled on Banach spaces, the isomorphism (5.1) holds iff M is a manifold of finite dimension.*

PROOF. Let us consider the following.

LEMMA 5.1 - *If F and E are Fréchet spaces, with E also nuclear, one has the canonical isomorphism*

$$\mathcal{L}_{\mathbf{K}}(E;F) \cong E'\widehat{\bigotimes}_{\mathbf{K}} F.$$

So, we can assume that the derivative Df of a map $f : E \to F$ of class C^1_{GL} between such spaces has values into $E'\widehat{\otimes}_{\mathbf{K}} F : Df : E \to E'\widehat{\otimes}_{\mathbf{K}} F$. This is exactly the case when E or F is of finite dimension.

Banach spaces are particular cases of Fréchet spaces. So we have that the k-derivative space $\mathcal{D}^k(M,N)$ between manifolds M, N modeled on Banach spaces, is isomorphic to $T'M\widehat{\otimes}_{\mathbf{K}} T\mathcal{D}^{k-1}(M,N)$ if the model of M is nuclear. On the other hand as a normable space is nuclear iff it is finite dimensional [144], it follows that for manifolds modeled on Banach spaces the isomorphism (5.1) holds iff M is a manifold of finite dimension. □

5.2 - DIFFERENTIAL GEOMETRY OF QUANTUM SITUS

In order to study the differential geometric structure of the quantum situs of a PDE, we shall, first resume some fundamental results about the space of mappings between two topological spaces. (See also refs.[15,82,126,144] and references quoted therein.)

DEFINITION 5.6 - *Let X and Y be topological spaces. The* **compact-open topology (KO-topology)** *on $C^0(X,Y)$ is obtained by taking as subspace for the topology all sets of the form*

$$N(K,U) \equiv \{f \in C^0(X,Y) | f(K) \subset U, K \subset X \quad \textit{compact}, U \subset Y \quad \textit{open}\}.$$

If $H \subset C^0(X,Y)$ is a subset of $C^0(X,Y)$ the topology induced on H by the KO-topology is also called KO-topology on H.

EXAMPLE 5.7 - The space of all continuous paths in Y:

$$C^0(I,Y), \quad I \equiv [0,1] \subset \mathbf{R},$$

is just the space with the KO-topology. A subset $H \subset C^0(I,Y)$ is for example the space of loops:

$$\Omega Y \equiv \{f \in C^0(I,Y), f(0) = f(1) = Y_0 \in Y\}.$$

REMARK 5.7 - Let X, Y be two sets. Set

$$Y^X \equiv \prod_{i \in X} Y_i, \quad Y_i = Y.$$

We have the identification $Y^X \equiv F(X,Y) =$ set of all maps $X \to Y$. If X, Y are topological spaces one has that $C^0(X,Y) \subset Y^X$. Furthermore, if Y^X is endowed with the product topology, it identifies on the subset $C^0(X,Y)$ a topology also called the **product topology** (**P-topology**) that is generated by all sets of the form $N(\{x\}, U)$, for $x \in X, U \subset Y$ open. So, this topology is coarser than the KO-topology: $P \subset KO$.[1]

THEOREM 5.5 - (PROPERTIES OF KO-TOPOLOGY ON $C^0(X,Y)$).

1) *Let R be a family of compact subsets of X such that for any compact subset L of X and any neighbourhood V of L, there exists a finite collection of sets $K_i \in R$ such that $L \subset U_i K_i \subset V$. Let G be a system of generators for the topology of Y. Then, the set $N(K,U), K \in R$ and $U \in G$ is a system of generators of the KO-topology on $C^0(X,Y)$.*

2) *Let X, Y, Z be topological spaces. If $f : X \times Y \to Z$ is a continuous map, the corresponding map $\tilde{f} : X \to C^0(Y,Z)$ given by $\tilde{f}(x)(y) = f(x,y)$ is continuous in the KO-topology. Vice versa, if $\tilde{f}$ is continuous and Y is locally compact, then f is continuous.*

3) *The* **evaluation function** *$e : C^0(X,Y) \times X \to Y$ defined by $e(f,x) = f(x)$ is continuous if X is locally compact.*

4) *Let X be a locally compact space, Y a topological space and $H \subset C^0(X,Y)$. Then, the KO-topology on H, is the coarsest such that the evaluation map $e : H \times X \to Y$ is continuous. (In fact from point 2. the continuity of this map means that the canonical injection $H \subset C^0(X,Y)$ is so.)*

5) *Let X be a Hausdorff space, Y a locally compact space and Z a topological space. One has the following commutative diagram:*

$$\begin{array}{ccc} Z^{X \times Y} & \cong & (Z^Y)^X \\ \uparrow & & \uparrow \\ C^0(X \times Y, Z) & \cong & C^0(X, C^0(Y,Z)) \end{array}$$

6) *Let X, Y, Z be topological spaces and suppose that Y is locally compact. The map $C^0(X,Y) \times C^0(Y,Z) \to C^0(X,Z)$ given by $(u,v) \mapsto v \circ u$, is continuous.*

---

[1] A subset of $C^0(X,Y)$ is uniformly closed if it contains the limit of every uniformly convergent sequence in it. A subset which is closed under pointwise convergence is uniformly closed, as is a subset which is closed in the KO-topology.

7) *Let each component of X be locally compact with a countable base (e.g. X compact). Let Y be a complete metric space. Then, $C^0(X,Y)$ endowed with the KO-topology has a complete metric.*

DEFINITION 5.7 - *Let X and Y be topological spaces. The* **Whitney topology** *(W^0-topology) on $C^0(X,Y)$ is generated by all sets of the form:*

$$N(f,W) \equiv \{g \in C^0(X,Y) | \Gamma_g \subset W | W \subset X \times Y \quad \textit{open containing} \quad \Gamma_f\},$$

where $f \in C^0(X,Y)$ and Γ_f, Γ_g are the graphs of the maps f and g respectively.

THEOREM 5.6 - (PROPERTIES OF W^0-TOPOLOGY ON $C^0(X,Y)$).

1) *When X is paracompact and Y is metric, the W^0-topology on $C^0(X,Y)$ is generated by all sets of the form*

$$N(f,\epsilon) \equiv \{g \in C^0(X,Y) | d(g(x),f(x)) < \epsilon, \forall\, x \in X, f \in C^0(X,Y), \epsilon \in C^0(X,\mathbf{R}^+)\}.$$

2) *If X is compact the KO-topology coincides with the W^0-topology.*

3) *Let X be a paracompact space and Y a complete metric space. Then, every* **uniformly closed subset** *$Q \subset C^0(X,Y)$ is a Baire space.*[2]

4) *If X and Y are C^0-manifolds, every KO-closed subset $Q \subset C^0(X,Y)$ is a Baire space in the W^0-topology.*

5) *Let X and Y be C^r-manifolds, $0 \le r \le \infty$. The image of the map*

$$D^r : C^r(X,Y) \to C^0(X, J\mathcal{D}^r(X,Y)), \quad f \mapsto D^r f,$$

is closed in the KO-topology.

DEFINITION 5.8 - *The W^r-***Whitney topology**
*(W^r-topology), resp. KO^r-***topology,** *on*

$$C^r(X,Y), 0 \le r \le \infty,$$

is the topology induced by D^r from the W^0-topology, resp. KO-topology, on

$$C^0(X, J\mathcal{D}^r(X,Y)).$$

THEOREM 5.7 - (PROPERTIES OF W^r-TOPOLOGY AND KO^r-TOPOLOGY ON $C^r(X,Y), 0 \le r \le \infty$).

---

[2] A topological space X is **Baire** if one the following equivalent condition is satisfied: (a) Any countable intersection of open sets, dense in X, is dense in X; (b) Any countable union of closed sets, without interior points in X is without interior points in X. Any open subspace Y of a Baire space X is so. A subset $W \subset Y^X$ is called open in the product topology if for each point $(y_i)_{i \in X} \in Y^X$ there are neighbourhoods U_i of y_i in Y such that $\prod_{i \in X} U_i \subset W$ in the W^0-topology.

1) $C^r(X,Y)$ endowed with the KO^r-topology has a complete metric.
2) Every KO^r-closed subspace of $C^r(X,Y)$ is a Baire space with respect to the W^r-topology.
3) The space $C^r(X,Y)$ is countable regular.
4) If X, Y are C^ω manifolds, then $C^\omega(X,Y)$ is W^r-dense in $C^r(X,Y)$.

DEFINITION 5.9 - *The* **fine-Whitney topology** *(FW^∞) on $C^\infty(X,Y)$ is the W^∞-topology refined in such a way that the following equivalence relation:*

$$f \sim g, f, g \in C^\infty(X,Y) \Leftrightarrow \text{the set } \{x \in X | f(x) \neq g(x)\} \text{ is relatively compact in } X,$$

becomes an open equivalence relation. One has $W^\infty \subset FW^\infty$.

DEFINITION 5.10 - *The* **D-topology** *on $C^\infty(X,Y)$ is given by taking all sets $N(m,U)$ as a basis for its open sets, where*

$$N(m,U) \equiv \{f \in C^\infty(X,Y) | D^{m_n} f(X - \overset{o}{K}_n) \subset U_n \, \forall n \in \mathbf{N}\},$$

where:
(a) $\{K_n\}_{n\in\mathbf{N}}$ is a fixed sequence of compact subset of X such that $K_0 = \emptyset$, $K_{n-1} \subset \overset{o}{K}_n \equiv$ interior of K_n, for each n and $X = \bigcup_n K_n$;
(b) m is a sequence of non-negative integers $m \equiv (m_n)$;
(c) $U = \{U_n\}_{n\in\mathbf{N}}$ is a family of open sets in $J\mathcal{D}^{m_n}(X,Y)$.
The D topology is independent of the choice of the sequence $\{K_n\}_{n\in\mathbf{N}}$. One has $W^\infty \subset D$.

DEFINITION 5.11 - *The* **FD-topology** *on $C^\infty(X,Y)$ is the weakest among all topologies on $C^\infty(X,Y)$ which are finer than the D-topology and for which all equivalence classes of the below equivalence relation are open: $f, g \in C^\infty(X,Y), f \sim g \Leftrightarrow$ the set $\{x \in X | f(x) \neq g(x)\}$ is relatively compact in X. One has*

$$FW^\infty \subset FD.$$

THEOREM 5.8 - 1) *$C^\infty(X,Y)$ is a Baire space in the topologies W^∞ and D, (but not in the topologies W^∞, FD).*
2) For each $k \geq 0$, the map

$$D^k : C^\infty(X,Y) \to C^\infty(X, J\mathcal{D}^k(X,Y))$$

is continuous in the topologies D and FD. Furthermore, $\mathrm{im}(D^k)$ is closed in the two topologies.
3) Embeddings, immersions, submersions, surjective submersions, diffeomorphism, proper mappings, (i.e. f^{-1}(compact) is compact), are all open subsets of $C^\infty(X,Y)$ in W and all finer topologies.

4)

$$C^\infty(X, Y\times Z)\cong C^\infty(X,Y)\times C^\infty(X,Z)$$

is a homeomorphism for all the above topologies.
5) *If* $f\in C^\infty(X,Y)$*, then*

$$f_*: C^\infty(X,Y)\to C^\infty(X,Z),\quad f_*(g)=f\circ g$$

is continuous for all topologies.
6) *Composition:*

$$C^\infty_{prop.}(X,Y)\times C^\infty(X,Z)\to C^\infty(X,Z)$$

is jointly continuous.
7) *Inversion:* $Diff(X)\to Diff(X)$ *is continuous.*
8) *A sequence* f_n *converges to* f *in* $C^\infty(X,Y)$*, (for all topologies) iff* f_n *equals* f *of some compact set (up to finitely many* n*'s) and* $f_n\to f$ *"uniformly in all derivatives" on this compact set.*

PROPOSITION 5.1 - *Let* $\pi: E\to X$ *be a smooth finite dimensional vector bundle. Let* $C^\infty_0(E)$ *denote the space of all smooth-section with compact support of this bundle. Then,* $C^\infty_0(E)$*, with the topology induced from the* FD*-topology on* $C^\infty(X,E)$ *is a locally convex topological linear space, in fact a fully nuclear* LF*-space. Furthermore, it is a Lindelöf space, hence paracompact and normal.*[3]

---

[3] The topology of a topological vector space E is determined by a base of neighbourhoods of the origin. A **balanced subset** $A\subset E$ is characterized by:

$$\lambda A\subseteq A,\forall\, \lambda\in \mathbf{K},|\lambda|\leq 1.$$

A **convex subset** $A\subset E$ is characterized by:

$$x,y\in A\Rightarrow \lambda x+(1+\lambda)y\in A, 0<\lambda<1.$$

A **locally convex spaces** is a topological vector space E characterized by the filter of neighbourhoods of zero with a base of convex neighbourhoods. The **jauge** of a convex C containing $0\in E$ is the numerical function

$$j_c: E\to\overline{\mathbf{R}}\equiv\mathbf{R}\cup\{+\infty\}=\begin{Bmatrix}\inf\{t|t>0,x\in tC\} \text{ if } x\in tC, t>0,\\ +\infty \text{ otherwise}\end{Bmatrix}$$

If $x\in C$ one has $j_c(x)\leq 1$; if $x\neq C$ one has $j_c(x)\geq 1$. Furthermore, the jauge of a convex set C verifies the following properties: a) $j_c(x)\geq 0,\forall x\neq 0; j_c(0)=0$; b) $j_c(\lambda x)=\lambda j_c(x)$ if $\lambda>0$; c) $j_c(x+y)\leq j_c(x)+j_c(y)$. Vice versa, to any function $j_c: E\to\overline{\mathbf{R}}^+$ verifying properties (a), (b) and (c), we can associate a convex set

$C^\infty(E)$ is FD-closed in the spaces $C^\infty(X,E)$ and $C_0^\infty(E)$ is closed and open in $C^\infty(E)$.

REMARK 5.8 - $C_0^\infty(E)$ as a fully nuclear LF-space is also reflexive: $C_0^\infty(E)'' \cong C_0^\infty(E)$.

REMARK 5.9 - Let us observe that the above structure of locally convex topological vector space induced on $C^\infty(E)$ and $C_0^\infty(E)$ can be recognized by a different approach. Let X be an n-dimensional manifold $n < \infty$, of class C^∞, Hausdorff, connected, paracompact. This implies that: 1) X is denumerable to infinity and admits differentiable partition of unity; 2) For any atlas $\mathcal{A}' \equiv (U'_j, \phi'_j)$ of X, there exists an equivalent atlas (**denumerable atlas**) $\mathcal{A} \equiv (U_j, \phi_j)$ such that: (i) the covering (U_i) of X is denumerable, locally finite, finer than (U'_j) and the U_i are relatively compact open sets; (ii) for any $j, \overline{U}_j$ is contained in the domain of a chart; (iii) there exists a sequence of compacts (K_i) of $\mathbf{R}^n$ such that $K_i \subset \phi(U_i)$, for each i and where the ${\phi_i}^{-1}(K_i)$ cover X. Then the structure of locally convex topological vector space on $C^\infty(E)$ can be defined by means of the following family of semi-norms:

$$N_{K,q}(s) \equiv \sup_{p\in K, |\alpha|\leq q, 1\leq i\leq m} |D^\alpha \bar{s}^i(\phi(p))|, \forall\, s \in C^\infty(E),$$

where $K =$ any compact of X contained in some domain $U, q \in \mathbf{N}, q \geq 0, m \equiv$ fiber dimension of E, $\bar{s}^i \equiv pr_i \circ \tau \circ s \circ \phi^{-1} : \phi(U) \to \mathbf{K}^m \to \mathbf{K}$, where (U, ϕ, τ) is a vector chart of E. The same topology can be obtained by considering as a base of topology the set $C^\infty(E|K_i)$, where the sequence K_i of compacts is subordinated to the denumerable atlas $\mathcal{A} \equiv (U_i, \phi_i, \tau_i)$ of E, i.e., $K_i \subset U_i, \cup_i K_i = X$. For the set $C_0^\infty(E)$ note first of all that it is a subspace of $C^\infty(E)$. In fact $s_1 + s_2$ has support compact if $s_i \in C_0^\infty(E), i = 1,2$. In fact, $\mathrm{supp}(s_1+s_2) \subseteq \mathrm{supp}(s_1) \cup \mathrm{supp}(s_2)$. On the other hand $\mathrm{supp}(s_1 + s_2)$ can be considered a closed subspace of the compact space $\mathrm{supp}(s_1) \cup \mathrm{supp}(s_2)$ and therefore it must be compact. Then the topology of $C_0^\infty(E)$

$C\equiv\{x|x\in E; j(x)\leq 1\}$ that admits j as gauge. C is called the **representing** of j. Furthermore, recall that a **F-space** (Fréchet space) is a complete metrizable locally convex topological vector space. Moreover, a **LF-space** is a space which can be represented as the topological inductive limit of F-spaces. A **Lindelöf space** is a topological vector space X such that every open cover of X has a countable subcover. Let E and F be Banach spaces. An operator $u\in\mathcal{L}(E,F)$ is **nuclear** if it belongs to the image of $j:E'\hat{\otimes}F\to\mathcal{L}(E,F)$. A **nuclear space** is a locally convex Hausdorff space E such that for every balanced, convex, open neighbourhood U of $0\in E$ there exists a balanced, convex open neighbourhood V of 0 in E contained in U and such that the canonical map $\widehat{E}_V\to\widehat{E}_U$ is nuclear. E_U (resp. E_V) is the normed spaces E/N, $N\equiv\cap_{\lambda>0}\lambda U\equiv\{x|q_U(x)=0\}$, where q_U is the gauge of U, equipped with the quotient of the seminorm q_U (resp. is the normed space E/M, $M=\cap_{\lambda>0}\lambda V\equiv\{x|q_V(x)=0\}$, where q_V is the gauge of V). $\widehat{E}_U$ (resp. $\widehat{E}_V$) is the completion of E_U (resp. E_V) which respect to q_U (resp. q_V).

as that of a locally convex topological vector sapce can be considered the inductive limit of $C_0^\infty(E|K)$, where K raises in the set of compact parts of X. Then, as X is denumerable to infinity, in order to define the topology of $C_0^\infty(E)$ we can use a sequence of compact (K_i) covering X.

THEOREM 5.9 - *$C^\infty(X,Y)$ with FW^∞ (resp. FD) topology has a natural structure of manifolds. Furthermore, $C^\infty(X,Y)$ is FD-paracompact. The tangent bundle $TC^\infty(X,Y)$ coincides with the open subset of $C^\infty(X,TY)$ of all C^∞ mappings $s: X \to TY$ with compact support; $TC^\infty(X,Y)$ is again a manifold (from $C^\infty(X,TY)$).*

PROOF. Let π_Y be the natural projection $\pi_Y : TY \to Y$. Fix $f \in C^\infty(X,Y)$ and consider the space $T_fC^\infty(X,Y)$ of all vector fields along f with compact support, (i.e., $s \in T_fC^\infty(X,Y)$ if $s \in C_0^\infty(X,TY), \pi_Y \circ s = f, s \sim 0_Y \circ f$). We have the isomorphism $T_fC^\infty(X,Y) \cong C_0^\infty(f^*TY)$. Let $\tau : TY \to Y$ be a local addition, (i.e. a mapping such that $(\pi_Y,\tau) : TY \to Y \times Y$ is a diffeomorphism and an open subset, $\tau(0_Y) = Y$). Let $U_f \subseteq C^\infty(X,Y)$ be the open subset of all g such that the image of (f,g) in $Y \times Y$ is contained in the open image of (π_Y,τ) and $g \sim f$. Now let

$$\left\{ \begin{array}{l} \phi_f : U_f \to C_0^\infty(f^*TY), \quad \phi_f(g) = (\pi_Y,\tau)^{-1} \circ (f,g), \\ \psi_f : C_0^\infty(f^*TY) \equiv T_fC^\infty(X,Y) \to U_f, \psi_f(s) = \tau \circ s \end{array} \right\} \tag{5.2}$$

Then ϕ_f, ψ_f are continuous and are inverse to each other. So, $C^\infty(X,Y)$ is a topological manifold, modeled on topological vector spaces $C_0^\infty(f^*TY)$ which bear the usual inductive limit topology:

$$C_0^\infty(f^*TY) = \lim_r C_0^r(f^*TY), r < \infty, C_0^r(f^*TY) = \lim_K C_K^r(f^*TY),$$

where $C_K^r(f^*TY)$ is the space of all sections of class C^r with support in $K \subset X$ compact, in case of FD. The change of chart in (5.2) looks like $\alpha_*(s) = \alpha \circ s$, where $\alpha : f^*TY \to g^*TY$, is a smooth fiber respecting mapping (defined on an open subset only). To make $C^\infty(X,Y)$ a differentiable manifold one has to choose a differential calculus on locally convex spaces such that the above mapping is of class C^∞. (See the above section.) More precisely $f : E \to F$ is C_c^1 if

$$Df(x)(y) = \lim_{t\to 0} \frac{1}{t}(f(x+ty) - f(x))$$

exists for all x, y with $Df(x) : E \to F$ a linear map such that $(x,y) \to Df(x)(y)$ is jointly continuous on $E \times E$. C_c^r and C_c^∞ are then defined by recursion. So, $C^\infty(X,Y)$ is a C_c^∞-manifold. $\square$

PROPOSITION 5.2 -

$$Diff(X) \subset C^\infty(X,X),$$

the open subset of all diffeomorphism of X*, is a* C_c^∞*-Lie group, i.e., composition and inversion are differentiable, of class* C_c^∞*. The corresponding Lie algebra is* $C_0^\infty(TX)$*. The canonical exponential mapping is not surjective on any neighbourhood of the identity in general.*

PROPOSITION 5.3 - 1) *$E(X,Y)$, the space of all embeddings from X into Y, is open in $C^\infty(X,Y)$, i.e. it is also a C_c^∞-manifold.*

2) *There is a right-action ρ of $Diff(X)$ onto $E(X,Y)$:*

$$\rho : Diff(X) \times E(X,Y) \to E(X,Y)$$

which is C_c^∞, and $E(X,Y)$ turns out to be a C^∞ principal fiber bundle with structure group $Diff(X)$.

PROPOSITION 5.4 - *Let $\pi : E \to X$ be a smooth finite dimensional vector bundle. Then, $C^\infty(E)$ is an affine manifold, modeled on the topological vector space $C_0^\infty(E)$. So, one has the identification*

$$TC^\infty(E) \equiv C^\infty(E) \times C_0^\infty(E).$$

2 THEOREM 5.10 - *If $\pi : W \to X$ is a fiber bundle, the space $C^\infty(W)$ has a natural of manifold induced by $C^\infty(X,W)$. The tangent bundle $TC^\infty(W)$ coincides with the open subset of $C^\infty(X,vTW)$ of all mappings $v : X \to vTW$ with compact support and one has*

$$TC^\infty(W) = \bigcup_{s\in C^\infty(W)} T_sC^\infty(W), \quad T_sC^\infty(W) = C_0^\infty(s^*vTW).$$

So, each $T_sC^\infty(W)$ is a locally convex topological vector space, fully nuclear, LF-space, Lindelöf, paracompact and normal. Therefore, $C^\infty(W)$ is a C_c^∞-manifold modeled on the topological vector space $C_0^\infty(s^*vTW)$.

PROPOSITION 5.5 - *The map $D^k : C^\infty(W) \to C^\infty(J\mathcal{D}^k(W))$ is continuous in all the above topologies and $\mathrm{im}(D^k)$ is closed.*

PROPOSITION 5.6 -

$$T(\mathrm{im}(D^k)) = \bigcup_{s\in C^\infty(W)} T_{D^ks}(\mathrm{im}(D^k)),$$

where $T_{D^ks}(\mathrm{im}(D^k))$ is the vector space given by

$$T_{D^ks}(\mathrm{im}(D^k) = \{X \in C_0^\infty(D^ks^*vTJ\mathcal{D}^k(W)) | \exists\, v \in T_sC^\infty(W); X = |_k \circ D^kv\},$$

where $|_k$ is the canonical isomorphism

$$|_k : J\mathcal{D}^k(s^*vTW) \cong D^ks^*vTJ\mathcal{D}^k(W).$$

PROOF. From the canonical map $D^k : C^\infty(W) \to C^\infty(J\mathcal{D}^k(W))$ we have, for functorial property:

$$T(D^k) : TC^\infty(W) \to TC^\infty(J\mathcal{D}^k(W)).$$

Note that the map D^k is injective; in fact one has

$$D^k s = D^k s' \Leftrightarrow s = s', \forall s, s' \in C^\infty(W).$$

So, we can identify $C^\infty(W)$ with its image into $C^\infty(J\mathcal{D}^k(W))$ and the tangent bundle $TC^\infty(W)$ with its image into $TC^\infty(J\mathcal{D}^k(W))$ by means of the map $T(D^k)$. This allows us to know the structure of the subset $T(\mathrm{im}(D^k)) \subset TC^\infty(J\mathcal{D}^k(W))$. In fact, one has:

$$\left\{ \begin{aligned} T_{D^k s}(\mathrm{im}(D^k)) &= T(D^k)(T_s C^\infty(W)) \\ &= \{X \in C_0^\infty(D^k s^* vTJ\mathcal{D}^k(W)) | \exists\, v \in T_s C^\infty(W), X = |_k \circ D^k v\} \end{aligned} \right\}$$

□

So, we have also proved the following.

THEOREM 5.11 - $\mathrm{im}(D^k)$ *(or $C^\infty(W)$) is a closed submanifolds of $C^\infty(J\mathcal{D}^k(W))$.*

THEOREM 5.12 - *Let $E_k \subset J\mathcal{D}^k(W)$ be a PDE on the fiber bundle $\pi : W \to M$. Then, $C^\infty(E_k)$ is a closed submanifold of $C^\infty(J\mathcal{D}^k(W))$ and the set of regular solutions of E_k, $Sol(E_k)$, is a closed submanifold of $C^\infty(E_k)$ (or $C^\infty(W)$).*

PROOF. In fact by means of the canonical inclusion $C^\infty(E_k) \subset C^\infty(J\mathcal{D}^k(W))$ we recognize a topological structure on $C^\infty(E_k)$. Furthermore, for the tangent bundle we have

$$TC^\infty(E_k) = \bigcup_{u \in C^\infty(E_k)} T_u C^\infty(E_k),$$

where $T_u C^\infty(E_k) = C_0^\infty(u^* vTE_k)$. So, $C^\infty(E_k)$ is a manifold modeled on the topological vector space $C_0^\infty(u^* vTE_k)$. Now, by means of the following commutative diagram

$$\begin{array}{ccc} Sol(E_k) & \xrightarrow{D^k} & C^\infty(E_k) \\ \cap & & \cap \\ C^\infty(W) & \xrightarrow[D^k]{} & C^\infty(J\mathcal{D}^k(W)) \end{array}$$

we induce a topological structure on $Sol(E_k)$. Furthermore, for the tangent bundle of $Sol(E_k)$ we have:

$$TSol(E_k) = \bigcup_{s \in Sol(E_k)} T_s Sol(E_k),$$

where

$$T_s Sol(E_k) = \{X \in C_0^\infty(D^k s^* vTE_k | \exists\, v \in T_s C^\infty(W); X = |_k \circ D^k v^k\}.$$

So, $Sol(E_k)$ is a closed submanifold of $C^\infty(W)$ (or $C^\infty(E_k)$) modeled on the topological vector spaces $T_s Sol(E_k), s \in C^\infty(W), D^k s \in C^\infty(E_k)$. $\square$

THEOREM 5.13 - *Let $E_k \subset J\mathcal{D}^k(W)$ be a PDE on the fiber bundle $\pi : W \to M$. Then, the quantum situs $\Omega(E_\infty)$ of E_∞ is a closed submanifold of $C^\infty(W)$ as well as $C^\infty(J\mathcal{D}^\infty(W))$.*

PROOF. First note that $\Omega(E_\infty)$ is a subset of the solution of the n-dimensional integral submanifolds of $J\mathcal{D}^\infty(W)$. More precisely, one has:

$$\Omega(E_\infty) = \left\{ \begin{array}{l} u \in C^\infty(J\mathcal{D}^\infty(W)), u = D^\infty s \\ \exists \text{ two } (n-1)\text{-dimensional subbundles}, N_i, i = 1, 2, \\ \text{of } E_\infty \subset J\mathcal{D}^\infty(W), \text{ such that } D^\infty s|_{V_i} = N_i, V_i \equiv \pi_\infty(N_i). \end{array} \right\}$$

So, on $\Omega(E_\infty)$ we recognize a topological structure, induced by $Sol(J\mathcal{D}^\infty(W))$. For the tangent bundle one has:

$$T\Omega(E_\infty) = \bigcup_{s \in \Omega(E_\infty)} T_s\Omega(E_\infty)$$

$$T_s\Omega(E_\infty) = \left\{ \begin{array}{l} X \in C_0^\infty(D^\infty s^* vTJ\mathcal{D}^\infty(W)) | \exists\, v \in T_s C^\infty(W); \\ X = |_k \circ D^\infty v; X|_V = 0, \text{for any Cauchy data} \\ N \text{on } E_\infty \text{ that } D^\infty s \text{ cobords}; V \equiv \pi_\infty(N) \subset M. \end{array} \right\}$$

So, $\Omega(E_\infty)$ is a closed submanifold of $Sol(J\mathcal{D}^\infty(W))$ modeled on the topological vector spaces $T_s\Omega(E_\infty)$. $\square$

COROLLARY 5.1 - *Any statistical set $\Omega \equiv \Omega(N_1, N_2)$ of a PDE $E_\infty \subset J\mathcal{D}^\infty(W)$ is a closed submanifold of $\Omega(E_\infty)$.*

Under suitable conditions we can recognize on $\Omega(E_\infty)$ a structure of Riemannian manifold. We shall use the following.

THEOREM 5.14 - *Let $\pi : W \to M$ be a fiber bundle over M, with paracompact fiber, and $W \subset E$ be canonically embedded as a closed subbundle of a C^∞ vector bundle $\pi : E \to M$. Then, the set $C^\infty(W)$ of smooth sections of π has a natural structure of Fréchet and Montel manifold immersed into the affine Fréchet and Montel space $C^\infty(E)$.*[4]

DEFINITION 5.12 - *Let $\pi : W \to M$ be a fiber bundle as in the above theorem. Let $H^r(W)$ be the set of sections of W such that the derivatives of order $\leq r$ are square integrable with respect to a system of charts on M and with respect to a volume form on M.*

---

[4] **A Montel space** is a reflexive locally convex topological vector space in which every closed bounded subset is relatively compact.

THEOREM 5.15 - 1) *If* $r > \frac{n}{2} + k, n = \dim M$, *one has the canonical continuous map* **(Sobolev lemma)**:

$$H^r(W) \to C^k(W).$$

2) $0 \leq k < \infty, C^k(W)$ *is a differentiable Fréchet manifold, in general of infinite dimension.*

3) $\frac{n}{2} < r < \infty, H^r(W)$ *is a differentiable manifold and the spaces of tangent vectors admit a Hilbert structure.*

4) *On* $H^r(W)$ *there exists a Riemannian structure.*

Then, we obtain the following result for the differential structure of the quantum situs of a PDE.

THEOREM 5.16 - *Let* $E_k \subset J\mathcal{D}^k(W)$ *be a PDE on a fiber bundle* $\pi : W \to M$ *with paracompact fiber embedded as a closed subbundle into a* C^∞ *vector bundle* $\pi : E \to M$ *over* M.

1) $\Omega(E_\infty)$ *is a closed submanifold of the Fréchet manifold* $C^\infty(W)$.

2) $\Omega^r(E_\infty) = \Omega(E_\infty) \cap H^r(W)$ *is a closed submanifold of* $H^r(W)$ *such that the spaces of tangent vectors admit a Hilbert structure. Furthermore, the Riemannian structure of* $H^r(W)$ *also identifies a Riemannian structure on* $\Omega^r(E_\infty)$.

3) *Same considerations hold for any statistical set* Ω *of* E_∞.

Under the hypotheses of the above theorem we recognize on $H^r(W)$ as well as on $\Omega^r(E_\infty)$ a metric structure.

PROPOSITION 5.7 - *On each component of* $H^r(W)$ *(resp.* $\Omega^r(E_\infty)$*), and* $\Omega^r \equiv \Omega \bigcap \Omega^r(E_\infty)$ *we can define a metric* d, *consistent with the manifold topology. So, each component* $H^r(W)$ *(resp.* $\Omega^r(E_\infty), \Omega^r$*) becomes a Hausdorff space.*

PROOF. In fact, one has the following metric function $d(\omega, \omega') \equiv \inf \int_{[0,1]} \|\dot{\underline{\omega}}(t)\| dt$, where $\underline{\omega}$ are C^1 paths,

$$\left\{ \begin{array}{l} \underline{\omega} : [0,1] \to H^r(W), \quad (\text{resp. } \underline{\omega} : [0,1] \to \Omega^r(E_\infty), \underline{\omega} : [0,1] \to \Omega^r) \\ \underline{\omega}(0) = \omega, \underline{\omega}(1) = \omega' \\ \|\dot{\underline{\omega}}(t)\| = \sqrt{< \dot{\underline{\omega}}(t), \dot{\underline{\omega}}(t) >} \\ \dot{\underline{\omega}}(t) \in T_{\underline{\omega}(t)} H^r(W), (\text{resp. } \dot{\underline{\omega}}(t) \in T_{\underline{\omega}(t)} \Omega^r(E_\infty), \dot{\underline{\omega}}(t) \in T_{\underline{\omega}(t)} \Omega^r) \end{array} \right\}$$

where $\dot{\underline{\omega}}(t)$ is the velocity of $\underline{\omega}$; $<,>_{\underline{\omega}(t)}$ is the inner product on the tangent space $T_{\underline{\omega}(t)} H^r(W)$, (resp. $T_{\underline{\omega}(t)} \Omega^r(E_k), T_{\underline{\omega}(t)} \Omega^r$). □

5.3 - MATHEMATICAL LOGIC AND QUANTIZATION OF PDEs

In this section we shall justify on the ground of the mathematical logic what we mean by the proposition: "*quantization of PDE*". Let us start with some definitions and fundamental results of mathematical logic. (See also e.g., ref.[11].)

DEFINITION 5.13 - 1) *An n-***ary operation** *on the set A is a function*

$$t : A^n \to A, \quad n \in \mathbf{N};$$

the number n is called **arity** *of t.*

2) *A* 0-**ary** *operation on a set A is a function $t : A^0 \equiv \emptyset \to A$. Hence, it can be regarded as a distinguished element of A. (For example, if A is a group, the 0-ary operation e gives the identity element of the group.)*

3) *A* **type** *$\mathcal{T}$ is a couple $\mathcal{T} \equiv (T, ar)$, where T is a set and ar is a map $ar : T \to \mathbf{N}$. We denote*

$$T_n \equiv \{t \in T \mid ar(t) = n\}.$$

4) *An* **algebra** *A* **of type** *$\mathcal{T}$, or $\mathcal{T}$-***algebra**, *is a triplet $(A, \mathcal{T}, \sigma)$, where A is a set, $\mathcal{T}$ is a type and σ is a section of the bundle*

$$\bigcup_{t \in T} Map(A^{ar(t)}; A) \to T, \quad i.e., \sigma(t) \equiv t_A : A^{ar(t)} \to A.$$

*The elements $t \in T$ are called n-***ary** *$\mathcal{T}$*-**algebra operations**.

5) *$\mathcal{T}$-algebras A and B are* **equal** *iff $A = B$ and $t_A = t_B, \forall t \in T$.*

6) *A $\mathcal{T}$*-**subalgebra** *of A is a subset $B \subset A$ such that it is a $\mathcal{T}$-algebra with respect to the operation restrictions to B of those in A.*

REMARK 5.10 - 1) Any intersection of subalgebras is a subalgebra.

2) $\emptyset$ is a subalgebra iff $T_0 = \emptyset$.

3) Given any subset $X \subset A$, there is a unique smallest subalgebra containing X- namely, the subalgebra $\cap\{U \mid U =$ subalgebra of A, $U \supseteq X\}$. We call this the **subalgebra generated by** X and denote it by $< X >_{\mathcal{T}}$ or simply $< X >$.

4) *For all T, every $\mathcal{T}$-algebra has a unique smallest subalgebra.*

EXAMPLE 5.8 - 1) Rings may be considered as algebras of type

$$\left\{ \begin{array}{l} \mathcal{T} \equiv (\{0, -, +, \cdot\}, ar) \\ ar(0) = 0, ar(-) = 1, ar(+) = 2, ar(\cdot) = 2 \end{array} \right\}$$

2) An R-module ($R \equiv$ ring) may be regarded as algebra of type

$$\left\{ \begin{array}{l} \mathcal{T} \equiv (\{0, -, +, \cdot\} \cup R, ar) \\ ar(0) = 0, ar(-) = 1, ar(+) = 2, ar(\lambda) = 1, \forall \lambda \in R \end{array} \right\}$$

3) Any group G can be considered as an algebra of type $\mathcal{T} \equiv (\{*\}, ar)$, with $ar(*) = 2$, or an algebra of type

$$\left\{ \begin{array}{l} \mathcal{T}' \equiv (\{e, i, *\}, ar) \\ ar(e) = 0, ar(i) = 1, ar(*) = 2 \end{array} \right\}$$

Then any $\mathcal{T}'$-subalgebra of a finite group is a group, but this is not true for any non-empty $\mathcal{T}$-subalgebra, if G is not a finite group.

DEFINITION 5.14 - 1) a **homomorphism** *between $\mathcal{T}$-algebras A and B is a function $\phi : A \to B$ such that the following diagram is commutative:*

$$\begin{array}{ccccl} & A^n & \xrightarrow{t_A} & A & \\ \phi^n & \downarrow & & \downarrow & \phi, \quad \forall t \in T, i.e., \phi(t_A(a_1, \cdots, a_n)) = t_B(\phi(a_1), \cdots, \phi(a_n)). \\ & B^n & \xrightarrow[t_B]{} & B & \end{array}$$

2) *An* **isomorphism** *of $\mathcal{T}$-algebras is an invertible homomorphism.*

2 REMARK 5.11 - The set of $\mathcal{T}$-algebras and homomorphisms of $\mathcal{T}$-algebras is a category.

DEFINITION 5.15 - *A* **free $\mathcal{T}$-algebra** *is a couple (F, σ), where F is a $\mathcal{T}$-algebra, and σ is a mapping $\sigma : X \to F$, such that for every $\mathcal{T}$-algebra A and function $\tau : X \to A$, there exists a unique homomorphism $\phi : F \to A$ such that $\tau = \phi \circ \sigma$, i.e., the following diagram is commutative:*

$$\begin{array}{cccc} & X & \stackrel{\sigma}{\to} & F \\ \tau & \downarrow & \swarrow & \\ & A & & \end{array}$$

Then, we say that X is a set of **free generators** *of the $\mathcal{T}$-algebra F (σ is necessarily injective).*

THEOREM 5.17 - 1) *For any set X and any type $\mathcal{T}$, there exists a free $\mathcal{T}$-algebra on X. This free $\mathcal{T}$-algebra on X is unique up to isomorphisms.*

2) *Let F be the free $\mathcal{T}$-algebra on the set $X_n \equiv \{x_1, \cdots, x_n\}$. Let A be any $\mathcal{T}$-algebra. Then, we have a canonical mapping (associated to $w \in F$)*

$$\left\{ \begin{array}{l} w_A : A^n \to A \\ w_A(a_1, \cdots, a_n) = w_A(a_1, \cdots, a_n) = \phi(w) \end{array} \right\}$$

where ϕ is the unique homomorphism

$$\left\{ \begin{array}{l} \phi : F \to A \\ \phi(x_i) = a_i, i = 1, \cdots, n \end{array} \right\}$$

☐ *In particular, if* $A = F, a_i = x_i, i = 1, \cdots, n$, *then* ϕ *is the identity and*

$$w_A(x_1, \cdots, x_n) = w.$$

DEFINITION 5.16 - 1) *A* $\mathcal{T}$**-algebra variable** *is an element of the free generating set of a free* $\mathcal{T}$*-algebra.*

2) *A* $\mathcal{T}$**-word** *in the variables* $x_1, \cdots, x_n$, *is an element of the free* $\mathcal{T}$*-algebra on the set*

$$X_n \equiv \{x_1, \cdots, x_n\}$$

of free generators.

3) *A* **word** *in the elements* $a_1, \cdots, a_n$ *of a* $\mathcal{T}$*-algebra* A, *is an element*

$$w_A(a_1, \cdots, a_n) \in A,$$

where w *is a* $\mathcal{T}$*-word in the variables* $x_1, \cdots, x_n$.

4) *Among the words in the variables* $x_1, \cdots, x_n$ *there are the words* $x_i, (i = 1, \cdots, n)$, *having the property that*

$$x_i(a_1, \cdots, a_n) = a_i.$$

These variables may also be regarded as coordinate functions.

DEFINITION 5.17 - 1) *An* **identical relation** *on* $\mathcal{T}$*-algebra* A *is a pair* $(u, v) \in F \times F$. *There is an* n *for which* u, v *are in the free algebra on* $X_n = \{x_1, \cdots, x_n\}$, *and we say that* (u, v) *is an* n**-variable identical relation** (u, v) *for any such* n.

2) *We say that* $(u, v) \in F \times F$ *is a* **law** *of the* $\mathcal{T}$*-algebra* A, *(or that* A **satisfies the** n**-variable identical relation** (u, v), *if* $\phi(u) = \phi(v)$ *for every homomorphism* $\phi : F \to A$, *(or equivalently* $u(a_1, \cdots, a_n) = v(a_1, \cdots, a_n)$ *for all* $a_1, \cdots, a_n$).

3) *Let* L *be a set of identical relations on* $\mathcal{T}$*-algebra. We call* **variety of** $\mathcal{T}$**-algebra defined by** L *the class* V *of all* $\mathcal{T}$*-algebras which satisfies all the identical relations satisfied by every algebra in* V. *(This set includes* L, *but may be larger.)*

EXAMPLE 5.9 - 1)

$$\left\{ \begin{array}{l} \mathcal{T} \equiv (\{*\}, ar), ar(*) = 2 \\ L \equiv \{x_1 * (x_2 * x_2), (x_1 * x_2) * x_3\} \end{array} \right\}$$

If A satisfies this identical relation, then $a * (b * c) = (a * b) * c$, for all $a, b, c \in A$. Thus the operation on A is associative and A is a semigroup. Hence, the variety defined by L in this case is the class of all semigroups.

2)

$$\left\{ \begin{array}{l} \mathcal{T} \equiv (\{e, i, *\}, ar), ar(e) = 0, ar(i) = 1, ar(*) = 2 \\ L \equiv \{x_1 * (x_2 * x_2), (x_1 * x_2) * x_3, (e * x_1, x_1), (i(x_1) * x_1, e)\} \end{array} \right\}$$

Thus, the first law ensures that $*$ is an associative operation in every algebra of the variety defined by L. The second law shows that the distinguished element e is always a left identity. The third law guarantees that $i(a)$ is a left inverse of the element e. Hence, the algebras of the variety are groups.

DEFINITION 5.18 - *Let V be the variety of $\mathcal{T}$-algebras defined by the set L of laws.*
1) A $\mathcal{T}$-algebra R in V is **(relatively) free algebra of** *V on the set X of* **(relatively) free generators**, *(where a function $\sigma : X \to R$ is given, usually an inclusion), if, for any algebra A in V and every function $\tau : X \to A$, there exists a unique homomorphism $\phi : R \to A$ such that $\phi \circ \tau$, i.e., one has the following commutative diagram:*

$$\begin{array}{ccc} X & \xrightarrow{\sigma} & R \\ & \searrow & \downarrow \phi \\ & & A \end{array}$$

2) An algebra is **relatively free** *if it is a free algebra of some variety.*

THEOREM 5.18 - *For any type $\mathcal{T}$, and any set L of laws, let V be the variety of $\mathcal{T}$-algebras defined by L. For any set X, there exists a free $\mathcal{T}$-algebra of V on X.*

DEFINITION 5.19 - *Let V be a variety of $\mathcal{T}$-algebras. Then, we call* **V-variables** *an element of the free generating set of a free algebra of V, and* **V-word** *in the V-variables $x_1, \cdots, x_n$ an element of the free algebra of V on the free generators $\{x_1, \cdots, x_n\}$.*

EXAMPLE 5.10 - 1)

$$\left\{ \begin{array}{l} \mathcal{T} \equiv (\{e, i, *\}, ar), ar(e) = 0, ar(i) = 1, ar(*) = 2 \\ L \equiv \{x_1 * (x_2 * x_2), (x_1 * x_2) * x_3, (e * x_1, x_1), (i(x_1) * x_1, e)\} \end{array} \right\}$$

Then, V is the variety of abelian groups. In this case, the relatively free algebra on $\{x_1, \cdots, x_n\}$ is the set of all

$$x_1^{r_1} x_2^{r_2} \cdots, x_n^{r_n}, \text{with} \quad r_i \in \mathbf{Z}.$$

2) Let κ be a field. Then, the vector spaces over κ form a variety V of algebras, and every vector space over κ is a free algebra of L.

3) Let R be a commutative ring with unity 1, and let V be the free variety of commutative rings S which contain R as a subring and in which 1_R is a multiplicative identity in S. Then, the free algebra of V on the set X of variables is the polynomial ring over R in the elements of X.

DEFINITION 5.20 - 1) *Let $\mathcal{T} \equiv (\{F, \Rightarrow\}, ar)$ be a type with $ar(F) = 0, ar(\Rightarrow) = 0$. Then, any $\mathcal{T}$-algebra is called a* **proposition algebra**.

2) The **proposition algebra $\mathcal{P}(X)$ of the proposition calculus on the set** *X* (**set of proposition variables**) *is the free $\mathcal{T}$-algebra on X.*

3) *In any proposition algebra, we introduce the following further operations:*

$$\begin{array}{ll} \sim: & \sim p = p \Rightarrow F \quad (ar(\sim) = 1) \\ \vee: & p \vee q = (\sim) \Rightarrow q \quad (ar(\vee) = 2) \\ \wedge: & p \wedge q = \sim (\sim p \vee \sim q) \quad (ar(\wedge) = 2) \\ \Leftrightarrow: & p \Leftrightarrow q = (p = q) \wedge (q \Rightarrow p) \quad (ar(\Leftrightarrow) = 2) \end{array}$$

☐ *We shall often omit brackets in writing* $\sim p \vee \sim q$ *for* $(\sim p) \vee (\sim q)$.

4) *A* **valuation** *of* $\mathcal{P}(X)$ *is a proposition algebra homomorphism* $v : \mathcal{P}(X) \to \mathbf{Z}_2$. *We say that* $p \in \mathcal{P}(X)$ *is* **true** *with respect to* v *if* $v(p) = 1$, *and that* p *is* **false** *with respect to* v *if* $v(p) = 0$.

☐ *Since* X *is a set of free generators of* $\mathcal{P}(X)$, *the values* $v(x)$ *for* $x \in X$ *may be assigned arbitrary. These values, once assigned, determine the homomorphism* v *uniquely and so determine* $v(p)$ *for all* $p \in \mathcal{P}(X)$.

5) *Let* $A \subseteq \mathcal{P}(X)$ *and* $q \in \mathcal{P}(X)$. *We say that* q *is a* **consequence** *of the set* A *of assumptions, or that* A **semantically implies** q, *if* $v(q) = 1$ *for every valuation* v *such that* $v(p) = 1$ *for all* $p \in A$. *We write:*

$$A \models q.$$

Set

$$Con(A) \equiv \{p \in \mathcal{P}(X) \mid A \models p\} \equiv \text{set of all } \textbf{consequences} \text{ of } A.$$

6) *Let* $p \in \mathcal{P}(X)$. *We say that* p *is* **valid**, *or is a* **tautology**, *if* $v(p) = 1$, *for every valuation* v *of* $\mathcal{P}(X)$.

PROPOSITION 5.8 - 1) *The proposition* p *is a tautology if* $\emptyset \models p$. *(We shall write this simply as* $\models p$.*)*

2) *Con is a* **closure operation** *on* $\mathcal{P}(X)$, *i.e., has the following properties:*

$$\left\{ \begin{array}{l} A \subseteq Con(A) \\ \text{If} \quad A_1 \subseteq A_2, \text{then} \quad Con(A_1) \subseteq Con(A_2) \\ Con(Con(A)) = Con(A) \end{array} \right\}$$

DEFINITION 5.21 - (PROOF IN PROPOSITIONAL CALCULUS). *For the propositional calculus on the set* X, *we take as axioms all elements of the subset*

$$\mathcal{A} \equiv \mathcal{A}_1 \cup \mathcal{A}_2 \cup \mathcal{A}_3 \subset \mathcal{P}(X),$$

where

$$\boxed{\begin{array}{l} \mathcal{A}_1 \equiv \{p \Rightarrow (q \Rightarrow p) \,|\, p, q \in \mathcal{P}(X)\} \\ \mathcal{A}_2 \equiv \{(p \Rightarrow (q \Rightarrow r)) \Rightarrow ((p \Rightarrow q) \Rightarrow (p \Rightarrow r)) \,|\, p, q, r \in \mathcal{P}(X)\} \\ \mathcal{A}_3 \equiv \{\sim\sim p \Rightarrow \,|\, p \in \mathcal{P}(X)\} \end{array}}$$

1) *Let* $q \in \mathcal{P}(X)$ *and let* $A \subseteq \mathcal{P}(X)$. *In the propositional calculus on the set* X, *a* **proof of** q **from the assumption** A *is a finite sequence* $p_1, p_2, \cdots, p_n$ *of elements* $p_i \in \mathcal{P}(X)$ *such that* $p_n = q$ *and for each* i, *either* $p_i \in \mathcal{A} \cup A$ *or for some* $j, k < i$, *we have*

$$p_k = (p_j \Rightarrow p_i).$$

2) *Let* $q \in \mathcal{P}(X)$ *and let* $A \subseteq \mathcal{P}(X)$. *We say that* q *is a* **deduction** *from* A, *or* q *is* **prevable** *from* A, *or that* A **syntactically implies** q, *if there exists a proof of* q *from* A. *We shall write this by*

$$A \vdash q.$$

Set

$$Ded(A) \equiv \text{set of all deductions from } A.$$

3) *Let* $p \in \mathcal{P}(X)$. *We say that* p *is a* **theorem** *of the propositional calculus on* X *if there exists a proof of* p *from* $\emptyset$ *and we write simply:*

$$p \quad \text{is a theorem if} \quad \vdash p.$$

THEOREM 5.19 - 1) *If* $q \in Ded(A)$, *then* $q \in Ded(A')$ *for some finite subset* A' *of* A.
2) *Ded is a closure operation on* $\mathcal{P}(X)$.
3) (**The substitution theorem**). *Let* X *and* Y *be any two sets, and let*

$$\phi : \mathcal{P}(X) \to \mathcal{P}(Y)$$

be a homomorphism of the (free) proposition algebra on X *into the (free) proposition algebra on* Y. *Let* $w = w(x_1, \cdots, x_n)$ *be any element of* $\mathcal{P}(X)$ *and let* A *be any subset of* $\mathcal{P}(X)$. *Put* $a_i = \phi(x_i)$. *Then, we have the following:*

$$\left\{\begin{array}{llll} (a) & A \vdash w, & \text{then} & \phi(A) \vdash w(a_1, \cdots, a_n) \\ (b) & A \models w, & \text{then} & \phi(A) \models w(a_1, \cdots, a_n) \end{array}\right\}$$

EXAMPLE 5.11 - 1) The algebra $\mathbf{Z}_2$ can be made into a proposition algebra by defining $F_{\mathbf{Z}_2} = 0$ and $w \Rightarrow u = 1 + m(1 + n)$. Furthermore, one can easily see the expressions $\sim, \vee, wedge$ in $\mathbf{Z}_2$ in terms of multiplications and addition.

2) $Ded(A)$ is the smallest subset D of $\mathcal{P}(X)$ such that $D \supseteq \mathcal{A}(X) \cup A$ and such that if $p, p \Rightarrow \in D$, then $q \in D$ also.

DEFINITION 5.22 - *A* **logic** $\mathcal{L}$ *is a triplet* $\mathcal{L} \equiv (P, \mathcal{V}, \mathcal{P})$, *where:*

(i) P *is a set of elements (called* **propositions***);*

(ii) $\mathcal{V}$ *is a set of functions (called* **valuations***)* $P \to W$ *(*$\equiv$ **value set***);*

(iii) $\mathcal{P} \equiv (\{p_1, \cdots, p_{nA}\})$. *For each subset* $A \subset P$, $\{p_1, \cdots, p_{nA}\}$ *is a set of finite sequences of elements of* P *(called* **proofs from the assumptions***).*

THEOREM 5.20 - *The propositional calculus on the set* X *is a logic:*

$$\mathcal{L} = Prop(X)$$

such that:

(i) $P = \mathcal{P}(X)$;

(ii) $\mathcal{V} \equiv$ *set of all homomorphisms* $\mathcal{P}(X) \to \mathbf{Z}_2$;

(iii) *For each subset* $A \subseteq \mathcal{P}(X)$, *the set of proofs are defined above.*

Now, we are ready to give the following important definitions that clarify the meaning of quantization of a physical theory.

DEFINITION 5.23 - 1) *The* **logic of a classical system**, *defined by a dynamic PDE* $E_k \subset J\mathcal{D}^k(W)$, *is the Boolean algebra of subsets of* $\Omega(E_k)_c \equiv$ *set of solutions of* E_k*:*

$$\mathcal{P}(\Omega(E_k)_c)$$

2) *A* **quantum logic** *is an algebra of self-adjoint operators on a Hilbert space* $\mathcal{H}$*:*

$$A \equiv \mathcal{A}d(\mathcal{H}) \subset L(\mathcal{H}).$$

3) **Quantize a classical system**, *defined by a PDE* $E_k \subset J\mathcal{D}^k(W)$, *means to define a homomorphism of algebras*

$$\mathcal{P}(\Omega(E_k)_c) \to A$$

or equivalently a pre-spectral measure on $\Omega(E_k)_c$*:*

$$\Omega(E_k)_c \circ\!\!\to L(\mathcal{H}).$$

(For definition and properties of pre-spectral measure see below and the following refs.[16, 124, 125, 127]*.)*

5.4 - FORMAL AND DIRAC QUANTIZATIONS OF PDEs

Formal quantization is, roughly speaking, a procedure that associates a measure space (quantum situs) to a PDE. This quantization becomes effective if on the quantum situs we recognize a (pre-)spectral measure (quantum spectral measure). Scope of this section is just to develop this linking between formal and Dirac quantizations. We will, also, recast in our geometric approach the concept of canonical quantization, that is the most used in the physical applications. In fact, in the next section we shall show, also, that canonical quantization of a PDE $E_k \subset J\mathcal{D}^k(W)$ is a phenomenon directly governed by singular solutions of an affine PDE associated to E_k (Jacobi equation of E_k) and that can be described in a purely geometric sense.

2 The following relates quantum cobordism to the axioms of "topological quantum field theory" as introduced by Atiyah [5,6].

THEOREM 5.21 - 1) *Let* $E_k \subset J\mathcal{D}^k(W)$ *be a formally integrable PDE on a fiber bundle* $\pi : W \to M, \dim M = n$. *Let* $\pi_k : H \to J\mathcal{D}^k(W)$ *be an hermitian vector bundle over* $J\mathcal{D}^k(W)$ *such that on the fibers an inner structure is defined such that* $(H, W; \mathbf{Q}\mathbb{B})$ *is a super-bundle of geometric objects over* M, *for some functor* $\mathbf{Q}\mathbb{B} : \mathcal{C}(W) \to \mathcal{C}(H)$. *We denote* $\mathcal{H} \equiv C_0^\infty(H)$. *Let us assume that* $L \subset J\mathcal{D}^k(W)$ *is an integral manifold such that* $\omega_1^{(k)}(L) = 0$. *Then, for any section* $\eta : J\mathcal{D}^k(W) \to \Lambda_n^0 J\mathcal{D}^k(W)$ *the fiber bundle* $H|_L \to L \subset J\mathcal{D}^k(W)$ *identifies a Hilbert space, that we shall denote by* $\mathcal{H}(L)$.

2) *If* $N \subset E_k$ *is a Cauchy hypersurface data on* E_k *such that* $N \subset L$ *then one also identifies a Hilbert space* $\mathcal{H}(N) \subset \mathcal{H}(L)$. *Furthermore, let* N^* *denote* N *with opposite orientation. Then, we set:*

(a) $\mathcal{H}(N^*) = \mathcal{H}(N)^*$;

(b) $\mathcal{H}(N_1 \dot{\bigcup} N_2) = Hom_{\mathbf{C}}(\mathcal{H}(N_1), \mathcal{H}(N_2))$. *Then, there exists a vector* $U(L) \in \mathcal{H}(N)$ *for each component oriented integral manifold with boundary* N. *Furthermore, if*

$$L = L_1 \bigcup_{N_2} L_2, \quad \partial L_1 = N_1^* \dot{\bigcup} N_2, \partial L_2 = N_2^* \bigcup N_3,$$

the vectors $U(L_1), U(L_2), U(L_3)$ *are related by*

$$U(L) = j(U(L_1), U(L_2)) = U(L_2) \circ U(L_1) \in Hom_{\mathbf{C}}(\mathcal{H}(N_1), \mathcal{H}(N_2))$$

where j *is the canonical mapping:*

$$\left\{ \begin{array}{l} j : Hom_{\mathbf{C}}(\mathcal{H}(N_1), \mathcal{H}(N_2)) \times Hom_{\mathbf{C}}(\mathcal{H}(N_2), \mathcal{H}(N_3)) \to Hom_{\mathbf{C}}(\mathcal{H}(N_1), \mathcal{H}(N_3)) \\ j(A, B) = A \circ B. \end{array} \right\}$$

The linear mappings $U(L), U(L^*)$ *are adjoint to each other. We also have:*

(c) $U(\emptyset) = \mathbf{C}$, ($\emptyset \equiv$ *empty hypersurface);*

(d) $U(N \times I) \in Hom_{\mathbf{C}}(\mathcal{H}(N_0), \mathcal{H}(N_1))$, *where* $I \equiv [0, 1] \subset \mathbf{R}, N_0 \equiv N \times \{0\}, N_1 \equiv \times\{1\}$. *Moreover, the correspondence that associates to each Cauchy hypersurface*

$N, \mathcal{H}(N)$ is of functorial nature and homotopy invariant, i.e., the group $Diff^+(N)$ of orientation preserving diffeomorphisms of N acts on $\mathcal{H}(N)$ via its group components $\Gamma(N)$.
3) To any closed integral manifold contained in $J\mathcal{D}^k(W)$ there corresponds a numerical invariant belonging to $\mathbf{C}$. This numerical invariant can be computed from any decomposition $L = L_1 \bigcup_N L_2$. One has $U(L^*) = \overline{U(L)} \in \mathbf{C}$.
PROOF. 1) Let us consider the following commutative diagram:

$$\begin{array}{ccc} j_L^* H & \rightarrow & H \\ \downarrow & & \downarrow \pi_H \\ L & \underset{j_L}{\rightarrow} & J\mathcal{D}^k(W) \end{array}$$

Let us consider the locally convex vector space $\mathcal{H}_L \equiv C_0^\infty(j_L^* H)$. Then, we can define the following inner product in $\mathcal{H}_L$:

$$< \psi_1 | \psi_2 >= \int_{L+} j_L^*((\psi_1, \psi_2)\eta) - \int_{L-} j_L^*((\psi_1, \psi_2)\eta),$$

where $(\psi_1, \psi_2) : J\mathcal{D}^k(W) \rightarrow \mathbf{C}, (\psi_1, \psi_2)(q) = (\psi_1(q), \psi_2(q))$, where the last bracket is the inner product defined on the fiber $H_q, \forall q \in J\mathcal{D}^k(W)$, and L^+ and L^- are the decompositions of L into orientable manifolds by removing the set of singular points $\Sigma(L)$. By completing $\mathcal{H}_L$ with respect to this inner product we get a Hilbert space that we denote by $\mathcal{H}(L)$.
2) By considering a section $\eta : J\mathcal{D}^k(W) \rightarrow \Lambda^0_{n-1} J\mathcal{D}^k(W)$, we can obtain similar results for each Cauchy hypresurface data $N \subset J\mathcal{D}^k(W)$. Now, let us show that it is possible to associate to any integral manifold L such that $\partial L = N_1^* \dot{\bigcup} N_2^*$, a vector $U(L) \in Hom_{\mathbf{C}}(\mathcal{H}(N_1), \mathcal{H}(N_2))$. For any probability measure μ on $\Omega_s(E_k)$ and physical observable $f : \Omega_s(E_k) \rightarrow \mathbf{C}$, one has that $U(L)$ is defined such that for any couple $|\psi_1 >\in \mathcal{H}(N_1), |\psi_2 >\in \mathcal{H}(N_2)$, one has

$$< \psi_2 | U(L) | \psi_1 >= \int_{\Omega(N_1, N_2)} f d\mu.$$

3) Let L be a closed integral manifold. Let us cut L along a Cauchy hypersurface data N. Then, $L = L_1 \bigcup_N L_2$. Then, one has $U(L_1) \in Hom_{\mathbf{C}}(\mathbf{C}, \mathcal{H}(N)) \cong \mathcal{H}(N)$, and $U(L_2) \in Hom_{\mathbf{C}}(\mathcal{H}(N), \mathbf{C} \cong \mathcal{H}(N)^*$. So, we get $U(L) =< U(L_1), U(L_2) >\in \mathbf{C}$.□

DEFINITION 5.24 - 1) *Let L be an integral manifold such that $\partial L = N_1^* \dot{\bigcup} N_2$. Let $|\psi_1 >\in \mathcal{H}(N_1), |\psi_2 >\in \mathcal{H}(N_2)$ be state vectors. Then,*

$$\boxed{< \psi_2 | U(L) | \psi_1 >}$$

is called the **transition amplitude between the two state vectors** $|\psi_1>$ *and* $|\psi_2>$.

2) *In particular, if* $N_2 = \emptyset, U(L) \in \mathcal{H}(N_1)^*$, *then* $U(L)|\psi_1> \in \mathbf{C}$ *and* $\Re(U(L)|\psi_1>)$ *is the* **transition amplitude between the state** $|\psi_1>$ **and the vacuum.**

3) *Conversely, if* $N_1 \equiv \emptyset, U(L) \in \mathcal{H}(N_2)$, *then* $<\psi_2|U(L)|\psi_1> \in \mathbf{C}$, *and* $\Re(<\psi_2|U(L)|\psi_1>)$ *is the* **transition amplitude between the vacuum and the state** $|\psi_2>$.

The following relates the concepts of (proper) Dirac quantization to a concept of algebras of local observables on a manifold similar to the one introduced in the literature (see e.g. ref.[132]).

DEFINITION 5.25 - 1) *Let M be a manifold. Let A be a natural sheaf of locally convex algebras over M, that is there exists a covariant functor $A : \mathcal{C}(M) \to \mathcal{C}_{l.c.}$, where $\mathcal{C}(M)$ is the category having as objects the open sets of M and whose morphisms are local diffeomorphisms of M, and $\mathcal{C}_{l.c.}$ is the category of locally convex algebras, such that:* (a) *$A(U)$ is a locally convex algebra for $U \subset M$; if $U \supset V$ one has the restriction map $\rho_{UV} : A(U) \to A(V)$ such that if $U \supset V \supset W \rho_{WV} = \rho_{WU} \circ \rho_{VU}$;* (b) *$A(M) = \bigcup A(U) \equiv A$ is an object of $\mathcal{C}_{l.c.}$;* (c) *for any covering (U_α) of an open $U \subset M$, and any family (s_α) of elements $s_\alpha \in A(U_\alpha)$ such that $\rho_{U_{\alpha\beta}U_\alpha}s_\alpha = \rho_{U_{\alpha\beta}U_\beta}s_\alpha$ for any couple (α, β) with $U_{\alpha\beta} \equiv U_\alpha \cap U_\beta \neq \emptyset$, there exists a unique $s \in A(U)$ such that $\rho_{U_\alpha U}s = s_\alpha$ for any α;* (d) *For any $f \in Hom_{\mathcal{C}(M)}(U, V)$, one has that $A(f) : A(U) \to A(V)$ is an isomorphism of locally convex algebras. We call A* **an algebra of local observables** *if any element $s \in A$ belongs to a subsheaf of C^*-algebras and there exists a faithful irreducible representation $A \to \mathcal{L}(\mathcal{H})$, where $\mathcal{H}$ is a locally convex topological vector space containing Hilbert spaces $\mathcal{H}_s$ and natural projections $\mathcal{H} \to \mathcal{H}_s$, so that $\mathcal{L}(\mathcal{H}_s)$ can be considered subspaces of $\mathcal{L}(\mathcal{H})$.*[5]

2) *Furthermore, if M is a 4-dimensional space-time with metric field g, we require that:* (**causality**) *if U is space-like separable from U' then $[A(U), A(U')] = 0$; (space-like separable means that there is no casual curve joining a point in U to a point in U'; a causal curve is a curve whose velocity is either time-like or null).*

PROPOSITION 5.9 - *Any Dirac quantization of a PDE $E_k \subset J\mathcal{D}^k(W)$, given on a*

---

[5] Let A be an algebra over $\mathbf{C}$. An **involution** on A is a bijection $^* : A \to A$ such that: (i) $(x^*)^* = x$; (ii) $(x+y)^* = x^* + y^*$; (iii) $(x.y)^* = y^* x^*$; (iv) $(\lambda x)^* = \bar{\lambda} x^*$, $\lambda \in \mathbf{C}$. One has $e^* = e$, where e is the unity of A and $(x^{-1})^* = (x^*)^{-1}$ if x is invertible. An **involutive algebra** is an algebra endowed with an involution. An **involutive Banach algebra** is a Banach algebra with an involution such that $\|x^*\| = \|x\|$. A C^***-algebra** is an involutive Banach algebra such that $\|x\|^2 = \|x^*.x\|$. Example. If $\mathcal{H}$ is a Hilbert space then $L_{\mathbf{C}}(\mathcal{H})$ is a C^*-algebra where the involution is to take the adjoint.

fiber bundle $\pi : W \to M$, i.e., any fiber bundle homomorphism

$$E_\infty \to J\mathcal{D}^\infty(J\mathcal{D}^h(H)^* \bigotimes H),$$

obtained by means of the prolongations

$$Q^{(q)} : J\mathcal{D}^{k+q}(W) \to J\mathcal{D}^q(J\mathcal{D}^h(H)^* \bigotimes H)$$

of some k-order differential operator

$$Q : J\mathcal{D}^k(W) \to J\mathcal{D}^h(H)^* \bigotimes H,$$

where $H \to J\mathcal{D}^p(H), -1 \leq p \leq \infty$, is a hermitian vector bundle, as defined in Theorem 5.21, and such that there is a map $\eta : C^\infty(W) \to C^\infty(\Lambda^0_n M), n = \dim M$, where $\eta(s)$ is a volume form for $s \in Sol(E_k)$, identifies an algebra of local observables over M.

PROOF. In fact, if $p = -1$, and η is independent of $s \in C^\infty(W)$ (proper Dirac quantization), i.e., it is a fixed volume form on M, we can define a Hilbert space $\mathcal{H}$ as the completion of $C^\infty_0(H)$ with respect to the following scalar product:

$$< \psi_1, \psi_2 >= \int_M (\psi_1, \psi_2)\eta,\ \psi_i \in C^\infty_0(H).$$

Furthermore, the sheaf Θ of solutions of E_k and that $\widehat{\Theta}$ of the corresponding quantized fields are related by means of the following exact sequence of sheaves and sheave homomorphisms:

$$\begin{array}{ccc} & & 0 & \\ & & \downarrow & \\ 0 \to K(\Theta) \to \Theta \overset{\wedge}{\to} & & \widehat{\Theta} & \to 0 \\ & & \downarrow & \\ & & \mathcal{L}(\mathcal{H}) & \end{array}$$

where $\wedge$ is the sheaf homomorphism induced by the Dirac quantization, i.e.,

$$s \mapsto \widehat{s},\quad \widehat{s}(\psi) =< Q \circ D^k s, D^h \psi >, \forall \psi \in \mathcal{H}.$$

As $A \equiv \Theta$ is a closed subalgebra of $\mathcal{L}(\mathcal{H})$, it is a sheaf of C^*-algebras, therefore we can consider A as an algebra of local observables.

If $p > -1$, we can consider the locally convex space $\mathcal{H} \equiv C^\infty_0(H)$ of sections $J\mathcal{D}(W) \to H$ with compact support. Therefore, for any section $s \in \Theta$ we can define a Hilbert space $\mathcal{H}(s) \subset \mathcal{H}$ as the completion of $C^\infty_0((D^\infty s)^* H)$ with respect to the scalar product

$$< \psi_1, \psi_2 >= \int_M (D^p s)^*(\psi_1, \psi_2)\eta(s),\ \forall \psi_1, \psi_2 \in C^\infty_0((D^p s)^* H).$$

Then one has yet a diagram like before, but of course, now $\mathcal{L}(\mathcal{H})$ is a sheaf of locally convex algebras only. Furthermore, $\mathcal{L}(\mathcal{H})$ contains as subsheaves, the sheaves of C^*-algebras $\mathcal{L}(\mathcal{H}(s))$. Then, we can take $A \equiv \widehat{\Theta}$ as a sheaf of locally convex algebras, where each element $\widehat{s}$ belongs to the sheaf of C^*-algebras $\mathcal{L}(\mathcal{H}(s))$, hence to the C^*-algebra $\mathcal{L}(\mathcal{H}(s))(M)$. □

DEFINITION 5.26 - *A* **quantum state** *of a PDE $E_k \subset J\mathcal{D}^k(W)$, quantized by means of a (proper) Dirac quantization $\wedge : C^\infty(W) \to \mathcal{L}(\mathcal{H})$ is an element $T \in \mathcal{L}(\mathcal{H})$ such that* $\mathrm{tr}(T) = 1$. *If $T = \psi' \otimes \psi$, with $\psi' \in \mathcal{H}', \psi \in \mathcal{H}$, T is called a* **pure quantum state**. *If the Dirac quantization is proper, we call* **proper quantum state** *a positive Hermitian nuclear operator $T \in \mathcal{L}_1(\mathcal{H})$ such that* $\mathrm{tr}(T) = 1$, *where $\mathcal{L}_1(\mathcal{H})$ is the space of nuclear operators on $\mathcal{H}$. If $T = \psi \otimes \psi, \psi \in \mathcal{H}$, and $\|\psi\| = 1$, T is called a* **proper pure quantum state**. *The space E of proper quantum states can be identified with a convex part of the unit bubble in $K(\mathcal{H})'$ where $K(\mathcal{H})$ is the space of compact operators on $\mathcal{H}$. In fact, from the canonical isomorphism $\mathcal{L}_1(\mathcal{H}) \cong K(\mathcal{H})'$ we can define a proper quantum state by using the following.*

DEFINITION 5.27 - 1) *A* **state** *of a C^* algebra $\mathcal{L}(\mathcal{H})$ is a self-adjoint complex-valued linear functional E on $\mathcal{L}(\mathcal{H})$ which is non-negative or non-negative self-adjoint, and normalized by the constraint:*

$$\sup_{\|A\| \leq 1} E(A^*A) = 1.$$

2) *A* **pure state** *is one which is not a linear combination with positive coefficients of two other states, otherwise it is called* **mixed**.

3) *A set of states is called* **complete** *if the only element of $\mathcal{L}(\mathcal{H})$ which vanishes in every state of the set is zero.*

REMARK 5.12 - Any vector $\psi \in \mathcal{H}$ identifies a pure state: $E(A) = (A(\psi), \psi)$ (called the **vertex state**). When $\mathcal{H}$ is finite-dimensional, all pure states have this form; furthermore in this case the general state has the form $E(A) = \mathrm{tr}(AD)$, where D is a positive semidefinite operator of unit trace (known as the "density operator" in statistical mechanics). Note that the normalized condition is equivalent to the condition $E(e) = 1$. The set of all pure states is always a Borel subset of state space of $\mathcal{L}(\mathcal{H})$.

DEFINITION 5.28 - 1) *The* **variance** *of a self-adjoint element A of $\mathcal{L}(\mathcal{H})$ in a state E is defined as*

$$E(A^2) - E(A)^2.$$

2) *A is said to* **have the exact value** *$E(A)$ in the state E in case its variance vanishes in this state.*

3) *A state of $\mathcal{L}(\mathcal{H})$ in which every self-adjoint element has an exact value is called an* **observation**.

PROPOSITION 5.10 - 1) *If F is a pure state of a C^*-subalgebra B, containing the identity of $\mathcal{L}(\mathcal{H})$, then there exists a pure state E of $\mathcal{L}(\mathcal{H})$, such that $E|_B = F$.*

2) *Any observation on $\mathcal{L}(\mathcal{H})$ is pure.*

3) *The variance of a self-adjoint operator of $A \in \mathcal{L}(\mathcal{H})$ in a state E of $\mathcal{L}(\mathcal{H})$ is identical to the variance in the usual probabilistic sense of the random variable which represents A on the regular compact measure space obtained by imposing on the spectrum of C^*-algebra $B(A) \subset \mathcal{L}(\mathcal{H})$ generated by A the measure representing $E|_{B(A)}$.*

PROOF. See e.g., ref.[132]. □

DEFINITION 5.29 - 1) *If G is a group of automorphisms of the C^*-algebra $\mathcal{L}(\mathcal{H})$, a* **stationary state** *(with respect to G) is defined as one with the property that*

$$E(A) = E(g(A)), \forall A \in \mathcal{L}(\mathcal{H}), g \in G.$$

2) *The state E is called* **ergodic** *in case it is an extreme point of the set of all stationary states.*

PROPOSITION 5.11 - *Every stationary state is w^*-limit of convex linear combinations of ergodic states.*

PROOF. See e.g., ref.[132]. □

REMARK 5.13 - (QUANTUM SPECTRAL CHARACTERIZATION OF FIELDS). As in a Dirac quantization a field $s \in C^\infty(W)$ is represented, in general, by means of element $\hat{s}$ of a locally convex algebra, $\hat{s} \in \mathcal{L}(\mathcal{H})$, (or in a proper Dirac quantization, by an element of a C^*-algebra), the quantum spectral properties of a field s can be obtained by studying the spectral properties of such algebras.

(See e.g. refs.[16,124,125,127].)

2 MEASURE FUNCTORS

As the relativistic uncertainty principle prohibits to representing quantum physical observables as punctual fields, the only right description it results into a global one by means of charges associated to physical entities. So, for any classical entity, that is a differential operator $K : J\mathcal{D}^k(W) \to \mathbb{K}$, relative to a dynamic system characterized by a PDE $E_k \subset J\mathcal{D}^k(W)$, we can identify a map (**charge map**)

$$Q : C^\infty(W) \times C_\bullet(M) \to C^\infty(\Lambda\mathbb{K})^*,$$

where $C_\bullet(M)$ is the space of chains on M and $C^\infty(\Lambda\mathbb{K})$ is the space of differential forms on $\mathbb{K}$, given by

$$Q(s, A) : \beta \mapsto \int_A (K \circ D^k s)^* \beta.$$

In fact, a charge

$$Q \equiv \int_A (K \circ D^k s)^* \beta$$

identifies a numerical function $f : \Omega \to \mathbf{R}$ on the statistical set Ω of a scattering process $c_{in} \to c_{out}$ given by

$$Q : \omega \equiv D^k s(A) \mapsto \int_A (K \circ D^k s)^* \beta.$$

More generally, we can say that on quantum level a procedure of measure of a physical entity is identified by means of a **random variable** f on $\Omega_s \supset \Omega$, even if Ω_s is endowed with a measure μ. Therefore, if μ is a definite measure on Ω_s, we can obtain the **quantum mean value** of f by considering the Feynman-like integral

$$\ll f \gg = \int_{\Omega_s} f d\mu / \int_{\Omega_s} d\mu$$

of the random variable f, i.e., the expectation of f with respect to the normalized measure ν on Ω_s defined by μ. In this way we can interpret the quantum fluctuations of a dynamic conservation law associated to $E_k \subset J\mathcal{D}^k(W)$. The generalized **Bohr Sommerfield quantization conditions** result in requiring the charge $Q.s$ corresponding to a configuration $s : M \to W$, to be multiple of integer. This imposes topological restrictions on the configurations [103]. More precisely we have the following.

THEOREM 5.22 - *Let $L \subset J\mathcal{D}^k(W)$ be an n-dimensional integral manifold solution of a PDE $E_k \subset J\mathcal{D}^k(W)$. Let L lead the charge $Q = \int_a i_L{}^* \beta$, where β is an $(n-1)$-differential form on $J\mathcal{D}^k(W)$; $i_L : L \hookrightarrow J\mathcal{D}^k(W)$ is the canonical inclusion and a is an $(n-1)$-chain on L. If $i_L{}^* \beta = k\theta$, where $k \in \mathbf{R}$ and θ belongs to the image of the canonical map*

$$\left\{ \begin{array}{ccc} \pi^{n-1}(L) & \overset{j_{n-1}}{\longrightarrow} & H^{n-1}(L) \\ \| & & \|\wr \\ [L; S^{n-1}] & \to Hom(H^{n-1}(S^{n-1}), H^{n-1}(L)) \cong & Hom(\mathbf{R}, H^{n-1}(L)) \\ [f] \mapsto [f^* \mu] & & \end{array} \right\}$$

where μ is the canonical volume form on S^{n-1}, then Q is **quantized**, *that is $Q = k.n, k \in \mathbf{R}, n \in \mathbf{Z}$.*

PROOF. Recall that any $(n-1)$-differential form α on S^{n-1} can be written as $\alpha = b\mu + d\gamma$, where $b \in \mathbf{R}$ and γ is an $(n-2)$-differential form on S^{n-1}. In fact, $H^{n-1}(S^{n-1}) \cong \mathbf{R}$. Then, we get

$$\int_a i_L{}^* \beta = \int_a \theta = k \int_a f^* \mu = k \deg(\bar{f}) \int_{S^{n-1}} \mu,$$

where $f : L \to S^{n-1}$ is the representative of the cohomology class $[f] \in \pi^{n-1}(L)$, $\bar{f} \equiv f|_\triangle$, where $\triangle$ is the support of a. So we get, $\int_a i_L{}^*\beta = \bar{k}.n$, where $\bar{k} \equiv k.vol(S^{n-1})$ and $n \equiv \deg(\bar{f}) \in \mathbf{Z}$ is the Brouwer degree of $\bar{f}$. □

REMARK 5.14 - Note that transition amplitudes for scattering processes in a PDE $E_k \subset J\mathcal{D}^k(W)$, can be considered as charges associated to E_k. More precisely, for any probability measure μ on $\Omega_s(E_k)$, and for any numerical function

$$f : \Omega_s(E_k) \to \mathbf{K}, (\mathbf{K} \equiv \mathbf{R}, \mathbf{C}),$$

we can define the (μ, f)**-quantum transition amplitude** between two Cauchy hypersurfaces data σ_1 and σ_2 of E_k as the integral

$$< \sigma_1, \sigma_2 > \equiv \int_{\Omega_s(\sigma_1,\sigma_2)} f d\mu,$$

where $\Omega_s(\sigma_1, \sigma_2) \subset \Omega_s(E_k)$ is just the set of (singular) quantum cobords cobording σ_1 and σ_2. Now, in order to extend the procedure of Dirac quantization to random variables associated to PDE, and so to relate the formal quantization with the Dirac quantization, we shall introduce the concepts of measure functor, stochastic process and spectral measure. So, we shall investigate under which conditions a Dirac quantization identifies a full quantization of a PDE, i.e. a quantization of the related random variables also. Here we shall give some fundamental results on the theory of measure. In order to represent mathematical objects with fully covariant behaviour more directly, we shall adopt the language of categories and functors.

Let $\mathcal{M}$ be the **category of measurable spaces**, that is the set of objects, $Ob(\mathcal{M})$, of $\mathcal{M}$ is given by all abstract measurable spaces $(X, \mathcal{T})$, where X is a set and $\mathcal{T}$ is a σ-algebra, that is a non-void collection of subset of X, closed for complementation and countable union (and intersection) and containing the empty set. (A σ-algebra on X results in a **monotone class** of sets of X, such that for every monotone sequence $\{A_k\}$ of sets in X, we have $\lim_k A_k \in X$.) The set of morphisms, $Hom(\mathcal{M})$, of $\mathcal{M}$ is given by all measurable transformations of $(X, \mathcal{T})$ into $(Y, \mathcal{R})$, $f : X \to Y$, that is $f^{-1}(B) \in \mathcal{T}, \forall B \in \mathcal{R}$. A subset $A \subset X$ of a measurable space $(X, \mathcal{T})$ is called **measurable** if A belongs to $\mathcal{T}$. If X is a topological space there is a canonical σ-algebra on X : that generated by the open (or closed) subset of X. This is called **Borel σ-algebra** on X and denoted by $\mathcal{B}(X) \equiv \mathcal{B}$.[6]

[6] Let $\mathcal{P}(X)$ denote the collection of all subsets of X. Then, for any subset $L \subset \mathcal{P}(X)$ the σ-algebra **generated** by L is the smallest σ-algebra, $\sigma(L)$, containing L. If $f{:}X \to Y$ is a mapping between two sets X and Y, and $\mathcal{T}$ is a σ-algebra on Y, then $f^{-1}(\mathcal{T}) \equiv \{A \subset X | A = f^{-1}(B), B \in \mathcal{T}\}$ is a σ-algebra on X. Of course $f{:}(X, f^{-1}(\mathcal{T})) \to (Y, \mathcal{T})$ is a measurable mapping. If $\mathcal{T} = \sigma(L)$, then $f^{-1}(\mathcal{T})$ is the σ-algebra generated by $f^{-1}(L)$. A **Borel mapping** $f{:}X \to Y$ between two topological spaces X and Y is a measurable map $(X, \mathcal{B}(X)) \to (Y, \mathcal{B}(Y))$. Any continuous map is Borel one.

DEFINITION 5.30 - *A* **measure** *is a covariant functor* $\mathbf{M}^+ : \mathcal{M} \to \mathcal{I}(\mathbf{R}) \equiv$ *category of* $\mathbf{R}$*-valued set functions given by:* (a) $\mathbf{M}^+(X, \mathcal{T}) \equiv \mu \equiv$ **measure** *on* $\mathcal{T}$, *that is a non-negative valued set function* $\mu : \mathcal{T} \to \overline{\mathbf{R}}$ *(possibly taking the value* $+\infty$*), equal to* 0 *on the empty set (monotone increasing), and countably additive (or* σ*-additive);* (b) *if*

$$f \in Hom((X, \mathcal{T}), (Y, \mathcal{R})), \mathbf{M}^+(f) : \mathbf{M}^+(X, \mathcal{T}) \equiv \mu \mapsto \mathbf{M}^+(f)(\mu) \equiv \nu = \mu \circ f^{-1}.$$

A measure μ *on* $(X, \mathcal{T})$ *is* **finite** *if* $\mu(X) < +\infty$. *A measure* μ *on* $(X, \mathcal{T})$ *is called* **complete** *if* $E \in \mathcal{T}, F \subset E, \mu(E) = 0 \Rightarrow F \in \mathcal{T}$. *A measure* μ *on* (X, T) *is called* σ**-finite** *if for all* $E \in \mathcal{T}$ *there is a uniquely sequence* $C_i (i = 1, 2, \cdots) \in \mathcal{T}$, *such that*

$$E = \bigcup_{i=1}^{\infty} C_i$$

and $\mu(E_i) < +\infty$. *Given a measure* μ *on* $(X, \mathcal{T})$ *there is a uniquely complete measure* $\bar{\mu}$ *on* $(X, \overline{\mathcal{T}})$, *given by:* (a) $\overline{\mathcal{T}}$ *is the* σ*-algebra of all sets of the form* $E \triangle N$, *with* $E \in \mathcal{T}$ *and* $N \subset F \in \mathcal{T}$ *with* $\mu(F) = 0$; (b) $\bar{\mu}(E \triangle N) = \mu(E)$.[7]
$\bar{\mu}$ *is called the* **completion** *of* μ. $\overline{\mathcal{T}}$ *is called the* **completion of** $\mathcal{T}$ **with respect to** μ. *A measure on a set* X *can also be defined with respect to a ring:* $\mu : R \to \mathbf{R}^+$ *such that* $X = \bigcup_{n=1}^{\infty} E_n$ *for some sequence of sets* $\{E_n\}$ *in* R. *Then, there is an extension of* μ *for a measure* ν *defined on* $\mathcal{S}(R)$, *the* σ*-ring generated by* R. *If* μ *is* σ*-finite on* R, *then the extension is unique, and is* σ*-finite on* $\mathcal{S}(R)$.
EXAMPLE 5.12 - 1) The **Lebesgue measure** on $\mathbf{R}$ is obtained by the measure μ on the ring E of elementary figures:

$$E \equiv \{E = \bigcup_{i=1}^{n}]a_i, b_i]\}$$

with $b_i < a_{i+1}$, putting $\mu(]a, b]) = b - a$. The corresponding σ-ring $\mathcal{S}(E)$ is the class of Borel sets $\mathcal{B}(\mathbf{R})$ in $\mathbf{R}$. The unique extension of μ to $\mathcal{B}(\mathbf{R})$ is called the **Lebesgue measure** on $\mathbf{R}$.
2) A similar example can be obtained considering the **Stieltjes measures** on $\mathbf{R}$ associated to a monotone increasing function $F : \mathbf{R} \to \mathbf{R}$, that is everywhere continuous on the right, putting $\mu(]a, b]) = F(b) - F(a)$. F is called **Stieltjes measure function**. This can be extended to $\mathbf{R}^n$ by considering an n-dimensional Stieltjes measure function $F : \mathbf{R}^n \to \mathbf{R}$ satisfying the following properties: (i) F is continuous

[7] $E \triangle N \equiv (E - N) \cup (N - E) \equiv$ **symmetric difference** of two sets E, N. Note that $E \triangle N = N \triangle E$.

on the right in each variable; (ii) $\mu_F(I) = \sum_{i=1}^{2^n} y_i F(v_i) \geq 0$, where v_i are the 2^n vertices of the set

$$I \in P^n \equiv \{\prod_{i=1}^{n}]a_i, b_i]\}$$

and $y_i = +1$ for the vertex in which each coordinate is the largest and $y_i = (-1)^r$ if the vertex v_i is such that r of its coordinates are at the lower bound; (iii) $F(x_1, \cdots, x_n) \to 0$ as any one of $x_1, \cdots, x_n \to -\infty$; $F(x_1, \cdots, x_n) \to 1$ as all of $x_1, \cdots, x_n \to +\infty$.

DEFINITION 5.31 - *A triplet $(X, \mathcal{T}, \mu)$ is called a* **measure space**. *The set of measures on $(X, \mathcal{T})$ is denoted by $M^+(X, \mathcal{T})$. The* **support of the measure** *$\mu = M^+(X, \mathcal{T})$ is the complement of the largest (open) set $V \subset X$ such that $\mu(V) = 0$. Let $\mathcal{M}.$ be the canonical sub-category of $\mathcal{M}$ of measurable spaces like $(X, \mathcal{P}(X))$, where $\mathcal{P}(X)$ is the collection of all subsets of X, (i.e., $\mathcal{M}.$ is the category of discrete topological spaces). Any measure μ on $(X, \mathcal{T}) \in Ob(\mathcal{M})$ identifies an* **outer measure** *μ^* (resp.* **inner measure** *μ_*) on $(X, \mathcal{P}(X)) \in Ob(\mathcal{M}.)$ given by $\mu^*(A) = \inf \mu(B)$ for all $B \supset A$ and $B \in \mathcal{T}$, (resp. $\mu_*(A) = \sup \mu(B)$ for all $B \subset A$ and $B \in \mathcal{T}$). Then, μ^* satisfies the* **Fatou property**: *if $A_n \uparrow A \Rightarrow \mu^*(A_n) \uparrow \mu^*(A)$.*

DEFINITION 5.32 - *If X is a topological space a* **Borel measure** *on X is a measure on $(X, \mathcal{B}(X))$. The set of all Borel measures on X is denoted by $M^+_{bor}(X)$. Let $\overline{\mathcal{M}}$ be the* **category of measure spaces**: *$f \in Hom_{\overline{\mathcal{M}}}((X, \mathcal{T}, \mu), (Y, \mathcal{R}, \nu))$ iff $f \in Hom_{\mathcal{M}}((X, \mathcal{T}), (Y, \mathcal{R}))$ and $\nu = \mu \circ f^{-1}$.*

DEFINITION 5.33 - *We have the following controvariant functor:*[8]

$$F : \overline{\mathcal{M}} \to \mathcal{B}_{an} \equiv \textit{category of Banach spaces}$$

such that $\forall (X, \mathcal{T}, \mu) \in Ob(\overline{\mathcal{M}})$, one has $F(X, \mathcal{T}, \mu) \equiv$ Banach space of $\overline{\mathbf{R}}$-valued measurable functions on $(X, \mathcal{T}, \mu)$; $\forall f \in Hom_{\overline{\mathcal{M}}}((X, \mathcal{T}, \mu), (Y, \mathcal{R}, \nu))$, one has $F(f) \in Hom_{\mathcal{B}an}(F(Y, \mathcal{R}, \nu), (X, \mathcal{T}, \mu))$, $F(f) = h \circ f, \forall h \in F(Y, \mathcal{R}, \nu)$. Then, $F(\overline{\mathcal{M}})$ is a sub-category of $\mathcal{B}_{an}$: $F(\overline{\mathcal{M}}) \subset \mathcal{B}_{an}$.

---

[8] On $\overline{\mathbf{R}} \equiv \mathbf{R} \cup \{-\infty, +\infty\}$ we say that a set $B \subset \overline{\mathbf{R}}$ is a Borel set in $\overline{\mathbf{R}}$ if it is the union of a set $B' \in \mathcal{B}(\mathbf{R})$ with any subset of $\overline{\mathbf{R}} - \mathbf{R} = \{-\infty, +\infty\}$. In practice in order to know if a function $f:(X, \mathcal{T}) \to \mathbf{R}$ is measurable, it is sufficient to check that $f^{-1}(E) \in \mathcal{T}$ for a suitable class of subset which generates the σ-field $\mathcal{B}(\mathbf{R})$. More precisely, the most important class is given by the following: (a) $\{x : f(x) \leq c\}, \forall c \in \mathbf{R}$; (b) $\{x : f(x) > c\}, \forall c \in \mathbf{R}$; (c) $\{x : f(x) \geq c\}, \forall c \in \mathbf{R}$; (d) $\{x : f(x) < c\}, \forall c \in \mathbf{R}$. Note that any continuous function $\Omega \to \mathbf{R}$ on a topological space Ω is Borel measurable.

DEFINITION 5.34 - 1) *The* **integral** *is a* **R**-*linear function*

$$\left\{\begin{array}{l}\int : F(X,\mathcal{T},\mu)\to\mathbf{R}\\ \int : f\mapsto \mu(f)\equiv\int f d\mu=\int f_{+}d\mu-\int f_{-}d\mu=\lim_{n\to\infty}(\int f_{+,n}d\mu-\int f_{-,n}d\mu)\end{array}\right\}$$

where $f = f_+ - f$, *and* $f_{+,n}, f_{-,n}$ *are monotone increasing sequences of simple functions such that* $f_{+,n} \to f_+$ *and* $f_{-,n} \to f_-$ *(that surely exist). Furthermore, if*

$$f_{\pm,n}(x)=\sum_{i=1}^{s_n}{}^{\pm}c_i^{(n)}\chi_{E_i}(x),\quad {}^{\pm}c_i^{(n)}\geq 0,$$

where χ_{E_i} *is the characteristic functions of* $\{E_i\}$, *that form a* $\mathcal{T}$-*dissection of* X; *we put*

$$\int f_{\pm,n}d\mu=\sum_{i=1}^{s_n}{}^{\pm}c_i^{(n)}\mu(E_i).$$

(This definition does not depend on the particular sequence $f_{+,n}$ *or dissection* $\{E_i\}$.*)* *A function* $f \in F(X, \mathcal{T}, \mu)$ *is called* μ-**integrable** *if* $|\int f d\mu| < +\infty$.
2) *The* **indefinite integral** *on* $F(X,\mathcal{T},\mu)$ *is a map*

$$\left\{\begin{array}{l}N : F(X,\mathcal{T},\mu)\to M(X,\mathcal{T})\\ N(f)\equiv\nu : A\to\nu(A)=\int_A f d\mu=\int\chi_A f d\mu\end{array}\right\}$$

One has that if

$$f\in Hom_{F(\overline{\mathcal{M}})}(F(X,\mathcal{T},\mu),F(Y,\mathcal{R},\nu))$$

then

$$\int_X(g\circ f)d\mu=\int_Y g d\nu,\quad\forall g\in F(Y,\mathcal{R},\nu).$$

3) *A set* $A \subset X$ *is said to be* μ-**integrable** *if its characteristic function* χ_A *is* μ-*integrable.*
4) *A set* $A \subset X$ *is said to be* μ-**measurable** *if its intersection with every* μ-*integrable set* C *is* μ-*integrable.*
5) *The* μ-*measurable sets form a* σ-*algebra* $\mathcal{T}_\mu$ *containing* T *and* μ *is a measure on* $\mathcal{T}_\mu$ **extending** μ.
EXAMPLE 5.13 - The set of **Lebesgue measurable sets** of **R** forms a class $\mathcal{L}$ that is larger than $\mathcal{B}(\mathbf{R}^n) \equiv \mathcal{B}$. If μ is the Lebesgue measure on **R** then $\mu^* : \mathcal{L} \to \mathbf{R}^+$ is also called Lebesgue measure on **R**. Similarly, the class $\mathcal{L}_F^n$ of sets in $\mathbf{R}^n$ which are Lebesgue-Stieltjes measurable, for a function F, is larger than $\mathcal{B}(\mathbf{R}^n)$. Then μ_F^* :

$\mathcal{L}_F^n \to \mathbf{R}$ is also called **Lebesgue-Stieltjes measure** on $\mathbf{R}^n$. This measure on the σ-algebra $\mathcal{L}_F^n$ satisfies the property $\mu_F(\mathbf{R}^n) = 1$ and it is complete.

PROPOSITION 5.12 - 1) *If*

$$f \in Hom_{\overline{\mathcal{M}}}((X, \mathcal{T}, \mu), (Y, \mathcal{R}, \nu)),$$

*then f is said to be $(\mu, \mathcal{R})$-***measurable** *if $f^{-1}(B)$ belongs to $\mathcal{T}_\mu$ for every $B \in \mathcal{R}$. So, in the category $\overline{\mathcal{M}}$ we can distinguish a sub-category*

$$\overline{\overline{\mathcal{M}}} \subset \overline{\mathcal{M}},$$

where

$$Ob(\overline{\overline{\mathcal{M}}}) = Ob(\overline{\mathcal{M}})$$

but

$$Hom_{\overline{\overline{\mathcal{M}}}}((X, \mathcal{T}, \mu), (Y, \mathcal{R}, \nu))$$

is the set of all

$$f \in Hom_{\overline{\mathcal{M}}}((X, \mathcal{T}, \mu), (Y, \mathcal{R}, \nu))$$

such that they are $(\mu, \mathcal{R})$-measurable.

2) *Let $\dot{\overline{\mathcal{M}}} \subset \overline{\mathcal{M}}$ be the* **sub-category of σ-finite measure spaces**. *Then, there is a canonical functor:*[9]

$$\left\{ \begin{array}{l} \otimes : \dot{\overline{\mathcal{M}}} \times \dot{\overline{\mathcal{M}}} \to \dot{\overline{\mathcal{M}}} \\ \otimes((X, \mathcal{T}, \mu), (Y, \mathcal{R}, \nu)) = (X \times Y, \mathcal{T} \otimes \mathcal{R}, \mu \otimes \nu) \end{array} \right\}$$

where $\mathcal{T} \otimes \mathcal{R}$ is the canonical σ-algebra in $X \times Y$ associated to $\mathcal{T}$ and $\mathcal{R}$, and $\mu \otimes \nu(A \times B) = \mu(A).\nu(B), A \in \mathcal{T}, B \in \mathcal{R}$, with the usual convention: $0\infty = \infty 0 = 0$. For each $E \in \mathcal{T} \otimes \mathcal{R}$ set $E_x = pr_1^{-1}(x) \cap E, E_y = pr_2^{-1}(y) \cap E$, where $pr_i, i = 1, 2$, are the canonical projections $X \times Y \to X, Y$. Then, one also has $\mu \otimes \nu(E) = \int \nu(E_x) d\mu(x) = \int \nu(E_y) d\nu(y)$. (For examples the Lebesgue measure on $\mathbf{R}^n$ is obtained by tensor product of the Lebesgue measure on $\mathbf{R}$.)

---

[9] Let $Y \equiv \prod_{i \in I} Y_i$, let $\mathcal{T}_i$ be the σ-algebra on $Y_i, i \in I$, and let $f \equiv (f_i)_{i \in I}: X \to Y$ be a map. Then the σ-algebra $\sigma(f_i)_{i \in I}$ on X generated by $\bigcup_{i \in I} f_i^{-1}(\mathcal{T}_i)$ is called the **σ-algebra generated by** $(f_i)_{i \in I}$. In particular, if $X = Y$ and $f_i \equiv pr_i \equiv$ projection on i-th factor, we call **product σ-algebra** the σ-algebra on Y generated by $(pr_i)_{i \in I}$ and we write $\otimes_{i \in I} \mathcal{T}_i \equiv \sigma(pr_i)_{i \in I}$. A map $f: (X, \mathcal{T}) \to (\prod_{i \in I} Y_i, \otimes_{i \in I} \mathcal{T}_i)$ between measurable spaces X and $Y \equiv \prod_{i \in I} Y_i$ is measurable iff the mappings $pr_i \circ f: (X, \mathcal{T}) \to (Y_i, \mathcal{T}_i)$ are measurable $\forall i \in I$. If $(X_n)_{n \in \mathbf{N}}$ is a countable family of topological spaces with countable basis then $\mathcal{B}(X) = \otimes_{n \in \mathbf{N}} \mathcal{B}(X_n)$, where $X = \prod_{n \in \mathbf{N}} X_n$.

3) *Let* $(X, \mathcal{T}, \mu) \in Ob(\overline{\mathcal{M}})$ *and let* $Y \subset X$ *be a* μ*-measurable set. We have*

$$B(Y) = \{A \cap Y; A \in B(X)\} \subset B(X)$$

and $\mu|B(Y)$ *is a measure on* Y *called* **induced by** μ.

DEFINITION 5.35 - *A* **signed measure** *is a covariant functor* $\mathbf{M} : \mathcal{M} \to \mathcal{I}(\mathbf{R})$ *such that* $\mu \equiv \mathbf{M}(X, \mathcal{T})$ *is a real valued, countably additive set function such that* $\mu(\emptyset) = 0$ *and assuming at most one of the values* $+\infty$ *or* $-\infty$. *Let us denote by* $\mathbf{M}(X, \mathcal{T})$ *the set of signed measures on* $(X, \mathcal{T})$. *If* $\nu \in \mathbf{M}(X, \mathcal{T})$ *there exist two disjoint sets* A *and* B *where* $A \cup B = X$ *such that* $\mu(A) > 0$ *and* $\mu(B) < 0$, *(***Hann decomposition***). Set* $\mu_+(E) \equiv \mu(E \cap A)$ *and* $\mu_-(E) = -\mu(E \cap B)$, *for every measurable set* E *of* X. *Then, we have* $\mu(E) = \mu_+(E) - \mu_-(E)$ *for* $\mu \in M(X, \mathcal{T})$, *(***Jordan decomposition***). If* μ *is finite then so are* μ_+ *and* μ_-, *at least one of the measures* μ_+ *and* μ_- *is always finite. We call* μ_+ *(resp.* μ_-*) the* **upper variation** *(resp.* **lower variation***) of* μ. *So, we have the canonical maps:* $p_\pm : M(X, \mathcal{T}) \to M^+(X, \mathcal{T}), p_\pm(\mu) = \mu_\pm$.

DEFINITION 5.36 - *Let* $\mathcal{M}_H$ *be the sub-category of* $\mathcal{M}$ *of measurable spaces* $(X, \mathcal{B})$ *where* X *is a Hausdorff space and* $\mathcal{B}(X) \equiv B$ *is the* σ*-algebra of all Borel sets of* X. *A* **Radon measure** *is a Borel measure* $\mathbf{M}_{Ra} : \mathcal{M}_H \to \mathcal{I}(\mathbf{K})$ *such that:*[10]

(a) $\mu(C) < \infty$ *for each compact subset* $C \subseteq X$;

(b) $\mu(B) = \sup\{\mu(C)|C \subseteq B, compact\}$, *for each* $B \in \mathcal{B}(X)$. *The set of all-Radon measures on* X *is denoted by* $M^+_{Ra}(X)$ *and the set of signed Radon measures is denoted by* $M_{Ra}(X)$. *A* **finite Radon measure** μ *(i.e.,* $\mu(X) < \infty$*) satisfies:*

$$\mu(B) = \inf\{\mu(G)|B \subseteq G, G \text{ open}\}, B \in \mathcal{B}(X).$$

A **molecular Radon measure** μ *has supp(*μ*) finite.*

DEFINITION 5.37 - *Let* X *be any infinite space and let* $(X, \mathcal{F})$ *be a measurable space where* $\mathcal{F}$ *contains all single points sets. Let* $x_1, x_2, \cdots$ *be an enumerable sequence of distinct points of* X *and assume that* $p_1, p_2, \cdots$ *is a sequence of real numbers such that* $\sum_{i=1}^{\infty} p_i$ *either converges absolutely or is properly divergent to* $+\infty$ *or* $-\infty$. *Put*

$$\mu(E) = \sum_{x_i \in E} p_i, \quad \forall E \in \mathcal{F}.$$

Then μ *is called a* **discrete measure**.

---

[10] A topological space is a **Hausdorff space** if every pair of distinct points have disjoint neighbourhoods. Two disjoint compact sets into a Hausdorff space have disjoint neighbouroods. A compact subset of a Hausdorff space is closed. If a locally compact space is a Hausdorff space or a separable space (i.e. it has a countable base), then so is its one-point compactification.

DEFINITION 5.38 - 1) *Let $\mathcal{C}$ be a set of subsets of X. The set $E_0 \in \mathcal{C}$ is said to be an* **atom** *of a set function $\phi : \mathcal{C} \to \overline{\mathbf{R}}$ if:* (a) $\phi(E_0) \neq 0$ and; (b) *for every*

$$E \subset E_0 \quad , \quad E \in \mathcal{C} \quad , \quad \phi(E) = \left\{ \begin{array}{c} 0 \\ \text{or} \\ \phi(E_0) \end{array} \right\}$$

2) *A set function $\phi : \mathcal{C} \to \overline{\mathbf{R}}$ is said to be* **non-atomic** *if it has no atom.*

PROPOSITION 5.13 - *Assume that $\phi : \mathcal{F} \to \overline{\mathbf{R}}$ is σ-additive on the σ-field $\mathcal{F}$, non-atomic, and finite valued. Then, for any $A \in \mathcal{F}$, ϕ takes every real value between $-\phi_-(A)$ and $\phi_+(A)$ for some subset $E \subset A$.*

DEFINITION 5.39 - *Given any measure space $(X, \mathcal{F}, \mu)$ in which $\mathcal{F}$ contains all single points sets, the point $p \in X$ is said to be an* **atom** *for the measure μ if $\mu(p) > 0$.*

2) *A measure μ with no atoms is said to be* **non-atomic**.

THEOREM 5.23 - 1) *If μ is σ-finite, the set of atoms of μ is countable.*

2) *In this case if we put*

$$\nu(E) = \sum_{p \in E, \mu(p) \neq 0} \mu(p),$$

we obtain a new measure ν defined on all subsets of X, and ν is a discrete measure.

3) *Given a σ-finite measure space $(X, \mathcal{F}, \mu)$ in which $\mathcal{F}$ contains all single point sets there is a unique decomposition of μ: $\mu = \nu + \tau$ for which ν is a discrete measure on X and τ is a non-atomic measure on $\mathcal{F}$.*

Note that the theory of Radon measures can be generalized to spaces which are not necessarily Hausdorff if we replace the word "compact" by "closed, quasi-compact". Recall that a subset E of a topological space X is said to be **quasi-compact** if every open covering of E has a finite subcovering. (A quasi-compact set in a non-Hausdorff space is not in general closed!)

DEFINITION 5.40 - *Let $\mathcal{K} \equiv \mathcal{K}(X)$ denote the family of all compact subsets of X. A* **Radon content** *is a set function*

$$\left\{ \begin{array}{l} \lambda : \mathcal{K} \to [0, \infty] \\ \lambda(C_2) - \lambda(C_1) = \sup\{\lambda(C) | C \subseteq C_2 - C_1, C \in \mathcal{K} \,|\, \forall C_1, C_2 \in \mathcal{K}, C_1 \subseteq C_2\} \end{array} \right\}$$

PROPOSITION 5.14 - 1) *If μ is a Radon measure, its restriction to $\mathcal{K}$ is a Radon content. Vice versa, any Radon content on a Hausdorff space has a unique extension to a Radon measure.*

2) *If X is a Hausdorff space and $B \in \mathcal{B}(X)$, then B is again a Hausdorff space and any Radon measure $\mu \in M^+_{Ra}(X)$ restricted to $\mathcal{B}(B)$ gives a Radon measure $\mu|_B$ on B. ($\mu|_B$ is called the* **restriction of** μ **to** *B.)*

3) *If* $\mu \in M^+_{Ra}(X)$ *and* $f : X \to [0,\infty]$ *is Borel measurable, then*

$$\int f d\mu = \sup_{K \in \mathcal{K}} \int_K f d\mu.$$

4) *If* $f : X \to [0,\infty[$ *is continuous, then* $\mathcal{B}(X) \to [0,\infty]$ *defined by* $\nu(B) = \int_B f d\mu$ *is again a Radon measure.*
5) *If* $\mu \in M^+_{Ra}(X), \nu \in M^+_{Ra}(Y)$, *with* $X, Y =$ *Hausdorff spaces, then there is a uniquely determined Radon measure on* $X \times Y$ *called* **product of** μ **and** ν *denoted by* $\mu \otimes \nu$, *with the property*

$$\mu \otimes \nu(K \times L) = \mu(K).\nu(L), \quad \forall K \in \mathcal{K}(X), L \in \mathcal{K}(Y).$$

For all Borel sets $A \subseteq X, B \subseteq Y$ *we have* $\mu \otimes \nu(A \times B) = \mu(A).\nu(B)$. *In particular the restriction of* $\mu \otimes \nu$ *to the product* σ*-algebra* $\mathcal{B}(X) \otimes \mathcal{B}(Y)$ *is a product measure of* μ *and* ν *in the usual sense. Furthermore, if* $f : X \times Y \to [0,\infty]$ *is lower semicontinuous, (resp. Borel measurable), one has* **(Fubini's theorem)**:

$$\int_{X\times Y} fd(\mu \otimes \nu) = \int_X \int_Y f(x,y)d\nu(y)d\mu(x) = \int_Y \int_X f(x,y)d\mu(x)d\nu(y).$$

6) *Let* $(G_\alpha)_{\alpha \in D}$ *be an open covering of the Hausdorff space of* X *and on each* G_α, *let a Radon measure* μ_α *be given such that* $\mu_\alpha(B) = \mu_\beta(B)$ *for each pair of indices* $\alpha, \beta \in D$ *and for each Borel set* $B \subseteq G_\alpha \cap G_\beta$. *Then there is a uniquely determined Radon measure* μ *on* X *such that* $\mu(B) = \mu_\alpha(B)$ *if* B *is a Borel set in* G_α.
7) *If* X *is strongly Lindelöf, every Radon measure* μ *on* X *is* σ*-finite.*
THEOREM 5.24 - (**Riesz representation theorem**). 1) *Let* X *be a Hausdorff space, let* C *denote a point separating convex cone of continuous functions* $f : X \to [0,\infty[$ *and let* $T : C \to [0,\infty]$ *together with* C *fulfil the following conditions:*
a) *if* $f, g \in C \Rightarrow f \wedge g \in C, f - g^+ \in C$ *and* $f \wedge 1 \in C$;
b) $T(f+g) = T(f) + T(g), \forall f, g \in C$;
c) *For each* $f \in C \Rightarrow T(f) = \sup\{T(g) | f \geq g \in C, g \text{ bounded}, T(g) < \infty\}$;
d) *For each compact set* $K \subseteq X$ *there is some* $f \in C$ *with* $1_K \leq f$ *and* $T(f) < \infty$.
e) *Given* $f \in C$ *such that* $T(f) < \infty$ *and given* $\epsilon > 0$ *there is a compact set* $K \subseteq X$ *such that* $g \in C, g \leq f$ *and* $g|_K = 0$ *implies* $T(g) < \epsilon$. *(As a consequence we have:* **monotonicity** *of* T: $T(f) \leq T(g)$ *if* $f \leq g$ *and* $T(\alpha f) = \alpha T(f)$ *for all* $f \in C$ *and* $\alpha \in \mathbf{R}^+$.)
Then, there is a uniquely determined Radon measure μ *on* X *such that* $T(f) = \int f d\mu$ *for all* $f \in C$. *The measure* μ *is furthermore locally finite.*

2) *If X is locally compact there is a bijection*

$$\left\{\begin{array}{l} C_0^0(X)' \to M_{Ra}(X) \\ T \mapsto \mu : T(f) = \int f d\mu = \int f d\mu_1 - \int f d\mu_2 \\ \mu = \mu_1 - \mu_2,\ \mu_i \in M_{Ra}^{+}(X) \end{array}\right\}$$

3) *If X is compact, $C^0(X)$ with the sup-norm is a Banach space and $C(X)'$ is also a Banach space with the norm*

$$\|T\| = \sup_{\|f\|\leq 1} |T(f)|.$$

The set $M_{Ra}(X)$ of all signed Radon measure on X is also a Banach space with norm

$$\|\mu\| = \sup\{|\mu(A)| + |\mu(CA)| \| A \in B(X)\}.$$

Then, one has the identification $M_{Ra}(X) \equiv C^0(X)'$ by means of the isometry

$$\left\{\begin{array}{l} C^0(X)' \to M_{Ra}(X) \\ T \mapsto \mu = \mu_1 - \mu_2, T(f) = \int f d\mu = \int f d\mu_1 - \int f d\mu_2, \mu_i \in M_{Ra}^{+}(X) \end{array}\right\}$$

DEFINITION 5.41 - *Let $\mathcal{CM}$ be the category of cylindrical measurable locally convex vector Hausdorff spaces over* **R**. *The set $Ob(\mathcal{CM})$ is given by all* **cylindrical measurable spaces** *that are couples $(E, \mathcal{T})$ where E is a convex Hausdorff vector space over the field* **R** *and $\mathcal{T}$ is the set of* **cylindrical subsets** *A of E, that is there exists a closed subspace of finite codimension $F \subset E$ and a subset $B \subset E/F$ such that*

$$A = \pi^{-1}{}_{E/F}(B),$$

where $\pi_{E/F}$ is the canonical projection $E \to E/F$. B is called the **base** *and F a* **generator subspace of the cylindrical set** *A. If $G \subset F$ has a finite codimension, here*

$$B_0 \equiv \pi^{-1}_{E/F,E/G}(B)$$

is also a base and G a generator subspace of A finer

$$A = \pi^{-1}_{E/F}(B) = \pi^{-1}_{E/G}(\pi^{-1}_{E/F,E/G}(B)) = \pi^{-1}_{E/G}(B)$$

where $\pi_{E/F,E/G}$ is the canonical map $E/G \to E/F$, such that

$$\pi_{E/F,E/G} \circ \pi_{E/G} = \pi_{E/F}.$$

Then, $f \in Hom_{\mathcal{CM}}((E,\mathcal{T}),(H,\mathcal{R}))$ *iff* $f: E \to H$ *is continuous and linear and* $f^{-1}(B) \in \mathcal{T}, \forall B \in \mathcal{R}$. *A* **cylindrical measure** *is a covariant functor*

$$\left\{\begin{array}{l}
\mathbf{CM}: \mathcal{CM} \to \mathcal{I}(\mathbf{R}) \\
(a) \quad \mathbf{CM}(E,\mathcal{T}) \equiv \mu \equiv \textbf{cylindrical measure on } E \textbf{ defined by} \\
\mu(A) = \mu_{E/F}(B) \forall A \in \mathcal{T} \\
\textit{with base } B \textit{ and generator subspace } F, \textit{ where} \\
\mu_{E/F} \textit{ is a probability on the finite dimensional space } E/F. \\
\textit{(This definition is independent of the particular choice of } F \textit{ and } B.) \\
(b) \quad \textit{If } f \in Hom_{\mathcal{CM}}((E,\mathcal{T}),(H,\mathcal{R})), \mathbf{CM}(f): \mu \mapsto \mathbf{CM}(f)(\mu) = \nu \equiv \mu \circ f^{-1}
\end{array}\right\}$$

Set $\overline{\mathcal{CM}}$ = **category of cylindrical measure spaces**; $\overline{\mathcal{M}}(X)$ = *subcategory of* $\overline{\mathcal{M}}$ *restricted to a set* X*, fixed.* $\overline{\mathcal{CM}}(E)$ = *subcategory of* $\overline{\mathcal{CM}}$ *restricted to a space* E.

PROPOSITION 5.15 - 1) *We have a canonical injective functor*

$$J: \overline{\mathcal{CM}}(F) \to \overline{\mathcal{M}}(\prod_F (E/F)), \quad \mu \mapsto \check{\mu},$$

where $(E/F)^{\vee}$ *is the Cech-compactification of* E/F. *(Note that we have a canonical injection* $j: E \to \prod_F (E/F)^{\vee}$ *but it can happen that* E *has measure zero in the measure* $\check{\mu}$.*)*

2) *Let* $\overline{\mathcal{M}}_{Ra}$ *be the subcategory of* $\overline{\mathcal{M}}$ *of Radon measure spaces and* $\mathcal{M}_{Ra}(X) = \overline{\mathcal{M}}_{Ra} \cap \overline{\mathcal{M}}(X)$. *We have a canonical functor:*

$$j: \overline{\mathcal{M}}_{Ra}(E) \to \overline{\mathcal{M}}(\prod_F (E/F)^{\vee}).$$

Then, we say that $j(\mu)$ **comes from the Radon measure** μ. *So, we can identify* $\overline{\mathcal{M}}_{Ra}(E)$ *with a subset of* $\overline{\mathcal{CM}}(E)$.

REMARK 5.15 - If E is a reflexive Banach space and μ is a cylindrical measure on E, concentrated cylindrically on the balls of E, then μ is a Radon measure on E concentrated on the strongly compact subsets of E. (Recall that a measure μ is said to be concentrated on a subset $A \subset E$ if $\mu(A) = 1$, i.e., if $\mu(CA) = 0$, where CA is the complementary of A. Furthermore, a cylindrical measure μ is cylindrically concentrated on a subset $A \subset E$ if $\mu^*(CA) = 0$.

THEOREM 5.25 - 1) (**Sazonov theorem**). *Let* E *and* F *be two Hilbert spaces and* $u: E \to F$ *a continuous linear map. The image* $\mu \circ u^{-1}$ *of every cylindrical measure* μ *scalarly concentrated on the balls of* E *is a Radon measure on* F *(i.e.,* u *is* **radonifying***) iff* u *is a Hilbert-Schmidt operator (that is belongs to the canonical* *image of* $\overline{E}\widehat{\bigotimes}F$ *into* $\mathcal{L}(E,F)$, *where* $\overline{E}$ *is the anti-space of* E *and* $\overline{E}\widehat{\bigotimes}F$ *is the Hilbert-Schmidt tensor product of* $\overline{E}$ *and* F*).*[11]

2) (**Minlos theorem**). *Let* E *be a locally convex Hausdorff vector space. Let* $\mathcal{G}$ *be a collection of convex, completing subsets of* E, *such that if* A *is a balanced, convex completing subsets of* E *contained in some* $B \in \mathcal{G}$, *then* $A \in \mathcal{G}$. *Assume that* E *is* $\mathcal{G}$-**conuclear**.[12]

If μ *a cylindrical measure on* E *concentrated scalarly on the sets of* $\mathcal{G}$, *then* μ *is a Radon measure on* E *concentrated on the sets of* $\mathcal{G}$.

DEFINITION 5.42 - *The* **Fourier transform of a Radon measure** *on a locally convex Hausdorff vector space* E *endowed with a finite Radon measure* μ *is a function* $\phi \equiv F(\mu) : E' \to \mathbf{K}$ *given by*

$$\phi(xi) = \int_E e^{-2i<x,\xi>} d\mu(x).$$

THEOREM 5.26 - (**Sazonov-Badrikian theorem**). *Let* E *be a quasi-complete locally convex Hausdorff vector space and let* E'_c *be its dual equipped with the topology of uniform convergence on compact subsets of* E. *Assume that every compact subset of E is contained in a balanced, convex, compact Hilbert set. A function of positive type* ϕ *on* E'_c *is the Fourier transform of a Radon measure on* E *iff it is continuous for Hilbert-Schmidt topology associated with the topology of* E'_c.

RANDOM VARIABLES AND RANDOM PROCESSES.

DEFINITION 5.43 - *A* **probability space** *(or* **sample space***) is a triplet* $(\Omega, \mathcal{T}, \mu)$ *consisting of a measurable space* $(\Omega, \mathcal{T})$ *and probability* μ, *i.e., a measure* $\mu : \mathcal{T} \to [0,1]$, *such that* $\mu(\Omega) = 1$. *An* **event** *is a* μ*-measurable subset* $E \subset \Omega$; *an* **elementary event** *is a point* $\omega \in \Omega$. *The* **probability** $P(E)$ *of an event* E, *is its measure:* $P(E) = \mu(E)$. *Two events* E *and* F *are said to be* **independent** *if* $P(E \cap F) = P(E).P(F)$. *The* **conditional probability** *of an event* $A \in \Omega$ *given the event* B *(with* $\mu(B) > 0$*), is the probability measure* μ^B *on* Ω *given by*

$$\mu^B(A) = P(A|B) = \mu(A \cap B)/\mu(B).$$

EXAMPLE 5.14 - 1) We recognize on $\mathbf{R}^n$ a structure of probability space

$$(\mathbf{R}^n, \mathcal{L}^n_F, \mu_F)$$

---

[11] Recall that a cylindrical measure μ is scalarly concentrated on $A \subset E$ if for every closed hyperplane $H \subset E$ we have $\mu_{E/H}(\pi_{E/H}(A)) \geq 1$.

[12] A $\mathcal{G}$-**conuclear space** is a locally convex Hausdorff space E such that there exists a subset collection $\mathcal{G}$ of balanced closed subsets of E, such that for every $A \in \mathcal{G}$ there exists $B \in \mathcal{G}$, $B \supset A$, such that the canonical map $\widehat{E}_A \to \widehat{E}_B$ is nuclear.

where F is a Stieltjes function and μ_F is the covariant measure on the σ-algebra $\mathcal{L}^n_F$.
2) A particular important example of probability measure on Hilbert spaces is the Gauss measure. Let $\mathcal{E}^\bullet(\mathbf{R})$ be the category of isometric finite dimensional real Euclidean vector spaces; $Hom_{\mathcal{E}^\bullet(\mathbf{R})}((E,g),(F,f)) = \emptyset$ if (E,g) and (F,f) are not isometric spaces. Then, there exists a canonical measure functor **(Gauss measure functor)** $\mathbf{M}^\bullet_G : \mathcal{E}^\bullet(\mathbf{R}) \to \mathcal{I}(\mathbf{R})$, such that $\gamma \equiv \mathbf{M}^\bullet_G(E,g)$ is the measure **(Gauss measure)** given by

$$\gamma(A) = \int_A e^{-\pi|x|^2} d\mu(x), \quad A \subseteq E,$$

where $d\mu(x) = *1$ is the canonical volume form associated to g. We simply write $\gamma = e^{-\pi|x|^2} d\mu(x)$.[13]
For example the Gauss measure on the real line $\mathbf{R}$ is $\gamma = e^{-\pi|x|^2} dx$ and one has $\gamma(\mathbf{R}) = \int_{\mathbf{R}} e^{-\pi|x|^2} dx = 1$. Let $\mathcal{E}(\mathbf{R})$ be the category of isometric Hilbert spaces. Then, there exists a canonical measure-functor **(Gauss measure functor)** $\mathbf{M}_G : \mathcal{E}(\mathbf{R}) \to \mathcal{I}(\mathbf{R})$ such that $\gamma \equiv \mathbf{M}_G(E)$ is the measure on E **(Gauss measure)** that is the cylindrical measure on E such that for any finite-dimensional subspace F of E one has $\gamma_F = \gamma \circ p^{-1}$, where $p : E \to F$ is the orthogonal projection and γ_F is the Gauss measure on F for the Euclidean structure induced on F by E. (The Gauss measure γ on E is scalarly concentrated on the balls of E since for any $\epsilon > 0$ there exists $R > 0$ such that

$$\int_{|x| \geq R} e^{-\pi|x|^2} d\mu(x) \leq \epsilon.)$$

Note also that the measure induced in the finite-dimensional Hilbert subspace $F \subset E$ by the Gauss measure γ on E coincides with the Gauss measure γ_F on F. In fact one has: (a) $\gamma \circ p^{-1} = \gamma_F$; (b) $p \circ j = id_F \Rightarrow \gamma \circ p^{-1} \circ p \circ j = \gamma_F \Rightarrow \gamma \circ j = \gamma_F$ i.e., $\gamma|F = \gamma_{F'}$ where $j : F \to E$ is the canonical injection. Furthermore, the Fourier transform $\phi = \mathcal{F}(\gamma)$ of the Gauss measure γ on E is given by $\phi(\xi) = e^{-\pi|\xi|^2}$, where $|\xi|$ is the norm corresponding to the canonical Euclidean structure of E'.
3) Let $\mathcal{H}_{l.c.}(\mathbf{R})$ be the category of the couples (E,H) where E is a locally convex Hausdorff $\mathbf{R}$-vector space and H is a Hilbert subspace of E.

$$Hom_{\mathcal{H}_{l.c.}(\mathbf{R})}((E,H),(\mathcal{F},K))$$

is the set of continuous maps $f : E \to F$ such that $f(H) = K$ and $f|H$ is an isometry between H and K. This set can be eventually empty. Then, there exists a canonical

[13] More generally a Gauss measure on (E,g) can be defined with respect to a vector $x_0 \in E$ and a positive definite isomorphism $B \in GL(E)$ as $\gamma_{x_0{}'B} = \frac{1}{\sqrt{2}} \frac{1}{n\sqrt{|B|}} \exp(-\frac{1}{2} < B^{-1}(x-x_0),(x-x_0) >) d\mu(X)$, where $n = \dim E$, $|B| = \det(B)$.

measure-functor **(relative Gauss measure-functor)** $\mathbf{M}_G^r : \mathcal{H}_{l.c.}(\mathbf{R}) \to \mathcal{I}(\mathbf{R})$, such that $\mu \equiv \mathbf{M}_G^r(E, H)$ is the image of the Gauss measure γ_H on the Hilbert subspace with respect to the canonical injection $j : H \to E, \mu = \gamma_H \circ j^{-1}$.
For example, the **Dirac measure** δ on any locally convex Hausdorff vector space E is a Gauss measure relative to the Hilbert subspace $H \equiv \{0\} \subset E$. One has $\delta(E) = 1$.[14]
Another example of relative Gauss measure is the **Wiener-Levy measure** of Brownian motion on $C^0([0,1])$ with respect to the Hilbert subspace $H \equiv H^1([0,1])$ of the real-valued continuous functions on $[0,1]$ which vanish at the origin and whose derivative in the sense of distributions in the open interval $[0,1]$ belongs to $L^2[0,1]$. The inner product in H is

$$(f, g) = \int_0^1 f'(t)g'(t)dt.$$

In fact, H is the image of the canonical continuous linear map $\mathcal{I} : L^2[0,1] \to C^0([0,1])$ given by $\phi \mapsto \mathcal{I}(\phi) \equiv \Phi =$ primitive of ϕ, i.e.,

$$\Phi(x) = \int_0^x \phi(t)dt.$$

Wiener's probability space is the couple $([0,1], \mu)$ where $[0,1] \subset \mathbf{R}$ is endowed with the class of all Lebesgue-measurable subsets of $[0,1]$ and μ is just the Lebesgue measure. This probability space describes the random choice of a point from the interval $[0,1]$. Let us quote the following theorems that relate Gauss measure and Radon measure.
THEOREM 5.27 - 1) *Let E be a separated locally convex vector space. Let H be a Hilbert space and let $u : H \to E$ be a linear continuous map. Let $\mu = \gamma_H \circ u^{-1}$ be the cylindrical probability image of γ_H by u. If μ is a Radon measure on E one has $supp(\mu) = u(H)$.*
2) *Let E be a Fréchet vector space and let μ be a Radon measure on E. There exists a Hilbert space H, a cylindrical measure ν on H scalar concentrated on the balls of H and a linear continuous map $u : H \to E$ such that $\nu \circ u^{-1} = \mu$.*

---

[14] The simplest type of probability on a measurable space $(\Omega, \mathcal{A})$ is the σ-measure μ_{δ_x} at $x \in \Omega$, defined as follows: for all $\bigtriangledown \in \mathcal{A}$:

$$\mu_{\delta_x}(\bigtriangledown) = \left\{ \begin{array}{ccc} 0 & \text{if} & x \notin \bigtriangledown \\ 1 & \text{if} & x \in \bigtriangledown \end{array} \right\}$$

where $\mathcal{A} =$ Boolean σ-algebra of subsets of Ω.

DEFINITION 5.44 - *Let*

$$(X, \mathcal{R})$$

be a measurable space. We say that two μ*-measurable maps*

$$\Omega \to X$$

are **equivalent** *if they coincide almost everywhere with respect to* μ.

1) *A* **random variable** *with value in* X*, on the probability space* $(\Omega, \mathcal{T}, \mu)$*, is an equivalent class of* μ*-measurable maps* $\Omega \to X$*. We shall denote by* $Mes(\Omega, \mu, X)$ *the set of all random variables* $\Omega \to X$*. If* $(X \equiv \mathbf{R}, \mathcal{R} \equiv \mathcal{B}(\mathbf{R}))$ *we shall speak about* **real random variables**.

2) *Let* $x : \Omega \to X$ *be a random variable. The* **probability distribution** *of* x *is the probability* $\mu \circ x^{-1} \equiv \nu$ *on* X *extended for completion. (In fact, it should not be, in general, complete!) If* $A \subset X$ *is a* $\nu \equiv \mu \circ x^{-1}$*-measurable subset of* X*, then its measure* $\nu(A)$ *will be denoted by* $P\{x \in A\}$ *and considered to be the* **probability** *that* $x(\omega)$ *should lie in* A.

DEFINITION 5.45 - *Two random variables* $x \in Mes(\Omega, \mu, X), y \in Mes(\Omega, \mu, Y)$ *define a random variable* $(x, y) \in Mes(\Omega, \nu; X \times Y)$*, where* $\nu \equiv \mu \circ (x \times y)^{-1}$*. Then,* x, y *are said to be* **independent** *if* $\nu \equiv (\mu \circ x^{-1}) \otimes (\mu \circ y^{-1})$*, or equivalently if* $P\{(x \times y) \in A \times B\} = P\{x \in A\} P\{y \in B\}$ *for arbitrary measurable* $A \subset X$ *and* $B \subset Y$.

REMARK 5.16 - This definition can be extended to any countable finite or infinite family $\{x_i\}_{i \in I}$ of random variables $x_i : \Omega \to X_i$. If the family is not countable, the definition is meaningless since the map $(x_i) : \omega \mapsto (x_i(\omega))$ from ω into $\prod_{i \in I} x_i$ is not necessarily μ-measurable. However, in this case we say that $(x_i)_{i \in I}$ is **independent** if every finite or countable subfamily is independent.

DEFINITION 5.46 - 1) *Let* $x \in Mes(\Omega, \mu, X), x' \in Mes(\Omega', \mu', X)$*. We say that* x *and* x' *are* **isonomous** *if they have the same probability distribution in* $X : \mu \circ x^{-1} = \mu' \circ x'^{-1}$.

2) *Let* $(X_i)_{i \in I}$ *be a family of completely regular spaces and let* $(x_i)_{i \in I}$ *be a family of random variables* $x_i : \Omega \to X$*. The family* (x_i) *is* **canonical** *(or* **regular***) if there exists a random variable*

$$x : \Omega \to X \equiv \prod_{i \in I} \check{X}_i, \quad (\check{X}_i \equiv \text{ Cech-compactification of } X_i),$$

such that $x_i = \pi_i \circ x$*, for each* $i \in I$*, where* π_i *denotes the canonical projection*

$$X \to \check{X}_i.$$

3) *Let* $x \in Mes(\Omega, \mu, E)$, *where* E *is a separable Banach space. The* **mean value** *of* x *is the vector of* E *(if it exists) given by*

$$E(x) \equiv < x >= (B) \int_{\Omega} x d\mu$$

where $(B) \int$ *denotes the Bochner integral.*
4) *Let* E *be a topological vector space, over* $\mathbf{K}(\equiv \mathbf{C}, \mathbf{R})$, *let* $x \in Mes(\Omega, \mu, E)$, *we define* **characteristic function** *of* x *the map* $\phi : E' \rightarrow \mathbf{C}$ *given by*

$$\phi(\theta) = E(\exp(i < \theta, x >)) = \int_{E} \exp(i < \theta, y >) d\nu, \nu \equiv \mu \circ x^{-1}, y \equiv id_{E}.$$

Here ϕ *is also called the* **characteristic functional** *of* x, *or characteristic functional of the measure* ν.

REMARK 5.17 - Let E be a separable Banach space and let $L : E \rightarrow E$ be a linear continuous operator. If $x \in Mes(\Omega, \mu, E)$, then $E(L \circ x) = L(E(x))$. Let $(H, <,>)$ be a separable Hilbert space. Let $x \in Mes(\Omega, \mu, H)$, then

$$< E(x), h' >= \int_{H} < h, h' > d\nu,$$

where $\nu \equiv \mu \circ x^{-1}$.

The following proposition gives a relation between random variables and random measures on a topological space X.

PROPOSITION 5.16 - *Let* X *be a uniform Hausdorff space. The map*

$$Mes(\Omega, \mu, X) \rightarrow M^{+1}_{Ra}(X)$$

where $M^{+1}_{Ra}(X) \equiv$ *set of finite positive Radon measures on* X, $x \mapsto \mu \circ x^{-1}$, *is continuous if* $Mes(\Omega, \mu, X)$ *is equipped with the topology of convergence in probability and* $M^{+1}_{Ra}(X)$ *with the narrow topology.*

In the following table we shall summarize some useful definitions and properties for real random variables.[15]

---

[15] Let $\xi \in Mes(\Omega,\mu,X)$ and $f : X \rightarrow \mathbf{R}^I$. Then $f \circ \xi \equiv f(\xi) : \Omega \rightarrow \mathbf{R}^I$ is a random variable and the expectation value $E(f(\xi))$ of $f(\xi)$ is defined by $E(f(\xi)) = \int_{\Omega} f(\xi) d\mu = \int_{X} f d\nu$, where $\nu \equiv \mu \circ f^{-1}$.

TAB.5.2 - **Useful definitions and properties for real random variables**

Name	Definitions and Properties
Expectation (†+)	$E(x) \equiv \int_\Omega x d\mu$ $E(x+y) = E(x) + E(y)$ $E(cx) = cE(x), c \in \mathbf{R}$
Generating function	$E(e^{tx}), t \in \mathbf{R}$ $E(xy) = E(x)E(y)$ if x and y are indep.
Moments	$E(x^n), (n = 1, 2, \cdots)$
Variance	$\sigma^2 \equiv var(x) = E[(x - E(x))^2]$ (Thebichef inequality)(+*): $P\{\lvert x - E(x)\rvert \geq \lambda\sigma\} \leq \lambda^{-2}, \forall \lambda \geq 1$
Covariance of two rand. var	$cov(x,y) = E\{(x - E(x))(y - E(y))\}$ $cov(x,x) = var(x)$ $var(x+y) = var(x) + var(y) + 2cov(x,y)$ If x, y are indep. $cov(x,y) = 0$ (Schwartz inequality): $(cov(x,y))^2 \leq var(x)var(y)$
Corrl. coeff. two rand.var.	$\rho(x,y) \equiv \frac{cov(x,y)}{(var(x)var(y))^{1/2}}$ $\rho(x,y) \in [-1,1]; \rho(x,x) = 1; \rho(x,-x) = -1$

(†+)($E(x)$ is also called **mean value** of x.)
((+*))(If $E(x)$ and $\sigma>0$ are finite.)

PROPOSITION 5.17 - (DISTRIBUTION FUNCTION OF REAL RANDOM VARIABLE).

1) *Assume that $x \in Mes(\Omega, \mu, \mathbf{R})$ is almost certainly finite, the probability distribution $\nu \equiv \mu \circ x^{-1}$ is a probability on* $\mathbf{R}$, *i.e.*, $\nu(\mathbf{R}) = 1$. *(In this case we say that the random variable x has probability* 1.*) Then, there exist:* (a) *monotone non-decreasing,* (b) *right-continuous function $F : \mathbf{R} \to [0,1]$ with* (c) $\lim_{x\to-\infty} F(x) = 0$, $\lim_{x\to\infty} F(x) = 1$, *such that $L_F \subset \mathcal{P}(\mathbf{R}) \equiv \sigma$-field of subsets on* $\mathbf{R}$; *(L_F contains all Borel subsets of* $\mathbf{R}$, *and such that ν coincides on L_F with the Lebesgue-Stieltjes measure μ_F determined by F). More precisely, one has:*

$$F(t) = P\{\omega; x(\omega) \leq t\} = \mu_F(-\infty, x] = \nu(-\infty, x].$$

Furthermore, for any Borel set $B \subseteq \mathbf{R}$,

$$P\{\omega; x(\omega \in B)\} = \int_B dF(t).$$

Then, F *is called* **distribution function** *of* $x \in Mes(\Omega, \mu, \mathbf{R})$. *Furthermore,* F *characterizes* ν *in the sense that any function* $F : \mathbf{R} \to [0,1]$ *satisfying the above three properties defines a probability measure on* $\mathbf{R}$. *(In fact we can put* $\nu((a,b]) = F(b) - F(a), a < b$, *this can be extended uniquely to a probability measure* ν *on* $(\mathbf{R}, \mathcal{B}(\mathbf{R}))$.

2) *If the random variable* $x \in Mes(\Omega, \mu, \mathbf{R})$ *is finite with probability one, and if* $f : \mathbf{R} \to \mathbf{R}$ *is Borel measurable, then*

$$E\{f \circ x\} = \int_{\mathbf{R}} f(t) dF(t).$$

Furthermore, the distribution function $f \circ x \in Mes(\Omega, \mu, \overline{\mathbf{R}})$ *is* $G(t) = F(\gamma(t))$, *where* $\gamma(t) = \sup\{t'; f(t') \leq t\}$.

3) *Let* x *be any (finite, real) random variable* $x \in Mes(\Omega, \mu, \mathbf{R})$. *Then, there exists a function* $\phi : \mathbf{R} \to \mathbf{R}$, *such that: (a) it is positive definite (for any finite sets* $\{t_1, \cdots, t_n\} \in \mathbf{R}$ *and* $\{\alpha_1, \cdots, \alpha_n\} \in \mathbf{C}, \sum_{j,k} \alpha_j \bar{\alpha}_k (z_j - z_k) \geq 0)$; *(b) uniformly continuous; (c)* $\phi(0) = 1$. ϕ *is uniquely determined by the distribution* F. *More precisely,*

$$\phi(t) = E(e^{itx}) = \int_{\mathbf{R}} e^{it\lambda} dF(\lambda) = \int_{\mathbf{R}} e^{it\lambda} d\nu$$

so ϕ *is the* **Fourier-Stieltjes transform** *of* F. ϕ *is called the* **characteristic function** *of* x. *If* $x, y \in Mes(\Omega, \mu, \mathbf{R})$ *are independent random variables, then the characteristic function* ϕ_{x+y} *of* $x + y$ *is*

$$\phi_{x+y} = \phi_x \phi_y.$$

4) (**P.Levy**). *Let* ϕ_n *be characteristic functions of distributions* F_n. *Let* $\phi_n \to \phi$ *uniformly on some neigbourhood of* $z = 0$. *Then,* ϕ *is also a characteristic function, and one has* $F_n \to F$, *where* F *is the distribution associated with* ϕ. *It is assumed that* $F_n \to F$ *means that for all* $f \in \widetilde{C}^0(\mathbf{R})$ = *Banach space of all continuous functions on* $\mathbf{R}$ *vanishing at* $+\infty$ *and* $-\infty$ *equipped with the topology of uniform convergence,*

$$\lim_{n \to \infty} \Big(\int_{-\infty}^{+\infty} f dF_n \Big) = \int_{-\infty}^{+\infty} f dF.$$

5) *The space R of distribution functions, $R \equiv \{F : \mathbf{R} \to [0,1]\}$, F monotone, non-decreasing, right-continuous function, such that*

$$\lim_{n \to -\infty} F(x) = 0 \quad , \quad \lim_{n \to \infty} F(x) = 1 \quad ,$$

is a complete metric space with respect to the **Levy's metric**:

$$d(F,G) \equiv \inf\{\epsilon > 0 | F(x-\epsilon) - \epsilon \leq G(x) \leq F(x+\epsilon) + \epsilon\}, \forall x \in \mathbf{R}, F, G \in R.$$

EXAMPLE 5.15 - 1) A **discrete random variable** is one that takes on at most countably many values. A such random variable is characterized by a jump function as distribution. Assume that x is a real random variable which takes on only (positive and negative) integer values and write $p_j = \mu(x = j), j = 0, \pm 1, \pm 2, \cdots$. Then x has the following characteristic function $\phi(t) = P(e^{it})$, where

$$P(z) = \sum_{j=-\infty}^{+\infty} p_j z^j, \quad z \in \mathbf{C}.$$

The series for $P(z)$ converges at least when $|z| = 1$, the resulting function is called the **probability generating function**. One has the following properties: (i) if $E(x) < \infty \Rightarrow E(x) = P'(1)$; (ii) if $E(x^2) < \infty \Rightarrow var(x) = P''(1) + P'(1) - \{P'(1)\}^2$.
2) A **continuous random variable** is one such that the distribution function F can be expressed as follows:

$$F(t) = \int_{-\infty}^{t} f(\lambda) d\lambda$$

where $f : \mathbf{R} \to \mathbf{R}$ is a measurable function such that

$$f(\lambda) \geq 0, \quad \int_{-\infty}^{+\infty} f(\lambda) d\lambda = 1.$$

Then, f is called **a probability density**. This is equivalent to saying that F is **absolutely continuous**, (which in turn occurs iff its probability distribution is absolutely continuous with respect to Lebesgue measure). Notice that if f_1 and f_2 are probability densities of the same random variable, then $f_1 = f_2$ at most everywhere. Furthermore $F'(t) = f(t)$ at most everywhere. The relation between probability density and characteristic function is the following:

$$\phi(\lambda) = \frac{\lambda}{2\pi} \int_{-\infty}^{+\infty} \phi(t) e^{-it\lambda} dt.$$

EXAMPLE 5.16 -

EXAMPLES OF REAL RANDOM VARIABLES: $x \in Mes(\Omega, \mu, \mathbf{R})$.

1) BINOMIAL: $B(n,p)$: (Discrete).

Distribution function: $\Sigma_{r\leq t}\binom{n}{r}p^r(1-p)^{n-r}$. Probability generating function: $(1-p+pz)^n$. Mean: np. Variance: $np(1-p)$.

☐ In the binomial case $B(n,p)$, the probability that x takes value r is

$$P(x=r) = \binom{n}{r}p^r(1-p)^{n-r} \equiv b(n,p;r).$$

2) POISSON: $P(\mu)$: (Discrete).

Distribution function: $\Sigma_{r\leq t}\mu^r e^{-\mu/r!}$. Characteristic function: $\exp[\mu(e^{it}-1)]$. Probability generating function: $e^{\mu(z-1)}$. Mean $= \mu =$ variance.

3) GEOMETRIC: $G(q), q \in [0,1]$: (Discrete).

Probability generating function: $\frac{1-q}{1-qz}$. Mean: $\frac{q}{1-q}$. Variance: $\frac{q}{(1-q)^2}$.

☐ In the geometric case $G(q)$, the probability that x takes value $r = 0, 1, 2, \cdots$ is

$$P(x=r) = (1-q)q^r.$$

4) GAUSSIAN:$N(0,1)$: (Continuous).

Characteristic function: $\exp(-\frac{1}{2}t^2)$. Probability density: $(2\pi)^{-\frac{1}{2}}e^{-\frac{1}{2}\lambda^2}$. Mean: 0. Variance: 1.

☐ In Gaussian case $N(0,1)$, the moments $E(x^k)$ of a Gaussian random variable are

$$E(x^k) = \left\{\begin{array}{l} 0 \quad \text{for} \quad k \quad \text{odd} \\ \text{related to integral gamma function for } k \text{ even} \end{array}\right\}$$

5) RANDOM VARIABLE WITH NORMAL DISTRIBUTION: $N(\mu, \sigma^2)$: (Continuous).

Charactersitic function: $\exp(i\mu t - \frac{1}{2}\sigma^2 t^2)$. Probability density:

$$\sigma^{-1}(2\pi)^{-\frac{1}{2}}\exp(-\frac{(\lambda-\mu)^2}{2\sigma^2}).$$

Mean: μ. Variance: σ^2.

6) RANDOM VARIABLE WITH UNIFORM DISTRIBUTION: $U(a,b)$: (Continuous).

Characteristic function: $\frac{e^{ibt}-e^{iat}}{i(b-a)t}$. Probability density:

$$f(\lambda) = \left\{\begin{array}{l} (b-a)^{-1} \quad \text{if} \quad a \leq \lambda \leq b \\ 0 \quad \text{otherwise} \end{array}\right\}$$

7) RANDOM VARIABLE WITH EXPONENTIAL DISTRIBUTION: $E(\xi)$: (Continuous)
Characteristic function: $\xi/(\xi - it)$. Probability density:

$$f(\lambda) = \left\{ \begin{matrix} \xi e^{-\xi\lambda} & \text{if} & \lambda \geq 0 \\ 0 \ \text{if} & \lambda < 0 & \end{matrix} \right\}$$

8) RANDOM VARIABLE WITH CAUCHY DISTRIBUTION: (Continuous).
Characteristic function: $e^{-|t|}$. Probability density: $[\pi(1+\lambda^2)]^{-1}$.
PROPOSITION 5.18 - (CONDITIONAL PROBABILITY). *Let $(\Omega, \mathcal{T}, \mu)$ be a probability space. Let A be a fixed element of $\mathcal{T}$. Let $\mathcal{G}$ be a sub-σ-field of $\mathcal{T}$, not necessarily containing A. Then, there exists a unique function $\mu|_{\mathcal{G}}$-measurable $\phi \equiv \mu(A|\mathcal{G}) : \Omega \to$ **R** such that*

$$\mu(A \cap E) \equiv \mu_A(E) = \int_E \phi d\mu|E = \int_E \phi d\mu, \forall E \in \mathcal{G}.$$

Then, $\mu(A|\mathcal{G})$ is called the **probability of A conditioned on the σ-field $\mathcal{G}$**. *(Note that, in general, $\mu(\bullet|\mathcal{G})$ is not a measure on $(\Omega, \mathcal{T})$.) One has:*
(a) *If $A \in \mathcal{G} \Rightarrow \mu(A|\mathcal{G}) = \chi_A =$ indicator function of A;*
(b) *If A and $\mathcal{G}$ are independent, then $\mu(A|\mathcal{G}) = \mu(A)$;*
(c) *If $\mathcal{S}$ is the σ-field containing only $\emptyset$ and $\Omega \Rightarrow \mu(A|\mathcal{S}) = \mu(A)$.*
DEFINITION 5.47 - *If $x \in Mes(\Omega, \mu, X)$ is a random variable on the probability space $(\Omega, \mathcal{T}, \mu)$, we call* **conditional probability** *of $A \in \mathcal{T}$ with respect to*

$$x\mu(A|\mathcal{G} \equiv x^{-1}(R)) \equiv \mu(A|x).$$

DEFINITION 5.48 - 1) *A* **stochastic process** *(***random function** *or* **random process***) is a map $f : T \to Mes(\Omega, \mu, X)$, where T is an arbitrary set. The map $f_\omega : T \to X$, for $\omega \in \Omega$ fixed, is called a* **sample function** *(or* **sample path** *if $T \equiv$* **R***). The map $\tilde{f} : T \times \Omega \to X, \tilde{f}(t, \omega) = f(t)(\omega)$ is called a* **random map.**
2) *Two stochastic processes $f : T \to Mes(\Omega, \mu, X), g : S \to Mes(\Omega, \mu, Y)$ corresponding to the same sample space (Ω, μ), are said to be* **independent** *if for arbitrary finite sequence, $t_1, \cdots, t_p \in T$ and $s_1, \cdots, s_p \in S$ the random variables $(f(t_1), \cdots, f(t_p)) : \Omega \to X \times \cdots_p \cdots \times X \equiv X^p$ and $(g(s_1), \cdots, g(s_p)) : \Omega \to Y \times \cdots_p \cdots \times Y \equiv Y^p$ are independent.*
3) *Let X be a completely regular space. We say that the stochastic process $f : T \to Mes(\Omega, \mu, X)$ is* **canonical** *(or* **regular***) if there exists a random variable $\check{f} : \Omega \to \check{X}^T \equiv F(T, \check{X})$, such that for every $t \in T, f(t)(\omega) = \check{f}(\omega)(t)$ for almost every $\omega \in \Omega$. (Every stochastic process is isonomous to a canonical stochastic process, so we shall often assume that a given random function is canonical.)*

4) *Let T be a topological space. A canonical stochastic process $f : T \to Mes(\Omega, \mu, X)$ is* **almost surely continuous** *if $\check{f}(\omega) : T \to X$ is an injective continuous map for almost every $\omega \in \Omega$.*

5) *If T and X are vector spaces, a canonical stochastic process $f : T \to Mes(\Omega, \mu, X)$ is* **linear** *if $\check{f}(\omega) : T \to X$ is a linear map for almost every $\omega \in \Omega$. f is one-to-one iff $\tilde{f}(t,\omega) = 0$ a.e. implies $t = 0$.*

6) *If $T \equiv \mathbf{Z}$ or $T \equiv \mathbf{N}$ the stochastic process is called a* **discrete stochastic process**.

REMARK 5.18 - If Y is a topological space and $\mathcal{B}$ is its Borellian σ-algebra, for any $n \in \mathbf{N}, B^n \equiv \bigotimes_{i=1}^{n} B_i, B_i \equiv \mathcal{B}$, is the product σ-algebra of Y^n. Furthermore, a **cylindrical set** in Y^X with **basis** $B_1 \times \cdots \times B_n \in B^n$, (over coordinates $x_1, \cdots, x_n \in X$), is the set $C_{x_1 \cdots x_n}(B_1 \times \cdots \times B_n) \equiv \{f \in Y^X | f(x_i) \in B_i, i = \overline{1,n}\}$. The class $\mathcal{C}$ of all cylindrical sets over a finite number of coordinates is an algebra of sets in Y^X. Denote by $\mathcal{A}$ the σ-algebra of sets in Y^X generated by $\mathcal{C}$. $\tilde{f} : X \times \Omega \to Y$ is a random map from X to Y iff the map $\check{f} : (\Omega, \mathcal{T}) \to (Y^X, \mathcal{A})$ defined by $f(\omega) = \tilde{f}_\omega$ is a measurable map. Then, $\nu \equiv \mu \circ \check{f}^{-1}$ is a probability measure on $(Y^X, \mathcal{A})$ called **distribution of the random process** $\tilde{f}$. For a random map

$$\tilde{f} : X \times \Omega \to (X, \mathcal{B}), n \in \mathbf{N}, x_1, \cdots, x_n \in T, B_1, \cdots, B_n \in \mathcal{B}$$

the probabilities

$$P\{\omega | \widetilde{f}_{x_1}(\omega) \in B_1, \cdots, \widetilde{f}_{x_n}(\omega) \in B_n\}$$

are called the **finite dimensional distributions** of $\tilde{f}$. One has:

$$P\{\omega | \widetilde{f}_{x_1}(\omega) \in B_1, \cdots, \widetilde{f}_{x_n}(\omega) \in B_n\} = \nu(C_{x_1 \cdots x_n}(B_1 \times \cdots \times B_n))$$

where $\nu \equiv \mu \circ \check{f}^{-1}$. The corresponding measures $\nu_{x_1 \cdots x_n}$ on (X^n, B^n) are probability measures, also called finite dimensional distributions of the stochastic process $\tilde{f}$. In particular if $X \equiv E$ is a separable Banach space, for any measurable map $h : (E^n, B^n) \to E$ one has

$$E(h(\widetilde{f}_{x_1}, \cdots, \widetilde{f}_{x_n})) = (B) \int_{E^n} h d\nu_{x_1 \cdots x_n}$$

The **characteristic functional** of $f : T \to Mes(\Omega, \mu, E)$ is the map

$$\left\{ \begin{array}{l} \phi : T \times E' \to \mathbf{C} \\ \phi(t, \theta) = \phi_t(\theta) = E(\exp(i < \theta, f_t >) = \int_E \exp(i < \theta, y >) d\nu_t \end{array} \right\}$$

where $\nu_t \equiv \mu \circ \widetilde{f}_t^{-1}, y \equiv id_E$.

THEOREM 5.28 - *Let (E, F) be a duality couple of vector spaces on* $\mathbf{R}$. *Then, there exists a bijection between cylindrical measures on E and the set of stochastic processes on F.*

EXAMPLE 5.17 - Set $T \equiv [0,1] \subset \mathbf{R}$. As $(C^0(T), M_{Ra}(T))$ is a duality couple of vector spaces on $\mathbf{R}$, we have that for a cylindrical measure w on $C^0(T)$ there corresponds a linear stochastic process on $M_{Ra}(T)$. For example, the stochastic process $f : M_{Ra}(T) \to Mes(C^0(T), w; \mathbf{R})$ to $\mu \in M_{Ra}(T)$ associates the w-class of functions

$$f(\mu) : C^0(T) \to \mathbf{R}, \quad x \mapsto \mu(x) \equiv \int_T x \, d\mu$$

is a canonical linear stochastic process associated to the Wiener (cylindrical) measure w.

DEFINITION 5.49 - *Let $X \equiv E$ be a topological vector space and $\{T, d\}$ a compact metric space. A map $f : T \to Mes(\Omega, \mu; E)$ is* μ-**Riemannian integrable** *if the base of the filter $\{B_\epsilon(t, \mu)\}_{\epsilon>0}$, where $B_\epsilon(t, \mu)$ is the set of all elements x of $Mes(\Omega, \mu; E)$ like*

$$x = \sum_{i \in I} \mu(A_i) x_i,$$

where I is finite, $\{A_i\}_{i \in I}$ is a partition of T in borelians with diameter ϵ and $x_i \in f(A_i)$, converges in $Mes(\Omega, \mu; E)$. This limit is called the μ-**Riemannian integral** *of f. (If f is continuous, f is μ-Riemannian integrable for any $\nu \in Ob(\overline{\mathcal{M}}(T))$.)*

THEOREM 5.29 - *Let (T, d) be a compact metric space and let $L : Ob(\overline{\mathcal{M}}_c(T)) \to Mes(\Omega, \mu; E)$ be a linear random process, $Ob(\overline{\mathcal{M}}_c(T)) \equiv$ vector space $Ob(\overline{\mathcal{M}}(T))$ endowed with the topology of uniform convergence on the compact parts of $C^0(T)$. Then, $L(\mu)$ is the μ-Riemannian integral of the function $f : T \to Mes(\Omega, \mu; E)$, $f(t) = L(\delta_t)$. Furthermore, the stochastic process f admits a version g with continuous trajectories $\omega \mapsto \int_T \tilde{f}(t, \omega)\mu(dt)$, is an element of the class $L(\mu)$.*

Of particular importance are the real random processes.

DEFINITION 5.50 - 1) *A* **real stochastic process** *is a map $f : T \to Mes(\Omega, \mu; \mathbf{R})$.*
2) *The* **covariance of a stochastic process** *$f : T \to Mes(\Omega, \mu; \mathbf{R})$ is the function*

$$\left\{ \begin{array}{l} \Gamma : T \times T \to \mathbf{R} \\ \Gamma(s,t) = E(f_s f_t) - E(f_s)E(f_t) \end{array} \right\}$$

3) *Let $\tilde{f} : T \times \Omega \to \mathbf{R}$ be a random function. The* **mean value** *of $\tilde{f}$ is the map*

$$\left\{ \begin{array}{l} <\tilde{f}>: T \to \mathbf{R} \\ <\tilde{f}>(t) = E(\tilde{f}_t) \end{array} \right\}$$

The **correlation function** *of* $\tilde{f}$ *is the map*

$$\left\{\begin{array}{l}\rho(\tilde{f}) : T \times T \to \mathbf{R} \\ \rho(\tilde{f})(t_1, t_2) = \rho(\tilde{f}_{t_1}, \tilde{f}_{t_2})\end{array}\right\}$$

The **variance** *of* $\tilde{f}$ *is the map*

$$\left\{\begin{array}{l}var(\tilde{f}) : T \to \mathbf{R} \\ var(\tilde{f})(t) = var(\tilde{f})\end{array}\right\}$$

PROPOSITION 5.19 - (PROPERTIES OF REAL STOCHASTIC PROCESSES).
1) *The covariance* Γ *of a real stochastic process* $f : T \to Mes(\Omega, \mu; \mathbf{R})$ *identifies a metric*

$$\left\{\begin{array}{l}d : T \times T \to \mathbf{R}^+ \\ d(s,t) = \|f_s - f_t\| = (\Gamma(t,t) + \Gamma(s,s) - 2\Gamma(s,t))^{1/2}.\end{array}\right\}$$

2) *The correlation function has the following properties:*
(a) $\rho(\tilde{f})(t_1, t_2) = \rho(\tilde{f})(t_2, t_1), \forall\, t_1, t_2 \in T$.
(b) $\sum_{i,j=1}^{n} \rho(\tilde{f})(t_i, t_j)\alpha^i \alpha^j \geq 0$, *for any* $n \in \mathbf{N}, t_i \in T, \alpha^i \in \mathbf{R}$.

THEOREM 5.30 - (FINITE DIMENSIONAL DISTRIBUTIONS OF REAL STOCHASTIC PROCESSES).
1) *Let* $T \equiv E$ *be a vector space and let* $f : E \to Mes(\Omega, \mu, \mathbf{R})$ *be a linear real stochastic process. Then* f *is characterized by a measurable map* $\tilde{f} : (\Omega, T) \to (E^*, A)$, *where* A *is the* σ*-algebra generated by cylindrical sets in* E^*. *More precisely, a cylindrical set in* E^* *is a subset* A *of* E^* *such that there exists a family of* n*-vectors of* E, $(x_1, \cdots, x_n), n \in \mathbf{N}$, *such that*

$$A = \{\alpha \in E^* | (< \alpha, x_1 >, \cdots, < \alpha, x_n >) \in B(\mathbf{R}^n)\}.$$

Furthermore, the finite dimensional distributions of $\tilde{f}$ *are probabilities measures* $\nu_{x_1 \cdots x_n}$ on $(\mathbf{R}^n, \mathcal{B}(\mathbf{R}^n)), n \in \mathbf{N}$.

☐ If E is a separable Banach space, for any measurable map $h : (E^n, B^n) \to (E, B)$ one has

$$E(h(\tilde{f}_{x_1}, \cdots, \tilde{f}_{x_n})) = (B) \int_{E^n} h d\nu_{x_1 \cdots x_n}.$$

☐ The mean value of $\tilde{f}$ is an element of $E^* :< \tilde{f} > \in E^*$.

☐ The characteristic functional of $\tilde{f}$ is:

$$\left\{\begin{array}{l}\phi : E \times \mathbf{R} \to \mathbf{C} \\ \phi(x,t) = E(\exp(it\tilde{f}_x)) = \int_{\mathbf{R}} \exp(it) d\nu_x, \quad \nu_x \equiv \mu \circ \tilde{f}_x^{-1}\end{array}\right\}$$

2) (**Inverse theorem**). *A system* $(\nu_{x_1\cdots x_n})$, *indexed by the set of finite families* $(x_1,\cdots,x_n)$ *of vectors belonging to a vector space* E, *is the finite dimensional distributions of a linear real stochastic process* $f : E \to Mes(\Omega,\mu,\mathbf{R})$ *iff the following conditions are satisfied:* (**Kolmogorov consistency conditions**)*:*
(a) *For any* $(x_1,\cdots,x_n), \nu_{x_1\cdots x_n}$ *is a probability measure on* $\mathbf{R}^n$;
(b) *For any* $(x_1,\cdots,x_n)$, *any linear map* $u \in L(\mathbf{R}^n,\mathbf{R}^m)$, *and any* $A \in \mathcal{B}(\mathbf{R}^m)$, *one has*

$$\nu_{u(x_1)\cdots u(x_n)}(A) = \nu_{x_1\cdots x_n}(u^{-1}(A)).$$

3) *Let* E *be a space, inductive limit of a sequence* $(E_n)_{n\in\mathbf{N}}$ *of Fréchet subspaces of countable type. A system* $(\nu_{x_1\cdots x_n})$ *like before, is the system of finite dimensional distributions of probability measure on* $(E',\mathcal{B}(E'))$ *iff the consistency conditions of Kolmogorov and the following are satisfied:*
(C) *For any* $\epsilon > 0$, *there exists a neighbourhood* U *of* 0 *in* E *such that, for any* $p \in \mathbf{N}$ *and any element* $(x_1,\cdots,x_p) \in U^p$, *one has* $\nu_{x_1\cdots x_p}(U^p) \geq 1-\epsilon$. *Furthermore, if* E_n *are nuclear spaces, then it is sufficient that the following are fulfilled: the Kolmogorov consistency conditions and that the map* $\phi_t : E \to \mathbf{C}, \phi_t(x) = \int_{\mathbf{R}} \exp(it)d\nu_x$, *should be continuous in* E *or* E_n.
EXAMPLE 5.18 - (REAL STOCHASTIC PROCESSES). 1) Let $T \equiv \mathcal{B}([0,1])$. Then, there exists a real stochastic process $f : T \to Mes(\Omega,\mu,\mathbf{R})$ such that: (a) $f(t)$ has a Poisson distribution whose mean is the Lebesgue measure of t; (b) if $t_1,\cdots,t_n$ are disjoint, $f(t_1),\cdots,f(t_n)$ are independent, and $f(t_1,\cdots,t_n) = f(t_1)+\cdots+f(t_n)$.
Let (Ω,T,μ) be a probability space. A vector subspace H of $Mes(\Omega,\mu,\mathbf{R})$ is **Gaussian** if each element of H is a Gaussian random variable: (We can see that the closure $\overline{H}$ of H in $Mes(\Omega,\mu,\mathbf{R})$ is also Gaussian and that $\overline{H}$ is a closed subspace of $Mes(\Omega,\mu,\mathbf{R})$.
A **Gaussian process** is a real stochastic process $f : T \to Mes(\Omega,\mu,\mathbf{R})$ such that the closed vector subspace generated by $\{f_t\}_{t\in T}$ is Gaussian.
For any Gaussian process $f : T \to Mes(\Omega,\mu,\mathbf{R})$ one has $E(f_t) = 0$ and the covariance Γ is a symmetric function $\Gamma : T\times T \to \mathbf{R}$ such that for any finite subset $\{t_1,\cdots,t_n\} \in T$ and any real numbers $z_1,\cdots,z_n$ not all zero

$$\sum_{1\leq i,j\leq n} \Gamma(t_i,t_j)z_iz_j > 0$$

(i.e., Γ is **positive definite**).
Furthermore, a Gaussian random function $\tilde{f} : T\times\Omega \to \mathbf{R}$ is characterized by definite dimensional distributions given by

$$F\{\omega|(\tilde{f}_{t_1}(\omega),\cdots,\tilde{f}_{t_k}(\omega)) \in A\} = \int_A \gamma_{x_0,B}$$

where $k \in \mathbf{N}, t_i \in T, i = \overline{1,k}$, A is Borel set in $\mathbf{R}^k, x_0 \equiv (<\tilde{f}>(t_1), \cdots, <\tilde{f}>(t_k))$. $B \equiv (a_{ij}), i,j = \overline{1,k}, a_{ij} \equiv \rho(\tilde{f})(t_i,t_j)$, and $\gamma_{x_0,B}$ is the Gaussian measure on $\mathbf{R}^k$ relative to the vector x_0 and the isomorphism B.

2) A **Wiener process** (or **Brownian motion**) is a real stochastic Gaussian process $f: T \to Mes(\Omega,\mu,\mathbf{R})$ such that: (a) $T \equiv [0,1]$; (b) The increments $f(s,t) = f(s) - f(t)$ are independent; (c) $f(0) = 0$; (d) the process $g: T \times T \to Mes(\Omega,\mu,\mathbf{R})$ given by $g(s,t) = f(s,t)$ is a Gaussian process with variance

$$var(g)(s,t) = |t-s|.$$

The distribution of Wiener process is a measure μ on $(\mathbf{R}^T, \mathcal{B}^T)$ supported by $C^0(T)$, (i.e., $\mu(C^0(T)) = 1$), so μ identifies a measure w on $C^0(T)$ endowed with the collection $\mathcal{B}'(C^0(T))$, trace of $\mathcal{B}^T$ on $C^0(T)$. w is just the Wiener measure. Further, one has: (a) $C^0(T) \neq \mathcal{B}^T$; (b) $\mathcal{B}(C^0(T)) = \mathcal{B}'(C^0(T))$. Furthermore, one can prove that almost all sample paths of a Brownian motion are nowhere differentiable.

3) A **Markov process** is a real stochastic process $f: T \to Mes(\Omega,\mu,\mathbf{R})$ where: (a) $T \subset \mathbf{R}$; (b) if for any subset $S \subset T, F_S$ denotes the smallest σ-field of event with respect to which the function $f(s): \Omega \to \mathbf{R}$ is measurable for any $s \in S$, then $(\mu(A \cap B)|f(t)) = \mu(A|f(t))\mu(B|f(t))$ with probability one with $A \in F_{\{s \in T; s<t\}}$, $B \in F_{\{s \in T; s>t\}}$. In the following we shall consider some fundamental properties of Markov process and a relation with Brownian motion.

□ Let $f: T \to Mes(\Omega,\mu,\mathbf{R})$ be a Markov process. Then, there exists a function $P: T\widehat{\times}T \times \Omega \times \mathcal{T} \to \mathbf{R}$, where $T\widehat{\times}T \equiv \{(s,t) \in T \times T | s < t\}$, such that: (i) $P(s,t,\bullet,A)$ is a Borel-measurable function on Ω and $P(s,t,\lambda,\bullet)$ is a probability measure on $\mathcal{T} \equiv \sigma$-algebra of the probability space; (ii) $P(s,s,\lambda,A) = \chi_A(\lambda)$; (iii) the following **Chapman-Kolmogorov equation** holds for almost all $\lambda \in \Omega$:

$$P(s,t,\lambda,A) = \int P(s,t,\lambda,dy)P(t,u,y,A), s<t<u.$$

Any such function $P(s,t,\lambda,A)$ is called a **transition probability** of the **Markov process** f.

□ The distribution of a Markov process f is uniquely determined by the initial distribution μ_0 of f_0 and the system $\{P(s,t,\lambda,A)\}$ of transition probabilities. Then finite dimensional distributions of a Markov process $f: T \to Mes(\Omega,\mu,\mathbf{R})$ may be defined so that to satisfy the relation

$$\left\{\begin{array}{l} E(\psi(f(t_0), f(t_1), \cdots, f(t_n))) \\ = \int_\Omega \cdots \int_\Omega \psi(\lambda_0,\lambda_1,\cdots \lambda_n) d\mu_0 \prod_{j-1}^{n} dP(t_{j-1},t_j,\lambda_{j-1},\bullet) \end{array}\right\}$$

for any measurable function $\psi : \Omega \times \cdots_{n+1} \cdots \times \Omega \to \mathbf{R}$.

☐ (**Relation between Brownian motion and Markov process**). Any Brownian motion $f : T \to Mes(\Omega, \mu, \mathbf{R})$ is a Markov process.

2 QUANTUM SPECTRAL MEASURES OF PDES.

DEFINITION 5.51 - 1) *A* **partially ordered system** $(E, \leq)$ *is a non-empty set* E, *together with a relation* $\leq$ *on* E, *such that:* (a) *if* $a \leq b$ *and* $b \leq c \Rightarrow a \leq c$; (b) $a \leq a$. *The relation* $\leq$ *is called an* **order relation** *in* E. *The notation* $y \geq x$ *is sometimes used in place of* $\leq$.

2) *A* **totally ordered subset** F *of partially ordered system* $(E, \leq)$ *is a subset of* E *such that for every pair* $x, y \in F$ *either* $x \leq y$ *or* $y \leq x$.

3) *If* F *is a subset of a partially ordered system* $(E, \leq)$ *then an element* x *in* E *is said to be an* **upper bound** *for* F *if every* f *in* F *has the property* $f \leq x$. *An upper bound for* F *is said to be a* **least upper bound** *of* F *(*$\sup F$*) if every upper bound* y *of* F *has the property* $x \leq y$.

4) *In a similar fashion, the terms* **lower bound** *and* **greatest lower bound** *(in* F*) may be defined.*

5) *An element* $x \in E$ *is said to be* **maximal** *if* $x \leq y$ *implies* $y \leq x$.

EXAMPLE 5.19 - Let X be a set and $\mathcal{P}(X)$ be the set of all subsets of X. The couple $(\mathcal{P}(X), \subseteq)$ is a partially ordered system where the order relation is the inclusion relation $\subseteq$ between the sets contained in X. An upper bound for a subfamily $B \subseteq \mathcal{P}(X)$ is any set containing $\cup B$, and $\cup B$ is the only least upper bound of B. Similarly $\cap B$ is the only greatest lower bound of B. The only maximal element of $\mathcal{P}(X)$ is X.

DEFINITION 5.52 - *Let* $(E, \leq)$ *be a non-void partially ordered system. A subset* $B \subset E$ *with the following three properties will be called* **admissible** *with respect to a function* $f : E \to E$ *such that* $f(x) \geq x$; *and with respect to an element* $a \in E$: (a) $a \in B$; (b) $f(B) \subseteq B$; (c) *Every least upper bound of a totally ordered subset of* B *is in* B.

THEOREM 5.31 - 1) (**Hausdorff maximality theorem**). *Every partially ordered system contains a maximal totally ordered subsystem.*

2) (**Zorn's lemma**). *A partially ordered system has a maximal element if every totally ordered subsystem has an upper bound.*

3) (**Zermelo well-ordering theorem**). *Every set* E *may be* **well-ordered** *that is a partially ordered system* $(E, \leq)$ *such that:* (*i*) $a \leq b, b \leq a \Rightarrow a = b$; (*ii*) *any non-void subset of* E *contains a lower bound for itself.*

EXAMPLE 5.20 - $\mathbf{N} \equiv$ the set of positive integers in the usual order is a familiar example of well-ordered system.

DEFINITION 5.53 - *A partially ordered system* $(E, \leq)$ *is said to be* **complete** *if:* (*i*) $a \leq b$ *and* $b \leq a \Rightarrow a = b$; (*ii*) *Every non-void subset has a least upper bound and* *a greatest lower bound.*

PROPOSITION 5.20 - (**Tarski**). *If $(E, \leq)$ is a complete partially ordered system , $f : E \to E$, and $x \leq y$ implies $f(x) \leq f(y)$, then f has a fixed element x_0, i.e., $f(x_0) = x_0$, and the set of all fixed elements contains its least upper bound and its greatest lower bound.*

DEFINITION 5.54 - *An element x in a ring R is said to be* **idempotent** *if $x^2 = x$, and to be* **nilpotent** *if $x^n = 0$ for some positive integer n.*

DEFINITION 5.55 - *A* **Boolean ring** *is one in which every element is idempotent.*

PROPOSITION 5.21 - 1) *Every Boolean ring is a commutative ring where the identity $x + x = 0$ (or equivalently $x = -x$) holds.*

2) *In a Boolean ring, any prime ideal is maximal.*

EXAMPLE 5.21 - 1) The smallest Boolean ring with unit is $\mathbf{Z}_2 \equiv \{0, 1\}$, i.e., the set of integers modulo 2. $\mathbf{Z}_2$ is actually a field.

2) Conversely every Boolean ring with unit which is also a field is necessarily isomorphic to the field $\mathbf{Z}_2$.

3) If a Boolean ring A does not contain zero divisor it is reduced to 0 or it is isomorphic to $\mathbf{Z}_2$.

4) The ring $\mathcal{P}(X)$ of subsets of a fixed set X is a Boolean ring with respect to the following definitions of multiplication and addition of arbitrary subsets of X: (a)**multiplication**: $EF = E \cap F, E, F \in \mathcal{P}(X))$; (b) **addition**: $E + F = (E \cap CF) \cup (CE \cup F) = (E \cup F) \cap (E \cap F) = (E - F) \cup (F - E) \equiv EF$ ($\equiv$ **symmetric difference**). X is the unity element and $\emptyset$ is the zero element.

4) The Boolean ring $\mathcal{P}(X)$ is isomorphic to the ring ${\mathbf{Z}_2}^X$ of the applications X into the ring $\mathbf{Z}_2$:

$$h : \mathcal{P}(X) \to {\mathbf{Z}_2}^X, h : A \mapsto h(A) \equiv \chi_A \equiv \text{characteristic function of } A.$$

5) Every Boolean ring is isomorphic to a sub-ring of the ring ${\mathbf{Z}_2}^X$.

6) Every finite Boolean ring is of the form ${\mathbf{Z}_2}^n$.

DEFINITION 5.56 - *A topological space is said to be* **totally disconnected** *if its topology has a base consisting of sets which are simultaneously open and closed.*

THEOREM 5.32 - (**Stone representation theorem**). *Every Boolean ring with unit is isomorphic to the Boolean ring of all open and closed subsets of a totally disconnected compact Hausdorff space.*

DEFINITION 5.57 - 1) *A* **lattice** *is a partially ordered set L such that every pair x, y of elements in L has a least upper bound and a greatest lower bound, denoted by $x \vee y$ and $x \wedge y$ respectively.*

2) *The lattice L has a* **unit** *if there exists an element 1 such that $x \leq 1, x \in L$.*

3) *The lattice L has a* **zero** *if there exists an element 0 such that $0 \leq x, x \in L$.*

4) *The lattice is called* **distributive** *if* $x \wedge (y \vee z) = (x \wedge y) \vee (x \wedge z), x, y, z \in L$.
5) *The lattice is called* **complemented** *if for every* $x \in L$*, there exists an* $x' \in L$ *such that* $x \vee x' = 1, x \wedge x' = 0$.
6) *A lattice* L *is said to be* **complete** *if every subset bounded above has a least upper bound, or equivalently, every subset bounded below has a greatest lower bound.*
7) *A lattice* L *is said to be* σ**-complete** *if this is true for denumerable subsets of* L.
DEFINITION 5.58 - *A* **Boolean algebra** *is a lattice with unit and zero which is distributive and complemented.*
PROPOSITION 5.22 - *The concepts of Boolean algebra and Boolean ring with unit are equivalent.*
PROOF. In fact let B be a Boolean algebra and define multiplication and addition as:

$$xy = x \wedge y, x + y = (x \wedge y') \vee (x' \wedge y).$$

Then, it may be verified that B is a Boolean ring with 1 as unit. On the other hand, if B is a Boolean ring with unit denoted by 1, then if $x \leq y$ is defined to mean $x = \lambda y$ and $x' = 1 + x$, then B is a Boolean algebra and $x \vee y = x + y, x \wedge y = xy$. □
DEFINITION 5.59 - 1) *If* B *and* C *are Boolean algebras and* $h : B \to C$*, then* h *is said to be* **homomorphism**, *or a* **Boolean algebra homomorphism**, *if*

$$h(x \wedge y) = h(x) \wedge h(y), \quad h(x \vee y) = h(x) \vee h(y), \quad h(x') = h(x)'.$$

2) *If* h *is one-to-one, it is called an* **isomorphism**.
3) *If* h *is an isomorphism and* $h(B) = C$*, then we say that* B *is* **isomorphic** *to* C.
EXAMPLE 5.22 - The following are examples of Boolean algebras.
1) $\mathbf{Z}_2 \equiv \{0, 1\}$.
2) $\mathcal{P}(X)$, where $\leq$ is taken to be the inclusion, and $\wedge, \vee$ are taken as intersection and union respectively.
THEOREM 5.33 - *Every Boolean algebra is isomorphic with the Boolean algebra of all open and closed subsets of a totally disconnected compact Hausdorff space.*
DEFINITION 5.60 - 1) *A* **projection** *in* V *means a linear operator* $E \in L(V)$ *with* $E^2 = E$.
2) *The* **intersection** $A \wedge B$ *and the* **union** $A \vee B$ *of two commutating projections* A *and* B *in* V *are projections: (a)* **intersection**: $A \wedge B = A \circ B$; *(b)* **union**: $A \vee B = A + B - (A \circ B)$. *The ranges of the intersection and union of two commutating projections are given by:*

$$(A \wedge B)(V) = A(V) \cap B(V); (A \vee B) = A(V) \oplus B(V) = S\bar{p}(A(V), B(V)),$$

where $S\bar{p}(A(V), B(V))$ means the closed linear manifold spanned by the sets $A(V)$ and $B(V)$.

3) *The* **natural ordering** $A \leq B$ *between two commutating projections A and B has the geometrical significance that $A \leq B$ is equivalent to $A(V) \subseteq B(V)$.*

4) *A* **Boolean algebra of projections** *in V is a set of projections in V which is a Boolean algebra under the operations $A \vee B$ and $A \wedge B$ which has for its zero and unit elements the operators 0 and $id_V \equiv I \in V$.*

5) *The notions of abstract σ-completeness, and completeness can be extended.*

2 DEFINITION 5.61 - *Let V be a locally convex topological vector space, X a set and $\mathcal{R}$ a ring of subsets of X. A* **pre-spectral measure** *on V is a map $E : \mathcal{R} \to \mathcal{L}(V)$ with the following properties:*

a) *E is countable additive in the weak operator topology, that is an $E(\bigcup_n A_n) = \sum_n E(A_n)$, whenever A_n are disjoint sets in $\mathcal{R}$.*

b) *E is* **continuous** *in the sense that $E(M_n) \to E(M)$ weakly, whenever M_n is an increasing sequence of sets in $\mathcal{R}$ whose union M is also in $\mathcal{R}$.*[16]

A pre-spectral measure $E : \mathcal{R} \to \mathcal{L}(V)$ is **normalized** *if $E(X) = id_V$, (if $X \in \mathcal{R}$). $E : \mathcal{R} \to \mathcal{L}(V)$ is said to be* **equicontinuous** *if the values of E on $\mathcal{R}$ are. $E : \mathcal{R} \to \mathcal{L}(V)$ is said to be* **Baire** *or* **Borel** *if X is a (compact) Hausdorff space and $\mathcal{R} \equiv \mathcal{B}_a(X)$ or $\mathcal{R} \equiv \mathcal{B}(X)$, its σ-algebra of Baire or Borel sets respectively. We call* **sub-spectral** *(or* **quasi-spectral***) a normalized pre-spectral measure. A* **spectral measure** *on V is a pre-spectral measure $E : \mathcal{R} \to \mathcal{L}(V)$ such that $E(A_1 \cap A_2) = E(A_1) \circ E(A_2)$ for any $A_1, A_2 \in \mathcal{R}$.*

REMARK 5.19 - For a spectral measure, normalization can always be achieved by throwing away an irrelevant part of the underlying locally convex space V. Furthermore, a spectral measure in V can be regarded as a homomorphic map of Boolean algebra of sets into a σ-complete or complete Boolean algebra of (equicontinuous) projections in V, which has the additional property that it maps the unit into the identity operator I.

DEFINITION 5.62 - *Let H be a nonzero complex Hilbert space. A* **pre-spectral** *(resp.* **sub-spectral, spectral***)* **measure** *on H is a pre-spectral (resp. sub-spectral, spectral) measure $E : \mathcal{R} \to \mathcal{L}(H)$, where H is considered as a locally convex space such that E is* **positive***, that is $E(M) \geq 0$ for each $M \in \mathcal{R}$; (so $E(M)$ is also hermitian).*

THEOREM 5.34 - 1) *Let $(X, \mathcal{R}, E)$ be a triplet where $\mathcal{R}$ is a ring of subsets of X and $E : \mathcal{R} \to \mathcal{L}(V)$ is a set function with V a locally convex space. Then, E is a*

---

[16] If $A, A_j \in \mathcal{L}(V)$, where V is a locally convex space, writing "$A_j \to A$ weakly", we mean that $< A_j(x), \alpha > = < A(x), \alpha >$, for each $\alpha \in V'$ ($\equiv$ topological dual of V), $x \in V$.

pre-spectral measure on V iff

$$E_{x,\alpha} : \mathcal{R} \to \mathbf{R}, \quad E_{x,\alpha}(A) = < E(A)(x), \alpha >,$$

is a signed measure on $(X, \mathcal{R})$ *for any* $x \in V, \alpha \in V'$.
2) *If* $V \equiv H$ *is a hermitian vector space, then E is a pre-spectral measure on H iff*

$$E_{x,x} : \mathcal{R} \to \mathbf{R} \quad , \quad E_{x,x}(A) = < E(A)(x), x >,$$

is a measure on $(X, \mathcal{R})$.
PROOF. 1) Note that a pre-spectral measure must take its values in a weakly normal cone in $\mathcal{L}(V)$, namely the convex cone generated by its values: if $x \in V$ and $\alpha \in V'$, then the set function $A \mapsto < E(A)(x), \alpha >$ is a finite-valued signed measure on $\mathcal{R}$ and hence there is a set $E \in \mathcal{R}$ for which $A \mapsto < E(A \cap E)(x), \alpha >$ is a non-negative finite valued measure and $A \mapsto < E(A \sim E)(x), \alpha >$ is a non-positive finite valued measure (**Hahn-decomposition**). Thus, for every $A \in \mathcal{R}$,

$$< E(A)(x), \alpha > = < E(A)(x), E(E)'(\alpha) > + < E(A)(x), E(A \sim E)'(\alpha) >,$$

where $'$ denotes the transposition, and $A \sim E \equiv C_A(A \cap E) = A \setminus (A \cap E)$.
2) The proof is similar. □

DEFINITION 5.63 - *A Borel spectral measure* $E : \mathcal{B}(X) \to \mathcal{L}(V)$ *is said to be* **regular** *if* $< E(\bullet)(x), \alpha >$ *is a regular Borel measure for each* $x \in V$ *and* $\alpha \in V'$.

THEOREM 5.35 - (**Naimark-Nagy**). *Let* $(X, \mathcal{R}, E)$ *be a quasi-spectral measure,* $E : \mathcal{R} \to \mathcal{L}(H)$, *where H is a Hermitian space and* $\mathcal{R}$ *is a* σ*-ring. Then, there exists an extended hermitian space* $\widetilde{H}$ *such that E can be extended to a spectral-measure* $\widetilde{E} : \mathcal{R} \to \mathcal{L}(\widetilde{H})$. *More precisely, the projections* $\widetilde{E}(\omega), \omega \in \mathcal{R}$, *are such that* $\widetilde{E}(\omega)(f), (f \in H, \omega \in \mathcal{R})$, *identify the space* $\widetilde{H}$.

PROOF. Let $E(\mathcal{R})$ be the space of complex numerical valued $\mathcal{R}$-sets functions and let $E(\mathcal{R}, H) \equiv E(\mathcal{R}) \bigotimes H$ be the space of H-valued, complex numerical $\mathcal{R}$-set functions. We can define a scalar semiproduct in $E(\mathcal{R}, H)$ by putting:

$$(\phi|\psi) = \sum_{k,e} (E(A_k \cap B_e)(x_k)|y_e); \text{if} \phi = \sum_k 1_{A_k} \otimes x_k, \quad \psi = \sum_e 1_{B_e} \otimes y_e.$$

We can prove that, thank to the additivity of E, $(\phi|\psi)$ is independent of the representation chosen for ϕ and ψ. Let $\widetilde{H} \equiv E(\mathcal{R}) \widehat{\bigotimes} H$ be the completion of $E(\mathcal{R}) \bigotimes H$ with respect to the above scalar semi-product. One has the isometric monomorphism $j : H \to \widetilde{H}$ given by $j : x \mapsto 1 \otimes x$. By means of j we can identify H with a closed subspace of $\widetilde{H}$. Furthermore, we have the orthogonal projection on

$$H : \pi_H : \widetilde{H} \to H$$

given by

$$\pi_H(1_A \bigotimes x) = E(A)(x), \quad \forall A \in \mathcal{R}, x \in H.$$

We define

$$\widetilde{E} : \mathcal{R} \to \mathcal{L}(E(\mathcal{R}, H))$$

as

$$\widetilde{E}(A)(\phi) = \widetilde{E}(A)(1_A \otimes x) = 1_A \otimes E(A)(x).$$

One verifies that

$$\|\widetilde{E}(A)(\phi)\| \leq \|\phi\|$$

and $\widetilde{E}(A)$ canonically extends to an operator $\widetilde{E}(A) : \widetilde{H} \to \widetilde{H}$. Then, we can see that $\widetilde{E}$ is a spectral measure on H such that

$$E(\bullet) = \pi_H \circ \widetilde{E}(\bullet) \circ \pi_H.$$

Finally, one can prove that $\widetilde{H}$ is a **minimal extension** of H, that is $\widetilde{H}$ is generated by the vectors $\widetilde{E}(A)(x), A \in \mathcal{R}, x \in H$. □

THEOREM 5.36 - *Any (pre-)spectral measure $E : (X, \mathcal{R}) \circ\!\!\to \mathcal{L}(V)$ identifies a map $\phi : M(X, \mathcal{R}) \to \mathcal{L}(V)$, where $M(X, \mathcal{R})$ is the algebra of measurable functions $X \to \mathbf{K}$ on $(X, \mathcal{R})$, such that*

$$< \phi(f)(x), y >= \int_x f dE_{x,\alpha}, \quad \forall\, x, \in V, \alpha \in V'.$$

If $V \equiv H$ is a hermitian space then one has

$$< \phi(f)(x), y >= \int_x f dE_x, \quad \forall\, x, y \in H.$$

If E is a spectral measure the map ϕ is a morphism of algebras.

DEFINITION 5.64 - 1) *A* **quantum (pre-)spectral measure** *on a PDE $E_k \subset J\mathcal{D}^k(W)$ is any (pre-)spectral measure*

$$E : (\Omega_s(E_k), \Sigma) \circ\!\!\to \mathcal{L}(\mathcal{H}),$$

where $\mathcal{H}$ is a locally convex topological vector space, obtained by means of the space $C_0^\infty(H)$ of the sections with support compact of a hermitian bundle

$$\pi_H : H \to J\mathcal{D}^p(W), (-1 \leq p \leq \infty),$$

satisfying axioms of Theorem 5.21, and such that there is a map

$$\eta : C^\infty(W) \to C^\infty(\Lambda_n^0 M), \quad (n = \dim M),$$

such that $\eta(s) : M \to \Lambda^0_n M$ is a volume form on the base manifold M of $\pi : W \to M$. So $\mathcal{H}(s) \subset \mathcal{H}$ is the completion of $C_0^\infty(D^p s^* H)$ with respect to the inner product

$$(\psi|\psi')_s = \int_M h(s)(\psi|\psi')\eta(s)$$

where $h(s)$ is the scalar product induced by the hermitian structure of $D^p s^* H$. Furthermore, $(H, W, \mathbf{Q}\mathbb{B})$ is a super-bundle of geometric objects over M, with respect to a covariant functor $\mathbf{Q}\mathbb{B} : \mathcal{C}(W) \to \mathcal{C}(H)$.
2) A **quantum weak (pre-) spectral measure** *on E_k is a quantum (pre-)spectral measure where H is not a hermitian bundle, but simply a vector bundle over*

$$J\mathcal{D}^p(W), \quad -1 \leq p \leq \infty.$$

So $\mathcal{H}(s)$ are not necessarily Hilbert spaces.
3) A **proper quantum (pre-) spectral measure** *of $E_k \subset J\mathcal{D}^k(W)$ is a (pre-) spectral measure*

$$E : (\Omega_s(E_k), \Sigma) \circ\!\!\to \mathcal{L}(\mathcal{H}),$$

where $\mathcal{H}$ is the Hilbert space obtained by means of the space $C_0^\infty(H)$ of sections with support compact of a hermitian bundle $\pi' : H \to M$, completed with respect to the inner product

$$(\psi|\psi') = \int_M h(\psi|\psi').\eta$$

where η is a volume form on M. Furthermore, $(H, W, \mathbf{Q}\mathbb{B})$ is a super-bundle of geometric objects over M, for some covariant functor

$$\mathbf{Q}\mathbb{B} : \mathcal{C}(W) \to \mathcal{C}(H).$$

THEOREM 5.37 - *Any* **(proper) quantum (pre-) spectral measure** *$E : (\Omega_s(E_k), \Sigma) \circ\!\!\to \mathcal{L}(\mathcal{H})$ of a differential equation $E_k \subset J\mathcal{D}^k(W)$ identifies a Dirac quantization of random variables of E_k.*

PROOF. In fact we have the following morphism (of $(*)$-algebras):

$$\phi : M(\Omega_s(E_k), \Sigma) \to \mathcal{L}(\mathcal{H}), \forall f \in M(\Omega_s(E_k), \Sigma)$$

such that $\phi(f)$ is the linear operator on $\mathcal{H}$ defined by the following:

$$< \phi(f)(\psi), \psi' > = \int_{\Omega_s(E_k)} f dE_{\psi,\psi'}, \forall \psi \in \mathcal{H}, \psi' \in \mathcal{H}'.$$

In the proper case we have:

$$< \phi(f)(\psi_1), \psi_2 >= \int_{\Omega_s(E_k)} f dE_{\psi_1,\psi_2}, \forall \psi_1 \in \mathcal{H}, \psi_2 \in \mathcal{H}.$$

$\square$

THEOREM 5.38 - *A (proper) quantum spectral measure on $E_k \subset J\mathcal{D}^k(W)$ identifies a (proper) quantum spectral measure on any differential equation $\overline{E}_{\bar{k}} \subset J\mathcal{D}^{\bar{k}}(\overline{W})$ equivalent to E_k by means of a fiber bundle diffeomorphism $f_W : W \to \overline{W}$ over a diffeomorphism $f_M : M \to \overline{M}$.*

PROOF. In fact, for functorial property, we have the following commutative diagram:

$$\begin{array}{cccccc} & (\Omega_s(E_k), \Sigma) & \overset{E}{\hookrightarrow} & \mathcal{L}(\mathcal{H}) & \\ \Omega_s(f) & \downarrow & & \downarrow & \mathcal{L}(Q(f)) \\ & (\Omega_s(\overline{E}_{\bar{k}}), \overline{\Sigma}) & \underset{\overline{E}}{\hookrightarrow} & \mathcal{L}(\overline{\mathcal{H}}) & \end{array}$$

that identifies $\overline{\Sigma}$ and $\overline{E}$ respectively. $\square$

COROLLARY 5.2 - (QUANTUM COVARIANCE). *Under the hypotheses of the above theorem we have that for any function*

$$\mathcal{F} : \Omega_s(\overline{E}_{\bar{k}}) \to \mathbf{K}$$

one has:

$$(\phi(\mathcal{F})(\bar{x}), \bar{y}) = \int_{\Omega_s(\overline{E}_{\bar{k}})} \mathcal{F} d\overline{E}_{\bar{x},\bar{y}} = \int_{\Omega_s(E_k)} \mathcal{F} \circ \Omega_s(f) dE_{\phi^{-1}(\bar{x}), \phi^{-1}(\bar{y})}$$

with $\phi \equiv Q(f)$.

THEOREM 5.39 - *Any (proper) Dirac quantization of a PDE $E_k \subset J\mathcal{D}^k(W)$ besides a measure μ on the singular quantum situs $\Omega_s(E_k)$ identifies a (proper) quantum pre-spectral measure on E_k.*

PROOF. In fact let $E : \Omega_s(E_k) \to \mathcal{L}(\mathcal{H})$ be the map induced by a proper Dirac quantization on E_k. Then, we get the following pre-spectral measure:

$$\check{E}(A) = \int_A E d\mu, (\check{E}(A)(\psi), \psi') = \int_A (E(\bullet)(\psi), \psi') d\mu, \forall \psi, \psi' \in \mathcal{H},$$

with $A \in \Sigma \equiv$ Borel σ-algebra on $\Omega_s(E_k)$. Furthermore, if the Dirac quantization is not proper, i.e., $\mathcal{H}$ is a locally convex space that contains for any section s of $W \to M$ a Hilbert space $\mathcal{H}(s)$, we have the quantum pre-spectral measure $\check{E} : \Sigma \to \mathcal{L}(\mathcal{H})$ such that

$$< \check{E}(A)(\psi), \psi' >= \int_A < E(\omega)(\psi_\omega), \psi'_\omega > d\mu,$$

for any $\psi \in \mathcal{H}, \psi' \in \mathcal{H}'$, where $\psi_\omega, {\psi'}_\omega$ represent the projections of ψ and ψ' into $\mathcal{H}(s)$ and $\mathcal{H}(s)'$ respectively, where s is the section of W such that $\omega = D^k s(U)$. In fact, we have a canonical projection $\mathcal{H} \to \mathcal{H}(s)$ given by the restriction of the section of $\pi_H : H \to J\mathcal{D}^p(W)$ on the submanifold $D^p s(M)$ of $J\mathcal{D}^p(W)$. Furthermore, we have a natural inclusion $\mathcal{H}(s) \to \mathcal{H}$ and therefore a natural projection $\mathcal{H}' \to \mathcal{H}(s)'$.
□

Therefore, we can give the following

DEFINITION 5.65 - *A PDE $E_k \subset J\mathcal{D}^k(W)$ is* **fully-quantizable** *if one canonically recognizes a (proper) Dirac quantization and a measure μ on the (singular) quantum situs $\Omega_s(E_k)$ or on any statistical set Ω_s of E_k. We say that E_k is* **μ-quantizable** *if for any measure μ on $\Omega_s(E_k)$ (or Ω_s) we recognize a (proper) quantum (pre-)spectral measure on E_k.*

THEOREM 5.40 - *Let $E_k \subset J\mathcal{D}^k(W)$ be a PDE on a fiber bundle W such that $C^\infty(W)$ is a C^∞-manifold modeled on a separable infinite dimensional Hilbert space E. Then, any Dirac quantization on E_k does not identify, in general, a fully-quantization of E_k, but only a μ-quantization.*

PROOF. In fact, even if there is a C^∞-embedding $j : C^\infty(W) \to E$ of $C^\infty(W)$ into an open set of E, and on the Hilbert space E there is the canonical Gauss measure γ, in general we cannot reproduce this measure on $\Omega_s(E_k)$ (or on each statistical set Ω_s) as $\Omega_s(E_k)$ (resp. Ω) is a closed submanifold of E and so the restriction of Ω_s to $\Omega_s(E_k)$ (resp. Ω_s) can be zero. □

REMARK 5.20 - 1) A **quantum value** of a random variable $f : \Omega_s(E_k) \to \mathbf{R}$ is a point λ of the spectrum of $\hat{f} \equiv \phi(f)$, where $\phi : M(\Omega_s, \Sigma) \to L(\mathcal{H})$ is the map induced by a quantum (pre-)spectral measure of $E_k \subset J\mathcal{D}^k(W) : \lambda \in S_p(\hat{f})$.

2) Furthermore, the **probability to find a quantum value** of f in the Borel set $\omega \subset \mathbf{R}$, if the system is in the quantum state T, is given by the following formula:

$$p(f; T, \omega) = \operatorname{tr}(E(\omega)T) = \operatorname{tr}(TE(\omega)).$$

(Here we have written $E = E \circ f^{-1}$.)

3) If $T = \psi \otimes \psi$ we have

$$p(f; T, \omega) = \int_\omega dE_\psi(\lambda).$$

4) The corresponding **mean value** is

$$\left\{ \begin{aligned} \ll f \gg_T &= \int_{S_p} (\hat{f})\lambda dp(f; T, \lambda) \\ &= \operatorname{tr} \int \lambda dE(\lambda)T = \operatorname{tr}(\hat{f}T) = \operatorname{tr}(T\hat{f}) = \Sigma\alpha_n(\hat{f}(e_n), e_n) \end{aligned} \right\}$$

(If $T = \psi \otimes \psi$ we have $\ll f \gg_T = \int \lambda dE_\psi(\lambda)$.)

5) The corresponding **variance** is:

$$\left\{ \begin{aligned} \sigma^2(f)_T &= \int_{S_p} (\widehat{f})(\lambda - \ll f \gg_T)^2 dp(f, T, \lambda) \\ &= \operatorname{tr}(\int \lambda^2 dE(\lambda) T) - \ll f \gg_T^2 \\ &= \operatorname{tr}(\widehat{f}^2 T) - [\operatorname{tr}(\widehat{f})] \end{aligned} \right\}$$

(If $T = \psi \otimes \psi$ one has:

$$\sigma^2(f)_T = \int (\lambda - \ll f \gg_T)^2 dE_\psi(\lambda) = \|\hat{f}(\psi)\|^2 - (\hat{f}(\psi)|\psi)^2 \text{ .})$$

Now, let $\hat{f}$, and $\hat{f}'$ be two elements of $L(\mathcal{H})$ corresponding to two charges f, f' of a continuum system $E_k \subset J\mathcal{D}^k(W)$. If $[\hat{f}, \hat{f}'] = 0$, we say that f and f' are **simultaneously observable**. Then, the probability to find the quantum value λ (resp. λ') of f (resp. f') in the Borel set $\omega \subset \mathbf{R}$ (resp. $\omega' \subset \mathbf{R}$), if the system is in the pure quantum state given by

$$p(f, f'; \psi, \omega, \omega') = \int_\omega \int_{\omega'} dG_\psi(\lambda, \lambda'),$$

where

$$G_\psi(\bullet, \bullet) = E_\psi(\bullet) \otimes F_\psi(\bullet)$$

is the spectral measure on $\mathbf{R}^2$ obtained by product of the spectral measures E and F of f and f' respectively.

THEOREM 5.41 - *Under the hypotheses of the above remark one has the following relation:* **Heisenberg uncertainty relation**:

$$\boxed{\sigma^2(f)_\psi . \sigma^2(f')_\psi \geq \frac{1}{4} | < [\hat{f}, \hat{f}'](\psi) | \psi > |^2}$$

REMARK 5.21 - The (μ, f)-quantum transition amplitude, between two Cauchy hypersurfaces data σ_1 and σ_2, of a PDE E_k is the expectation value of the random variable

$$f \in Mes(\Omega_s(\sigma_1, \sigma_2), \mu, \mathbf{K}).$$

REMARK 5.22 - In order to recognize a canonical measure on a statistical set we can, sometimes, restrict these sets by adding some further requirements. More precisely, we say that a PDE $E_k \subset J\mathcal{D}^k(W)$ has a **quantum-variational constraint** if the bundle $\pi : W \to M$ is such that we can identify a submanifold $H^r(W)$ of the manifold of C^k sections, $0 \leq k \leq \infty$, of π such that:

(a) The canonical map $H^r(W) \to C^k(W)$ is continuous, for $r > n/2 + k, n = \dim M$, (**Sobolev lemma**);

(b) The space $\Gamma(TH^r(W))$ of tangent vectors to $H^r(W)$ admits a Hilbert structure;

(c) On $H^r(W)$ there is a Riemannian structure γ;

(d) A numerical function $f : J\mathcal{D}^k(W) \to \mathbf{R}$ on $J\mathcal{D}^k(W)$ is assigned. This identifies a functional $I_f : H^r(W) \to \mathbf{R}$ given by

$$I_f : s \mapsto \int_A (f \circ D^k s).\eta$$

for any volume form η on M and compact n-dimensional submanifold $A \subset M$ with smooth boundary, then, we call **Schroedinger equation** of E_k the equation for the lines in $H^r(W)$ with the most rapid decreasing for the functional I_f, that is

$$(5.3)(S) \qquad \boxed{\partial t.\psi = \zeta_f.\psi}$$

where:

(a) ζ_f is the Euler-Lagrange field corresponding to the functional

$$I_f : \zeta_f = \gamma' \circ \epsilon(f) : H^r(W) \to TH^r(W),$$

where γ' is the isomorphism $T^*H^r(W) \cong TH^r(W)$ induced by the metric γ on $H^r(W)$ and $\epsilon(f) : H^r(W) \to T^*H^r(W)$ is the Euler-Lagrange operator associated to the Lagrangian f;

(b) $\psi : \mathbf{R} \times H^r(W) \to H^r(W)$ is a 1-parameter semigroup of deformations of sections $H^r(W)$, i.e., for any section $s \in H^r(W)$ we have a path $\psi_s : \mathbf{R} \to H^r(W)$ through s, given by $\psi_s(t) = \widetilde{s}_t$, where $\widetilde{s} : \mathbf{R} \times M \to W$ is a deformation of the section s.

Of course one also has for any $t \in \mathbf{R}$ the map $\psi_t : H^r(W) \to H^r(W)$ given by $\psi_t(s) = \widetilde{s}_t$.

So, the **quantum cobordism problem**, characterized by a singular quantum situs $\Omega_s(E_k)$, is to find the solutions ψ of the equation (S) which satisfy the internal constraint ${\Omega_s}^r(E_k) \equiv \Omega_s(E_k) \cap H^r(W)$.

We have the following important theorem.

THEOREM 5.42 - *Let $E_k \subset J\mathcal{D}^k(W)$ be a PDE endowed with a quantum variational constraint*

$$(H^r(W), \gamma, f).$$

Then, any solution of the corresponding quantum cobordism problem, besides a Dirac quantization of E_k, identifies a full quantization of E_k.

PROOF. In fact, let $\psi_\omega : \mathbf{R} \to \Omega_s(E_k)$ be a solution of the quantum cobordism problem for a fixed quantum cobord $\omega \in \Omega(E_k)$. Let μ be the Gauss measure on $\mathbf{R}$. Then, we recognize a probability measure ν on $\Omega_s(E_k)$. More precisely, one has $\nu(U) = \mu(\psi_\omega^{-1}(U_\omega))$, where $U_\omega \equiv U \cap \psi_\omega(\mathbf{R})$, for any $U \in B(\Omega_s(E_k))$. □

REMARK 5.23 - Note, also, that in this context we have a direct interpretation of any physical entity as a stochastic process. In fact let $f : \Omega_s(E_k) \to \mathbf{R}$ be a random variable that translates on quantum level a differential operator associated to a PDE $E_k \subset J\mathcal{D}^k(W)$. Then, any solution $\psi : \mathbf{R}\times\Omega_s(E_k) \to \Omega_s(E_k)$ of a quantum cobordism problem of E_k identifies a stochastic process $\phi : \mathbf{R} \to Mes(\Omega_s(E_k), \nu; \mathbf{K})$ if Ω_s is endowed with the probability measure ν as we have stressed in the proof of the above theorem. More precisely ϕ is defined by: $\phi(t, \omega) = f(\psi(t, \omega)), \forall t \in \mathbf{R}, \omega \in \Omega_s(E_k)$. Furthermore, for any ν-measurable function $\phi : \Omega_s(E_k) \to \mathbf{K}$, we can define the ϕ-quantum transition amplitude between two Cauchy hypersurfaces data σ_1 and σ_2 of E_k as the number

$$< \sigma_1, \sigma_2 >\equiv \int_{\Omega_s(\sigma_1,\sigma_2)} \phi d\nu.$$

We can also obtain a full quantization by a procedure of discretization of statistical sets. In fact we have the following

THEOREM 5.43 - *Let $E_k \subset J\mathcal{D}^k(W)$ be a PDE such that for any statistical set Ω_s we can recognize a discrete subset $\Omega_{s,d} \subset \Omega_s$. Then, to any Dirac quantization $E : \Omega_s(E_k) \to \mathcal{L}(V)$, where V is a locally convex topological vector space, we can associate a quantization of any random variable, therefore a full-quantization.*

PROOF. In fact E identifies, with respect to $\Omega_{s,d}$ a discrete measure on Ω_s. So, the quantization $\phi(f)$ of any random variable f on Ω_s is obtained as follows:

$$(\phi(f)(x, \alpha) = \int_\Omega f d\nu_{x,\alpha} = \sum_{\sigma\in\Omega_{s,d}} f(\sigma)E_{x,\alpha}(\sigma),$$

for any $x \in V, \alpha \in V'$. □

The problem that now arises is: how to identify the discrete set $\Omega_{s,d} \subset \Omega_s$? In the following we will list two methods. The first consists of using a Morse function while the second is to adopt a particular type of spectral measure.

THEOREM 5.44 - *Let $E_k \subset J\mathcal{D}^k(W)$ be a PDE endowed with a quantum-variational constraint*

$$(H^r(W), \gamma, f).$$

Let $I_f : H^r(W) \to \mathbf{R}$ satisfy the following conditions:

(C) (**Palais-Smale**). *If A is a subset of $H^r(W)$ on which $|I_f|$ is bounded but on which $\|\text{grad}\, I_f\|$ is not bounded away from zero, then there is a critical point of I_f in the closure of A.*

(D) *All critical points of* I_f *are non degenerate. Then, for any statistical set of* E_k *we recognize a discrete subset* $\Omega_{s,d}$*. So, any Dirac quantization on* E_k *identifies a fully-quantization of* E_k *for the admissible configurations of* $\Omega_{s,d}$*.*[17]

PROOF. In fact for any real numbers $a < b$ there are only finitely many critical points of I_f in $I_f^{-1}[a,b]$. (See also ref.[89].) □

THEOREM 5.45 - *Under the hypotheses of the above theorem there exists a set* $\Omega_{s,d}$*, for any statistical set* Ω_s*, for which Theorem 5.44 is satisfied.*

PROOF. In fact in these cases there is a **Morse function** $\phi : \Omega_s \to \mathbf{R}$, i.e., a C^∞-function of which all its critical points are non-degenerate and which satisfies conditions (C) of Palais-Smale, relative to a complete Riemannian metric on Ω_s. (Given any sequence $(\omega_i)_{i\geq 1}$ of points of Ω_s such that $(\phi(\omega_i))_{i\geq 1}$ is bounded and for which

$$\|D\phi(\omega_i)\|_{\omega_i} \to 0$$

then there is a convergent subsequence. □

THEOREM 5.46 - *Let* $E_k \subset J\mathcal{D}^k(W)$ *be a PDE and let* $E : \mathcal{B}_a(\Omega(E_k)) \to \mathcal{L}(V)$ *be an equicontinuous Baire spectral measure in a bounded complete convex space* V*, and let*

$$\widetilde{E} : \mathcal{B}(\Omega_s(E_k)) \to \mathcal{L}(V)$$

be its unique extension to a regular Borel spectral measure. Suppose that for each

---

[17] Let X be a C^2-Riemannian manifold without boundary modeled on a separable Hilbert space. A **critical point** of a C^2 function $f{:}X\to\mathbf{R}$ is a stationary point p of the maximum local one-parameter group ϕ_t generated by

$$-\text{grad } f\equiv v{:}X\to TX, v\equiv\partial\phi, v(p)=0.$$

Let $H(f)$ be the Hessian of f. Then $H(f)(p)$ identifies a self-adjoint operator $A{:}T_pX\to T_pX$ such that $(A(u),v)=H(f)(p)(u,v)$. The dimension of subspaces on which $H(f)(p)<0$ (resp. >0) is called **index** (resp. **coindex**) of f at p. A critical point p is **non-degenerate** if A has a bounded inverse. In this case p is isolated in the set of critical points. A **critical value** of f is a point $c\in\mathbf{R}$ such that $f^{-1}(c)$ contains a critical point of f. Otherwise c is a **regular point** of f. Set

$$X_{a,b}\equiv f^{-1}[a,b] \quad , \quad X_b\equiv X_{-\infty,b}.$$

If c is a regular value of f, X_c is a closed submanifold of X with boundary

$$f^{-1}(c) \quad , \quad \partial X_c=f^{-1}(c).$$

Furthermore, from the inverse function theorem it follows that, $f^{-1}(c)$ is a (possibly empty) smooth, codimension one submanifold of X.

idempotent $e \neq 0$ in the strong closure of $\{E(A)\}, A \in \mathcal{B}_a(\Omega_s(E_k))$, there exists a non-zero $x \in e(V)$ for which the order interval $[-x, x] \subseteq M_{\mathbf{R}}(x)$ ($\equiv$ smallest closed real subspace of V which contains $\{E(A)(x)\}, A \in \mathcal{B}_a(\Omega_s(E_k))$), is compact. Then, in each statistical set Ω_s of E_k we can identify a discrete subset $\Omega_{s,d}$. So, any Dirac quantization on E_k identifies a full quantization of E_k for the admissible configurations of $\Omega_{s,d}$.

PROOF. In fact we can prove [127] that $E : \mathcal{B}(\Omega_s(E_k)) \to \mathcal{L}(V)$ possesses atoms hence point atoms, and their supremum (in the sense of [127]) is the identity element of $\mathcal{L}(V)$. So if $\Omega_s(E_k)_d$ is the set of point atoms of E we recognize $\Omega_{s,d} = \Omega_s \cap \Omega_s(E_k)_d$ as a discrete subset for any statistical set Ω_s of E_k. □

REMARK 5.24 - Same results can be obtained by using a equicontinuous Baire spectral measure in $\mathcal{L}(V)$ for E_k, where (V, T) is a boundedly complete quasi-montel space (namely its weakly compact subsets are T compact). (See ref. [127]).

EXAMPLE 5.23 - (LINEAR PDEs).

1) KLEIN-GORDON EQUATION. (KG) admits a natural proper Dirac quantization

$$Q : J\mathcal{D}^2(E) \to J\mathcal{D}(H)^* \bigotimes H,$$

where $H \equiv Clif^c(M)$ is the complex Clifford bundle over M endowed with the hermitian structure h, natural extension on $Clif^c(M)$ of the hermitian structure $\bar{h}$ on $T^cM \equiv \mathbf{C} \bigotimes TM$, defined for real vector fields by:

$$\bar{h}(i\xi, \nu) = ig(\xi, \nu), \ \underline{h}(\xi, i\nu) \equiv -ig(\xi, \nu)$$

and extended by linearity to complex vector fields. The corresponding Hilbert space $\mathcal{H}$ is the space of sections of H with support compact, completed with respect to the scalar product $(\psi, \psi') = \int_M h(\psi, \psi').\eta$, where $\eta \equiv *1$ is the canonical volume form on (M, g). More precisely Q is given by:

$$Q.f \equiv Q \circ D^2 f = \check{\nabla}_{\mathrm{grad}\, f} + \frac{1}{2}\kappa f \tag{5.4}$$

for any $f : M \to \mathbf{R}$, where $\check{\nabla}$ is the absolute differential on $Clif^c(M)$ canonically associated to the Levi-Civita connection on (M, g). (For more details see ref. [104].) Resuming we have the following.[18]

THEOREM 5.47 - *Klein-Gordon equation admits a natural proper Dirac quantization $Q : J\mathcal{D}^2(E) \to J\mathcal{D}(H)^* \bigotimes H$, with $H \equiv Clif^c(M)$, and Q given in (5.4). (KG) is μ-quantizable.*

[18] Compare also with Crumeyrolle [21], where a similar quantification has been produced by considering instead of $Clif^c(M)$ the fiber bundle of pure spinors $S(M)$.

The commutator corresponding to this Dirac quantization is the following:

$$(5.5) \qquad \boxed{[\widehat{f_1},\widehat{f_2}]_- = R(X_1,X_2) - \nabla_{[X_1,X_2]} + \frac{1}{2}[X_1(\kappa f_2) - X_2(\kappa f_1)]}$$

where $\widehat{f_i} \equiv Q.f_i$, $X_i \equiv \text{grad } f_i, i = 1,2$, *and* R *is the curvature associated to the Levi-Civita connection. Of course* $[\hat{f},\hat{f}]_- = 0$!
Furthermore, as the Klein-Gordon equation (KG) *is the Euler-Lagrange equation of the following Lagrangian:*

$$L \equiv \frac{1}{2}\sqrt{|g|}\{g^{\mu\nu}y_\mu y_\nu - \kappa y\} : J\mathcal{D}(E) \to \mathbf{R},$$

we can consider the conjugate moments

$$p^\alpha \equiv (\partial y^\alpha.L) : J\mathcal{D}(E) \to \mathbf{R}.$$

For any section $\phi \equiv (id_M, f) \in C^\infty(E)$, p^α *identify functions* $\pi^\alpha \equiv p^\alpha \circ D\phi : M \to \mathbf{R}$ *locally defined on* M. *More precisely one has*

$$\pi^\alpha = \sqrt{|g|}g^{\alpha\nu}(\partial x_\nu.f).$$

So we can quantize π^α *by using expression* (5.4). *In particular with respect to a physical frame on* M, *we distinguish a* **time coordinate** x^0 *in* $\{x^\alpha\}$ *and we can consider the conjugate field* $\pi \equiv \pi^0$ *to* ϕ:

$$\pi = \sqrt{|g|}g^{o\nu}(\partial x_\nu.f).$$

(If $\{x^\alpha\}$ *is an orthonormal coordinate system one has* $\pi = (\partial x_0.f)$.*) Then by using expression* (5.5), *we can calculate the commutator* $[\hat{f},\widehat{\pi}]$.
2) MAXWELL EQUATION. (M) admits a canonical proper Dirac quantization

$$Q : J\mathcal{D}^2(E) \to J\mathcal{D}(H)^* \bigotimes H,$$

where $H \equiv Clif^c(M)$ and Q is given by the following:

$$(5.6) \qquad Q.A \equiv Q \circ D^2 A = \check{\nabla}_{X(A)} + \frac{1}{2} < \nabla A,_A \overline{F} >,$$

where $\check{\nabla}$ is like in the above example, $X(A) : M \to TM$ is the vector field canonically associated to the differential 1-form A given by

$$X(A) = (A_\alpha F^{\alpha\beta})\partial x_\beta$$

and ${}_A\overline{F}$ is the controvariant expression of

$$ {}_AF \equiv dA = F_{\alpha\beta} dx^\alpha \wedge dx^\beta = [(\partial x_\alpha . A_\beta) - (\partial x_\beta . A_\alpha)] dx^\alpha \wedge dx^\beta . $$

Therefore, we have the following.

2 THEOREM 5.48 - *Maxwell equation admits a natural proper Dirac quantization* $Q : J\mathcal{D}^2(E) \to J\mathcal{D}(H)^* \bigotimes H$, *with* $H \equiv Clif^c(M)$, *and* Q *given in* (5.6). (M) *is* μ*-quantizable.*

The commutator corresponding to this Dirac quantization is given by:

$$ \boxed{\begin{aligned} &[\widehat{A}_1, \widehat{A}_2]_- = R(X_1, X_2) + \nabla_{[X_1, X_2]} \\ &+ \frac{1}{2}[(\nabla_{X_1} . < \nabla A_2, F_2 >) - (\nabla_{X_2} < \nabla A_1, F_1 >)] \end{aligned}} \tag{5.7} $$

where $\widehat{A}_i \equiv Q.A_i, X_i = X(A_i), F_i =_{A_i} \overline{F}, i = 1, 2.$

3) DIRAC EQUATION. (D) admits a natural proper Dirac quantization:

$$ Q : J\mathcal{D}(E) \to J\mathcal{D}(H)^* \bigotimes H, \quad Q.\phi \equiv Q \circ D\phi = \check{\nabla}_{X(\phi)} + [-m.h(\phi, \phi) + f(\phi)], \tag{5.8} $$

where H is like in the above example 1); $X(\phi) : M \to TM$ is a vector field canonically associated to the field $\phi : M \to Clif^c(M)$, given by $X(\phi) = Y(\phi) + \bar{y}(\phi)$, $\bar{y}(\phi) \equiv$ complex conjugate of $Y(\phi)$,

$$ Y(\phi) \equiv g^{\mu\nu} \phi^A \gamma_{\nu A}^B \bar{\phi}_B \partial x_\mu . $$

Moreover,

$$ f(\phi) \equiv \phi^A \nabla_\mu (g^{\mu\nu} \gamma_{\nu A}^B \bar{\phi}_B) : M \to \mathbf{C}. $$

Therefore we have the following.

2 THEOREM 5.49 - *Dirac-equation* (D) *admits a natural proper Dirac quantization* $Q : J\mathcal{D}(E) \to J\mathcal{D}(H)^* \bigotimes H$, *with* $H \equiv Clif^c(M)$, *and* Q *given in* (5.8). (D) *is* μ*-quantizable.*

The commutator corresponding to (5.8) *is the following:*

$$ \boxed{[\widehat{\phi}_1, \widehat{\phi}_2]_- = R(X_1, X_2) + \nabla_{[X_1, X_2]} + [(X_1 . f_2) - (X_2 . f_1)]} \tag{5.9} $$

where

$$ \left\{ \begin{array}{l} \widehat{\phi}_i \equiv Q.\phi_i, X_i = X(\phi_i) \\ f_i \equiv [-m.h(\phi_i, \phi_i) + f(\phi_i)], i = 1, 2. \end{array} \right\} $$

2 EXAMPLE 5.24 - (NON-LINEAR PDEs)

1) EINSTEIN EQUATION. Let us first consider Einstein equation on a compact n-dimensional manifold M. Then, (Ein) admits a canonical Dirac quantization

$$Q : J\mathcal{D}^2(E) \to J\mathcal{D}(H)^* \bigotimes_E H,$$

where

$$H \equiv \bigcup_{p\in M} \left\{ \bigcup_{g(p)\in E_p} C(T_pM, g(p))^c \right\} \quad (\clubsuit)$$

where $C(T_pM, g(p))^c$ is the complexificated Clifford algebra on the space T_pM with inner product $g(p)$. (We also write $H \equiv Clif^c(M)$.) More precisely, for any metric g, $Q.g \equiv Q \circ D^2g$ identifies a linear operator $Q.g$ with domain $D(Q.g) = \mathcal{H}(g) \subset \mathcal{H}$, where $\mathcal{H}$ is the locally convex topological vector space of sections with support compact of $H \to E$ and $\mathcal{H}(g)$ is the subspace of $\mathcal{H}$ defined by sections restricted to the submanifold $g(M) \subset E$. The operator $Q.g$ is defined by

$$Q.g = \check{\nabla}_{X(g)} - f(g) \tag{5.10}$$

where $X(g) : M \to TM$ is a vector field on M canonically associated to g as the dual of the differential 1-form $\overline{X}(g) : M \to T^*M$ such that

$$-nR(g) = \delta\overline{X}(g) + \phi(g), \tag{5.11}$$

where $\phi(g) : M \to \mathbf{R}$ is a harmonic function on M. Note, that $X(g)$ is unique, up to divergence free vector fields, by means of Hodge's decomposition theorem. Furthermore,

$$f(g) = -\frac{1}{2}[R(g)_{\alpha\beta}\, g^{\alpha\beta} + \chi T_{\alpha\beta}\, g^{\alpha\beta} + \phi(g)]\,.$$

So, we have the following.

THEOREM 5.50 - *Einstein equation (Ein) admits a natural Dirac quantization Q : $J\mathcal{D}^2(E) \to J\mathcal{D}(H)^* \bigotimes_E H$, with H as given in $(\clubsuit)$ and Q given in (5.10), and $\eta(g) = *_g 1$. (Ein) is μ-quantizable.*

The commutator corresponding to (5.10) is given by:

$$\boxed{[\widehat{g}_1, \widehat{g}_2]_- = [\overset{(1)}{\nabla}_{X_1}, \overset{(2)}{\nabla}_{X_2}]_- - [(X_1.f_2) - (X_2.f_1)]}$$

where

$$\widehat{g}_i \equiv Q.g_i, \quad X_i \equiv X(g_i), \quad f_i \equiv f(g_i), i=1,2.$$

Here ∇_{X_i} *is the covariant derivative induced by the metric* g_i.
If $\mathcal{H}(g_1) \cap \mathcal{H}(g_2) = \emptyset$ *one has* $[\widehat{g}_1, \widehat{g}_2]_- = 0$.

From the physical point of view the interest of Einstein equation on compact manifolds is not neglectable. For examples compact gravitational instantons fit this class. However, there are serious obstructions in order to consider hyperbolic structures on compact manifolds. In facts, as it is well known, a connected, compact orientable differentiable manifold M can be endowed with a hyperbolic structure if its Euler Poincaré characteristic is zero:

$$\chi(M) \equiv \sum_{p=0}^{n} (-1)^p b_p = 0.$$

For odd dimensional manifolds this circumstance is always verified, but for even-dimensional manifolds can be $\chi(M) \neq 0$. For example, $\chi(S^{2n}) = 2$, so on S^{2n} we cannot define a hyperbolic structure. However, Theorem 5.50 can be generalized to non-compact manifolds. In fact, the compacticity of M intervenes only when one uses Hodge's decomposition theorem, in order to construct the Dirac quantization of (Ein). In fact, we always use decomposition (5.11) in compact regions of M, as $Q.g$ acts on fields that have compact support. So, Hodge's decomposition works well in (5.11) also for non compact manifolds. (Note also that restricting the class of metric fields on M, we have a Hodge's decomposition theorem over non-compact manifolds, (see e.g.ref.[22].)
2) YANG-MILLS EQUATION. Let us consider Yang-Mills equation on an n-dimensional manifold. Then, (YM) admits a canonical proper Dirac quantization

$$Q : J\mathcal{D}^2(E) \to J\mathcal{D}(H)^* \bigotimes H,$$

where H is similar to above examples and Q is given by:

$$Q.A \equiv Q \circ D^2 A = \check{\nabla}_{X(A)} - f(A), \tag{5.12}$$

where $X(A) : M \to TM$ is a vector field on M canonically associated to A as follows:

$$X(A) \equiv (A_{\alpha H} F^{\alpha\beta H}) \partial x_\beta = < A, F >: M \to TM.$$

Note that the pairing $<,>$ is taken with respect to the metric isomorphism $T^*M \cong TM$ and the isomorphism $A(P)^* \cong A(P)$ induced by means of the Killing metric on

G. The Hilbert space $\mathcal{H}$ of wave functions is similar to the above examples. Moreover, we have:

$$f(A) \equiv -\frac{1}{2} < A, {}_A F >= \frac{1}{2}(\nabla_\beta A_{\alpha H})F^{\alpha\beta H} : M \to \mathbf{R}.$$

Therefore, we have the following.

THEOREM 5.51 - *Yang-Mills equation* (YM) *admits a canonical proper Dirac -quantization* $Q : J\mathcal{D}^2(E) \to J\mathcal{D}(H)^* \bigotimes H$, *with* $H \equiv Clif^c(M)$, *and* Q *given in* (5.12). *(YM) is* μ*-quantizable.*

The commutator corresponding to (5.11) *is given by:*

$$[\widehat{A}_1, \widehat{A}_2]_- = R(X_1, X_2) + \nabla_{[X_1,X_2]} - [(X_1.f_2) - (X_2.f_1)] \tag{5.13}$$

where

$$X_i \equiv X(A_i), \quad f_i \equiv f(A_i), \quad \widehat{A}_i \equiv Q.A_i, \quad i = 1, 2.$$

2 *Quantization* (5.12) *reduces to that for the Maxwell-equation when* $G = U(1)$.

3) EINSTEIN-YANG-MILLS EQUATION. We consider the situation of a gravitational field coupled also with a gauge field. Let $(P, M, \pi; G)$ be a G-principal bundle over an n-dimensional manifold M, such that G is a semi-simple compact Lie group. Let $E \equiv E_0 \bigotimes C(P)$ where E_0 is the open subbundle of non-degenerate metrics on M, and $C(P)$ is the fiber bundle of principal connections on E. Einstein-Yang-Mills equation is a second order PDE on the fiber bundle P given as the kernel (EYM) of the following epimorphism of constant rank

$$\left\{\begin{array}{l} \mathcal{EYM} : J\mathcal{D}^2(E) \to S^0_2 M \times_M T^*M \bigotimes A(P) \\ \times_M TM \times_M A(P) \times_M \Lambda^0_3 M \bigotimes A(P) \\ \mathcal{EYM} \circ D^2(g, A) = \begin{pmatrix} (R..(g) - \frac{1}{2}R(g)g - \chi T(g, F), \\ {}_A\delta_A dA + J(A, g), \\ \mathrm{div}\, T(g, F), \\ \mathrm{div}\, J(A, g), \\ {}_A d_A F \end{pmatrix} \end{array}\right\}$$

where $R..(g)$ is the Ricci curvature of g, χ is the gravitational constant, $T(g, F)$ is the energy momentum-tensor of the field $F \equiv {}_A F \equiv {}_A dA$, and $J(A, g)$ is the interaction current corresponding to the configuration (A, g). The local expression of (EYM) is the following:

$$(EYM)\left\{\begin{array}{l} F^{\alpha\beta H}\equiv g^{\alpha\delta}g^{\beta\gamma}F^H_{\delta\gamma}\equiv g^{\alpha\delta}g^{\beta\gamma}[(\partial x_\delta.A^H_\gamma)-(\partial x_\gamma.A^H_\delta)+C^H_{BC}A^B_\delta A^C_\gamma] \\ g_{\alpha\omega}[(\partial x_\beta.F^{\alpha\beta H})+C^H_{BC}A^C_\beta.F^{\alpha\beta B}]-J^H_\omega=0 \\ C^H_{BC}\equiv \text{structure constants of } G \\ (\partial x_\lambda.F^H{}_{\alpha\beta})+C^H_{AB}A^A_\lambda F^B_{\alpha\beta}=0 \\ R_{\alpha\beta}-\frac{1}{2}R_{\gamma\delta}g^{\gamma\delta}g_{\alpha\beta}=T_{\alpha\beta}\chi \\ R_{\alpha\beta}\equiv R^\gamma_{\alpha\gamma\beta}=-(\partial x_\beta.\Gamma^\lambda_{\lambda\alpha})+(\partial x_\lambda.\Gamma^\lambda_{\beta\alpha})+\Gamma^\lambda_{\lambda\omega}\Gamma^\omega_{\beta\alpha}-\Gamma^\omega_{\lambda\beta}\Gamma^\lambda_{\omega\alpha} \\ \Gamma^\lambda_{\nu\omega}=\frac{1}{2}g^{\lambda\alpha}[(\partial x_\nu.g_{\omega\alpha})+(\partial x_\omega.g_{\alpha\nu})-(\partial x_\alpha.g_{\nu\omega})] \\ \nabla_\alpha T^\alpha_\beta=0 \\ \nabla_\alpha J^{\alpha H}=0. \end{array}\right\}$$

This analytic equation is involutive, formally integrable and completely integrable. Furthermore, it is of Cauchy type, that is one has the existence and unicity for the Cauchy problem. (EYM) admits a natural Dirac quantization

$$Q: J\mathcal{D}^2(E)\to J\mathcal{D}(H)^*\bigotimes H,$$

where H is similar to the above examples and

$$(5.14)\quad\left\{\begin{array}{l} Q.(g,A)\in\mathcal{L}(\mathcal{H}(g)) \\ Q.(g,A)=\check{\nabla}_{X(g,A)}-f(g,A) \\ X(g,A):M\to TM,\quad X(g,A)=X(g)+X(A) \\ X(g):M\to TM, \text{dual of } \quad \overline{X}(g):M\to T^*M \\ -nR(g)=\delta\overline{X}(g)+\phi(g) \\ \phi(g):M\to\mathbf{R}: \text{harmonic function on } M \\ X(A)=((A_{\alpha H}F^{\alpha\beta H})+J^{\beta H}F^\alpha_{\alpha H})\partial x_\beta:M\to TM \\ f(g,A):M\to\mathbf{R} \\ f(g,A)\equiv -\frac{1}{2}(R_{\alpha\beta}g^{\alpha\beta}-\chi T_{\alpha\beta}g^{\alpha\beta}+\phi(g)) \\ -\frac{1}{2}(\nabla_\beta A_{\alpha H})F^{\alpha\beta H}+\frac{1}{2}\nabla_\alpha T^\alpha_\beta X^\beta+\frac{1}{2}J^{\alpha H}A_{\alpha H} \\ +\frac{1}{2}(J^{\beta H}\nabla_\beta F^\alpha_{\alpha H})-\frac{1}{2}[(\partial x_\lambda.F^H_{\alpha\beta})+C^H_{AB}A^A_\lambda F^B_{\alpha\beta}]F^{\alpha\beta}_H X^\lambda \end{array}\right\}$$

In fact, $Q.(g,A)$ is antihermitian iff $\delta\overline{X}(g,A)-2f(g,A)=0$, that is iff

$$\begin{aligned}&(R_{\alpha\beta}-Rg_{\alpha\beta}-\chi T_{\alpha\beta})g^{\alpha\beta}-(\nabla_\alpha T^\alpha_\beta)X^\beta+[(\partial x_\lambda.F^H_{\alpha\beta})+C^H_{AB}A^A_\lambda F^B_{\alpha\beta}]F^{\alpha\beta}_H X^\lambda\\ &-(\nabla_\beta F^{\alpha\beta H}-J^{\alpha H})A_{\alpha H}+(\nabla_\beta J^{\beta H})F^\alpha_{\alpha H}=0.\end{aligned}$$

But the above equation is certainly satisfied on the set of solutions of (EYM). Therefore, we have the following.

THEOREM 5.52 - *Einstein-Yang-Mills equation* (EYM) *admits a canonical Dirac quantization* $Q : J\mathcal{D}^2(E) \to J\mathcal{D}(H)^* \bigotimes H$, *with* $H \equiv Clif^c(M)$, *in the sense specified above, and* Q *given in* (5.14). (EYM) *is* μ*-quantizable.*

4) NEWTON-LAGRANGE EQUATION. The classical mechanics of one particle (in the presence of a potential V), is identified by a second order differential equation (e.g., Newton equation, Lagrange equation), $E_2 \subset J\mathcal{D}^2(G)$, on the (non-trivial) fiber bundle $\mathcal{G} \equiv \{\tau : M \to T\}$, where M is the 4-dimensional affine space-time, T is the time-axis and τ is an affine surjective map of constant rank 1. We shall assume that E_2 is formally integrable. A "field", in this context, is a **motion**, that is a section (of class C^∞) of $\mathcal{G}$, or, equivalently, a time-like smooth curve in M. The formal quantization of E_2 consists in the identification of the quantum situs, or statistical sets $\Omega(a, b)$, of E_2, where a and b belong to $E_{2+1} = E_3$. In fact, in this context, a Chauchy hypersurfaces data reduces to a point on E_3, so, a statistical set $\Omega \equiv \Omega(a, b)$ between two points $a, b \in E_3$ is the set of 3-jet-derivative curves joining a and b. Now, $C^\infty(\mathcal{G})$ is an affine space modeled on the infinite dimensional vector space $C^\infty(T, \mathbf{S})$, where $\mathbf{S}$ is the vector subspace of the space of free vectors $\mathbf{M}$ of M given as $\mathbf{S} = \ker \underline{\tau}$, where $\underline{\tau} : \mathbf{M} \to \mathbf{T}$ the linear map associated to τ, ($\mathbf{T} \equiv$ the space of free vectors of T). We can also see that $C^\infty(\mathcal{G})$ is a differential manifold modeled to the Hilbert space $\mathcal{P}_{\mathcal{H}}$ that is the completion of $C_0^\infty(T, \mathbf{S})$ with respect to the scalar product

$$(v, v') = \int_T \bar{g}(v, v')dt,$$

where $\bar{g}$ is the Euclidean structure of $\mathbf{S}$. In this way $C^\infty(\mathcal{G})$ becomes a Riemannian manifold with metric field

$$\left\{ \begin{array}{l} g : C^\infty(\mathcal{G}) \to S_2^0(C^\infty(\mathcal{G})) \\ g(m)(v_m, v'_m) = \int_{[0,1]} \bar{g}(v_m, v'_m)dt. \end{array} \right\}$$

Then, on $C^\infty(\mathcal{G})$ we recognize a metric

$$d(m, m') = \inf \int_{[0,1]} \|\dot{g}(t)\| dt,$$

where $\gamma : [0, 1] \to C^\infty(\mathcal{G})$ is a C^1 path such that $\gamma(0) = m, \gamma(1) = m'$ and $\|\dot{g}(t)\| = \sqrt{g(\gamma(t))(\dot{g}(t), \dot{g}(t))}$, where $\dot{g}(t) \in T_{\gamma(t)}C^\infty(\mathcal{G}) \equiv \mathcal{P}_{\mathcal{H}}$, which is the velocity of γ. Therefore, $C^\infty(\mathcal{G})$ is also a Hausdorff metric space. The statistical set $\Omega(a, b)$ is a closed set of $C^\infty(\mathcal{G})$ (in the FW^∞-topology), and also results in a closed submanifold of $C^\infty(\mathcal{G})$. More precisely, $\Omega(a,b)$ is an affine space modeled on the Hilbert space $\mathcal{H}_{[a,b]}$ that is the completion of the infinite dimensional vector space:

$$\left\{\begin{array}{l} X \in C^\infty(J\mathcal{D}^3([a_0,b_0],\mathbf{S})) \\ X = |_3 \circ D^3 v \\ v \in C^\infty([a_0,b_0],\mathbf{S}) \\ X(a_0) = X(b_0) = 0 \end{array}\right\}$$

where $a_0 \equiv \pi_3(a), b_0 \equiv \pi_3(b)$, with respect to the scalar product

$$(v,v') = \int_{[a_0,b_0]} \bar{g}(v,v')dt.$$

So $\Omega(a,b)$ is a metric space and as a consequence it is also Hausdorff. Furthermore, if the statistical set $\Omega(a,b)$ is the quantum fluctuation of some dynamic cobord ω that represents the unique classical solution cobording the points a and b, (e.g., if the equantion E_2 is Lagrangian, ω is represented by the particular trajectory that makes the action integral

$$I_{[t_a,t_b]} \equiv \int_{t_a}^{t_b} (L \circ Dm)dt$$

stationary, where $t_a \equiv \pi_3(a) \in T, t_b \equiv \pi_3(b) \in T$), taking into account the affine structure of $\Omega(a,b)$, we can canonically identify $\Omega(a,b)$ with the vector Hilbert space $\mathcal{H}_{[a,b]}$:

$$\Omega(a,b) \equiv \mathcal{H}_{[a,b]}.$$

Then, on $\Omega(a,b)$ we can induce the Gauss measure of $\mathcal{H}_{[a,b]}$. So, we have proved the following.

THEOREM 5.53 - *Let $E_2 \subset J\mathcal{D}^2(\mathcal{G})$ be the Newton-Lagrange equation of classical particle on the Galilean space-time fiber bundle $\mathcal{G} = \{\tau : M \to T\}$. Then we have:*

1) The quantum situs $\Omega(E_2)$ is a Hausdorff closed submanifold of the Hausdorff space $C^\infty(\mathcal{G})$.

2) Any statistical set $\Omega(a,b), a,b \in E_{2+1}$, of E_2 is a Hausdorff, metric affine manifold.

3) If $\Omega(a,b)$ is the quantum fluctuation of some dynamic cobord of E_2, on $\Omega(a,b)$ we recognize a canonical probability measure.

The relation between Kostant-Souriau quantization [58] and formal quantization is given by the following.

THEOREM 5.54 - *Let $L : J\mathcal{D}(G) \to \mathbf{R}$ be the Lagrangian characterizing a classical particle. Then, there exists a canonical pre-spectral measure on any statistical set Ω of $E_2 \equiv$ Euler-Lagrange equation of L.*

PROOF. Let $\Omega \equiv d\theta = dp_i \wedge dx^i - dH \wedge dx^0$, be the closed 2-form, where θ is the Cartan form, associated to L and $H \equiv (\partial \dot{x}_i.L)\dot{x}^i - L$ the corresponding Hamiltonian. For any $t \in T$ we have that

$$JD(G)_t \equiv M_t \times \mathbf{I}, M_t \equiv \tau^{-1}(t), \mathbf{I} \equiv \{v \in BM < \underline{\tau}, v >= 1\},$$

is the 6-dimensional manifold endowed with a symplectic structure $\Omega_t = i_t^* d\theta = i_t^* \Omega$, where i_t is the canonical inclusion $i_t : J\mathcal{D}(\mathcal{G})_t \to J\mathcal{D}(\mathcal{G})$. The space $F(J\mathcal{D}(\mathcal{G})_t)$ of real-valued differential function on $J\mathcal{D}(\mathcal{G})_t$ together with the usual Poisson bracket defined in terms of Ω_t is a Lie algebra. Let $\mathcal{P}$ be the Poisson algebra of 0-forms f on $J\mathcal{D}(\mathcal{G})$ such that $\xi \rfloor \Omega = df$, for some vector field ξ on $J\mathcal{D}(\mathcal{G})$, ($\xi$ is unique and vertical). The restriction map $|_t : \mathcal{P} \to F(J\mathcal{D}(\mathcal{G})_t)$ is a homomorphism of Lie algebra. Furthermore, if ζ is the unique vector field on $J\mathcal{D}(\mathcal{G})$ determined by the conditions: $\zeta \rfloor \Omega = 0, < \zeta, dx^0 >= 1$, (**Euler-field**) then ζ generates a one-parameter group of diffeomorphism ϕ_t on $J\mathcal{D}(\mathcal{G})$ such that

$$\phi_s : J\mathcal{D}(\mathcal{G})_s \to J\mathcal{D}(\mathcal{G})_{s+t}, \forall s, t \in \mathbf{R}.$$

The map $|_t$ is an isomorphism of Lie algebras. A function $f \in F(J\mathcal{D}(\mathcal{G}))$ also belongs to $\mathcal{P}$ iff $\zeta.f = 0$.
Let us consider, now, the following vector fiber bundle on $J\mathcal{D}(\mathcal{G})$:

$$H \equiv \Lambda_6^0 vT^c J\mathcal{D}(\mathcal{G}) \bigotimes_{J\mathcal{D}(\mathcal{G})} L(\mathbb{E}^c),$$

where vT^c denotes complexificated vertical tangent functor and

$$\mathbb{E}^c \equiv \mathbb{E} \bigotimes \mathbf{C},$$

where $\mathbb{E}$ is a vertical distribution on $J\mathcal{D}(\mathcal{G})$ belonging to the kernel of Ω (**polarization**); $L(\mathbb{E}^c)$ denotes the super-bundle of vertical half-forms on $J\mathcal{D}(\mathcal{G})$. We have a canonical connection on the complex line bundle

$$K \equiv \Lambda_6^0 vT^c J\mathcal{D}(\mathcal{G})$$

induced by the canonical connection on Galilean space-time M. Furthermore, by using the vertical metric field g on M we can consider a Hermitian structure h on K. So we can build a Hilbert space $\mathcal{H}$, completed with respect to the following scalar product:

$$< s \otimes \rho, s' \otimes \rho' >= \int_{J\mathcal{D}(\mathcal{G})} h(s, s') \rho \bar{\rho}'.$$

Recall that any "observables" of the classical mechanics is a real function

$$f : J\mathcal{D}(\mathcal{G}) \to \mathbf{R},$$

so that the corresponding random variable over a statistical set Ω is a function

$$\left\{ \begin{array}{l} \tilde{f} : \Omega \to \mathbf{R} \\ \tilde{f}(\omega) \equiv \tilde{f}(D^2 m(I)) = \int_I (f \circ D^2 m)\eta_0, \end{array} \right\}$$

where $I \subset T$ is a closed interval and η_0 is the canonical volume form on T. Now, for any $f \in F(J\mathcal{D}(\mathcal{G}))$, we define the following linear operator on $\mathcal{H}$: (**K-S quantization**)

$$\hat{f} : s \otimes \rho \mapsto [\frac{1}{2\pi i}(\nabla_{\mathrm{sg}(f)} s) + f.s] \otimes \rho + \frac{1}{2\pi i}(s \otimes L_{\mathrm{sg}(f)}\rho)$$

where $\mathrm{sg}(f) \equiv \Omega' \circ vdf \equiv$ symplectic gradient of f, Ω' is the symplectic isomorphism $vT^* J\mathcal{D}(\mathcal{G}) \cong vTJ\mathcal{D}(\mathcal{G})$ and ∇ is the absolute differential associated to the connection on K. If f is such that $\mathrm{sg}(f)$ is adapted to the super-bundle structure identified by $L(\mathbb{E}^c)$, then f is a Hermitian operator. This condition is equivalent to the following:

$$[\mathrm{sg}(f), \xi] \rfloor \Omega = 0 = L_{\mathrm{sg}(f)}(\xi \rfloor \Omega), \forall \xi \in \mathbb{E}^c.$$

Then, the quantization of the random variable $\tilde{f}$ associated to f is defined by using as Hilbert space just the space $\mathcal{H}$ and putting

$$(\phi(\tilde{f})(x), y) = \int_\Omega \tilde{f} dE_{x,y} = (\hat{f}(x), y).$$

In this way we identify a pre-spectral measure

$$E : (\Omega, \Sigma) \circ\!\!\to \mathcal{L}(\mathcal{H})$$

with respect to the Borel σ-algebra Σ on Ω in the FW^∞-topology.

2 If $\psi : T \times M \to M$ is an inertial frame we can represent M as a 3-dimensional affine Euclidean space $S_\psi \equiv M/\sim$, by using the canonical projections $\psi/ : M \to S_\psi$. (If one fixes a point on S_ψ we get the identification $S_\psi \equiv \mathbf{S}$.) One has the diffeomorphism $M \cong S_\psi \times T$. So, we can represent $\Omega(a,b)$ on S_ψ as $\Omega(a,b)^\psi = \psi/_*(\Omega(a,b))$. This is a subset of the set $\Omega[a,b]^\psi$ of all C^∞ curves in S_ψ connecting the points $\bar{a}_0 \equiv \psi/(a_0)$ and $\bar{b}_0 \equiv \psi/(b_0)$. We call $\Omega(a,b)^\psi$ the **observed statistical set** of $\Omega(a,b)$ by means of the inertial frame ψ. Note, that $\Omega(a,b)^\psi$ is contained in the Brownian bridge

$$B(\bar{a}_0, \bar{b}_0) \equiv \{\gamma \in C^0([0,1], S_\psi) | \gamma(0) = \bar{a}_0, \gamma(1) = \bar{b}_0\},$$

which one usually considers in order to calculate the amplitude of probability (**propagator**) to the transition $\bar{a}_0 \to \bar{b}_0$ into a time t:

$$P(\bar{a}_0, 0|\bar{b}_0, t) = E(\exp \int_{[0,t]} V(\gamma(t'))dt'|^{\gamma(0)=\bar{a}_0}_{\gamma(1)=\bar{b}_0}), \tag{5.15}$$

where E is the expectation associated to the Wiener measure w on $B(\bar{a}_0, \bar{b}_0)$. In fact, (5.15) results in the propagator of the heat equation

$$\hbar(\partial t.\phi) - \left(\frac{\hbar^2}{2m} \triangle + V\right) \phi = 0, \tag{5.16}$$

e.g.,

$$\phi(\bar{b}_0, t) = \int_{S_\psi} P(\bar{a}_0, 0|\bar{b}_0, t)\phi(\bar{a}_0, 0) * 1.$$

Equation (5.16) gives the classical Schroedinger equation

$$i\hbar(\partial t.\phi) + \left(\frac{\hbar^2}{2m} \triangle - V\right) \phi = 0$$

if the bar-Planck-constant $\hbar$ is replaced by $i\hbar$.[19]
Unfortunately, $\Omega(a,b)^\psi$ has zero measure with respect to the Wiener measure. So, this later is not suitable if we want to stay in the class of differentiable maps. However,

[19] Note that the classical Schroedinger equation can be written intrinsically in Galilean space-time, with respect to a frame ψ as follows:

$$i\hbar\mathcal{L}_{\dot\psi}\phi + (\tfrac{\hbar^2}{2}\triangle - V)\phi = 0$$

where $\mathcal{L}_{\dot\psi}$ denotes the Lie derivative with respect to the velocity $\dot\psi$ of ψ and $\triangle$ is the Laplacian given by $\triangle\phi \equiv \mathrm{div}\,\mathrm{grad}\,\phi$, with $\mathrm{grad}\,\phi = {}'g \circ vd\phi$, where vd is the vertical differential and $'g$ is the vertical isomorphism $vT^*M = vTM$ induced by the vertical metric on M. With respect to a coordinate system $(x^0 \equiv t, x^k)$ on M adapted to ψ, the above equation can be written:

$$i\hbar(\partial t.\phi) + \tfrac{\hbar^2}{2}|g|^{-1/2}\partial x_i.[|g|^{1/2}g^{ij}(\partial x_i.\phi)] - V\phi = 0.$$

Remark also that in the Schroedinger equation ϕ is usually a $\mathbf{C}$-valued function on $T \times S_\psi$ such that for any $t \in T, \phi_t \in L^2(S_\psi)$. In order to recover such a classical meaning of quantum state, it is necessary to restrict the above Hilbert space $\mathcal{H}$ to the following subspace

$$\mathcal{H}_0 \equiv \{\phi \in \mathcal{H} | \nabla_\xi \phi = 0, \forall \xi \in \mathbf{E}^c\} = \textbf{polarized Hilbert space}.$$

if $\Omega(a,b)$ is the quantum fluctuation of some dynamic cobord of E_2, we can consider on $\Omega(a,b)^{\psi}$ the measure ν induced by the Gauss measure γ on $\Omega(a,b)$ by means of the surjection

$$\psi/* : \Omega(a,b) \to \Omega(a,b)^{\psi} : \nu \equiv \gamma \circ \psi/*^{-1} : \Omega(a,b)^{\psi} \multimap \mathbf{R}.$$

So $\Omega(a,b)^{\psi}$ becomes a probability space with respect to ν.[20]

5.5 - CANONICAL QUANTIZATION OF PDEs

The "canonical quantization" of a classical mechanical system (system of particles or field theory), consists into reformulate the classical dynamical PDE (or ODE) as a differential equation of operators on a suitable vector space, with an additional constraint on the initial data, given by "canonical" commutation relations. Even if in the linear cases this approach is satisfactory enough, for non-linear equations a direct interpretation of the classical equations as operatorial ones is, in general, very problematic.

In the context of the formal quantization of PDEs we have, instead, a natural way to consider classical entities as operatorial ones, whether in linear or non linear cases. Scope of this section is, just, to show that the formal quantization of PDEs is able to interpret the usual concept of canonical quantization. (See also refs.[106,108,109,111,113,114] and compare with the approaches by Lychnerowicz [65,66] and Maslov [54].)

REMARK 5.25 - In this section we will use the name "physical observables", associated to some PDE $E_k \subset J\mathcal{D}^k(W)$, to mean random variables

$$A : \Omega(E_k)_c \to \mathbf{R}$$

given as

$$A(s) = \int_M f \circ s d\mu,$$

where μ is a suitable measure on M and $f : W \to \mathbf{R}$ is a numerical function on W. Local coordinates on W identify random variables of this type.

DEFINITION 5.66 - *Let $\pi : W \to M$ be a fiber bundle. Let $K : J\mathcal{D}^k(W) \to \mathbb{K}$ be a k-order differential operator, where $\underline{\pi} : \mathbb{K} \to M$ is a fiber bundle over M. Let us assume also that $\underline{\pi}_{\bar{h}} : \mathbb{K} \to J\mathcal{D}^{\bar{h}}(W)$ is a fiber bundle such that $\underline{\pi} = \pi_{\bar{h}} \circ \underline{\pi}_{\bar{h}}$. Furthermore, let us assume that K is a fiber bundle morphism over $\pi_{h,\bar{h}} : J\mathcal{D}^h(W) \to J\mathcal{D}^{\bar{h}}(W)$,*

[20] Compare with the approach by B.Lascar [62] which considers the representation of the kernel for the half-wave equation by means of geodesic paths of class H^1.

with $-1 \leq \bar{h} \leq h \leq k$. *Let* $\chi : J\mathcal{D}^{\bar{h}}(W) \to \mathbb{K}$ *be a section of* $\underline{\pi}_{\bar{h}}$. *Let* $E_k = \ker_\chi K \subset J\mathcal{D}^k(W)$ *be a PDE defined by the following exact sequence:*

$$0 \to E_k \hookrightarrow J\mathcal{D}^k(W) \underset{\chi \circ \pi_{k,\bar{h}}}{\overset{K}{\rightrightarrows}} \mathbb{K}$$

So any solution $s \in Sol(E_k)$ *satisfies the following equation:*

$$K \circ D^k s = \chi \circ D^{\bar{h}} s. \tag{5.17}$$

We say that E_k *is a* **dynamic PDE** *if its Jacobi operator* $J[s]$, *associated at a solution* s, *is a self-adjoint differential operator, and "past" and "future" can be defined with respect to any foliation of the base manifold* M *into "space-like" hypersurfaces. Then, we have two kernels* $G^{\pm}$ *such that* ${}_\phi G^+$ *has support in the future of* $\operatorname{supp}\phi$ *and* ${}_\phi G^-$ *has support in the past of* $\operatorname{supp}\phi$. *So we have* $(J[s] \otimes \mathbf{1})G^{\pm} = (\mathbf{1} \otimes J[s])G^{\pm} = \mathbb{D}$. *Let the retarded and advanced solutions of the Jacobi equation* $E_k[s]$ *be denoted by* ν^- *and* ν^+ *respectively. Then we have*

$$\nu^{\pm} =_{j[\chi.s]} G^{\pm}.$$

We have the following important theorem.

THEOREM 5.55 - *Let* $E_k = \ker_\chi K \subset J\mathcal{D}^k(W)$ *be a dynamic PDE such that* $K = vT^*W$. *Then the set of physical observables defined on the classic limit* $\Omega(E_k)_c$ *of* E_k *has a natural structure of Lie algebra.*

PROOF. Let us be given a function $f : W \to \mathbf{R}$ on the fiber bundle W. This identifies a physical observable characterized by a function $A : C^\infty(W) \to C^\infty(M, \mathbf{R}) \to \mathbf{R}$. To A we can associate a **current**

$$\left\{ \begin{array}{l} jA : C^\infty(W) \to \bigcup_{s \in C^\infty(W)} C^\infty(s^* vT^*W) \\ jA(s) \equiv (vdf) \circ s \in C^\infty(s^* vT^*W). \end{array} \right\}$$

If $B : C^\infty(W) \to \mathbf{R}$ is another physical observable, we can consider the corresponding current of B calculated in correspondence of a solution of E_k for any vector field $\nu \in T_s C^\infty(W)$ belonging to the set of solutions of the Jacobi equation of E_k corresponding to the current $j[\chi.s] = -jA(s) : J[s].\nu = -jA(s)$. Then, we get

$$< \nu^{\pm}, jB(s) \otimes \eta >= -_{jA(s)} G^{\pm}(jB(s) \otimes \eta) = G^{\pm}(jA(s) \otimes jB(s) \otimes \eta).$$

Set

$$(D_A B)(s) \equiv < \nu^-, jB(s) \otimes \eta >= G^-(jA(s) \otimes jB(s) \otimes \eta)\,.$$

Thus, in $\mathcal{P}_{hys}(E_k) \equiv$ set of physical observables on $\Omega(E_k)_c$, we can define the following bracket:

$$\left\{\begin{array}{l} (A,B)(s) \equiv (D_A B - D_B A)(s) \\ = G^-(jB(s) \otimes jA(s) \otimes \eta) - G^-(jA(s) \otimes jB(s) \otimes \eta) \\ = G^+(jA(s) \otimes jB(s) \otimes \eta) - G^-(jA(s) \otimes jB(s) \otimes \eta) \\ \equiv \tilde{G}(jA(s) \otimes jB(s) \otimes \eta). \end{array}\right\}$$

We can easily see that this bracket satisfies the following relations:

$$(A,B) = -(B,A); \quad ((A,B),C) + ((B,C),A) + ((C,A),B) = 0.$$

So $\mathcal{P}_{hys}(E_k)$ becomes a Lie algebra. □

REMARK 5.26 - The structure of $\tilde{G}$ assures the local causality. In other words, for any two observable fields A and B, if the support of $jA(s)$ is separated from every point of the support of $jB(s)$ by spacelike interval, we have $(A,B)(s) = 0$.

The above considerations allow us to point out a general criterion to quantize physical observables.

DEFINITION 5.67 - (GENERAL CRITERION TO QUANTIZE PHYSICAL OBSERVABLES). *$\mathcal{P}_{hys}(E_k)$, with the above Lie algebra structure, suggests the following general criterion to obtain a canonical quantization of physical observables defined on $\Omega(E_k)_c$, for dynamic PDEs:*

$$\boxed{[\widehat{A},\widehat{B}](s) = (\widehat{A} \circ \widehat{B} - \widehat{B} \circ \widehat{A})(s) = i\hbar\tilde{G}(jA(s) \otimes jB(s) \otimes \eta)\widehat{\otimes} id_{\mathcal{H}(s)} \ (\triangle)}$$

where $\mathcal{H}(s)$ is a suitable locally convex topological vector space. The corresponding expectation value for any $\psi \in \mathcal{H}(s), \psi' \in \mathcal{H}(s)'$ is as follows:

$$\boxed{< \psi' | [\widehat{A},\widehat{B}](s) | \psi >= i\hbar\tilde{G}(jA(s) \otimes jB(s) \otimes \eta) < \psi', \psi >}$$

REMARK 5.27 - The choice of the quantum commutator allows us to also recognize quantum spectral measures

$$E : (\Omega(E_k)_c, \Sigma) \circ\!\!\rightarrow \mathcal{L}(\mathcal{H})$$

associated to random functions $\Omega(E_k)_c \to \mathbf{R}$.

EXAMPLE 5.25 - 1) In ref.[111] the canonical quantization of unharmonic oscillator is shown. More generally it also produces the canonical quantization of Newton-equation. Furthermore, it is stressed that the above geometric method allows us to reproduce the canonical quantizations of the usual linear PDEs of field theories, like Klein-Gordon, Dirac, Maxwell equations.

2) QUANTUM FLUIDS: (CANONICAL QUANTIZATION OF EULER AND NAVIER-STOKES EQUATIONS).

In this last example we shall simply state that also for the classic nonlinear equations (E) and (NS) we can recognize a procedure of canonical quantization and to obtain a description of quantum fluids.

Let us write the Jacobi equations $(E)[s]$ and $(NS)[s]$ respectively of (E) and (NS) that govern the canonical quantization.

$(E)[s]$: JACOBI EQUATION OF (E) (at the section $s = (v, p)$)

Dependent variables: $(\nu^j, \mu), 1 \leq j \leq 3$.

$$\left\{\begin{array}{ll} (a) & G^j_{jk}\nu^k + \partial x_i.\nu^i = J[s] \\ (b) & \rho[2G^j_{ik}v^k + (\partial x_i.v^j) - (\partial v_i.(\rho B^j))]\nu^i + [\rho\delta^j_k - (\partial v^0_k.(\rho B^j))](\partial x_0.\nu^k) \\ & + [\rho v^i\delta^j_k - (\partial v^i_k.(\rho B^j))](\partial x_i.\nu^k) + (\partial x_i.\mu)g^{ij} = J^j[s] \end{array}\right\}$$

$(NS)[s]$: JACOBI EQUATION OF (NS) (at the section $s = (v, p, \theta)$)

Dependent variables: $(\nu^j, \mu, \epsilon), 1 \leq j \leq 3$.

$$\left\{\begin{array}{ll} (a) & G^j_{jk}\nu^k + (\partial x_i.\nu^i) = J[s] \\ (b) & \nu^s\{2\rho G^i_{sk}v^k + (\partial x_s.v^j) - 2\chi[G^i_{ik}(\partial x_s.g^{kj}) + G^j_{ik}(\partial x_s.g^{ki}) \\ & (\partial v_s.\rho B^j) + (\partial x_i.\partial x_s.g^{ij})]\} + (\partial x_0.\nu^i)\{\rho\delta^j_i - (\partial v_0.\rho B^j)\} \\ & + (\partial x_s..\nu^k)\{v^s\delta^j_k - 2\chi[-G^i_{ik}g^{js} - G^i_{ip}g^{ps}\delta^j_k - 2G^i_{ik}g^{is} - (\partial x_i.g^{is})\delta^j_k] \\ & - (\partial v^s_k.\rho B^j)\} + (\partial x_i.\mu)g^{ij} + (\partial x_i.\partial x_s.v^k)\{2\chi[g^{js}\delta^i_k - g^{is}\delta^j_k] = J[s]^j \\ (c) & \nu^s\{\rho C_p(\partial x_s.\theta) - 2\chi[G^j_{ik}(\partial x_s.g^{pi})v^k g_{jp} - G^j_{is}(\partial x_p.v^k)g_{kq}g^{qp} \\ & - G^j_{is}(\partial x_j.v^i) + (\partial x_i.v^k)(\partial x_s.g^{pi})g_{kp}]\} \\ & + (\partial x_s.\nu^k)\{-2\chi[-G^j_{iq}v^q g_{kj}g^{is} - 2(\partial x_p.v^i)g_{ki}g^{sp} - 2(\partial x_k.v^s) \\ & - G^s_{ik}v^i + v^p(\partial x_p.g^{is})g_{ki}]\} + (\partial x_0.\nu^i)\rho C_p \\ & + (\partial x_k.\epsilon)\{2\rho C_p v^k - \phi[G^j_{ji}g^{ki} + (\partial x_i.g^{ik})]\} \\ & - (\partial x_i.\partial x_s.\epsilon)\phi g^{is} = \overline{J}[s]. \end{array}\right\}$$

Then, the canonical quantization of physical observables representing the components of a section s is obtained by means of the following quantum bracket:

$$[\widehat{s}^i(x,t), \widehat{s}^j(x',t')] = -i\hbar\widetilde{G}^{ij}(x,t;x',t';s)$$

where $\widetilde{G}^{ij}(x,t;x',t';s)$ is the propagator of $(E)[s]$ or $(NS)[s]$. The derivation with respect to t' and taking $t'=t$ gives

$$[\widehat{s}^i(x,t),\dot{\widehat{s}}^j(x',t)]=-i\hbar(\partial t'.\widetilde{G}^{ij}(x,t;x',t';s))|_{t'=t}.$$

Furthermore, taking into account that the propagators satisfy the boundary conditions:

$$(\partial t'.\widetilde{G}^{ij}(x,t;x',t';s))|_{t'=t}=\delta(x,x')\delta^{ij},$$

we get

$$[\widehat{s}^i(x,t),\dot{\widehat{s}}^j(x',t)]=-i\hbar\delta(x,x')\delta^{ij}$$

that is the desired canonical quantization of (E) and (NS) respectively.

REMARK 5.28 - Note, that the method of canonical quantization given in this section, by means of Jacobi equation clarifies the linking between quantification and deformation of original structure (i.e., PDE). Furthermore, taking into account the relation between generalized Green kernels and singular solutions, we see that multi-valued solutions of Jacobi equation govern the canonical quantization. Finally, note that the knowledge of Bäcklunds of PDEs [111] can aid the procedure of canonical quantization. (For details see ref.[111].)

In ref.[111] a categorial interpretation of Dirac quantization is also given. There we have proved that Dirac quantizations are reconducible to coverings in the category of PDEs.

6 - GEOMETRY OF QUANTUM PDEs

In this chapter we formulate a geometrical theory of PDEs in a suitable category of noncommutative manifolds. So this chapter gives first mathematical foundation to a research program that aims to build a theory of quantum supergravity by translating the point of view of General Relativity in microphysics. In other words, we shall develop a theory of quantum physics by introducing special structured manifolds (quantum manifolds) and dynamic equations for such quantum manifolds. These should describe the microphysics worlds, similarly as pseudo-Riemannian manifolds, satisfying Einstein equations, describe macroscopic worlds.

So the first step for such a theory is to define quantum manifolds. These are seen as locally convex manifolds, where the model is A^{m+n}, with A a suitable noncommutative $\mathbf{K}$-algebra ($\mathbf{K} = \mathbf{R}, \mathbf{C}$). (Such an algebra is not, in general, a superalgebra, as, in general, gradiation is required.) Then, for such manifolds we develop a geometry and a differential calculus. In particular, we emphasize that, similarly to what happens in General Relativity, where the metric field describes the structure of pseudo Riemannian manifolds, we recognize for quantum manifolds particular fields of geometric objects that fully describe on microlevel the physical world. We call such fields the quantum structure of a quantum manifold. A dynamic theory for quantum physics should also formulate a PDE for such a structure. So, for this purpose, we should be able to formulate a geometric theory of PDEs in the category of quantum manifolds. From the similarity between quantum manifolds and supermanifolds, our previous formulation of geometry of super PDEs [112,113,114] has been of great help. In fact, many steps of that theory can be translated in the present theory of quantum PDEs.

So, we formulate a theory of nonholonomic connections for quantum manifolds, as these objects remain fundamental also in our mathematical description of microscopic world. Then, a quantum gauge theory is formulated and a quantum supergravity model is described as an application.

Note that in this framework we are able to recognize also the meaning of canonical quantization of our theory by using a non-local propagator, canonically associated to the quantum dynamic equation. In fact, if we start from a classical theory (i.e. classical PDE), canonical quantization produces a quantum algebra of observables, by means of propagators that depend on a classical field [111]. Instead, if we start from a quantum context (i.e. quantum PDE), the canonical quantization produces yet a quantum algebra of observables, but, now, by means of propagators depending on quantum fields. So, in both cases, canonical quantization can simply be seen as a changing of the commutation rules between observables, by means of (quantum) dynamic equations. Furthermore, singular solutions of QPDEs and cobordism theory in QPDEs give us a full geometric meaning of the canonical quantization and quantum tunnel effects. The interaction between macrolevel and microlevel, i.e., the observation of microscopic phenomena at macroscopic level, is geometrically described by means of the fiber bundle structure that we impose on our quantum manifolds. The base manifold for such a structure is an ordinary manifold (m-dimensional space-time). Sections of this fiber bundle embed the ordinary manifold in the quantum one. The pullback of the quantum geometric structures on these ordinary manifolds just interpreted the meaning of observations of the microscopic world at macroscopic level. In fact quantum fields, i.e. sections of quantum fiber bundles, identify on the classic limit of the base manifolds, e.g., space-time, A-valued distributions. So, if A is an algebra of linear operators on some Hilbert space, one recovers the usual interpretation of "quantum fields" [111]. Remark that in our formulation of quantum supergravity we recognize that the quantum structure defines on the space-time curvature and torsion, as well as gravitino and electromagnetic fields, as A-valued distributions. This clarifies that such macroscopic objects have their origin at the quantum level.

Finally, observe that our previous theory of quantum geometry in PDEs [109], as well as super PDEs [113,114] can be considered as the first necessary step in a process of **canonical quantization** of PDEs. In fact, in this chapter we develop what it is necessary when a quantum algebra has been chosen. In practice, we start with a classical theory, i.e., a classic (super) PDE, $E_k \subset JD^k(W)$, where $\pi : W \to M$ is a fiber bundle in the category of (super) manifolds. So, after the process of canonical quantization of E_k, (as illustrated in refs.[111], and the above chapter), we are able to build a quantum algebra A, i.e., the algebra of quantized observables of E_k. As a consequence we also have a way to build quantum manifolds over A, (as illustrated in this chapter), and to formulate a quantum (super) PDE $\hat{E}_k$ as the quantized of E_k. ($\hat{E}_k$ can almost be formally written as E_k, but in the category of quantum manifolds over A). In this chapter we will show, also, that it is possible to further "quantize" $\hat{E}_k$, by means of a procedure of canonical quantization. This will have the effect to further change the product in the algebra of obervables of $\hat{E}_k$.

6.1 - DIFFERENTIAL GEOMETRY OF QUANTUM MANIFOLDS

Non-commutative geometry started with some work on graded commutative manifolds (supermanifolds), (see e.g., the books quoted in refs.[23,162,163]) and continued with non-commutative structures in geometry, (see e.g., the pioneer work by A.Connes [20]). The approach that we present here from one side uses non-commutative algebra in order to build suitable locally convex topological vector spaces as models for non-commutative manifolds, and from the other side characterizes such manifolds from the differential-algebraic point of view. In the last decade efforts have been made in order to study non-commutative algebras. (See e.g., refs.[1,49,52,79,88] and references quoted therein). Also attention has been devoted to the study of infinite dimensional manifolds. (See e.g., refs.[61,90,91,121].) The new approach that we suggest here is to combine these two different informations in order to obtain a new concept of non-commutative manifold (quantum manifolds) suitable to interpret physical phenomena in microworlds.

DEFINITION 6.1 - *A* **quantum algebra** *is a triplet* (A, ϵ, c), *where:*

(i) *A is a metrizable, complete, Hausdorff locally convex topological* $\mathbf{K}$*-vector space, i.e., a Fréchet space, which is also a unit ring;*

(ii) *$\epsilon : \mathbf{K} \to A_0 \subset A$ is a ring homomorphism, where A_0 is the centre of*

$$A: \quad A_0 \equiv \{a \in A | ab = ba, \forall\, b \in A\}.$$

The homomorphism ϵ is necessarily injective. So A is a $\mathbf{K}$*-algebra i.e., A is a ring, and a* $\mathbf{K}$*-vector space satisfying*

$$\lambda(ab) = a(\lambda b) = (\lambda a)b, \forall \lambda \in \mathbf{K}, a, b \in A.$$

(iii) *$c : A \to \mathbf{K}$ is a* $\mathbf{K}$*-algebra morphism, with $c(e) = 1, e =$ unit of A. For any $a \in A$ we call $a_C \equiv c(a) \in \mathbf{K}$ the* **classic-limit** *of a.*

(iv) *A is an associative* $\mathbf{K}$*-algebra.*

DEFINITION 6.2 - *We call* **quantum extension** *(resp.* **full quantum extension**), *of a quantum algebra A, $\hat{A} \equiv Hom_{\mathbf{K}}(A, A)$ (resp. $\widehat{A} \equiv Hom_{A_0}(A, A) \equiv AOA$) . $\hat{A}$ and $\widehat{A}$ have natural structures of* $\mathbf{K}$*-algebra containing A as a subalgebra. $(\hat{a} + \hat{b})(a) = \hat{a}(a) + \hat{b}(a), (\lambda\hat{a})(a) = \lambda\hat{a}(a), (\hat{a}\hat{b})(a) = \hat{a}(a)\hat{b}(a)$, or $(\hat{a}\hat{b}) = \hat{a} \circ \hat{b}, \forall \hat{a}, \hat{b} \in \hat{A}, \widehat{A}, a \in A, \lambda \in \mathbf{K}$. Any element $a \in A$ identifies the element $\hat{a} \in \hat{A}$, or $\widehat{A}$, given by $\hat{a}(b) = ab, \forall b \in A$. $\hat{A}$ and $\widehat{A}$ are also two-sided A-modules.*

DEFINITION 6.3 - *We call s*-**quantum coextension** *(resp. s*-**full quantum coextension**) *of a quantum algebra A :*

$$\left\{\begin{array}{l} \overset{s}{\hat{A}} \equiv Hom_{\mathbf{K}}(A \bigotimes_{\mathbf{K}} \cdots_s \cdots \bigotimes_{\mathbf{K}} A, A) \equiv Hom_{\mathbf{K}}(T^s_0(A), A), s \in \mathbf{N}\, \overset{0}{\hat{A}} \equiv A; \overset{1}{\hat{A}} \equiv \hat{A} \\ (\text{resp. } \overset{s}{\widehat{A}} \equiv Hom_{A_0}(A \bigotimes_{A_0} \cdots_s \cdots \bigotimes_{A_0} A, A) \equiv \dot{T}^s_0(A) O A, s \in \mathbf{N}, \overset{0}{\widehat{A}} \equiv A, \overset{1}{\widehat{A}} \equiv \widehat{A}) \end{array}\right\}$$

Then $\overset{s}{\hat{A}}$ and $\overset{s}{\widehat{A}}$ have natural structures of $\mathbf{K}$-algebras with the following product:

$$\left\{\begin{array}{l} (\hat{b}\hat{a})(a_1 \otimes \cdots \otimes a_s) = \hat{b}(a_1 \otimes \cdots \otimes a_s)\hat{a}(a_1 \otimes \cdots \otimes a_s), \forall \hat{a}, \hat{b} \in \overset{s}{\hat{A}} \\ \text{or } \overset{s}{\widehat{A}}, a_1 \otimes \cdots \otimes a_s \in \dot{T}^s_0(A), \text{ or } T^s_0(A) \end{array}\right\}$$

Then, $\overset{s}{\hat{A}}$ contains $T^s_0(A)$ as a subalgebra . More precisely, any $b_1 \otimes \cdots \otimes b_s \in T^s_0(A)$ identifies the element of $\overset{s}{\hat{A}}$ given by : $b_1 \otimes \cdots \otimes b_s(a_1 \otimes \cdots \otimes a_s) = b_1 a_1 \cdots b_s a_s$. Set

$$\overset{\infty}{\hat{A}} \equiv \bigoplus_{q \geq 0} \overset{q}{\hat{A}}; \quad {}_q\hat{A} \equiv \bigoplus_{0 \leq p \leq q} \overset{p}{\hat{A}}.$$

One has the natural filtration:

$$A = {}_0\hat{A} \subset {}_1\hat{A} \subset {}_2\hat{A} \subset \cdots \subset \overset{\infty}{\hat{A}}.$$

Furthermore, $\overset{\infty}{\hat{A}}$ has a natural structure of associative $\mathbf{Z}$-graded -algebra. (Similar considerations hold for $\widehat{A}$).

DEFINITION 6.4 - *Given a quantum algebra A, we define its* **symmetric** s-**full quantum coextension,** *(resp.* **skew-symmetric***)* s-**full quantum coextension***:*

$$\left\{ \begin{array}{l} \overset{(q)}{\widehat{A}} \equiv \dot{S}^q_0(A) O A, \quad q \geq 1; \quad \overset{(1)}{\widehat{A}} \equiv \widehat{A}; \quad \overset{(0)}{\widehat{A}} \equiv A \\ (resp. \ \overset{[q]}{\widehat{A}} \equiv \dot{\Lambda}^q_0(A) O A, \quad q \geq 1; \quad \overset{[1]}{\widehat{A}} \equiv \widehat{A}; \quad \overset{[0]}{\widehat{A}} \equiv A) \end{array} \right\}$$

where $\dot{S}^q_0(A)$ *(resp.* $\dot{\Lambda}^q_0(A)$*) is the symmetric (resp. skew-symmetric) part of* $\dot{T}^q_0(A)$. *Then* $\overset{(q)}{\widehat{A}}$, *and* $\overset{[q]}{\widehat{A}}$, *have natural structures of associative* $\mathbf{K}$*-algebras.*

$$\overset{(\infty)}{\widehat{A}} \equiv \bigoplus_{q \geq 0} \overset{(q)}{\widehat{A}}, \quad \text{and} \quad \overset{[\infty]}{\widehat{A}} \equiv \bigoplus_{q \geq 0} \overset{[q]}{\widehat{A}}$$

have natural structures of associative $\mathbf{Z}$*-graded* $\mathbf{K}$*-algebras with unit, that we call* **symmetric** ∞-**full quantum coextension,** *(resp.* **skew-symmetric***)* ∞-**full quantum coextension,** *of A. The graded product*

$$\overset{(r)}{\widehat{A}} \overset{(s)}{\widehat{A}} \subset \overset{(r+s)}{\widehat{A}}, (resp. \overset{[r]}{\widehat{A}} \overset{[s]}{\widehat{A}} \subset \overset{[r+s]}{\widehat{A}})$$

is defined by:

$$\left\{ \begin{array}{l} (\widehat{b}\widehat{a})(a_1 \odot \cdots \odot a_{r+s}) \\ = \sum_{\sigma \in S_{s+r}} \frac{1}{(r+s)!} \widehat{b}(a_{\sigma(1)} \otimes \cdots \otimes a_{\sigma(r)}) \widehat{a}(a_{\sigma(r+1)} \otimes \cdots \otimes a_{\sigma(r+s)}) \\ \forall \widehat{a} \in \overset{(r)}{\widehat{A}}, \widehat{b} \in \overset{(s)}{\widehat{A}} \end{array} \right\}$$

(resp.

$$\left\{\begin{array}{l}(\widehat{b}\widehat{a})(a_1\wedge\cdots\wedge a_{r+s})\\ \sum_{\sigma\in S_{s+r}}\frac{\epsilon(\sigma)}{(r+s)!}\widehat{b}(a_{\sigma(1)}\otimes\cdots\otimes a_{\sigma(r)})\widehat{a}(a_{\sigma(r+1)}\otimes\cdots\otimes a_{\sigma(r+s)})\\ \forall\widehat{a}\in\overset{[r]}{\widehat{A}},\widehat{b}\in\overset{[s]}{\widehat{A}})\end{array}\right\})$$

Set:

$$_{(q)}\widehat{A}\equiv\bigoplus_{0\leq p\leq q}\overset{(p)}{\widehat{A}}\quad,\quad(resp.\ _{[q]}\widehat{A}\equiv\bigoplus_{0\leq p\leq q}\overset{[p]}{\widehat{A}}).$$

Thus one has the natural filtration:

$$A={}_{(0)}\widehat{A}\subset\widehat{A}\subset{}_{(2)}\widehat{A}\subset\cdots\subset\overset{(\infty)}{\widehat{A}}\quad,\quad(resp.\ A={}_{[0]}\widehat{A}\subset\widehat{A}\subset{}_{[2]}\widehat{A}\subset\cdots\subset\overset{[\infty]}{\widehat{A}}).$$

Of course one has the natural inclusion $\overset{(\infty)}{\widehat{A}}\subset\overset{\infty}{\widehat{A}}$, *(resp.* $\overset{[\infty]}{\widehat{A}}\subset\overset{\infty}{\widehat{A}}$), *and the natural projection* $\overset{(\infty)}{\widehat{A}}\to\overset{\infty}{\widehat{A}}$. *(resp.* $\overset{[\infty]}{\widehat{A}}\to\overset{\infty}{\widehat{A}}$.) *(Similar considerations hold for* $\hat{A}$, *then the term "full" is omitted.)*

REMARK 6.1 - The product in A is a **K**-bilinear mapping $\mu : A\times A\to A$. This induces the **K**-linear mapping $\hat{\mu} : A\bigotimes_{\mathbf{K}}A\to A$ (resp. A_0-linear mapping $\widehat{\mu} : A\bigotimes_{A_0}A\to A$) by means of the following commutative diagrams:

$$\begin{array}{ccc}A\times A&\overset{\mu}{\to}&A\\ \downarrow&&\|\\ A\bigotimes_{\mathbf{K}}A&\underset{\hat{\mu}}{\to}&A\end{array}\qquad\qquad\begin{array}{ccc}A\bigotimes_{\mathbf{K}}A&\overset{\hat{\mu}}{\to}&A\\ \downarrow&&\|\\ A\bigotimes_{A_0}A&\underset{\widehat{\mu}}{\to}&A\end{array}$$

REMARK 6.2 - A quantum algebra identifies a **K**-coalgebra [1] (A,Δ,c), where $\Delta : A\to A\bigotimes_{\mathbf{K}}A,\Delta(a)=1\otimes a+a\otimes 1$. Then $(A,\mu,c,\Delta,\epsilon)$ is a **K**-bialgebra [1].

REMARK 6.3 - Given a **K**-coalgebra C, the set $R\equiv Hom_{\mathbf{K}\text{-}module}(C,A)$ has a natural structure of **K**-algebra with structure maps : $\mu_R(f\otimes g)=f\star g\equiv\mu\circ(f\otimes g)\circ\Delta$, (**convolution** of f and g) ; $\epsilon_R(\lambda)=\lambda\epsilon\circ c_C$. One has $(f\star g)(a)=f(a)g(a),\forall a\in A$. A **K**-algebra (resp. **K**-coalgebra) A is said to be **commutative** (resp. **cocommutative**) if $\mu=\sigma\circ\mu$ (resp. $\Delta=\sigma\circ\Delta$), where σ is a linear mapping $\sigma : A\bigotimes_{\mathbf{K}}A\to A\bigotimes_{\mathbf{K}}A$ (**braiding**) such that $\sigma(a\otimes b)=b\otimes a,\forall a,b\in A$. So a quantum algebra is not commutative but can be considered cocommutative with respect to the natural structure of coalgebra. When the identity map 1_A of A is a regular element of R with respect to multiplication on R, the inverse s of 1_A is called the **antipode** of A. More precisely s satisfies the following equivalent conditions: $\mu \circ (s \otimes 1) \circ \Delta = \mu \circ (1 \otimes s) \circ \Delta = \epsilon \circ c; \quad s \star 1 = 1 \star s = \epsilon \circ c$. Furthermore, if A is a cocommutative algebra one also has $\sigma \circ (\sigma \otimes s) \circ \Delta = \Delta \circ s$. In such a case the quantum algebra becomes a $\mathbf{K}$-Hopf algebra [1].

REMARK 6.4 - A quantum algebra is called a $*$-**quantum algebra** (or **involutive quantum algebra**) if A is a $*$-algebra, i.e., A is equipped with a $*$-operation (**involution**) $* : A \to A$ with the properties : (i) $(\alpha a + \beta b)^* = \bar{\alpha} a^* + \bar{\beta} b^*$, where $\bar{\gamma}$ denotes complex conjugation for any $\gamma \in \mathbf{K}$; (ii) $(a^*)^* = a$; (iii) $(ab)^* = b^* a^*$. One has $e^* = e$, and $(a^{-1})^* = (a^*)^{-1}$ if a is invertible. Then an element $a \in A$ such that $a^* = a$ is called **hermitian**; if $aa^* = a^*a$, a is called **normal**, and if $aa^* = a^*a = e$, a is called **unitary**. An **involutive Banach quantum algebra** is an involutive Banach algebra that is also a quantum algebra, i.e. $\|a^*\| = \|a\|$. A C^*-**quantum algebra** is an involutive Banach quantum algebra that is also a C^*-algebra, i.e., (iv) $\|a\|^2 = \|a^* a\|$. If A is a $*$-quantum algebra that is also an $*$-Hopf algebra, then one has: (v) $s((s(a^*))^*) = a$; (vi) Δ is a $*$-homomorphism : $\Delta(a^*) = \sum_i b_i^* \otimes c_i^*$, where $\Delta(a) = \sum_i b_i \otimes c_i$; (vii) c is a $*$-homomorphism $c(a^*) = \overline{c(a)}, \forall a \in A$. A **von Neumann quantum algebra** is a C^*-quantum algebra A that is a sub-algebra of $\mathcal{L}(\mathcal{H}) \equiv$ algebra of continuous authomorphisms of a complex Hilbert space $\mathcal{H}$, such that A is stable by involution $*$ and closed with respect to the weak topology of $\mathcal{L}(\mathcal{H})$. [Any C^*-quantum algebra A is isomorphic to a closed subalgebra B (with respect to the norm topology) of $\mathcal{L}(\mathcal{H})$ with respect to a suitable Hilbert space $\mathcal{H}: \quad A \cong B \subset \mathcal{L}(\mathcal{H})$. The closure $\overline{B}$ of B (with respect to the weak topology) is a von Neumann quantum algebra.]

REMARK 6.5 - The underlying $\mathbf{K}$-vector space structure of a quantum algebra A is determined by a family $\{p_\alpha\}_{\alpha \in B}$ of seminorms, i.e., a family of functions $p_\alpha : A \to \mathbf{K}$, such that $p_\alpha(a+b) \leq p_\alpha(a) + p_\alpha(b), p_\alpha(\lambda a) = |\lambda| p_\alpha(a), p_\alpha(a) > 0, \forall a, b \in A, \lambda \in \mathbf{K}$. As A is metrizable, B is a countable set, and hence $\{p_\alpha\}_{\alpha \in B}$ is a sequence of seminorms which is increasing. Furthermore, as A is a locally convex topological vector space, we have $A \subseteq \Pi_{\alpha \in B} A_\alpha$, where A_α are Banach spaces. Moreover, as A is complete, the above inclusion is more precisely an equality. Finally, as A is metrizable, B is a countable set.

THEOREM 6.1 - *Set:* $A(\mathbf{K}) \equiv$ *category of associative* $\mathbf{K}$*-algebras;* $Q(\mathbf{K}) \equiv$ *category of quantum algebras,* $\mathcal{L}A(\mathbf{K}) \equiv$ *category of Lie algebras over* $\mathbf{K}$*,* $\mathcal{J}A(\mathbf{K}) \equiv$ *category of Jordan algebras over* $\mathbf{K}$*. Then, one has two natural functors:* $Q_- : Q(\mathbf{K}) \subset A(\mathbf{K}) \to \mathcal{L}A(\mathbf{K}), \qquad Q_+ : Q(\mathbf{K}) \subset A(\mathbf{K}) \to \mathcal{J}A(\mathbf{K})$*. More precisely, if* $A \in Ob(A(\mathbf{K}))$*, the algebras* $Q_-(A)$ *and* $Q_+(A)$ *have the same underlying* $\mathbf{K}$*-vector space of* A*, but the products are given respectively by* $[a,b]_- \equiv ab - ba, [a,b]_+ \equiv \frac{1}{2}(ab+ba)$*. These satisfy* *the following properties:*

$$\begin{aligned}[a,b]_- + [b,a]_- &= 0,\\ [a,[b,c]_-]_- + [b,[c,a]_-]_- + [c,[a,b]_-]_- &= 0,\\ [a,b]_+ - [b,a]_+ &= 0,\\ [a^2,[b,a]_+]_+ - [[a^2,b]_+,b]_+ &= 0,\\ \frac{1}{2}[a,b]_- + [a,b]_+ - ab &= 0.\end{aligned}$$

We call $Q_-(A)$ *(resp.* $Q_+(A)$*) the* **Lie algebra** *(resp.* **Jordan algebra***), associated to* A*. We call* $[a,b]_-$*, (resp.* $[a,b]_+$*) the,* **commutator***, (resp.* **anticommutator***), of* $a,b\in A$*.*

NOTATION. In the following we shall also use the notation:

$$[b,a]_- \equiv [b,a]; \quad \{b,a\} \equiv 2[b,a]_+.$$

DEFINITION 6.5 - *Let* A *be a quantum algebra. A* **representation** *of* A *is a* $\mathbf{K}$*-algebra morphism*

$$\pi : A \to Hom_{\mathbf{K}}(E,E) \equiv B_\pi,$$

where: E *is a locally convex topological* $\mathbf{K}$*-vector space, such that there exists a* $\mathbf{K}$*-algebra morphism* $c_E : B_\pi \to \mathbf{K}$ *with the following commutative diagram*

$$\begin{array}{ccc}\mathbf{K} & \xrightarrow{\epsilon_E} & Hom_{\mathbf{K}}(E,E)\\ \downarrow \epsilon & & \|\\ A & \xrightarrow{\pi} & Hom_{\mathbf{K}}(E,E)\\ \downarrow c & & \downarrow c_E\\ \mathbf{K} & = & \mathbf{K}\end{array}$$

where ϵ_E *is the canonical mapping given by* $\epsilon_E(k) : v \mapsto kv$*.* B_π *is called* **model algebra** *and* $E \equiv E_\pi$ **model space** *of the representation* π *of the quantum algebra* A*.*

DEFINITION 6.6 - *By using the following commutative diagram*

$$\begin{array}{ccc}Hom_{\mathbf{K}-algebra}(A;Hom_{\mathbf{K}}(E,E)) & \xrightarrow{j} & Hom_{\mathbf{K}}(A\bigotimes_{\mathbf{K}}E,E)\\ \bigcap & & \|\\ Hom_{\mathbf{K}}(A;Hom_{\mathbf{K}}(E,E)) & \cong & Hom_{\mathbf{K}}(A\bigotimes_{\mathbf{K}}E,E)\end{array}$$

we can identify any representation of π *of* A *with an element*

$$j(\pi) \in Hom_{\mathbf{K}}(A\bigotimes_{\mathbf{K}}E,E).$$

We say that a vector subspace $F \subset E$ is **invariant** *with respect to a representation π if $j(\pi)(A \bigotimes_{\mathbf{K}} F) \subset F$. We say that a representation is* **irreducible** *if it only has trivial invariant subobjects.*

DEFINITION 6.7 - *We call* **spectrum** *of A the set $Spec(A)$ of all irreducible representations of A. The space $Spec(A)$ is topologized by making the closure $\overline{X}$ of $X \in Spec(A)$ the following subset of $Spec(A)$:*

$$\overline{X} \equiv \{q \in Spec(A) \mid q|_{\ker(\pi)} = 0, \forall \pi \in X\}.$$

The topology obtained in this way is called **Jacobson topology**. *One has the following mapping*

$$\left\{ \begin{array}{l} A \to Map(Spec(A), \mathcal{M}_A) \equiv \mathcal{M}_A{}^{Spec(A)} \\ a \mapsto \hat{a} : \pi : A \to B_\pi, \quad \pi(a) \in B_\pi \end{array} \right\}$$

where $\mathcal{M}_A \equiv$ set of all the model algebras of A. We call $\hat{a}$ **Fourier image** *(or* **Fourier transform***) of $a \in A$.*

REMARK 6.6 - The spectrum is related to the ideals of A. In fact, any irreducible representation $\pi \in Spec(A)$ identifies a closed two-sided ideal (**primitive ideal**) $A_\pi \subset A$ given by : $A_\pi \equiv \ker(\pi)$. Denote by $Prim(A)$ the set of primitive ideals in A. However, $Prim(A)$ is not suitable to parametrize the space of irreducible representations. In fact, e.g. if A is a simple C^*-algebra, i.e., has no ideals closed under the $*$-operation, $Prim(A)$ consists of a single point. On the other hand, there are simple quantum algebras having many inequivalent irreducible representations (all having kernel 0). A particular important subset of $Spec(A)$ is

$$Sp(A) \equiv Hom_{\mathbf{K}\text{-}algebra}(A, \mathbf{K}) \subset Spec(A).$$

(Here $E \equiv \mathbf{K}$). Of course, on $Sp(A)$ we recognize the Jacobson topology induced by $Spec(A)$. As $\hat{A} = Hom_{\mathbf{K}}(A, A)$, we can consider the natural immersion $A \subset \hat{A}$ as a representation of A. This representation is irreducible iff A has no trivial ideals. In such cases the immersion $A \subset \hat{A}$ belongs to the spectrum $Spec(A)$ of A. Therefore, if A has no trivial ideals, the product of A identifies an element of $Spec(A)$.

NOTE. In the literature quantum groups and quantum algebras are often obtained as deformations of some classic groups and algebras respectively. Representations of such objects are, e.g., considered in ref.[151], where relations with the theory of special functions are also obtained.

DEFINITION 6.8 - *We call* **quantum vector space** *of dimension $m, m \in \mathbf{N}$, built on the quantum algebra A, a locally convex topological $\mathbf{K}$-vector space E isomorphic to $A^m \equiv A \times \cdots_m \cdots \times A$:*

$$E \cong A^m \equiv A \times \cdots_m \cdots \times A.$$

REMARK 6.7 - A^m , and so E, has a natural structure of A_0-module.
DEFINITION 6.9 - *We say that a mapping $f : U \to A^p$ is of* **class** $Q^s_w, 0 \leq s \leq \infty, \omega$, *on $U \subset A^m$, where U is an open subset of A^m, if it is a weakly s-times differentiable map (in the sense of* [56] *and above chapter) (or of class C^s_w) such that*

$$Df(x) \in Hom_{A_0}(A^m, A^p) \equiv A^m O A^p \cong \widehat{A}^{mp}, \forall x \in U.$$

DEFINITION 6.10 - *A* **quantum manifold** *of* **dimension** m *(over a quantum algebra A) (of class $Q^k_w, 0 \leq k \leq \infty, \omega$) is a* $\mathcal{Q}$-**quantum structure**, *i.e., $\mathcal{Q}$ is a pseudogroup of Q^k_w-homeomorphisms of A^m. So a quantum manifold of dimension m is a locally convex manifold M, modeled on A^m and with a Q^k_w atlas Ξ of local coordinate mappings such that the transition functions define a pseudogroup $\mathcal{Q}$ of local Q^k_w-homeomorphisms on A^m. So for each open coordinate set $U \subset M$ we have a set of m coordinate functions,* (**quantum coordinates**), $(x^A) : U \to A^{m+n}$. *We denote by $\mathcal{A}_p(M)$ the algebra of germs of A-valued Q^∞_w- functions at $p \in M$.*
REMARK 6.8 - (RELATION BETWEEN QUANTUM MANIFOLD AND SPECTRUM OF THE UNDERLYING QUANTUM ALGEBRA). One has a canonical mapping

$$j : M \to Sp(\hat{\mathcal{A}}) \subset Spec(\hat{\mathcal{A}}), p \mapsto j(p) \in Sp(\hat{\mathcal{A}}), j(p)(f) = f(p),$$

where $\hat{\mathcal{A}} \equiv Q^\infty_w(M; A)$, and M is a quantum manifold, modeled on the quantum algebra A.
Let M_C be the classic limit of M, then one has the following commutative and exact diagram:

$$\begin{array}{ccccc} 0 & \to & \mathcal{A}_C & \to & \mathcal{A} \\ & & \| & & \| \\ 0 & \to & C^\infty_w(M_C; A) & \to & C^\infty_w(M; A) \end{array}$$

As a consequence we get the following exact commutative diagram

$$\begin{array}{ccccc} 0 & \leftarrow & M_C & \leftarrow & M \\ & & \downarrow & & \downarrow \\ 0 & \leftarrow & Sp(\mathcal{A}_C) & \leftarrow & Sp(\mathcal{A}) \end{array}$$

One has the canonical short exact sequences : $0 \to \hat{\mathcal{A}} \to \mathcal{A}$; $\quad Sp(\mathcal{A}) \to Sp(\hat{\mathcal{A}}) \to 0$.
DEFINITION 6.11 - *Let M be a quantum manifold of dimension m. The* **tangent space** *at $p \in M$ is the vector space T_pM defined in the following way: $T_pM \equiv$ space of equivalent classes $v = [f]$ of C^1_w (or equivalently C^1) curves $f : I \to M, I =$open neighbourhood of $0 \in \mathbf{R}, f(0) = p$; two curves f, f' are equivalent if for each (equivalently, for some) coordinate system μ around p, the functions $\mu \circ f, \mu \circ f' : I \to A^m$, have the same derivative at $0 \in \mathbf{R}$.*

REMARK 6.9 - T_pM results in a **K**-vector space or a topological locally convex **K**-vector space, so there exists the **dual**: $(T_pM)^* \equiv L_{\mathbf{K}}(T_pM;\mathbf{K})$, of T_pM, or the **topological dual**: $(T_pM)' \equiv \mathcal{L}_{\mathbf{K}}(T_pM;\mathbf{K})$. On the other hand one also has the following noncanonical isomorphism $T_pM \cong A^m$, so on T_pM one recognizes also a structure of m-dimensional quantum vector space. If $v \in T_pM$ has two different local representations (v^A) and $(\bar{v}^A)$ in coordinates respectively (x^A) and $(\bar{x}^A)$, then the relations between (v^A) and $(\bar{v}^A)$ are the following : $\bar{v}^A = \sum_{1\leq B\leq m}(\partial x_B.\bar{x}^A)(v^B)$, where $(\partial x_B.\bar{x}^A) \in \widehat{A}$. As we must also have $v^C = \sum_{1\leq A\leq m}(\partial\bar{x}_A.x^C)(\bar{v}^A)$, we get $\sum_{1\leq A\leq m}(\partial\bar{x}_A.x^C)\circ(\partial x_B.\bar{x}^A) \equiv \sum_{1\leq A\leq m}(\partial\bar{x}_A.x^C)(\partial x_B.\bar{x}^A) = \delta^A_B$. Similarly, we must also have $\sum_{1\leq A\leq m}(\partial x_B.\bar{x}^A)(\partial\bar{x}_C.x^B) = \delta^A_C$. Hence T_pM is an A_0-module.
So we can give the following.

DEFINITION 6.12 - *Let M be a quantum manifold of dimension m. For any $p \in M$, we have*

(a)

$$\hat{T}_pM \equiv Hom_{\mathbf{K}}(A, T_pM), \quad (\textbf{quantum tangent space} \text{ at } p\in M);$$

(b)

$$\widehat{T}_pM \equiv AOT_pM \equiv Hom_{A_0}(A, T_pM), \quad (\textbf{full quantum tangent space} \ \textit{at } p\in M);$$

(c)

$$(T_pM)^\star \equiv Hom_{\mathbf{K}}(T_pM, A), \quad (\textbf{quantum dual} \ \textit{of } T_pM);$$

(d)

$$(T_pM)^+ \equiv T_pMOA \equiv Hom_{A_0}(T_pM, A), \quad (\textbf{full quantum dual} \ \textit{of } T_pM);$$

(e)

$$\hat{T}M \equiv \bigcup_{p\in M}\hat{T}_pM, \quad (\textbf{quantum tangent bundle});$$

(f)

$$\widehat{T}M \equiv \bigcup_{p\in M}\widehat{T}_pM, \quad (\textbf{full quantum tangent bundle});$$

(g)

$$(TM)^\star \equiv \bigcup_{p\in M}(T_pM)^\star, \quad (\textbf{quantum dual bundle} \ \textit{of } TM);$$

(h)

$$(TM)^+ \equiv \bigcup_{p\in M}(T_pM)^+, \quad (\textbf{full quantum dual bundle} \ \textit{of } TM).$$

REMARK 6.10 - In the following, in order not to make the exposition too heavy, we will almost consider only full quantum bundles. In fact, similar considerations for quantum bundles can be obtained. $\widehat{T}_pM$ and $(T_pM)^+$ are A_0-modules. Furthermore, $\widehat{T}_pM$ is a right $\widehat{A}$-module with action given by : $v\hat{a} = v \circ \hat{a}, \forall v \in \widehat{T}_pM, \hat{a} \in \widehat{A}$. $(T_pM)^+$, has, instead, a left $\widehat{A}$-module structure : $\hat{a}\alpha = \hat{a} \circ \alpha, \forall \hat{a} \in \widehat{A}, \alpha \in (T_pM)^+$. Furthermore, $(T_pM)^+$ also has a two-sided A-module structure. One has the following exact sequence of A_0-modules: $\widehat{T}_pM \to T_pM \to 0$ induced by the canonical inclusion $A_0 \to A$.

NOTATION. By considering $T_pM, \forall p \in M$, as a $\mathbf{K}$-vector space, we will write:

$$T^r_sM \equiv TM \bigotimes_{\mathbf{K}} \cdots r \cdots \bigotimes_{\mathbf{K}} TM \bigotimes_{\mathbf{K}} T^*M \bigotimes_{\mathbf{K}} \cdots_s \cdots \bigotimes_{\mathbf{K}} T^*M$$

S^r_sM and Λ^r_sM will denote the corresponding symmetric and skew-symmetric subfiber bundles respectively. Furthermore, we shall denote by $\dot{T}^r_sM$ the tensor bundle of type (r, s) if each tangent space T_pM is considered as A_0-module. More precisely we put:

$$\dot{T}^r_sM \equiv TM \bigotimes_{A_0} \cdots_r \cdots \bigotimes_{A_0} TM \bigotimes_{A_0} (TM)^+_0 \bigotimes_{A_0} \cdots_s \cdots \bigotimes_{A_0} (TM)^+_0,$$

with $(TM)^+_0 \equiv Hom_{A_0}(TM, A_0)$. The corresponding symmetric and skew-symmetric bundles will be denoted by $\dot{S}^r_sM$ and $\dot{\Lambda}^r_sM$ respectively. Similarly we put:

$$\widehat{T}^r_sM \equiv \widehat{T}M \bigotimes_{A_0} \cdots_r \cdots \bigotimes_{A_0} \widehat{T}M \bigotimes_{A_0} (TM)^+ \bigotimes_{A_0} \cdots_s \cdots \bigotimes_{A_0} (TM)^+.$$

The corresponding symmetric and skew-symmetric bundles will be denoted by $\widehat{S}^r_sM$ and $\widehat{\Lambda}^r_sM$ respectively.

REMARK 6.11 - (REPRESENTATION OF QUANTUM p-FORMS AS DUAL QUANTUM p-VECTORS).

One has a canonical A_0-homomorphism

$$\widehat{T}^0_pM \to (\dot{T}^p_0M)^+ \equiv Hom_{A_0}(\dot{T}^p_0M, A)$$

induced by composition

$$\widehat{T}^0_pM \to Hom_{A_0}(\dot{T}^p_0M; \dot{T}^p_0A) \to Hom_{A_0}(\dot{T}^p_0M, A),$$

where the last homomorphism is induced by the product of A. One has the following commutative diagram:

$$\begin{array}{ccc} \widehat{T}^0_pM & \to & (\dot{T}^p_0M)^+ \\ \uparrow & & \downarrow \\ \widehat{\Lambda}^0_pM & \to & (\dot{\Lambda}^p_0M)^+ \end{array}$$

Furthermore, one can write

$$\overset{k}{\widehat{A}} = (\dot{T}^k_0 A)^+ \quad , \quad \overset{(k)}{\widehat{A}} = (\dot{S}^k_0 A)^+ \quad , \quad \overset{[k]}{\widehat{A}} = (\dot{\Lambda}^k_0 A)^+$$

i.e., the k-full quantum coextension, (resp. symmetric k-full quantum coextension, skew-symmetric k-full quantum coextension) of A can be interpreted as the full quantum dual space of the k-vectors (resp. symmetric k-vectors, skew-symmetric k-vectors) of A.

In the following we will denote by $Q^s_w(W)$ the set of sections of class Q^s_w of a fiber bundle $\pi : W \to M$.

REMARK 6.12 - (REPRESENTATION OF TENSOR PRODUCTS OF OPERATOR VALUED MAPS INTO ALTERNATING MAPS).

Let N be another quantum manifold, of dimension p. One has the following commutative diagram of A_0-modules:

$$\begin{array}{ccc} \dot{T}^q_0(Hom_{A_0}(T_pM, T_qN)) & \to & Hom_{A_0}(\dot{T}^q_0(T_pM); \dot{T}^q_0(T_qN)) \\ j\downarrow & & \downarrow \\ Hom_{A_0}(\dot{\Lambda}^q_0(T_pM); \dot{\Lambda}^q_0(T_qN)) & = & Hom_{A_0}(\dot{\Lambda}^q_0(T_pM); \dot{\Lambda}^q_0(T_qN)) \, . \end{array}$$

We put

$$j(h_1, \dots, h_q) \equiv \dot{\Lambda}^q(h_1, \dots, h_q), \quad \forall\, h_i \in Hom_{A_0}(T_pM, T_qN).$$

Note that

$$\dot{\Lambda}^q(h_1, \dots, h_q) \neq h_1 \wedge \cdots \wedge h_q.$$

One has,

$$\dot{\Lambda}^q(h_1, \dots, h_q)(v_1 \wedge \cdots \wedge v_q) = \sum_{\sigma \in S_q} h_{\sigma(1)}(v_1) \wedge \cdots \wedge h_{\sigma(q)}(v_q), \quad \forall\, v_i \in T_pM.$$

REMARK 6.13 - (δ-QUANTUM COMPLEX). Let $\alpha : N \to M$ be a fiber bundle between quantum manifolds M and N, and let $\beta : E \to N$ be a fiber bundle of A_0-modules over N. We have a natural morphism of fiber bundles of A_0-modules over M, $\delta : \dot{\Lambda}^j_0 M \to \dot{\Lambda}^{j-1}_0 M \bigotimes_{A_0} TM$, that extends to a morphism of A_0-modules $\delta : \dot{\Lambda}^j_0 M \bigotimes_{A_0} \dot{S}^i_0 M \to \dot{\Lambda}^{j-1}_0 M \bigotimes_{A_0} \dot{S}^{i+1}_0 M$. One can easily see that $\delta^2 = 0$, so we get the following complex:

$$0 \leftarrow \dot{S}^p_0 M \overset{\delta}{\leftarrow} TM \bigotimes_{A_0} \dot{S}^{p-1}_0 M \overset{\delta}{\leftarrow} \dot{\Lambda}^2_0 M \bigotimes_{A_0} \dot{S}^{p-2}_0 M \overset{\delta}{\leftarrow} \cdots \overset{\delta}{\leftarrow} \dot{\Lambda}^r_0 M \bigotimes_{A_0} \dot{S}^{p-r}_0 M \overset{\delta}{\leftarrow} \cdots,$$

where $\dot{S}^l_0 M = 0$, for $l < 0$. Now, taking the pull-back of the above complex by means of α, and applied the functor $Hom_{A_0}(-, E)$, we get the following one

$$0 \to \alpha^* \dot{S}^p_0 MOE \overset{\delta}{\to} \alpha^* TMO(\alpha^* \dot{S}^{p-1}_0 MOE) \overset{\delta}{\to} \cdots \overset{\delta}{\to} \alpha^* \dot{\Lambda}^r_0 MO(\alpha^* \dot{S}^{p-r}_0 MOE) \overset{\delta}{\to} \cdots,$$

where, for abuse of notation, we have denoted by δ the induced boundary homomorphism too. We call the above complex **δ-quantum complex** of $E \stackrel{\beta}{\to} N \stackrel{\alpha}{\to} M$. Such a complex will be useful in order to characterize quantum PDEs. (See Chapter 2.)

DEFINITION 6.13 - *We will call* **classic regular quantum manifold** *a quantum manifold M of dimension $m+n$ such that it admits a projection $c : M \to M_C$ on an m-dimensional manifold M_C. We will call M_C the* **classic limit** *of M.*[1]

[This is equivalent to saying that for any $p \in M$ we have a system of quantum coordinates adapted to the projection $c : M \to M_C$.]

In the following we shall only consider classic regular quantum manifolds.

DEFINITION 6.14 - *Set :*

$$\widehat{A}_0 \equiv AOA_0 \quad ; \quad A^{r|s} \equiv A^r \times \widehat{A}^s;$$

$$A^{s|(1,q_1)\dots(k,q_k)} \equiv A^s \times \widehat{A}^{q_1} \times \cdots \times (\overset{(k)}{\widehat{A}})^{q_k} \quad ; \quad A^{s|(1,q_1)} \equiv A^{s|q_1};$$

$$A^{s|(\infty)} \equiv \lim_k A^{s|(1,q_1)\dots(k,q_k)} = A^s \times \overset{(\infty)}{\widehat{A}} \quad ; \quad A^{s|(k,q_k)} \equiv A^s \times (\overset{(k)}{\widehat{A}})^{q_k};$$

$$A^{s|\{1,q_1\}\dots\{k,q_k\}} \equiv A^s \times \widehat{A}^{q_1} \times \cdots \times (\overset{k}{\widehat{A}})^{q_k} \quad ; \quad A^{s|\{1,q_1\}} \equiv A^{s|q_1};$$

$$A^{s|\infty} \equiv \lim_k A^{s|\{1,q_1\}\dots\{k,q_k\}} = A^s \times \overset{\infty}{\widehat{A}} \quad ; \quad A^{s|\{k,q\}} \equiv A^s \times (\overset{k}{\widehat{A}})^{q_k}.$$

By using standard results of algebra [14] we can prove the following theorems.

THEOREM 6.2 - *One has the noncanonical isomorphisms of A_0-modules :*

$$(6.1) \qquad \widehat{T}_pM \cong \widehat{A}^{m+n}; \quad (T_pM)^+ \cong \widehat{A}^{m+n}.$$

Furthermore, the first is also an isomorphism of right $\widehat{A}$-modules, and the second of left $\widehat{A}$-modules. Thus, $\widehat{T}_pM$ and $(T_pM)^+$ are respectively right and left free $\widehat{A}$-modules of finite type. Moreover, one has the canonical isomorphism of left $\widehat{A}$-modules :

$$j : (T_pM)^+ \cong Hom_{\widehat{A}}(\widehat{T}_pM.\widehat{A}), \quad j(\gamma)(v) = \gamma \circ v, \quad \forall v \in \widehat{T}_pM, \gamma \in (T_pM)^+.$$

So $(T_pM)^+$ can be considered the dual of $\widehat{T}_pM$ if these are considered as left $\widehat{A}$-module and right $\widehat{A}$-module respectively. Let M be an $(m+n)$-dimensional quantum manifold:

(a) *$\widehat{T}M$ is a quantum manifold of dimension $m+n|m+n$, i.e., a locally convex manifold modeled on $A^{m+n|m+n}$ of class Q^{s-1}_w , if M is of class Q^s_w.*

[1] If we are not interested to emphasize the projection $M \to M_C$, or if $\dim_{\mathbf{K}} M_C = \dim_A M = s$, we will simply say that M is a quantum manifold of dimension s.

(b) *Similarly* $(TM)^+$ *is a quantum manifold of dimension* $m+n|m+n$.
(c) *One has the following noncanonical isomorphisms of* A_0*-modules:*

$$\widehat{T}^r_s(T_pM) \cong \dot{T}^{r+s}_0(\widehat{A})^{(m+n)^{r+s}}, \quad \dot{T}^r_0(T_pM) \cong \dot{T}^r_0(A)^{(m+n)^r}, \quad \forall p \in M.$$

THEOREM 6.3 - *One has a mapping:*

$$<,>: \widehat{T}_pM \times (T_pM)^+ \to \widehat{A},$$

given by

$$< \alpha, \zeta >= \alpha \circ \zeta.$$

It is a A_0*-bilinear mapping, and identifies a* A_0*-linear mapping* $\widehat{T}_pM \bigotimes_{A_0}(T_pM)^+ \to \widehat{A}$.

THEOREM 6.4 - *Let* M *and* N *be quantum manifolds of dimensions* $m+n$ *and* $p+q$ *respectively. One has the following canonical isomorphism of* A_0*-modules:*

$$\begin{aligned} &Hom_{A_0}(T_pM, T_qN) \cong Hom_{\widehat{A}}(\widehat{T}_pM, \widehat{T}_qN) \\ &\cong \widehat{T}_qN \bigotimes_{\widehat{A}} Hom_{\widehat{A}}(\widehat{T}_pM, \widehat{A}) \cong \widehat{T}_qN \bigotimes_{\widehat{A}} (T_pM)^+ . \end{aligned}$$

If $h \in Hom_{A_0}(T_pM, T_qN)$ *we get that the corresponding* $\widehat{h}$ *in* $Hom_{\widehat{A}}(\widehat{T}_pM.\widehat{T}_qN)$ *is defined by* $\widehat{h}(v) = h \circ v$, *for any* $v \in \widehat{T}_pM$. *We call* $\widehat{h}$ **full quantum extension** *of* h. *Conversely, for any* $\widehat{h} \in Hom_{\widehat{A}}(\widehat{T}_pM, \widehat{T}_qN)$ *there exists a unique* $h \in Hom_{A_0}(T_pM, T_qN)$ *such that the following diagram is commutative:*

$$\begin{array}{ccc} \widehat{T}_pM & \xrightarrow{\widehat{h}} & \widehat{T}_qN \\ \downarrow & & \downarrow \\ T_pM & \xrightarrow[h]{} & T_qN \end{array}$$

THEOREM 6.5 - *Let* M *be a quantum manifold of dimension* $m+n$. *For any section* i *of* $c: M \to M_C$ *we have a canonical* A_0*-homomorphism:*

$$i_* : (T_{i(p)}M)^* \to A \bigotimes_{\mathbf{K}} (T_pM_C)^*$$

given by means of the following commutative diagram:

$$\begin{array}{ccc} (T_{i(p)}M)^* & \xrightarrow{i_*} & A \bigotimes_{\mathbf{K}} (T_pM_C)^* \\ \| & & \|\wr \\ Hom_{\mathbf{K}}(T_{i(p)}M, A) & \xrightarrow[h_i]{} & Hom_A(A \bigotimes_{\mathbf{K}} T_pM_C, A) \end{array}$$

Here h_i is the A_0-homomorphism defined by :

$$h_i(\alpha)(a \otimes v) = a < \alpha, T(i)(v) >, \forall \alpha \in (T_{i(p)}M)^+, a \otimes v \in A \bigotimes_{\mathbf{K}} T_p M_C.$$

Thus we can calculate the pull-back on M_C of any quantum p-differential form $\alpha : M \to (\Lambda_0^p M)^\star$ on M, by means of any section i of $c : M \to M_C$:

$$i^*\alpha \equiv \hat{i}_* \circ \alpha \circ i : M_C \to A \bigotimes_{\mathbf{K}} \Lambda_p^0 M_C,$$

by using the natural extension $\hat{i}_*$ of i_*. Taking into account the A_0-homomorphism $(\dot{\Lambda}_0^p M)^+ \to (\Lambda_0^p M)^\star$, similar considerations can be made for quantum p-differential forms $\alpha : M \to (\dot{\Lambda}_0^p M)^+$ too.

DEFINITION 6.15 - *We call* **quantum (product) bundle** *the following one*

$$\dot{T}_0^2 MOTM \equiv Hom_{A_0}(TM \bigotimes_{A_0} TM, TM).$$

DEFINITION 6.16 - *A* **quantum (product) structure** *is a Q_w^s-section $\widehat{\mu}$ of this bundle over M.*

PROPOSITION 6.1 - *The local representation of $\widehat{\mu}(p)$ is the following:*

$$\widehat{\mu}(p) = (\mu_{HK}^R) \in (\overset{2}{\widehat{A}})^{(m+n)^3}, 1 \leq H, K, R \leq m+n.$$

Furthermore, $\dot{T}_0^2 MOTM$ is a quantum manifold of dimension $m+n|\{2,(m+n)^3\}$, i.e., modeled on $A^{m+n|\{2,(m+n)^3\}}$.

THEOREM 6.6 - *For any section i of*

$$i : M_C \to M, \quad c \circ i = id_{M_C},$$

one has a natural A_0-homomorphism

$$i_* : Hom_{A_0}(T_{i(p)}M \bigotimes_{A_0} T_{i(p)}M, T_{i(p)}M) \to T_2^1(T_p M_C) \bigotimes_{\mathbf{K}} A.$$

PROOF. One has, in fact, the following sequence of A_0-homomorphisms and isomorphisms

$$Hom_{A_0}(T_{i(p)}M \bigotimes_{A_0} T_{i(p)}M, T_{i(p)}M)$$

$$\downarrow$$

$$Hom_{A_0}((A \bigotimes_{\mathbf{K}} T_p M_C) \bigotimes_{A_0} (A \bigotimes_{\mathbf{K}} T_p M_C), A_0 \bigotimes_{\mathbf{K}} T_p M_C)$$

$$\cong Hom_{A_0}(A \bigotimes_{\mathbf{K}} (T_p M_C \bigotimes_{\mathbf{K}} T_p M_C), A \bigotimes_{\mathbf{K}} T_p M_C)$$

$$\cong T_p^* M_C \bigotimes_{\mathbf{K}} T_p^* M_C \bigotimes_{\mathbf{K}} T_p M_C \bigotimes_{\mathbf{K}} A$$

$$\equiv T_2^1(T_p M_C) \bigotimes_{\mathbf{K}} A$$

The first isomorphism is induced from the fact that T_pM_C is a projective **K**-module of finite type. The same is true for the last two. □

COROLLARY 6.1 - *As a consequence, for any section $i : M_C \to M$, a quantum structure $\widehat{\mu}$ identifies an A-valued tensor field of type $(1,2)$ on M_C:*

$$i^*\widehat{\mu} : M_C \to T_2^1(T_pM_C) \bigotimes_{\mathbf{K}} A.$$

COROLLARY 6.2 - *For any section $i : M_C \to M$ one has the following canonical A_0-homomorphisms*

$$i_* : Hom_{A_0}(T_{i(p)}M \bigotimes_{A_0} T_{i(p)}M, A) \to T_2^0(T_pM_C) \bigotimes_{\mathbf{K}} A$$

$$\bar{i}_* : Hom_{A_0}(\widehat{T}_{i(p)}M \bigotimes_{A_0} \widehat{T}_{i(p)}M, A) \to T_2^0(T_pM_C) \bigotimes_{\mathbf{K}} A$$

and the following commutative diagram

$$\begin{array}{ccc} Hom_{A_0}(\widehat{T}_{i(p)}M \otimes_{A_0} \widehat{T}_{i(p)}M, A) & \to & T_2^0(T_pM_C) \otimes_{\mathbf{K}} A \\ \uparrow j_* & & \| \\ Hom_{A_0}(T_{i(p)}M \otimes_{A_0} T_{i(p)}M, A_0) & \to & T_2^0(T_pM_C) \otimes_{\mathbf{K}} A \end{array}$$

where $j_*(\phi)(\zeta \otimes \xi) = \phi(j(\zeta) \otimes j(\xi))$, where j is the A_0-homomorphism

$$j : \widehat{T}_{i(p)}M \to T_{i(p)}M,$$

given by

$$f \mapsto f(1).$$

DEFINITION 6.17 - *A* **full quantum metric** *on a quantum manifold is a Q^∞_w-section*

$$\widehat{g} : M \to Hom_{A_0}(\dot{T}_0^2 M, A).$$

PROPOSITION 6.2 - *Taking into account the canonical isomorphism*

$$Hom_{A_0}(\dot{T}_0^2 M, A) \cong Hom_{A_0}(\dot{S}_0^2 M, A) \bigoplus Hom_{A_0}(\dot{\Lambda}_0^2 M, A),$$

we get that a full quantum metric, $\widehat{g}$ splits into symmetric and skew-symmetric part: $\widehat{g} = (\widehat{g}) + [\widehat{g}]$. Furthermore, a quantum metric gives two representations:

(left representation) $'\widehat{g} : TM \to (TM)^+ \quad , \quad '\widehat{g}(v) = {}_v\widehat{g} \quad , \quad {}_v\widehat{g}(u) \equiv \widehat{g}(v \otimes u)$;
(right representation) $\widehat{g}' : TM \to (TM)^+, \quad \widehat{g}'(v) = \widehat{g}_v \quad , \quad \widehat{g}_v(u) \equiv \widehat{g}(u \otimes v)$,

for all $v, u \in T_pM, p \in M$.

PROPOSITION 6.3 - *For any section $i : M_C \to M$ a full quantum metric $\widehat{g}$ identifies an A-valued metric on M_C :*

$$i^*\widehat{g} : M_C \to T_2^0(M_C) \bigotimes_{\mathbf{K}} A.$$

DEFINITION 6.18 - *The* **quantum derivative space** *$\hat{D}(M,N)$ for quantum manifolds M and N of dimension $m+n$ and $p+q$ respectively is defined by*

$$\hat{D}(M,N) \equiv TMOTN \equiv \bigcup_{(p,q)\in M\times N} T_pMOT_qN,$$

where $T_pMOT_qN \equiv Hom_{A_0}(T_qN, T_pM)$ is the space of A_0-linear mappings $T_pM \to T_qN$. Let us denote by $Q^s_w(M,N)$ the space of mappings $M \to N$ of class Q^s_w. Then if

$$f \in Q^s_w(M,N) \Rightarrow D^k f(p) \in \hat{D}(M,N), \quad \forall\, p \in M.$$

DEFINITION 6.19 - *If $\pi : W \to M$ is a fiber bundle between quantum manifolds, the* **quantum derivative space of sections** *of π, $\hat{D}(W)$, is defined by $\hat{D} \equiv d^*\hat{D}(M,W)$, where d is the diagonal mapping $d : M \to M \times M$.*

DEFINITION 6.20 - **Quantum derivative spaces of higher orders** *between quantum manifolds can be defined by*

$$\hat{D}^k(M,N) \equiv \hat{D}(\hat{D}^{k-1}(M,N)), \quad \hat{D}^k(W) \equiv \hat{D}(\hat{D}^{k-1}(W)).$$

REMARK 6.14 - One has a natural injection over $M\times N$: $\hat{D}(M,N) \hookrightarrow D(M,N)$, where $D(M,N)$ denotes the derivative space considering M and N as locally convex manifolds. $\hat{D}(M,N)$ results a quantum manifold of dimension $m+n+p+q|(m+n)(p+q)$. If (x^A) is a quantum coordinate system on M and (y^B) is a quantum coordinate system on N then $(x^A, y^B, y^B_A), 1 \leq A \leq m+n, 1 \leq B \leq p+q$, is a quantum coordinate system on $\hat{D}(M,N)$. The first **derivative** of a quantum differentiable mapping $f : M \to N$ is a section $Df : M \to \hat{D}(M,N)$ of the fiber bundle $\pi_1 : \hat{D}(M,N) \to M$. The local expression of Df is as follows: $x^A \circ Df = x^A, y^B \circ Df = f^B \equiv y^B \circ f, y^B_A \circ Df \equiv (\partial x_A f^B)$. The $(m+n)\times(p+q)$ matrix $(\partial x_A f^B)(p), \forall p \in M$, has entries in $\widehat{A}$:

$$(\partial x_A f^B)(p) \in \widehat{A}.$$

So the local representation of $Df(p)(v)$ is the following:

$$Df(p)(v) = (Df(p)(v)^B) \equiv (\sum_{1\leq A\leq m+n} (\partial x_A f^B)(p)(v^A)) \in A^{p+q},$$

where $v \equiv (v^1, \ldots, v^{m+n}) \in A^{m+n}$ is the local representation of $v \in T_pM$. Let $\pi : W \to M$ be a fiber bundle, $\dim M = m+n$, $\dim W = m+n+p+q$. Let us consider the following system of quantum coordinates:

(a)

$$M : (x^H), 1 \leq H \leq m+n;$$

(b)

$$W : (x^H, y^K), 1 \leq H \leq m+n, 1 \leq K \leq p+q;$$

(c)

$$\hat{D}(M,W) : (x^H, \bar{x}^H, y^K, \bar{x}^H_{\bar{H}}, y^K_{\bar{H}}), 1 \leq H, \bar{H} \leq m+n, 1 \leq K \leq p+q.$$

Then, $\hat{D}(W)$ is a quantum submanifold of $\hat{D}(M,W)$ characterized by means of the following equations :

$$x^H = \bar{x}^H, \quad x^{\bar{H}}_H = \delta^{\bar{H}}_H, 1 \leq H, \quad \bar{H} \leq m+n.$$

So, (x^H, y^K, y^K_H) are good quantum coordinates on $\hat{D}(W)$. Then

$$\dim \hat{D}(W) = m+n+p+q|(m+n)(m+n+p+q).$$

The **velocity** of a quantum differentiable curve in M, $f : A \to M$, is the map $\dot{f} \equiv pr_2 \circ Df : A \to \widehat{T}M$, where pr_2 is the canonical projection $pr_2 : \hat{D}(A,M) \cong A \times \widehat{T}M \to \widehat{T}M$. The **differential** of a quantum differentiable map $f : M \to A$ is a map $df \equiv pr_2 \circ Df : M \to (TM)^+$, where pr_2 is , now, the canonical projection $pr_2 : \hat{D}(M,A) \cong A \times (TM)^+ \to (TM)^+$. For a quantum differentiable map $f : A \times M \to N$ we can define its **infinitesimal variation** $\partial f : M \to f_0^*\widehat{T}N$ as $\partial f = D_1 f \circ j$, where j is the inclusion $j : M \to A \times M$ given by $p \mapsto (0,p)$. So, if f is a quantum deformation of a quantum differentiable section s of the fiber bundle $\pi : W \to M$, such that $f_0 = s$, then $\partial f : M \to s^* v\widehat{T}W$, where the symbol v denotes "vertical". From the above considerations we get that $\widehat{T}M$ can be identified with the space of vectors tangent to quantum differentiable curves $A \to M$ of M, and $(TM)^+$ can be identified with the space of differentials of quantum differentiable A-valued functions on M.

More precisely we have the following.

DEFINITION 6.21 - *A C^∞_w vector field $v : M \to TM$, (resp. Q^∞_w quantum vector field $v : M \to \widehat{T}M$) is called* **integrable** *if there exists a unique (local)* **flow** $\{\phi_t\}_{t \in J \subset \mathbf{R}}$, *(resp. $\{\phi_a\}_{a \in B \subset A}$), such that $\partial\phi = v$, where ϕ_t (resp. ϕ_a) are local diffeomorphisms of M of class C^∞_w (resp. Q^∞_w) such that $\phi_0 = id_M, \phi_{-t} = \phi_t^{-1}, \phi_{t+t'} = \phi \circ \phi_{t'}, t, t' \in J$, (resp. $\phi_0 = id_M, \phi_{-a} = \phi_a^{-1}, \phi_{a+a'} = \phi \circ \phi_{a'}, a, a' \in B$). Here, J is a neighbourhood of $0 \in \mathbf{R}$, (resp. $0 \in A$).*

2 THEOREM 6.7 - *Let M be a quantum manifold of dimension m, over a Banach quantum algebra (or von Neumann quantum algebra) A. Let $v \in C^\infty_w(TM)$ be a vector field over M. Then, v is an integrable vector field.*

PROOF. We shall consider the following lemmas (see also ref.[56]).

LEMMA 6.1 - *Let $[a,b] \equiv \{t \in \mathbf{R} | a \leq t \leq b\}$ be a compact interval on $\mathbf{R}$. Let F be a l.c.s. To any function $f : [a,b] \to F$ we associate a filter on F, (**Riemannian filter** of f), generated by the following basis $(S(\delta))_{\delta>0}, \delta \in \mathbf{R}$, where $S(\delta) \equiv \bigcup_{|\epsilon|\leq\delta} S_\epsilon \subset F$, where $\epsilon \equiv a = t_0 < t_1 < \cdots < t_n \equiv b$ is any subdivision of $[a,b]$ and $S_\epsilon \equiv \sum_{1\leq i\leq n}(t_i - t_{i-1})f([t_{i-1}, t_i]) \subset F$, and $|\epsilon| \equiv \max_{1\leq i\leq n} |t_i - t_{i-1}| \leq \delta$. If f is a continuous function $S(\delta)$ is a Cauchy filter and then has a well-defined limit in the completion $\tilde{F}$ of F. We call this limit the **Riemannian integral** of f and we will denote it by the symbol $\int_{[a,b]} f(t)dt$. We also define $\int_{[b,a]} f(t)dt = -\int_{[a,b]} f(t)dt$. One has the following properties: (a) The mapping $\int_{[a,b]} C^0([a,b],F) \to \tilde{F}$ is linear; (b) Let a,b,c be real numbers such that $a < b < c$. Let $f \in C^0([a,c],F)$. Then $\int_{[a,c]} f(t)dt = \int_{[a,b]} f(t)dt + \int_{[b,c]} f(t)dt$; (c) Let $\phi \in C^0([a,b],\mathbf{R}), y_0 \in F$. Then $\int_{[a,b]}(\phi(t)y_0)dt = (\int_{[a,b]} \phi(t)dt)y_0$; (d) Let $f \in C^0([a,b],F)$ and $u \in \mathcal{L}(F,G)$, where G is a locally convex space, with completion $\tilde{G}$. Let $\tilde{u} : \tilde{F} \to \tilde{G}$ denote the continuous extension of u. Then one has $\int_{[a,b]}(u \circ f)(t)dt = \tilde{u}(\int_{[a,b]} f(t)dt)$; (e)(Mean value theorem). Let $f \in C^0([a,b],F)$. Assume that M is a convex subset of F such that $f([a,b]) \subset M$. Then $\int_{[a,b]} f(t)dt \in (b-a)\tilde{M}$, where $\tilde{M}$ denotes the completion of M, i.e., the closure of M in F; (f) Let $f \in C^0([a,b],F)$. For every continuous seminorm β on F, we have*

$$(6.2) \qquad |\int_{[a,b]} f(t)dt|_\beta \leq \int_{[a,b]} |f(t)|_\beta \leq (b-a) \sup_{a\leq t\leq b} |f(t)|_\beta.$$

Here, for abuse of notation, we have denoted the continuous extension of β to $\tilde{F}$ by the same symbol β. If one introduces the topology of uniform convergence on the vector space $C^0([a,b],F)$, the l.c. topology is defined by the semi-norms $|f|_\beta \equiv \sup_{a\leq t\leq b} |f(t)|_\beta, \beta \in \Gamma_F \equiv$set of all continuous seminorms on F directed by its natural order. Then, formula (6.2) can also be written

$$(6.3) \qquad |\int_{[a,b]} f(t)dt|_\beta \leq (a,b)|f(t)|_\beta.$$

(g) From points (e) and (f) we get that if $[a,b]$ is a compact interval on $\mathbf{R}$ and F is a locally convex space then one has a continuous linear mapping $C^0([a,b],F) \to \tilde{F}, f \mapsto \int_{[a,b]} f(t)dt$, where $C^0([a,b],F)$ is considered a locally convex space with respect to the topology of uniform convergence.

LEMMA 6.2 - (**Relations between integrals and derivatives**). *(a) Let $J \subset \mathbf{R}$ be an open interval, F a l.c.s. and $f \in C^0(J,F)$. For any $a \in J$ the function*

$\int_a f : J \to \tilde{F}, (\int_a f)(t) = \int_{[a,t]} f(t')dt', \forall t \in J$ *is of class* C^1 *and* $\frac{d}{dt}(\int_a f)(t) = f(t), \forall t \in J$. *(b) If* $f \in C^r(J,F), \forall r \in \mathbf{N}$, *then* $\int_a f \in C^{r+1}(J,F)$ *and* $\frac{d^k}{dt^k}(\int_a f)(t) = \frac{d^{k-1}}{dt^{k-1}} f(t), 1 \leq k \leq r+1$. *(c) Let* $J \subset \mathbf{R}$ *be an open interval,* F *a l.c.s. and* $f \in C^1(J,F)$. *Then for any* $(a,b) \in J \times J$, *one has* $f(b) - f(a) = \int_{[a,b]} \frac{df}{dt} dt$. *The integral on the right hand side is always an element of* F *regardless of whether* F *is sequentially complete or not. Furthermore, for any continuous semi-norm* $\beta \in \Gamma_F$, *one has* $|f(b) - f(a)|_\beta \leq \int_{[a,b]} |f'(t)|_\beta |dt| \leq |b-a| \sup_{t \in I} |f'(t)|_\beta$, *where* $I \equiv [a,b]$ *if* $a \leq b$ *and* $I \equiv [b,a]$ *if* $b \leq a$. *Finally if* $\frac{df}{dt} = 0$, *then there exists* $y_0 \in F$ *such that* $f(t) = y_0, \forall t \in J$.

LEMMA 6.3 - *If* a, b *are points in a vector space* E, *we denote by* $[a,b] \equiv \{a + t(b-a) | t \in [0,1] \subset \mathbf{R}\}$, *the* **closed segment** *in* E *joining* a *to* b. *Let* E *and* F *be l.c.s. and* X *an open set in* E. *Let* $f : X \to F$ *be of class* C^1_Λ *with respect to some convergence structure* Λ *in* $\mathcal{L}(E,F)$[56]. *For each* $x \in X$ *and each* $h \in E$ *such that* $x + h \in X$, *the* **remainder** *of* f *at* x *in the direction* h *is defined by* $Rf_x(h) \equiv Rf(x,h) \equiv f(x+h) - f(x) - Df(x)(h)$. *Then we have a mapping* $Rf : \Omega_X \equiv \{(x,h) \in X \times E | x + h \in X\} \to F$. *Furthermore, for fixed* $x \in X$ *define* $\theta Rf_x : \Omega_x \equiv \{(t,h) \in \mathbf{R} \times E | x + th \in X\} \to F, \theta Rf_x(t,h) = t^{-1} Rf_x(th)$, *if* $t \neq 0, \theta Rf_x(0,h) = 0$. *If* $f : X \to F$ *is of class* C^1_s, *then for every* $(x,h) \in E \times E$ *the functions* $\phi : J_{x,h} \equiv \{t \in \mathbf{R} | x + th \in X\} \subset \mathbf{R} \to F, \phi(t) = f(x+th)$, *and* $\psi : J_{x,h} \to F, \psi(t) = Rf_x(th)$ *are of class* C^1 *and one has*

$$(6.4) \quad \frac{d\phi}{dt}(t) = Df(x+th)(h), \qquad \frac{d\psi}{dt}(t) = \big(Df(x+th) - Df(x)\big)(h), \forall t \in J_{x,h}.$$

Furthermore, for every $(x,h) \in X \times E$ *such that* $[x, x+h] \subset X$, *we have the following representations by Riemannian integrals:*

$$(6.5) \quad \left\{ \begin{aligned} f(x+h) - f(x) &= \int_{[0,1]} Df(x+th)(h)dt \\ Rf_x(h) &= \int_{[0,1]} \big(Df(x+th) - Df(x)\big)(h)dt \end{aligned} \right\}$$

For every $(x,h) \in X \times E$ *and every* $t \in \mathbf{R}, t \neq 0$, *such that* $[x, x+th] \subset X$, *we have* $\theta Rf_x(t,h) = \frac{1}{t} \int_{[0,1]} (Df(x+\tau h) - Df(x))(h) d\tau$. *For every* $\beta \in \Gamma_F$ *we have the inequalities:*

$$(6.6) \quad |f(x+h) - f(x)|_\beta \leq \sup_{0 \leq t \leq 1} |Df(x+th)(h)|_\beta$$

$$(6.7) \quad |Rf_x(h)|_\beta \leq \sup_{0 \leq t \leq 1} |\big(Df(x+th) - Df(x)\big)(h)|_\beta$$

$$(6.8) \qquad |\theta R f_x(t,h)|_\beta \leq \sup_{0\leq\tau\leq 1} |(Df(x+\tau h)-Df(x))(h)|_\beta .$$

LEMMA 6.4 - *Let $p_0 \in X$. If the vectors $v(p) \in E$, satisfy the Lipschitz-conditions $|v(p_1)-v(p_2)| \leq k|p_1-p_2|, p_1,p_2 \in X_J \subset X$, a suitable neighbourhood of $p_0 \in X$, where $k \geq 0$, then there exists a neighbourhood J of $0 \in \mathbf{R}$ such that the Cauchy system*

$$(6.9) \qquad \frac{d\phi_p}{dt}(t) = v(\phi_p(t)), \phi_p(0) = p, \forall t \in J \subset \mathbf{R}, p \in X,$$

admits a unique solution $\phi : J \times X_J \to E$.

PROOF OF LEMMA 6.4 - The proof can be found in refs.[56]. □

Now, if $v \in C^\infty_w(TM)$ we get that locally v is representable by a mapping $\bar{v} : X \subset E \to E$, where X is an open set in a Banach space E, and satisfies the Lipschitz conditions. Therefore, v is integrable. □

The following also gives a criterion to recognize if a vector field on a quantum manifold is integrable.

LEMMA 6.5 - (GENERALIZED HILL-YOSIDA THEOREM)[168]. *Let E be a locally convex, sequentially complete, linear topological space. A linear operator $K \in L(D(E),E)$, with domain $D(E)$ dense in E such that the resolvent $R(n;K) \equiv (r\mathbf{1}-K)^{-1} \in L(E,E)$ exists for $r = 1,2,\cdots$ is the infinitesimal generator of a uniquely determined equi-continuous semi-group of class (C_0) iff the operators $\{(\mathbf{1}-r^{-1}K)^{-s}\}$ are equi-continuous in $r = 1,2,\cdots$ and $s = 0,1,\cdots$.*

THEOREM 6.8 - (CRITERION FOR INTEGRABILITY OF VECTOR FIELDS ON QUANTUM MANIFOLDS).

Let A be a quantum algebra and let M be a quantum manifold over A. Every vector field $v \in C^\infty_w(TM)$ is integrable if the following condition is satisfied.

(i) The mapping $\bar{v} : U \subset E \to E$ locally characterizing v, with E a Fréchet space, and U an open subset in E, is such that $D\bar{v}(p), p \in U$, satisfies the generalized Hille-Yosida lemma. So, the linear operators $D\bar{v}(p) \in L(E;E)$ have flows.

PROOF. The proof follows directly from the above Lemma by taking into account that the flows associated to $D\bar{v}(p)$ vary smoothly with p. □

REMARK 6.15 - Note that if a quantum vector field $v \in Q^\infty_w(\widehat{T}M)$ is such that the corresponding vector field $v(a) \in C^\infty_w(TM)$, given by $v(a)(p) = v(p)(a), \forall p \in M, a \in A$, is integrable, then, we can write a mapping $\phi_a : [0,\epsilon_a a] \to M$, given by

$$\phi_p(ta) = p + \int_{[0,t]} D\phi_p(t'a)(a)dt',$$

with $D\phi_p(t'a)(a) = v(\psi_p(t'))(a)$, where $\psi : J_a \times U \to M$ is the local flow such that $\partial\psi = v(a), J_a \equiv]-\epsilon_a, \epsilon_a[\subset \mathbf{R}$.

2 THEOREM 6.9 - (CRITERION OF INTEGRABILITY OF QUANTUM VECTOR FIELDS). *Let M be a quantum manifold, of dimension $m+n$, over a Banach quantum algebra A. Then, a quantum vector field $v \in Q^{\infty}_{w}(\widehat{T}M)$ is integrable if it is of Lipschitz type, i.e., v is locally represented by a mapping $\bar{v}$: $U \subset M \to \mathcal{L}_{A_0}(A; A^{m+n})$, where $\mathcal{L}_{A_0}(A; A^{m+n}) \equiv \mathcal{L}(A; A^{m+n}) \cap Hom_{A_0}(A, A^{m+n})$, where $\mathcal{L}(E; F)$ is the linear space of all continuous linear maps between the normed linear spaces E and F.*

PROOF. In fact, it is well known [61] that a linear operator $u : E \to F$ between Banach spaces E and F is continuous iff for some $\beta > 0, |u(x)| \leq \beta|x|, x \in E$. The smallest such β is called the norm of u and written $|u|$, that is

$$|u| = \sup_{x \in E} \left\{ \frac{|u(x)|}{|x|}, x \neq 0 \right\}.$$

This quantity serves as Lipschitz constant for u. So $\mathcal{L}(E; F)$ (that is also called the space of **bounded linear operators** $E \to F$) results a normed space. Now, if v is of Lipschitz type, $\bar{v} : U \to \widehat{A}^{m+n}$ has domain and range into normed spaces, and, if we can complete $\mathcal{L}_{A_0}(A; A^{m+n})$ into Banach spaces. Then, for such a situation we can simulate the proof for the integrability of vector fields. More precisely we have the following. Let $B_b(0)$ (resp. $D_a(x_0)$) be the open ball of center $0 \in A$ and radius b into A, (resp. open ball of center $x_0 \in U \subset A^{m+n}$ and radius a into U). Let $a > 0, a > 1$, a real number such that the closed ball $\overline{D}_{3a}(x_0)$ lies in U. Assume that the map $\bar{v} : U \to \widehat{A}^{m+n}$ is bounded by a constant $L \geq 1$ on U and satisfies a Lipschitz condition on U with Lipschitz constant $K \geq 1 :\quad |\bar{v}(x) - \bar{v}(y)| \leq K|x-y|, \forall x, y \in U$. If $b < a/LK$, then for each $x \in \bar{D}_a(x_0)$ there exists a unique flow

$$\phi : B_b(0) \times D_a(x_0) \to U.$$

In fact, set $F \equiv \{\phi \in C^0(\bar{B}_b(0), \bar{D}_{2a}(x_0) \,|\, \phi(0) = x\}$. Then, F is a complete metric space if we define the distance

$$d(\phi, \psi) \equiv \sup_{\lambda \in \bar{B}_b(0)} |\phi(\lambda) - \psi(\lambda)|, \phi, \psi \in F.$$

Set

$$S(\phi)(\lambda) = x + \int_{[0,1]} \bar{v}(\phi(t\lambda))(\lambda) dt, \forall \phi \in F.$$

As $S(\phi)(0) = x, |S(\phi)(\lambda) - x| < a$, and $S(\phi)$ is a continuous mapping, we conclude that S identifies a mapping $S \in L(F;F)$. Furthermore, we have

$$\begin{aligned}|S(\phi)-S(\psi)| &= \sup_{\lambda\in\bar{B}_b(0)} |S(\phi)(\lambda)-S(\psi)(\lambda)| \\ &\leq \sup_{\lambda\in\bar{B}_b(0)} \int_{[0,1]} |\bar{v}(\phi(t\lambda))(\lambda)-\bar{v}(\psi(t\lambda))(\lambda)|dt \\ &\leq \sup_{\lambda\in\bar{B}_b(0)} \sup_{0\leq t\leq 1} |\bar{v}(\phi(t\lambda))(\lambda)-\bar{v}(\psi(t\lambda))(\lambda)| \\ &\leq Kb|\phi-\psi|\,.\end{aligned}$$

Let us, now, consider the following lemma.

LEMMA 6.6 - (THE CONTRACTION THEOREM)[126]. *Let F be a complete metric space with distance function d and let $f: F \to F$ be a mapping of F into itself. Assume that there is a constant $K, 0 < K < 1$, such that for any $(x,y) \in F \times F$ we have $d(f(x), f(y)) \leq Kd(x,y)$. (In this case we say that f is a* **contraction**.*) Then, f has a unique fixed point ($f(x) = x$). Given the point $x_0 \in F$, the fixed point x can be obtained as the following limit $\lim_{n\to\infty} f^n(x_0) = x$, where f^n means iterating f n times.*

Thus, as S is a contraction, we can apply the above lemma and say that S has a unique fixed mapping ϕ that satisfies the integral equation

$$\phi(\lambda) = x + \int_{[0,1]} \bar{v}(\phi(t\lambda))(\lambda)dt.$$

This equation is, of course, locally equivalent to the differential system $\partial\phi = v$. □

REMARK 6.16 - (QUANTUM COORDINATES AND NATURAL BASES). To any quantum coordinate system $(x^A), 1 \leq A \leq m+n$, on $U \subset M$, we can associate a system of quantum coordinate lines

$$x_A : I \times U \to M, x_A(0,p) = p, I \equiv \text{neighbourhood of} \quad 0 \in A, 1 \leq A \leq m+n.$$

Then, $\{\partial x_A(p)\}_{1\leq A\leq m+n}$ is a basis for the right $\widehat{A}$-module, $\widehat{T}_pM$. So any quantum vector field $v \in Q^\infty_w(\widehat{T}M)$ can be locally written as follows:

$$v = \sum_{1\leq A\leq m+n} \partial x_A \circ v^A \equiv \sum_{1\leq A\leq m+n} \partial x_A v^A, \qquad v^A \in Q^\infty_w(U;\widehat{A}).$$

The canonical projection of $v \in \widehat{T}_pM$ on T_pM is the restriction $v_0 \equiv v|_{A_0} : A_0 \to T_pM$. Thus, if $v = \sum_{1\leq A\leq m+n} \partial x_A(p)v^A$ is the representation of v in coordinates (x^A) , we get the following representation of v_0 : $v = \sum_{1\leq A\leq m+n} \partial x_A(p)(v_0^A)$, where $v_0^A \in A$. (But this does not mean that T_pM is a free A-module). Furthermore, any quantum vector field $v : M \to \widehat{T}M$ is a **quantum derivation** between $\mathcal{A}_p(M)$, and $\widehat{\mathcal{A}}_p(M), \forall p \in M$ i.e., a A_0-derivation [1]. In particular, the corresponding vector field $v_0 : M \to TM$ identifies a quantum derivation $\mathcal{A}_p(M) \to \mathcal{A}_p(M)$. Furthermore, to $(x^A), 1 \leq A \leq m+n$, we can also associate a natural basis $\{dx^A(p)\}_{1 \leq A \leq m+n}$ on $(T_pM)^+$ considered as a left $\widehat{A}$-module. So any quantum differential 1-form, $\omega \in Q^\infty_w((TM)^+)$ has the following local representation:

$$\omega = \sum_{1 \leq A \leq m+n} \omega_A \circ dx^A \equiv \sum_{1 \leq A \leq m+n} \omega_A dx^A, \quad \omega_A \in Q^\infty_w(U; \widehat{A}).$$

$\{dx^A(p)\}_{1 \leq A \leq m+n}$ is a basis dual of $\{\partial x_A(p)\}_{1 \leq A \leq m+n}$ in the sense that

$$< \partial x_B(p), dx^A(p) >= \delta^A_B.$$

Moreover, by using Theorem 6.3 one has:

$$\begin{aligned} < v, \omega > &=< \sum_{1 \leq A \leq m+n} \partial x_A \bar{v}^A, \sum_{1 \leq B \leq m+n} \omega_B dx^B > \\ &= \sum_{1 \leq A, B \leq m+n} \omega_B \circ dx^B \circ \partial x_A \circ v^A \\ &= \sum_{1 \leq B \leq m+n} \omega_B \circ v^B \\ &\equiv \sum_{1 \leq B \leq m+n} \omega_B v^B \end{aligned}$$

REMARK 6.17 - Taking into account the local isomorphism

$$\widehat{T}^0_k(T_pM) \cong (\dot{T}^k_0(\widehat{A}))^{(m+n)^k}$$

we get the following local representation of any section $\alpha \in Q^\infty_w(\widehat{T}^0_k M)$:

$$(6.10) \qquad \left\{ \begin{aligned} \alpha = & \sum_{1 \leq A_1, \ldots, A_k \leq m+n} \alpha_{A_1 \ldots A_k} \circ dx^{A_1} \otimes \cdots \otimes dx^{A_k} \\ \equiv & \sum_{1 \leq A_1, \ldots, A_k \leq m+n} \alpha_{A_1 \ldots A_k} dx^{A_1} \otimes \cdots \otimes dx^{A_k} \end{aligned} \right\}$$

where $\alpha_{A_1 .. A_k} \in Q^\infty_w(U, \dot{T}^k_0(\widehat{A})), U \subset M$. In particular, if

$$\alpha = \alpha_1 \otimes \cdots \otimes \alpha_k = (\sum \alpha_{A_1} dx^{A_1}) \otimes \cdots \otimes (\sum \alpha_{A_k} dx^{A_k}),$$

we have $\alpha_{A_1 \ldots A_k} = \underbrace{\alpha_{A_1} \otimes \cdots \otimes \alpha_{A_k}}_{k}$. Furthermore, taking into account the local isomorphism $(\dot{T}^k_0(T_pM))^+ \cong (\widehat{A})^{(m+n)^k}$, we get a local representation of any section

$\alpha \in Q^\infty_w((\dot{T}^k_0 M)^+)$, similar to equation (6.10) but with $\alpha_{A_1..A_k} \in Q^\infty_w(U, \overset{k}{\widehat{A}})$. For the corresponding skew-symmetric and symmetric fields one has respectively the following local representations:

(a)

$$\alpha \in Q^\infty_w(\widehat{\Lambda}^0_k M), \quad \alpha = \sum_{1 \le i_1, \ldots, i_k \le m+n} \alpha_{i_1 \ldots i_k} \circ dx^{i_1} \otimes \cdots \otimes dx^{i_k},$$

with

$$\alpha_{i_1 \ldots i_k} = \frac{1}{k!} \sum_{\sigma \in S_k} \epsilon(\sigma) \alpha^{\sigma(1)}_{i_1} \otimes \cdots \otimes \alpha^{\sigma(k)}_{i_k} \in Q^\infty_w(U, \dot{T}^k_0(\widehat{A})).$$

(b)

$$\alpha \in Q^\infty_w(\widehat{S}^0_k M), \quad \alpha = \sum_{1 \le i_1, \ldots, i_k \le m+n} \alpha_{i_1 \ldots i_k} \circ dx^{i_1} \otimes \cdots \otimes dx^{i_k},$$

with

$$\alpha_{i_1 \ldots i_k} = \frac{1}{k!} \sum_{\sigma \in S_k} \alpha^{\sigma(1)}_{i_1} \otimes \cdots \otimes \alpha^{\sigma(k)}_{i_k} \in Q^\infty_w(U, \dot{T}^k_0(\widehat{A})).$$

(c)

$$\alpha \in Q^\infty_w((\dot{\Lambda}^k_0 M)^+), \quad \alpha = \sum_{1 \le i_1, \ldots, i_k \le m+n} \alpha_{i_1 \ldots i_k} \circ dx^{i_1} \triangle \cdots \triangle dx^{i_k},$$

with

$$\left\{ \begin{array}{l} \alpha_{i_1 \ldots i_k} \in Q^\infty_w(U, \overset{k}{\widehat{A}}) \\ dx^{i_1}(p) \triangle \cdots \triangle dx^{i_k}(p) \equiv dx^{i_1}(p) \otimes \cdots \otimes dx^{i_k}(p)|_{\dot{\Lambda}^k_0(T_p M)}, \quad \forall p \in M \end{array} \right\}$$

(d)

$$\alpha \in Q^\infty_w((\dot{S}^k_0 M)^+), \quad \alpha = \sum_{1 \le i_1, \ldots, i_k \le m+n} \alpha_{i_1 \ldots i_k} \circ dx^{i_1} \bullet \cdots \bullet dx^{i_k},$$

with

$$\left\{ \begin{array}{l} \alpha_{i_1 \ldots i_k} \in Q^\infty_w(U, \overset{k}{\widehat{A}}) \\ dx^{i_1}(p) \bullet \cdots \bullet dx^{i_k}(p) \equiv dx^{i_1}(p) \otimes \cdots \otimes dx^{i_k}(p)|_{\dot{S}^k_0(T_p M)}, \quad \forall p \in M \end{array} \right\}$$

Similarly we also get the following local representations:

(e)

$$\alpha \in Q^\infty_w(\widehat{T}^k_0 M), \quad \alpha = \sum_{1 \le A_1, \ldots, A_k \le m+n} \partial x_{A_1} \otimes \cdots \otimes \partial x_{A_k} \circ \alpha^{A_1 \ldots A_k},$$

where $\alpha^{A_1..A_k} \in Q^\infty_w(U, \dot{T}^k_0(\widehat{A}))$.

(f)

$$\alpha \in Q^{\infty}_w(\widehat{\Lambda}^k_0 M), \quad \alpha = \sum_{1\le i_1,\dots,i_k\le m+n} \partial x_{i_1}\otimes\cdots\otimes\partial x_{i_k}\circ\alpha^{i_1\dots i_k},$$

with

$$\alpha_{i_1\dots i_k} = \frac{1}{k!}\sum_{\sigma\in S_k}\epsilon(\sigma)\alpha^{i_1}_{\sigma(1)}\otimes\cdots\otimes\alpha^{i_k}_{\sigma(k)} \in Q^{\infty}_w(U,\dot{T}^k_0(\widehat{A})).$$

(g)

$$\alpha \in Q^{\infty}_w(\widehat{S}^k_0 M), \quad \alpha = \sum_{1\le i_1,\dots,i_k\le m+n} \partial x_{i_1}\otimes\cdots\otimes\partial x_{i_k}\circ\alpha^{i_1\dots i_k},$$

with $\alpha_{i_1\dots i_k} = \frac{1}{k!}\sum_{\sigma\in S_k}\alpha^{i_1}_{\sigma(1)}\otimes\cdots\otimes\alpha^{i_k}_{\sigma(k)} \in Q^{\infty}_w(U,\dot{T}^k_0(\widehat{A}))$.

REMARK 6.18 - Let N be a quantum manifold modeled on $A^r\times E^s$, where $E\equiv \mathbb{E}(A)$ is a quantum algebra built starting from A by means of some functor $\mathbb{E}$. We say that N is a **quantum manifold of dimension** $r|(E;s)$. For example $\widehat{T}^k_0(M), \dim M = m+n$, is a quantum manifold of dimension $m+n|\big(\dot{T}^k_0(\widehat{A});(m+n)^k\big)$. More precisely, let (x^r) and $(\bar{x}^r)$ be quantum coordinates on M in open neighbourhoods U and $\bar{U}$ respectively of a point $p\in M$, with $x^r = x^r(\bar{x}^s)$ as quantum coordinates transformations. Let $\pi:\widehat{T}^0_k(M)\to M$ be the canonical projection. Then, if $(x^r,\dot{x}^{\alpha_1\dots\alpha_k})$ and $(\bar{x}^r,\dot{\bar{x}}^{\alpha_1\dots\alpha_k})$ are the quantum coordinates on $\pi^{-1}(U)$ and $\pi^{-1}(\bar{U})$ respectively, their quantum coordinates trasformations are

$$\begin{Bmatrix} x^r = x^r(\bar{x}^s) \\ \dot{x}^{\alpha_1\dots\alpha_k} = (\partial\bar{x}_{\beta_1}.x^{\alpha_1})\otimes\cdots\otimes(\partial\bar{x}_{\beta_k}.x^{\alpha_k})\dot{\bar{x}}^{\beta_1\dots\beta_k}\end{Bmatrix}$$

where $\dot{x}^{\alpha_1\dots\alpha_k},\dot{\bar{x}}^{\beta_1\dots\beta_k}$ are $\dot{T}^k_0(\widehat{A})$-valued functions on $\pi^{-1}(U)$ and $\pi^{-1}(\bar{U})$ respectively. In the following we shall also meet more complex situations where quantum manifolds are modeled on $A^r\times E_1^{s_1}\times\cdots\times E_q^{s_q}$, with each $E_q\equiv\mathbb{E}_q(A)$ as before. In such cases we will say that the quantum manifolds are of dimension $r|(E_1;s_1),\dots,(E_q;s_q)$. Then we will also talk about A-valued coordinates (x^r), and E_i-valued ones E_i. Examples of such manifolds are the higher order (jet-) derivative spaces. (See below.) We identify a quantum manifold (over A) of dimension r with one of dimension $r|(E_1;0),\dots,(E_q;0)$. If M is a quantum manifold of dimension $r|(E_1;s_1),\dots,(E_q;s_q)$ the definition of full quantum tangent space at $p\in M$ is, of course, the following:

$$\begin{aligned}\widehat{T}_pM &\equiv Hom_{A_0}(A\times E_1\times\cdots\times E_q, T_pM)\\ &\cong Hom_{A_0}(A,T_pM)\times\Pi_{1\le i\le q}Hom_{A_0}(E_i,T_pM).\end{aligned}$$

Similarly, the definition of full quantum dual space is

$$\begin{aligned}(T_pM)^+ &\equiv Hom_{A_0}(T_pM, A\times E_1\times\cdots\times E_q)\\ &\cong Hom_{A_0}(T_pM,A)\times\Pi_{1\le i\le q}Hom_{A_0}(T_pM,E_i).\end{aligned}$$

REMARK 6.19 - Let M and N be quantum manifolds of dimension $m+n$ and $p+q$ respectively. Let $\{x^A\}$ and $\{y^B\}$ be quantum coordinates on M and N respectively. Let us consider the following canonical A_0-isomorphism:

$$\begin{aligned} j : \widehat{T}_q N \otimes_{\widehat{A}} Hom_{A_0}(\dot{T}^k_0(T_pM), A) &\equiv \widehat{T}_q N \otimes_{\widehat{A}} (\dot{T}^k_0(T_pM))^+ \\ &\cong Hom_{A_0}(\dot{T}^k_0(T_pM), T_qN) \end{aligned}$$

given by

$$v \otimes \alpha \mapsto j(v \otimes \alpha), \quad j(v \otimes \alpha)(u) = v(\alpha(u)).$$

Then, by using the non-canonical isomorphism:

$$Hom_{A_0}(\dot{T}^k_0(T_pM), T_qN) \cong \widehat{A}^{(m+n)^k(p+q)},$$

we get that any vector $X \in Hom_{A_0}(\dot{T}^k_0(T_pM), T_qN)$ can be locally represented by an $(m+n)^k(p+q)$ matrix $X^B_{A_1 \dots A_k} \in \overset{k}{\widehat{A}}$. So we can write:

$$X = \sum_{1 \leq A_i \leq m+n, 1 \leq B \leq p+q} \partial y_B(q) \otimes_{\widehat{A}} X^B_{A_1 \dots A_k} dx^{A_1} \otimes_{A_0} \cdots \otimes_{A_0} dx^{A_k},$$

where

$$X^B_{A_1 \dots A_k} \equiv dy^B \circ X \circ \partial x_{A_1} \otimes \cdots \otimes \partial x_{A_k}.$$

In particular, for any $v = \partial x_{A_1} \otimes \cdots \otimes \partial x_{A_k} v^{A_1 \dots A_k} \in \dot{T}^k_0(T_pM)$ we get

$$X(v) = \partial y_B(q) X^B_{A_1 \dots A_k} v^{A_1 \dots A_k} \equiv \partial y_B(q) \circ X^B_{A_1 \dots A_k} \circ v^{A_1 \dots A_k} \in T_qN.$$

Similar considerations hold for symmetric and skew-symmetric restrictions, i.e., the following isomorphisms of A_0-modules hold:

$$\begin{aligned} \widehat{T}_q N \bigotimes_{\widehat{A}} Hom_{A_0}(\dot{S}^k_0(T_pM), A) &\equiv \widehat{T}_q N \bigotimes_{\widehat{A}} (\dot{S}^k_0(T_pM))^+ \\ &\cong Hom_{A_0}(\dot{S}^k_0(T_pM), T_qN) \\ \widehat{T}_q N \bigotimes_{\widehat{A}} Hom_{A_0}(\dot{\Lambda}^k_0(T_pM), A) &\equiv \widehat{T}_q N \bigotimes_{\widehat{A}} (\dot{\Lambda}^k_0(T_pM))^+ \\ &\cong Hom_{A_0}(\dot{\Lambda}^k_0(T_pM), T_qN), \end{aligned}$$

that allow us to locally represent any

$$Y \in Hom_{A_0}(\dot{S}^k_0(T_pM), T_qN)$$

and

$$Z \in Hom_{A_0}(\dot{\Lambda}_0^k(T_pM), T_qN)$$

respectively as follows:

$$\begin{aligned} Y &= \sum_{1\le A_i\le m+n, 1\le B\le p+q} \partial y_B(q) \otimes_{\widehat{A}} Y^B_{A_1\dots A_k} dx^{A_1} \bullet_{A_0} \cdots \bullet_{A_0} dx^{A_k}, \\ Z &= \sum_{1\le A_i\le m+n, 1\le B\le p+q} \partial y_B(q) \otimes_{\widehat{A}} Z^B_{A_1\dots A_k} dx^{A_1} \triangle_{A_0} \cdots \triangle_{A_0} dx^{A_k} \end{aligned}$$

where

$$\left\{ \begin{array}{l} Y^B_{A_1\dots A_k} \equiv dy^B \circ Y \circ \partial x_{A_1} \otimes \cdots \otimes \partial x_{A_k} \\ Z^B_{A_1\dots A_k} \equiv dy^B \circ Z \circ \partial x_{A_1} \otimes \cdots \otimes \partial x_{A_k} \in Q^\infty_w(U, \overset{k}{\widehat{A}}) \end{array} \right\}$$

and

$$dx^{A_1} \bullet_{A_0} \cdots \bullet_{A_0} dx^{A_k} \equiv dx^{A_1} \otimes_{A_0} \cdots \otimes_{A_0} dx^{A_k}|_{\dot{S}^k_0(T_pM)}$$

$$dx^{A_1} \triangle_{A_0} \cdots \triangle_{A_0} dx^{A_k} \equiv dx^{A_1} \otimes_{A_0} \cdots \otimes_{A_0} dx^{A_k}|_{\dot{\Lambda}^k_0(T_pM)}$$

Note that for abuse of notation we have written $dx^{A_1} \otimes_{A_0} \cdots \otimes_{A_0} dx^{A_k}$ and $\partial x_{A_1} \otimes_{A_0} \cdots \otimes_{A_0} \partial x_{A_k}$ instead of $j(dx^{A_1} \otimes_{A_0} \cdots \otimes_{A_0} dx^{A_k})$ and $j(\partial x_{A_1} \otimes_{A_0} \cdots \otimes_{A_0} \partial x_{A_k})$ respectively, where j is the canonical A_0-homomorphism

$$j : (TM)^+ \otimes_{A_0} \cdots_k \cdots \otimes_{A_0} (TM)^+ \to Hom_{A_0}(\dot{T}^k_0 M, \dot{T}^k_0 A),$$

or

$$j : \widehat{T}M \otimes_{A_0} \cdots_k \cdots \otimes_{A_0} \widehat{T}M \to Hom_{A_0}(\dot{T}^k_0 A, \dot{T}^k_0 M).$$

REMARK 6.20 - As a consequence, for any Q^∞_w-mapping $f : M \to N$ we have the following local representation of the derivative Df :

$$Df = \partial y_B \circ f \otimes_{\widehat{A}} (\partial x_A f^B) dx^A, \quad \text{with} \quad (\partial x_A f^B) \in Q^\infty_w(U, \widehat{A}).$$

For example, if $f \in Q^\infty_w(\widehat{\Lambda}^0_0 M) = Q^\infty_w(M, A)$, we get the following representation for its differential

$$df : M \to \widehat{\Lambda}^0_1 M \cong J\hat{D}(\widehat{\Lambda}^0_0 M), df = \sum_{1\le A\le m+n} (\partial x_A f) dx^A, (\partial x_A f) : U \to \widehat{A}.$$

In particular, if $u : M \to W$ is a section of a quantum fiber bundle $\pi : W \to M, \dim W = (m+n) + (p+q), \dim M = m+n$, we get

$$(6.11) \qquad Du = \partial x_B \otimes (\delta^B_A) dx^A + \partial y_B \circ (\partial x_A u^B) dx^A, (\partial x_A u^B) \in Q^\infty_w(U, \widehat{A}).$$

REMARK 6.21 - Similarly, we can represent a full quantum product structure $\widehat{\mu}$, and a full quantum metric $\widehat{g}$ respectively, in the following way:

$$\widehat{\mu} = \partial x_R \circ \mu \otimes \mu^R_{HK} dx^H, \quad \widehat{g} = g_{HK} dx^H \otimes dx^K, \quad \mu^R_{HK}, g_{HK} \in Q^\infty_w(U, \overset{2}{\widehat{A}}).$$

REMARK 6.22 - Let $u : M \to W$ be a section of a quantum fiber bundle $\pi : W \to M$, where the fiber F is a quantum manifold modeled on E^s, i.e., W is a quantum manifold of dimension $m+n|(E;s)$. One has the following local isomorphism :

$$\begin{aligned} Hom_{A_0}(T_pM, T_{u(p)}W) &\cong Hom_{A_0}(T_pM, T_pM) \times Hom_{A_0}(T_pM, T_qF) \\ &\cong Hom_{A_0}(T_pM, T_pM) \times Hom_{A_0}(T_pM, E)^s \\ &\cong \widehat{A}^{(m+n)^2} \times \widehat{E}^{(m+n)s} \end{aligned}$$

where $\widehat{E} \equiv Hom_{A_0}(A; E)$ is the **full quantum extension** of E. Now, taking into account the canonical isomorphism : $\widehat{T}_qF \bigotimes_{\widehat{A}} (T_pM)^+ \cong Hom_{A_0}(T_pM, T_qF)$, and (x^A, y^B) a fibered quantum coordinate system on W, we see that the derivative of u can be locally written with the same formula (6.11), but with $(\partial x_A u^B) \in Q^\infty_w(U; \widehat{E})$.
REMARK 6.23 - We get natural structures of fiber bundles on $\hat{D}^k(M,N)$ as in the case of (super)manifolds [97,111,112,113]:

$$\pi_{k,h} : \hat{D}^k(M,N) \to \hat{D}^h(M,N), 0 \leq h \leq k$$
$$\pi_k : \hat{D}^k(M,N) \to M, \bar{\pi}_k : \hat{D}^k(M,N) \to N$$

If $\pi : W \to M$ is a fiber bundle between quantum manifolds then into $\hat{D}^k(W)$ we can distinguish some further subbundles homonymous to classic case ones. For example the **holonomic k-jet quantum derivative space** :

$$J\hat{D}^k \equiv \left\{ \begin{array}{l} u \in \hat{D}^k(W) \\ \exists \quad \text{quantum differentiable sections } s \text{ of} \quad \pi \\ D^k s(p) = u, p = \pi_k(u) \end{array} \right\}$$

THEOREM 6.10 - *$J\hat{D}^k(W)$ is a quantum manifold of dimension*

$$m+n+p+q|\{1,(m+n)(p+q)\}\{(2,(m+n)^2(p+q)\}\cdots\{k,(m+n)^k(p+q)\}.$$

THEOREM 6.11 - *$J\hat{D}^k(W)$ is an affine bundle over $J\hat{D}^{k-1}(W)$ with associated vector bundle*

$$\pi^*_{k,0}(\dot{S}^k_0 MOvTW).$$

So we have the following exact sequence of vector bundles over $J\hat{D}^k(W)$:

$$0 \to \pi^*{}_{k,0}(\dot{S}^k_0 MOvTW) \to vTJ\hat{D}^k(W) \to \pi^*{}_{k,k-1} vTJ\hat{D}^{k-1} \to 0.$$

The relation with k-jet derivative spaces is given by the following exact commutative diagram:

$$\begin{array}{ccccccc} & & 0 & & 0 & & 0 & \\ & & \downarrow & & \downarrow & & \downarrow & \\ 0 \to & & \pi^*_{k,0}(\dot{S}^k_0 MOvTW) & \to & vTJ\hat{D}^k(W) & \to & \pi^*_{k,k-1} vTJ\hat{D}^{k-1}(W) & \to 0 \\ & & \downarrow & & \downarrow & & \downarrow & \\ 0 \to & & \pi^*_{k,0} Hom_{\mathbf{K}}(S^k_0 M, vTW) & \to & vTJD^k(W) & \to & \pi^*_{k,k-1} vTJD^{k-1}(W) & \to 0 \end{array}$$

PROOF. A system of quantum coordinates on a k-jet quantum derivative space can be written:

$$\{x^A, y^B, y^B_A, y^B_{A_1A_2}, \dots, y^B_{A_1\dots A_k}\}, 1 \leq A, A_1, \dots, A_k \leq m+n, 1 \leq B \leq p+q,$$

where

$$y^B_{A_1\dots A_r}(D^k s(p)) \equiv (\partial x_{A_r} \dots \partial x_{A_1}.s^B)(p).$$

It is a matrix with entries in $\overset{r}{\widehat{A}}$. Locally we can write:

$$\begin{aligned} D^r s(p) = {} & \partial x_B(p) \otimes_{\widehat{A}} \delta^B_A dx^H(p) \\ & + \partial y_B(s(p)) \otimes_{\widehat{A}} (\partial x_H.s^B)(p) dx^H(p) \\ & + \partial y_B(s(p)) \otimes_{\widehat{A}} (\partial x_{H_1} \partial x_{H_2}.s^B)(p) dx^{H_1}(p) \bullet_{A_0} dx^{H_2}(p) \\ & + \cdots + \partial y_B(s(p)) \otimes_{\widehat{A}} (\partial x_{H_1} \cdots \partial x_{H_r}.s^B)(p) dx^{H_1}(p) \bullet_{A_0} \cdots \bullet_{A_0} dx^{H_r}(p). \end{aligned}$$

In general,

$$(\partial x_{H_1} \cdots \partial x_{H_r}.s^B)(p) \neq (\partial x_{H_{\sigma(1)}} \cdots \partial x_{H_{\sigma(r)}}.s^B)(p),$$

$\forall \sigma \in S_r \equiv$ group of permutations of r objects. (The equality, instead, holds if the restriction on $\dot{T}^r_0(A_0) \cong A_0$ is taken.) Furthermore, one has:

$$\begin{aligned} & (\partial x_{H_r} \cdots \partial x_{H_1}.s^B)(p)(v^{H_1} \otimes \cdots \otimes v^{H_r}) \\ & = (\partial x_{H_{\sigma(r)}} \cdots \partial x_{H_{\sigma(1)}}.s^B)(p)(v^{H_{\sigma(1)}} \otimes \cdots \otimes v^{H_{\sigma(r)}}), \end{aligned}$$

$\forall \sigma \in S_r, (v^{H_i}) \in A^{m+n}$. □

REMARK 6.24 - Let $\pi : W \to M$ be a fiber bundle such that $\dim W = m+n|(E;s), \dim M = m+n$. Then $J\hat{D}^k(W)$ is a quantum manifold of dimension

$$(6.12) \qquad \left\{ \begin{aligned} & m+n|(E;s) + (\overset{1}{\widehat{A}}(E);(m+n)s) + \cdots + (\overset{k}{\widehat{A}}(E);(m+n)^k s) \\ & \equiv m+n|\, \overset{k}{\widehat{A}}(E)(s, m+n) \end{aligned} \right\}$$

i.e., $J\hat{D}^k(W)$ is modeled on

$$A^{m+n} \times E^s \times (\overset{1}{\widehat{A}}(E))^{(m+n)s} \times \cdots \times (\overset{k}{\widehat{A}}(E))^{(m+n)^k s} \equiv A^{m+n} \times \overset{k}{\hat{A}}(E)(s, m+n),$$

where $\overset{k}{\widehat{A}}(E) \equiv Hom_{A_0}(\dot{T}^k_0(A), E), \overset{0}{\widehat{A}}(E) \equiv E$. For example, $J\hat{D}^k(\hat{T}^0_p M)$ is a quantum manifold of dimension of the type (6.12), with $s = (m+n)^p$ and $E = \dot{T}^p_0(\widehat{A})$. Furthermore, $J\hat{D}^\infty(W)$ is modeled on $A^{m+n} \times \overset{\infty}{\widehat{A}}(E)(s, m+n)$, where $\overset{\infty}{\widehat{A}}(E)(s, m+n) \equiv$

$\bigoplus_{k\geq 0} \overset{k}{\widehat{A}}(E)(s, m+n)$. In particular, if $E \equiv A, s = p+q$, we get that $J\hat{D}^k(W)$ has dimension

$$\begin{aligned} &m+n|\overset{k}{\widehat{A}}(E)(p+q, m+n) \\ &= m+n+p+q|\{1,(m+n)(p+q)\}\{(2,(m+n)^2(p+q)\}\cdots\{k,(m+n)^k(p+q)\} \end{aligned}$$

as $\overset{k}{\widehat{A}}(A) = \overset{k}{\widehat{A}}$. We also have $\overset{\infty}{\widehat{A}}(E) \equiv \bigoplus_{k\geq 0} \overset{k}{\widehat{A}}(E)$. Set ${}_k\widehat{A}(E) \equiv \bigoplus_{0\leq i\leq k} \overset{i}{\widehat{A}}(E)$. Then, one has the natural filtration:

$$E \subset \widehat{A}(E) \subset \ldots \subset {}_k\widehat{A}(E) \subset \cdots \subset \overset{\infty}{\widehat{A}}(E).$$

REMARK 6.25 - (LIE QUANTUM DERIVATIVE AND ALGEBRA OF QUANTUM VECTOR FIELDS.) Let $\pi : W \to M$ be a bundle of geometric objects in the category of quantum manifolds. Let $s \in Q^\infty_w(W)$ and $X = \partial\phi : M \to \widehat{T}M$, where $\phi_a, a \in A$, is a (local) group of (local) Q^∞_w-diffeomorphisms of M. Then, similarly to that in the ordinary differential geometry [97], we can define **Lie quantum derivative** of s with respect to $X, \mathcal{L}_X s \equiv \partial\tilde{s} : M \to s^* v\widehat{T}W$, where $\tilde{s}$ is the deformed of s by means of ϕ. One has $\mathcal{L}_X s = \partial(s\circ\phi) + (\partial\hat{\phi})\circ s$, where $\hat{\phi} : A \times M \to W$ is the lifting of ϕ on W.

REMARK 6.26 - In particular, if $s = Y : M \to \widehat{T}M$ is a quantum vector field on M, we get

$$\partial\tilde{Y} : M \to Hom_{A_0}(\dot{T}^2_0(A), TM) \cong \widehat{T}M \bigotimes_{\widehat{A}} (\dot{T}^2_0(A))^+.$$

So $\partial\tilde{Y}$ is a $(\dot{T}^2_0(A))^+$-valued quantum vector field on M. In quantum coordinates one has

$$\partial\tilde{Y} = \partial x_\beta \otimes [(\partial x_\alpha.Y^\beta)X^\alpha - (\partial x_\alpha X^\beta)Y^\alpha],$$

where

$$\left\{ \begin{array}{l} X = \partial x_\alpha X^\alpha , Y = \partial x_\beta Y^\beta , X^\alpha , Y^\alpha \in Q^\infty_w(U, \widehat{A}) \\ [(\partial x_\alpha.Y^\beta)X^\alpha - (\partial x_\alpha X^\beta)Y^\alpha] \in Q^\infty_w(U, \overset{2}{\widehat{A}}), U \subset M \end{array} \right\}$$

Moreover, if we put

$$\partial_\bullet\tilde{Y}(p) \equiv \partial\tilde{Y}(p)|_{A_0} \in Hom_{A_0}(A_0, \widehat{T}_pM) \cong \widehat{T}_pM , \forall p \in M,$$

we see that $\partial_\bullet\tilde{Y} : M \to \widehat{T}M$. In coordinates we have:

$$\left\{ \begin{array}{l} \partial_\bullet\tilde{Y} = \partial x_\beta((\partial x_\alpha.Y^\beta)X^\alpha_\bullet - (\partial x_\alpha X^\beta)Y^\alpha_\bullet) \\ X^\alpha_\bullet(p) \equiv X^\alpha|_{A_0}, Y^\alpha_\bullet(p) \equiv Y^\alpha|_{A_0}, \forall p \in M \end{array} \right\}$$

As a consequence if we define **Lie quantum bracket**, $[X, Y]$, between the quantum vector fields $X = \partial\phi, Y : M \to \widehat{T}M$, as

$$[X,Y] \equiv \mathcal{L}_X Y = \partial\tilde{Y} : M \to \widehat{T}M \bigotimes_{\widehat{A}} (\dot{T}^2_0(A))^+,$$

we see that the **restricted quantum Lie bracket**,$[X,Y]_\bullet \equiv \partial_\bullet \tilde{Y} : M \to \widehat{T}M$, gives a quantum vector field yet. One has: $[X,Y] = -[Y,X]$, $[X,Y]_\bullet = -[Y,X]_\bullet$. Note that the bracket $[X,Y]_\bullet$ satisfies the Jacobi identity too. In fact , a simple calculation in quantum coordinates gives

$$\begin{aligned}
[[X,Y]_\bullet, Z]^\gamma_\bullet + [[Y,Z]_\bullet, X]^\gamma_\bullet + [[Z,X]_\bullet, Y]^\gamma_\bullet =& (\partial x_\omega \partial x_\alpha . X^\gamma)(Y^\alpha_\bullet \otimes Z^\omega_\bullet) \\
&- (\partial x_\alpha \partial x_\omega . X^\gamma)(Z^\omega_\bullet \otimes Y^\alpha_\bullet) \\
&+ (\partial x_\omega \partial x_\alpha . Y^\gamma)(Z^\alpha_\bullet \otimes X^\omega_\bullet) \\
&- (\partial x_\alpha \partial x_\omega . Y^\gamma)(X^\omega_\bullet \otimes Z^\alpha_\bullet) \\
&+ (\partial x_\omega \partial x_\alpha . Z^\gamma)(X^\alpha_\bullet \otimes Y^\omega_\bullet) \\
&- (\partial x_\alpha \partial x_\omega . Z^\gamma)(Y^\omega_\bullet \otimes X^\alpha_\bullet) \\
=& 0 ,
\end{aligned}$$

where $X^\alpha, Y^\alpha, Z^\alpha \in Q^\infty_w(U, \widehat{A})$, and $X^\alpha_\bullet \equiv X^\alpha|_{A_0}, Y^\alpha_\bullet \equiv Y^\alpha|_{A_0}, Z^\alpha_\bullet \equiv Z^\alpha|_{A_0} \in Q^\infty_w(U, A)$.

REMARK 6.27 - It is obvious that $C^\infty_w(TM)$ is a Lie algebra with respect to the Lie bracket $[X,Y] \equiv \mathcal{L}_X Y = \partial\tilde{Y}$, where $X = \partial\phi, Y \in C^\infty_w(TM)$, and $\phi : \mathbf{R} \times M \to M$. Furthermore, $Q^\infty_w(TM)$ is a sub-Lie algebra of $C^\infty_w(TM)$. Denote $\mathcal{L}(M)$ this Lie algebra on $C^\infty_w(TM)$, and by $\hat{\mathcal{L}}(M)$ the corresponding Lie algebra on $Q^\infty_w(TM)$. Let us denote by $\mathcal{D}er(M)$ the algebra of $\mathbf{R}$-derivations of $C^\infty_w(M, \mathbf{R})$. Any vector field $X \in C^\infty_w(TM)$ can be identified with an element of $\mathcal{D}er(M)$. Furthermore, $\hat{\mathcal{L}}(M)$ is a Lie subalgebra of $\mathcal{D}er(M)$, i.e, $[X,Y] = XY - YX, \forall X, Y \in C^\infty_w(TM)$. It is important to note that any quantum vector field $X \in Q^\infty_w(\widehat{T}M)$ is also a derivation, and one has $[X,Y]_\bullet = XY - YX, \forall X, Y \in Q^\infty_w(\widehat{T}M)$. In fact, for any $f \in Q^\infty_w(M, A)$ one has : $[X,Y]_\bullet . f = (\partial x_\alpha . f)[(\partial x_\beta Y^\alpha) X^\beta_\bullet - (\partial x_\beta . X^\alpha) Y^\beta_\bullet]$, and $(XY - YX).f = [X,Y]_\bullet . f + (\partial x_\beta \partial x_\alpha . f)(X^\beta_\bullet \otimes Y^\alpha_\bullet) - (\partial x_\alpha \partial x_\beta . f)(Y^\alpha_\bullet \otimes X^\beta_\bullet) = [X,Y]_\bullet . f$. So we can identify the bracket $[X,Y]_\bullet$ with $XY - YX$. Therefore, we can adopt the definition $[X,Y]_\bullet \equiv XY - YX$ even if X and Y are not integrable quantum vector fields. Then, the bracket $[,]_\bullet$ defines on $Q^\infty_w(\widehat{T}M)$ an anticommutative algebra that we shall denote by $\widehat{\mathcal{L}}(M)$. One has the following commutative diagrams of homomorphisms of A_0-modules and Lie algebras:

$$\begin{array}{ccc} Q^\infty_w(\widehat{T}M)\bigotimes_{A_0} Q^\infty_w(\widehat{T}M) & \stackrel{[,]}{\to} & Q^\infty_w(\widehat{T}M)\bigotimes_{\widehat{A}}(\dot{T}^2_0(A))^+ \\ \| & & \downarrow \\ Q^\infty_w(\widehat{T}M)\bigotimes_{A_0} Q^\infty_w(\widehat{T}M) & \stackrel{[,]_\bullet}{\to} & Q^\infty_w(\widehat{T}M) \\ & & \downarrow \\ & & 0 \end{array}$$

(In the above diagram the last column is exact).

$$\begin{array}{ccccc} & & 0 & & \\ & & \uparrow & & \\ 0 & \to & \hat{\mathcal{L}}(M) & \to & \mathcal{L}(M) \\ & & \uparrow & & \\ & & \widehat{\mathcal{L}}(M) & & \end{array}$$

REMARK 6.28 - A **Lie quantum group** G, over A, of dimension $m+n$ and class Q^∞_w, is defined to be simultaneously a quantum manifold over A of dimension $m+n$ and a topological group such that the mapping $G \times G \to G, (f,g) \mapsto f \circ g^{-1}$ is quantum differentiable of class Q^∞_w. The set of left-invariant C^∞_w vector fields on G forms the **Lie algebra** $A(G)$ of G, i.e., a Lie algebra with respect to the **K**-bilinear operation $[X,Y]$ induced by the Lie algebra $\mathcal{L}(G)$. Furthermore, we denote by $\hat{A}(G)$ the Lie sub-algebra of $A(G)$ of left-invariant Q^∞_w vector fields on G. We call $\hat{A}(G)$ the **quantum Lie algebra** of G. Moreover, we will denote by $\widehat{A}(G)$ the Lie algebra of left-invariant Q^∞_w quantum vector fields on G induced by $\widehat{\mathcal{L}}(G)$. We call $\widehat{A}(G)$ the **full quantum Lie algebra** of G. If $\{Z_k\}_{1\le k\le m+n}$ is a basis for $\widehat{A}(G)$, one has $[Z_j,Z_k]_\bullet = Z_i C^i_{jk}$, where $C^i_{jk} \in \widehat{A}$ satisfy the following relations :

$$\begin{gathered} C^i_{jk} + C^i_{kj} = 0, \\ C^r_{ij}C^s_{hk} + C^r_{hj}C^s_{ki} + C^r_{kj}C^s_{ih} = 0, \end{gathered}$$

and are called **quantum structure constants** of G. One has the following commutative and exact diagram of A_0-modules and Lie algebras:

$$\begin{array}{ccccc} & & 0 & & \\ & & \uparrow & & \\ 0 & \to & \hat{A}(G) & \to & A(G) \\ & & \uparrow & & \\ & & \widehat{A}(G) & & \end{array}$$

Let M and P be quantum manifolds, G a Lie quantum group acting quantum differentially on P , and $\pi : P \to M$ a surjective quantum differentiable mapping. The **quantum principal bundle** $(P,M,\pi;G)$ is defined similarly as on ordinary manifolds, but taking into account that now we are working in the category of quantum manifolds. A **quantum principal connection** is a G-compatible quantum linear (i.e., $\widehat{A}$-linear) splitting of the following exact sequence of bundles of right $\widehat{A}$-modules over P: $0 \to v\widehat{T}P \to \widehat{T}P \to \pi^*\widehat{T}M \to 0$. A principal connection gives a splitting $\widehat{T}_pP \cong v\widehat{T}_pP \bigoplus \widehat{H}_p$, where $\widehat{H}_p$ is the horizontal subspace at $p \in P$ such that the following conditions hold: 1) $(\widehat{H}_p)_a = (R_a)^*(\widehat{H}_p)$, for any $p \in P$ and $a \in G$, where R_a is the transformation in P induced by the element $a \in G$, namely $R_ap = pa$; 2) $\widehat{H}_p$ depends quantum differentially on p. Each vector field $X \in \widehat{A}(G)$ induces a fundamental quantum vector field $\sigma(X)$ on P, and $\sigma : X \to \sigma(X)(p)$ is an isomorphism of right $\widehat{A}$-modules $\widehat{A}(G) \cong v\widehat{T}_pP$. For any $Y \in \widehat{T}_pP$, let $\hat{\omega}(Y)$ be the unique $X \in \widehat{A}(G)$ such that $\sigma(X)(p)$ is the vertical component for Y. We call the map $\hat{\omega} : P \to Hom_{\widehat{A}}(\widehat{T}P, \widehat{A}(G))$ **quantum Ehresmann connection** of the given principal quantum connection on P. Similarly we can define a **quantum principal connection** on a quantum principal bundle $(P, M, \pi; G)$ as a G-compatible quantum linear-splitting of the following exact sequence of bundles of A_0-modules over P: $0 \to vTP \to TP \to \pi^*TM \to 0$. Then, the corresponding **quantum Ehresmann connection** $\dot{\omega}$ is a A_0-homomorphism $TP \to A(G)$ over $P : \dot{\omega} : P \to TPOA(G)$. Standing the canonical isomorphisms of A_0-modules $Hom_{\widehat{A}}(\widehat{T}P, \widehat{A}(G)) \cong \widehat{A}(G) \bigotimes_{\widehat{A}} (\widehat{T}P)^+ \cong Hom_{A_0}(TP, A(G))$, we get that any principal quantum connection $\hat{\omega}$ identifies a principal quantum connection $\dot{\omega}$ on P and vice versa.

REMARK 6.29 - (QUANTUM SUBMANIFOLDS AND IMPLICIT QUANTUM FUNCTION THEOREM). Let M be a smooth quantum manifold (i.e., a quantum manifold of class Q^∞_w) of dimension $m+n$. A subset $S \subset M$ is called a **quantum submanifold** of M if for every point $p \in M$ there exists a chart $(W, \chi, A^{p+r} \times A^{q+s})$ on $M, p \in W$, with quantum coordinates $(x^A, y^B), (x^A) : W \to A^{p+r}, (y^B) : W \to A^{q+s}, p+q = m, r+s = n$, such that in chart

$$\chi(S \cap W) = \{(u, v) \in \chi(W) \subset A^{p+r} \times A^{q+s} : (v^B = 0) \in A^{q+s}\}.$$

In this case there is the natural smooth quantum manifold structure on S induced from M. Namely, to every chart (W, χ, A^{m+n}) on M there corresponds the chart (U, ϕ, A^{p+r}) on S, with $U \equiv S \cap W, \phi = \pi \circ \chi$, where π is the natural projection $\pi : A^{m+n} \to A^{p+r}$. We can locally define a quantum submanifold S of M as a solution set of a system of equations:

$$S \equiv \{p \in M : F^\alpha(p) = 0, (F^\alpha) : M \to A^{a+b}\}.$$

In a chart (W, χ, A^{m+n}) on M with quantum coordinates $\{\xi^C\}_{1 \le C \le m+n}$, the above equations take the form

$$S \equiv \{w \in \chi(W) \subset A^{m+n} : \Phi^\alpha(w) = 0\},$$

where

$$(\Phi^\alpha) \equiv \left(F^\alpha \circ \chi^{-1}|_{\chi(W)} : \chi(W) \to A^{a+b}\right).$$

Such set will be a quantum submanifold of M if in a neighbourhood of each point $w \in W$ the coordinates ξ^C can be split into two classes $(\xi^C) = (x^A, y^B)$, in a way that the system $\{\Phi^\alpha = 0\}$ will be solvable, in this neighbourhood, with respect to the coordinates (y^B), i.e., if it can be written in the following form

$$S = \{y^B = f^B(x^A)\}.$$

If

$$\Phi : U_1 \times U_2 \subset A^{p+r} \times A^{q+s} \to A^{a+b}, \Phi(u_0, v_0) = 0, (u_0, v_0) \in U_1 \times U_2, a+b = q+s$$

and the matrix $(\partial y_B.\Phi^\alpha)(u,v)$ is continuous and invertible in $\chi(W) = U_1 \times U_2 \subset A^{p+r} \times A^{q+s}$, there exist a neighbourhood $U' \subset U_1$ of u_0 and a (unique) continuous function $f : U' \to A^{q+s}$ such that $v_0 = f(u_0)$ and $\Phi(u, f(u)) = 0$. So, if there is a derivative $(\partial x_A.\Phi^\alpha)(u,v)$ in $U_1 \times U_2$, then the explicit function $\left(y^B = f^B(x^A)\right)$ is differentiable at U' and we have:

$$(\partial x_A.f^B)(u) = -(\partial y_S.\Phi^{-1\,B})(\partial x_A.\Phi^S)(u,v)(u,v), v = f(u).$$

Generalizations of the above considerations to quantum manifolds of dimension $m + n|(E_1; s_1)\cdots(E_q; s_q)$ are easily obtained.

REMARK 6.30 - (QUANTUM INVOLUTIVE QUANTUM DISTRIBUTIONS). Let M be a smooth quantum manifold. A **quantum distribution** of dimension $(m+n)$ on M is a smooth subbundle $\widehat{\mathbb{D}} \subset \widehat{T}M$ on M, full quantum extension of some distribution $\mathbb{D} \subset TM$ of dimension $(m+n)$ on M:

$$\widehat{\mathbb{D}} = Hom_{A_0}(A, \mathbb{D}) \equiv A O \mathbb{D}.$$

A **(maximal) integral quantum submanifold** of $\widehat{\mathbb{D}}$ is a connected quantum submanifold S of M such that $\widehat{T}_pM \subseteq \widehat{\mathbb{D}}_p, \forall p \in S$, (equality implies maximality). Then, one has $\dim S = m+n$, and the quantum submanifold S defined by equations

$$S \equiv \{p \in M : F^\alpha(p) = 0, \alpha \in I\}$$

is integral iff

$$\zeta.F^\alpha \equiv <\zeta, dF^\alpha> = (\partial x_\beta.F^\alpha) \circ \zeta^\beta = 0 \quad \text{on} \quad S, \forall \zeta \in Q^\infty_w(\widehat{\mathbb{D}}), \alpha \in I.$$

A quantum distribution $\widehat{\mathbb{D}} \subset \widehat{T}M$ is **quantum involutive** if

$$[X,Y]_{\bullet} \in Q^{\infty}_{w}(\widehat{\mathbb{D}}), \forall X, Y \in Q^{\infty}_{w}(\widehat{\mathbb{D}}).$$

An $(m+n)$-dimensional quantum distribution $\widehat{\mathbb{D}} \subset \widehat{T}M$ can be defined as the quantum extension of the common null subbundle $\mathbb{D}$ of some subbundle $\Theta \subset (TM)^{+}$:

$$\widehat{\mathbb{D}} = \bigcup_{p\in M} \widehat{\mathbb{D}}_p, \widehat{\mathbb{D}}_p \equiv A \otimes \mathbb{D}_p, \mathbb{D}_p \equiv \{\zeta \in T_pM :< \zeta, \omega(p) >= 0, \forall \omega \in Q^{\infty}_{w}(\Theta)\}.$$

$\widehat{\mathbb{D}}$ is quantum involutive iff

$$d\omega(\zeta,\eta) = 0, \forall \omega \in Q^{\infty}_{w}(\Theta) \quad \text{and} \quad \zeta, \eta \in Q^{\infty}_{w}(\mathbb{D}),$$

where d is the exterior quantum differential of ω (see next section for the definition).
THEOREM 6.12 - (QUANTUM FROBENIUS THEOREM). *If the quantum distribution $\widehat{\mathbb{D}} \subset \widehat{T}M$ is quantum involutive and if its Q^{∞}_{w}-vector fields are integrable, then for each point $p \in M$ passes a unique maximal integral quantum manifold of $\widehat{\mathbb{D}}$. In this case a set $S \subset W$ is an integral quantum manifold iff*

$$S = \{x, y : y^B = c^B, B \in I'\},$$

where c^B are some constants. In other words, integral quantum submanifolds are ***folios*** *of some chart.*

PROOF. In fact, under our hypotheses we can reproduce analogous standard proof for distributions on ordinary manifolds. □

REMARK 6.31 - (QUANTUM DIFFIETIES). Quantum diffieties are diffieties [152] in the sense of quantum. These structures are strictly related to quantum PDEs. In fact infinite quantum prlongations of quantum PDEs are quantum diffieties (see Chapter 2). More precisely, a **quantum diffiety** is a pair $\mathcal{M} \equiv (M, \widehat{C}(M))$, where M is a smooth quantum manifold and $\widehat{C}(M) \subset \widehat{T}M$ is an involutive quantum distribution on M, of dimension $m+n$, called **Cartan quantum distribution**. We call **Cartan quantum dimension** of $\mathcal{M}$ the dimension $m+n$ of $\widehat{C}(M)$. We call **Cartan quantum fields** local quantum smooth sections of the bundle $\widehat{C}(M) \to M$. Let $\{x^A, y^i\}_{1\le A\le m+n, i\in I}$ be a quantum coordinate system on $U \subset M$. Let $(\zeta_1, \dots, \zeta_{m+n})$ be basis vector fields of $\widehat{C}(M)$ on $U \subset M$. The system of quantum coordinates $\{x^A, y^i\}$ is called **natural** if

$$\zeta_A = \partial x_A + \sum_{i\in I} \partial y_i A^i_A \equiv \partial_A, 1 \le A \le m+n,$$

where $A^i_A : U \to \widehat{A}$ are Q^∞_w local functions.

THEOREM 6.13 - *Let $C(M)$ be a distribution of dimension $m+n$ on M. Then $\widehat{C}(M)$ is a quantum diffiety, i.e. involutive quantum distribution, iff*

$$(\partial_B . A^i_A)(Y^B_\bullet \otimes X^A_\bullet) - (\partial_A . A^i_B)(X^A_\bullet \otimes Y^B_\bullet) = 0, \qquad A, B = 1, \dots, m+n, i, j \in I.$$

This implies $(\partial_B . A^i_A)|_{A_0} - (\partial_A . A^i_B)|_{A_0} = 0$.

PROOF. We shall use the following two lemmas.

LEMMA 6.7 - *Let X, Y, Z quantum manifolds of class Q^∞_w. Let $\mu : Q^\infty_w(Y, Z) \times Q^\infty_w(X, Y) \to Q^\infty_w(X, Z)$ the composition mapping, i.e., $\mu(g, f) = g \circ f$, for any $g \in Q^\infty_w(Y, Z), f \in Q^\infty_w(X, Y)$. Then μ is a quantum differentiable mapping and one has*

$$T(\mu) : Q^\infty_w(Y, TZ) \times Q^\infty_w(X, TY) \to Q^\infty_w(X, TZ), \qquad D\mu(g, f)(t, s) = T(g) \circ s + t \circ f.$$

PROOF OF LEMMA 6.7 - It is a natural extension of similar situation in ordinary smooth manifolds [29]. □

LEMMA 6.8 - (LEIBNITZ RULE IN $Q^\infty_w(M, \widehat{A})$). *If $f : M \to \widehat{A}$ is a Q^∞_w-mapping such that the following diagram is commutative*

$$\begin{array}{cccccc} & M & \xrightarrow{f} & \widehat{A} & \\ (f_2, f_1) & \downarrow & & \uparrow & \mu \\ & \widehat{A} \times \widehat{A} & = & \widehat{A} \times \widehat{A} & \end{array}$$

where μ is the product given by composition, i.e., $\mu(\hat{b}, \hat{a}) = \hat{b} \circ \hat{a}$, for any $\hat{b}, \hat{a} \in \widehat{A}$, we get that the derivative of f is given by the following formula:

$$Df(p)(v) = f_2(p) \circ Df_1(p)(v) + Df_2(p)(v) \circ f_1(p).$$

In particular, if the local representation of v is $v = \partial x_\mu(v^\mu), v^\mu \in A$, we get the following local representation of $Df(p)(v)$:

$$Df(p)(v) = f_2(p) \circ (\partial x_\mu . f_1)(p)(v^\mu) + (\partial x_\mu . f_2)(p)(v^\mu) \circ f_1(p)$$

where $(\partial x_\mu . f_i)(p) \in Hom_{A_0}(A, \widehat{A}), f_i(p) \in \widehat{A}, i = 1, 2$.

PROOF OF LEMMA 6.8 - It is a direct consequence of the above lemma taking into account that $f_i(p), i = 1, 2$, are homomorphisms of A_0-modules, so that their derivatives coincide with $f_i(p), i = 1, 2$. □

Now, let us consider $X, Y \in \widehat{C}(M)$. Let us calculate the bracket $[X, Y]_\bullet$ applied to any $f \in Q^\infty_w(M, A)$. By using the above lemma we can write

$$\left\{ \begin{array}{l} [X, Y]_\bullet . f = (\partial_B \partial_A . f)(Y^B_\bullet \otimes X^A_\bullet) + (\partial_B . f) \circ (\partial_A . Y^B)(X^A_\bullet) \\ \qquad - (\partial_A \partial_B . f)(X^A_\bullet \otimes Y^B_\bullet) \\ \qquad - (\partial_A . f) \circ (\partial_B . X^A)(Y^B_\bullet) \end{array} \right\}$$

So $\widehat{C}(M)$ is involutive iff $[X,Y]_{\bullet}.f = (\partial_C.Z^C).f = (\partial_C.f)Z^C \Leftrightarrow (\partial_B\partial_A.f)(Y_{\bullet}^B \otimes X_{\bullet}^A) - (\partial_A\partial_B.f)(X_{\bullet}^A \otimes Y_{\bullet}^B) = 0$. Then a simple calculation show that this is equivalent to saying that $(\partial_B.A_A^i)(Y_{\bullet}^B \otimes X_{\bullet}^A) - (\partial_A.A_B^i)(X_{\bullet}^A \otimes Y_{\bullet}^B) = 0$. □

An $m+n$-dimensional smooth quantum submanifold $S \subset M$ is integral iff it has a local representation

$$S \equiv \{x, y : y^i = s^i(x^1, \cdots, x^{m+n}), i \in I\} \subset M$$

where the quantum smooth function $s^i(x^1, \ldots, x^{m+n})$ satisfy the contact equations

$$\zeta.(y^i - s^i) = 0, \forall \zeta \in Q_w^\infty(\widehat{C}(M)), i \in I, \quad \text{i.e. iff} \quad (\partial x_A.s^i) - A_A^i = 0.$$

A quantum diffiety

$$\mathcal{F} \equiv (F, \widehat{C}(F))$$

is called **quantum subdiffiety** of $\mathcal{M} \equiv (M, \widehat{C}(M))$ if F is a quantum submanifold of M and $\widehat{C}(F) = \widehat{T}F \cap \widehat{C}(M)|_F$. The imbedding $i : \mathcal{F} \to \mathcal{M}$ is a Lie-immersion. If the Cartan quantum dimension of $\mathcal{F}$ and $\mathcal{M}$ coincide, then $\mathcal{F}$ is called a **reduction** of the quantum diffiety $\mathcal{M}$. In this case any integral quantum submanifold of $\mathcal{F}$ is so for all $\mathcal{M}$, and conversely every integral quantum submanifold of $\mathcal{M}$, lying in F, is an integral quantum submanifold of $\mathcal{F}$. Let $\mathcal{M} \equiv (M, \widehat{C}(M))$ be a quantum diffiety, then $\mathcal{F} \equiv (F, \widehat{C}(M)|_F)$ will be a reduction of $\mathcal{M}$ iff $\zeta : \mathcal{I}_F \to \mathcal{I}_F$ for every Cartan field $\zeta \in Q_w^\infty(\widehat{C}(M))$, where $\mathcal{I}_F \equiv$ defining two-sided ideal of F, i.e., the set of all functions $f \in Q_w^\infty(M; A)$ such that $f|_F = 0$.

REMARK 6.32 - Every quantum vector field $\zeta \in Q_w^\infty(\widehat{T}M)$ can be locally represented as follows:

$$\zeta = \sum_{1 \le \mu \le m+n} \partial_\mu \circ \xi^\mu + \sum_{i \in I} \partial y_i \circ \eta^i$$

where $\xi^\mu, \eta^i : U \to \widehat{A}$ are local Q_w^∞-functions. Set

$$\delta y^i \equiv dy^i - \sum_{1 \le \mu \le m+n} A_\mu^i \circ dx^\mu, i \in I$$

Then every 1-form $\omega \in Q_w^\infty((TM)^+)$ has the unique representation

$$\omega = \sum_{1 \le \mu \le m+n} \phi_\mu \circ dx^\mu + \sum_{i \in I} \chi_i \circ \delta y^i$$

where $\phi_\mu, \chi_i : U \to \widehat{A}$ are local Q_w^∞-functions. One has

$$< \partial_\mu, \delta y^i >= 0.$$

The common null subbundle of the 1-form $\delta y^i, i \in I$, is the restriction of the distribution $C(M)$ on U. Let $f : M \to A$ be a Q^∞_w-function on M. Then we can write

$$df = \sum_{1\le\mu\le m+n} (\partial_\mu . f)dx^\mu + \sum_{i\in I}(\partial y_i . f)\delta y^i$$

where we have forgotten the symbol of composition.

6.2 - COHOMOLOGY OF QUANTUM MANIFOLDS

Scope of this section is to explicitly characterize the (co)homological structures of quantum manifolds. This is necessary in order to develop a formal theory of quantum PDEs.

Let M be a quantum manifold modeled on A^{m+n}. Set

$$\Omega^\bullet(M) \equiv \bigoplus_{p\ge 0} \Omega^p(M), \text{where} \quad \Omega^p(M) \equiv C^\infty_w((\Lambda^p_0 M)^*).$$

The **de Rham complex** on M is given by

$$0 \to \mathbf{R} \to \Omega^0(M) \xrightarrow{d} \Omega^1(M) \xrightarrow{d} \Omega^2(M) \to \cdots$$

where d is the **exterior differential** defined by:

$$(6.13) \quad \left\{ \begin{aligned} d\omega(\zeta_0,\dots,\zeta_q) = & \frac{1}{q+1}\sum_{\lambda=0}^{q}(-1)^\lambda\omega(\zeta_0,\dots,\hat{\zeta}_\lambda,\dots,\zeta_q) \\ & + \frac{1}{q+1}\sum_{\lambda<\mu}(-1)^{\lambda+\mu}\omega([\zeta_\lambda,\zeta_\mu],\zeta_0,\dots,\hat{\zeta}_\lambda,\dots,\hat{\zeta}_\mu,\dots,\zeta_q) \end{aligned} \right\}$$

for all fields $\zeta_0,\dots,\zeta_q : M \to TM$. Here the terms $\hat{\zeta}_\lambda$ and $\hat{\zeta}_\mu$ are omitted.
This is a first order differential operator, i.e., one has the following commutative diagram

$$\begin{array}{ccc} J\hat{D}((\Lambda^q_0 M)^*) & \longrightarrow & (\Lambda^{q+1}_0 M)^* \\ {\scriptstyle D\alpha}\uparrow & & \uparrow{\scriptstyle d\alpha} \\ M & \cong & M \end{array}$$

for any $\alpha \in \Omega^q(M)$. We will denote by $H^\bullet(M)$ the homology (**de Rham cohomology**) corresponding to de Rham complex of M. One has the following properties for the exterior product and exterior derivative.

(a)

$$\omega \wedge \theta = (-1)^{qp}\theta \wedge \omega, \forall\omega \in \Omega^q(M), \theta \in \Omega^p(M);$$

(b)

$$d(\omega \wedge \theta) = (d\omega) \wedge \theta + (-1)^q \omega \wedge (d\theta);$$

(c)

$$\zeta\rfloor(\omega \wedge \theta) = (\zeta\rfloor\omega) \wedge \theta + (-1)^q \omega \wedge (\zeta\rfloor\theta), \forall \zeta \in C^\infty(TM);$$

(d)

$$\zeta_1(\zeta_2\rfloor\omega) + \zeta_2(\zeta_1\rfloor\omega) = 0, \forall \zeta_1, \zeta_2 \in C^\infty(TM);$$

(e)

$$\left\{ \begin{array}{l} (\mathcal{L}_\zeta\omega)(\zeta_1, \ldots, \zeta_q) = \zeta.\omega(\zeta_1, \ldots, \zeta_q) - \sum\limits_{1\le\lambda\le q} \omega(\zeta_1, \ldots, [\zeta, \zeta_\lambda], \ldots, \zeta_q) \\ \forall \zeta_i \in C^\infty(TM), i = 1, \ldots, q \end{array} \right\}$$

(f)

$$\mathcal{L}_\zeta(\omega \wedge \theta) = (\mathcal{L}_\zeta\omega) \wedge \theta + \omega \wedge (\mathcal{L}_\zeta\theta);$$

(g)

$$\mathcal{L}_{\zeta_1}(\mathcal{L}_{\zeta_2}\omega) - \mathcal{L}_{\zeta_2}(\mathcal{L}_{\zeta_1}\omega) = \mathcal{L}_{[\zeta_1,\zeta_2]}\omega;$$

(h)

$$\mathcal{L}_{\zeta_1}(\zeta_2\rfloor\omega) = [\zeta_1, \zeta_2]\rfloor\omega + \zeta_2\rfloor(\mathcal{L}_{\zeta_1}\omega);$$

(i)

$$\mathcal{L}_\zeta\omega = d(\zeta\rfloor\omega) + \zeta\rfloor(d\omega);$$

(l) For any smooth mapping $f : M \to M$ between quantum manifolds, one has

$$(f^*\omega)(\zeta_1, \ldots, \zeta_q) = \omega(f_*\zeta_1, \ldots, f_*\zeta_q), \quad f^*(\omega\wedge\theta) = (f^*\omega)\wedge(f^*\theta), \quad d(f^*\omega) = f^*(d\omega);$$

(m) For any local one-parameter transformation group $\{\phi_t\}_{t\in\mathbf{R}}$ on M, a smooth vector field $\zeta = \partial\phi$ is defined. The converse is generally a false statement. (See Theorems 6.7 and 6.8.)

REMARK 6.33 - One has the following commutative diagram

$$\begin{array}{ccccc}
0 & & (\dot\Lambda^q_0 M)^+ & \equiv & Hom_{A_0}(\dot\Lambda^q_0 M, A) \\
\uparrow & & & & \\
(\Lambda^q_0 M)^* \bigotimes_{\mathbf{K}}(A/\mathbf{K}) & & \downarrow & & \downarrow \\
\uparrow & & & & \\
(\Lambda^q_0 M)^* \bigotimes_{\mathbf{K}} A & \to & (\Lambda^q_0 M)^\star \equiv Hom_{\mathbf{K}}(\Lambda^q_0 M, A) & \leftarrow & Hom_{A_0}(\Lambda^q_0 M, A) \\
\uparrow & & \uparrow & & \\
(\Lambda^q_0 M)^* & = & (\Lambda^q_0 M)^* & & \\
\uparrow & & \uparrow & & \\
0 & & \Lambda^0_q M & &
\end{array}$$

(ATTENTION: Do not confuse the symbol "star" with "asterisk"!!).
Set:

$$\hat{\Omega}^q(M) \cong C^\infty_w(Hom_{\mathbf{K}}(\Lambda^q_0 M, A))$$
$$\overset{\bullet}{\Omega}{}^q(M) \cong C^\infty_w(Hom_{\mathbf{K}}(\Lambda^q_0 M, A/\mathbf{K})), \widehat{\Omega}^q(M) \cong Q^\infty_w((\dot{\Lambda}^q_0 M)^+)$$

Then the spaces $\Omega^q(M), \overset{\bullet}{\Omega}{}^q(M), \widehat{\Omega}^q(M)$ can be represented in $\hat{\Omega}^q(M)$. One has the following cochain complex (**de Rham quantum complex** of M):

$$0 \to A \to \hat{\Omega}^0(M) \xrightarrow{d} \hat{\Omega}^1(M) \xrightarrow{d} \hat{\Omega}^2(M) \to \cdots$$

where d is the **quantum exterior differential**, formally defined as in (6.13). One has the following commutative diagram:

$$\begin{array}{ccccccccc} 0 & \to & \Omega^q(M) & \to & \hat{\Omega}^q(M) & \to & \overset{\bullet}{\Omega}{}^q(M) & \to & 0 \\ & & d\downarrow & & d\downarrow & & d\downarrow & & \\ 0 & \to & \Omega^{q+1}(M) & \to & \hat{\Omega}^{q+1}(M) & \to & \overset{\bullet}{\Omega}{}^{q+1}(M) & \to & 0 \end{array}$$

that defines an exterior differential $d : \overset{\bullet}{\Omega}{}^q(M) \to \overset{\bullet}{\Omega}{}^{q+1}(M)$. Furthermore, for differential forms of $\hat{\Omega}^\bullet(M)$ the same formulas like those reported in equations (a),...,(m) for differential forms belonging to $\Omega^\bullet(M)$ hold. Another important complex is obtained taking into account that $d : \widehat{\Omega}^q(M) \to \widehat{\Omega}^{q+1}(M)$. So we get the following commutative diagram

$$\begin{array}{ccccccccccc} 0 & \to & A & \to & \widehat{\Omega}^0(M) & \xrightarrow{d} & \widehat{\Omega}^1(M) & \xrightarrow{d} & \widehat{\Omega}^2(M) & \xrightarrow{d} & \cdots \\ & & \| & & \downarrow & & \downarrow & & \downarrow & & \\ 0 & \to & A & \to & \hat{\Omega}^0(M) & \xrightarrow{d} & \hat{\Omega}^1(M) & \xrightarrow{d} & \hat{\Omega}^2(M) & \xrightarrow{d} & \cdots \end{array}$$

We will call the above upper complex the **de Rham full quantum complex** of M.
REMARK 6.34 - Taking into account that a quantum manifold M , of dimension $m+n$, is locally a Fréchet space, we can always consider $\mathbf{K}$-valued local coordinates $\{x^i\}_{i\in J}$, where J is a (finite or infinite) countable set. So if $\alpha \in \Omega^p(M)$ we can locally represent α in the form

$$\alpha = \sum_{i_1<\ldots<i_p, i_s\in J} \alpha_{i_1\ldots i_p} dx^{i_1} \wedge \cdots \wedge dx^{i_p} \tag{6.14}$$

where $\alpha_{i_1\ldots i_p} \in C^\infty_w(U, \mathbf{K})$, are skew-symmetric in the indexes. Thus the differential $d\alpha$ can be written

$$d\alpha = \sum_{i_1<\ldots<i_p, i_s\in J} (\partial x_{[\alpha}.\alpha_{i_1\ldots i_p]}) dx^\alpha \wedge dx^{i_1} \wedge \cdots \wedge dx^{i_p}. \tag{6.15}$$

Furthermore, if $\alpha \in \hat{\Omega}^p(M)$ we can also represent α in the above form but with $\alpha_{i_1 \dots i_p} \in C^\infty_w(U, A)$, skew-symmetric in the indexes. Finally, if $\alpha \in \widehat{\Omega}^p(M)$ we have similar representations like before, but with functions of class Q^∞_w. However, a more interesting local representation of such differential forms can be obtained by means of quantum coordinates $\{x^A\}_{A \in \{1,\dots,m+n\}}$, on M. So if $\alpha \in \widehat{\Omega}^p(M)$, one has

$$(6.16) \qquad \alpha = \sum_{1 \leq A_1, \dots, A_p \leq m+n} \alpha_{A_1 \dots A_p} dx^{A_1} \triangle \cdots \triangle dx^{A_p}$$

where $\alpha_{A_1 \dots A_p} \in C^\infty_w(U, \overset{p}{\hat{A}})$. Then $d\alpha$ is locally written in the following form

$$(6.17) \qquad d\alpha = \sum_{1 \leq B, A_1, \dots, A_p \leq m+n} (\partial x_B . \alpha_{A_1 \dots A_p}) dx^B \triangle dx^{A_1} \triangle \cdots \triangle dx^{A_p}$$

with $(\partial x_B . \alpha_{A_1 \dots A_p}) \in C^\infty_w(U, \overset{p+1}{\hat{A}})$. Of course, if $\alpha \in \widehat{\Omega}^p(M)$ one has a local representation in quantum coordinates similar to (6.16) but with

$$\alpha_{A_1 \dots A_p} \in Q^\infty_w(U, \overset{p}{\widehat{A}}).$$

Furthermore, $d\alpha$ has a local representation like in (6.17) but with

$$(\partial x_B . \alpha_{A_1 \dots A_p}) \in Q^\infty_w(U, \overset{p+1}{\widehat{A}}).$$

Finally, if $\alpha \in \hat{\Omega}^p(M)$, then the pull-back $i^*\alpha$ of α by means of any section $i : M_C \to M$, has the following local representation

$$i^*\alpha = \sum_{1 \leq j_1 < \cdots < j_p \leq m} \beta_{j_1 \dots j_p} d\xi^{j_1} \wedge \cdots \wedge d\xi^{j_p}$$

with $\beta_{j_1 \dots j_p} \in C^\infty_w(c(U), A)$, given by

$$\beta_{j_1 \dots j_p} = \sum_{1 \leq A_1, \cdots, A_p \leq m+n} \alpha_{A_1 \dots A_p} \circ i(\partial \xi_{[j_1} . i^{A_1}) \otimes \cdots \otimes (\partial \xi_{j_p]} . i^{A_p}).$$

Here c is the canonical projection $M \to M_C$ and $(\xi^j)_{1 \leq j \leq m}$ are coordinates on M_C induced by the quantum coordinates $(x^A)_{1 \leq A \leq m+n}$ on M.
REMARK 6.35 - Let $\mathcal{M} \equiv (M, \widehat{C}(M))$ be a quantum diffiety of dimension $m + n$. Then the exterior derivatives $d\delta y^i$ can be written:

$$d\delta y^i = -(\partial_\nu . A^i_\mu) dx^\nu \triangle dx^\mu - \sum_{1 \leq A \leq m+n, j \in I} (\partial y_j . A^i_A) \delta y^i \triangle dx^A, i \in I.$$

For any vector field ζ,η we get

$$\begin{aligned} d\delta y^i(\zeta,\eta) =& \frac{1}{2}[(\partial_n u A^i_\mu)(\eta^\nu \otimes \zeta^\mu) - (\partial_n u A^i_\mu)(\zeta^\nu \otimes \eta^\mu)] \\ &+ \frac{1}{2}[(\partial y_j A^i_A)(\eta^j \otimes \zeta^A) - (\partial y_j A^i_A)(\zeta^j \otimes \eta^A)] \\ =& \frac{1}{2}[(\partial y_j A^i_A)(\eta^j \otimes \zeta^A) - (\partial y_j A^i_A)(\zeta^j \otimes \eta^A)] \,. \end{aligned}$$

In particular,

$$d\delta y^i(\zeta,\eta) = 0, \quad \forall i \in I, \zeta,\eta \in Q^\infty_w(C(M)).$$

DEFINITION 6.22 - *The* **de Rham quantum cohomology** *(resp.* **de Rham full quantum cohomology***) of the quantum manifold M is the homology $\hat{H}^\bullet(M)$, (resp. $\widehat{H}^\bullet(M)$), of the de Rham quantum complex, (resp. de Rham full quantum complex) on M.*

In the following table we also define some further useful cochain complexes and the corresponding cohomology spaces.

TAB.6.2 - **Cohomologies of quantum manifolds**		
Cochain Complex	C.Spaces	Cohomology Name
$\{\hat{\Omega}^\bullet(M), d\}$	$\hat{H}^\bullet(M)$	de Rham q.coh.
$\{\widehat{\Omega}^\bullet(M), d\}$	$\widehat{H}^\bullet(M)$	de Rham fullq.coh.
$\{\Omega^\bullet(M), d\}$	$H^\bullet(M)$	de Rham coh.
$\{\overset{\bullet}{\Omega}{}^\bullet(M), d\}$	$\overset{\bullet}{H}{}^\bullet(M)$	bullet de Rham q.coh.
$\{A \bigotimes_{\mathbf{K}} \Omega^\bullet(M), d\}$	$H^\bullet(M;A)$	de Rham coh.coefs.in A
$\{A_0 \bigotimes_{\mathbf{K}} \Omega^\bullet(M), d\}$	$H^\bullet(M;A_0)$	de Rham coh.coefs.in A_0

NOTE. $(\dot{\Lambda}^0_0 M)^+ = \hat{\Lambda}^0_0 M \equiv M \times A = Hom_{\mathbf{K}}(\Lambda^0_0 M, A)$; $\dot{\Lambda}^0_0 M \equiv M \times A_0$; $\Lambda^0_0 M \equiv M \times \mathbf{K}$. As a consequence one has: $\hat{\Omega}^0(M) = C^\infty_w(M;A)$, $\dot{\Omega}^0(M) \equiv C^\infty_w(\dot{\Lambda}^0_0 M) = C^\infty_w(M;A_0)$, $\overset{\bullet}{\Omega}{}^0(M) = C^\infty_w(M;A/\mathbf{K})$, $\widehat{\Omega}^0(M) = Q^\infty_w(M;A)$.

REMARK 6.36 - The above construction of quantum de Rham complex is a geometric variant to previous algebraic constructions of noncommutative de Rham complex that have been related to Hoschild cohomology and cyclic homology [52].

By using standard results of algebra and homological algebra [49,88] we can prove the following theorems.

THEOREM 6.14 - *One has the canonical isomorphisms of* **K***-modules:*

$$H^q(M;A) \cong A \bigotimes_{\mathbf{K}} H^q(M); H^q(M;A_0) \cong A_0 \bigotimes_{\mathbf{K}} H^q(M);$$
$$\hat{H}^q(M) \cong H^q(M) \bigoplus \overset{\bullet}{H}{}^q(M),$$

THEOREM 6.15 - *If the quantum manifold M is the product of two quantum manifolds $M_i, i = 1,2, M = M_1 \times M_2$, then we can also consider the bicomplex*

$$\{\Omega^{\bullet,\bullet}(M), d_1, d_2\}, \Omega^{r,s}(M) \equiv \Omega^r(M_1) \bigotimes_{\mathbf{K}} \Omega^s(M_2).$$

In general one has the monomorphism:

$$\bigoplus_{r+s=p} \Omega^{r,s}(M) \to \Omega^p(M)\,. \tag{6.18}$$

This induces the morphism

$$\bigoplus_{r+s=p} H^r(M_1) \bigotimes_{\mathbf{K}} H^s(M_2) \to H^p(M).$$

This morphism becomes an isomorphism if at least one of the quantum manifolds $M_i, i = 1,2$ is modeled on a finite dimensional **K***-vector space , as in that situation the morphism (6.18) becomes an isomorphism.*

THEOREM 6.16 - *One has the following morphism :*

$$\bigoplus_{r+s=p} H^r(M_1; A) \bigotimes_{\mathbf{K}} H^s(M_2; A) \to H^p(M_1 \times M_2; A).$$

This becomes an isomorphism if at least one of the quantum manifolds $M_i, i = 1,2$ is modeled on a finite dimensional **K***-vector space.*

THEOREM 6.17 - *Let M be a classic regular $(m+n)$-dimensional quantum manifold. Then one has the canonical* **K***-isomorphisms :*

$$H^p(M) \cong H^p(M_C); \quad H^p(M; A) \cong A \bigotimes_{\mathbf{K}} H^p(M_C); \quad H^0(M; A) \cong A^s,$$

where s is the number of connected components of M.

COROLLARY 6.3 - *One has the following canonical homomorphism:*

$$H^p(M; A) \to H^p(M_C).$$

REMARK 6.37 - In the following we shall assume $\mathbf{K} = \mathbf{R}$. Let X be a quantum manifold. Then we have the usual **singular homology** of X, with coefficients in $\mathbf{K}$, as the homology $H_\bullet(X; \mathbf{K})$ of the chain complex $\{C_p(X; \mathbf{K}), \partial\}$ of the groups of p-chains of X. $(H_p(X; \mathbf{K})$ is a $\mathbf{K}$-module. Let X and Y be quantum manifolds. Then one has the following isomorphism of $\mathbf{K}$-modules:

$$H_p(X \times Y; \mathbf{K}) \cong \bigoplus_{r+s=p} H_r(X; \mathbf{K}) \bigotimes_{\mathbf{K}} H_s(Y; \mathbf{K}).$$

The **singular cohomology** of X (with coefficients in $\mathbf{K}$, $H^p(X;\mathbf{K})$, is the cohomology of the cochain complex $\{C^p(X;\mathbf{K}),\delta\}$, where

$$C^p(X;\mathbf{K}) \equiv Hom_{\mathbf{K}}(C_p(X,\mathbf{K}),\mathbf{K}) \equiv C_p(X;\mathbf{K})^*,$$

and the coboundary is defined by $< \delta\alpha, c > +(-1)^p < \alpha, \partial c >= 0$. One has the canonical isomorphisms:

$$H^p(X;\mathbf{K}) \cong Hom_{\mathbf{K}}(H_p(X,\mathbf{K}),\mathbf{K}) \cong H^p(X).$$

Let X and Y be quantum manifolds. Then one has the following canonical $\mathbf{K}$-isomorphism:

$$H^p(X \times Y;\mathbf{K}) \cong \bigoplus_{r+s=p} H^r(X;\mathbf{K}) \bigotimes_{\mathbf{K}} H^s(Y;\mathbf{K}).$$

The **group of p-chains**, of the quantum manifold X, with coefficients in A, is the set

$$C_p(X;A) \equiv A \bigotimes_{\mathbf{K}} C_p(X;\mathbf{K}).$$

$C_p(X;A)$ is a free two-sided A-module: $C_p(X;A) = \{c = \sum_i c^i u_i | c^i \in A\}$. For $p < 0$ the group $C_p(X;A)$ is defined to be zero. Let $\partial : C_p(X;A) \to C_{p-1}(X;A)$ be the homomorphism defined by

$$\partial u = \sum_{0 \leq i \leq p} (-1)^i \partial_i u : \Delta^{p-1} \to X;$$

∂ is called the **boundary homomorphism**. As $\partial \circ \partial = 0$, one has that $\{C_p(X;A)$ $,\partial\}$ is a chain complex. (For 0-chains one has $\partial \equiv 0$.) So $C_p(X;A)$ identifies a covariant functor from the category of quantum manifolds to the category of free two-sided A-modules. The **singular homology** of X, with coefficients in A, is the homology $H_\bullet(X;A)$ of the chain complex $\{C_p(X;A),\partial\}$. $H_p(X;A)$ is a two-sided A-module, for $p \geq 0$. One has the following canonical isomorphism of $\mathbf{K}$-modules:

(a) $H_p(X;A) \cong A \bigotimes_{\mathbf{K}} H_p(X;\mathbf{K})$

(b) $H_p(X \times Y;A) \cong \bigoplus_{r+s=p} H_r(X;A) \bigotimes_{\mathbf{K}} H_s(X;A)$.

DEFINITION 6.23 - *The p-**quantum cochain group** $\hat{C}^p(X;A)$ is defined by :*

$$\hat{C}^p(X;A) \equiv Hom_A(C_p(X;A),A).$$

REMARK 6.38 - $\hat{C}^p(X;A)$ is a two-sided A-module. The following pairing: $<,>$: $\hat{C}^p(X;A)\times C_p(X;A)\to A$ can be defined. The corresponding cochain complex $\{\hat{C}^p(X;A),\delta\}$ is defined by considering the **coboundary** of a p-quantum cochain $\alpha\in\hat{C}^p(X;A)$ as the p-quantum cochain $\delta\alpha\in\hat{C}^{p+1}(X;A)$ such that

$$<\delta\alpha,c>+(-1)^p<\alpha,\partial c>=0.$$

DEFINITION 6.24 - *The* **singular quantum cohomology** *of X with coefficients in A, is the cohomology $\hat{H}^\bullet(X;A)$ of the cochain complex $\{\hat{C}^p(X;A),\delta\}$.*
In the following table we summarize some further useful cochain complexes and the corresponding cohomology spaces.

TAB.6.3 - **Cohomologies of quantum manifolds**		
Cochain Complex	C.Spaces	Cohomology Name
$\{\hat{C}^\bullet(X,A)\,\delta\}$	$\hat{H}^\bullet(X;A)$	Sing.q.coh.
$\{C^p(X,A)\equiv Hom_{\mathbf{K}}(C_p(X;\mathbf{K}),A),\delta\}$	$H^\bullet_s(X;A)$	Sing.coh.coefs.in A

THEOREM 6.18 - (RELATION BETWEEN SINGULAR COHOMOLOGY WITH COEFFICIENTS IN A, AND SINGULAR HOMOLOGY).
One has the canonical isomorphisms of $\mathbf{K}$*-modules :*

$$H^p_s(X;A)\cong Hom_{\mathbf{K}}(H_p(X;\mathbf{K}),A).$$

THEOREM 6.19 - (RELATION BETWEEN SINGULAR QUANTUM COHOMOLOGY AND SINGULAR COHOMOLOGY WITH COEFFICIENTS IN A).
One has the canonical isomorphisms:

$$\hat{H}^p(X;A)\cong Hom_A(H_p(X;A),A);\quad \hat{H}^p(X;A)\cong H^p{}_s(X;A);$$
$$\hat{H}^p(X;A)\cong Hom_{\mathbf{K}}(H_p(X;\mathbf{K}),A).$$

COROLLARY 6.4 - *One has the following canonical homomorphism:*

$$\hat{H}^p(X;A)\to H^p(X;\mathbf{K}).$$

THEOREM 6.20 - (RELATION BETWEEN SINGULAR (QUANTUM-) COHOMOLOGY AND DE RHAM COHOMOLOGY). *If X is a classic regular quantum manifold one has the following canonical exact sequence of* $\mathbf{K}$*-modules:*

$$0\to A\bigotimes_{\mathbf{K}}H^p(X)\cong H^p(X;A)\to\hat{H}^p(X;A).$$

In particular, $H^p(X;A) \cong \hat{H}^p(X;A)$ *iff* $H_p(X_C;\mathbf{K})$ *is a finitely generated* $\mathbf{K}$*-space.*

THEOREM 6.21 - (RELATION BETWEEN SINGULAR QUANTUM COHOMOLOGY AND DE RHAM QUANTUM COHOMOLOGY).

One has the canonical morphisms of two-sided A-modules :

$$\widehat{H}^p(X) \to \hat{H}^p(X) \to H^p_s(X;A).$$

THEOREM 6.22 - *If X is a contractible quantum manifold, then*

$$H^p_s(X;A) \cong \hat{H}^p(X;A) \cong H^p(X;A) \cong \hat{H}^p(X) \cong \widehat{H}^p(X) = 0, \text{ for } \quad p \geq 1.$$

THEOREM 6.23 - *If X has s connected components one has*

$$H^0_s(X;A) \cong \hat{H}^0(X;A) \cong H^0(X;A) \cong \hat{H}^0(X) \cong \widehat{H}^0(X) \cong A^s.$$

THEOREM 6.24 - (QUANTUM STOKES THEOREM).

$$< d\phi, c > +(-1)^{\dim \phi} < \phi, \partial c >= 0, \text{for any} \quad \phi \in \hat{\Omega}^{p-1}(X) \quad \text{and} \quad c \in C_p(X;A).$$

In the following table we summarize the isomorphic cohomology spaces as results from the above theorems.

TAB.6.4 - **Cohomologies of quantum manifolds**	
Cohomology Name	(Quantum) Cohomology Spaces
de Rham coh.coeffs.in A	$H^\bullet(X;A) \cong A \bigotimes_{\mathbf{K}} H^\bullet(X) \cong A \bigotimes_{\mathbf{K}} H^\bullet(X_C)$
Singular q.coh.	$\hat{H}^\bullet(X;A) \cong Hom_A(H_\bullet(X;A),A)^{(*)}$
Singular coh.with coeffs.in A	$H^\bullet_s(X;A) \cong Hom_{\mathbf{K}}(H_\bullet(X;A),\mathbf{K})$

* ($\hat{H}^\bullet(X;A) \cong H^\bullet(X;A)$ iff $H_p(X_C;\mathbf{K})$ is a finite generated $\mathbf{K}$-space,$\forall\, p \in \mathbf{N}$.)

6.3 - FORMAL THEORY OF QUANTUM PDEs

Scope of this section is to generalize PDEs defined in the category of quantum manifolds (QPDEs), some fundamental definitions and results on the geometry of PDEs. In particular, we give a formal theory of QPDEs by emphasizing the role of a criterion to formal integrability for QPDEs. Furthermore, we show that it is possible to give a geometric theory of singular solutions for QPDEs and we prove as these can be characterized by means of characteristic classes. Spectral sequences for QPDEs are introduced in order to characterize conservation laws, Euler-Lagrange quantum equations, as well as wholly cohomological properties for QPDEs. Symmetry properties of QPDEs are characterized by means of Lie quantum equations. Quantum distributions are related to singular solutions. Finally, quantum cobordism in QPDEs is related to canonical quantization and quantum tunnel effects. A general method to calculate integral and quantum cobordism groups of QPDEs is given.

DEFINITION 6.25 - *A* **quantum PDE** *(QPDE) of order k on the fiber bundle $\pi : W \to M$, defined in the category of quantum manifolds, is a subfiber bundle $\hat{E}_k \subset J\hat{D}^k(W)$ of the jet quantum derivative space $J\hat{D}^k(W)$ over M. The* r-**quantum prolongation** *of $\hat{E}_k \subset J\hat{D}^k(W)$ is*

$$\hat{E}_{k+r} \equiv J\hat{D}^r(\hat{E}_k) \cap J\hat{D}^{k+r}(W) \subset J\hat{D}^{k+r}(W).$$

The **quantum symbol** *of $\hat{E}_{k+r}, r \geq 0$, is a family of A_0-modules over $\hat{E}_k$ given as the kernel $\dot{g}_{k+r}$ of the morphism*

$$\dot{S}_0^{k+r} MOvTW \to \dot{S}_0^r MOF_0,$$

with

$$F_0 \equiv vTJ\hat{D}^k(W)/vT\hat{E}_k.$$

We can also say that $\dot{g}_{k+r}$ is the kernel of the following composition :

$$\begin{aligned}\dot{S}_0^{k+r} MOvTW &\to (\dot{S}_0^r M \bigotimes_{A_0} \dot{S}_0^k M) OvTW \cong \dot{S}_0^r MO(\dot{S}_0^k MOvTW) \\ &\to \dot{S}_0^r MO(\dot{S}_0^k MOvTW/\dot{g}_k).\end{aligned}$$

Another way to characterize $\dot{g}_{k+r}$ is by means of the following short exact sequence of A_0-modules:

$$0 \to \pi^*_{k+r}\dot{g}_{k+r} \to vT\hat{E}_{k+r} \to \pi^*_{k+r,k+r-1} vT\hat{E}_{k+r-1}\,.$$

We can formally write

$$\dot{g}_{k+r} = vT\hat{E}_{k+r} \cap \pi^*_{k+r,0}(\dot{S}_0^{k+r} MOvTW).$$

We say that $\hat{E}_k$ is of **finite type** *if $\dot{g}_{k+r} = 0$ for some (and hence all large) r. Otherwise we say that $\hat{E}_k$ is of* **infinite type**.

DEFINITION 6.26 - *A QPDE $\hat{E}_k \subset J\hat{D}^k(W)$ is* **formally quantum integrable** *if $\forall r \geq 0$ one has that $\dot{g}_{k+r+1} \to \hat{E}_k$ is a fiber bundle, and $\hat{E}_{k+r+1} \to \hat{E}_{k+r}$ is surjective.*

Let us consider the δ-quantum complex of $E \equiv \pi^*_{k,0} vTW \xrightarrow{\beta} N \equiv \hat{E}_k \xrightarrow{\alpha \equiv \pi_k} M$. Then taking into account the following commutative diagram:

$$\begin{array}{ccc} \dot{S}_0^{k+r} M & \to & \dot{S}_0^r M \bigotimes \dot{S}_0^k M \\ {\scriptstyle\delta}\uparrow & & \uparrow{\scriptstyle\delta} \\ TM \bigotimes \dot{S}_0^{k+r-1} M & \leftarrow & TM \bigotimes \dot{S}_0^{r-1} M \bigotimes \dot{S}_0^k M \end{array}$$

we get that $\delta(\dot{g}_{k+r}) \subset \pi_k^* TMO\dot{g}_{k+r-1}$. So the above δ-quantum complex can be restricted to the symbols $\dot{g}_m \subset \pi_k^* \dot{S}_0^m MO\pi_{k,0}^* vTW$. Hence, associated to $\hat{E}_k$ one has the following complexes of A_0-modules over $\hat{E}_k$ (**δ-quantum sequence**):

$$\begin{array}{llll}
0 \rightarrow & \dot{g}_m & \rightarrow TMO\dot{g}_{m-1} & \cdots \rightarrow \delta(\dot{\Lambda}_0^{m-k} MO\dot{g}_k) \rightarrow 0 \\
 & \| & \|\wr & \|\wr \\
0 \rightarrow & \dot{g}_m & \rightarrow \widehat{g}_{m-1} \otimes_{\widehat{A}} (TM)^+ & \cdots \rightarrow \delta(\widehat{g}_k \otimes_{\widehat{A}} (\dot{\Lambda}_0^{m-k} M)^+) \rightarrow 0 \\
 & \uparrow & \| & \| \\
0 \rightarrow & \widehat{g}_m & \rightarrow \widehat{g}_{m-1} \otimes_{\widehat{A}} (TM)^+ & \cdots \rightarrow \delta(\widehat{g}_k \otimes_{\widehat{A}} (\dot{\Lambda}_0^{m-k} M)^+) \rightarrow 0
\end{array}$$

(For the sake of simplicity, we have eliminated the pull-back symbols). The last line is exact iff the original δ-quantum sequence is exact. In fact, we have the following commutative diagram:

$$\begin{array}{ccccc}
 & & \widehat{g}_m & & \\
 & \nearrow & \downarrow & \searrow & \\
0 & \rightarrow & \dot{g}_m & \rightarrow & \widehat{g}_{m-1} \bigotimes_{\widehat{A}} (TM)^+ \\
 & & \downarrow & & \\
 & & 0 & &
\end{array}$$

where the vertical line is exact. We call the corresponding cohomology $\{H_q^{m-j,j}\}_{q\in\hat{E}_k}$ at $(\dot{\Lambda}_0^j MO\dot{g}_{m-j})_q$ the **Spencer quantum cohomology** of $\hat{E}_k$. We say that $\hat{E}_k$ is **r-quantum acyclic** if $H_q^{m,j} = 0, m \geq k, 0 \leq j \leq r, \forall q \in \hat{E}_k$. We say that $\hat{E}_k$ is **quantum involutive** if $H_q^{m,j} = 0, m \geq k, j \geq 0$, (e.g. the δ-quantum sequences are exact for $m \geq k, \forall q \in \hat{E}_k$). Note that each $H_q^{m-j,j}$ can be considered an $\widehat{A}$-module. In fact, $\widehat{g}_{m-1} \bigotimes_{\widehat{A}} (\dot{\Lambda}_0^j M)^+$ has a natural structure of $\widehat{A}$-module: $a(v \otimes \alpha) = v \bigotimes a \circ \alpha = v \circ a \otimes \alpha = (v \otimes \alpha)a, \forall a \in \widehat{A}, v \in \widehat{g}_{m-1}, \alpha \in (\dot{\Lambda}_0^j M)^+$. We say that E_k is **quantum regular** if E_{k+r} are subbundles of $J\hat{D}^{k+r}(W), \forall r \geq 0$.

REMARK 6.39 - If $\hat{E}_k$ is a formally quantum integrable QPDE, then one has the following sequence of A_0-bundles over $\hat{E}_k$:

$$0 \rightarrow \pi_{k+r,k}^* \dot{g}_{k+r} \rightarrow vT\hat{E}_{k+r} \rightarrow \pi_{k+r,k+r-1}^* vT\hat{E}_{k+r-1}$$

that means that $\hat{E}_{k+r}$ is an affine bundle over $\hat{E}_{k+r-1}$ with the associated vector bundle

$$\pi_{k+r-1,k}{}^* \dot{g}_{k+r} \rightarrow \hat{E}_{k+r-1}.$$

The following theorem gives a relation between the formal structure of quantum PDEs and spectral sequences.

THEOREM 6.25 - (FORMAL PROPERTIES AND SPECTRAL SEQUENCES OF QPDEs). *Let $\hat{E}_k \subset J\hat{D}^k(W)$ be a quantum regular QPDE, $\dim W = n + m, \dim M = m, \pi : W \rightarrow M$. Let us consider the following filtration $\{X_q\}$ of $\hat{E}_k$:*

$$(6.19) \qquad \emptyset = X_{-1} \subset X_0 \subset X_1 \subset \cdots \subset X_N = \hat{E}_k,$$

where $X_q \equiv \pi_{k+q,k}((\hat{E}_k)_{+q})$. *Then we can associate to such a filtration a spectral sequence* $\{E^r_{p,q}, d^r\}$ *(**homological formal spectral sequence** of* $\hat{E}_k$*), such that it converges to* $H_\bullet(\hat{E}_k)$. *Furthermore, to the filtration (6.19) of* $\hat{E}_k$ *there also corresponds a spectral sequence* $\{E^{p,q}_r, d_r\}$ *(**cohomological formal spectral sequence** of* $\hat{E}_k$*), such that it converges to* $H^\bullet(\hat{E}_k)$. *Finally, if* $\hat{E}_k$ *is quantum formally integrable, then one has:*[2]

$$H_\bullet(\hat{E}_k) \cong H_\bullet(\hat{E}_\infty) \quad , \quad H^\bullet(\hat{E}_k) \cong H^\bullet(\hat{E}_\infty).$$

PROOF. It is a direct consequence of the well known result of algebraic topology that to filtrations of topological spaces correspond spectral sequences. Note that the above filtration is stationary. In fact, as $\dim \hat{E}_k \geq \dim X_{N-1} \geq \dim X_{N-2} \geq \cdots$, then there must exist an integer $s = s(k)$ such that $X_{s(k)} = X_{s(k)-1} = \cdots = X_0$. This means that one has the following exact commutative diagram, for all $r \geq 0$:

$$\begin{array}{ccccccc}
 & & 0 & & 0 & & \\
 & & \uparrow & & \uparrow & & \\
 & & \hat{E}_{k+s+r+1} & \to & \hat{E}_{k+s+r} & \to 0 & \\
 & \nearrow & \uparrow & \nwarrow & \uparrow & & \\
0 \to a^*\dot{g}_{k+s+r+1} & \to & vT\hat{E}_{k+s+r+1} & \to & b^*vT\hat{E}_{k+s+r} & \to 0 &
\end{array}$$

where $a \equiv \pi^*_{k+s+r+1,k}$ and $b \equiv \pi^*_{k+s+r+1,k+s+r}$. So, assuming that $\hat{E}_k$ is quantum regular one has that $\hat{E}_{k+s+r+1}$ are affine bundles over $\hat{E}_{k+s+r}$, with associated vector bundles $\pi^*_{k+s+r,k}\dot{g}_{k+s+r+1} \to \hat{E}_{k+s+r}$. Finally, the last part of the theorem follows directly from the fact that $\hat{E}_\infty \to \hat{E}_k$ is an affine bundle, when $\hat{E}_k$ is quantum formally integrable. □

We have the following important criterion of formal quantum integrability that extends to QPDEs previously given in [112,113,114] for (super) PDEs.

THEOREM 6.26 - (CRITERION OF FORMAL QUANTUM INTEGRABILITY).

Let $\hat{E}_k \subset J\hat{D}^k(W)$ *be a quantum regular QPDE. Then, there exists a number* $s = s(k)$ *such that if* $\hat{E}_{k+r+1} \to \hat{E}_{k+r}$ *is surjective for* $0 \leq r \leq s$, *then* $\hat{E}_k$ *is formally quantum integrable.*

PROOF. It is a direct consequence of the above theorem and the definition of formally quantum integrable QPDE. □

DEFINITION 6.27 - *We say that a QPDE* $\hat{E}_k \subset J\hat{D}^k(W)$ *is* **δ-regular** *if there exists an integer* $k_0 \geq k$, *such that* $\dot{g}_{k_0}$ *is quantum involutive or 2-quantum acyclic.*

---

[2] Here $H_\bullet$ (resp. $H^\bullet$) can be any homology (resp. cohomology) theory satisfying the wedge and WHE axioms, (see e.g., ref.[141]). In particular, we can take $H_\bullet(\hat{E}_k)$=$H_\bullet(\hat{E}_k;G)$, and $H^\bullet(\hat{E}_k)$=$H^\bullet(\hat{E}_k;G)$, with G=$\mathbf{K}$ or G=A.

THEOREM 6.27 - (CRITERION OF FORMAL QUANTUM INTEGRABILITY).

Let $\hat{E}_k \subset J\hat{D}^k(W)$ be a δ-regular QPDE. Then, if $\dot{g}_{k+r+1}$ is a bundle of A_0-modules, over $\hat{E}_k$, and $\hat{E}_{k+r+1} \to \hat{E}_{k+r}$ is surjective for $0 \leq r \leq h$, for some suitable h, then $\hat{E}_k$ is formally quantum integrable.

PROOF. Let us consider the following lemmas.

LEMMA 6.9 - *The sequences*

$$0 \to \dot{g}_{k+r} \xrightarrow{\delta} TMO\dot{g}_{k+r-1} \xrightarrow{\delta} \dot{\Lambda}_0^2 MO\dot{g}_{k+r-2} \tag{6.20}$$

are exact for $r \geq 1$, that is, $\dot{g}_k$ is 1-quantum acyclic.

PROOF OF LEMMA 6.9 - From the above commutative diagram we also get the commutativity of the following diagram:

$$\begin{array}{cccccc}
 & 0 & & 0 & & 0 \\
 & \downarrow & & \downarrow & & \downarrow \\
0 \to & \dot{g}_{k+r} & \to & \dot{S}_0^{k+r} MOvTW & \to & \dot{S}_0^r MOR \\
 & \downarrow \delta & & \downarrow \delta & & \downarrow \delta \\
0 \to & TMO\dot{g}_{k+r-1} & \to & TMO(\dot{S}_0^{k+r-1} MOvTW) & \to & TM \otimes \dot{S}_0^{r-1} MOR \\
 & \downarrow \delta & & \downarrow \delta & & \\
0 \to & \dot{\Lambda}_0^2 MO\dot{g}_{k+r-2} & \to & \dot{\Lambda}_0^2 MO(\dot{S}_0^{k+r-2} MOvTW) & &
\end{array}$$

where the rows are exact, for $r \geq 1$, and $R \equiv \dot{S}_0^k MOvTW/\dot{g}_k$. Then, taking into account the δ-quantum sequence we get that the last two columns of this diagram are also exact, and hence so is the first. □

LEMMA 6.10 - *The m-th prolongation $\dot{g}_{(k+r)+m}$ of the family of subspaces $\dot{g}_{k+r} \subset \dot{S}_0^{k+r} MOvTW$ is the same as the $(r+m)$-th prolongation of $\dot{g}_k$.*

PROOF OF LEMMA 6.10 - We proceed by induction on m. For $m = 0$, the lemma is trivial. Assume that $m \geq 1$ and that $\dot{g}_{(k+r)+m} = \dot{g}_{k+r+m}$, for $q < m$. By Lemma 6.9 we have that the following diagram

$$\begin{array}{ccccccc}
0 & \to & \dot{g}_{k+r+m} & \xrightarrow{\delta} & TMO\dot{g}_{k+r+m-1} & \xrightarrow{\delta} & \dot{\Lambda}_0^2 MO\dot{g}_{k+r+m-2} \\
 & & & & \downarrow & & \downarrow \\
0 & \to & \dot{g}_{(k+r)+m} & \xrightarrow{\delta} & TMO\dot{g}_{(k+r)+m-1} & \xrightarrow{\delta} & \dot{\Lambda}_0^2 MO\dot{g}_{(k+r)+m-2}
\end{array}$$

is exact and commutative, so that $\dot{g}_{(k+r)+m} = \dot{g}_{k+r+m}$. □

Let us, now, apply to the exact sequence (2.1) the functor $Hom_{A_0}(\dot{\Lambda}_0^l M, -)$, with $l \geq 1, k+r \equiv m > k$. We get the following commutative diagram:

$$\begin{array}{ccccccc}
0 & \to & \dot{\Lambda}_0^l MO\dot{g}_m & \to & \dot{\Lambda}_0^l MO(TMO\dot{g}_{m-1}) & \to & \dot{\Lambda}_0^l MO(\dot{\Lambda}_0^2 MO\dot{g}_{m-2}) \\
 & & \| & & \|\wr & & \|\wr \\
0 & \to & \dot{\Lambda}_0^l MO\dot{g}_m & \to & \dot{\Lambda}_0^l M \bigotimes_{A_0} TMO\dot{g}_{m-1} & \to & \dot{\Lambda}_0^l M \bigotimes_{A_0} \dot{\Lambda}_0^2 MO\dot{g}_{m-2} \\
 & & \| & & \downarrow & & \downarrow \\
0 & \to & \dot{\Lambda}_0^l MO\dot{g}_m & \to & (\dot{\Lambda}_0^{l+1} M)O\dot{g}_{m-1} & \to & \dot{\Lambda}_0^{l+2} MO\dot{g}_{m-2}
\end{array}$$

There the upper (and the middle) row is exact, as $Hom_{A_0}(\dot{\Lambda}^l_0 M, -)$ is a left exact covariant functor. But, it is important to note that this does not imply that the row on the bottom is, in general, exact: it is simply a 0-sequence. So, in general, the Spencer cohomology is not zero.

LEMMA 6.11 - *Let $\hat{E}_k \subset J\hat{D}^k(W)$ be a QPDE of order k on $\pi : W \to M$, such that $\pi_k : \hat{E}_{k+1} \to \hat{E}_k$ is surjective. If $\dot{g}_k$ is 2-quantum acyclic, and $\dot{g}_{k+1}$ is a vector bundle over E_k, then E_k is formally quantum integrable.*

PROOF OF LEMMA 6.11 - Similarly to ordinary and super cases [111,112,113, 114], we can prove that $\dot{g}_{k+r}$ is a vector bundle for $r \geq 1$. Let us now proceed by induction on r. Assume that $\pi_{k+m+1,k+m} : \hat{E}_{k+m+1} \to \hat{E}_{k+m}$ is a surjective map for $0 \leq m \leq r$. Then, let us consider the equation $\hat{E}_{k+r} \subset J\hat{D}^{k+r}(W)$. Then, similarly to what happens in PDEs for ordinary manifolds, there is a morphism

$$\hat{\kappa}(\hat{E}_{k+r}) : \hat{E}_{k+r+1} \to H^{k+r,2} \equiv \frac{\ker\left\{\delta : \dot{\Lambda}^2_0 M O \dot{g}_{k+r} \to \dot{\Lambda}^3_0 M O \dot{g}_{k+r-1}\right\}}{\delta(TMO\dot{g}_{k+r+1})}.$$

over $\hat{E}_{k+r}$ such that one has the following exact sequence:

$$\hat{E}_{k+r+2} \to \hat{E}_{k+r+1} \xrightarrow{\hat{\kappa}(\hat{E}_{k+r})} \pi^*_k H^{k+r,2}_q. \tag{6.21}$$

We call $\hat{\kappa}(\hat{E}_{k+r})$ the **quantum curvature map** of $\hat{E}_{k+r}$. On the other hand as $\dot{g}_k$ is 2-quantum acyclic, $H^{k+r,2}_q = 0$. Hence, the map $\pi_{k+r+2,k+r+1} \to: \hat{E}_{k+r+2} \to \hat{E}_{k+r+1}$ is surjective. □

Now, as we have assumed that $\hat{E}_k$ is δ-regular, we are in a situation similar to the classic case ones, where we can apply the δ-Poincaré lemma. So there exists an integer $k_0 > k$ such that $\dot{g}_{k_0}$ is involutive or 2-quantum acyclic. Then, from the above lemma we conclude that $\hat{E}_k$ is formally quantum integrable. □

COROLLARY 6.5 - *In the quantum analytic case, and assuming that $\hat{E}_k$ is formally quantum integrable, we have that, for any initial condition $q \in \hat{E}_k \subset J\hat{D}^k(W)$, there exists a Q^ω_w solution of $\hat{E}_k$. In fact, we can choose points $q_{k+t} \in \hat{E}_{k+t}, t \geq -k$, where for $t < 0, \hat{E}_{k+t} = \pi_{k,k+t}(\hat{E}_k)$, such that $\pi_{k+t,k+t-1}(q_{k+t}) = q_{k+t-1}$. Then a local solution can be represented in a suitable neighbourhood U of $p \equiv \pi_k(q)$, by the following functions:*

$$\left\{\begin{array}{l} s^C(x) = \sum_{0\leq r\leq\infty} \frac{1}{r!}(\partial x_{A_r} \cdots \partial x_{A_1}.s^C)(p)((x-p)^{A_1} \otimes \cdots \otimes (x-p)^{A_r}) \\ 1 \leq C \leq p+q, (\equiv \textit{fiber dimension of } W) \\ x \in U \\ (\partial x_{A_r} \cdots \partial x_{A_r}.s^C)(p) = y^C_{A_1\ldots A_r}(q_{r+s}) \end{array}\right\}$$

where $y^C_{A_1\dots A_r}$ are vertical quantum coordinates on $J\hat{D}^{r+s}(W)$.
Similarly to classic (super) case [112,113,114], we can develop a geometric theory of singular quantum solutions for QPDEs. In fact, one recognizes on $J\hat{D}^k(W)$ a canonical quantum distribution, **Cartan quantum distribution**, $\mathbb{E}_k \subset TJ\hat{D}^k(W)$, that is the distribution on $J\hat{D}^k(W)$ spanned by tangent spaces to graphs of k-quantum derivatives of sections of the bundle $\pi : W \to M$. If $\dim M = m$, we simply call **integral quantum manifolds** the m-dimensional integral quantum manifolds of $\mathbb{E}_k$. We call also **regular integral quantum manifolds** the m-dimensional integral quantum manifolds of $\mathbb{E}_k$ that can locally be represented as the image of the k-quantum derivative of some section of π. Then a **singular quantum solution** of $\hat{E}_k \subset J\hat{D}^k(W)$ is a closed quantum submanifold $V \subset J\hat{D}^k(W)$ with the property that for each $q \in V$ there exists a smooth integral quantum manifold $\bar{V} \subset J\hat{D}^{k+r}(W), r \geq 0$, such that V coincides with $\pi_{k+r,k}(\bar{V})$ in some neighbourhood of q and $\bar{V} \cap U \subset \hat{E}_{k+r}$, for some neighbourhood $\bar{U}$ of some point $\bar{q} \in {\pi_{k+r,k}}^{-1}(q) \cap \bar{V}$. By using the "blow up" $\bar{V}$ at each point it is possible to construct a smooth quantum manifold $\bar{V}$, called the **blow-up** of V. The points $q \in V$ for which $q \in \Sigma(\bar{V}) \equiv$ set of singular points of the map $\pi_{k+r}|\bar{V} : \bar{V} \to M$, will be called the **geometric singularities** of the quantum solution V. A **non-singular** quantum solution of $\hat{E}_k$ is a regular integral quantum manifold $V \subset J\hat{D}^k(W)$ contained also in $\hat{E}_k$. We define a QPDE $\hat{E}_k \subset J\hat{D}^k(W)$ **singular formally quantum integrable at** $q \in \hat{E}_k$ if there exists an integral quantum manifold L tangent to $\hat{E}_k$ at this point with infinite order. $\hat{E}_k$ is **formally quantum integrable** at $q \in \hat{E}_k$ if it is singular formally quantum integrable at q and L is diffeomorphically projected onto M. A QPDE $\hat{E}_k \subset J\hat{D}^k(W)$ is formally quantum integrable iff the projections $\pi_{k+s+1,k+s} : (\hat{E}_k)_{+(s+1)} \to (\hat{E}_k)_{+s}$ are smooth quantum bundles, $s \geq 0$. An equation $\hat{E}_s \subset J\hat{D}^s(W)$ is called an **intermediate quantum integral** of the QPDE $\hat{E}_k$ if its $(k-s)$-quantum prolongation is contained in $\hat{E}_k : (\hat{E}_s)_{+(k-s)} \subset \hat{E}_k$. The number s is the **order** of the intermediate quantum integral. An equation $\hat{E}_k \subset J\hat{D}^k(W)$ is called V-**quantum involutive** if: (a) its first quantum prolongation $\hat{E}_k$ is **quantum regular**, (i.e., at each point $q \in (\hat{E}_k)_{+1}$ the subspaces $T_q((\hat{E}_k)_{+1})$ and $\dot{S}^{k+1}_0(T_xM)OvT_pW, p \equiv \pi_{k+1,0}(q)$, intersect transversally); (b) the map $\pi_{k+1,k} : (\hat{E}_k)_{+1} \to \hat{E}_k$ is surjective; (c) $\hat{E}_k$ is quantum acyclic.

DEFINITION 6.28 - *We call* **regular quantum plane** *at the point $q \in J\hat{D}^k(W)$ the linear subspace in $T_qJ\hat{D}^k(W)$ of the form, $T_qD^ks(M)$, for some Q^∞_w-section s of π such that $D^ks(x) = q, x = \pi_k(q)$.*

Any point $q \in J\hat{D}^k(W)$ identifies a unique regular quantum plane

$$\left\{\begin{array}{ll} L_q \quad \text{at} \quad q' \equiv \pi_{k,k-1}(q) \in J\hat{D}^{k-1}(W) \\ L_q \subset T_{q'}J\hat{D}^{k-1}(W) \end{array}\right\}$$

If $P \subset T_{\bar{q}}J\hat{D}^{k-1}(W)$ is some subspace, we set

$$S_i(P) \equiv \{q \in J\hat{D}^k(W) | \bar{q} \equiv \pi_{k,k-1}(q), \dim(L_q \cap P) = i\}.$$

We call $S_i(q) \equiv S_i(L_q)$ **characteristic quantum cones**. These are affine subspaces of ${\pi^{-1}}_{k,k-1}(\bar{q})$. Furthermore, as $S_i(q)$ can be considered pointed at q, it can be identified with its tangent space $T_q S_i(q)$ that is a subspace of

$$T_q \pi^{-1}_{k,k-1}(\bar{q}) \cong S^k_0(T_x M) Ov T_p W \ , \ p \equiv \pi_{k,0}(q) \, .$$

The **characteristic quantum manifold**,

$$Char_i(\hat{E}_k)_q,$$

of all i-dimensional characteristic subspaces for a QPDE $\hat{E}_k \subset J\hat{D}^k(W)$ at $q \in \hat{E}_k$ is defined by:

$$Char_i(\hat{E}_k)_q \equiv \bigcup_{q' \in \dot{g}_{k,q} \cap S_{(m-i)}(q)} Char_i(\hat{E}_k)_{q,q'},$$

where

$$Char_i(\hat{E}_k)_{q,q'} \equiv \left\{ \begin{array}{l} \xi \in (T_x M)^+ - \{0\} \\ <\xi, T(\pi_{k-1}(L_q \cap L_{q'}) >= 0 \\ x = \pi_k(q) = \pi_k(q') \end{array} \right\}$$

$T(\pi_{k-1}(L_q \cap L_{q'}))$ is the $(m-i)$-dimensional subspace of $T_x M$ identified by q'. A vector ξ contained in a 1-dimensional characteristic subspace is called a **characteristic quantum covector** of $\hat{E}_k$ at q.
Let us now characterize singularities by means of characteristic classes. Let $\pi : W \to M$ be a fiber bundle over an m-dimensional quantum manifold M. Set:

$$I_k(W) \equiv \bigcup_{u \in J\hat{D}^k(W)} I_k(W)_u$$

$I_k(W)_u \equiv$ Grassmannian of integral quantum planes at q. An **integral quantum plane** at a point $u \in J\hat{D}^k(W)$ is defined to be an m-dimensional subspace of $(\mathbb{E}_k)_u$ tangent to some integral quantum manifold of the Cartan distribution

$$\mathbb{E}_k \subset TJ\hat{D}^k(W)).$$

It is also useful to consider the quantum extension counterpart. Let us denote by $\widehat{I}_k(W)_u$ Grassmannian of full quantum extension integral quantum planes at u and by $\widehat{I}_k(W)$ the corresponding bundle over $J\hat{D}^k(W)$. Let $\widehat{I}(\hat{E}_{k+s})$ be the fiber bundle of

Grassmannian $m|m$-dimensional integral quantum planes of the Cartan quantum distribution $\widehat{\mathbb{E}}(\hat{E}_{k+s})$ on $\hat{E}_{k+s}$ where $\hat{E}_k \subset J\hat{D}^k(W)$ is a QPDE. If $\hat{E}_{k+s} = J\hat{D}^{k+s}(W)$ one has

$$\widehat{I}(J\hat{D}^{k+s}(W)) = \widehat{I}_{k+s}(W).$$

DEFINITION 6.29 - *Each m-dimensional integral quantum manifold $V \subset J\widehat{D}^k(W)$ determines a* **tangential map** $i_V : V \to \widehat{I}_k(W)$ *given by* $i_V(q) = \widehat{T}_q V \in \widehat{I}_k(W)_q$. *Then, any integral quantum manifold $V \subset J\hat{D}^k(W)$ determines a map*

$$\left\{\begin{array}{l} H^i(\widehat{I}_k(W); A) \to H^i(V; A) \\ \omega \mapsto i_V^* \omega \end{array}\right\}$$

where

$$i_V^*\omega \equiv \textbf{characteristic quantum class} \textit{ on } V$$
$$\textit{corresponding to } \omega \in H^i(\widehat{I}_k(W); A).$$

Then we have the following important theorem that extends the previous one for manifolds and supermanifolds.

THEOREM 6.28 - *Let $\hat{E}_k \subset J\hat{D}^k(W)$ be a QPDE. Each cohomology quantum class*

$$\omega \in H^i(\widehat{I}(\hat{E}_{k+s}); A)$$

defines characteristic quantum class $\omega_V \equiv i_V^*\omega$ on the singular solutions $V \subset \hat{E}_{k+s}$ of $\hat{E}_k$. $H^\bullet(\widehat{I}(\hat{E}_{k+s}); A)$ is an algebra over $H^\bullet(\hat{E}_{k+s}; A)$.

PROPOSITION 6.4 - *Let $\hat{E}_k$ be formally quantum integrable QPDE then the fiber bundles*

$$\pi_{k+s,k+s-1} : \hat{E}_{k+s} \to \hat{E}_{k+s-1}, \quad s \geq 1$$

are affine subbundles in $\pi_{k+s,k+s-1} : J\hat{D}^{k+s}(W) \to J\hat{D}^{k+s-1}(W)$ and, hence

$$H^\bullet(\hat{E}_{k+s}; A) \cong H^\bullet(\hat{E}_k; A).$$

If V is a non-singular integral quantum manifold, then all its quantum characteristics classes are zero in dimension ≥ 1.

PROOF. We shall prove that if V is a non-singular integral quantum manifold, then the mapping $i_V{}^* : H^i(\widehat{I}_k(W); A) \to H^i(V; A)$ is the zero map. In fact, one has the following commutative and exact diagram:

$$\begin{array}{ccccccc}
 & & & & 0 & & \\
 & & & & \downarrow & & \\
 & & H^i(\widehat{I}_k(W); A) & \xrightarrow{i_V{}^*} & H^i(V; A) & & \\
 & & {\scriptstyle i_V{}^*}\downarrow & & \downarrow & & \\
0 & \leftarrow & A \bigotimes_{\mathbf{K}} H^i(V_C) & \leftarrow & A \bigotimes_{\mathbf{K}} H^i(V) & \leftarrow & 0 \\
 & & & & \downarrow & & \\
 & & & & 0 & &
\end{array}$$

So any characteristic quantum class $i_V{}^*\omega$ on V can be written as

$$i_V{}^*\omega = \sum_i a^i \otimes \beta_i \tag{6.22}$$

where $a^i \in A$, and β_i are characteristic classes on V_C. On the other hand if V is non-singular, V_C must also be non-singular; hence all β_i must be zero [111]. Then, from (6.22) it follows that $i_V{}^*\omega = 0, \forall \omega \in H^i(\widehat{I}_k(W); A)$. □

THEOREM 6.29 - *Lie quantum transformations, that are local Q^∞_w-diffeomorphisms*

$$\phi : J\hat{D}^k(W) \to J\hat{D}^k(W)$$

which preserves the Cartan quantum distributions,

$$\hat{T}(\phi)(\widehat{\mathbf{E}}_k)_u = (\widehat{\mathbf{E}}_k)_{\phi(u)}, \forall u \in J\hat{D}^k(W),$$

preserve the cohomology space $\hat{H}^i(\widehat{I}_k(W); A)$ and $H^i(\widehat{I}_k(W); A)$ (if they are not just isomorphic). Characteristic Lie quantum transformations, that are Lie quantum transformations that transform each integral quantum manifold in itself, preserve the space of characteristic quantum classes of any integral quantum manifold. In particular, any characteristic Lie quantum transformation that preserves also the classes of $H^i(\widehat{I}_k(W); A)$ preserves characteristic quantum classes of any integral quantum manifold.

THEOREM 6.30 - (CAUCHY PROBLEM FOR QPDEs AND CHARACTERISTIC QUANTUM NUMBERS). *The Cauchy problem for (singular) quantum solutions of QPDE can be formulated in the following way. A Cauchy data for the system $\hat{E}_k \subset J\hat{D}^k(W)$ is defined by an $(m-1)$-dimensional integral quantum manifold $\sigma \subset \hat{E}_{k+r}$ of the Cartan quantum distribution $\widehat{\mathbf{E}}_{k+r}$, besides a section h of the bundle $\widehat{I}(\hat{E}_{k+r})|\sigma \to \sigma$, such that $h(\bar{q}) \supset T_{\bar{q}}\sigma$, for all $\bar{q} \in \sigma$. A quantum solution of the Cauchy problem with initial data, (σ, h), is defined to be an m-dimensional integral quantum manifold V, with boundary, such that $V \subset \hat{E}_{k+r}, \partial V = \sigma$ and $T_{\bar{q}}V = h(\bar{q}), \forall \bar{q} \in \sigma$. With each quantum cohomology class $\omega \in H^{m-1}(\widehat{I}(\hat{E}_{k+r}); A)$ and each Cauchy data (σ, h) we associate the characteristic quantum number $\chi_\omega(\sigma, h) =< h^*\omega, Z_\sigma >$, where $Z_\sigma \in H_{m-1}(\sigma; A)$ is a fundamental cycle of σ. For solvability of the Cauchy problem (σ, h) for QPDEs it is necessary that all the characteristic quantum numbers $\chi_\omega(\sigma, h)$ be equal to zero.*[3]

[3] Here and in the following for p-dimensional integral quantum manifold $V, 0 \leq p \leq m$, with boundary ∂V, (or eventualy with $\partial V = \emptyset$), we mean an element $V \in C_p(\hat{E}_{k+h}, A), h \geq 0$, such that $\widehat{T}V \subset \widehat{\mathbf{E}}_{k+r}$. So, if $V = \sum_i s^i u_i, a^i \in A$, one has $\partial V = \sum_i (-1)^i a^i \partial_i u$. Then, the canonical homomorphism $c: A \to \mathbf{K}$ induces the canonical mapping $V \to V_C = \sum_i \lambda^i u_i$, with $\lambda^i \equiv c(a^i) \in \mathbf{K}$, and one has $\partial V_C = \sum_i (-1)^i \lambda^i \partial_i u$, where $V_C \in C_p(\hat{E}_{k+h}; \mathbf{K})$ is the classic limit of V. Finally note that by means of the monomorphism $\epsilon: \mathbf{K} \to A$, we can induce an injection $\epsilon: V_C \to V$, such that $c \circ \epsilon = id_{V_C}$, hence ϵ is a section of $V \to V_C$. In other words V canonically contains its classic limit V_C.

PROOF. As quantum manifolds involve classic regular, we have the following canonical isomorphisms:

$$H^{m-1}(\sigma;A)\cong A\bigotimes_{\mathbf{K}} H^{m-1}(\sigma_C);H_{m-1}(\sigma;A)\cong A\bigotimes_{\mathbf{K}} H_{m-1}(\sigma_C;\mathbf{K}),$$

where σ_C is the classic-limit of σ. (In this way it makes sense to talk of fundamental cycle of σ.) Hence, the characteristic quantum numbers can be written as follows:

$$\chi_\omega(\sigma,h)=<h^*\omega,Z_\sigma>=<a^i\otimes\beta_i,1\otimes Z_{\sigma_C}>=a^i<\beta_i,Z_C>, \tag{6.23}$$

where $a^i\in A,\beta_i\in H^{m-1}(\sigma_C)$, and Z_{σ_C} is a fundamental cycle of σ_C. On the other and, if $V\subset\hat{E}_{k+r}$ is an integral quantum manifold such that $\partial V=\sigma$, we also get $\partial V_C=\sigma_C$. Then, taking into account that for ordinary manifolds, a necessary condition for solvability of Cauchy problem is that all characteristic numbers $<\beta_i,Z_C>$ must be zero, from (6.23) we get that, in order to solve the Cauchy problem for QPDEs, all the characteristic quantum numbers $\chi_\omega(\sigma,h)$ must also be zero. □

6.4 - CARTAN SPECTRAL SEQUENCES OF QUANTUM PDEs

We shall, now, consider a spectral sequence that can be associated to QPDEs and that is useful in order to characterize the space of conservation laws of QPDEs. The wholly cohomological character of QPDEs will be considered and a general criterion to recognize conservation laws for QPDEs will be given. These results are similar to the previous ones formulated for PDEs as well as for super PDEs.
Let us assume that $\pi:W\to M$ is a quantum fiber bundle, where $\dim W=m+n$, and $\dim M=m$. Following our previous considerations in section 1.1, the full quantum dual tangent space of $T_qJ\hat{D}^\infty(W)$ is

$$(T_qJ\hat{D}^\infty(W))^+\equiv Hom_{A_0}(T_qJ\hat{D}^\infty(W),B)$$

where $B\equiv\overset{\infty}{\widehat{A}}$. One has the local isomorphisms:

$$(T_qJ\hat{D}^\infty(W))^+\cong Hom_{A_0}(A,B)^m\bigoplus_{r>0} Hom_{A_0}(\overset{r}{\widehat{A}},B)^{s_r}$$

where $s_t\equiv m^tn$, so the differential of a quantum function $f:J\hat{D}^\infty(W)\to B$ has the following local representation:

$$df=(\partial x_\alpha f)dx^\alpha+\sum_{1\le\alpha_1,\dots,\alpha_r\le m+n;1\le i\le p+q}(\partial y_i^{\alpha_1\dots\alpha_r}.f)dy^i_{\alpha_1\dots\alpha_r}.$$

Furthermore, if W is of dimension $m|(E;s)$. So the quantum dual tangent space of $T_qJ\hat{D}^\infty(W)$ is

$$(T_qJ\hat{D}^\infty(W))^+ \equiv Hom_{A_0}(T_qJ\hat{D}^\infty(W);B),$$

where $B \equiv \overset{\infty}{\hat{A}}(E)$. One has the local isomorphisms:

$$(T_qJ\hat{D}^\infty(W))^+ \cong Hom_{A_0}(E,B)^m \bigoplus_{r>0} Hom_{A_0}(\overset{r}{\widehat{A}}(E),B)^{s_r}$$

where $s_t \equiv m^t s$.

REMARK 6.40 - In the following we shall assume that $\pi : W \to M$ is a quantum fiber bundle with $\dim W = m|(E;s)$. Then we can extend Theorem 1.5 to quantum differential forms on $J\hat{D}^\infty(W)$. So we get the following canonical A_0-homomorphisms

$$\begin{aligned}(T_{D^\infty s(i(p))}J\hat{D}^\infty(W))^+ &\to (T_{D^\infty s(i(p))}J\hat{D}^\infty(W))^* \\ &\to E\bigotimes_{\mathbf{K}}(T_qM_C)^*, s\in Q^\infty_w(W), i\in C^\infty_w(M).\end{aligned}$$

As a consequence we obtain the following:

$$\begin{array}{ccccc}(\dot{\Lambda}^p_0(J\hat{D}^\infty(W)))^+ & \to & (\Lambda^p_0(J\hat{D}^\infty(W)))^* & \to & E\bigotimes_{\mathbf{K}}(\Lambda^0_pM_C) \\ \| & & \| & & \| \\ Hom_{A_0}(\dot{\Lambda}^p_0(J\hat{D}^\infty(W),B) & \to & Hom_{\mathbf{K}}(\Lambda^p_0(J\hat{D}^\infty(W),B) & \to & E\bigotimes_{\mathbf{K}}\Lambda^0_p(M_C)\end{array}$$

So any quantum differential form $\alpha\in\hat{\Omega}^p(J\hat{D}^\infty(W))$, or $\alpha\in\widehat{\Omega}^p(J\hat{D}^\infty(W))$, can be represented by a E-valued p-differential form on the classic limit M_C of M, by means of pull-back of s and i:

$$((D^\infty s)\circ i)^*\alpha : M_C \to E\bigotimes_{\mathbf{K}}\Lambda^0_p(M_C).$$

In particular, if there exists a canonical mapping $\gamma : E\to A$ we also get the following representation of α :

$$((D^\infty s)\circ i)^*\alpha : M_C \to A\bigotimes_{\mathbf{K}}\Lambda^0_p(M_C).$$

$J\hat{D}^\infty(W)$ has a canonical connection, that is a canonical A_0-splitting

$$TJ\hat{D}^\infty(W)\cong vTJ\hat{D}^\infty(W)\bigoplus H_W,$$

where H_W is the canonical horizontal subbundle $H_W\subset TJ\hat{D}^\infty(W)$ given by

$$H_W \equiv \bigcup_{q\in J\hat{D}^\infty(W)} (H_W)_q$$

where $(H_W)_q \equiv \mathrm{im}(D(D^\infty s)(p)), p = \pi_\infty(q), D^\infty s(p) = q$, and $D(D^\infty s)(p) : T_pM \to T_qJ\hat{D}^\infty(W)$. Associated to H_W we have the quantum bundle

$$\left\{\begin{array}{l} H_W^\perp \equiv \bigcup_{q\in J\hat{D}^\infty(W)} (H_W^\perp)_q \subset T^+J\hat{D}^\infty(W) \\ (H_W^\perp)_q \equiv \{\beta \in T_q^+J\hat{D}^\infty(W) \ \mid \ \beta|_{H_q} = 0\} \end{array}\right\}$$

Furthermore, H_W is **quantum flat** in the sense that $dQ_w^\infty(H_W^\perp)|_{H_W} = 0$. A section $u : M \to J\hat{D}^\infty(W)$ is called **quantum flat** if $Tu(M) \subset H_W$. The requirement of quantum flatness for a section $u : M \to J\hat{D}^\infty(W)$ of π_∞ is equivalent to the following equivalent conditions:

(a)

$$u^*Q_w^\infty(H_W^\perp) = 0;$$

(b)

$$u = D^\infty s, s = \pi_{\infty,0} \circ u : M \to W.$$

Let $\hat{E}_k \subset J\hat{D}^k(W)$ be a formally quantum integrable QPDE on a fiber bundle $\pi : W \to M$. Then, the canonical connection on $J\hat{D}^\infty(W)$ identifies a canonical connection on $\hat{E}_\infty \equiv \lim_r \hat{E}_{k+r}$. More precisely, we have a canonical A_0-splitting

$$T\hat{E}_\infty \cong vT\hat{E}_\infty \bigoplus H_{\hat{E}_k}, \text{where} \quad H_{\hat{E}_k} \equiv H_W|_{\hat{E}_\infty} \subset T\hat{E}_\infty.$$

$H_{\hat{E}_k}$ is **quantum flat** in the sense that $dQ_w^\infty(H_{\hat{E}_k}^\perp|_{H_{\hat{E}_k}}) = i^*dQ_w^\infty(H_W^\perp|_{H_{\hat{E}_k}}) = 0$, where i is the canonical embedding $i : \hat{E}_\infty \subset J\hat{D}^\infty(W)$. A section $u \in Q_w^\infty(\hat{E}_k)$ of $\pi_\infty|_{\hat{E}_\infty} : \hat{E}_\infty \to M$ is a solution of $\hat{E}_k$ iff one of the following equivalent conditions holds:

$$\text{(a)} \quad u \quad \text{is quantum flat;} \quad \text{(b)} \quad u^*Q_w^\infty(H_{\hat{E}_k}^\perp) = 0.$$

Let $\hat{E}_k \subset J\hat{D}^k(W)$ be a quantum PDE. Let $\mathbb{E}(\hat{E}_k)$ be the Cartan distribution of E_k, i.e., $\mathbb{E}(\hat{E}_k) = \mathbb{E}_k|_{\hat{E}_k}$. We have a spectral sequence $\{\hat{E}_r^{p,q}, d_r\}$, associated to the following filtration

(6.24) $\quad \hat{\Omega}^q(\hat{E}_\infty) = C^0\hat{\Omega}^q(\hat{E}_\infty) \supset C^1\hat{\Omega}^q(\hat{E}_\infty) \supset C^2\hat{\Omega}^q(\hat{E}_\infty) \supset \cdots C^p\hat{\Omega}^q(\hat{E}_\infty) \supset \dots$

for $q = 0, 1, 2, \dots$, where

$C^1\hat{\Omega}^q(\hat{E}_\infty)$

$\|$

$C\hat{\Omega}^q(\hat{E}_\infty) \equiv \{$space of all Cartan quantum q-forms on $\hat{E}_\infty\}$

$\equiv \{\omega \in \hat{\Omega}^q(\hat{E}_\infty) : \omega(p)(\zeta_1, \dots, \zeta_q) = 0, \forall p \in \hat{E}_\infty, \forall \zeta_1, \dots, \zeta_q \in C(\hat{E}_\infty), q = 0, 1, 2, \dots\}$;

$C^p\hat{\Omega}^q(\hat{E}_\infty) \equiv \big\{$space of all p-Cartan quantum q-forms on $\hat{E}_\infty, p = 0, 1, 2, \dots\big\}$

$\equiv \big\{\omega \in \hat{\Omega}^q(\hat{E}_\infty) : \omega(p)(\zeta_1, \cdots, \zeta_q) = 0, \forall p \in \hat{E}_\infty$, if at least $q - p + 1\, \zeta_i \in \mathbb{E}(\hat{E}_\infty)_p\big\}$

One has

$$C^p\hat{\Omega}^q(\hat{E}_\infty) = \left\{ \begin{array}{ll} 0 \text{ if } p > q & \\ \hat{\Omega}^q(\hat{E}_\infty) \text{ if } p < 0 \end{array} \right\}$$

$$\hat{\Omega}^q(\hat{E}_\infty) = 0 \text{ if } q < 0.$$

If $\omega \in C^p\hat{\Omega}^q(\hat{E}_\infty) \Rightarrow d\omega \in C^p\hat{\Omega}^{q+1}(\hat{E}_\infty)$. Thus, the de Rham quantum complex

$$0 \to B \to \hat{\Omega}^0(\hat{E}_\infty) \to \hat{\Omega}^1(\hat{E}_\infty) \to \hat{\Omega}^2(\hat{E}_\infty) \to \cdots$$

of $\hat{E}_\infty$ has the filtration (6.24). Here $B \equiv \overset{\infty}{\widehat{A}}(E)$.

DEFINITION 6.30 - *The above spectral sequence is called* **Cartan quantum spectral sequence** *of* $\hat{E}_k$.

REMARK 6.41 - Similar considerations can of course be made for the quantum forms in $\widehat{\Omega}^q(\hat{E}_\infty)$. So we also have the **Cartan full quantum spectral sequence** $\{\widehat{E}_r^{p,q}, d_r\}$ of $\hat{E}_k$, and the de Rham **full quantum complex** of $\hat{E}_k$. One has the following commutative diagrams:

$$\begin{array}{ccccccccc} \widehat{\Omega}^q(\hat{E}_\infty) & = & C^0\widehat{\Omega}^q(\hat{E}_\infty) & \supset & \cdots & \supset & C^p\widehat{\Omega}^q(\hat{E}_\infty) & \supset & \cdots \\ \downarrow & & \downarrow & & & & \downarrow & & \\ \hat{\Omega}^q(\hat{E}_\infty) & = & C^0\hat{\Omega}^q(\hat{E}_\infty) & \supset & \cdots & \supset & C^p\hat{\Omega}^q(\hat{E}_\infty) & \supset & \cdots \end{array}$$

$$\begin{array}{ccccccccccc} 0 & \to & B & \to & \widehat{\Omega}^0(\hat{E}_\infty) & \overset{d}{\to} & \widehat{\Omega}^1(\hat{E}_\infty) & \overset{d}{\to} & \widehat{\Omega}^2(\hat{E}_\infty) & \overset{d}{\to} & \cdots \\ & & \| & & \downarrow & & \downarrow & & \downarrow & & \\ 0 & \to & B & \to & \hat{\Omega}^0(\hat{E}_\infty) & \overset{d}{\to} & \hat{\Omega}^1(\hat{E}_\infty) & \overset{d}{\to} & \hat{\Omega}^2(\hat{E}_\infty) & \overset{d}{\to} & \cdots \end{array}$$

In the following in order not to make the expression too heavy, we shall explicitly consider Cartan quantum spectral sequences only. Similar results for Cartan full quantum spectral sequences can be easily obtained.

Any quantum form $\omega \in \hat{\Omega}^q(\hat{E}_\infty)$ has the unique local representation

$$\omega = \omega_{\mu_1 \ldots \mu_q} dx^{\mu_1} \triangle \cdots \triangle dx^{\mu_q} \bmod C\hat{\Omega}^q(\hat{E}_\infty),$$

where $\omega_{\mu_1 \ldots \mu_q}$ is a $\overset{q}{\hat{B}}$-valued local function of class C^∞_w on $\hat{E}_\infty$, where

$$\overset{q}{\hat{B}} \equiv Hom_{\mathbf{K}}(T^q_0 B, B),$$

where $B \equiv \overset{\infty}{\widehat{A}}(E)$. Furthermore, one has

$$d\omega = \partial\omega \bmod C\hat{\Omega}^{q+1}(\hat{E}_\infty),$$

where

$$\partial\omega = (\partial_{\mu_0}\omega_{\mu_1\dots\mu_q})dx^{\mu_0}\triangle\cdots\triangle dx^{\mu_q}.$$

One has $i^*\partial\omega = 0$ for any $\omega \in \hat{\Omega}^{m+n}(\hat{E}_\infty)$, and section $i : M_C \to M$.

THEOREM 6.31 - (CARTAN QUANTUM SPECTRAL SEQUENCE). *The Cartan quantum spectral sequence converges to $\hat{H}^\bullet(\hat{E}_\infty)$. If $\hat{E}_k$ is a path-connected formally quantum integrable QPDE one has:*

$$\hat{H}^\bullet(\hat{E}_k) \cong \hat{H}^\bullet(\hat{E}_\infty).$$

The following properties hold:

$$\hat{E}_r^{p,q}(\hat{E}_\infty) = \frac{\hat{Z}_r^{p,q}}{\hat{Z}_{r-1}^{p+1,q-1} \oplus \hat{B}_{r-1}^{p,q}}, p, q, r = 0, 1, 2, \dots$$

with

$$\begin{aligned}
\hat{Z}_r^{p,q} &= C^p\hat{\Omega}^{p+q}(\hat{E}_\infty) \cap d^{-1}C^{p+r}\hat{\Omega}^{p+q-1}(\hat{E}_\infty),\\
\hat{B}_r^{p,q} &= C^p\hat{\Omega}^{p+q}(\hat{E}_\infty) \cap dC^{p-r}\hat{\Omega}^{p+q-1}(\hat{E}_\infty);\\
d_r^{p,q} &: \hat{E}_r^{p,q}(\hat{E}_\infty) \to \hat{E}_r^{p+r,q-r+1}(\hat{E}_\infty), p, q, = 0, 1, 2, \dots .
\end{aligned}$$

One has the following convergence condition:

$$\hat{E}_r^{p,q}(\hat{E}_\infty) = \hat{E}_\infty^{p,q}(\hat{E}_\infty), p, q, = 0, 1, 2, \dots, r > \max\{p, q+1\},$$

with

$$\hat{E}_\infty^{p,q}(\hat{E}_\infty) = \frac{\hat{Z}_\infty^{p,q}}{\hat{Z}_\infty^{p+1,q-1} \oplus \hat{B}_\infty^{p,q}}, p, q = 0, 1, 2, \dots$$

where

$$\begin{aligned}
\hat{Z}_\infty^{p,q} &= C^p\hat{\Omega}^{p+q}(\hat{E}_\infty) \cap d^{-1}(0)\\
&= \{\omega \in C^p\hat{\Omega}^{p+q}(\hat{E}_\infty) : d\omega = 0\};\\
\hat{B}_\infty^{p,q} &= C^p\hat{\Omega}^{p+q}(\hat{E}_\infty) \cap d\hat{\Omega}^{p+q-1}(\hat{E}_\infty)\\
&= \{\omega \in C^p\hat{\Omega}^{p+q}(\hat{E}_\infty) : \omega = d\theta, \theta \in \hat{\Omega}^{p+q-1}(\hat{E}_\infty)\}.
\end{aligned}$$

Furthermore,

$$\hat{E}_r^{p,q}(\hat{E}_\infty) = 0, \text{ for } q > m$$

because

$$C^p\hat{\Omega}^{p+q}(\hat{E}_\infty) = C^{p+q-(m+n)}\hat{\Omega}^{p+q}(\hat{E}_\infty) \text{ for } q > m.$$

One has the following:

(a)

$$\hat{E}_0^{0,0}(\hat{E}_\infty) = C_w^\infty(\hat{E}_\infty; B);$$

(b)

$$\hat{E}_1^{0,0}(\hat{E}_\infty) = C_w^\infty(\hat{E}_\infty; B) \cap d^{-1}(C\hat{\Omega}^1(\hat{E}_\infty))$$
$$= \{f \in C_w^\infty(\hat{E}_\infty; B) : \zeta.f = 0, \forall \zeta \in C_w^\infty(\widehat{\mathbb{E}}(\hat{E}_\infty))\}$$
$$= \textit{constants for the Lie quantum algebra}$$
of differentiations of the quantum algebra
$$C_w^\infty(\hat{E}_\infty; B)$$

(c)

$$\hat{E}_r^{0,0}(\hat{E}_\infty) = C_w^\infty(\hat{E}_\infty; B) \cap d^{-1}\{0\}$$
$$= \{f \in C_w^\infty(\hat{E}_\infty; B) : df = 0\}, r = 2, 3, \cdots$$

(d) *In particular, if* $\hat{E}_\infty$ *is connected one has:*

$$\hat{E}_r^{0,0}(\hat{E}_\infty) \cong B.$$

PROOF. The proof can be conducted by using the standard techniques of homological algebra and algebraic topology. □

DEFINITION 6.31 - *A* **quantum current** *of dimension* $q = 0, 1, 2, \ldots$ *on* $\hat{E}_k \subset J\hat{D}^k(W)$ *is a* q*-quantum form* $\omega \in \hat{\Omega}^q(\hat{E}_\infty)$ *such that* $d\omega \in C\hat{\Omega}^{q+1}(\hat{E}_\infty)$. *Denote by* $\hat{C}^q_{urr}(\hat{E}_\infty)$ *the space of all quantum currents on* $\hat{E}_k$ *of dimension* q. *Exact quantum forms are called* **trivial quantum currents**.

DEFINITION 6.32 - *The* **space of conservation quantum laws** *of dimension* q *is the following quotient:*

$$\hat{C}^q_{ons}(\hat{E}_\infty) = \frac{\hat{\Omega}^q(\hat{E}_\infty) \cap d^{-1}(C\hat{\Omega}^{q+1}(\hat{E}_\infty))}{C\hat{\Omega}^q(\hat{E}_\infty) \oplus d\hat{\Omega}^{q-1}(\hat{E}_\infty)}, q = 0, 1, \ldots$$

PROPOSITION 6.5 -

1) *Let* $\omega \in \hat{C}^q_{urr}(\hat{E}_\infty)$. *Then for any compact domain* D *identifiable with a* $(q+1)$*-chain, with coefficients in* $B \equiv \overset{\infty}{\widehat{A}}(E)$, *of an integral quantum submanifold* $S \subset \hat{E}_\infty$, *one has:*

$$\int_{\partial D} \omega|_S = 0.$$

2) *One has*

$$\hat{C}^0_{ons}(\hat{E}_\infty) = \{f \in C_w^\infty(\hat{E}_\infty; B) : \partial f = 0\};$$
$$\hat{C}^q_{ons}(\hat{E}_\infty) = 0, \text{ for } q > m+n, \text{as} \quad C\hat{\Omega}^q(\hat{E}_\infty) = \hat{\Omega}^q(\hat{E}_\infty) \quad \textit{for} \quad q > m+n.$$

3) *One has*

$$\hat{E}_1^{0,q}(\hat{E}_\infty) = \frac{\hat{\Omega}^q(\hat{E}_\infty) \cap d^{-1}(C\hat{\Omega}^{q+1}(\hat{E}_\infty))}{C\hat{\Omega}^q(\hat{E}_\infty) \oplus d\hat{\Omega}^{q-1}(\hat{E}_\infty)} = \hat{C}^q_{ons}(\hat{E}_\infty), q = 0, 1, \ldots.$$

4) *Let* $\omega \in \hat{C}^q_{ons}(\hat{E}_\infty)$. *Then, for any integral quantum submanifold* $S \subset \hat{E}_\infty$, *and for any two cobording* $(q-1)$*-chains, with coefficients in* B, N, N' *on* $S, N = N' + \partial P, P \subset S$, *we have*

$$\int_N \omega|_S = \int_{N'} \omega|_S.$$

PROOF.1) In fact, as $\omega|_S$ is an exact q-quantum form, we can write

$$\int_{\partial D} \omega|_S = \int_D d(\omega|_S) = \int_D (d\omega)|_S = 0.$$

2) and 3) are trivial consequences of definitions.
4) In fact, for any integral quantum submanifold $S \subset \hat{E}_\infty$ one has that ω is nontrivially closed on S. Then we get

$$\int_N \omega|_S = \int_{N'} \omega|_S + \int_{\partial P} \omega|_S = \int_{N'} \omega|_S + \int_P (d\omega)|_S = \int_{N'} \omega|_S.$$

□

COROLLARY 6.6 - *If* S_C *denotes the classic-limit of a quantum submanifold of* $\hat{E}_\infty$ *and* $i : S_C \to S$ *is a section, then any* q*-quantum current or* q*-conservation quantum law* ω *with* $q > m = \dim S_C$ *is zero on* $i(S_C)$*:*

$$i^*(\omega|_S) = 0, \ \textit{if} \quad \deg \omega > m.$$

DEFINITION 6.33 - *The set* $\hat{\mathcal{L}}_{ie}(\hat{E}_\infty)$ *of* **quantum Lie-fields** *of* $\hat{E}_\infty$ *is defined by*

$$\hat{\mathcal{L}}_{ie}(\hat{E}_\infty) \equiv \{\zeta \in C^\infty_w(T\hat{E}_\infty) : [\zeta, \eta] \in C^\infty_w(\mathbb{E}(\hat{E}_\infty)), \forall \eta \in C^\infty_w(\mathbb{E}(\hat{E}_\infty))\}.$$

DEFINITION 6.34 - *The set* $\hat{\mathcal{S}}_{ym}(\hat{E}_\infty)$ *of* **infinitesimal quantum symmetries** *of* $\hat{E}_\infty$ *is defined by*

$$\hat{\mathcal{S}}_{ym}(\hat{E}_\infty) \equiv \frac{\hat{\mathcal{L}}_{ie}(\hat{E}_\infty)}{C^\infty_w(\mathbb{E}(\hat{E}_\infty))}.$$

DEFINITION 6.35 - *The mapping*

$$d_1 : \hat{E}^{0,q}_1(\hat{E}_\infty) \to \hat{E}^{1,q}_1(\hat{E}_\infty)$$

is called **quantum characteristic** *when* $q = m-1$, *and* **quantum Euler-Lagrange** *(or* **quantum variational***) when* $q = m$, *where* $\dim M = m$.
PROPOSITION 6.6 - *Any Lie-quantum field* $\zeta \in C^\infty_w(T\hat{E}_\infty)$ *generates a local 1-parameter group of Lie quantum transformations. Conversely, a 1-parameter group of Lie quantum transformations* $\{\phi_t\}_{t\in\mathbf{R}}$, *identifies a Lie-quantum field. The local*

1-parameter group $\{\phi_t\}_{t\in\mathbf{R}}$, generated by a Cartan quantum field $\zeta \in C^\infty_w(\mathbb{E}(\hat{E}_\infty))$ is trivial, i.e.,

$$y^i \circ \phi = y^i_0 = const., \forall i \in I.$$

Conversely, a trivial 1-parameter group of trivial quantum transformations of $\hat{E}_\infty$ generates a Cartan quantum field ζ. Let $f : \hat{E}_\infty \to \hat{E}_\infty$ be a Lie mapping between QPDEs. Then one has the following induced mappings:

$$\left\{\begin{array}{l} f^* : C\hat{\Omega}^q(\hat{E}'_\infty) \to C\hat{\Omega}^q(\hat{E}_\infty) \\ f^* : \hat{C}^q_{ons}(\hat{E}'_\infty) \to \hat{C}^q_{ons}((\hat{E}_\infty) \\ f^* : \hat{E}^{p,q}_r(\hat{E}'_\infty) \to \hat{E}^{p,q}_r(\hat{E}_\infty) \end{array}\right\}$$

for $r,p,q = 0,1,2\dots$. In particular f^* are isomorphisms if f is a Lie isomorphism. If $\zeta \in Q^\infty_w(\hat{T}\hat{E}_\infty)$, the Lie derivative $\mathcal{L}_\zeta$ identifies the following mapping:

$$\mathcal{L}_\zeta : \hat{\Omega}^q(\hat{E}_\infty) \to \hat{\Omega}^q(\hat{E}_\infty); \quad \mathcal{L}_\zeta : d\hat{\Omega}^q(\hat{E}'_\infty) \to d\hat{\Omega}^q(\hat{E}_\infty).$$

one has:

$$\begin{array}{l} \mathcal{L}_\zeta\omega = \zeta\rfloor d\omega + d(\zeta\rfloor\omega) \\ \mathcal{L}_\zeta(\omega\wedge\alpha) = (\mathcal{L}_\zeta\omega)\wedge\alpha + \omega\wedge(\mathcal{L}_\zeta\alpha). \end{array}$$

If $\zeta \in \hat{\mathcal{L}}_{ie}(\hat{E}_\infty) \Rightarrow [\zeta,\eta]$ is Cartan if η is Cartan and one has the following mapping:

$$\mathcal{L}_\zeta : C\hat{\Omega}^q(\hat{E}_\infty) \to C\hat{\Omega}^q(\hat{E}_\infty).$$

If $\zeta \in \hat{S}_{ym}(\hat{E}_\infty)$, then one has the quotient mapping:

$$\mathcal{L}_{[\zeta]} : \hat{C}^q_{ons}(\hat{E}_\infty) \to \hat{C}^q_{ons}(\hat{E}_\infty).$$

REMARK 6.42 - (EVOLUTIONARY DERIVATION). Let $\pi' : \widehat{E} \to M$ be a bundle, full quantum extension of a natural bundle $\pi : E \to M$ of vector quantum spaces over $M : \widehat{E} \equiv AOE$. Set $\hat{\Phi} \equiv Q^\infty_w(\pi^*_\infty\widehat{E})$, where $\pi_\infty : \hat{E}_\infty \subset J\hat{D}^\infty(W) \to M$ is a QPDE. $\hat{\Phi}$ is called **horizontal** if it is of the form

$$\hat{\Phi}^0 \equiv Q^\infty_w(\hat{E}_\infty; A) \bigotimes_{Q^\infty_w(M;A)} Q^\infty_w(\widehat{V})$$

where $\rho : \widehat{V} \to M$ is the full quantum extension of some bundle of vector quantum spaces over M. Set

$$C\hat{\Phi} \equiv \{\phi \in \hat{\Phi} | (D^\infty s)^*\phi(x) = 0 \text{ if } \quad D^\infty s(x) \in \hat{E}_\infty\}.$$

Then, there exists a unique operation called **horizontalization** $\rceil : \hat{\Phi} \to \hat{\Phi}^0$, which splits the sequence

$$0 \to C\hat{\Phi} \to \hat{\Phi} \xrightarrow{\rceil} \hat{\Phi}^0 \to 0$$

and has the following property

$$(D^\infty s)(\rceil(\phi))(x) = ((D^\infty s)^* \phi)(x), \forall \phi \in \hat{\Phi}, \forall s \in Q^\infty_w(W), D^\infty s(x) \in \hat{E}_\infty.$$

Set $\hat{U}_k(\phi) = \hat{\mathbb{D}}_{0,k}(\rceil(D^k\phi))$, where $\hat{\mathbb{D}}_{0,k}$ is the linear Spencer quantum operator defined by means of the following exact sequence:

$$0 \to \hat{\Phi} \to J\hat{D}^k(\hat{\Phi}) \overset{\hat{\mathbf{D}}_{0,k}}{\to} J\hat{D}^{k-1}(\hat{\Phi}) \bigotimes_{\mathbf{K}} \widehat{\Omega}^1(M) \overset{\hat{\mathbf{D}}_{1,k-1}}{\to} J\hat{D}^{k-2}(\hat{\Phi}) \bigotimes \widehat{\Omega}^2(M) \to 0\,,$$

i.e.,$\hat{\mathbb{D}}_{s,t}(D^t\phi \otimes \omega) = D^{t-1}\phi \otimes d\omega$, with $\phi \in \hat{\Phi}, \omega \in \widehat{\Omega}^s(M)$. In particular, $\hat{\mathbb{D}}_{o,1} : J\hat{D}(\hat{\Phi}) \to \hat{\Phi} \otimes \widehat{\Omega}^1$. For any $\phi \in \hat{\Phi}$ we have the operator

$$l_\phi : \hat{S}_{ym}(\hat{E}_\infty) \to \hat{\Phi}, \quad \zeta \mapsto \zeta \rfloor \hat{U}_1(\phi).$$

The linear operator

$$\Xi_\zeta : \hat{\Phi} \to \hat{\Phi}, \Xi_\zeta(\phi) = l_\phi(\phi)$$

is called **evolutionary derivation** and satisfies

$$\Xi_\zeta(f\phi) = \Xi_\zeta(f).\phi + f\Xi_\zeta(\phi), \forall f \in Q^\infty_w(\hat{E}_\infty : A).$$

THEOREM 6.32 - *The infinite prolongation $\hat{E}_\infty \subset J\hat{D}^\infty(W)$ of a QPDE $\hat{E}_k \subset J\hat{D}^k(W)$ is a reduced quantum diffiety with respect to $J\hat{D}^\infty(W)$. So if $\mathcal{I}(\hat{E}_\infty)$ is the defining ideal of $\hat{E}_\infty$, then*

$$\zeta : \mathcal{I}(\hat{E}_\infty) \to \mathcal{I}(E_\infty)$$

for every Cartan quantum field $\zeta \in C^\infty_w(\mathbb{E}(J\hat{D}^\infty(W)))$. In other words every $\zeta \in C^\infty_w(\mathbb{E}(J\hat{D}^\infty(W)))$ is such that $\zeta|_{\hat{E}_\infty} \in C^\infty_w(T\hat{E}_\infty)$.

THEOREM 6.33 - *The natural imbedding $i : \hat{E}_\infty \hookrightarrow J\hat{D}^\infty(W)$ is a Lie immersion, and the induced mappings*

$$i_* : T\hat{E}_\infty \hookrightarrow TJ\hat{D}^\infty(W), i_* : \mathbb{E}(\hat{E}_\infty) \hookrightarrow \mathbb{E}(J\hat{D}^\infty(W))$$

are respectively monomorphism and isomorphism of bundles of modules.

PROPOSITION 6.7 - (LOCAL PROPERTIES). *One has the following local representations.*

$$\hat{E}_1^{0,m}(\hat{E}_\infty) = \frac{\{\omega = Ldx^1\triangle\cdots\triangle dx^m | L \in C^\infty_w(\hat{E}_\infty, \overset{m}{\hat{B}})\}}{\{\omega = \partial\theta = \sum_{1\leq\mu\leq m}(\partial_\mu J^\mu) | J^\mu \in C^\infty_w(\hat{E}_\infty, \overset{m-1}{\hat{B}})\}}.$$

$$\hat{E}_1^{1,m}(\hat{E}_\infty) = \{\Omega = \sum_{1\leq\alpha\leq s} \Omega_\alpha \delta y^\alpha \wedge dx^1\triangle\cdots\triangle dx^m; \Omega_\alpha \in C^\infty_w(\hat{E}_\infty; \overset{m+1}{\hat{B}})\}.$$

The mapping $d_1^{0,m}$ *is acting by the rule:*

$$\left\{\begin{array}{l} \hat{E}_1^{0,m}(\hat{E}_\infty) \to \hat{E}_1^{1,m}(\hat{E}_\infty) \\ [Ldx^1\triangle\cdots\triangle dx^m] \mapsto [\sum_{1\leq\alpha\leq s}\frac{\delta L}{\delta y^\alpha}\delta y^\alpha \triangle dx^1\triangle\cdots\triangle dx^m] \\ \frac{\delta L}{\delta y^\alpha} = \sum_{i\in I}(-1)^{[i]}\partial_i(\frac{\partial L}{\partial y^\alpha_i}), \alpha = 1,\cdots,s \end{array}\right\}$$

where $[i]$ *is the length of the multi-index*

$$i : [i] \equiv i_1 + \cdots + i_m,$$

and

$$\partial_i \equiv (\partial_1)^{i_1}\cdots(\partial_{m+n})^{i_m}.$$

The A_0*-linear mapping*

$$L \mapsto \frac{\delta L}{\delta y} \equiv (\frac{\delta L}{\delta y^\alpha})$$

is the Euler-Lagrange quantum operator, $\mathcal{E}[-] \equiv \frac{\delta}{\delta y}$, *of the variational quantum calculus. The kernel of* $\mathcal{E}[-]$ *is equal to*

$$\{L \equiv \sum_{1\leq\mu\leq m}(\partial_\mu J^\mu), J^\mu \in Q^\infty_w(\hat{E}_\infty; \overset{m+1}{\hat{B}})\} = \ker\frac{\delta}{\delta y}.$$

In order to interpret the meaning of the above structures with respect to classic limit manifolds, let us define a **solution quantum manifold** of $\hat{E}_k$ as a map $\beta : X \to Sol(\hat{E}_k)$, where X is a d-dimensional manifold ($d \geq 0$) such that the $\underline{\beta} : X \times M \to E, \underline{\beta}(x,y) = D^\infty(\beta(x))(y), x \in X, y \in M$, is quantum smooth. Let $\beta^\bullet : \hat{\Omega}^\bullet(\hat{E}_\infty) \to \hat{\Omega}^\bullet(X\times M)$ be the induced map. $\hat{\Omega}^\bullet(X\times M)$ has a bicomplex structure $\{\hat{\Omega}^{\bullet,\bullet}(X \times$

$M), d_X, d_M\}$ and $\beta^\bullet$ is a bicomplex map (it preserves filtration). Let $\{'\hat{E}_r^{p,q}, 'd_r\}$ be the spectral sequence induced by the filtration $'F^p \equiv \bigoplus_{p'\geq p} \hat{\Omega}^{p',\bullet}(X \times M)$. One has:

$$'\hat{E}_1^{p,q} \cong \hat{\Omega}^p(X) \bigotimes_{A_0} \hat{H}^q(M); \qquad '\hat{E}_2^{p,q} \cong \hat{H}^p(X) \bigotimes_{A_0} \hat{H}^q(M).$$

So $\{'\hat{E}_1^{\bullet,q}, 'd_1\}$ is isomorphic to the $\hat{H}^q(M)$-valued **de Rham quantum complex** on X. Furthermore, $\beta^\bullet$ identifies a spectral sequence homomorphism $\{\beta_r\} : \{\hat{E}_r^{p,q}\} \to \{'\hat{E}_r^{p,q}\}$ that gives a representation of the quantum spectral sequence of $\hat{E}_k$, associated to any solution quantum manifold β. In particular one has the following homomorphisms:

$$\beta_1 : \hat{E}_1^{p,q} \to \hat{\Omega}^p(X) \bigotimes_{A_0} \hat{H}^q(M); \beta_2 : \hat{E}_2^{p,q} \to \hat{H}^p(X) \bigotimes_{A_0} \hat{H}^q(M).$$

If the dimension of X is 0 then a solution quantum manifold is identified with a solution $s \in Sol(\hat{E}_k)$, and we have the homomorphisms:

$$s_r : \hat{E}_r^{0,q} \to \hat{H}^q(M), [\mu] \mapsto s_r([\mu]) = [D^\infty s^* \mu], r = 1, 2.$$

EXAMPLE 6.1 - If $\hat{E}_k$ is an ordinary quantum differential equation then $\hat{E}_1^{0,0}$ is the space of **quantum constants of motions**.

EXAMPLE 6.2 - If $\dim W = m + n + p + q, \dim M = m + n$, and $\pi : W \to M$, is a fiber bundle in the category of quantum manifolds, then $\hat{E}_1^{0,m-1}$ is the space of **classic-conserved quantum laws** of $\hat{E}_k$. Note that the physical interpretation of any classic-conserved quantum law $\mu \in \hat{E}_1^{0,m-1}$ is made by considering the $(m-1)$-differential quantum form $(D^\infty s)^* \mu \in \hat{\Omega}^{m-1}(M)$, where s is a Q_w^∞-solution of $\hat{E}_k$, restricted to any $(m-1)$-dimensional submanifold N of M obtained as the image of a section i of $c : M \to M_C$:

$$i^*(D^\infty s)^* \mu : M_C \to E \bigotimes_{\mathbf{K}} \Lambda_{m-1}^0 M_C.$$

In fact, $\alpha \equiv i^*(D^\infty s)^* \mu$ results in a closed E-valued $(m-1)$-differential form on M_C so that for any two cobording $(m-1)$-chains B, B' on $M_C : B = B' + \partial C$, we have

$$\int_B \alpha = \int_{B'} \alpha + \int_{\partial C} \alpha = \int_{B'} \alpha + \int_C d\alpha = \int_{B'} \alpha.$$

$\hat{E}_1^{0,m}$ is the space of **classic-Lagrangian quantum densities**. Then $d_1^{0,m}[\mu] = 0$ is the **classic-Euler-Lagrange quantum equation** of the Lagrangian quantum density $[\mu] \in \hat{E}_1^{0,m}$ and $d_1^{0,m}$ is the **classic-Euler-Lagrange quantum operator**.

The effective construction of conservation laws for PDEs defined in the category of ordinary manifolds can be made by a fully covariant method given in refs.[100,102]. That method has also been directly extended to the category of supermanifolds [107,112,113,114]. Taking into account previous considerations we can extend that method also to this quantum framework. More precisely we have the following.

THEOREM 6.34 - (CONSERVATION QUANTUM LAWS AND QUANTUM SYMMETRIES FOR QPDEs). *Let $\hat{E}_k \subset J\hat{D}^k(W)$ be a k-order QPDE on a fiber bundle $\pi : W \to M$ between quantum manifolds W and M respectively. Let μ be an m-quantum form (resp. $(m+n)$-quantum form) on the $(m+n)$-dimensional quantum manifold M. Let $\hat{\mathbb{E}}_H$ be the quantum involutive quantum distribution generated by the infinitesimal generators of the quantum symmetry group H of $\hat{E}_k$. Then for any differential invariant*

$$\left\{\begin{array}{l} f : J\hat{D}^k(W) \to \overset{m}{\hat{B}}(E) \\ (\text{resp.} f : J\hat{D}^k(W) \to \overset{m+n}{\hat{B}}(E)) \end{array}\right\}$$

of $\hat{\mathbb{E}}_H$ we have some configuration $s : M \to W$ that leads a conservation quantum law if $Sol(\hat{E}_k) \cap Ext(\beta) \neq \emptyset$, where $Ext(\beta)$ denotes the set of extremals for the following classic- Lagrangian quantum density:

$$\beta \equiv f \circ \pi_k^* \mu \in \hat{\Omega}^m(J\hat{D}^k(W))$$

(resp. Lagrangian quantum density

$$\beta \equiv f \circ \pi_k{}^* \mu \in \hat{\Omega}^{m+n}(J\hat{D}^k(W)).$$

THEOREM 6.35 - *Furthermore, if $\hat{E}_k$ is formally quantum integrable and $\hat{E}_{2k} \equiv \hat{E}_{k+k}$ is a quantum submanifold of $(EL)_f \subset J\hat{D}^k(W)$, where $(EL)_f$ is the classic-Euler-Lagrange (resp. Euler-Lagrange) quantum equation of f, then any quantum solution s of $\hat{E}_k$ brings a dynamic conservation laws of class Q^∞_w associated to f. For any C^∞_w section i of b we have the following map:*

$$\left\{\begin{array}{l} j_i : Sol(\hat{E}_k) \times \hat{E}_1^{0,m-1} \to H^{m-1}(E \bigotimes_{\mathbf{K}} \Omega^\bullet(M_C)) \cong E \bigotimes_{\mathbf{K}} H^{m-1}(M_C) \\ j_i(s,\beta) = [((D^\infty s) \circ i)^* \beta] \end{array}\right\}.$$

PROOF. The proof is similar to the ordinary case. □

In general the partial map $j_{i,s} : \hat{E}_1^{0,m-1} \to E \bigotimes_{\mathbf{K}} H^{m-1}(M_C)$ is not surjective for any s belonging to $Sol(\hat{E}_k)$. This suggests the following.

DEFINITION 6.36 - *For any C^∞_w section $i : M_C \to M$ of b let $[E \bigotimes_{\mathbf{K}} H^{m-1}(M_C)]_i$ denote the image of the homomorphism*

$$(a_i)^* : \hat{H}^{m-1}(M) \to E \bigotimes_{\mathbf{K}} H^{m-1}(M_C).$$

$[E \bigotimes_{\mathbf{K}} H^{m-1}(M_C)]_i$ represents the space of classic conserved quantum laws led by the even m-dimensional manifold $i(M_C)$. A QPDE $\hat{E}_k \subset J\hat{D}^k(W)$ is **wholly quantum cohomologic** *if in any cohomological class $[\mu] \in [E \bigotimes_{\mathbf{K}} H^{m-1}(M_C)]_i$ there exists an $(m-1)$-differential form (nontrivial) obtained by means of the above theorem, for a suitable Q^∞_w solution s of $\hat{E}_k$.*

THEOREM 6.36 - (CRITERIA OF WHOLLY QUANTUM COHOMOLOGICITY FOR QPDEs). *Let $\hat{E}_k$ be a k-order formally quantum integrable QPDE on a fiber bundle $\pi : W \to M$, in the category of quantum manifolds, where M is an $(m+n)$-dimensional quantum manifold. Let $\hat{E}_k$ be complete with respect to a conservation law $\beta \in \hat{\Omega}^{m-1}(J\hat{D}^k(W))$, which is of the type considered in the above theorem. Then we have the following criteria of wholly quantum cohomologicity: $\hat{E}_k$ is wholly quantum cohomologic if $\hat{H}^{m-1}(M) \cong \pi_0(Sol(\hat{E}_k))$, where π_0 denotes the 0-th Hurewicz homotopy group functor. If $\hat{E}_k$ is also a conic QPDE, it is wholly quantum cohomologic if*

$$\hat{E}_2^{0,m-1} \cong \pi_0(Sol(\hat{E}_k)).$$

PROOF. Taking into account that any section $i : M_C \to M$ identifies a natural epimorphism

$$u(a_i)^* : \hat{H}^{m-1}(M) \to [E \bigotimes_{\mathbf{K}} H^{m-1}(M_C)]_i$$

we get that $\hat{E}_k$ will be surely wholly quantum cohomologic if in any class of $\hat{H}^{m-1}(M)$ there is a differential form of the type $(D^k s)^*\beta$ for some quantum solution s of $\hat{E}_k$. But this is equivalent to saying that $\hat{H}^{m-1}(M) \cong \pi_0(Q^\infty_{adm}(\hat{E}_k))$, where $Q^\infty_{adm}(\hat{E}_k)$ is the set of Q^∞_w-section u of $\hat{E}_k$ that can be represented as $u = D^k s$ for some section s of π. For functorial property we also have that $\pi_0(Q^\infty_{adm}(\hat{E}_k)) \cong \pi_0(Sol(\hat{E}_k))$.

THEOREM 6.37 - *Let $\hat{E}_k \subset J\hat{D}^k(W)$ be a linear QPDE, that is complete with respect to a classic conserved quantum law, then $\hat{E}_k$ is wholly quantum cohomologic if*

$$\hat{H}^{m-1}(M) = 0.$$

PROOF. Taking into account that for any section $s \in Sol(\hat{E}_k)$, there exists the following homomorphism

$$s_2 : \hat{E}_2^{0,q} \to \hat{H}^q(M)$$

and that for conic QPDE the above homomorphism s_2 becomes an isomorphism ([78]), we get that if

$$\hat{E}_2^{0,m-1} \cong \hat{H}^{m-1}(M) \cong \pi_0(Sol(\hat{E}_k))$$

$\hat{E}_k$ is surely wholly quantum cohomologic. □

EXAMPLE 6.3 - All the linear quantum equations of the field theory on the Minkowski quantum space-time (that is a quantum manifold of dimension $4+4$), are wholly quantum cohomologic.

6.5 - QUANTUM DISTRIBUTIONS AND SINGULAR SOLUTIONS OF QUANTUM PDEs

In this section we shall consider quantum distributions for fiber bundles of quantum spaces on quantum manifolds and as these are related to solutions of QPDEs. Furthermore, we will show that any singular solution L of an affine QPDE is associated to a quantum distributive kernel (**quantum Green kernel** of L).

DEFINITION 6.37 - *Let M be an $(m+n)$-dimensional quantum manifold. Let $\pi : E \to M$ be a fiber bundle of $(m'+n')$-dimensional vector quantum spaces. E can be considered whether as a fiber bundle of A_0-modules or fiber bundle of $\mathbf{K}$-vector spaces over M. In the following table we define some structures, associated to $\pi : E \to M$. (M_C is the classic limit of M and i is any section of $M \to M_C$.)*

TAB.6.5 - **Spaces associated to** $\pi : E \to M$	
Definition	*Space Name*
$C_0^\infty(i^*E)$	$C_w^\infty(i^*E)$ *compact supp.* (*)
$(i^*E)' \equiv (i^*E)^* \bigotimes_{\mathbf{K}} \Lambda_m^0 M_C$	**formal adjoint** *of* $i^*E((i^*E)'' \cong i^*E)$
$(C_0^\infty(i^*E))' \equiv \mathcal{L}(i^*E, \mathbf{K})$	**sp.dist.sects.** i^*E
$(C_0^\infty((i^*E)'))'$	**sp.dist.sects.** $(i^*E)'$
$E' \equiv E^* \bigotimes_{\mathbf{K}} \Lambda_m^0 M$	**formal adjoint** *of* E
$E^{\blacksquare} \equiv E^+ \bigotimes_A Hom_{\mathbf{K}}(\Lambda_0^m M, A)$	**formal q.adjoint** *of* E
$C_0^\infty(E)^* \equiv Hom_{\mathbf{K}}(C_0^\infty(E), \mathbf{K})$	**alg.dual** *of* $C_0^\infty(E)$
$C_0^\infty(E)^+ \equiv Hom_{A_0}(C_0^\infty(E), A)$	**sp.q.dists.** *on* E

*(Nuclear, reflexive space)

REMARK 6.43 - Note that $C_0^\infty(E)$ is the space of smooth sections of $\pi : E \to M$ with compact support on some m-dimensional submanifold of M, identified by means of some section i of $c : M \to M_C$.

The following theorems can be proved by simulating analogous theorems for super PDEs given in ref.[112,113,114].

THEOREM 6.38 - (INTERPRETATION OF QUANTUM FIELDS AS QUANTUM DISTRIBUTIONS AND A-VALUED DISTRIBUTIONS).

1) *One has the following exact sequences of* $\mathbf{K}$*-vector spaces:*

$$0 \to C^\infty_w((i^*E)') \xrightarrow{\gamma} (C^\infty_0(i^*E))', \gamma : \sigma \mapsto \gamma(\sigma) : f \mapsto \int_{M_C} < \sigma, f >;$$

$$0 \to C^\infty_w(i^*E) \xrightarrow{\gamma} (C^\infty_0((i^*E)'))', \omega : f \mapsto \omega(f) : \sigma \mapsto \int_{M_C} < \sigma, f > .$$

One has the following exact sequence of vector bundles over M_C*:*

$$i^*E' \to (i^*E)' \to 0.$$

As a consequence we also have the following:

$$C^\infty_0(i^*E') \to C^\infty_0((i^*E)') \to 0; (C^\infty_0(i^*E'))' \leftarrow (C^\infty_0((i^*E)'))' \leftarrow 0.$$

2) *For any section* i *of* $b : M \to M_C$ *we have the following mappings:*

$$l[i] : C^\infty_w(E) \times C^\infty_0(E') \to A; \qquad l[i] : C^\infty_w(E) \times C^\infty_0(E^{\blacksquare}) \to A$$

given by

$$l[i](s, \alpha) = \int_{M_C} i^* < \alpha, s >, \forall \alpha \in C^\infty_0(E'), C^\infty_0(E^{\blacksquare}), s \in C^\infty(E)$$

where the integral belongs to $\mathbf{K}$ *in the first case and to* A *in the second one. Recall, (see Theorem 6.5), that* $i^*\eta : M_C \to A \bigotimes_{\mathbf{K}} \Lambda^0_m M_C$, *for any quantum* m*-form* η *on* M. *So, for any section* i *of* b *we have the following natural homomorphisms of* $\mathbf{K}$*-vector spaces and* A_0*-modules respectively :*

$$C^\infty_w(E) \xrightarrow{\underline{l}[i]} (C^\infty_0(E'))^+; C^\infty_w(E) \xrightarrow{\underline{l}[i]} (C^\infty_0(E^{\blacksquare}))^+$$

given by

$$\underline{l}[i](s)(\alpha) = \int_{M_C} i^* < \alpha, s > .$$

So for any of such sections i *we can represent a quantum field* $s : M \to E$ *as a quantum distribution.*

3) *If the vector bundle* $\pi : E \to M$ *is such that for any section* $i : M_C \to M$ *there exists a canonical mapping* $\mathbb{B}(i) : E \to A \bigotimes_{\mathbf{K}} E_C$ *where* $\pi_C : E_C \to M_C$ *is a vector bundle over* M_C *such that the following diagram*

$$\begin{array}{cccccc} & M & \leftarrow & E & \\ i & \uparrow & & \downarrow & \mathbf{B}(i) \\ & M_C & \leftarrow & A \bigotimes_{\mathbf{K}} E_C & \end{array}$$

is commutative, then one has the following natural **K***-homomorphism:*

$$\left\{\begin{array}{l} C^\infty_w(E) \to C^\infty_0(E'_C)'\bigotimes_{\mathbf{K}} A \\ \hat{s}[i](\phi) = \int_{M_C} < \phi, i^* s >, \quad \forall \phi \in C^\infty_0(E'_C) \end{array}\right\}$$

where $i^* s \equiv \mathbb{B}(i) \circ s \circ i : M_C \to A \bigotimes_{\mathbf{K}} E_C$. *In this way any quantum field* $s : M \to E$ *can be represented as a* A*-valued distribution.*

THEOREM 6.39 - (LOCAL REPRESENTATION). *Let* $\{g_U\}$ *be a partition of unity on* M_C *subordined on the coordinate-covering* $\{U\}$. *Then the local representation of* $\hat{s}[i]$ *is the following:*

$$< \hat{s}[i], \phi >= \sum_{1\le j\le p} \sum_U \int_U g_U \phi_{U,j} s^j_U \otimes dx^1 \wedge \cdots \wedge dx^m$$

where $s^j_U : U \to A$, *and* $\{e_{U,j}\}_{1\le j\le p}$ *is a local basis for sections of* $E_C \to M_C$, *where* $\phi = \sum_{1\le j\le p} \phi_{U,j} \theta^j_U \otimes dx^1 \wedge \cdots \wedge dx^m$ *is the local expression of* ϕ, *with* $\phi_{U,j} : U \to \mathbf{K}$. *Furthermore,* $\{\theta^j_U\}$ *is the local basis dual of* $\{e_{U,j}\}$. *Note that* E_C *is a* **K***-vector bundle over* M_C *of fiber dimension* p.

THEOREM 6.40 - *The above considerations can also be extended to any chiral quantum field* $s : M \to W$, *i.e., a section of the fiber bundle* W *that is not a vector bundle on* M. *In fact, it is enough to substitute* $E \equiv s^* vTW$, *and fix an infinitesimal variation* $\nu \equiv \partial \tilde{s}$ *of some deformation* $\tilde{s}$ *of* s.

$$\left\{\begin{array}{l} \hat{s}[i|\nu] \in C^\infty_0((s^* vTW)')' \bigotimes_{\mathbf{K}} A \\ \hat{s}[i|\nu](\phi) = \int_{M_C} < \phi, i^* \nu > \quad , \forall \phi \in C^\infty_0((s^* vTW)') \end{array}\right\}$$

Of course if W *is a bundle of geometric objects on* M, *and 1-parameter group of diffeomorphisms of* M *is introduced on* M, $f : \mathbf{R} \times M \to M$, *such that it induces one* $f_C : \mathbf{R} \times M_C \to M_C$ *on* M_C, *then one has a canonical deformation* $\tilde{s}$ *of* s *and so a canonical vertical vector field* $\dot{s} \equiv \partial \tilde{s} : M \to s^* vTW \equiv E$ *can be recognized. Then we put*

$$\hat{s}[i] \equiv \hat{s}[i|\dot{s}] .$$

The local representation is the following:

$$< \hat{s}[i], \phi >= \sum_{1\le j\le p} \sum_U \int_U g_U \phi_{U,j} \dot{s}^j_U \otimes dx^1 \wedge \cdots \wedge dx^m .$$

THEOREM 6.41 - *Let M and N be $(m+n)$-dimensional quantum manifolds. Let $h \equiv (h_F, h_N)$ be a proper homomorphism of fiber bundles of A_0- modules, i.e. the following diagram:*

$$\begin{array}{ccccc} & F & \stackrel{h_F}{\rightarrow} & E & \\ \pi_F & \downarrow & & \downarrow & \pi_E \\ & N & \stackrel{h_N}{\rightarrow} & M & \end{array}$$

is commutative. "Proper" in the sense that the inverse image of a compact in M is also a compact. Then h identifies the following homomorphism for any section i of $b: N \to N_C$, where N_C is the classic limit of N:

$$\omega_h[i] : C^\infty(F) \to (C_0^\infty(E^\blacksquare))^+, \omega_h[i](f) : \sigma \mapsto \int_{N_C} i^* < h^*\sigma, f > .$$

So, for any section $f \in C^\infty(F)$ we get a quantum distribution of $E^\blacksquare$: $\omega_h[i](f) \in (C_0^\infty(E^\blacksquare))^+$.

THEOREM 6.42 - *Let L be an $(m+n)$-dimensional integral quantum manifold of the Cartan quantum distribution on $J\hat{D}^k(E)$, such that:*
(i) *$\pi_k|L : L \to M$ is a proper map;*
(ii) *${\omega_1}^{(k)}(L_C) = 0$, where L_C is the classic limit of L, and*

$${\omega_1}^{(k)}(L_C) \equiv i^*_{L_C}{\omega_1}^{(k)} \in H^1(L_C; \mathbf{Z}_2)$$

where

$${\omega_1}^{(k)} \in H^1(I_k(E_C); \mathbf{Z}_2)$$

is the first generator of

$$H^1(I_k(E_C); \mathbf{Z}_2) \cong \mathbf{Z}_2[{\omega_1}^{(k)}, \dots, {\omega_m}^{(k)}]$$

where $I_k(E_C)$ is the bundle of Grassmannian of oriented integral planes on $JD^k(E_C)$, where E_C is the classic limit of E. Then L identifies an element $F[L|i]$ of $C_0^\infty(E^\blacksquare)^+$ for any section i of $L \to L_C$, where L_C is the classic limit of L.

THEOREM 6.43 - *Let E and F be bundles of vector quantum spaces over an $(m+n)$-dimensional quantum manifold M. Let $\kappa : C_w^\infty(E) \to C_w^\infty(F)$ be a quantum differential operator over M of order $\leq k$. Then to κ we can associate its formal quantum adjoint $\kappa^\blacksquare : C_w^\infty(F^\blacksquare) \to C_w^\infty(E^\blacksquare)$ that is a quantum differential operator of order $\leq k$ such that*

$$(6.25) \qquad \int_{M_C} i^* < \kappa(e), \alpha > = \int_{M_C} i^* < e, \kappa^\blacksquare(\alpha) >, \forall e \in C_0^\infty(E), \alpha \in C_w^\infty(F^\blacksquare)$$

and for any section i of $c : M \to M_C$. More precisely, one has:

$$(6.26) \qquad < \kappa^{\blacksquare}(\alpha), e >=< \alpha, \kappa(e) > -dG(e \otimes \alpha), \forall e \in C_0^\infty(E), \alpha \in C_w^\infty(F^{\blacksquare})$$

where G is a $(k-1)$-order quantum differential operator (Green quantum operator)

$$(6.27) \qquad G : C_w^\infty(E \bigotimes F^{\blacksquare}) \to \hat{\Omega}^{m-1}(M),$$

THEOREM 6.44 - *One has the following properties: (a) The mapping $K \mapsto K^{\blacksquare}$ is A_0-linear. (b) Let κ_1, κ_2 be two quantum differential operators then*

$$(\kappa_2 \circ \kappa_1)^{\blacksquare} = \kappa_1^{\blacksquare} \circ \kappa_2^{\blacksquare}.$$

(c) In particular, for trivial bundles $E \cong F \cong M \times A_0$ one has

$$E^+ \cong M \times A, E^{\blacksquare} \cong F^{\blacksquare} \cong Hom_{\mathbf{K}}(\Lambda_0^m M, A) \equiv (\Lambda_0^m M)^\star$$

and the operation $K \mapsto K^{\blacksquare}$ transforms scalar differential operators into quantum differential operators between quantum m-differential forms on M.

THEOREM 6.45 - (GREEN's FORMULA). *Let N be a compact domain of M. Endow N (and hence ∂N) with the induced orientation from M. Let κ be a quantum differential operator on M between E and F, of order $\leq k$. Let G be a Green quantum operator for κ. Let $e \in C_0^k(E)$ and $\alpha \in C_0^k(F^{\blacksquare})$ such that* $\mathrm{supp}(e) \cap N \neq \emptyset \neq$ $\mathrm{supp}\, \alpha \cap N$. *Then one has*

$$(6.28) \qquad \int_{N_C} i^* < \kappa(e), \alpha > - \int_{N_C} i^* < e, \kappa^{\blacksquare}(\alpha) >= \int_{\partial N_C} i^* G(e \otimes \alpha)$$

where N_C is the classic limit of N, and i is any section of $b : N \to N_C$. In particular, if $\partial N = \emptyset$, one has

$$(6.29) \qquad \int_{N_C} i^* < \kappa(e), \alpha >= \int_{N_C} i^* < e, \kappa^{\blacksquare}(\alpha) > .$$

EXAMPLE 6.4 - (INFINITESIMAL TRANSFORMATIONS). Let κ be the quantum differential operator of order ≤ 1 defined by means of the infinitesimal variation induced by a vector field ξ on M: $\kappa(e) = \partial\tilde{e}$, for $u : M \to \widehat{E} \equiv AOE$ a section of a natural bundle of A_0-modules $\pi : E \to M$ over $M, \tilde{e}_t \equiv \mathbb{B}({\phi_t}^{-1}) \circ e \circ \phi_t, \mathbb{B} \equiv$ functor of the natural structure of E. Then, the Green quantum operator of K is $G(e \otimes \alpha) = \xi \rfloor < e, \alpha >$. In fact, one has the following Green's formula:

$$< \partial\tilde{e}, \alpha > + < e, \partial\tilde{\alpha} >= d(\xi \rfloor < e, \alpha >).$$

EXAMPLE 6.5 - (EXTERIOR DIFFERENTIAL). If

$$E \equiv Hom_{\mathbf{K}}(\Lambda_0^p M, A) \equiv (\Lambda_0^p M)^\star$$

one can identify $E^\blacksquare$ with $Hom_{\mathbf{K}}(\Lambda_0^{m-p-1} M, A) \equiv (\Lambda_0^{m-p-1} M)^\star$. So, if $\kappa \equiv d$ one has $\kappa^\blacksquare = (-1)^{p+1} d$. The Green quantum operator is the exterior product:

$$G : \hat{\Omega}^p(M) \bigotimes \hat{\Omega}^{m-p-1}(M) \to \hat{\Omega}^{m-1}(M)\,.$$

THEOREM 6.46 - *Let $\kappa : C_w^\infty(E) \to C_w^\infty(F)$ be a k-order A_0-linear differential operator, where E and F are fiber bundles of A_0-modules over an $(m+n)$-dimensional quantum manifold M. Then for any section i of $b : M \to M_C$ we have a natural extension of K to quantum distributional spaces :*

$$\begin{array}{ccc} C_0^\infty(E^\blacksquare)^+ & \xrightarrow{\hat{\kappa}[i]} & C_0^\infty(F^\blacksquare)^+ \\ \bigcup & & \bigcup \\ C_w^\infty(E) & \xrightarrow{\kappa} & C_w^\infty(F)\,. \end{array}$$

THEOREM 6.47 - (DISTRIBUTIONAL EXTENSION FOR GREEN FORMULA ON COMPACT QUANTUM MANIFOLDS WITH BOUNDARY). *One has the following formula for linear differential operators of first order*

$$\hat{\kappa}[i](\omega)(\alpha) = \omega(\kappa[i]^\blacksquare(\alpha)) + \hat{\sigma}_1(\kappa)[i]|_{\partial M}(\omega)(\alpha), \tag{6.30}$$

where $\omega \in (C_0^\infty(E^\blacksquare))^+, \alpha \in C_0^\infty(F^\blacksquare), \hat{\kappa}$ is the distributive extension of $\kappa. : C_w^\infty(E) \to C_w^\infty(F)$, and $\hat{\sigma}_1(\kappa)[i]|_{\partial M}(\omega)$ is the distribution supported on ∂M identified by the symbol $\dot{\sigma}_1(\kappa)$ of κ.

THEOREM 6.48 - *Let $\kappa : C_w^\infty(E) \to C_w^\infty(F)$ be a k-order A_0-linear differential operator. Let $\hat{E}_k \equiv \ker_f \kappa \subset J\hat{D}^k(E)$ be an affine quantum equation, where $f \in C_w^\infty(F)$. Suppose that $L \subset J\hat{D}^{k+s}(E)$ is an $(m+n)$-dimensional integral quantum manifold such that $\pi_{k+s}|L : L \to M$ is a proper map, and $\omega_1^{(k+s)}(L_C) = 0$. Then, for any section i of $b : M \to M_C, F[L|i] \in (C_0^\infty(E^\blacksquare))^+$ is a quantum distribution-solution of $\hat{E}_k$ iff L is a multivalued quantum solution. Furthermore, for any section i of b we have that the quantum distributional solution of $\hat{E}_k$ can be written as follows: $\nu = G[i]_f$, where $G[i]$ is the Green kernel of K at the section i, that is $G[i]$ is the quantum distributional kernel*

$$G[i] \in C_0^\infty(E^\blacksquare \boxed{\times} F)^+ \cong C_0^\infty(F \boxed{\times} E^\blacksquare)^+$$

such that

$$(\hat{\kappa}[i] \otimes \mathbf{1})(G[i]) = (\mathbf{1} \otimes \hat{\kappa}[i])(G[i]) = \hat{\mathbb{D}}[i],$$

where $\hat{\mathbb{D}}[i] \in C_0^\infty(F \boxtimes F^\blacksquare)^+$ *is the Dirac quantum kernel of* F*, at the section* i*, i.e.,*

$$\hat{\mathbb{D}}[i](f \otimes \alpha) = \int_{M_C} i^* < f, \alpha >, \forall f \in C_0^\infty(F), \alpha \in C_0^\infty(F^\blacksquare).$$

THEOREM 6.49 - *In the particular case that* K *is a first order differential quantum operator we get:*

(a) *As for the Green's kernel* $G[i]$ *of the operator* $\hat{\kappa}[i]$

$${}_fG[i] \in (C_0^\infty(E^\blacksquare))^+$$

for any $f \in C^\infty(F)$, *we have*

$$\hat{\kappa}[i]({}_fG[i])(\alpha) =_f G[i](\kappa^\blacksquare(\alpha)) + \hat{\sigma}_1(\kappa)[i]|_{\partial M}({}_fG[i])(\alpha), \forall \alpha \in C_0^\infty(F^\blacksquare).$$

(b) *In the case where the Green-kernel* $G[i]$ *is such that it can be identified with a section* $G[i] \in C^\infty(E \boxtimes F)$*, then we can write:*

$$\int_{M_C} i^* < \tilde{\kappa}[i]({}_fG[i]), \alpha >= \int_{M_C} i^* <_f G[i], \kappa^\blacksquare(\alpha) > + \int_{\partial M_C} i^* < \dot{\sigma}_1(\kappa)({}_fG[i]), \alpha > .$$

THEOREM 6.50 - *The Green kernel satisfies the following equivalent equations*

$$\hat{\kappa}[i]({}_fG[i]) =_f \mathbb{D}[i], \forall f \in C_w^\infty(F)$$

$${}_fG[i](\kappa^\blacksquare(\alpha)) + \hat{\sigma}_1(\kappa)[i]|_{\partial M}({}_fG)(\alpha) = \int_{M_C} i^* < f, \alpha >, \forall f \in C_w^\infty(F), \alpha \in C_0^\infty(F^\blacksquare).$$

If $G[i] \in C^\infty(E \boxtimes F)$*, the above equations become*

$$\hat{\kappa}[i]({}_fG[i]) = {}_f\mathbb{D}[i], \forall f \in C_w^\infty(F) \tag{6.31}$$

and

$$\int_{M_C} i^* <_f G[i], \kappa^\blacksquare(\alpha) > + \int_{\partial M_C} i^* < \dot{\sigma}_1(\kappa)({}_fG[i]), \alpha >= \int_{M_C} i^* < f, \alpha > . \tag{6.32}$$

Finally, we are ready to give the following important theorem that relates singular solutions of affine QPDEs with quantum distributional kernels.

THEOREM 6.51 - *Any singular solution* L *of an affine QPDE* $\hat{E}_k \equiv \ker_f \kappa \subset J\hat{D}^k(E)$*, that satisfies some boundary conditions, i.e.,* $\partial L = X_1 \dot{\bigcup} X_2$*, where* X_1 *and* X_2 *are some fixed hypersurfaces Cauchy data, (*$\dot{\bigcup}$ *denotes disjoint union), and such that:*

(i) $\pi_{k+s}|L : L \to M$ *is proper map;*

(ii) $\omega_1^{(k+s)}(L_C) = 0$*, identifies, for any section* $i : M_C \to M$*, a quantum distributional kernel* $G[L|i] \in C_0^\infty(E^\blacksquare \boxtimes F^\blacksquare)^+$.

PROOF. Let $F[L|i]$ be the quantum distribution associated to L (and i) by means of the above theorem. Then, $G[L|i] \in C_0^\infty(i^*E^{\blacksquare}\boxed{\times}i^*F^{\blacksquare})^+$ is defined by means of the following:

$$G[L|i](\alpha \otimes \phi) = F[L|i](\alpha) \int_{M_C} i^* < \phi, f > .$$

□

DEFINITION 6.38 - *We call* $G[L|i]$ *the* **quantum Green kernel of the singular solution** $L \subset \hat{E}_{k+s}$ *satisfying the boundary condition* $\partial L = X_1 \dot{\bigcup} X_2$. *In particular, if L is the singular solution corresponding to the Green kernel* $G[i]$ *of* κ, *i.e.,* $F[L|i] = G[i]_f$, *then the corresponding quantum Green kernel is defined by*

$$G[L|i](\alpha \otimes \phi) = G[i]_f(\alpha) \int_{M_C} i^* < \phi, f > .$$

In this case, we call L **Green singular solution of the boundary value problem** $\partial L = X_1 \dot{\bigcup} X_2$.

6.6 - TUNNEL EFFECTS IN QUANTUM PDEs

In this section we consider cobordism in QPDEs and we will show how this is related to the canonical quantization.

Let $\Omega(\hat{E}_k)$ denote the set of integral quantum submanifolds of $\hat{E}_k$ cobording two Cauchy hypersurfaces data of $\hat{E}_k$.

THEOREM 6.52 - *Let* $\hat{E}_k = \ker_\chi \kappa \subset J\hat{D}^k(W)$ *be a QPDE obtained as the kernel of a k-order quantum differential operator* $\kappa : J\hat{D}^k(W) \to K$, *with respect to a* Q_w^∞ *section* $\chi : M \to K$, *of the fiber bundle* $\underline{\pi} : K \to M$, *defined in the category of quantum manifolds. Then for any section* $s \in Q_w^\infty(W)$, *solution of* $\hat{E}_k$, *and* $j[\chi] = \partial\tilde{\chi}$, *where* $\tilde{\chi}$ *is a quantum deformation of* χ, *we can associate to* $\hat{E}_k$ *an affine quantum differential equation*

$$\hat{E}_k[s] = \ker_{j[\chi]} J[s] \subset J\hat{D}^k(s^*vTW)$$

where $J[s] : C_w^\infty(i^*s^*vTW) \to C_w^\infty(i^*\chi^*vTK)$ *is a* A_0*-linear map. We call* $\hat{E}_k[s]$ **Jacobi quantum equation** *of* $\hat{E}_k$.

THEOREM 6.53 - *For any section* i *of* $b : M \to M_C$, *we have the following natural extension of* $J[s]$*:*

$$\begin{array}{ccc} (C_0^\infty(s^*(vTW)^{\blacksquare})^+ & \xrightarrow{J[s|i]} & (C_0^\infty(\chi^*vTK)^{\blacksquare})^+ \\ \uparrow & & \uparrow \\ Q_w^\infty(s^*vTW) & \xrightarrow[J[s]]{} & Q_w^\infty(\chi^*vTK) \end{array}$$

So, the distributional solution of $\hat{E}_k[s]$ *can be written as follows:* $\nu = G[s|i]_{j[s]}$, *where* $G[s|i]$ *is the Green's kernel of* $J[s]$, *(associated to the section* i *of* c*).*

DEFINITION 6.39 - *For any function*

$$f[i] : \Omega(\hat{E}_\infty) \to B \,,$$

the physical quantum observable for $\hat{E}_k \subset J\hat{D}^k(W)$ *is defined as*

$$f[i](s) = \int_{M_C} i^*(\underline{f} \circ D^p s) d\mu$$

where μ *is a suitable measure on* M_C, *where*

$$\underline{f} : J\hat{D}^p(W) \to B$$

is a function on $J\hat{D}^k(W), p \geq 0$, *and* i *is a section of* $c : M \to M_C$. *Let us denote by* $\mathcal{P}_{hys}(\hat{E}_k)$ *the set of physical quantum observables of* $\hat{E}_k$. *Then, we have the following structure of Lie quantum algebra, (for any section* $i : M_C \to M$*):*

$$\left\{ \begin{array}{l} (f_1, f_2)[i](s) = G^+[s|i](jf_1(s) \otimes jf_2(s) \otimes \eta) - G^-[s|i](jf_1(s) \otimes jf_2(s) \otimes \eta) \\ \qquad \equiv \hat{G}[s|i](jf_1(s) \otimes jf_2(s) \otimes \eta) \end{array} \right\}$$

where

$$jf : C^\infty_w(W) \to \bigcup_{s \in C^\infty_w(W)} C^\infty_0(s^* vT^+ W)$$

is the current associated to f *as follows:* $jf(s) = (vdf) \circ s$, *where* vdf *is the vertical differential of the quantum differentiable function* $f : W \to E$. $G^\pm[s|i]$ *are advanced and retarded Green-kernels respectively, with respect to the space-time foliation on* $i(M_C)$. *Furthermore* η *is a volume form on* M_C.

DEFINITION 6.40 - *We call* $\hat{G}[s|i]$ *the* **quantum propagator at the sections** s *and* i. *Furthermore, we call* **canonical quantization** *of* $\hat{E}_k$ *the quantum algebra defined by means of the following commutator:*

$$[\hat{f}_1, \hat{f}_2][i](s) = i\hbar \hat{G}[s|i](jf_1(s) \otimes jf_2(s) \otimes \eta)\mathbf{1}.$$

REMARK 6.44 - This quantum algebra has been associated to a quantum PDE by using a way similar to that previously introduced in [108,111], and previous chapter, in order to quantize PDEs. The big difference between the canonical quantization of a classical theory and the present one is that now the propagator $\hat{G}$ is a function of a quantum field instead of classical one.

2 THEOREM 6.54 - (TOPOLOGY TRANSITIONS AS QUANTUM TUNNEL EFFECTS IN QPDEs).
Let $\hat{E}_k \subset J\hat{D}^k(W)$ be a k-order QPDE on the fiber bundle $\pi : W \to M, \dim W = m+n, \dim M = m$. Let Y_0 and Y_1 be Cauchy hypersurfaces data of $\hat{E}_k$. Let us assume that there are closed domains $N_i \subset Y_i, i = 0, 1$, such that there exists a (singular) cobordism $L \in \Omega(\hat{E}_k)$ between N_0 and $N_1 : \partial L = N_0 \times 0 \bigcup N_1 \times 1$. Then there exists an admissible Morse function $f : L_C \to [a, b]$, where L_C is the classic limit of L, such that $(N_0)_C = f^{-1}(a), (N_1)_C = f^{-1}(b)$, such that:
(A)(SIMPLE QUANTUM TUNNEL EFFECT). *If f has a critical point q of index k, then there exists a k-cell $e^k \subset L_C - (N_1)_C$ and a $(m-k)$-cell $e_*^{(m-k)} \subset L_C - (N_0)_C$ such that:*
(i) $e^k \cap (N_0)_C = \partial e^k$;
(ii) $e_*^{(m-k)} \cap (N_1)_C = \partial e_*^{(m-k)}$;
(iii) *There is a deformation retraction of L_C onto $(N_0)_C \cup e^k$;*
(iv) *There is a deformation retraction of L_C onto $(N_1)_C \cup e_*^{(m-k)}$;*
(v) $e_*^{(m-k)} \cap e^k = q$; $\quad e_*^{(m-k)} \pitchfork e^k$. *[Here $\pitchfork$ denotes transverse].*
(B)(MULTI QUANTUM TUNNEL EFFECT). *If f is of type $(\nu_0, \dots, \nu_m)$, such that f has just one critical value $c, a < c < b$, then there are disjoint k-cells $e_i^k \subset L_C - (N_1)_C$, and disjoint $(m-k)$-cells $(e_*)_i^{(m-k)} \subset L_C - (N_0)_C, 1 \le i \le \nu_k, k = 0, \dots, m$, such that:*
(i) $e_i^k \cap (N_0)_C = \partial e_i^k$;
(ii) $(e_*)_i^{(m-k)} \cap (N_1)_C = \partial (e_*)_i^{(m-k)}$;
(iii) *There is a deformation retraction of L_C onto $(N_0)_C \cup \{\bigcup_{i,k} e_i^k\}$;*
(iv) *There is a deformation retraction of L_C onto $(N_1)_C \cup \{\bigcup_{i,k} (e_*)_i^{(m-k)}\}$;*
(v) $(e_*)_i^{(m-k)} \cap e_i^k = q_i \in L_C; (e_*)_i^{(m-k)} \pitchfork e_i^k$.
In the cases (A) and (B), L changes topology by passing from N_0 to N_1.
(C)(NO TOPOLOGY TRANSITION). *If f has no critical point then $L_C \cong (N_0)_C \times I$, where $I \equiv [0, 1]$. In this case L does not change topology by passing from N_0 to N_1.*
PROOF. The proof can be conducted on L_C by using Theorem 23 in ref.[111]. Then taking into account that $b : L \to L_C$ is a vector fiber bundle the conclusions on L get out from the fact that $H^\bullet(L; A) \cong A \bigotimes_{\mathbf{K}} H^\bullet(L_C)$; (see tab 6.4). □

2 THEOREM 6.55 - *Let $\hat{E}_k \subset J\hat{D}^k(W)$ be k-order QPDE on the fiber bundle $\pi : W \to M, \dim M = m + n$. Let (N_i, h_i), $i = 0, 1$, be a couple of Cauchy data for the system $\hat{E}_k$, defined on $\hat{E}_{k+r}$. Then one has a canonical mapping*

$$\phi_{(N,h)} : H^{m-1}(\hat{I}_{k+r}(W); A) \to A$$

where $N \equiv N_0 \times 0 \bigcup N_1 \times 1$, and h is the natural extension to N of h_0 and h_1. We call $H^{m-1}(\hat{I}_{k+r}(W); A)$ the $(m-1)$- **Leray-Serre quantum cohomology** *of*

$J\hat{D}^{k+r}(W)$. *Then, in order for an integral cobord V cobording N_0 and N_1 to exist it is necessary that the mapping $\phi_{(N,h)}$ should be zero.*
PROOF. In fact $\phi_{(N,h)}(\omega), \forall\omega \in H^{m-1}(\hat{I}_{k+r}(W); A)$, is the characteristic quantum number:

$$\phi_{(N,h)}(\omega) = a^i < \beta_i, Z_{N_C} >= a^i[\int_{Z_{N_{0C}}} \beta_i - \int_{Z_{N_{1C}}} \beta_i] = a^i[\int_{V_C} d\beta_i] = 0$$

as $d\beta_i = 0$. So we can conclude that $\phi_{(N,h)}$ is a zero mapping on $H^{m-1}(\hat{I}_{k+r}(W); A)$.□
THEOREM 6.56 - *Furthermore, in order for the topology to remain the same passing from N_0 to N_1 it is necessary that all the quantum characteristic classes $i_V^*\omega$ should be zero, $\forall\omega \in H^i(\hat{I}_{k+r}(W); A), i \geq 1$.*
PROOF. In fact, the topology does not change passing from N_0 to N_1 iff the integral quantum manifold V that cobords N_0 to N_1 is a cylinder. But this requires all the quantum characteristic classes $i_V^*\omega$ to be zero, in dimension $i \geq 1$. □
The following theorem gives a constructive method to obtain integral cobords, for fixed boundary $X = X_1 \dot{\bigcup} X_2$, also for non-linear QPDEs.
2 THEOREM 6.57 - (RELATION BETWEEN INTEGRAL COBORDISM AND QUANTUM PROPAGATORS). *Let us assume that $\hat{E}_k \equiv \ker_f K \subset J\hat{D}^k(W)$ is a QPDE as given in Theorem 6.55. Then, the quantum propagator $\hat{G}[s|i]$ identifies an integral cobord $L \subset \Omega(\hat{E}_k)$, such that if $\hat{G}[s|i]$ satisfies the boundary conditions*

$$X_1 \dot{\bigcup} X_2, \quad \textit{then} \quad \partial L = X_1 \dot{\bigcup} X_2.$$

PROOF. The proof can be conducted directly translating in the present quantum context one given for ordinary manifolds [73,111] or supermanifolds [112,113,114]. □
2 The following definitions and results are extensions to the present non-commutative context of similar ones given for quantum and integral cobordism in PDEs. (See Chapter 2.)
DEFINITION 6.41 - *Let $\hat{E}_k \subset J\hat{D}^k(W)$ be a quantum PDE, $\pi : W \to M, \dim W = n+m, \dim M = m$. We denote by $\hat{\Omega}_{p,s}(\hat{E}_k)$ the singular quantum p-cobordism groups of $\hat{E}_k$ and by $\hat{\Omega}_{p,s}^{\hat{E}_k}$ the corresponding singular integral cobordism groups, with $0 \leq p \leq m-1$. We denote by $[N]^s_{\bar{\hat{E}}_k}$ and $[N]^s_{\hat{E}_k}$ the corresponding equivalence classes.*
THEOREM 6.58 - *One has the following isomorphisms:*

$$(6.33) \qquad \left\{ \begin{array}{l} \hat{\Omega}_{p,s}^{\hat{E}_k} \cong \bar{H}_p(\hat{E}_k; A) \cong \Omega_{p,s}(\hat{E}_k) \bigotimes_{\mathbf{K}} A \\ \\ \hat{\Omega}_{p,s}(\hat{E}_k) \cong \bar{H}_p(J\hat{D}^k(W), \hat{E}_k; A) \cong \Omega_{p,s}(\hat{E}_k) \bigotimes_{\mathbf{K}} A \end{array} \right\}$$

COROLLARY 6.7 - *One has the following isomorphisms:*

$$(6.34)\qquad \left\{\begin{array}{l}\hat{\Omega}_{p,s}(\hat{E}_k)\cong\hat{\Omega}_{p,s}^{J\hat{\mathcal{D}}^k(W))}\times\hat{\Omega}_{p-1,s}^{\hat{E}_k}\\ \hat{\Omega}_{p,s}^{J\hat{\mathcal{D}}^k(W))}\times\hat{\Omega}_{p,s}(\hat{E}_k)\end{array}\right\}$$

PROOF OF COROLLARY 6.7 - The proof can be directly obtained by using the isomorphisms (6.33) and the corresponding isomorphisms in theory of PDEs for ordinary manifolds. □

THEOREM 6.59 - (CHARACTERIZATION OF COBORDISM GROUPS OF $\hat{E}_k$ BY MEANS OF CHARACTERISTIC NUMBERS).

1) *We give the following definitions.*

(a) **Quantum Stiefel-Whitney characteristic numbers** *of $\hat{E}_k$:*

$$\hat{w}[N]^s_{\hat{E}_k}=w[N_C]^s_{\hat{E}_k}\otimes a\in A\,.$$

(b) **Quantum Euler characteristic numbers** *of $\hat{E}_k$:*

$$\hat{\chi}[N]^s_{\hat{E}_k}=\chi[N_C]^s_{\hat{E}_k}\otimes a\in A\,.$$

(c) **Quantum integral characteristic numbers** *of $\hat{E}_k$:*

$$\hat{i}[N]^s_{\hat{E}_k}=i[N_C]^s_{\hat{E}_k}\otimes a\in A.$$

Now, one has the following:

$$N'\in[N]^s_{\hat{E}_k}\Leftrightarrow\left\{\begin{array}{l}\hat{w}(N')=\hat{w}(N)\\ \hat{\chi}[N']=\hat{\chi}[N]\\ \hat{i}[N']=\hat{i}[N]\end{array}\right\}$$

2) *We define*

(d) **full quantum characteristic numbers** *of $\hat{E}_k$:*

$$\hat{q}[N]^s_{\hat{E}_k}=q[N_C]^s_{\hat{E}_k}\otimes a\in A.$$

Then one has the following:

$$N'\in[N]^s_{\overline{\hat{E}_k}}\Leftrightarrow\left\{\begin{array}{l}\hat{w}(N')=\hat{w}(N)\\ \hat{\chi}[N']=\hat{\chi}[N]\\ \hat{q}[N']=\hat{q}[N]\end{array}\right\}$$

PROOF. It is a direct consequence of isomorphisms (6.33) and the characterization of cobordism groups of PDEs. □

6.7 - QUANTUM PDEs NON-HOLONOMIC CONNECTIONS

In order to formulate the quantum gauge theory, by using the tools developed in the previous chapters, we need also to formulate a theory for non-holonomic quantum connections. So the present section will be devoted to this subject. We will extend to quantum manifolds our previous results on ordinary manifolds and supermanifolds [113,114,117].

Let M be an $(m+n)$-dimensional quantum manifold with quantum coordinates $\{x^A\}$ and let $\pi : W \to M$ be a fiber bundle over M with quantum coordinates $\{x^A, y^B\}, 1 \leq A \leq m+n, 1 \leq B \leq p+q$. We define non-holonomic jet-quantum derivative spaces by formally reproducing homonymous definitions given for ordinary manifolds or for supermanifolds. So we have:

(A) k-**Semiholonomic jet-quantum derivative spaces**: $\bar{J}\hat{D}^k(W)$. The fiber bundle $\bar{\pi}_{k,k-1} : \bar{J}\hat{D}^k(W) \to \bar{J}\hat{D}^{k-1}(W)$ has an affine structure on the fibers. The associated vector bundle is the following:

$\bar{\pi}^*_{k-1,0}(\dot{T}^k_0 MOvTW)$. The induced quantum coordinates on $\bar{J}\hat{D}^k(W)$ are

$$\{x^A, y^B, y^B_A, \dots, y^B_{A_1\dots A_k}\}$$

where no symmetries in the indexes are involved.

(B) k-**Sesquiholonomic jet-quantum derivative spaces**: $\breve{J}\hat{D}^k(W)$. The fiber bundle $\breve{\pi}_{k,k-1} : \breve{J}\hat{D}^k(W) \to \breve{J}\hat{D}^{k-1}(W)$ is an affine fiber bundle with the associated vector bundle $\pi^*_{k-1,0}(TMO(\dot{S}^{k-1}_0 MOvTW))$. The induced quantum coordinates on $\breve{J}\hat{D}^k(W)$ are $\{x^A, y^B, \dots, y^B_{A_1\dots A_{k-1}}, y^B_{A_1\dots A_{k-1}C}\}$, where no symmetries between the indexes are involved.

If $\pi : E \to M$ is a vector bundle over M, then

$$\bar{\pi}_k : \bar{J}\hat{D}^k(E) \to M \quad \text{and} \quad \breve{\pi}_k : \breve{J}\hat{D}^k(E) \to M, \quad \text{as well as} \quad \pi_k : J\hat{D}^k(E) \to M,$$

are vector bundles over M.

If $\pi : E \to M$ is an affine fiber bundle with associated vector bundle $\rho : \mathbf{E} \to M$, then $\pi_k : \bar{J}\hat{D}^k(E) \to M$ (resp. $\breve{J}\hat{D}^k(E)$, resp. $J\hat{D}^k(E)$) has a canonical structure of affine fiber bundle with associated vector bundle $\bar{\rho}_k : \quad J\hat{D}^k(\mathbf{E}) \to M$, (resp. $\breve{\rho}_k : \breve{J}\hat{D}^k(\mathbf{E}) \to M$, resp. $\rho_k : J\hat{D}^k(\mathbf{E}) \to M$). The first quantum prolongation

$$\rceil^{(1)} : J\hat{D}^k(W) \to J\hat{D}(J\hat{D}^k(W))$$

of a k-quantum connection on $\pi : W \to M$, i.e., a Q^∞_w-section

$$\rceil : J\hat{D}^{k-1}(W) \to J\hat{D}^k(W)$$

of the fiber bundle $J\hat{D}^k(W) \to J\hat{D}^{k-1}(W)$ has values into the $(k+1)$-sesquiholonomic space $\breve{J}\hat{D}^{k+1}(W)$:

$$\rceil^{(1)} : J\hat{D}^k(W) \to \breve{J}\hat{D}^{k+1}(W) \hookrightarrow J\hat{D}(J\hat{D}^k(W).$$

Note that the second superprolongation $\rceil^{(2)}$ of $\rceil$ does not take values into $\breve{J}\hat{D}^{k+2}(W)$ but into a different subspace of $\bar{J}\hat{D}^{k+2}(W)$.
Let $\overline{C}_k \subset \bar{J}\hat{D}^k(W)$ be a semiholonomic k-quantum connection on $\pi : W \to M$, i.e., a sub-fiber bundle

$$\overline{C}_k \subset \bar{J}\hat{D}^k(W)$$

given as image of a Q^∞_w-section

$$\bar{\rceil} : \bar{J}\hat{D}^{k-1}(W) \to \bar{J}\hat{D}^k(W)$$

of

$$\bar{\pi}_{k,k-1} : \bar{J}\hat{D}^k(W) \to \bar{J}\hat{D}^{k-1}(W) : \qquad \overline{C}_k = \bar{\rceil}(\bar{J}\hat{D}^{k-1}(W)) \subset \bar{J}\hat{D}^k(W).$$

Then, $\overline{C}_k$ identifies (and it is identified by) an affine morphism

$$\bar{\lceil} : \bar{J}\hat{D}^k(W) \to \bar{\pi}^*_{k-1,0}(\dot{T}^k_0 MOvTW)$$

over $\bar{J}\hat{D}^{k-1}(W)$, whose fiber derivative is $vD\bar{\lceil} = 1$. One has

$$\bar{\lceil} \circ \bar{\rceil} = 0, \bar{\lceil} = id_{\bar{J}\hat{D}^k(W)} - \bar{\rceil} \circ \hat{\pi}_{k,k-1}.$$

Let

$$\{x^A, y^B, \ldots, y^B_{A_1\ldots A_{k-1}}, \bar{y}^B_{A_1\ldots A_k}\}$$

be the naturally associated linear quantum coordinates on $\bar{\pi}^*_{k-1,0}(\dot{T}^k_0 MOvTW)$. Then, one has the following local expression for $\bar{\lceil}$:

$$x^A \circ \bar{\lceil} = x^A, \quad \ldots \quad , y^B_{A_1\ldots A_{k-1}} \circ \bar{\lceil} = y^B_{A_1\ldots A_{k-1}}$$

$$\bar{y}^B_{A_1\ldots A_k} \circ \bar{\lceil} = y^B_{A_1\ldots A_k} + \lceil^B_{A_1\ldots A_k} \circ \bar{\pi}_{k,k-1}, \quad -\lceil^B_{A_1\ldots A_k} = y^B_{A_1\ldots A_k} \circ \bar{\rceil} : \bar{J}\hat{D}^{k-1}(W) \to \overset{k}{\widehat{A}}.$$

So the local expression of $\overline{C}_k$ is the following:

$$y^B_{A_1\ldots A_k} + \lceil^B_{A_1\ldots A_k} = 0. \tag{6.35}$$

(The proofs of these propositions are similar to supercase ones [114].)
DEFINITION 6.42 - *If $\overline{C}_k$ is a semiholonomic k-quantum connection on $\pi : W \to M$, then we can define* **semiholonomic covariant map of order k of a Q^∞_w-section** $u : M \to \bar{J}\hat{D}^k(W)$ *the following section:*

$$\tilde{\nabla} u = \bar{\lceil} \circ u : M \to \dot{T}^k_0 MOs^* vTW$$

where $s \equiv \overline{\pi}_{k-1,0} \circ u : M \to W$.
If $W \equiv E$ is a vector bundle one has $\tilde{\nabla} u : M \to \dot{T}^k_0 MOE$. If $u = D^k s : M \to J\hat{D}^k(W) \subset \bar{J}\hat{D}^k(W)$, we obtain the **semiholonomic covariant derivative, of order k, of the section** s:

$$\overline{\overset{k}{\nabla}} s \equiv \tilde{\nabla} D^k s = \bar{\lceil} \circ D^k s : M \to \dot{T}^k_0 MOs^* vTW.$$

Furthermore, if $\bar{\lceil}$ is such that there exists a (holonomic) k-quantum connection $\rceil$: $J\hat{D}^{k-1}(W) \to J\hat{D}^k(W)$ such that the following diagram is commutative

$$\begin{array}{ccc} \bar{J}\hat{D}^{k-1}(W) & \overset{\bar{\rceil}}{\to} & \bar{J}\hat{D}^k(W) \\ \bigcup & & \bigcup \\ J\hat{D}^{k-1}(W) & \underset{\rceil}{\to} & J\hat{D}^k(W) \end{array}$$

then we say that $\bar{\rceil}$ is the **semiholonomic blow-up** of the (holonomic) k-quantum connection $\rceil$. In this case one has that the semiholonomic covariant derivative of order k of a section $s : M \to W$ coincides with the **covariant derivative of order** k of s :

$$\overline{\overset{k}{\nabla}} s = \overset{k}{\nabla} s : M \to \dot{S}^k_0 MOs^* vTW.$$

Local expressions:

$$\tilde{\nabla} u : \bar{y}^B_{A_1\ldots A_k} \circ (\tilde{\nabla} u) \equiv (\tilde{\nabla} u)^B_{A_1\ldots A_k} = u^B_{A_1\ldots A_k} + \lceil^B_{A_1\ldots A_k} \circ \pi_{k,k-1} \circ u; \tag{6.36}$$

$$\overline{\overset{k}{\nabla}} s : \bar{y}^B_{A_1\ldots A_k} \circ (\overline{\overset{k}{\nabla}} s) \equiv (\overline{\overset{k}{\nabla}} s)^B_{A_1\ldots A_k} = (\partial x_{A_1} \ldots \partial x_{A_k} . s^B) + \lceil^B_{A_1\ldots A_k} \circ D^k s;$$

(6.37)(Blow up case).

$$\bar{y}^B_{(A_1\ldots A_k)} \circ (\overset{k}{\nabla} s) \equiv (\overset{k}{\nabla} s)^B_{(A_1\ldots A_k)} = (\partial x_{A_1} \ldots \partial x_{A_k} . s^B) + \lceil^B_{(A_1\ldots A_k)} \circ D^{k-1} s.$$

DEFINITION 6.43 - *The* **semiholonomic s-quantum prolongation** *of the semiholonomic k-quantum connection* $\bar{\rceil}:\bar{J}\hat{D}^{k-1}(W)\to\bar{J}\hat{D}^{k}(W)$ *is*

$$\overset{(s)}{\bar{\rceil}}\equiv J\hat{D}(\bar{\rceil}^{(s-1)})|_{\bar{J}\hat{D}^{k+s-1}(W)}$$

where

$$\bar{\rceil}^{(1)}=J\hat{D}(\bar{\rceil})|_{\bar{J}\hat{D}^{k}(W)}:\bar{J}\hat{D}^{k}(W)\to J\hat{D}(\bar{J}\hat{D}^{k}(W)).$$

PROPOSITION 6.8 - $\bar{\rceil}^{(s)}$ *has values into* $\bar{J}\hat{D}^{k+s}(W)$*:*

$$\bar{\rceil}^{(s)}:\bar{J}\hat{D}^{k-1+s}(W)\to\bar{J}\hat{D}^{k+s}(W).$$

PROPOSITION 6.9 - *If* $\overline{C}_k$ *is a semiholonomic quantum connection one has that its s-order semiholonomic quantum prolongation* $\overline{C}_{k+s}$ *is given by*

$$\overline{C}_{k+s}=\bar{\rceil}^{(s)}(\overline{C}_{k+s-1}).$$

PROPOSITION 6.10 - *Let* $\pi:E\to M$ *be a quantum vector bundle (i.e., the fibers are quantum vector spaces). Then, there is a monomorphism of vector bundles over* $\bar{J}\hat{D}^{k-1}(E)$*:*

$$\beta^{(k)}:\pi^{*}_{k-1}\dot{\Lambda}^{k}_{0}MOE\cong\bar{J}\hat{D}^{k-1}(E)\times_M(\dot{\Lambda}^{k}_{0}MOE)\to\bar{J}\hat{D}^{k}(E)$$

defined by means of the following commutative diagram, where the lines are exact sequence of vector bundles over M*:*

$$\begin{array}{ccccccc}
0 & \to & \dot{T}^{k}_{0}MOE & \overset{\bar{\mu}^{(k)}}{\to} & \bar{J}\hat{D}^{k}(E) & \overset{\pi_{k,k-1}}{\to} & \bar{J}\hat{D}^{k-1}(E) \to 0\\
 & & \uparrow & & & & \\
 & & \dot{\Lambda}^{k}_{0}MOE & & & & \\
 & & \uparrow & & & & \\
 & & 0 & & & &
\end{array}$$

PROPOSITION 6.11 - *A semiholonomic k-quantum connection* $\bar{\rceil}:\bar{J}\hat{D}^{k-1}(E)\to\bar{J}\hat{D}^{k}(E)$ *which is A_0-linear over M, identifies an isomorphism*

$$\bar{J}\hat{D}^{k}(E)\cong\dot{T}^{k}_{0}MOE\times_M\bar{J}\hat{D}^{k-1}(E),$$

obtained by means of the following split exact sequence of bundles of A_0-modules over M*:*

$$0\to\dot{T}^{k}_{0}MOE\underset{\bar{\rceil}}{\rightleftarrows}\bar{J}\hat{D}^{k}(E)\underset{\bar{\rceil}}{\rightleftarrows}\bar{J}\hat{D}^{k-1}(E)\to 0\,.$$

More precisely, one has the A_0-isomorphism: $u \mapsto (\bar{\daleth}(u), \bar{\pi}_{k,k-1}(u)), \forall u \in \bar{J}\hat{D}^k(E)$, and its inverse: $(v,q) \mapsto (v + \bar{\daleth}(q)), (v,q) \in \dot{T}^k_0 MOE \times_M \bar{J}\hat{D}^{k-1}(E)$. One has the following commutative diagram:

$$
\begin{array}{ccc}
\bar{J}\hat{D}^k(E) & \stackrel{\bar{\daleth}}{\rightarrow} & \bar{\pi}^*_{k-1}(\dot{T}^k_0 MOE) \\
\bar{\daleth}\uparrow & & \uparrow 0 \\
\bar{J}\hat{D}^{k-1}(E) & = & \bar{J}\hat{D}^{k-1}(E)
\end{array}
$$

In particular one has the following. Any E-valued quantum differential k-form μ : $\dot{\Lambda}^k_0 MOE$, on M, identifies a semiholonomic k-quantum connection $\bar{\daleth}$ on E. More precisely, one has: $\bar{\daleth}(q) = \beta^{(k)}(q, \mu(\bar{\pi}_{k-1}(q))), \forall q \in \bar{J}\hat{D}^{k-1}(E)$. As $\bar{J}\hat{D}(E) = J\hat{D}(E)$, a semiholonomic 1-quantum connection $\bar{\daleth} : E \to \bar{J}\hat{D}(E)$ on a fiber bundle $p : E \to M$ coincides with a (holonomic) 1-quantum connection on E. So, if $\bar{\daleth}$ is A_0-linear over M, it identifies a **covariant exterior differential**

$$
(6.38) \qquad {}_{\bar{\daleth}}d : J\hat{D}(\dot{\Lambda}^\bullet MOE) \to \dot{\Lambda}^\bullet MOE
$$

defined by means of the following commutative diagram:

$$
\begin{array}{ccc}
J\hat{D}(\dot{\Lambda}^\bullet MOE) & \stackrel{{}_{\bar{\daleth}}d}{\rightarrow} & \dot{\Lambda}^\bullet MOE \\
{}_{\bar{\daleth}}j \,\|\wr & & \uparrow \\
TMO(\dot{\Lambda}^0 MOE) \times \dot{\Lambda}^\bullet MOE) & \rightarrow & TMO(\dot{\Lambda}^\bullet MOE)
\end{array}
$$

where ${}_{\bar{\daleth}}j$ is the isomorphism induced by $\bar{\daleth}$. In local quantum coordinates we have, for any E-valued graded quantum differential form

$$
\beta = \sum_p \partial y_B \otimes \beta^B{}_{A_1 \dots A_p} dx^{A_1} \triangle \cdots \triangle dx^{A_p}
$$

the following expression of ${}_{\bar{\daleth}}d\beta$:

$$
{}_{\bar{\daleth}}d\beta = \sum_p \partial y_B \otimes [(\partial x_A . \beta^B_{A_1 \dots A_p}) + \daleth^B_{AA_1 \dots A_p}] \circ dx^{A_1} \triangle \cdots \triangle dx^{A_p}.
$$

PROPOSITION 6.12 - *Let G be a Lie quantum group. Let us consider the following trivial fiber bundle over G: $\pi : E \equiv G \times A(G) \to G$. One has the canonical isomorphism over G: $\bar{J}\hat{D}^k(E) \cong \bar{J}\hat{D}^{k-1}(TGOA(G)) \times A(G)$. So, one has the folowing commutative diagram with exact lines of bundles of A_0-modules over G:*

$$
\begin{array}{ccccccc}
0 \to \dot{T}^p_0 GOA(G) & \to & \bar{J}\hat{D}^{p-1}(R) \times A(G) & \to & \bar{J}\hat{D}^{p-2}(R) \times A(G) \to 0 \\
\| & & \| & & \| \\
0 \to \dot{T}^p_0 GOA(G) & \to & \bar{J}\hat{D}^p(E) & \to & \bar{J}\hat{D}^{p-1}(E) \to 0 \\
\uparrow & & \uparrow & & \\
\dot{\Lambda}^p_0 GOA(G) & = & \dot{\Lambda}^p_0 GOA(G) & & \\
\uparrow & & \uparrow & & \\
0 & & 0 & &
\end{array}
$$

where $R \equiv TGOA(G)$.

DEFINITION 6.44 - *A* p-**quantum pseudoconnection on a Lie quantum group** G *is a semiholonomic* p*-quantum connection*

$$\overline{C}_p \equiv \rceil(\bar{J}\hat{D}^{p-1}(E)) \subset \bar{J}\hat{D}^p(E)$$

on the trivial fiber bundle $\pi : E \equiv G \times A(G) \to G$, *such that* $\overline{C}_p$ *is freely generated by* $\widetilde{\overline{C}}_p$ *and* $\bar{J}\hat{D}^{p-1}(E)$, *that is*

$$\overline{C}_p = \widetilde{\overline{C}}_p \times \bar{J}\hat{D}^{p-1}(E)$$

where

$$\widetilde{\overline{C}}_p = \mu(G) \subset \dot{\Lambda}^p_0 GOA(G)$$

for some $A(G)$*-valued even quantum differential* p*-form* μ *on* G.

DEFINITION 6.45 - *A* **quantum pseudoconnection** *is a* p*-quantum pseudoconnection with* $p = 1$. *(It also results a* $A(G)$*-principal quantum connection on the* $A(G)$*-principal fiber bundle* $E \equiv G \times A(G) \to G$.*)*

DEFINITION 6.46 - *A* 0-**quantum pseudoconnection** *is by definition a section of* $E \to G$.

DEFINITION 6.47 - *A* **reduction of** p-**quantum pseudoconnection** $\rceil\mu : G \to \dot{\Lambda}^p_0 GOA(G)$ *with respect to a closed subgroup (= Lie quantum subgroup)* $H \subset G$, *is the fixing of a* $A(H)$*-valued quantum differential form* $_\rceil\mu : G \to \dot{\Lambda}^p_0 GOA(H)$ *besides a* H*-principal quantum connection* $_\rceil\beta : G \to TGOA(H)$ *such that one has the following commutative diagram:*

$$\begin{array}{ccc} G & \stackrel{\rceil\mu}{\to} & \dot{\Lambda}^p_0 GOA(G) \\ \downarrow {}_\rceil\mu & & \downarrow {}_\rceil j \\ \dot{\Lambda}^p_0 GOA(H) & = & \dot{\Lambda}^p_0 GOA(H) \end{array}$$

where $_\rceil j$ *is the projection induced by* $_\rceil\beta$.

DEFINITION 6.48 - *If* $p = 1$ *and* $_\rceil\beta = {}_\rceil\mu$ *we say that* $\rceil\mu$ *is a* **reductif quantum pseudoconnection.**

DEFINITION 6.49 - *A reductif* $A(H)$*-valued quantum differential* p*-form on* $G, p \geq 1$, *is called a* **fully reductif** p**-quantum pseudoconnection**.

DEFINITION 6.50 - *If* $p = 1$ *we call* **fully reductif quantum pseudoconnection** *on* G *any* H*-principal quantum connection on* G *with* $H \subset G$.

The following propositions can be proved simulating the proofs reported in ref.[110] for similar propositions for super PDEs.

PROPOSITION 6.13 - $_{\rceil}\mu : G \to TGOA(G)$ *is a reductif quantum pseudoconnection with respect to the H-principal quantum connection*

$$_{\rceil}\beta : G \to TGOA(H)$$

iff

$$< f_a^* {}_{\rceil}\beta(e) \circ {}_{\rceil}\mu, X >= ad(a^{-1} <_{\rceil} \beta(e) \circ {}_{\rceil}\beta, X >)$$

for all $a \in H$, *and* $f : H \times G \to G$ *is the action map of* H *on* G.

PROPOSITION 6.14 - *The local expression of a p-quantum pseudoconnection on a Lie quantum group G is the following:*

$$\overline{C}_p \equiv \overline{\rceil}(\bar{J}\hat{D}^{p-1}(E)) \subset \bar{J}\hat{D}^p(E) : \quad y^B_{A_1\ldots A_p} - \mu^B_{A_1\ldots A_p} = 0$$

where $\mu^B_{A_1\ldots A_p}$ *are the components of* $_{\rceil}\mu$ *in the system of quantum coordinates* $\{x^A, \dot{x}^B_{A_1\ldots A_p}\}$ *on* $\dot{\Lambda}^p_0 GOA(G)$*:*

$$\left\{\mu^B_{A_1\ldots A_p} \equiv \dot{x}^B_{A_1\ldots A_p} \circ {}_{\rceil}\mu\right\}.$$

PROPOSITION 6.15 - *The semiholonomic covariant map (of p-order) of* $_{\rceil}\mu$ *is zero:* $\tilde{\nabla}_{\rceil}\mu = \lceil \circ_{\rceil}\mu = 0$.

PROPOSITION 6.16 - *If*

$$\overline{\rceil} = ({}_{\rceil}\mu, id_{A(G)}) : E = G \times A(G) \to J\hat{D}(E) \cong (TGOA(G)) \times A(G)$$

is a quantum pseudoconnection, for the semiholonomic p-quantum prolongation $\overline{\rceil}^{(p)}$ *of* $\overline{\rceil}$ *we have:*

$$\begin{array}{ccc} \bar{J}\hat{D}^p(E) & \xrightarrow{\overline{\rceil}^{(p)}} & \bar{J}\hat{D}^{p+1}(E) \\ \| & & \| \\ \bar{J}\hat{D}^{p-1}(TGOA(G)) \times A(G) & \xrightarrow{(D^p_{\rceil}\mu \circ \pi_p) \times id_{A(G)}} & \bar{J}\hat{D}^p(TGOA(G)) \times A(G) \end{array}$$

DEFINITION 6.51 - *A* **Z-graded quantum pseudoconnection on a Lie quantum group** *G is a subbundle* $\overline{C}$ *of the vector bundle*

$$\bigoplus_{p\geq 0} \bar{J}\hat{D}^p(E), \pi : E \equiv G \times A(G) \to G$$

over G, that can be written as

$$\overline{C} = \prod_{p\geq 0} \overline{C}_p = \prod_{p\geq 0} \overline{\rceil}^{(s)}(\bar{J}\hat{D}^{p-1}(E))$$

where each $\overline{C}_p$ is a p-quantum pseudoconnection on G. This is equivalent to assigning a sequence of sections

$$G \underset{{}_0\rceil}{\longrightarrow} E \underset{{}_1\rceil}{\longrightarrow} \bar{J}\hat{D}(E) \underset{{}_2\rceil}{\longrightarrow} \bar{J}\hat{D}^2(E) \longrightarrow \cdots . \underset{{}_{m-1}\rceil}{\longrightarrow} \bar{J}\hat{D}^{m-1}(E) \underset{{}_m\rceil}{\longrightarrow} \bar{J}\hat{D}^m(E)$$

where each ${}_p\rceil$ is a p-quantum pseudoconnection on $G, p \geq 0$.

DEFINITION 6.52 - *A* **natural Z-graded quantum pseudoconnection** *on G is a* **Z**-*graded quantum pseudoconnection such that $\overline{C}_p = \bar{J}\hat{D}^{p-1}(E) \times \{0\}, p \neq 1$, that is $\overline{C}_p$ is identified with a quantum pseudoconnection.*

PROPOSITION 6.17 - *A* **Z**-*graded quantum pseudoconnection $\overline{C} = \prod_{p\geq 0} \overline{C}_p$ is identified with a section*

$$\mu = \sum_{p\geq 0} {}_p\mu : G \to \dot{\Lambda}^\bullet GOA(G)$$

of the **Z**-*graded bundle $\dot{\Lambda}^\bullet GOA(G)$ over G. So, $\dot{\Lambda}^\bullet GOA(G)$ can be considered the fiber bundle of* **Z**-*graded quantum pseudoconnections.*

PROPOSITION 6.18 - *A natural* **Z**-*graded quantum pseudoconnection is characterized by means of a section $\mu = \sum_{p\geq 0} {}_p\mu$, such that ${}_p\mu = 0$ for $p \neq 1$.*

DEFINITION 6.53 - *Let $\mu : G \to \dot{\Lambda}^\bullet GOA(G)$ be a section characterizing a* **Z**-*graded quantum pseudoconnection. Let $\beta : G \to \dot{\Lambda}^\bullet GOA(G)$ be any $A(G)$-valued differential form on G. Then, we can define* **exterior covariant differential** *of β by means of the following formula:*

$$\rceil d\beta = d\beta + \frac{1}{2}[\mu, \beta],$$

where, by means of the isomorphism $\dot{\Lambda}^\bullet GOA(G) \cong \widehat{A}(G) \bigotimes_{\widehat{A}} (\dot{\Lambda}^\bullet G)^+$, we get

$$\begin{aligned}
[\mu, \beta] &= [\oplus_p {}_p\mu, \oplus_q {}_q\beta] \\
&= [\oplus_p Z_K \otimes {}_p\overline{\mu}_p^K, \oplus_q Z_J \otimes {}_q\beta_q^J] \\
&= \oplus_{p,q}[Z_K, Z_J]_\bullet \otimes \overline{\mu}^K \wedge_q \overline{\beta}^J \\
&= \oplus_{p,q} Z_I C^I_{KJ} \otimes \overline{\mu}^K \wedge_q \overline{\beta}^J
\end{aligned}$$

where $C^I_{KJ} \in \widehat{A}$ are the quantum structure constants of G.

DEFINITION 6.54 - *The* **Z-graded curvature** *of a* **Z**-*graded quantum pseudoconnection μ on G is by definition*

$$\rceil R \equiv \rceil d\mu = d\mu + \frac{1}{2}[\mu, \mu].$$

Let μ be a natural **Z**-graded quantum pseudoconnection on G; then for any $A(G)$-valued quantum differential form β on G one has

$$ {}_{\rceil}d^2\beta = {}_{\rceil}R\triangle\beta \quad ; \quad {}_{\rceil}d\,{}_{\rceil}R = 0 \quad \text{(Bianchi identity).} $$

Let

$$ {}_{\rceil}\mu = \sum_{p\geq 0} Z_K \otimes \mu^K_{\alpha_1\dots\alpha_p} d\xi^{\alpha_1}\triangle\cdots\triangle d\xi^{\alpha_p} $$

with

$$ \mu^K_{\alpha_1\dots\alpha_p} \in Q^\infty_w(U, \overset{p}{\widehat{A}}), $$

and

$$ \beta = \sum_{q\geq 0} Z_H \otimes \beta^H_{\beta_1\dots\beta_q} d\xi^{\beta_1}\triangle\cdots\triangle d\xi^{\beta_q} $$

with $\beta^H_{\beta_1\dots\beta_q} \in Q^\infty_w(U, \overset{q}{\widehat{A}})$ local representations of

$$ {}_{\rceil}\mu, \beta \in Q^\infty_w(\dot{\Lambda}^\bullet GOA(G)), U \subset G. $$

Then one has the following local representation of the exterior covariant differential:

$$ \begin{aligned} {}_{\rceil}d\beta = & \sum_{q\geq 0} Z_H \otimes (\partial\xi_\gamma.\beta^H_{\beta_1\dots\beta_q}) d\xi^\gamma\triangle d\xi^{\beta_1}\triangle\cdots\triangle d\xi^{\beta_q} \\ & + \frac{1}{2}\sum_{p,q\geq 0} [Z_K, Z_J]_\bullet \otimes \mu^K_{\alpha_1\dots\alpha_p}\beta^J_{\beta_1\dots\beta_q} d\xi^{\alpha_1}\triangle\cdots\triangle d\xi^{\alpha_p}\triangle d\xi^{\beta_1}\triangle\cdots\triangle d\xi^{\beta_q} \end{aligned} $$

with $\mu^K_{\alpha_1\dots\alpha_p}\beta^J_{\beta_1\dots\beta_q} \in Q^\infty_w(U, \overset{p+q}{\widehat{A}})$, defined by the following commutative diagram:

$$ \begin{array}{ccc} \dot{T}^p_0(A)\bigotimes_{A_0}\dot{T}^q_0(A) & \xrightarrow{\mu^K_{\alpha_1\dots\alpha_p}\otimes\beta^J_{\beta_1\dots\beta_q}} & A\otimes_{A_0} A \\ \uparrow & & \downarrow \widehat{\mu} \\ \dot{T}^{p+q}_0(A) & \xrightarrow[\mu^K_{\alpha_1\dots\alpha_p}\beta^J_{\beta_1\dots\beta_q}]{} & A \end{array} $$

The above definition of **Z-graded quantum pseudoconnection** can be generalized by using vector-valued quantum differential forms. Let V be a vector quantum space and let $r : A(G) \to L(V)$ be a representation of $A(G)$ on V. Then we can define a **Z-graded quantum pseudoconnection** on G associated to the representation r, as a subbundle $\overline{C}$ of the vector bundle $\bigoplus_{p\geq 0} \bar{J}\hat{D}^p(E)$, where $\pi : E \equiv G\times V \to G$ is the trivial vector bundle over G associated to r, such that

$$ \overline{C} = \prod_{p\geq 0}\overline{C}_p = \prod_{p\geq 0} {}_p\rceil(\bar{J}\hat{D}^{p-1}(E)) $$

where each of $\overline{C}_p$'s is a p-quantum pseudoconnection on G. This is equivalent to assigning a section

$$\mu = \sum_{p\geq 0} {}_p\mu : G \to \dot{\Lambda}^{\bullet}GOV \text{ of the graded bundle } \dot{\Lambda}^{\bullet}GOV \to G.$$

Then, the above definitions of exterior covariant differential and **Z**-graded curvature, can be extended directly. Note, also, that in some applications the vector quantum space V can be structured in a product $V = V_1 \times \cdots \times V_s$ and the representation r is just obtained as a product of representations $r_i : A(G) \to L(V_i), r = (r_1, \ldots, r_s) : A(G) \to L(V_1 \times \cdots \times V_s)$. Then a **Z**-graded quantum pseudoconnection associated to r ${}_{\urcorner}\mu : G \to \dot{\Lambda}^{\bullet}GOV$ splits into a **Z**-graded quantum pseudoconnections on G associated to the representations

$$r_i : {}_{\urcorner}\mu = ({}_{\urcorner}\mu^{(1)}, \ldots, {}_{\urcorner}\mu^{(s)}) : G \to \dot{\Lambda}^{\bullet}GOV_1 \times \cdots \times \dot{\Lambda}^{\bullet}GOV_s.$$

The above definitions can be further generalized with respect to a generic vector bundle $p : E \to B$ (where the fibers are vector quantum spaces). More precisely, we define a **p-quantum pseudoconnection** on E a semiholonomic p-quantum connection $\overline{C}_p = \urcorner(\bar{J}\hat{D}^{p-1}(E)) \subset \bar{J}\hat{D}^p(E))$ on the fiber bundle $p : E \to B$ such that $\overline{C}_p$ is freely generated by $\tilde{\overline{C}}_p$ and $\bar{J}\hat{D}^{p-1}(E)$, that is $\overline{C}_p = \tilde{\overline{C}}_p \times \bar{J}\hat{D}^{p-1}(E)$, where $\tilde{\overline{C}}_p = \mu(B) \subset \dot{\Lambda}^{\bullet}BOE$ for some E-valued quantum differential p-form μ on B. Then a **Z-graded quantum pseudoconnection** on E is a subbundle $\overline{C}$ of the vector bundle $\bigoplus_{p\geq 0} \bar{J}\hat{D}^p(E)$ over B that can be written as $\overline{C} = \bigoplus_{p\geq 0} \overline{C}_p$, where $\overline{C}_p$ is a p-quantum pseudoconnection on E. So a **Z**-graded quantum pseudoconnection on E is characterized by a section $\mu = \sum_{p\geq 0} {}_p\mu$ of $\dot{\Lambda}^{\bullet}BOE$ over B.

6.8 - GAUGE QUANTUM PDEs

Scope of the present section is first to give a unitary intrinsic description of gauge theories in the category of quantum manifolds. We will refer to our previous geometrical setting for gauge theory [112,113,114,117] in order to extend that framework to the present quantum situation.

REMARK 6.45 - (GEOMETRIC STRUCTURES AND QUANTUM CONNECTIONS ON PRINCIPAL FIBER BUNDLES).

Before defining a quantum gauge system let us emphasize some geometrical structures about quantum connections on principal fiber bundles. Any principal quantum connection $k : M \to \hat{C}(P) \cong J\hat{D}(P)/G$(= fiber bundle of principal quantum connections) on the G-principal fiber bundle $(P, M, \pi; G)$ (in the category of quantum manifolds), identifies a section $\underline{\mathbf{k}}$ of $\pi^*\hat{C}(P) \to P$ such that the following diagram is commutative:

$$\begin{array}{ccccccccc} TPO_{A_0}A(G) & \overset{j}{\leftarrow} & \pi^*\hat{C}(P) & = & \pi^*\hat{C}(P) & \rightarrow & \hat{C}(P) & = & \hat{C}(P) \\ {\scriptstyle\omega_{\mathbf{k}}}\uparrow & & \downarrow & & \uparrow{\scriptstyle\underline{\mathbf{k}}} & & \downarrow & & \uparrow{\scriptstyle\mathbf{k}} \\ P & = & P & = & P & \underset{\pi}{\rightarrow} & M & = & M \end{array}$$

More precisely one has $\underline{\mathbf{k}} = (id_P, \breve{\mathbf{k}})$, where $\breve{\mathbf{k}} = \mathbf{k} \circ \pi$, and $\omega_{\mathbf{k}}$ is the Ehresmann quantum connection associated to $\mathbf{k}$. j is the canonical embedding over P given by : $j(p, \mathbf{k}(x)) = \omega_{\mathbf{k}}(p)$, with $x = \pi(p)$. Set

$$< \pi^*\hat{C}(P) >\equiv j(\pi^*\hat{C}(P)) \subset TPOA(G)$$

Of course one has the following isomorphism: $< \pi^*\hat{C}(P) >\cong \pi^*\hat{C}(P)$ and the following exact sequences of quantum vector bundles over P:

$$0 \to \pi^*\hat{C}(P) \to TPOA(G).$$

The morphism j induces a morphism

$$j^k : \pi^* J\hat{D}^k(\hat{C}(P)) \to J\hat{D}^k(TPOA(G))$$

over P given by

$$j^k(p, u = D^k\mathbf{k}(x)) = D^k\omega_{\mathbf{k}}(p), x = \pi(p).$$

Set

$$< \hat{C}(P) >^k\equiv j^k(\pi^* J\hat{D}^k(\hat{C}(P))) \subset J\hat{D}^k(TPOA(G)).$$

THEOREM 6.60 - *Let $(P, M, \pi; G)$ be a principal fiber bundle in the category of quantum manifolds. Let $\rho : E \to P$ be a vector fiber bundle over P. Let C denote the fiber bundle $\dot{\Lambda}^\bullet POE$ over M with projection π_C and let $\underline{E}$ denote the fiber bundle $\dot{\Lambda}^\bullet POE$ over P with projection ρ. One has a canonical epimorphism of fiber bundles over $\dot{\Lambda}^\bullet POE$:*

$$\mathbf{n} : \pi^*_{k,0} J\hat{D}^k(\underline{E}) \to J\hat{D}^k(C), \qquad (6.39)$$

given by

$$\mathbf{n}(D^k s(x), D^k{}_1\mu(s(x))) = D^k({}_1\mu \circ s)(x) = D^k c(x),$$

where $\pi_{k,0} : J\hat{D}^k(P) \to P$ is the canonical epimorphism.

DEFINITION 6.55 - *A k-order* **quantum gauge system** *with blow-up structure is a k-order gauge continuum system $G(M) \equiv (\mathcal{B}, \hat{E}_k)$, in the category of quantum manifolds, such that: $\mathcal{B}$ is a bundle of geometric objects $\mathcal{B} \equiv (P, C; \mathbb{B})$, where:* (a) *$P$ is a* *principal fiber bundle* $(P, M, \pi; G)$, *G is a Lie quantum group of dimension* $(m'' + n'')$ *called* **gauge structure quantum group** *and M is an* $(m+n)$*-dimensional quantum manifold called* **base quantum manifold**. *The classic limit* M_C *of M is called (m-dimensional)* **space-time**; (b) *C is a fiber bundle over M (called* **configuration bundle**) *that coincides with the fiber bundle of* **Z**-*graded quantum connections over a quantum vector fiber bundle* $\rho : E \to P$ *over P*; (c) $\mathbb{B}$ *is the natural functor* $\mathbb{B} : \mathcal{C}(P) \to \mathcal{C}(C)$, *where* $\mathcal{C}(P)$, *(resp.* $\mathcal{C}(C)$) *is the category whose objects are open subbundles of P (resp. C), and whose morphisms are the local fiber bundle authomorphisms among these objects. (We shall denote C by* $\underline{E}$ *when it is considered as a fiber bundle over P.) The* **blow-up structure** *is given by an embedding morphism* $j : \pi^* \hat{C}(P) \to \underline{E}$ *over P given by means of the following commutative diagram with exact lines*

$$\begin{array}{ccccc} & & \underline{E} & = & \dot{\Lambda}^{\bullet} POE \\ & & {\scriptstyle j}\uparrow & & \uparrow \\ 0 & \to & \pi^* \hat{C}(P) & \to & TPOE \\ & & & & \uparrow \\ & & & & 0 \end{array}$$

Furthermore, $\hat{E}_k$ *is a k-order formally quantum integrable QPDE on C,*

$$\left\{ \begin{array}{l} \hat{E}_k \subset J\hat{D}^k(C) \\ \hat{E}_k = \mathbf{n}(\pi^*_{k,0} \hat{E}_k) \end{array} \right\}$$

where $\underline{E}_k$ *is a k-order formally quantum integrable QPDE on* $\overline{\rho} : \underline{E} \to P, \underline{\hat{E}}_k \subset J\hat{D}^k(\underline{E})$, *such that* $\mathbf{b}^k(\underline{\hat{E}}_k) \subset < \hat{C}(P) >^k$, *where* $\mathbf{b}^k : J\hat{D}^k(\underline{E}) \to J\hat{D}^k(TPOE)$ *is the canonical epimorphism over P induced by the canonical epimorphism* $\mathbf{b} : \underline{E} \to TPOE$ *over P.*

REMARK 6.46 - The set of regular Q^∞_w-solutions of $\hat{E}_k$, $Sol(\hat{E}_k)$, is characterized by

$$Sol(\hat{E}_k) = \left\{ \begin{array}{l} c \in Q^\infty_w(C) \\ c = {}_1\mu \circ s \\ s = \overline{\rho} \circ c \\ {}_1\mu \equiv \textbf{Z-graded quantum} \\ \text{pseudoconnection on } P \\ \text{such that } D^k{}_1\mu(s(x)) \in \underline{\hat{E}}_k \\ s \in Q^\infty_w(P) \\ \text{the corresponding quantum pseudoconnection} \\ \mathbf{b} \circ {}_1\mu : P \to TPOE \\ \text{is a Ehresmann quantum connection on } P \end{array} \right\}$$

DEFINITION 6.56 - *A k-order* **FDA-quantum gauge system,** *(FDA = free differential algebra) is a k-order quantum gauge system with blow up structure, $G(M) \equiv (\mathcal{B}, \hat{E}_k)$, such that: The configuration bundle C in $\mathcal{B} \equiv (P, C; \mathbf{B})$ is the fiber bundle of graded quantum pseudoconnections on the trivial vector bundle $\rho : E \equiv P \times V \to V$, where V is a vector quantum space containing $A(G) : V \equiv A(G) \times V_1 \times \cdots \times V_s$, such that there is a representation of $A(G)$ on V of the type*

$$r = (\mathbf{1}, r_1, \ldots, r_s) : A(G) \to L(A(G) \times V_1 \times \cdots \times V_s)$$

where $\mathbf{1}$ is the representation identity on $A(G)$ and $r_i : A(G) \to L(V_i), i = 1, \ldots, s$, are irreducible representations of $A(G)$ on V_i. The blow up structure

$$j : \pi^* \hat{C}(P) \to \underline{E} \equiv \dot{\Lambda}^\bullet POV$$

is induced by the following morphism of fiber bundles over P defined by means of the following commutative diagram of embedding mappings:

$$\begin{array}{ccc} \pi^* \hat{C}(P) & \xrightarrow{j} & \dot{\Lambda}^\bullet POV \equiv \underline{E} \\ \downarrow & & \bigcup \\ TPOA(G) & \to & TPOV \end{array}$$

The QPDE $\hat{\underline{E}}_k \subset J\hat{D}^k(\underline{E})$ is induced by means of a variational principle by using a classic-Lagrangian quantum density $\hat{\Omega}^m(J\hat{D}^h(\underline{E}))$, ($m$ = space-time dimension).

DEFINITION 6.57 - *A* **quantum symmetry** *of a quantum gauge system with blow up structure, $G(M) \equiv (\mathcal{B}, \hat{E}_k)$, is a diffeomorphism $f : P \to P$, (that does not come from the action of the group structure of G on P), such that the natural diffeomorphism $\mathbb{B}(f)$, induced on C, identifies a diffeomorphism of $J\hat{D}^k(C)$ that preserves $\hat{E}_k$. (It is similar to classic case that a quantum symmetry f transforms any regular solution $c = {}_\rceil\mu \circ s$ of $\hat{E}_k$ into a new regular solution $c' = {}_\rceil\mu' \circ s'$ such that f is a symmetry of $\hat{\underline{E}}_k$ too.)*

DEFINITION 6.58 - *The* **curvature map** *of a quantum gauge system with blow up structure is a differential operator $R : J\hat{D}(\underline{E}) \to \dot{\Lambda}^\bullet POE$ such that*

$${}_\rceil R \equiv R \circ D_\rceil \mu = {}_\rceil d_\rceil \mu + P({}_\rceil \mu)$$

where $P({}_\rceil\mu)$ is an element of the polynomial algebra $A[Q^\infty_w(\underline{E})]$, such that on the canonical constraint $TPOA(G) \subset \underline{E}$, R reduces to the corresponding curvature for quantum pseudoconnections:

$$R \circ D_\rceil \mu = d_\rceil \mu + \frac{1}{2}[{}_\rceil\mu, {}_\rceil\mu].$$

REMARK 6.47 - Note that the **Z**-graded curvature

$$_\rceil R = R \circ D_\rceil \mu = {}_\rceil d({}_\rceil \mu)$$

is a particular curvature map. Note also that, in order to define $_\rceil d$, we shall assume a bilinear morphism $[,] : \underline{E} \times \underline{E} \to \underline{E}$ on the vector bundle $\underline{E}$.

6.9 - SUPERGRAVITY QUANTUM PDEs

DEFINITION 6.59 - *A* **super-Poincaré group** *is an extension, G, of the Poincaré group such that on the tangent space in the unit, T_eG, one recognizes a structure of superalgebra.*

REMARK 6.48 - As a super-Poincaré group is not compact, their unitary irreducible representations are infinite-dimensional, and decompose into a finite number of unitary irreducible (infinite-dimensional) representations of the Poincaré subgroup. So every irreducible unitary representation of a super-Poincaré group contains a finite number of particles (**supermultiplets**) parametrized by **mass** and **spin**, eigenvalues of the two Casimir operators.

EXAMPLE 6.6 - In the following table we report the **N-extended super-Poincaré algebra with electric charge** e.

TAB.6.6 - **N-Extended Super-Poincaré Algebra with electric charge** e

$$[M_{ab}, M_{cd}] = \tfrac{1}{2}(\eta_{bc}M_{ad} + \eta_{ad}M_{bc} - \eta_{ac}M_{bd} - \eta_{bd}M_{ac})$$
$$[P_a, P_b] = -8e^2 M_{ab}$$
$$[M_{ab}, P_c] = -\tfrac{1}{2}(\eta_{ac}P_b - \eta_{bc}P_a)$$
$$[M_{ab}, \overline{Q}_{A\beta}] = \tfrac{1}{4}\overline{Q}_{A\alpha}(\gamma_{ab})_{\alpha\beta}$$
$$[P_a, \overline{Q}_{A\beta}] = -ie\overline{Q}_{A\alpha}(\gamma_a)_{\alpha\beta}$$
$$[T_{A\beta}, \overline{Q}_{C\alpha}] = -\tfrac{1}{2}e(\delta_{CA}\overline{Q}_{B\alpha} - \delta_{CB}\overline{Q}_{A\alpha})$$
$$\{\overline{Q}_{A\alpha}, \overline{Q}_{B\beta}\} = i(C\gamma^a)_{\alpha\beta}\delta_{AB}{}^p_a - 2e(C\gamma_{ab})_{\alpha\beta}\delta_{AB}M^{ab} - 4C_{\alpha\beta}T_{AB}$$
$$[T_{AB}, T_{CD}] = \tfrac{e}{2}(\delta_{BC}T_{AD} + \delta_{AD}T_{BC} - \delta_{BD}T_{AC} - \delta_{AC}T_{BD})$$

The Casimir operator (square mass) $M^2 \equiv P_a P^a$ commute with all generators, so all particles in the same supermultiplet have the same mass.

EXAMPLE 6.7 - The case $e \to 0$ is called **N-extended super-Poincaré algebra**.

DEFINITION 6.60 - *A* **supergravity quantum gauge system** *is a quantum gauge system such that the Lie quantum algebra $\widehat{A}(G)$ of the structure quantum group G is a super-Poincaré algebra* (**supergravity quantum algebra**)*. In such a case we say that G is a* **super-Poincaré quantum group**.

REMARK 6.49 - The points of the spectrum of a supergravity quantum algebra are identified with supermultiplets, or with the quantum numbers that characterize them.

EXAMPLE 6.8 - An example of quantum algebra with a structure of Lie superalgbra is the superalgebra of conservation superlaws of a super-PDE $\tilde{E}_k$, quantized by means of the canonical quantization. In such a case $\widehat{A}(G)$ has as generators the quantized conserved charges $\widehat{Q}_k$ associated to a system of generators Z_k of the superalgebra $\tilde{A}(G)$ of the symmetry supergroup G of $\tilde{E}_k$.

In the following we shall recast on microlevel an example of supergravity by using our unified geometric model of quantum gauge theory. We shall specialize on some model for $N = 2, d = 4$ supergravity.

$$\left\{\begin{array}{l} A = \text{quantum algebra} \\ M = (4+0)\text{-dimensional quantum manifold} = M_C \\ (P, M, \pi; G) = \text{principal bundle in the category of quantum manifolds} \\ \text{with structure group}\, G, U(1) - \text{central extension} \\ \text{of the super-Poincaré quantum group} \\ \widehat{A}(G) : \text{generators} : \{Z_B\}_{1\le B\le 19} \equiv \{J_{ab}, P_c, Q_{i\beta}, Z\} \\ 0 \le a, b, c, \beta \le 3, i = 1, 2, J_{ab} = -J_{ba} \end{array}\right\}$$

In other words any $v \in \widehat{A}(G)$ has the following linear representation :

$$v = J_{ab}v^{ab} + P_c v^c + Q_{i\beta}v^{i\beta} + Zu, \text{ where } v^{ab}, v^c, v^{i\beta}, u \in \widehat{A}.$$

So $\dim G = 19$, and $\widehat{A}(G) \cong \widehat{A}^{19}, \dim P = 4 + 19 = 23$. The nonzero commutation relations are resumed in the following table.

TAB.6.7 - **Poincaré Quantum Group**

$$\begin{array}{|l|} \hline [J_{ab}, J_{cd}] = C^{ef}_{ab\ \ cd} J_{ef} = (\eta_{bc}\delta^{ef}_{ad} + \eta_{ad}\delta^{ef}_{bc} - \eta_{ac}\delta^{ef}_{bd} - \eta_{bd}\delta^{ef}_{ac})J_{ef} \\ [J_{ab}, P_c] = C^{\nu}_{ab\ \ \mu} P_{\nu} = (\eta_{bc}\delta^{\nu}_{a} - \eta_{ac}\delta^{\nu}_{b})P_n u \\ \{Q_{\beta i}, Q_{\mu j}\} = C^{\nu}_{\beta i\ \ \mu j} P_{\nu} + C_{\beta i\ \ \mu j} Z = (C\gamma^{\nu})_{\beta\mu}\delta_{ij}P_{\nu} + C_{\beta\mu}\epsilon_{ij}Z \\ [J_{ab}, Q_{\beta i}] = C^{\mu j}_{ab\ \ \beta i} Q_{\mu j} = (\sigma_{ab})^{\mu j}_{\beta i} Q_{\mu j} \\ \hline \end{array}$$

$\eta_{\beta\mu} = \mathrm{diag}(-1, 1, 1, 1), \sigma_{\beta\mu} = \frac{1}{4}[\gamma_{\beta}, \gamma_{\mu}], \gamma^{\mu} =$ Dirac-matrices, $C =$ charge conjugation matrix.

$\overline{\rho} : \underline{E} \equiv TPO_{A_0}A(G) \cong \widehat{A}(G) \bigotimes_{\widehat{A}} (TP)^+ \to P$: fiber bundle of quantum pseudoconnections on the trivial vector bundle $\rho : E \equiv P \times A(G) \to P$. Configuration bundle:

$\pi_C : C \equiv TPO_{A_0}A(G) \cong \widehat{A}(G) \bigotimes_{\widehat{A}} (TP)^+ \to M$. A configuration is a section c of π_C such that $c = {}_{\rceil}\mu \circ s$, where $s \in Q^\infty_w(P)$ and ${}_{\rceil}\mu \in Q^\infty_w(\underline{E})$ is locally given by:

$$ {}_{\rceil}\mu = \sum_{1 \le k \le 19} Z_k \otimes \mu^k_\alpha d\xi^\alpha = J^b_a \otimes \omega^a_{b\beta} d\xi^\beta + P_\mu \otimes \theta^\mu_\beta d\xi^\beta + Q_{\mu j} \otimes \psi^{\mu j}_\beta d\xi^\beta + Z \otimes A_\beta d\xi^\beta $$

where (ξ^β) is a quantum coordinate system on P fibered on (x^α) on M with respect to π. Furthermore,

$$ \omega^a_{b\beta}, \theta^\mu_\beta, \psi^{\mu j}_\beta, A_\beta \in Q^\infty_w(U, \widehat{A}) $$

where $U \subset P$ is an open set. Note that on $\underline{E}$ we have a A_0-bilinear morphism $[,] : \underline{E} \times \underline{E} \to \underline{E}$ over P defined by means of the following commutative diagram:

$$ \begin{array}{ccc} \underline{E} \times \underline{E} & \xrightarrow{[,]} & \widehat{A}(G) \bigotimes_{\widehat{A}} (\dot{\Lambda}^2_0 P)^+ \\ \| & & \|\wr \\ TPOA(G) \times TPOA(G) & \to & \dot{\Lambda}^2_0 POA(G) \\ \downarrow & & \uparrow \\ TPOA(G) \bigotimes TPOA(G) & \to & \dot{T}^2_0 POA(G) \bigotimes_{A_0} A(G) \end{array} $$

such that if $\psi = Z_\beta \otimes \psi^\beta, \phi = Z_\beta \otimes \phi^\beta \in TPOA(G) \cong \widehat{A}(G) \bigotimes_{\widehat{A}} (TP)^+$, $[\phi, \psi]$ is given by

$$ [\phi, \psi] = \sum_{A,B,C} Z_C \otimes C^C_{AB} (\phi^A \Delta \psi^B) $$

where C^C_{AB} are the quantum structure constants of G. They satisfy the following relations:

$$ \left\{ \begin{array}{l} C^C_{AB} = -(-1)^{|A||B|} C^C_{BA} \\ (-1)^{|A||C|} C^E_{AD} C^D_{BC} + (-1)^{|B||A|} C^E_{BD} C^D_{CA} + (-1)^{|C||B|} C^E_{CD} C^D_{AB} = 0 \end{array} \right\} $$

Then, the set $Q^\infty_w(\underline{E})$ becomes a $\mathbf{Z}_2$-graded Lie algebra, as for any $\phi, \psi \in Q^\infty_w(\underline{E})$ one has :

$$ \left\{ \begin{array}{l} [\phi, \psi] = -(-1)^{|\phi||\psi|} [\psi, \phi] \\ (-1)^{|\phi||\rho|} [[\phi, \psi], \rho] + (-1)^{|\rho||\psi|} [[\rho, \phi], \psi] + (-1)^{|\phi||\psi|} [[\psi, \rho], \phi] = 0 \end{array} \right\} $$

The curvature ${}_{\rceil}R = {}_{\rceil}d_{\rceil}\mu$ has the following representation:

$$ {}_{\rceil}R = Z_K \otimes ((\partial\xi_\beta \mu^K_\alpha) + \frac{1}{2} C^K_{IJ} \mu^I_\beta \mu^J_\alpha) d\xi^\beta \Delta d\xi^\alpha = Z_K \otimes R^K_{\beta\alpha} d\xi^\beta \Delta d\xi^\alpha $$

with $R^K_{\beta\alpha} \in Q^\infty_w(U, \overset{2}{\widehat{A}})$, $U \subset P$. The corresponding Bianchi identities locally read

$$ (\partial\xi_\gamma R^K_{\beta\alpha}) + \frac{1}{2} C^K_{IJ} \mu^I_\gamma R^J_{\beta\alpha} = 0. $$

Furthermore, for any section s of $P \to M$ a quantum pseudoconnection ${}_{\rceil}\mu$ and its curvature ${}_{\rceil}R$ identify $A(G)$-valued differential 1-form and 2-form respectively on M:

$$s^*{}_{\rceil}\mu : M \to A(G) \bigotimes_{\mathbf{K}} (TM)^*, \quad s^*{}_{\rceil}\mu = Z_K(\overline{\mu}^K_\gamma) \otimes dx^\gamma;$$

$$s^*{}_{\rceil}R : M \to A(G) \bigotimes_{\mathbf{K}} \Lambda^0_2 M, \quad s^*{}_{\rceil}R = Z_K(\overline{R}^K_{\gamma\delta}) \otimes dx^\gamma \wedge dx^\delta;$$

with $\overline{\mu}^K_\gamma, \overline{R}^K_{\gamma\delta} \in C^\infty_w(\overline{U}, A), \overline{U} \equiv \pi(U), U \subset P$. The blow up structure: $\pi^*\hat{C}(P) \hookrightarrow \underline{E}$ implies that only acceptable quantum pseudoconnections can be written in the form ${}_{\rceil}\mu = (\partial x_\alpha + Z_K \mu^K_\alpha) \otimes dx^\alpha$, where $\mu^K_\alpha \in Q^\infty_w(\pi(U), \widehat{A})$. The corresponding curvature can be written in the form ${}_{\rceil}R^K_{\beta\alpha} = (\partial x_{[\beta}\mu^K_{\alpha]}) + \frac{1}{2} C^K_{IJ}\mu^I_{[\beta}\mu^J_{\alpha]}$. The local expression of the **dynamic equation** $\hat{E}_2 \subset J\hat{D}^2(C)$ is given by the following equations.

$TAB.6.8$ – Dynamic Equation $\hat{E}_2 \subset J\hat{D}^2(C)$

(Bianchi identities)
$$\left\{\begin{array}{r} (\partial x_{[\gamma}.R^{ab}_{\beta\alpha]}) + 2\omega^a_{e[\gamma}R^{eb}_{\beta\alpha]} = 0 \\ (\partial x_{[\gamma}.R^\alpha_{\beta\omega]} + \omega^{ab}_{[\gamma}R_{\beta\omega]b} + \frac{1}{2}(C\gamma^\alpha)_{\delta\mu}\psi^\delta_{j[\gamma}\rho^{\mu j}_{\beta\omega]} = 0 \\ (\partial x_{[\gamma}.\rho^{\beta i}_{\omega\alpha]} + \frac{1}{2}(\sigma_{ab})^{\beta i}_{\delta j}\omega^{ab}_{[\gamma}\rho^{\delta j}_{\omega\alpha]} = 0 \\ (\partial x_{[\gamma}F_{\beta\alpha]}) + \frac{1}{2}C_{\delta\mu}\epsilon_{ij}\psi^{\delta i}_{[\gamma}\rho^{\mu j}_{\beta\alpha]} = 0 \end{array}\right.$$

(Fields)
$$\left\{\begin{array}{l} R^{ab}_{\mu\nu} = (\partial x_{[\mu}.\omega^{ab}_{\nu]}) + 2\omega^a_{e[\mu}\omega^{eb}_{\nu]} \text{ (\underline{curvature})} \\ R^\alpha_{\mu\nu} = (\partial x_{[\mu}.\theta^\alpha_{\nu]}) + \omega^\alpha_{\beta[\mu}\theta^\beta_{\nu]} + \frac{1}{2}(C\gamma^\alpha)_{\beta\delta}\psi^\beta_{j[\mu}\psi^{\delta j}_{\nu]} \text{ (\underline{torsion})} \\ \rho^{\beta i}_{\mu\nu} = (\partial x_{[\mu}.\psi^{\beta i}_{\nu]}) + \frac{1}{2}(\sigma_{ab})^{\beta i}_{\gamma j}\omega^{ab}_{[\mu}\psi^{\gamma j}_{\nu]} \text{ (\underline{gravitino})} \\ F_{\mu\nu} = (\partial x_{[\mu}.A_{\nu]}) + \frac{1}{2}C_{\beta\gamma}\epsilon_{ij}\psi^{\beta i}_{[\mu}\psi^{\gamma j}_{\nu]} \text{ (\underline{electromagnetic field})} \end{array}\right.$$

(Fields equations)
$$\left\{\begin{array}{c} (\partial\omega^\gamma_{ab}.L) - \partial_\mu(\partial\omega^{\gamma\mu}_{ab}.L) = 0 \text{ (\underline{curvature equation})} \\ (\partial\theta^\gamma_\alpha.L) - \partial_\mu(\partial\theta^{\gamma\mu}_\alpha.L) = 0 \text{ (\underline{torsion equation})} \\ (\partial\psi^\gamma_{\beta i}.L) - \partial_\mu(\partial\psi^{\gamma\mu}_{\beta i}.L) = 0 \text{ (\underline{gravitino equation})} \\ (\partial A^\gamma.L) - \partial_\mu(\partial A^{\gamma\mu}.L) = 0 \text{ (\underline{Maxwell's equation})} \end{array}\right.$$

where $L : J\hat{D}(\underline{E}) \rightarrow \overset{4}{\widehat{A}}$ is a Lagrangian function of first order.

REMARK 6.50 - Underline that all the above local functions on M are A-valued. Thus it is clear that the quantum structure identifies curvature, torsion, gravitino and electromagnetic fields as A-valued distributions on the classic space-time.

The canonical quantization of $\hat{\underline{E}}_2$ is given by means of the following quantum bracket:

$$[\hat{c}^I(x), \hat{c}^J(x')] = i\hbar \hat{G}[c]^{IJ}(x, x')$$

where $(c^I) \equiv (\urcorner \mu^I \circ s)$, and $\tilde{G}[c]$ is the quantum propagator of $\hat{\underline{E}}_2$ solution of the following Cauchy problem

$$(6.40) \qquad \left\{ \begin{array}{l} J[c].\hat{G}[c] = 0 \\ \hat{G}[c]^{IJ}(x^0, \overline{x}, x^0, \overline{x}') = 0 \\ (\partial x_0.\hat{G}[c]^{IJ})(x, x')|_{x^0 = x^{0\prime}} = \delta^{IJ}\delta(x, x')|_{x^0 = x^{0\prime}} \end{array} \right\}$$

where $J[c]$ is the Jacobi operator of $\hat{\underline{E}}_2$. As a consequence we get the following commutation relations:

$$[\hat{c}^I(x^0, \overline{x}), \hat{c}^J(x^0, \overline{x})] = 0; \qquad [\hat{c}^I(x^0, \overline{x}), \dot{\hat{c}}^J(x^0, \overline{x}')] = i\hbar\delta^{IJ}\delta(x, x')|_{x^0 = x^{0\prime}}.$$

REMARK 6.51 - Note that the fundamental difference between this canonical quantization of supergravity and the one given in refs.[113,114] is that now the quantum propagator is a function of a quantum field instead of classic one.

REFERENCES

[1] ABE, E., *Hopf algebras*, Cambridge Univ. Press, Cambridge, 1980.

[2] ABRHAM, R. & MARSDEN, J., *Foundations of Mechanics*, Benjamin, 1978.

[3] ARNOLD, V.I., GUSEIN-ZADES, S.M. & VARCEHENKO, A.N., *Singularities of differentiable maps.I,II*, Birkhauser, Boston, Basel, Stuttgart, 1985, 1988.

[4] ASTARITA, G., *Mass transfer with chemical reaction*, Elsevier, 1967.

[5] ATIYAH, M., **Publ. Math. I.H.E.S. 68** (1988), 175-186.

[6] ATIYAH, M., *The geometry and physics of knots*, Cambridge Univ. Press, Cambridge, 1990.

[7] ATIYAH, M.F. & McDONALD, I.G., *Introduction to commutative algebra*, Addison-Wesley, London, 1969.

[8] AVIS, S.J. & ISHAM, C.J., **Comm. Math. Phys. 72** (1980), 103.

[9] AWADA, M.A., **Comm. Math. Phys. 91** (1983), 53-64.

[10] BACK, A., FREUND, P.G. & FORGER, M., **Phys. Lett. 77 B** (1978), 181.

[11] BARNES, D. & MACK, J.M., *An algebraic introduction to mathematic logic*, Springer-Verlag, 1975.

[12] BORDMAN, J.M., **Publ. Math. I.H.E.S. 37** (1967), 383-419.

[13] BOTT, R. & TU, L.W., *Differential forms in algebraic topology*, Springer-Verlag, Berlin, 1982.

[14] BOURBAKI, N., *Algèbre.*, Hermann, Paris, 1970.

[15] BOURBAKI, N., *Topologie générale*, Hermann, Paris, I(1971), II(1974).

[16] BOURBAKI, N., *Théorie spectrales*, Hermann, Paris, 1967.

[17] BROWN, K.S., *Cohomology of groups*, Springer-Verlag, Berlin, 1982.

[18] CHOQUET-BRUHAT, Y., *Géométrie différentielle et systèmes extérieures*, Dunod, Paris, 1968.

[19] COLEMAN, S., Harvard Preprint HUTP/AOO4.

[20] CONNES, A., **Publ. Math. I.H.E.S.** (1986), 44-144.

[21] CRUMEYROLLE, A., **Ann. Inst. H. Poincaré (29)(2)** (1978), 217-231.

[22] DE RHAM, G., *Variétés differentiables*, Hermann, Paris, 1973.

[23] DE WITT, B., *Supermanifolds*, Cambridge Univ. Press, Cambridge, 1984.

[24] DIEUDONNÉ, J., *Éléments d'analyse*, Gauthier-Villars, Paris, III.(1974);VII. (1978).

[25] DUBROVIN, B.A., NOVIKOV, S.P. & FOMENKO, A., *Modern geometry, vols 1,2,3*, Springer-Verlag, Berlin, 1987,88,89.

[26] EGUCHI, T. & FREUND, P.G.O., **Phys. Rev. Lett. 37** (1976), 1251.

[27] EHRENPREIS, L., GUILLEMIN, V.W. & STERNBERG, S., **Ann. Math. 82** (1963), 128-138.

[28] EELLS, J. & SAMPSON, J.H., **Ann. Inst. Fourier, Grenoble 14(1)** (1964), 61-70.

[29] EELLS, J. & ELWORTHY, K.D., **Ann. Math. 91** (1970), 465-485.

[30] ELIASHBERG, J., *Cobordism des solutions de relations différentielles*, Traveaux en cours, Hermann, Paris, 1986.

[31] ERDELYI, A., *Higher transcendental functions, vols.1,2,3*, McGraw-Hill, N.Y., 1953-55.

[32] ERINGEN, A., *Continuum Physics, vols.1,2,3* Academic Press.

[33] GEL'FAND, I.M., **Sov. Math. Dokl. 33(3)** (1986), 573-577.

[34] GEL'FAND, I.M. & GEL'FAND, S.I., **Sov. Math. Dokl. 33(3)** (1986), 643-546.

[35] GEORGI, M. & GLASHOW, L., **Phys. Rev. Lett. 32** (1974), 438.

[36] GIBBONS, G.W. & HAWKING, S.W., **Comm. Math. Phys. 66**(1979), 291.

[37] GIBBONS, G.W. & POPE, G.M., **Comm. Math. Phys. 61** (1978), 231.

[38] GINZBURG, V.L., **Topology 31(4)** (1992), 767-773.

[39] GOLDSCHMIDT, H., **J. Diff. Geometry 1** (1967), 269-307.

[40] GOLDSCHMIDT, H. & SPENCER, D., *Submanifolds and over-determined differential operators*, Complex Analysis & Algebraic Geometry, Cambridge (1977), 319-356.

[41] GOLDSCHMIDT, H. & SPENCER, D., **Acta Math. 136** (1976), 103-239.

[42] GOLDSCHMIDT, H. & STERNBERG, S., **Ann. Inst. Fourier, Grenoble 27(1)** (1973), 203-267.

[43] GREUB, W., *Multilinear algebra*, Springer-Verlag, Berlin, 1978.

[44] GROTHENDIECK, A. & DIEUDONNÉ, J., *Eléments de géométrie algébrique*, Springer-Verlag, Berlin, 1971.

[45] GROMOV, M., *Partial differential relations*, Springer-Verlag, Berlin, 1986.

[46] GUILLEMIN, V. & STERNBERG, S., **Bull. Am. Math. Soc. 70** (1964), 16-47.

[47] HAWKING, S.W., **Phys. Lett. 60A** (1977), 84.

[48] HAWKING, S.W. & POPE, C.N., **Nucl. Phys. B146** (1978), 381.

[49] HILTON, J. & STANMBACH, V., *A course in homological algebra*, Springer-Verlag, Berlin, 1971.

[50] HIRSCH, M.W., *Differential topology*, Springer-Verlag, Berlin, 1976.

[51] HUILGOL, R.R., *Continuum mechanics of viscoelastic liquids*, J.Wiley, 1974.

[52] HUSEMOLLER, D., *Cyclic homology*, Tata Inst. Fund. Res., Bombay, Springer-Verlag, Berlin, 1991.

[53] JACBSON, N., *The theory of rings*, **Amer. Math. Soc. Surv. 2**, Providence 1943.

[54] KARASEV, M.V. & MASLOV, V.P., **Russian Math. Surv. 39(6)** (1984), 133-205.

[55] KASTLER, D., *Introduction a l'électrodynamique quantique*, Dunod, Paris, 1961.

[56] KELLER, H. H., **Lect. Notes Math. 417** (1974), Springer-Verlag, Berlin.

[57] KOBAYASHI, S. & NOMIZU, K., *Foundations of differential geometry, vols.1,2.*, J.Wiley 1963, 1969.

[58] KOSTANT, B., **Lect. Notes Math. 570** (1977), 127-306, Springer-Verlag, Berlin.

[59] KÖTHE, G., *Topological vector spaces, vols.1,2*, Springer-Verlag, Berlin, 1969,1979.

[60] KRASILISHCHIK, I.S., LYCHAGIN, V.V., & VINOGRADOV, A.M., *Geometry of jet spaces and nonlinear partial differential equations*, Gordon & Breach, N.Y. 1986.

[61] LANG, S., *Introduction to differential manifolds*, Interscience, N.Y., 1962.

[62] LASCAR, B., **J. Anal. Math. 37** (1980), 1-31.

[63] LEWY, H., **Ann. Math. 65(1)** (1957), 155-158.

[64] LIBERMANN, P. & MARLE, C.M., *Symplectic geometry and analytical mechanics*, D.Reidel, Dordrecht, 1987.

[65] LICHNEROWICZ, A., **Bull. Soc. Math. France 92** (1964), 11-100.

[66] LICHNEROWICZ, A., **Lect. Notes Phys. 106** (1979), 209-219, Springer-Verlag, Berlin.

[67] LODGE, A.S., *Elastic liquids*, Academic Press, 1964.

[68] LUGO, G.G., **Lect. Notes Phys. 135** (1980), 450.

[69] LYCHAGIN, V.V., **Russian Math. Surv. 30(1)** (1975), 105-175; **34(1)** (1979), 149-180.

[70] LYCHAGIN, V.V., **Sov. Math. Dokl. 24(3)** (1981), 680-685; **25(3)** (1982), 685-689; **28(1)** (1983), 275-279.

[71] LYCHAGIN, V.V., **Lect. Notes Math. 1214** (1986), 102-116, Springer-Verlag, Berlin.

[72] LYCHAGIN, V.V., **J. Sov. Math. 51(6)** (1990), 2735-2757.

[73] LYCHAGIN, V.V. & PRÁSTARO, A., **Differential Geom. Appl. 4** (1994), 283-300.

[74] MACLANE, S. & BIRKOFF, G., *Algebra*, Mursia, 1975.

[75] MADSEN, I. & MILGRAN, R.J., *The classifying spaces for surgery and cobordism of manifolds*, Princeton Univ. Press, Princeton, 1979.

[76] MALGRANGE, B., **Ann. Inst. Fourier, Grenoble 6** (1955-56), 271-355.

[77] MARINO, V. & PRÁSTARO, A., **Lect. Notes Math. 1214** (1986), 101-116, Springer-Verlag, Berlin.

[78] MARINO, V. & PRÁSTARO, A., **Rep. Math. Phys. 26(2)** (1987/8), 211-225.

[79] McCONNEL, J. & ROBSON, J., *Noncommutative Noetherian rings*, Wiley 1987.

[80] McCLEARY, J., *User's guide to spectral sequences*, Publish or Perish Inc., Delaware, 1985.

[81] McLANE, S., *Category theory for the working mathematicians*, Springer-Verlag, Berlin 1971.

[82] MICHOR, P., **Cahiers Top. Geom. Diff. 19(1)** (1078), 47-78; **21(1)** (1980), 63-86; **21(3)** (1980), 325-337; **24(1)** (1983), 57-86.

[83] MILNOR, J., *Morse theory*, Princeton Univ. Press, Princeton, 1963.

[84] MILNOR, J. & STASHETF, J.D., *Characteristic classes*, Princeton Univ. Press, Princeton, 1974.

[85] MISNER, C., THORNE, K. & WHEELER, J.A., *Gravitation*, W.H.Freeman & Co., San Francisco, 1973.

[86] NAIMARK, M.A. & STERN, A.F., *Theory of groups representations*, Springer-Verlag, Berlin, 1982.

[87] NAKANISHI, N.& OJIMA, I., *Covariant operator formalism of gauge theories and quantum gravity*, World Scientific, Singapore, 1990.

[88] NORTHCOTT, D.G., *An introduction to homological algebra*, Cambridge, 1962.

[89] PAGE, D.H., **Phys. Lett. 79 B** (1978), 235.

[90] PALAIS, R.S., *Seminar on the Atiyah-Singer index theorem*, Princeton Univ. Press, Princeton, 1967.

[91] PALAIS, R.S. & SMALE, S., **Bull. Amer. Math. Soc. 70** (1967), 167-172.

[92] POMMARET, J.F., *Lie pseudogroups and mechanics*, Gordon & Breach, Paris, 1990.

[93] PONTRJAGIN, L.S., **Trans. Amer. Math. Soc. 11** (1959), 1-14.

[94] PRÁSTARO, A. & PARRINI, P., **Tex. Res. J. 15** (1975), 118-127.

[95] PRÁSTARO, A., **Riv. Mat. Univ., Parma (4)2** (1976), 295-313.

[96] PRÁSTARO, A., **Ing. Chim. Ital., Suppl. 3-4** (1977), 37-44.

[97] PRÁSTARO, A., **Stochastica 3(2)** (1979), 15-31.

[98] PRÁSTARO, A., **Boll. Un. Mat. Ital. (5)17-B** (1980), 704-726; **(5)S.-FM** (1981),107-129; **(5)S.-FM** (1981), 107-129.

[99] PRÁSTARO, A., *Geometry and existence theorems for incompressible fluids*, Geometrodynamics Proceedings 1983, Pitagora Ed., Bologna (1984), 65-90.

[100] PRÁSTARO, A., **Rend. Sem. Mat., Torino 40(2)** (1982), 89-117; **43(1)** (1985), 89-116.

[101] PRÁSTARO, A., **Boll. Un. Mat. Ital. 6(1-B)** (1982), 1015-1028.

[102] PRÁSTARO, A., **Riv. Nuovo Cimento 5(4)** (1982), 1-122.

[103] PRÁSTARO, A., *A geometric point of view for the quantization of non-linear field theories*, Atti VI Conv.Naz.Relativitá, Firenze 1984, Pitagora Ed. Bologna, 1986, 289-292.

[104] PRÁSTARO, A., *Dynamic conservation laws*, Geometrodynamics Proceedings 1985, World Scientific, Singapore.

[105] PRÁSTARO, A., *Quantum gravity and group-model gauge theory*,Journéés Relativistes de Toulouse, Univ.P.Sabatier, Toulouse (1986), 213-222.

[106] PRÁSTARO, A., *On the quantization of Newton equation*, Atti IX Congresso AIMETA, Bari 1988, AIMETA(1988), 13-16.

[107] PRÁSTARO, A., *Wholly cohomological PDEs*, International Conference on Differential Geometry and Applications, Dubrovnik 1988, Univ. Beograd & Univ. Novi Sad (1989), 305-314.

[108] PRÁSTARO, A., *Geometry of quantized PDEs*, Differential Geometry and Applications, Eds. J. Janyska & D. Krupka, World Scientific, Singapore (1990), 392-404.

[109] PRÁSTARO, A., *On the singular solutions of PDEs*, Atti X Congresso AIMETA, Pisa 1990, AIMETA(1990), 17-20.

[110] PRÁSTARO, A., **Boll. Un. Mat. Ital. 7(5-B)** (1991), 977-1001.

[111] PRÁSTARO, A., **Rep. Math. Phys. 30(3)** (1991), 273-354.

[112] PRÁSTARO, A., *Geometry of super PDEs*, Geometry in Partial Differential Equations, Eds.A. Prástaro & Th. Rassias, World Scientific, Singapore, (1994), 259-315.

[113] PRÁSTARO, A., *Geometry of quantized super PDEs*, **Amer. Math. Soc. Transl. 167(2)** (1995), 165-192.

[114] PRÁSTARO, A., *Quantum geometry of super PDEs*, **Rep. Math. Phys. 37(1)** (1996), 23-140.

[115] PRÁSTARO, A. & RASSIAS, Th., *On a geometric approach to an equation of J.D'Alembert*, Geometry in partial differential equations, Eds. A. Prástaro & Th. Rassias, World Scientific, Singapore (1994), 316-322.

[116] PRÁSTARO, A. & RASSIAS, Th., **Proc. Amer. Math. Soc. 123(5)** (1995), 1597-1606.[1]

[117] PRÁSTARO, A. & REGGE, T., **Ann. Inst. H. Poincaré, 4(1)** (1986), 39-89.

[118] QUAN, P.M., *Introduction à la géométrie des variétés différentiables*, Dunod, 1969.

[119] QUILLEN, D., **Adv. Math. 7** (1971), 29-56.

[120] RICCI, G. & LEVI-CIVITA, T., **Math. Ann. 54** (1901), 125.

[121] ROUX, A., **Publications Dep. Math., Lyon 7-4** (1970), 1-43.

[122] SALAM, A. & STRATHDEE, J., **Nucl. Phys. B76** (1974), 477-482.

[123] SANSONE, G., *Equazioni differenziali, vols 1, 2*, Zanichelli, Bologna, 1965.

[124] SCHAEFER, H., **Acta Math. 107** (1962), 125-173.

[125] SCHAEFER, H., **Trans. Amer. Math. Soc. 120** (1965), 295-326.

[126] SCHAEFER, H., *Topological vector spaces*, Springer-Verlag, Berlin, 1971.

[127] SCHREIBER, M., **Trans. Amer. Math. Soc. 87** (1958), 108-118.

---

[1] See also chapter 8 in the following book: Th. M. Rassias & J. Šimša, Finite sums decompositions in mathematical analysis, J. Wiley & Sons Ltd., England, 1995.

[128] SCHWARTZ, L., *Radon measures on arbitrary topological spaces and cylindrical measures*, Tata Inst. Fund. Res., Oxford Univ. Press, 1973.

[129] SCHWARTZ, L., *Theorie des distributions*, Hermann, Paris, 1978.

[130] SCHWARZ, A.S., **Comm. Math. Phys. 87** (1982), 37-63; **95** (1984), 161-184; **96** (1984), 161-184.

[131] SEELEY, R.T., **Trans. Amer. Math. Soc. 117** (1965), 108-118.

[132] SEGAL, I.E. & KUNZE, R.A., *Integrals and operators*, Springer-Verlag, New York, 1970.

[133] SPENCER, D., **Acta Math. 76** (1062) 306-445.

[134] SPENCER, D., **Bull. Amer. Math. Soc. 75** (1969), 179-239.

[135] SPIVAK, M., *A comprensive introduction do Differential Geometry, vols.1-5*, Publish or Perish, Houston, Texas, 1970.

[136] STEENROD, N., *Topology of fibre bundles*, Princeton Univ. Press, Princeton, 1974.

[137] STERNBERG, S., *Lectures of Differential Geometry*, Englewood Cliffs, N.J., 1964.

[138] STONG, R.F., *Notes on cobordism theories*, Princeton Univ. Press, Princeton, 1968.

[139] SWEENEY, W.J., **Acta Math. 120** (1965), 223-277.

[140] SWEENEY, W.J., **Ann. Math. 90** (1969), 353-360.

[141] SWITZER, R., *Algebraic topology-homotopy and homology*, Springer-Verlag, Berlin, 1975.

[142] THOM, R., **Comm. Math. Helv. 28** (1954), 17-86.

[143] THOM, R., **Ann. Inst. Fourier, Grenoble 6** (1956) 43-87.

[144] TREVES, F., *Topological vector spaces, distributions and kernels*, Academic Press, N.Y., 1967.

[145] TRICOMI, F., *Funzioni ellittiche*, Zanichelli, Bologna, 1951.

[146] TRICOMI, F., *Funzioni ipergeometriche confluenti*, Cremonese, Roma, 1954.

[147] TRICOMI, F., *Integral equations*, Interscience, 1957.

[148] TRUESDELL, C., *The elements of continuum mechanics*, Springer-Verlag, Berlin, 1966.

[149] VAISMAN, I., *Cohomology and differential forms*, N.Y., 1973.

[150] VAISMAN, I., *Symplectic geometry and secondary characteristic classes*, Birbrek khauser, Boston, Basel, 1987.

[151] VILENKIN, N.J., & KLIMYK, A.J., *Representation of Lie groups and spacial functions*, Kluwer Academic Press, Dordrecht, 1992.

[152] VINOGRADOV, A.M., **Lect. Notes Math. 1108** (1984), 77-102, Springer-Verlag, Berlin.

[153] WALL, C.T.C., **Ann. Math. 72** (1960), 292-311.

[154] WALL, C.T.C., **Ann. Math. 84** (1966), 217-276.

[155] WALL, C.T.C., *Surgery on compact manifolds*, Academic Press, London, 1970.

[156] WALL, C.T.C., **Topology 15** (1976), 189-210.

[157] WARNER, F.W., *Foundations of differential manifolds and Lie groups*, Glenview, Illinois, 1971.

[158] WALSH, B., **Trans. Amer. Math. Soc. 120** (1965), 295-326.

[159] WALSH, B., **Proceedings Amer. Math. Soc. 17** (1966), 1267-1271.

[160] WEISS, J. & ZUMINO, B., **Nucl. Phys. 70B** (1974), 39-50.

[161] WELLS, R.O., *Differential analysis on complex manifolds*, Springer-Verlag, Berlin, 1980.

[162] WESS, S. & BAGGER, J., *Supersymmetry and supergarvity*, Princeton Univ. Press, Princeton, 1983.

[163] WEST, P., *Introduction to supersymmetry and supergravity*, World Scientific, Singapore, 1986.

[164] YATES, R.G., **Comm. Math. Phys. 76** (1980), 255-268.

[165] YOSIDA, K., *Functional analysis*, Springer-Verlag, Berlin, 1965.

[166] ZHARINOV, V.V., *Geometrical aspects of partial differential equations*, World Scientific, Singapore, 1992.

INDEX

A

B

C

D

E

F

G

H

I

J

K

L

M

N

O

P

Q

R

S

T

U

V

W

X

Y

Z

LIST OF SYMBOLS

$Ann(-)$ annulator of $(-)$.

$Aut(-)$ authomorphism of $(-)$.

CA, $C_X A$ complement of subset $A \subset X$.

$C^\alpha(X,Y)$ space of functions $X \to T$ of class C^α.

$Clif(M)$, $Clif^c(M)$, $Clif^G(M)$ Clifford fiber bundles.

div divergence.

$D^k f$ k-derivative of a map f.

Df derivative of a map f.

$\mathbf{R},\mathbf{C},\mathbf{N},\mathbf{Q},\mathbf{Z},\mathbf{Z_2},\mathbf{H}$ sets of real, complex, natural, rational, integers, integers module 2, and quaternionic respectively.

$\mathbf{R}_+$ set of positive real numbers.

$E_k \subset J^k_n(W)$ PDE.

$E_k \subset J\mathcal{D}^k_n(W)$ PDE.

$\mathbb{E}^k_n(W) \subset TJ^k_n(W)$ Cartan distribution.

$\mathbb{E}^k(W) \subset T\mathcal{D}J^k(W)$ Cartan distribution.

$\mathbb{E}^k_n \subset TE_k$ Cartan distribution.

$\mathbf{G}$ Galilean group.

$G_{l,l+p}(E)$ Grassmann manifold.

$G^+_{l,l+p}(E)$ oriented Grassmann manifold.

grad gradient.

$\mathbb{P}$ projectivization functor.

RP real projective space.

$\mathbf{RP}^n = \mathbb{P}(\mathbf{R}^{n+1})$.

CP complex projective space.

rot , rotor.

$R_{..}$ Ricci tensor.

$S(M)$, $S(M)^c$, $S(M)^G$ spinor bundles.

$\mathcal{S}ol(E_k)$ set of solutions of PDE E_k.

$Sol(E_k)$ set of regular solutions of PDE E_k.

T tangent functor; time-axis.

vd vertical differential.

vT vertical tangent functor.

$v_{/k}$ $v^{/k}$ covariant derivative of v.

$T^r_s M$, $\Lambda^r_s M$, $S^r_s M$ tensor fiber bundles of type (r,s) and (skew-)symmetric subbundles.

$\underline{W}$ sheaf of sections of the fiber bundle $\pi: W \to M$.

$\Re$ real part of complex number.

$\Im$ imaginary part of complex number.

$\Omega_p(E_k)$, $\Omega_{p,s}(E_k)$ quantum p-cobordism groups of PDE E_k.

$\Omega_p^{E_k}$, $\Omega_{p,s}^{E_k}$ integral p-cobordism group of PDE E_k.

$\Omega(E_k)$ quantum situs of PDE E_k.

Ω_s s-cobordism group.

$^+\Omega_s$ oriented s-cobordism group.

$\otimes$ tensor product.

$\widehat{\otimes}$ completed tensor product.

$\boxed{\times}$ two points tensor product.

∂ partial derivative, infinitesimal derivative, boundary-symbol.

$\pitchfork$ transverse.

$\triangleleft$ subgroup.

$\overline{\wedge}$, special wedge product.

$\bullet_{A_0}$, $\triangle_{A_0}$ particular tensor products in quantum manifolds.

$\emptyset$, $\emptyset$ empty set.

$\square$ end of proof.

$\boxed{} \equiv \delta d$. It marks also some propositions.

☡ "dangerous bend".

▪ end of example or remark.

$\circ\!\!\rightarrow$ set-arrow.

$\varinjlim$, $\lim\limits_{n}$ direct (or inductive) limit.

$\varprojlim$, $\lim\limits_{n}$ invers (or projective) limit.